Unit Systems and Conversion

SI Units

Fundamental units	mass	kg	kilogram
	length	m	meter
	time	s	second
	electric current	A	ampere
Supplementary unit	angle	rad	radian
Some derived units	force	$N = kg \cdot m/s^2$	newton
	energy	$J = N \cdot m$	joule
	power	$W = J/s$	watt
	frequency	$Hz = 1/s$	hertz
	pressure	$Pa = N/m^2$	pascal
	charge	$C = A \cdot s$	coulomb
	electric potential	$V = J/C$	volt
	magnetic field	$T = N/A \cdot m$	tesla
	magnetic flux	$Wb = T \cdot m^2$	weber
	capacitance	$F = C/V$	farad
	resistance	$\Omega = V/A$	ohm

Selected British Units

length	1 inch $\equiv$ 2.540 cm
	1 foot = 0.3048 m
	1 mile = 1.609 km
mass	1 pound mass (used in the U.K.) = 0.4536 kg
	1 slug (mass unit used in the U.S.) = 14.59 kg
	1 ton = 2240 lb mass (British or long ton)
	1 U.S. ton = 2000 lb mass (short ton)
energy	1 British thermal unit = 1.055×10^3 J
power	1 horse power = 745.7 W
force	1 pound (called pound-weight in the U.K.) = 4.448 N
pressure	1 lb/in.^{2} = 6.895×10^3 Pa

Selected cgs/Gaussian Units

length	1 cm = 10^{-2} m
mass	1 g = 10^{-3} kg
energy	1 erg = 10^{-7} J
force	1 dyne = 10^{-5} N
pressure	1 dyn/cm^2 = 0.1 Pa
magnetic field	1 gauss corresponds to 10^{-4} T

Selected Units Used in Astronomy

length	1 astronomical unit = 1.50×10^{11} m
	1 light-year = 9.46×10^{15} m
	1 angstrom = 10^{-10} m
mass	1 solar mass = 1.99×10^{30} kg
power	1 solar luminosity = 3.90×10^{26} W

Miscellaneous Units

time	1 y $\approx \pi \times 10^7$ s
	1 d = 86400 s
length	1 nautical mile = 1.852 km
speed	1 mph = 0.4470 m/s
	1 knot = 0.5145 m/s
mass	1 u (atomic mass unit) = 1.660×10^{-27} kg
	1 metric ton = 1000 kg
energy	1 calorie = 4.18 J
	1 electron volt = 1.60×10^{-19} J
	1 kilowatt-hour = 3.60×10^6 J
	1 kiloton of TNT = 4.2×10^{12} J
pressure	1 atmosphere = 1.013×10^5 Pa
	1 torr = 133.3 Pa
	1 cm Hg = 1.333×10^3 Pa
	1 in. Hg = 3.386×10^3 Pa
volume	1 liter = 10^{-3} m^3
	1 U.S. gallon = 3.785×10^{-3} m^3
area	1 acre = 4.05×10^3 m^2
	1 barn = 10^{-28} m^2
angle	$1^\circ = 1.745 \times 10^{-2}$ rad
	$1' =$ 1 minute of arc $= \frac{1}{60}{}^\circ$
	$1'' =$ 1 second of arc $= \frac{1}{60}{}'$

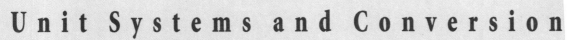

"This dread and darkness of the mind cannot be dispelled by the sunbeams, the shining shafts of day, but only by an understanding of the outward form and inner workings of nature."

LUCRETIUS

Volume Two

Physics: The Nature of Things

Susan M. Lea
San Francisco State University

John Robert Burke
San Francisco State University

Brooks/Cole Publishing Company
I(T)P® An International Thomson Publishing Company

Pacific Grove • Albany • Belmont • Bonn • Boston • Cincinnati • Detroit
Johannesburg • London • Madrid • Melbourne • Mexico City
New York • Paris • Singapore • Tokyo • Toronto • Washington

SPONSORING EDITOR: Richard W. Mixter
DEVELOPMENTAL EDITOR: Keith Dodson
MARKETING TEAM: Ann Hillstrom, Ellen Stanton
INTERIOR DESIGN: Gerri Davis, Quadrata, Inc.
COVER IMAGE: Leo de Wys Inc./De Wys/Sipa/Fritz
INTERIOR ILLUSTRATION: Scientific Illustrators, Inc.
PROOFREADING: Elliot Simon, Simon and Associates
Production, Prepress, Printing and binding by West Publishing Company
Photo Credits follow the index.

For more information, contact:

Brooks/Cole Publishing Company
511 Forest Lodge Road
Pacific Grove, CA 93950
USA

International Thomson Editores
Campos Eliseos 385, Piso 7
Col. Polanco
11560 México D. F., México

International Thomson Publishing Europe
Berkshire House 168-173
High Holborn
London WC1V 7AA
England

International Thomson Publishing GmbH
Königswinterer Strasse 418
53227 Bonn
Germany

Thomas Nelson Australia
102 Dodds Street
South Melbourne, 3205
Victoria, Australia

International Thomson Publishing Asia
221 Henderson Road
#05-10 Henderson Building
Singapore 0315

Nelson Canada
1120 Birchmount Road
Scarborough, Ontario
Canada M1K 5G4

International Thomson Publishing Japan
Hirakawacho Kyowa Building, 3F
2-2-1 Hirakawacho
Chiyoda-ku, Tokyo 102
Japan

Printed in the United States of America

10 9 8 7 6 5 4 3 2 1

The Library of Congress has cataloged the combined volume as follows:

Lea, Susan.
Physics : the nature of things / Susan Lea, John Burke.
p. cm.
Includes bibliographical references and index.
ISBN 0-314-05273-9 student edition (alk. paper)
ISBN 0-314-07012-5 annotated instructor's edition (alk. paper)
1. Physics. I. Burke, John (John Robert) II. Title.
QC21.2.L43 1997
530—dc20 96-13354
CIP

British Library Cataloging in Publication Data
A catalogue record for this book is available from the British Library.
ISBN this volume: 0-534-35735-0

Systematic Use of Color in this Text

(Components of vectors are shown in a lighter shade of the same color)

Part I: Newtonian Mechanics

- Path of a particle
- Unit vector
- Position vector
- Displacement vector
- Velocity vector
- Acceleration vector
- Force vector
- Moving frame box

Part II: Conservation laws

- Momentum vector
- Gravitational field vector
- Angular momentum vector
- Torque vector

Part III: Continuous systems

- Angular velocity vector
- Angular acceleration vector
- Streamlines

Part IV: Oscillatory and Wave Motion

- Sound wave phasors
- Light wave phasors
- Light rays

Part V: Thermodynamics

- Adiabat
- Isotherm
- Isochor
- Isobar

Part VI: Electromagnetic Fields

- Positive charge
- Negative charge
- Electric field vector
- Magnetic field vector
- Equipotential surface
- Electric dipole
- Magnetic dipole
- Electric current
- Electric displacement vector
- Gaussian surface
- Amperian curve

Part VII: Electrodynamics

- Poynting vector

Part VIII: Twentieth Century Physics

- World line
- Photon
- α-decay
- β-decay

To the special people in my life:

my father and mother, my husband Michael and my daughter Jennifer.

Thank you.

Susan Lea

To my father, whose thirst for knowledge was an inspiration.

John Burke

ABOUT THE AUTHORS

Susan Lea is a professor of Physics and Astronomy at San Francisco State University, where she has taught since 1981. Born in Wales, she received her undergraduate degree from Cambridge University, with 1st class honors in applied mathematics and theoretical physics. She did her graduate work at the University of California, Berkeley, receiving a Ph.D in Astrophysics. She worked extensively with data from x-ray satellite missions, including Uhuru, HEAO 1 and the Einstein Observatory. She and her husband own and operate a software company offering optical ray tracing software. She has published extensively in the astronomical journals, but her first refereed paper (in an engineering journal) was on the theory of loudspeaker design! She began teaching physics at the age of 16 (in high school), and hasn't stopped since.

Dr. Lea's interests include flying (she holds a flight instructor certificate with airplane and instrument ratings), horse riding and music.

Professor of Physics at San Francisco State University since 1972, Dr. Burke has enjoyed sharing his love of science with young people deciding on their careers. As a voracious young consumer of science fiction and serious studies of space exploration, Dr. Burke's own path was set by visits to dad's job at then new particle accelerators and by Fred Hoyle's popular astronomy books. "It was so cool to know we could explore atoms or picture the Earth four billion years ago, melted by meteorite bombardment, its core forming from liquid iron dribbling inward." Undergraduate work at Caltech and graduate work in astrophysics at Harvard led to a research specialty in physics of the interstellar medium, with occasional forays into acoustics, economics, and relativity. It was also at Harvard that Dr. Burke's interest in physics eduation bloomed. "I had the opportunity to study teaching with outstanding masters of the craft. Concern for how people come to understand, how they fit science into their lives, and how they learn to think with precision have since guided all my work."

Of course, it's not all work. On occasion "J.R.B." can be caught taking in an early music concert, trekking a wilderness, climbing the odd mountain, or taking his plane into some out-of-the-way airport.

PREFACE

To the Instructor

Our book's title is taken from De Rerum Natura, a work by Lucretius, a Roman writer of the first century AD[1] who tried to persuade his readers by using logical arguments based on observation and experience. This approach is still in style—modern physicists employ the same methods. Like Lucretius, both research physicists and physics students struggle to understand "the nature of things."

Goals

A primary goal of this book is to help science students develop the kinds of logical thinking that they will need to understand physics. These skills are useful in physics and other disciplines as well. Students often find physics the most difficult of the sciences because, even in the introductory courses, it demands much more than the memorization of facts. To study physics successfully, students need to learn to think like physicists. Students must move beyond being hunters and gatherers of formulae to solve problems—they must become, like physicists, creative problem solvers. In this book we have tried to help students develop the logical reasoning and analytical skills that enable a physicist to practice his or her art.

Citizens in a modern technological society need to be scientifically literate. That only a small, elite group of bright students survives introductory physics and goes on to become aerodynamic engineers or physics professors is no longer acceptable. We hope to make physics accessible to all those who choose to take a physics course. We make it accessible not by watering it down, but by giving students the tools they need to grab hold of the subject and make it their own. Physics is fascinating and fun—at least we think so—and we have tried to convey some of our own enthusiasm for the subject. Examples such as the motion of a "hot-dog" skier (Example 3.5, Exercise 3.2) show the power of physics as a tool for understanding the world and, at the same time, spark students' interest.

Intended for a course that requires calculus as a prerequisite or co-requisite, this text uses calculus throughout, in derivations, examples, and problems. In the first few chapters calculus is used sparingly, mostly in optional sections, so that those students who are just starting calculus will not be overwhelmed. Later in the book, more familiarity is assumed. An interlude following Chapter 7 discusses the use of integration in physics and presents a five-step plan for setting up integrals. Basic knowledge of algebra, geometry, and trigonometry is assumed. Appendix I includes some basic relations from these disciplines as a reminder and reference for students.

This book can be used by students with widely varying levels of ability. Each chapter stresses the basic concepts first. By including or excluding the *Digging Deeper boxes,* the optional *Math Topic boxes, Optional sections* (marked with an ✲), and the *Advanced and Challenge Problems,* the instructor can tailor the text to her or his own students. *Instructor marginal notes (in blue)* indicate which optional topics are used later in the book, and also explain the reason for some of our choices of topic and organization. We have also given references for some of our sources.

Organization

The order of topics in the book is largely traditional, but is organized to allow a large range of sequencing options. For example, introducing angular momentum of a particle in Part II offers the option of foregoing rigid body dynamics in favor of a faster move to the twentieth century. The chapter on oscillatory motion could be used any time after the discussion of energy (Chapter 8). We have included optics in the section on wave motion, to stress the unity of such wave phenomena as interference. However, Chapters 16–18 on optics could easily be covered after E&M if desired. Part V on thermodynamics is self contained and could be studied any time after basic mechanics. The first three sections of Chapter 34 (relativity) could be covered after Chapter 3, and Section 34.4 could be introduced after Chapter 8. The chapters on modern physics tend to be more qualitative, because of the level of mathematical sophistication required for a detailed treatment. They are designed to serve as the culmination of a two- or three-semester survey or as an impedance-matching introduction to a standard course on modern physics. These chapters emphasize the conservation principles developed in Part II.

[1]Lucretius based his book on earlier work by the Greek philosopher Democritus.

Throughout the book we stress *two major themes: conceptual understanding and a consistent approach to problem solving.* The material in the book is divided into eight parts, each introducing a unified body of concepts: Newtonian Mechanics; Conservation Laws; Continuous Systems; Oscillations and Waves; Thermodynamics; Electromagnetic Fields; Electrodynamics; and Twentieth Century Physics. This division helps the students organize their knowledge. The introduction to each part explains the theme to be covered and provides some historical perspective. We begin each chapter with a discussion of the opening photograph, frequently raising a question that we answer within the chapter. Just as each chapter begins with a physical situation to introduce the concepts of the chapter, each topic within the chapter is introduced with a conceptual discussion before the mathematics is presented. In this way we emphasize that working with the concepts is the first essential step in solving a problem. Then the mathematics is used to complete the solution. Similarly, we place a great deal of emphasis on using diagrams to help conceptualize problems and plan their solution. We encourage students to use diagrams as graphical tools to aid their understanding and to help make the transition from a verbal presentation to a mathematical model. Unlike many texts, we not only tell students to use diagrams, we *always* do it ourselves.

PROBLEM SOLVING

Two *Interludes* in the early parts of the text help lay the groundwork for a systematic approach to problem solving. In the first interlude, following Chapter 3, we lay out *our basic four-part problem-solving strategy.* The major stages of each problem solution— **MODEL** , **SETUP** , **SOLVE** , and **ANALYZE** — are identified and discussed at this point. These steps are used and labelled in every example throughout the book. Seeing the method at work in each example better enables students to apply a similar approach in their own solutions.

The second Interlude, following Chapter 7, shows students how to set up problem solutions using *integration.* The method involves five steps. The first four steps are a procedure for describing a physical process or system in terms of differential elements and transforming a sum over such elements to a standard mathematical form. Only at the final step does the actual evaluation of an integral occur. This final step is the one that students learn in their calculus classes. In each example requiring integration we use this method, with the steps clearly labelled.

Throughout the book we present *Solution Plans.* These are problem-solving strategies that show the logical steps necessary in certain specific classes of problems. Each plan is explicitly laid out in flow-diagram form. The method for analyzing dynamical systems with Newton's laws (Chapter 5, p. 167) provides a good example of a Solution Plan. A table in the appendix lists all the plans for easy reference. These Solution Plans will help the students develop the skills they need to solve problems in physics, and help them to go beyond that hunter-gatherer, "find the right equation and stuff in," stage. As students become more proficient they will be able to adapt these problem-solving strategies to their personal style.

The Solution Plans can also be valuable teaching tools, allowing you to identify precisely where students have difficulties. For example, using the plan in Chapter 5, we found that an astonishingly large number of students are convinced that they can't analyze a system with strings unless they know the value of the tension before carrying out the algebra. Once these difficulties have been identified, it is much easier to confront them and, ultimately, eliminate them.

The careful use of *vectors* is stressed throughout. In particular we introduce vectors as the primary descriptive tool in kinematics, using geometrical addition (Sections 1.4–1.6), and then solve one-dimensional problems as a special case of one-component vectors (Section 2.3). Not only does this approach stress the importance of vectors from the beginning, but it makes the meaning of signs in one-dimensional motion obvious. (An instructor's marginal note on page 52 explains how this material can be presented in other sequences.) In addition to boldface type, we have used the *"arrow-over" notation* so that equations in the book will look the same as the equations you write on the blackboard, or the students write in their notes. We have avoided the use of "magic" minus signs (as in the spring force) that are not explicitly tied to a coordinate choice or stated sign convention.

Beginning students often focus on finding "the answer" without first framing any expectation of what the magnitude, units or other characteristics of the answer might be. As scientists, instructors know the importance of estimation as a problem-solving strategy. It can be difficult to integrate this strategy into teaching, however, especially if students don't see it used regularly in their text. We introduce students to these valuable skills by using *back-of-the-envelope calculations* to estimate results, or to decide what is or is not important in a given situation. These methods are also used to estimate the reasonableness of an answer or to figure out the basic physics behind a complicated event like a thunderstorm. The envelope symbol ( ) alerts the students whenever we use these techniques in examples or discussions. Some problems show this symbol to indicate that the students should use these techniques in their solution, and that an exact answer is not expected.

EXAMPLES, QUESTIONS, AND PROBLEMS

Each chapter starts by emphasizing the basic concept, then developing it through a carefully graded series of *Examples.* All Examples consistently use the four-part problem-solving strategy presented in the first Interlude, and show the appropriate free-body diagram or other illustration at each step. While we have attempted to keep the introductory examples straightforward, and to assure that they demonstrate a steady gradual increase in difficulty throughout a chapter or part, twenty *Study Problems,* spread throughout the book, emphasize the use of the problem-solving method in detail with interesting and sometimes intricate problems. The inclusion of these problems should help to alleviate the complaint that the "examples didn't prepare me to do the problems."

The text offers many opportunities for students to test their knowledge and their ability to use the material. Within each chapter, *Exercises* allow students to practice with ideas they have just learned. Abbreviated solutions—not just answers—are given at the end of each chapter, so students can get real feedback after they work an exercise.

The end-of-chapter material includes a carefully structured array of problems for student review or assignment by an instructor. *Review Questions* emphasize conceptual understanding and can be answered by a quote or paraphrase of material from the chapter; *Basic Skill Drill* is a set of problems that test student's knowledge of fundamental mathematical relations and the meaning of terms introduced in the chapter; an extensive set of *Questions and Problems* include practical applications and conceptual questions as well as the usual "textbook exercises." Symbols preceding each problem identify the level of difficulty, and also indicate the conceptual problems. Many of the problems are sorted by chapter sections, but numerous *Additional Problems* are included that may require use of material from several sections, or even from previous chapters. *Computer problems* give the students an opportunity to hone their computer skills—an increasingly important component of education. Most of these problems can be solved using a spreadsheet program, or one of the simple programs on the supplementary computer disk available with the text. Students with more advanced computer skills will have an opportunity to incorporate these skills into their physics problem solving. *Challenge problems* introduce the more capable students to interesting and stimulating exercises that require advanced problem-solving skills. *Part Problems,* found at the end of each of the eight parts of the text, give students an opportunity to synthesize their understanding and to see how each topic builds on and enhances what went before.

Other Helpful Features

Math Toolboxes appear throughout the text. Each one presents a set of techniques that are necessary tools for doing physics. They are located in the text where the techniques are first needed. For examples, see the Math Toolbox on the properties of the scalar product (p. 229) or the one called *How to Solve a Differential Equation* on p. 1005.

Digging Deeper boxes and *Math Topic* boxes present ideas that are not essential but that provoke interest, give greater depth to a point in the text, or simply point out a delightful consequence of the physical principles. See for example *More on Cyclotrons* (p. 928), *Use of Calculus in Circular Motion* (p. 98), and *How Do Fish Survive the Winter?* (p. 686).

Essays, some by guest authors, address interesting sidelights or more advanced topics. We happily remember the student who suddenly remarked "I get it!" after reading the bicycle essay. By applying Newton's laws to a subject he enjoyed, he finally made sense of it all.

Definitions and equations are color-coded to help students recognize their level of importance. Despite the emphasis throughout the text on problem-solving as a reasoning process, some things must be memorized to be used efficiently. Anything in a gold box is fundamental and should be memorized!!! Level 2 equations, in tan boxes, are important and will often be useful in solving problems. Level 3 equations, unboxed and unnumbered, are intermediate or less important results that need not be memorized. Occasionally we need to refer to intermediate results in order to guide students through a problem solution or derivation. Such results are given lower case Roman numerals. Any reference to these equations is local (within a page or so of the original statement).

Marginal notes (in black) alert the students to common errors, point out important features and special cases, give additional references, refer to previously discussed, related issues, and add clarifying commentary.

Instructor's Marginal Notes (in blue) appear throughout the *Instructor's Annotated Edition*. In this special version of the text, these marginal notes signal the location of related material, explain why a particular approach is used, cite references to the physics literature, provide suggestions and comments on possible changes in the sequence of topics and so forth. Many of the *Instructor's Marginal Notes* in the text are the result of "dialogues" that are carried on between reviewers of the manuscript and ourselves through many drafts of the text. In the *Instructor's Annotated Edition*, the Contents (on page *xi*) includes instructor's notes that comment on various organizational and content features of the text.

Our book has *more art* than other texts presenting the same material. We don't just tell students that good problem solving starts by making a drawing as conceptual link from the physical situation to the correct mathematical model, we consistently follow this practice. To help reinforce the importance of using and understanding graphical models, *color is consistently used in all illustrations* throughout the book. Acceleration is always blue, for example. See the color key that appears on page ii in the front of the book.

Accuracy

The authors and publisher recognize that errors in quantitative material can undermine the effectiveness of a text. A great deal of attention and effort has been invested to assure that all of the quantitative material in the text (and the solutions manuals) is correct and accurate. Accuracy checking went on throughout preparation of the manuscript, as well as during production of the physical book. During the years that manuscript was being written and developed, many people were involved in assuring accuracy.

- Dozens of physics professors reviewed numerous drafts of the manuscript. All were asked to review the examples, exercises and problems. Many reviewers focused specifically on this quantitative material, at the publisher's request.
- The authors solved every end-of-chapter problem and checked each to be sure that it did not make unstated assumptions or rely upon unstated information.
- Jon Celesia of San Francisco State University carefully reviewed the final manuscript, checking for unstated assumptions, unclear explanations, and any possible inaccuracies.

During the year-long process of drawing all the art and setting all the type, numerous checks were performed.

- The authors proofread every syllable and symbol through two (in some cases three) stages of proof.

- One independent proofreader was hired to read the entire first stage proof. Another professional proofreader checked both stages of proof, and all subsequent revisions.
- Barbara Uchida of the College of San Mateo checked every one of the in-chapter Examples and Exercises for accuracy and clarity.
- Every end-of-chapter problem was solved by a team of physics professors and graduate students. These solutions were then reviewed by an independent accuracy checker before going to the authors for final approval.

Because students remember best what they learned first, we have taken pains to keep the concept discussions accurate. Even if topics must be expanded on later, students should never have to unlearn anything. Extensive review of the manuscript by dozens of teaching colleagues and consultations with several authorities on specific topics have helped to ensure that all concepts are correctly presented.

A final source of quality assurance for both quantitative and non-quantitative material has been students. During the development of the manuscript, many of the Examples, Exercises and end-of-chapter problems were tested with students in class and in homework assignments. The first half of the book has been used by students at San Francisco state University and at the University of California at Davis. The results have been very gratifying, both for the instructors and the students.

SUPPLEMENTS

A carefully prepared package of supplements has been created to support both the instructor and the student. Contact your local sales representative for information about the complete list of print and electronic supplements available. Among them are the following:

The Solutions Manuals have been written by the authors in conjunction with a team of graduate students and fellow physics professors. The Solutions Manuals have been carefully checked by the authors at least three times and by an independent accuracy checker for clarity, consistency, and accuracy. A clearly designed format featuring accurate, professionally rendered art makes the Solutions Manuals more accessible and useful. The Solutions Manuals also include a section of recommended readings ("For Further Reading") for reference.

> *Student's Solutions Manual*—provides complete solutions to selected odd-numbered end-of-chapter problems including solutions for every odd-numbered Basic Skill Drill problem.
>
> *Instructor's Solutions Manual*—contains complete solutions to all of the odd-numbered end-of-chapter problems and answers to all of the even-numbered problems.

The Test Bank, prepared by Darry S. Carlstone of the University of Central Oklahoma, includes over 3000 questions in multiple-choice format. The test bank is available in hard copy and on disk with a computerized test generator that allows instructors to modify, write, and display test questions. The testing program has outstanding graphics capability and a full range of physics symbols. IBM and Macintosh versions available.

The Optional Student Program Disk, created by Susan Lea and Michael Lampton in conjunction with Donnelly Software, contains data files for use with computer problems in the text (see above). In addition, some small programs for demonstrations and for problems on specific topics are also included. The files are in ASCII and can be imported into any spreadsheet program. The text may be ordered with or without the Student Program Disk.

Acetate Transparencies of numerous important illustrations from the text are available in full color.

ACKNOWLEDGEMENTS

This textbook represents not only the work of the authors, but also the extraordinary efforts and contributions of a number of people to whom we owe thanks.

This project would not have been possible without the support, encouragement, and help of our editor, Richard Mixter. His guidance and insight have been invaluable. He has kept us going through the difficult times, and shared our joy in the happy times. He has held our hands as we learned the ins and outs of publishing. Without him, this book would never have been completed.

Keith Dodson, the developmental editor, has also provided important guidance through his perceptive and thoughtful analyses of reviews and manuscript. We are grateful to the publishing team in Eagan, Minnesota, especially to Tamborah Moore and Emily Autumn for their outstanding work in the production of this complex book, and to Ann Hillstrom and Ellen Stanton for their insight and effort in marketing. Our thanks go to George Morris of Scientific Illustrators for creatively rendering the hundreds of pieces of art in the text, beautifully, clearly and correctly, and to Denee Reiton Skipper for her elegant and effective page layouts. Patricia Burke deserves recognition for her invaluable help with the illustration program. We should also like to acknowledge the contributions of those photographers and scientists who have allowed us to use their pictures, and especially to Tom Pantages who took many photographs to our specifications. We appreciate the help of Chuck and Janet Donnelly of Donnelly Software with the optional student program disk. (Any errors in the programs, however, are our responsibility.)

We also wish to thank our colleagues for their help: Jon Celesia for checking every word of the manuscript, Barbara Uchida for checking all of the examples and exercises, J. David Jackson for helpful discussions on topics in E&M, Edwin F. Taylor for brilliant thought experiments that clarified our thinking and inspired several end-of-chapter problems, Jim Lockhart and Shirley Chiang for searching out errors in the preliminary edition and the solutions manual, Alma Zook for taking data on vibrating strings especially for this book, Peter Linde for compiling the gas data in Figure 19.6, and our colleagues at SFSU for their continuing support. We especially appreciate the contributions of our guest essayists and their commitment to giving students a perspective on a wide range of careers in applied physics.

An important group of people has been helpful in working end-of-chapter problems to check both problems and solutions for accuracy and clarity. Special thanks go to Chris Kelly, Shuleen Martin, Russ Patrick, Peter Salzman, and Ladye Wilkinson. We appreciate the herculean efforts of Jeremy Hayhurst and his team at Chrysalis Productions, who turned a mountain of manuscript into the two Solutions Manuals. Lauren Fogel, at West, also deserves our thanks for her steadfast work on coordinating the entire Solutions Manual project, and keeping the authors going when we thought no more was possible. And thanks for the brownies, Lauren!

The development of this book has greatly benefitted from the contributions of numerous reviewers who offered their various perspectives and insights. We are sincerely grateful for their ideas and suggestions and offer each of them our thanks: Bill Adams, Baylor University; Edward Adelson, Ohio State University; Clifton Albergotti, University of San Francisco; S. N. Antani, Edgewood College; Paul Baum, CUNY, Queen's College; David Boness, Seattle University; Peter Border, University of Minnesota; Nick Brown, California State Polytechnic University, San Luis Obispo; Michael Browne, University of Idaho; Joseph J. Boyle, Miami-Dade Community College; Anthony Buffa, California State Polytechnic University, San Luis Obispo; Lou Cadwell, Providence College; Bob Camley, University of Colorado at Colorado Springs; D. S. Carlstone, University of Central Oklahoma; Colston Chandler, University of New Mexico; Edward Chang, University of Massachusetts; William Cochran, Youngstown State University; James R. Conrad, Contra Costa College; Roger Crawford, LA Pierce College; John E. Crew, Illinois State University; Gordon Emalbie, University of Alabama, Huntsville; Lewis Ford, Texas A&M University; David Gavenda, University of Texas at Austin; Edward F. Gibson, California State University, Sacramento; Gerald Hart, Moorhead State University; Scott Hildreth, Chabot College; Richard Hilt, Colorado College; Stanley Hirschi, Central Michigan University; Laurent Hodges, Iowa State University; C. Gregor Hood, Tidewater Community College; Ruth H. Howes, Ball State University; John Hubisy, North Carolina State University; Alvin W. Jenkins, North Carolina State University; Darrell Huwe, Ohio University; Larry Johnson, Northeast Louisiana University; Karen Johnston, North Carolina State University; John King, University of Central Oklahoma; Leonard Kleinman, University of Texas at Austin; Claude M. Laird, University of Kansas; Robert Larson, St. Louis Community College; Michael Lieber, University of Arkansas; David Markowitz, University of Connecticut; L. C. McIntyre, Jr., University of Arizona; Howard Miles, Washington State University; Lewis Miller, Canada College; David Mills, College of the Redwoods; Matthew J. Moelter, California State University, Sacramento; Richard Mould, SUNY Stony Brook; Raymond Nelson, U.S. Military Academy; Jack Noon, University of Central Florida; Aileen O'Donoghue, St. Lawrence University; Harry Otteson, Utah State University; Rob Parsons, Bakersfield College; Eric Peterson, Highland Community College; R. Jerry Peterson, University of Colorado; Ronald Poling, University of Minnesota; Richard Reimann, Boise State University; Charles W. Scherr, University of Texas at Austin; Arthur Schmidt, Northwestern University; Achin Sen, Eastern Washington University; John Shelton, College of Lake County; Stanley J. Shepherd, Pennsylvania State University; Gregory Snow, University of Michigan; Kevork Spartalian, University of Vermont; Richard Swanson, Sandhills Community College; Chuck Taylor, University of Oregon; Carl T. Tomizuka, University of Arizona; Sam Tyagi, Drexel University; Gianfranco Vidali, Syracuse University; James S. Walker, Washington State University; Arthur West, Shoreline Community College; Gary Williams, University of California, Los Angeles; John G. Wills, University of Indiana at Bloomington; Lowell Wood, University of Houston.

We also thank all the students at SFSU on whom we have tried out our ideas over the years, and who have used various early editions of this book.

To Andy Crowley, who started with us on this project many years ago: we thank you for your faith in us, and wish you well in your present ventures.

Finally, we owe an enormous debt of thanks to our families, who have endured the enormous piles of paper that have littered our homes for years and who have tolerated our long hours and grouchiness for lack of sleep. We would especially like to thank Jennifer Lampton who gracefully consented to be the guinea-pig on whom we tested our ideas and explanations to see if they made sense. To those friends we haven't seen for ages, perhaps we'll see you soon, and we thank you, too, for your patience.

In Conclusion

In writing this book we have been guided by our students. We have listened to their complaints, watched how they work and noted where they have difficulties. We have also been cognizant of recent research on physics education, which, for the most part, supports our own observations. Thus, this book is written for the student. No book can make physics easy for everyone, but we can show students an approach that works. Our problem-solving strategy has been tested and approved by hundreds of our students, and it has increased their exam scores dramatically. We are confident it can work for your students as well.

CONTENTS IN BRIEF

CONTENTS

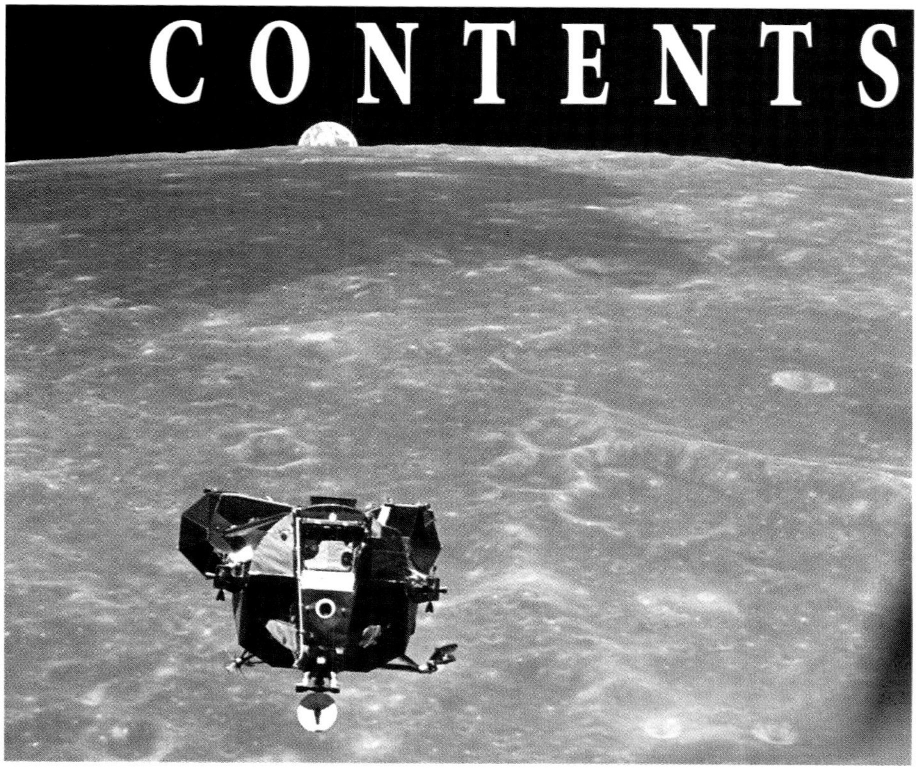

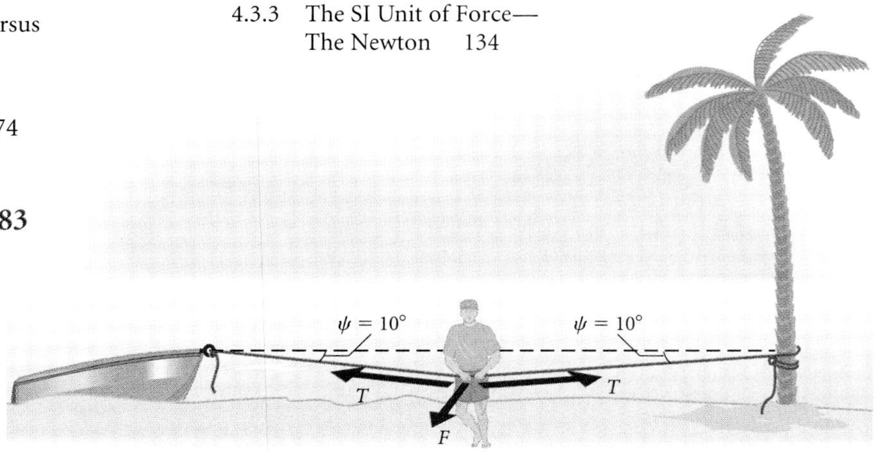

$\psi = 10°$
$\psi = 10°$
T
T
F

BOEING 777
777

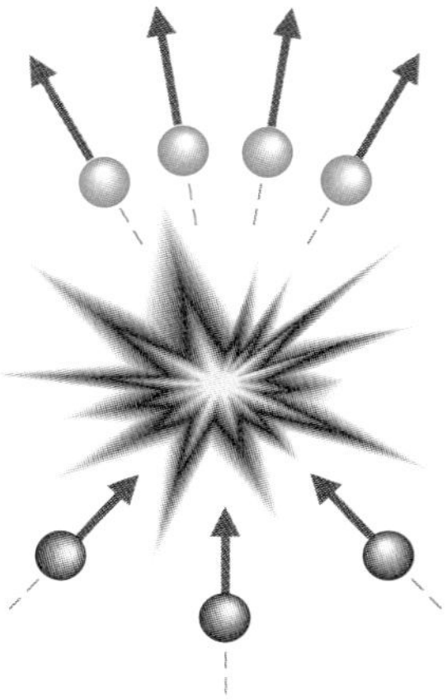

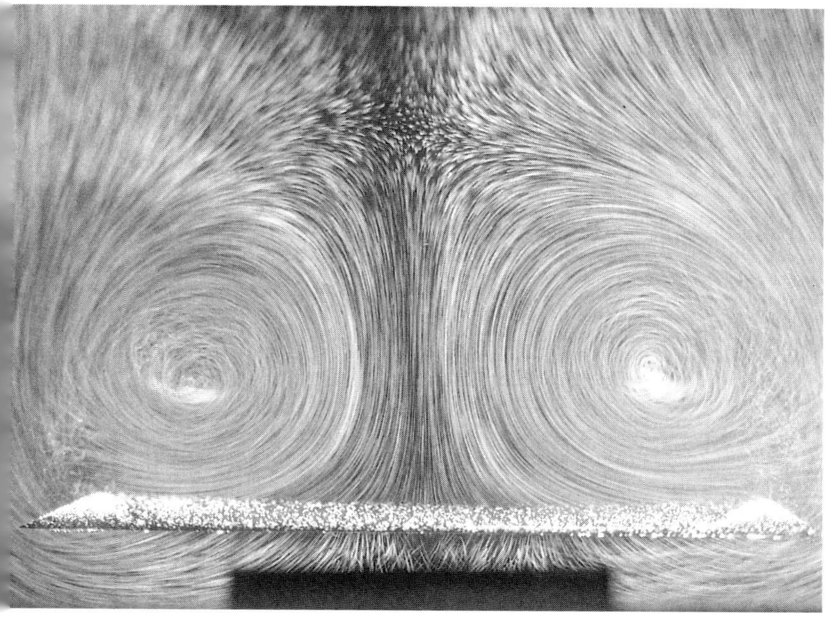

PART THREE CONTINUOUS SYSTEMS

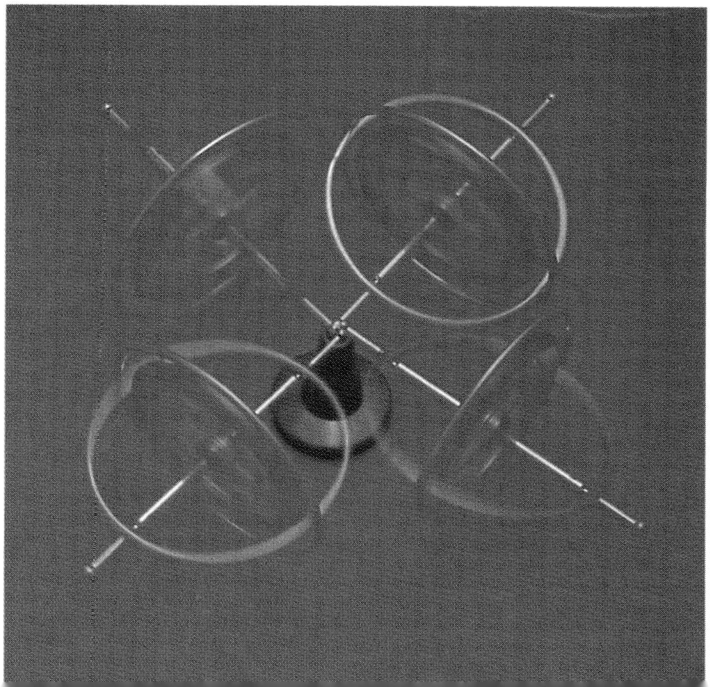

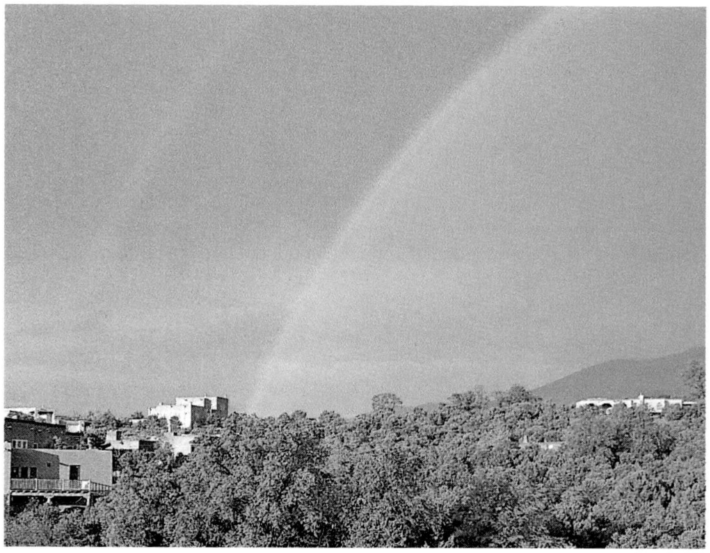

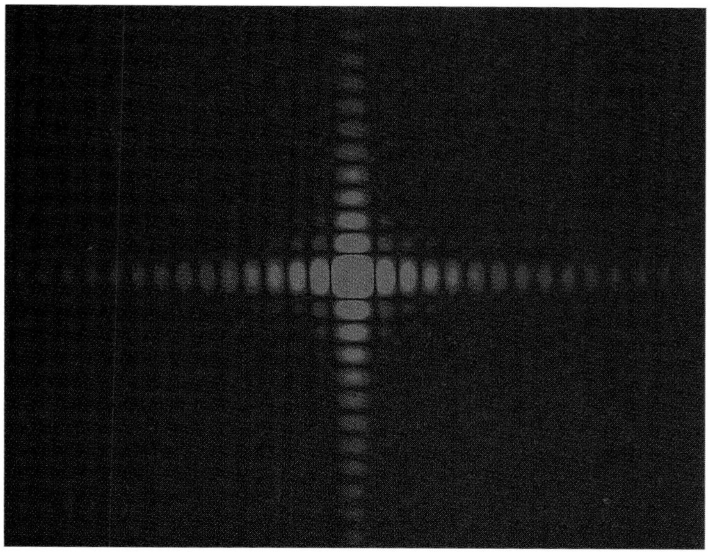

PART FIVE THERMODYNAMICS

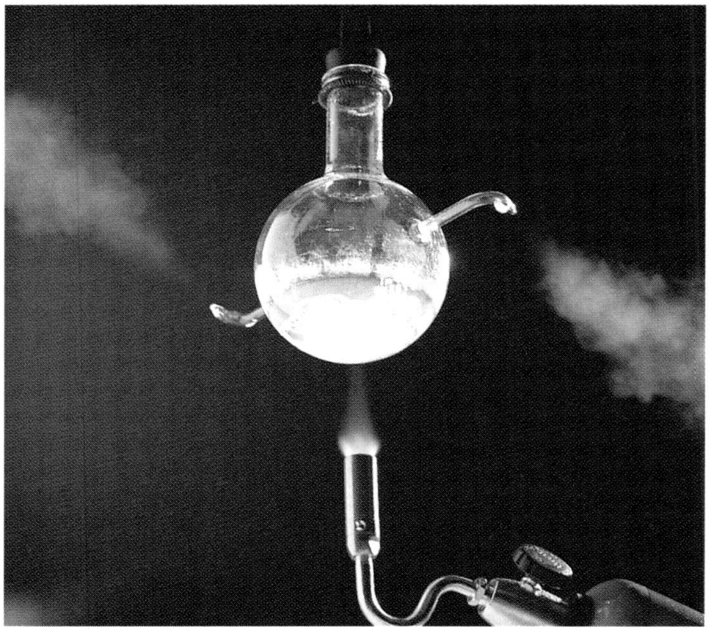

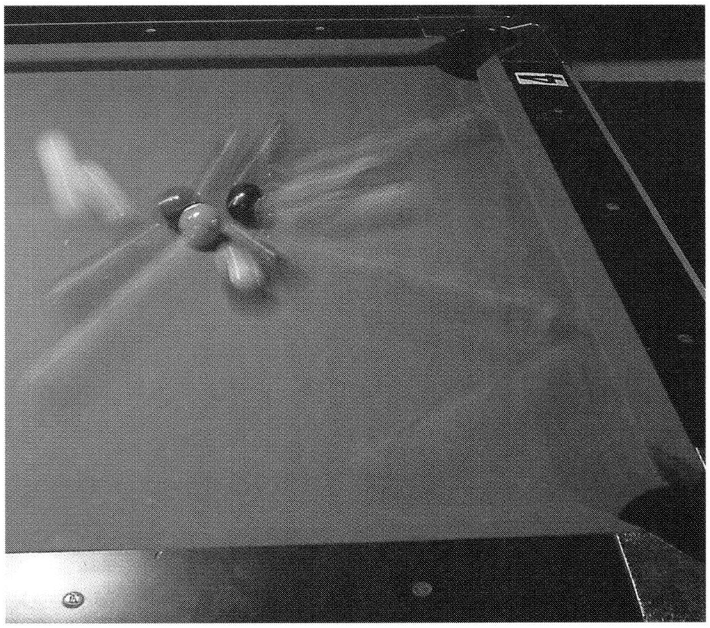

PART SIX ELECTROMAGNETIC FIELDS

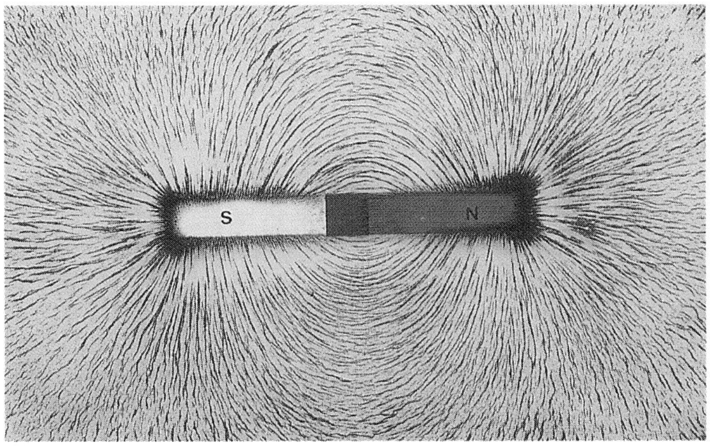

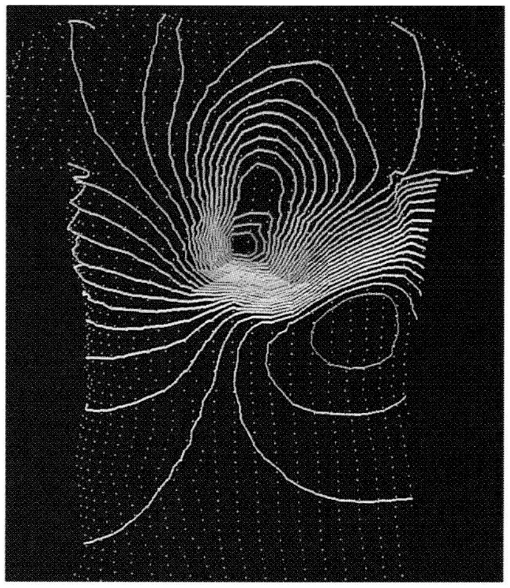

PART EIGHT TWENTIETH-CENTURY PHYSICS

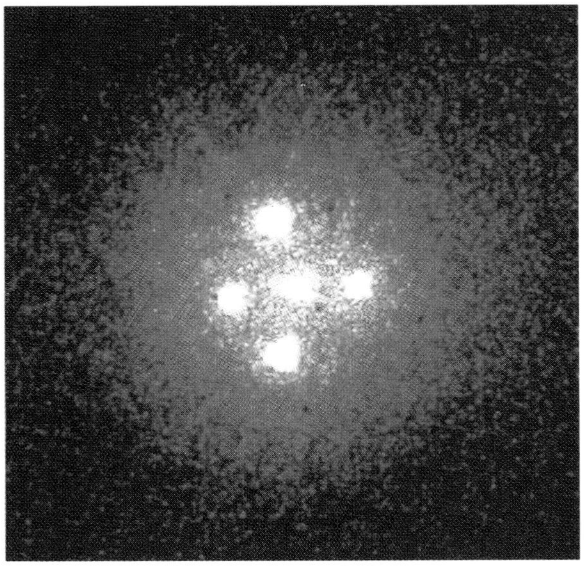

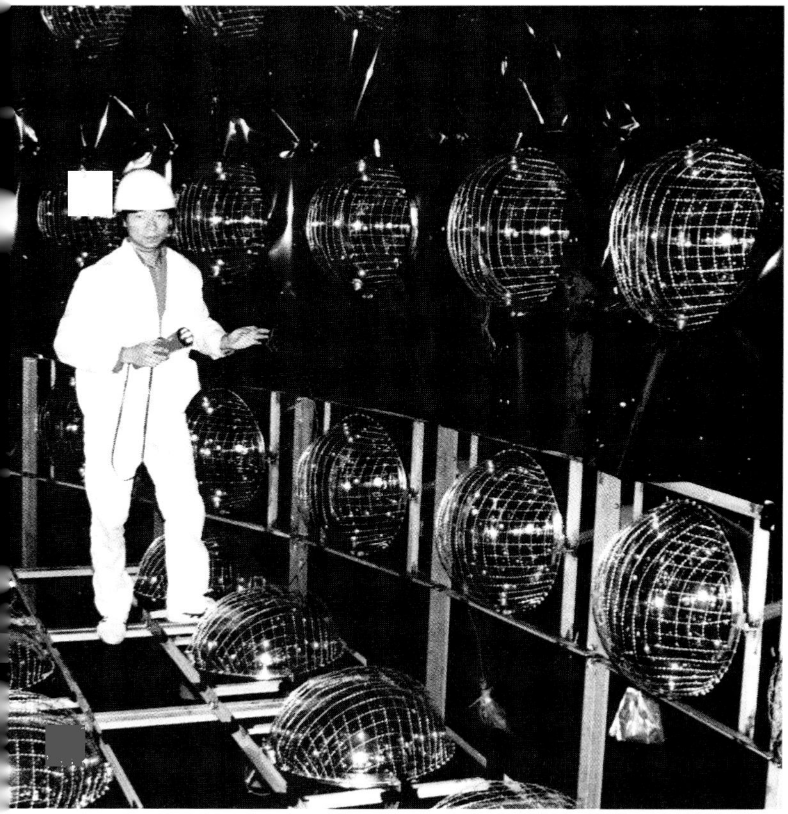

. . . just for the fun of doing Physics.

MARIA GOEPPERT-MEYER

To see the world for a moment as something rich and strange is the private reward of many a discovery.

EDWARD M. PURCELL

. . . If there turn out to be any practical applications, that's fine and dandy. But we think it's important that the human race understands where sunlight comes from.

WILLIAM FOWLER

PROLOGUE

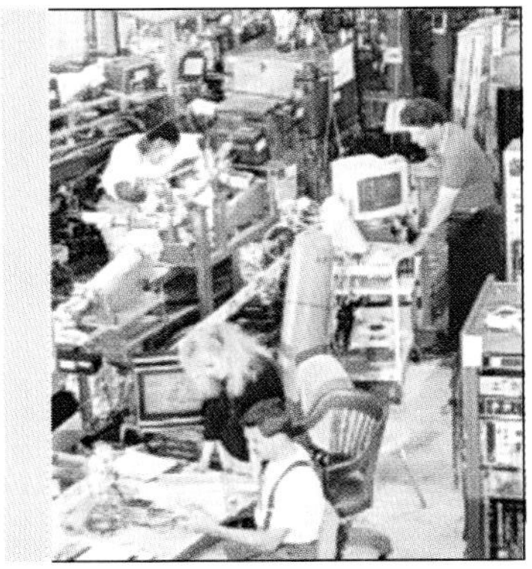

WHY DO PHYSICS?

Reasons for doing physics are nearly as diverse as the people who do it. For the professional, the challenge of teasing secrets from nature is a calling, an exciting occupation, and often a source of profound personal satisfaction. Physicists often view their discoveries as major additions to human culture, not unlike great symphonies or epic poems. Physics is, at the same time, a very practical science—basic to the design of your alarm clock, the computer that handles your bank account, and whatever transportation system gets you to work and school. Most students take a physics course because of this practical aspect.

Every citizen in a modern society needs to have some scientific understanding. The scientific way of thinking about our world has become an integral part of modern culture, interwoven with theories of politics and justice and with the economic structure of our society. Most scientists believe that a scientific worldview liberates the mind and that technological progress will continue to be beneficial. Critics of science argue that it diminishes traditional, humanistic ways of thinking without offering a valid, alternative view, and that technology has left us with problems of pollution, atomic bombs, global warming, to name a few. As both an individual and a citizen, you will need to judge these issues for yourself, and this physics course offers an introduction to the necessary scientific reasoning.

Whatever your reason for studying physics, you will acquire powerful skills that you can use professionally as well as in developing your personal philosophy. Perhaps you will also come to share some of the physicist's deep fascination with the beauty and logic of the universe *and* to enjoy solving its puzzles. Welcome to the enterprise!

SO, WHAT IS PHYSICS?

The name is derived from an ancient Greek word meaning *the nature of things that move of themselves.* Through physics, we strive to discover the fundamental structure of the universe and the rules by which it operates. This structure turns out to be both *simple* and *complex!* It is *simple* because only a small number of rules are needed to explain the world around us. It is *complex* because of the large numbers of objects that interact. We have a good set of rules for the behavior of everyday objects and can understand those rules in terms of atoms and only two kinds of interaction. For atoms we have yet deeper *levels of description* that involve three kinds of interaction. We're pretty sure we haven't reached bottom yet!

Occasionally someone (who should know better) declares to the world that we now know it all . . . then someone else will discover nuclear energy, or semiconductors, or lasers! Physics is a dynamic subject, and physicists continually test the limits of current ideas, probe for exciting new phenomena, attempt to explain puzzling phenomena when they are discovered, and strive to create new ideas that provide deeper or more wide-ranging explanations. The fun of science is in this dynamic quest.

New scientific ideas, as Einstein put it, are "free creations of the human mind," as fresh and unpredictable as any other creative endeavor. But any theoretical picture must be consistent with the actual behavior of the world. So, scientific ideas experience evolutionary pressure as intense as do biological species in a jungle—with similar results: some stable, well-adapted, broad-ranging ideas thrive, while certain variant ideas test the limits of survival. Most of the variants become extinct, but occasionally one proves highly adaptive, takes over the whole environment, and establishes a new level of description. Physics research is the process of creativity, skepticism, and competition that drives this evolution.

Good ideas, unlike dinosaurs, don't always become fossils when a new one takes over their habitat. Most often, a long-lived old idea remains the easiest to learn and use where it is valid, even though it is recognized as a *special case* of the newer and more penetrating idea. For example, a mechanical engineer works almost entirely with mechanical principles obeyed by everyday objects; a metallurgist uses atomic physics to develop stronger metals. Neither would probably ever work with subatomic physics or the theory of relativity.

So, what *is* physics? It is at least three things: a set of ideas describing the universe at various levels of detail; a set of methods for using these ideas to understand the world about us; and a dynamic, evolutionary process for testing, extending, and refining those ideas and methods. The study of physics calls on us to employ a peculiar way of thinking—that of viewing familiar events as the sum of many parts, each governed by the principles of physics and interacting with one another. The term *natural philosophy,* used until recently in Britain, describes physics well: it is a method that has evolved for thinking successfully about the natural world.

What Are the Aims of this Text?

Fortunately, you don't need to master the whole of physics to achieve your purposes in this introductory course. Our main aim here is to help you learn how to become a *natural philosopher*—to understand the structure of physics and to be able to apply it to the world. Like most introductory physics texts, we shall work primarily with classical physics. These ideas, developed largely before 1900, describe most systems on the everyday scale of existence and still find broad application. Though everyday events are familiar and we can study them at a level consistent with your mathematical experience, don't make the error of thinking them trivial. It took 2000 years to get *everyday* physics right, and you will find it a challenge to figure out just how the basic rules work. Once you've met the challenge though, you'll have a method for using physics, for further study of science, or for deciding whether a political candidate takes sound positions on technical issues.

At the beginning of the twentieth century, physicists discovered that phenomena involving strong gravity, objects moving near the speed of light, small numbers of atoms, or low temperature are not well described by classical ideas. The last part of the text introduces you to the modern ideas that have resolved these difficulties and provides a framework for appreciating discoveries at the current frontiers of physics.

We know you will find your study of physics challenging. We hope you will also find it fascinating and rewarding. Good luck!

Suggestions for Using the Text

We have divided the text into eight parts. The chapters in each part form a conceptual unit that will prove useful in organizing your knowledge. We suggest that you read each chapter before attending a lecture on the material. You will understand the lecture better and also be able to ask your instructor about anything that was not clear. Be sure to work the exercises.

Complete solutions are given at the end of the chapter. Peek for hints, but don't just copy them; that doesn't do you much good. The chapter summaries review the major ideas.

The lists of concepts and goals indicate the ideas and methods you should understand after reading the chapter. A wise way to use them is to scan the list as you begin reading so that you know which terms to look for. When you have finished the chapter, go back and be sure you know what each item is about. Then you are ready to tackle the problem set.

The problem set is divided into two parts: *Basic Skills* and *Questions and Problems.* The *Basic Skills* section includes review questions and a basic skill drill. The review questions bring out the main points of the chapter and should be answered with a short quote or paraphrase. The skill drill tests your knowledge of the most fundamental concepts in the chapter. We suggest that you answer all the questions in *Basic Skills,* whether or not your instructor assigns them.

We have provided questions and problems for each chapter section, as well as additional problems for the whole chapter. They are rated according to the following scheme:

CONCEPTUAL ❖
These questions involve primarily verbal and/or graphical discussion. These questions are not necessarily easy!

BASIC ♦
These problems are mostly calculations (more than 10% of the total effort), but ones that involve only a single physical principle from the current chapter.

INTERMEDIATE ♦♦
These problems (except those in the *Additional Problems* category) rely on ideas from the current chapter or ideas encountered so frequently before that they are now taken for granted.

ADVANCED ♦♦♦
Advanced problems may involve subtleties that go beyond the examples and exercises, require more difficult mathematics, take more than one page to complete, or involve ideas from previous chapters. These problems usually involve more than one physical principle.

COMPUTER PROBLEMS are intended to be used with a simple computer program or spreadsheet. Some may be solved graphically, or with a calculator and patience.

CHALLENGE PROBLEMS, at the end of each problem set, require an intricate or subtle argument and/or an expert level of computational skill.

The *Additional Problems* may involve concepts from one or more sections of the chapter, or even from different chapters. The text is divided into eight parts, and you will find a problem set at the end of each part. These problem sets involve material such as might be asked on comprehensive examinations.

THE UNIVERSE: AN OVERVIEW

Small children quickly learn that the world is made up of definite objects with identifiable properties: soft blankets, hard floors, hot water, cold ice. They also learn that certain behaviors are predictable: push your cup off the table and it falls to the floor! As adults, we notice that changes occur because the objects interact with each other. To model this world, we need to classify the kinds of objects that exist and the ways in which they interact. Physicists do this systematically, distilling intuitive experience into a precise and succinct set of ideas, then probing far beyond common experience with carefully designed experiments.

In daily life, we interact with a wide variety of objects more or less similar in size to our own bodies. A description on this scale is completely adequate for a study of mechanics and yields precise methods for problems as diverse as the design of machines or the maneuvering of spacecraft. However, on the everyday scale we find no explanation of why such a huge variety of objects exists or of the reasons for their interactions. Better understanding comes

SEE APPENDIX IA FOR A DISCUSSION OF SCIENTIFIC NOTATION.

from looking at different size scales—different magnifications. For both very large and very small systems, we find a simpler, if weirder, description, although the everyday description, which we shall study first, remains an important and useful approximation. Physicists can now shed light on phenomena with size scales ranging from 10^{-37} meter to 10^{26} meters and can discuss events that occurred as early as 10^{-45} second after a beginning some 10^{10} years ago or that will occur as late as some 10^{100} years in the future. Touring the universe on different length scales will allow us to sample the ideas physicists now use.

The Everyday Scale

■ **New York City.** *We are very familiar with size scales ranging from 1 millimeter to about 10 kilometers—that is, from roughly the size of a grain of sand to the size of a city.*

■ *We are familiar with the sensation of force. Your muscles ache after carrying your physics books around all day. In Part I we'll begin to study the forces we experience in our daily lives.*

■ *A jet aircraft is a good example of modern technology. To build one, you must understand mechanics (Part I), to understand how it flies is an exercise in fluid mechanics (Part III), its engine is a thermodynamic machine (Part V), and plotting its course is an exercise in kinematics (Part I). This picture also shows interesting optical effects due to refraction of sunlight through the jet engine exhaust (Part IV).*

The Solar System

■ *At a size scale of 10^7 meters, things begin to look different. The Earth's surface now appears curved. Gravitational attraction by the Earth is the dominant interaction, accelerating the space shuttle in its orbit around the Earth. The Hubble telescope (Part IV), being refurbished on this shuttle mission, offers us a view of the universe we live in.*

■ *On a scale of 10^9 meters, the Earth's spherical shape is obvious. This view is from Earth's closest natural companion, the Moon, a rocky body similar to Earth but without atmosphere or native life-forms. Both the Earth and the Moon exert gravitational forces on each other and on the spacecraft used to reach the Moon.*

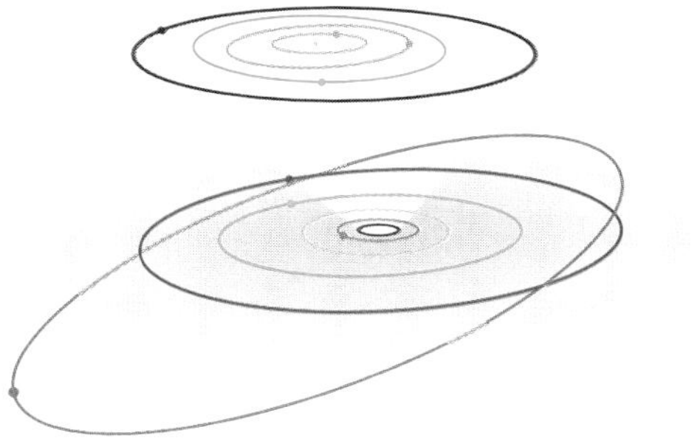

■ *At a scale of 10^{12} meters, the Earth has faded into insignificance, and there is no evidence of the magnificent detail we are so familiar with on Earth. The Sun's gravitational pull holds planets, comets, and sundry other debris in orbits that form the solar system. Isaac Newton's study of the solar system led him to discover that every object exerts a gravitational attraction on every other object (Part I).*

The Universe of Stars

Outside the solar system, there is no trace of human existence. On very large scales, the interactions between objects are simplified, and a single force—the gravitational force—dominates.

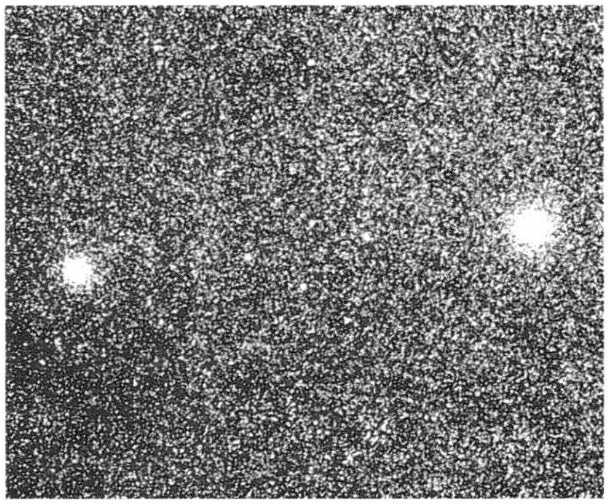

■ *The distance to the nearest star is some 30 000 times the size of the solar system. A cube around the Sun with sides of 10^{18} meters contains about 10 000 stars so distant from one another that their individual gravitational attractions have negligible influence on the motions of the other stars.*

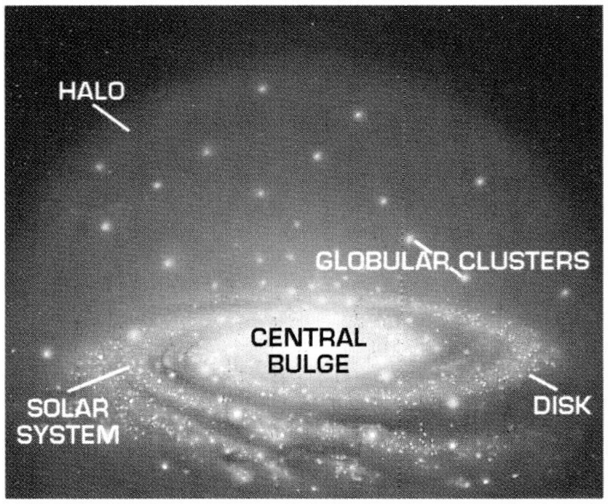

■ *We belong to a galaxy of roughly a thousand billion stars called the Milky Way, a system that is about 10^{21} meters in size. All the stars in a galaxy (plus about ten times as much material that we don't see visually) together exert enough gravitational force to hold individual stars in orbit around the center of the galaxy.*

■ *On a scale of 10^{24} meters, galaxies cluster together, moving under their mutual gravitational attraction.*

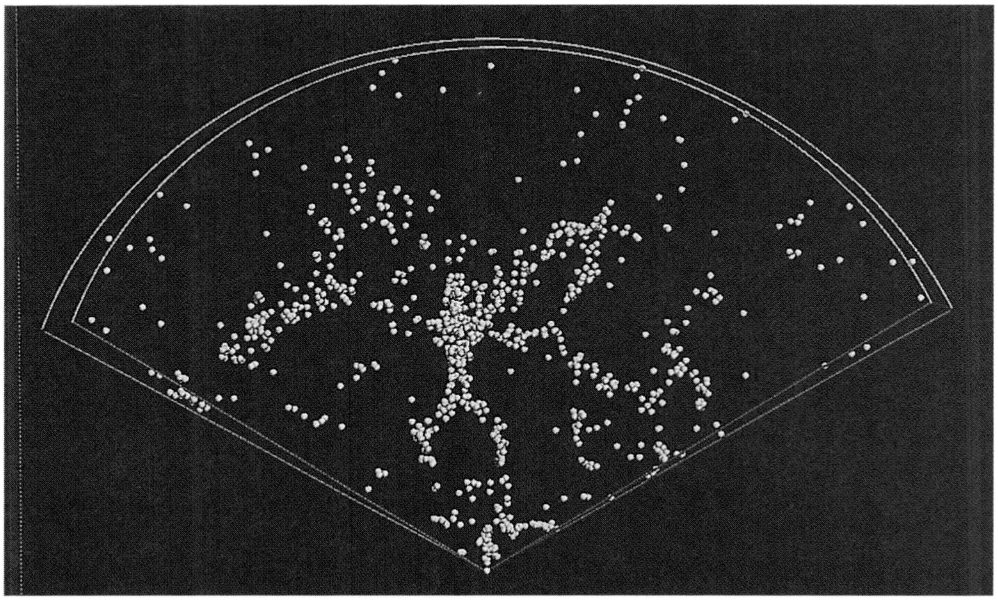

■ *Here we see a plot of galaxy positions within a slice of the universe. The distribution is clumpy, but on size scales larger than about 10^{27} meters, the structure seems to average out. This uniform universe is expanding; all the galaxies are rushing away from each other. Albert Einstein's concept of gravity as variations in the geometry of space and time (Part VIII) explains this expansion, but current observations cannot yet determine whether the expansion will stop or continue forever. In the past, the part of the universe we can see must have been much smaller. Cosmological theories suggest that, between 10 and 20 billion years ago, all the stars and galaxies we can see were squeezed into a volume the size of a single atomic nucleus.*

The World as Atoms

As with the large scales of astronomy, our description of the world changes radically when we look at very small size scales. Again, we find odd and wonderful things and a small number of fundamental interactions.

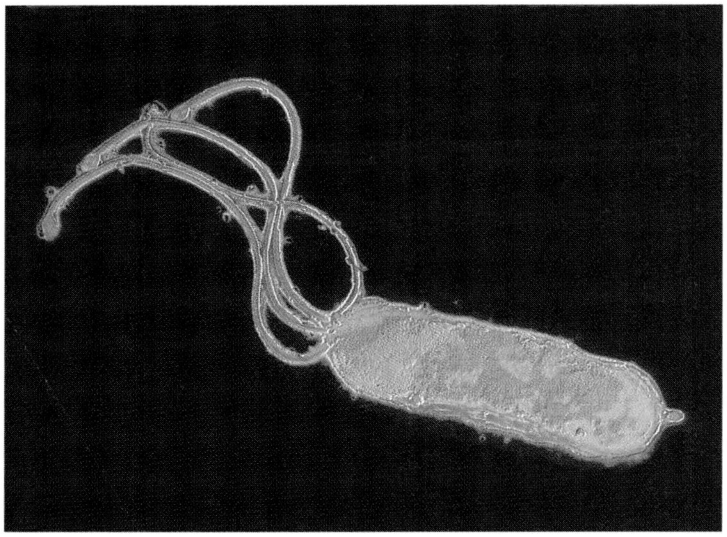

■ *Single cells of living creatures are several micrometers (10^{-6} meter) in size. Although we cannot see them with our own eyes, we can still comprehend their behavior with concepts from the everyday world.*

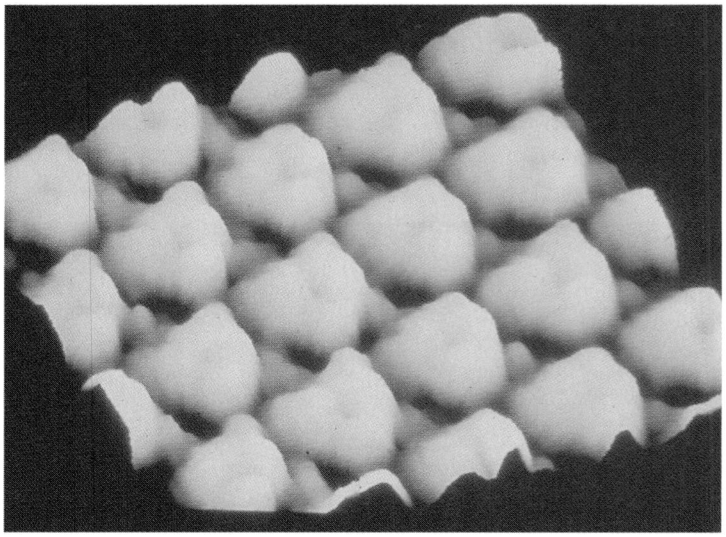

■ *At a scale of 10^{-9} meter, we observe the atomic nature of matter. Fluid forces result from collisions between rapidly moving atoms of gas or liquid. Forces between solid bodies in contact result from the forces between individual atoms in the surfaces of the bodies. The atoms consist of electrons, with negative electric charge, surrounding small, positively charged nuclei. Interatomic forces are electromagnetic forces between these charged pieces of the atoms. All of the kinds of force we experience on the everyday scale are either gravitational or result from electromagnetic interactions between atoms (Part VI). This photo shows benzene molecules.*

The Subatomic World

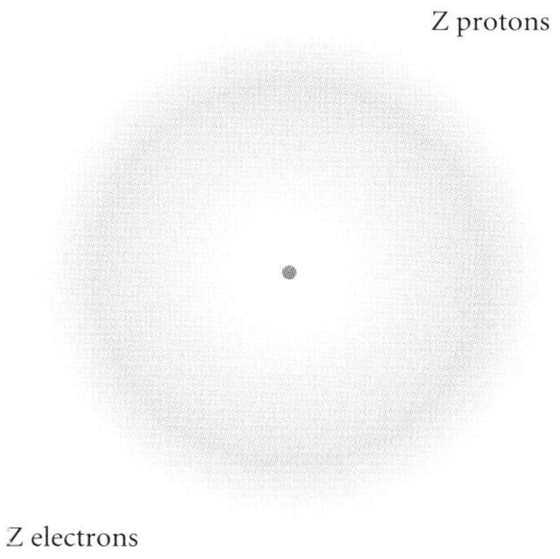

■ *Nearly all the mass of an atom is concentrated in the nucleus, pointlike compared with the size of the atom, 10^{-10} meter. The volume of the atom is filled by much less massive electrons, which form a cloud around the nucleus. (Part VIII)*

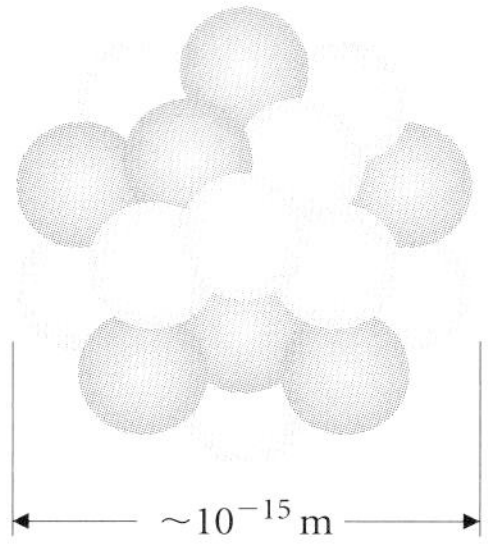

■ *On a scale of 10^{-15} meter, we become aware of the electrically neutral neutrons and positively charged protons that comprise the nucleus. The protons repel each other electrically and are held together by the strong nuclear force. A second nuclear force, the weak nuclear force, causes some nuclei to change form.*

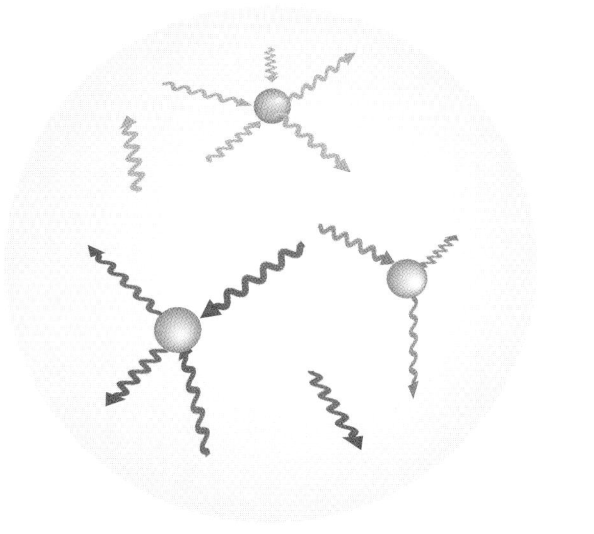

■ *Nuclear particles themselves have structure. A proton consists of three particles called quarks, which exert strong nuclear forces on each other by exchanging particles called gluons, and exert electromagnetic forces by exchanging photons. The quarks also exert weak forces through the exchange of particles. In the 1970s, these particles were shown to be cousins of the photons. In this sense, there is but one "electroweak" kind of force, rather than separate electromagnetic and weak nuclear forces. Theorists are now trying to show that the electroweak and strong nuclear forces are just different aspects of one force. Yet more intriguing is the possibility that gravity and this unified force may be aspects of a single interaction. An experimental test of these ideas is far beyond current techniques. Because only these most fundamental particles could exist at the beginning of the universe, the way they behave may be responsible for the way the universe is today. In this way, the smallest and largest scales are intimately connected.*

Summary Chart

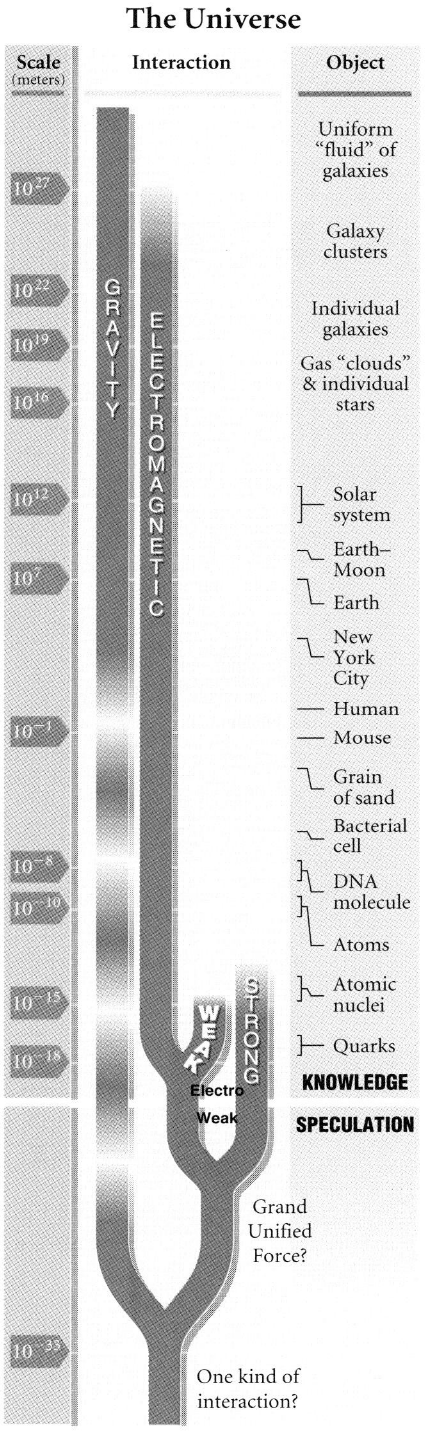

■ *The most important forms of material substance are listed for each size scale. For the fundamental types of force, solid lines denote scales at which the force is of major importance. Fuzzy lines indicate scales at which a particular kind of force is present but relatively unimportant.*

Newtonian Mechanics

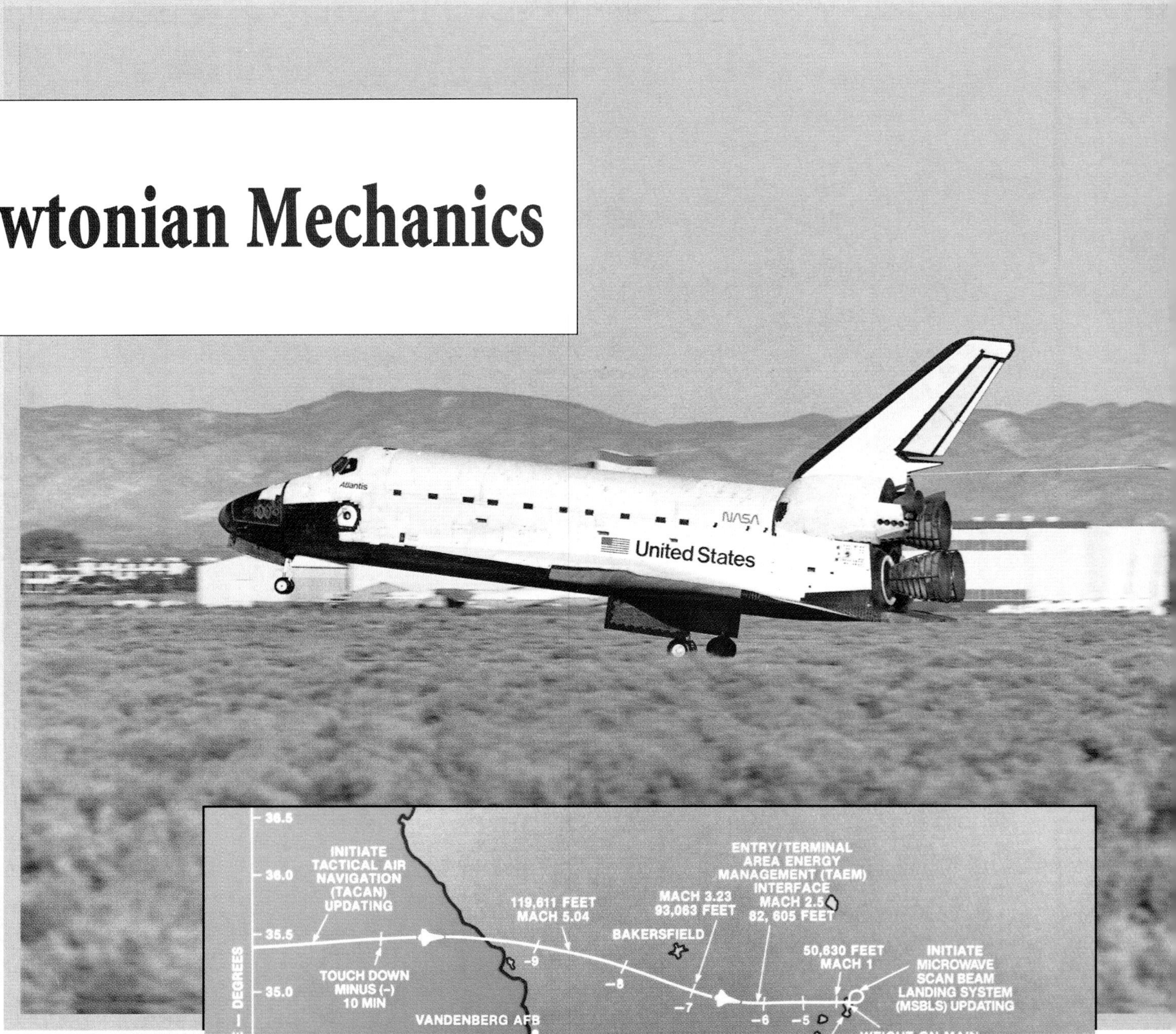

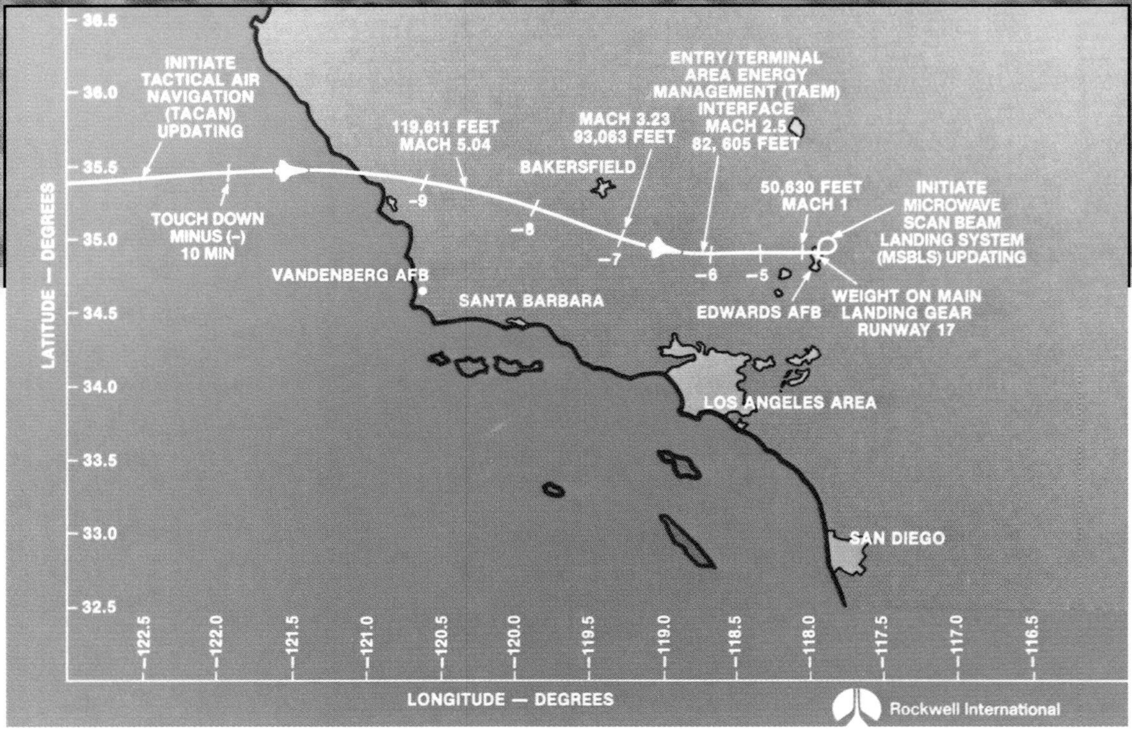

Volume Two

Physics: The Nature of Things

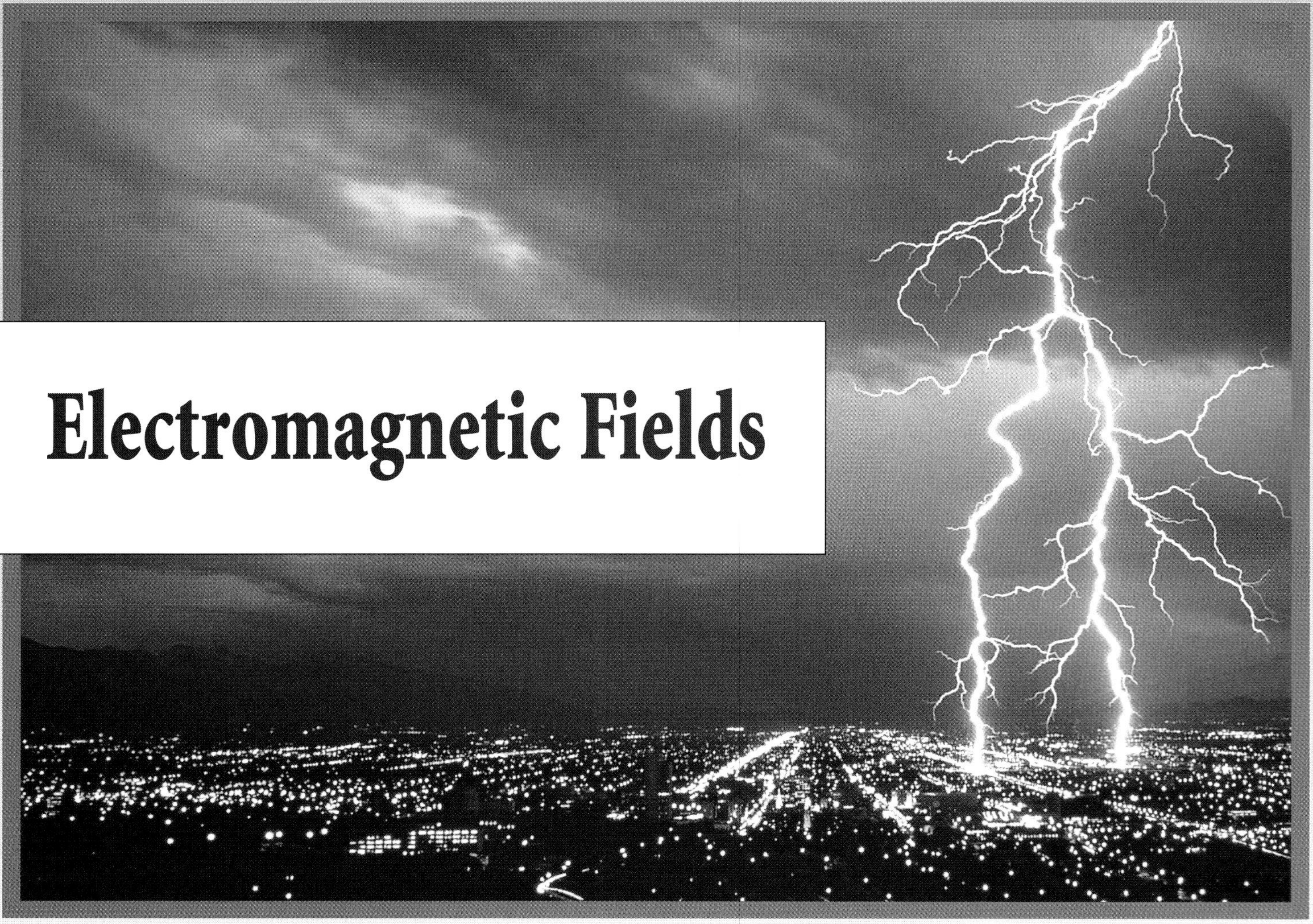

Electromagnetic Fields

Fear no more the lightning flash,
Nor the all-dreaded thunder stone;
WILLIAM SHAKESPEARE

The mechanical systems we studied in the first five parts are governed by Newton's laws of motion and the laws of thermodynamics. In Newtonian mechanics, the primary attribute of stuff that determines its behavior is *mass,* and the quantity that changes behavior is force. In Part I we discussed various different kinds of forces, such as friction and tension, but we presented a fundamental explanation only for gravity. Mechanical forces like tension in a rope result from electromagnetic interactions between the atoms that make up the rope. The laws that govern those interactions are the subject of the next two parts.

Until the last half of the eighteenth century, electricity was known as a curiosity, capable of producing entertaining effects. Magnetism, important because of the use of magnetic compasses in navigation, seemed to be quite distinct from electricity. But in the early nineteenth century, electric and magnetic phenomena were found to be closely related, and in the 1860s James Clerk Maxwell presented a unified theory of electricity, magnetism, and optics. Electricity and magnetism are different aspects of a single interaction that not only accounts for the nature of mechanical forces like friction but is also important for determining the laws of chemistry and understanding light. Applications of Maxwell's theory have caused profound revolutions in technology and communications and, through the work of Einstein, the theory has led to a fundamental revision of our concepts of space and time.

The unification of apparently diverse phenomena under a single theoretical umbrella is a major theme of modern physics. To study these phenomena, however, we have to break the theory into small pieces and learn them one at a time. Traditionally you first learn about static electricity, and weeks later you learn about magnetism. The coherence is lost. Our goal in this essay is to give a sense of the unity of electromagnetic phenomena.

Like mass in Newtonian mechanics, another fundamental property of matter—*electric charge*—is the basis of electromagnetism. Charge both produces electromagnetic (EM) fields and determines how objects respond to EM fields. Unlike particles, which are localized, the fields they produce extend throughout space. Fields play an essential role in electromagnetic phenomena, and their nature and description are major themes of this part.

SEE THE DISCUSSION OF THE GRAVITATIONAL FIELD IN ESSAY 2.

PART SIX

OVERVIEW OF ELECTROMAGNETISM

VI.1 Magnetic Field

Most people are familiar with the magnetic compass and its use in navigation. The word *magnetism* derives from the province of Magnesia, a region in Asia Minor with substantial deposits of the mineral magnetite. The ancient Greeks had noticed that pieces of this mineral, *lodestones,* would attract iron. Furthermore, a lodestone suspended from a string orients itself along a north-south line, providing a practical navigational device. The first documented use of a magnetic compass was in China in the eleventh century A.D., though both Chinese and Viking explorers probably used compasses as early as the sixth century. The magnetic compass remains a primary navigational tool (■ Figure VI.1).

By the thirteenth century it was known that magnets can exert both attractive and repulsive forces on each other (■ Figure VI.2). The end of a magnet that points north is called a *north-seeking, north,* or *positive* pole. The opposite end is a *south-seeking, south,* or *negative* pole. If we try to hold two magnets with their north poles or with

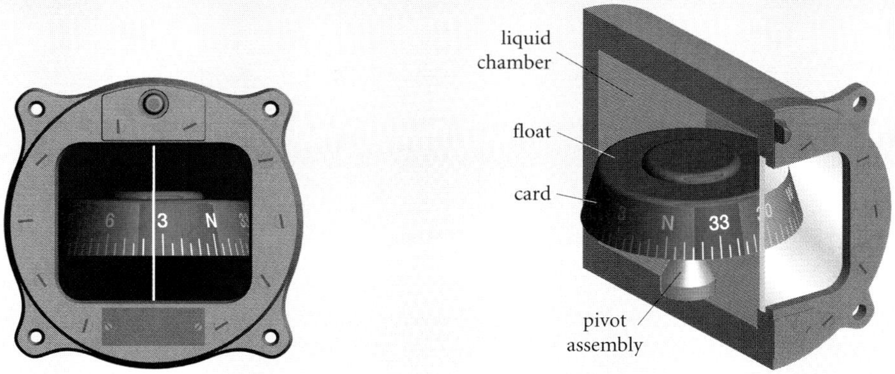

■ FIGURE VI.1
Navigational compass used in an aircraft. A small magnet floats in liquid and rotates so that its north-seeking pole points toward magnetic north.

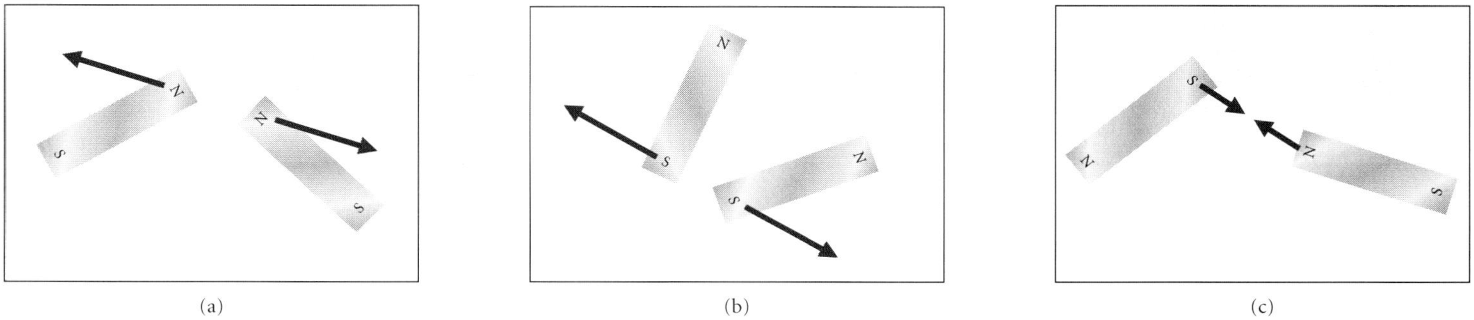

■ FIGURE VI.2
Magnetic forces exerted by bar magnets. Like poles repel (a and b); unlike poles attract (c). You can easily verify these results for yourself with magnets from your refrigerator door.

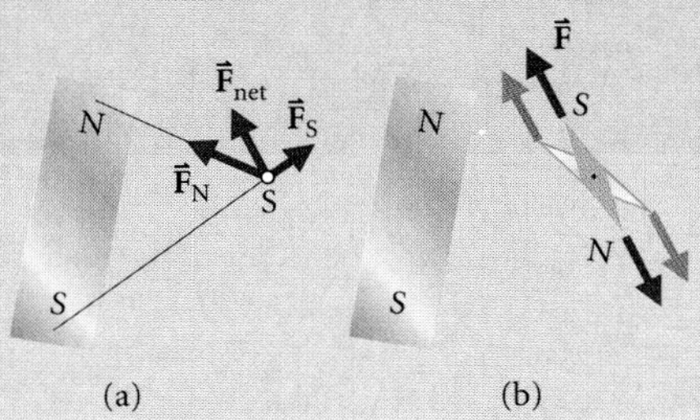

■ FIGURE VI.3
A compass needle placed near a magnet indicates the direction of the magnetic field at that point. (a) The needle's south(-seeking) pole is attracted to the magnet's north pole and repelled by the magnet's south pole. (b) Forces on the compass needle's north and south poles are equal and opposite. The resulting torque causes an angular acceleration of the needle. Once the resulting oscillations damp out, the compass needle is aligned with the magnetic forces.

their south poles together, they repel each other, while a north and a south pole attract each other.

Like poles repel; unlike poles attract.

When a small compass needle is placed near a magnet, its south pole is attracted to the north pole of the magnet and repelled by the south pole (■ Figure VI.3). Opposite forces act on the north pole. The needle rotates until the net torque on it is zero, and it is parallel to the direction of the net force on each pole. Iron filings scattered on a sheet of paper near the magnet act like small compass needles that show the influence of the magnet on its surroundings—the magnetic field (■ Figure VI.4). The filings trace out lines from the north to the south pole of the magnet, forming a pattern called a *dipole field* pattern.

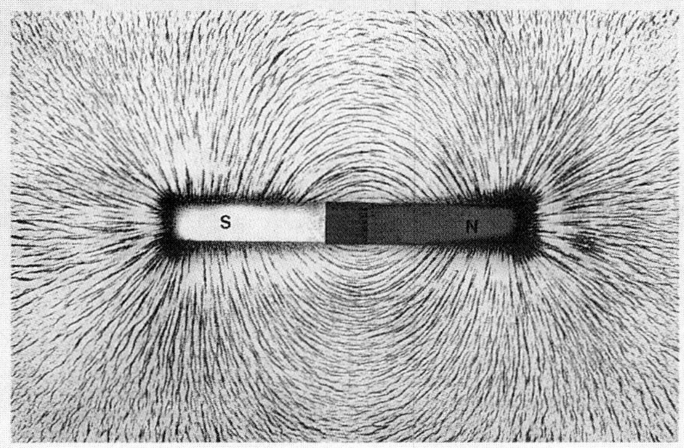

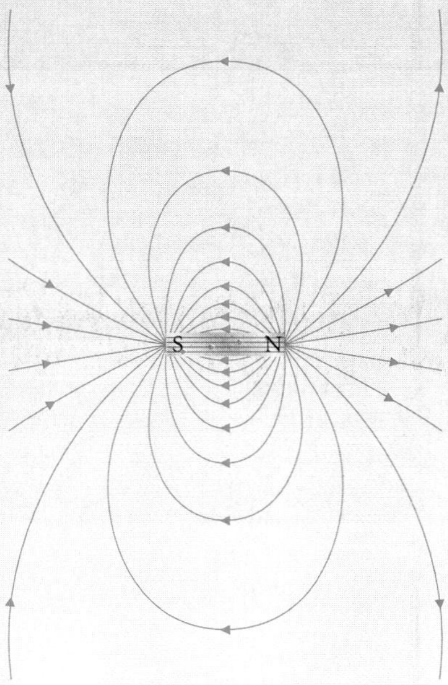

■ FIGURE VI.4
(a) Iron filings act like tiny compass needles. Each aligns with the magnetic forces acting on it. Together they trace out the pattern of the magnetic field.

(b) Computer-constructed drawing of the field lines of a bar magnet. Compare with the distribution of iron filings in Figure VI.4a. Each field line loop is closed by a line running from S to N within the magnet.

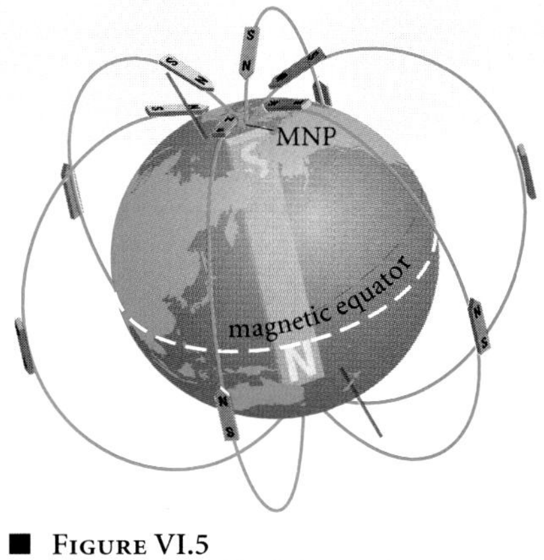

■ FIGURE VI.5
The magnetic field of the Earth is very nearly a dipole field. Magnetic rocks produce deviations from dipole behavior in local regions.

The first systematic investigations of electromagnetic phenomena were conducted by the Englishman William Gilbert (1540–1603), who distinguished electric and magnetic forces and described the behavior of a compass needle near another magnet. Gilbert realized that the Earth itself behaves like a great magnet, thus explaining how a compass senses north. A dipole pattern of field lines originates deep within the Earth (■ Figure VI.5), and a compass suspended freely aligns itself with those field lines. The magnetic field isn't aligned with the Earth's rotation axis, so a compass doesn't point exactly toward geographic north. Instead it points toward magnetic north, currently a point in Canada near Baffin Bay. The naming of magnetic poles arises historically from the use of the compass. The point that cartographers call the *North Magnetic Pole* of the Earth is the point where magnetic field lines converge as they enter the Earth. A compass placed parallel to the surface there would not point in any particular direction. If the Earth's field were exactly a dipole field, this point would actually mark the location of a south-seeking, or negative, magnetic pole, since a compass' north-seeking, or positive, magnetic pole points toward it.

The Earth's magnetic field arises from motion of the liquid rock in the core of the Earth. The field is constant over short time scales, but over decades it changes in both magnitude and direction. Over geologic time scales the poles wander about, and the field reverses direction (■ Figure VI.6). The last reversal occurred about 100 000 years ago.

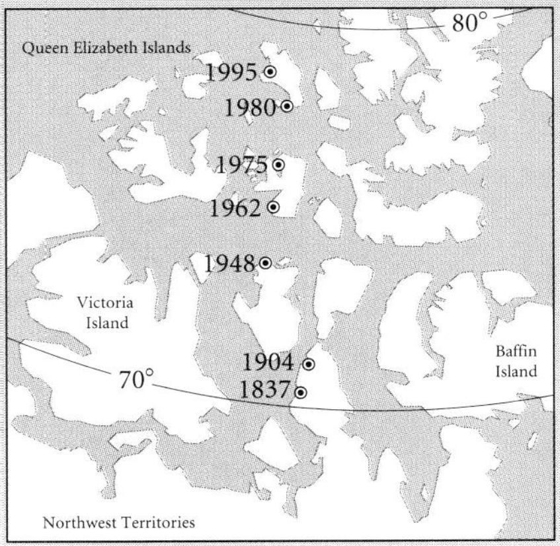

■ **Figure VI.6**
The Earth's magnetic poles wander about with time. Magnetic signatures in certain kinds of rock allow us to reconstruct the path of the poles over time.

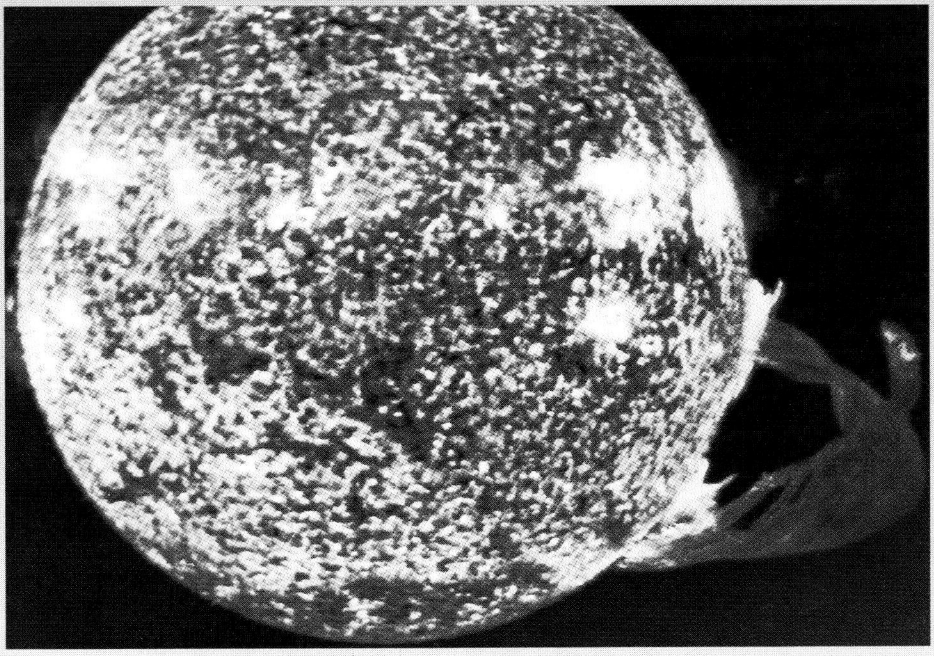

■ **Figure VI.7**
In a solar prominence we may observe the paths of charged particles that follow the magnetic field lines (see Chapter 29). The magnetic field emerges from the Sun in great loops.

The Sun displays some of the most visually dramatic magnetic phenomena. In a solar prominence (■ Figure VI.7) great loops of magnetic field erupt from the surface, and material flows along the field lines. Even more energetic outbursts on the Sun expel particles that would prove fatal if we were not shielded by the Earth's magnetic field. Most of the particles approaching the Earth are deflected away from the surface by the Earth's field, which is itself swept out into a long tail (■ Figure VI.8). Two bands of trapped particles, called the Van Allen belts, pose a radiation hazard to astronauts and leak particles into the atmosphere to produce the aurora.

What causes magnetic fields? Magnetic poles would have been William Gilbert's answer, but that answer cannot be correct. If you cut a magnet in half to isolate a magnetic pole, each piece ends up with both a north and south pole (■ Figure VI.9). It appears that a dipole is the simplest magnetic field source. So what produces a dipole? To answer this question, we must first understand some electric phenomena.

VI.2 Electric Charge

Lightning is a common and spectacular electric phenomenon that results from an electric field set up between the thundercloud and the ground. The electric spark that leaps from your finger to the doorknob after you shuffle over the rug on a dry day differs from lightning

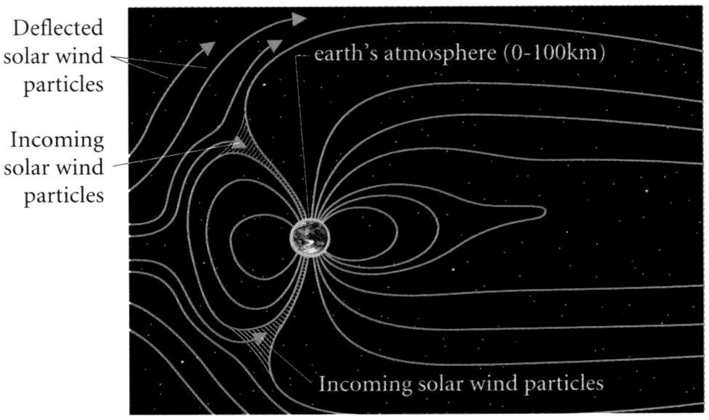

■ **Figure VI.8**
The Earth's magnetic field, which looks like a dipole near the surface, is drawn into a long tail by particles streaming from the Sun. Some of these particles are trapped in the field and contribute to phenomena such as the Van Allen belts and aurora.

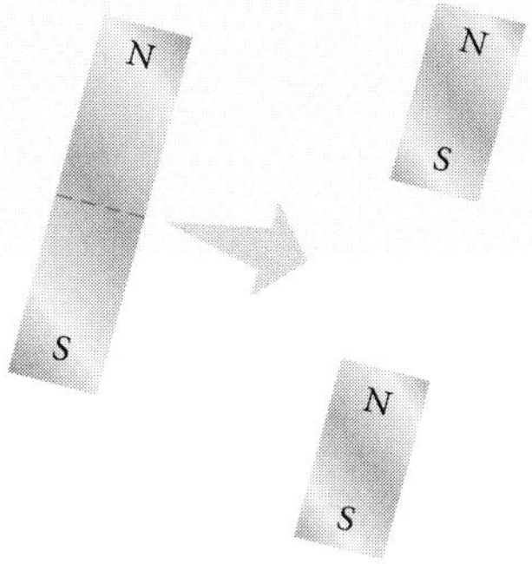

■ **Figure VI.9**
If we attempt to determine the source of magnetic fields by cutting a bar magnet in half, we obtain two smaller bar magnets. Magnetic poles always come in north-south pairs.

■ FIGURE VI.10
Benjamin Franklin (1706–1790). Born in Boston in 1706, Franklin's early years were spent as an assistant in his father's soap and candle factory, and as an apprentice in his brother's printing shop. By the age of 40 he was able to retire with a comfortable income from his own printing business in Philadelphia. Fascinated by a public lecture, he bought the lecturer's equipment and began his own research in electricity. His immense success as a scientist made him a natural choice as a representative of the American colonies in London and later as ambassador to France. His rustic American ways charmed the French court, but his suggestion for saving candles by getting up at dawn did not prove popular!

■ FIGURE VI.11
A lightning rod. The sharply pointed rod is connected to ground. As a thundercloud approaches, charge leaks from the ground into the air through the rod's point. The discharge is rapid enough to reduce the risk of a lightning strike. If the connection to ground is broken, the rod becomes a hazard. Instead of reducing the nearby concentration of electric charge, it becomes a relatively easy path for the lightning discharge.

primarily in the amount of energy released. Benjamin Franklin, ■ Figure VI.10, suggested the first experiments that showed the similarity between lightning and electric sparks. Franklin transformed the study of electricity from a display of freakish and amusing phenomena to a scientific discipline with great practical importance. His invention of the lightning rod, ■ Figure VI.11, was the first major practical application of the new discipline.

ABOUT 600 B.C., CF. CHAPTER 0.

The root *electr* derives from the Greek word *elektron,* meaning amber. Thales of Miletus observed that when amber is rubbed, it attracts other small bodies to itself. A few simple experiments illustrate the important features of this electric force. If we stroke a rubber rod with cat fur and hold it near a small ball suspended by a piece of thread, the ball is attracted to the rod (■ Figure VI.12). If the rod and ball are allowed to touch, the ball is then repelled. Afterwards, the ball is attracted by the cat fur. These experiments show that electric forces can be either attractive or repulsive and suggest that something is transferred between the rod and ball to cause the change from attraction to repulsion. That something is the quantity that allows us to understand electric phenomena: *electric charge.* There are two kinds of electric charge, the kind on the rod and the kind on the cat's fur.

Benjamin Franklin first named the two kinds of charge. As a result of his experiments, Franklin developed a theory of electric fluid that could flow from place to place: an object with excess fluid he called *electrised positively,* and one with a lack of fluid he called *electrised negatively.* The fluid was thought to attract ordinary matter but repel itself. His theory could not explain the repulsive forces between two negatively charged objects. Now we recognize two kinds of charge, still called positive and negative. Negative charge is associated with electrons that move more easily than the more massive, positively charged atomic nuclei. Usually objects are charged by transfer of negatively charged electrons, the opposite of Franklin's conjecture.

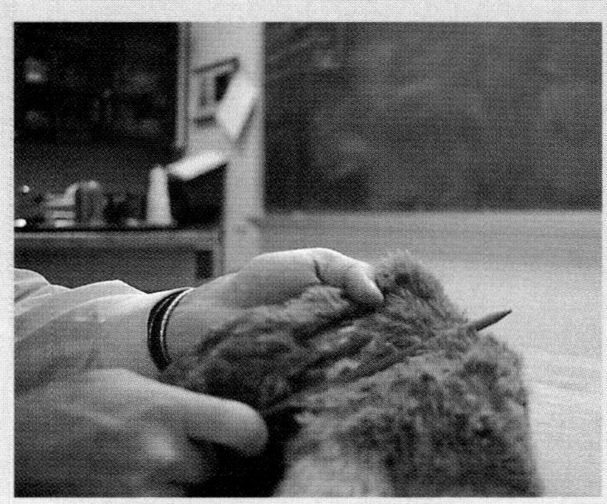

(a)

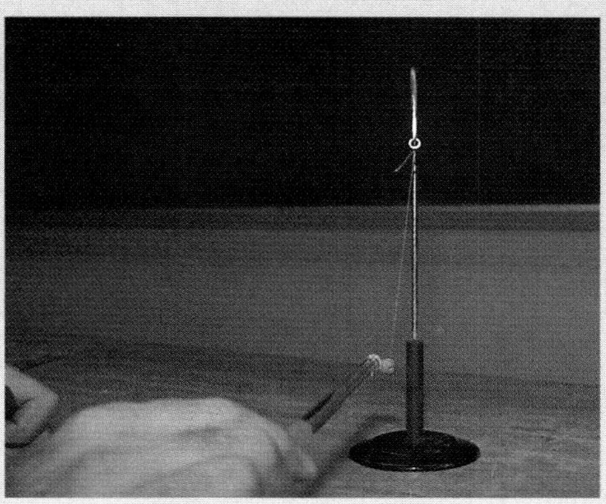

(b)

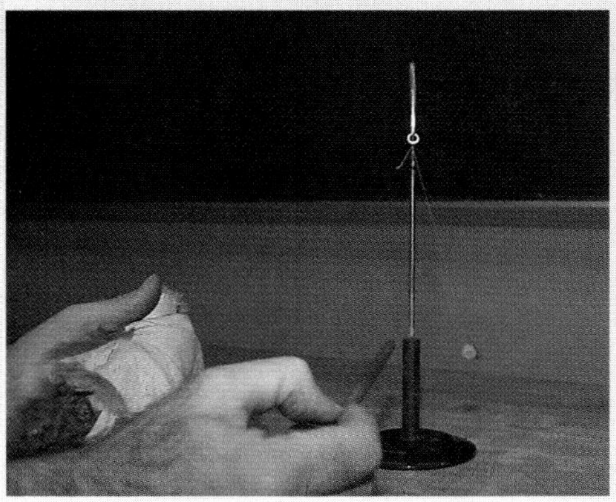

(c)

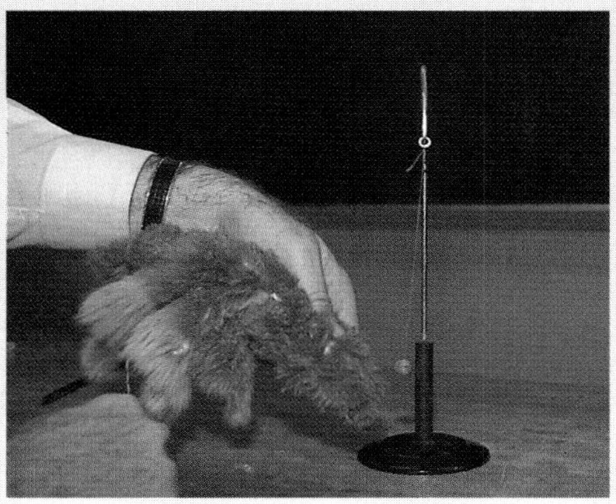

(d)

■ FIGURE VI.12
Experiment that demonstrates the existence and nature of electric charge. (a) A rubber rod is stroked with cat fur. (b) The rod attracts a small pith ball. (Pith is a light material found in the stems of certain plants.) (c) If the rod is allowed to touch the ball and is then removed, the ball is subsequently repelled by the rod. (d) The ball is also attracted to the fur. When the rod is rubbed on the fur, charge is transferred between them, leaving the fur and rod oppositely charged. When the charged rod touches the ball, some charge is transferred to the ball. Then the ball and rod have the same kind of charge and repel each other. The ball is attracted to the oppositely charged fur. (The reason why the ball is initially attracted to the rod is discussed in Chapter 27.)

VI.3 The Electrical Structure of Matter

To most of us, electric charge is an elusive concept. We never come across any sizable amount of it in the natural course of things, and so we have little intuition about it; yet it does fantastic things—light lamps, move trolleys, and shoot thunderbolts across the sky. It is astonishing to learn that electric interactions are 10^{40} times stronger than gravity! The elusive behavior of something so powerful arises from its occurrence in balancing positive and negative forms. We notice gravitational force because we live on an astronomical-sized chunk of mass. In stark contrast, it takes ingenuity to tease apart even a tiny amount of the Earth's electric charge. Ironically, the very strength of the electric force binding positive and negative charges together ensures that electric forces play a less obvious role than gravity in our everyday lives.

EVEN A SINGLE ELECTRON IN A HYDROGEN ATOM FORMS A *CLOUD*. WE'LL DISCUSS THIS BIZARRE NOTION IN CHAPTER 35.

There is a close connection between charge and matter. Just how close only became clear early in the twentieth century with the triumph of the atomic theory. An atom consists of a nucleus surrounded by a comparatively large cloud of electrons (■ Figure VI.13). The nucleus is made up of particles called protons and neutrons. Each electron has a negative charge, given the traditional symbol e.

Charge of an electron $= -e$.

Each proton carries a positive charge of the same magnitude.

Charge of a proton $= +e$.

Neutrons are electrically *neutral;* they carry no net electric charge.

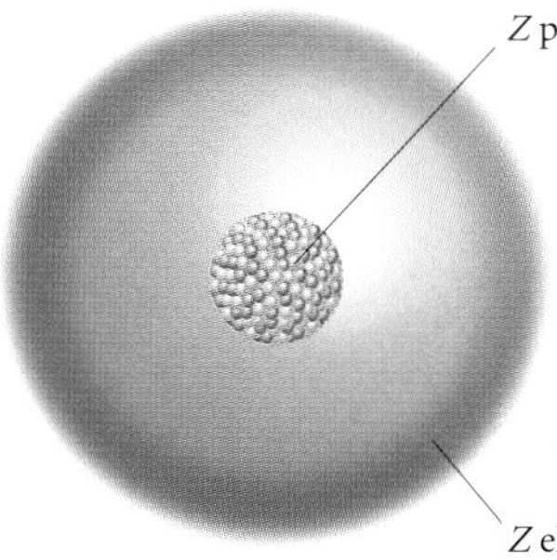

■ FIGURE VI.13
Schematic of an atom. Positively charged protons and uncharged neutrons crowd together in the nucleus. The less massive, negatively charged electrons form a cloud around the nucleus. The size of the electron cloud relative to the nucleus is much larger than we can show here: $r_{\text{cloud}}/r_{\text{nucleus}}$ is about 10^5. (This diagram does not attempt to show the electron distribution accurately. For more details, see Chapter 35.)

Z IS AN INTEGER.

A typical atom consists of *Z* protons and *Z* electrons, which determine the chemical nature of the atom (e.g., if $Z = 8$, it is an oxygen atom). The atom also contains a number of neutrons. Different isotopes of the same element have different numbers of neutrons (e.g., ^{16}O has 8 neutrons, while ^{18}O has 10). The patterns of the electrons around atomic nuclei, and the ways electrons can be added to, stripped from, or shared between atoms, determine the chemical properties of matter.

Each proton or neutron has roughly 2000 times the mass of an electron, so an atom's mass is concentrated in a tiny volume (its nucleus). All the *solid* objects that surround us are mostly vacuum. It is the electrical interactions that bind the structure together and produce its solid feeling.

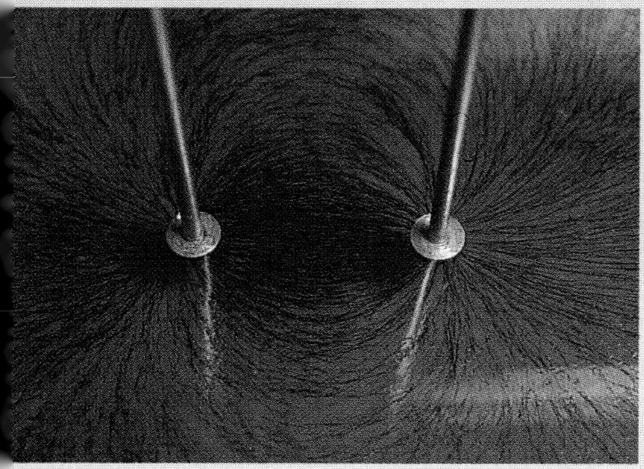

■ FIGURE VI.14
Small fibers suspended in an electric field behave like iron filings in a magnetic field: they align themselves with the field. The field line pattern produced by a positive charge and an equally large negative charge with a small separation looks like the pattern of a magnetic field produced by a bar magnet. This charge configuration is called an electric dipole. Charges in the individual fibers are slightly separated, so that each fiber itself behaves like a tiny electric dipole. A field line diagram for this system is also shown in Figure 23.14.

VI.4 Electric Field

In ■ Figure VI.14 we see small fibers suspended in a fluid surrounding two concentrations of electric charge, one positive and the other negative. This charge configuration is called an electric dipole. The fibers lie along a pattern of lines strikingly similar to the pattern of magnetic field lines surrounding a magnet. The dipole influences its environment by producing an *electric field.* The electric field tugs on the electric charges in the atoms of the fibers, separating them slightly and forming small electric dipoles that respond to the electric field just as a compass needle responds to a magnetic field. Electric forces arise from a two-step process. One electrically charged object produces an electric field, while a second charged object, exposed to that field, experiences a force. The gravitational forces exerted by the Earth may be modeled similarly: the Earth produces a gravitational field described at each point by the acceleration due to gravity $\vec{\mathbf{g}}$. The force exerted on an object is given by its mass multiplied by the gravitational field vector at the object's location:

$$\vec{\mathbf{F}}_{\text{grav}} = m\vec{\mathbf{g}}.$$

The electric field is also described by a vector at each point in space, called the electric field vector $\vec{\mathbf{E}}$. The force exerted on an object with charge q is the charge multiplied by the electric field vector $\vec{\mathbf{E}}$ at the object's location:

$$\vec{\mathbf{F}}_{\text{elec}} = q\vec{\mathbf{E}}.$$

Gravitational forces are always attractive, but electric forces may be attractive or repulsive because there are two kinds of charge.

Electric field lines indicate the direction of the electric field vector at each point in space. A diagram of the field lines is a useful tool for visualizing the electric field (■ Figure VI.15). The field near a single charge looks quite different from the pictures of magnetic field loops. Field lines emerge from isolated positive charges, which are the simplest sources of electric field. To produce an isolated electric charge, however, we have to separate equal amounts of positive and negative charge. When we look at a region large enough to include charges of both signs, the familiar looped field patterns reappear (Figure VI.14).

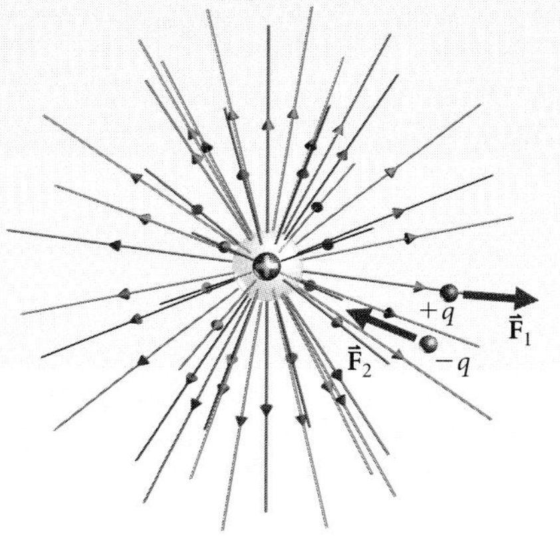

■ FIGURE VI.15
The electric field produced by a single positive charge. Positive charges are repelled by and negative charges are attracted toward the central positive charge.

VI.5 Moving Charge as the Source of Magnetic Field

When electric charges move, they produce magnetic field as well as electric field. Christian Oersted (1777–1851) demonstrated this relationship in 1820 during a public lecture. While performing a demonstration to illustrate the similarities between electricity and magnetism, he noticed that charges moving through a wire (*electric current*) influenced a compass needle. His experiment showed that the magnetic field lines form circles around the wire (■ Figure VI.16). The audience apparently was not impressed!

There is no obvious flow of charge in a bar magnet. Nonetheless, moving charge is the source of its magnetic field. In some atoms, the electrons' motions result in a net circulation around the nucleus, forming a miniature current loop, and each atom acts as a small magnetic dipole. In most materials, these atomic dipoles are oriented randomly and produce no net magnetic field. In iron, cobalt, nickel, and some alloys the atomic dipoles can be aligned in the same direction, producing large total magnetic field strengths.

THE LODESTONES DISCOVERED BY THE CHINESE AND THE ANCIENT GREEKS ARE COMPOSED OF AN IRON ORE NOW KNOWN AS MAGNETITE (Fe_2O_3).

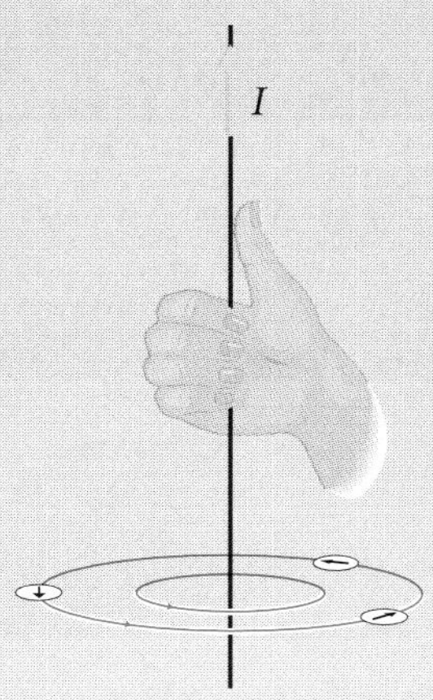

■ FIGURE VI.16
Oersted discovered that magnetic field lines form circles around a wire carrying electric current. The direction of the field lines follows a right-hand rule: with your right thumb in the direction of the current, the field lines run around the wire in the direction that your fingers curl. Oersted demonstrated this rule using compass needles.

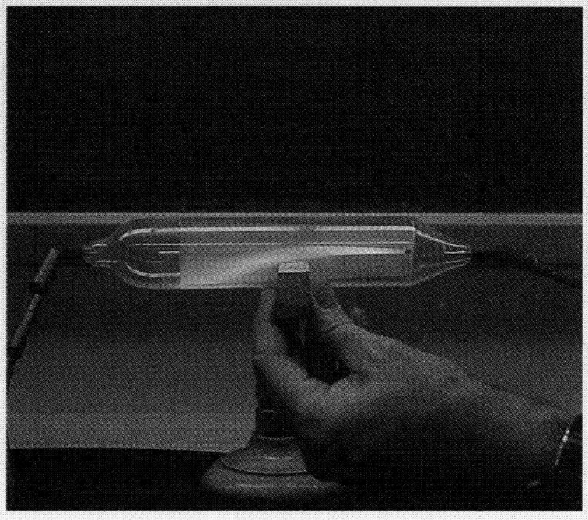

■ FIGURE VI.17
A phosphor screen indicates the acceleration of moving electrons in response to an applied magnetic field. The force is perpendicular both to the electrons' velocity and to the applied magnetic field.

VI.6 *Magnetic Force on Moving Charges*

Magnetic fields are produced by moving charges, and only moving charges experience magnetic force. ■ Figure VI.17 shows an electron beam moving from right to left in an evacuated tube. A phosphor screen (like a TV screen) intercepts a portion of the beam and makes its position visible. We may investigate the magnetic forces on the electrons by placing a bar magnet near the beam. The force is perpendicular both to the magnetic field $\vec{\mathbf{B}}$ and to the velocity of the electrons.

Electrons moving through a wire form an electric current. If the wire passes through a region containing a magnetic field (■ Figure VI.18), the field exerts a force on each of the moving electrons. The total force on a piece of wire is the vector sum of the forces on all the individual electrons. The magnitude of the force is proportional to the magnetic field strength, the magnitude of the current, and the length of wire. Electric motors, ■ Figure VI.19, are a major technological application of magnetic forces on wires.

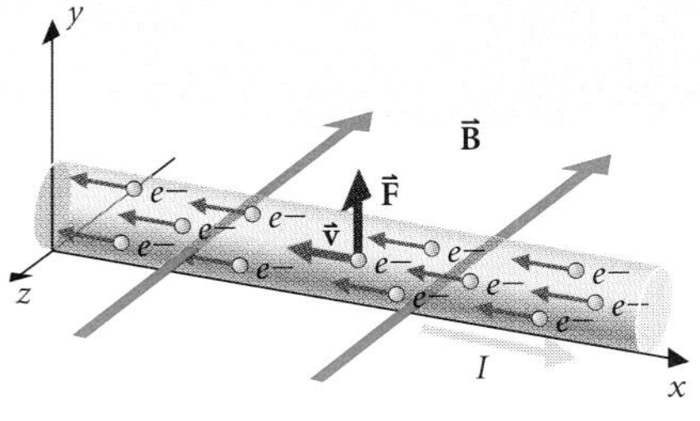

■ FIGURE VI.18
The electric current in a metal wire consists of moving electrons. When a magnetic field is applied, each electron feels a force. The total force on the wire is the vector sum of the forces on all the moving electrons. Thanks to Franklin's sign conventions, we describe negatively charged electrons moving to the left as an electric current I toward the right.

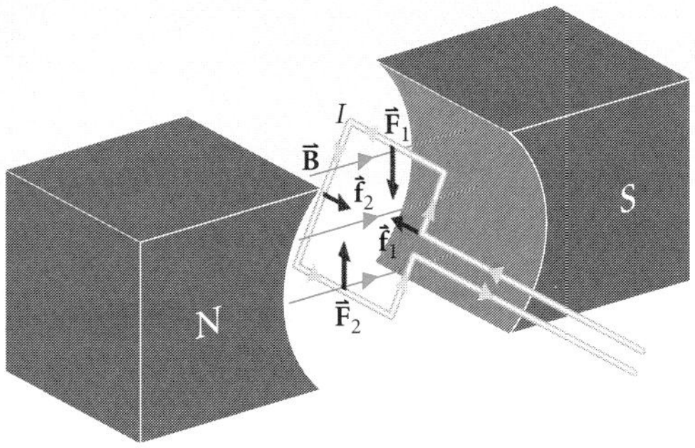

■ FIGURE VI.19
Electric motors make use of the forces on current-carrying wires. In this simple version, the current is in opposite directions on opposite sides of the square wire loop. The forces on opposite sides are equal and opposite. The resulting torque does mechanical work as the loop turns.

VI.7 SI Units for Charge and Current

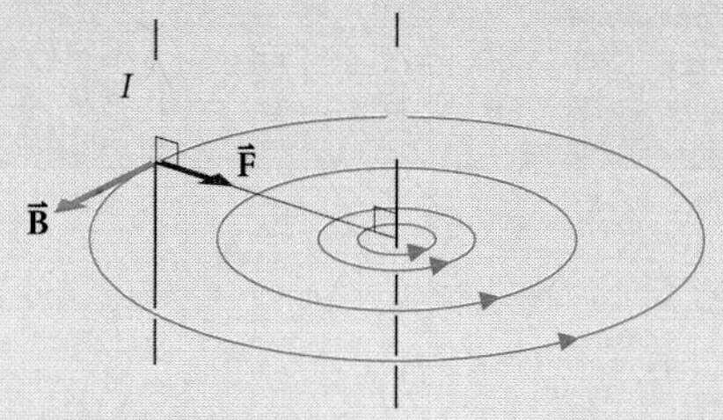

■ FIGURE VI.20
When each of two wires carries a current, the magnetic field produced by each wire causes the moving electrons in the other wire to experience a force. This observation forms the basis for the definition of the SI unit of current: the ampere.

The magnetic field produced by the moving charges in a current-carrying wire influences the charges in a nearby wire (■ Figure VI.20). The force between two such wires is used to define the SI unit of electric current:

> If the same current is maintained in two long parallel wires 1 m apart that repel each other with a force of 2×10^{-7} N per meter of wire, then the current in each wire is 1 ampere (abbreviation A).

The definition of the charge unit follows from the ampere:

> When a current of 1 ampere exists in a wire, the charge flowing past any point of the wire in 1 second is 1 coulomb, (abbreviation C).
>
> $$1\ \text{C} \equiv 1\ \text{A·s}.$$

VI.8 Unity of the Electromagnetic Field

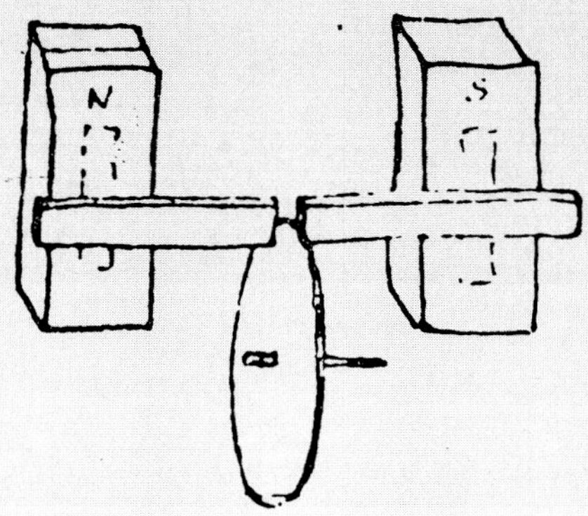

■ FIGURE VI.21
Diagram from Faraday's diary, August 29, 1831. Michael Faraday's experiment with a rotating metal disk showed some of the links between electric and magnetic fields (see Chapter 30) and paved the way for Maxwell's synthesis.

A compass seems unrelated to lightning, yet magnetic and electric phenomena are closely connected. Electric and magnetic fields are both produced by electric charge, and both exert forces on electric charge. In fact the two have an even deeper unity. In 1831 Michael Faraday discovered an electric current in a metal disk rotating between the poles of a magnet (■ Figure VI.21). Faraday described this effect as a production of electric field by the *cutting* of magnetic field lines. His image is illustrated in ■ Figure VI.22. When a wire *cuts* a field line, it retains a small loop of field corresponding to the current in the wire. Faraday's rotating disk continuously cuts through the field lines.

Magnetic field is also produced by changing the electric field. James Maxwell deduced this effect in 1864 from theoretical studies and predicted the existence of electromagnetic waves that move at the speed of light. In 1887 Heinrich Hertz produced and detected such waves, providing experimental verification of Maxwell's theory.

Albert Einstein showed in 1905 that there is a still closer relationship between electric and magnetic fields. His paper on special relativity established that the laws of electromagnetism are the same in every uniformly moving reference frame. Moving charge produces a magnetic field, but whether a charge is moving depends on the reference frame of the observer (■ Figure VI.23). A pure electric field in one reference frame is a mixture of electric and magnetic fields in another. Whether the field is electric, magnetic, or some combination of the two depends on whom you ask! Both kinds of field are aspects of a single electromagnetic field.

IN §3.3 WE DISCUSSED HOW THE REFERENCE FRAME AFFECTS THE DESCRIPTION OF MOTION.

VI.9 Electromagnetic Waves

Maxwell's electromagnetic theory of 1865 was the first to incorporate the symmetry between $\vec{E}$ and $\vec{B}$. According to this theory, EM waves are emitted whenever electric charges are accelerated. ■ Figure VI.24 illustrates how a wave is generated when a charge is displaced rapidly

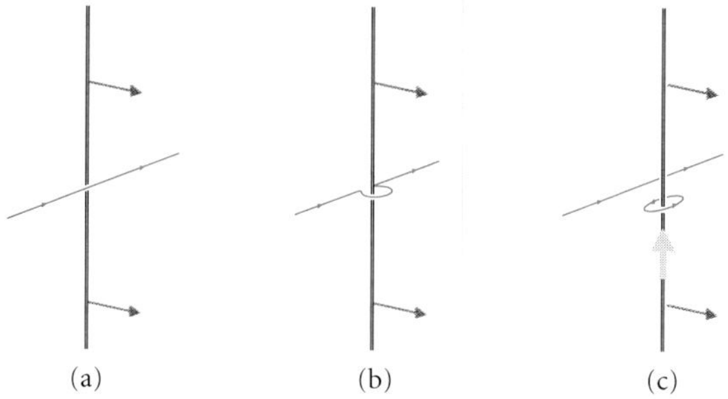

■ FIGURE VI.22
Michael Faraday used field lines as vivid images for his experimental results. In one such image, he imagined a moving metal wire passing through a magnetic field (a) and cutting off a loop (b). If the wire is part of a complete electrical circuit, a current arises as a result of the wire's passage through the field (c). The field line surrounding the wire indicates the presence of that electric current.

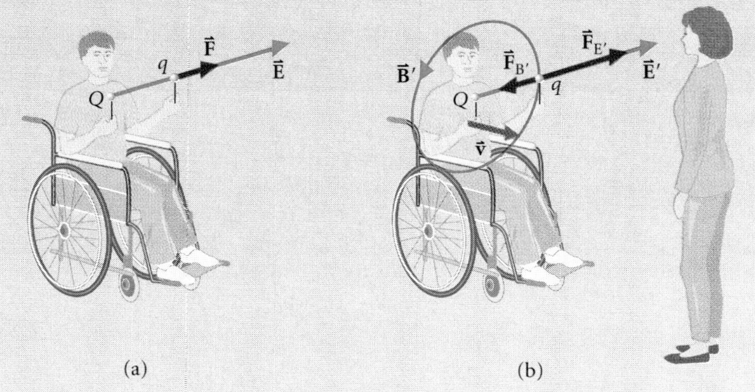

■ **Figure VI.23**
(a) Martin is holding two positive charges, one in each hand. He notices that the charge q experiences an electric field due to the charge Q and is repelled. According to Martin, the interaction between Q and q is purely electric. (b) If Martin is sitting in a wheelchair moving forward, Elizabeth notices that the charge Q moving with velocity $\vec{v}$ produces both electric and magnetic fields. According to Elizabeth, the repulsion of the moving charge q results from both electric and magnetic forces.

Martin and Elizabeth are both right. A pure electric field in Martin's reference frame is a combination of electric and magnetic fields in Elizabeth's frame. There is a single electromagnetic field. How we describe it is a matter of perspective—our reference frame.

and then left at rest. When the charge moves, its field lines cannot change everywhere at one instant. They respond somewhat like stretched rubber bands attached to the charge. By shaking the charge repeatedly, we can send a series of kinks along the field lines at the speed of light.

With Maxwell's identification of light as electromagnetic waves, electromagnetism grew from an obscure branch of physics into a fundamental theory of nature that describes a broad range of apparently disparate phenomena as the interaction of electric charge with fields. This unity inspired Einstein to search for a combined theory of electromagnetism and gravity. He didn't succeed, but electromagnetism has since been unified with the weak nuclear force, and grand unified theories are being constructed that link together all forces except gravity. Einstein's goal of unifying these forces with gravity has not yet been achieved.

As you study the chapters in Part VI, remember that these are the first steps along the road to understanding the unified structure of electromagnetism that does not become apparent until Part VII. So let's begin!

See Chapter 33 for further discussion of EM waves.

Maxwell also briefly considered a possible unification of EM theory and gravity.

The weak nuclear force controls decay of certain atomic nuclei.

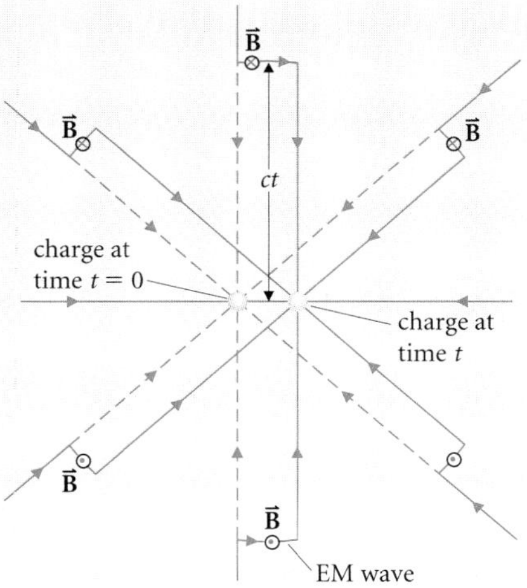

■ **Figure VI.24**
An idealized example shows how electromagnetic waves are generated. A charge initially at the origin is suddenly displaced to the right and abruptly brought to rest again. The field lines cannot move rigidly with the charge. It takes time for the lines to respond to the displacement. Dashed lines show the electric field before the displacement. The solid lines illustrate the field at time t after the displacement. The information that the charge has moved has traveled outward a distance ct. The field lines are centered at the charge's new position out to a distance ct but remain centered on the old position beyond that distance. At distance ct, the field lines are connected by a kink resulting from the charge's displacement.

During the displacement, the motion of the charge constituted an electric current in the negative x-direction that produced a magnetic field according to Oersted's rule (see Figure VI.16). This magnetic field, together with the kink in the electric field, is the wave disturbance that propagates away from the charge. The diagram shows that the electric and magnetic fields of the wave are perpendicular to each other and to the direction of travel of the disturbance. See Chapter 33 for a more detailed explanation.

CHAPTER 23
Charge and the Electric Field

Concepts

Electric charge
Point charge
Test charge
Coulomb's law
Electric field vector
Field line diagram
Superposition
Flux
Gauss' law

Goals

Understand:

The concept of field.

The relation between electromagnetic phenomena and the microscopic structure of matter.

Be able to:

Use Coulomb's law to find the force on a charged particle.

Find the electric field produced by a number of point charges.

Draw quantitative field line diagrams.

Use Gauss' law to relate flux to net charge for a set of point charges.

What objects are the fountains
Of thy happy strain?
What fields, or waves, or mountains?
What shapes of sky or plain?

Percy Bysshe Shelley

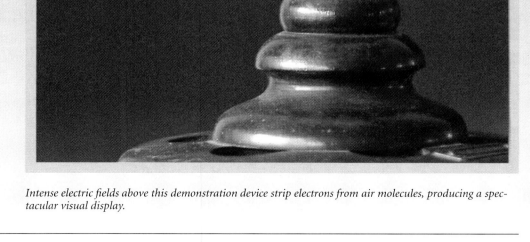

Intense electric fields above this demonstration device strip electrons from air molecules, producing a spectacular visual display.

If you walk across a wool carpet on a dry winter day, you may get a nasty surprise when you reach for the door knob: an electric spark jumps out and your finger feels a sharp snap. While walking, you generate an electric field around your body that disrupts air molecules and causes the spark. The same effect produces an eerie purple glow hovering above demonstration equipment designed to maintain a large field continuously. Electric fields also hold the plastic wrap around your dish of leftovers, stick toner to paper in copy machines, and remove solid particles from exhaust gases in smokestacks. In this chapter we shall begin to study the electric fields that make these devices work.

SEE §23.1.4 FOR AN EXPLANATION OF THIS EFFECT.

BEFORE STARTING THIS CHAPTER, READ THE PART INTRODUCTION, WHERE WE SET THE STAGE FOR THE MATERIAL IN THIS PART AND THE NEXT.

As in mechanics, very small particles—*point charges*—obey the simplest rules. We use the particle model to introduce the ideas of electric charge and electric field and to develop field line diagrams as a tool for visualizing the electric field. Most practical electrical systems employ distributions of charge whose geometric shape is important. An electrostatic microphone, for example, uses a thin charged sheet of plastic to convert sound energy into an electric signal. In the next chapter we shall find that the rules and techniques for point charges provide a firm foundation for the study of such continuous systems.

ELEMENTARY PARTICLES COME CLOSEST TO THE THEORETICAL IDEAL OF A POINT CHARGE, BUT THEIR BEHAVIOR IS OFTEN COMPLICATED BY ADDITIONAL PHYSICAL PRINCIPLES.

23.1 ELECTRIC CHARGE

23.1.1 *Charge and Matter*

In mechanics, the most significant property of a particle is its mass, which determines its acceleration when forces act on it. In electromagnetic theory, there is another significant property: charge. There are two kinds of charge, called positive and negative. Ordinary matter is electrically neutral. Each atom contains a positively charged nucleus and enough negatively charged electrons to make its net charge zero. To study the properties of electric charge, we must first separate the positive from the negative. One way to do this is to stroke a rubber rod with cat fur. Before stroking, the fur and rod show no electric effects at all. In contact, they acquire electric charges of opposite kinds. Negative charge is removed from the fur and added to the rod, leaving the fur positively charged.

THE CHARGE ON THE NUCLEUS IS Ze, WHERE Z IS THE ATOMIC NUMBER. YOU CAN FIND THIS NUMBER IN THE PERIODIC TABLE OF THE ELEMENTS, INSIDE THE BACK COVER.

THIS EXPERIMENT IS DESCRIBED IN THE PART INTRODUCTION.

In all such experiments, we find that charge is neither created nor destroyed; it is simply transferred from one place to another. To produce a positively charged object means transferring some negative charge to one or more other objects. We may summarize this idea in a principle called the *law of conservation of charge:*

THUS WE SHOULD ADD CHARGE TO THE LIST OF CONSERVED QUANTITIES IN PART II.

> Electric charge may be transported, but it may not be created or destroyed. The total amount of charge in the universe is constant.

The SI unit of charge is the coulomb (C). The charge on an electron is $-e$: $e = 1.602 \times 10^{-19}$ C. Ordinary matter is composed of whole numbers of electrons and protons and is charged by addition or removal of particles (usually the electrons). We now believe that protons are composed of smaller particles called quarks, whose charges are multiples of $e/3$. Apparently, single, isolated quarks cannot exist outside the particles they compose, so for all practical purposes the smallest observable unit of charge is e. Thus charge is quantized: the charge on any object is an integer number N times e.

A PROTON HAS A CHARGE $+e$. SEE THE PART INTRODUCTION.

FOR AN INTRODUCTION TO THE THEORY OF QUARKS, SEE CHAPTER 37.

Because e is so small, N is usually very large in laboratory-scale phenomena. Just as air may be considered to be a continuous fluid even though it is a collection of individual molecules, a charge distribution may often be considered continuous even though it is made up of individual elementary charges.

EXAMPLE 23.1 ♦ Estimate the total amount of positive electric charge in your body.

MODEL We model a human body as approximately 100 kg of water.

SETUP Each water molecule contains two hydrogen nuclei, each with positive charge e, and one oxygen nucleus with positive charge $8e$. The total positive charge in each

Remember: The atomic mass unit u is 1.66×10^{-27} kg.

molecule is $10e$. We find the number of molecules by dividing the mass of the body by the mass of one molecule, which is 18 u.

SOLVE The number of molecules is:

$$n = \frac{M_{\text{body}}}{M_{H_2O}} = \frac{100 \text{ kg}}{18(1.7 \times 10^{-27} \text{ kg})} = 3 \times 10^{27}.$$

Thus: $Q = Ne = n(10e) = (3 \times 10^{27})(10)(1.6 \times 10^{-19} \text{ C}) = 5 \times 10^{9} \text{ C}.$

ANALYZE Of course, the body contains an equal amount of negative charge as well. ∎

23.1.2 The Forces Between Charges

$F = Gm_1m_2/r^2$. See §5.4.

The law governing forces between charged objects is similar to the law of gravitation. Newton found that the expression for the gravitational force between two objects is simplest if they can be modeled as particles—objects whose size and structure are unimportant. A particle with a net electric charge also exerts and experiences electric force.

A *point charge* is a particle that carries a net electric charge.

The electric force law was discovered by Cavendish between 1771 and 1781, and independently by Coulomb between 1785 and 1789. Cavendish did not publish his results, so the law was named after Coulomb.

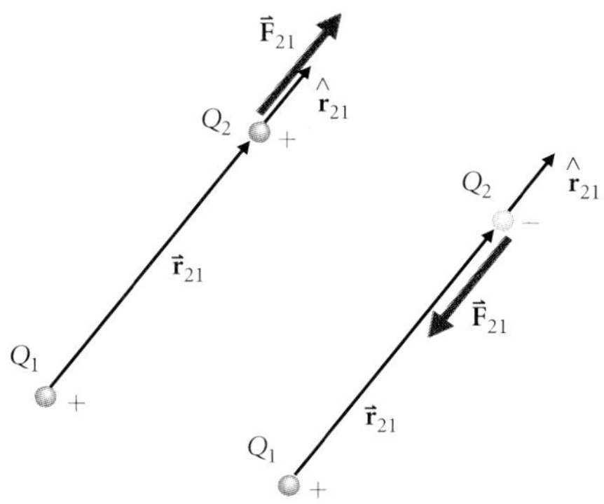

■ Figure 23.1
The force $\vec{\mathbf{F}}_{21}$ on particle 2 due to particle 1 lies along the line joining them. The force is repulsive if the two particles have charges of the same sign and attractive if the particles have charges with opposite signs. The unit vector $\hat{\mathbf{r}}_{21}$ points from particle 1 toward particle 2.

Coulomb's Law

Charges of like sign repel each other, while charges of unlike sign attract. The forces between point charges are proportional to the product of their charges and inversely proportional to the square of the distance between them. (■ Figure 23.1)

$$\vec{\mathbf{F}}_{21} = k\frac{Q_1Q_2}{r^2}\hat{\mathbf{r}}_{21}. \quad \textbf{(23.1)}$$

Here $\vec{\mathbf{F}}_{21}$ is the force exerted on Q_2 by Q_1. The direction of the unit vector $\hat{\mathbf{r}}_{21}$ is from Q_1 toward Q_2. Notice how the formula works: if Q_1 and Q_2 have the same sign, their product is positive, and the force is repulsive. Conversely, if they have opposite signs, their product is negative, so the force is opposite $\hat{\mathbf{r}}_{21}$ and the force is attractive.

The constant of proportionality k, the Coulomb constant, is determined by measurement. Its value to three significant figures is:

$$k = 8.99 \times 10^{9} \text{ N}\cdot\text{m}^2/\text{C}^2. \quad (23.2)$$

We should say it was *originally* determined by measurement. In fact k is related to the speed of light (Chapter 33), whose value is now considered a fixed constant. The value of k is therefore also fixed.

The Coulomb constant is often expressed in terms of the quantity ϵ_0, which is most often used in conjunction with Gauss' law (§23.4) and capacitance (Chapter 27).

$$k \equiv \frac{1}{4\pi\epsilon_0}, \quad (23.3)$$

$$\epsilon_0 = 8.85 \times 10^{-12} \text{ C}^2/\text{N}\cdot\text{m}^2. \quad (23.4)$$

EXAMPLE 23.2 ♦ A small spherical ball with charge 5.6 C is 2.5 m from a second ball with charge −3.5 C. How large a force do the balls exert on each other?

MODEL We model the two balls as particles. The force between them is given by Coulomb's law.

SETUP Since the charges have opposite signs, the force between them is attractive and has magnitude:

$$|\vec{\mathbf{F}}| = k\frac{|Q_1 Q_2|}{r^2}.$$

SOLVE
$$F = 9.0 \times 10^9 \frac{\text{N}\cdot\text{m}^2}{\text{C}^2}\frac{(5.6\ \text{C})(3.5\ \text{C})}{(2.5\ \text{m})^2} = 2.8 \times 10^{10}\ \text{N}.$$

NOTICE THAT WE ROUNDED k TO TWO SIGNIFICANT FIGURES, SINCE THE OTHER NUMBERS ARE NOT KNOWN TO GREATER ACCURACY.

ANALYZE This is an enormous force! No known heavy machinery could hold the two "small balls" apart. ∎

23.1.3 The Strength of the Electric Force

The coulomb is a very large amount of electric charge. The force between charges of a few coulombs separated by a few meters (Example 23.2) is about 10^{10} N, equal to the weight of a million metric tons. If you were within a meter of a 1-C point charge, it would exert forces on the electrons and protons in your body strong enough to pull you apart. In the following example, we investigate the amount of charge involved in a laboratory exercise.

EXAMPLE 23.3 ♦♦ Two charged pith balls, each of mass $m = 1.0$ g, are suspended from strings $\ell = 21$ cm long (■ Figure 23.2). The angle between the strings is $2\theta = 12°$, and the balls carry equal charges Q. What is Q?

MODEL We model each ball as a point charge in static equilibrium under three forces: weight, string tension, and electric repulsion.

SETUP We find the distance s between the balls using geometry, and the electric force by analyzing a free-body diagram (Figure 23.2b). Then we apply Coulomb's law to find the charge Q.

In the free-body diagram, choose the x-axis horizontal and y up. Then:

$$\sum F_x = T\sin\theta - F_e = 0$$

and
$$\sum F_y = T\cos\theta - mg = 0.$$

From Coulomb's law:
$$F_e = kQ^2/s^2.$$

We eliminate the unknown tension by dividing the first two equations:

$$\tan\theta = \frac{F_e}{mg} = k\frac{Q^2}{mgs^2}.$$

The distance between the two charges is:

$$s = 2\ell\sin\theta,$$

so:
$$\tan\theta = \frac{kQ^2}{4mg\ell^2\sin^2\theta}.$$

SOLVE
$$Q = \pm\sqrt{\frac{4mg\ell^2}{k}\sin^2\theta\tan\theta} = \pm 2\ell\sin\theta\sqrt{\frac{mg}{k}\tan\theta}$$

$$= \pm 2(0.21\ \text{m})\sin(6°)\sqrt{\frac{(1.0\times10^{-3}\ \text{kg})(9.8\ \text{m/s}^2)}{9.0\times10^9\ \text{N}\cdot\text{m}^2/\text{C}^2}\tan(6°)}$$

$$= \pm 1.5\times10^{-8}\ \text{C} = \pm 15\ \text{nC}.$$

ANALYZE Since $F \propto Q^2$, we cannot determine the sign from the given information. Either two positive or two negative charges would produce the same observed behavior. ∎

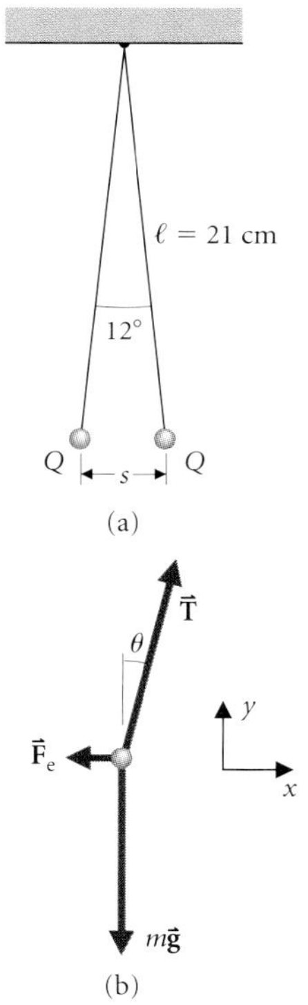

■ FIGURE 23.2
(a) Two charged pith balls suspended by strings from the same point are pushed apart by the Coulomb force between them. The angle between the strings is $2\theta = 12°$. (b) Free-body diagram for the pith ball on the left. Three forces act on each ball: string tension, Coulomb force, and weight.

In the spirit of a back-of-the-envelope estimate, we take the average mass of a pith ball molecule to be 10 u.

The net charge on the pith balls in Example 23.3 results from the addition (or removal) of about $(1.5 \times 10^{-8}\,\mathrm{C})/(1.6 \times 10^{-19}\,\mathrm{C}) = 100$ billion electrons. But each ball contains roughly

$$N \approx \frac{m_{\text{ball}}}{m_{\text{molecule}}} \approx \frac{m_{\text{ball}}}{10\ \mathrm{u}} = \frac{10^{-3}\ \mathrm{kg}}{10^{-26}\ \mathrm{kg}} = 10^{23}\ \text{molecules},$$

and several electrons per molecule. So only about one part in 10^{13} of the ball's electrons are transferred. Even spectacular electric effects involve only a minuscule fraction of the charge present.

EXAMPLE 23.4 ♦ Compare the electrical and gravitational forces between a proton and an electron.

MODEL Both the electrical and gravitational forces vary in the same way with distance, so their ratio is independent of distance.

The power of 10 is the important result of this calculation, not the 2. That is why we have not used more figures.

SETUP

$$\frac{|\vec{\mathbf{F}}_{\text{elec}}|}{|\vec{\mathbf{F}}_{\text{grav}}|} = \frac{ke^2/r^2}{Gm_p m_e/r^2} = \frac{ke^2}{Gm_p m_e}$$

SOLVE

$$= \frac{(9 \times 10^9\ \mathrm{N{\cdot}m^2/C^2})(1.6 \times 10^{-19}\ \mathrm{C})^2}{(6.7 \times 10^{-11}\ \mathrm{m^3/kg{\cdot}s^2})(1.7 \times 10^{-27}\ \mathrm{kg})(9 \times 10^{-31}\ \mathrm{kg})}$$

$$= 2 \times 10^{39}.$$

ANALYZE Gravitational forces are completely unimportant to the structure of atoms. ▮

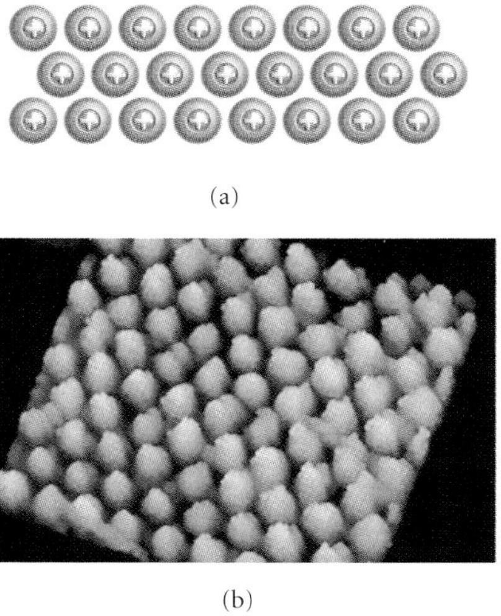

■ **Figure 23.3**
(a) A crude model of solid matter. The nuclei are surrounded by electron clouds. (b) A scanning tunneling microscope image of a gold surface shows the electron clouds as bumps separated by about 0.3 nm.

23.1.4 *Triboelectricity*

We can understand many electrical phenomena by using Coulomb's force law qualitatively together with a crude model of a piece of solid matter (■ Figure 23.3). Heavy positive nuclei form a nearly rigid framework. Each nucleus is surrounded by a cloud of electrons held by the attractive electrical force between the electrons' negative charge and the positive charge on the nucleus. The outermost electrons are held by the weakest force for two reasons: They are farthest from the positive nucleus, and they are also repelled by the electrons closer in. The electrical properties of a solid depend on the relative freedom of outer electrons to move between atoms, so they are extremely sensitive to the strength of this binding force. The electrons are more mobile than the nuclei because of their much smaller mass.

EXERCISE 23.1 ♦ If the same force F acts on a proton and an electron, what is the ratio of their accelerations?

The atoms in rubber attract their outer electrons more strongly than do atoms in cat fur. When cat fur and rubber are put in very close contact with each other (■ Figure 23.4), some electrons near the common surface are more strongly attracted to the rubber than to the cat fur, and so they move across the surface. When enough electrons have moved into the rubber, their repulsion overcomes the excess attraction of the rubber and stops further electrons from switching sides. When the two materials are separated, the rubber has taken a small fraction of the cat fur's electrons and so is negatively charged. The cat fur ends up positively charged. This phenomenon is called *tribo-* or contact electricity. In the triboelectric sequence (● Table 23.1) materials listed early in the sequence obtain positive charge from contact with those later in the list.

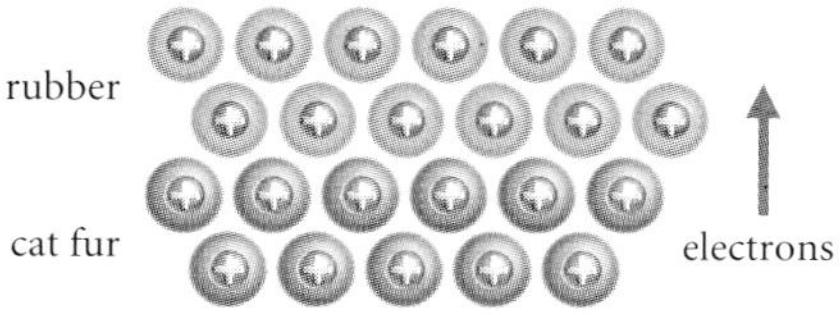

■ **Figure 23.4**
Triboelectricity. When the rubber and fur are put in contact, electrons move from the cat fur into the rubber, and the rubber becomes negatively charged. Migration stops when electrons in the fur are repelled as strongly by the excess electrons in the rubber as they are attracted by the rubber nuclei.

23.1.5 *Conductors and Insulators*

On a dry day, a rubber rod remains charged for several minutes, while a charged metal rod loses its charge immediately when you pick it up. The difference lies in the ability of charge to move through the bodies of the two rods. The metal is called a *conductor,* and the rubber is an *insulator.* Each molecule in the rubber holds its electrons strongly, while outer electrons in the metal are attracted as strongly by neighboring atoms as by their *own* atom and are thus free to wander throughout the material. Any charge placed on a conductor may move about freely or even pass through the body of the conductor. The most common conducting materials are metals. The same free electrons that sustain current in copper wire also transport heat efficiently through the bottom of a cooking pot and reflect light to produce its shiny appearance. (Values of conductivity are listed in Chapter 26.)

The degree to which electrons are free to move in a material is described by its *conductivity.* Conductivity varies enormously among materials. Metals allow charge to move some 10^{23} times more freely than does a typical insulator such as glass. Germanium and silicon have an intermediate number of mobile electrons per atom. Controlling conductivity of such materials via clever doping with impurities is the essence of the semiconductor industry.

Table 23.1 The Triboelectric Sequence

Positive
Rabbit's fur
Glass
Mica
Wool
Cat's fur
Silk
Cellulose acetate
Cotton
Wood
Amber
Resins
Natural rubber
Metals (Cu, Ni, Co, Ag)
Sulfur
Metals (Pt, Au)
Celluloid
Negative

Adapted from Gaylord P. Harnwell, *Principles of Electricity and Electromagnetism* (New York: McGraw-Hill, 1949), 3, and *Physics Today,* May 1986:51.

23.2 The Electric Field of a Point Charge

The interaction between charges may be described as a two-step process. A charge creates an *electric field* in the region around it, while the electric field exerts force on any charge placed in it. The electric field exists at each point whether or not any charge is present to experience a force. To measure the field at a point, however, we place a charge there and measure the force on it. So as not to disturb the system significantly, we introduce as small a *test charge q* as possible. From the measured force on the test charge we find the value of the *electric field vector* at the location of the test charge:

$$\vec{\mathbf{E}} \equiv \lim_{q \to 0} \frac{\vec{\mathbf{F}}}{q}. \tag{23.5}$$

The magnitude of the electric field vector $|\vec{\mathbf{E}}|$ is the *electric field strength.*

The vector $\vec{\mathbf{E}}$ describes the electric field in the same way the vector $\vec{\mathbf{g}}$ describes the gravitational field at a point.

$$\vec{\mathbf{g}} = \lim_{m \to 0} \frac{\vec{\mathbf{F}}_g}{m}.$$

Even objects as large as airplanes or supertankers experience the gravitational field of the Earth without themselves contributing significantly to it. We are surrounded by test masses!

See the part introduction, §VI.4, for further discussion of the field concept.

■ Figure 23.5 illustrates electric field vectors produced by a point charge Q_1 at two nearby points P and S. A second charge Q_2 experiences a force exerted on it by the electric field (■ Figure 23.6).

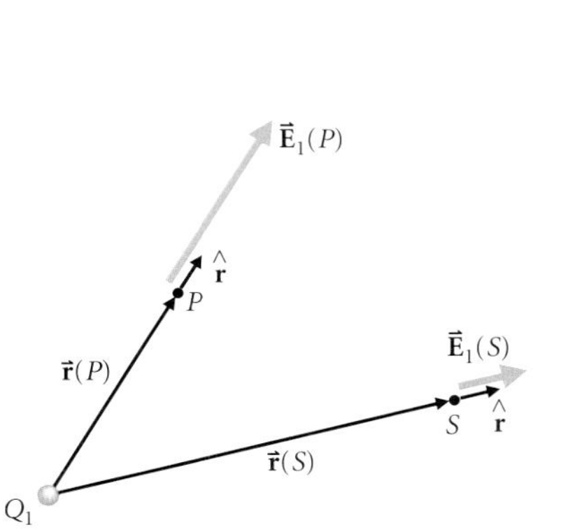

■ Figure 23.5
Electric field vectors due to the positive charge Q_1 at two points P and S. In each case the electric field points directly away from Q_1.

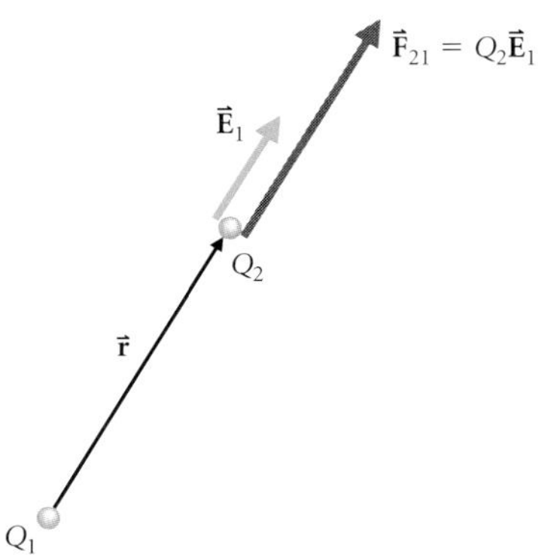

■ Figure 23.6
If another point charge Q_2 is placed at P, it experiences a force $\vec{\mathbf{F}}_{21} = Q_2\vec{\mathbf{E}}_1$.

Coulomb's experimental results (§23.1) for the field and force may be restated as follows:

Remember: The electric field vector is a function of position. It may be different at each point of space.

> **Coulomb's Law for a Point Charge**
>
> A point charge Q_1 produces an electric field vector $\vec{\mathbf{E}}_1(\vec{\mathbf{r}})$ at each point *P*. If *P* has position $\vec{\mathbf{r}}$ with respect to Q_1, then:
>
> $$\vec{\mathbf{E}}_1 = \frac{kQ_1\hat{\mathbf{r}}}{r^2}. \tag{23.6}$$
>
> The force exerted by the field on the point charge Q_2 at *P* is its charge multiplied by the electric field vector at *P*.
>
> $$\vec{\mathbf{F}}_{21} \equiv Q_2\vec{\mathbf{E}}_1. \tag{23.7}$$

In Chapter 25 we shall introduce another unit for measuring field: volt/meter.

In eqn. (23.6), the unit vector $\hat{\mathbf{r}}$ points *from* the point charge *to* the field point *P*. Thus the electric field vector points away from a positive charge and toward a negative charge. The unit of electric field vector is the newton per coulomb (N/C) and has no special name.

Combining eqns. (23.6) and (23.7), with $\hat{\mathbf{r}} = \hat{\mathbf{r}}_{21}$, the unit vector from Q_1 to Q_2, we get back eqn. 23.1:

$$\vec{\mathbf{F}}_{21} = Q_2\frac{kQ_1}{r^2}\hat{\mathbf{r}}_{21} = \frac{kQ_2Q_1}{r^2}\hat{\mathbf{r}}_{21}.$$

EXAMPLE 23.5 ♦ A point charge $Q = 10.0\ \mu$C is at the origin. Find the electric field vector at the point *P* with coordinates $x = 2.0$ m, $y = 1.0$ m (■ Figure 23.7).

MODEL The electric field vector is given by Coulomb's law in the form of eqn. (23.6).

SETUP The distance from the origin to *P* may be found from Pythagoras' theorem. Then $\vec{\mathbf{E}}\,(P)$ has magnitude:

$$|\vec{\mathbf{E}}| = 9.0 \times 10^9\,\frac{\text{N}\cdot\text{m}^2}{\text{C}^2}\,\frac{10.0 \times 10^{-6}\ \text{C}}{[(2.0\ \text{m})^2 + (1.0\ \text{m})^2]} = 1.8 \times 10^4\ \text{N/C}.$$

The direction of $\vec{\mathbf{E}}$ is directly away from the origin (Figure 23.7b), along:

$$\hat{\mathbf{r}} \equiv (\cos\theta)\hat{\mathbf{i}} + (\sin\theta)\hat{\mathbf{j}} = \frac{2\hat{\mathbf{i}} + \hat{\mathbf{j}}}{\sqrt{5}} = [(0.894)\hat{\mathbf{i}} + (0.447)\hat{\mathbf{j}}].$$

SOLVE $\vec{\mathbf{E}} = (1.8 \times 10^4\ \text{N/C})\hat{\mathbf{r}} = [1.6\hat{\mathbf{i}} + 0.80\hat{\mathbf{j}}] \times 10^4\ \text{N/C}.$

The field vector makes an angle $\theta = \tan^{-1}(0.50) = 27°$ with the positive *x*-axis, so we may also write:

$\vec{\mathbf{E}} = (1.8 \times 10^4$ N/C, in the *x*-*y*-plane at $\theta = 27°$ counterclockwise from the *x*-axis).

ANALYZE *Remember:* The electric field vector points away from the positive charge and lies in the plane containing both the charge *Q* and the point *P*. Thus in this example, the *z*-component is zero. ▮

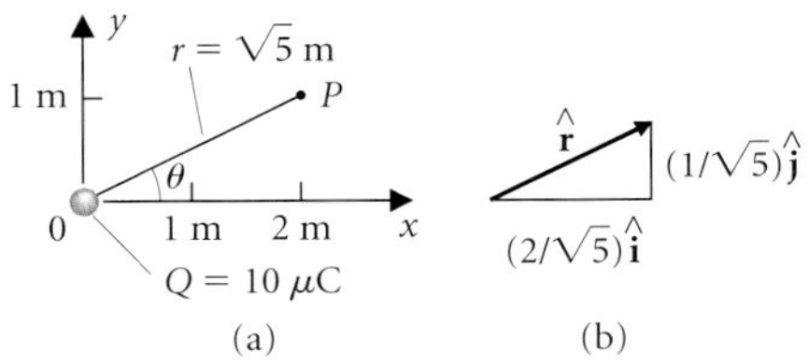

■ **Figure 23.7**
(a) A point charge $Q = 10.0\ \mu$C is at the origin. The point *P* is $\sqrt{5}$ m away. (b) The unit vector $\hat{\mathbf{r}}$ at *P* equals $(2/\sqrt{5})\hat{\mathbf{i}} + (1/\sqrt{5})\hat{\mathbf{j}}$. All the vectors, $\vec{\mathbf{r}}$, $\hat{\mathbf{r}}$, and $\vec{\mathbf{E}}$, lie in the *x*-*y*-plane.

EXAMPLE 23.6 ♦ What is the force acting on a charge $Q_2 = -1.0\ \mu$C at point *P* in Example 23.5?

MODEL The force on the charge is $\vec{\mathbf{F}} = Q_2\vec{\mathbf{E}}$, where $\vec{\mathbf{E}}$ is the electric field vector at *P*, the position of Q_2.

SETUP The force is: $\vec{\mathbf{F}}_{21} = Q_2\vec{\mathbf{E}} = (-1.0 \times 10^{-6}\ \text{C})(10^4\ \text{N/C})[1.6\hat{\mathbf{i}} + 0.80\hat{\mathbf{j}}].$

SOLVE $\vec{\mathbf{F}}_{21} = [-1.6\hat{\mathbf{i}} - 0.80\hat{\mathbf{j}}] \times 10^{-2}$ N.

$1.8 = \sqrt{1.6^2 + 0.80^2}$; see also Example 23.5.

The force has magnitude 1.8×10^{-2} N and is directed toward the origin.

ANALYZE Since Q_2 is negative, it is attracted toward Q_1. Notice that this is automatically described by the mathematical expression for $\vec{F}_{21}$.

EXERCISE 23.2 ♦ In Figure 23.7, the 10.0 μC charge is moved to the point R with coordinates $x = -1.0$ m, $y = -1.0$ m. What is the electric field vector at P then?

If an electric force acts on the point charge Q_2, then an equal and opposite electric force acts on the charge Q_1. This force arises in the same way as its pair: Q_2 produces an electric field $\vec{E}_2$ (■ Figure 23.8) that exerts the force on Q_1. According to Coulomb's law, $\vec{E}_2$ at the location of Q_1 is:

$$\vec{E}_2 = \frac{kQ_2}{r^2}\hat{r}_{12},$$

and the force on Q_1 is: $\vec{F}_{12} \equiv Q_1\vec{E}_2 = Q_1\frac{kQ_2}{r^2}\hat{r}_{12} = -\frac{kQ_1Q_2}{r^2}\hat{r}_{21} = -\vec{F}_{21}.$

The Coulomb force satisfies the strong form of Newton's third law.

Taking Coulomb's law at face value, the field at a point charge's own location is both infinite in magnitude and undefined in direction. It wouldn't make much sense to compute with such a thing, and indeed, if we tried, we would find momentum not conserved. The field of a point charge does not act on the charge that produces it. The force acting on any point charge is computed from the field produced by *other* charges.

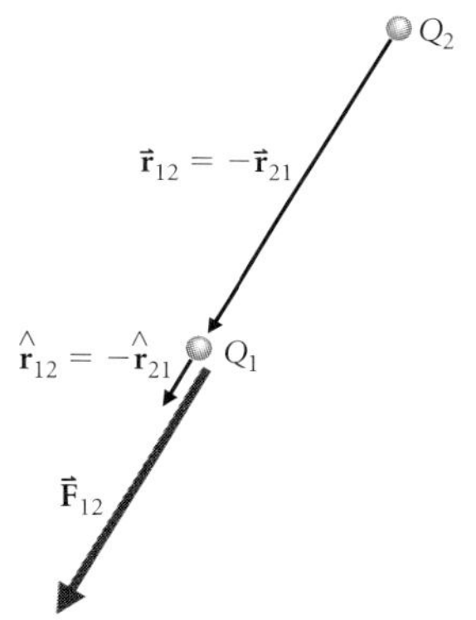

■ FIGURE 23.8
Coulomb forces on point charges obey Newton's third law. The charge Q_2 in Figure 23.6 exerts an equal and opposite force $\vec{F}_{12}$ on Q_1. The vector $\hat{r}_{12}$ points from 2 toward 1 and equals $-\hat{r}_{21}$.

WE HAVE TO BE CAREFUL APPLYING NEWTON'S THIRD LAW WHEN DEALING WITH SYSTEMS OF PARTICLES *AND FIELDS*. WE'LL HAVE MORE TO SAY ABOUT THIS IN CHAPTER 29.

23.3 THE PRINCIPLE OF SUPERPOSITION

23.3.1 Superposition of Fields at a Point

Coulomb's law describes the electric field produced by a single charge. If several charges exist in a region of space, each contributes to the net electric field. Experimentally, we find that the total electric field is the vector sum of the individual contributions. The presence of one charge does not affect the contribution due to another. This rule is called *the principle of superposition.*

> If several point charges Q_i exist in a region of space, the total electric field vector $\vec{E}$ at a point P is the sum of the electric field vectors produced by the individual charges.
>
> $$\vec{E} = \sum_i \vec{E}_i = \sum_i \frac{kQ_i}{r_i^2}\hat{r}_i. \quad \textbf{(23.8)}$$

EXAMPLE 23.7 ♦ A point charge $Q_1 = +10.0$ μC is at the origin, and a second charge $Q_2 = -5.0$ μC is placed on the y-axis at $y = 1.0$ m. What is the total electric field at the point P with coordinates $x = 2.0$ m, $y = 1.0$ m?

MODEL We use Coulomb's law to find the field due to each charge, and then we find the vector sum according to the principle of superposition (■ Figure 23.9).

SETUP We found the field $\vec{E}_1$ due to Q_1 in Example 23.5:

$$\vec{E}_1 = [1.6\hat{i} + 0.80\hat{j}] \times 10^4 \text{ N/C}.$$

The field at P due to the negative charge Q_2 is in the $-x$-direction:

$$\vec{E}_2 = k\frac{Q_2}{r_2^2}\hat{r}_2 = \left(9.0 \times 10^9 \frac{\text{N}\cdot\text{m}^2}{\text{C}^2}\right)\frac{(-5.0 \times 10^{-6}\text{ C})}{(2.0\text{ m})^2}\hat{i} = (-1.1 \times 10^4\text{ N/C})\hat{i}.$$

SOLVE

$$\vec{E} \equiv \vec{E}_1 + \vec{E}_2 = [1.6\hat{i} + 0.80\hat{j} - 1.1\hat{i}] \times 10^4 \text{ N/C}$$
$$= [0.5\hat{i} + 0.80\hat{j}] \times 10^4 \text{ N/C},$$

or $\vec{E} = (0.94 \times 10^4$ N/C, making an angle of 58° counterclockwise from the x-axis).

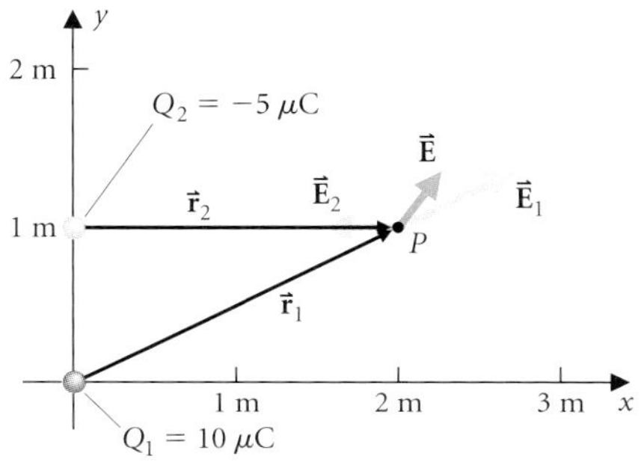

■ FIGURE 23.9
Two charges Q_1 and Q_2 lie on the y-axis. The electric field $\vec{E}$ at P is the sum of the electric fields $\vec{E}_1$ and $\vec{E}_2$ due to Q_1 and Q_2.

Digging Deeper

Measuring the Electron Charge

During the period 1910–1913, Robert A. Millikan developed the following experiment for measuring very small amounts of electric charge (■ Figure 23.9). An atomizer sprays a cloud of small oil droplets into the experimental chamber. As they are formed, the droplets acquire a small charge. Droplets occasionally fall through a small hole into a region between two metal plates. There the experimenter can observe the illuminated droplets against a dark background.

With no electric field between the plates, a drop falls at constant speed, its weight balanced by air resistance. When an electric field is applied by connecting a battery to the plates (see Chapters 24 and 25), the drop moves at a different speed, since air resistance now balances the vector sum of the drop's weight and the electric force $q\vec{\mathbf{E}}$. (Air resistance is proportional to speed.) By observing the drop's speed in the two cases, the experimenter can deduce the electric charge q on the drop.

To 1% accuracy, Millikan found that all his measured charges were integer multiples of a basic unit that he identified as the electron charge e.

$$q = ne = \pm e, \pm 2e, \pm 3e, \ldots$$

Because the speed of an individual drop is very small, a drop may be followed for hours by varying the electric field. In such extended observations, Millikan could detect the transfer of charge between a drop and the surrounding air as the value of the drop's measured charge changed abruptly by one or two units. Millikan received the Nobel prize for physics in 1923 for this work and also for his measurements of the photoelectric effect (Chapter 35).

During the 1980s, in one of several attempts to detect free quarks, physicists at San Francisco State University used a modernized Millikan apparatus (■ Figure 23.10) to measure charges on droplets of a wide variety of substances, from mercury to mineral water. All the experiments confirmed Millikan's result:

> All objects larger than subatomic particles have electric charges that are integer multiples of e.

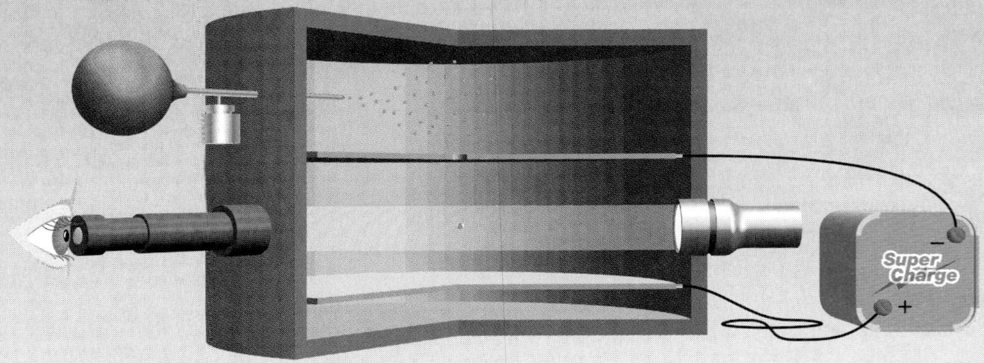

■ **Figure 23.10**
Millikan oil drop experiment. The speed of small oil drops moving through air varies as the applied electric field varies. By observing the drops' speeds, the experimenter can determine the charges on the drops.

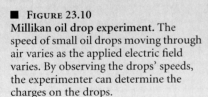

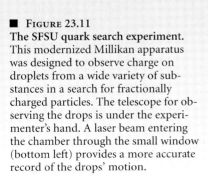

■ **Figure 23.11**
The SFSU quark search experiment. This modernized Millikan apparatus was designed to observe charge on droplets from a wide variety of substances in a search for fractionally charged particles. The telescope for observing the drops is under the experimenter's hand. A laser beam entering the chamber through the small window (bottom left) provides a more accurate record of the drops' motion.

ANALYZE Remember that we obtain the total electric field at a point by adding the field vectors *at the same spatial position.* There is no physical meaning to the sum of field vectors at different positions.

EXAMPLE 23.8 ♦♦ Two positive and two negative point charges are at the corners of a square $\ell = 1.0$ m on a side (■ Figure 23.12). If $Q = 1.0\ \mu$C, find the electric field at point P produced by the charges at the other three corners. What electric force acts on the charge at P?

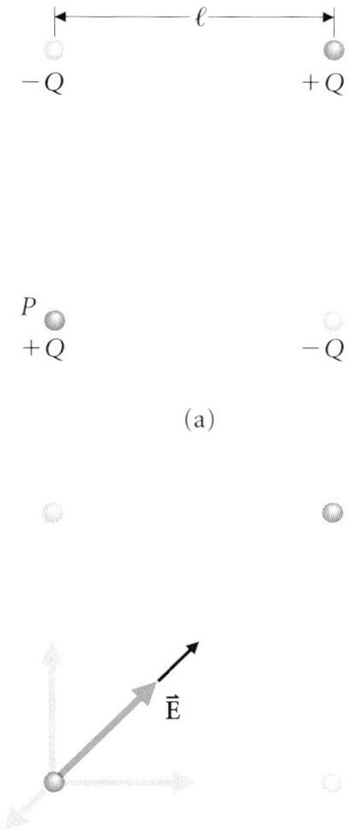

■ **Figure 23.12**
(a) Four point charges are at the corners of a square. Two are positive and two are negative. (b) The electric field at point P, the lower left-hand corner, is due to the three charges at the other three corners. The sum of the three vectors lies along the diagonal of the square.

MODEL Using the principle of superposition, the electric field is the sum of three contributions (Figure 23.12b). *Remember:* The charge at P does not contribute to the force acting on itself.

SETUP The contributions to the field at P due to the two negative charges have equal magnitude, $E_1 = E_2 = kQ/\ell^2$, and are at right angles. Their vector sum points toward the center of the square. The field due to the other positive charge points away from the center of the square and has magnitude $E_3 = kQ/(2\ell^2)$.

SOLVE The sum of all three vectors points toward the center and has magnitude:

$$\begin{aligned}|\vec{\mathbf{E}}| &= 2|\vec{\mathbf{E}}_1|\cos 45^\circ - |\vec{\mathbf{E}}_3| \\ &= 2\frac{kQ}{\ell^2}\frac{\sqrt{2}}{2} - \frac{kQ}{2\ell^2} \\ &= \frac{kQ}{\ell^2}\left(\sqrt{2} - \tfrac{1}{2}\right) \\ &= \frac{(9.0 \times 10^9\ \mathrm{N\cdot m^2/C^2})\ (1.0 \times 10^{-6}\ \mathrm{C})}{(1.0\ \mathrm{m})^2}(1.41 - 0.50) \\ &= 8.2 \times 10^3\ \mathrm{N/C}.\end{aligned}$$

The force is: $\vec{\mathbf{F}} = Q\vec{\mathbf{E}} = (8.2 \times 10^{-3}$ N, toward the center of the square).

ANALYZE We could have argued that the force is toward or away from the center of the square from the charges' symmetry with respect to the diagonal through P.

THE USE OF FIELD LINES IN STUDYING ELECTROMAGNETIC PHENOMENA WAS INSTITUTED BY MICHAEL FARADAY IN THE 1830S.

23.3.2 Field Line Diagrams

A good way to visualize the electric field due to any charge distribution is to draw a field line diagram. The direction of the electric field vector at each point is tangent to the field line at that point. In other words, at any point the field line has the same direction as the electric field vector.

Electric field lines diverge from positive charges and converge into negative charges. We may imagine creating a positively charged object by pulling negative charge out. The electric field lines stretch between the two pieces of charge as we pull them apart. In ■ Figure 23.13 the field lines surrounding a single positive charge disappear at *infinity,* the final location of the negative charge that was removed.

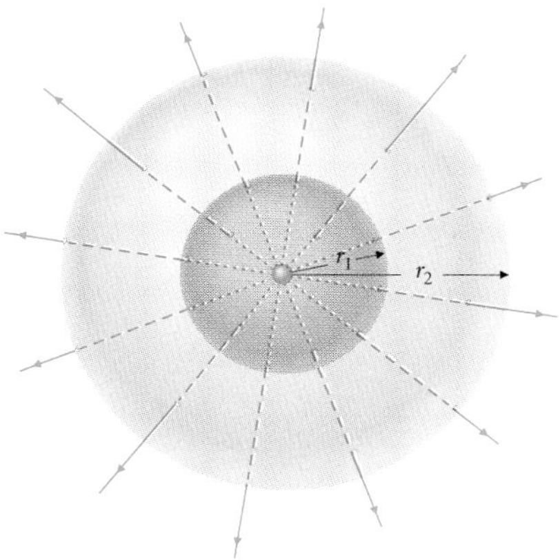

■ **Figure 23.13**
Field line diagram for a single positive point charge. If we draw a sequence of spheres centered on the charge, each field line passes through all of them. The number of lines per unit area of surface decreases with distance from the charge.

In principle, a field line passes through every point. However, we obtain a useful picture by drawing some finite number N of field lines. These lines pass through ever-larger spheres centered on the object (Figure 23.13). The field lines that pass through the sphere at r_1 also pass through the sphere at r_2. Thus:

$$\frac{\text{number of lines per unit area at } r_1}{\text{number of lines per unit area at } r_2} = \frac{N/(4\pi r_1^2)}{N/(4\pi r_2^2)} = \left(\frac{r_2}{r_1}\right)^2.$$

According to Coulomb's law, this is also the ratio of the electric field strengths at the two positions.

$$\frac{|\vec{\mathbf{E}}_1|}{|\vec{\mathbf{E}}_2|} = \frac{kQ/r_1^2}{kQ/r_2^2} = \left(\frac{r_2}{r_1}\right)^2.$$

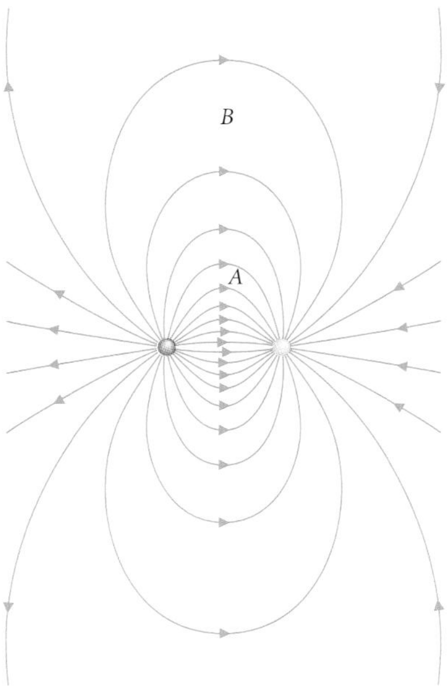

■ FIGURE 23.14
(a) Field line diagram for two equal and opposite point charges. (See also Figure VI.14.) The positive charge is at the left. All the lines emerging from the positive charge converge into the negative charge. Some of the lines form very large loops, much larger than the diagram. The lines that leave the left border of the diagram reenter at the right. The system of field lines is rotationally symmetric about the line joining the two charges; that is, the field line diagram is the same in any plane containing the two charges.

The distance between field lines at A, d_A, is about one-fourth that at B, $d_B \approx 4d_A$. The relative field strengths are determined by the separation of field lines.

The field line diagram describes the electric field according to the following rules:

> The direction of the electric field vector $\vec{\mathbf{E}}$ at a point is tangent to the field line through that point.
>
> The strength of an electric field $|\vec{\mathbf{E}}|$ at any location is proportional to the relative number of field lines passing through a unit area perpendicular to the lines.

The total number of field lines drawn in any diagram is arbitrary. However, the number of field lines drawn emerging from or converging into a point charge is proportional to the amount of charge, since electric field is proportional to the amount of charge that produces it. With two equal and opposite point charges (■ Figure 23.14), the same number of lines emerge from the positive charge as converge into the negative charge. None extend to infinity.

Field lines never cross each other. If two field lines were to cross at any point, the electric field vector would have two different directions. A test charge placed there would have to accelerate in both directions at the same time. That can't happen.

There are five rules for constructing field line diagrams for a collection of point charges:

1. Field lines begin at positive charge and end at negative charge.
2. The number of field lines shown diverging from or converging into a point charge is proportional to the magnitude of the charge.
3. Field lines are spherically symmetric near a point charge.
4. If the system has a net charge, the field lines are spherically symmetric at great distances.
5. Field lines never cross.

EXAMPLE 23.9 ♦♦ Estimate the relative strengths of the electric field at points A and B in Figure 23.14.

MODEL The strength of the electric field at a point is proportional to the number of field lines per unit area perpendicular to the lines.

SETUP The relative number of field lines per unit area (and thus the strength of the field) at a point is inversely proportional to the area surrounding each field line at that point. That area is roughly the square of the distance between two lines in a two-dimensional diagram (Figure 23.14).

SOLVE We measure the distance between lines near the two points and compute the ratio of their squares.

$$\frac{E_B}{E_A} \approx \left(\frac{d_A}{d_B}\right)^2 = \left(\frac{3\text{ mm}}{13\text{ mm}}\right)^2 \approx 0.05.$$

ANALYZE Field lines spread apart as the strength of the field decreases. ∎

EXAMPLE 23.10 ♦ Two positive charges are at opposite corners of a square, and two negative charges occupy the other two corners. All four charges have the same magnitude. Sketch the field lines that lie in the plane of the square.

MODEL We choose to draw eight field lines for each charge.

SETUP Since the total charge of the system is zero, no field lines go to infinity (rule 4). Each line begins at a positive charge and ends at a negative charge (rule 1).

SOLVE We complete the diagram using rules 2 and 3 (■ Figure 23.15). Remember rule 5! The lines bend away from the center of the square; they do not cross there.

ANALYZE The field lines share the symmetry of the system about the diagonals of the square. ∎

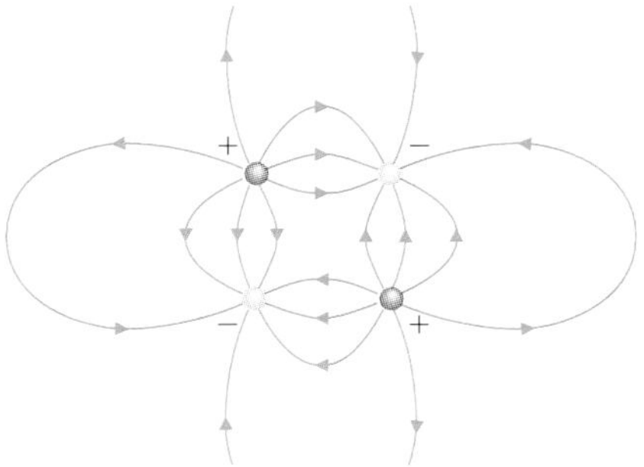

■ FIGURE 23.15
Field line diagram for a system of four charges, two positive and two negative, in the plane containing all four charges. (The diagram in other planes would look different.)

EXERCISE 23.3 ♦ A positive charge $2Q$ has two negative charges $-Q$ at equal distances on either side of it. Sketch a field line diagram for this system. ∎

23.3.3 *Calculation of Fields as a Function of Position*

The electric field due to a system of charges exists at each point in space. Using Coulomb's law and the principle of superposition, we can express the electric field vector as a function of the point's coordinates. For an arbitrary point the calculation can be quite intricate. In this section we shall explore the methods and concepts involved in such calculations, but we shall minimize mathematical complexity by obtaining values of $\vec{\mathbf{E}}$ for specially chosen points. Since the field lines are everywhere tangent to the electric field vector $\vec{\mathbf{E}}$, it is usually helpful to begin by sketching the field line diagram.

EXAMPLE 23.11 ♦♦ Two equal positive charges Q are at positions $x = -a$ and $x = a$ on the x-axis of a coordinate system (■ Figure 23.16). Compute $\vec{\mathbf{E}}$ at each point on the y-axis.

MODEL We begin with the field line diagram. Since the system is rotationally symmetric about the x-axis, we need to draw the field lines in only a single plane through the line of the charges. Showing six field lines per charge gives sufficient detail. We use the five rules given in §23.3.2.

1. There are no negative charges in the system, so the field lines end at *infinity*.
2. Six field lines emerge from each charge.
3. The field lines emerge symmetrically from the charges.
4. From a large distance, the separation of the charges is negligible, and the system appears as a point charge of magnitude $2Q$. The total of 12 lines diverge symmetrically to *infinity*.
5. The field lines do not cross between the charges, but bend away from each other.

The resulting diagram is shown in Figure 23.16a.

Now we calculate $\vec{\mathbf{E}}$ at an arbitrary point P on the y-axis, as a function of the y-coordinate. According to the principle of superposition (Figure 23.16b), the total electric field is the sum of the two vectors $\vec{\mathbf{E}}_1$ and $\vec{\mathbf{E}}_2$ produced by the two charges. The two equal charges are at the same distance r from P.

SETUP We use Coulomb's law to find $\vec{\mathbf{E}}_1$ and $\vec{\mathbf{E}}_2$. The magnitudes are equal:

$$|\vec{\mathbf{E}}_1| = |\vec{\mathbf{E}}_2| = \frac{kQ}{r^2} = \frac{kQ}{a^2 + y^2}.$$

Since both vectors make equal angles with the y-axis, their vector sum lies along the y-axis. The x-components of $\vec{\mathbf{E}}_1$ and $\vec{\mathbf{E}}_2$ cancel, while their y-components add.

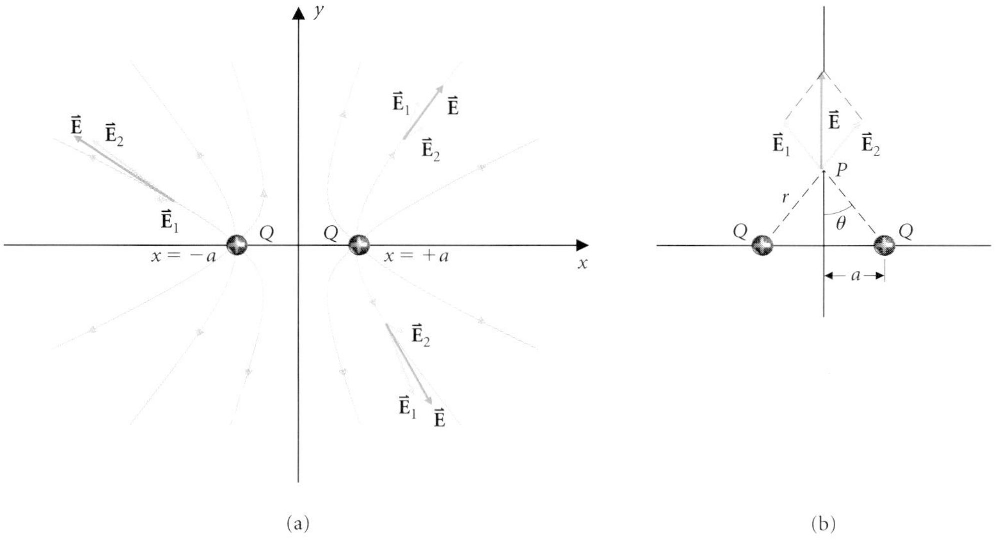

■ FIGURE 23.16
(a) Field line diagram for two equal positive point charges at $x = -a$ and $x = a$ on the x-axis. (b) The electric field at a point P on the y-axis is the sum of the fields due to each of the two charges. The vector sum lies along the y-axis.

SOLVE $$\vec{\mathbf{E}} = 2|\vec{\mathbf{E}}_1| \cos\theta\ \hat{\mathbf{j}} = 2\frac{kQ}{r^2}\frac{y}{r}\hat{\mathbf{j}} = \frac{2kQy}{(a^2 + y^2)^{3/2}}\hat{\mathbf{j}}. \tag{23.9}$$

ANALYZE The direction we computed for $\vec{\mathbf{E}}$ is consistent with the field line diagram. At the origin, halfway between the charges, the electric field vectors produced by the two charges are equal and opposite, so $\vec{\mathbf{E}} = 0$. The field lines bend away from the origin and none passes through the origin. Along the y-axis, the field lines crowd together before beginning to diverge, indicating that the electric field increases to a maximum strength before decreasing again at large distances. At points where $y \gg a$, the separation of the two charges becomes insignificant. In this limit we have:

$$\lim_{y \gg a} |\vec{\mathbf{E}}| = \lim_{y \gg a} \frac{2kQy}{y^3(1 + a^2/y^2)^{3/2}} = \frac{2kQ}{y^2}.$$

This is just what we would expect for a point charge of magnitude $2Q$ at the origin.

The calculation actually gives the value of $\vec{\mathbf{E}}$ at any point in the y-z-plane. Since the system is symmetric about the x-axis, the electric field at a point in the y-z-plane a distance $d = \sqrt{(y^2 + z^2)}$ from the origin points directly away from the x-axis and has magnitude $|\vec{\mathbf{E}}| = 2kQd/(a^2 + d^2)^{3/2}$. ■

EXERCISE 23.4 ♦♦ Find the electric field at each point on the x-axis with $x > a$.

Study Problem 15 ♦♦♦ *Two Unequal Charges*

Two charged objects $Q_1 = +3Q$ and $Q_2 = -Q$ are separated by a distance D (■ Figure 23.17a). Find a point near the objects where $\vec{\mathbf{E}} = 0$.

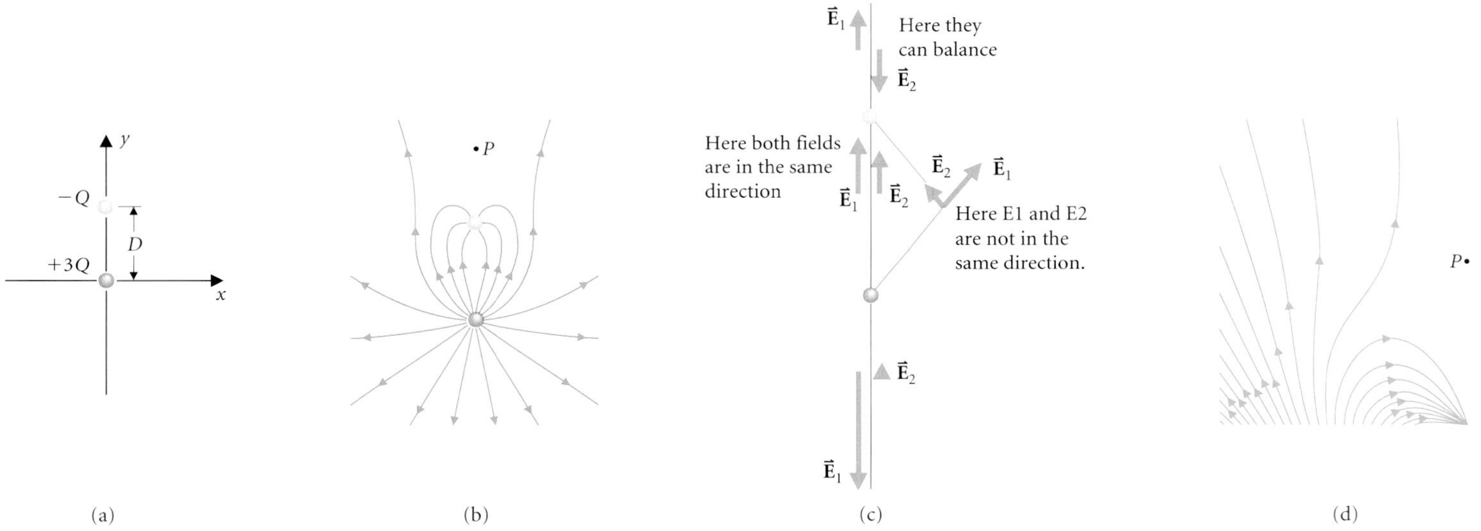

■ FIGURE 23.17
(a) Two unequal charges are on the y-axis: $+3Q$ at $y = 0$ and $-Q$ at $y = D$. (b) Field line diagram. One-third of the field lines emerging from the positive charge converge at the negative charge; the rest go off to infinity. (c) The net field can be zero only on the y-axis where the fields $\vec{\mathbf{E}}_1$ and $\vec{\mathbf{E}}_2$ are in opposite directions, that is, above the negative charge, at $y > D$. (d) An enlargement of the field line diagram in the region $y > D$. Many more field lines are shown. For $x > 0$, the diagram is the mirror image of that shown for $x < 0$. Notice how the lines bend away from P, where the field is zero.

Modeling the System

We begin by drawing the field line diagram (Figure 23.17b). The algebraic sum of the two charges is $+2Q$, so when viewed from a large distance (much greater than the separation D of the two charges), the system looks like a single point charge $+2Q$.

Applying the rules for field line diagrams from §23.3.2, we choose to represent a charge Q with 6 field lines; then 18 lines emerge from the positive charge $Q_1 = 3Q$, and 6 converge into the negative charge. The remaining 12 field lines are distorted by the presence of Q_2 but flare out symmetrically at large distances.

The total electric field $\vec{\mathbf{E}}$ is the vector sum of contributions $\vec{\mathbf{E}}_1$ and $\vec{\mathbf{E}}_2$ produced by the charges $Q_1 = 3Q$ and $Q_2 = -Q$. The total field is zero at any point where $\vec{\mathbf{E}}_1$ and $\vec{\mathbf{E}}_2$ balance each other; that is, when $\vec{\mathbf{E}}_1$ and $\vec{\mathbf{E}}_2$ are (1) in opposite directions and (2) equal in magnitude (Figure 23.17c). The two vectors have opposite directions only at points on the line through the two charges, either below the positive charge or above the negative charge. The field of a point charge increases with the magnitude of the charge and decreases with distance from it. So the magnitudes $|\vec{\mathbf{E}}_1|$ and $|\vec{\mathbf{E}}_2|$ can be equal only at points closer to Q_2 than to Q_1, that is, above Q_2.

Setup

We put the origin at Q_1 and the y-axis along the line joining the two charges. We derive expressions for the electric field vectors at a point P on the y-axis due to Q_1 and Q_2, add them to find an expression for the total field vector $\vec{\mathbf{E}}$, set this equal to zero, and solve for y.

From Coulomb's law:

$$\vec{\mathbf{E}}_1 = k\frac{3Q}{y^2}\hat{\mathbf{j}} \qquad (y > 0),$$

$$\vec{\mathbf{E}}_2 = -\frac{kQ}{(y - D)^2}\hat{\mathbf{j}} \qquad (y > D),$$

$$\vec{\mathbf{E}} = \vec{\mathbf{E}}_1 + \vec{\mathbf{E}}_2 = kQ\left[\frac{3}{y^2} - \frac{1}{(y - D)^2}\right]\hat{\mathbf{j}} \qquad (y > D).$$

Solution of Equations

The field is zero where:

$$\frac{3}{y^2} - \frac{1}{(y - D)^2} = 0. \qquad \text{(i)}$$

Setting the two terms equal and taking their square roots:

$$\frac{3}{y^2} = \frac{1}{(y - D)^2} \Rightarrow \sqrt{\frac{3}{y^2}} = \pm\sqrt{\frac{1}{(y - D)^2}}.$$

$$\frac{\sqrt{3}}{y} = \frac{\pm 1}{y - D} \Rightarrow (\sqrt{3} \mp 1)y = D\sqrt{3}.$$

$$y = \frac{\sqrt{3}}{\sqrt{3} \mp 1}D = \frac{3 \pm \sqrt{3}}{2}D.$$

Don't forget that there are two solutions, corresponding to the positive and negative square root.

Are both of these solutions valid? No. In setting up the problem we established that $y > D$. For $y < D$, the direction of $\vec{\mathbf{E}}_2$ is reversed, and eqn. (i) does not hold. The + sign gives the correct result:

$$y = (3 + \sqrt{3})D/2 = 2.37D.$$

Analysis

The point P lies above the negative charge, as shown on the diagram. Notice that below this point, all field lines curve inward and terminate at the negative charge, while above this point the field lines terminate at infinity (Figure 23.17d). There is no well-defined direction for the field at P, corresponding to its zero magnitude.

23.4 Gauss' Law

23.4.1 *The Relation Between Charge and Field Lines*

Coulomb's law describes the electric field created by a charge distribution: given a point, you can calculate the field there. By drawing the field line diagram, we get a sense of how the charge distribution affects all space: it gives a more global perspective. The values of electric field vectors at different places are related because they are produced by the same charge distribution. In this section we introduce Gauss' law, which expresses this global relation of charge to field. It does not contain any new information, but it expresses the same information as Coulomb's law in a different way.

After Karl Friedrich Gauss (1777–1855).

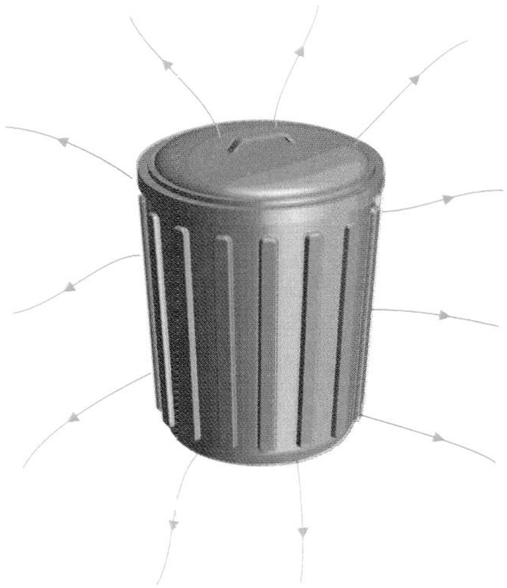

■ FIGURE 23.18
If field lines emerge from a plastic garbage pail, we may conclude that there is charge inside to produce the field. In this case the charge is positive, since the field lines point out of the pail.

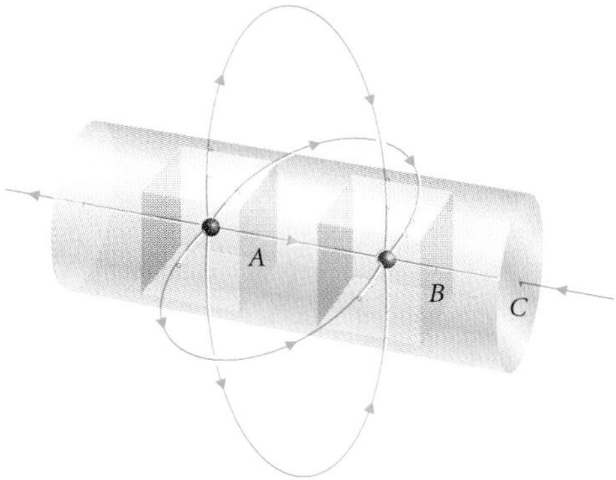

■ FIGURE 23.19
A system of two equal and opposite charges. The field line along the symmetry axis forms a very large loop seen leaving to the left and returning from the right.

Electric field lines begin and end on electric charges. Thus the field lines emerging from the pail in ■ Figure 23.18 indicate that there is charge inside to produce them. That is the main idea contained in Gauss' law. To begin developing a more quantitative statement, let's consider a specific system: two equal and opposite point charges (■ Figure 23.19). The six field lines we have drawn emerge from the positive charge and converge into the negative charge. The field lines that emerge from or pass into any volume indicate the charge contained *inside* that volume. For example, volume A contains a charge $+Q$, and six field lines emerge from its surface. Volume B contains charge $-Q$, and six field lines pass inward through its surface. If we describe a field line passing inward as a negative line coming out, then:

GAUSS' LAW: PRELIMINARY STATEMENT.

> The net number of field lines emerging from any volume is proportional to the net charge inside.

This result holds for a volume of arbitrary shape containing an arbitrary number of charged objects. Volume C contains both charges, with a total inside of $Q - Q = 0$. One of the field lines emerging from the positive charge doesn't intersect the surface of volume C at all. The rest of the lines emerge and then reenter the volume; they each count positive once and negative once, for a net contribution of zero.

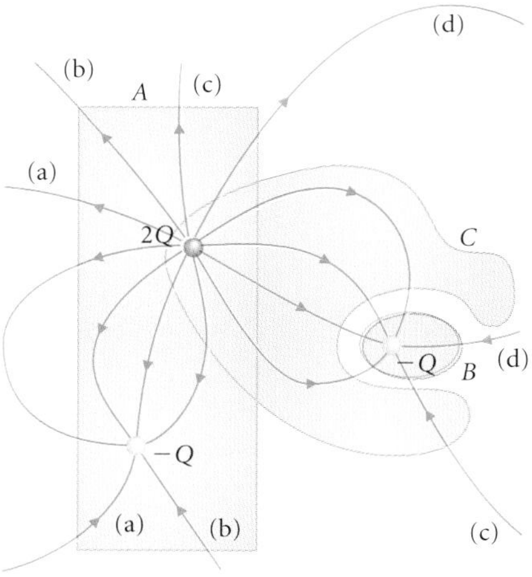

■ FIGURE 23.20
Gauss' law in a two-dimensional world. The number of field lines leaving each *volume* is proportional to the charge inside.

EXAMPLE 23.12 ❖ It becomes increasingly difficult to draw accurate three-dimensional diagrams for more complicated charge distributions. So, imagine for the moment that we live in a two-dimensional world. Sketch the field line diagram for the system of three "point" charges shown in ■ Figure 23.20, and verify the relation between charge and field lines for volumes B and C.

MODEL If we draw six field lines per charge unit Q, 12 lines diverge from the positive charge and six lines converge into each of the negative charges. Four lines, labeled (a), (b), (c), and (d), form large loops: they leave and reenter the diagram.

SETUP Six field lines enter the elliptical volume B, which contains a total net charge $-Q$. Volume C contains a net charge $2Q$. Even though the volume wraps around the negative charge on the right, the charge is *not* inside the volume. All 12 field lines leaving the positive point charge leave volume C. At the lower right, field line (c) reenters the volume but then leaves again, so its net contribution is still one line leaving the volume.

SOLVE The number of field lines entering each volume (-6 versus $+12$) is in the same ratio as the charge contained ($-Q$ versus $+2Q$).

ANALYZE A two-dimensional slice does not always give a true representation of a three-dimensional situation. However, a slice like this is an accurate representation of the field pattern if each of the "points" is actually an infinite *line* of charge lying perpendicular to the page.

SEE §24.2.

EXERCISE 23.5 ♦ Compare the number of field lines emerging from volume A in Figure 23.20 with the charge inside the volume.

23.4.2 *Electric Flux*

Field lines are a useful visual aid, but at best they offer a rough quantitative description of the electric field strength. Our next step is to find a precise definition of the quantity *electric flux* that we have just modeled as "number of field lines." The strength of the electric field is represented by the relative number of field lines passing through a unit area (§23.3.2):

$$|\vec{\mathbf{E}}| \propto \frac{\text{number of field lines}}{\text{area}}.$$

■ Figure 23.21 shows a bundle of field lines passing through an area A_o perpendicular to the lines, and a second area A inclined at angle θ to the lines. The number of field lines in the bundle is represented by:

$$N \propto |\vec{\mathbf{E}}|A_o = |\vec{\mathbf{E}}|A \cos\theta = A\vec{\mathbf{E}} \cdot \hat{\mathbf{n}}.$$

The quantity on the right-hand side of this expression is the *electric flux* through the area A.

The *electric flux* $\Delta\Phi_E$ passing through an element of area ΔA with normal $\hat{\mathbf{n}}$ is:

$$\Delta\Phi_E \equiv (\vec{\mathbf{E}} \cdot \hat{\mathbf{n}})\Delta A. \tag{23.10}$$

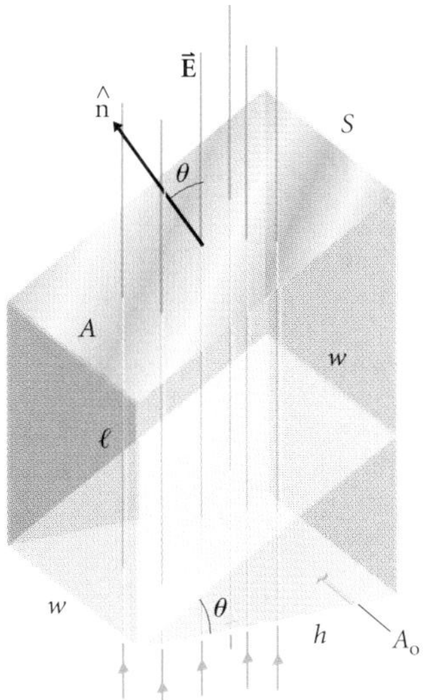

■ **FIGURE 23.21**
(a) The area of the tilted surface $A = w\ell = wh/\cos\theta = A_o/\cos\theta$. (b) The flux of electric field through the surface S is $\vec{\mathbf{E}} \cdot \hat{\mathbf{n}}A = |\vec{\mathbf{E}}|A\cos\theta = |\vec{\mathbf{E}}|A_o$.

Digging Deeper

FLUX

The word *flux* derives from the Latin for flow, and electric flux is mathematically very similar to the flow of liquid in a pipe. Imagine a pipe carrying water at constant velocity $\vec{\mathbf{v}}$ (■ Figure 23.22). The volume of water flowing past any cross section A_o of the pipe per unit time is:

$$\frac{dV}{dt} = \frac{A_o|\vec{\mathbf{v}}|\,dt}{dt} = A_o|\vec{\mathbf{v}}|.$$

The same volume flows out of the diagonal surface at the end of the pipe. The *flux* out of the pipe is due to the velocity component $\vec{\mathbf{v}} \cdot \hat{\mathbf{n}}$ normal to the surface; motion parallel to the surface does not transport water across it.

$$\frac{dV}{dt} = A\vec{\mathbf{v}} \cdot \hat{\mathbf{n}} = A|\vec{\mathbf{v}}|\cos\theta = A_o|\vec{\mathbf{v}}|.$$

A similar expression (eqn. 23.10) describes electric flux.

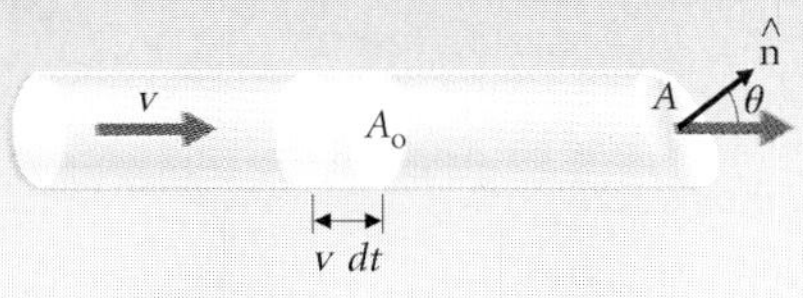

■ **FIGURE 23.22**
Flux of water through a pipe is
$A\vec{\mathbf{v}} \cdot \hat{\mathbf{n}} = A_o|\vec{\mathbf{v}}|$.

EXERCISE 23.6 ♦ Find the electric flux through an element of area $\Delta A = 1.0\ \text{m}^2$ lying in the x-y-plane with normal in the positive z-direction (■ Figure 23.23) if the electric field is uniform:

$$\vec{\mathbf{E}} = (1.0 \times 10^{-6}\ \text{N/C})\,\frac{(\hat{\mathbf{i}} + \hat{\mathbf{k}})}{\sqrt{2}}.$$

A CLOSED SURFACE IS THE BOUNDARY OF A VOLUME. FOR EXAMPLE, A CLOSED CYLINDRICAL SURFACE SUCH AS A SOUP CAN INCLUDES THE TWO FLAT FACES AT THE ENDS. NO MATTER WHICH WAY YOU ORIENT THE SURFACE, THE SOUP WON'T RUN OUT.

A surface that completely encloses a volume is called a *closed surface.* The total flux emerging from a volume is found by dividing the closed surface bounding the volume into elements of area and adding the flux through each of them (■ Figure 23.24). The normal for each element always points *outward* from the volume. (This is equivalent to counting emerging field lines as positive and entering field lines as negative.)

$$\Phi_E = \oint d\Phi_E = \oint \vec{\mathbf{E}} \cdot \hat{\mathbf{n}}\, dA. \qquad (23.11)$$

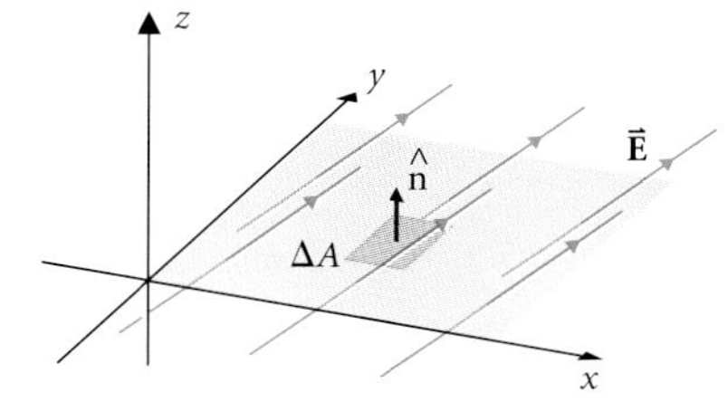

■ FIGURE 23.23
What is the electric flux through the area shown?

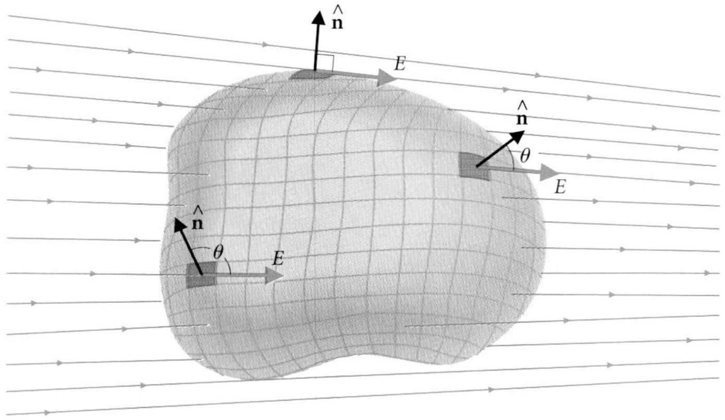

■ FIGURE 23.24
To find the electric flux through a closed, curved surface, divide the surface up into differential elements and sum the flux through all the elements. The normal to each element points outward from the volume enclosed by the surface.

EXAMPLE 23.13 ♦ Find the electric flux through a cubical box placed in a uniform field so that two of the faces are perpendicular to $\vec{\mathbf{E}} = E\hat{\mathbf{i}}$ (■ Figure 23.25).

MODEL Since $\vec{\mathbf{E}}$ and $\hat{\mathbf{n}}$ are each constant on any one side, we may calculate the flux through each surface of the box and add the results.

SETUP The electric field is parallel to the top and bottom of the box and also to two of the sides. On these faces, $\vec{\mathbf{E}} \cdot \hat{\mathbf{n}} = 0$, and the flux through them is zero. No field lines pierce these faces. Since normal vectors point outward from the box, $\hat{\mathbf{n}} = \hat{\mathbf{i}}$ on the side at $x = L$, and $\hat{\mathbf{n}} = -\hat{\mathbf{i}}$ on the side at $x = 0$. The flux through the side at $x = L$ is:

$$\Phi_L = (\vec{\mathbf{E}} \cdot \hat{\mathbf{n}})a^2 = \vec{\mathbf{E}} \cdot \hat{\mathbf{i}}\,a^2 = Ea^2.$$

The flux through the side at $x = 0$ is:

$$\Phi_0 = (\vec{\mathbf{E}} \cdot \hat{\mathbf{n}})a^2 = \vec{\mathbf{E}} \cdot (-\hat{\mathbf{i}})a^2 = -Ea^2.$$

SOLVE The net flux through the box is:

$$\oint \vec{\mathbf{E}} \cdot \hat{\mathbf{n}}\, dA = \Phi_L + \Phi_0 = Ea^2 + (-Ea^2) = 0.$$

ANALYZE There is no charge inside the box and the net flux through its surface is zero. All the field lines that pass in at $x = 0$ pass out again at $x = L$.

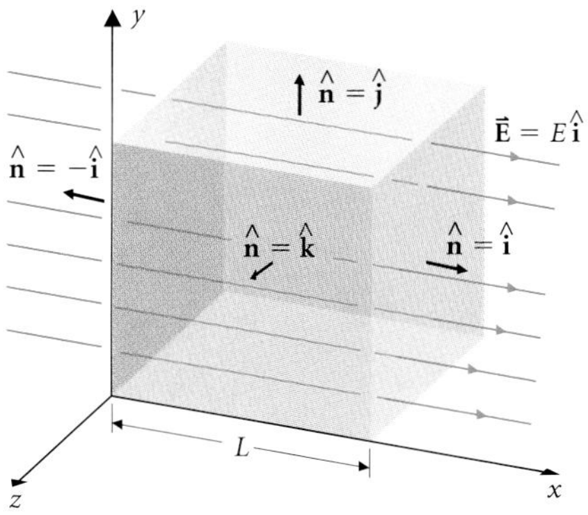

■ FIGURE 23.25
Flux of a uniform electric field through a cube. Notice that each field line that enters the cube at the left side leaves at the right. The net flux is zero.

EXAMPLE 23.14 ♦♦ Find the electric flux through a sphere of radius $r = a$ surrounding a point charge Q (■ Figure 23.26).

MODEL Field lines emerge from the point charge and exit through the surface of the sphere. Since there is charge inside, the flux is not zero.

SETUP The normal to the surface at any point is along the outward radius, $\hat{\mathbf{n}} = \hat{\mathbf{r}}$. The electric field is also along the outward radius:

$$\vec{\mathbf{E}} = \frac{kQ}{r^2}\hat{\mathbf{r}}.$$

SOLVE The dot product $\vec{\mathbf{E}} \cdot \hat{\mathbf{n}} = kQ/r^2$ has the same value kQ/a^2 at each point on the surface $r = a$. Thus:

$$\Phi_E = \oint \frac{kQ}{a^2}\,dA = \frac{kQ}{a^2}\oint dA = \frac{kQ}{a^2}4\pi a^2 = 4\pi kQ = \frac{Q}{\epsilon_0}.$$

ANALYZE The flux does not depend on the specific value $r = a$ of the sphere's radius. Once again the flux through the surface is proportional to the charge inside. More generally, outside any spherically symmetric distribution of charge:

$$\vec{\mathbf{E}} \cdot \hat{\mathbf{n}} = E_r \quad \text{and} \quad \Phi_E = \oint E_r\,dA = E_r \oint dA = 4\pi r^2 E_r.$$

The radial component of the field E_r may be factored from the integral because it is constant over the surface. This simplification occurs because the surface has the same symmetry—spherical symmetry—as the field line pattern. ∎

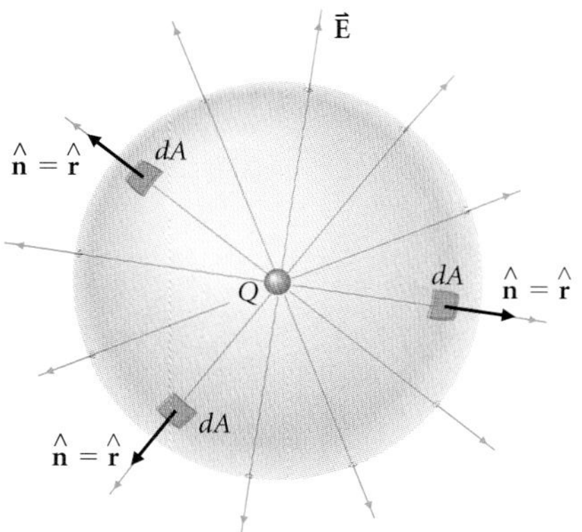

■ FIGURE 23.26
Electric flux through a sphere centered on a point charge Q. Here Q is positive. Everywhere on the surface of the sphere $\hat{\mathbf{n}} = \hat{\mathbf{r}}$ and is parallel to $\vec{\mathbf{E}}$. The product $\vec{\mathbf{E}} \cdot \hat{\mathbf{n}} = |\vec{\mathbf{E}}| = kQ/r^2$ has the same value everywhere on the surface of the sphere.

23.4.3 *Gauss' Law for the Electric Field*

A formal statement of the relation between charge and flux is known as Gauss' law.

> **GAUSS' LAW**
>
> The total electric flux emerging from an arbitrary volume equals the net charge enclosed within the volume divided by ϵ_0.
>
> $$\Phi_E \equiv \oint \vec{\mathbf{E}} \cdot \hat{\mathbf{n}}\,dA = \frac{Q_{\text{enclosed}}}{\epsilon_0}. \qquad \textbf{(23.12)}$$
>
> The integral is taken over the closed surface surrounding the volume, and $\hat{\mathbf{n}}$ is the outward normal to the surface element dA.

Karl Friedrich Gauss (1777–1855). A child prodigy, Gauss corrected his father's account books at age three. His family was not wealthy. Luckily his mathematical talents came to the attention of the Duke of Brunswick, who provided financial support until his career was established. Gauss received his Ph.D. in mathematics from the University of Helmstedt in 1799 and published his major work on arithmetic in 1801. He made contributions in number theory, non-Euclidean geometry, statistics, and the determination of cometary orbits. Most of his work in electricity and magnetism was accomplished between 1830 and 1840. He was one of the founders of the German Magnetic Union, which had the objective of making continuous observations of the Earth's magnetic field. He also developed an early version of the telegraph. The cgs-Gaussian unit system and its unit of magnetic induction both bear his name.

Examples 23.13 and 23.14 illustrate this statement. Notice that the total electric field, including that due to charges outside, is used to calculate the flux, but only the charge inside the surface contributes to Q_{enclosed}. Gauss' law is extremely useful for computing the electric field due to symmetric charge distributions.

EXAMPLE 23.15 ♦ A cubical box is known to contain a net charge of 6 μC. The measured flux through one face of the box is $8 \times 10^5\ \text{N}\cdot\text{m}^2/\text{C}$. What is the total flux through the other five sides?

MODEL Gauss' law relates the total flux to the known charge inside. Since we are given the flux through one side, the flux through the other five sides equals the total flux minus the given flux.

SETUP From Gauss' law, the total flux through the walls of the cubical box is:

$$\Phi_E = \frac{Q}{\epsilon_0} = \frac{6\ \mu\text{C}}{8.9 \times 10^{-12}\ \text{C}^2/\text{N}\cdot\text{m}^2} = 7 \times 10^5\ \text{N}\cdot\text{m}^2/\text{C}.$$

SOLVE The flux through the remaining five sides is:

$$\Phi_E = 7 \times 10^5\ \text{N}\cdot\text{m}^2/\text{C} - 8 \times 10^5\ \text{N}\cdot\text{m}^2/\text{C}$$
$$= -1 \times 10^5\ \text{N}\cdot\text{m}^2/\text{C}.$$

Digging Deeper

Formal Proof of Gauss' Law

The formal proof of Gauss' law uses the concept of *solid angle.*

The solid angle $d\Omega$ subtended by a surface of area dA at point P (■ Figure 23.28) is defined by:

$$d\Omega = \hat{\mathbf{r}} \cdot \hat{\mathbf{n}} \frac{dA}{r^2} = \frac{dA_\perp}{r^2}.$$

The unit of solid angle is the steradian; it is dimensionless. Qualitatively, the solid angle is the cone defined by the area dA and the point P. Any other area dA' perpendicular to $\vec{\mathbf{r}}$ and bounded by the cone also defines the solid angle: $d\Omega = dA'/r'^2$ (Figure 23.28b). In this way, a closed surface surrounding P can be replaced by a sphere centered on P. The sphere has area $4\pi r'^2$ and subtends solid angle $A/r'^2 = 4\pi r'^2/r'^2 = 4\pi$ at P. The solid angle subtended by any closed surface at a point within it is 4π steradians.

Consider a point charge q located at O inside the surface shown in ■ Figure 23.29. The element of surface shown has normal $\hat{\mathbf{n}}$, and the flux through the surface element due to the single charge is:

$$d\Phi_E = \vec{\mathbf{E}} \cdot \hat{\mathbf{n}}\, dA = \frac{kq}{r^2} \hat{\mathbf{r}} \cdot \hat{\mathbf{n}}\, dA.$$

But $dA\, \hat{\mathbf{r}} \cdot \hat{\mathbf{n}}/r^2$ is the element of solid angle $d\Omega$ subtended at O by dA.

$$d\Phi_E = kq\, d\Omega.$$

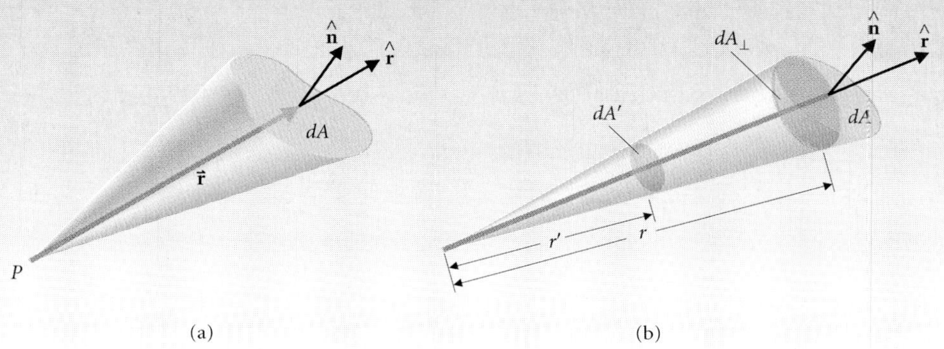

■ **Figure 23.28**
(a) The solid angle subtended at point P by the surface element of area dA is defined to be $d\Omega = \hat{\mathbf{r}} \cdot \hat{\mathbf{n}}/r^2$, where $\hat{\mathbf{r}}$ points from P to the surface element, and $\hat{\mathbf{n}}$ is the normal to the surface element. Qualitatively, $d\Omega$ is the angle of the cone formed by the area element with apex at P. (b) Any two areas bounded by the same cone define the same solid angle at P. Thus a closed surface and any sphere within it subtend the same solid angle at the center of the sphere. That solid angle is 4π steradians.

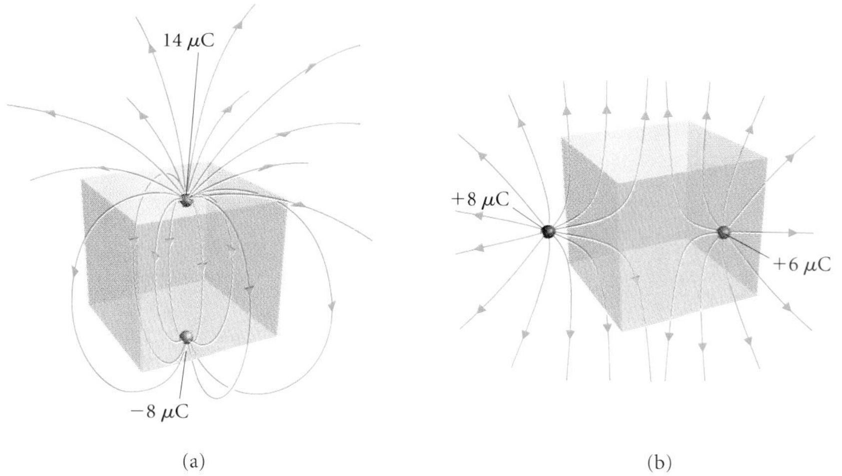

■ **Figure 23.27**
(a) The net charge in the box is 14 μC − 8 μC = 6 μC. However, the flux through the top surface is positive, while the net flux through the rest of the box is negative. (b) Field lines from the positive charge outside the box pass inward, contributing negative flux through some of the faces. However, all these field lines also leave the box, contributing positive flux through other faces. The net flux through the entire surface corresponds to the 6-μC charge *inside.*

ANALYZE How could this situation arise? The box could contain both positive and negative charges, since the net flux through the other five sides is negative (■ Figure 23.27a). Alternatively, negative flux through one or more surfaces could be due to positive charge outside the box (Figure 23.27b).

EXERCISE 23.7 ♦ A spherical volume surrounds two point charges $Q_1 = 5$ μC and $Q_2 = -3$ μC. What is the electric flux through the surface of the sphere?

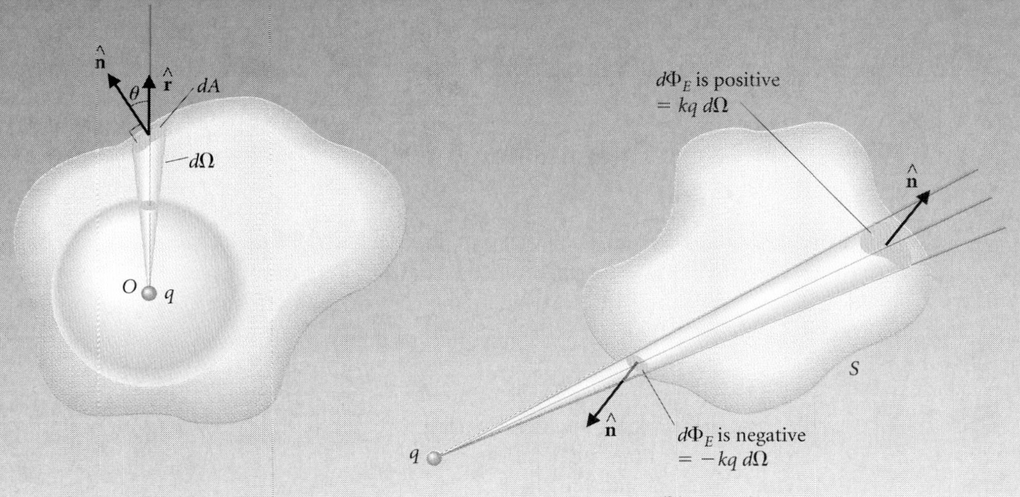

(a) (b)

■ FIGURE 23.29
(a) We may express the electric flux due to a point charge through an element of surface in terms of the solid angle subtended by the surface at the position of the charge. (b) A cone of field lines from any charge outside a closed surface intersects the surface twice. The flux contributed by the element of area closest to the charge is negative, since $\hat{\mathbf{n}}$ and $\hat{\mathbf{r}}$ are almost opposite each other: $d\Phi_E = -kq\,d\Omega$. The element of area on the far side of the surface contributes a positive flux $kq\,d\Omega$, and the total is zero.

To find the total flux we sum over all the surface elements:

$$\Phi_E = \oint kq\,d\Omega = kq \oint d\Omega = 4\pi kq,$$

where $\oint d\Omega = 4\pi$ is the solid angle subtended by the whole surface at O. If there is more than one charge inside, we use the principle of superposition to find the flux they contribute:

$$\vec{\mathbf{E}} = \sum_i \vec{\mathbf{E}}_i.$$

Then: $$\Phi_E = \oint \vec{\mathbf{E}} \cdot \hat{\mathbf{n}}\,dA = \oint \sum_i \vec{\mathbf{E}}_i \cdot \hat{\mathbf{n}}\,dA = \sum_i \oint \vec{\mathbf{E}}_i \cdot \hat{\mathbf{n}}\,dA$$

$$= \sum_i 4\pi k q_i = 4\pi k \sum_i q_i.$$

If a charge is outside the surface (Figure 23.29b), the cone that defines the solid angle subtended at O by an element of the surface also cuts the surface at another place. The contributions to the flux through the surface contributed by the pair of surface elements are equal and opposite. So any charge outside the surface contributes zero net flux. Then the net flux due to all charges, both inside and outside, is:

$$\Phi_E = 4\pi k \sum_{\substack{\text{charges}\\\text{inside}\\\text{surface}}} q_i.$$

This is Gauss' law.

Chapter Summary

Where Are We Now?

We introduced electric charge as the fundamental property that produces electromagnetic fields and upon which those fields act. We learned how to use Coulomb's law to find the electric field due to a system of point charges, and we developed field line diagrams as a tool for visualizing electric fields. Gauss' law provides us a quantitative statement of global relations between charge and field. These ideas should form the basis for your thinking as we develop the concepts more fully in the chapters that follow.

What Did We Do?

There are two kinds of charge, labeled positive and negative. Charge can neither be created nor destroyed: electric phenomena occur when positive and negative charges are separated from each other. The SI unit of charge is the coulomb (C). Ordinary matter is composed of atoms, each with a positively charged nucleus surrounded by negatively charged electrons. A proton has charge $+e$ and an electron has charge $-e$: $e = 1.6 \times 10^{-19}$ C. The electrical behavior of everyday materials is determined by the forces the atoms exert on their outermost electrons.

A point charge Q produces an electric field vector at P according to Coulomb's law:

$$\vec{\mathbf{E}} = k\frac{Q}{r^2}\hat{\mathbf{r}},$$

where $\hat{\mathbf{r}}$ is a unit vector that points from Q to P. The constant $k = 8.99 \times 10^9\ \mathrm{N\cdot m^2/C^2} = 1/(4\pi\epsilon_0)$, where $\epsilon_0 = 8.85 \times 10^{-12}\ \mathrm{C^2/N\cdot m^2}$. The force exerted on a test charge q placed at P is

$$\vec{\mathbf{F}} = q\vec{\mathbf{E}}.$$

The principle of superposition states that the electric field due to a system of charges is the sum of the electric field vectors produced by each charge in the system.

Field line diagrams are useful for visualizing electric fields. There are five rules for drawing diagrams:

1. Field lines begin at positive charge and end at negative charge.
2. The number of field lines shown diverging from or converging into a point charge is proportional to the magnitude of the charge.
3. Field lines are spherically symmetric near a point charge.
4. If the system has a net charge, the field lines are spherically symmetric at great distances.
5. Field lines never cross.

The electric flux Φ_E through a surface S is a quantitative measure of field lines passing through that surface:

$$\Phi_E = \int \vec{\mathbf{E}} \cdot \hat{\mathbf{n}}\, dA.$$

Gauss' law states the relation between charge and flux:

> The electric flux Φ_E through a closed surface S enclosing a volume V is equal to the net charge inside the volume divided by ϵ_0.

Practical Applications

Charge buildup in clouds produces strong electric fields and lightning. The topics discussed in this chapter form the basis for a classical description of the structure of matter. Field line diagrams are a useful tool for estimating how different components of a semiconductor circuit interfere with each other.

Solutions to Exercises

23.1 From Newton's second law, the acceleration of a particle is $a = F/m$. If forces of equal magnitude act on two different particles, their accelerations are in the ratio:

$$\frac{a_1}{a_2} = \frac{F/m_1}{F/m_2} = \frac{m_2}{m_1}.$$

A proton is roughly 2000 times as massive as an electron, so

$$a_e/a_p \approx 2000.$$

23.2 See ■ Figure 23.30. The distance r of point P from the charge is given by:

$$r^2 = [2.0\ \mathrm{m} - (-1.0\ \mathrm{m})]^2 + [1.0\ \mathrm{m} - (-1.0\ \mathrm{m})]^2 = 13.0\ \mathrm{m}^2.$$

Then the electric field at P has magnitude:

$$E = \frac{kQ}{r^2} = \frac{(8.99 \times 10^9\ \mathrm{N\cdot m^2/C^2})(10.0 \times 10^{-6}\ \mathrm{C})}{13.0\ \mathrm{m}^2}$$

$$= 6.92 \times 10^3\ \mathrm{N/C}.$$

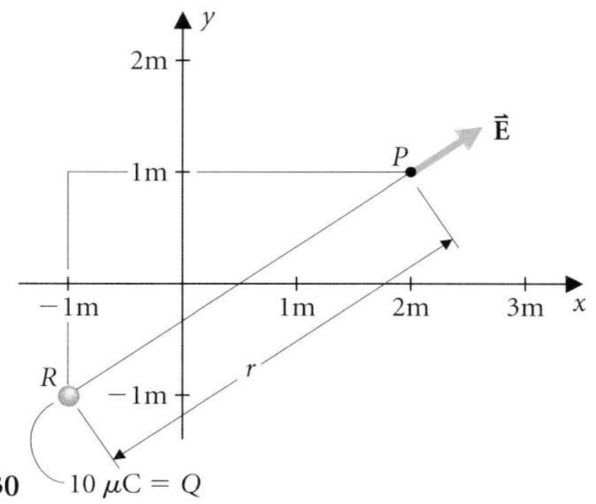

■ FIGURE 23.30

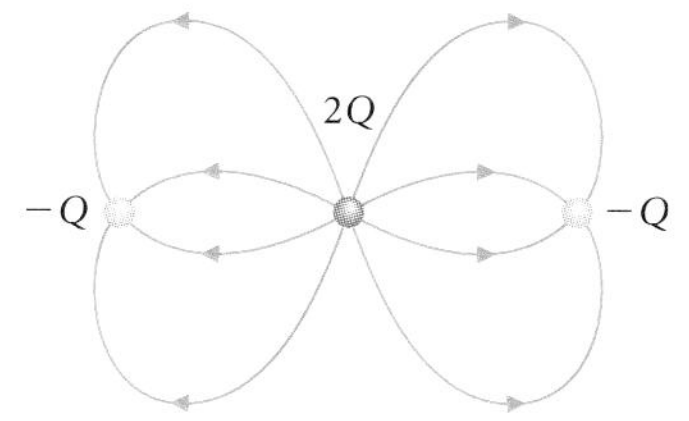

■ FIGURE 23.31

The direction of the field is:

$$\hat{\mathbf{r}} = \frac{3}{\sqrt{13}}\,\hat{\mathbf{i}} + \frac{2}{\sqrt{13}}\,\hat{\mathbf{j}} = 0.832\hat{\mathbf{i}} + 0.555\hat{\mathbf{j}}.$$

Finally we have: $\vec{\mathbf{E}} = (5.8\hat{\mathbf{i}} + 3.8\hat{\mathbf{j}}) \times 10^3$ N/C.

23.3 Since the positive charge has twice the magnitude of each negative charge, we show twice as many field lines emerging from it as converging into each of the negative charges (■ Figure 23.31). The lines have spherical symmetry near each charge.

23.4 Each charge produces an electric field that points along the x-axis:

$$\vec{\mathbf{E}}_1 = \hat{\mathbf{i}}\frac{kQ}{(x-a)^2} \quad \text{and} \quad \vec{\mathbf{E}}_2 = \hat{\mathbf{i}}\frac{kQ}{(x+a)^2}.$$

Using the principle of superposition:

$$\vec{\mathbf{E}} = \vec{\mathbf{E}}_1 + \vec{\mathbf{E}}_2 = \hat{\mathbf{i}}kQ\left[\frac{1}{(x-a)^2} + \frac{1}{(x+a)^2}\right] = \hat{\mathbf{i}}kQ\frac{2(x^2+a^2)}{(x^2-a^2)^2}.$$

The field has magnitude $\approx 2kQ/x^2$ at large x, as expected from Coulomb's law.

23.5 Six field lines (the ones at the right) leave volume A. Three lines at the left (including those labeled (a) and (b)) leave and then reenter. Three lines pass directly to the negative charge inside the volume and do not leave at all. The net number of lines leaving is six, consistent with the net charge $2Q - Q = Q$ inside volume A.

23.6 We need the component of $\vec{\mathbf{E}}$ normal to the surface:

$$\vec{\mathbf{E}}\cdot\hat{\mathbf{n}} = \frac{(1.0 \times 10^{-6}\ \text{N/C})}{\sqrt{2}}(\hat{\mathbf{i}} + \hat{\mathbf{k}})\cdot\hat{\mathbf{k}} = 7.1 \times 10^{-7}\ \text{N/C}.$$

Since the field is uniform across the area, the flux is:

$$\Phi_E = \vec{\mathbf{E}}\cdot\hat{\mathbf{n}}A = (7.1 \times 10^{-7}\ \text{N/C})(1.0\ \text{m}^2) = 7.1 \times 10^{-7}\ \text{N}\cdot\text{m}^2/\text{C}.$$

23.7 From Gauss' law, the charge inside the volume determines the flux through the surface. The net charge inside is $5\ \mu\text{C} - 3\ \mu\text{C} = 2\ \mu\text{C}$. Thus the flux through the surface of the sphere is:

$$\Phi_E = \frac{Q}{\epsilon_0} = \frac{2 \times 10^{-6}\ \text{C}}{8.85 \times 10^{-12}\ \text{C}^2/\text{N}\cdot\text{m}^2} = 2 \times 10^5\ \text{N}\cdot\text{m}^2/\text{C}.$$

Basic Skills

Review Questions

§23.1 ELECTRIC CHARGE

- Describe an experiment that indicates there are two kinds of charge.
- State the *law of conservation of charge.*
- What is the charge on an electron?
- What is a *point charge?*
- State *Coulomb's law.*
- What is the *triboelectric sequence?*
- What is the difference between a *conductor* and an *insulator?*
- In what sense is the coulomb a large unit of charge?

§23.2 THE ELECTRIC FIELD OF A POINT CHARGE

- What is the operational definition of *electric field?*
- Describe the electric field due to a point charge Q. State how its magnitude and direction vary with position.

§23.3 THE PRINCIPLE OF SUPERPOSITION

- State the *principle of superposition.*
- What is a *field line diagram?* Explain how the field lines describe the magnitude and direction of $\vec{\mathbf{E}}$.
- State the five rules for drawing field line diagrams for a collection of point charges.

§23.4 GAUSS' LAW

- Explain how a field line diagram illustrates Gauss' law.
- What is the electric flux through a plane surface of area A?
- What is a closed surface?
- State Gauss' law (a) in words and (b) as an equation. Carefully define all the symbols that appear in your equation.

Basic Skill Drill

§23.1 ELECTRIC CHARGE

1. A neutron is an unstable particle that decays into a proton, an electron, and an antineutrino. What is the charge on the antineutrino?
2. (a) A repulsive force of 1 N is exerted by a charge Q on a charge q. What force is exerted on Q by q? **(b)** If q is replaced by a charge $2q$ at the same location, what force is exerted by Q on the $2q$ charge? **(c)** If q is moved to twice its original distance from Q, what force is exerted on q by Q?
3. A person wearing a wool sweater and a cotton T-shirt steps onto an insulating platform and then removes the sweater. What sign of charge remains on the person? On the sweater?
4. Find the force on a charge $q = 1.0 \times 10^{-10}$ C placed 1.0 m from a charge $Q = -5.0 \times 10^{-3}$ C. (Give magnitude and direction.)

§23.2 THE ELECTRIC FIELD OF A POINT CHARGE

5. A point P is a distance $r = 1.0$ mm from a charge $Q = -1.0\ \mu$C. Compute the magnitude of the electric field vector at P. Describe its direction.
6. A particle with a charge of 1.0×10^{-9} C experiences a force of 3.0×10^{-5} N when placed at P. What is the magnitude of the electric field at P?

§23.3 THE PRINCIPLE OF SUPERPOSITION

7. Two identical positive charges, each of magnitude Q, are held fixed on the x-axis at $x = +1.0$ m and $x = -1.0$ m. A tiny test charge $+q$ is used to test the field of the two charges. Where, in addition to the points at infinite distance, is the electrostatic force on the test charge zero?
(a) Only at the exact midpoint between the two charges.
(b) At two points, $x = +2.0$ m and $x = -2.0$ m on the x-axis.
(c) At two points, $y = +1.0$ m and $y = -1.0$ m on the y-axis.
(d) At any point on the y-axis.
(e) Nowhere.
Explain your reasoning.
8. Two point charges of 15 μC each are on the x-axis at $x = 1.0$ m and $x = 1.5$ m. Find the electric field vector at the origin.
9. Three equal charges are at the corners of an equilateral triangle. Sketch the field line diagram in the plane of the triangle.

§23.4 GAUSS' LAW

10. An electric dipole (two equal and opposite point charges separated by a small distance) lies at the center of a spherical container. What is the electric flux through the sphere? Does your answer change if the dipole is not at the exact center? Explain your reasoning.
11. The point charges shown in ■ Figure 23.32 are held stationary. What are the values of electric flux through surfaces A, B, and C? If $Q = 4.425\ \mu$C, give numerical values for the three values of flux.

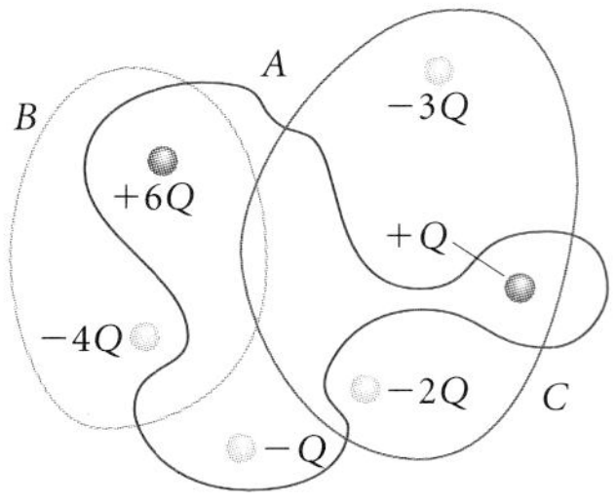

■ FIGURE 23.32

Questions and Problems

§23.1 ELECTRIC CHARGE

12. ❖ The distance between two charges is reduced by half. The force between them:
(a) becomes insignificant compared with gravity.
(b) becomes twice as large.
(c) becomes four times as large.
(d) becomes half as large.
(e) becomes one-fourth as large.
Explain how you chose your answer.
13. ❖ A rubber rod is stroked with wool and then allowed to come in contact with a pith ball. A glass rod is rubbed with silk and then placed in contact with a second pith ball. Is the force between the two charged pith balls attractive or repulsive? Explain your reasoning.
14. ❖ In the early universe and in stars, deuterium is produced from the combination of a proton and a neutron, with the release of a gamma ray. What is the charge on a deuterium nucleus?
15. ♦ Two electrons in a helium atom are separated by about 10^{-10} m. Estimate the magnitude of the repulsive force between them.
16. ♦ Two particles with equal mass have charges of 1.0 μC and 1.0 nC, respectively. What must be their mass if their mutual gravitational attraction is to overcome their Coulomb repulsion?
17. ♦ Two point charges of 1.0×10^{-6} C each are held together by a string 15 cm long. What is the tension in the string?
18. ♦♦ Two point charges 2.0 m apart exert repulsive forces of magnitude 2.7×10^{-2} N on each other. If one charge is known to be twice as large as the other, what are the two charges? Another pair of charges exert attractive forces of magnitude 2.7×10^{-2} N when 2.0 m apart. If the total charge of the pair is $+1.0\ \mu$C, what are the two charges?
19. ♦♦ Two objects, each carrying charge Q, are connected by a spring with force constant k_s and relaxed length zero. When the system is in equilibrium, what is the distance between the two charges?
20. ♦♦ In the cgs-Gaussian unit system, Coulomb's law takes the form $F = q_1q_2/r^2$, with no constant. Find the dimensions of charge in the Gaussian unit system.

21. ♦♦♦ Show that the force between two objects that share a total charge Q is largest when they share the charge equally, that is, when each has charge $Q/2$.

§23.2 THE ELECTRIC FIELD OF A POINT CHARGE

22. ♦ A point charge $Q = 16.5\ \mu C$ is at the origin. Find the electric field vector at the point P with coordinates $x = 15$ cm, $y = 25$ cm, $z = 0$.

23. ♦ The electric field strength 17 cm from a point charge is measured to be 56 N/C. What is the value of the charge?

24. ♦♦ A point charge $Q = -15$ nC is at the point $x = 1.0$ m, $y = 2.0$ m, $z = 0$. Find the electric field vector at the point $x = 2.0$ m, $y = -1.0$ m, $z = 1.0$ m.

25. ♦♦ A point charge $Q = 33$ nC is located at $x = 55$ mm, $y = 27$ mm, $z = 15$ mm. Find the electric field vector at the point P with coordinates $x = 15$ mm, $y = 35$ mm, $z = -6$ mm.

26. ♦♦ A pith ball with mass $m = 1.0$ g and charge 150 nC hangs from a thread in a region with uniform horizontal electric field of strength $E = 1.0 \times 10^4$ N/C. At what angle from the vertical does the ball hang?

27. ♦♦ What electric field vector is required to support an electron (mass 9×10^{-31} kg) against gravity?

28. ♦♦ A small oil drop contains exactly one excess electron. If the mass of the drop is 3×10^{-15} kg, what electric field vector is needed to support the drop against gravity?

29. ♦♦ On a windless day, when the electric field strength above the Earth's surface is 100 N/C, a fog droplet with a net charge of 250 electrons remains motionless. What is the mass of the droplet? What is its radius?

§23.3 THE PRINCIPLE OF SUPERPOSITION

30. ❖ A positive and a negative charge of equal magnitude are separated by a distance ℓ (■ Figure 23.33). What is the direction of the electric field vector at R and at P?

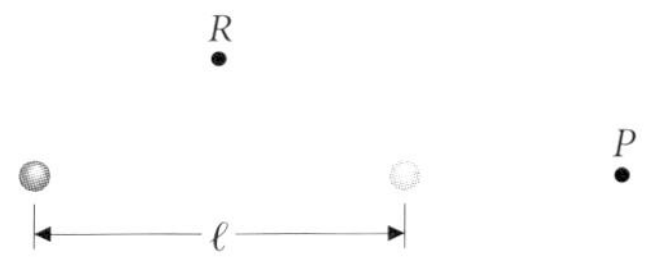

■ FIGURE 23.33

31. ❖ What is the direction of the electric field vector at R and at P in Figure 23.33 if the negative charge is replaced by a positive charge of equal magnitude?

32. ❖ A positive charge $2Q$ is at one vertex and two charges $-Q$ are at the other two vertices of an equilateral triangle. Sketch the field line diagram in the plane of the triangle.

33. ❖ Criticize the following statement: "An object cannot produce an electric field unless it has a net electric charge."

34. ❖ A charge of $+3Q$ is at the apex of a tetrahedron. Three negative charges, each $-Q$, are at the corners of the base. Sketch the field line diagram.

35. ❖ Positive charges $+Q$ and negative charges $-Q$ are spaced along a line with a separation d between neighboring charges, so that charges alternate in sign. Sketch a field line diagram for this system.

36. ❖ Six equal positive charges are equally spaced around a circle. Sketch a field line diagram in the plane of the circle and in a plane through the center and perpendicular to the circle.

37. ❖ Six equal charges Q are placed at the corners of a hexagon. What is the force on a test charge q placed at the center of the hexagon?

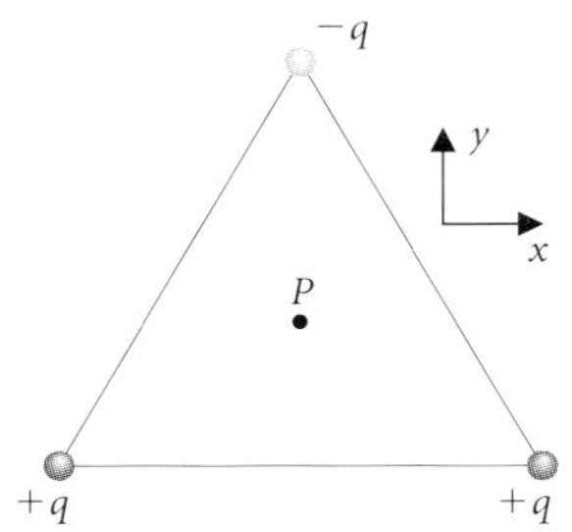

■ FIGURE 23.34

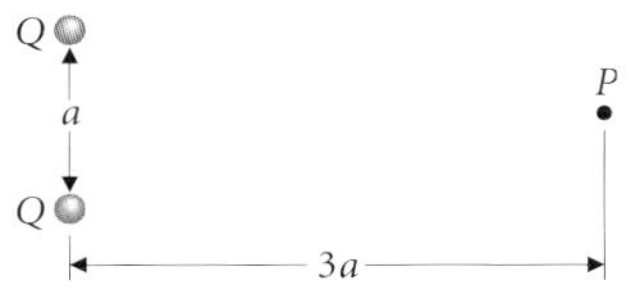

■ FIGURE 23.35

38. ❖ Three point charges (two $+q$, one $-q$) are arranged at the points of an equilateral triangle with sides of length L (■ Figure 23.34). The electric field at the center of the triangle:
(a) is zero. **(b)** is infinite.
(c) is in the y-direction. **(d)** is in the x-direction.
(e) is equal to q/L^2.
Explain what is wrong with the answers you don't choose.

39. ❖ Five equal negative charges are placed at the corners of a pentagon. A small positive test charge q is placed at the center of the pentagon. What is the force on the test charge? If one of the charges on the pentagon is removed, in which direction does the test charge accelerate?

40. ❖ Three charges, one $+Q$ and two with charge $-4Q$, are to be arranged so that the total electric force on each charge is zero. How can this be done? Is the arrangement stable?

41. ❖ If the distance between the two point charges Q in ■ Figure 23.35 is reduced by half, the magnitude of the electric field at P:
(a) decreases slightly. **(b)** is reduced by a factor 2.
(c) is increased by a factor 2. **(d)** is increased by a factor 4.
(e) increases slightly.
Explain what is wrong with the answers you don't choose.

42. ❖ Four equal positive charges are at the corners of a square. Sketch a field line diagram for this system in a plane perpendicular to the square and through a diagonal. Describe how the diagram is similar to that for two equal charges, and how it differs.

43. ♦ A point charge $Q = 3.0\ \mu C$ is located at the origin. **(a)** What is the magnitude of the electrostatic force on a positive charge $q = 1.0\ \mu C$ placed at point P with coordinates $x = 2.0$ m, $y = 2.0$ m, $z = 0$? **(b)** A second charge $-Q$ is placed at position R with coordinates $x = 0$, $y = 4.0$ m, $z = 0$. What is the total electrostatic force on q?

44. ♦ A point charge of 20.0 nC is at the origin. A second charge of -30.0 nC is on the z-axis at $z = 1.00$ m. Find the electric field vector at the point $x = 0$, $y = 2.00$ m, $z = 1.00$ m.

45. ♦ Three charges $+Q$ and one charge $-Q$ are at the corners of a square of side a. Find the magnitude of the electric field at the center of the square produced by each charge. Find the magnitude of the total electric field at the center. What direction is the force on a small negative test charge at the center?

46. ♦♦ Two point charges, each of 15 nC, are on the x-axis at $x = 1.0$ m and $x = 1.5$ m. Find the electric field as a function of position on the x-axis.

47. ♦♦ Four equal charges $+Q$ are placed at the corners of a square of side a. Find the electric field strength **(a)** at the center of the square, **(b)** at one corner, and **(c)** at the midpoint of one side. In each case, state which charges contribute to the field you compute, and why.

48. ♦♦ A charge $2Q$ and two equal charges $-Q$ lie on a line with each negative charge a distance d from the positive charge (see Exercise 23.3). What is the electric field vector: **(a)** at the position of the

positive charge? **(b)** at the position of one of the negative charges? **(c)** at an arbitrary point on the positive y-axis? **(d)** on the x-axis, at a point with $x > d$? In each case, state which charges contribute to the field you compute, and why.

49. ◆◆ Two point charges are held fixed on the x-axis. One with charge $Q_1 = -2.0\ \mu C$ is at $x = -3.0$ m, and the other with charge $Q_2 = +4.0\ \mu C$ is at $x = +1.0$ m. Find a point where a third charge may be placed and experience no net electric force.

50. ◆◆ Three positive charges of 1.0×10^{-5} C are each at the corner of an equilateral triangle of side 10.0 cm. Find the force on a fourth charge of 1.0×10^{-8} C placed at the midpoint of one side of the triangle.

51. ◆◆ Find the electric field at the origin due to the system of charges in ■ Figure 23.36.

52. ◆◆◆ Find the electric field produced by the system of charges in ■ Figure 23.36 as a function of position on the z-axis.

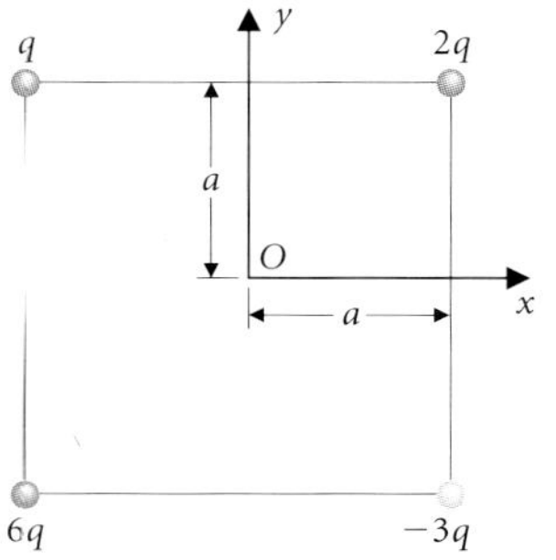

■ FIGURE 23.36

§23.4 GAUSS' LAW

53. ❖ If the electric flux through a closed surface is zero, is the field identically zero everywhere on the surface? Discuss.

54. ❖ An opaque cubical box is observed to have a net number of electric field lines emerging from it. What can you say about the net charge in the box?

(a) Nothing. **(b)** Its sign but not its magnitude.
(c) Its magnitude but not its sign.
(d) Both its sign and its magnitude.

Explain how you chose your answer, and if you chose (b), (c), or (d), give the sign or magnitude, as appropriate.

55. ❖ A point charge Q is in the exact center of a cubical surface of side a. What is the electric flux emerging through one face of the cube? Can you answer the question easily if the charge is within the cube but not at the exact center? Why or why not?

56. ❖ You observe that field lines are emerging from a closed box, but every field line that leaves the box reenters. Which of the following statements are true? Explain your answers.

(a) There is no charge in the box.
(b) The net charge in the box is positive.
(c) The net charge in the box is negative.
(d) There are both positive and negative charges in the box, but the net charge is zero.
(e) None of the statements (a)–(d) is true: it is impossible to say anything about the charge in the box.

57. ◆ A tetrahedron contains a point charge $Q = 65$ nC. Find the electric flux through the surface of the tetrahedron. Can you find the flux through each face separately? Under what conditions?

58. ◆ A cube of side $\ell = 2.0$ m is centered at the origin, with the coordinate axes perpendicular to its faces. Find the flux of the electric field $\vec{E} = (15\ N/C)\hat{\imath} + (27\ N/C)\hat{\jmath} + (39\ N/C)\hat{k}$ through each face of the cube.

59. ◆◆ The electric field in a certain region of space is described by the function:

$$\vec{E} = (6.0\ N/C)\hat{\imath} + (7.0\ N/C)\hat{\jmath}.$$

Find the electric flux through the surface $x = 6y$, $0 < x < 6.0$ m, $0 < z < 1.0$ m.

60. ◆◆ A tetrahedron of side ℓ has one face in the x-y-plane. The electric field in the region has magnitude E_o and is in the positive z-direction. What is the electric flux through the surface of the tetrahedron? Find the flux through each surface of the tetrahedron.

61. ◆◆◆ Find the flux of the electric field $\vec{E} = (250\ N/C)\hat{\jmath}$ through the surface $x^2 + y^2 + z^2 = 1.0\ m^2$, $x > 0$, $y > 0$, $z > 0$. (*Hint:* The easiest way is to use Gauss' law.)

Additional Problems

62. ❖ How can you use a field line diagram to estimate the electric field strength at a point? Use this result to explain why the electric flux through a surface is proportional to the number of field lines crossing that surface.

63. ❖ Three equal charges are in a line with equal separations between them. The outer two are fixed; the center one is free. Show that the force on the central charge is zero. Is the equilibrium stable or unstable? What if the central charge is negative? Is the magnitude of the central charge important for its stability?

64. ❖ A positive test charge q is in equilibrium at the center of a square that has four equal charges $+Q$ at its corners. Is the equilibrium stable against small displacements: **(a)** perpendicular to the plane of the square? **(b)** in the plane of the square and perpendicular to one side? **(c)** in the plane of the square and toward one corner?

65. ◆◆ What is the total amount of negative charge contained in a raindrop 3.0 mm in diameter? If by magic we could suddenly remove all this negative charge, leaving the positive charge behind, what would be the acceleration of two raindrops 1 m apart?

66. ◆◆ Suppose an electron is removed from this book and placed 1 mm above the page. Calculate its acceleration, and *estimate* how long it would take to return to its proper place.

67. ◆◆◆ In the system of Study Problem 15, describe how the strength of the electric field varies along the y-axis with $y > D$. At what point is the y-component of the field maximum? What feature of the field line diagram shows the location of this point?

68. ❖ While standing vertically, brushing dust from your formal suit, you notice a piece of lint moving at constant speed toward your coat at an angle of 45° to the vertical. Why does the lint not fall vertically? Compare the weight of the lint, the electrical force on the lint, and the air resistance to its motion.

69. ◆◆◆ Two equal point charges Q are on the x-axis at $x = \pm a$ (cf. Example 23.11). Where on the y-axis is the electric field magnitude a maximum?

Computer Problems

70. Set up a spreadsheet program to calculate the x-, y-, and z-components of the electric field due to the collection of point charges described in the file CH23P70 on the supplementary computer disk. Calculate the field components at points P_1 and P_2 with coordinates $(-3.0\ m, -2.0\ m, 2.3\ m)$ and $(1.5\ m, 2.0\ m, 1.3\ m)$.

71. A point charge $+Q$ is at the origin, and two charges $-Q$ are on the x-axis at $x = +L$ and $x = -L$. Locate qualitatively two points where $\vec{E} = 0$. Using 12 lines per charge, sketch the electric field line diagram for the system in the x-y-plane. Discuss the relation of your diagram to the $\vec{E} = 0$ points. Use Coulomb's law to find the exact location of the two points.

72. Two point charges of $1.00\ \mu C$ and $-5.00\ \mu C$ are at the ends of a spring of unstretched length $\ell = 5.00$ cm and $k = 5.00 \times 10^4$ N/m. What is the separation of the charges in equilibrium?

73. Use the program "FIELDLINE" on your supplementary computer disk to draw the field line diagram for the following sets of charges: **(a)** A negative charge of -1 unit at $x = 200$, $y = 250$; a charge of -1 unit at $x = 300$, $y = 175$; and a charge of $+3$ units at $x = 200$, $y = 200$. Use a step size of 1.0, 10 lines per charge unit, and starting angle $\pi/4$. The program works best if you enter the largest charge first. **(b)** Four charges of ± 1 unit at the corners of a square, arranged so that charges of the same sign are at opposite ends of each diagonal. Comment on the diagrams. In particular note how many lines go off to infinity, where the field is weak, and where it is strong.

Challenge Problems

74. Show that a displacement along a field line lying in the x-y-plane satisfies the equation:

$$\frac{dy}{dx} = \frac{E_y}{E_x}.$$

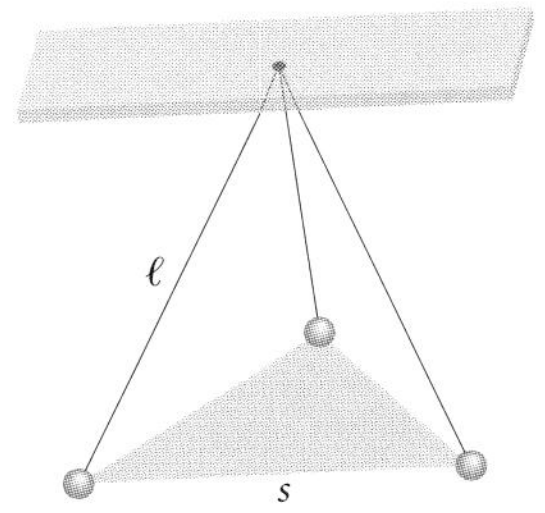

■ FIGURE 23.37

75. Find an analytic expression for the x- and y-components of the electric field due to two positive charges on the x-axis at $x = a$ and $x = -a$. Use the result of Problem 74 to find a differential equation for the field lines. (There is no need to solve the equation.)

76. Three identical charged objects of mass m and charge Q are suspended from a common point by massless strings as shown in ■ Figure 23.37. Show that the dimension s satisfies the equation:

$$s^3 = \frac{3kQ^2}{mg}\sqrt{\ell^2 - s^2/3}.$$

Solve this equation approximately in the two limits **(a)** $s \ll \ell$ and **(b)** $s \approx \sqrt{(3)}\ell$. Under what conditions are the approximate solutions valid?

Computer problem: For the cases $m = 1.0$ g, $\ell = 0.25$ m, **(a)** $Q = 55$ nC and **(b)** $Q = 450$ nC, use a numerical method to solve this equation exactly for $x = s/\ell$, and compare each result with the appropriate approximate solution.

In a television set, electric fields accelerate electrons in the television tube. Magnetic forces (Chapter 29) move the electrons across the end of the tube to form the picture. To understand how the television works, we must be able to compute the electric field in the tube and to determine how the electron responds. In this chapter we begin to develop the necessary techniques. The electron motion is discussed in Example 24.8.

CHAPTER 24
Static Electric Fields

Concepts

Field line symmetry
Charge density
Dipole

Goals

Be able to:

Find the electric field produced by a continuous distribution of charge.

Use symmetry arguments to simplify electric field calculations.

Compute the motion of a test charge in a given electric field.

And I have tried to apprehend the Pythagorean power
by which number holds sway above the flux.
A little of this, but not much, I have achieved.

Bertrand Russell

The picture on your television screen is produced by a beam of electrons accelerated down the tube by an electric field. To design and build a good TV set, the manufacturer needs to control the electric field in the tube precisely and to understand the electron's response to that field. Photocopiers and ink-jet printers require a similarly well-tuned electric field if they are to produce high-quality documents and graphics. In this chapter we'll study how to calculate such fields and the behavior of charged particles exposed to them.

BEFORE BEGINNING THIS CHAPTER, REVIEW THE STEPS IN THE INTEGRATION METHOD FROM INTERLUDE 2.

The electric field in a television set or a photocopier is due to a distribution of charge spread over surfaces rather than to a few isolated point charges. Of course, the charge is a property of individual electrons and atomic nuclei, but their number is so immense that we may consider their distribution as continuous. We may model any distribution as a collection of differential elements, each containing many electrons and nuclei but small enough to be treated as a point charge. Then the principle of superposition gives $\vec{E}$ as a physical integral of the kind described in Interlude 2. If the charge distribution has appropriate symmetry, we may use Gauss' law to calculate the field.

REMEMBER: MODELING THE SYSTEM AS A COLLECTION OF DIFFERENTIAL ELEMENTS IS THE FIRST STEP IN THE INTEGRATION METHOD. AN EXAMPLE IS ILLUSTRATED IN FIGURE 24.8.

24.1 USING GAUSS' LAW TO CALCULATE ELECTRIC FIELD

Before beginning to calculate the electric field of any charge distribution, it is wise to sketch a field line diagram that shows the field's overall behavior. When the charge distribution has appropriate symmetry, the resulting field pattern is also symmetric, and Gauss' law provides the simplest method for computing the field. The general method is outlined in ■ Figure 24.1.

THERE ARE THREE IMPORTANT CASES: SPHERICAL, CYLINDRICAL, AND PLANE SYMMETRY. WE'LL ENCOUNTER EXAMPLES OF EACH.

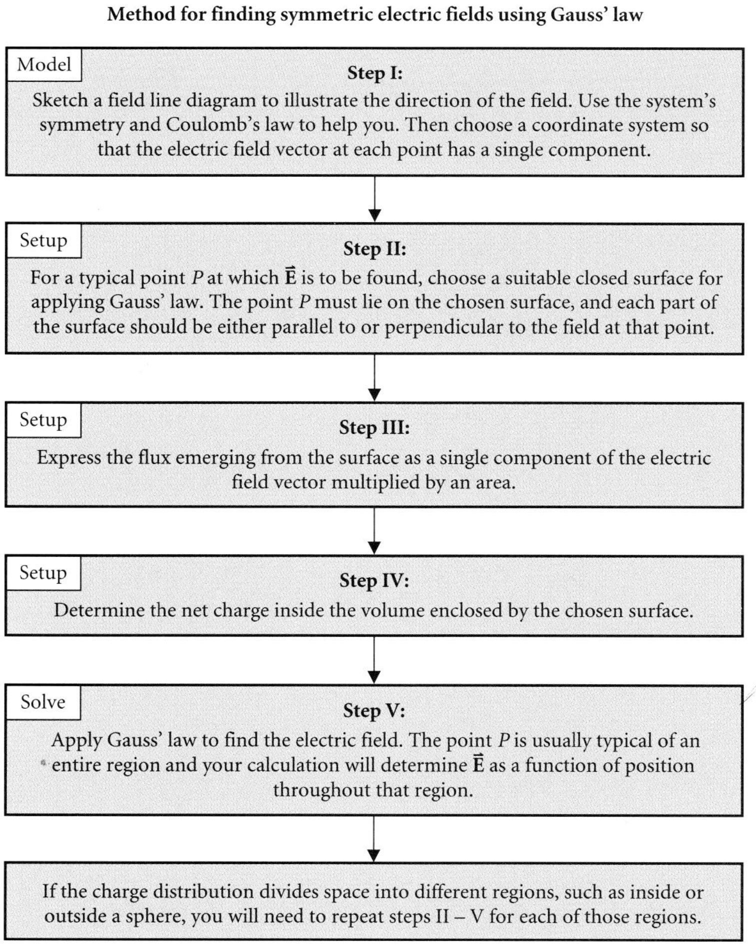

Method for finding symmetric electric fields using Gauss' law
Model
Step I:
Sketch a field line diagram to illustrate the direction of the field. Use the system's symmetry and Coulomb's law to help you. Then choose a coordinate system so that the electric field vector at each point has a single component.
Setup
Step II:
For a typical point P at which E is to be found, choose a suitable closed surface for applying Gauss' law. The point P must lie on the chosen surface, and each part of the surface should be either parallel to or perpendicular to the field at that point.
Setup
Step III:
Express the flux emerging from the surface as a single component of the electric field vector multiplied by an area.
Setup
Step IV:
Determine the net charge inside the volume enclosed by the chosen surface.
Solve
Step V:
Apply Gauss' law to find the electric field. The point P is usually typical of an entire region and your calculation will determine E as a function of position throughout that region.
If the charge distribution divides space into different regions, such as inside or outside a sphere, you will need to repeat steps II – V for each of those regions.

■ FIGURE 24.1

REMEMBER: A GAUSSIAN SURFACE COMPLETELY ENCLOSES A VOLUME.

IF THERE IS NOT ENOUGH SYMMETRY TO REDUCE THE FLUX INTEGRAL IN THIS WAY, GAUSS' LAW WON'T HELP YOU.

To find the electric field vector at a typical point P, construct an imaginary closed surface that passes through the point. This surface, which is called a *Gaussian surface,* should share the symmetry of the charge distribution and at each point be either parallel or perpendicular to $\vec{\mathbf{E}}$. Then, with *properly chosen coordinates,* $\vec{\mathbf{E}}$ will have only one component. This component will be constant on all parts of the surface where $\vec{\mathbf{E}} \cdot \hat{\mathbf{n}}$ is not zero. The flux integral $\oint \vec{\mathbf{E}} \cdot \hat{\mathbf{n}}\, dA$ for the Gaussian surface then reduces to an area multiplied by this constant component. We use the known charge distribution to calculate the total charge enclosed by the surface, apply Gauss' law to find the one component of $\vec{\mathbf{E}}$, and so determine $\vec{\mathbf{E}}$ at each point on the surface.

EXAMPLE 24.1 ♦♦ Find the value of $\vec{\mathbf{E}}$ at any point either inside or outside a uniformly charged, thin, spherical shell with radius R and total charge Q.

MODEL The charge distribution has spherical symmetry, so we may use Gauss' law to find the field. We'll need to do two calculations: one for the region outside the shell and one for the region inside.

STEP I The electric field lines share the spherical symmetry of the charge distribution (■ Figure 24.2). Outside the shell the field lines extend radially outward, making the same pattern that would be produced by a point charge located at the center of the shell.

Inside the shell, any field lines leaving the positive charge on the shell would have to point radially inward, converging on the center. Field lines can converge only on electric charge, and there is none at the center, so field lines cannot exist inside the shell.

SETUP We choose a spherical surface of radius r, concentric with the charged shell.
STEP II The normal to the surface is $\hat{\mathbf{n}} = \hat{\mathbf{r}}$ and is parallel to the electric field vector everywhere on the surface.

STEP III Because of the spherical symmetry, the radial component of the field has the same value at each point on the shell. We calculate the flux through the surface as follows:

Definition of flux:	$\Phi_E = \oint \vec{\mathbf{E}} \cdot \hat{\mathbf{n}}\, dA,$
normal $\hat{\mathbf{n}} = \hat{\mathbf{r}}$:	$= \oint \vec{\mathbf{E}} \cdot \hat{\mathbf{r}}\, dA,$
$\vec{\mathbf{E}} \cdot \hat{\mathbf{r}} = r$ component of $\vec{\mathbf{E}}$:	$= \oint E_r\, dA,$
E_r is constant over the surface:	$= E_r \oint dA,$
Surface area of a sphere is $4\pi r^2$:	$= 4\pi r^2 E_r.$

STEP IV The charge inside the Gaussian sphere is the total charge Q on the shell.

SOLVE Applying Gauss' law:
STEP V

$$\Phi_E = 4\pi r^2 E_r = Q/\epsilon_0. \tag{24.1}$$

Thus anywhere outside the shell:

$$\vec{\mathbf{E}} = E_r\hat{\mathbf{r}} = \frac{Q}{4\pi r^2 \epsilon_0}\hat{\mathbf{r}} \qquad (r > R).$$

SETUP AND SOLVE The field inside the shell is found in exactly the same way, except that there is no charge inside any Gaussian sphere within the shell. Setting $Q = 0$ in eqn. (24.1), we find that the electric field is zero anywhere inside the shell.

$$\vec{\mathbf{E}} = 0 \qquad (r < R).$$

ANALYZE The electric field outside the shell is identical to the field produced by a point charge Q at the center of the shell, but the field inside the shell is zero. The need to make separate arguments for separate spatial regions is typical of calculations with Gauss' law. ■

STUDY HINT: A SIMILAR LOGIC CHAIN APPLIES WHENEVER GAUSS' LAW IS USED.

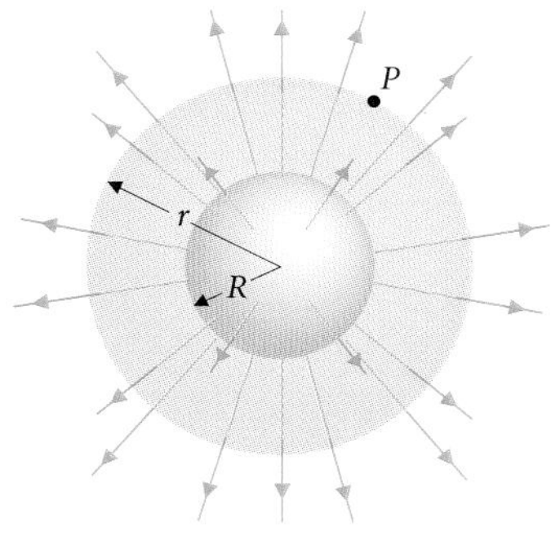

■ FIGURE 24.2
Electric field due to a uniformly charged spherical shell. Outside the shell, the field lines have spherical symmetry: they diverge from the origin. The field line pattern is the same as that due to a point charge at the origin, and the mathematical expression for the electric field is the same, too. Inside the shell, there are no field lines at all, and $\vec{\mathbf{E}} = 0$. We choose a sphere as the volume to which we apply Gauss' law.

EXERCISE 24.1 ❖ What is the electric field outside a ball with total charge Q uniformly distributed throughout its volume?

24.2 The Electric Field Due to a Linear Charge Distribution

Next we examine the electric field produced by charge uniformly distributed on a long straight filament such as the central electrode of a Geiger counter (■ Figure 24.3). This example illustrates the interplay between approximation and exact calculation in obtaining useful answers.

At large distances from any finite charge distribution, no matter how complicated, the system's structure is not noticeable. If the distribution has nonzero net charge, its field line diagram looks like that of a point charge. The electric field is approximately given by Coulomb's law with Q equal to the total net charge in the distribution.

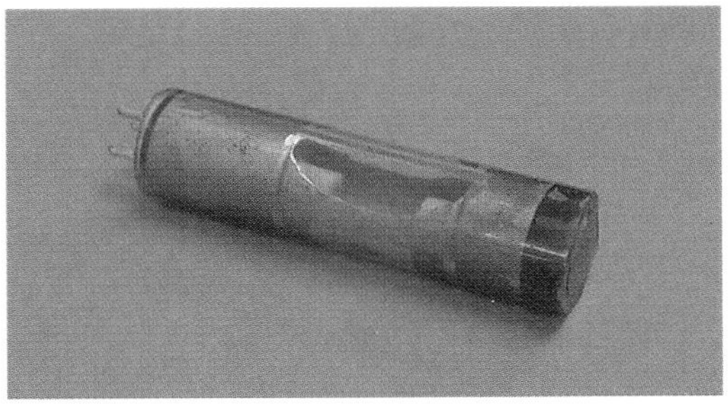

■ Figure 24.3
The central electrode of a Geiger counter is an example of a charged filament.

From nearby points, the distribution appears large. Often we may model it as infinite and find enough symmetry to apply Gauss' law. Even though no real distribution is truly infinite, this method usually provides an excellent approximation in a region close to the charge distribution. At intermediate distances, the electric field vector is found using Coulomb's law and integration. Because $\vec{\mathbf{E}}$ and each differential contribution $d\vec{\mathbf{E}}$ are vectors, we must take particular care with direction when evaluating the integral. It is usually best to calculate the x-, y-, and z-components of $\vec{\mathbf{E}}$ separately. These calculations are often intricate. The approximations at near and far distances provide a valuable check on the results.

24.2.1 *An Infinitely Long, Uniformly Charged Filament*

We model the filament as having negligible diameter. The amount of charge on the filament is described by a linear charge density:

> The *linear charge density* λ at a point P of a filament is the amount of charge per unit length along the filament. The charge dQ on a length $d\ell$ of the filament is given by:
>
> $$dQ = \lambda \; d\ell. \qquad (24.2)$$

A *uniformly* charged filament with length ℓ and total charge Q has the same value of λ at each point, and $\lambda = Q/\ell$.

Before calculating, we draw the field line diagram for the filament to see what we should expect. If the filament is very long, or if we consider points very close to it, it appears *infinitely* long. Such an infinitely long, uniformly charged filament is a highly symmetric object with a symmetric electric field. Its properties—charge density and diameter—don't depend on position along its length, and so the field lines do not bend along the direction of the filament but lie in planes perpendicular to it (■ Figure 24.4).

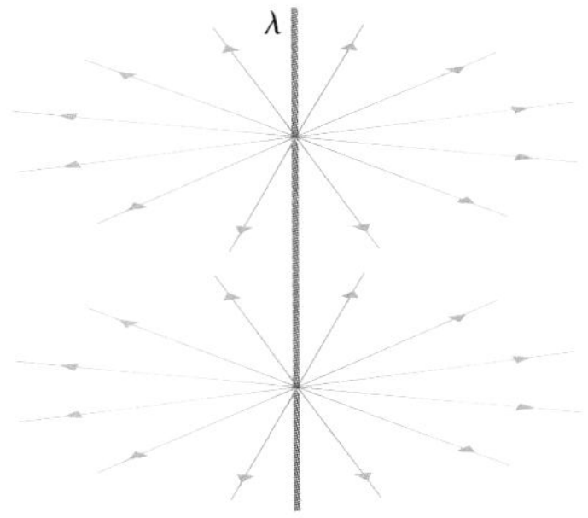

■ Figure 24.4
Field line diagram for an infinitely long, uniformly charged filament. Since the properties of the filament are the same everywhere along it, the field lines do not bend but emerge perpendicular to the filament, like the spokes of a wheel.

EXAMPLE 24.2 ♦♦ Find the electric field due to an infinite filament with uniform linear charge density λ.

MODEL The charge density has cylindrical symmetry, and we may use Gauss' law to compute the field it produces.

Compare with the reasoning in Example 24.1.

Be careful to distinguish the unit vector $\hat{\mathbf{r}}$ in cylindrical coordinates from that in spherical coordinates. Some books use the Greek letter $\hat{\rho}$ for the unit vector in cylindrical coordinates.

STEP I The field lines radiate from the filament, perpendicular to it and uniformly distributed around it (Figure 24.4). In cylindrical coordinates, the electric field has a single component E_r.

SETUP
STEP II For the Gaussian surface we choose a cylinder of height h and radius r centered on the filament. The normal to this surface is parallel to $\vec{\mathbf{E}}$ at points on the curved surface and perpendicular to $\vec{\mathbf{E}}$ on the flat ends (■ Figure 24.5).

STEP III The flux through the flat ends is zero (no field lines pass through the ends; $\vec{\mathbf{E}} \cdot \hat{\mathbf{n}} = 0$). The flux through the curved surface ($\hat{\mathbf{n}} = \hat{\mathbf{r}}$) is:

$$\Phi_E = \int d\Phi_E = \oint \vec{\mathbf{E}} \cdot \hat{\mathbf{n}}\, dA = \oint \vec{\mathbf{E}} \cdot \hat{\mathbf{r}}\, dA = \oint E_r\, dA.$$

Because of the rotational symmetry, the electric field component E_r is constant along the curved surface, and so it may be pulled out of the integral:

$$\Phi_E = E_r \oint dA = E_r 2\pi rh. \qquad (24.3)$$

This is the expression for flux through a cylindrical Gaussian surface in any example where the charge distribution has cylindrical symmetry.

STEP IV The charge inside the cylinder is $Q = \lambda h$.

SOLVE
STEP V Applying Gauss' law:

$$\Phi_E = \oint \vec{\mathbf{E}} \cdot \hat{\mathbf{n}}\, dA = 2\pi rhE_r = \lambda h/\epsilon_0$$

$$\Rightarrow \vec{\mathbf{E}} = \left(\frac{\lambda}{2\pi r\epsilon_0}, \text{ radially outward from the filament}\right). \qquad (24.4)$$

ANALYZE Notice that the arbitrarily chosen height h of the Gaussian cylinder cancels.

The electric field due to an infinite line decreases as 1/(distance from the line). Compare with the $1/(\text{distance})^2$ behavior of the field due to a point charge. ■

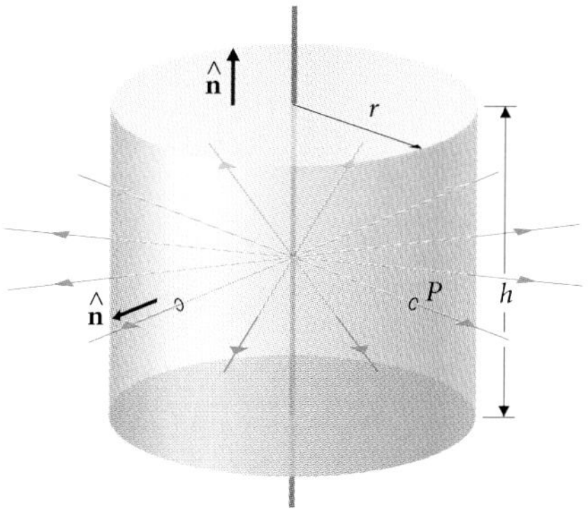

■ **Figure 24.5**
To find the field due to an infinite filament, we choose a cylindrical surface of height h and radius r. The electric field is parallel to $\hat{\mathbf{n}}$ on the curved surface and perpendicular to $\hat{\mathbf{n}}$ on the flat ends. Since a length h of the filament is inside the cylinder, charge $Q = \lambda h$ is enclosed inside.

24.2.2 *A Finite, Uniformly Charged Filament*

If the filament has a finite length ℓ, it has a finite charge $Q = \lambda\ell$. At very large distances from the filament, the field line diagram looks like that for a point charge Q (■ Figure 24.6) and we expect $|\vec{\mathbf{E}}| \approx kQ/r^2$. Very close to the filament, the field lines emerge at right angles, and the field is given by Gauss' law (§24.2.1). At intermediate distances, we evaluate the field using Coulomb's law and the principle of superposition.

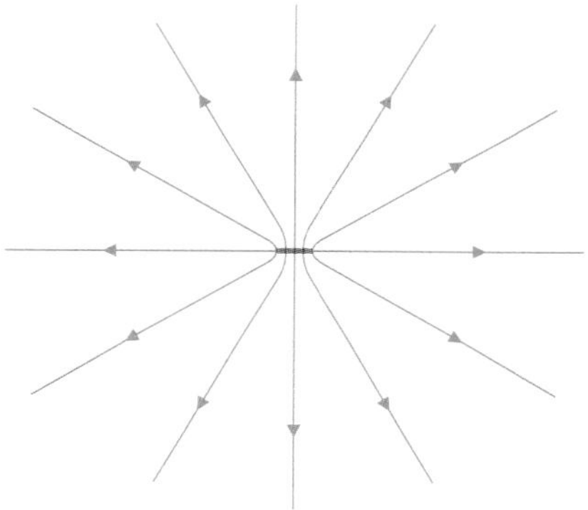

■ **Figure 24.6**
Field line diagram for a real charged filament viewed from a large distance away. The field lines diverge as from a point charge.

EXAMPLE 24.3 ♦♦♦ A filament of length ℓ has total charge Q distributed uniformly along it. Find the electric field at any point P in a plane bisecting the filament.

MODEL ■ Figure 24.7 is the field line diagram. We model the filament as a collection of differential pieces and use Coulomb's law for the field due to each piece. We sum the contributions $d\vec{\mathbf{E}}$ from each piece using the principle of superposition.

Here the steps refer to the integration method in Interlude 2.

SETUP
STEP I We choose the plane containing the filament and point P to be the x-y-plane, with the x-axis along the filament and the y-axis along its bisector, through P (■ Figure 24.8). Each differential piece of the filament has length dx.

STEP II A typical element is a piece of the filament with coordinate x, length dx, and charge $dQ = \lambda\, dx$. Since the filament is uniformly charged, λ is independent of x and equals Q/ℓ.

STEP III We wish to find the electric field at the arbitrary point P as a function of its y-coordinate. The contribution of the element dQ to this field is:

$$d\vec{\mathbf{E}}_1 = \frac{k\, dQ}{r^2}\hat{\mathbf{r}} = \frac{k\lambda\, dx}{r^2}\hat{\mathbf{r}} \qquad (\text{where } r^2 = x^2 + y^2) \text{ (Figure 24.8).}$$

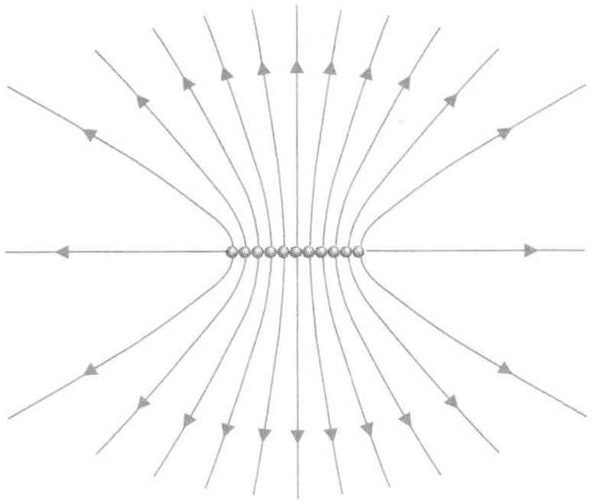

■ FIGURE 24.7
Field line diagram for a finite, uniformly charged filament at intermediate distances. The field lines emerge perpendicular to the filament and diverge at large distances.

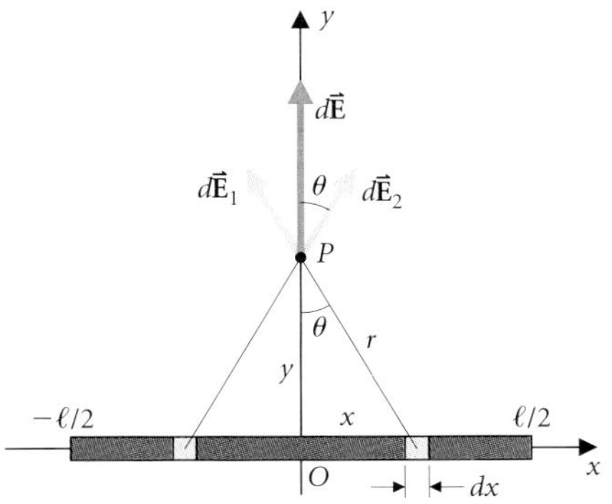

■ FIGURE 24.8
To calculate $\vec{E}$ at a point on the y-axis, we look at the field $d\vec{E}_1$ produced by a differential element of the filament at x. By summing the contributions from elements located symmetrically at x and $-x$, we determine that the electric field at P is in the y-direction. The element $d\vec{E}_1$ makes an angle θ with the y-axis, with $\tan\theta = x/y$.

We may simplify the calculation by using the symmetry about the y-axis. A second element of charge at coordinate $-x$ has an equal charge dQ, since λ is the same at both positions. The contribution $d\vec{E}_2$ from this element has the same magnitude as that due to the first element and makes an equal angle with the y-axis, on the opposite side. Thus the x-components of the two contributions sum to zero. For the y-component of the sum we have (cf. eqn. 23.9):

$$d\vec{E} = 2\hat{j}\ dE_{1,y} = 2\hat{j}|d\vec{E}_1|\cos\theta = 2\hat{j}\frac{k\ dQ}{r^2}\frac{y}{r} = 2\hat{j}\frac{ky\lambda\ dx}{r^3}.$$

THIS IS THE FIELD ON THE PERPENDICULAR BISECTOR OF TWO EQUAL POINT CHARGES (EXAMPLE 23.11).

We obtain the final result by adding the contributions of all such pairs of charge elements. Remember that r is a function of x and y.

STEP IV We add contributions from charge elements with coordinates ranging from $x = -\ell/2$ to $x = +\ell/2$. However, we add them as symmetrical pairs labeled by the coordinate of their right-hand element, which varies from $x = 0$ to $x = +\ell/2$.

IF YOU PREFER, YOU MAY CALCULATE EACH COMPONENT OF $\vec{E}$ SEPARATELY AND INTEGRATE OVER INDIVIDUAL ELEMENTS BETWEEN $-\ell/2$ AND $+\ell/2$. THE CALCULATION IS LONGER, BUT THE ANSWER IS THE SAME, OF COURSE.

SOLVE
STEP V

$$\vec{E}(y) \equiv \int d\vec{E} = \int 2\hat{j}\frac{ky\lambda\ dx}{r^3} = 2k\lambda y\hat{j}\int_0^{\ell/2}\frac{dx}{(x^2+y^2)^{3/2}}.$$

Since $x = y\tan\theta$, we may convert the calculation to an integral over θ as the independent variable:

$$dx \equiv d\theta\left(\frac{dx}{d\theta}\right) = y\sec^2\theta\ d\theta,$$

THIS IS A USEFUL TRICK THAT WE'LL WANT TO USE OFTEN.

and $x^2 + y^2 = y^2(1+\tan^2\theta) = y^2\sec^2\theta$. So:

$$\int\frac{dx}{(x^2+y^2)^{3/2}} = \int\frac{y\sec^2\theta}{y^3\sec^3\theta}d\theta = \frac{1}{y^2}\int\cos\theta\ d\theta = \frac{\sin\theta}{y^2}.$$

REMEMBER: WE ARE INTEGRATING OVER x, HOLDING y CONSTANT.

From Figure 24.8, $\sin\theta = x/r = x/(x^2+y^2)^{1/2}$, so:

$$\vec{E}(y) = \frac{2k\lambda\hat{j}}{y}\frac{x}{(x^2+y^2)^{1/2}}\Bigg|_0^{\ell/2} = \frac{k\lambda\ell\hat{j}}{y(y^2+\ell^2/4)^{1/2}} = \frac{kQ}{y\sqrt{y^2+\ell^2/4}}\hat{j}.$$

ANALYZE It is the symmetry of the *charge* distribution that we used here. If the left half of the filament were negatively charged, its contributions $d\vec{E}_2$ would reverse direction. The y-components would then sum to zero, and the resulting field would be in the negative x-direction. Always remember to sketch a field line diagram first. ■

24.2.3 Comparison of Exact and Approximate Calculations

NOTICE HOW THE SYMBOLS $\hat{\mathbf{j}}$ AND y ARE READ AS "VECTOR RADIALLY OUTWARD FROM THE FILAMENT" AND "DISTANCE FROM THE FILAMENT" WHEN COMPARING WITH EQN. (24.4).

In the near limit $y \ll \ell$, the exact expression for the finite filament becomes:

$$\lim_{y\ll\ell} \vec{\mathbf{E}}(y) = \frac{k\lambda\ell}{y\sqrt{\ell^2/4}}\,\hat{\mathbf{j}} = \frac{2k\lambda}{y}\,\hat{\mathbf{j}},$$

which verifies our conclusions from Gauss' law and the field line diagram for the infinite filament.

In the far limit $y \gg \ell$, the exact expression should agree with Coulomb's law for a point charge. Indeed, if we neglect $\ell^2/4$ compared with y^2:

$$\lim_{y\gg\ell} |\vec{\mathbf{E}}(y)| = \frac{kQ}{y\sqrt{y^2}} = \frac{kQ}{y^2}.$$

THESE ESTIMATES APPLY TO THE ELECTRIC FIELD OF THE FILAMENT. THEY ARE NOT GENERAL STATEMENTS ABOUT ANY CALCULATION.

Comparing the graphs of these two approximations with the exact result (■ Figure 24.9), we obtain rough rules for estimating $\vec{\mathbf{E}}$: if we want answers accurate to 10%, then *near* means closer than about 0.2ℓ and *far* means more distant than 0.8ℓ.

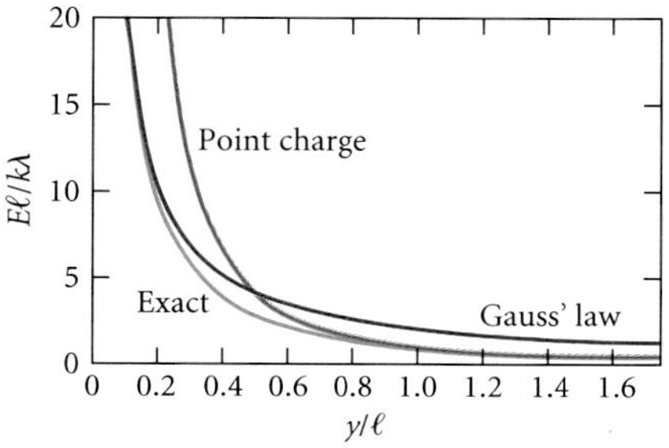

■ **FIGURE 24.9**
Comparison of exact and approximate solutions for the electric field due to a finite filament. The near solution, found using Gauss' law (red line) is close to the exact solution (green line) when $y \lesssim 0.2\ell$. The far (point charge) solution (blue line) is close to the exact solution for $y \gtrsim 0.8\ell$.

24.2.4 More Complicated Linear Charge Distributions

Once we know the field due to a uniformly charged filament, such filaments can be used as elements in more complicated problems.

EXAMPLE 24.4 ♦♦ Two parallel, infinitely long filaments lying in the x-y-plane are separated by a distance $d = 1.0$ mm. Each has uniform charge density $\lambda = 1.0$ nC/m (■ Figure 24.10). What is the electric field vector at point P on the z-axis?

MODEL The electric field at P is the sum of the two field vectors produced by the individual filaments, each found using Gauss' law, and given by eqn. (24.4).

SETUP Figure 24.10b shows a view along the filaments, with the field vectors due to each and their sum. The x-components cancel, and the z-components add.

SOLVE We substitute $2k = 1/(2\pi\epsilon_0)$ into eqn. (24.4):

$$\vec{\mathbf{E}} = \vec{\mathbf{E}}_1 + \vec{\mathbf{E}}_2 = 2|\vec{\mathbf{E}}_1|(\cos\theta)\hat{\mathbf{k}} = 2\frac{2k\lambda}{r}(\cos\theta)\hat{\mathbf{k}}$$

$$= 4k\lambda\frac{d}{r^2}\hat{\mathbf{k}} = \frac{4k\lambda d}{d^2 + d^2/4}\hat{\mathbf{k}} = \frac{16k\lambda}{5d}\hat{\mathbf{k}}.$$

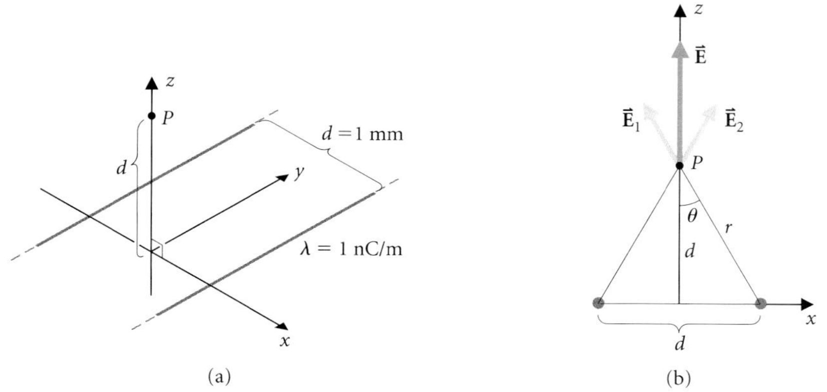

■ **FIGURE 24.10**
(a) Electric field due to two parallel, infinite filaments separated by a distance d. The filaments lie in the x-y-plane, and we find the field at P, on the z-axis at $z = d$.

(b) View in the x-z-plane of the two contributions to $\vec{\mathbf{E}}$ at P. The filaments are perpendicular to the page, and $\cos\theta = d/r$.

$$|\vec{\mathbf{E}}| = \frac{16(9.0 \times 10^9\ \mathrm{N \cdot m^2/C^2})(1.0 \times 10^{-9}\ \mathrm{C/m})}{5.0 \times 10^{-3}\ \mathrm{m}} = 2.9 \times 10^4\ \mathrm{N/C}.$$

Thus: $$\vec{\mathbf{E}} = (2.9 \times 10^4\ \mathrm{N/C})\hat{\mathbf{k}}.$$

ANALYZE Since λ has dimensions [charge]/[length], λ/d has dimensions of [charge]/[length]^2, and $k\lambda/d$ has the correct dimensions for electric field.

EXERCISE 24.2 ♦♦ Point P in ■ Figure 24.11 is a distance z above the center of a uniformly charged circular filament with radius a and total charge Q. Show that the electric field vector at P is given by:

$$\vec{\mathbf{E}} = \frac{kQz}{(z^2 + a^2)^{3/2}}\hat{\mathbf{k}}. \qquad (24.5)$$

(*Hint:* Use a symmetry argument to show that the field vector $\vec{\mathbf{E}}$ is in the z-direction.)

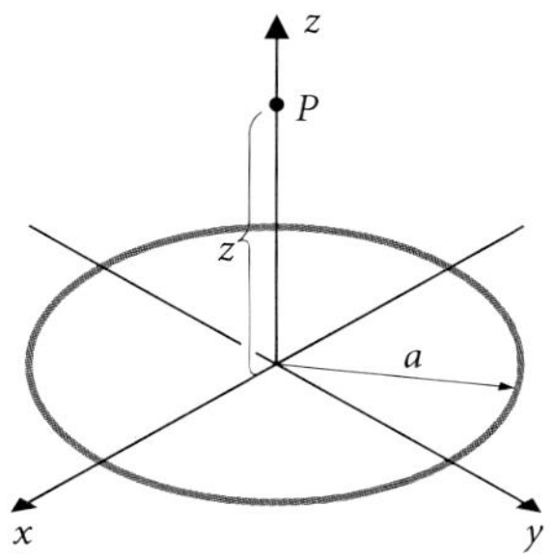

■ **FIGURE 24.11**
The circular filament lying in the x-y-plane has a uniform charge density.

24.3 ELECTRIC FIELD DUE TO SURFACE AND VOLUME CHARGE DISTRIBUTIONS

24.3.1 *Surface Charge Distributions*

When charge is distributed over a surface, we describe it with a surface charge density:

The *surface charge density* σ at a point is the amount of charge per unit area on a surface. The charge dQ on an area dA surrounding the point is given by:

$$dQ = \sigma\ dA. \qquad (24.6)$$

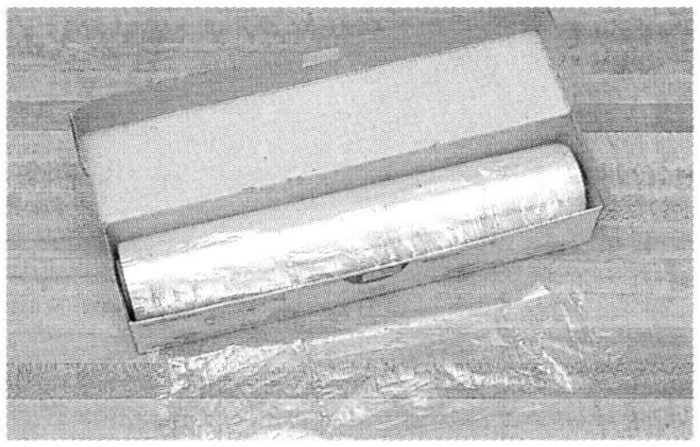

Pulling plastic wrap from the roll produces a charged sheet.

In a *uniform* distribution, σ has the same value at every point on the surface. Our first example is a uniformly charged infinite plane sheet. It has sufficient symmetry that we may use Gauss' law to find the field.

EXAMPLE 24.5 ♦♦ Find the electric field due to an infinite plane sheet that carries a uniform surface charge density σ (■ Figure 24.12).

MODEL STEP I Since the sheet has the same charge density everywhere in the plane, the field lines emerging from it do not bend at all. They are perpendicular to the surface and parallel to each other. The separation of field lines, and hence $|\vec{\mathbf{E}}|$, doesn't change. The magnitude of the field is the same on both sides and remains finite at the sheet. We choose the plane of the sheet as the x-y-plane; then only the z-component of $\vec{\mathbf{E}}$ is nonzero, and E_z changes sign at $z = 0$. We use Gauss' law to find E_z.

IN CONTRAST, FIELD LINES CONVERGE AT POINT CHARGES AND FILAMENTS, WHERE THE FIELD MAGNITUDE BECOMES INFINITELY LARGE.

SETUP STEP II We choose as our Gaussian surface a box with vertical sides (parallel to $\vec{\mathbf{E}}$) and with top and bottom parallel to the sheet (perpendicular to $\vec{\mathbf{E}}$). The shape and area of the horizontal faces are unimportant. The box extends as far below the sheet as above, to make use of the symmetry.

ACTUALLY, THE VALUE OF E_z TURNS OUT TO BE INDEPENDENT OF DISTANCE FROM THE SHEET, AS WE ARGUED IN OUR MODELING SECTION. LOCATING THE GAUSSIAN SURFACE SYMMETRICALLY IS ALWAYS A SAFE APPROACH. THE AREA A CANCELS FROM THE FINAL RESULT.

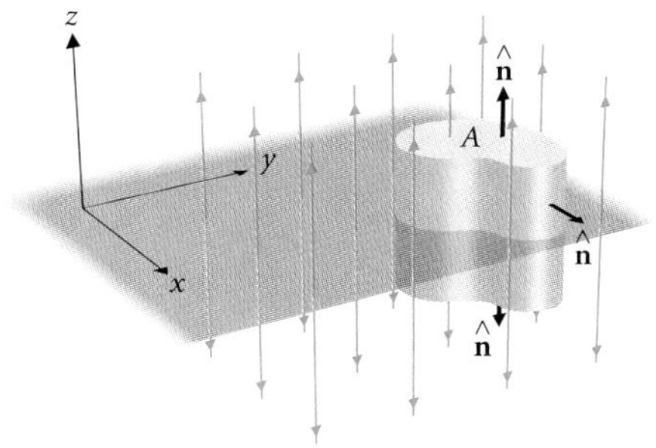

■ **FIGURE 24.12**
An infinite plane sheet has a uniform surface charge density σ. We choose the z-axis perpendicular to the sheet. The field lines emerge perpendicular to the sheet and remain parallel as they go off to infinity. Since the separation of fields lines indicates the field strength, the field due to the sheet is uniform.

STEP III Since the field is perpendicular to $\hat{\mathbf{n}}$ at points on the vertical sides of the box, there is no flux through the sides. On the top and bottom $\vec{\mathbf{E}}$ is parallel to $\hat{\mathbf{n}} = \pm\hat{\mathbf{k}}$, and $|E_z|$ is constant over each surface, so the net flux is twice that emerging from the top surface:

$$\Phi_E = 2AE_{z,\text{top}}.$$

STEP IV The box contains a charge σA.

SOLVE Applying Gauss' law:
STEP V

$$2AE_{z,\text{top}} = \frac{\sigma A}{\epsilon_0}$$

$$\Rightarrow \vec{\mathbf{E}}(z) = \frac{\sigma}{2\epsilon_0}\hat{\mathbf{k}} \qquad (z > 0). \qquad (24.7)$$

Below the sheet the electric field changes direction: $\vec{\mathbf{E}} = (\sigma/2\epsilon_0)(-\hat{\mathbf{k}})$ for $z < 0$.

ANALYZE The surface charge density has dimensions [charge]/[length]^2, so σ/ϵ_0 has the correct dimensions for electric field. The uniform field lines can't go on forever. In a real system, the lines terminate on distant charges of the opposite sign that were removed to create the charged sheet. ∎

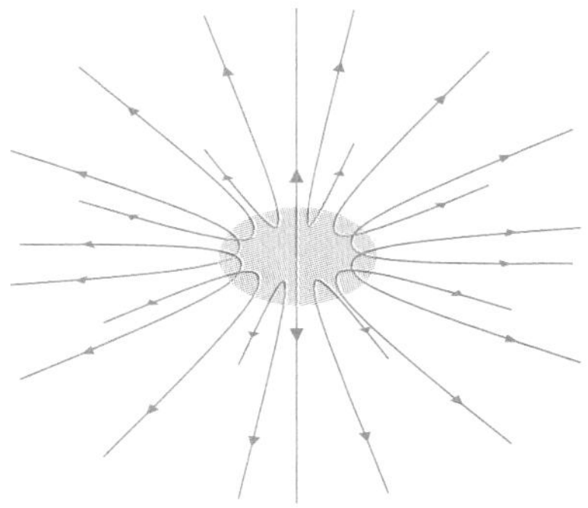

(a)

■ FIGURE 24.13 (a)
Electric field due to a uniformly charged disk. The field lines emerge perpendicular to the disk (cf. Figure 24.12), and at large distances they diverge as from a point charge.

EXAMPLE 24.6 ♦♦♦ An element of an electrostatic microphone is a circular plastic disk of radius a that carries a total charge Q uniformly distributed over its surface. Find the electric field $\vec{\mathbf{E}}$ at point P a distance z above the center of the disk.

MODEL As always, we begin by sketching a field line diagram for the system (■ Figure 24.13a). Far away from the disk, the field lines diverge as from a point charge. Since the disk is uniformly charged, the same number of field lines emerge from each unit of area. The lines do not diverge from a single point or line. Near the surface they resemble the field of an infinite sheet, and $|\vec{\mathbf{E}}|$ has a finite limit at the surface of the disk. When $z \ll a$, $\vec{\mathbf{E}}$ is perpendicular to the disk and approaches the value found in Example 24.5 using Gauss' law. To calculate the field at intermediate distances, we use the principle of superposition.

HERE THE STEP NUMBERS REFER TO THE INTEGRATION METHOD IN INTERLUDE 2.

SETUP Since we know the electric field produced by a circular filament (Exercise 24.2),
STEP I such filaments are appropriate differential elements to use in analyzing the disk. We use polar coordinates r and θ in the plane of the disk and z perpendicular to the disk (Figure 24.13b).

THAT IS, WE USE CYLINDRICAL COORDINATES.

STEP II A typical element is a ring of radius r and thickness dr, which has area $dA = 2\pi r\,dr$. The charge density on the disk is $\sigma = Q/(\pi a^2)$, and the charge on the element is:

$$dQ = \sigma\,dA = \frac{Q}{\pi a^2}2\pi r\,dr = \frac{2Qr\,dr}{a^2}.$$

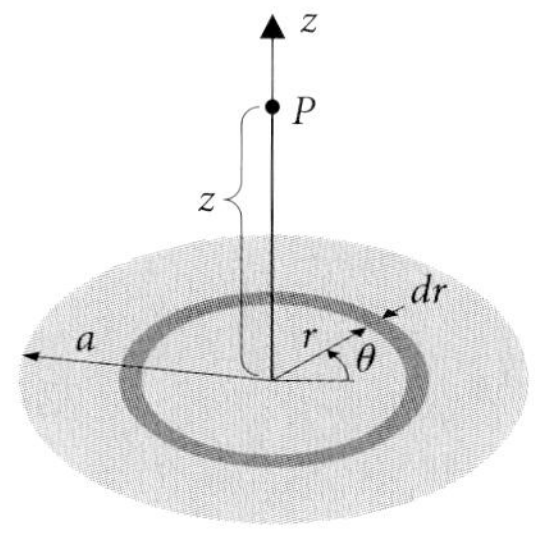

(b)

■ FIGURE 24.13 (b)
We model the disk as a collection of rings, each with thickness dr and charge $dQ = \sigma 2\pi r\,dr$. The field at P due to each ring is in the z-direction, perpendicular to the disk.

STEP III The electric field at P produced by this ring is given by eqn. 24.5:

$$d\vec{\mathbf{E}} = \frac{k(dQ)z}{(z^2 + r^2)^{3/2}}\hat{\mathbf{k}} = \frac{kQz}{a^2}\frac{2r\,dr}{(z^2 + r^2)^{3/2}}\hat{\mathbf{k}}.$$

STEP IV Since we expressed a positive ring thickness as dr, we sum the rings from the center toward the edge. That is, the radius r ranges from 0 to a.

SOLVE Each element of field $d\vec{\mathbf{E}}$ is in the z-direction, so the sum $\vec{\mathbf{E}}$ is, too. The unit
STEP V vector $\hat{\mathbf{k}}$ is constant and may be brought out of the integral:

$$\vec{\mathbf{E}} = \hat{\mathbf{k}}\frac{kQz}{a^2}\int_0^a \frac{2r\,dr}{(z^2 + r^2)^{3/2}}.$$

We change variables to $u = z^2 + r^2$. Then $du = 2r\,dr$, and we have:

$$\int_0^a \frac{2r\,dr}{(z^2 + r^2)^{3/2}} = \int_{z^2}^{z^2+a^2} \frac{du}{u^{3/2}} = \frac{u^{-1/2}}{-1/2}\Bigg|_{z^2}^{z^2+a^2}$$

$$= -2\left(\frac{1}{\sqrt{z^2 + a^2}} - \frac{1}{z}\right).$$

Thus:

$$\vec{\mathbf{E}} = \frac{2kQ\hat{\mathbf{k}}}{a^2}\left(1 - \frac{z}{\sqrt{z^2 + a^2}}\right).$$

ANALYZE Checking the limits, we find for $z \ll a$:

$$|\vec{\mathbf{E}}| \approx \frac{2kQ}{a^2} = 2\pi k\sigma, \tag{24.8}$$

which agrees with eqn. (24.7) from Gauss' law. For $z \gg a$, we expand the square root using the binomial expansion:

$$|\vec{\mathbf{E}}| \approx \frac{2kQ}{a^2}\left[1 - \left(1 - \frac{a^2}{2z^2}\right)\right] = \frac{kQ}{z^2},$$

which is the expected result for a point charge located at the origin. ∎

NOTICE THAT WE CANNOT IGNORE THE a^2 TERM ENTIRELY BECAUSE WE WOULD GET A LIMIT OF ZERO. ALTHOUGH CORRECT, THIS IS NOT VERY INFORMATIVE.

24.3.2 *Volume Charge Distributions*

Charge distributed throughout a volume is described by a volume charge density:

> The volume *charge density* ρ at a point is the amount of charge per unit volume. The charge dQ in a volume element dV surrounding the point is given by:
>
> $$dQ = \rho\, dV. \tag{24.9}$$

WHEREAS σ IS CALLED "SURFACE CHARGE DENSITY," ρ IS USUALLY CALLED JUST "CHARGE DENSITY."

The techniques for finding the resulting electric field are the same ones that we used for linear and surface distributions: Gauss' law, or Coulomb's law and the principle of superposition.

EXAMPLE 24.7 ♦♦ Find the electric field due to an infinite plane slab of thickness w that carries a uniform charge density ρ (■ Figure 24.14).

MODEL The system is sufficiently symmetric that we may use Gauss' law to calculate the electric field. We choose coordinates with the x-y-plane as the midplane of the slab.

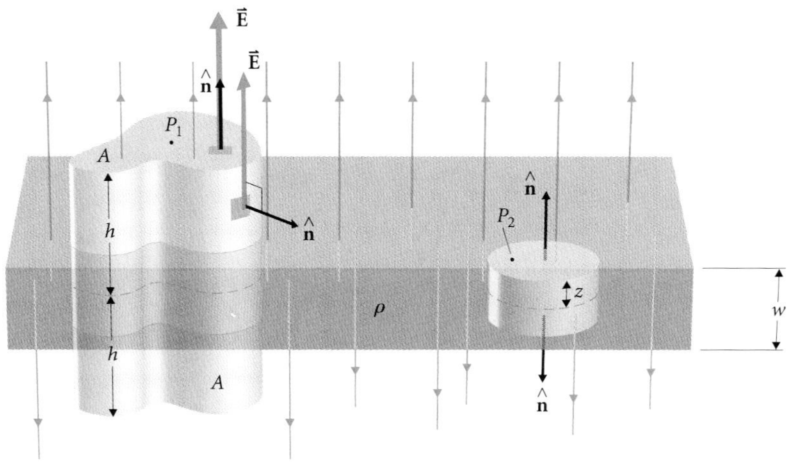

■ FIGURE 24.14
Electric field due to a uniformly charged slab of thickness *w*. We choose the z-axis perpendicular to the slab. The field lines are everywhere perpendicular to the sides of the slab; however, inside the slab, new field lines originate at the charge within each layer. The field strength increases with distance from the central plane. To apply Gauss' law, we choose a box with sides parallel and perpendicular to the slab.

STEP I The field lines outside the slab have the same symmetry as the lines from the plane sheet (Example 24.5). At points outside the slab ($|z| > w/2$), it looks like a charged sheet with surface charge density $\sigma = \rho w$. The calculation of the electric field proceeds exactly as in Example 24.5, and the result is the same:

$$\vec{\mathbf{E}} = \frac{\sigma}{2\epsilon_0}\hat{\mathbf{k}} = \frac{\rho w}{2\epsilon_0}\hat{\mathbf{k}} \qquad (z > w/2).$$

Since the charge distribution within the slab also has plane symmetry, the field lines there remain parallel to the z-axis. However, in each successive layer field lines begin on the charges in that layer. Thus the magnitude of the field increases with distance from the center of the slab, and E_z is a function of z. Because the slab is symmetric about its midplane, we expect $E_z(-z) = -E_z(z)$.

REMEMBER THAT $E(z)$ MEANS "E AT COORDINATE z."

SETUP STEP II We choose the Gaussian surface to be a box with vertical sides and with top and bottom faces inside and parallel to the slab. The box extends as far below the midplane of the slab as above, to make use of the symmetry, and its top is at $z < w/2$. The volume of the box is $V = 2zA$.

STEP III Since the field is perpendicular to $\hat{\mathbf{n}}$ on the vertical sides of the box, there is no flux through the sides. On both the top and bottom $\vec{\mathbf{E}}(z) = E_z(z)\hat{\mathbf{k}}$ is parallel to $\hat{\mathbf{n}} = \pm\hat{\mathbf{k}}$. Since $E_z(-z) = -E_z(z)$, the total flux through the box is:

$$\Phi_E = 2E_z(z)A.$$

STEP IV The total charge inside the box is $Q = \rho V = \rho 2Az$.

SOLVE STEP V Applying Gauss' law:

$$\Phi_E = 2E_z(z)A = \frac{2\rho Az}{\epsilon_0}$$

$$\Rightarrow \vec{\mathbf{E}}(z) = \frac{\rho z}{\epsilon_0}\hat{\mathbf{k}} \qquad (|z| < w/2).$$

ANALYZE The results for the fields inside and outside give the same expression at the edge of the slab, where $z = w/2$. The electric field at the midpoint of the slab is zero, as expected from the symmetry.

EXERCISE 24.3 ❖ Why is it necessary for the ends of the box to be equidistant from the midplane when computing $\vec{\mathbf{E}}$ for $|z| < w/2$? Is this requirement necessary for computing the field outside the slab? Can you find another box that works for the field inside?

EXERCISE 24.4 ♦♦ Find the electric field within a uniformly charged sphere with charge Q and radius R.

24.4 MOTION OF CHARGES IN AN ELECTRIC FIELD

If a particle with charge q is in a region with electric field $\vec{\mathbf{E}}$, it experiences a force $\vec{\mathbf{F}} = q\vec{\mathbf{E}}$. Frequently, this electric force is much stronger than any other force acting on the particle and we may neglect the other forces. For example, because it is impossible to produce an environment completely free of electric fields, it has only recently become possible to detect the gravitational force acting on an electron (see ■ Figure 24.15).

Neglecting other forces, the acceleration of a particle in an electric field $\vec{\mathbf{E}}$ is:

$$\vec{\mathbf{a}} = \frac{\vec{\mathbf{F}}}{m} = \frac{q}{m}\vec{\mathbf{E}}. \qquad (24.10)$$

Measuring the acceleration of a particle in a known electric field is one method of determining its *charge-to-mass ratio* q/m.

■ FIGURE 24.15
Apparatus for measuring the gravitational force on an electron. Electrons are contained in the cylinder just left of center, which extends into the room below. Great pains must be taken to minimize electric and magnetic fields inside the cylinder and to correct for the effects of any fields that do remain.

Electric fields along the oscilloscope tube accelerate the electrons in the beam. Electric fields across the tube control their sideways deflection to produce the image on the screen.

EXAMPLE 24.8 ♦ In the *electron gun* of a color TV set the electric field $|\vec{E}| = 2.5 \times 10^6$ N/C in a region 1.0 cm long (■ Figure 24.16). Electrons enter the region at low speed and are accelerated into a beam that produces light when it strikes phosphors on the tube face. What is the acceleration of each electron? What is the speed of an electron as it emerges from the gun?

MODEL We model the electric field as uniform, so the electron (charge $-e$) has a constant acceleration while it is in the 1.0-cm-long region where $|\vec{E}|$ is nonzero. We may find the electron's speed using the kinematic equations from Chapter 2.

SETUP The electron's acceleration has magnitude:

$$a = \frac{F}{m} = \frac{e}{m}E.$$

SOLVE The electron's speed after moving $s = 1.0$ cm from rest at this acceleration is (eqn. 2.13):

$$v = \sqrt{2as} = \sqrt{2\frac{e}{m}Es}$$

$$= \sqrt{(2)\frac{1.60 \times 10^{-19}\ \text{C}}{9.11 \times 10^{-31}\ \text{kg}}(2.5 \times 10^6\ \text{N/C})(1.0 \times 10^{-2}\ \text{m})}$$

$$= 9.4 \times 10^7\ \text{m/s}.$$

ANALYZE These very energetic electrons are moving at about a third the speed of light. In addition to exciting the phosphors on the TV screen, they can produce x rays. TV manufacturers have to put x-ray-absorbent coatings on the front of the TV screen to avoid exposing viewers to hazardous radiation.

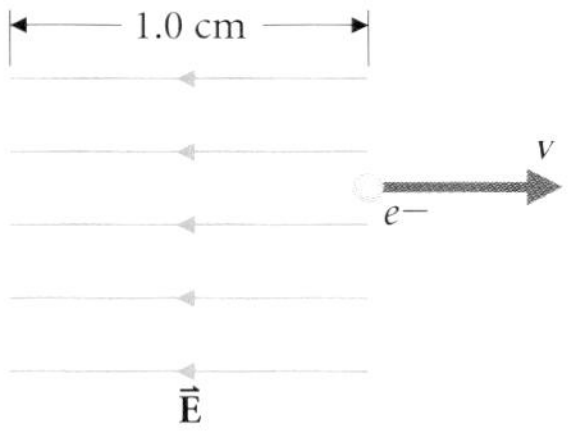

■ FIGURE 24.16
In a television set, electrons are accelerated by an electric field, modeled here as uniform. Remember that the force on a negatively charged electron is opposite $\vec{E}$.

When a calculation using Newtonian mechanics leads to an answer for v that is a large fraction of the speed of light, we should check the result using relativistic mechanics (see Chapter 34). In this example, the Newtonian calculation overestimates the electron speed by about 4%. ∎

EXAMPLE 24.9 ♦♦♦ An electron is observed to move in a circle around a long, uniformly charged filament. If the speed of the electron is 6.0×10^{6} m/s, what is the charge density on the filament?

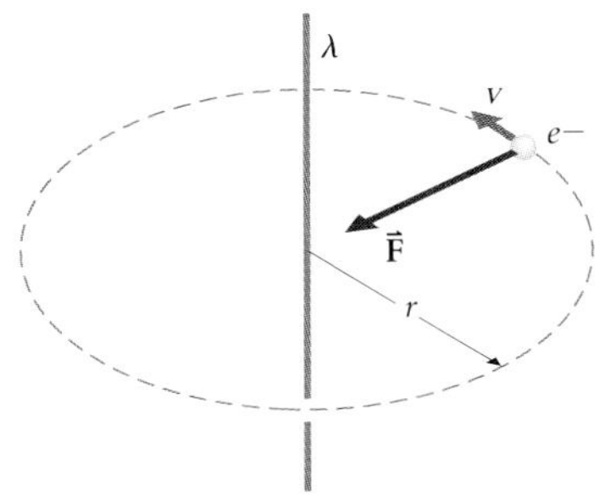

■ FIGURE 24.17
An electron orbits a charged filament. To produce an inwardly directed force, the filament must have a positive charge.

MODEL The electric field of a positively charged filament is radially outward. The electron moves in a circle around the filament (■ Figure 24.17) because the force $\vec{\mathbf{F}} = q\vec{\mathbf{E}} = -e\vec{\mathbf{E}}$ acting on it is always directed toward the center of the circle.

SETUP The electric field due to the filament is given by eqn. (24.4). Letting r be the radius of the observed circular path and λ be the charge density on the filament, we calculate the magnitude of the electric force on the electron:

$$F = eE = e\,(2k\lambda/r).$$

This force causes the observed centripetal acceleration:

$$F = ma.$$
$$2k\lambda e/r = mv^{2}/r.$$

SOLVE The distance of the electron from the filament cancels:

$$\lambda = mv^{2}/(2ke) = \frac{(9.11 \times 10^{-31}\ \text{kg})(6.0 \times 10^{6}\ \text{m/s})^{2}}{2(9.0 \times 10^{9}\ \text{N}\cdot\text{m}^{2}/\text{C}^{2})(1.60 \times 10^{-19}\ \text{C})} = 1.1 \times 10^{-8}\ \text{C/m}.$$

ANALYZE Remember that the electron has a negative charge, so the electric force on it is opposite the electric field vector. ∎

24.5 THE DIPOLE

An electric dipole—two equal but opposite charges separated by a small distance—is one of the most important and widely applicable physical models. For example, the forces between uncharged molecules of vapor condensing into a liquid occur because the molecules act like electric dipoles. Since it has no net charge, the dipole is a useful model for predicting the behavior of neutral matter exposed to electric fields. The dipole field pattern describes the simplest magnetic field that can occur and approximates the magnetic field produced by a complex system like the Earth. In this section, we investigate the behavior of electric dipoles.

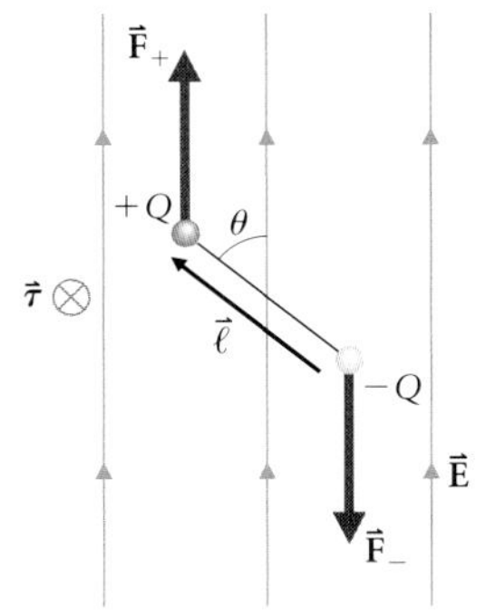

■ FIGURE 24.18
A dipole placed in a uniform field. The two charges in the dipole experience equal and opposite forces, forming a couple whose torque accelerates the dipole toward a configuration with $\vec{\mathbf{p}}$ parallel to $\vec{\mathbf{E}}$.

MOLECULAR FORCES AND THEIR CONSEQUENCES WERE DISCUSSED IN CHAPTER 20.

SEE CHAPTER 27 FOR MORE ON THIS TOPIC.

EXAMPLE 24.10 ❖ Two point charges $+Q$ and $-Q$, each with mass m, are at the ends of a massless rod of length ℓ. The object is placed in a region where there is a uniform electric field $\vec{\mathbf{E}}$ (■ Figure 24.18). Describe the motion of the dipole.

MODEL Each of the charges in the dipole experiences an electric force. The net force and net torque on the dipole determine its linear and angular accelerations.

THE TWO FORCES FORM A *COUPLE* (CHAPTER 11). *REMEMBER:* THE TORQUE EXERTED BY A COUPLE ABOUT ANY POINT IS THE SAME.

SETUP The positive charge experiences a force $\vec{\mathbf{F}}_{+} = Q\vec{\mathbf{E}}$, and the negative charge experiences a force $\vec{\mathbf{F}}_{-} = -Q\vec{\mathbf{E}} = -\vec{\mathbf{F}}_{+}$. The total force:

$$\vec{\mathbf{F}}_{t} = \vec{\mathbf{F}}_{+} + \vec{\mathbf{F}}_{-} = Q\vec{\mathbf{E}} - Q\vec{\mathbf{E}},$$

is zero, and the dipole's center of mass does not accelerate. With the origin at the negative charge, the torque on the dipole is:

$$\vec{\boldsymbol{\tau}} = \vec{\mathbf{r}} \times \vec{\mathbf{F}} = \vec{\ell} \times Q\vec{\mathbf{E}} = QE\ell\,\sin\theta\ \hat{\otimes} \quad \text{(into the plane of the figure).}$$

This torque causes the dipole to begin rotating toward alignment with the electric field. When the dipole is parallel to $\vec{\mathbf{E}}$, the torque is zero, but the dipole has gained angular momentum. It continues to rotate beyond alignment, and the torque reverses direction. The dipole oscillates about an equilibrium position parallel to the field direction.

ANALYZE The example gives no mechanism for the dipole to lose energy, so it will continue to oscillate indefinitely. Real dipoles, such as the particles in Figure VI.14 in the overview, experience friction. Consequently their oscillations die out, and they align with the field. A compass needle is a small *magnetic* dipole that points north as a result of a similar alignment process. ■

THERE ARE OTHER DAMPING MECHANISMS TOO. FOR EXAMPLE, AN OSCILLATING ELECTRIC DIPOLE RADIATES EM WAVES (CHAPTER 33).

The torque exerted on the dipole depends on the magnitude of the point charges Q, their separation ℓ, and their orientation with respect to the field. Together these define a vector called the dipole moment.

> The *dipole moment* $\vec{\mathbf{p}}$ of a pair of equal but opposite charges has magnitude equal to the product of the charge magnitude and the separation of the pair.
>
> $$|\vec{\mathbf{p}}| = Q\ell. \tag{24.11}$$
>
> The direction of $\vec{\mathbf{p}}$ is from the negative charge toward the positive charge.

In terms of the dipole moment, the torque on the dipole is

$$\vec{\tau} = \vec{\mathbf{p}} \times \vec{\mathbf{E}}. \tag{24.12}$$

EXERCISE 24.5 ♦♦♦ Find the force on a dipole placed at a distance r from a point charge and oriented with $\vec{\mathbf{p}}$ along a radial line from the charge. Assume $\ell \ll r$.

EXAMPLE 24.11 ♦♦♦ Find an expression for the electric field at a point P on the x-axis due to a dipole at the origin if the distance to the point is much greater than the size of the dipole and $\vec{\mathbf{p}} = p\hat{\mathbf{i}}$.

MODEL The electric field is the sum of the field vectors $\vec{\mathbf{E}}_+$ and $\vec{\mathbf{E}}_-$ due to the two charges in the dipole. When the separation of the two charges is small, their field vectors nearly cancel, and the sum is relatively small.

SETUP We choose the origin to be at the negatively charged end of the dipole (■ Figure 24.19). From the principle of superposition and Coulomb's law:

USING AN ORIGIN SOMEWHERE ELSE ON THE DIPOLE DOESN'T CHANGE THE EXPRESSION WE OBTAIN FOR $\vec{\mathbf{E}}$.

$$\vec{\mathbf{E}} = \vec{\mathbf{E}}_+ + \vec{\mathbf{E}}_- = \frac{kQ}{(x-\ell)^2}\hat{\mathbf{i}} + \frac{k(-Q)}{x^2}\hat{\mathbf{i}} = kQ\hat{\mathbf{i}}\left[\frac{x^2-(x-\ell)^2}{x^2(x-\ell)^2}\right]$$
$$= kQ\hat{\mathbf{i}}\left[\frac{2\ell x-\ell^2}{x^2(x-\ell)^2}\right] = kQ\ell\hat{\mathbf{i}}\left[\frac{2x-\ell}{x^2(x-\ell)^2}\right].$$

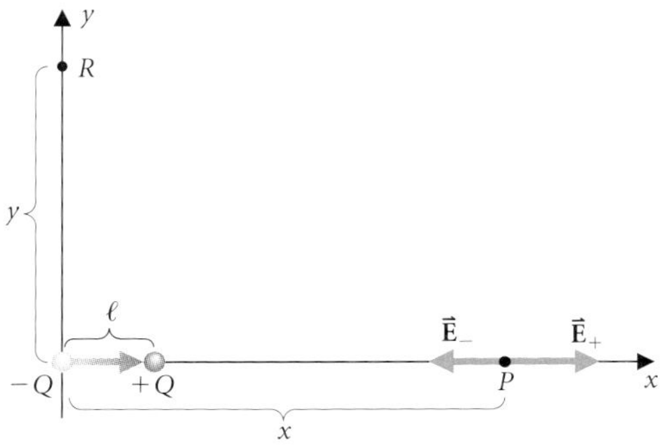

■ FIGURE 24.19
A dipole lies along the x-axis. The electric field at P is the superposition of the fields due to each of the charges that make up the dipole. Their magnitudes differ slightly, because of the different distances of the two charges from P. (The diagram is not drawn to scale: $x \gg \ell$.) In Exercise 24.6, you are asked to find the electric field at point R, with $y \gg \ell$.

This is an example of the rule in Interlude 1.

SOLVE Having simplified as much as possible, we may now use the fact that $\ell \ll x$. We neglect ℓ with respect to x whenever they are added or subtracted but not when they are multiplied. Thus:

$$\vec{\mathbf{E}} \approx kQ\ell\hat{\mathbf{i}}\,\frac{2}{x^3}.$$

$$\vec{\mathbf{E}} = \frac{2kp}{x^3}\,\hat{\mathbf{i}}. \tag{24.13}$$

ANALYZE *Remember:* This result is correct for points on the dipole axis, with $x \gg \ell$. However, the decrease of $|\vec{\mathbf{E}}|$ with distance cubed is a general feature of the dipole field. ∎

EXERCISE 24.6 ♦♦♦ Show that the electric field at a point on the y-axis in Figure 24.19 is: $\vec{\mathbf{E}} = -k\vec{\mathbf{p}}/|y|^3$. ∎

EXAMPLE 24.12 ♦♦♦ Water molecules have a dipole moment p of approximately 6×10^{-30} C·m. Two water molecules are separated by 10^{-7} m, and their dipole moments are in the same direction along their line of separation (■ Figure 24.20a). What electric force acts on each molecule?

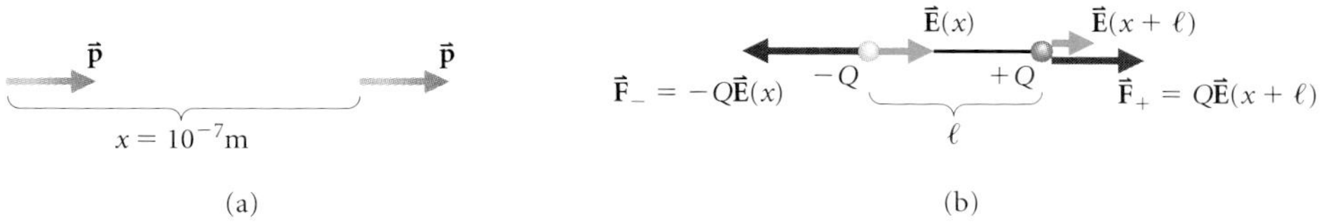

■ Figure 24.20
(a) The force that one dipole exerts on another depends on their relative orientation. Here the two dipoles are parallel and colinear, separated by a distance x.

(b) Enlargement of the dipole on the right, showing the electric field due to the other dipole and the forces exerted on the two charges.

MODEL We model each dipole as two point charges $\pm Q$ separated by a distance ℓ, with $p = Q\ell$. One dipole produces an electric field (eqn. 24.13) at the location of the other. As a result, each charge in the second dipole experiences a force, and the net force on the second dipole is the sum of the forces acting on its two charges (Figure 24.20b).

SETUP With the origin at the first dipole, the net force on the second is:

$$\begin{aligned}\vec{\mathbf{F}} &= Q\vec{\mathbf{E}}(x + \ell) - Q\vec{\mathbf{E}}(x)\\ &= Q\left[\frac{2kp\hat{\mathbf{i}}}{(x + \ell)^3} - \frac{2kp\hat{\mathbf{i}}}{x^3}\right]\\ &= 2kpQ\hat{\mathbf{i}}\left[\frac{x^3 - (x + \ell)^3}{(x + \ell)^3 x^3}\right]\\ &= 2kpQ\hat{\mathbf{i}}\left[\frac{-3x^2\ell - 3x\ell^2 - \ell^3}{(x + \ell)^3 x^3}\right].\end{aligned}$$

Note also: $\ell^3/(3x^2\ell) = \frac{1}{3}(\ell/x)^2 \ll 1$.

SOLVE Since the dimension ℓ of a molecule is approximately 10^{-9} m $\ll 10^{-7}$ m $= x$, we ignore ℓ compared with x whenever they are added or subtracted:

$$\vec{\mathbf{F}} = \hat{\mathbf{i}}2kQp\ell\left[\frac{-3x(x + \ell) - \ell^2}{(x + \ell)^3 x^3}\right] \approx \hat{\mathbf{i}}2kQp\ell\frac{-3x^2}{x^6} = -\hat{\mathbf{i}}\,\frac{6kp^2}{x^4}.$$

The minus sign indicates that the force is attractive. Its magnitude is:

$$|\vec{\mathbf{F}}| = \frac{6(9 \times 10^{9} \text{ N}\cdot\text{m}^2/\text{C}^2)(6 \times 10^{-30} \text{ C}\cdot\text{m})^2}{(10^{-7} \text{ m})^4}$$

$$= 2 \times 10^{-20} \text{ N}.$$

The force on the other dipole is equal and opposite.

ANALYZE The mass of each molecule (two hydrogen atoms plus one oxygen atom) is roughly 18 u, or about 3×10^{-26} kg, so each molecule has a huge acceleration:

$$|\vec{\mathbf{a}}| = |\vec{\mathbf{F}}|/m \approx 10^{6} \text{ m/s}^2,$$

or about 10^5 *g*! This acceleration occurs in a gas when two molecules come very close together and the influence of other molecules can be ignored: that is, during a *collision.* In liquid water the molecules are influenced by the fields of many other nearby molecules as well. ∎

Chapter Summary

Where Are We Now?

We have shown how to use Coulomb's law and Gauss' law to compute the electric field due to a distribution of charge. We have discussed the motion of charges under the influence of electric fields. We described a configuration of charge called a dipole.

What Did We Do?

The principle of superposition states that the electric field vector due to a charge distribution is the sum of the electric field vectors produced by each charge element in the distribution. The vector sum is evaluated by integration taking proper account of the directions of the vectors. When the system has appropriate symmetry, Gauss' law provides an efficient way to compute the field. A solution plan for using Gauss' law is given in Figure 24.1. We used these two principles to find the electric field due to a continuous charge distribution. Two important results are the field due to a long filament ($E = 2k\lambda/r$, eqn. 24.4) and the field due to an infinite plane sheet ($E = \sigma/2\epsilon_0$, eqn. 24.7).

A particle's acceleration when placed in an electric field depends on its charge-to-mass ratio, $\vec{\mathbf{a}} = (q/m)\vec{\mathbf{E}}$. A dipole comprises a positive charge Q and negative charge $-Q$ separated by a distance ℓ. The dipole moment $\vec{\mathbf{p}}$ has magnitude $Q\ell$ and points from the negative charge toward the positive charge. A dipole placed in an external electric field feels a torque $\vec{\boldsymbol{\tau}} = \vec{\mathbf{p}} \times \vec{\mathbf{E}}$ that rotates it toward alignment with that field. The electric field of a dipole decreases as the inverse cube of distance from the dipole.

Practical Applications

Static electric fields accelerate particles inside the cathode ray tubes used in oscilloscopes and television sets, and in linear accelerators for high-energy particle research. Electrostatic fields are also used in some designs of high-frequency loudspeakers and microphones, and in photocopiers. Dipoles are a good model for molecules exposed to electric fields (Chapter 27) and form the basis for the $1/r^6$ behavior of molecular forces in the van der Waals model of gases (Chapter 20).

Solutions to Exercises

24.1 The field line diagram outside the ball looks just the same as the field line diagram outside the spherical shell. The flux through any sphere surrounding and centered on the ball is the same as we calculated in Example 24.1: $\Phi_E = 4\pi r^2 E_r$, and the charge enclosed is Q. From Gauss' law, the electric field outside the ball is the same as for a point charge Q at the center of the ball.

24.2 Consider the ring to be made up of differential elements with coordinate ϕ and length $a\,d\phi$ (■ Figure 24.21). Each element has charge $dQ = \lambda a\,d\phi = (Q/2\pi a)a\,d\phi$. The electric field vectors contributed by two such elements on opposite ends of a diameter are shown in Figure 24.21. Once again we have two equal charges, and the electric field vector produced by the pair is along the z-axis and has magnitude (eqn. 23.9):

$$dE = \frac{2k(dQ)z}{(a^2 + z^2)^{3/2}} = \frac{kQz(d\phi)}{\pi(a^2 + z^2)^{3/2}}.$$

Now we add up the contributions of all such pairs. We need to consider the coordinate of only one charge in each pair, which ranges from $\phi = 0$ to $\phi = \pi$.

$$\vec{\mathbf{E}} = \int d\vec{\mathbf{E}} = \hat{\mathbf{k}}\,\frac{kQz}{\pi(a^2 + z^2)^{3/2}}\int_0^{\pi} d\phi = \hat{\mathbf{k}}\,\frac{kQz}{(a^2 + z^2)^{3/2}},$$

as required. The integrand is independent of ϕ because all the differential elements are the same distance from P.

24.3 Since the electric field outside the slab is independent of z, we could put the box ends anywhere. So long as one end is on each side of the slab, the net flux is the same. The field inside the slab does depend on z; here we need to make full use of the symmetry and put the ends of the box at the same distance from the midplane so as to have the same magnitude of $\vec{\mathbf{E}}$ on each end.

Another method would be to put one end of the box outside the slab and one end inside. Then the known value of the field outside could be used to find the flux through that end of the box. Alternatively, we could put one end of the box at the midplane itself, where $\vec{\mathbf{E}}$ is zero due to the symmetry.

24.4 We choose a spherical Gaussian surface of radius r, concentric with the sphere of charge. The flux through the surface is $4\pi r^2 E_r$, and the charge enclosed in the volume within the surface is $\frac{4}{3}\pi r^3 \rho$. The charge density $\rho = Q/(\frac{4}{3}\pi R^3)$. Applying Gauss' law:

$$4\pi r^2 E_r = \left(\frac{r}{R}\right)^3 \frac{Q}{\epsilon_0} \Rightarrow \vec{\mathbf{E}} = \frac{Qr}{4\pi R^3 \epsilon_0}\,\hat{\mathbf{r}}.$$

At the surface of the charged sphere, where $r = R$, this expression gives the same result that we found in Exercise 24.1 for the electric field vector outside the sphere.

24.5 In ■ Figure 24.22 the point charge is Q, and the dipole moment is $p = q\ell$. The electric field produced at the negative end of the dipole is:

$$\vec{\mathbf{E}}_- = \frac{kQ}{r^2}\,\hat{\mathbf{r}},$$

so the force on the dipole's negative charge is:

$$\vec{\mathbf{F}}_- = -\frac{kQq}{r^2}\,\hat{\mathbf{r}}.$$

At the positive end:

$$\vec{\mathbf{E}}_+ = \frac{kQ}{r_+^2}\,\hat{\mathbf{r}} \quad \text{and} \quad \vec{\mathbf{F}}_+ = \frac{kQq}{r_+^2}\,\hat{\mathbf{r}} = \frac{kQq}{(r + \ell)^2}\,\hat{\mathbf{r}}.$$

Therefore:

$$\vec{\mathbf{F}} = \vec{\mathbf{F}}_+ + \vec{\mathbf{F}}_- = kQq\left[\frac{1}{(r + \ell)^2} - \frac{1}{r^2}\right]\hat{\mathbf{r}}.$$

The term in brackets is:

$$\frac{-2r\ell - \ell^2}{r^2\,(r + \ell)^2} \approx \frac{-2\ell}{r^3},$$

where, since $\ell \ll r$, we have dropped ℓ whenever it is added to r. Thus:

$$\vec{\mathbf{F}} = -\frac{2kQq\ell}{r^3}\,\hat{\mathbf{r}} = -\frac{2kQ\vec{\mathbf{p}}}{r^3}.$$

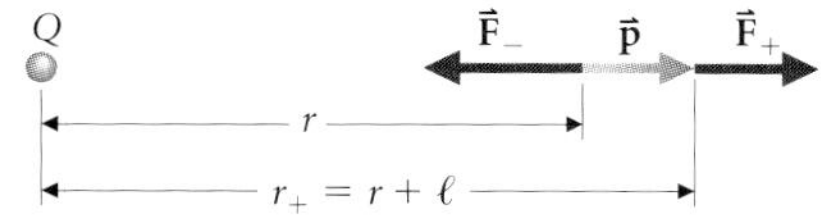

■ FIGURE 24.22
The electric forces acting on each charge of the dipole are not quite equal in magnitude because of the different distances from the point charge Q.

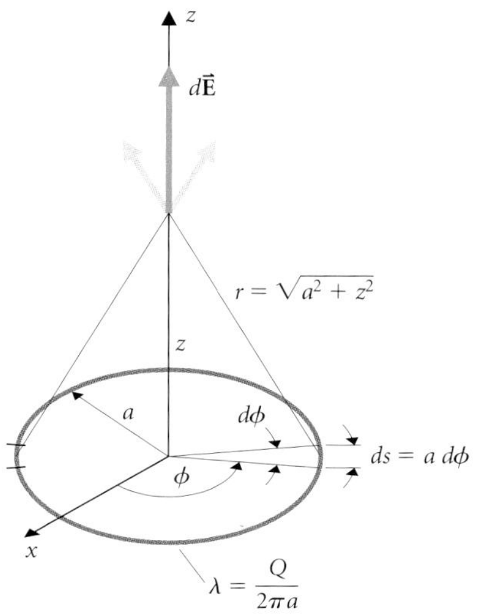

■ FIGURE 24.21
Two charge elements at opposite ends of a diameter form a pair of equal charges, as in Example 23.11.

You can check the result by using eqn. 24.13 to compute the force exerted by the dipole on the point charge, then applying Newton's third law.

Since

$$-2kQ/r^3 = (d/dr)(kQ/r^2),$$

the force may also be written:

$$\vec{\mathbf{F}} = p\frac{d\vec{\mathbf{E}}}{dr}.$$

Only in a nonuniform field does a dipole experience a net force.

24.6 We compute the fields produced by each charge at point R:

$$\vec{\mathbf{E}}_- = \frac{kQ}{y^2}(-\hat{\mathbf{j}})$$

and

$$\vec{\mathbf{E}}_+ = \frac{kQ}{(y^2 + \ell^2)^{3/2}}\left(y\hat{\mathbf{j}} - \ell\hat{\mathbf{i}}\right).$$

If we ignore ℓ with respect to y when they are added, the y-components cancel. The net electric field is:

$$\vec{\mathbf{E}} = \vec{\mathbf{E}}_- + \vec{\mathbf{E}}_+ = \frac{-kQ\ell}{(y^2)^{3/2}}\hat{\mathbf{i}} = \frac{-k}{|y|^3}\vec{\mathbf{p}}.$$

Basic Skills

Review Questions

§24.1 USING GAUSS' LAW TO CALCULATE ELECTRIC FIELDS

- Under what conditions may Gauss' law be used to find the electric field?
- How should the Gaussian surface be matched to the symmetry of the field line diagram?
- What is the electric field inside a uniformly charged spherical shell?

§24.2 THE ELECTRIC FIELD DUE TO A LINEAR CHARGE DISTRIBUTION

- When evaluating $\vec{\mathbf{E}}$ using integration, what feature of the calculation requires particular care?
- Define *linear charge density.*
- How does the electric field of a uniformly charged infinite filament depend on distance from the filament?

§24.3 ELECTRIC FIELDS DUE TO SURFACE AND VOLUME CHARGE DISTRIBUTIONS

- Define *surface charge density.*
- What is the electric field due to an infinite, uniformly charged plane sheet?
- Define *volume charge density.*
- Describe a volume that you could use to find the electric field inside a uniformly charged slab using Gauss' law.

§24.4 MOTION OF CHARGES IN AN ELECTRIC FIELD

- What feature of a particle describes its response to an applied electric field?

§24.5 THE DIPOLE

- What is an *electric dipole?* Define its electric dipole moment $\vec{\mathbf{p}}$.
- How does a dipole behave when placed in a uniform electric field?
- How does the electric field at a point P due to a dipole depend on the distance of P from the dipole?

Basic Skill Drill

§24.1 USING GAUSS' LAW TO CALCULATE ELECTRIC FIELDS

1. Find the electric field 16 cm from the surface of a uniformly charged sphere of radius 5.0 cm carrying a total charge of 16 nC.

§24.2 THE ELECTRIC FIELD DUE TO A LINEAR CHARGE DISTRIBUTION

2. A uniformly charged rod 12 cm long carries a charge of 16 μC. Find the linear charge density on the rod.
3. An infinitely long filament carries a uniform charge density $\lambda = 175\ \mu$C/m. What is the electric field 0.50 m from the filament?

§24.3 ELECTRIC FIELDS DUE TO SURFACE AND VOLUME CHARGE DISTRIBUTIONS

4. A uniformly charged square plate 11.6 cm on a side carries a total charge of 98.7 μC. What is the surface charge density on the plate?
5. What is the electric field near the center of and 5.5 mm above a large flat plate carrying a uniform charge density 4.6 mC/m^2?
6. A uniformly charged rectangular solid measuring 15 cm $\times$ 25 cm $\times$ 35 cm carries a total charge of 750 μC. What is the charge density?

§24.4 MOTION OF CHARGES IN AN ELECTRIC FIELD

7. A proton is released from rest 10.0 cm above an infinite plane with a charge density of $-175\ \mu$C/m^2. What is its acceleration? How long does it take to reach the plane?
8. A helium nucleus with charge $2e$ and mass 4.0 u is placed 3.0 μm from an infinite uniformly charged filament with $\lambda = 5.0$ nC/m. Find the magnitude and direction of its initial acceleration.

§24.5 THE DIPOLE

9. An object 1.5 μm long has a dipole moment of 25×10^{-14} C·m. What is the magnitude of each charge in the dipole?
10. A dipole of moment 0.85×10^{-15} C·m is placed in a uniform field $E = 4.3$ N/C, with $\vec{\mathbf{p}}$ perpendicular to $\vec{\mathbf{E}}$. What torque acts on the dipole?

Questions and Problems

§24.1 USING GAUSS' LAW TO CALCULATE ELECTRIC FIELDS

11. ❖ One half of a sphere carries a charge $+Q$ uniformly distributed on its surface. The other half of the sphere has a charge $-Q$. Can you use Gauss' law to find the electric field produced by the sphere? Why or why not?

12. ❖ A cube carries a charge Q uniformly distributed throughout its volume. Can you use Gauss' law to find the electric field produced by the cube? Why or why not?

13. ♦ In good weather, the electric field near the surface of the Earth is about 100 N/C and points inward. Estimate the net charge on the Earth's surface. If the charge is uniformly distributed, how much is there on each square meter of surface? Assume the Earth is a perfect sphere.

14. ♦ Two concentric spherical surfaces with radii R_1 and R_2 each carry a total charge Q. What is the electric field between the two shells?

15. ♦♦ Two concentric plastic spherical shells carry uniformly distributed charges, Q on the inner shell and $-Q$ on the outer shell. Find the electric field **(a)** inside the smaller shell, **(b)** between the shells, and **(c)** outside the larger shell.

16. ♦♦ Three concentric glass spherical shells each carry a charge Q, uniformly distributed over the shell. The radii of the shells are $a < b < c$. Find the electric field everywhere. (*Hint:* There are four distinct regions.)

§24.2 THE ELECTRIC FIELD DUE TO A LINEAR CHARGE DISTRIBUTION

17. ❖ What is the electric field at the center of an equilateral triangle made out of three uniformly charged rods, each carrying the same charge density λ?

18. ❖ An infinite, uniformly charged filament is perpendicular to the plane of a uniformly charged circular filament and through its center. What is the total force each filament exerts on the other? Does the answer depend on the signs or magnitudes of their charge densities?

19. ♦ A long glass filament carries a charge density $\lambda = +1.34\ \mu\text{C/m}$. If two 1.0-mm-long pieces are cut from the filament, how much charge does each piece have? What force would the segments exert on each other if they were held 1.0 m apart?

20. ♦ A long glass filament carries a charge density $\lambda = -2.7$ nC/m. What is the magnitude of the electric field 0.67 mm from the filament?

21. ♦ A circular filament of radius $R = 0.65$ m has a total charge of 6.3 μC. What electric field does it produce on its axis a distance R from its center? Compare with the field produced on its bisector at a distance $\sqrt{2}\,R$ by a straight filament of length $2\pi R$ with the same total charge.

22. ♦ The tube of a Geiger counter is 15 cm long and has a radius of 1.2 cm. The electric field near the outer wall is 1.5×10^4 N/C. If electrons are attracted toward the central filament, what is the linear charge density on the filament? What is the total charge on the filament?

23. ♦♦ Two uniformly charged infinite filaments are parallel and separated by a distance d. If each has charge density λ, find the force per unit length each exerts on the other.

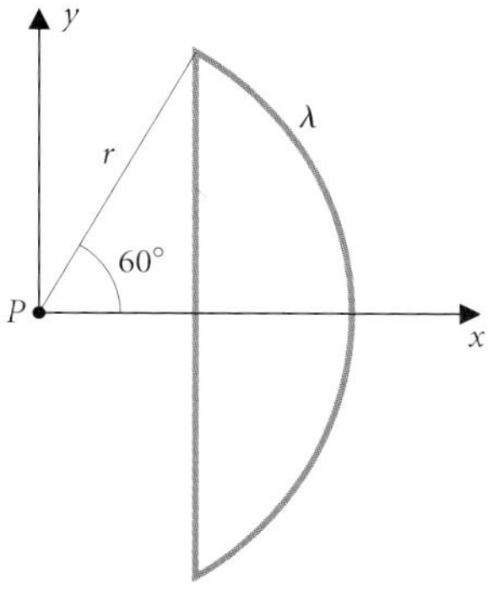

■ FIGURE 24.23

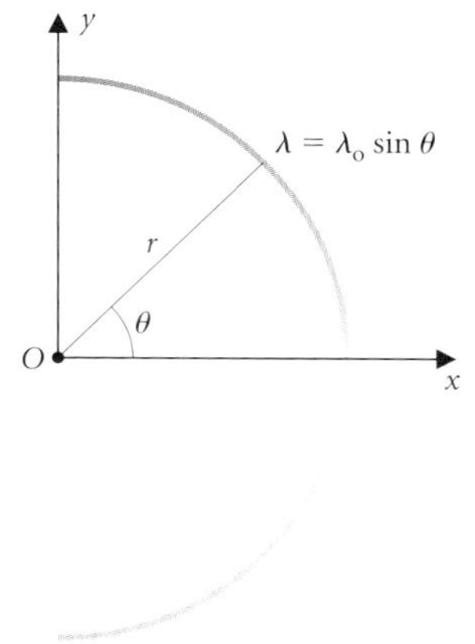

■ FIGURE 24.24

24. ♦♦ A point charge Q is on the axis of a uniformly charged circular filament with charge density λ and radius a. If the point charge is a distance z from the center of the filament, show that the force it exerts on the filament is:

$$\vec{\mathbf{F}} = \left(\frac{2\pi k a \lambda Q z}{[z^2 + a^2]^{3/2}}, \text{ along the axis of the filament} \right).$$

25. ♦♦ A uniformly charged rod of length ℓ with charge density λ lies along the x-axis, with its midpoint at the origin. Find the electric field at a point on the x-axis, with $x > \ell/2$.

26. ♦♦ A semi-infinite filament with uniform charge density λ lies along the positive x-axis. Find the electric field at points on the y-axis.

27. ♦♦♦ A uniformly charged filament is in the shape of a circular arc and a chord (■ Figure 24.23). Find the electric field at the center of the circle (point P).

28. ♦♦♦ Two semicircular filaments have a common diameter. One lies in the x-y-plane and one in the x-z-plane. The curved portion of each carries a uniform line charge density λ. Find the electric field at the common center of the semicircles.

29. ♦♦♦ A semicircular piece of wire carries a charge density $\lambda = \lambda_o \sin\theta$, where θ is zero at the midpoint of the wire (■ Figure 24.24). Find the electric field at the center of the circle (point O).

30. ♦♦♦ Two semi-infinite filaments lie along the positive x- and y-axes, respectively. Each carries the same charge density λ. Find the electric field at a point on the diagonal (■ Figure 24.25).

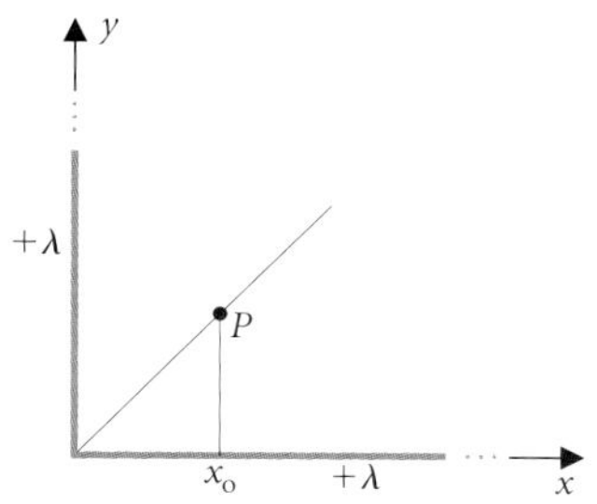

■ FIGURE 24.25

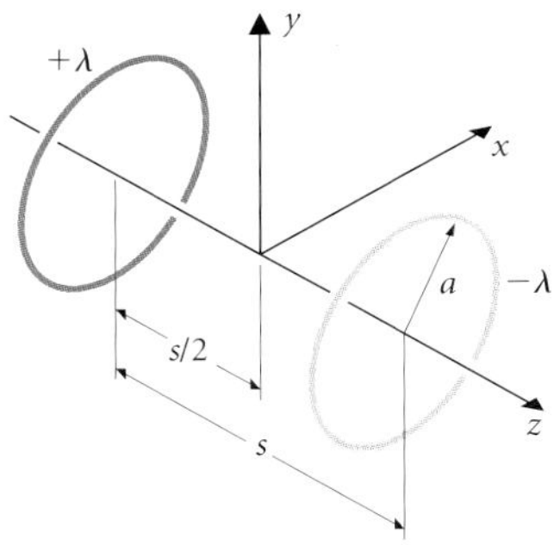

■ FIGURE 24.26

31. ♦♦♦ Two uniformly charged circular filaments with charge densities $+\lambda$ and $-\lambda$ are parallel, have a common axis, and are separated by a distance s (■ Figure 24.26). Find the electric field vector at points on the axis between the filaments as a function of z. Show that $d\vec{E}/dz = 0$ at $z = 0$.

§24.3 ELECTRIC FIELDS DUE TO SURFACE AND VOLUME CHARGE DISTRIBUTIONS

32. ❖ A regular tetrahedron of side L carries a total positive charge Q uniformly distributed over its surface. At point P, at very large distance from the object ($D \gg L$), the electric field due to the object is: **(a)** kQ/L^2. **(b)** kQ/D^2. **(c)** kQD/L^2. **(d)** kQL/D. **(e)** kQD/L^3. Explain what is wrong with the answers you don't choose.

33. ❖ A regular tetrahedron carrying a total positive charge Q uniformly distributed over its surface has one of its faces in the x-z-plane (■ Figure 24.27). At point R just below the center of this face, the direction of the electric field is:
(a) undetermined, since the field is zero.
(b) in the negative y-direction.
(c) in the positive x-direction.
(d) in the negative z-direction.
(e) impossible to determine without a detailed calculation.
Explain why you chose your answer.

34. ❖ Find the electric field on the axis near the center of a very long, uniformly charged cylindrical tube.

35. ❖ Two uniformly charged spherical shells have radii R_1 and R_2, with $R_1 - R_2 \ll R_1$. At a point P between the shells they look like infinite planes, but the electric field is not the same as that between two infinite charged planes. Why? Find a model for the sphere, using infinite planes, that correctly describes the field between the shells near point P.

36. ♦ A uniformly charged glass plate carries surface charge density $\sigma = 4.3\ \text{nC/m}^2$. If a circular disk of radius $R = 3.2$ cm is cut from the plate, what charge does the disk carry?

37. ♦ What is the electric field 0.35 cm above the surface of the glass plate described in Problem 36?

38. ♦ A glass sphere of radius 0.025 cm is bombarded with ions so as to have a total charge of 7.6 pC uniformly distributed throughout its volume. What is the sphere's charge density? What is the electric field 0.05 mm above the surface of the sphere?

39. ♦♦ The cylindrical drum of a copy machine is 40 cm long and has a radius of 5 cm. If charge is uniformly distributed over the curved surface of the drum, and the electric field near the surface is 2×10^5 N/C, estimate the amount of charge on the drum.

40. ♦♦ Two parallel, infinite planes are separated by a distance d. Find the electric field everywhere **(a)** if both planes carry a surface charge density σ and **(b)** if one plane has charge density σ and the other $-\sigma$. (This problem has application in studying capacitors, Chapter 27.)

41. ♦♦ Two uniformly charged infinite planes with equal charge densities σ intersect at right angles. Find the electric field everywhere.

42. ♦♦ A thick spherical shell with inner radius R_1 and outer radius R_2 has a uniform charge density ρ. What is the total charge on the shell? Use Gauss' law to find the electric field everywhere. Sketch the magnitude of the field as a function of distance r from the center of the spheres.

43. ♦♦ A very long cylindrical shell with inner radius a and outer radius b carries a uniform charge density ρ. Use Gauss' law to find the electric field produced by the shell in the regions $r < a$, $a < r < b$, and $r > b$.

44. ♦♦♦ Find the electric field due to an infinite plane by treating the plane as a collection of infinite, uniformly charged filaments, and using superposition.

45. ♦♦♦ Find the electric field on the z-axis due to a disc of radius a lying in the x-y-plane with center at the origin and carrying a charge density $\sigma = \sigma_0(r/a)^2$.

46. ♦♦♦ An infinite plane slab parallel to the x-y-plane occupies the region $-w < z < w$. The charge density in the slab varies with z according to the formula $\rho = \rho_0|z|/w$, for $|z| < w$. Find the electric field produced by the slab for $|z| > w$ and for $|z| < w$.

47. ♦♦♦ Find the electric field everywhere due to the slab of charge in Problem 46 if the charge density is given by $\rho = \rho_0(z/w)$, for $|z| < w$.

48. ♦♦♦ A sphere of radius a has a variable charge density $\rho(r) = \rho_0(r/a)^2$. Find the electric field everywhere inside and outside the sphere. Plot $|\vec{E}(r)|$ versus r. Make a sketch of the field line diagram, showing where field lines begin.

49. ♦♦♦ A sphere of radius R has uniform charge density ρ except within a spherical hole of radius $R/2$, where $\rho = 0$ (■ Figure 24.28). Use the result of Exercise 24.4 to show that $\vec{E}$ is uniform inside the hole, and find its magnitude.

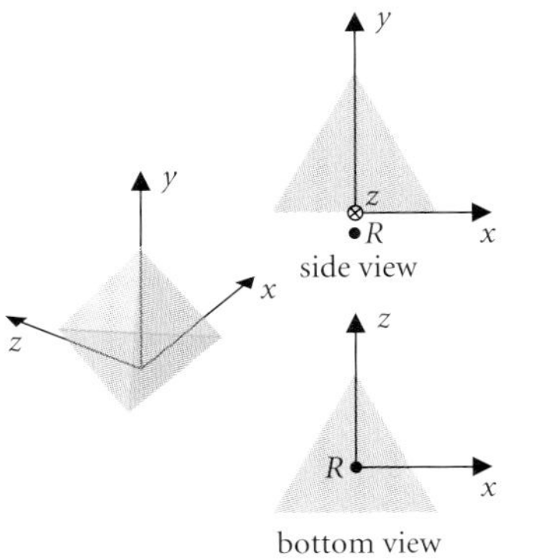

■ FIGURE 24.27

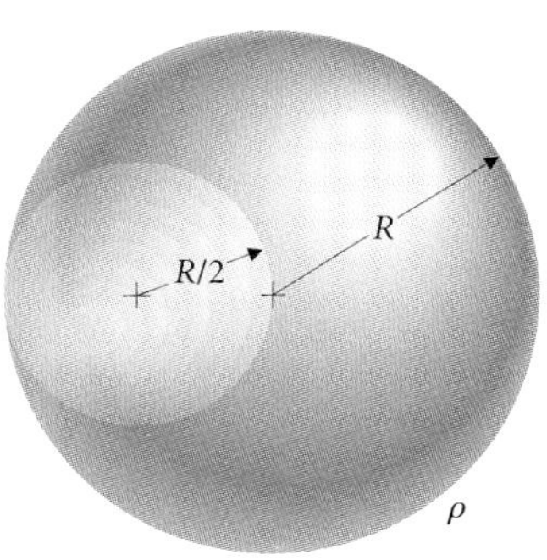

■ FIGURE 24.28

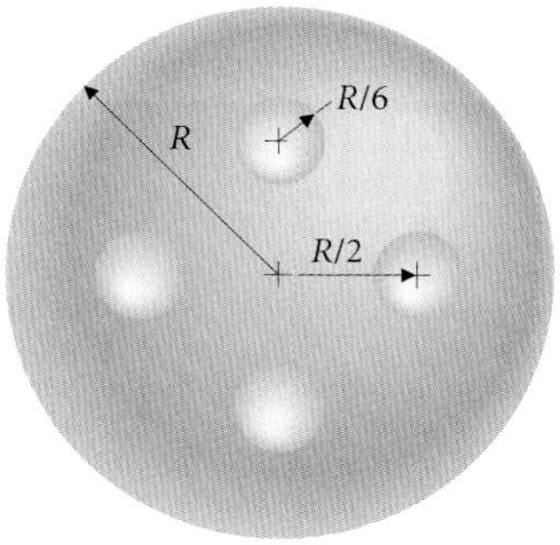

■ FIGURE 24.29

50. ♦♦♦ A sphere of radius R has uniform charge density ρ except within four spherical holes of radius $R/6$. The centers of all five spheres lie in the same plane (■ Figure 24.29). What is the electric field at the center of the large sphere? What is the electric field at the center of one of the holes? (*Hint:* Use the result of Exercise 24.4.)

51. ♦♦♦ An infinitely long cylindrical object with radius R has a charge distribution that depends on distance from the axis according to the formula:

$$\rho = Ar + Br^2 \qquad (r \leq R).$$

Apply Gauss' law to find the electric field produced by the object. What ratio A/B results in zero field outside the object? For this ratio, sketch plots of ρ and $|\vec{\mathbf{E}}|$ as functions of r.

§24.4 MOTION OF CHARGES IN AN ELECTRIC FIELD

52. ❖ If a charged particle is released in a region with an electric field $\vec{\mathbf{E}}$, will the particle move along a field line? Why or why not?

53. ❖ An electron is set in motion in the x-direction at a distance d above an infinite sheet with surface charge density $+\sigma$ (■ Figure 24.30). The electron will:
(a) continue to move in the x-direction with a uniform speed.
(b) be accelerated in the x-direction.
(c) be decelerated and finally move off in the $-x$-direction.
(d) be accelerated in the $+y$-direction.
(e) be accelerated in the $-y$-direction.
Explain what is wrong with the answers you don't choose.

54. ❖ An electron is originally at point P, 1.0 m away from a very large uniformly charged plane surface. The acceleration of the electron is (a, away from the plane). Later, the electron is at point R, 2.0 m away from the surface. Its acceleration has magnitude:
(a) $2a$. **(b)** $a/2$. **(c)** $4a$. **(d)** a. **(e)** $a/4$.
Explain how you chose your answer.

55. ❖ Three charges of equal magnitude are held at the corners of a triangle as shown in ■ Figure 24.31. If the negative charge is released, describe its subsequent motion qualitatively.

56. ♦ A charged particle of mass 5.0 g is placed in an electric field of 10.0 N/C. It is observed to accelerate antiparallel to the field at 8.0 m/s^2. What is the charge on the particle?

57. ♦ Two identical charged particles of mass 5.0 g and charge 17 nC are held 0.10 m apart and released. What is the initial acceleration of each particle?

58. ♦ If the mass of an electron is 9.11×10^{-31} kg, find the electric field necessary to accelerate the electron at $a = 1.76 \times 10^{10}$ m/s^2. The mass of a proton is 1.67×10^{-27} kg. What is its acceleration in the same electric field?

59. ♦ At a certain time an electron is observed to accelerate to the right at $a_e = 1.6 \times 10^5$ m/s^2. At the same time a charge is known to be 1.0 m to the left of the electron. If the unknown charge produces the only force on the electron, what is the value of the charge?

60. ♦♦ What is the acceleration of an electron in an electric field of 10^3 N/C? What distance does the electron travel in such a field before its speed equals one-tenth the speed of light? How far would a proton have to travel to reach the same speed? [At faster speeds the laws of special relativity (Chapter 34) are necessary for accurate results.]

61. ♦♦ An object with unknown charge q and mass $m = 1.0$ g moves in a circle about a very much more massive object with charge $Q = 1.0\ \mu$C. The distance between the two is $r = 2.0$ cm, and the smaller object moves at speed $v = 5.0$ m/s. What is q?

62. ♦♦ An infinite plane surface carries a uniform charge density $\sigma = +20.0$ nC/m^2. An electron emerges from a small hole in the surface with a speed of 2.7×10^7 m/s at an angle of 60° from the normal. Describe the motion of the electron. What time interval passes before the electron returns to the surface? At what distance from the hole does it strike the surface?

63. ♦♦ If a proton emerges from the hole in Problem 62 with the same initial velocity, find the position of the proton as a function of time.

64. ♦♦ An electron is released 1.0 cm above a uniformly charged infinite plate with charge density 1.0×10^{-9} C/m^2. What is the speed of the electron when it hits the plate?

65. ♦♦ A particle of mass 1.5 g is moving in a circular path around a filament carrying a charge density of 1.0 μC/m. The charge on the particle is -15 nC. Find the speed of the particle.

66. ♦♦ Three parallel, infinite, uniformly charged planes are arranged as shown in ■ Figure 24.32. A small hole passes through the middle plane. At $t = 0$ an electron emerges from the hole moving perpendicular to the planes with speed v_0. Assuming v_0 is small enough that the electron does not collide with the negative plate, show that its motion is periodic, and find the period.

67. ♦♦ In an ink-jet printer, drops of ink are given a controlled amount of charge before they pass through a uniform electric field generated by two charged plates (■ Figure 24.33). If the electric field between the plates is 1.5×10^6 N/C over a region 1.7 cm long, and drops of mass 0.12 ng enter the electric field region moving horizon-

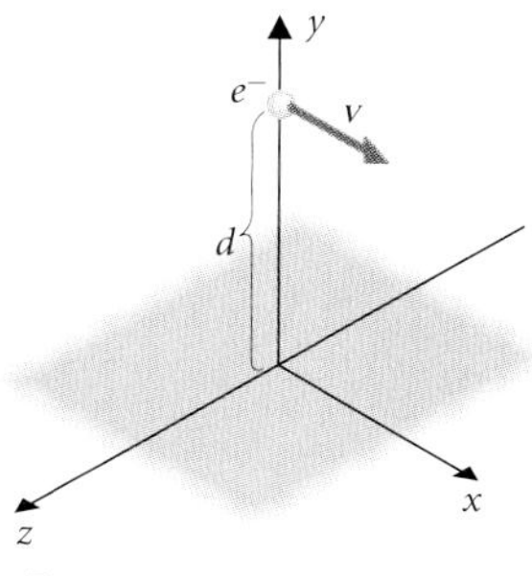

■ FIGURE 24.30

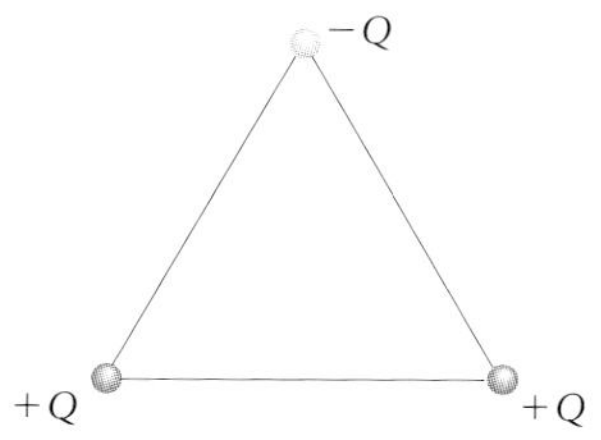

■ FIGURE 24.31

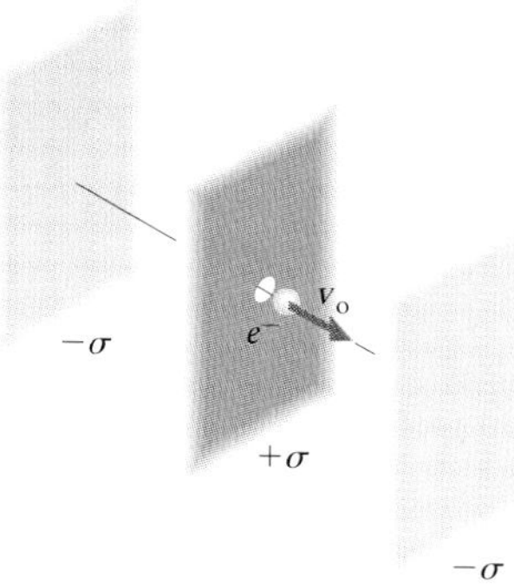

■ FIGURE 24.32

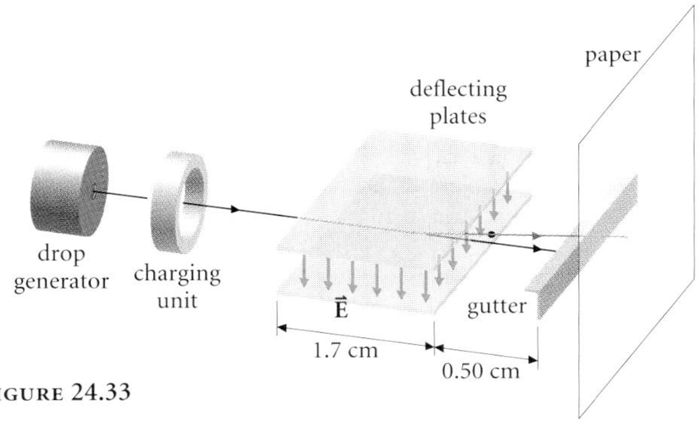

■ FIGURE 24.33

tally at 17 m/s, how much charge on a drop is necessary to produce a vertical deflection of **(a)** 0.50 mm and **(b)** 0.75 mm at the end of the electric field region? What is the deflection at the position of the paper, 0.50 cm from the end of the charged plates, in each case? Ignore other forces such as gravity and air resistance, and assume the field drops abruptly to zero at the edge of the plates.

68. ♦♦♦ In a cathode-ray tube used in an oscilloscope, an electric field $\vec{E}_0$ parallel to the tube accelerates the electrons to a speed v (cf. Example 24.8). A second electric field $\vec{E}$ perpendicular to the tube deflects them sideways (■ Figure 24.34). Find the deflection angle θ in terms of v, E, and the length of the electric field region, ℓ. For a tube with $\ell = 10$ cm, and $v = 9 \times 10^7$ m/s, how large an electric field is required to give a deflection of 40°?

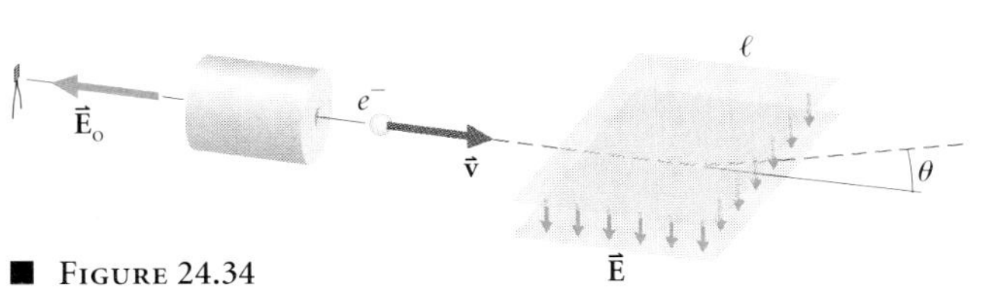

■ FIGURE 24.34

§24.5 THE DIPOLE

69. ❖ A dipole is placed near a long, uniformly charged filament. Describe the orientation the dipole achieves if it can rotate about its center, and the resulting oscillations damp.

70. ❖ A charge configuration consists of three equally spaced charges on a line. The central charge is $+2Q$; the other two have charge $-Q$ and are at a distance a on either side of the center. Draw a diagram showing the forces acting on the charge configuration in a uniform external electric field. Show that the total force and total torque on the configuration are both zero.

71. ❖ A circular filament lying in the x-y-plane has charge density $\lambda = \lambda_0 \sin\theta$ (■ Figure 24.35). What is the direction of the electric field at each of the points A, B, C, and D?

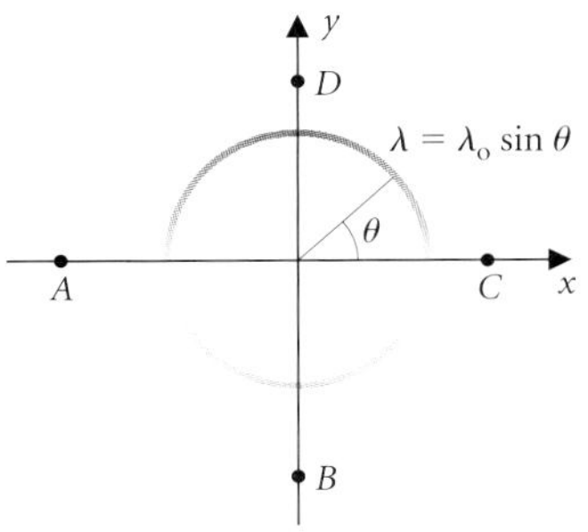

■ FIGURE 24.35

72. ♦ A needle 2.0 mm long has a charge of 1.0 nC at one end and -1.0 nC at the other. Find the dipole moment of the needle.

73. ♦ The needle described in Problem 72 lies along the x-axis with its center at the origin. Find the electric field on the x-axis at $x = 33$ cm.

74. ♦ A needle suspended from a string hangs horizontally. The electric field at the needle's location is horizontal with magnitude 3.7×10^3 N/C and is at an angle of 30° with the needle. There is no net electrical force acting on the needle, but the string exerts a torque of 3.7×10^{-4} N·m to hold the needle in equilibrium. What is the needle's dipole moment?

75. ♦♦ A dipole of moment 9.0 nC·m is placed in a region where the electric field is 1.5 N/C. If the dipole may be modeled as a rod 5.0 mm long with a mass of 1.0 g, find the period of small oscillations about the equilibrium position.

76. ♦♦♦ A dipole of moment $\vec{p}$ is placed near a point charge Q. Find the force and torque on the dipole as a function of distance r from the charge and of the orientation of the dipole.

Additional Problems

77. ❖ Two uniformly charged spherical plastic shells of diameter D each carry charge Q. They are separated by a distance $3D$ center to center. What is the electric field halfway between the shells? At the center of each shell?

78. ♦♦ A uniformly charged filament with charge density λ is parallel to a uniformly charged plane, with density σ, a distance s away. At what positions would an electron be in static equilibrium? Is the equilibrium stable? Would a proton be in stable equilibrium?

79. ♦♦ A thin, square sheet of side ℓ carries a total charge Q uniformly spread along its surface. Use Gauss' law to find the electric field *very near* the center of the square. From Coulomb's law obtain the magnitude of the electric field at *very large* distances from the square. By comparing these two approximations, estimate the distance from the square at which the infinite plane approximation becomes invalid.

80. ♦♦ In an experiment to measure charge, a small water droplet 8.41 μm in diameter falls through an evacuated region in which there is an electric field 1.10×10^6 N/C downward. A laser beam system measures the times when the drop passes successive slits 2.00 cm apart to be 0.0000 s, 0.0179 s, 0.0330 s, and 0.0463 s. What is the acceleration of the drop? What is the charge on the drop in units of e? What would the acceleration of the drop be if the field were reversed?

81. ♦♦♦ A classical model of an atom has electrons in orbit about a positively charged nucleus. Consider an electron in a circular orbit about a proton. Find the radius of the orbit in terms of the angular momentum of the electron.

82. ♦♦♦ The machine shown in ■ Figure 24.36 uses two oppositely charged sheets to support the block against gravity. Using the

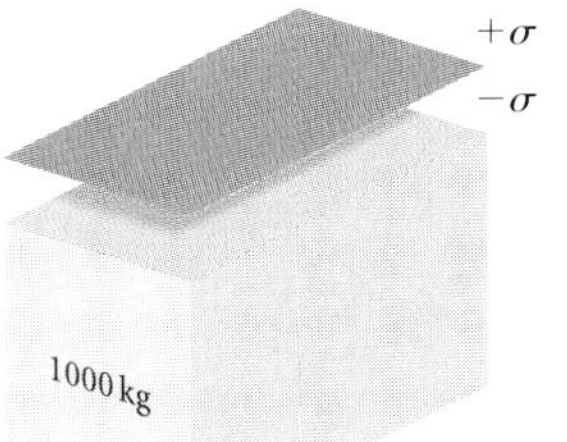

■ FIGURE 24.36

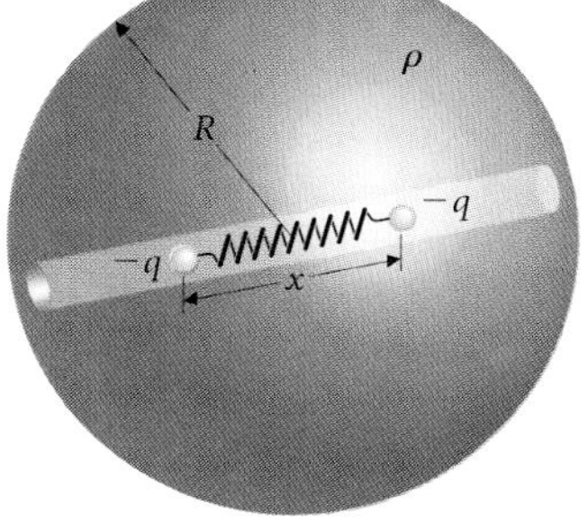

■ FIGURE 24.37

result for an electric field due to an infinite plane to approximate the electric field due to each sheet, what charge density is required if the sheet is 1 m on a side and the block has a mass of 1000 kg?

83. ♦♦♦ A small hole is drilled through a sphere with uniform charge density ρ. Two charges $-q$ are attached to a spring of constant k and zero relaxed length (■ Figure 24.37). Find the equilibrium separation x of the charges. Consider both $x < 2R$ and $x > 2R$.

84. ♦♦♦ An infinite plane slab of material of thickness $2s$ contains a variable charge density $\rho = \rho_o(x^2/s^2)$ (■ Figure 24.38). Apply Gauss' law to find the electric field due to the slab.

85. ♦♦♦ A uniform rod with mass $M = 0.75$ kg, charge density $\lambda = 1.5\ \mu\text{C/m}$, and length $L = 1.2$ m is attached with a frictionless hinge to an infinite, vertical plane surface with uniform charge density $\sigma = 85\ \mu\text{C/m}^2$. At what angle with the vertical does the rod hang? What force does the hinge exert?

86. ♦♦♦ A very thin hole is cut along a diameter of an otherwise uniformly charged sphere with radius $R = 0.20$ m and total charge $Q = 1.0\ \mu\text{C}$. If an electron is released from rest at one end of the hole, show that its motion is simple harmonic, and find the period.

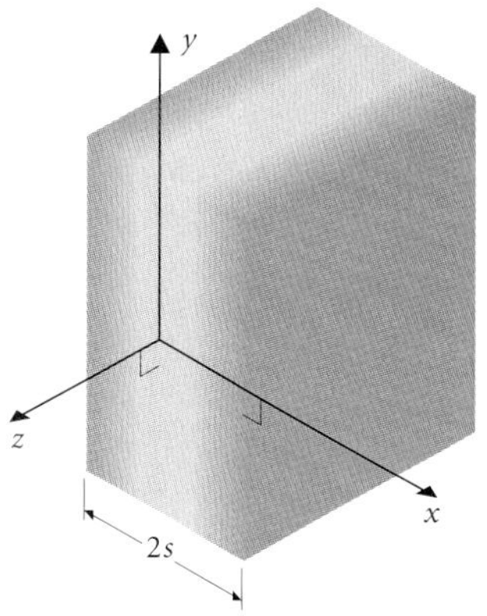

■ FIGURE 24.38

Computer Problems

87. A line charge of length 1.00 m with $\lambda = 1.00$ nC/m lies along the x-axis with its center at the origin. Divide the line into 20 segments each of length 0.05 m. Compute the x- and y-components of the electric field produced by each segment at a point P_1 on the y-axis at $y_1 = 0.50$ m. Use the trapezoidal rule to sum the contributions to find the electric field at P_1, and compare your result with the result of Example 24.3. Now compute the x- and y-components of the electric field at points P_2 and P_3 with coordinates $x_2 = 0.25$ m, $x_3 = 0.50$ m, $y_2 = y_3 = 0.50$ m.

88. A circular filament of radius 1.0 m with $\lambda = 1.0$ nC/m lies in the x-y-plane with its center at the origin. Calculate the components of the electric field at the point P with coordinates $x = -1.0$ m, $y = 0.0$ m, $z = +1.0$ m.

89. Two identical dipoles with $\vec{\mathbf{p}} = (1.0\ \mu\text{C}\cdot\text{m})\hat{\mathbf{j}}$ are located at the origin and at a distance $r = 5.0$ m from the origin. Calculate the force exerted on the second dipole by the first, when the line from the first to the second makes an angle of **(a)** 30°, **(b)** 60°, and **(c)** 90° with the x-axis. *Hint:* Model each dipole as two point charges of magnitude 10 μC separated by a distance 0.10 m. Compare your result for (c) with the result of Example 24.12.

Challenge Problems

90. A rod of length 2ℓ lies along the x-axis with its center at the origin. The rod carries a charge density $\lambda = \lambda_o(x/\ell)$. Find the electric field at a point on the y-axis.

91. Show that the electric field at an arbitrary point P with polar coordinates r and θ due to a dipole at the origin is:

$$\vec{\mathbf{E}} = \frac{kp}{r^3}\,[\sin\theta\ \hat{\boldsymbol{\theta}} + 2\cos\theta\ \hat{\mathbf{r}}].$$

92. Show that a displacement along a field line satisfies the relation:

$$\frac{dr}{d\theta} = \frac{rE_r}{E_\theta},$$

where r and θ are polar coordinates in the plane containing the field line. Using the result of Problem 91, obtain a differential equation for the field lines due to a dipole at distances $r \gg$ the length ℓ of the dipole. Show that each field line satisfies the equation:

$$r = r_m \sin^2\theta,$$

where r_m is the maximum distance of the field line from the origin.

93. A hemispherical shell has a uniform charge density σ spread across its surface. What is the value of the electric field at point P (■ Figure 24.39)? If the electric force on a point charge q placed at P just balances gravity, what is the mass of the particle? Is the equilibrium stable?

94. A quadrupole consists of four charges at the corners of a small square of side ℓ. Two positive charges are at the ends of one diagonal, and two negative charges are at the ends of the other diagonal. Find the electric field due to this charge distribution **(a)** at a point on the perpendicular through the center of the square and **(b)** at a point in the plane of the square on the bisector of one side. Can you find a reasonable definition of *quadrupole moment?*

95. Find the electric field at a point P at distance y above the center line of a uniformly charged, infinitely long strip of width L (■ Figure 24.40). The strip has charge density σ.

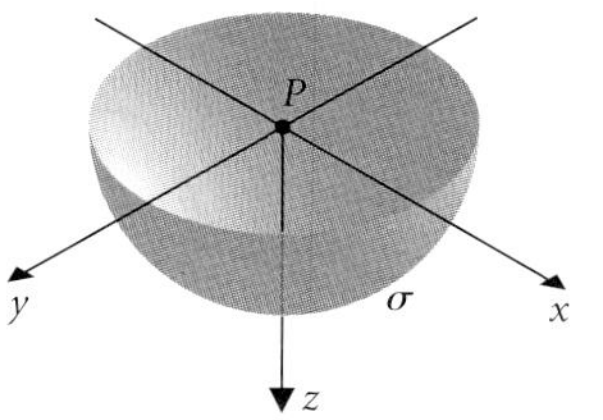

■ FIGURE 24.39

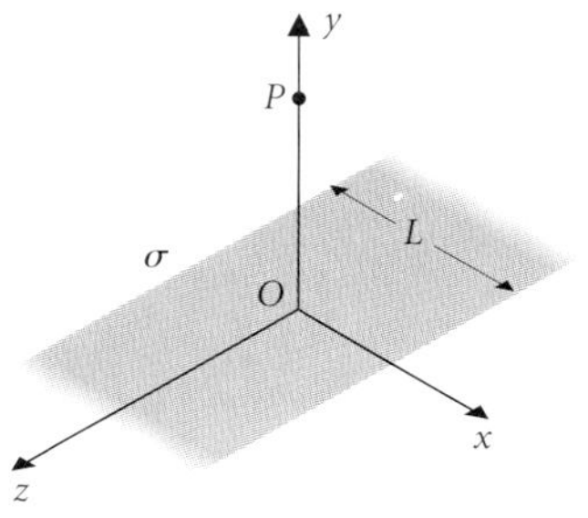

■ FIGURE 24.40

CHAPTER 25
Electric Potential Energy

In the Boston Museum of Science, operator Don Salvatore sits in a metal cage. The giant towers behind him are Van de Graaff generators. The large spheres at their tops are charged up to 2.5 million volts. The electric fields that surround them are great enough to ionize the air, providing current paths that allow sparks to jump to nearby wires and smaller metal spheres. Inside his cage, Mr. Salvatore is protected from the artificial lightning storm that surrounds him. In this chapter we'll investigate properties of conductors that explain how the cage can protect its occupant (§25.6). We introduce electric potential—measured in volts—and see how it helps us to understand situations like this. (Robert J. Van de Graaff built this model in 1931.)

Concepts

Electric potential energy
Electric potential
Equipotential surface
Conductor
Electron volt

Goals

Be able to:

Compute the potential energy of a system of point charges and use the results in energy conservation problems.

Calculate the potential due to a discrete or continuous charge distribution.

Sketch the equipotential surfaces for a given charge distribution.

Understand:

The relation between electric potential and the electric field vector.

The use of potential in explaining the behavior of conductors.

I sometimes fancy that such nearness to nature as I have described keeps the spirit sensitive to impressions not commonly felt, and in touch with unseen powers.

Ohiyesa

Electric energy provides spectacular entertainment in museum displays that snap and crackle like miniature lightning storms. Huge sparks look and sound scary, but the operator is quite safe in his cage. The display relies on the ability of electrons to flow within the bars of the metal cage and on the fact that the sparks transmit very little total charge though impressive amounts of energy per electron.

Study hint: Review the four properties of potential energy in §8.1.

We have already discovered energy in several different forms: kinetic energy, gravitational and elastic potential energy, and thermal energy. Electric potential energy—our next addition to this list—is the key to understanding such diverse phenomena as chemical interactions and high-voltage power transmission as well as sparks in museums. In this chapter, we shall introduce electric potential energy and explore its relation to the electric field vector $\vec{\mathbf{E}}$.

25.1 Potential Energy of a Pair of Point Charges

25.1.1 *Work Done by the Coulomb Force*

In Chapters 23 and 24 we discussed the electric field, described by the vector $\vec{\mathbf{E}}$, and the force $\vec{\mathbf{F}} = Q\vec{\mathbf{E}}$ it exerts on an electrically charged particle. The force is a function of position only and is similar in form to the gravitational force. Also like the gravitational force, the Coulomb force is conservative (cf. §8.3), and the system of charged particles and electric field possesses electric potential energy. If a particle is released, the electric force causes it to accelerate and gain kinetic energy at the expense of the system's potential energy.

These are the decay products of radon.

EXAMPLE 25.1 ♦♦ A polonium nucleus (charge $Q_1 = 1.3 \times 10^{-17}$ C) and an alpha particle (charge $Q_2 = 3.2 \times 10^{-19}$ C) are a distance $d_1 = 9.1 \times 10^{-15}$ m apart. Assume the polonium is fixed and the alpha particle is free to move. Find the work done on the alpha particle when it moves to a new position at distance $d_2 = 2d_1$ from the polonium nucleus (■ Figure 25.1). If the alpha particle is initially at rest, find its speed at its new position.

See Interlude 2 and §8.2.2.

MODEL The alpha particle accelerates away because the electric field $\vec{\mathbf{E}}_1$ of the polonium nucleus exerts a repulsive force on it. Since the force depends on position, we find the work done as an integral of the differential amounts of work done during each differential displacement of Q_2. The speed of the alpha particle can be found by setting the work done on it equal to the change in its kinetic energy.

SETUP The force exerted by $\vec{\mathbf{E}}_1$ when Q_2 is a distance r from Q_1 is:

$$\vec{\mathbf{F}}(r) = \frac{kQ_2Q_1}{r^2}\,\hat{\mathbf{r}}.$$

STEPS I AND II The path of Q_2 between its initial and final positions is along a radial line. The appropriate coordinate is the distance r, and a differential displacement is described by: $d\vec{\mathbf{r}} = dr\,\hat{\mathbf{r}}$.

STEP III The work done on Q_2 during a differential displacement is:

$$dW = \vec{\mathbf{F}} \cdot d\vec{\mathbf{r}} = \frac{kQ_2Q_1}{r^2}\,\hat{\mathbf{r}} \cdot (dr\,\hat{\mathbf{r}}) = \frac{kQ_2Q_1\,dr}{r^2}.$$

STEPS IV AND V The particle moves from $r = d_1$ to $r = d_2$, so:

$$W \equiv \int dW = \int_{d_1}^{d_2} \frac{kQ_2Q_1\,dr}{r^2} = kQ_2Q_1\left(-\frac{1}{r}\right)\Bigg|_{d_1}^{d_2} = kQ_2Q_1\left(-\frac{1}{d_2} + \frac{1}{d_1}\right). \quad (25.1)$$

SOLVE

$$W = \frac{(9.0 \times 10^{9}\ \mathrm{N \cdot m^2/C^2})(1.3 \times 10^{-17}\ \mathrm{C})(3.2 \times 10^{-19}\ \mathrm{C})}{(9.1 \times 10^{-15}\ \mathrm{m})}\left(-\frac{1}{2} + \frac{1}{1}\right)$$

$$= 2.1 \times 10^{-12}\ \mathrm{J}.$$

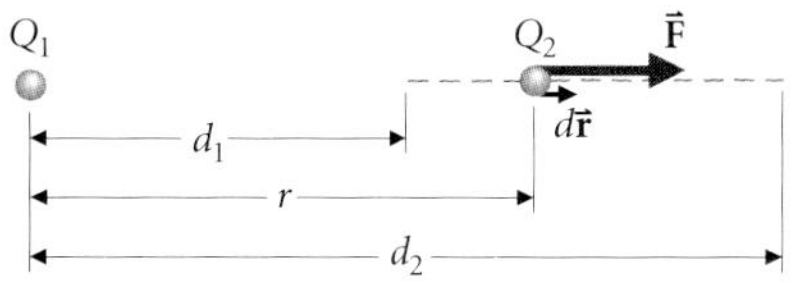

■ Figure 25.1
The electric field due to a fixed polonium nucleus exerts a force on the nearby alpha particle. The polonium nucleus does work on the alpha particle, which accelerates away.

SETUP Using the work-energy theorem:

$$\tfrac{1}{2}mv_f^2 - \tfrac{1}{2}mv_i^2 = W.$$

With $v_i = 0$, $\quad v_f = \sqrt{2W/m}$.

SOLVE The mass m of the alpha particle $= 4.0$ u $= 6.6 \times 10^{-27}$ kg, so:

$$v_f = \sqrt{\frac{2(2.1 \times 10^{-12}\ \text{J})}{6.6 \times 10^{-27}\ \text{kg}}} = 2.5 \times 10^7\ \text{m/s}.$$

ANALYZE The final velocity is 8% of the speed of light. This is another indication of the strength of the electric force. This example provides an approximate model for the nuclear decay of radon.

25.1.2 The Electric Force as a Conservative Force

The Coulomb force is conservative. That is, the work done moving a test charge between two points is independent of the path. ■ Figure 25.2a shows an arbitrary path between two points at distances d_1 and d_2 from a point charge Q. The work done by the electric field on a test charge q when it undergoes a displacement $d\vec{\ell}$ is:

$$dW = \vec{\mathbf{F}}_e \cdot d\vec{\ell} = q\frac{kQ}{r^2}\hat{\mathbf{r}} \cdot d\vec{\ell}.$$

From Figure 25.2b, $\hat{\mathbf{r}} \cdot d\vec{\ell} = dr$, and $dW = q\dfrac{kQ\,dr}{r^2}$.

This is the same expression for dW that we obtained in Example 25.1, where we calculated the work done along a radial path. Thus eqn. (25.1) also gives the work done in separating two charges from d_1 to d_2 along an *arbitrary* path.

$$W = kqQ\left(\frac{1}{d_1} - \frac{1}{d_2}\right). \qquad (25.2)$$

25.1.3 Potential Energy of a Pair of Charges

The work done by a conservative force decreases the *potential* energy of the system:

REMEMBER: ONLY *CHANGES* OF POTENTIAL ENERGY ARE MEASURABLE.

$$W = U(d_1) - U(d_2).$$

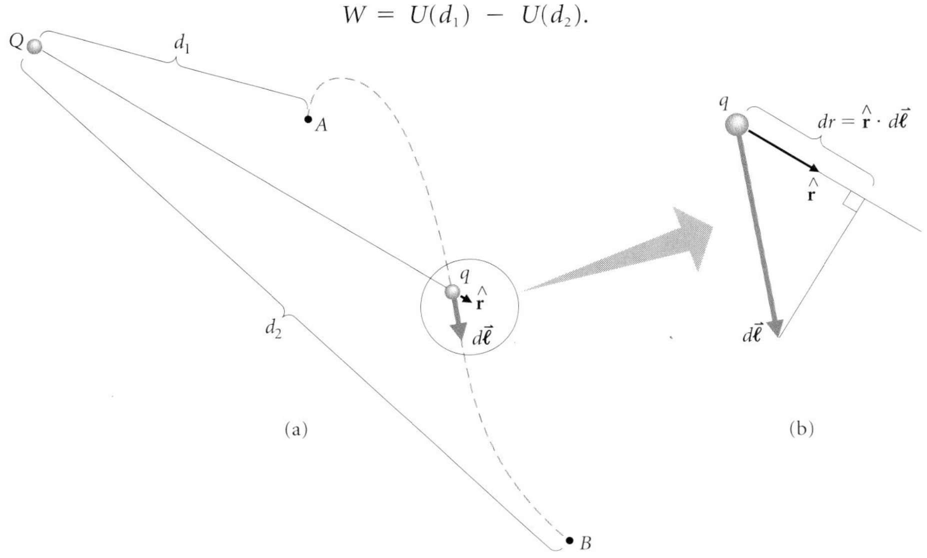

■ FIGURE 25.2
The Coulomb force is conservative; that is, the work done by the Coulomb force on a charge that moves from A to B is independent of the path taken between the two points. Since the field points radially outward, the work done in a differential displacement $d\vec{\ell}$ depends only on the dot product $\hat{\mathbf{r}} \cdot d\vec{\ell} = dr$, the change in distance from Q_1.

Comparing this equation with eqn. (25.2), we find the potential energy as a function of the distance between a pair of charges:

$$U(d) \equiv \frac{kqQ}{d} + \text{arbitrary constant.}$$

THE SAME CONVENTION IS USED FOR GRAVITATIONAL POTENTIAL ENERGY. SEE §8.2.

As usual, the potential energy function contains an arbitrary constant because we may choose where to set the function equal to zero for *our convenience* (cf. §8.2). It is customary to set the energy function to zero when the two particles are infinitely far apart. In that case the arbitrary constant is zero, and the potential energy function is:

$$U(d) = \frac{kqQ}{d}. \tag{25.3}$$

Recall that a *system* possesses potential energy, not its individual particles. The system here is the pair of charges, and the potential energy we have calculated is that of a pair of charges separated by distance d. If instead of only one charge moving, both are free to move, they both accelerate and share the energy of the system.

EXAMPLE 25.2 ♦♦ In the first stage of a controlled nuclear fusion reaction, nuclei of deuterium (D, mass 2.0 u) and tritium (T, mass 3.0 u) react to form a helium nucleus and a neutron. The D and T nuclei each have charge $+e$, so the Coulomb force between them is repulsive. Assume the center of mass of the two particles is at rest in the lab frame. How fast must each particle be moving as they approach each other from a great distance if they are to reach a minimum separation of 2.0×10^{-15} m?

MODEL We are given an initial state of the system and asked about a final state, without being asked how the system arrives at the final state or how much time it requires. These are the features that identify a conservation law problem (cf. §6.2). Both particles come to rest in their CM frame as they reach minimum separation. The initial kinetic energy of the system is converted to potential energy. The linear momentum of the system is zero throughout. The Coulomb force lies along the line between the particles, so the angular momentum of each particle about the CM remains zero and we need not consider it further. The final state of the system is illustrated in ■ Figure 25.3.

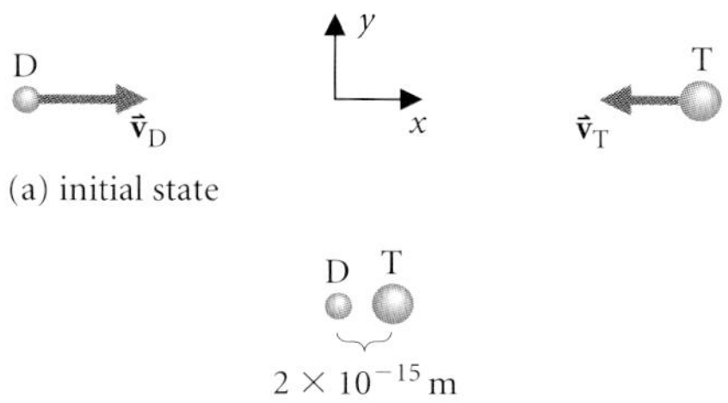

(a) initial state

(b) final state:

■ FIGURE 25.3
In nuclear fusion, a deuterium nucleus and a tritium nucleus combine to form helium. Since both reacting nuclei are positively charged, they repel each other. They must move toward each other at high speed to get close enough to react. We choose the x-axis along the line joining the two nuclei. (a) Nuclei separated by a large distance. (b) Nuclei at rest with a separation of 2 fm.

SETUP

	BEFORE		AFTER	
Energy:	Potential	Kinetic	Potential	Kinetic
	zero	$\frac{1}{2}M_D v_D^2 + \frac{1}{2}M_T v_T^2$	$\frac{kQ_D Q_T}{d}$	zero
Momentum (x-component):		$M_D v_D - M_T v_T$	zero	

Set values *before* equal to values *after*:

$$\tfrac{1}{2}M_D v_D^2 + \tfrac{1}{2}M_T v_T^2 = \frac{kQ_D Q_T}{d}. \tag{i}$$

$$M_D v_D - M_T v_T = 0. \tag{ii}$$

SOLVE From eqn. (ii):

$$v_D = (M_T/M_D)v_T.$$

Substituting in eqn. (i):

$$\frac{kQ_D Q_T}{d} = \tfrac{1}{2}M_T v_T^2 + \tfrac{1}{2}M_D\left(\frac{M_T v_T}{M_D}\right)^2 = \tfrac{1}{2}M_T v_T^2\left(1 + \frac{M_T}{M_D}\right).$$

Then:

$$v_T^2 = \frac{2kQ_D Q_T}{M_T d}\,\frac{M_D}{(M_T + M_D)}.$$

So: $v_T = \sqrt{\dfrac{2kQ_DQ_T}{M_Td}\dfrac{M_D}{(M_T + M_D)}}$ and $v_D = \sqrt{\dfrac{2kQ_DQ_T}{M_Dd}\dfrac{M_T}{(M_D + M_T)}}$.

$$v_T = \sqrt{\frac{2(9.0 \times 10^9\ \mathrm{N\cdot m^2/C^2})(1.6 \times 10^{-19}\ \mathrm{C})^2}{(2.0 \times 10^{-15}\ \mathrm{m})(3.0)(1.66 \times 10^{-27}\ \mathrm{kg})}\frac{(2.0\ \mathrm{u})}{(2.0\ \mathrm{u} + 3.0\ \mathrm{u})}}$$

$$= 4.3 \times 10^6\ \mathrm{m/s}.$$

$$v_D = \frac{M_T}{M_D}v_T = \frac{3.0\ \mathrm{u}}{2.0\ \mathrm{u}} 4.3 \times 10^6\ \mathrm{m/s} = 6.5 \times 10^6\ \mathrm{m/s}.$$

ANALYZE Notice that we can use any unit for the mass in a ratio such as M_T/M_D, provided that we use the same unit in the numerator and the denominator. Since the speed of the particles in a gas increases as the square root of the temperature (Chapter 19), the deuterium/tritium gas must be very hot (at least 10^7 K) for the nuclei to achieve the necessary speeds.

25.2 ELECTRIC POTENTIAL

25.2.1 *Potential Energy of a Charge in an Arbitrary Electric Field*

In an oscilloscope, charges on the electron gun and the deflection plates produce electric fields that accelerate and guide the electron beam. The forces between different electrons in the beam are comparatively small, and we may think of the individual electrons as test charges moving in the fields of the tube. As each electron accelerates, it gains kinetic energy, and the system of electron plus electric field loses potential energy.

To find the potential energy of a charge in an arbitrary electric field, we begin by calculating the work done on the charge by the field. As the charge moves through a displacement $d\vec{\ell}$ the work done is:

$$dW = \vec{\mathbf{F}} \cdot d\vec{\ell} = q\vec{\mathbf{E}} \cdot d\vec{\ell}.$$

If the charge moves from P_1 to a new position P_2 (■ Figure 25.4), the work done on it by electric forces is:

$$W = q\int_{P_1}^{P_2} \vec{\mathbf{E}} \cdot d\vec{\ell}. \tag{25.4}$$

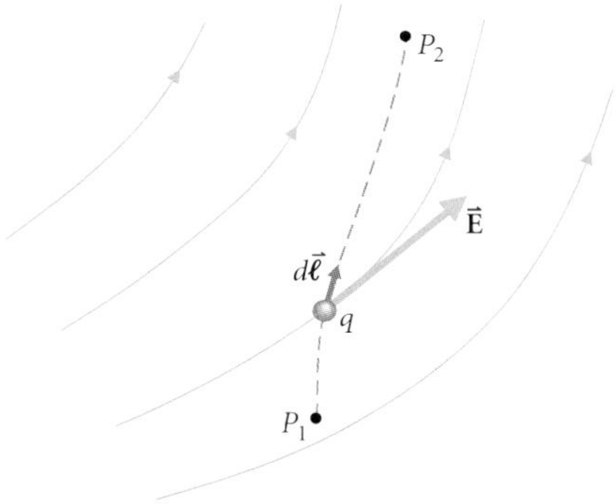

■ FIGURE 25.4
A charged particle moves from point P_1 to P_2 in a region where the electric field is $\vec{\mathbf{E}}$, illustrated here by the field line diagram. The field does work on the particle, corresponding to a change in its potential energy.

The electric force is conservative, so W is independent of the path the electron travels between P_1 and P_2. This work is done at the expense of potential energy belonging to the system of the electron and the charges that produce the field:

$$-\Delta U \equiv -(U_2 - U_1) = W = q\int_{P_1}^{P_2} \vec{\mathbf{E}} \cdot d\vec{\ell}.$$

Thus:

$$\Delta U = U_2 - U_1 = -q\int_{P_1}^{P_2} \vec{\mathbf{E}} \cdot d\vec{\ell} = q\int_{P_2}^{P_1} \vec{\mathbf{E}} \cdot d\vec{\ell}. \tag{25.5}$$

THIS DEVELOPMENT FOLLOWS THAT IN CHAPTER 8. COMPARE EQN. (25.5) WITH EQN. (8.3).

In the second form of this integral, we move along the path from P_2 to P_1 rather than from P_1 to P_2. The direction of each differential displacement $d\vec{\ell}$ is reversed, and therefore so is the sign of the integral. You may use whichever form is more convenient. The two expressions have slightly different physical interpretations: doing work *on* the system increases its potential energy; when the *system* does work, its potential energy decreases.

25.2.2 *Electric Potential*

IN CHAPTER 23, EQN. (23.5).

We defined the electric field vector at any point as the force per unit charge exerted on a test charge placed at that point. Similarly, we may measure the change in energy of the system as a test charge moves between two points and interpret it in terms of a field quantity, called *electric potential V*, produced by the other charges. Like the field vector $\vec{\mathbf{E}}$, potential describes the electric field itself and is independent of the test charge that moves through the field.

> The *difference in electric potential* ΔV between two points equals the change in potential energy of the system when a test charge is moved between the same two points, divided by the value of the charge.
>
> $$\Delta V = \lim_{q \to 0} \frac{\Delta U}{q}. \tag{25.6}$$

Using eqn. (25.5) to express ΔU in terms of $\vec{\mathbf{E}}$, we have:

$$\Delta V \equiv V_{21} = V(P_2) - V(P_1) = \frac{U_2 - U_1}{q}.$$

$$\Delta V = -\int_{P_1}^{P_2} \vec{\mathbf{E}} \cdot d\vec{\ell}. \tag{25.7}$$

THE DIRECTION OF $\vec{\mathbf{E}}$ NEED NOT BE ALONG A STRAIGHT LINE BETWEEN THE TWO POINTS. THE RULE GIVES THE GENERAL TREND.

A useful rule for remembering the signs in eqn. (25.7) is:

The electric field vector points toward decreasing potential.

Just as we have the freedom to choose an arbitrary constant in defining potential energy, we can also choose an arbitrary constant in defining potential. Relative to a reference point R where we choose to define the potential as zero, the electric potential is given by:

$$V(P) = -\int_{R}^{P} \vec{\mathbf{E}} \cdot d\vec{\ell}. \tag{25.8}$$

The value of the potential at P is independent of the path taken between R and P. In deriving the expression for the potential energy of a pair of point charges, we chose to assign zero potential energy to a pair separated by an infinite distance. Similarly, a common choice for the reference point R is at *infinity*, particularly when dealing with a system of point charges. In laboratory practice, electrical *ground* is taken as the reference point. Connecting a conducting wire to the water pipes is a satisfactory way to construct a ground reference for simple electric circuits. The wiring in modern buildings includes a ground wire that is connected to a distant buried metal plate.

AFTER ALESSANDRO VOLTA (1745–1827), THE INVENTOR OF THE BATTERY.

Since potential is energy per unit charge, its SI unit is the joule per coulomb. This unit is given the name volt (symbol V):

$$1 \text{ V} \equiv 1 \text{ J/C}.$$

Combining the definition (eqn. 25.6) with eqn. (25.3) for the potential energy of a pair of point charges with R at infinity, we find the electric potential at a point P a distance d from a single point charge Q:

SPECIAL CASE: POTENTIAL AT DISTANCE d FROM A SINGLE POINT CHARGE, WITH REFERENCE POINT AT INFINITY.

$$V = k\frac{Q}{d}. \tag{25.9}$$

EXERCISE 25.1 ♦♦ Derive the formula $V = kQ/d$ for the potential produced by a single point charge from eqn. (25.8).

EXAMPLE 25.3 ♦ Find the potential V at a point P a distance $d = 30.0$ cm from a point charge $Q = -1.00\ \mu$C.

MODEL We choose the reference point at infinity and use eqn. (25.9).

SETUP AND SOLVE

$$V = k\frac{Q}{d} = (8.99 \times 10^{9}\ \text{N}\cdot\text{m}^2/\text{C}^2)\frac{(-1.00 \times 10^{-6}\ \text{C})}{(0.300\ \text{m})}$$

$$= -3.00 \times 10^{4}\ \text{N}\cdot\text{m/C} = -3.00 \times 10^{4}\ \text{V}.$$

ANALYZE It would require $qV = 3 \times 10^{-5}$ J to bring a -1-nC charge to this position from infinity. ∎

EXAMPLE 25.4 ♦♦ An electron moves from rest at P_1 with potential $V_1 = 9.0$ V to a point P_2 with potential $V_2 = 10.0$ V. What is the electron's speed when it reaches P_2?

MODEL Since the electron has negative charge, its motion results in a decrease of the system's potential energy and a corresponding increase in the electron's kinetic energy.

SETUP

	BEFORE		AFTER	
Energy:	Kinetic	Potential	Kinetic	Potential
	zero	$-eV_1$	$\frac{1}{2}mv^2$	$-eV_2$

Set the total energy *before* equal to its value *after:*

$$-eV_1 = \tfrac{1}{2}mv^2 - eV_2.$$

SOLVE

$$\tfrac{1}{2}mv^2 = e(V_2 - V_1).$$

$$v = \sqrt{2\frac{e(V_2 - V_1)}{m}} = \sqrt{\frac{2(1.6 \times 10^{-19}\ \text{C})(10.0\ \text{V} - 9.0\ \text{V})}{9.1 \times 10^{-31}\ \text{kg}}}$$

$$= 5.9 \times 10^{5}\ \text{m/s}.$$

ANALYZE Since the potential decreases between P_2 and P_1, according to the rule following eqn. (25.7), the electric field points generally from P_2 toward P_1. The negatively charged electron experiences a force opposite $\vec{\mathbf{E}}$, that is, toward P_2. ∎

The kinetic energy gained by the electron in Example 25.4 is a very convenient energy unit in electronics and atomic physics.

An *electron volt* (eV) is the work done on an electron that moves through a potential difference of one volt. Equivalently, 1 eV is the change in an electron's potential energy when it moves through a potential difference of one volt.

$$1.00\ \text{eV} = 1.60 \times 10^{-19}\ \text{J}. \quad (25.10)$$

BECAUSE THEY ARE RELATED BY A FACTOR OF 1 V, THE NUMERICAL VALUE OF THE ELECTRON VOLT IN JOULES EQUALS THE NUMERICAL VALUE OF THE ELECTRON CHARGE IN COULOMBS.

EXAMPLE 25.5 ♦♦♦ A battery maintains a potential difference of 10.0 V between two large parallel metal plates separated by a distance of 1.0 mm (■ Figure 25.5). Electrons emerge in all directions from a small hole in the positive (high-potential) plate. If the electrons all have the same initial speed $v_0 = 2.0 \times 10^{6}$ m/s, but they emerge at different angles θ with the normal to the plates, for what range of angles θ can electrons reach the negative plate?

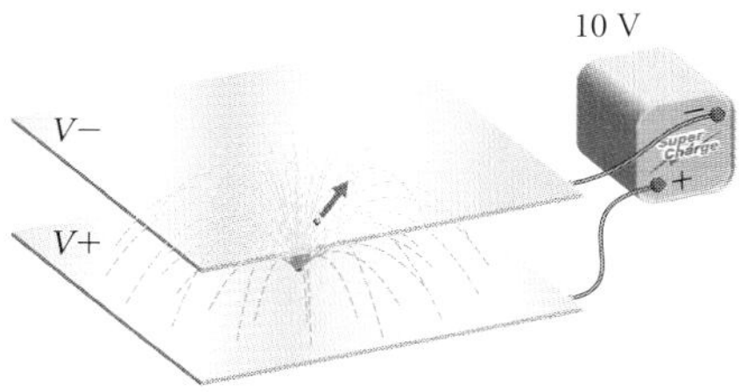

■ FIGURE 25.5 (a) A battery hooked up to two metal plates maintains a potential difference of 10.0 V between them.

MODEL The battery is designed to maintain a constant potential difference between the two plates. In this case, the negative plate is held at 10.0 V lower potential than the positive plate. If we choose the reference point at the positive plate, $V_+ \equiv 0$, then the potential at the negative plate $V_- = -10.0$ V.

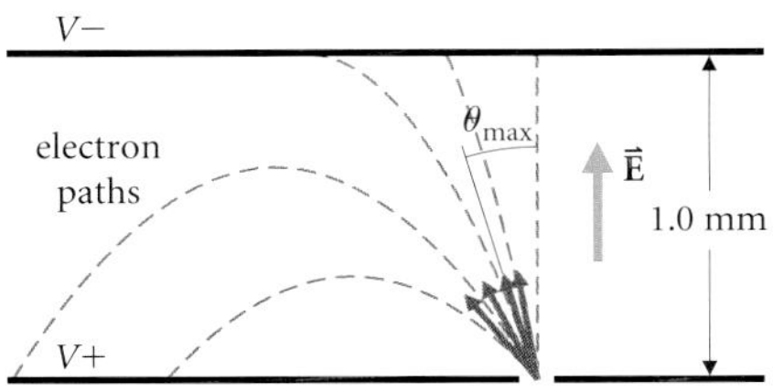

■ FIGURE 25.5 (b)
Electrons passing through a hole in the plate at higher potential follow parabolic paths as the electric field accelerates them back toward that same plate. Electrons that emerge at angle θ_{max} to the normal just make it to the other plate.

The electrons move on parabolic paths similar to that of a baseball accelerated by gravity (Figure 25.5b). The electric field between the plates decelerates the electrons, reducing the component of velocity perpendicular to the plates while leaving the velocity parallel to the plates unchanged. We choose the x-axis parallel to the plates and y perpendicular. Electrons leave the hole with $v_y = v_o \cos\theta$. An electron that just reaches the second plate arrives there with v_y equal to zero. This electron has the maximum possible exit angle, θ_{max}. Electrons with smaller exit angles leave the hole with a larger v_y and are able to reach the negative plate; those with larger exit angles do not. We may find θ_{max} with the conservation law technique.

SETUP The x-component of velocity, $v_o \sin\theta$, remains constant during the motion.

	BEFORE (AT POSITIVE PLATE)		AFTER (AT NEGATIVE PLATE) $v_y = 0$	
Energy:	Kinetic	Potential	Kinetic	Potential
	$\frac{1}{2}mv_o^2$	$-eV_+ = 0$	$\frac{1}{2}m(v_o \sin\theta_{max})^2$	$-eV_-$

Set value *before* equal to value *after:*

$$\tfrac{1}{2}mv_o^2 = \tfrac{1}{2}mv_o^2 \sin^2\theta_{max} - eV_- .$$

SOLVE

$$\tfrac{1}{2}mv_o^2(1 - \sin^2\theta_{max}) = -eV_- .$$

$$\cos^2\theta_{max} = \frac{-2eV_-}{mv_o^2} = \frac{-2(1.6\times10^{-19}\ \text{C})(-10.0\ \text{V})}{(9.1\times10^{-31}\ \text{kg})(2.0\times10^{6}\ \text{m/s})^2} = 0.88.$$

$$\theta_{max} = 20°.$$

ANALYZE A 14% increase in the potential difference stops electrons from reaching the negative plate altogether ($\cos^2\theta_{max}$ increases to 1.00, and θ_{max} becomes zero). A similar reduction in the potential difference more than doubles the number of electrons reaching the plate. This principle is used to control the intensity of electron beams. The brightness control of a TV tube is one example.

Notice how we used the freedom to choose a reference point to simplify the calculation. Here we chose $V = 0$ at the positive plate; the choice $R = \infty$ would not make much sense in this example. ■

25.2.3 *Calculation of Field from Potential*

ACTUALLY WE CAN ALWAYS CALCULATE THE FIELD IF WE KNOW ENOUGH MATH. SEE THE MATH TOPIC ON GRADIENTS.

Equation 25.8 allows us to calculate the potential difference between two points from a known electric field. Conversely, we can calculate the field from the potential. To use eqn. (25.8), we need to know the direction of the field and how it varies along the path.

COMPARE WITH EXAMPLE 24.5. THE ELECTRIC FIELD HERE IS THE SUPERPOSITION OF THE FIELD DUE TO TWO UNIFORM SHEETS.

EXAMPLE 25.6 ♦♦ Given that the electric field between the plates in Example 25.5 is uniform and perpendicular to the plates, find its magnitude.

MODEL We begin with the relation between the given potential difference and the electric field (eqn. 25.7). The integral may be carried out along any path between the plates, but it is simplest along a straight line perpendicular to the plates (■ Figure 25.6).

SETUP The electric field vector is antiparallel to each displacement $d\vec{\ell}$ along the chosen path and has constant magnitude. The integral simplifies accordingly:

RECALL HOW WE SIMPLIFY FLUX INTEGRALS WHEN APPLYING GAUSS' LAW. THE SAME KIND OF REASONING IS USED HERE.

$$\Delta V = -\int_-^+ \vec{E}\cdot d\vec{\ell} = -\int_-^+ -|\vec{E}|\,d\ell = |\vec{E}|\int_-^+ d\ell = |\vec{E}|d,$$

where $d = 1.0$ mm is the distance between the plates.

SOLVE $|\vec{\mathbf{E}}| = \dfrac{\Delta V}{d} = \dfrac{V_+ - V_-}{d} = \dfrac{10.0\text{ V}}{1.0 \times 10^{-3}\text{ m}} = 1.0 \times 10^4\text{ V/m}.$

ANALYZE These units look unfamiliar, but notice that:

$$\frac{\text{V}}{\text{m}} = \frac{\text{J/C}}{\text{m}} = \frac{\text{N}\cdot\text{m}}{\text{C}\cdot\text{m}} = \frac{\text{N}}{\text{C}}.$$

Volts per meter is a more common unit for electric field strength than newtons per coulomb. ∎

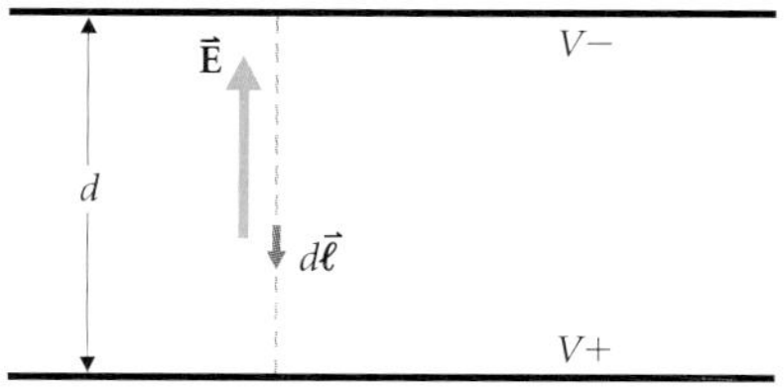

■ FIGURE 25.6
The electric field between the plates is known to be uniform and perpendicular to the plates. An integral of $-\vec{\mathbf{E}} \cdot d\vec{\ell}$ along any path between the plates gives the potential difference. If we choose a path with $d\vec{\ell}$ antiparallel to $\vec{\mathbf{E}}$, that is, perpendicular to the plates, the integral simplifies to $|\vec{\mathbf{E}}|d = \Delta V$.

25.3 Potential Energy of a System of Charges

25.3.1 *The Principle of Superposition*

Using the definition of potential in terms of the electric field vector $\vec{\mathbf{E}}$, we may show that the principle of superposition (§23.3) may be restated as an addition law for potential. Suppose $\vec{\mathbf{E}}$ is the sum of a finite number N of different contributions $\vec{\mathbf{E}}_i$, $i = 1 \ldots N$. (For example, each contribution $\vec{\mathbf{E}}_i$ could be the field due to a point charge, a line of charge, or a sheet of charge.)

$$\vec{\mathbf{E}} = \sum_{i=1}^{N} \vec{\mathbf{E}}_i.$$

Then: $$V(P) = -\int_R^P \vec{\mathbf{E}} \cdot d\vec{\ell} = -\int_R^P \sum_{i=1}^{N} \vec{\mathbf{E}}_i \cdot d\vec{\ell} = \sum_{i=1}^{N}\left(-\int_R^P \vec{\mathbf{E}}_i \cdot d\vec{\ell}\right).$$

The ith term in this sum is a separate contribution V_i to the potential at P due to the electric field $\vec{\mathbf{E}}_i$ alone (eqn. 25.8). Thus:

$$V(P) = \sum_{i=1}^{N} V_i(P). \qquad \textbf{(25.11)}$$

For a set of point charges, this result becomes:

$$V(P) = \sum_{i=1}^{N} V_i(P) = \sum_{i=1}^{N} \frac{kQ_i}{r_i}, \qquad (25.12)$$

where r_i is the distance of P from Q_i.

> The potential at a point due to an arbitrary, finite set of point charges is the sum of the potentials at that point due to the individual charges.

EXAMPLE 25.7 ♦♦ Find the potential at P, the midpoint of one side of a square of side s with a charge Q at each corner (■ Figure 25.7).

MODEL We use the principle of superposition. The potential at P is the sum of potentials produced by each of the charges. With the charges labeled as shown in the figure:

$$V(P) = V_1(P) + V_2(P) + V_3(P) + V_4(P).$$

SETUP To find the individual potential contributions, we need only the distances of the charges from P. For Q_1 and Q_3, the distances are $r_1 = r_3 = s/2$. For the other two charges:

$$r_2 = r_4 = \sqrt{s^2 + (s/2)^2} = \sqrt{5}s/2.$$

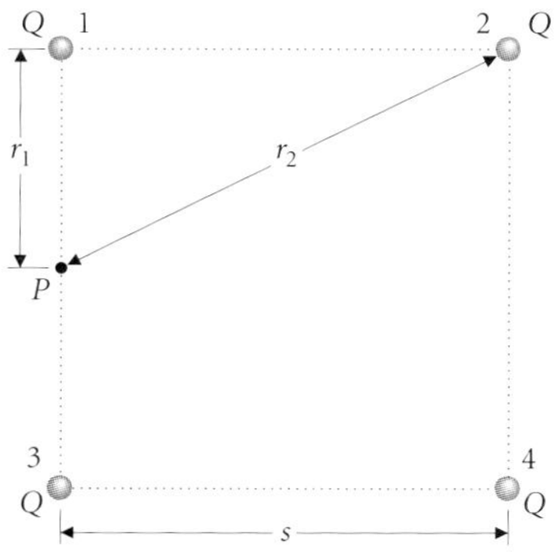

■ FIGURE 25.7
To find the potential at point P, we use the principle of superposition to sum the potentials due to the four charges at the corners of the square.

SOLVE $$V(P) = kQ\left(\frac{2}{r_1} + \frac{2}{r_2}\right) = \frac{kQ}{s}\left[2(2) + 2\frac{2}{\sqrt{5}}\right] = \frac{4kQ}{s}\left(\frac{5 + \sqrt{5}}{5}\right) = 5.8\frac{kQ}{s}.$$

ANALYZE The principle of superposition for potential is easier to use than that for the electric field vector because potential is a scalar quantity.

EXERCISE 25.2 ♦♦ A dipole (§24.5) with moment $\vec{\mathbf{p}} = p\hat{\mathbf{i}}$ is at the origin. Find the potential it produces at an arbitrary point on the x-axis.

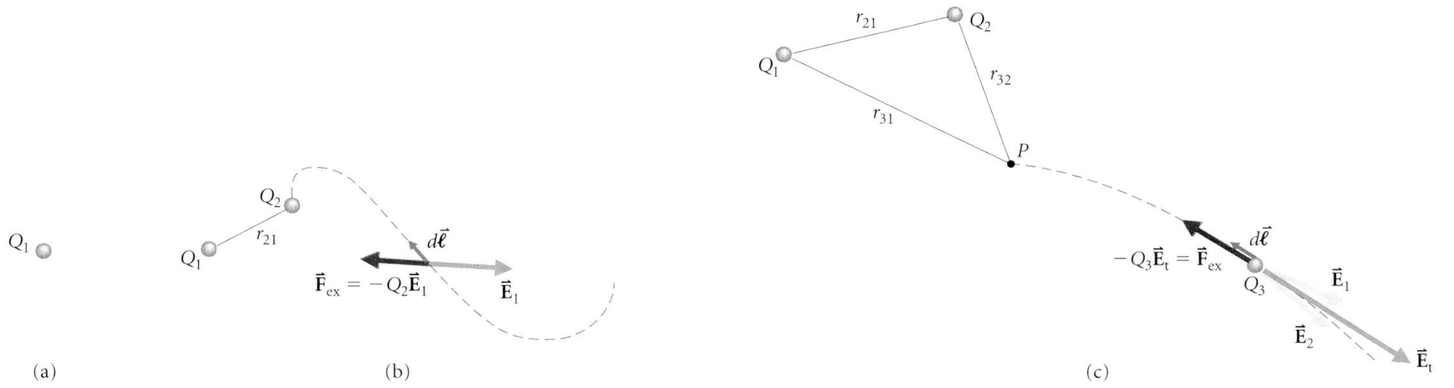

■ FIGURE 25.8
To find the potential energy of this system of three charges, we calculate the work done to assemble the system. (a) No work is needed to put the first charge in place. (b) Next we bring up the second charge, doing work kQ_1Q_2/r_{21}. (c) Finally, we bring up the third charge and have to do work against the electric force exerted by both the other charges. The work done may be expressed as the sum of the potential energies of the two new charge pairs formed. The two new pairs are Q_1Q_3 and Q_2Q_3.

25.3.2 Potential Energy of Systems of Charges

The potential energy of any physical system equals the work done to assemble the system (§8.4) and is independent of the procedure we choose. By convention the potential energy of a system of point charges is taken to be zero when the charges are very far apart. First, let us look at an arbitrary system of three charges (■ Figure 25.8).

We imagine assembling the system one charge at a time, beginning with them all very far apart. No work is required to put the first charge Q_1 in place, since no electrical forces act on it: $W_1 = 0$. To place the second charge in position (Figure 25.8b), we do work against the Coulomb force exerted by the first charge. This work equals the final potential energy of the system of two charges (cf. eqn. 25.3 and Example 25.1):

ACTUALLY IN EXAMPLE 25.1 WE CALCULATED THE WORK DONE *BY THE FIELD* TO EXPEL THE CHARGE TO INFINITY. THAT EQUALS THE WORK DONE *BY US* TO ASSEMBLE THE PAIR.

$$W_2 = Q_2V_1(r_{21}) = \frac{kQ_2Q_1}{r_{21}}.$$

The force $\vec{\mathbf{F}}_{ex}$ we exert on the third particle balances the electric force exerted by the first two charges. From the principle of superposition, that electric force is due to the sum of the electric field vectors produced by the two individual charges:

$$\vec{\mathbf{F}}_{ex} = -Q_3\vec{\mathbf{E}} = -Q_3(\vec{\mathbf{E}}_1 + \vec{\mathbf{E}}_2).$$

Now we find the work needed to put Q_3 in place at P:

$$W_3 = \int_\infty^P \vec{\mathbf{F}}_{ex} \cdot d\vec{\ell} = \int_\infty^P [-Q_3(\vec{\mathbf{E}}_1 + \vec{\mathbf{E}}_2)] \cdot d\vec{\ell}$$
$$= \int_\infty^P (-Q_3\vec{\mathbf{E}}_1) \cdot d\vec{\ell} + \int_\infty^P (-Q_3\vec{\mathbf{E}}_2) \cdot d\vec{\ell}.$$

REMEMBER: EACH OF THE WORK INTEGRALS IS INDEPENDENT OF PATH.

Each integral is the work required to bring Q_3 into place near a single other charge:

$$W_3 = Q_3V_1(r_{31}) + Q_3V_2(r_{32}) = \frac{kQ_3Q_1}{r_{31}} + \frac{kQ_3Q_2}{r_{32}}.$$

Notice that $W_3 = Q_3\,\Delta V = Q_3[V(P) - 0]$, where $V(P)$ is the potential at P due to the other two charges.

As Q_3 is put in place it forms two new charge pairs, one with each charge already in position, and two new terms are added to our expression for the potential energy of the system. This argument extends readily to a system containing any finite number of point charges. Bringing in a fourth charge creates three new charge pairs and adds three new terms to the system energy, and so on. The potential energy of an arbitrary system of charges is (■ Figure 25.9):

$$U = W = \sum_{\substack{\text{all}\\ \text{distinct}\\ \text{pairs}}} \frac{kQ_iQ_j}{r_{ij}}. \qquad (25.13)$$

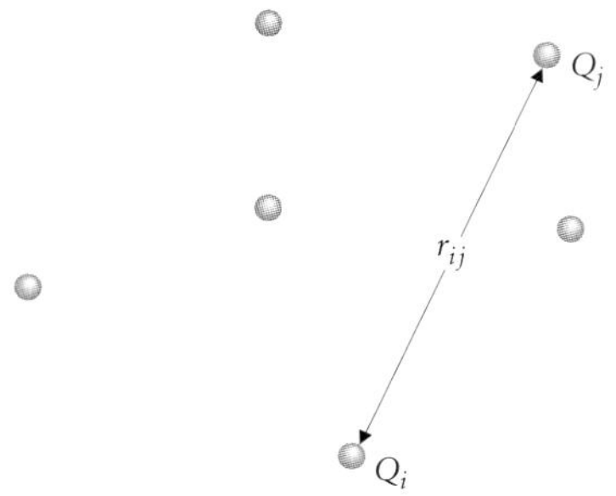

■ FIGURE 25.9
The energy of an arbitrary system of charges is stored in charge pairs. One such pair, Q_i and Q_j separated by distance r_{ij}, is shown here.

EXAMPLE 25.8 ♦♦ Find the potential energy of a system of three identical charges Q placed at the corners of an equilateral triangle of side s.

MODEL We imagine assembling the system one charge at a time.

SETUP No work is required to place the first charge at one corner of the triangle, $W_1 = 0$. To bring up the second charge, we do work against the Coulomb force exerted by the first charge:

$$W_2 = QV_1(s) = kQ^2/s.$$

Finally we bring up the third charge, doing work against the Coulomb force exerted by the first two (■ Figure 25.10):

$$W_3 = QV_1(s) + QV_2(s) = 2kQ^2/s.$$

SOLVE The potential energy of the system equals the total work done to assemble it:

$$U = W_1 + W_2 + W_3 = 0 + \frac{kQ^2}{s} + \frac{2kQ^2}{s} = \frac{3kQ^2}{s}.$$

ANALYZE To compare with eqn. (25.13), count up the number of distinct pairs in the system and add the potential energies of the pairs. Here we have three pairs (1-2, 2-3, and 3-1) and the potential energy of each pair is kQ^2/s, so the potential energy of the system is $3kQ^2/s$.

IN PROBLEM 93 WE ASK YOU TO VERIFY THIS RESULT BY DIRECT INTEGRATION OF THE WORK DONE, USING EQN. (23.9) FOR THE ELECTRIC FIELD.

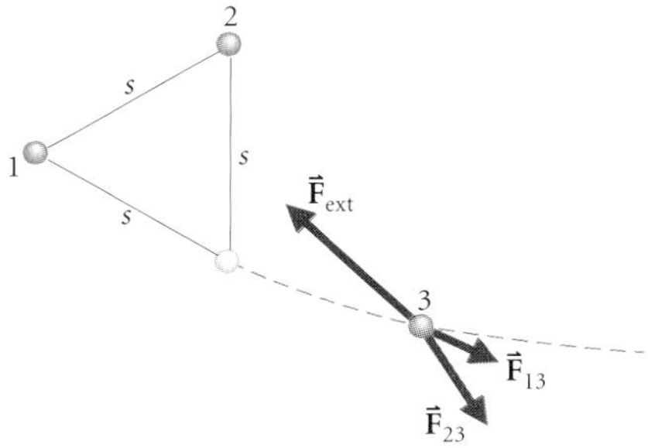

■ FIGURE 25.10
Bringing up the third charge to form an equilateral triangle.

EXERCISE 25.3 ♦ Use eqn. (25.13) to show that the potential energy of a system of four identical charges at the corners of a square of side s is: $U = (4 + \sqrt{2})kQ^2/s$.

If we bring a fifth charge q up from infinity and place it at P in Figure 25.7, the potential energy of the system of five charges increases by an amount $qV(P)$. It is tempting to think of this quantity as energy *possessed by* the charge q and to calculate the energy of the system by adding up energies *possessed by* each charge in the system. That procedure is faulty because it implicitly assumes that each charge is brought up with all the others already in place. Potential energy is unavoidably a property of an *entire system,* not of its individual particles.

The sum $\Sigma\, Q_iV_i$ overestimates the system energy by a factor of 2:

$$\sum_i Q_iV_i = \sum_i Q_i \sum_{j\neq i} \frac{kQ_j}{r_{ij}} = \sum_{j\neq i}\sum_i \frac{kQ_iQ_j}{r_{ij}}.$$

Let's compare this expression with eqn. (25.13). Each term in the double sum is the potential energy of a pair of charges i and j. However, each pair appears twice in the sum. For example, the pair of charges labeled 2 and 3 appears once with $i = 2$ and $j = 3$, and once with $i = 3$ and $j = 2$. Thus the sum is exactly twice the total potential energy of the system. For a system of point charges then:

$$U_{\text{system}} = \tfrac{1}{2}\sum_i Q_iV_i. \qquad (25.14)$$

IN THIS EXPRESSION V_i IS THE POTENTIAL AT THE POSITION OF CHARGE Q_i DUE TO ALL THE OTHER CHARGES.

EXAMPLE 25.9 ♦ Compute the potential energy of the system of four charges Q at the corners of a square of side s using eqn. (25.14).

MODEL To use eqn. 25.14 we first calculate the potential at the position of each of the four charges due to the other three. Because the system is symmetric, the result is the same at all four corners.

SETUP The potential at one corner of the square produced by charges at the other three corners is:

$$V_{\text{corner}} = \frac{2kQ}{s} + \frac{kQ}{s\sqrt{2}} = \frac{kQ}{s}\left(2 + \frac{1}{\sqrt{2}}\right) = \frac{kQ}{s}\left(\frac{4 + \sqrt{2}}{2}\right).$$

SOLVE Equation (25.14) becomes the sum of four equal terms:

$$U_{\text{system}} = \tfrac{1}{2}\sum_{i=1}^{4} Q_i V_i = \tfrac{1}{2}(4QV_{\text{corner}}) = \frac{kQ^2}{s}(4 + \sqrt{2}).$$

ANALYZE You should have obtained the same result in Exercise 25.3. ▮

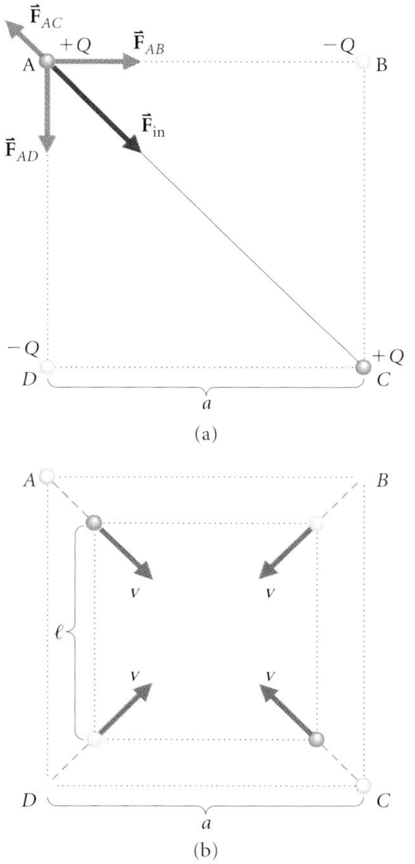

■ **FIGURE 25.11**
(a) Four charges of alternating sign are placed at the corners of a square. Each charge feels forces from the other three. The two attractive forces $\vec{F}_{AB} + \vec{F}_{AD}$ sum to give a force $\vec{F}_{\text{in}}$ inward across the diagonal. The net force $\vec{F}_{\text{in}} + \vec{F}_{AC}$ is toward the center of the square. (b) As the charges accelerate inward, each has the same speed at any time. The square shrinks while maintaining its shape.

Study Problem 16 ♦♦♦ *The Collapsing Square*

Four particles, two with charge $+Q$ and two with charge $-Q$, are placed at the corners of a square of side a (■ Figure 25.11a). Each has mass m, and they are released from rest. What is the speed of each charge when the side of the square has decreased to ℓ?

Modeling the System

Each particle is attracted to its two oppositely charged neighbors and repelled by the identical charge across the diagonal of the square. The two attractive forces have the same magnitude and sum to give an attractive force directly opposite the repulsive force. The net force on each charge is inward along the diagonal (cf. Example 23.8), so the square shrinks while maintaining its shape (Figure 25.11b). At any one time, the speed of each charge is the same. No *external* forces act, so the total energy of the system is conserved. The charges gain kinetic energy as the electric potential energy of the system decreases. We compute the system's potential energy as a function of the square's size. Conservation of energy gives the final speed of each charge.

Setup

The potential energy of the system is the potential energy of the six pairs of charges AB, AC, AD, BC, BD, and CD. For any given pair, the energy $U = kQ_1Q_2/d$, where d is the separation of the pair. When the square has sides of length s, the energy of each pair is:

PAIR	Q_1	Q_2	d	U
AB	Q	$-Q$	s	$-kQ^2/s$
AC	Q	Q	$s\sqrt{2}$	$kQ^2/(s\sqrt{2})$
AD	Q	$-Q$	s	$-kQ^2/s$
BC	$-Q$	Q	s	$-kQ^2/s$
BD	$-Q$	$-Q$	$s\sqrt{2}$	$kQ^2/(s\sqrt{2})$
CD	$-Q$	Q	s	$-kQ^2/s$

The total energy of the system is thus:

$$U_{\text{total}} = \frac{kQ^2}{s}\left(-1 + \frac{1}{\sqrt{2}} - 1 - 1 + \frac{1}{\sqrt{2}} - 1\right)$$

$$= \frac{kQ^2}{s}\left(-4 + \frac{2}{\sqrt{2}}\right) = -\frac{kQ^2}{s}(4 - \sqrt{2}) = -2.6\frac{kQ^2}{s}.$$

Once again, we apply the standard method for using conservation laws.

	Before (side of square $s = a$)		After (side of square $s = \ell$)	
Energy:	Kinetic	Potential	Kinetic	Potential
	zero	$-\dfrac{kQ^2}{a}(4 - \sqrt{2})$	$4(\frac{1}{2}mv^2)$	$-\dfrac{kQ^2}{\ell}(4 - \sqrt{2})$

In Problem 104 we ask you to calculate the time it takes for the square to collapse.

Solution of Equations

Set total energy *before* equal to total energy *after:*

$$-\frac{kQ^2}{a}(4 - \sqrt{2}) = 4(\tfrac{1}{2}mv^2) - \frac{kQ^2}{\ell}(4 - \sqrt{2}).$$

$$2mv^2 = kQ^2(4 - \sqrt{2})\left(\frac{1}{\ell} - \frac{1}{a}\right).$$

So:

$$v = Q\sqrt{\frac{k(4 - \sqrt{2})}{2m}\left(\frac{1}{\ell} - \frac{1}{a}\right)}.$$

Analysis

The particles give up potential energy as they move. Since we chose $U = 0$ for $s \to \infty$, the particles move toward smaller values of s, where the potential energy is less (U is a larger negative number). With $U = 0$ at infinity, a negative potential energy indicates a bound system, that is, one that tends to collapse rather than fly apart. With all the charges positive as in Example 25.9, the potential energy of the system is positive: the system is unbound and tends to expand to infinity.

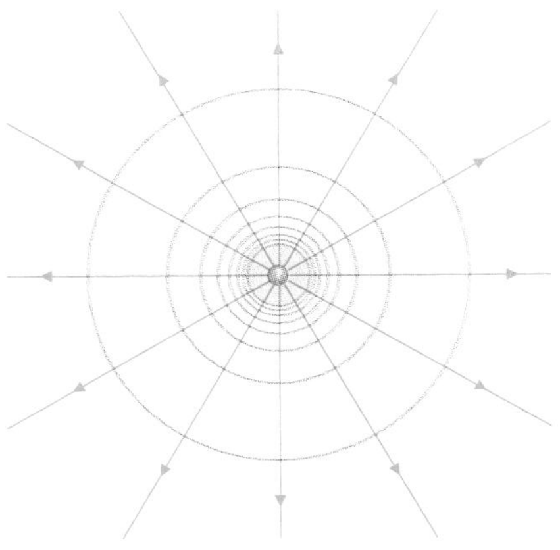

■ **Figure 25.12**
A positive point charge illustrates the relation between field lines and equipotential surfaces. Since $V = kQ/r$, the equipotential surfaces are spheres with $r =$ constant. The field lines are radial, so they are perpendicular to the equipotential surfaces. Note that if we draw surfaces with a constant potential difference, they are closer together near the charge. (In the central shaded region the surfaces are too close to be distinguished in this diagram.) The field lines are also closer together at the center. With a central charge of 3 nC, the potential difference between the surfaces shown is 1 kV.

25.4 Equipotential Surfaces

25.4.1 The Relation Between Equipotential Surfaces and Field Lines

Field line diagrams give one method of visualizing the electric field. An alternate picture of the field is provided by equipotential surfaces.

> An *equipotential surface* is a surface on which the electric potential has a constant value.

The potential due to a point charge depends only on distance from the charge; the equipotential surfaces are spheres centered on the charge (■ Figure 25.12). If the charge is positive, the value of the potential decreases outward. The field lines are radial, cross the spherical equipotential surfaces at right angles, and point in the direction of lower potential. In Figure 25.12, the potential difference between successive surfaces is the same, so they concentrate near the charge. The field lines are also closest together near the charge, indicating that the field strength is greatest where the potential is changing most rapidly.

EXAMPLE 25.10 ♦♦ A constant electric field $\vec{\mathbf{E}} = E_0\hat{\mathbf{i}}$ exists in a region of space. Describe the equipotential surfaces.

MODEL The relation between potential and field is given by eqn. (25.8). We may use this relation to find the potential V at an arbitrary point as a function of the coordinates. Equipotential surfaces are defined by the relation $V(x, y, z) =$ constant.

SETUP We choose the x-axis to be parallel to $\vec{\mathbf{E}}$ and put the reference point R on the x-axis with coordinate $x = r$ (■ Figure 25.13). Then the potential at a point P with coordinates (x, y, z) is:

$$V(P) = -\int_R^P \vec{\mathbf{E}} \cdot d\vec{\ell} = -\int_R^P E_0\hat{\mathbf{i}} \cdot d\vec{\ell}.$$

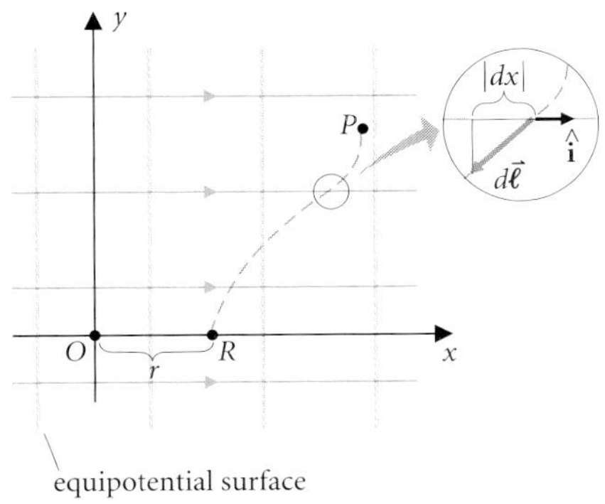

■ **Figure 25.13**
When the electric field is uniform, the equipotential surfaces are planes perpendicular to $\vec{\mathbf{E}}$. Here $\vec{\mathbf{E}} = E_0\hat{\mathbf{i}}$.

For every path element, $\hat{\mathbf{i}} \cdot d\vec{\ell} = dx$, so:

$$V(P) = V(x, y, z) = -\int_r^x E_o\, dx = E_o(r - x).$$

SOLVE The function $V(x, y, z)$ depends only on x. The equipotential surfaces are surfaces of constant x, that is, planes perpendicular to the x-axis.

ANALYZE The field lines are again perpendicular to the equipotential surfaces and point toward smaller values of V. ■

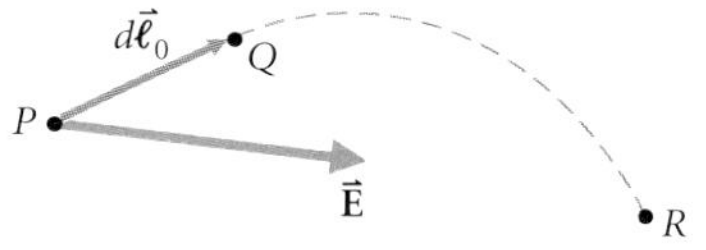

■ FIGURE 25.14
The potential difference between P and Q is $dV = -\vec{E} \cdot d\vec{\ell}_o$. If P and Q are on the same equipotential surface, $dV = 0$ and so $d\vec{\ell}_o$ is perpendicular to $\vec{E}$. Note also that the electric field points toward lower potential.

In the case of a point charge or a uniform field, the field lines are perpendicular to the equipotential surfaces. This is always true (■ Figure 25.14). The potential difference between two points P and Q separated by a differential displacement $d\vec{\ell}_o$ is (eqn. 25.7):

$$dV_{QP} \equiv V(Q) - V(P) = -\int_P^Q \vec{E} \cdot d\vec{\ell} = -\vec{E} \cdot d\vec{\ell}_o. \qquad (25.15)$$

If P and Q are on the same equipotential surface, $dV_{QP} = 0$, so $\vec{E} \cdot d\vec{\ell}_o = 0$ and $\vec{E}$ is perpendicular to $d\vec{\ell}_o$.

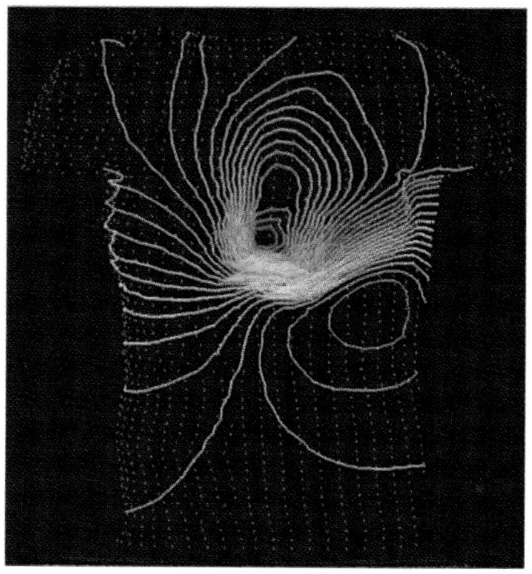

The contour lines in this picture show where equipotential surfaces cross the skin on a person's body. The lines are color coded from blue = most negative to red = most positive. They indicate an electric dipole. Compare with the field line diagram for a dipole (Figure 23.14).

25.4.2 *Equipotential Surfaces for a System of Point Charges*

As we did for field line diagrams, we may determine a set of rules for drawing equipotential surfaces. Near a point charge, the equipotential surfaces are spheres with the point charge at the center (Figure 25.12). At a large distance from a system of point charges, the equipotential surfaces are spheres surrounding the whole system. In between, some rather odd shapes can result. There is usually a transition surface that may be double-lobed or may touch itself. At such connections there is no definite direction perpendicular to the surface, and the direction of $\vec{E}$ is undefined. Thus these surface connections can occur only at points where $\vec{E} = 0$.

Avoid confusing $\vec{E} = 0$ and $V = 0$! That $\vec{E} = 0$ means only that V is not changing and says nothing about its value. (The point is a maximum, minimum, or saddle point.) Conversely, $V = 0$ by itself says nothing about the value of $\vec{E}$. For example, between two point charges of opposite sign, the potential varies continuously from a positive value near the positive charge to a negative value near the negative charge. It is zero somewhere between them, where $\vec{E}$ points toward the negative charge. The following example illustrates these ideas.

Math Topic

ELECTRIC FIELD AS THE GRADIENT OF POTENTIAL

A vector field like $\vec{E}$ assigns a vector to each point. A scalar field like V assigns a number to each point. In mathematics the *gradient* of a scalar field, such as electric potential, is a vector that points in the direction in which the scalar field varies most rapidly. Its magnitude is the maximum rate of change of the scalar field. Perpendicular to the gradient, the scalar field does not change at all. The relation between the electric field vector and the electric potential is summarized by the statement:

$$\vec{E} = -\text{gradient } V.$$

In Cartesian coordinates:

$$\text{gradient } V = \frac{\partial V}{\partial x}\hat{\mathbf{i}} + \frac{\partial V}{\partial y}\hat{\mathbf{j}} + \frac{\partial V}{\partial z}\hat{\mathbf{k}} = -\vec{E}$$

The uniform field in Example 25.10 has potential $V(x, y, z) = E_o(r - x)$, so $\partial V/\partial x = -E_o$, $\partial V/\partial y = \partial V/\partial z = 0$, and

$$\vec{E} = -\text{gradient } V = E_0\hat{\mathbf{i}},$$

as expected.

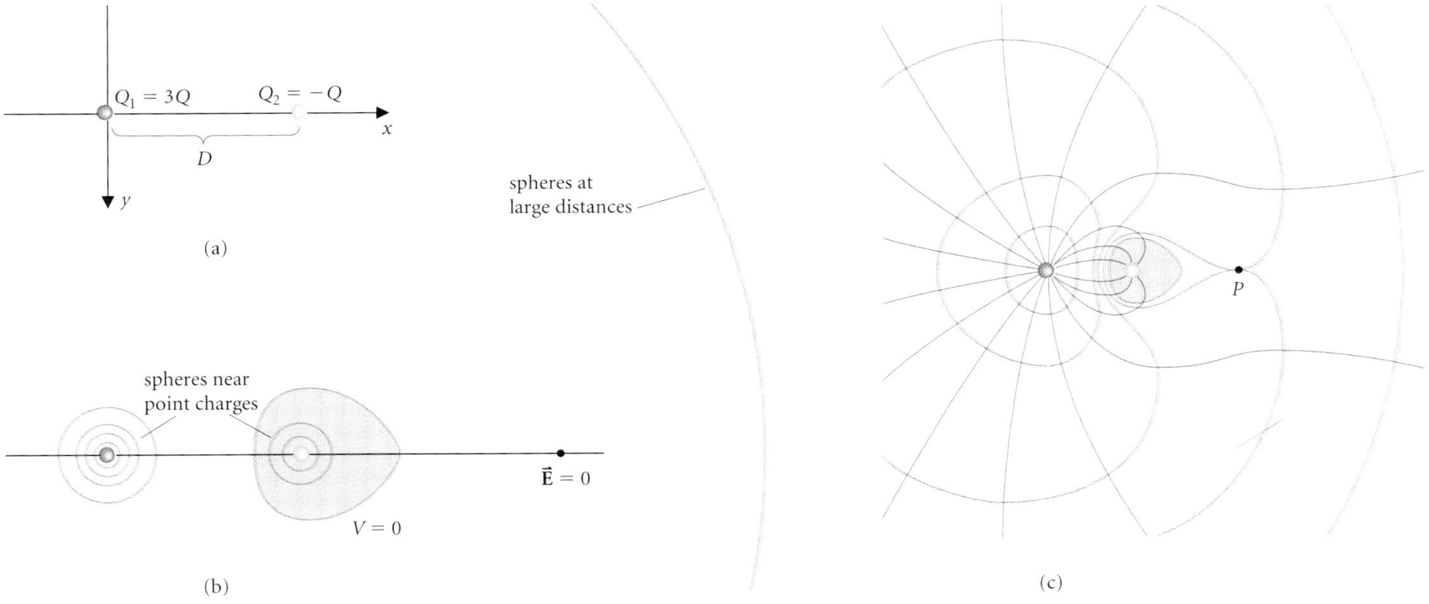

■ FIGURE 25.15
(a) Two point charges, $3Q$ and $-Q$, are on the x-axis separated by distance D. (b) Since $V \to +\infty$ at the positive charge and $V \to -\infty$ at the negative charge, it is zero somewhere in between. Also, since the *net* charge is $+2Q$, the potential is positive at large distances from the system. Thus the potential is also zero somewhere to the right of the negative charge. The surface of zero potential bounds the shaded region, where the potential is negative. (c) The equipotential surfaces are perpendicular to the field lines. Since the equipotential surface through P touches itself there, $\vec{E}$ is zero at P. The potential decreases away from P along the x-axis and increases away from P perpendicular to the x-axis.

EXAMPLE 25.11 ♦♦♦ Two particles with charges $Q_1 = 3Q$ and $Q_2 = -Q$ are on the x-axis separated by a distance D (■ Figure 25.15a). Sketch the equipotential surfaces in the x-y-plane, and find two points on the x-axis, other than $\pm\infty$, where $V = 0$.

MODEL Near each charge the equipotential surfaces are spheres surrounding the charge. We adopt the usual convention that $V \to 0$ as $r \to \infty$. The potential is $+\infty$ at the positive charge and is $-\infty$ at the negative charge. Far from the system, at a distance much greater than D but less than infinity, the potential is positive, since the system's net charge $(+2Q)$ is positive. Separating the regions of positive and negative potential is an equipotential surface on which $V = 0$ (Figure 25.15b).

SETUP Since the $V = 0$ surface surrounds the negative charge, one of the two required points lies between the two charges $(x < D)$, and one is to the right of the negative charge $(x > D)$.

$$\text{For } x > D: \qquad V(x) = \frac{k(3Q)}{x} + \frac{k(-Q)}{x - D} = kQ\left(\frac{3}{x} - \frac{1}{x - D}\right).$$

SOLVE The potential is zero where:

$$\frac{3}{x} = \frac{1}{x - D} \qquad \text{or} \qquad 3x - 3D = x \;\Rightarrow\; x = \frac{3D}{2}.$$

SETUP For $0 < x < D$: $\quad V(x) = \dfrac{k(3Q)}{x} + \dfrac{k(-Q)}{D - x} = kQ\left(\dfrac{3}{x} - \dfrac{1}{D - x}\right).$

SOLVE This function is zero where:

$$\frac{3}{x} = \frac{1}{D - x} \qquad \text{or} \qquad 3D - 3x = x \;\Rightarrow\; x = \frac{3D}{4}.$$

ANALYZE In order for the potential to be zero at a point, the point must be closer to the smaller, negative charge so that the closeness can overcome the factor of 3 in the sizes of the charges. Check the ratio of distances in the two cases:

$$\frac{\text{Distance to + charge}}{\text{Distance to − charge}} = \begin{cases} \dfrac{3D/2}{(3D/2) - D} = 3 & \text{for } x = 3D/2. \\[2ex] \dfrac{3D/4}{D - (3D/4)} = 3 & \text{for } x = 3D/4. \end{cases}$$

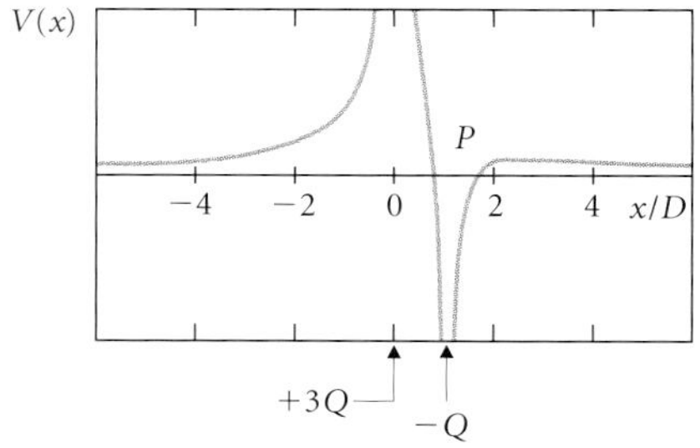

■ FIGURE 25.16
Graph of $V(x)$ along the x-axis through the two charges. The potential has a local maximum at point P, where the electric field vector is zero.

Notice that we need two different expressions for $x > D$ and $x < D$ because the potential at a point depends on the distance of that point from each charge, which is by definition positive. The denominator $x - D$ becomes negative for $x < D$, so it is replaced with $D - x$. We need yet a third expression for $x < 0$.

The system of charges in this problem is the same as that in Study Problem 15. The equipotential surfaces are everywhere normal to the field lines (Figure 25.15c). In Study Problem 15 we found that $\vec{\mathbf{E}} = 0$ on the x-axis at point P where $x = (3 + \sqrt{3})D/2$. Note that one of the equipotential surfaces in Figure 25.15c touches itself on the x-axis at this point, which corresponds to a maximum in the function $V(x)$ (■ Figure 25.16). ■

25.5 Potential Due to a Continuous Distribution of Charge

There are two ways to find the electrical potential produced by charge distributed continuously on an object. The first way is to compute the electric field vector and then integrate it along an appropriate path, using eqn. (25.8). The second way is to model the charge distribution as a collection of differential elements and use the principle of superposition to express the potential function as an integral. The second method fails for infinite objects, because the interchange of sum and integral in the derivation of eqn. (25.12) is valid only if both sum and integral converge. However, for symmetric infinite objects we may use Gauss' law to find the electric field, and the first method usually works well.

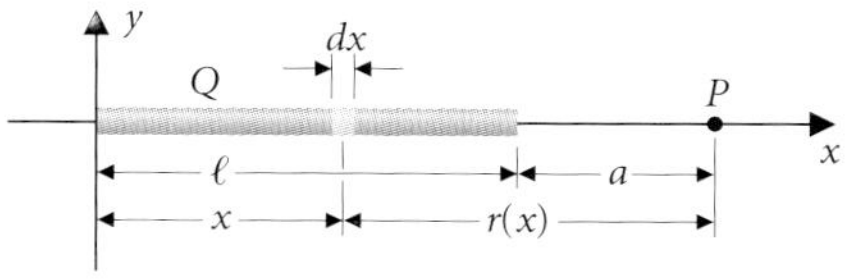

■ FIGURE 25.17
A finite, uniformly charged rod. A differential element with coordinate x is a distance $r = a + \ell - x$ from P.

EXAMPLE 25.12 ♦♦♦ Find the potential at distance a from one end of a uniformly charged rod of length ℓ and total charge Q, as shown in ■ Figure 25.17.

MODEL We model the rod as a collection of differential elements, and integrate.

SETUP We choose the x-axis along the rod, with the origin at its far end. A typical
STEPS I AND II element is a differential piece of rod with coordinate x, length dx, and charge:

$$dQ = \frac{Q}{\ell}\,dx.$$

STEP III The distance of the element from P is $r = a + \ell - x$, and it produces potential dV at P:

$$dV = k\frac{dQ}{r} = \frac{(kQ/\ell)dx}{a + \ell - x}.$$

STEP IV The rod extends from $x = 0$ to $x = \ell$, and we have chosen dx to represent a positive length, so we integrate from 0 to ℓ.

SOLVE
STEP V

$$V = \int_0^{\ell} \frac{kQ\,dx}{\ell(a + \ell - x)} = -\frac{kQ}{\ell}\left[\ln(a + \ell - x)\right]\Big|_0^{\ell}.$$

$$V = \frac{kQ}{\ell}\ln\left(\frac{a + \ell}{a}\right).$$

ANALYZE If $a \gg \ell$, we may use the approximation $\ln(1 + \epsilon) \approx \epsilon$ (Appendix IB) to obtain:

$$V \approx \frac{kQ}{\ell}\frac{\ell}{a} = \frac{kQ}{a} \qquad (a \gg \ell).$$

As expected, this is the result for a point charge. ■

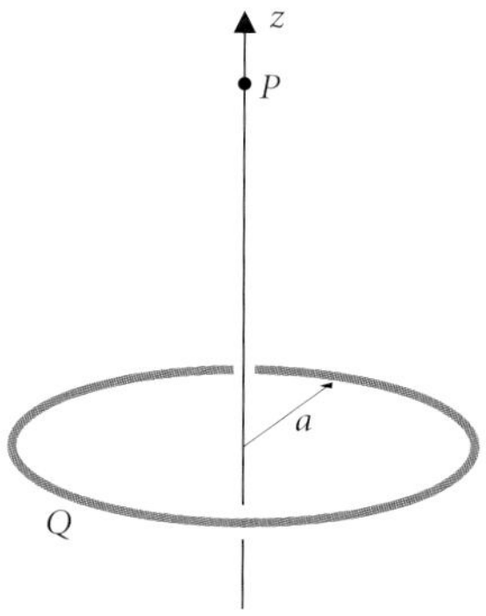

■ FIGURE 25.18

EXERCISE 25.4 ♦♦ Find the potential at points on the axis of a uniformly charged ring of radius a and total charge Q (■ Figure 25.18).

EXERCISE 25.5 ♦♦♦ Use the result of Exercise 25.4 to find the potential on the axis of a uniformly charged disk of radius R.

EXAMPLE 25.13 ♦♦♦ Find the potential at a perpendicular distance a from an infinite, uniformly charged filament.

MODEL The method used in Example 25.12 may not be used here. The filament is infinitely long: its total charge is infinite, and the integral is again a logarithm that diverges as $\ell \to \infty$. Instead, we obtain the potential as an integral of the field vector (eqn. 25.8).

SETUP The electric field vector due to the infinite filament is (eqn. 24.4):

Using Gauss' law is the easiest way to get this result; λ is the linear charge density on the filament, cf. §24.2.

$$\vec{\mathbf{E}} = \frac{2k\lambda}{r}\hat{\mathbf{r}}.$$

SOLVE The potential difference between P and R is:

$$V(P) - V(R) = -\int_R^P \vec{\mathbf{E}} \cdot d\vec{\ell} = -\int_R^P \frac{2k\lambda}{r}\hat{\mathbf{r}} \cdot d\vec{\ell}.$$

For any path we choose between P and R, $\hat{\mathbf{r}} \cdot d\vec{\ell} = dr$, the change in radial distance from the line. Let $r = r_0$ at the reference point R, where we choose $V(R) \equiv 0$. Then:

$$V(P) = -\int_{r_0}^{a} \frac{2k\lambda}{r}dr = -2kr\lambda\ \ln(r)\Bigg|_{r_0}^{a}$$

$$V(P) = 2k\lambda\ \ln(r_0/a). \qquad (25.16)$$

ANALYZE There is no obvious place for the reference point. We cannot choose the reference point R at $r_0 = 0$ or $r_0 = \infty$, since the logarithm is infinite at both places. We must choose a position at some intermediate distance r_0 from the filament. Of course, no charge distribution is truly infinite. Using such a model in a practical situation, you would choose the reference point by considering the relation of the filament to other objects. The potential *difference* between any two points at finite distance is well defined:

$$V(P_1) - V(P_2) = 2k\lambda\ \ln(a_2/a_1).$$

See Problem 66.

25.6 The Behavior of Conductors

25.6.1 Response of a Conductor to an Electric Field

Conductors are materials in which some of the electrons move relatively freely. The copper wire inside an electrical cable is a typical, efficient conductor. Any net electric field within the body of such a material exerts a force on electrons that causes them to move. For example, if we try to produce an electric field within a piece of conducting material by bringing a positively charged object near it, electrons accelerate toward the surface of the conductor closest to the object (■ Figure 25.19). The resulting deficit of electrons on the opposite side of the

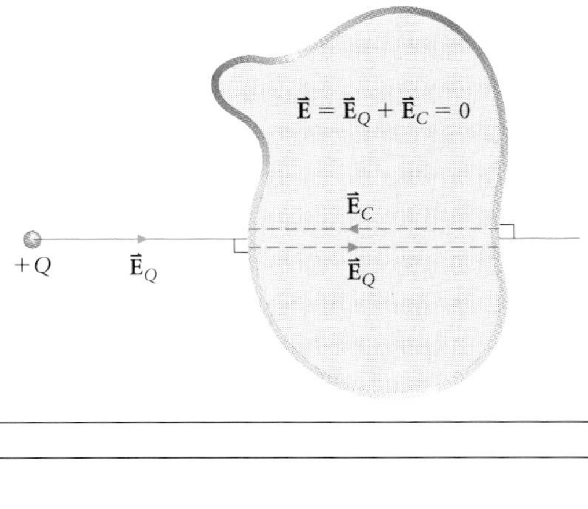

■ Figure 25.19
When a charged object is brought near a conductor, equilibrium is established rapidly. At each point within the conductor the electric field $\vec{\mathbf{E}}_c$ produced by the surface charge distribution is equal to and opposite the field $\vec{\mathbf{E}}_Q$ produced by the point charge. The surface charges also distort the field outside the conductor (see Figure 25.22).

conductor leaves a positive charge there. These surface charges produce an electric field within the conductor that opposes the electric field due to the charged object. As long as any electric field remains inside the conductor, electrons continue to respond until the distribution of charge on the conductor surface exactly neutralizes the external field within the conductor and leaves it in equilibrium. This process happens extremely rapidly. Since all the electrons move, each electron travels only a very short distance in creating the necessary charge distribution.

THE TIME REQUIRED IS APPROXIMATELY THE LIGHT TRAVEL TIME ACROSS THE CONDUCTOR (SEE PROBLEM 82).

> In static equilibrium, the electric field $\vec{\mathbf{E}}$ is zero at every point within the body of a conducting material.

Any net charge within the conductor would be the source of electric field. If the field is zero, then the charge must be also. To see why, we apply Gauss' law to a closed surface within the metal. Since $\vec{\mathbf{E}} \equiv 0$, the flux through *any* such surface is zero, and thus the surface encloses zero net charge. We may choose a very small Gaussian surface, surrounding an arbitrary point P, or one so large that it coincides with the surface of the conductor (■ Figure 25.20). Thus the charge density is zero at every point inside the conductor, and any net charge on the conductor must be on its surface.

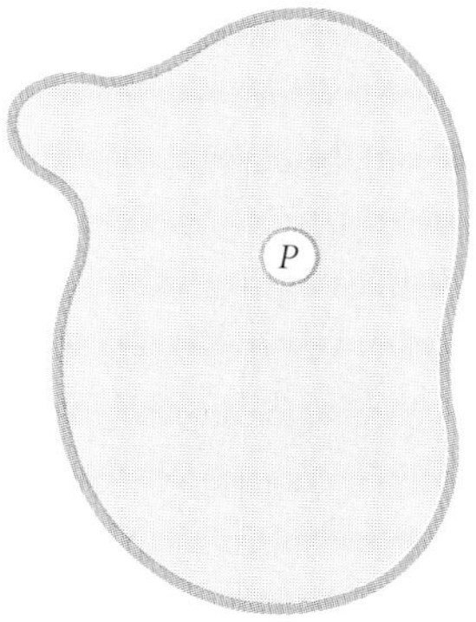

■ FIGURE 25.20
There is no flux through any Gaussian surface that is entirely within the conductor. Here we see two such surfaces. A small surface surrounding the arbitrary point P shows that $\rho = 0$ at P. A large surface reaching to the boundary of the conductor shows that all the charge on the conductor must be on its surface.

> In static equilibrium, net electric charge can exist only on the surfaces of a conductor, not within its body.

CHARGES ON THE SURFACE ARE ACTUALLY CONCENTRATED WITHIN ROUGHLY AN ATOMIC DIAMETER OF THE SURFACE.

A charged object exerts a substantial force on a neutral piece of conductor, because the surface charge is not all at the same distance from the object.

EXAMPLE 25.14 ♦♦ A point charge $Q = +1\ \mu\text{C}$ is 10 cm from the middle of a thin metal needle 1 cm long (■ Figure 25.21). Estimate the charges on the ends of the needle required to maintain $\vec{\mathbf{E}} \equiv 0$ within the metal. Estimate the net attractive force on the needle.

MODEL We model the charges on the needle as point charges $\pm Q_c$ at its ends and estimate their magnitude by requiring $\vec{\mathbf{E}} = 0$ at the middle of the needle. Since negative charge is attracted toward the positive charge Q, and positive charge is repelled, the end of the needle closest to Q is negatively charged. This model is only approximate, since it does not make $\vec{\mathbf{E}}$ zero throughout the needle.

IN REALITY THE CHARGE IS DISTRIBUTED OVER THE ENTIRE SURFACE AS WELL AS THE ENDS OF THE NEEDLE.

SETUP The field at the center of the needle is the superposition of the fields due to the charges at the two ends, plus the field due to the external point charge Q:

$$\vec{\mathbf{E}}_Q = \frac{kQ}{D^2}\hat{\mathbf{i}}, \qquad \vec{\mathbf{E}}_{c+} = \frac{kQ_c}{(\ell/2)^2}(-\hat{\mathbf{i}}), \qquad \text{and} \qquad \vec{\mathbf{E}}_{c-} = \frac{kQ_c}{(\ell/2)^2}(-\hat{\mathbf{i}}).$$

The net electric field is zero:

$$0 = \vec{\mathbf{E}}_Q + \vec{\mathbf{E}}_{c+} + \vec{\mathbf{E}}_{c-} = \frac{kQ}{D^2}\hat{\mathbf{i}} + \left[-\frac{2kQ_c}{(\ell/2)^2}\hat{\mathbf{i}}\right].$$

SOLVE

$$Q_c = \frac{Q\ell^2}{8D^2} = \frac{10^{-6}\ \text{C}}{8}\left(\frac{1\ \text{cm}}{10\ \text{cm}}\right)^2 = 1\ \text{nC}.$$

SETUP The force exerted by Q on the needle is the sum of the forces on the two charges $\pm Q_c$. Using Coulomb's law:

SOLVE

$$\vec{\mathbf{F}} = \frac{k(-Q_c)Q}{(D - \ell/2)^2}\hat{\mathbf{i}} + \frac{kQ_cQ}{(D + \ell/2)^2}\hat{\mathbf{i}}.$$

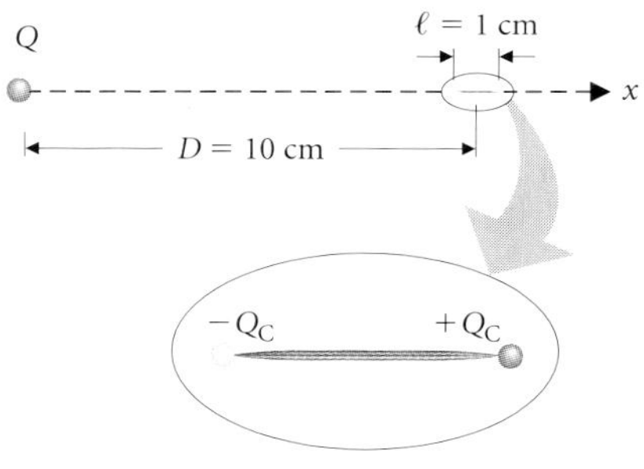

■ FIGURE 25.21
An external point charge induces surface charge on a metal needle, here modeled with two point charges $\pm Q_c$ on the ends. The external point charge exerts a net attractive force on the needle because the negative charge $-Q_c$ is closer to Q than the positive charge $+Q_c$.

We simplify this expression by factoring the common quantities:

$$\vec{\mathbf{F}} = -\frac{kQ_c Q\hat{\mathbf{i}}}{D^2}\left[\frac{1}{(1 - \ell/2D)^2} - \frac{1}{(1 + \ell/2D)^2}\right].$$

Then we combine terms over a common denominator:

$$\vec{\mathbf{F}} = -\frac{kQ_c Q\hat{\mathbf{i}}}{D^2}\left[\frac{(1 + \ell/2D)^2 - (1 - \ell/2D)^2}{(1 - \ell^2/4D^2)^2}\right].$$

Now, since $\ell^2/4D^2 = 0.002$, we neglect it compared with 1:

$$\vec{\mathbf{F}} \approx -\frac{kQ_c Q\hat{\mathbf{i}}}{D^2}\left(\frac{2\ell/D}{1}\right) = \frac{-2kQ_c Q\ell\hat{\mathbf{i}}}{D^3}.$$

Then substituting the value we found for Q_c gives:

$$\vec{\mathbf{F}} \approx \frac{-kQ^2\ell^3\hat{\mathbf{i}}}{4D^5} = \frac{-\hat{\mathbf{i}}(9 \times 10^9\ \mathrm{N\cdot m^2/C^2})(10^{-6}\ \mathrm{C})^2(10^{-2}\ \mathrm{m})^3}{4(10^{-1}\ \mathrm{m})^5}$$
$$= -\hat{\mathbf{i}}(2 \times 10^{-4}\ \mathrm{N}).$$

ANALYZE The direction $(-\hat{\mathbf{i}})$ confirms that the force is attractive. A steel needle 1 mm in diameter has a mass m of approximately 6×10^{-5} kg and would accelerate at $|\vec{\mathbf{a}}| = |\vec{\mathbf{F}}|/m \approx 3\ \mathrm{m/s^2}$, or at about $\frac{1}{3}\,g$.

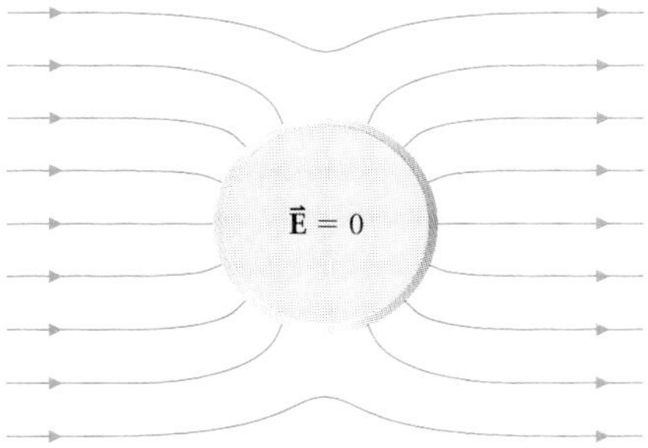

■ FIGURE 25.22
A conducting sphere is placed in a uniform field. The applied electric field creates a surface charge distribution on the sphere that distorts the field outside. The field lines meet the sphere perpendicular to the surface. The electric field within the sphere is zero.

25.6.2 *Conductors and Electric Potential*

Since the electric field vanishes within a conductor, there is no potential difference between any two points within the conductor (eqn. 25.7). The entire volume is at a constant potential, and its surface is an equipotential surface. Electric field lines are always perpendicular to equipotential surfaces (§25.4.1), so the field lines that emanate from or converge into charges on the conductor are perpendicular to the surface.

■ Figure 25.22 shows a conducting sphere placed in a region where the electric field is initially uniform and parallel to the x-axis. The external electric field points to the right, pushing electrons within the sphere to its left surface. A net positive charge remains at the right surface. The field produced by these charges neutralizes the applied field within the sphere and also modifies the field outside. External field lines connect to these charges and meet the sphere at right angles to its surface. There is no electric field within the sphere.

Now we can see why Don Salvatore is safe within his metal cage at the Boston Museum of Science. Imagine trying to make an electric field within a cavity inside a conductor (■ Figure 25.23a). Surface charges keep $\vec{\mathbf{E}}$ out of the conducting material, and since the surface is an equipotential, field lines may not run from one point on the surface to another across the cavity. There is no field at all inside the cavity. The metal cage isn't a perfect shield because of the holes between bars (Figure 25.23b), but it is good enough.

EXERCISE 25.6 ❖ Is it safe to sit in your car during a lightning storm? What happens when you get out?

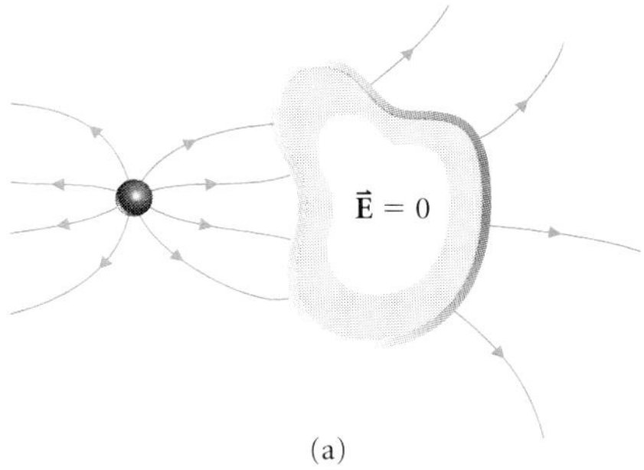

(a)

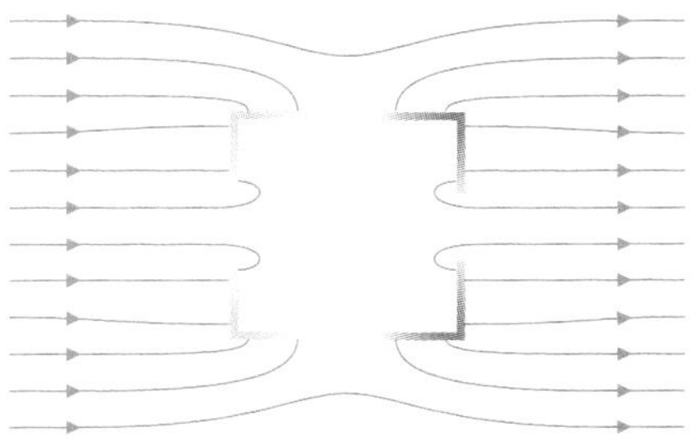

(b)

■ FIGURE 25.23
(a) Since the whole conductor is at the same potential, no field lines pass from one point of the conductor to another. That includes field lines inside the cavity. Since there is no charge inside the cavity, it is not possible to have any static electric field in the cavity. The surface charge induced by the external charge is all on the outer surface of the conductor. (b) Even if there are holes in the conducting surface, the electric field does not penetrate very far into the cavity.

25.6.3 *The Relation Between Field and Surface Charge Density on a Conductor*

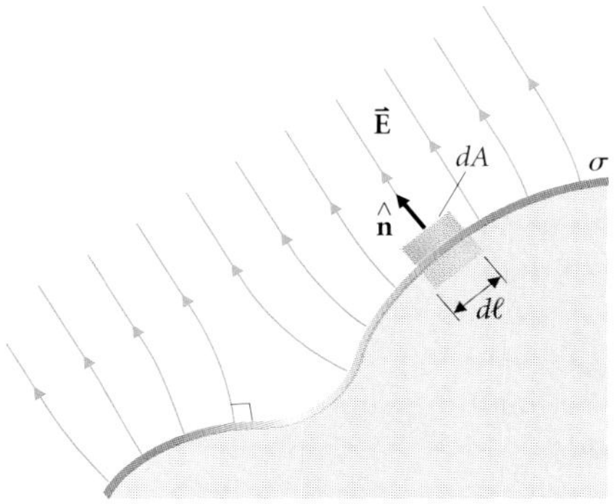

■ FIGURE 25.24
To find the relation between electric field and the surface charge density on a conductor, we apply Gauss' law to the small pillbox shown here. It is constructed with its curved sides perpendicular to the conductor surface and thus parallel to the field just outside the surface. The flat faces are parallel to the conductor surface and perpendicular to $\vec{E}$.

We may use Gauss' law to find the relation between the surface charge density on a conductor and the electric field vector just outside. We choose a small volume (a *pillbox*) with one face just outside the conductor and the pillbox side walls perpendicular to the conductor surface (■ Figure 25.24). If the conductor surface is not flat, we choose the pillbox area dA small enough that the curvature is negligible. That is, we take $d\ell \sim \sqrt{dA} \ll R$, where R is the radius of curvature of the surface at that point. Since there is no electric field inside the conductor, there is no electric flux through the pillbox walls inside the conductor. With $\vec{E}$ perpendicular to the surface, there is no flux through the side walls outside the conductor either. The remaining face is parallel to the conductor surface, $\vec{E}$ is parallel to $\hat{n}$ on this face, and the flux through it is $E\ dA$. The charge inside the pillbox is all on the surface of the conductor: $Q = \sigma\ dA$. Applying Gauss' law, we find:

$$\Phi_E = \oint \vec{E} \cdot \hat{n}\ dA = E\ dA = \frac{Q}{\epsilon_0} = \frac{\sigma\ dA}{\epsilon_0}.$$

So:

$$\vec{E} = \frac{\sigma}{\epsilon_0}\hat{n} = 4\pi k\sigma\hat{n}. \tag{25.17}$$

IF THE FIELD LINES POINT TOWARD THE SURFACE, σ IS NEGATIVE; IF THEY POINT AWAY, σ IS POSITIVE.

Equivalently, the charge density is:

$$|\sigma| = \epsilon_0|\vec{E}|. \tag{25.18}$$

The electric field near a large, uniformly charged plane with charge density σ has magnitude $\sigma/(2\epsilon_0)$ (eqn. 24.7). Electric field lines emerge from both sides perpendicular to the surface (■ Figure 25.25a). If a conducting surface has the same charge density, the same number of field lines come out, but all on one side (Figure 25.25b). The field outside is double the value for a single charged plane (Figure 25.25c).

EXERCISE 25.7 ♦ The charge density at a point on the surface of a conductor is 1.2 μC/m^2. What is the electric field just outside the conductor at that point?

25.6.4 *The Relation Between the Shape of a Conductor and the Electric Field at its Surface*

Benjamin Franklin was the first to note that sharply pointed objects leak charge into the air much more rapidly than do blunt objects. Electric fields are strongest near points, and sufficiently intense fields can strip electrons from molecules in the air, increasing its conductivity. We can determine the relationship between electric field and the shape of a conductor by considering how conductors in contact share charge.

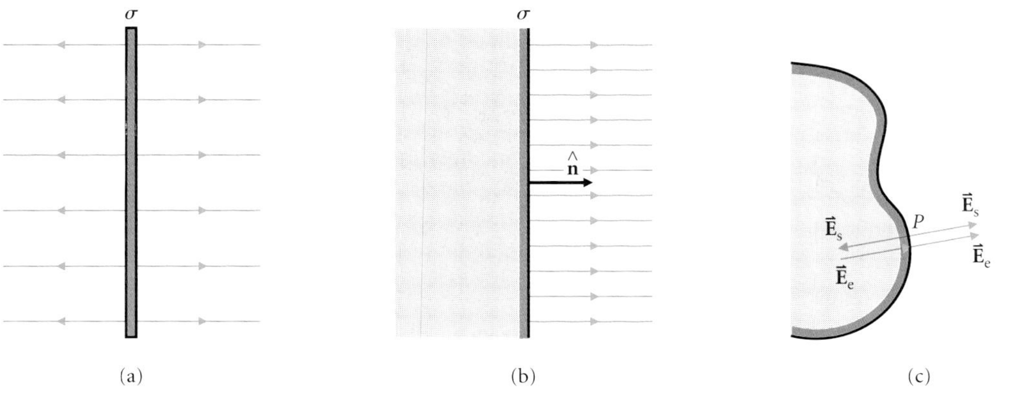

■ FIGURE 25.25
(a) A plane sheet of charge with surface charge density σ. The electric field is normal to the sheet and has magnitude $\sigma/(2\epsilon_0)$. (b) The same charge density on the surface of a plane conductor. All the field lines emerge outward from the surface, and the field is twice as large: $|\vec{E}| = \sigma/\epsilon_0$. (c) The field near a conducting surface at P is the superposition of the field due to the nearby surface charge density [$|\vec{E}_s| = \sigma/(2\epsilon_0)$] and the field $\vec{E}_e$ due to charges elsewhere on the surface of the conductor. In order that the field inside the conductor be zero, the electric field $\vec{E}_e$ must be equal to and opposite the field *within* the conductor produced by the surface charge at P, and thus its magnitude is also $\sigma/(2\epsilon_0)$. Outside the conductor, $\vec{E}_s$ reverses direction, but $\vec{E}_e$ does not. The total field outside has magnitude σ/ϵ_0.

EXAMPLE 25.15 ♦♦ Two conducting spheres of radii R_1 and R_2 are far apart but connected by a thin conducting wire (■ Figure 25.26). They share a total charge Q. Find the charge density on each sphere and the electric field strength at each sphere's surface.

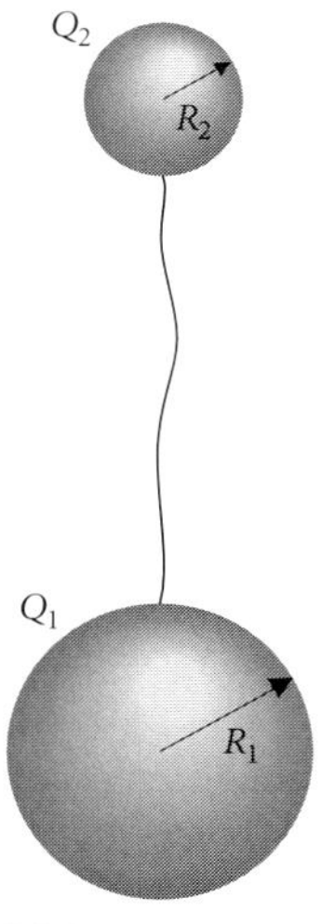

■ **FIGURE 25.26**
Two spheres sharing a total charge Q are connected by a long, thin wire. The spheres and wire form a single conductor with a uniform potential. The charge on each sphere is proportional to its radius.

MODEL The two spheres connected by a wire form a single conducting object. No electric field can exist within this object, so the electric potential is constant everywhere on it. Since the two spheres are far apart, they act like isolated spherical charges with electric fields and potentials given by Coulomb's law (cf. Example 24.1 and Exercise 25.1). The charge on the wire is negligible so long as it is thin.

SETUP The potential at distance R from the center of a sphere carrying charge Q is kQ/R. This formula holds even at the surface of the sphere. If the two spheres have charges Q_1 and Q_2, they have potentials $V_1 = kQ_1/R_1$ and $V_2 = kQ_2/R_2$. Since the spheres are connected by a wire, they are at the *same* potential. Thus:

$$Q_1/R_1 = Q_2/R_2.$$

Since the total charge $Q = Q_1 + Q_2 = Q_1 + (R_2/R_1)Q_1$, we have:

$$Q_1 = \frac{QR_1}{R_1 + R_2} \quad \text{and} \quad Q_2 = \frac{QR_2}{R_1 + R_2}.$$

Since the spheres are far apart, they do not much influence each other's charge distributions, which we may assume uniform. Then:

$$\sigma_1 = \frac{Q_1}{4\pi R_1^2} \quad \text{and} \quad \sigma_2 = \frac{Q_2}{4\pi R_2^2}.$$

THE WIRE CHANGES THE CHARGE DISTRIBUTION SLIGHTLY WHERE IT TOUCHES THE SPHERES. WE ALSO IGNORE THIS EFFECT.

SOLVE Substituting the values of Q_1 and Q_2 into these equations, we find:

$$\sigma_1 = \frac{QR_1}{4\pi R_1^2(R_1 + R_2)} = \frac{Q}{4\pi R_1(R_1 + R_2)},$$

and similarly:
$$\sigma_2 = \frac{Q}{4\pi R_2(R_1 + R_2)}.$$

The electric field strength at the surface of one of the spheres is $|\vec{\mathbf{E}}| = 4\pi k\sigma$ (eqn. 25.17). So:

$$|\vec{\mathbf{E}}_1| = \frac{kQ}{R_1(R_1 + R_2)},$$

and:
$$|\vec{\mathbf{E}}_2| = \frac{kQ}{R_2(R_1 + R_2)}.$$

ANALYZE $|\vec{\mathbf{E}}| \propto 1/R$, that is:
$$\frac{|\vec{\mathbf{E}}_1|}{|\vec{\mathbf{E}}_2|} = \frac{R_2}{R_1}.$$

This is a general relation; the electric field is large near a sharp point because the radius of curvature is small (■ Figures 25.27, 25.28).

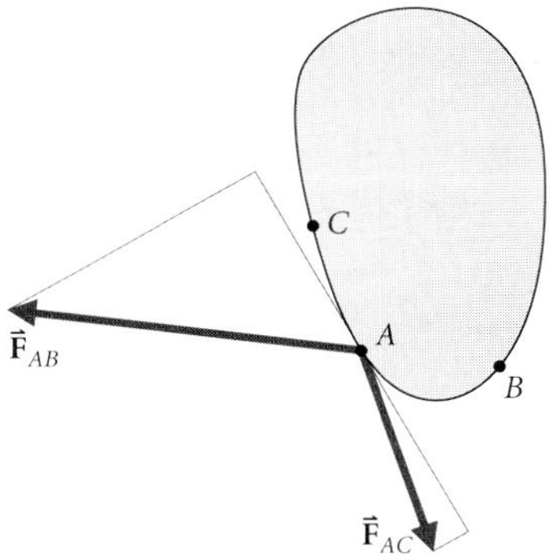

■ **FIGURE 25.27**
Charge is in equilibrium on a conducting surface when the net force parallel to the surface acting on any element of charge is zero. Two charge elements at points B and C exert very different forces on the element at A, though they are the same distance away. Because of the greater curvature between B and A, the force $\vec{\mathbf{F}}_{AB}$ is more nearly perpendicular to the surface than is $\vec{\mathbf{F}}_{AC}$. Since the tangential components balance, $|\vec{\mathbf{F}}_{AB}| > |\vec{\mathbf{F}}_{AC}|$, and the charge element at B is greater than the charge element at C. The charge density, and hence the electric field, is greater at B than at C. ∎

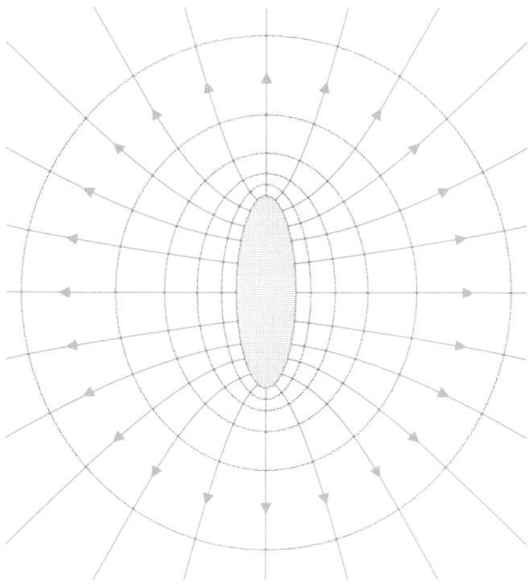

■ **FIGURE 25.28**
Field lines and equipotential surfaces for an object with varying curvature. The surfaces evolve toward spherical sphape and are more closely spaced where their initial curvature is greatest. Thus electric field is greatest where curvature is greatest.

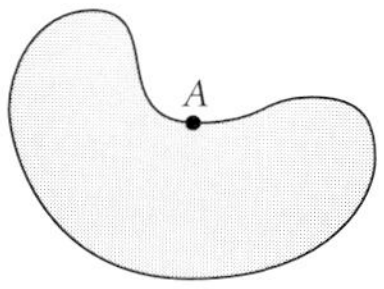

■ FIGURE 25.29

EXERCISE 25.8 ❖ The object in ■ Figure 25.29 is a charged conductor. Sketch a field line diagram for this object, and convince yourself that the surface charge density is small at point A, where the surface has a dimple, that is, a region of negative curvature.

Chapter Summary

Where Are We Now?

We have discussed the electric potential energy of systems of charges. This is the third form of potential energy we have studied, the others being elastic and gravitational energy. We used potential energy to introduce the idea of electric potential. Electric potential is a scalar quantity that describes the effect of a charge on its surroundings and is closely related to the electric field vector. We studied the behavior of conductors. We now have enough tools to begin a serious study of electric systems.

What Did We Do?

Electric potential describes the work done to move a charge between two points in an electric field: $\Delta U = q\,\Delta V$. The potential difference between two points P and R is:

$$V(P) - V(R) = -\int_R^P \vec{\mathbf{E}} \cdot d\vec{\ell}.$$

If R is the reference point where we choose to assign the potential a value of zero, this equation defines the potential $V(P)$. Potential is measured in volts. One volt is 1 joule per coulomb.

When the charge distribution is a collection of point charges, the reference point R is usually chosen at infinity. Then the potential due to a point charge is kQ/r, where r is the distance from the charge (r is always positive). The potential due to a finite, continuous distribution of charge may be found by integrating the potentials due to each differential charge element.

We found the potential energy of a system of charges by computing the work done to assemble the system. The energy of such a system is the sum of the energies kQ_iQ_j/r_{ij} of each distinct pair of charges Q_i and Q_j in the system. The energy may also be computed from the potential at the position of each charge in the system:

$$U = \tfrac{1}{2}\Sigma Q_i V_i.$$

When a charge q is moved from the reference point R to P, the potential energy of the system changes by

$$\Delta U = qV(P).$$

Surfaces of constant potential are always perpendicular to field lines, and the value of the potential decreases along field lines. In static equilibrium, the electric field is zero within a conductor, and the surface of a conductor is an equipotential surface. The charge density on the surface of the conductor is related to the electric field at the surface:

$$|\sigma| = \epsilon_0|\vec{\mathbf{E}}|.$$

Practical Applications

Potential is important for understanding electric circuits (Chapter 26). Many electronic devices require carefully designed patterns of electric field to operate properly. Since conductors are equipotential surfaces, the field pattern can be controlled by properly shaping the metal electrodes used in a device. Electric potential is an essential concept in the design of television sets, photocopiers, x-ray machines for medical diagnosis, and particle accelerators for physics research.

Solutions to Exercises

25.1 We put the origin at the point charge and integrate along a radial line, which we'll call the x-axis. The reference point R is at ∞ and P is at $x = d$.

$$V(P) = -\int_{\infty}^{P} \vec{\mathbf{E}} \cdot d\vec{\ell} = -\int_{\infty}^{d} \frac{kQ}{x^2}\hat{\mathbf{i}} \cdot (dx\,\hat{\mathbf{i}}) = -\int_{\infty}^{d} \frac{kQ\,dx}{x^2} = \left.\frac{kQ}{x}\right|_{\infty}^{d} = \frac{kQ}{d}.$$

25.2 Model the dipole as two point charges, $-Q$ at the origin and $+Q$ at $x = a$ (■ Figure 25.30). The potential at $x > a$ is:

$$V_a(x) = kQ\left(\frac{1}{x-a} - \frac{1}{x}\right) = kQ\frac{a}{x(x-a)}.$$

The potential produced by an ideal dipole of moment $p = Qa$ is:

$$V(x) = \lim_{a \to 0} V_a(x) = \frac{kp}{x^2}.$$

■ FIGURE 25.30

25.3 (■ Figure 25.31) There are six charge pairs, AB, AC, BD, CD, AD, and BC. The first four have separation s, and the other two have separation $s\sqrt{2}$. Thus:

$$U_{\text{system}} = 4\frac{kQ^2}{s} + 2\frac{kQ^2}{s\sqrt{2}} = \frac{kQ^2}{s}(4 + \sqrt{2}).$$

25.4 STEPS I AND II Take the z-axis as the axis of the ring, with the origin at its center (■ Figure 25.32). A typical element of the ring at angle θ has length $a\,d\theta$ and charge $dQ = Q\,d\theta/(2\pi)$.

STEP III The potential at P due to the element is:

$$dV = k\,dQ/s.$$

The distance $s = \sqrt{z^2 + a^2}$ is the same for each point on the ring.

STEPS IV AND V The angle θ runs from 0 to 2π, so:

$$V = \int_0^{2\pi} \frac{kQ}{2\pi}\frac{d\theta}{\sqrt{z^2 + a^2}} = \frac{kQ}{\sqrt{z^2 + a^2}}\int_0^{2\pi}\frac{d\theta}{2\pi} = \frac{kQ}{\sqrt{z^2 + a^2}}.$$

Note: Since all the charge is at the same distance:

$$V = k(\text{total charge})/(\text{distance}).$$

25.5 The disk is a collection of differential rings of radius r and width dr (cf. Example 24.6). The charge on such a ring is:

$$dQ_{\text{ring}} = \sigma\,dA = \frac{Q}{\pi R^2}(2\pi r\,dr) = \frac{2Q}{R^2}r\,dr.$$

From Exercise 25.4 the potential due to the ring is:

$$dV = \frac{k\,dQ_{\text{ring}}}{\sqrt{z^2 + r^2}} = \frac{2kQr\,dr}{R^2\sqrt{z^2 + r^2}}$$

Thus:

$$V = \int dV = \int_0^R \frac{kQ}{R^2}\frac{2r\,dr}{\sqrt{z^2 + r^2}} = \frac{kQ}{R^2}\int_{z^2}^{z^2+R^2}\frac{du}{\sqrt{u}}, \quad \text{where } u = z^2 + r^2.$$

So:

$$V = \frac{kQ}{R^2}\left(\frac{\sqrt{u}}{\frac{1}{2}}\right)\Bigg|_{z^2}^{z^2+R^2} = \frac{2kQ}{R^2}\left(\sqrt{z^2 + R^2} - z\right).$$

25.6 Your car acts like the metal cage in the Boston Museum of Science, so you'll be relatively safe. If a lightning bolt does strike the car, wait several minutes before getting out so any charge can leak away, equalizing potentials of car and ground. This happens faster if you connect the car frame to ground with something metallic. Use gloves! You don't want your body to be the path the current takes!

25.7 From eqn. (25.17), the field is normal to the surface and has magnitude:

$$E = 4\pi k\sigma = 4\pi(9.0 \times 10^9\ \text{N}\cdot\text{m}^2/\text{C}^2)(1.2 \times 10^{-6}\ \text{C/m}^2) = 1.4 \times 10^5\ \text{N/C} = 1.4 \times 10^5\ \text{V/m}.$$

25.8 See ■ Figure 25.33. The field lines are perpendicular to the surface and as distance from the object increases, they bend toward the radial pattern expected for a point charge. To accomplish this, they spread apart near the surface at A. The increased distance between the lines indicates a reduced field magnitude and a reduced surface charge density.

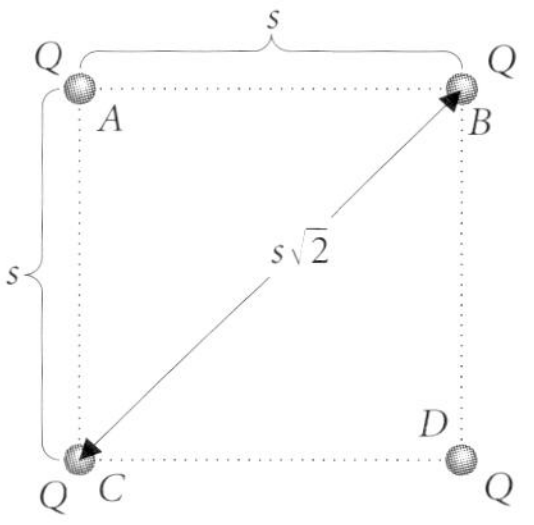

■ FIGURE 25.31

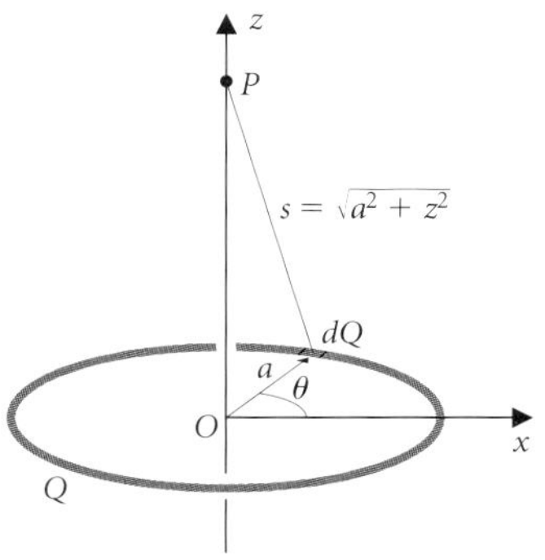

■ FIGURE 25.32

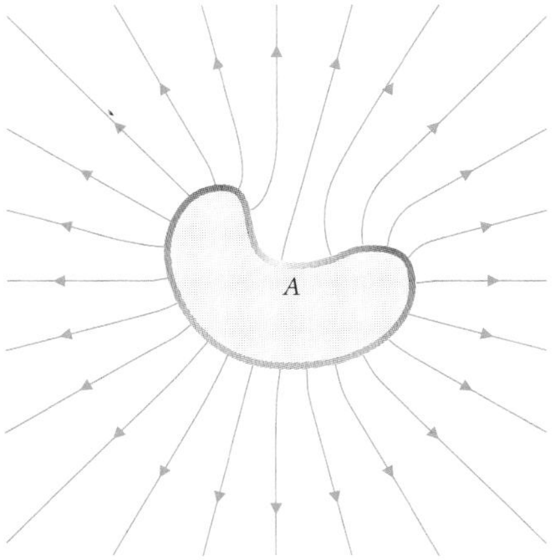

■ FIGURE 25.33

Basic Skills

Review Questions

§25.1 POTENTIAL ENERGY OF A PAIR OF POINT CHARGES

- What feature of the Coulomb force allows us to define electric potential energy?
- How does the potential energy of a pair of charges depend on the distance between them?

§25.2 ELECTRIC POTENTIAL

- What is the operational definition of electric potential difference?
- State a useful rule for remembering how potential difference is related to electric field.
- When the electric field is due to a system of point charges, where do we usually choose the reference point R such that $V(R)$ is zero?
- What is the unit of electric potential?
- Write two versions of the SI units for electric field.
- What is the electric potential at distance d from a point charge Q if the reference point is at infinity?
- What is an *electron volt?* What physical quantity is measured in electron volts?

§25.3 POTENTIAL ENERGY OF A SYSTEM OF CHARGES

- State the principle of superposition for electric potential.
- Describe a procedure for calculating the potential energy of a system of point charges.
- How is the potential energy of a system related to the energies of its charge pairs?
- How is the potential energy of a system of charges expressed in terms of the potential at the location of each charge?

§25.4 EQUIPOTENTIAL SURFACES

- How are equipotential surfaces related to field lines?
- What can you say about the electric field vector at a position where an equipotential surface crosses itself?
- Is the electric field necessarily zero where the potential is zero? Why or why not?

§25.5 POTENTIAL DUE TO A CONTINUOUS DISTRIBUTION OF CHARGE

- Can the principle of superposition be used to find the potential due to an infinite plane sheet of charge? Why or why not?

§25.6 THE BEHAVIOR OF CONDUCTORS

- What is the static electric field within a conductor?
- How does electric potential vary within a conducting object? Does it make a difference whether the conductor is solid or hollow?
- What is the direction and magnitude of the electric field vector at the surface of a conductor?
- How does the surface charge density on a conductor's surface depend on the shape of the conductor?

Basic Skill Drill

§25.1 POTENTIAL ENERGY OF A PAIR OF POINT CHARGES

1. Two protons are separated by a distance of 5.0 nm. What is the potential energy of the system?
2. Two protons are initially separated by 5.0 nm and are at rest. What is the speed of each proton after they have moved far apart?

§25.2 ELECTRIC POTENTIAL

3. How much work is required to transfer a charge of 6 C against a potential difference of 100 V?
4. What is the potential difference between two points $x = 0$, $y = 0$ and $x = 1.30$ m, $y = 0$ in a region where the field is uniform and equal to $(1.50 \times 10^4 \text{ V/m})\hat{\mathbf{i}}$?

§25.3 POTENTIAL ENERGY OF A SYSTEM OF CHARGES

5. Three identical charges are fixed at the corners of an equilateral triangle. At the center of the triangle:
(a) only the electric field is zero.
(b) only the electric potential is zero.
(c) both the electric field and the potential are zero.
(d) neither the electric field nor the potential is zero.
Explain why you chose your answer.
6. Three equal charges Q lie in a line, with the distance between two neighboring charges equal to s. What is the potential energy of the system?

§25.4 EQUIPOTENTIAL SURFACES

7. Sketch the equipotential surfaces due to a system of two equal point charges Q separated by a distance d.
8. ■ Figure 25.34 shows several equipotential surfaces in a certain region of space. Each surface is labeled with the value of the potential on the surface. In which region is the magnitude of the electric field greatest? In which direction does $\vec{\mathbf{E}}$ point? Explain your reasoning.

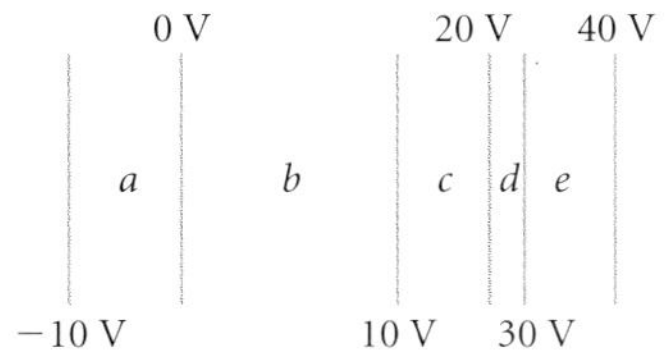

■ FIGURE 25.34

§25.5 POTENTIAL DUE TO A CONTINUOUS DISTRIBUTION OF CHARGE

9. Find the potential at the center of a hemispherical shell of radius a carrying a total charge Q uniformly distributed over it.
10. A filament of length $L = 3.0$ cm carries a uniform charge density $\lambda = 6.7\ \mu$C/m. Estimate the potential at a point on a perpendicular bisector of the filament and 1.8 m away.

§25.6 THE BEHAVIOR OF CONDUCTORS

11. A large point charge Q is held near a metal shell and causes the charge distribution shown in ■ Figure 25.35. Which choice best describes the magnitude of the electric field at point A within the shell? **(a)** kQ/D^2 **(b)** $2\pi k\sigma$ **(c)** zero **(d)** $kQ\sigma/(Dd)$ **(e)** $k\sigma/d^2$
Explain why you chose your answer.

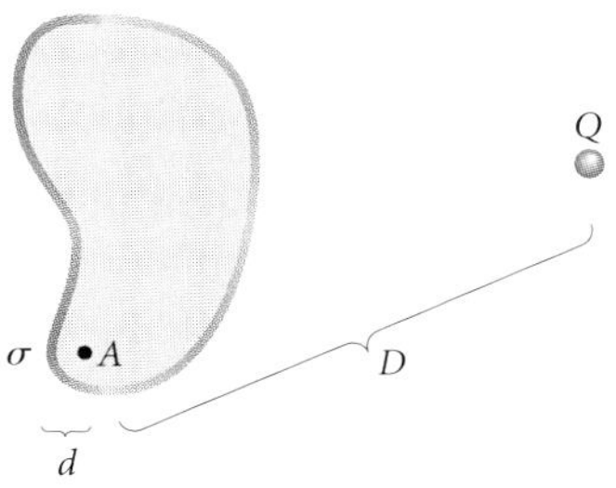

■ FIGURE 25.35

12. A point charge Q is initially a distance R from the center of a *solid* metal sphere of radius $a \ll R$. If the charge is moved to a distance $R/2$, the magnitude of the electric field at the center of the sphere:
(a) increases by a factor 2. **(b)** decreases by a factor 2. **(c)** stays the same.
Explain why you chose your answer.

13. A conducting sphere of radius 10 cm carries a charge Q. A second conducting sphere of radius 20 cm is held at the same potential as the first sphere. The charge on the second sphere is:
(a) Q. **(b)** $2Q$. **(c)** $Q/2$. **(d)** $4Q$. **(e)** $Q/4$.
Explain your reasoning.

14. A solid conducting object is placed between two infinite, uniformly charged planes (■ Figure 25.36). Which arrow best describes the direction of the electric field at point A? What is wrong with the answers you don't choose?

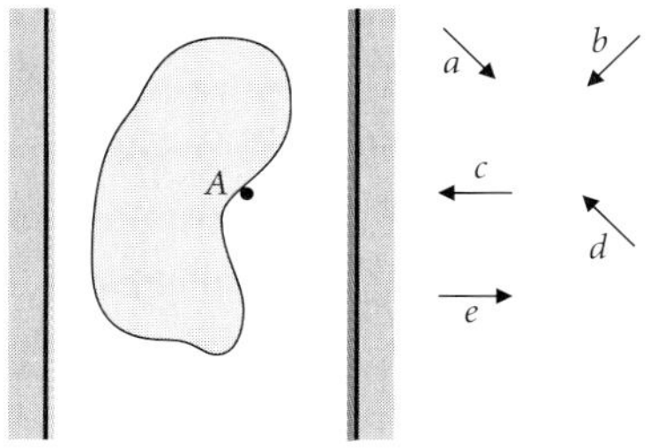

■ FIGURE 25.36

Questions and Problems

§25.1 POTENTIAL ENERGY OF A PAIR OF CHARGES

15. ❖ A point charge Q is at the center of a square. How much work is required to move a second charge q slowly along one side of the square?

16. ♦ **(a)** Find the potential energy of two point charges of $+3.0\ \mu C$ and $-0.20\ \mu C$ separated by 5.0 cm. **(b)** How much work would be required to increase their separation to 10.0 cm?

17. ♦ A system of two equal charges separated by 7.5 μm has a potential energy of 25 eV. What is the value of the charge?

18. ♦♦ Two identical point charges, each with charge $Q = 1.00 \times 10^{-6}$ C and mass $m = 1.00 \times 10^{-3}$ g, are released from rest when separated by $d = 0.100$ m. What is their speed when they are very far apart?

19. ♦♦ A point charge Q is fixed in position, and a second object with charge q and mass m moves directly toward it from a great distance. If the initial speed of the object is v, compute the minimum distance between the two objects. If $Q = 1.0\ \mu C$, $q = 1.0$ nC, $m = 1.0 \times 10^{-5}$ kg, and $v = 3.0 \times 10^5$ m/s, what is the minimum distance of approach?

20. ♦♦ Two α-particles (diameter 1.9×10^{-15} m) are headed directly toward each other with equal speeds. Compute the minimum energy in electron volts each particle must have if they are to collide. What initial speed must each particle have?

21. ♦♦ A helium nucleus has radius $r_{He} = 1.9$ fm, mass $m = 6.6 \times 10^{-27}$ kg, and charge $+2e$. A gold nucleus has charge $+87e$ and radius 7.0 fm. What initial speed must a helium nucleus have if it is to come into contact with a fixed gold nucleus in a head-on collision? (Ernest Rutherford studied this kind of collision in 1910 to establish the small size of atomic nuclei. An accurate calculation requires the use of special relativity, but this answer is good to about 10%.)

22. ♦♦ A uranium nucleus in a reactor captures a slow neutron and divides, or fissions, into two smaller *daughter* nuclei. Assuming the nucleus divides into two equal daughters with charge $Q = 46e$ and diameter $d = 2 \times 10^{-14}$ m, calculate their electric potential energy. (The electrostatic energy is only about 1% of the particles' total energy.)

§25.2 ELECTRIC POTENTIAL

23. ❖ At point C in ■ Figure 25.37, is the potential greater than, less than, or the same as the potential at B? Explain your reasoning.

24. ❖ Is there any physical significance to a uniform change in the value of the electric potential everywhere in space? Why or why not?

25. ❖ What is the potential difference between two points on the x-axis at $x = 1$ cm and $x = 10$ cm if the electric field in the region is $(1.0 \times 10^3\ \text{V/m})\hat{\jmath}$?

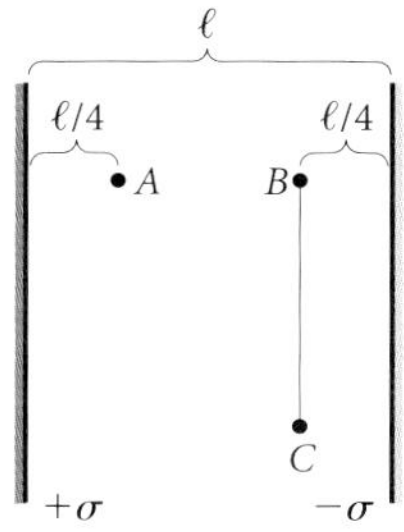

■ FIGURE 25.37

26. ❖ Points A, B, and C are distinct points in space. If it is known that $V_A - V_B = 10.0$ V, and $V_C - V_B = 2.0$ V, calculate:

$$\int_A^C \vec{\mathbf{E}} \cdot d\vec{\ell}.$$

27. ♦ Show from the formula for potential due to a point charge that the units of the Coulomb constant must be $[k] = \text{V}\cdot\text{m/C}$. Derive this same expression from the known units $[k] = \text{N}\cdot\text{m}^2/\text{C}^2$.

28. ♦ A proton is accelerated from rest through a potential difference of 0.100 MV. What is the final speed of the proton? Why don't we need to know the electric field strength? ($m_p = 1.67 \times 10^{-27}$ kg)

29. ♦ If $\vec{\mathbf{E}} = (16\text{ V/m})\,\hat{\mathbf{i}} + (8.5\text{ V/m})\,\hat{\mathbf{j}}$, and the potential is zero at the origin, find the potential at point P with coordinates $x = 1.5$ m, $y = 3.5$ m.

30. ♦ Compute the potential difference between points $O(x = 0, y = 0)$ and $P(x = 3.00\text{ cm}, y = 2.00\text{ cm})$ if the electric field in the region is $\vec{\mathbf{E}} = (2.50\text{ kV/m})(0.300\hat{\mathbf{i}} + 0.500\hat{\mathbf{j}})$.

31. ♦ Point P is 6.6 cm from a uniformly charged concrete wall. Point Q is 27.0 cm from the wall. The potential difference between P and Q is $V_{PQ} = -75.0$ V. Find the electric field vector in the region between P and Q. What is the surface charge density on the wall?

32. ♦♦ An infinite plane sheet of charge has uniform surface density $\sigma = 1.00\text{ nC/m}^2$. What is the separation between equipotential surfaces differing in potential by 10.0 V?

33. ♦♦ Two concentric metal spherical shells have radii $R_1 = 0.200$ m and $R_2 = 0.300$ m. The shells carry equal and opposite charges $Q_1 = -Q_2 = 3.27\ \mu$C. What is the potential difference $\Delta V = V_2 - V_1$ between the shells?

34. ♦♦ An electric field $\vec{\mathbf{E}}(x) = E_o(x/a)\hat{\mathbf{i}}$ exists in a certain region of space. If $E_o = 5.0 \times 10^3$ V/m and $a = 1.0$ cm, find the potential difference between the planes $x = 1.5$ cm and $x = 2.8$ cm.

35. ♦♦ A Geiger counter (Figure 24.3) is operated at a potential difference of 11 kV. If the inner wire has a radius of 25 μm and the chamber radius is 1.05 cm, find the electric field at the edge of the wire and at the outer radius of the chamber.

36. ♦♦♦ Two equal point charges $Q = 1.5$ nC are on the x-axis at $x = -2.5$ cm and $x = +2.5$ cm. Using eqn. (23.9) for the electric field on the y-axis, compute the potential difference between the points $x = 0$, $y = 1.0$ cm and $x = 0$, $y = 10.0$ cm.

§25.3 POTENTIAL ENERGY OF A SYSTEM OF CHARGES

37. ❖ A system consists of five point charges. How many distinct charge pairs are there in the system?

38. ❖ If the charge Q in the upper right-hand corner of the square in ■ Figure 25.38 is moved to the center of the square, the energy of the system:
(a) decreases. **(b)** increases. **(c)** remains the same.
(d) depends on the path the charge travels on its trip to the center.
Explain how you arrive at your answer.

39. ♦ Compute the electrostatic energy of the system in ■ Figure 25.38.

40. ♦ A positive charge Q and two negative charges $-Q$ are at the corners of an equilateral triangle (■ Figure 25.39). What is the potential at point P?

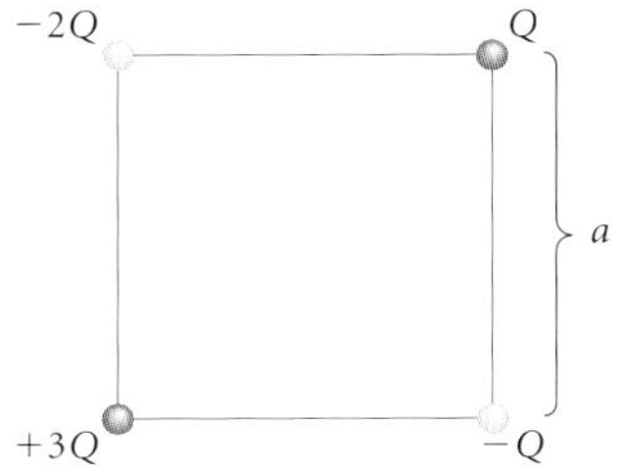

■ FIGURE 25.38

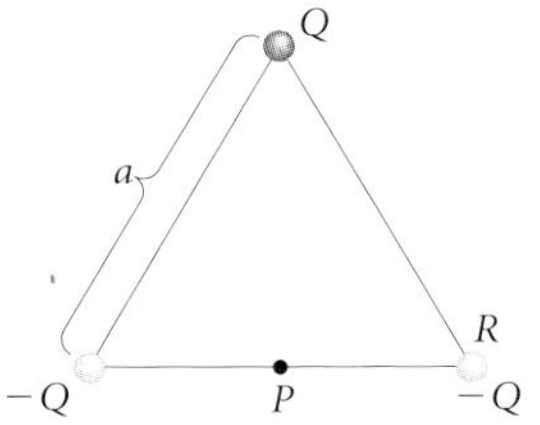

■ FIGURE 25.39

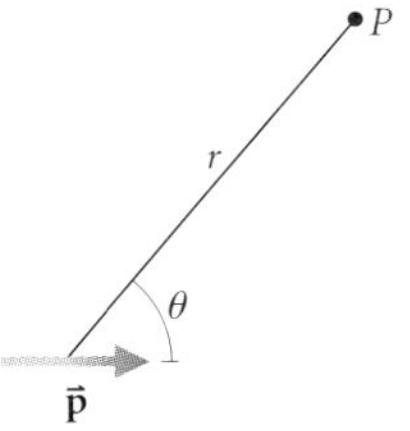

■ FIGURE 25.40

41. ♦ What is the potential energy of the system of charges in ■ Figure 25.39?

42. ♦ What is the potential at point R in ■ Figure 25.39? Which charges do you include in your calculation of potential? Why?

43. ♦♦ Four identical particles, each with mass m and charge Q, are initially at the four corners of a square with side a. If all four particles are simultaneously released from rest, what is their speed when they are a large distance apart? Take $m = 1.0 \times 10^{-3}$ g, $Q = 1.0\ \mu$C, and $a = 1.0$ cm.

44. ♦♦ A point charge $Q = +3.0\ \mu$C is fixed at the origin, and two particles with mass $m = 0.010$ kg and charge $q = +1.0\ \mu$C are on the x-axis at $a = +1.0$ mm and $x = -1.0$ mm. What is the electrical potential energy of the system? If the two particles are simultaneously released from rest, what is their speed when they are a very large distance apart?

45. ♦♦ Two identical charges $+Q$ are fixed at two corners of an equilateral triangle of side ℓ. A third charge q is placed at the third corner and then released. Find the kinetic energy of the third charge when it has reached a great distance from the other two.

46. ♦♦ Two equal positive charges with $Q = 120$ nC are separated by a distance $D = 15$ cm. A particle of mass $m = 1.5$ g and charge $q = 1.0$ nC is placed midway between them. Is the particle in equilibrium? If it is displaced slightly, determine the direction of its motion and find its speed when it is a large distance from the system of $2Q$.

47. ♦♦♦ Find the potential due to a dipole at an arbitrary point with polar coordinates r and θ with respect to the dipole (■ Figure 25.40).

§25.4 EQUIPOTENTIAL SURFACES

48. ❖ Two charges $+Q$ and $-2Q$ are placed as shown in ■ Figure 25.41. In which of the three regions A, B, or C is it possible for the potential to be zero? Explain your reasoning.

49. ❖ Sketch the equipotential surfaces for two equal and opposite charges a distance s apart.

50. ❖ The equipotential surfaces in ■ Figure 25.42 are spaced at equal intervals of 10^{-3} V. Using the distance scale at the bottom of the figure, estimate the magnitude and show the direction of the electric field at points A and B.

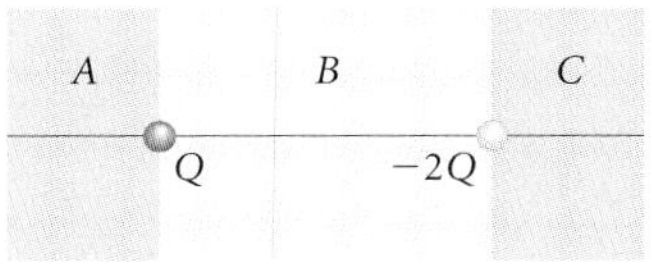

■ FIGURE 25.41

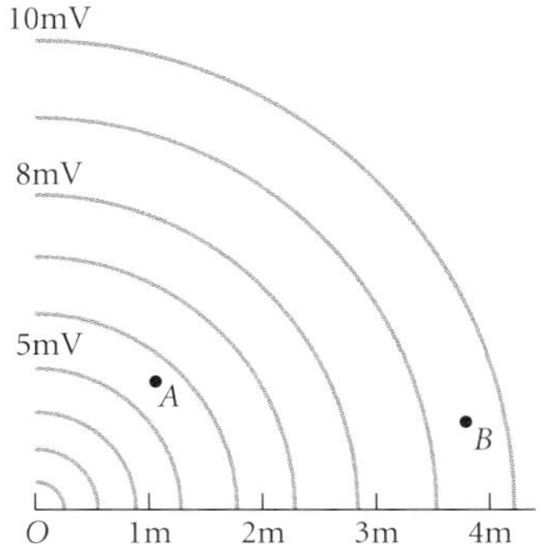

■ FIGURE 25.42

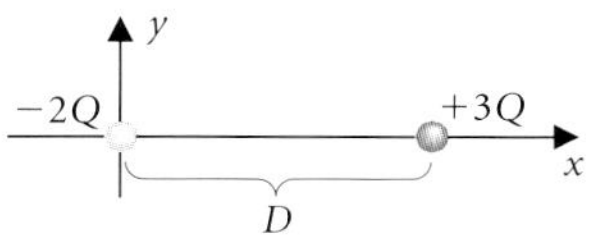

■ FIGURE 25.43

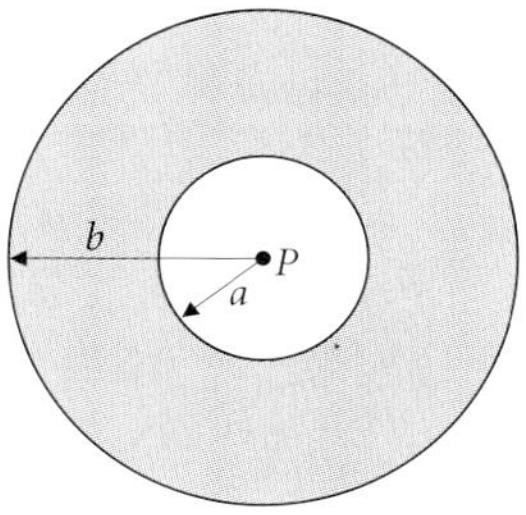

■ FIGURE 25.44

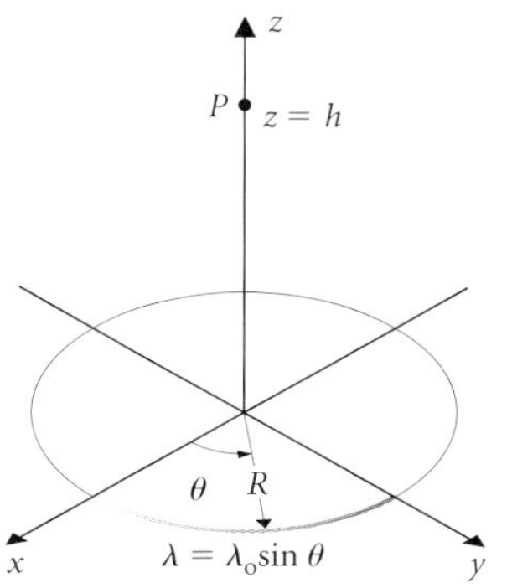

■ FIGURE 25.45

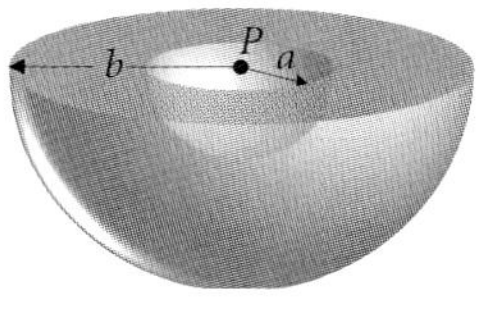

■ FIGURE 25.46

51. ❖ Three equal positive charges are fixed at the corners of an equilateral triangle. Sketch the pattern of equipotential surfaces for the system where they intersect the plane of the triangle. Using eight lines per charge, sketch the electric field line diagram in the plane of the triangle and compare with the equipotential surfaces.

52. ❖ A point charge $+Q$ is at the origin, and two negative charges $-Q$ are on the x-axis at $x = -a$ and $x = +a$. Sketch the equipotential surfaces for the system where they intersect the x-y-plane. Compare with the electric field line diagram in the x-y-plane.

53. ❖ A point charge Q is a distance D above a uniformly charged plane sheet with surface charge density σ. Both Q and σ are positive. Sketch the field line diagram and the equipotential surfaces for this system.

54. ♦♦ Two point charges $-2Q$ and $+3Q$ are on the x-axis at the origin and at $x = D$ (■ Figure 25.43). Find two points on the x-axis (not at infinity) where the electric potential is zero.

55. ♦♦ Two point charges of $+1.5\ \mu$C and $-3.0\ \mu$C are on the y-axis at $y = 0$ and $y = 2.0$ cm. Find all the points on the y-axis (not at infinity) where **(a)** $V = 0$ and **(b)** $\vec{\mathbf{E}} = 0$. What is the value of the potential where $\vec{\mathbf{E}} = 0$?

56. ♦♦ A point charge $3Q/4$ is at the origin, and two charges $-Q/2$ are on the x-axis at $x = -L$ and $x = +L$. Find points on the x-axis and on the y-axis where $V = 0$.

§25.5 POTENTIAL DUE TO A CONTINUOUS DISTRIBUTION OF CHARGE

57. ❖ A filament of length L carries a uniform charge density λ. Is the potential at P larger **(a)** when the filament is formed into a circle with its center at P or **(b)** if the filament is straightened out and lies along a tangent to the original circle?

58. ❖ Sketch the equipotential surfaces produced by a uniformly charged ring. Make the sketch in a plane perpendicular to the plane of the ring and through its center.

59. ♦ Model yourself as a sphere 30 cm in diameter receiving charge from the rug as you walk over it. How much charge do you need to reach a potential of 10^4 V?

60. ♦ A charge of $1.0\ \mu$C is uniformly distributed on a thin spherical shell of radius 0.20 m. Find the potential at the center of the shell. What is the potential at points outside the shell? Draw a graph showing the value of the electric potential as a function of distance from the center of the shell.

61. ♦♦ Points A and B lie between two infinite, uniformly charged planes with surface charge densities $\pm\sigma$ (Figure 25.37). The potential difference $\Delta V = V_A - V_B$ is:
(a) 0. **(b)** $+4\pi k\sigma\ell$. **(c)** $-4\pi k\sigma\ell$. **(d)** $+2\pi k\sigma\ell$. **(e)** $-2\pi k\sigma\ell$.
What is incorrect about the answers you didn't choose?

62. ♦♦ A test particle with charge q and mass m is projected from a large distance along the axis of a uniformly charged ring of radius a and charge Q. What initial speed is required if the particle is to pass through the center of the ring?

63. ♦♦♦ A rod lies between $x = -L/2$ and $+L/2$. It carries a linear charge density λ C/m. Find expressions for the potential at a point on the y-axis by three methods. **(a)** For $y \ll L$, treat L as infinite. **(b)** For $y \gg L$, use Coulomb's law. **(c)** Do an exact calculation using integration. Compare the results of (a), (b), and (c) graphically as functions of y and determine the ranges of y in which (a) and (b) are reasonable approximations. Compare your results with those from §24.2.3.

64. ♦♦♦ Find the potential at the center of a plane annular ring of inner radius a and outer radius b carrying a surface charge density σ (■ Figure 25.44).

65. ♦♦♦ An object in the shape of a quarter circle with radius R lies in the x-y-plane (■ Figure 25.45). If the object carries a charge per unit length given by $\lambda = \lambda_o \sin\theta$, find the electric potential at a point P at $z = h$ on the z-axis.

66. ♦♦♦ **(a)** Find the potential at an arbitrary point due to two parallel, oppositely charged infinite filaments (linear charge densities λ and $-\lambda$) separated by a distance d. **(b)** Explain why the mathematical difficulties that occur for a single infinite filament do not occur for this pair.

67. ♦♦♦ A uniformly charged hemisphere of radius b and charge density ρ has a hemispherical depression of radius a cut from its center (■ Figure 25.46). Find the potential at point P, the center of the depression.

§25.6 THE BEHAVIOR OF CONDUCTORS

68. ❖ Each of the infinitely long, cylindrical, conducting objects shown in ■ Figure 25.47 carries the same total charge. Sketch a field line diagram in the x-y-plane for each.

69. ❖ Two lumps of metal, one uncharged and the other carrying a net charge $-Q$, are placed near each other. Make a sketch of the situation showing qualitatively the charge densities on the surfaces of the conductors. Is there any net electric force acting between the conductors? Explain your reasoning.

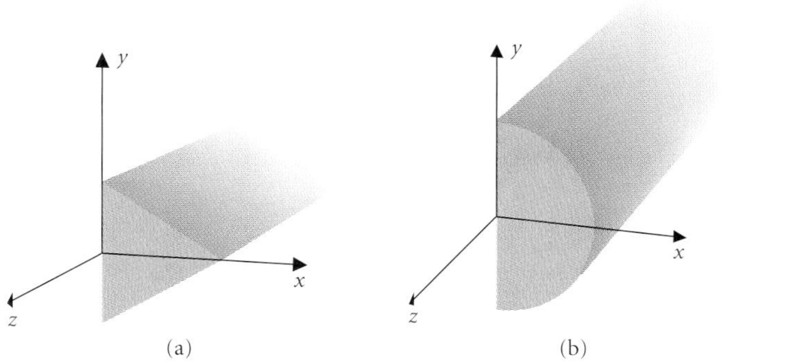

■ FIGURE 25.47

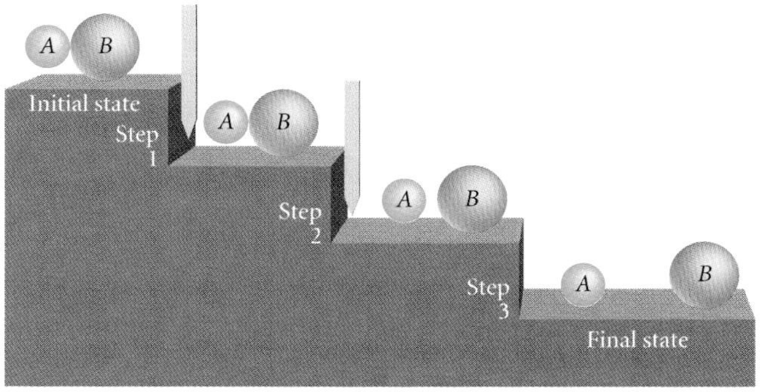

■ FIGURE 25.48

70. ❖ Two conducting spheres are initially in contact and uncharged. They undergo the process depicted in ■ Figure 25.48. Step 1: A negatively charged object is brought up and held close to the smaller sphere. Step 2: With the charged object held fixed, the two spheres are separated using insulated gloves. Step 3: The charged object is removed. Let Q_A and Q_B be the charges on the spheres in the final state. Then Q_A is:
(a) greater in magnitude than Q_B and of the same sign.
(b) greater in magnitude than Q_B and of the opposite sign.
(c) lesser in magnitude than Q_B and of the same sign.
(d) lesser in magnitude than Q_B and of the opposite sign.
(e) equal in magnitude to Q_B and of the opposite sign.
Explain your reasoning. If the two spheres are then separated by a great distance, which will be at the greater potential? Why? Which potential will be of greater magnitude? Why?

71. ❖ **(a)** A spherical conducting shell and a solid conducting ball are a great distance apart (■ Figure 25.49a). The shell has a positive charge Q_s, and the ball has a negative charge Q_b. $|Q_b| > |Q_s|$. With $V = 0$ at infinity, which of the diagrams in Figure 25.49b correctly represents the relation between the potential of the shell V_s and of the ball V_b? Explain your reasoning. **(b)** The ball is placed inside the shell without touching it (Figure 25.49c). Which of the diagrams in Figure 25.49b now represents the potentials of the two objects?

72. ❖ Explain why the electric field between the plates in Example 25.5 should be very nearly uniform.

73. ❖ A coaxial cable consists of a long, straight metal wire at the center of a thin cylindrical sheath of woven metal fibers (■ Figure 25.50). If the wire has a uniform charge density λ, describe the charge density (if any) that resides on the interior and exterior surfaces of the sheath.

74. ❖ A lump of conducting material with no net charge has an irregular hole within it, inside of which is a point charge Q. What charges reside on the inner and outer surfaces of the lump? *Approximately* what is the electric field at a great distance r from the lump?

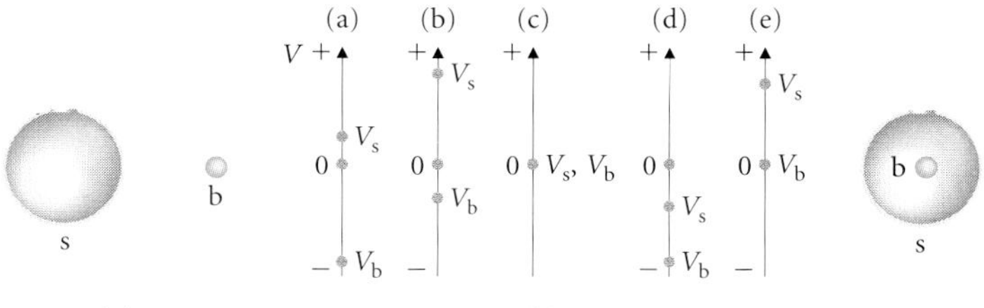

■ FIGURE 25.49

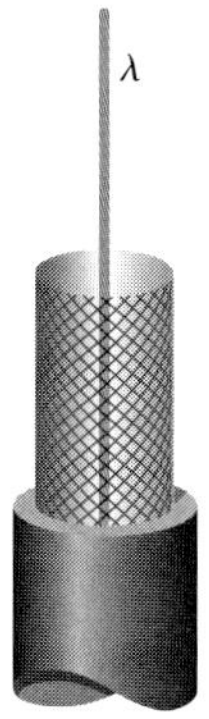

■ FIGURE 25.50

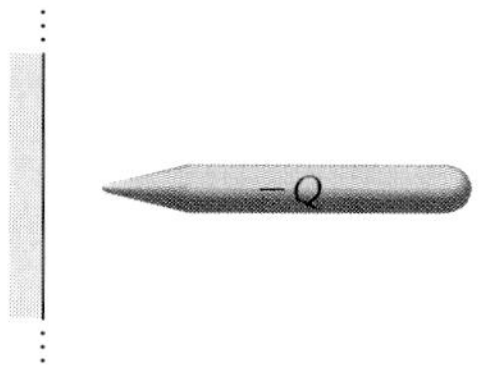

■ FIGURE 25.51

75. ❖ Sketch a field line and equipotential surface diagram for the configuration of conductors shown in ■ Figure 25.51, if the object carries a charge $-Q$ and the plane is grounded.

76. ❖ A hollow conducting bowl, ■ Figure 25.52, carries a large amount of charge. The location of the charge is tested by touching the bowl with a proof plane—a small metal disk mounted on an insulating rod. Compare the charges received by a plane that touches the bowl at B and one that touches at A. Explain your reasoning.

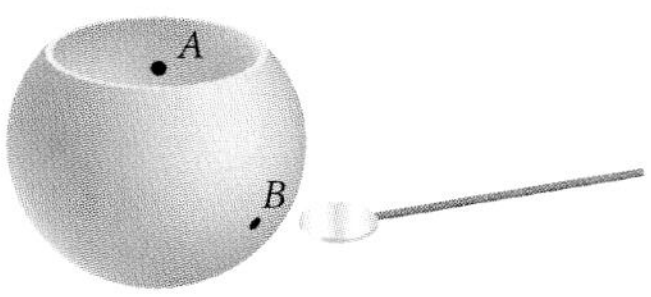

■ FIGURE 25.52

77. ❖ An electroscope (■ Figure 25.53) consists of a metal ball mounted on a rod that connects to a metal plate and a thin metal foil. When there is a net charge on the plate and foil, repulsive electric force causes the foil to hang away from the plate at an angle. When a charged object is brought near the electroscope ball, the foil rises. Why? If you touch the ball of the electroscope with your finger while the charged object is held near, the foil falls. Why? If you remove first your finger and then the charged object, the foil lifts once again. Explain this behavior. If the charge on the object is positive, what is the sign of charge remaining on the electroscope?

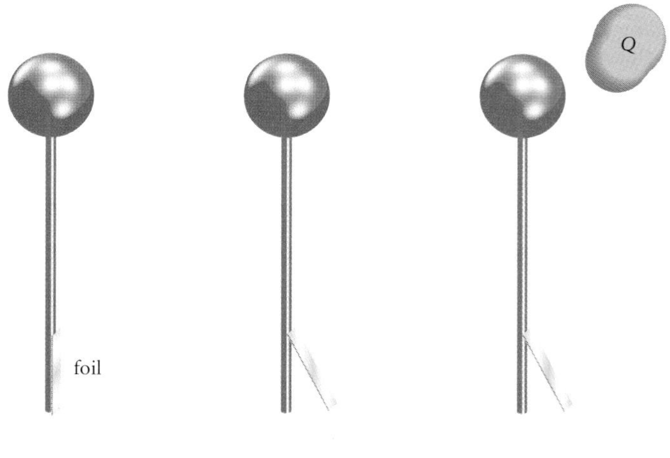

(a) electroscope uncharged (b) electroscope charged (c) charged object brought near the electroscope

■ FIGURE 25.53

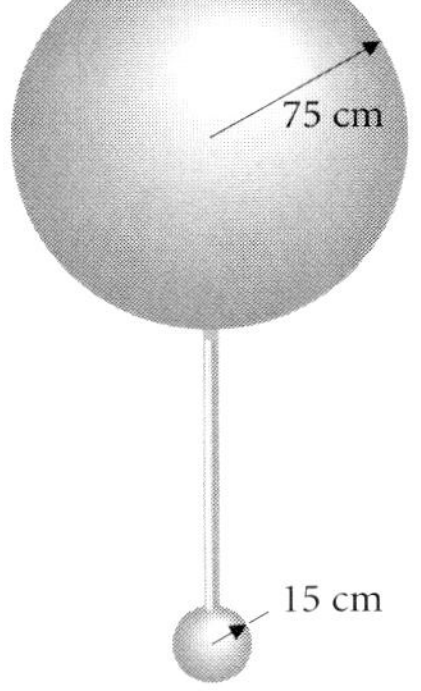

■ FIGURE 25.54

78. ♦ The electric field at point P on the surface of a conductor is measured to be 157 kV/m. What is the surface charge density on the conductor at P?

79. ♦ A copper cube has a charge density of 0.78 μC/m^2 at the center of each face. What is the electric field outside the cube near the center of the face? Discuss *qualitatively* how the field strength changes near the edges and corners of the cube.

80. ♦♦ A dumbbell-shaped conductor has spherical ends of radius 15 cm and 75 cm, respectively (■ Figure 25.54). The electric field at the surface of the small end is measured to be 98 kV/m. What is the electric field at the other end? What is the potential at the surface? Estimate the total charge on the object.

81. ♦♦ A steel needle 0.7 cm long hangs from a string in front of a plastic sheet that has a uniform surface charge density of 300 μC/m^2. Estimate how large a point charge at each end of the needle models the induced charge distribution. What is the net force on the needle?

82. ♦♦♦ Assume that the 1-cm-long needle in Example 25.14 is made of copper (density 9×10^3 kg/m^3). A copper atom has a mass of approximately 10^{-25} kg and contributes two free electrons to the metal. What fraction f of the needle's electrons must be displaced to balance the external charge? Rather than some electrons moving the length of the needle, all the electrons are displaced a small distance $\Delta\ell$. Show that $\Delta\ell/\ell \approx f$, where ℓ is the length of the needle. Assuming constant acceleration, calculate the time required for an electron to move the same fraction of the needle's length if it is exposed to the unbalanced electric field of the point charge. (Use the value of the electric field at the center of the needle.) A change in the electric field propagates at roughly the speed of light. At that speed how long would it take for an abrupt change in the electric field to propagate through the needle? Which calculation provides the better estimate of the time required for the needle to reach equilibrium?

Additional Problems

83. ♦ The chemical energy stored in a gallon of gasoline is about 130 MJ. How many coulombs would have to be stored at a potential of 12 V to provide the same energy as a 15-gallon tank of gas? If one battery can deliver 5 C/s for 10 h, how many would be needed to match the energy in the gas tank?

84. ♦ Use the fact that charge multiplied by potential difference gives potential energy to express the volt in terms of the meter, the coulomb, the kilogram, and the second.

85. ♦♦ Two concentric spherical metal shells of radii 0.350 m and 0.150 m are maintained at a potential difference of 175 kV. What is the electric field at a radius of 0.250 m?

86. ♦♦ In the Bohr model of the hydrogen atom, a single electron orbits a single proton in a circle of radius $a_0 = 5.29 \times 10^{-11}$ m. Find the energy needed to ionize the atom, that is, to remove the electron to infinity. Express your result in electron volts.

87. ♦♦ Two parallel plastic sheets are separated by a distance $d =$ 1.3 mm. If they carry equal and opposite surface charge densities $\sigma = 1.3$ μC/m^2, what is the potential difference between them? Describe what happens if a metal slab of thickness $t = 1.0$ mm is inserted between the plastic sheets. What is the new potential difference between the sheets?

88. ♦♦ A proton with initial kinetic energy of 5 MeV is headed directly toward a distant positively charged sphere of total charge $Q =$ 5×10^{-5} C and radius $R = 0.1$ m. Does the proton hit the sphere? If so, what is the kinetic energy of the proton as it hits? With what speed does the proton hit the sphere?

89. ♦♦ The probe tip for a scanning tunneling microscope has radius of curvature $r \approx 10^{-9}$ m. If the electrode is held at a potential of 100 mV with respect to the surface to be scanned, estimate the electric field at the tip of the electrode.

90. ♦♦ The breakdown field for dry air is about 10^6 V/m. What charge on an aluminum sphere 20 cm in radius will result in a spark discharge from the sphere?

91. ♦♦ A Van de Graaff machine has a dome of radius 2.0 m (■ Figure 25.55). A motor causes a charged insulating belt to travel continuously between grounded spikes at the bottom of the device and another set of spikes connected to the dome at the top. Charge is transferred through the upper spikes to the outer surface of the dome at a rate of 17 μC/s. How long does it take to charge the dome to 3.0 MV? How much power is required to operate the machine when the dome potential is 3.0 MV?

92. ♦♦ At what temperature do two protons have enough energy on average to reach a separation of 2 fm in a head-on collision? How can the Sun (central temperature about 1.3×10^7 K) succeed in generating energy by nuclear fusion? (*Hint:* Do all the protons have the same energy in thermal equilibrium?)

93. ♦♦♦ Verify the result of Example 25.8 by calculating the work done to bring up the third charge using eqn. (23.9) for the electric field of the first two charges.

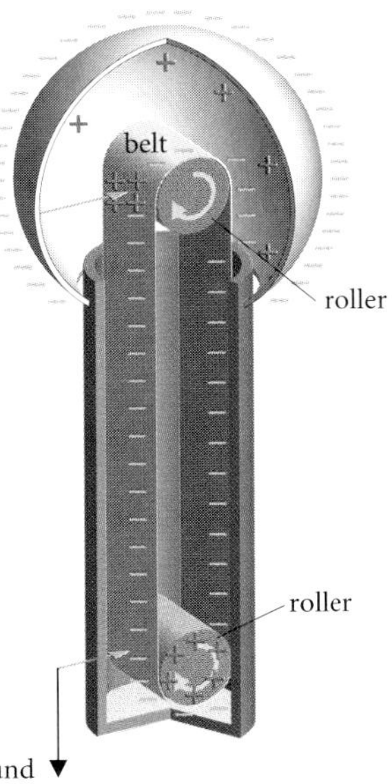

■ FIGURE 25.55

Computer Problems

94. Three point charges are placed on the x-axis. Two, each $1.0\ \mu C$, are at $x = -1.0$ m and $x = 0.0$. The third, $-1.0\ \mu C$, is at $x = +1.0$ m. Find all the points on the x-axis, not at infinity, where **(a)** $V = 0$ and **(b)** $\vec{E} = 0$. What is the value of the potential where $\vec{E} = 0$?

95. Set up a spreadsheet program to calculate the potential due to a set of point charges. Find the potential at points P_1 (coordinates $x = 1.5$ m, $y = 2.5$ m, $z = 0.3$ m) and P_2 ($x = -0.45$ m, $y = 1.7$ m, $z = -2.2$ m) due to the set of charges listed in file CH25P95 on the supplementary computer disk.

96. Set up a spreadsheet program to calculate the potential energy of a system of charges. Find the potential energy of the system in file CH25P95 (cf. Problem 95).

Challenge Problems

97. A filament extends from $x = -a$ to $x = +a$. Its charge density is given by: $\lambda(x) = \lambda_0(x/a)$. Find the potential produced by the filament at a point on the x-axis with $|x| > a$. Compare your result for $|x| \gg a$ with the result for a dipole (cf. Exercise 25.2), and determine the dipole moment of the filament.

98. A point charge $+3Q$ is at the origin, and two negative charges $-Q$ are on the x-axis at $x = d$ and $x = 2d$. Sketch the equipotential surfaces for the system where they intersect the x-y-plane. Compare your sketch with the field line diagram in the x-y-plane.

99. **(a)** Find the potential due to a uniformly charged sphere of radius R and charge density ρ, at points both outside and inside the sphere. **(b)** Use the result of part (a) to find the electric potential energy stored in the sphere. Compare your result with the work necessary to assemble the sphere by adding successive thin spherical shells of charge.

100. Find the potential on the axis of a disk of radius a, a distance d from it, if the disk has a charge density $\sigma = \sigma_0(r/a)^2$.

101. A hollow cylindrical object of radius a and length $2L$ (■ Figure 25.56) is uniformly charged with density σ. Find the electric potential at an arbitrary point P on the z-axis.

102. Three point charges, one $+Q$ and two $-Q$, lie at the vertices of an equilateral triangle. Draw two diagrams, one showing electric field lines and the other showing equipotential surfaces in the plane of the triangle. *State* how you use symmetry, behavior of the fields at large distances, and behavior of the fields near the charges. Show qualitatively the locations of any points where $\vec{E} = 0$ and the $V = 0$ equipotential surfaces.

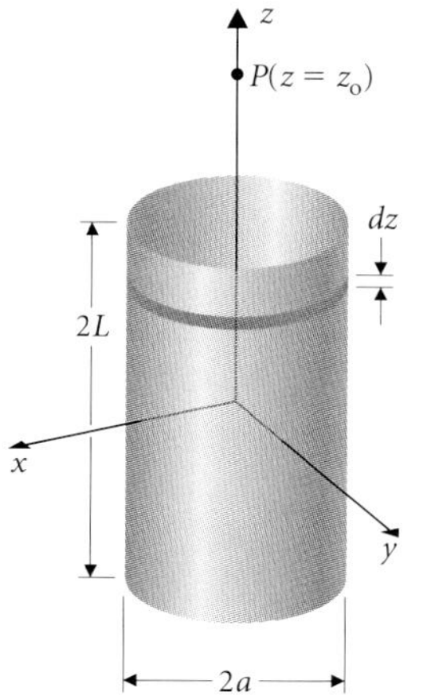

■ FIGURE 25.56

Hint for problems 103 and 104: Show in each case that the problem reduces to solving the differential equation:

$$\frac{d\ell}{\sqrt{\dfrac{1}{\ell} - \dfrac{1}{\ell_0}}} = -(\text{constant})\ dt,$$

where ℓ is a characteristic dimension of the system, ℓ_0 is its value when the system is at rest, and the constant is of the form $\sqrt{2kq^2/m}$ times a factor that depends on the specific system. Then show that the equation can be solved using the substitution $\ell/\ell_0 = \cos^2\theta$.

103. Two particles each with mass m but with equal and opposite charges q and $-q$ are released from rest when separated by distance d. Find the time required for them to collide.

104. Find the time taken for the square of charges in Study Problem 16 to collapse to zero size. Take $a = 1.0$ m, $Q = 1.0\ \mu C$, and $m = 1.0$ mg. *Hint:* When the square side shrinks by ds, how far does each charge move?

CHAPTER 26

Introduction to Electric Circuits

This person is using a simple electric circuit. The car battery supplies the power to operate the light. Wires from the battery form a closed current path. In this chapter we'll discuss some basic circuit designs and the principles that allow power transfer from electrical sources to practical devices.

Concepts

Current
Electromotive force
Resistance
Conductivity/resistivity
Ohm's law
Ohmic losses
Series and parallel circuits
Equivalent circuit
Kirchhoff's rules

Goals

Be able to:

Compute the resistance of a simple object given its resistivity.

Analyze simple circuits using series/parallel relations or Kirchhoff's rules.

Compute the power requirements for simple circuits.

A smell of burning fills the startled air—
The electrician is no longer there!

Hilaire Belloc

A person reading a road map in a car late at night is using one of the simplest and most common applications of electric current. Charge flows through wires to the roof lamp and back to the battery. Energy is supplied by the car's battery and is converted to heat and light in the lamp. This system illustrates the three basic elements in all circuits: an energy source, a closed current path, and a practical device that uses energy (■ Figure 26.1). In this chapter we shall study the principles on which these elements operate and then discuss how to apply these principles in more complex situations. For the moment we consider only circuits in which current is constant in time.

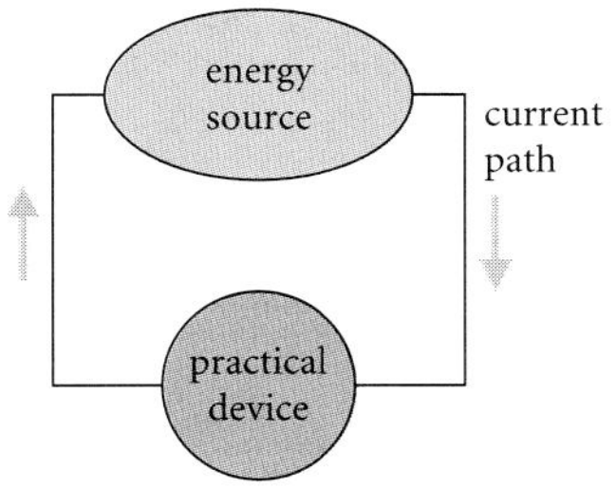

■ FIGURE 26.1
Schematic of an electric circuit. The purpose of any circuit is to transfer power from the electric source to the practical device.

26.1 BASIC CIRCUIT BEHAVIOR

26.1.1 Electric Current

Some electrons in a conductor can move in response to an applied electric field. Electrons quickly build up at the surface of the conductor and neutralize the applied field (§25.6). But if we remove electrons from the conductor as they arrive at the surface and put them back in at the other end, the result is a continuous flow of electrons—an electric current. You can make a device that does this by sticking an iron nail and a copper spike into a lemon. Chemical reactions at the spike remove electrons from the copper, while reactions at the nail push electrons into the iron. The device drives a small current through a wire connected between its iron and copper *electrodes* (■ Figure 26.2).

Benjamin Franklin envisioned electric fluid as positively charged. This historical fact has left us with the sign convention illustrated in ■ Figure 26.3. Negatively charged electrons flowing through a conducting wire toward the left constitute a current that is described as an equivalent flow of positive charge to the right. (The actual positive charges—atomic nuclei—oscillate about their equilibrium positions and do not contribute to the current.)

THIS DEFINITION REFERS TO THE MAGNITUDE OF THE CURRENT. SIGN CONVENTIONS ARE IMPORTANT IN ANALYZING CIRCUITS. SEE §26.4.

The current in a wire is the rate at which charge flows through any cross section of the wire:

$$I \equiv \pm dQ/dt. \qquad \textbf{(26.1)}$$

REMEMBER: THE AMPERE IS THE SI BASE UNIT: 1 C = 1 A·S. SEE PART VI INTRODUCTION AND CHAPTER 29.

The SI unit of current is the ampere: 1 A = 1 C/s.

26.1.2 Batteries and Electromotive Force

We can make an analogy between the flow of electrons in a wire and fluid flow. Imagine a large pile of rocks on a hillside that is porous enough for water to flow through it, over and around the rocks (■ Figure 26.4). Water starts at the top of the hill and flows down through

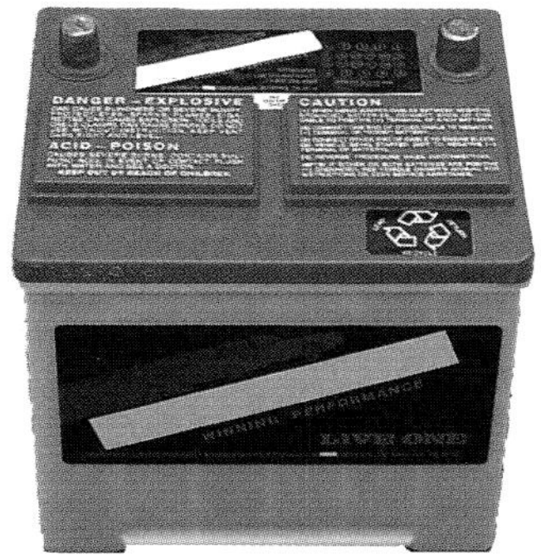

■ FIGURE 26.2
A simple battery is easily constructed from a lemon, a nail, and a piece of copper wire. The voltmeter indicates the potential difference between the terminals.

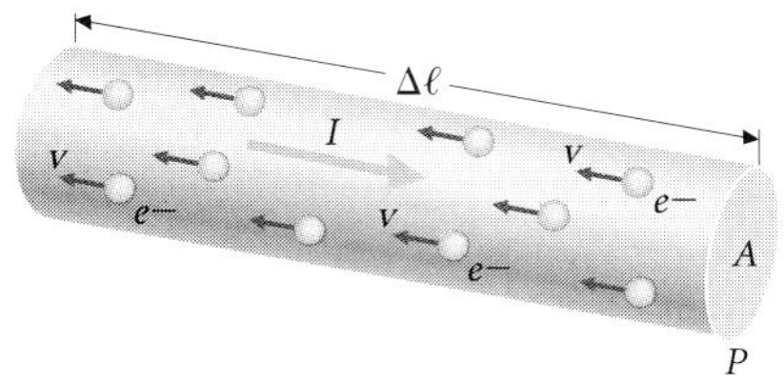

■ FIGURE 26.3
Definition of electric current. Negatively charged electrons moving to the left constitute a rightward electric current *I*. The current is the rate at which charge flows through any cross section of the wire.

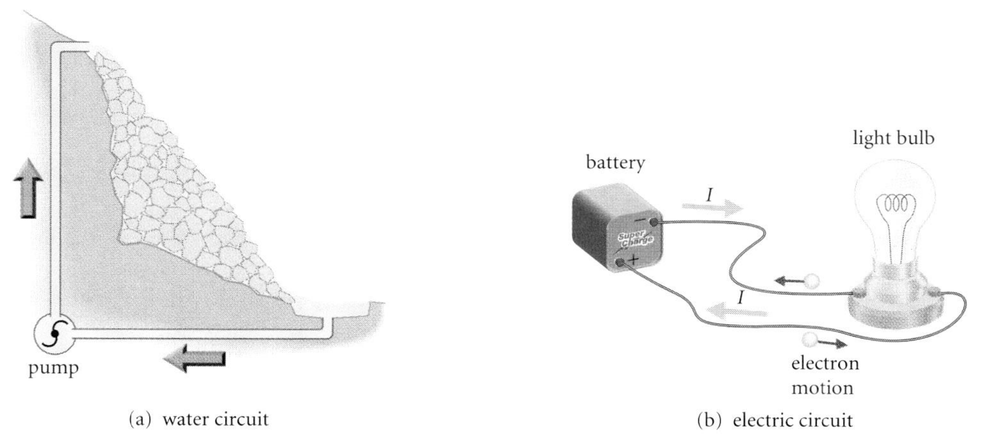

■ FIGURE 26.4
(a) A water circuit consists of a pump and a continuous stream of water that flows downhill through a pile of rocks. (b) In an electric circuit, the battery raises the electric potential of charge in the same way that the pump raises the gravitational potential of water. Here the arrow shows the direction of electric current I. The electrons move in the opposite direction, indicated by the brown arrows.

the pile. The gravitational potential energy of the water is less at the bottom than at the top, and the difference is used up by *friction* as the water flows through the pile. If the water is to circulate continuously, a pump is needed to lift it from the bottom to the top of the hill. In an electrical system an energy source such as a battery acts as a pump, and a device such as a lamp corresponds to the pile of rocks. The electrical "pump" is called a source of *electromotive force,* or emf.

Batteries use stored chemical energy to replace energy lost by the electrons moving through the conducting circuit. Chemical energy is another form of electric potential energy. Electrons arrange themselves around atomic nuclei so as to have the minimum possible potential energy. If two atoms come close together and can share their electrons so as to reduce their combined potential energy, they bind together to form a chemical compound. In a car battery (■ Figure 26.5) reactions occur at each terminal. At the positive terminal, lead oxide reacts with positive ions in the sulfuric acid solution, and at the negative terminal metallic lead reacts with negative sulfate ions. As a result of these reactions, charges of opposite sign build up on the terminals. The electric field arising from this charge separation creates a potential difference between the two terminals. A single lead-acid battery cell develops a 2-V potential difference. Six such cells are connected together in a standard 12-V battery. The potential difference developed by the battery is called its electromotive force. In spite of its name, emf is *not* a force but is the energy transferred to each unit of charge that moves between the terminals.

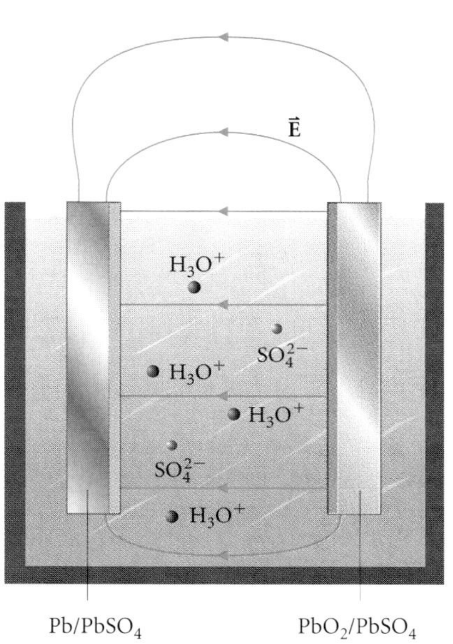

■ FIGURE 26.5
At the negative terminal of the battery, lead atoms react with sulfate ions (SO_4^{2-}) in the solution. Each reaction delivers a charge $-2e$ to the negative terminal. At the positive terminal, lead oxide PbO_2 reacts with hydronium ($H_3O_2^+$ = water plus an extra proton) and sulfate ions to produce lead sulfate and water. Each reaction delivers a charge $+2e$ to the positive terminal. The chemical reactions release approximately 2 eV for each electron removed from the positive plate and added to the negative plate and deplete the acid in the battery. The reactions reach equilibrium when the potential difference between the plates is about 2 V. When the battery is used, electrons flow from the negative plate to the positive plate through the external circuit, disturbing the chemical equilibrium. Inside the battery, more electrons flow to the negative terminal to compensate. To charge the battery, current is forced through it in the opposite direction, reversing the chemical reactions and replenishing the battery acid.

> The *electromotive force* (emf) $\mathcal{E}$ of a battery or other electric power source is the value of the potential difference it maintains between its terminals in the absence of current.

Generators, solar power cells, and the like are also sources of emf that convert energy in other forms to electric potential energy. ■ Figure 26.6 shows the symbol used to represent an emf in a circuit. The longer bar represents the positive terminal.

Electrons flowing through a battery collide with electrons and molecules in the battery fluid. Some of the electrical energy they gain is dissipated as heat, and the potential difference

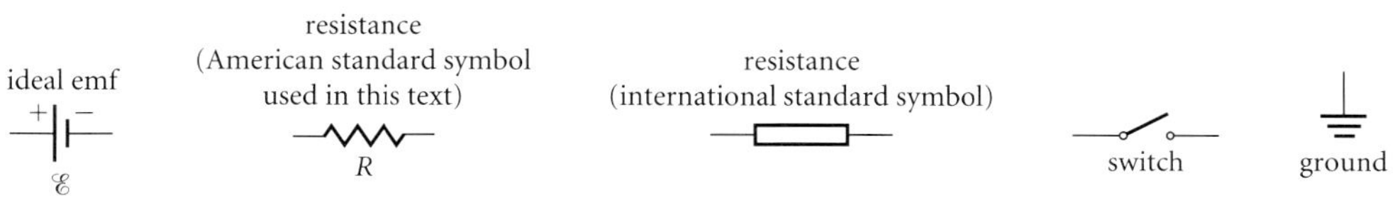

■ FIGURE 26.6
Standard circuit symbols for an ideal emf or battery, a resistance, a switch, and connection to ground. The positive battery terminal is indicated by the longer line. Batteries are sometimes indicated by a series of these symbols. In this text we use the American standard symbol for resistance. The international standard symbol is also shown here. Potential is taken to be zero at ground.

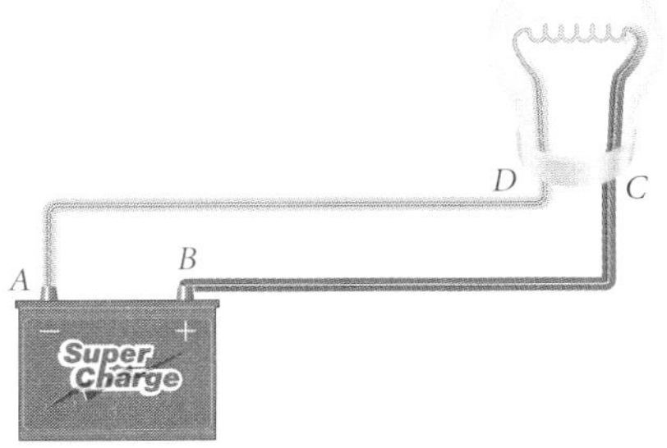

■ FIGURE 26.7
The circuit connecting a car dome light to the battery. The charge distribution in the circuit is set up immediately after the switch is closed (Problem 25.82) to produce the necessary electric fields. The field in the wires is small compared with that in the light bulb filament. The potential difference across the lamp equals that across the battery.

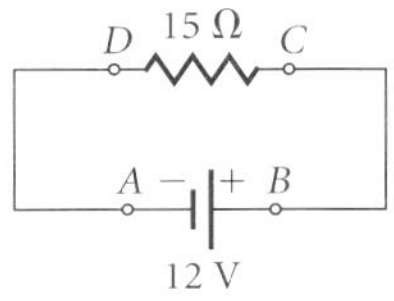

■ FIGURE 26.8
A circuit diagram shows the circuit elements and the pattern of their connections. The potential increases by 12 V across the battery ($A \to B$), and decreases by 12 V across the resistor ($C \to D$).

■ FIGURE 26.9
Georg Simon Ohm (1787–1854). Based on his experiments with batteries, Ohm derived eqn. (26.2) in 1826. He was motivated by the connection between temperature difference and heat flow (see Chapter 21). Ohm's theory was confirmed experimentally by Fechner (1801–1887) and Wheatstone (1802–1875) (see §26.5). However, Ohm did not receive official recognition for his discovery until 1841, when he was awarded the Copley medal of the Royal Society. He was finally promoted to a university professorship 22 years after his discovery, 5 years before his death.

across the battery terminals is reduced. When a battery provides current to an external circuit, the potential difference across its terminals is less than its emf. The measured potential difference is called the battery's *terminal voltage.*

26.1.3 Resistance

We may use an ideal battery with no internal resistance to illustrate how the ideas of electric potential that we developed in Chapter 25 are applied to electric circuits. Closing the switch on the car's dome light completes a conducting path from the battery through the light and back to the battery (■ Figure 26.7). The complex chemical reactions within the battery produce a simple result: a fixed potential difference across the battery terminals. An equal potential difference exists across the light bulb. Immediately after the switch is closed, a very small amount of charge flows through the battery and is distributed in the circuit to produce the necessary electric field. A negligible electric field is sufficient to maintain current through the thick connecting wires; a much larger field is required in the light bulb's thin filament. The bulb is an example of an electrical *resistance.* ■ Figure 26.8 is an abstract diagram of the circuit using the standard symbol for resistance.

To see why the potential difference across the light equals that across the battery, recall that the potential difference between any two points is independent of the path used to compute it:

$$\Delta V = V_B - V_A = V_{BA} = -\int_A^B \vec{\mathbf{E}} \cdot d\vec{\ell}.$$

If the two points A and B are the terminals of the ideal battery (Figure 26.8), the potential difference V_{BA} is the electromotive force of the battery. Along a path from A to B through the connecting wires and the light bulb:

$$V_{BA} = V_{BC} + V_{CD} + V_{DA} = (V_{BC} + V_{DA}) + V_{CD}.$$

$$V_{BA} = -\underbrace{\left(\int_C^B + \int_A^D\right) \vec{\mathbf{E}} \cdot d\vec{\ell}}_{\text{connecting wires}} - \underbrace{\int_D^C \vec{\mathbf{E}} \cdot d\vec{\ell}}_{\text{light bulb}}.$$

The first term may usually be ignored because it is much smaller than the second. The second term is the potential difference across the light bulb.

The potential difference V_{CD} and the resulting current I through the light bulb are the two quantities that describe the bulb's behavior in the circuit. Their ratio gives an operational definition of the bulb's resistance to charge flow:

> The *electrical resistance* of a circuit component is the potential difference between its terminals divided by the current that is established in response to that potential difference:
>
> $$R \equiv \Delta V / I. \qquad \textbf{(26.2)}$$

The unit of resistance is the ohm (symbol Ω), after Georg Simon Ohm (■ Figure 26.9). One ohm equals 1 volt per ampere:

$$1\ \Omega \equiv \frac{1\ \text{V}}{1\ \text{A}} = \frac{1\ \text{J/C}}{1\ \text{C/s}} = 1\ \text{J}\cdot\text{s/C}^2.$$

EXAMPLE 26.1 ♦ The potential difference between a light bulb's terminals is 12 V when it carries a current of 0.80 A. What is the resistance of the bulb?

MODEL We apply the definition of resistance, eqn. (26.2):

SETUP AND SOLVE $$R = \frac{\Delta V}{I} = \frac{12\ \text{V}}{0.80\ \text{A}} = 15\ \Omega.$$

THE SYMBOL Ω IS THE GREEK CAPITAL OMEGA (APPENDIX IIA).

ANALYZE Another light bulb carrying a current of 1.60 A with the same potential difference across it would have half the resistance, or 7.5 Ω. ∎

A useful model for a real battery is an ideal, lossless emf together with an *internal resistance.* The resistance accounts for the difference between emf and terminal voltage as well as for the energy required to move charge through the battery (■ Figure 26.10). Efficient batteries have very small internal resistance. A *dead* battery has a large internal resistance.

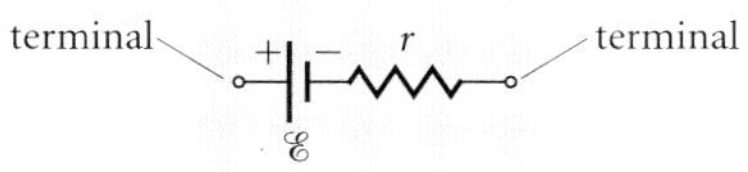

■ FIGURE 26.10
A model for a real battery. Power loss within the battery is represented by the internal resistance r connected to the ideal emf. This end-to-end connection is called a series connection (see §26.3.1).

EXAMPLE 26.2 ♦♦ A 12.0-V car battery has a terminal voltage of 11.7 V when supplying a current of 4.8 A to the automobile circuit. What is the battery's internal resistance?

MODEL The terminal voltage is the emf minus the potential drop across the internal resistance.

SETUP We use eqn. (26.2) to express the potential difference ΔV_{int} across the internal resistance r in terms of the current I:

$$\Delta V_{\text{term}} = \mathscr{E} - \Delta V_{\text{int}} = \mathscr{E} - Ir.$$

SOLVE $$r = \frac{\mathscr{E} - \Delta V_{\text{term}}}{I} = \frac{12.0\ \text{V} - 11.7\ \text{V}}{4.8\ \text{A}} = 0.06\ \Omega.$$

ANALYZE A good car battery has an internal resistance of less than 0.02 Ω. This internal resistance is much larger than normal; the battery is in bad shape. ∎

26.1.4 Energy Relations in a Simple Circuit

As electrons flow through a circuit, chemical reactions in the battery raise their potential energy. In the rest of the circuit, this potential energy is converted to thermal energy. In ordinary conductors, the electrons are accelerated by the electric field but collide frequently in the material. After a collision, an electron's velocity is in a random direction, unrelated to the direction of the applied force. Its kinetic energy is converted to energy of random motion (thermal energy), and a rise in temperature is observed. Electric current nearly always dissipates energy as heat. This is a nuisance in a computer but is the purpose of an electric toaster.

ELECTRIC CURRENT IN SUPERCONDUCTORS IS AN IMPORTANT EXCEPTION. SEE *DIGGING DEEPER.*

WE'LL DISCUSS THE COLLISIONS FURTHER IN §26.2.

The potential difference between the ends of the resistor in Figure 26.8 is $\Delta V = V_{CD} = V_C - V_D$. An amount of charge δq passing through this component loses potential energy $\delta U = V_{CD}\,\delta q$. If the charge δq passes through the resistor in time δt, the energy is dissipated as heat at a rate:

$$P_\Omega \equiv \frac{\delta U}{\delta t} = \frac{V_{CD}\,\delta q}{\delta t} = I\,\Delta V. \qquad (26.3a)$$

WE USED EQN. (26.1) FOR I. SEE §7.2 FOR THE DEFINITION OF POWER.

This power is called rate of ohmic loss or Joule heating. The former term emphasizes loss of electrical energy, and the latter emphasizes conversion to thermal energy.

Using eqn. (26.2), we may express the power in terms of resistance:

$$P_\Omega = I(IR) = I^2R, \qquad (26.3b)$$

and

$$P_\Omega = \frac{\Delta V}{R}\,\Delta V = \frac{(\Delta V)^2}{R}. \qquad (26.3c)$$

An equal amount of power is generated by the battery to maintain the current in the resistor. Charge δq passing through the battery gains potential energy $\delta U = \mathscr{E}\delta q$, so the power output of the battery is:

$$P_{\mathscr{E}} \equiv \frac{\delta U}{\delta t} = \frac{\mathscr{E}\delta q}{\delta t} = I\mathscr{E}.$$

THIS RESULT IS REQUIRED BY CONSERVATION OF ENERGY.

Since the current is the same in each case, and $\Delta V = \mathscr{E}$, $P_{\mathscr{E}} = P_{\Omega}$.

EXAMPLE 26.3 ♦ A toaster draws a current $I = 6.7$ A and has resistance $R = 18\ \Omega$. How much power does the toaster require?

MODEL Since we are given I and R, we use eqn. (26.3b) to compute the power dissipated.

SETUP AND SOLVE $P_{\Omega} = I^2R = (6.7\ \text{A})^2\ (18\ \Omega) = 0.81\ \text{kW}.$

ANALYZE Check the units:

$$\text{A}^2\cdot\Omega = \text{A}^2\ \frac{\text{V}}{\text{A}} = \frac{\text{C}}{\text{s}}\ \text{V} = \frac{\text{C}\cdot\text{J}}{\text{s}\cdot\text{C}} = \text{W}.$$

REMEMBER: EMF IS A SOURCE OF ENERGY. IT IS NOT ACTUALLY A FORCE AT ALL.

EXERCISE 26.1 ❖ A simple circuit contains an emf connected to a resistance. If the resistance is increased while the emf remains the same, does the dissipated power increase, decrease, or stay the same?

26.1.5 Safety Considerations

Despite the warning sign (■ Figure 26.11a), high voltage alone is not necessarily dangerous. The student in Figure 26.11b is at a potential of 100 000 V and is smiling! She is not in danger because she is standing on an insulating mat, and her whole body is at the same potential. It *is* very dangerous to send large electric currents through the body. The human body is a conductor, so a current is established in response to a potential *difference* maintained across different parts of the body. Currents in the human body can cause burns and can disrupt the function of muscles such as those in the heart and around the lungs. Even small currents, 5 mA or less, cause the sensation of shock; 100 mA can be fatal. Thus it is important to avoid any situation in which the human body forms part of a conducting circuit.

Old electrical equipment becomes dangerous if the insulation wears thin and *live* parts (those at high potential) can be touched. Electrical appliances in kitchens and bathrooms can be particularly dangerous because water is a reasonably good conductor, and charge flows more easily into a wet hand than a dry one. Proper ground connections greatly improve the safety of any appliance. Current takes the path of least resistance, through the ground wire rather than your body.

WE'LL DEMONSTRATE THIS IN §26.3.

■ FIGURE 26.11
High voltage itself is not dangerous. It is a large potential *difference* that can be dangerous if it is applied across part of a human body. Always exercise care when handling electrical equipment.

26.2 A Model for Current and Resistance

To demonstrate how current depends on potential difference, we use a simple model of electron flow in a conducting substance. An accurate analysis requires quantum theory, but the model we present allows us to understand important qualitative features of resistance and obtain rough quantitative estimates.

In a conductor, a large number of electrons, usually one or two per atom, can move through the material with speeds of the order of 10^6 m/s. As they travel, the electrons collide with the surrounding material. The electrons accelerate in response to an electric field, and the collisions amount to a resistive force akin to friction. A reasonable model is that the electrons move with a constant velocity, with the resistive force balancing the applied force.

According to quantum mechanics, the electron speed is determined by the properties of the material.

The atoms in a solid are arrayed in a regular pattern called a lattice. In a cold, perfectly regular lattice, electrons move almost unimpeded. But in a real substance the lattice has imperfections (missing atoms, atoms slightly out of place, even atoms of a different kind), and as temperature rises, the atoms oscillate with increasing frequency. An electron's motion is disturbed by these imperfections. For convenience we shall refer to these interactions as collisions with atoms, but you should remember that the electron actually interacts with the lattice as a whole.

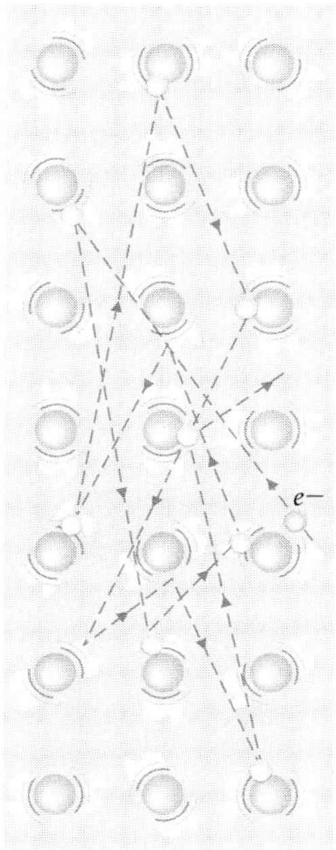

■ **Figure 26.12**
Classical model of electron motion in a solid. A rapidly moving electron suffers frequent collisions with the lattice of atoms in the solid that redirect its motion. On average, it goes nowhere.

The instantaneous velocity of an electron is redirected in each collision, so the electrons move on random paths that on average go nowhere (■ Figure 26.12). When an electric field is applied, each electron is accelerated between collisions and gains a small additional velocity, always in the direction opposite the electric field (■ Figure 26.13). An electron that collided at time t_o and obtained a velocity $\vec{v}_o$ as a result of that collision, has a velocity $\vec{v}$ at time t, where:

$$\vec{v} = \vec{v}_o + \vec{a}(t - t_o).$$

The acceleration $\vec{a}$ is $\vec{F}/m = -e\vec{E}/m$, so:

$$\vec{v} = \vec{v}_o - \frac{e}{m}\vec{E}(t - t_o).$$

The average velocity of the electrons at any time is:

$$\langle\vec{v}\rangle = \left\langle \vec{v}_o - \frac{e}{m}\vec{E}(t - t_o)\right\rangle.$$

The average velocity after a collision is random, so $\langle\vec{v}_o\rangle = 0$. Thus:

$$\langle\vec{v}\rangle = -\frac{e}{m}\vec{E}\left\langle (t - t_o)\right\rangle.$$

The average time since the last collision, $\langle(t - t_o)\rangle$, equals the average time between collisions, τ. Thus the free electrons drift with an average velocity $\vec{v}_d$ given by:

The argument proving this result is quite subtle. See, for example, *The Feynman Lectures on Physics*, v. 1, Ch. 43.

$$\vec{v}_d = -\frac{e}{m}\vec{E}\tau. \tag{26.4}$$

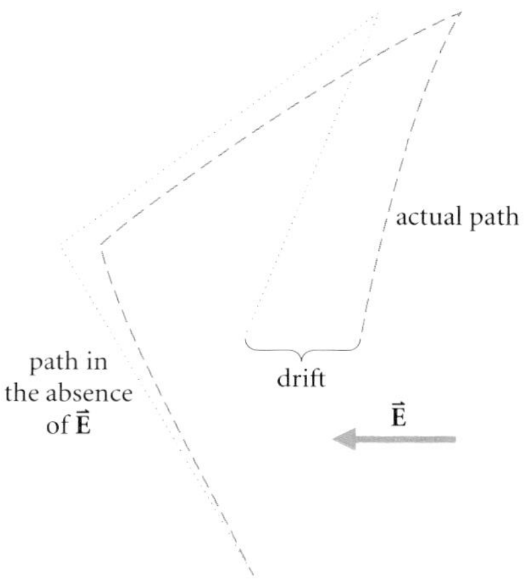

■ **Figure 26.13**
When an electric field is applied, the electron accelerates in the direction opposite $\vec{E}$ between collisions and travels on a parabolic path. The result is a net drift in the direction opposite $\vec{E}$. The size of the drift is vastly exaggerated in this drawing, to make it visible. ($v_d/v = \frac{1}{10}$ instead of the actual 10^{-10}!)

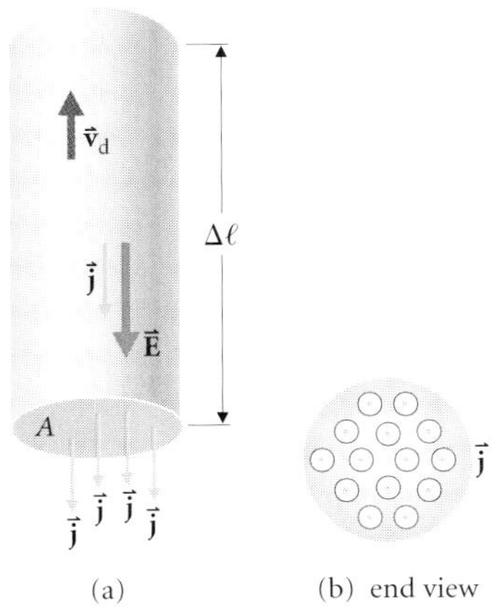

■ FIGURE 26.14
(a) Electrons, n_e per unit volume, drifting with velocity $\vec{v}_d$, produce a current density $\vec{j} = -en_e\vec{v}_d$ parallel to $\vec{E}$. (b) End view of the wire. The total current $I = jA$.

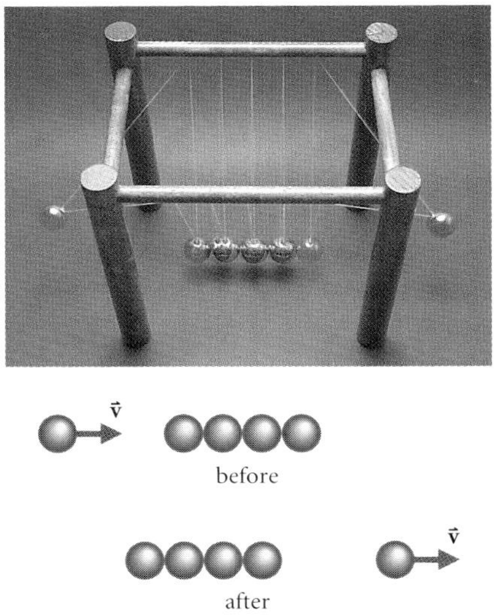

■ FIGURE 26.15
When a ball collides with a set of identical balls, initially in contact, one ball is rapidly ejected from the end of the row. The signal is transmitted by compression waves in the middle three balls, which hardly move at all. This system forms a mechanical analogue of signal transmission in electric circuits. No one electron moves very far; the signal is transmitted by the electric field.

THE CURRENT $I = \int \vec{j} \cdot \hat{n}\, dA$ IS THE FLUX OF THE CURRENT DENSITY.

REMEMBER THE SIGN CONVENTION: CURRENT IS IN THE DIRECTION OPPOSITE ELECTRON VELOCITY.

A piece of wire of length $\Delta\ell$ and cross-sectional area A with n_e free electrons per unit volume contains $N_e = n_e A\, \Delta\ell$ free electrons (■ Figure 26.14). These electrons move out of the piece of wire in a time $\Delta t = \Delta\ell / v_d$, so the resulting current is:

$$I = \frac{|dQ|}{dt} = \frac{en_e A\, \Delta\ell}{\Delta t} = n_e e A v_d. \tag{26.5}$$

Electron drift speeds are remarkably small compared with the speed of the electrons between collisions, or the speed of electrical signals along a wire. Signals are transmitted by the electric field, not by the electrons themselves (■ Figure 26.15).

EXAMPLE 26.4 ♦♦ A 12-gauge copper wire has a diameter of 2.0 mm. If the wire carries a 15 A current, find the drift velocity of the electrons in the copper. (The number of atoms per cubic meter of copper is 8.5×10^{28} and there is one free electron per atom.) If the electric field in the wire is 77 mV/m, and the electrons have a speed of 2×10^6 m/s, estimate the average distance an electron travels between collisions.

MODEL The drift velocity is related to the current in the wire through eqn. (26.5).

SETUP The cross-sectional area of the wire is πr^2, with $r = d/2 = 1.0$ mm.

SOLVE

$$v_d = \frac{I}{n_e e A} = \frac{15\ \text{A}}{(8.5 \times 10^{28}\ \text{m}^{-3})(1.6 \times 10^{-19}\ \text{C})\pi(1.0 \times 10^{-3}\ \text{m})^2} = 3.5 \times 10^{-4}\ \text{m/s}.$$

SETUP The distance traveled between collisions, λ_c, is the collision time multiplied by the speed of the electrons. Using eqn. (26.4), we find:

$$\tau = \frac{mv_d}{eE} = \frac{(9.1 \times 10^{-31}\ \text{kg})(3.5 \times 10^{-4}\ \text{m/s})}{(1.6 \times 10^{-19}\ \text{C})(77 \times 10^{-3}\ \text{V/m})} = 2.6 \times 10^{-14}\ \text{s}.$$

SOLVE $\lambda_c = v\tau = (2 \times 10^6\ \text{m/s})(2.6 \times 10^{-14}\ \text{s}) = 5 \times 10^{-8}$ m.

ANALYZE The mean free path λ_c is roughly 150 atomic spacings.
Check the units of τ:

$$\frac{1\ \text{kg}\cdot\text{m/s}}{1\ \text{C}\cdot\text{V/m}} = \frac{1\ \text{N}\cdot\text{s}}{1\ \text{C}\cdot\text{V/m}} = \frac{1\ (\text{N/C})\cdot\text{s}}{1\ \text{V/m}} = 1\ \text{s}.$$ ■

The total current depends on the wire's cross section. The quantity that relates charge flow directly to electric field and the properties of the material is the current per unit area, or *current density:*

$$\vec{j} = -n_e e \vec{v}_d. \tag{26.6}$$

Using eqn. (26.4) for $\vec{v}_d$, we have:

$$\vec{j} = \frac{n_e e^2 \tau}{m} \vec{E} = \sigma \vec{E}, \tag{26.7}$$

where the *conductivity,*

$$\sigma = \frac{n_e e^2 \tau}{m}, \tag{26.8}$$

TABLE 26.1 Resistivities of Common Materials (at 20°C)

Conductors

SUBSTANCE	ρ ($\Omega\cdot$m)	TEMPERATURE COEFFICIENT α (per °C)
Silver	1.6×10^{-8}	3.8×10^{-3}
Copper	1.7×10^{-8}	3.9×10^{-3}
Gold	2.4×10^{-8}	3.4×10^{-3}
Aluminum	2.8×10^{-8}	3.9×10^{-3}
Magnesium	4.6×10^{-8}	4×10^{-3}
Tungsten	5.6×10^{-8}	4.5×10^{-3}
Platinum	1.0×10^{-7}	3×10^{-3}
Iron	1.0×10^{-7}	5.0×10^{-3}
Steel (piano wire)	1.2×10^{-7}	4×10^{-3}
Lead	2.2×10^{-7}	3.9×10^{-3}
Constantan	4.9×10^{-7}	1×10^{-5}
Invar 36	8.2×10^{-7}	1.1×10^{-3}
Mercury	9.6×10^{-7}	8.9×10^{-4}
Nichrome	1.15×10^{-6}	4.0×10^{-4}

Semiconductors

SUBSTANCE	ρ ($\Omega\cdot$m)
Carbon (graphite)	1.375×10^{-5}
Germanium	0.46
Silicon	0.03

Insulators

SUBSTANCE	ρ ($\Omega\cdot$m)
Wood	10^{8}–10^{11}
Glass	10^{10}–10^{14}
Sulfur	2×10^{17}
Quartz	7.5×10^{17}

Data from *CRC Handbook*, 64th ed. Data on Invar courtesy of Carpenter Technology Corporation.

is a property of the conducting material. The relation is often written as:

$$\vec{\mathbf{E}} = \rho \vec{\mathbf{j}}, \tag{26.9}$$

where

$$\rho \equiv \frac{1}{\sigma} = \frac{m}{n_e e^2 \tau} \tag{26.10}$$

is the *resistivity* of the material. Its unit is the ohm-meter ($\Omega\cdot$m). Resistivity is small if there are many electrons per unit volume and they collide infrequently. Resistivities of several common materials are listed in ● Table 26.1.

DO NOT CONFUSE σ = CONDUCTIVITY WITH σ = SURFACE CHARGE DENSITY OR ρ = RESISTIVITY WITH ρ = CHARGE DENSITY.

EXERCISE 26.2 ♦ Use eqn. (26.10) to show that the units of ρ are $\Omega\cdot$m.

EXERCISE 26.3 ♦ Use the data in Example 26.4 to estimate the resistivity of copper. Compare your answer with the value listed in Table 26.1.

The resistance of a wire depends on its length ℓ, its area A, and the resistivity of the material. For a uniform wire, eqns. (26.5) and (26.9) give a relation between electric field and total current:

$$I = jA = (A/\rho)E.$$

For calculating field from potential see, for instance, Example 25.6.

Thus when I, A, and ρ are constant along a wire, E is also constant. Since $\vec{\mathbf{E}}$ is parallel to $\vec{\mathbf{j}}$, and hence to $d\vec{\ell}$, $E = \Delta V/\ell$, where ΔV is the potential difference between the ends of the wire. So:

$$I = \left(\frac{A}{\rho \ell}\right)\Delta V.$$

Comparing with the definition of resistance (eqn. 26.2), we obtain:

$$R = \frac{\rho \ell}{A}. \tag{26.11}$$

Ohm discovered that in many conductors current is directly proportional to applied voltage; that is, resistance is independent of current.

> A material obeys *Ohm's law* if its resistivity is constant, independent of the applied potential difference and the resulting current.

Materials that obey Ohm's law are called *ohmic materials.* If its temperature is held fixed, metal wire is an excellent example of an ohmic material.

In our model, resistivity depends on a number of well-defined physical quantities all of which, except the collision time τ, are constants for any given material. In most substances, τ decreases as temperature increases, and resistance is observed to increase linearly with temperature. If ρ_0 is the resistivity at temperature T_0, then:

$$\rho = \rho_0[1 + (T - T_0)\alpha]. \tag{26.12}$$

Table 26.1 also lists the temperature coefficients α for several substances.

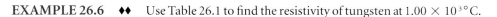

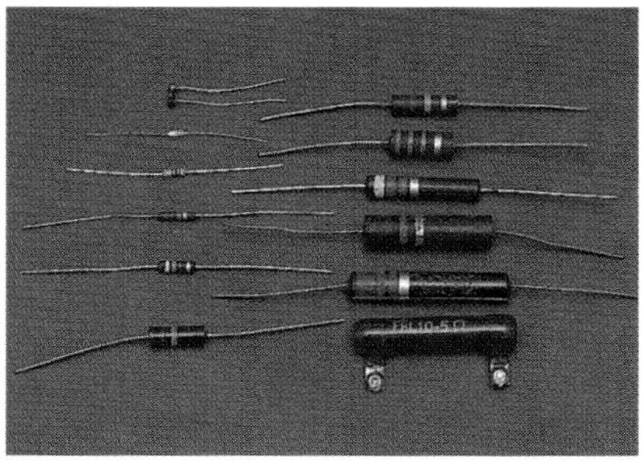

■ Figure 26.16
Commercial resistors are usually made of graphite in a clay substrate. The colored bands identify the magnitude of the resistance and the accuracy of that number as specified by the manufacturer.

EXAMPLE 26.5 ♦ An automobile dome light is connected to the positive terminal of the battery with copper wire of length 1.0 m, radius 0.50 mm, and resistivity $1.6 \times 10^{-8}\ \Omega\cdot$m. What is the resistance of the wire?

MODEL The resistance is given by eqn. (26.11).

SETUP The area of the wire is $\pi r^2 = \pi(5.0 \times 10^{-4}\ \text{m})^2$.

SOLVE From eqn. (26.11), the resistance is:

$$R = \frac{\rho \ell}{A} = \frac{(1.6 \times 10^{-8}\ \Omega\cdot\text{m})(1.0\ \text{m})}{\pi(25 \times 10^{-8}\ \text{m}^2)} = 0.020\ \Omega.$$

ANALYZE The resistance of ordinary wires is much less than that of most circuit components and is usually ignored in analyzing circuits. When a circuit requires a measured amount of resistance, a *resistor* is used. Resistors are usually made of graphite in a clay substrate and have resistances ranging from about 1 Ω to 22 MΩ (■ Figure 26.16). ▮

EXAMPLE 26.6 ♦♦ Use Table 26.1 to find the resistivity of tungsten at $1.00 \times 10^3\,^\circ$C.

MODEL We assume the resistivity increases linearly with temperature up to 1000°C.

SETUP Table 26.1 lists values of resistivity and temperature coefficient at 20°C. Using eqn. (26.12), we find the resistivity $\rho(T)$ at temperature T:

$$\rho(T) = \rho(20^\circ\text{C})[1 + (T - 20^\circ\text{C})\alpha].$$

SOLVE

$$\rho(1000^\circ\text{C}) = (5.6 \times 10^{-8}\ \Omega\cdot\text{m})[1 + (980\ \text{K})(4.5 \times 10^{-3}\ /\text{K})]$$
$$= 3.0 \times 10^{-7}\ \Omega\cdot\text{m}.$$

Remember: A temperature *difference* on the Celsius scale is expressed in kelvin.

ANALYZE Ordinary metals have small values of α and obey Ohm's law quite well over a moderate range of temperatures.

EXERCISE 26.4 ♦♦ The filament of a light bulb is to have a resistance of 16 Ω and is to be made from 10.0 cm of tungsten wire wound into a helical coil. If its operating temperature is 1000°C, what diameter wire should be used?

26.3 Series and Parallel Combinations of Resistors

Nearly all interesting electric circuits consist of more than a single energy source and a single energy-consuming device. When several elements are combined in a circuit, each influences the behavior of the others. Electric potential and the conservation of charge are the key physical ideas that allow us to understand and predict these influences. In this section, we apply these ideas to the simplest combinations of resistive elements—series and parallel connections.

26.3.1 Resistors in Series

The indicator (mounted in the panel) and the sensor (in the tank) of an automobile fuel gauge are connected in *series* (end to end) (■ Figure 26.17a). Once a steady state is established, charge does not build up anywhere in the circuit, since the changing electric fields would alter the electric current. Thus charge passes point A in the circuit at the same rate as it passes points B and C. The current is the same through both components.

This is the same argument that we used in discussing steady state flow of fluids (Chapter 13) and heat (Chapter 21).

Two or more resistors are in *series* if there is a single current path through all of them; that is, the current in each element is the same.

Potential differences are the key to analyzing the circuit. The potential difference between points A and C equals the sum of potential differences across the two resistors (Figure 26.17b):

$$\begin{aligned} V_{AC} &= V_A - V_C \\ &= (V_A - V_B) + (V_B - V_C) \\ &\equiv V_{AB} + V_{BC}. \end{aligned}$$

The potential differences are related to the current and the individual resistances (eqn. 26.2):

$$V_{AB} = IR_1$$

and

$$V_{BC} = IR_2.$$

Thus

$$V_{AC} = IR_1 + IR_2 = I(R_1 + R_2),$$

and the *equivalent resistance* for the series combination is:

$$R_s \equiv \frac{V_{AC}}{I} = R_1 + R_2.$$

Here, *equivalent* means that the actual circuit and the *equivalent circuit,* Figure 26.17c, draw the same current from the battery and have the same potential difference between points A and C. The only differences between the two circuits are within the series combination itself.

The method for finding R_s applies to any number of components connected in series.

The equivalent resistance for any number of components connected in series is the sum of their individual resistances:

$$R_s = \sum_i R_i. \tag{26.13}$$

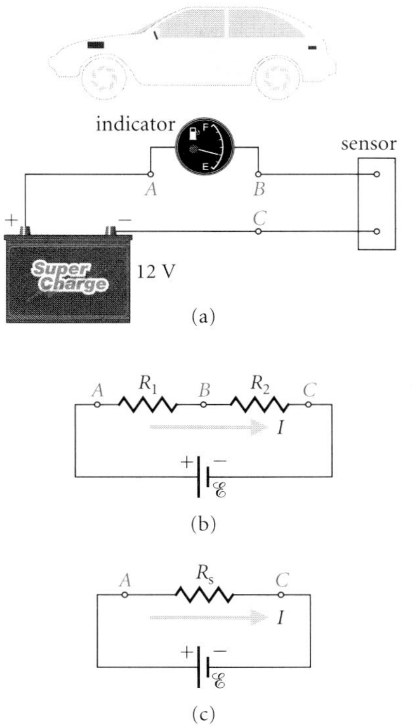

■ **Figure 26.17**
(a) An automobile fuel gauge. There are two resistances in series: the indicator in the panel and the sensor in the tank. The resistance of the tank sensor depends on the quantity of fuel in the tank. (b) Circuit diagram for the fuel gauge. (c) Equivalent circuit. The current I is the same as in the actual circuit, and the potential difference between points A and C is also the same.

Digging Deeper

Nonohmic Devices

I: The Light Bulb

The resistance of a metal wire is independent of the current flowing through it only so long as the metal's temperature is held constant. Increase of temperature increases the thermal speed of the electrons and the vibration of the atoms, decreasing the collision time and increasing the resistivity. ■ Figure 26.18 is a graph of current in a tungsten filament versus potential difference across it. When the filament's temperature is held constant, the graph is a straight line whose slope $1/R$ measures the constant resistance of the filament. When the filament is used in a light bulb, its resistance increases with temperature, and the corresponding graph curves downward. This contrast emphasizes that Ohm's law is a practical relation rather than a fundamental one. It works well for some systems in specific circumstances. The graph of voltage versus current, or *characteristic curve,* for a device provides a useful tool for describing its behavior.

II: Superconductivity

In some materials resistivity decreases abruptly to zero at very low temperatures. Electrons form pairs that avoid collisions altogether, allowing current to flow with no resistance at all. For mercury, this transition occurs at 4 K (■ Figure 26.19). Since 1986 it has been found that certain metallic oxides become superconducting at temperatures as high as 125 K = −148° C. Since this transition temperature is above the temperature of liquid nitrogen, it is relatively easy to achieve. Such materials are potentially very important for making practical superconducting devices. However, the materials tend to be brittle, so it is not easy to manufacture flexible wires from them. Understanding the mechanism by which superconductivity occurs in these materials, and learning how to build practical devices with them, remain fertile fields of research.

III: Semiconductors

■ Figure 26.19 shows the characteristic curve of a semiconductor diode, whose resistance depends strongly on the direction of current through it. In silicon, the major material of such a device, four valence electrons are shared among neighboring

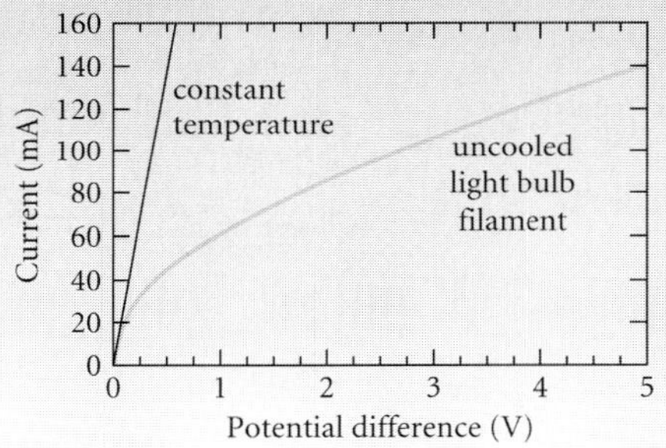

■ **Figure 26.18**
Current in a tungsten light bulb filament versus applied voltage. As the current increases, the power dissipated increases and the light bulb heats up. Its resistance increases, and so the rate at which current changes with ΔV decreases.

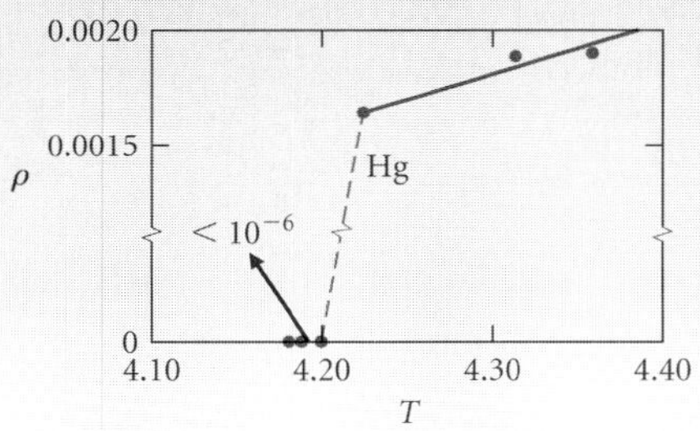

■ **Figure 26.19**
Superconducting transition for mercury, as observed by H. Kamerlingh Onnes in 1911. The transition begins at between 4.2 K and 4.3 K (about −269°C). When the transition is complete, the resistivity is truly zero. Current in such material persists forever without any electric field to drive it. Experiments have demonstrated current persistence over periods of years.

EXAMPLE 26.7 ♦ In the fuel gauge shown in Figure 26.17, the resistance of the indicator is 11 Ω. The resistance of the sensor is 75 Ω when the tank is half-full. What is the current in the circuit when the indicator reads half-full? How much power does the fuel gauge use then?

MODEL The two resistances are connected in series.

SETUP The equivalent resistance in the circuit is given by eqn. (26.13).

$$R_s = R_1 + R_2 = 11\ \Omega + 75\ \Omega = 86\ \Omega.$$

SOLVE The current in the circuit is given by eqn. (26.2):

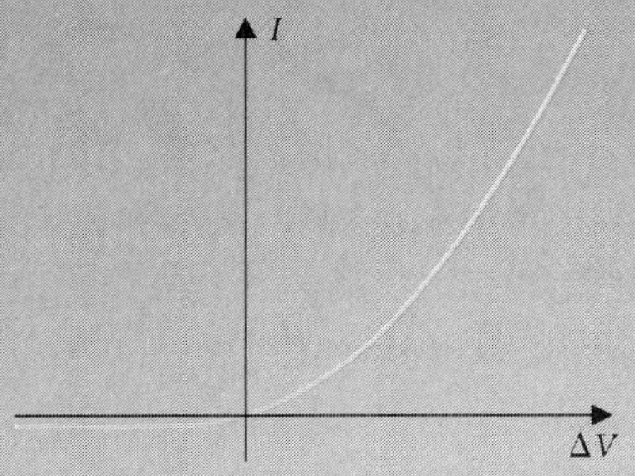

■ FIGURE 26.20
Characteristic curve for a semiconductor diode. The device permits current in only one direction and so acts like an electric "valve."

atoms but usually are not free to conduct current through the material. Silicon is a *semi*conductor because at normal temperature only a small fraction of its electrons have enough energy to become conduction electrons. Each electron that does so enhances conductivity, both by its own motion and by the curious ability of the "hole" (or lack of an electron) it leaves behind to move from atom to atom.

The conductivity of semiconductors is enhanced by *doping*. Replacement of a silicon atom with gallium, which has only three valence electrons, immediately produces a hole due to the lack of a valence electron. Thus silicon doped with gallium has a supply of positive charge carriers and so has a larger conductivity than pure silicon. It is called a *p-type* semiconductor. Conversely, doping silicon with arsenic, which has five valence electrons, gives a conduction electron for each arsenic atom, producing an *n-type* semiconductor. When p-type and n-type semiconductors are put in contact (■ Figure 26.21), the excess of free electrons on one side of the junction and holes on the other side causes a net flow of charge, until equilibrium is established by a buildup of negative charge in the p-type side and positive charge on the n-type side. In equilibrium a small number of thermal electrons are produced in the p-type material and flow toward the n-type region. These are balanced by electrons from the n-type region with enough kinetic energy to cross the potential barrier. A similar balance occurs between the much smaller flows of holes. An external electric field in the "forward-bias" direction reduces the internal potential barrier and allows electrons to flow from the n-type region. In contrast, an electric field in the "reverse-bias" direction suppresses the n-type flow. The resulting asymmetry of current can be very dramatic (Figure 26.21). Diodes are often used in circuits to ensure that current occurs in only one direction.

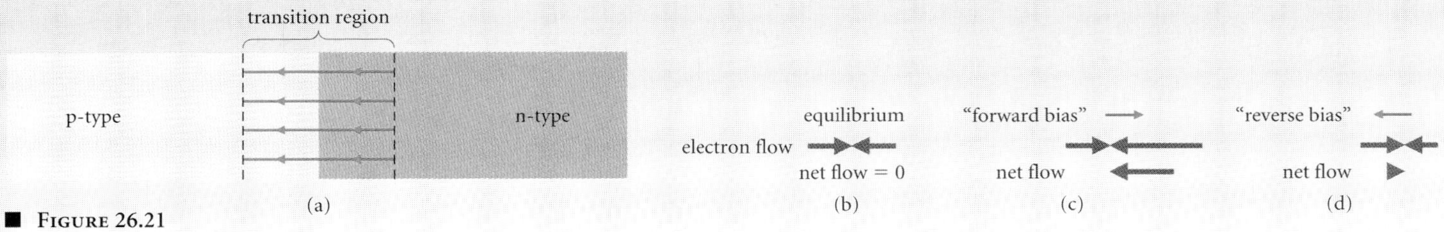

■ FIGURE 26.21
(a) Semiconductor junction diode. Silicon doped with gallium has a shortage of valence electrons in its atomic lattice—it is called p-type. Silicon doped with arsenic is n-type; it has an excess of valence electrons. Electrons migrate from the n-type material to the p-type until the electric field produced by the resulting charge separation stops the migration. (b) A small number of electrons in the p-type region escape from silicon atoms and flow toward the right. Meanwhile a few electrons in the n-type region have sufficient kinetic energy to cross the transition region. In equilibrium, the potential difference between the two regions is just right to keep these flows equal. (c) An applied field in the *forward bias* direction greatly increases the number of electrons able to cross the transition region. (d) Reverse bias decreases the number of electrons able to cross the transition region. The current is almost zero.

$$I = \frac{\Delta V}{R_s} = \frac{12\ \text{V}}{86\ \Omega} = 0.14\ \text{A}.$$

The power used is (eqn. 26.3c):

$$P = \frac{(\Delta V)^2}{R_s} = \frac{(12\ \text{V})^2}{86\ \Omega} = 1.7\ \text{W}.$$

ANALYZE The additional resistance of the indicator actually reduces the power used by the sensor. Connecting additional resistance in series reduces the current in the circuit and hence reduces the power I^2R used by each resistance. ∎

26.3.2 Resistors in Parallel

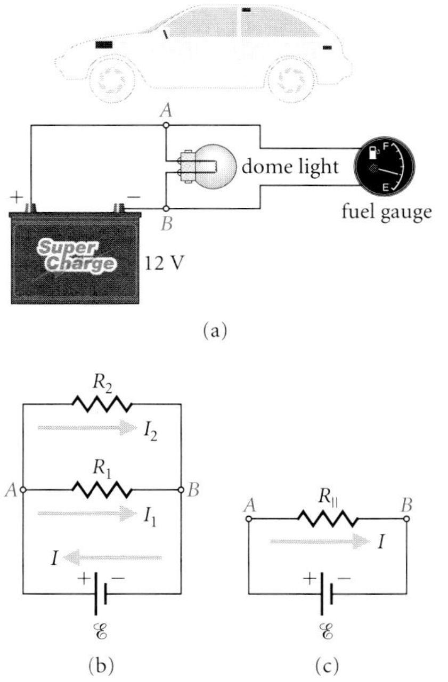

■ **Figure 26.22**
(a) A car dome light and fuel gauge are connected in parallel across the car battery. This configuration allows each to be turned on independently of the other and provides the full battery emf across each device.
(b) Circuit diagram for the two devices connected to the battery. The current divides at point A and recombines at point B. (c) Equivalent circuit.

If two components are connected to a single power source, but you want the option of turning on the power to one at a time, a series connection is not appropriate. In addition, neither component operates at full power when they are connected in series. The appliances in a car are actually connected so that each has the full emf of the battery across its terminals; they are connected in *parallel.* As you stop to read a map at night, the dome light and the fuel gauge are connected in parallel across the battery (■ Figure 26.22). The current I from the battery divides at point A into a current I_1 through the dome light and a current I_2 through the fuel gauge. Since charge is neither created nor destroyed, in a steady state the current arriving at A from the battery equals the total current leaving A and passing through the appliances:

$$I = I_1 + I_2.$$

Two or more resistors are in *parallel* when they are all connected between the same two points in a circuit, and the potential difference across each is the same.

The currents in parallel resistors derive from a common source and recombine into a common outgoing current.

The equivalent resistance for a parallel combination is defined by the potential difference and the total current:

$$R_\| \equiv \frac{V_{AB}}{I}.$$

The individual currents are: $I_1 = \dfrac{V_{AB}}{R_1}$ and $I_2 = \dfrac{V_{AB}}{R_2}$.

To use the relation between the currents, we invert the expression for $R_\|$:

$$\frac{1}{R_\|} = \frac{I}{V_{AB}} = \frac{I_1 + I_2}{V_{AB}} = \frac{1}{R_1} + \frac{1}{R_2}.$$

A similar result holds for any number of resistors in parallel.

For a set of parallel resistors, the reciprocal of their equivalent resistance equals the sum of reciprocals of their individual resistances:

$$\frac{1}{R_\|} = \sum_i \frac{1}{R_i}. \qquad (26.14)$$

EXAMPLE 26.8 ♦ Find the equivalent resistance of an automobile fuel gauge ($R = 86\ \Omega$) and dome light ($R = 18\ \Omega$) connected in parallel across a 12-V battery (Figure 26.22). Determine the current in each device and the total current through the battery. What is the ratio of the currents in the two devices?

MODEL The devices are connected in parallel.

SETUP From eqn. (26.14):

$$\frac{1}{R_\|} = \frac{1}{R_1} + \frac{1}{R_2} = \frac{R_2 + R_1}{R_1 R_2}.$$

$$R_\| = \frac{R_1 R_2}{R_1 + R_2} = \frac{(86.0\ \Omega)(18.0\ \Omega)}{86.0\ \Omega + 18.0\ \Omega} = 14.9\ \Omega.$$

SOLVE The currents are:

$$I_1 = \frac{V_{AB}}{R_1} = \frac{12\ \text{V}}{86\ \Omega} = 0.14\ \text{A}.$$

$$I_2 = \frac{V_{AB}}{R_2} = \frac{12\ \text{V}}{18\ \Omega} = 0.67\ \text{A}.$$

The total current is: $$I = \frac{V_{AB}}{R_{\|}} = \frac{12\ \text{V}}{14.9\ \Omega} = 0.81\ \text{A} = I_1 + I_2.$$

Since $V_{AB} = I_1R_1 = I_2R_2$: $$\frac{0.14\ \text{A}}{0.67\ \text{A}} = \frac{I_1}{I_2} = \frac{R_2}{R_1} = \frac{18\ \Omega}{86\ \Omega} = 0.21.$$

ANALYZE The currents are inversely proportional to the resistances; the dome light draws five times as much current as the fuel gauge.

EXERCISE 26.5 ♦♦ Find the power used by each of the devices in Example 26.8.

EXERCISE 26.6 ♦♦ The resistance of the two devices in parallel is 15 Ω, less than the resistance of either alone. Show that for any parallel combination of resistors the equivalent resistance is less than any of the individual resistances.

26.3.3 *Combined Series and Parallel Circuits*

Because a battery has internal resistance, a parallel connection of two components to a car battery is actually a series-parallel connection. The circuit diagram is shown in ■ Figure 26.23a. Such circuits can be simplified by replacing each parallel or series combination of resistors with its equivalent resistance.

EXAMPLE 26.9 ♦♦ A 12.0-V auto battery has an internal resistance of $r = 0.02\ \Omega$. The 0.200-Ω starter motor and a 1.20-Ω headlight are connected in parallel across it. What is the current in the battery? What is the total power required from the emf? At what rate is energy dissipated as heat within the battery itself?

MODEL Two resistors R_1 and R_2 in parallel can be replaced with their equivalent resistance, as shown in Figure 26.23b. The equivalent resistance forms a series combination with the internal resistance of the battery.

SETUP $$\frac{1}{R_{\|}} = \frac{1}{R_1} + \frac{1}{R_2}$$
$$= \frac{1}{0.200\ \Omega} + \frac{1}{1.20\ \Omega} = (5.00 + 0.833)\ \Omega^{-1} = 5.83\ \Omega^{-1}.$$

So: $$R_{\|} = 0.171\ \Omega.$$

The equivalent resistance $R_{\|}$ is in series with the internal resistance r. Together they are equivalent to a single resistor $R_s = r + R_{\|} = 0.02\ \Omega + 0.171\ \Omega = 0.19\ \Omega$ (Figure 26.23c).

SOLVE The total current in circuit (c) is:

$$I = \frac{\mathscr{E}}{R_s} = \frac{12.0\ \text{V}}{0.19\ \Omega} = 63\ \text{A}.$$

This is the current through the battery in the actual circuit. The battery supplies power:

$$P_{\mathscr{E}} = \mathscr{E}I = (12.0\ \text{V})(63\ \text{A}) = 760\ \text{W}.$$

The current is the same through the internal resistance of the battery, resulting in an ohmic loss:

$$P_r = I^2r = (63\ \text{A})^2(0.02\ \Omega) = 80\ \text{W},$$

which is dissipated as heat in the battery.

ANALYZE The remaining 680 W is delivered to the starter and the headlight. Since the power dissipated in each resistor is $(\Delta V)^2/R$, the amounts of power delivered to each component are inversely proportional to their resistances. The starter uses six times as much as the headlight, or $\frac{6}{7} = 86\%$ of the total.

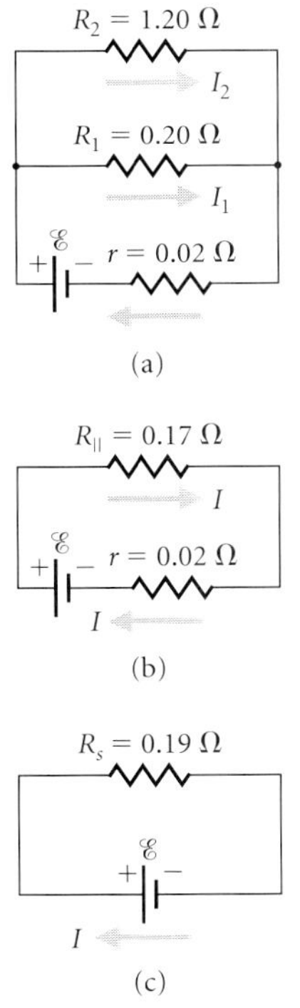

■ FIGURE 26.23
(a) Circuit diagram for a real battery, with internal resistance, connected to a starter motor and a headlight. (b) The two parallel resistances are reduced to one equivalent resistance. (c) The two series resistances combine to form this equivalent circuit.

EXERCISE 26.7 ♦ Find the equivalent resistance of the three resistors in the circuit shown in ■ Figure 26.24.

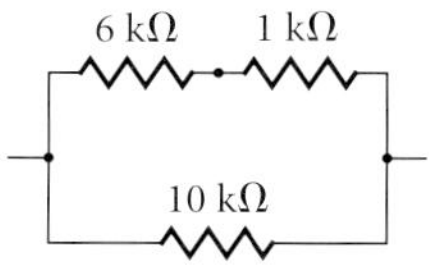

■ FIGURE 26.24

Circuits containing any number of resistors with series and parallel connections may be analyzed by this technique of sequentially replacing parts of the circuit with equivalent single resistors to obtain an equivalent circuit that contains a single resistor and a single emf. Often, complex resistor combinations may be reduced to series/parallel circuits using symmetry arguments.

EXAMPLE 26.10 ♦♦♦ Six equal 15-kΩ resistors are connected to form a tetrahedron (■ Figure 26.25a). A 6.0-V battery is connected to points A and B. How much power is drawn from the battery?

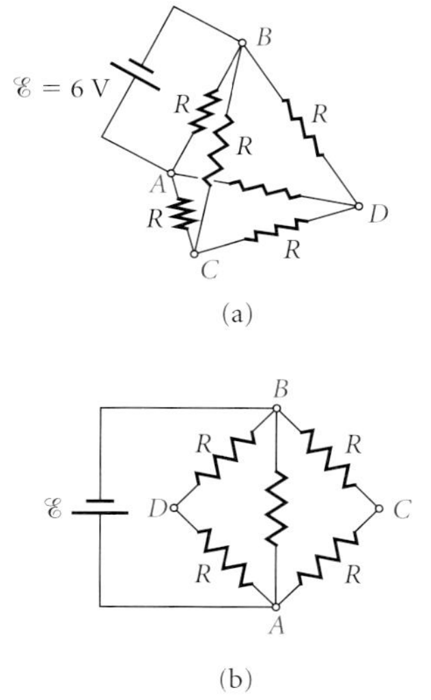

■ FIGURE 26.25
(a) This tetrahedron is not a simple series/parallel connection but may be reduced to one once we realize that there is no current between C and D. (b) The equivalent series/parallel circuit diagram.

MODEL To find the power, we need the equivalent resistance of the object between points A and B. At point A, current divides into three currents along paths AB, ACB, and ADB. The system of resistors is symmetric. If we imagine standing the tetrahedron on edge AB, we have mirror symmetry about AB. Thus the currents along paths ACB and ADB are the same, and so are the potential differences $V_{AC} = V_{AD}$. Thus C and D are at the same potential. With no potential difference between C and D, there is no current along CD, and eliminating it leaves an equivalent circuit. The circuit simplifies to the series/parallel combination in Figure 26.25b.

SETUP Each of the series combinations R_{ACB} and R_{ADB} equals $R + R = 2R$. The parallel combination between A and B is then:

$$\frac{1}{R_\parallel} = \frac{1}{R_{AB}} + \frac{1}{R_{ACB}} + \frac{1}{R_{ADB}} = \frac{1}{R} + 2\left(\frac{1}{2R}\right) = \frac{2}{R}.$$

Thus:
$$R_\parallel = R/2.$$

SOLVE The power used by the tetrahedron is:

$$P = \frac{\mathscr{E}^2}{R_\parallel} = \frac{2\mathscr{E}^2}{R} = \frac{2(6.0\ \text{V})^2}{15 \times 10^3\ \Omega} = 4.8\ \text{mW}.$$

ANALYZE Current is zero along CD provided that the resistances are *exactly* equal. For this to be true, the connecting wires should be exactly the same length on each side, too, to make the small resistances of the wires themselves equal along AC, AD, CB, and DB. ■

26.4 KIRCHHOFF'S RULES

Circuits containing only series and parallel combinations of resistors may be analyzed with a sequence of equivalent circuits using the methods of §26.3. However, these rules are not sufficient for a circuit with two batteries in parallel connected to a load (■ Figure 26.26). The internal resistances of the two batteries are *not* a parallel combination, because the emfs can make the potentials differ at points B and H, with the result that $V_{BC} \neq V_{HD}$. In this section, we shall outline a procedure that works for any circuit. The two basic principles are those we have been using all along: charge conservation and path independence of potential difference (or equivalently, energy conservation). Their formal statements are called Kirchhoff's rules, after Gustav Kirchhoff (1824–1887). We begin with a few definitions and then state the rules.

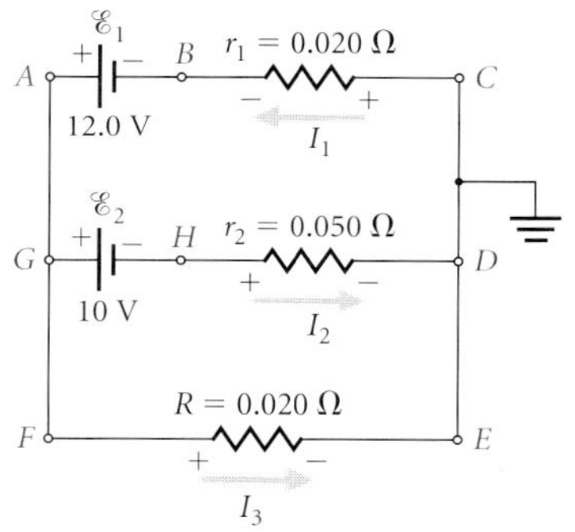

■ FIGURE 26.26
This circuit, used to jump-start a car, is not a series/parallel connection. The two different emfs make the potentials at B and H different. The negative terminal of the battery is connected to the car chassis, which serves as *ground* ($V = 0$), indicated by the multiple bar symbol. (*Remember:* Resistances r_1 and r_2 are *inside* the batteries.)

An *arm* of a circuit is a portion in which there is a single current. Examples are the arms $GABCD$, GHD, and $GFED$ in Figure 26.26.

A *junction* is a point where three or more arms join. Examples are points G and D in the figure.

A *loop* is a path through the circuit that returns to its starting point. Examples are paths $ABCDHGA$ and $ABCDEFGA$ in the figure.

The fact that charge does not continuously build up at any junction requires

Kirchhoff's junction rule: The sum of currents entering a junction equals the sum of currents leaving the junction.

$$\sum I_{in} = \sum I_{out}.$$

THE JUNCTION RULE IS A STATEMENT OF *CHARGE CONSERVATION.*

The fact that the potential has a definite value at any point in the circuit requires

Kirchhoff's loop rule: The algebraic sum of the potential changes around any loop in a circuit is zero.

$$\sum_{\substack{\text{closed}\\ \text{loop}}} \Delta V = 0.$$

THE LOOP RULE IS A STATEMENT OF *ENERGY CONSERVATION.*

In the following examples and study problem we'll apply these rules, using the plan shown in ■ Figure 26.27.

EXAMPLE 26.11 ♦ In a flashlight two 1.5-V batteries are connected in series with a light bulb ($R = 1.0\ \Omega$). How much power does the light bulb draw from the batteries?

MODEL We can use Kirchhoff's loop rule to determine the potential difference across the light bulb. ■ Figure 26.28 is the circuit diagram.

SETUP The circuit has only one loop, and we traverse it clockwise:

$$\Delta V = -\mathscr{E}_1 - \mathscr{E}_2 + \Delta V_{bulb} = 0 \Rightarrow \Delta V_{bulb} = \mathscr{E}_1 + \mathscr{E}_2 = 3.0\ \text{V}.$$

SOLVE The power used by the bulb is (eqn. 26.3c):

$$P = (\Delta V)^2/R = (3.0\ \text{V})^2/(1.0\ \Omega) = 9.0\ \text{W}.$$

ANALYZE When batteries are connected in series, their emfs add. ■

EXAMPLE 26.12 ♦♦ In an automobile starter circuit, a battery is connected to a starter with resistance $R = 0.20\ \Omega$. When your battery is low, your friend Jane can help out by attaching jumper cables from her battery to yours. If Jane's battery has an emf of 12.0 V and internal resistance of $0.020\ \Omega$, and your battery has an internal resistance of $0.050\ \Omega$ and emf of 10.0 V, how much current is delivered to the starter?

MODEL Figure 26.26 is the circuit diagram. The current variables for the three arms are I_1, I_2, and I_3; the arrows define their positive directions. Here we can guess the actual directions of the currents and define the directions so as to get positive values for the current variables. Notice that current I_2 through your battery (#2) is in the direction opposite that expected in normal operation. Jane's battery is not only powering the starter but is charging your battery too.

SETUP We apply the junction rule at G:

$$I_1 = I_2 + I_3. \tag{i}$$

There are three currents and they all appear in this one equation, so we do not need to look at any more junctions. (The junction rule at D generates the same equation.)

We may apply the loop rule to any loop and count potential changes in either direction around the loop. There are three possible choices of loop: *ABCDHGA*, *ABCDEFGA*, or *GHDEFG*. Any two include all the arms of the circuit. We choose to use loops *ABCDHGA* and *GHDEFG*, both traversed clockwise as indicated by the sequence of letters. Potential drops across a resistor to drive current through it. The + and − signs drawn on the resistors are a convenient shorthand for keeping track of the signs in loop equations. The plus sign is always where the current arrow "enters" the resistor.

Loop *ABCDHGA* $\quad \Delta V = V_{AB} + V_{BC} + V_{DH} + V_{HG}.$

$$0 = -\mathscr{E}_1 + I_1 r_1 + I_2 r_2 + \mathscr{E}_2. \tag{ii}$$

Solution plan for circuit problems using Kirchhoff's rules

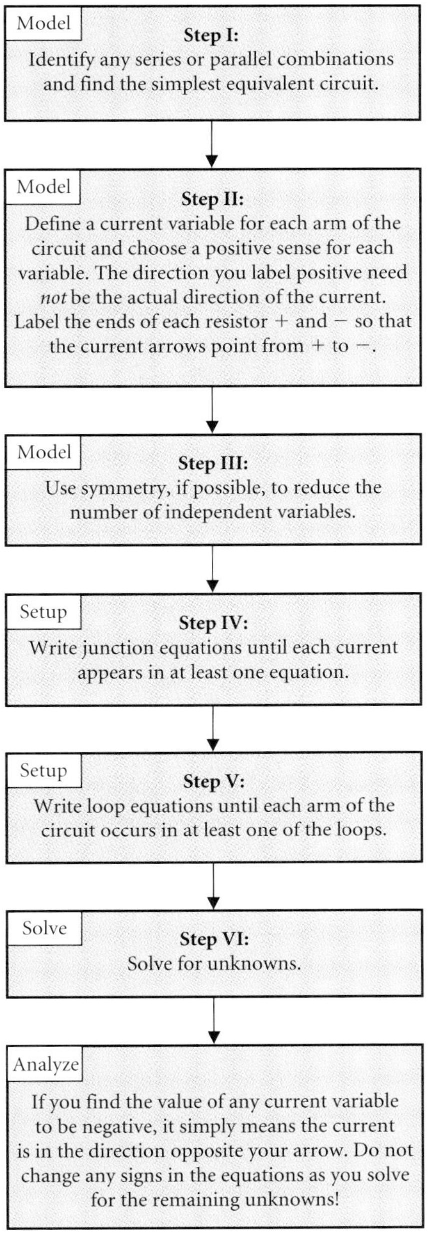

Model
Step I:
Identify any series or parallel combinations and find the simplest equivalent circuit.
Model
Step II:
Define a current variable for each arm of the circuit and choose a positive sense for each variable. The direction you label positive need *not* be the actual direction of the current. Label the ends of each resistor + and − so that the current arrows point from + to −.
Model
Step III:
Use symmetry, if possible, to reduce the number of independent variables.
Setup
Step IV:
Write junction equations until each current appears in at least one equation.
Setup
Step V:
Write loop equations until each arm of the circuit occurs in at least one of the loops.
Solve
Step VI:
Solve for unknowns.
Analyze
If you find the value of any current variable to be negative, it simply means the current is in the direction opposite your arrow. Do not change any signs in the equations as you solve for the remaining unknowns!

■ FIGURE 26.27

IT IS NOT NECESSARY TO GUESS THE ACTUAL DIRECTIONS. WE SHALL EXPLORE THIS POINT FURTHER IN THE STUDY PROBLEM.

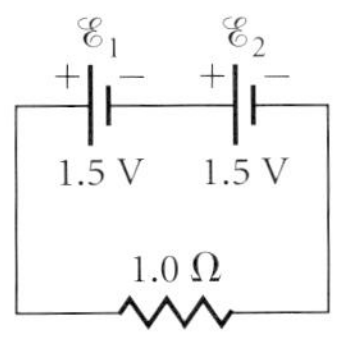

$\mathscr{E}_1$
$\mathscr{E}_2$
+ −
+ −
1.5 V
1.5 V
1.0 Ω

■ FIGURE 26.28
The circuit diagram for a flashlight with two batteries.

If we traversed the loops counterclockwise, each term would change sign, but the resulting equations would be the same.

Loop *GHDEFG* $\Delta V = V_{GH} + V_{HD} + V_{EF}$.

$$0 = -\mathscr{E}_2 - I_2 r_2 + I_3 R. \qquad \text{(iii)}$$

We now have three equations in three unknowns (I_1, I_2, and I_3). ■ Figures 26.29a and b are graphs of potential versus position around these two loops.

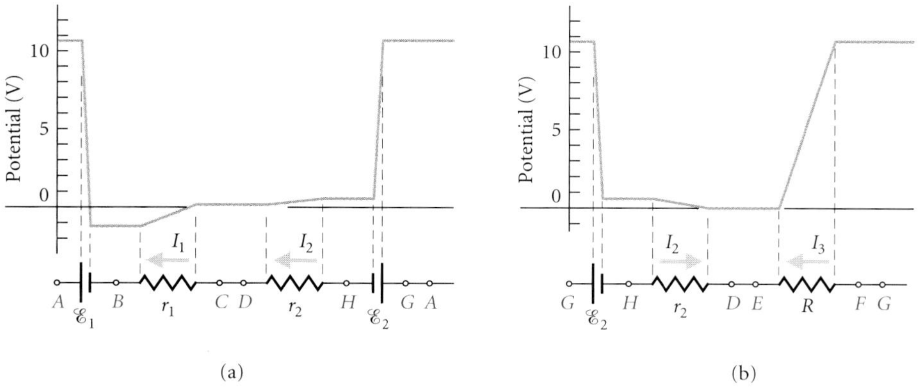

■ **Figure 26.29**

Potential changes around the loops in the circuit diagramed in Figure 26.26. We chose the zero level at ground, points *C*, *D*, and *E*. The potential is almost constant in the connecting wires and increases or decreases across batteries and resistors according to the polarity and direction of current. You could sketch a curve like this before solving the problem. The numerical values shown here are taken from the solution. (a) Loop *ABCDHGA*. (b) Loop *GHDEFG*.

SOLVE Using eqns. (i) and (ii), we eliminate I_1:

$$-\mathscr{E}_1 + r_1(I_2 + I_3) + I_2 r_2 + \mathscr{E}_2 = 0,$$

or:

$$I_2 = \frac{\mathscr{E}_1 - \mathscr{E}_2 - r_1 I_3}{r_1 + r_2}.$$

Substituting this result into eqn. (iii) gives:

$$-\mathscr{E}_2 - \frac{(\mathscr{E}_1 - \mathscr{E}_2 - r_1 I_3)}{(r_1 + r_2)} r_2 + I_3 R = 0,$$

from which we obtain:

$$I_3 = \frac{(\mathscr{E}_1 - \mathscr{E}_2) r_2 + \mathscr{E}_2 (r_1 + r_2)}{R(r_1 + r_2) + r_1 r_2} = \frac{r_2 \mathscr{E}_1 + r_1 \mathscr{E}_2}{R(r_1 + r_2) + r_1 r_2}$$

$$= \frac{(0.050\ \Omega)(12.0\ \text{V}) + (0.020\ \Omega)(10.0\ \text{V})}{(0.20\ \Omega)(0.020\ \Omega + 0.050\ \Omega) + (0.020\ \Omega)(0.050\ \Omega)}$$

$$= \frac{0.80\ \text{V}}{0.015\ \Omega} = 53\ \text{A}.$$

ANALYZE The results for the other currents are $I_2 = [2.0\ \text{V} - (0.020\ \Omega)(53\ \text{A})]/(0.070) = 13$ A, and $I_1 = I_2 + I_3 = 66$ A. All the values are positive, indicating that the currents are in the directions we expected. Note the nice symmetry in the expression for I_3. ■

EXERCISE 26.8 ♦♦ Compare the current delivered to the starter by the parallel batteries in Example 26.12 with that delivered by the low battery alone.

Study Problem 17 ♦♦ *A Compound Circuit*

Find the current in each arm of the circuit diagramed in ■ Figure 26.30a. What total power is required of $\mathscr{E}_2$ if $\mathscr{E}_1 = 3\mathscr{E}_2 = 30.0$ V, and $R = 10.0\ \Omega$?

Modeling the System

First, we notice that the two resistors between points A and B are in parallel, so the circuit reduces to the equivalent circuit shown in Figure 26.30b. The equivalent circuit has five arms, and we define five current variables. Here it is not easy to guess the actual current directions, so some of our current variables may have negative values. We still label the + and − ends of the resistors *as if* all the current values were positive. These conventions ensure that we will obtain correct equations from Kirchhoff's rules regardless of the choice of directions for the current variables.

FOR A DIFFERENT CHOICE, SEE EXERCISE 26.10.

Setup of Solution

Applying the junction rule, we find:

$$\text{At } A \qquad I_1 + I_3 = I_2. \tag{i}$$

$$\text{At } B \qquad I_4 = I_3 + I_5. \tag{ii}$$

All five current variables appear in these two equations, so an equation for junction C is unnecessary.

Applying the loop rule, we find:

AS USUAL, THE SEQUENCE OF LETTERS INDICATES THE DIRECTION AROUND THE LOOP, CLOCKWISE HERE.

$$\text{Loop } ACDA \qquad \Delta V = V_{AC} + V_{DA} = -I_2R + \mathscr{E}_1 = 0. \tag{iii}$$

$$\text{Loop } ABCA \qquad \Delta V = V_{AB} + V_{BC} + V_{CA}.$$

$$0 = I_3\frac{R}{2} - \mathscr{E}_2 + I_2R. \tag{iv}$$

$$\text{Loop } BEFCB \qquad \Delta V = V_{EF} + V_{CB} = -I_5R + \mathscr{E}_2 = 0. \tag{v}$$

We do not need any more loop equations because each arm of the circuit has appeared in at least one of the loops. We have five equations for the five unknowns.

Solution

Equations (iii) and (v) give I_2 and I_5 immediately in terms of the known emfs:

$$I_2 = \frac{\mathscr{E}_1}{R} \quad \text{and} \quad I_5 = \frac{\mathscr{E}_2}{R}.$$

Substituting for I_2 in eqn. (iv) gives I_3:

$$I_3\frac{R}{2} - \mathscr{E}_2 + \mathscr{E}_1 = 0 \Rightarrow I_3 = \frac{2(\mathscr{E}_2 - \mathscr{E}_1)}{R}.$$

Then from eqn. (ii):

$$I_4 = I_3 + I_5 = \frac{3\mathscr{E}_2 - 2\mathscr{E}_1}{R},$$

and from eqn. (i):

$$I_1 = I_2 - I_3 = \frac{3\mathscr{E}_1 - 2\mathscr{E}_2}{R}.$$

Substituting the known values into the equations gives:

$$I_1 = \frac{3(30.0\ \text{V}) - 2(10.0\ \text{V})}{10.0\ \Omega} = 7.00\ \text{A}, \qquad I_2 = \frac{30.0\ \text{V}}{10.0\ \Omega} = 3.00\ \text{A},$$

$$I_3 = \frac{2(10.0\ \text{V} - 30.0\ \text{V})}{10.0\ \Omega} = -4.00\ \text{A},$$

$$I_4 = \frac{3(10.0\ \text{V}) - 2(30.0\ \text{V})}{10.0\ \Omega} = -3.00\ \text{A}, \quad \text{and} \quad I_5 = \frac{10.0\ \text{V}}{10.0\ \Omega} = 1.00\ \text{A}.$$

The power output from battery #2 is:

$$P_2 = \mathscr{E}_2I_4 = (10.0\ \text{V})(-3.00\ \text{A}) = -30.0\ \text{W}.$$

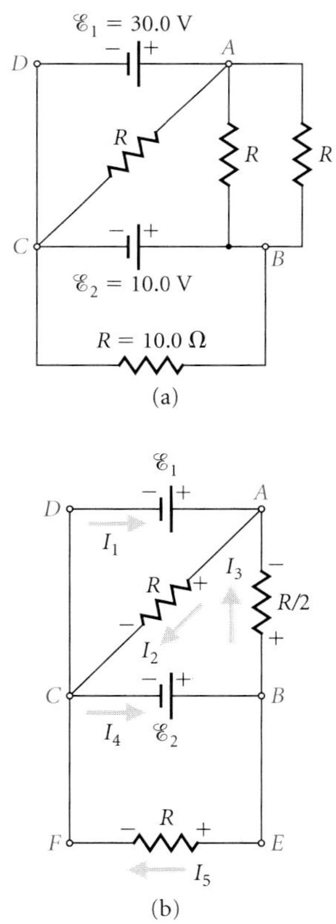

■ FIGURE 26.30
(a) Circuit diagram for a compound circuit. (b) The two parallel resistors may be replaced with a single equivalent resistance to form this circuit. We define the current variables as shown. *Remember:* The arrows are not necessarily in the actual directions of the currents.

BATTERIES ARE NOT CHARGED REALLY. THEY ARE ENERGIZED. ORDINARY SPEECH CAN BE MISLEADING HERE.

MANY COMPUTER MATH PACKAGES OFFER ROUTINES FOR SOLVING SUCH SETS OF EQUATIONS. MATRIX METHODS MAY ALSO BE USED. SEE PROBLEM 85.

Analysis

As a quick check on the arithmetic, note that the answers do indeed satisfy eqns. (i) and (ii). Two of the variables, I_3 and I_4, have negative values; those currents are in the direction opposite the arrows in the diagram. The negative value of P_2 means that battery #2 is absorbing energy from the circuit: it is being charged.

Do not be discouraged by the number of equations that result from applying Kirchhoff's rules. They are linear equations and always succumb to a systematic process of elimination.

EXERCISE 26.9 ♦♦ The junction equation for point C in Figure 26.30 would give redundant information. Show that it can be derived from eqns. (i) and (ii). Similarly, derive the loop equation for loop $ABCDA$ from eqns. (iii) and (iv).

EXERCISE 26.10 ♦♦ Do the problem again, choosing the current variable in arm AB of Figure 26.30 pointing from A to B, and the current variable in arm CB pointing from B to C. Show that the physical results are the same.

* 26.5 ELECTRICAL MEASUREMENT

The art of electrical measurement relies on clever circuit design. Precision measurements now use sophisticated solid-state circuitry with designs beyond the scope of an introductory text. However, many of these designs rely on principles we have studied, and we can get a sense of electronic design by studying a few simple systems.

MAGNETIC FORCE ON A CURRENT-CARRYING WIRE IS DISCUSSED IN §29.2.

In a *galvanometer* (■ Figure 26.31) magnetic force on current in a wire coil results in a deflection of the meter needle proportional to the current. Carefully constructed meters of this type are sensitive to currents as small as 10^{-8} A. Through careful design, such current meters may be used to measure potential difference or resistance.

Modern solid-state devices can be constructed to measure potential differences of a few volts with great sensitivity. These devices are the basis of practical multimeters for use in analyzing electric circuits. In this section we'll explore several aspects of their design.

Whenever a meter is used to measure voltage or current in a circuit, the circuit that is being measured is the one containing the meter, not the original circuit. Thus an important aspect of meter design is that it disturb the circuit as little as possible.

26.5.1 An Ammeter

To measure the current in an arm of a circuit, an ammeter is inserted into the arm, in series with the circuit components (■ Figure 26.32a). The resistance of the ammeter increases the total resistance in that arm and thus decreases the current. Thus an ammeter should have as small a resistance as possible. To make an ammeter, we may connect a small shunt resistance r_s in parallel with a galvanometer (Figure 26.32b). Most of the current in the circuit passes through the shunt, while a small amount passes through the galvanometer. This design has the additional benefit of protecting the galvanometer, since any other than a very small current would damage it.

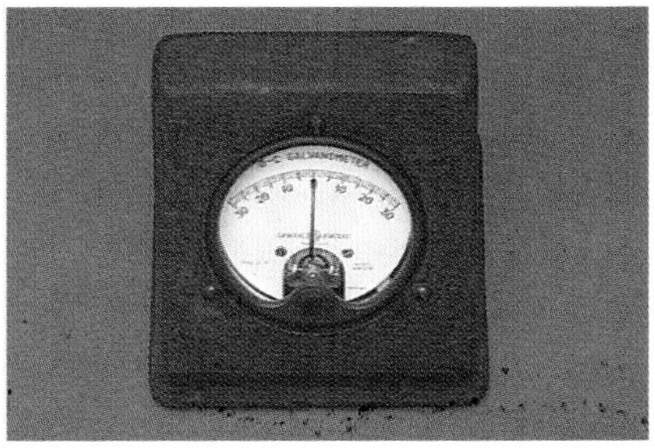

■ FIGURE 26.31
A galvanometer is an accurate instrument for measuring current. Magnetic force on current in a wire coil results in a deflection of the meter needle that is proportional to current. Since even moderate currents can damage the mechanism, meters using a galvanometer are designed so that only a small current passes through the galvanometer itself.

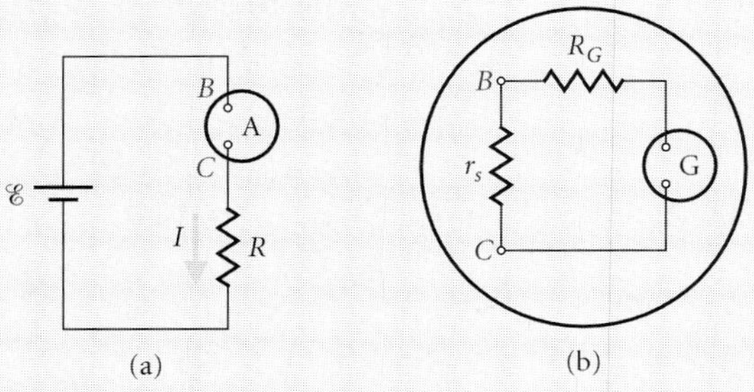

■ FIGURE 26.32
(a) To measure current through a resistor, connect an ammeter in series with it. (b) In an ammeter, the galvanometer (resistance R_G) is in parallel with a shunt resistance $r_s \ll R_G$ that carries most of the current. The small shunt resistance disturbs the current in the circuit only slightly.

EXAMPLE 26.13 ♦♦ In the circuit whose diagram is shown in Figure 26.32, $\mathscr{E} = 110$ V, $R = 1.0$ kΩ, $R_G = 35$ Ω, and $r_s = 1.0$ Ω. Find the current in the galvanometer.

MODEL Since the resistance of the ammeter is much less than the 1.0-kΩ resistance in the circuit, we may ignore any change of current caused by the ammeter.

SETUP The resistance of the ammeter is R_A, where:

$$R_A = \left(\frac{1}{R_G} + \frac{1}{r_s}\right)^{-1} = \left(\frac{1}{35\ \Omega} + \frac{1}{1.0\ \Omega}\right)^{-1} = 0.97\ \Omega.$$

The current through the ammeter is $I \approx \mathscr{E}/R = (110\ \text{V})/(1.0\ \text{k}\Omega) = 0.11$ A, and the potential difference across the ammeter is $V_{BC} = IR_A = 0.11$ V.

THE AMMETER IS CALIBRATED TO READ THE TOTAL CURRENT THROUGH IT, NOT THE CURRENT IN THE GALVANOMETER ITSELF.

SOLVE The current through the galvanometer is then:

$$I_G = \frac{V_{BC}}{R_G} = \frac{0.11\ \text{V}}{35\ \Omega} = 3.1\ \text{mA}.$$

ANALYZE Connecting the ammeter into the circuit increases the total resistance (and decreases the current through R) by less than 0.1%. ■

26.5.2 *Voltmeters*

To measure the potential difference between two points in a circuit, a voltmeter is connected between the points, in parallel with the circuit components (■ Figure 26.33). To disturb the circuit as little as possible, the meter should have a large resistance. One voltmeter design (■ Figure 26.34) has a sensitive galvanometer and a large resistance R_m in series with it.

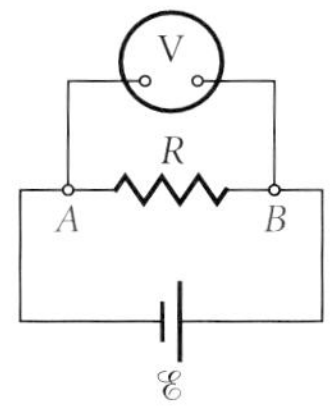

■ FIGURE 26.33
A voltmeter is connected in parallel with a circuit component to measure potential difference across it.

EXAMPLE 26.14 ♦♦ A voltmeter has a resistance $R_m = 1.0 \times 10^5$ Ω, and the galvanometer can measure current with an uncertainty of $\delta I = \pm 5.0 \times 10^{-7}$ A. What is the minimum potential difference the meter can measure to 10% accuracy?

MODEL If the uncertainty in the measurement is to be 10%, the current in the galvanometer should be at least $I_G = 10\,\delta I = 5.0 \times 10^{-6}$ A, so that $\delta I/I \lesssim \frac{1}{10}$.

SETUP AND SOLVE The required potential difference is

$$\Delta V = R_m I_G = (1.0 \times 10^5\ \Omega)(5.0 \times 10^{-6}\ \text{A}) = 0.50\ \text{V}.$$

ANALYZE We ignored the galvanometer's resistance, which is much less than 10^5 Ω. ■

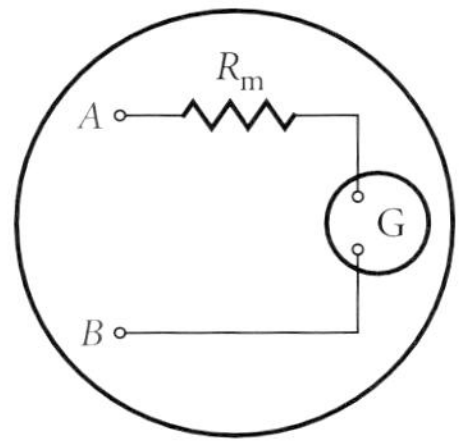

■ FIGURE 26.34
In one design of voltmeter, a galvanometer is in series with a large resistance. The meter draws only a small amount of current from the original circuit.

Modern solid-state voltmeters use semiconductor devices that have extremely large resistance ($\geq 10^{12}$ Ω). However, they are limited to potential differences of at most a volt or two. To reduce the potential difference across the device, it is connected in parallel with a set of resistors in series that are used as a *voltage divider* (■ Figure 26.35). A typical voltmeter

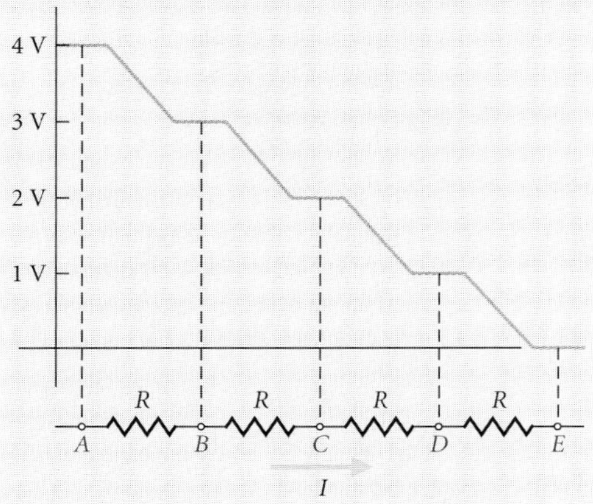

■ FIGURE 26.35
Resistors in series act as a voltage divider. The sum of potential differences across each resistor equals the total potential difference across the entire series. Thus the resistors *divide* the total potential difference. With the four equal resistors shown here, $V_{AB} = \frac{1}{2}V_{AC} = \frac{1}{3}V_{AD} = \frac{1}{4}V_{AE}$.

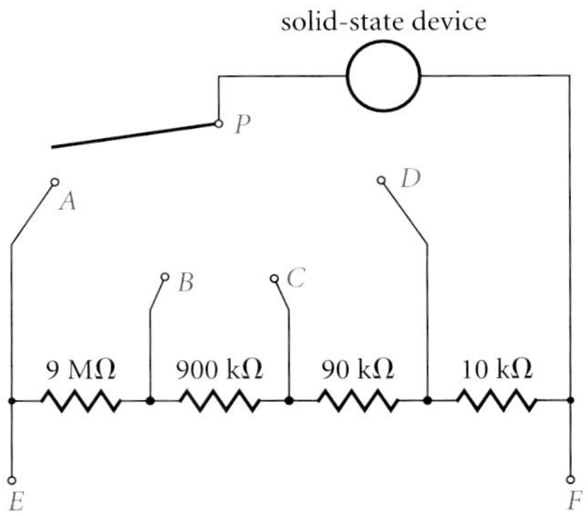

■ FIGURE 26.36

A modern voltmeter uses a solid-state device to measure the potential difference. The series of resistors allows the voltmeter to operate over four different ranges, depending on the connection of P to one of the four points A, B, C, or D. The solid-state device itself has an extremely high resistance. For all practical purposes we may assume it draws no current at all. The 10-MΩ resistance of this meter is typical.

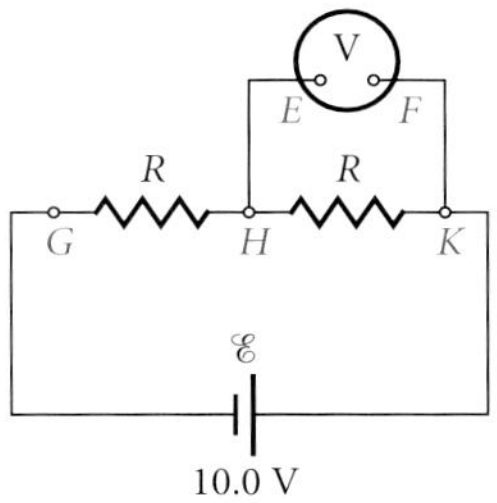

■ FIGURE 26.37

The voltmeter in Figure 26.36 is connected across one of two identical resistors R in a circuit.

<small>SUCH ACCURACY IS NEEDED WHEN MEASURING EXTREMELY LOW TEMPERATURES, FOR EXAMPLE. (SEE ESSAY 6.)</small>

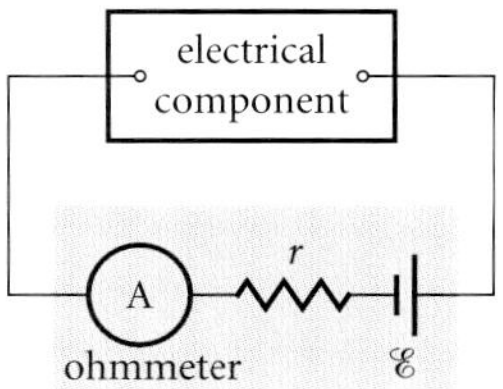

■ FIGURE 26.38

An ohmmeter contains an emf and an ammeter. The component to be measured is disconnected from its circuit and then connected across the ohmmeter.

(■ Figure 26.36) has a 10 MΩ resistance, divided into 9 MΩ, 900 kΩ, 90 kΩ, and 10 kΩ. A switch moves connection P to points A, B, C, or D depending on the magnitude of the voltage to be measured. Connected at D, the solid-state device measures the potential across the 10-kΩ resistor, which is $\frac{1}{1000}$ of the total potential drop across the series combination. Similarly, when connected at C, the device measures the potential drop across 100 kΩ of resistance, or $\frac{1}{100}$ of the total. Connected at B, the device measures $\frac{1}{10}$ of the total potential difference.

EXAMPLE 26.15 ◆◆ A voltmeter is used to measure the potential difference across one of two identical resistors connected in series with a 10.0-V emf (■ Figure 26.37). If the solid-state device is limited to 0.2 V, which connection, A, B, C, or D (Figure 26.36), should be used?

MODEL The voltage across the solid-state device cannot exceed 0.2 V, so we use the voltage divider to reduce the measured voltage to less than 0.2 V.

SETUP Since we are measuring the potential difference across one of two identical resistors in series, the result should be half of the battery emf, or 5 V.

SOLVE We need to reduce the measured voltage by at least a factor of 5/0.2, or 25. Ten won't do, so we choose to reduce by a factor of 100 and use the connection at point C. ($R_{CF} = 100$ k$\Omega = \frac{1}{100} R_{EF}$, so $V_{CF} = \frac{1}{100} V_{EF}$.)

ANALYZE Always err on the side of caution. You don't want to blow out the meter. ■

26.5.3   The Wheatstone Bridge

A simple ohmmeter for measuring resistance is a battery in series with an ammeter and a resistor r (■ Figure 26.38). A component to be measured is disconnected from any circuit, and the ohmmeter is placed across it. Then the equivalent resistance R_{eq} is the battery emf divided by the ammeter current, and the component's resistance is $R_{eq} - r$. An ohmmeter is very convenient to use, but the accuracy of its measurements is limited by the internal resistance of the battery and the accuracy of the galvanometer. When extremely accurate measurements are needed, a Wheatstone bridge may be used (■ Figure 26.39, after Charles Wheatstone, 1802–1875).

The Wheatstone bridge uses four resistances. Resistor R_1 and the ratio R_4/R_3 have known values, and the ratio may be varied. R_x is the resistance to be determined. In use, the bridge is balanced by adjusting the known ratio R_4/R_3 until the current through the galvanometer is zero. Thus the potential difference between points A and B is zero.

The potential at A is related to the potentials at C and D and to the potential differences across the resistors:

$$V_A \equiv V_C - V_{CA} \qquad \text{or} \qquad V_A \equiv V_D + V_{AD}.$$

Similarly:

$$V_B \equiv V_C - V_{CB} \qquad \text{or} \qquad V_B \equiv V_D + V_{BD}.$$

Since $V_A = V_B$, we may subtract these equations to find:

$$V_{CA} = V_{CB} \qquad \text{and} \qquad V_{AD} = V_{BD}.$$

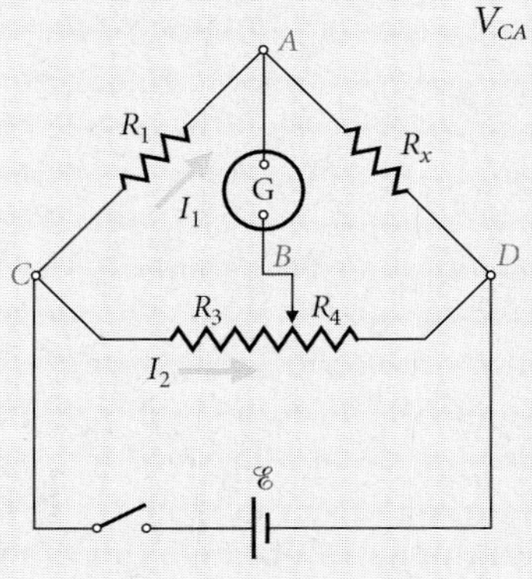

■ FIGURE 26.39

A Wheatstone bridge uses a *null measurement* to determine the unknown resistance R_x. One of the other resistors, say R_1, is precisely known. The ratio of the other two is variable. Here we show a single coil that is divided by the tap B. The ratio of the resistances R_3 and R_4 equals the ratio of the corresponding lengths of coil. This device is called a potentiometer. With no current in the galvanometer, the potential at A equals the potential at B. When the bridge is balanced, the galvanometer reading does not change when the switch is opened and closed. Such null measurements are the basis of the most accurate instruments.

Since there is no current between A and B, both R_1 and R_x have the same current I_1, while R_3 and R_4 have the same current I_2. So:

$$I_1R_1 = I_2R_3 \quad \text{and} \quad I_1R_x = I_2R_4.$$

Dividing these equations, we obtain:
$$\frac{R_x}{R_1} = \frac{R_4}{R_3}. \tag{26.15}$$

The unknown resistance is determined from the values of R_1, R_3, and R_4.

Since the measurement uses a zero reading from the galvanometer, most uncertainties are eliminated. You can easily find the galvanometer reading for zero current by disconnecting it from the circuit. Since no current or potential difference is measured, battery stability is unimportant. Resistance R_1 should be manufactured with great precision, but only the ratio R_4/R_3 needs to be accurately known. If both are made of the same material, temperature variations do not affect the ratio, for example. Similar use of *null* results is essential for many precision measurement techniques.

Chapter Summary

Where Are We Now?

We have developed methods for analyzing direct current circuits. These methods are the starting point for understanding and designing electrical devices.

What Did We Do?

In an ideal battery, chemical reactions at each terminal maintain a constant potential difference between the terminals. This potential difference is the electromotive force of the battery. Charge flowing through a circuit gains energy from a source of emf and delivers that energy to practical devices. The resistance of a circuit component is the ratio of the potential difference between its ends to the current through it: $R = \Delta V/I$. Metal wire at constant temperature is a material whose resistance does not vary with the current. Such a material follows Ohm's law: voltage is proportional to current. Ohm's law is widely used to describe circuits, but a large variety of interesting and useful devices show deviations from this simple behavior.

The power output of a battery equals the current in the battery multiplied by its emf. Similarly, the power dissipated by a resistor equals the current multiplied by the potential difference across the resistor:

$$P_{\mathcal{E}} = I\mathcal{E} \quad \text{and} \quad P_{\Omega} = I\,\Delta V = I^2R = (\Delta V)^2/R.$$

Two or more resistors in *series* are connected end to end and have the same current in each of them. Replacing resistors in series with an equivalent resistance, equal to the sum of the individual resistances, produces an equivalent circuit, one that behaves in the same way as the original circuit:

$$R_s = \sum_i R_i.$$

Two or more resistors in *parallel* are connected across the same two points in a circuit and have the same potential difference across each of them. Current in parallel resistors divides and then recombines after passing through the resistors. The reciprocals of the individual parallel resistances add to give the reciprocal of the equivalent resistance:

$$\frac{1}{R_{\parallel}} = \sum_i \frac{1}{R_i}.$$

The equivalent resistance $R_{\|}$ is less than each individual resistance R_i. Combined series and parallel networks of resistors may be analyzed by using these rules to construct a sequence of simplified equivalent circuits.

A network of batteries and resistances may be analyzed by a systematic procedure based on Kirchhoff's rules. Kirchhoff's junction rule—the sum of currents entering a junction equals the sum of currents leaving—embodies conservation of charge in circuits. Kirchhoff's loop rule—the algebraic sum of potential changes around any loop in a circuit equals zero—embodies conservation of energy. Each point in the circuit has a definite value of potential.

Electrical measurement involves careful design of measuring devices. The Wheatstone bridge illustrates the use of null techniques to achieve precision in measurement. Ammeters and voltmeters necessarily disturb the circuit they measure. Simple models of such meters allow estimation of and correction for such disturbance.

Practical Applications

The circuit theory developed in this chapter applies directly to power supply systems in aircraft and the space shuttle as well as in cars, and to most battery-powered devices, such as flashlights. These are direct current (DC) circuits. The current in your house oscillates in time; nevertheless, at any instant the currents and potential differences have the same relations they would have in a direct current system. DC circuit analysis with a proper averaging over time gives an adequate description of your household circuitry. (We'll investigate this topic further in Chapter 32.) Because the electric signals can travel across the chip much faster than the input signal changes, direct current techniques also apply to micron-sized solid-state circuit elements used in microwave technology, where signals oscillate at frequencies of about 10^{11} Hz!

Solutions to Exercises

26.1. From eqn. (26.3c), $P = (\Delta V)^2/R$, so if R is increased with ΔV held constant, P decreases.

26.2. Replacing each algebraic quantity on the right-hand side of eqn. (26.10) with its appropriate unit gives us the units of ρ:

$$\frac{\text{kg}\cdot\text{m}^3}{\text{C}^2\cdot\text{s}} = \frac{(\text{kg}\cdot\text{m}^2/\text{s}^2)\cdot\text{m}}{\text{C}\cdot(\text{C/s})} = \frac{\text{J}\cdot\text{m/C}}{\text{A}} = \frac{\text{V}\cdot\text{m}}{\text{A}} = \Omega\cdot\text{m}.$$

26.3. From eqn. (26.10):

$$\rho = \frac{1}{\sigma} = \frac{m}{ne^2\tau} = \frac{9.11 \times 10^{-31}\ \text{kg}}{(8.5 \times 10^{28}\ \text{m}^{-3})(1.6 \times 10^{-19}\ \text{C})^2(2.6 \times 10^{-14}\ \text{s})} = 1.6 \times 10^{-8}\ \Omega\cdot\text{m}.$$

Table 26.1 gives $\rho = 1.7 \times 10^{-8}\ \Omega\cdot$m for copper at 20°C.

26.4. From eqn. (26.11):

$$A = \frac{\pi d^2}{4} = \frac{\rho \ell}{R}$$

Inserting the value of ρ from Example 26.6, we obtain:

$$d = \sqrt{\frac{4\rho\ell}{\pi R}} = \sqrt{\frac{4(3.0 \times 10^{-7}\ \Omega\cdot\text{m})(0.10\ \text{m})}{\pi(16\ \Omega)}} = 4.9 \times 10^{-5}\ \text{m}.$$

26.5. The potential difference across each device is 12 V. Using eqn. (26.3c), we find the power used by each:

$$P_1 = \frac{\mathscr{E}^2}{R_1} = \frac{(12\ \text{V})^2}{86\ \Omega} = 1.7\ \text{W}$$

and

$$P_2 = \frac{(12\ \text{V})^2}{18\ \Omega} = 8.0\ \text{W}.$$

26.6. Suppose a set of resistors are connected in parallel across an emf $\mathscr{E}$. Adding another resistor doesn't change the current through the resistors already in place but does increase the total current by the amount that now exists in the new resistor. The equivalent resistance, $R_{\|} \equiv \mathscr{E}/I_{\text{total}}$, is reduced by the increase in current.

We may also give a mathematical proof:

$$R_{\|} = \frac{1}{\sum_i (1/R_i)}.$$

For any individual resistor, labeled j,

$$\frac{R_{\|}}{R_j} = \frac{1/R_j}{\sum_i (1/R_i)} < 1.$$

26.7. The two resistors in series are equivalent to a 7-kΩ resistor. This combination is in parallel with the 10-kΩ resistor. The equivalent of the parallel combination is:

$$R_{\|} = \frac{(7\ \text{k}\Omega)(10\ \text{k}\Omega)}{7\ \text{k}\Omega + 10\ \text{k}\Omega} = 4\ \text{k}\Omega.$$

26.8. The current delivered by the single battery is:

$$I' = \frac{\mathcal{E}_2}{R + r_2} = \frac{10.0\ \text{V}}{0.25\ \Omega} = 40\ \text{A} \qquad \text{(2 significant figures)}.$$

This current is only 75% of the current delivered when Jane helps.

26.9. The junction equation for point C is:

$$I_2 + I_5 = I_1 + I_4.$$

This is the sum of eqns. (i) and (ii). The loop equation for loop $ABCDA$ is:

$$\Delta V = +I_3R/2 - \mathcal{E}_2 + \mathcal{E}_1 = 0.$$

This is the sum of eqns. (iii) and (iv).

26.10. ■ Figure 26.40 shows the circuit with the revised sign conventions. The resulting changes in the Kirchhoff's equations are underlined:

At A $\qquad I_1 = \underline{I_2 + I_3'}.$

At B $\qquad \underline{I_3'} = \underline{I_4'} + I_5.$

Loop $ACDA$ $\qquad \Delta V = -I_2R + \mathcal{E}_1 = 0.$

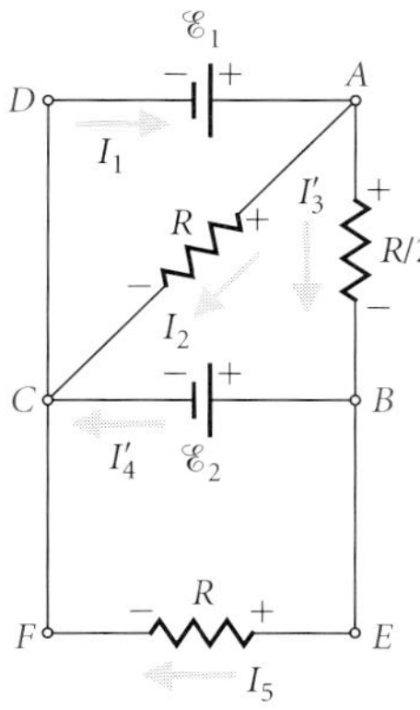

■ FIGURE 26.40

Loop $ABCA$ $\quad \Delta V = \underline{-I_3'R/2} - \mathcal{E}_2 + I_2R = 0.$

Loop $BEFCB$ $\Delta V = -I_5R + \mathcal{E}_2 = 0.$

Following the same sequence of steps as in the study problem, we obtain:

$$I_3' = \frac{2(\mathcal{E}_1 - \mathcal{E}_2)}{R} = +4.00\ \text{A},$$

and

$$I_4' = \frac{2\mathcal{E}_1 - 3\mathcal{E}_2}{R} = +3.00\ \text{A}.$$

Both sets of results describe the same currents. The signs of the results for the current variables are reversed ($I_3 = -4.00$ A, $I_3' = +4.00$ A), but so are the meanings of "positive" and "negative." This is why you don't have to guess actual directions of currents before defining your current variables.

Basic Skills

Review Questions

§26.1 BASIC CIRCUIT BEHAVIOR

- Define the *electric current* in a wire.
- How is current related to electron motion in the wire?
- What is the *electromotive force* of a battery? How does it differ from *terminal voltage?*
- Define the *electrical resistance* of a circuit component.
- Why is the resistance of connecting wires in a circuit usually ignored?
- Give three expressions for the rate of energy dissipation in a resistor.
- What situations should you avoid in order to use electrical equipment safely?

§26.2 A MODEL FOR CURRENT AND RESISTANCE

- What is the electron drift velocity in a conductor? How is it related to the average time between collisions?
- Define *current density.* How is it related to electric field?
- What is *resistivity?* Which properties of a material determine its resistivity?
- Which properties of a wire determine its resistance?
- State *Ohm's law.*
- How does resistance depend on temperature?
- What is the order-of-magnitude value of resistivity for a conductor? A semiconductor? An insulator?

§26.3 SERIES AND PARALLEL COMBINATIONS OF RESISTORS

- What does it mean for resistors to be connected in *series?*
- What is an *equivalent circuit?*
- What is the equivalent resistance of a set of resistors connected in series? Is it greater than or less than the value of each individual resistance?
- What does it mean for resistors to be connected in *parallel?*
- What is the equivalent resistance of a set of resistors connected in parallel? Is it greater than or less than the value of each individual resistance?

§26.4 KIRCHHOFF'S RULES

- Define an *arm*, a *junction*, and a *loop* in a circuit.
- State Kirchhoff's two rules of circuit behavior. Which physical principles do they express?
- When should you stop writing junction equations? Loop equations?
- What does it mean if one of the current variables in your solution has a negative value?

✲ §26.5 ELECTRICAL MEASUREMENT

- How is an ammeter connected into a circuit?
- What are the components of a simple ammeter? How are they connected?
- Is a small or a large resistance needed in an ammeter? Why?
- How is a voltmeter connected into a circuit?
- What are the components of a simple voltmeter? Draw a circuit diagram indicating how they are connected.
- How does a modern solid-state voltmeter differ from this simple design?
- How would you use an ohmmeter to measure the resistance of a circuit component?
- How does a Wheatstone bridge allow accurate measurement of resistance? Why is it more accurate than an ohmmeter?

Basic Skill Drill

§26.1 BASIC CIRCUIT BEHAVIOR

1. A current of 5.0 A exists in a 10-Ω resistor for 4.0 min. How much charge passes through each cross section of the wire in this time?
2. A 12-V battery is connected to a 7.5-Ω light bulb. What is the current in the circuit?
3. A simple circuit has an ideal emf of 5.0 V and contains a 15-W lamp. What is the current in the circuit? What is the resistance of the lamp?

§26.2 A MODEL FOR CURRENT AND RESISTANCE

4. A copper wire 2.0 mm in diameter extends 4800 km from New York to Bristol, England, as part of an undersea cable. What is the resistance of the wire?
5. One wire is 1.0 m long and 2.0 mm in diameter. A second wire of the same material is 0.75 m long and 1.0 mm in diameter. Find the ratio R_1/R_2 of the resistances of the two wires.

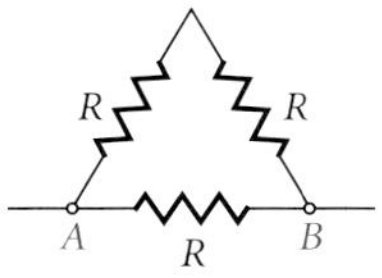

■ FIGURE 26.41

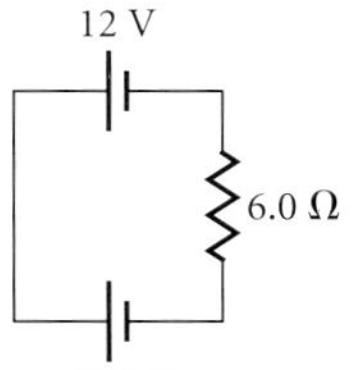

■ FIGURE 26.42

§26.3 SERIES AND PARALLEL COMBINATIONS OF RESISTORS

6. A 36-kΩ resistor and a 5.0-kΩ resistor are connected in series. What is the equivalent resistance?
7. A 36-kΩ resistor and a 5.0-kΩ resistor are connected in parallel. What is the equivalent resistance?
8. What is the equivalent resistance between points A and B in ■ Figure 26.41?

§26.4 KIRCHHOFF'S RULES

9. What terminal voltage is required of an emf used to charge a battery of emf 10.0 V and internal resistance 0.60 Ω with a current of 2.0 A?
10. Refer to the circuit diagram in ■ Figure 26.42. What is the current in the circuit?

✲ §26.5 ELECTRICAL MEASUREMENT

11. A Wheatstone bridge is used to measure an unknown resistance. The bridge balances when configured as shown in ■ Figure 26.43. What is the value of the unknown resistance?
12. An ammeter contains a galvanometer of resistance 15.00 Ω and a shunt resistance $r = 0.25\ \Omega$. It is connected into a circuit with a 6.00-V power supply and a 25.00-Ω resistor (■ Figure 26.44). What current does it measure? What was the current before the ammeter was inserted in the circuit?

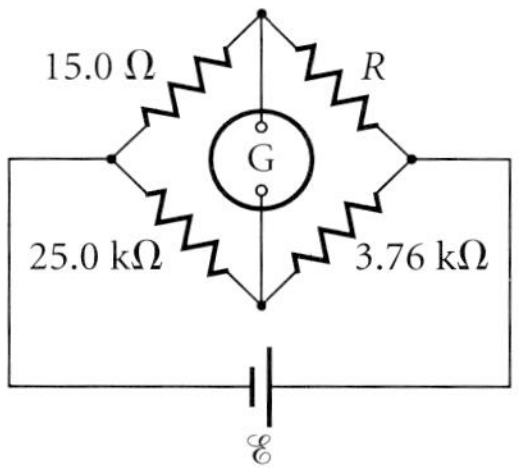

■ FIGURE 26.43

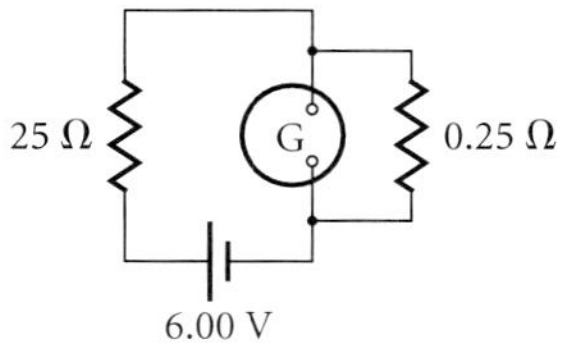

■ FIGURE 26.44

Questions and Problems

§26.1 BASIC CIRCUIT BEHAVIOR

13. ❖ What current is produced by a ring of radius r and uniform charge density λ rotating about its axis at angular frequency ω?
14. ♦ An average lightning flash lasts 0.19 s and transfers 25 C of charge. What is the average electric current during the flash?
15. ♦ A 12-V car battery is rated at 120 A·h. Assuming that the terminal voltage remains 12 V, for how long can this battery supply power at 29 W? How much charge flows through the battery in this time? How much energy is initially stored in the battery?
16. ♦ The electron beam in a cathode-ray tube carries 5.0×10^{15} electrons per second to a phosphor screen. What is the electric current in the tube?
17. ♦ A 1.50-V battery has an internal resistance of 0.025 Ω. What is the battery's terminal voltage when it is delivering 2.33 A current?

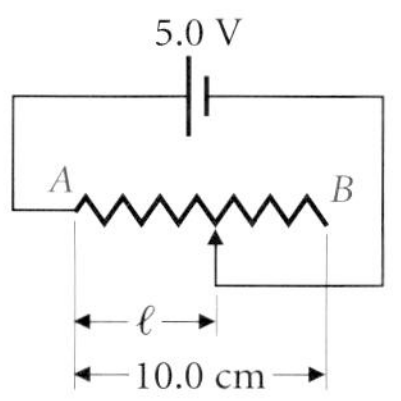

■ FIGURE 26.45

18. ♦♦ A single electron in uniform circular motion constitutes an average current around the circle. If the electron moves clockwise with angular speed ω in a circle of radius r, find the current. Evaluate I if $\omega = 2 \times 10^{16}$ rad/s and $r = 5$ nm.

19. ♦♦ ■ Figure 26.45 shows a circuit containing a potentiometer (variable resistance). The resistance of the potentiometer is proportional to the length connected into the circuit. If the total resistance of the potentiometer when the slide wire is at end B is 1.0 kΩ, and the length AB is 10.0 cm, find the current in the circuit as a function of the position of the slidewire from end A.

20. ♦♦ A simple circuit consists of an ideal emf of 12 V and a resistance of 20 Ω. The negative pole of the battery is grounded (this is usually the case in an automobile). Plot the potential as a function of position around the circuit.

21. ♦♦ What size (voltage) battery is required to operate a 25-W bulb with a resistance of 0.15 Ω?

§26.2 A MODEL FOR CURRENT AND RESISTANCE

22. ❖ Which has the larger resistance, a copper kettle with electrical connections at spout and handle, or a thin copper wire just long enough to stretch from spout to handle along the surface of the kettle?

23. ❖ Which has the larger resistance, a gold wire or a copper wire of the same length and diameter?

24. ❖ Which causes the greatest change in the resistance of a wire: **(a)** changing its diameter by a factor of 2, **(b)** changing its length by a factor of 2, or **(c)** raising its temperature from 300 K to 600 K. Which causes the least change? Explain your reasoning.

25. ♦ An aluminum transmission line extends 150 km from a power plant to a nearby city. What diameter should the line have if its resistance is to be less than 0.50 kΩ?

26. ♦ Eighteen-gauge copper wire (diameter 1.024 mm) is rated for a maximum current of 5 A. What is the maximum current density? What electric field in the copper is required to maintain the maximum current?

27. ♦♦ A potential difference of 10.0 V is applied to a wire 1.0 m long and 3.0 mm in diameter. The resistance of the wire is found to be 0.017 Ω. Compute the current density in the material and the resistivity of the wire. Using Table 26.1, can you identify the material?

28. ♦♦ The toaster in Example 26.3 ($R = 18\ \Omega$) uses a nichrome strip of 1.0 mm × 0.25 mm rectangular cross section for its heating elements. If the operating temperature is 310°C, what is the length of the strip?

29. ♦♦ A copper wire has a diameter of 0.80 mm. What diameter tungsten wire has the same resistance per unit length? Find the electric field in each wire when it is carrying a current of 5.0 mA.

30. ♦♦ A mercury-filled rubber tube is 1.00 m long and has 1.00 mm inside diameter. The tube is wrapped around a person's arm and is stretched 0.04 cm by expansion of the arm each time the person's heart beats. What is the fractional change in resistance of the mercury during a heartbeat? (Assume the cross section remains circular, and the volume of mercury remains constant.)

31. ♦♦ Aluminum has three free electrons per atom. Using data from Tables 26.1, 13.1, and the periodic table, determine the drift speed of conduction electrons in an aluminum wire 1.0 mm in diameter at 20°C when the current in the wire is 1.0 mA. What is the average time between collisions?

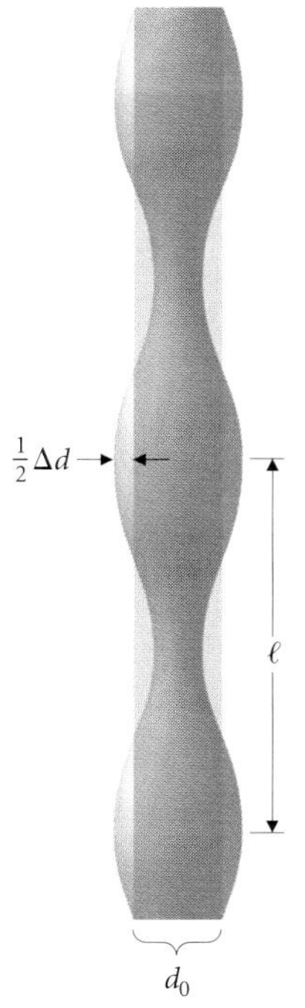

■ FIGURE 26.46

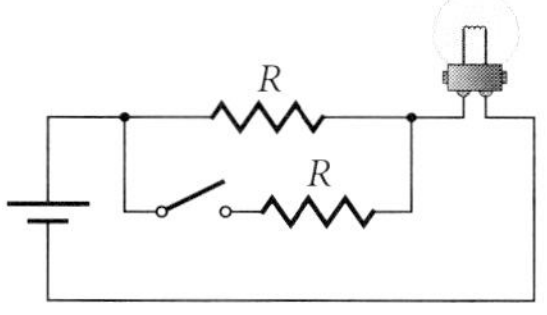

■ FIGURE 26.47

32. ♦♦♦ Because of a manufacturing error the diameter of a strand of wire varies periodically along its length (■ Figure 26.46).

$$d(z) = 1.00 \text{ mm} + (0.01 \text{ mm})\{\sin[2\pi z/(1 \text{ cm})]\}.$$

Calculate *approximately* the difference between the average resistance per unit length of this wire and that of a wire of the same material with a constant diameter of 1.0 mm.

§26.3 SERIES AND PARALLEL COMBINATIONS OF RESISTORS

33. ❖ Refer to the circuit diagram in ■ Figure 26.47. Does the lamp burn more brightly with the switch open or closed?

34. ❖ A power transmission line has a resistance r. Is it better to transmit power with high emf and low current or low emf and high current? Why?

35. ❖ A string of decorative lights could be connected in series or in parallel. What simple experiment, requiring no equipment, allows you to determine how they are connected? Explain your reasoning.

36. ♦ A wire of length L and resistance R is cut in half, and the two halves are connected in parallel. What is the resistance of the parallel combination?

37. ♦ Find the equivalent resistance of the circuit diagramed in ■ Figure 26.48.

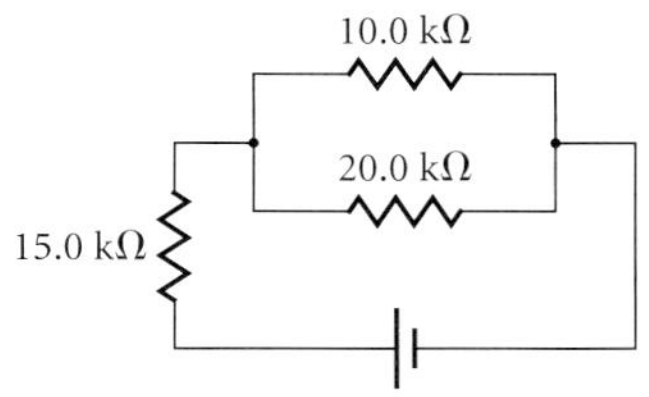

■ FIGURE 26.48

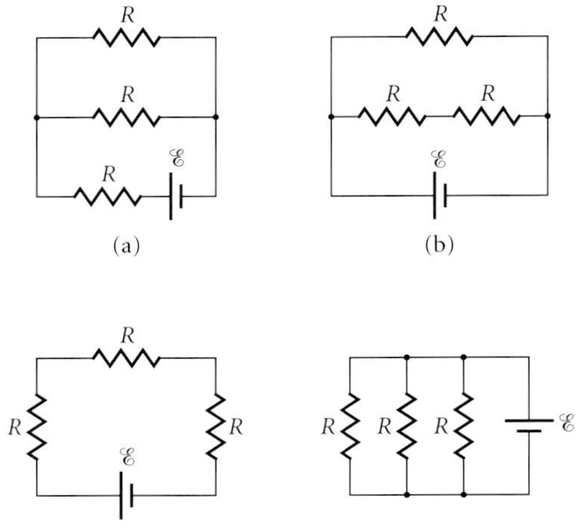

■ FIGURE 26.49

38. ♦♦ If a total current I passes through resistors R_1 and R_2 connected in parallel, show that the currents in the individual resistors are:

$$I_1 = \frac{R_2}{R_1 + R_2} I \quad \text{and} \quad I_2 = \frac{R_1}{R_1 + R_2} I.$$

39. ♦♦ In ■ Figure 26.49 are diagrams of four different circuits that may be constructed with an ideal battery and three equal resistors. Which has the greatest current through the battery? Which has the least? If $\mathscr{E} = 10.0$ V and $R = 5.00\ \Omega$, what are the greatest and least currents?

40. ♦♦ Find the power used by each resistor in the circuit diagramed in ■ Figure 26.49a. How much less power does each resistor use than it would if it were the only resistor in the circuit?

41. ♦♦ A real battery has emf $\mathscr{E} = 50.0$ V. When it is connected to a 20.0-Ω resistor, the current is 2.45 A. What are the internal resistance of the battery and the potential difference between its terminals?

42. ♦♦ A 5.0-Ω and a 20.0-Ω resistor are connected in parallel across an ideal 10.0-V emf. What is the current through each circuit element? Find the power output of the battery and the power dissipated by each resistor.

43. ♦♦ In the circuit shown in ■ Figure 26.50, find the current in each resistor and the values of the potential at points A, B, and C. (Take $V = 0$ at ground.)

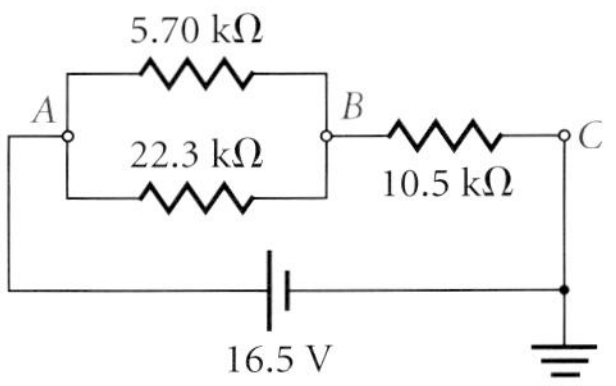

■ FIGURE 26.50

44. ♦♦ There is a total current I through resistors R_1, R_2, and R_3 connected in parallel. Derive expressions for the currents in the individual resistors.

45. ♦♦♦ Find the equivalent resistance of the combination shown in ■ Figure 26.51. Find the current through the battery and through each resistor. Find the potential V at points A, B, and C.

46. ♦♦♦ In the circuit diagramed in ■ Figure 26.52, the current through the 50-Ω resistor is the same whether the switch is open or closed. What is R?

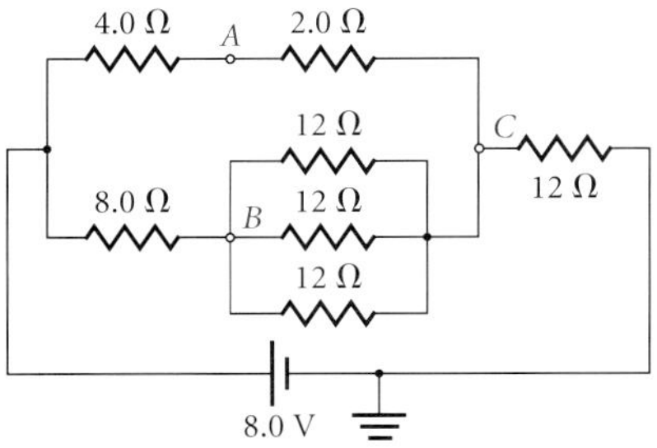

■ FIGURE 26.51

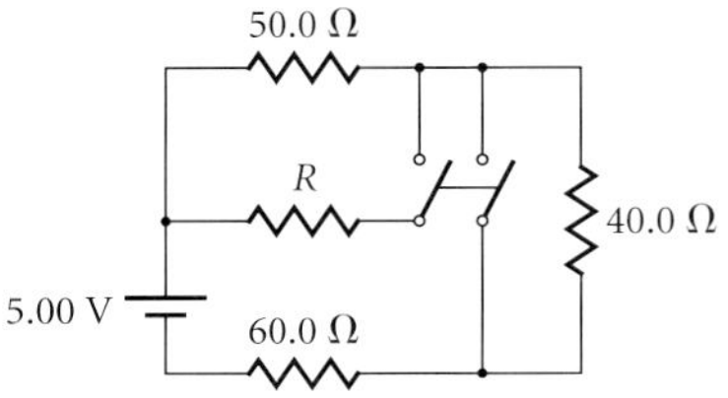

■ FIGURE 26.52

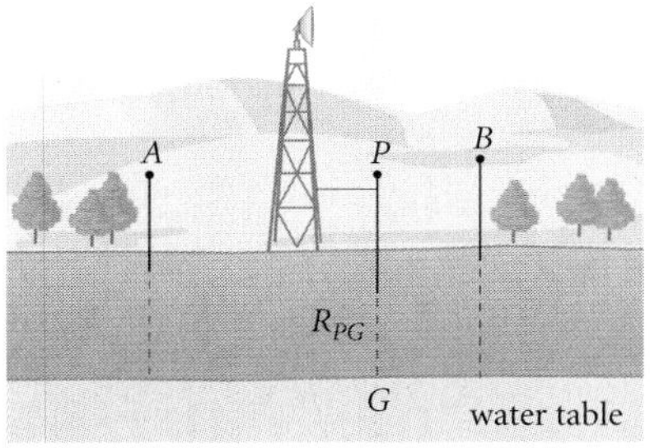

■ FIGURE 26.53

47. ♦♦♦ To protect a radio transmitter against lightning, it is connected to a metal post in the ground (■ Figure 26.53). Proper protection requires that the resistance between the post and the water table be less than 10 Ω. To test for this, the station engineer drives two spikes into the ground at A and B. The resistance between the two spikes is measured to be $R_{AB} = 40\ \Omega$. The resistance between A and the post (R_{AP}) is measured as 28 Ω, and the resistance between B and the post (R_{BP}) is 30 Ω. Is the radio tower adequately grounded?

§26.4 KIRCHHOFF'S RULES

48. ❖ In a compound circuit with more than one battery, does current always circulate around a loop from the positive battery terminal to the negative terminal? Why or why not?

49. ❖ If the 100-Ω resistor in ■ Figure 26.54 is increased to 101 Ω, does the current through the resistor increase, decrease, or stay the same? Does the potential difference across it increase, decrease, or stay the same?

50. ♦ A flashlight uses two 1.5-V batteries to light a bulb. The current in the circuit is 0.53 A. What is the resistance of the bulb? How much power does it consume?

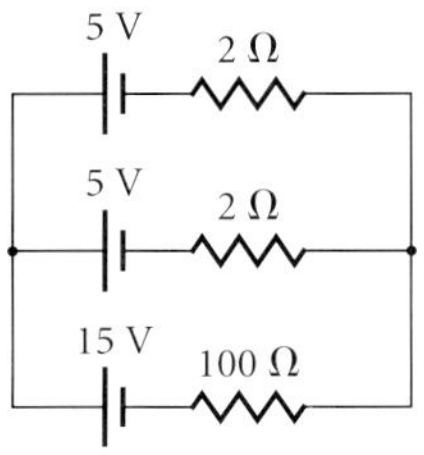

■ FIGURE 26.54

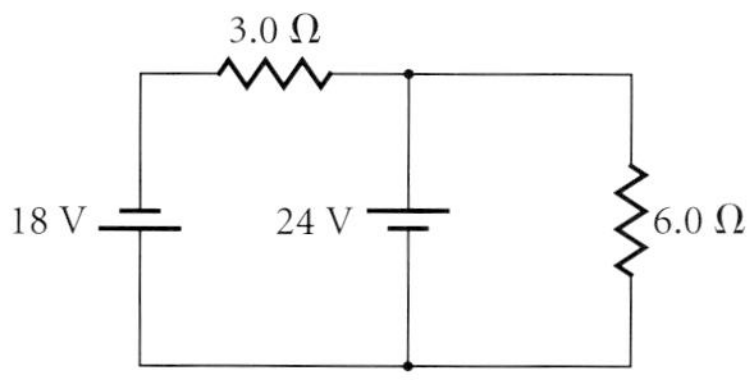

■ FIGURE 26.55

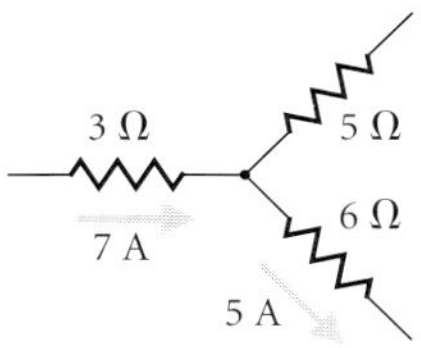

■ FIGURE 26.56

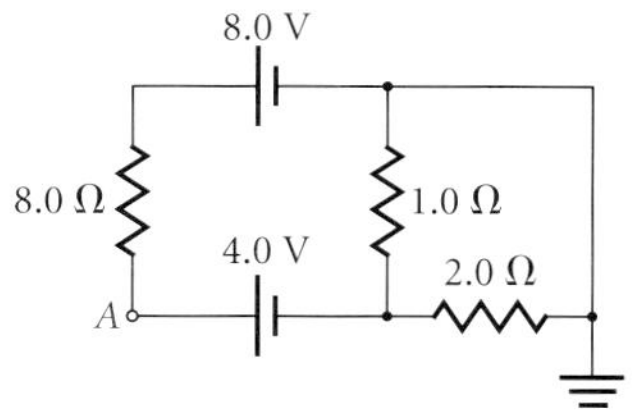

■ FIGURE 26.58

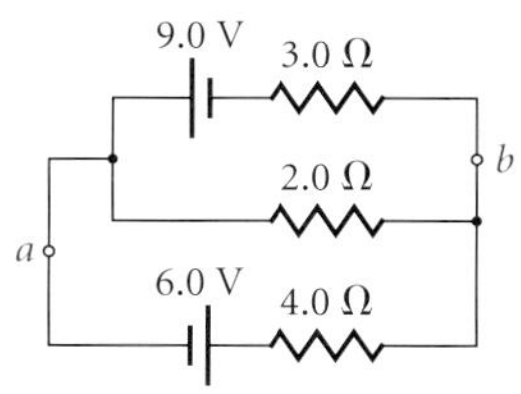

■ FIGURE 26.59

51. ♦ A 12.0-V battery with internal resistance of 0.02 Ω is used to charge a battery with an emf of 10.5 V and an internal resistance of 0.15 Ω. What is the charging current?
52. ♦ A flashlight contains two batteries in series, each with an emf of 1.50 V and internal resistance of 0.17 Ω. The bulb has a resistance of 25.0 Ω. Draw a circuit diagram, and find the current in the circuit. How much power does the bulb draw?
53. ♦♦ Find the current through each resistor in ■ Figure 26.55.
54. ♦♦ What value of the emf $\mathscr{E}_2$ in the circuit of Figure 26.30 is necessary if that battery is to deliver power to the circuit rather than absorbing power from the circuit?
55. ♦♦ What is the potential difference across the 100-Ω resistor in the circuit of ■ Figure 26.54?
56. ♦♦ Two automobile batteries are connected in parallel to power a wheelchair. If each of the batteries has an emf $\mathscr{E} = 12.0$ V and internal resistance $r = 0.020\ \Omega$, and the wheelchair motor has resistance $R = 1.00\ \Omega$, find the current provided to the motor. What would be the current delivered to the motor if the batteries were connected in series? What are the relative advantages of series and parallel connections?
57. ♦♦ Show that the currents I_1 and I_2 in Figure 26.26 (Example 26.12) are given by:

$$I_1 = \frac{(\mathscr{E}_1 - \mathscr{E}_2)R + \mathscr{E}_1 r_2}{R(r_1 + r_2) + r_1 r_2}$$

and

$$I_2 = -\frac{(\mathscr{E}_2 - \mathscr{E}_1)R + \mathscr{E}_2 r_1}{R(r_1 + r_2) + r_1 r_2}.$$

58. ♦♦ In the circuit segment diagramed in ■ Figure 26.56, find the potential difference across each resistor.
59. ♦♦ Three ideal batteries and three identical resistors are connected in the circuit diagramed in ■ Figure 26.57. Find the current in each arm of the circuit.
60. ♦♦ A 3-Ω resistor, a 1-Ω resistor, a 3-V emf, and a 9-V emf are all connected in series. Draw diagrams of two circuits that fit this description but that carry different currents. Apply Kirchhoff's loop rule to each circuit. Find the current and draw a graph showing the variation of potential around the circuit.
61. ♦♦ What would be the current if the good battery in Figure 26.26 (Example 26.12) were connected in series with the dead battery? What would you have to do to make the connection? Would it be worth it?

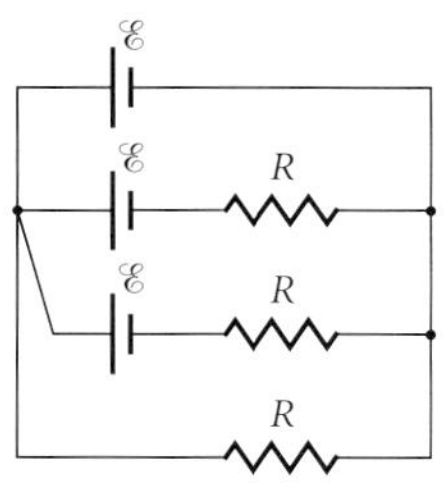

■ FIGURE 26.57

62. ♦♦ Find all the currents in the circuit diagramed in ■ Figure 26.58. What is the potential at point *A*?
63. ♦♦ Find all the currents in the circuit shown in ■ Figure 26.59 and the potential difference between points *a* and *b*.
64. ♦♦ Find the resistance of the combination in ■ Figure 26.60.
65. ♦♦ Twelve identical resistors are connected together to form a cube (■ Figure 26.61). What is the cube's equivalent resistance when connected into a circuit at points *A* and *D*? *Hint:* Use symmetry to equate the currents in some of the arms before applying Kirchhoff's rules.

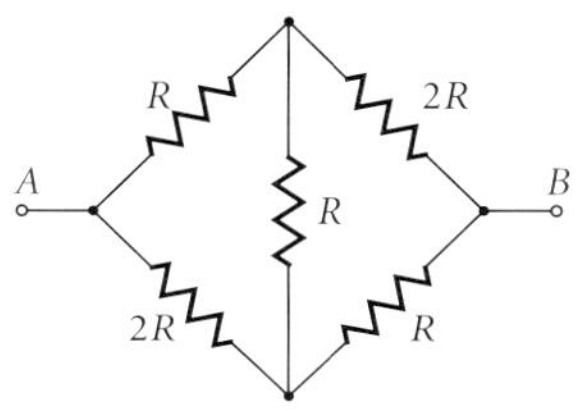

■ FIGURE 26.60

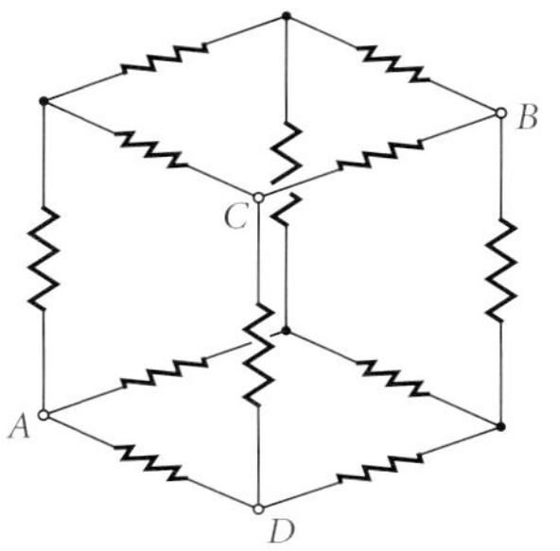

■ FIGURE 26.61

§26.5 ELECTRICAL MEASUREMENT

66. ♦ A Wheatstone bridge is used to determine the resistance of a piece of steel wire. With $R_1 = 4.00\ \Omega$, the ratio of R_4 to R_3 is found to be 0.092. What is R_x? If the wire is 45.0 cm long and 0.50 mm in diameter, what is its resistivity?
67. ♦ A voltmeter with a resistance $R_m = 5.00 \times 10^5\ \Omega$ is used to measure the potential difference across one of two 3.00-kΩ resistors connected in series to a 5.00-V battery (■ Figure 26.62). What potential difference does the voltmeter measure? What is the true potential difference before the voltmeter is connected?
68. ♦♦ A multirange ammeter uses a series of shunt resistances and a solid-state voltage detector (■ Figure 26.63). The solid-state device is sensitive to potential differences of 0 to 0.2 V. The ammeter is con-

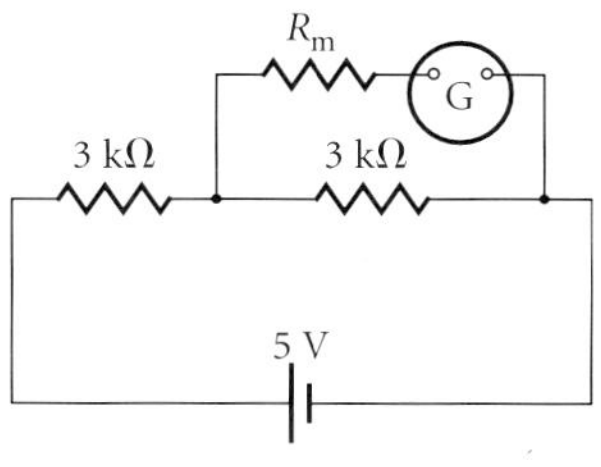

■ FIGURE 26.62

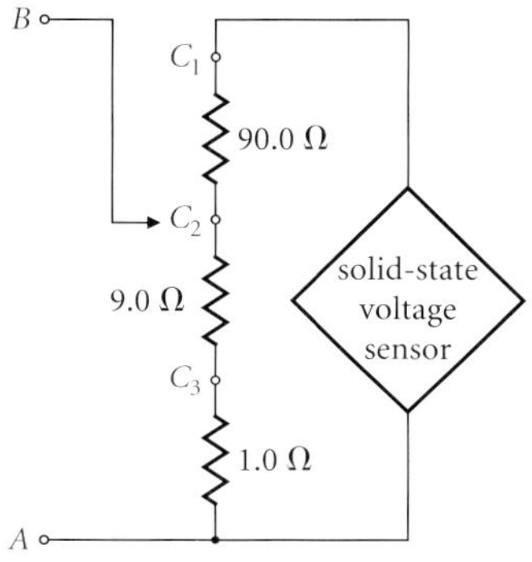

■ FIGURE 26.63

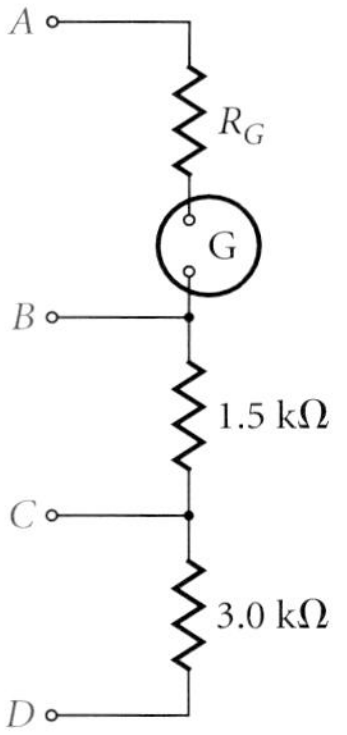

■ FIGURE 26.64

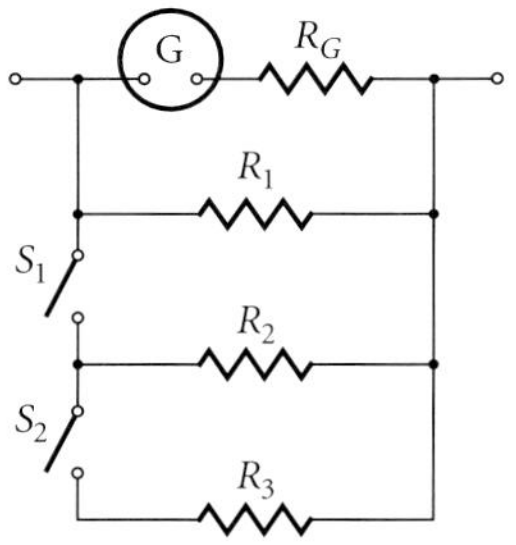

■ FIGURE 26.65

nected into the circuit between points A and B, with B connected to one of points C_1, C_2, or C_3. If the current to be measured is 3.00 mA, which connection should be used? If the current is produced in a circuit with an emf of 1.50 V and a resistance of 0.500 kΩ, what is the measured current? By what percentage does the ammeter disturb the circuit?

69. ♦♦ A voltmeter is made using a galvanometer of internal resistance 1.5 kΩ and two resistances of 1.5 kΩ and 3.0 kΩ (■ Figure 26.64). The galvanometer gives a full-scale deflection when there is a current of 5 μA in it. What voltages across terminals AB, AC, and AD produce a full-scale deflection?

70. ♦♦♦ An ohmmeter is constructed by connecting a known emf in series with a galvanometer of resistance 15 Ω. The circuit is completed by connecting the unknown resistance in series. If the galvanometer sensitivity is 50.0 μA at full scale and a 2.0-V cell is used as the known emf, what is the minimum resistance that can be measured? If the uncertainty in meter reading is $\frac{1}{100}$ of full scale, what is the largest resistance that can be measured to 10% accuracy? What known resistance, in series with the galvanometer, will protect it against damage when the ohmmeter leads are touched together?

71. ♦♦♦ An ammeter is constructed from a galvanometer with resistance $R_G = 18.0\ \Omega$ and three resistances connected in parallel (■ Figure 26.65). $R_1 = 2.0\ \Omega$. The galvanometer shows a full-scale deflection when a 50.0-μA current passes through it. What is the maximum current the ammeter can measure when the switches are both open? If switch S_1 is closed, does the ammeter become more or less sensitive? (A more sensitive ammeter can measure a smaller current.) What resistances R_2 and R_3 are required to change the sensitivity by a factor of 10 when switch S_1 is closed, and another factor of 10 when both switches are closed?

72. ♦♦♦ A potentiometer may be used to measure potential differences without requiring any current (■ Figure 26.66). The emf $\mathscr{E}_s$ is accurately known. To measure a potential difference V the switch is first set at position a and the sliding contact is moved until $\ell = \ell_s$, and the galvanometer detects no current. The switch is then set at position b, and the contact is moved again until $\ell = \ell_x$, and the galvanometer detects no current. Show that $V = \mathscr{E}_s(\ell_x/\ell_s)$.

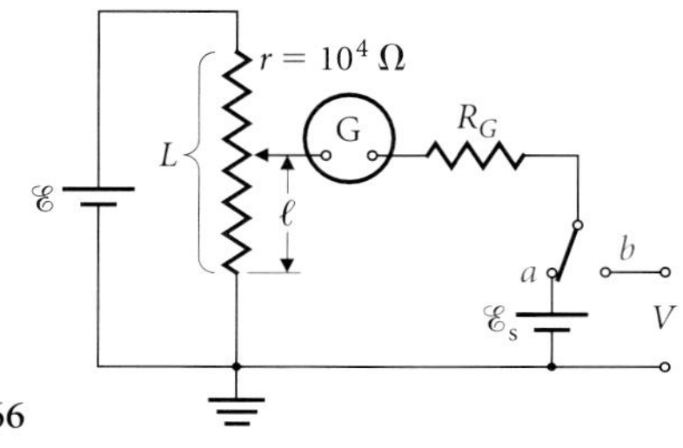

■ FIGURE 26.66

Additional Problems

73. ❖ A disk of radius R and uniform charge density ρ rotates with angular speed ω about an axis perpendicular to the disk at its center. Describe the resulting distribution of current.

74. ♦ Electric companies charge for power in units of kW·h. How many joules equal 1 kilowatt-hour? If the power company charges 12¢/kW·h, and supplies power at 120 V, what is the cost per coulomb?

75. ♦♦ A 1.50-V battery delivers a current of 0.235 A when connected to a 6.35-Ω load. What is the internal resistance of the battery?

76. ♦♦ A battery delivers a current of 1.98 A when connected to a 6.00-Ω resistance and 2.96 A when connected to a 4.00-Ω resistance. Find the emf and internal resistance of the battery.

77. ♦♦ A 12.00-V battery is connected to two identical resistors, each of resistance R. With the resistors connected in series, the measured current is 0.5985 A. With the resistors connected in parallel, the measured current is 2.376 A. Find the internal resistance of the battery and the resistance R.

78. ♦♦ A whimsical physics instructor connects 12 identical 10.0-Ω resistors together to make the cubical resistor shown in Figure 26.61. An ideal battery of emf 4.0 V is connected between points A and B (across a body diagonal). Find the current through the battery. Use symmetry arguments to simplify the problem as much as possible before applying Kirchhoff's rules. What is the current through a real battery with an internal resistance of 0.30 Ω?

79. ♦♦♦ The resistor network in ■ Figure 26.67 extends infinitely to the right. What is the resistance between points A and B? *Hint:* If you remove the two leftmost resistors, the network is unchanged.

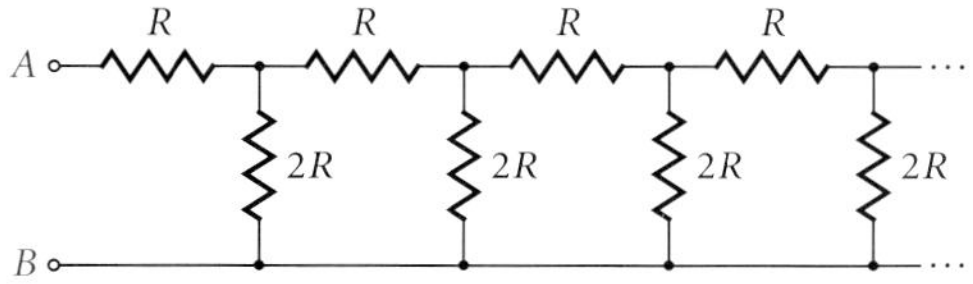

■ FIGURE 26.67

80. ♦♦♦ A 55-kg woman pedaling a bicycle wheel powers an electric generator (source of emf). The generator maintains a current of 5.0 A through a 20.0-Ω resistance. Using this power more enjoyably, how rapidly may she gain altitude while riding a 15-kg bicycle?

81. ♦♦♦ Show that maximum power is delivered to a load when the load resistance is equal to the internal resistance of the power supply.

82. ♦♦♦ The circuit in ■ Figure 26.68 has geometric symmetry. Make use of the symmetry to find the currents. Solve the problem using

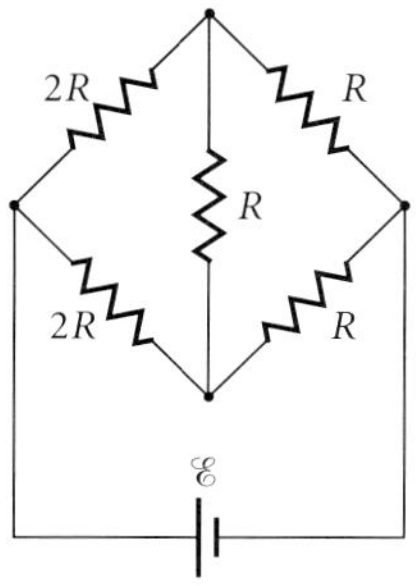

■ FIGURE 26.68

Kirchhoff's rules with five current variables and show that the first method gives correct results.

83. ♦♦♦ A battery with $\mathscr{E} = 1.0$ V is connected between points A and C (across a face diagonal) of the cubical resistor in Figure 26.61. Each resistance is 10.0 Ω. Find the current through the battery. (*Hint:* Use geometric symmetry to reduce the circuit to a series/parallel combination.)

Computer Problems

84. File CH26P84 on the supplementary computer disk contains measured values of current when a potential difference ΔV is applied across a light-emitting diode. Use a spreadsheet program to plot the characteristic curve (I vs. ΔV). The slope of the characteristic curve is called the dynamic resistance. Use the spreadsheet to estimate the dynamic resistance of the diode at 1.50 V and 1.70 V.

85. Applying Kirchhoff's rules to a circuit, we obtain a set of linear equations for the currents that may be described by a matrix equation:

$$MI = J,$$

where I is the vector of current values with components $I_1, I_2, \ldots, I_n$, and the matrix M contains the coefficients of the currents in each of the equations. The equations in Example 26.12 are:

$$\begin{aligned}
&\text{(i)} && I_1 - I_2 - I_3 = 0.\\
&\text{(ii)} && (0.02\ \Omega)I_1 + (0.05\ \Omega)I_2 = 2.0\ \text{V}.\\
&\text{(iii)} && -(0.05\ \Omega)I_2 + (0.20\ \Omega)I_3 = 10.0\ \text{V}.
\end{aligned}$$

The matrix M contains the coefficients of the Is:

$$M = \begin{pmatrix} 1 & -1 & -1 \\ 0.02 & 0.05 & 0 \\ 0 & -0.05 & 0.20 \end{pmatrix} \Omega,$$

and the vector $J = (0, 2.0, 10.0)$ V. The problem is solved by inverting the matrix:

$$I = M^{-1}J.$$

Use the matrix inversion program on the supplementary computer disk to show that the inverse of the matrix is:

$$M^{-1} = \begin{pmatrix} 2/3 & 50/3 & 10/3 \\ -4/15 & 40/3 & -4/3 \\ -1/15 & 10/3 & 14/3 \end{pmatrix} \Omega^{-1}.$$

So then:

$$\begin{pmatrix} I_1 \\ I_2 \\ I_3 \end{pmatrix} = \begin{pmatrix} 2/3 & 50/3 & 10/3 \\ -4/15 & 40/3 & -4/3 \\ -1/15 & 10/3 & 14/3 \end{pmatrix} \begin{pmatrix} 0 \\ 2.0 \\ 10.0 \end{pmatrix} \text{A}.$$

Each element in the product matrix is the sum of the products of the numbers in a row of the matrix times the numbers in the vector J:

$$\begin{pmatrix} (2/3)\times 0 + (50/3)\times(2.0) + (10/3)\times(10.0) \\ (-4/15)\times 0 + (40/3)\times(2.0) + (-4/3)\times(10.0) \\ (-1/15)\times 0 + (10/3)\times(2.0) + (14/3)\times(10.0) \end{pmatrix}\text{A} = \begin{pmatrix} 66.7 \\ 13.3 \\ 53.3 \end{pmatrix}\text{A}.$$

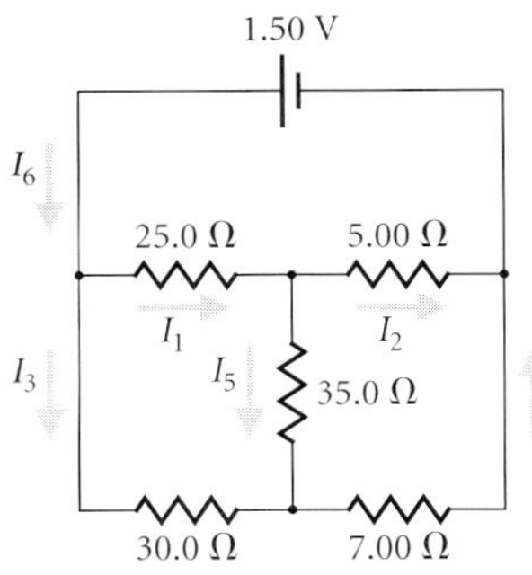

■ FIGURE 26.69

Use this method to find the currents in the circuit in ■ Figure 26.69. (Spreadsheet programs and calculators often have the ability to do matrix inversion and matrix multiplication.)

Challenge Problems

86. A superposition theorem states that one may solve a circuit by first solving a number of simplified circuits, each containing only a single emf but otherwise the same as the original circuit. The current in any arm of the actual circuit is the sum of the currents in the same arm of each simplified circuit. Use this method to verify the results of Example 26.12. Find currents I_1, I_2, and I_3 in Figure 26.26 if $\mathscr{E}_1 = 12.0$ V, $\mathscr{E}_2 = 10.0$ V, $r_1 = 1.5\ \Omega$, $r_2 = 1.2\ \Omega$, and $R = 36.0\ \Omega$.

87. Loop currents: An alternative method for using Kirchhoff's rules defines one loop current for each loop used in the solution. The actual current in any arm of the circuit is the algebraic sum of the loop currents associated with that arm. **(a)** Using the circuit in ■ Figure 26.70 as an example, show that the loop current method automatically satisfies Kirchhoff's junction rule. **(b)** Use Kirchhoff's loop rule to write a set of equations for the loop currents, and solve for i_1 and i_2. **Warning:** Never use a combination of loop currents and arm currents in one circuit. Pick one method and stick to it.

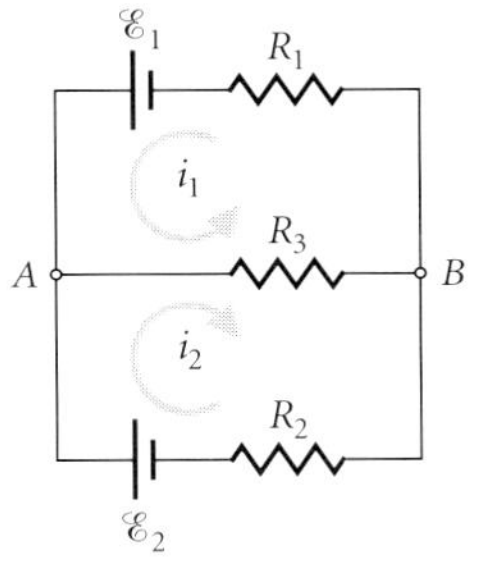

■ FIGURE 26.70

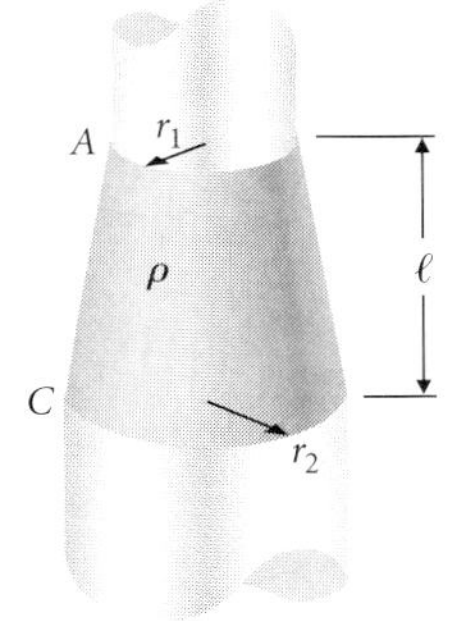

■ FIGURE 26.71

88. If the galvanometer in a Wheatstone bridge circuit (Figure 26.39) has 20.0 Ω of resistance and an uncertainty of ±1 μA in its reading, what is the uncertainty in the measured value of R_x? Take $R_1 = 5.00\ \Omega$, $R_3 = 10.0\ \Omega$, $R_4 = 20.0\ \Omega$, and $\mathscr{E} = 5.00$ V. Is the uncertainty significant?

89. A device designed to connect two wires of different radii has the shape of a conical frustum of height ℓ, smaller radius r_1, and larger radius r_2 (■ Figure 26.71). If the device is made of material with resistivity ρ, compute its resistance approximately. Assume $r_2 - r_1 \ll \ell$.

90. Thevenin's theorem states that any portion of a circuit containing only emfs and resistors and with two junctions that connect to the rest of the circuit is equivalent to a single battery with emf $\mathscr{E}_{TH}$ and internal resistance R_{TH} (■ Figure 26.72). **(a)** Show that E_{TH} and R_{TH}

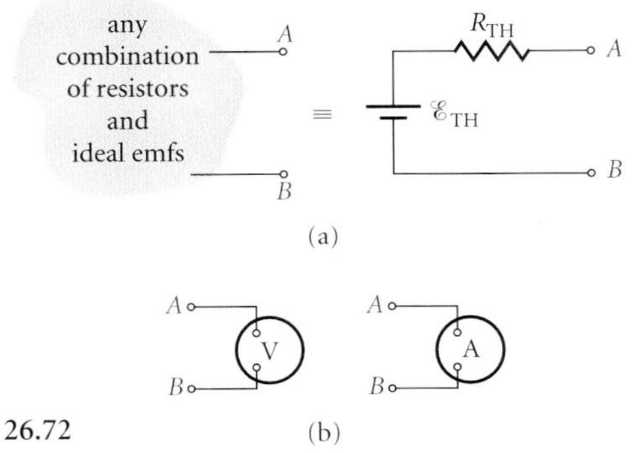

■ FIGURE 26.72

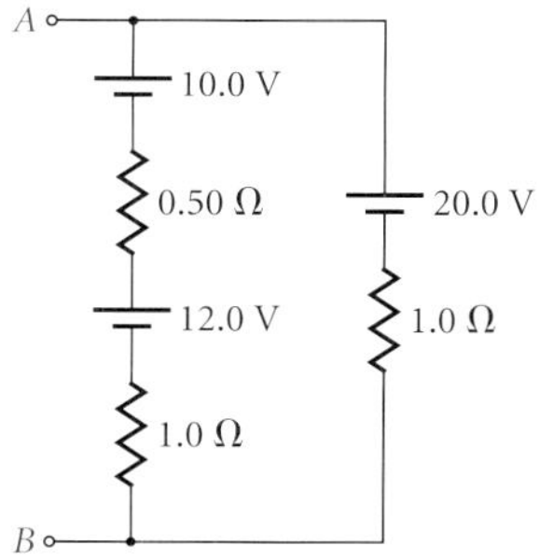

■ FIGURE 26.73

can be found by measuring two quantities: (i) the open-circuit potential difference between points B and A and (ii) the short-circuit current between A and B (Figure 26.72b). **(b)** Find the Thevenin equivalent of the power supply in ■ Figure 26.73. What is the current through a 1.00-kΩ load? **(c)** Use Thevenin's theorem to show that if any single resistance in a circuit is increased, the current through it decreases and the potential difference across it increases.

91. Norton's theorem states that any combination of emfs and resistors with two junctions that connect to the rest of the circuit is equivalent to a current source I_N in parallel with a resistance R_N (■ Figure 26.74). **(a)** Show that I_N and R_N may be found by measuring (i) the open-circuit potential difference between points A and B and (ii) the short-circuit current from A to B. (A current source is a device that adjusts its emf to maintain a constant current.) **(b)** Find the Norton equivalent of the circuit in ■ Figure 26.75.

92. Refer to Problems 90 and 91. Prove that the Norton and Thevenin resistances for any combination of emfs and resistors are the same.

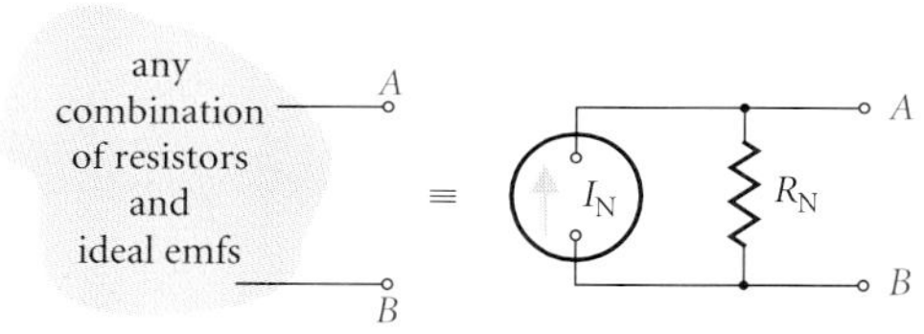

■ FIGURE 26.74

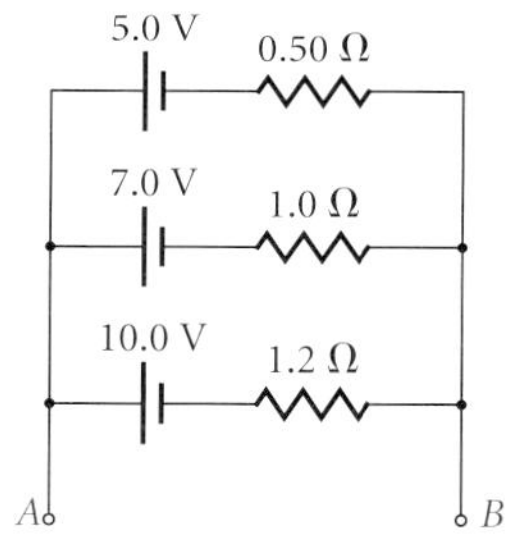

■ FIGURE 26.75

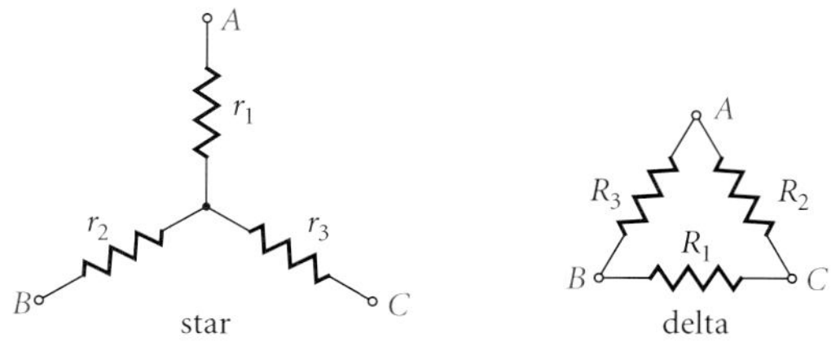

■ FIGURE 26.76

93. *Star-Delta Transformation* ■ Figure 26.76 shows a *star* and a *delta* resistor combination. Each connects to the rest of a circuit at three points. Show that replacing the star with the delta (or vice-versa) produces an equivalent circuit if and only if:

$$r_1 = \frac{R_2 R_3}{R_1 + R_2 + R_3}$$

and

$$r_2 = \frac{R_1 R_3}{R_1 + R_2 + R_3}$$

and

$$r_3 = \frac{R_1 R_2}{R_1 + R_2 + R_3},$$

or:

$$R_1 = \frac{r_1 r_2 + r_2 r_3 + r_3 r_1}{r_1},$$

with similar expressions for R_2 and R_3. A battery with $\mathscr{E} = 10.0$ V is connected between points A and D of the cubical resistor (Figure 26.61) with $R = 10.0\ \Omega$. Use star-delta transformations to reduce the problem to a series-parallel circuit, and find the total current through the battery.

94. In a lead-acid battery, the reactions at each terminal are reversible. When a charged battery is idle, the reactions in fact proceed in both directions at equal rates with no net energy use or charge transfer. Explain how connecting the battery in a circuit and allowing charge to flow from the terminals causes the reactions to replace the charge on the terminals.

95. An unknown resistance is connected to an unknown ideal battery in a simple circuit. A voltmeter and ammeter are to be used to measure current and potential difference simultaneously. Both the voltmeter and ammeter use galvanometers with 30.0-Ω resistance. The voltmeter has a 1.00-kΩ resistor in series with the galvanometer, and the ammeter has a 0.500-Ω shunt. When the voltmeter is connected in parallel with the resistor, the meters read $\Delta V = 4.76$ V and $I = 0.481$ A. What are the emf and resistance in the circuit? What would the meter readings be if the voltmeter were connected in parallel with the battery? If you take the voltmeter reading as the emf and the voltmeter reading divided by the ammeter reading as the resistance, which position gives the more accurate measurements?

Bank of capacitors designed at the Ernest O. Lawrence Berkeley National Laboratory and used to power a neutral beam injector. Each bank consists of ten modules of 272 capacitors. (See also Figure 27.11).

CHAPTER 27
Capacitance and Electrostatic Energy

CONCEPTS

Capacitor
Capacitance
Dielectric
Bound charge
Free charge
Electrostatic energy density

GOALS

Be able to:
Calculate the capacitance of symmetric objects.

Compute the equivalent capacitance of series and parallel combinations.

Compute the energy stored in a capacitor.

Compute the electrostatic energy density at a point.

They were hoarding and dispersing energy
while the inanimate universe was running down around us.

LOREN EISELEY

Plentiful energy for a billion years from the hydrogen in seawater—this is the hope of nuclear fusion research. The energy is available; the problem is how to extract it. All the experiments striving to achieve fusion need large pulses of energy either to inject the fuel into the experiment chamber or to compress and heat it. For example, a neutral beam injector designed at the University of California, Berkeley, produces pulses lasting 50 ms that deliver nearly 1.6 MJ of energy. Such pulses greatly exceed the capability of the local power company and require a method of storing electrical energy for quick release. A bank of 8160 individual *capacitors* stores the energy. In this chapter we shall study capacitors and give an introduction to their use in circuits. We shall also gain a deeper understanding of electrostatic energy and its relation to the electric field.

A MORE EXTENSIVE DISCUSSION OF CAPACITOR CIRCUITS WILL BE GIVEN IN CHAPTERS 31 AND 32.

27.1 CAPACITANCE

27.1.1 The Parallel Plate Capacitor

RECALL: σ = CHARGE/AREA.

WE USED EQN. (25.7) WITH E = CONSTANT.

Any object that can store charge is a capacitor. The simplest capacitor to analyze is made from two parallel conducting plates, each of area A (■ Figure 27.1). In use, one plate holds a positive charge density σ while the other has an equal negative charge density. We model the system as two infinite uniformly charged sheets. Each sheet produces an electric field of magnitude $E = \sigma/(2\epsilon_0)$ (eqn. 24.7) perpendicular to the plates. The two fields reinforce each other to give $E = \sigma/\epsilon_0$ between the plates and cancel to give zero outside (Figure 27.1b). The potential difference between the plates is $|\Delta V| = Ed = \sigma d/\epsilon_0$. The total charge on one plate is $Q = \sigma A$, so $\sigma = Q/A$. We find that the potential difference ΔV between the plates is proportional to the charge Q:

$$|\Delta V| = \frac{d}{\epsilon_0 A} Q. \tag{27.1}$$

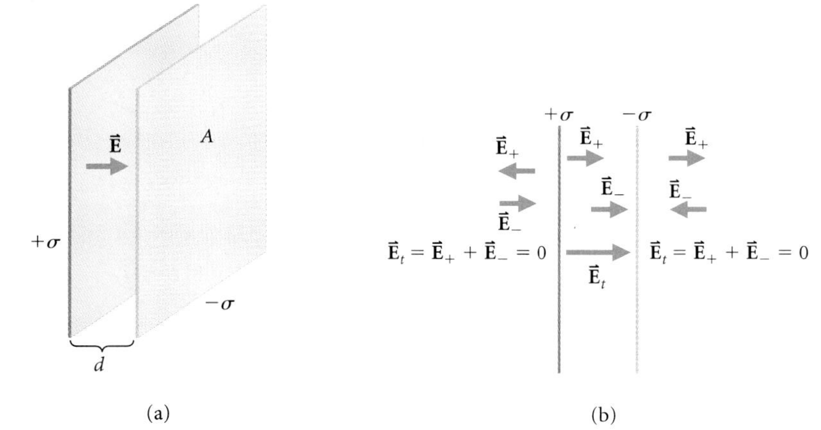

■ FIGURE 27.1
(a) A charged parallel plate capacitor. The two conducting plates are separated by a distance d. Each plate has area A; one has a uniform charge density σ, and the other, $-\sigma$. (b) The electric field vector is the sum of the vectors produced by the two plates. The electric field vector produced by a uniform sheet of charge has magnitude $\sigma/(2\epsilon_0)$ and is constant. Between the plates, the vectors are in the same direction, and $E = \sigma/\epsilon_0$. On either side, the vectors are in opposite directions and sum to zero.

In fact, for *any* pair of conductors, charge and potential difference are proportional (■ Figure 27.2). Their *ratio* depends only on the shapes, sizes, and relative positions of the conductors and gives an operational definition of capacitance:

> The *capacitance* of two conductors is the magnitude of the ratio of the charge on one of them to the potential difference between them, when they are oppositely charged:
>
> $$C \equiv \left|\frac{Q}{\Delta V}\right|. \tag{27.2}$$

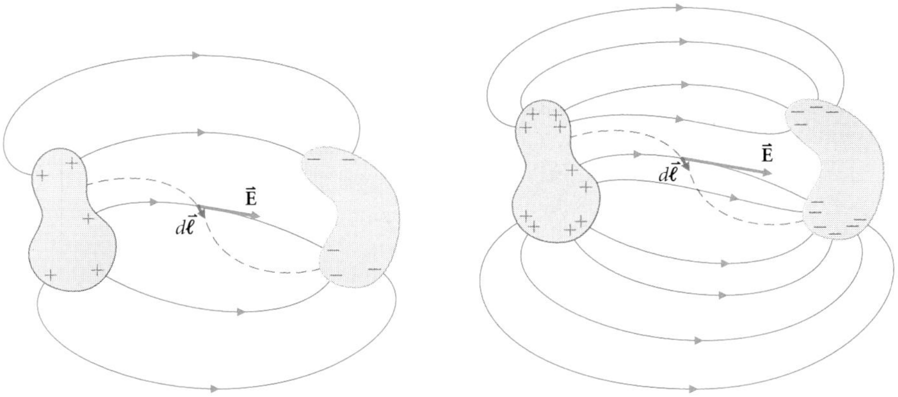

■ FIGURE 27.2
The potential difference between any two conductors with equal and opposite charge is proportional to that charge. Charge is distributed on the surface of each conductor so that $\vec{E}$ is zero within the conductor, and just outside the conductor, $\vec{E}$ is perpendicular to the surface. Only one pattern of charge density accomplishes this. If the total charge is doubled, the charge density at each point doubles, leaving the pattern unchanged. As a result, the electric field strength doubles, whereas the direction at each point is unchanged. Thus the potential difference $\int \vec{E} \cdot d\vec{\ell}$ between the conductors also doubles.

We find the capacitance of the two parallel plates by combining eqns. (27.1) and (27.2):

$$C = \frac{\epsilon_0 A}{d}. \tag{27.3}$$

STRICTLY SPEAKING THIS EXPRESSION IS CORRECT WITH VACUUM BETWEEN THE PLATES. AIR MAKES LITTLE DIFFERENCE (BUT SEE §27.3).

Once the capacitor is made, the plate area A and separation d are set, and the capacitance does not change. The amount of charge on the plates does change as the potential difference between the plates is varied.

The unit of capacitance is the farad (F). Conductors with a capacitance of 1 farad at a potential difference of 1 volt store 1 coulomb of charge:

$$1\ \text{F} \equiv \frac{1\ \text{C}}{1\ \text{V}}.$$

Just as a coulomb is a large unit of charge, a farad is a very large unit of capacitance.

NAMED AFTER MICHAEL FARADAY (1791–1867). FARADAY MADE FUNDAMENTAL CONTRIBUTIONS TO A WIDE RANGE OF FIELDS. HIS FULL NAME IS USED FOR A UNIT OF CHARGE IN ELECTROCHEMISTRY. THE PASSAGE OF 1 FARADAY OF CHARGE THROUGH A SOLUTION DEPOSITS 1 MOLE OF A UNIVALENT ION ON THE ELECTRODES.

EXAMPLE 27.1 ♦ Two parallel metal plates are separated by 2.0 mm. What area should they have to form a 1.0-F capacitor?

MODEL We use eqn. (27.3) for a parallel plate capacitor.

SETUP AND SOLVE

$$A = \frac{dC}{\epsilon_0} = \frac{(2.0 \times 10^{-3}\ \text{m})(1.0\ \text{F})}{8.85 \times 10^{-12}\ \text{C}^2/\text{N}\cdot\text{m}^2} = 2.3 \times 10^{8}\ \text{m}^2.$$

ANALYZE Square plates 15 km on a side would do the job! Check the units:

$$\frac{\text{N}\cdot\text{m}^3\cdot\text{F}}{\text{C}^2} = \frac{\text{J}\cdot\text{F}\cdot\text{m}^2}{\text{C}^2} = \frac{(\text{C}\cdot\text{V})\cdot(\text{C/V})\cdot\text{m}^2}{\text{C}^2} = \text{m}^2.$$

The units of ϵ_0, $\text{C}^2/\text{N}\cdot\text{m}^2$, are equivalent to F/m. ■

EXERCISE 27.1 ♦ What area plates, separated by 2.0 mm, would form a 1.0-nF capacitor? How much charge would this capacitor store with a potential difference of 2000 V? ■

27.1.2 Calculating Capacitance

The definition of capacitance reflects a capacitor's practical use: we apply a known potential difference ΔV to the capacitor, which then stores charge. To calculate the capacitance of a given pair of conductors we work in the opposite direction and compute the potential difference between the conductors when a charge Q is stored.

FOR EXAMPLE, A BATTERY IS CONNECTED ACROSS THE PLATES AND HOLDS CHARGE ON THEM.

EXAMPLE 27.2 ♦♦ A spherical capacitor consists of two concentric metal spheres of radii R_1 and R_2 (■ Figure 27.3). The inner sphere carries a charge Q and the outer sphere, $-Q$. What is the potential difference between the spheres? What is their capacitance?

MODEL A uniformly charged spherical shell produces zero net field in its interior. Outside, it produces the same field strength as a point charge at its center (cf. Example 24.1). Thus the electric field between the two shells of the capacitor is due only to the inner shell, and $|\vec{\mathbf{E}}| = kQ/r^2$.

OUTSIDE BOTH SHELLS, THE NET FIELD IS ZERO.

SETUP Field lines run from the inner shell to the outer, so the inner shell is at the higher potential. The potential difference between the shells is (eqn. 25.7):

SINCE WE ARE INTEGRATING FROM R_2 TO R_1, EACH dr IS NEGATIVE. DON'T BE TEMPTED TO ADD EXTRA MINUS SIGNS!

$$\Delta V = V_1 - V_2 = -\int_{R_2}^{R_1} \vec{\mathbf{E}} \cdot d\vec{\ell} = -\int_{R_2}^{R_1} \frac{kQ}{r^2}\hat{\mathbf{r}} \cdot dr\,\hat{\mathbf{r}} = -\int_{R_2}^{R_1} \frac{kQ}{r^2}\,dr = \left.\frac{kQ}{r}\right|_{R_2}^{R_1}$$

$$= kQ\left(\frac{1}{R_1} - \frac{1}{R_2}\right) = kQ\,\frac{R_2 - R_1}{R_1 R_2}.$$

SOLVE The capacitance of the object is:

$$C = \frac{Q}{\Delta V} = \frac{R_1 R_2}{k(R_2 - R_1)} = \frac{4\pi\epsilon_0 R_1 R_2}{R_2 - R_1}. \qquad (27.4)$$

ANALYZE Again, the capacitance depends only on the physical dimensions of the capacitor. ■

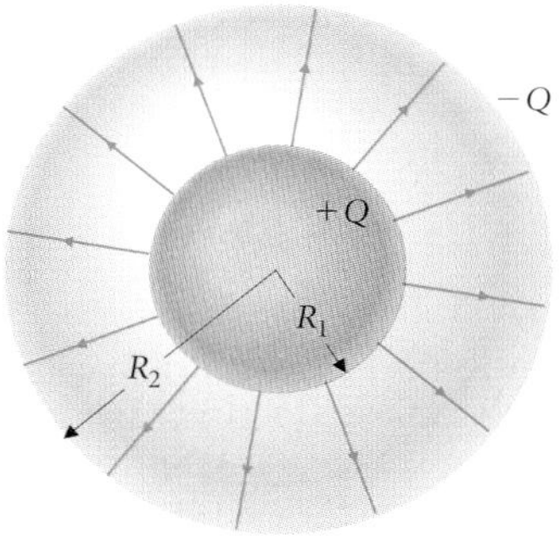

■ FIGURE 27.3
A spherical capacitor consists of two concentric spherical shells. The electric field due to the outer sphere is zero between the shells.

EXERCISE 27.2 ♦ What is the capacitance of a pair of spherical shells with $R_1 = 2.0$ cm and $R_2 = 3.0$ cm?

EXERCISE 27.3 ♦♦ Show that when $R_2 - R_1 \ll R_1$, the capacitance of two nested spheres is the same as that of parallel plates with the same surface area and separation. ■

An isolated conductor may be viewed as one of a pair of conductors, the other of which is at infinity; then we may speak of the capacitance of an isolated object.

EXAMPLE 27.3 ♦♦ Find the capacitance of an aluminum ball with a radius of 10.0 cm.

MODEL This is a special case of the capacitor in Example 27.2, in which $R_2 \to \infty$.

SETUP When taking the limit, it is easiest to use the intermediate result in Example 27.2:

$$C = \frac{Q}{\Delta V} = \lim_{R_2 \to \infty} \frac{1}{k(1/R_1 - 1/R_2)} = \frac{R_1}{k} = 4\pi\epsilon_0 R_1.$$

SOLVE With $R_1 = 10.0$ cm:

$$C = \frac{1.00 \times 10^{-1}\ \text{m}}{8.99 \times 10^9\ \text{N}\cdot\text{m}^2/\text{C}^2} = 11.1\ \text{pF}.$$

ANALYZE The ball would have to be huge to make a 1-F capacitor. The Earth is good for about 0.7 mF. ■

Our result for a sphere gives a useful rule for estimating the capacitance of any isolated object:

$$\text{Capacitance} \sim (\text{linear dimension})/k.$$

For the sphere, the linear dimension is the radius of the sphere, and the rule gives an exact result. With more complicated systems, the result is cruder. For parallel plates, the only combination of A and d with dimensions of length is A/d. The rule gives $C \sim A/kd$, which is too large by a factor 4π.

EXAMPLE 27.4 ♦♦ Find the capacitance per unit length of coaxial cable if its inner wire has a radius of 1.5 mm and its outer conductor has an inside radius of 4.5 mm. (■ Figure 27.4).

MODEL We calculate the potential difference between the two conductors when they carry charges per unit length $+\lambda$ and $-\lambda$; then we apply the definition of capacitance.

SETUP A charged cylindrical shell, like a spherical shell, produces zero net field inside itself. The electric field the shell produces outside itself is (eqn. 24.4):

$$\vec{\mathbf{E}} = \left(\frac{\lambda}{2\pi\epsilon_0 r}, \quad \text{radially outward}\right).$$

Between the conductors in the cable, the net field is that due to the charge on the inner cylinder. The potential difference is:

$$\Delta V = V_1 - V_2 = -\int_{r_2}^{r_1} \vec{\mathbf{E}} \cdot d\vec{\ell} = -\int_{r_2}^{r_1} \frac{\lambda}{2\pi\epsilon_0 r}\, dr = \frac{\lambda}{2\pi\epsilon_0} \ln(r_2/r_1).$$

SOLVE The capacitance per unit length is:

$$\frac{dC}{d\ell} = \frac{dQ/d\ell}{\Delta V} = \frac{\lambda}{\Delta V} = \frac{2\pi\epsilon_0}{\ln(r_2/r_1)} = \frac{1}{2k\ln(r_2/r_1)}. \tag{27.5}$$

Putting in the numbers, we get $r_2/r_1 = (4.5\text{ mm})/(1.5\text{ mm}) = 3.0$. Then:

$$\frac{dC}{d\ell} = \frac{1}{2(9.0 \times 10^9\ \text{N}\cdot\text{m}^2/\text{C}^2)\ln(3.0)} = 0.051\ \text{nF/m}.$$

ANALYZE The capacitance of a finite length of cable is approximately its length times the capacitance per unit length. The appropriate dimension of a cylinder to use in our rule for estimating capacitance is the length. ■

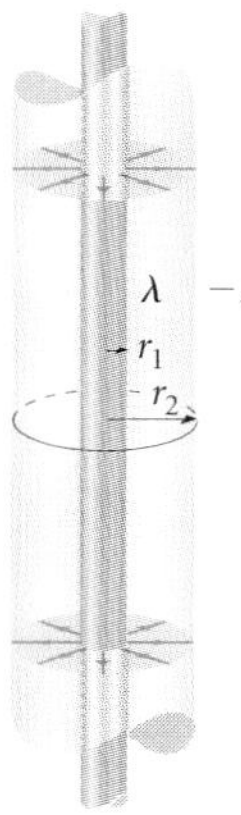

■ FIGURE 27.4
A coaxial cable has an inner conducting cylinder surrounded by a conducting cylindrical shell. Each carries charge per unit length $\pm\lambda$. Here we assume the inner conductor is positively charged. As in the spherical capacitor, the electric field due to the outer conductor is zero inside the cable.

EXERCISE 27.4 ♦ Estimate the capacitance of a person.

Digging Deeper

GROUNDING A DC CIRCUIT

When a DC circuit is grounded, why isn't the circuit disrupted by a current through the ground connection? The circuit in ■ Figure 27.5 operates disconnected from ground when the switch is open, and carries a current of 7 mA. The circuit can be grounded at either terminal of the battery by moving the switch to A or to B. When the switch is moved from A to B the potential of every point in the circuit increases by $\Delta V = 14$ V. To produce this change, a charge ΔQ flows through the ground wire, where $\Delta Q = C\,\Delta V$, and C is the capacitance of the circuit. (The other conductor is at infinity, cf. Example 27.3). Using the estimation rule with a linear dimension of 10 cm for the circuit, we obtain $C \approx (10^{-1}\text{ m})/k \approx 10$ pF, and $\Delta Q = C\,\Delta V \approx 0.14$ nC.

The current of 7 mA in the circuit can distribute this charge in a time $\tau \approx \Delta Q/I \approx 20$ ns. Charge does flow from ground—for this very short time—but once equilibrium is established, no further charge transfer occurs.

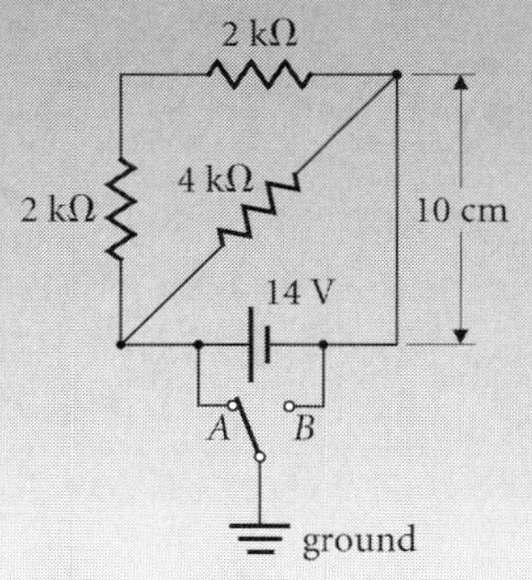

■ FIGURE 27.5
This DC circuit may be grounded at either terminal of the battery by placing the switch at point A or at point B. When the switch setting is changed, the circuit requires about 20 ns to adjust.

27.2 Energy Storage in Capacitors

27.2.1 Charging a Capacitor

We'll investigate this circuit more carefully in Chapter 31.

To charge a capacitor, we connect it to an emf that supplies energy to separate charge into equal positive and negative amounts on the two conductors. Unavoidably the circuit contains some resistance r (■ Figure 27.6a). At $t = 0$, the switch is closed in position A, and charge begins to flow. As time passes and charge accumulates on the capacitor, the potential difference across the capacitor becomes equal to the emf of the battery, and the current decreases to zero. Each of the two conductors then has charge of magnitude:

$$Q = C\,\Delta V = C\mathcal{E}.$$

The emf does work both to charge the capacitor and to drive current through the resistor. The work done on the resistor is dissipated as heat. What happens to the work done on the capacitor? If the switch is subsequently moved to position B (Figure 27.6b), the potential difference across resistor R drives current in the right-hand loop as the capacitor discharges. The energy dissipated in resistor R must have been stored in the capacitor.

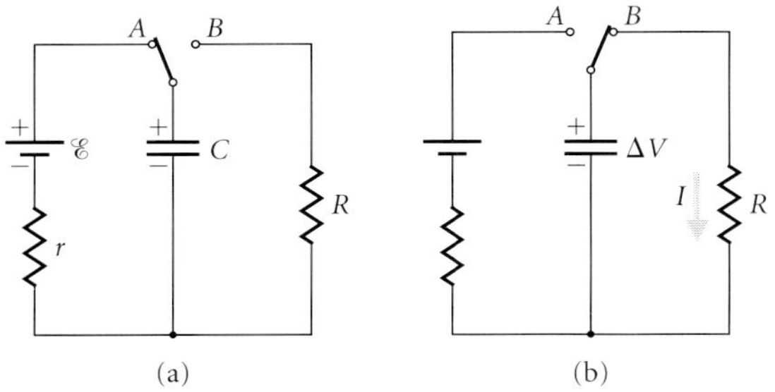

■ Figure 27.6
Circuit for charging and discharging a capacitor. (a) With the switch in position A, as shown, the capacitor stores charge as the potential difference across it approaches the emf of the battery. (b) When the switch is moved to position B, the capacitor discharges through the resistor. Energy is stored in the capacitor when it is charged and is subsequently dissipated as heat by the resistor.

Remember: This is an *imaginary* process!

We use a capital D for the plate separation here to avoid confusion with the differential quantities dq and dW.

The electrostatic energy stored in the capacitor equals the work done to charge it. The energy depends only on the amount of charge, not how it was put in place. So we may compute the energy by analyzing any imaginary process that would charge the capacitor (cf. §25.3). So we imagine a process that is easy to compute: carrying charge across the gap from one plate of a parallel plate capacitor to the other (■ Figure 27.7). When the capacitor is partially charged, the electric field strength between the plates is $E = 4\pi k\sigma$. The surface charge density σ is related to the amount of charge q already on the plate, $\sigma = q/A$, and the potential difference between the plates is:

$$\Delta V = ED = 4\pi k\sigma D = 4\pi kqD/A.$$

To increase the charge on the plate by dq we do work:

$$dW = \Delta V\,dq = \frac{4\pi kqD}{A}\,dq.$$

Since the work done depends on the charge already transferred, we compute the total work done by integrating. The physical process is a sum of differential charge transfers dq, each involving work dW. The coordinate q ranges from 0 (uncharged) to Q (fully charged). Thus the total work done to charge the capacitor is:

$$W = \int_0^Q \Delta V\,dq = \frac{4\pi kD}{A}\int_0^Q q\,dq = \frac{4\pi kD}{A}\frac{Q^2}{2} = \frac{Q^2}{2C}.$$

We set the work done equal to the stored energy:

$$U = Q^2/2C. \tag{27.6a}$$

Since $Q = C\,\Delta V$, we may also write this as:

$$U = \tfrac{1}{2}C(\Delta V)^2, \tag{27.6b}$$

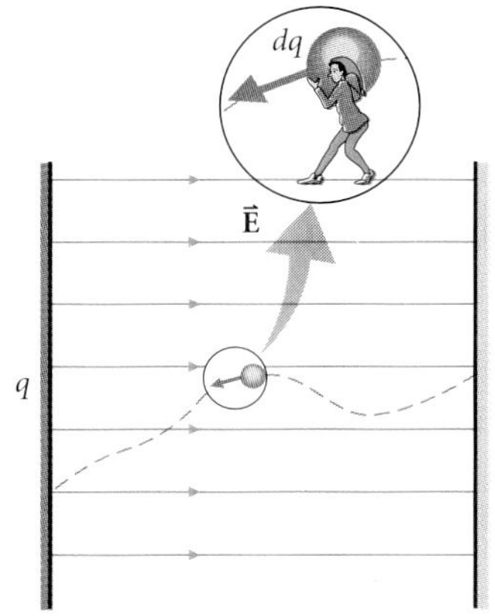

■ Figure 27.7
An imaginary process for charging a capacitor. An elf carries positive charge from the negative plate to the positive plate. The elf has to do work against the electric field produced by the charge already in place.

or: $$U = \tfrac{1}{2}Q\,\Delta V. \quad (27.6c)$$

COMPARE THE EXPRESSION $U = \frac{1}{2}Q\,\Delta V$ WITH EQN. (25.14).

These expressions apply to any capacitor. If we replace the capacitance $A/4\pi kD$ with the arbitrary capacitance C, the argument goes through unchanged.

EXERCISE 27.5 ♦ A parallel plate capacitor has plates of area 10.0 m^2 separated by a gap 0.10 mm wide. How much energy does it store when connected to a 5.0-V battery?

According to eqn. (27.6b), to store more energy in a given capacitor we should increase the potential difference across it. But the potential difference is limited: too large an electric field destroys the capacitor. For example, an electric field greater than a few million volts per meter ionizes air, and charge can flow from one plate to another across the air gap. The largest individual capacitors used in the Berkeley neutral beam injector have a capacitance of 6.0 mF and a limiting voltage of 450 V. Large numbers of these capacitors are connected together to increase both the total potential difference and the total energy stored in the system. Next we'll investigate how these connections work.

27.2.2 *Capacitors in Parallel*

Stored energy is increased with no change in voltage by using a *parallel* combination.

Two or more capacitors are in *parallel* when they are connected between the same two points in a circuit, and the potential difference across each is the same.

The two parallel capacitors in ■ Figure 27.8 are connected across the same potential difference $\Delta V = \mathcal{E}$. Thus:

$$\Delta V = \frac{Q_1}{C_1} = \frac{Q_2}{C_2}.$$

The total amount of charge stored is $Q \equiv Q_1 + Q_2$. Thus the equivalent capacitance of the combination is:

$$C_\parallel \equiv \frac{Q}{\Delta V} = \frac{Q_1 + Q_2}{\Delta V} = C_1 + C_2.$$

The same argument extends to any number of capacitors in parallel, so the equivalent capacitance of any parallel combination of capacitors is the sum of the individual capacitances:

$$C_\parallel = \sum_i C_i. \quad (27.7)$$

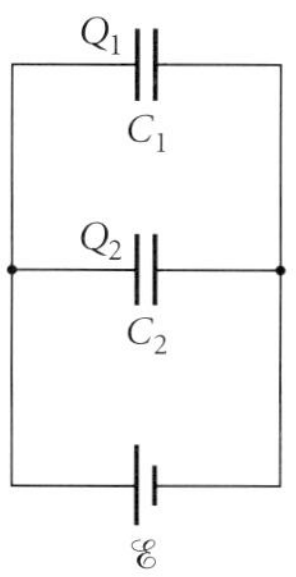

■ FIGURE 27.8
Two capacitors in parallel. The potential difference across each capacitor is the same. The total charge is the sum of the amounts of charge on each capacitor: $Q = Q_1 + Q_2$.

EXAMPLE 27.5 ♦ A 1.0-μF and a 6.0-μF capacitor are connected in parallel. What is the capacitance of the combination? How much energy is stored when they are connected to a 10.0-V power supply?

MODEL The capacitance of the parallel combination is the sum of the individual capacitances.

SETUP AND SOLVE $$C_\parallel = C_1 + C_2 = 7.0\ \mu\text{F}.$$

The stored energy is (eqn. 27.6b):

$$W = \tfrac{1}{2}C_\parallel(\Delta V)^2 = \tfrac{1}{2}(7.0 \times 10^{-6}\ \text{F})(10.0\ \text{V})^2 = 3.5 \times 10^{-4}\ \text{J}.$$

ANALYZE Check the units:

$$\text{F}\cdot\text{V}^2 = \frac{\text{C}}{\text{V}}\text{V}^2 = \text{C}\cdot\text{V} = \text{C}\frac{\text{N}\cdot\text{m}}{\text{C}} = \text{J}.$$

The total energy stored is the sum of the energies stored in the individual capacitors.

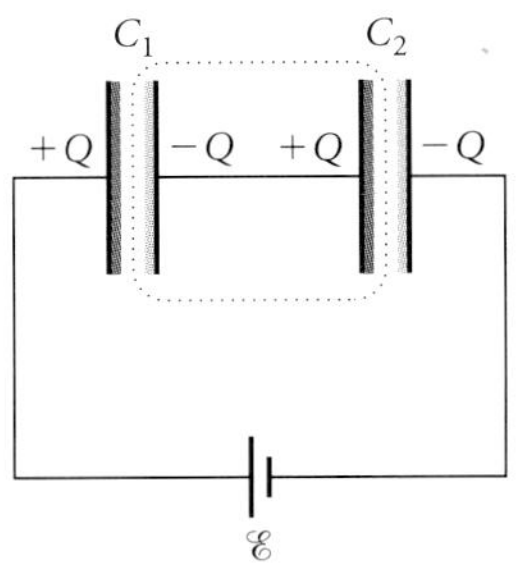

■ FIGURE 27.9
Two capacitors in series. The dotted curve encloses an electrically isolated piece of the circuit whose total charge must remain constant and equal to zero. The charge on the negative plate of C_1 and that on the positive plate of C_2 have the same magnitude. The capacitors store equal amounts of charge.

27.2.3 *Capacitors in Series*

A series connection produces a compound capacitor able to withstand a greater total potential difference.

> Two or more capacitors are in *series* if they are connected in sequence in one arm of a circuit.

The capacitors in ■ Figure 27.9 are in series. The dotted curve in the figure includes one conducting surface from each capacitor and the wire used to connect them. Together these elements form one unified conducting body that is electrically isolated from the rest of the circuit and so contains zero net charge. Any electrons removed from the positive plate of C_2 are deposited on the negative plate of C_1. The two capacitors store equal charges: $Q_1 = Q_2 = Q$. The total potential difference across the two capacitors is the sum of the potential differences across each one:

$$\mathscr{E} = \Delta V = \Delta V_1 + \Delta V_2 = \frac{Q_1}{C_1} + \frac{Q_2}{C_2} = Q\left(\frac{1}{C_1} + \frac{1}{C_2}\right).$$

In terms of the total stored charge Q and the equivalent capacitance of the combination, C_s, the potential difference is: $\Delta V = Q/C_s$. So:

$$\frac{1}{C_s} = \frac{1}{C_1} + \frac{1}{C_2} \quad \text{or} \quad C_s = \frac{C_1 C_2}{C_1 + C_2}.$$

Once again, the argument extends to an arbitrary number of capacitors.

> For any number of capacitors in series, the reciprocal of the equivalent capacitance equals the sum of the reciprocals of the individual capacitances:
>
> $$\frac{1}{C_s} = \sum_i \frac{1}{C_i}. \tag{27.8}$$

EXERCISE 27.6 ♦ What is the equivalent capacitance when the 1.0-μF and 6.0-μF capacitors in Example 27.5 are connected in series? What is the stored energy when they are connected across the 10.0-V power supply?

A QUANTITY CALLED *REACTANCE* MEASURES VOLTS PER AMPERE, AND REACTANCES COMBINE LIKE RESISTANCES (SEE CHAPTER 32).

The rules for combining capacitors in series are similar to those for combining resistors in parallel; capacitors in parallel add in the same way as resistors in series. This reversal arises from the role of potential difference in the definitions of capacitance and resistance: capacitance is coulombs per volt, but resistance is volts per ampere.

27.2.4 *Series and Parallel Combinations*

A capacitor network containing both series and parallel combinations of capacitors may be analyzed in the same way as a resistor network. We find a sequence of equivalent circuits that reduces to a single equivalent capacitor.

EXAMPLE 27.6 ♦♦ What is the equivalent capacitance of the network shown in ■ Figure 27.10a?

MODEL We proceed to simplify each series or parallel combination, as we did in Chapter 26 with resistor combinations.

SETUP The two series combinations have the following equivalents:

$$C_1 = \frac{(12\ \mu\text{F})(12\ \mu\text{F})}{12\ \mu\text{F} + 12\ \mu\text{F}} = 6.0\ \mu\text{F} \quad \text{and} \quad C_2 = \frac{(24\ \mu\text{F})(6.0\ \mu\text{F})}{24\ \mu\text{F} + 6.0\ \mu\text{F}} = 4.8\ \mu\text{F}.$$

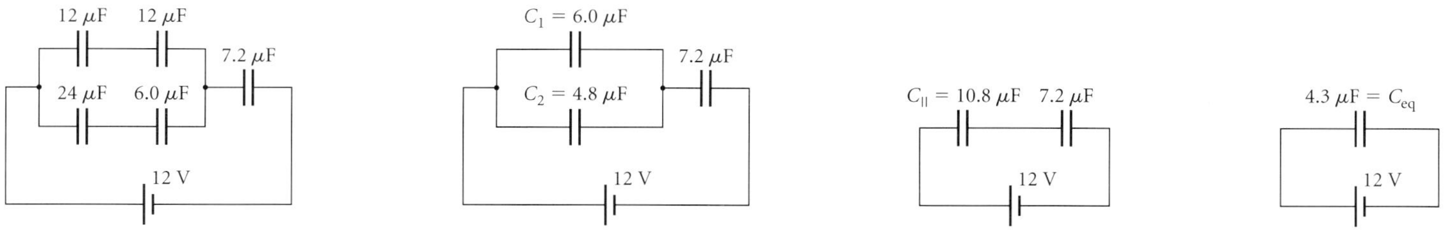

■ FIGURE 27.10
(a) A series-parallel capacitor combination is simplified in steps.
(b) The series combinations are replaced with their series equivalents.
(c) The parallel combination is replaced with its equivalent.
(d) The final series combination is replaced with its equivalent.

Figure 27.10b is the equivalent circuit with these substitutions. Next, the parallel combination of C_1 and C_2 is equivalent to:

$$C_{\parallel} = 6.0\ \mu\text{F} + 4.8\ \mu\text{F} = 10.8\ \mu\text{F}.$$

SOLVE The 7.2-μF capacitor is in series with $C_{\parallel}$ (Figure 27.10c). The equivalent capacitance of the whole combination is:

$$C_{eq} = \frac{(7.2\ \mu\text{F})(10.8\ \mu\text{F})}{7.2\ \mu\text{F} + 10.8\ \mu\text{F}} = 4.3\ \mu\text{F}.$$

ANALYZE The capacitance of a series combination is less than any of the individual capacitances. The capacitance of a parallel combination is greater than each individual capacitance. Again these are the opposite of the rules for resistors. ■

EXAMPLE 27.7 ♦♦ What is the charge on the 24-μF capacitor in Figure 27.10a?

MODEL We proceed by using the sequence of equivalent circuits found in the previous example, working backward toward the individual capacitors.

SETUP The total charge stored by the combination (Figure 27.10d) is:

$$Q = C_{eq}\mathcal{E} = (4.3\ \mu\text{F})(12\ \text{V}) = 52\ \mu\text{C}.$$

Since series capacitors have the same charge, both the 7.2-μF capacitor and the 10.8-μF parallel combination of C_1 and C_2 hold equal charges Q (Figure 27.10c). Capacitors C_1 and C_2 share this charge so as to have the same potential difference: $Q_1 + Q_2 = Q$ (Figure 27.10b). Also:

$$\Delta V_{\parallel} = \frac{Q_1}{C_1} = \frac{Q_2}{C_2} = \frac{Q}{C_{\parallel}} = \frac{Q}{C_1 + C_2}.$$

Thus the parallel capacitors share the charge in proportion to their capacitances. So:

$$Q_2 = \frac{C_2}{C_1 + C_2}Q = \frac{4.8}{10.8}(52\ \mu\text{C}) = 23\ \mu\text{C}.$$

SOLVE Capacitor C_2 is actually the series combination of the 6.0-μF and 24-μF capacitors (Figure 27.10a), each of which holds $Q_2 = 23$ μC.

ANALYZE If we replaced the 7.2-μF capacitor with a larger one, the charge on the 24-μF capacitor would increase. Do you see why? ■

Individual capacitors in the Lawrence Berkeley facility have capacitance $C_0 = 2.0$ mF. Each side of the module shown in ■ Figure 27.11 holds a series string of 136 capacitors (17 rows of 8). The two strings are connected in parallel to make the module, and ten modules are connected in parallel to form a *bank*.

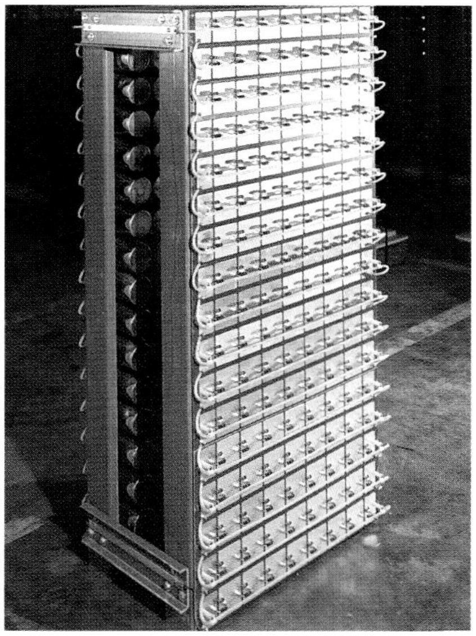

■ FIGURE 27.11
Module of capacitors used at Lawrence Berkeley Laboratory. Each side of the module holds a string of 136 2-mF capacitors (17 rows of 8). The two strings on each side are connected in parallel.

RECALL: EACH CAPACITOR IN A PARALLEL COMBINATION HAS THE SAME POTENTIAL DIFFERENCE ACROSS IT.

RECALL: CAPACITORS IN SERIES HAVE THE SAME Q.

EXERCISE 27.7 ♦♦ The capacitor banks are charged to a potential difference of 54.4 kV. Find the potential difference across an individual capacitor and the total energy stored in the bank.

IN PROBLEM 33 WE ASK YOU TO PROVE THIS RESULT.

To achieve a greater potential difference, the three banks of capacitors are charged while connected in parallel and then switched to a series connection. This increases the potential difference across the entire array of capacitors to three times the emf of the charging supply. Since the three banks act as identical capacitors, there is no current during the switching process, and no energy is needed to achieve this increase of potential difference!

27.3 DIELECTRICS AND PRACTICAL CAPACITORS

Design of a practical capacitor to store large amounts of energy presents several problems. The device needs conducting surfaces with large surface area and small separation that fit into a reasonably small volume. A design with parallel plates separated by air meets none of these needs. Charged capacitor plates attract each other and so they must be held apart mechanically (■ Figure 27.12). The electric field between the plates can produce electrical breakdown (ionization) of the air, which leaks the charge across the gap, or, at high enough voltage, can produce sparks that ruin the device. All these problems are solved by placing a layer of insulating material or *dielectric* between the conducting plates (■ Figure 27.13). In this section we shall study the electrical properties of dielectrics.

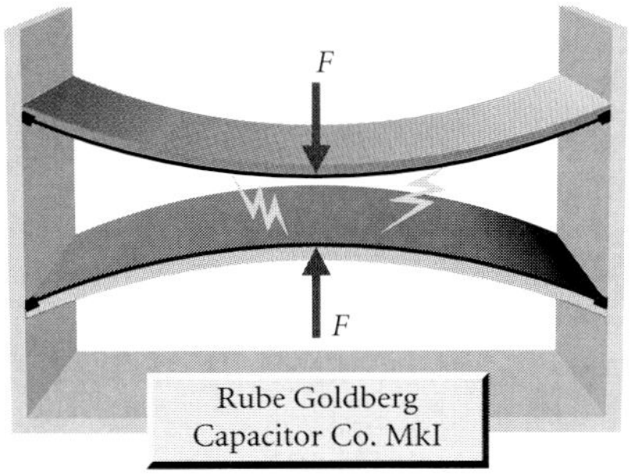

■ FIGURE 27.12
Faulty design for a large capacitor. The conductors in a capacitor are oppositely charged, so the Coulomb force between them is attractive. The device must be mechanically strong enough to resist this force. If the electric field between the plates is large enough, charge flow between the plates discharges the capacitor. For air, the limiting field is a few million volts per meter, and for dielectrics it is typically five to ten times that for air.

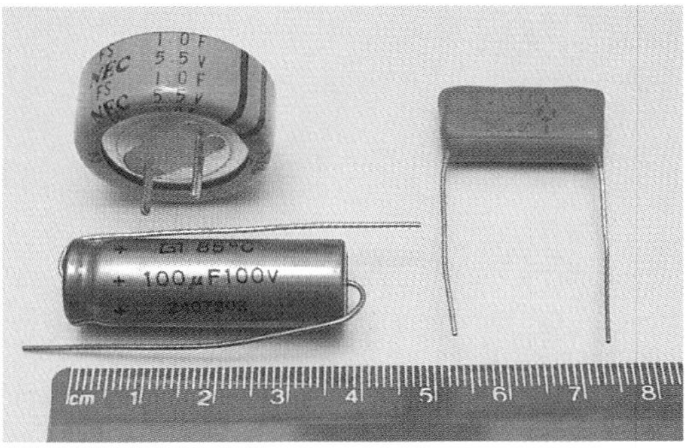

■ FIGURE 27.13
Commercially available capacitors. In the red capacitor, Mylar plastic and aluminum foil conductors form layers that are rolled into a compact cylinder. The blue capacitor was made by electrodepositing a thin dielectric layer on one conductor. This electrolytic capacitor is *polar* (one-way!)—charging the wrong conductor positive removes the dielectric layer. Compact capacitors as large as 1.0 F are now available.

27.3.1 *The Dielectric Constant*

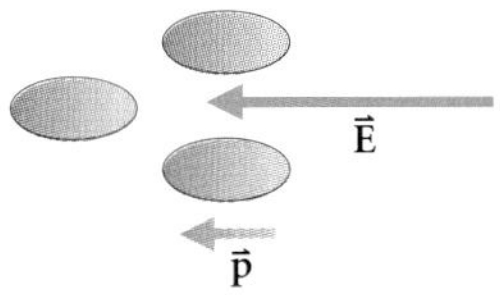

■ FIGURE 27.14
Electric field acting on individual molecules of an insulating substance polarizes them. Each molecule becomes a small dipole $\vec{\mathbf{p}}$ oriented in the same direction as the electric field vector $\vec{\mathbf{E}}$.

In an insulating substance, electrons are not free to move from atom to atom when an electric field is applied. Instead, the electric force produces *polarization* (■ Figure 27.14). A small displacement of the electrons in each atom or molecule of the substance forms a small dipole, oriented in the direction of the electric field. Any volume in the interior of the dielectric remains electrically neutral, but on the dielectric's surface the ends of these atomic dipoles produce a net charge density, σ_B, called *bound* charge. Like the free charge density on the surface of conductors, this bound charge occurs within the outermost atomic layer of the dielectric surface. Unlike the free charge, however, bound charge is attached to its parent atoms. It cannot flow throughout the volume of the dielectric or into the external circuit.

Bound charge on the surface of a dielectric material, like surface charge on a conductor, produces a net electric field that opposes the applied field (■ Figure 27.15). But unlike surface charge on a conductor, bound charge is not large enough to eliminate the net electric field within the dielectric material entirely. The polarization of individual atoms is proportional to the field acting on them, so the amount of bound charge and the internal field it produces are both proportional to the applied field. Consequently, the actual net field $\vec{\mathbf{E}}$ that exists within the material is proportional to the external applied field. The ratio between the two is a property of the particular material.

IN SOME MATERIALS, PARTICULARLY CRYSTALS, THE DIPOLE ORIENTATION DEPENDS ON THE STRUCTURE OF THE MATERIAL AS WELL AS THE DIRECTION OF $\vec{\mathbf{E}}$. THIS EFFECT IS NOT IMPORTANT FOR OUR DISCUSSION OF CAPACITORS.

SEE §27.3.2 FOR DETAILS.

The *dielectric constant* κ of a material is the ratio of the applied electric field to the net electric field within the material:

$$\kappa = \frac{|\vec{\mathbf{E}}_a|}{|\vec{\mathbf{E}}|}. \tag{27.9}$$

κ IS THE GREEK LETTER KAPPA (APPENDIX IIA).

• Table 27.1 lists dielectric constants of some materials commonly used in capacitors. The dielectric constant of air is very close to unity, and for most purposes we may ignore the difference.

When the applied electric field is very large, the force on individual molecules becomes large enough to strip off electrons, and the material begins to conduct. The maximum electric field that a material can withstand before it loses its insulating properties is called its *dielectric strength.* The value of dielectric strength is quite variable, because it depends on temperature, presence of impurities, and so forth. Values vary from about 3 MV/m for air to a few tens of MV/m for ceramics and plastics.

TABLE 27.1 Dielectric Constants of Some Common Materials

Material	Dielectric Constant
Air (1 atm)	1.0006
Water	80
Ruby mica	5–7
Glass	4–10
Class I ceramic	5–450
Class II ceramic	200–12000
Oil-impregnated paper	4–6
Mylar	3
Polytetrafluoroethylene (Teflon)	2.1
Polyethylene	2.4
Polystyrene	2.5
Polycarbonate	3
Plexiglas	3.1
Epoxy	3.6
Polyvinyl chloride	4.6
Aluminum oxide	11
Tantalum oxide	28
Diamond	5.7

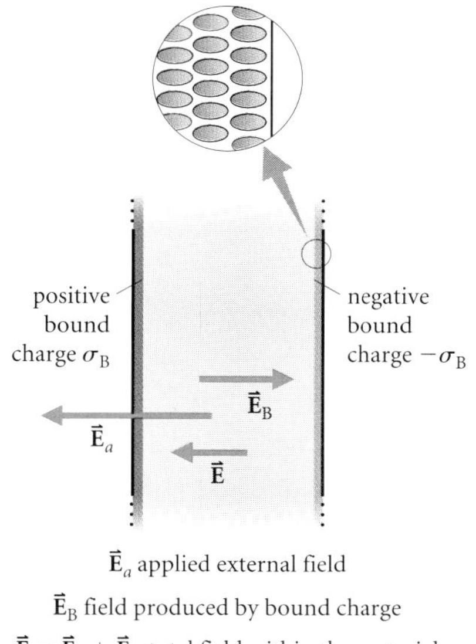

■ FIGURE 27.15
At the surface of a dielectric material the molecular dipoles form an unbalanced layer of charge known as *bound charge.* The bound charge produces an electric field that opposes the externally applied field within the material.

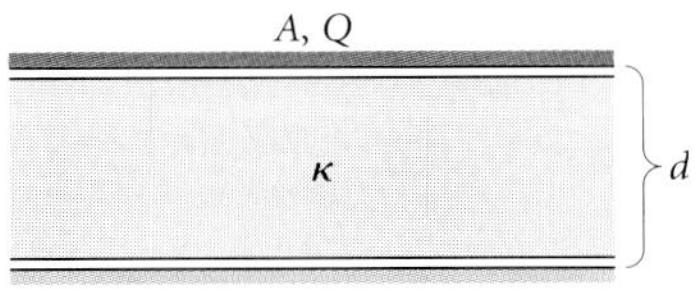

■ FIGURE 27.16
A parallel plate capacitor filled with dielectric. Because the potential difference between the plates for a given charge Q is reduced, the dielectric increases the capacitance by a factor κ.

EXAMPLE 27.8 ♦♦ A material with dielectric constant κ fills a parallel plate capacitor with plate area A and separation d (■ Figure 27.16). When the magnitude of charge on each plate is Q, what is the electric field between the plates? What is the potential difference between the plates? What is the capacitance of the device?

MODEL The applied field is produced by the charged plates:

$$|\vec{\mathbf{E}}_a| = \frac{\sigma}{\epsilon_0} = \frac{Q}{\epsilon_0 A}.$$

The field within the dielectric is reduced by the induced bound charge, so the potential difference is reduced, and the capacitance (eqn. 27.2) is increased.

SETUP From eqn. (27.9):

$$|\vec{\mathbf{E}}| = \frac{|\vec{\mathbf{E}}_a|}{\kappa} = \frac{Q}{\kappa\epsilon_0 A}.$$

SOLVE The potential difference is:

$$\Delta V = |\vec{\mathbf{E}}|d = \frac{Qd}{\kappa\epsilon_0 A},$$

and the capacitance is:

$$C \equiv \frac{Q}{\Delta V} = \frac{Q\kappa\epsilon_0 A}{Qd} = \kappa\frac{\epsilon_0 A}{d}.$$

ANALYZE The capacitance of the plates is increased by a factor of κ over its value with vacuum between the plates (eqn. 27.1). An alternative definition of the dielectric constant is the factor by which a material increases the capacitance of a system. ■

EXERCISE 27.8 ♦ The red device shown in Figure 27.13 has a capacitance $C = 1.0\ \mu$F. If the dielectric material is a 2.5×10^{-5}-m-thick sheet of mylar, what is the surface area of each aluminum foil conductor? ■

EXAMPLE 27.9 ♦♦ A parallel plate capacitor with plates of area $A = 16\ \text{cm}^2$ and separation $d = 0.50$ cm is connected to a 12.0-V battery. How much energy is stored in the capacitor? A slab of polyvinyl chloride (PVC) that just fits between the plates is inserted into the capacitor. What is the energy stored then? What difference does it make if the battery is disconnected before the PVC slab is inserted?

MODEL The battery charges the capacitor, and the stored energy is given by eqn. (27.6). Inserting the PVC slab increases the capacitance. With the battery connected, the potential difference remains fixed, and the stored energy increases. If the battery is disconnected first, the *charge* remains fixed. Using eqn. (27.6a), we see that the stored energy decreases as the capacitance is increased.

SETUP With air between the plates, the capacitance is (eqn. 27.3):

$$C_{\text{air}} = \frac{\epsilon_0 A}{d} = \frac{(8.85 \times 10^{-12}\ \text{F/m})(16 \times 10^{-4}\ \text{m}^2)}{5.0 \times 10^{-3}\ \text{m}} = 2.83\ \text{pF}.$$

Connected to the 12.0-V battery, the stored charge is:

$$Q = C\,\Delta V = (2.83\ \text{pF})(12.0\ \text{V}) = 34.0\ \text{pC}.$$

SOLVE The stored energy at 12.0 V is:

$$U_{\text{air}} = \tfrac{1}{2}C_{\text{air}}(\Delta V)^2 = \tfrac{1}{2}(2.83\ \text{pF})(12.0\ \text{V})^2 = 0.20\ \text{nJ}.$$

SETUP After insertion of the PVC slab ($\kappa = 4.6$ from Table 27.1), the capacitance becomes:

$$C_{\text{PVC}} = \kappa C_{\text{air}} = (4.6)(2.83\ \text{pF}) = 13.0\ \text{pF}.$$

SOLVE The stored energy at 12.0 V is then:

$$U_1 = \tfrac{1}{2}C_{PVC}(\Delta V)^2 = \kappa U_{air} = 4.6(0.204\ \text{nJ}) = 0.94\ \text{nJ}.$$

Remember: Keep an extra figure in intermediate results.

If instead the battery is disconnected before the slab is inserted, the new energy is:

$$U_2 = \frac{Q^2}{2C_{PVC}} = \frac{(3.40 \times 10^{-11}\ \text{C})^2}{2(13.0 \times 10^{-12}\ \text{F})} = 0.044\ \text{nJ}.$$

ANALYZE When it remains connected to the battery, the capacitor's stored energy increases as the slab enters. That means work is *done on* the system of capacitor and slab. The charge on the capacitor after the slab is inserted is (eqn. 27.6c) $2U_1/\Delta V = 0.16$ nC. The extra charge is drawn from the battery, which also supplies the required energy.

If the battery is disconnected first, the energy decreases as the slab is inserted. That means the *system does work;* it sucks the slab in! The final potential difference is $Q/C_{PVC} = 2.6$ V. ∎

✻ 27.3.2 Polarization and Susceptibility

■ Figure 27.17 illustrates the connection between the bound charge at the surface of a dielectric and the polarization of the molecules. The bound charge per unit area equals the number of molecules per unit area at the surface times the bound charge at the end of each molecule:

$$\begin{aligned}\text{Bound charge per unit area} &= \text{molecules per unit area} \times \text{charge per molecule}\\ &= \text{molecules per unit area} \times \frac{\text{dipole moment}}{\text{length of molecule}}\\ &= \text{molecules per unit volume} \times \text{dipole moment.}\end{aligned}$$

$$\sigma_B = np,$$

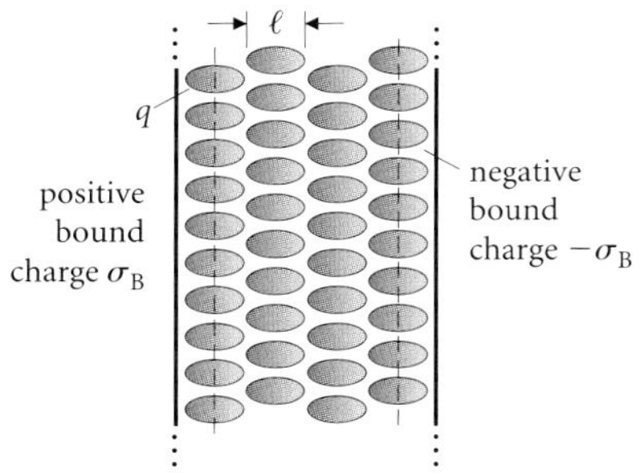

■ Figure 27.17
The bound charge arises because the end of each molecule at the surface has the same sign of charge. The charge on its other end is balanced by molecules one layer in from the surface. The amount of bound charge contributed by each molecule equals its dipole moment $p = q\ell$ divided by its length ℓ.

where n is the number of molecules per unit volume and p is the dipole moment per molecule. The product np is the dipole moment per unit volume in the material. Together with the direction of the dipoles this quantity forms a vector called the *polarization vector* $\vec{\mathbf{P}} = n\vec{\mathbf{p}}$. The electric field produced by the bound charge has magnitude $E_B = \sigma_B/\epsilon_0$ and is in the direction opposite the polarization of the material:

$$\vec{\mathbf{E}}_B = -\frac{n\vec{\mathbf{p}}}{\epsilon_0} = \frac{-\vec{\mathbf{P}}}{\epsilon_0}.$$

The net electric field $\vec{\mathbf{E}}$ is the sum of the applied field $\vec{\mathbf{E}}_a$ and the field $\vec{\mathbf{E}}_B$ produced by the bound charge:

$$\vec{\mathbf{E}} = \vec{\mathbf{E}}_a + \vec{\mathbf{E}}_B = \vec{\mathbf{E}}_a - \vec{\mathbf{P}}/\epsilon_0.$$

The polarization vector $\vec{\mathbf{P}}$ results from the dipole moments of individual molecules, each proportional to the net field:

A greater field produces a greater charge separation and thus a greater dipole moment.

$$\vec{\mathbf{P}} \equiv n\vec{\mathbf{p}} \equiv \epsilon_0\chi\vec{\mathbf{E}}.$$

The proportionality constant χ is called the *dielectric susceptibility* of the material. Combining these relations, we find:

$$\vec{\mathbf{E}}_a = \vec{\mathbf{E}} + \frac{\vec{\mathbf{P}}}{\epsilon_0} = \vec{\mathbf{E}} + \frac{\epsilon_0\chi\vec{\mathbf{E}}}{\epsilon_0} = \vec{\mathbf{E}}(1 + \chi). \tag{27.10}$$

Comparing eqn. (27.10) with eqn. (27.9), we find that the dielectric constant is given by:

$$\kappa = 1 + \chi. \tag{27.11}$$

SEE §24.5, ESPECIALLY EXAMPLE 24.10.

In some materials, such as water, each molecule has a dipole moment even without an applied field. An applied field acts to align the otherwise randomly oriented molecular dipoles. Because the degree of alignment is proportional to $\vec{E}_a$, eqns. (27.10) and (27.11) still apply.

✲ 27.3.3 *Electric Displacement*

Electric field within a dielectric material is produced by both free and bound charge. It is sometimes more convenient to work with a vector called electric *displacement* $\vec{D}$, which is related to the free charge alone. The displacement is defined by its relation to the electric field vector and the polarization:

$$\vec{D} \equiv \epsilon_0\vec{E} + \vec{P} = \epsilon_0\vec{E} - \epsilon_0\vec{E}_B = \epsilon_0\vec{E}_a.$$

$$\vec{D} = \epsilon_0\kappa\vec{E}. \tag{27.12}$$

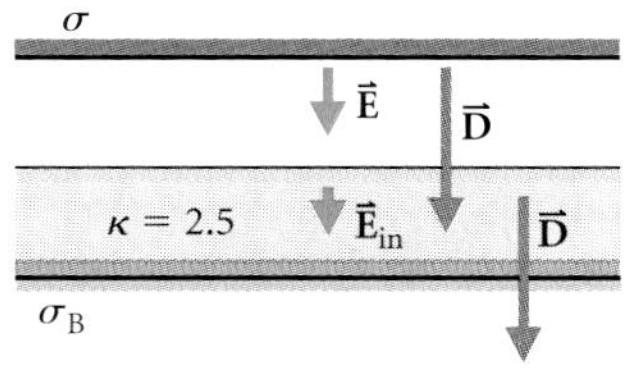

■ FIGURE 27.18
A capacitor half-filled with dielectric. The electric field $\vec{E}$ is reduced within the dielectric, whereas the displacement $\vec{D}$ is uniform throughout the capacitor's interior.

EXAMPLE 27.10 ◆◆ A plastic slab with dielectric constant $\kappa = 2.50$ fills half the volume between the plates of a capacitor (■ Figure 27.18). The plates have area $A = 25\ \text{cm}^2$ and separation 1.50 mm. The electric field in the air is $|\vec{E}_a| = 1.00\ \text{kV/m}$. Find $|\vec{E}_{in}|$ in the dielectric and $|\vec{D}|$ both in the air and in the dielectric. What is the capacitance?

MODEL Outside the dielectric, the two bound charge layers produce canceling fields, so the electric field in the air gap is just the applied field due to the free charges on the capacitor plates. Within the dielectric, the applied field is reduced by $\vec{E}_B$.

SETUP In the air gap: $|\vec{E}| \equiv |\vec{E}_a| = \sigma/\epsilon_0 = 1.00\ \text{kV/m}$.

SOLVE The electric field in the dielectric is reduced by a factor κ, so:

$$|\vec{E}_{in}| = |\vec{E}_a|/\kappa = (1.00\ \text{kV/m})/(2.50) = 0.400\ \text{kV/m}.$$

In air, $\kappa \approx 1$, so that $|\vec{D}| = \epsilon_0|\vec{E}| = \epsilon_0\sigma/\epsilon_0 = \sigma$.

This expression shows directly the relation between D and the free charge. Numerically:

$$D = (8.85 \times 10^{-12}\ \text{F/m})(1.00\ \text{kV/m}) = 8.85 \times 10^{-9}\ \text{C/m}^2.$$

In the dielectric the magnitude of $\vec{D}$ is:

$$|\vec{D}_{in}| = \epsilon_0\kappa|\vec{E}_{in}| = \epsilon_0\kappa\frac{|\vec{E}_a|}{\kappa} = \epsilon_0\frac{\sigma}{\epsilon_0} = \sigma,$$

the same as in the air.

ANALYZE The displacement vector has the same value both inside and outside the dielectric and depends only on σ. Note that its units are C/m^2—the units of surface charge density.

SETUP To find the capacitance we need the potential difference:

$$|\Delta V| = \int \vec{E}\cdot d\vec{\ell} = E_a\frac{d}{2} + E_{in}\frac{d}{2} = E_a\frac{d}{2}\left(1 + \frac{1}{\kappa}\right).$$

SOLVE Then: $$\frac{1}{C} = \frac{\Delta V}{Q} = \frac{\sigma}{\epsilon_0}\frac{d}{2}\left(1 + \frac{1}{\kappa}\right)\frac{1}{\sigma A} = \frac{d}{2\epsilon_0 A}\left(1 + \frac{1}{\kappa}\right).$$

So: $$C = \frac{2(8.85 \times 10^{-12}\ \text{F/m})(25 \times 10^{-4}\ \text{m}^2)}{(1.5 \times 10^{-3}\ \text{m})(1 + 1/2.50)} = 21\ \text{pF}.$$

ANALYZE This system behaves like two capacitors in series, with capacitances

$$C_1 = 2\epsilon_0 A/d \qquad \text{and} \qquad C_2 = 2\kappa\epsilon_0 A/d$$

combining according to eqn. (27.8).

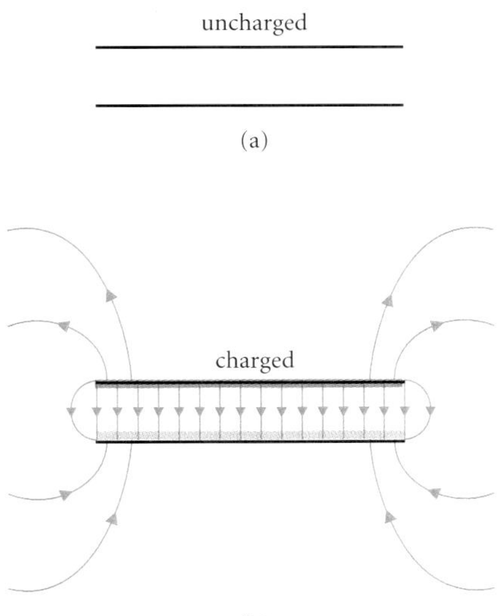

■ FIGURE 27.19
A charged capacitor differs from an uncharged one because of the separated charge on the plates and the electric field between the plates. Thinking of the field lines as stretched rubber bands gives an image for the electrostatic energy stored in the field.

27.4 ENERGY IN THE ELECTRIC FIELD

27.4.1 Electrostatic Energy Density in Vacuum

A capacitor stores electrostatic potential energy, but we don't yet have a picture of how or where it is stored. ■ Figure 27.19 shows a parallel plate capacitor before and after being charged. The important difference between the two states is the presence of the charges and their electric field. The electric field exerts forces on the charged plates, much as if the field lines were a collection of stretched rubber bands. Thinking of the field lines in this way gives us an image of the energy stored *in the field.* We shall use this capacitor to derive an expression for the amount of energy stored in a unit volume containing electric field.

The work done to charge the capacitor is (eqn. 27.6b) $W = \frac{1}{2}C(\Delta V)^2$. The capacitance is given by eqn. (27.3) ($C = \epsilon_0 A/d$), and the potential difference between the plates is $|\Delta V| = Ed$. So:

$$W = \frac{1}{2}\left(\frac{\epsilon_0 A}{d}\right)E^2 d^2 = \left(\frac{\epsilon_0 E^2}{2}\right)(Ad).$$

The stored energy is the product of the volume Ad between the capacitor plates and the uniform density u_E of electrostatic energy in vacuum:

$$u_E \equiv \frac{E^2}{8\pi k} = \frac{1}{2}\epsilon_0 E^2. \qquad \textbf{(27.13)}$$

ENERGY DENSITY IS ENERGY PER UNIT VOLUME.

Although we obtained this expression using a special model—a parallel plate capacitor—it is a completely general result for electric energy density in vacuum.

SEE PROBLEM 75 AND §27.4.2 FOR FURTHER DISCUSSION OF THIS POINT.

If dielectric materials are present, energy is stored both in the average electric field within the material and in the increased potential energy of the polarized molecules. For practical purposes we need an expression for the total energy density in terms of the electric field and the material's dielectric constant.

We derived eqn. (27.6a) for stored energy in terms of charge and capacitance by calculating the work required to charge the capacitor. A similar derivation works for capacitors with dielectrics. For a parallel plate capacitor, the stored energy is:

$$U = \frac{1}{2}\left(\frac{Q^2}{C}\right) = \frac{1}{2}\left(\frac{Q^2}{\kappa\epsilon_0 A/d}\right).$$

WE USED EQN. (27.6a) AND THE RESULT OF EXAMPLE 27.8.

The electric field magnitude is $E = Q/(\kappa\epsilon_0 A)$ (cf. Example 27.8), so:

$$Q = \kappa\epsilon_0 AE,$$

and:

$$U = \frac{1}{2}\frac{(\kappa\epsilon_0 AE)^2 d}{\kappa\epsilon_0 A} = \tfrac{1}{2}\kappa\epsilon_0 E^2(Ad).$$

Thus the energy density in a dielectric medium is the energy divided by the volume:

$$u_E = \tfrac{1}{2}\kappa\epsilon_0 E^2. \qquad (27.14)$$

COMPARE WITH EQN. (27.13); $\kappa = 1$ IN VACUUM.

EXAMPLE 27.11 ♦♦ A spherical high-voltage terminal with a radius of 2.5 mm is insulated with a layer of epoxy. If the terminal potential is 10.0 kV, what is the electric field energy density in the epoxy 1.0 mm from the surface of the terminal?

MODEL Using the known potential, we may find the electric field in vacuum at any distance from the terminal. The field in the epoxy is reduced by the dielectric constant. Once we know the value of E, eqn. (27.14) gives the energy density.

SETUP The potential due to a charged spherical conductor is $V = kQ/r$. At the surface, $r = r_s$, we set V equal to the given value of $V_s = 10.0$ kV. The applied electric field has magnitude:

$$E_a = \frac{kQ}{r^2} = \frac{kQ}{r_s}\frac{r_s}{r^2} = \frac{V_s r_s}{r^2}.$$

Within the epoxy, the electric field strength is:

$$E = \frac{E_a}{\kappa} = \frac{V_s r_s}{\kappa r^2}.$$

SOLVE From Table 27.1, $\kappa = 3.6$ for epoxy. The energy density at distance d from the center of the terminal, $r = r_s + d = 2.5\text{ mm} + 1.0\text{ mm} = 3.5\text{ mm}$, is:

$$\begin{aligned} u &= \tfrac{1}{2}\kappa\epsilon_0 E^2 = \kappa\epsilon_0 \frac{(V_s r_s)^2}{2(\kappa r^2)^2} = \frac{\epsilon_0 (V_s r_s)^2}{2\kappa(r_s + d)^4} \\ &= \frac{(8.85 \times 10^{-12}\text{ F/m})(1.0 \times 10^4\text{ V})^2(2.5 \times 10^{-3}\text{ m})^2}{2(3.6)(3.5 \times 10^{-3}\text{ m})^4} \\ &= 5.1\text{ J/m}^3. \end{aligned}$$

ANALYZE Check the units:

$$\frac{(\text{F/m})\cdot\text{V}^2\cdot\text{m}^2}{\text{m}^4} = \frac{\text{F}\cdot\text{V}^2}{\text{m}^3} = \frac{\text{C}\cdot\text{V}^2}{\text{V}\cdot\text{m}^3} = \frac{\text{C}\cdot\text{V}}{\text{m}^3} = \frac{\text{J}}{\text{m}^3}.$$

The net effect of the dielectric is to reduce the energy density. The increase of potential energy due to formation of dipoles is outweighed by the reduction of $|\vec{\mathbf{E}}|$. ■

✲ 27.4.2 *Electrostatic Energy of Two Point Charges*

In Chapter 25 we computed the potential energy of a pair of point charges Q_1 and Q_2: $U = kQ_1Q_2/r$ (eqn. 25.3). Now we have another way of computing electrostatic energy, using the energy density u_E. These two descriptions ought to agree.

The electric field vector at a point P due to the two charges is the sum of their individual fields:

$$\vec{\mathbf{E}} = \vec{\mathbf{E}}_1 + \vec{\mathbf{E}}_2.$$

Thus at P:

$$\begin{aligned} u_E &= \tfrac{1}{2}\epsilon_0 E^2 = \tfrac{1}{2}\epsilon_0(\vec{\mathbf{E}}_1 + \vec{\mathbf{E}}_2)\cdot(\vec{\mathbf{E}}_1 + \vec{\mathbf{E}}_2) \\ &= \tfrac{1}{2}\epsilon_0(E_1^2 + E_2^2 + 2\vec{\mathbf{E}}_1\cdot\vec{\mathbf{E}}_2). \end{aligned}$$

The total electrostatic energy of the two charges is expressed as the sum of three terms:

THE INTEGRAL IS OVER "ALL SPACE" BECAUSE THE ELECTRIC FIELD EXTENDS TO INFINITY.

$$U = \int_{\substack{\text{all}\\\text{space}}} \tfrac{1}{2}\epsilon_0 E_1^2\,dV + \int_{\substack{\text{all}\\\text{space}}} \tfrac{1}{2}\epsilon_0 E_2^2\,dV + \int_{\substack{\text{all}\\\text{space}}} \epsilon_0\vec{\mathbf{E}}_1\cdot\vec{\mathbf{E}}_2\,dV \equiv U_1 + U_2 + U_{\text{int}}.$$

The first two terms, U_1 and U_2, are the same energies we would calculate for the individual charges if they were isolated point charges. They are called the *self-energies* of the two charges.

The *interaction energy* U_{int} results from the combined fields and is the only part of the energy that changes as the particles move. By evaluating the integral, it can be shown that:

$$U_{\text{int}} = \frac{kQ_1 Q_2}{r}.$$

This interaction energy is the potential energy we described in Chapter 25.

The self-energy term for an individual charge is the energy stored throughout all space outside the particle. The electrostatic energy outside a spherical distribution of charge may be found by integrating the energy density (■ Figure 27.20). The answer is intriguing.

■ FIGURE 27.20
The self-energy of a charged shell is the total electrostatic energy stored in the region $r > R$.

STEPS I AND II We model the region outside a sphere of radius R as a set of spherical shells with radius r and thickness dr.

STEP III The electric field at radius $r > R$ has magnitude $|\vec{\mathbf{E}}| = kQ/r^2$. The corresponding energy density is:

$$u_E = \tfrac{1}{2}\epsilon_0 E^2 = \frac{1}{2}\epsilon_0\left(\frac{kQ}{r^2}\right)^2 = \frac{kQ^2}{8\pi r^4}.$$

REMEMBER: $k = 1/(4\pi\epsilon_0)$.

The volume dV of a spherical shell of radius r and thickness dr is:

$$dV = 4\pi r^2\, dr,$$

and the electrostatic energy stored in the shell is the energy density times the volume:

$$dU = u_E\, dV = \frac{kQ^2}{8\pi r^4} 4\pi r^2\, dr = \frac{kQ^2\, dr}{2r^2}.$$

STEPS IV AND V The total energy stored outside radius R is then:

$$U = \int_R^\infty dU = \int_R^\infty \frac{kQ^2}{2r^2}\, dr = -\frac{kQ^2}{2r}\bigg|_R^\infty = \frac{kQ^2}{2R}.$$

What inner radius R should we use? With $R = 0$ the self-energy is infinite! Clearly, point charges cannot exist—or can they?

✲ 27.4.3 *The Classical Electron Radius and Renormalization*

In his special theory of relativity Einstein argued that mass is a form of energy, according to the famous relation $E = mc^2$. By extending this idea, he supposed that the observed mass of an electron might be none other than its electrical self-energy!

EXAMPLE 27.12 ♦♦ Model an electron as a thin spherical shell of charge. If its mass-energy equals its electrical self-energy, what is the radius r_s of the shell?

MODEL We choose to model the electron as a spherical shell, since then all its charge is at the same potential, $V = ke/r_s$.

IF WE MODEL THE ELECTRON AS A SOLID SPHERE, ONLY THE NUMERICAL FACTOR $\frac{1}{2}$ DIFFERS IN THE EXPRESSION FOR U. SEE PROBLEM 73.

SETUP The potential energy of the shell is (eqn. 25.14) $U = \frac{1}{2}eV$. Setting the mass-energy and the electrical self-energy equal, we have:

$$mc^2 = \frac{1}{2}\left(\frac{ke^2}{r_s}\right).$$

SOLVE

$$r_s = \frac{1}{2}\left(\frac{ke^2}{mc^2}\right)$$

$$= \frac{(9.0 \times 10^9\ \text{N}\cdot\text{m}^2/\text{C}^2)(1.6 \times 10^{-19}\ \text{C})^2}{2(9.1 \times 10^{-31}\ \text{kg})(3.0 \times 10^8\ \text{m/s}^2)^2} = 1.4 \times 10^{-15}\ \text{m}.$$

ANALYZE The quantity $r_e \equiv ke^2/mc^2 = 3 \times 10^{-15}$ m is known as the classical electron radius. Unfortunately, this beautiful idea is not correct. Modern experiments show that the electron radius is much less than 10^{-15} m. ∎

The self-energy of elementary particles, even if not truly infinite, is enormous. Since the self-energy is constant, we may simply agree to ignore it and work with finite interaction energies only! The subtraction of infinite constants from physical properties of particles is a common feature of current theories and is called *renormalization.* The process produces consistent theories that predict phenomena with remarkable accuracy. Some recent, highly speculative, and as yet untested ideas—the superstring theories—do not require renormalization. It remains to be seen if infinite constants are a property of nature or an inadequacy of our ideas.

Chapter Summary

Where Are We Now?

We have studied capacitors as devices for storage of charge and energy. Electrostatic energy is stored throughout space wherever there is electric field.

What Did We Do?

Two conducting objects form a capacitor. The potential difference between the conductors when they carry equal and opposite charges is proportional to the amount of charge. Their capacitance is the ratio of charge to potential difference:

$$C \equiv \left| \frac{Q}{\Delta V} \right|.$$

The unit of capacitance is the farad (F). The capacitance of any conductor is roughly ϵ_0 times an appropriate linear dimension. For a parallel plate capacitor,

$$C = \epsilon_0 A/d.$$

Work must be done to store charge on a capacitor and increase its electric potential energy. The amount of stored energy is proportional to the square of the stored charge: $U = \frac{1}{2}Q^2/C = \frac{1}{2}C(\Delta V)^2$ (eqns. 27.6a, b, c).

To store large amounts of energy, capacitors are connected into banks. Capacitors connected across the same potential difference (in parallel) have an equivalent capacitance equal to the sum of their individual capacitances.

$$C_{\|} = \sum_i C_i. \tag{27.7}$$

Each capacitor in a series connection stores the same amount of charge. The reciprocal of the equivalent capacitance is the sum of the reciprocals of the individual capacitances.

$$\frac{1}{C_s} = \sum_i \frac{1}{C_i}. \tag{27.8}$$

Dielectric insulators provide mechanical support and improve the electrical properties of practical capacitors. Polarization of molecules in the dielectric results in bound charge at the surface of the dielectric that reduces the electric field within the dielectric material. The dielectric constant κ is the ratio of the applied electric field to the field in the dielectric medium (eqn. 27.9, Table 27.1). Dielectrics increase the capacitance of a device by a factor κ, since decreasing the electric field results in a lower potential difference for the same stored charge.

Electric potential energy is actually stored throughout space wherever there is electric field. The energy density is proportional to the square of the field vector:

$$u_E = \tfrac{1}{2}\epsilon_0 E^2.$$

⁂ The electric displacement vector $\vec{\mathbf{D}} = \kappa\epsilon_0\vec{\mathbf{E}}$ is useful for analyzing systems with dielectrics, because it is related directly to the free charge. The polarization vector $\vec{\mathbf{P}}$, or dipole moment per unit volume, is related to the bound charge.

Practical Applications

Capacitors are used in electrical circuits whenever temporary storage of small amounts of charge is necessary. Flash units on cameras are a familiar example. We'll explore these applications further in Chapter 31. Large banks of capacitors can store and quickly release energy in amounts that would be very inconvenient or very expensive to draw directly from commercial power systems. Such systems are used to power lasers, electromagnets, and neutral beam accelerators used in nuclear fusion research.

Solutions to Exercises

27.1. From eqn. (27.3), the area is:

$$A = \frac{dC}{\epsilon_0} = \frac{(2.0 \times 10^{-3}\ \text{m})(1.0 \times 10^{-9}\ \text{F})}{(8.85 \times 10^{-12}\ \text{F/m})} = 0.23\ \text{m}^2.$$

The charge stored is:

$$Q = CV = (10^{-9}\ \text{F})(2 \times 10^3\ \text{V}) = 2\ \mu\text{C}.$$

27.2. We use eqn. (27.4):

$$C = \frac{(2.0 \times 10^{-2}\ \text{m})(3.0 \times 10^{-2}\ \text{m})}{(9.0 \times 10^9\ \text{N}\cdot\text{m}^2/\text{C}^2)(3.0 \times 10^{-2}\ \text{m} - 2.0 \times 10^{-2}\ \text{m})}$$
$$= 6.7\ \text{pF}.$$

27.3. When $R_1 \approx R_2$, both inner and outer spheres have area:

$$A \approx 4\pi R_1^2 \approx 4\pi R_2^2 \approx 4\pi R_1 R_2.$$

Their separation is $d = R_2 - R_1$, so from eqn. (27.4):

$$C = \frac{4\pi\epsilon_0 R_1 R_2}{(R_2 - R_1)} \approx \frac{\epsilon_0 A}{d},$$

which is the same expression as for parallel plates.

27.4. A typical person has dimension of the order of 1 m and a capacitance:

$$C \sim (1\ \text{m})/(9 \times 10^9\ \text{N}\cdot\text{m}^2/\text{C}^2) = 0.1\ \text{nF}.$$

27.5. Using eqn. (27.6b) relating stored energy to potential difference, we have:

$$U = \tfrac{1}{2}C(\Delta V)^2 = \frac{\epsilon_0 A}{2d}(\Delta V)^2$$
$$= \frac{(8.85 \times 10^{-12}\ \text{F/m})(10.0\ \text{m}^2)(5.0\ \text{V})^2}{2(1.0 \times 10^{-4}\ \text{m})} = 11\ \mu\text{J}.$$

27.6. The series capacitance is:

$$C_s = \frac{C_1 C_2}{C_1 + C_2} = \frac{(1\ \mu\text{F})(6\ \mu\text{F})}{(1\ \mu\text{F} + 6\ \mu\text{F})} = \frac{6}{7}\ \mu\text{F}.$$

Then the energy stored is:

$$U = \tfrac{1}{2}C_s(\Delta V)^2$$
$$= \frac{1}{2}\left(\frac{6}{7}\ \mu\text{F}\right)(10.0\ \text{V})^2 = 4.3 \times 10^{-5}\ \text{J}.$$

27.7. The capacitor bank consists of 20 parallel strings, each of 136 capacitors in series. Each string is connected across the full 54.4-kV potential difference. The potential difference across each individual capacitor is:

$$V_1 = (54.4\ \text{kV})/136 = 0.400\ \text{kV}.$$

Each individual capacitor in the bank contains the same energy, so the total stored is:

$$U = (20)(136)\tfrac{1}{2}(2.0 \times 10^{-3}\ \text{F})(400\ \text{V})^2 = 0.44\ \text{MJ}.$$

27.8. The dielectric constant for Mylar is 3 (Table 27.1). The capacitance is $C = \kappa\epsilon_0 A/d$, so the area is:

$$A = \frac{dC}{\kappa\epsilon_0}$$
$$= \frac{(2.5 \times 10^{-5}\ \text{m})(1.0 \times 10^{-6}\ \text{F})}{3(8.85 \times 10^{-12}\ \text{F/m})} = 0.94\ \text{m}^2.$$

Basic Skills

Review Questions

§27.1 CAPACITANCE

- Define *capacitance.*
- How does the capacitance of a parallel plate capacitor depend on the plate area? On the plate separation?
- What procedure is used to calculate the capacitance of any pair of conductors?
- Describe a *back-of-the-envelope* method for estimating capacitance.

§27.2 ENERGY STORAGE IN CAPACITORS

- How does the energy stored in a capacitor depend on its charge? On the potential difference?
- Describe a parallel connection of capacitors.
- What is the equivalent capacitance of a set of capacitors connected in parallel?
- What is the advantage of a parallel connection?
- Describe a series connection of capacitors.
- What is the equivalent capacitance of a set of capacitors connected in series?
- Why might you use a series connection?

§27.3 DIELECTRICS AND PRACTICAL CAPACITORS

- What is *polarization?*
- What is the *dielectric constant* of a material?
- What is *bound charge?* Describe how it results from polarization of a material.
- If a capacitor is filled with material whose dielectric constant is $\kappa > 1$, how does its capacitance change?
- How does inserting a dielectric into a capacitor change its stored energy **(a)** if it is kept connected to a battery and **(b)** if it is charged but insulated from its surroundings? Describe the energy flow in each of these processes.
- ✲ What are the *polarization vector* $\vec{\mathbf{P}}$ and the *displacement vector* $\vec{\mathbf{D}}$?

§27.4 ENERGY IN THE ELECTRIC FIELD

- How is the *energy density* in the electric field in vacuum related to the field strength?
- How is the expression for energy density modified to account for energy of polarization?
- ✲ What is *interaction energy?* How does it differ from *self-energy?*

Basic Skill Drill

§27.1 CAPACITANCE

1. A 10.0-pF capacitor is connected to a 12-V battery. What is the charge on the capacitor?
2. A capacitor connected to a 1.5-V battery carries a 27-μC charge. What is its capacitance?
3. You want to make a parallel plate capacitor from two soup-can ends of radius 4.25 cm. If the capacitance is to be 16 pF, what should be the plate separation?
4. What is the capacitance of a steel ball bearing of radius 0.75 cm?

§27.2 ENERGY STORAGE IN CAPACITORS

5. A 3.9-nF capacitor is connected to a 6.0-V battery. How much energy does it store?
6. How large a capacitor would you need to store 16 mJ at 120 V?
7. A 1.8-pF and a 4.7-pF capacitor are connected in parallel. What is the equivalent capacitance?
8. A 1.8-μF capacitor and a 0.56-μF capacitor are connected in series. What is the equivalent capacitance of the combination?

§27.3 DIELECTRICS AND PRACTICAL CAPACITORS

9. Find the capacitance of a parallel plate capacitor filled with polystyrene, with $A = 11$ cm^2 and $d = 0.74$ mm.
10. A parallel plate capacitor has square plates of side ℓ with separation d. If the capacitor is filled with plexiglas and d is unchanged, how much smaller is the value of ℓ for the same capacitance? If ℓ were unchanged, how much larger would d have to be?

§27.4 ENERGY IN THE ELECTRIC FIELD

11. What is the electrostatic energy density inside a parallel plate capacitor with a plate separation of 2.2 mm and connected to a 1.5-V battery?
12. What is the electrostatic energy density 6.0 cm from a point charge of 3.0 μC?

Questions and Problems

§27.1 CAPACITANCE

13. ❖ Estimate the capacitance of a car.
14. ❖ Two parallel plate capacitors have the same separation d, but one has twice the area of the other. If both have the same potential difference ΔV between their plates, what is the ratio of **(a)** their stored charges and **(b)** their charge densities?
15. ❖ A parallel plate capacitor is charged and then disconnected from the battery. The plates are then moved apart using insulated handles. Which of the following results?
(a) The charge on the capacitor increases.
(b) The charge on the capacitor decreases.
(c) The capacitance of the plates increases.

(d) The voltage between the plates remains the same.
(e) The voltage between the plates increases.
Explain what is wrong with the answers you didn't choose.

16. ♦ What is the radius of an isolated conducting sphere with a capacitance of 1 F? Compare your result with the radius of the Moon's orbit about the Earth.

17. ♦ A parallel plate capacitor with $A = 125\ \text{cm}^2$ and $d = 1.25$ mm is charged to a potential difference of 10.0 V. What is the charge density on each plate?

18. ♦ A coaxial cable consists of a cylindrical metal sheath of inner radius $b = 0.3$ cm surrounding an inner wire of outer radius $a = 0.1$ cm. Find the capacitance per unit length of the cable.

19. ♦♦ A 28.0-pF parallel plate capacitor has plates of area 420 cm^2. What is the electric field between its plates when it is connected across a potential difference of 12 V?

20. ♦♦ A parallel plate capacitor has plate area A and separation d. A metal slab of thickness $d/3$ and area A is inserted between the plates (■ Figure 27.21). The capacitor is given a charge Q_0. Explaining your reasoning at each step, find: **(a)** the charge density on each plate, **(b)** the electric field between the top plate and the slab, within the slab, and between the slab and the lower plate, **(c)** the potential difference between the plates, and **(d)** the capacitance of the system.

21. ♦♦♦ Two long parallel wires, each of radius a, are separated by a distance d (■ Figure 27.22). One wire carries a uniform charge density $+\lambda$, and the other carries $-\lambda$. Calculate the electric field due to the wires by superposing contributions from the two filaments. Use the result to estimate the capacitance per unit length of the wires.

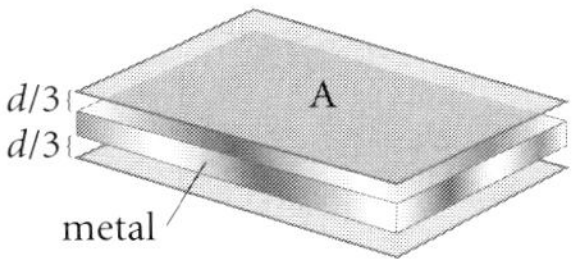

■ FIGURE 27.21

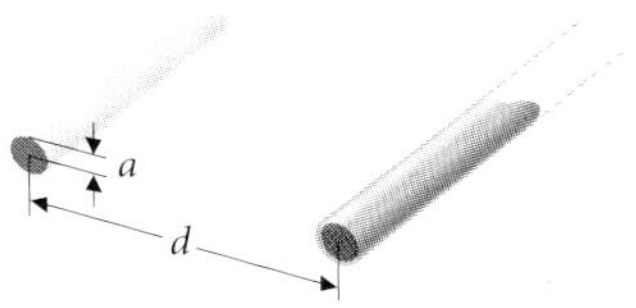

■ FIGURE 27.22

§27.2 ENERGY STORAGE IN CAPACITORS

22. ❖ Two capacitors are charged to the same potential difference. The second has twice the capacitance of the first. What is the ratio of the energies stored in each?

23. ❖ Two capacitors carry the same charge Q. If $C_1 = 3C_2$, what is V_1/V_2? What is the ratio of the stored energies?

24. ❖ If the charge stored in a capacitor is doubled, by what factor is the stored energy increased?

25. ❖ You have six identical 5-μF capacitors. How can you connect them together to form a 6-μF module?

26. ♦ What is the equivalent capacitance of the combination in ■ Figure 27.23 if it is connected into a circuit at points a and b?

27. ♦ What is the equivalent capacitance of the combination in ■ Figure 27.24 if it is connected into a circuit at points a and b?

28. ♦ What is the equivalent capacitance of the combination in ■ Figure 27.25 if it is connected into a circuit at points a and b?

29. ♦♦ Three capacitors have capacitances of 2.0 μF, 4.0 μF, and 8.0 μF. How would you connect them across a battery in order to

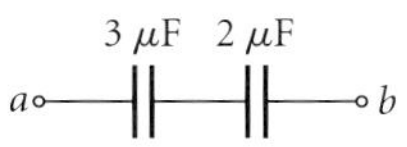

■ FIGURE 27.23

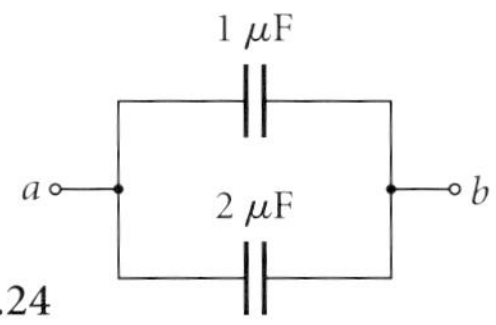

■ FIGURE 27.24

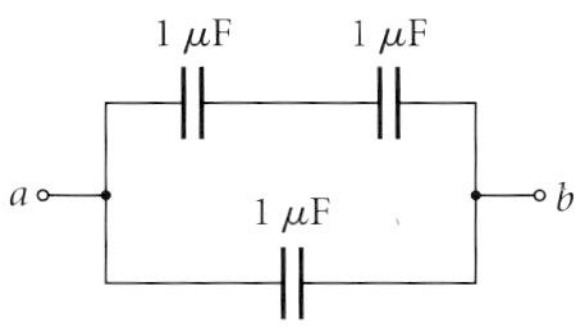

■ FIGURE 27.25

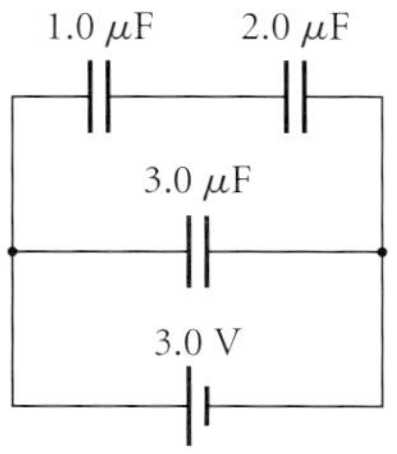

■ FIGURE 27.26

minimize the potential difference across each capacitance? What is the equivalent capacitance in this configuration?

30. ♦♦ How would you connect three capacitors (2.0 μF, 4.0 μF, and 8.0 μF) across a battery to store maximum charge? What is their equivalent capacitance in this configuration?

31. ♦♦ What values of capacitance can be produced by various combinations of a 5.0-mF, a 10.0-mF, and a 30.0-mF capacitor? Draw a circuit diagram for each case.

32. ♦♦ Find the charge on each capacitor in the system in ■ Figure 27.26. What is the total energy stored?

33. ♦♦ Three identical capacitors with capacitance C are connected in parallel with an emf $\mathcal{E}$. Find the charge on each capacitor and the total stored energy. A person using insulating gloves then disconnects the emf and connects the capacitors in series. Find the final values of the charge on each capacitor, the total stored energy, and the potential difference across the combination.

34. ♦♦♦ Capacitors C_1 and C_2 are initially connected in parallel and carry a combined charge Q_0 (■ Figure 27.27). Find the charge on each capacitor and the energy stored in the system. The switches are opened, C_2's terminals are reversed, and then the switches are closed again. Find the stored charge and the energy in the final state.

35. ♦♦♦ What is the capacitance of the semi-infinite network in ■ Figure 27.28? (*Hint:* Removing the leftmost two capacitors leaves an identical semi-infinite network!)

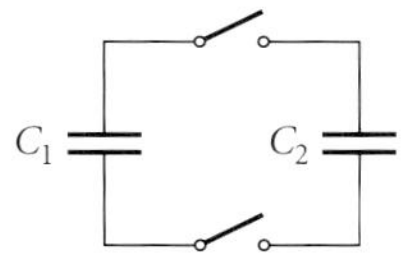

■ FIGURE 27.27

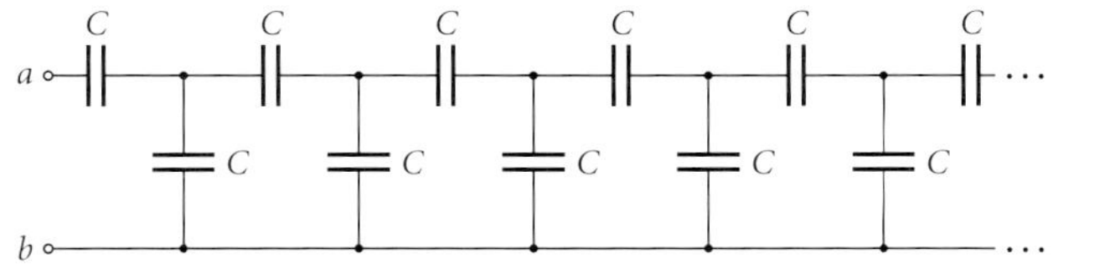

■ FIGURE 27.28

§27.3 DIELECTRICS AND PRACTICAL CAPACITORS

36. ❖ Each plate of a parallel plate capacitor has a charge density of magnitude σ_{free}, and is insulated from its surroundings. If a thin conducting plate the same shape as the capacitor plate is inserted into the capacitor, the induced charge density on the conductor:
(a) is greater than σ_{free}. **(b)** is less than σ_{free}. **(c)** equals σ_{free}.
(d) can be either greater than or less than σ_{free} depending on the conductivity of the conductor.
Explain why you chose your answer.

37. ❖ Each of the plates of a parallel plate capacitor has a charge density of magnitude σ_{free} and is insulated from its surroundings. If a thin

dielectric slab the same shape as the capacitor plate is inserted into the capacitor, the induced bound charge density on the dielectric:
(a) is greater than σ_{free}. **(b)** is less than σ_{free}. **(c)** equals σ_{free}.
(d) can be either greater or less than σ_{free} depending on the dielectric constant of the slab.
Explain why you chose your answer.

38. ❖ A parallel plate capacitor consists of two conducting plates of area A separated by an air space of thickness d. If a plastic slab of dielectric constant $\kappa > 1$ is placed between the plates:
(a) the electric field between the plates increases.
(b) the potential difference between the plates increases.
(c) the capacitance of the capacitor increases.
(d) the charge density on the plates decreases.
Explain what is wrong with the answers you didn't choose. Does your answer change if the capacitor is connected to a battery?

39. ♦ Two capacitors have identical metal plates and carry the same charge. One has no dielectric; the other is filled with epoxy. If the electric field in the empty capacitor has magnitude 7.5×10^4 V/m, find the electric field in the epoxy.

40. ♦ A demonstration capacitor has circular plates 15 cm in radius that are separated by $d = 0.30$ cm. A dielectric with constant $\kappa = 2.0$ is inserted between the plates. What is the capacitance of the device?

41. ♦ Find the capacitance of a Pyrex shell ($\kappa = 4.7$) of 1.0-cm inner radius and 1.3-cm outer radius with copper plated on its inner and outer surfaces.

42. ♦♦ A rectangular block of Teflon with sides 3 cm × 2 cm × 1 cm is to be plated with metal on two opposite sides to produce a capacitor. Use a parallel plate capacitor model to decide which choice of coated sides results in the greatest capacitance. What is that capacitance? For each pair of sides, comment on whether parallel plates or parallel wires are a better model for the conductors.

43. ♦♦ Find the capacitance per unit length of a coaxial cable filled with polystyrene. The inner conductor has a radius of 0.75 mm, and the outer conductor has a radius of 0.75 cm.

44. ♦♦ You need a 6.0-pF capacitor. What length of coaxial cable should you use if its inner wire has a radius of 2.3 mm, its outer conductor has an inner radius of 1.1 cm, and it is filled with epoxy?

45. ♦♦ Find the capacitance of the parallel plate capacitor in ■ Figure 27.29 if the plate area is $A = 25\text{ cm}^2$, $d = 2.0$ mm, $\kappa_1 = 1.5$, and $\kappa_2 = 2.5$.

46. ♦♦ A spherical capacitor with inner radius of 2.5 cm and outer radius of 3.5 cm is filled with epoxy. It is connected to a 1.5-V battery. What is the charge on the capacitor? Find the surface charge density on the inner and outer conductors. What is the bound charge density on the inner and outer surfaces of the epoxy?

47. ♦♦♦ Two capacitors are connected in parallel with a 12-V battery. Each has plates of area 16 cm². The plate separations are $d_1 = 2.0$ mm and $d_2 = 3.0$ mm. The first capacitor has a 2.0-mm slab of plexiglas between the plates. Find the two capacitances, the stored charge, and the stored energy. The capacitors are now disconnected from the battery but remain connected in parallel with each other. The plexiglas slab is moved from the first to the second. What are the two capacitances then? Calculate the charge on each and the total energy stored. How much work is done to move the slab?

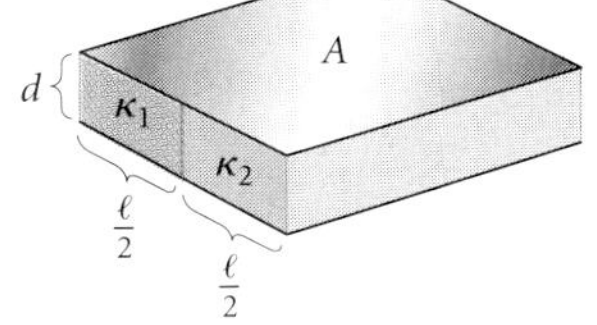

■ FIGURE 27.29

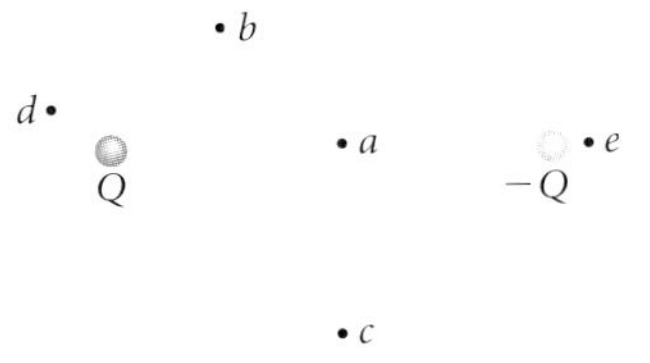

■ FIGURE 27.30

§27.4 ENERGY IN THE ELECTRIC FIELD

48. ❖ You have measured the energy density at a distance 1 m from a certain point charge. If the charge is increased by a factor of 2, by what factor does the energy density change? If the charge is replaced with an equal and opposite charge, does the energy density change?

49. ❖ Two point charges are located as shown in ■ Figure 27.30. At which of points a, b, c, d, or e is the electric field energy density greatest? Why?

50. ❖ Which charge configuration has the greater electrostatic energy, a charge Q on the surface of a metallic conductor with radius R or the same charge Q distributed uniformly throughout a sphere of radius R?

51. ♦ What is the electrostatic energy density 10 cm from a 6-μC point charge?

52. ♦ Find the energy density: **(a)** midway between two point charges $+Q$ a distance D apart. **(b)** midway between two point charges $+Q$ and $-Q$ a distance D apart. **(c)** 1.0 cm from a filament carrying 1.0 μC/m.

53. ♦♦ What is the energy density in vacuum: **(a)** 5 cm from a point charge of 3 μC? **(b)** 5 cm from a charge of −3 μC? **(c)** midway between a 3-μC and a −3-μC charge separated by 10 cm? **(d)** midway between two +3-μC charges or two −3-μC charges separated by 10 cm? Are the third and fourth answers consistent with the first two? Discuss.

54. ♦♦ A total charge of 5.0 μC is spread over **(a)** a sphere of radius $R = 2.0$ cm and **(b)** a square plate with the same area (side $4\sqrt{\pi} = 7.1$ cm). Find the electrostatic energy density at a point 1.0 mm from the surface in each case. How do the values compare? (For the square, use the infinite plane result for $\vec{\mathbf{E}}$.)

55. ♦♦ A charged metal ball has radius R. Find the radius r of a spherical surface that contains half the electrostatic energy.

56. ♦♦ Three equal 5.3-nC charges are at the corners of an equilateral triangle of side 2.5 cm. Find the electric energy density **(a)** at the center of the triangle and **(b)** at the center of one side.

57. ♦♦ Two point charges $\pm Q$ are on the x-axis at $x_Q = \pm \ell$. Find the electrostatic energy density at a point P on the x-axis with **(a)** $|x_P| < \ell$ and **(b)** $|x_P| > \ell$. In particular, find the energy density at $x = 2\ell$.

58. ✲ ❖ Show that the interaction energy density produced by two equal but opposite point charges is positive within a sphere with the two charges at opposite ends of a diameter, and negative outside that sphere.

Additional Problems

59. ❖ ■ Figure 27.31 shows a combination of capacitors that is neither series nor parallel. Kirchhoff's loop rule applies without change to such a circuit. How would you restate Kirchhoff's junction rule to relate the charges on the capacitors? (Remember to state carefully the meaning of the algebraic symbol Q_i describing the charge on a capacitor.)

60. ❖ How would you expect the dielectric constant of a dilute gas to vary with the density of the gas?

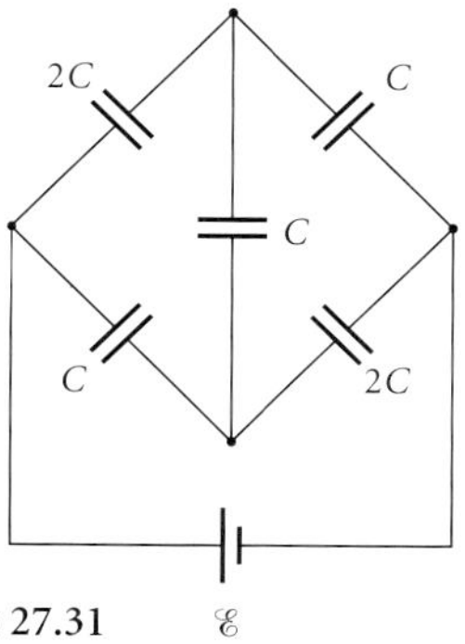

■ FIGURE 27.31

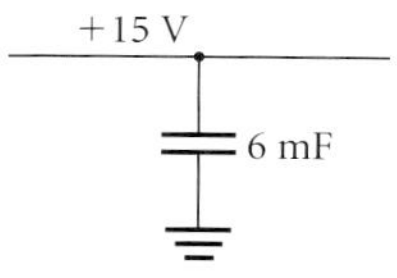

■ FIGURE 27.32

61. ♦♦ Two parallel plate capacitors have the same dimensions, but one is filled with forsterite ($\kappa = 6.2$, dielectric strength $= 9.4$ MV/m), and the other is filled with paraffin ($\kappa = 2.2$, dielectric strength $=$ 9.8 MV/m). Which can hold more charge? How much more? Explain your reasoning carefully.

62. ♦♦ ■ Figure 27.32 shows part of the circuit diagram for the LZ-1 stereo preamplifier. What is the charge on the capacitor when the amplifier is in use, with +15 V on the positive terminal of the capacitor? How much energy is stored?

63. ♦♦ After the switch is moved to position A in Figure 27.6, some energy from the battery is stored in the capacitor, and some is dissipated as heat by the resistor. Show the two amounts of energy are equal. (The easiest method is to calculate the work done by the battery.)

64. ♦♦ You want to construct a parallel plate capacitor with capacitance C and volume V. Find the required values of A and d in terms of V and C.

65. ♦♦ If the maximum dielectric strength of any material is of order 10 MV/m, estimate the minimum volume of a capacitor that could store 1 MJ of energy.

66. ♦♦♦ (■ Figure 27.33) With the switch at position f, a total charge $Q_0 = C\mathcal{E}$ resides on the isolated circuit segment between the series capacitors. Find the charge on each capacitor, the energy stored in the capacitors, and the potential at point h. The switch is moved to position g. Find the new values of these quantities, a long time after the switch is moved. How much energy is dissipated as heat in the resistor during this process?

67. ♦♦♦ A tunable capacitor for use in a radio has two sets of semicircular conducting plates (■ Figure 27.34). One set can be rotated with respect to the other set. If there are n positive and n negative plates, how many capacitors are there? How are they connected? Find the capacitance of the set of plates when the negative ones make an angle θ with the positive ones.

68. ♦♦♦ Starting from the electrostatic energy density, find the energy stored in a spherical capacitor with inner radius r_1 and outer radius r_2. Use the expression $U = \frac{1}{2}Q^2/C$ to obtain a value for C. Compare with eqn. (27.4).

Computer Problems

69. A -1.0-μC charge is at the origin, and a 2.0-μC charge is at $x = 1.0$ cm. Either analytically or numerically, compute the energy density at points on the x-axis for both $x < 1.0$ cm and $x > 1.0$ cm. Plot $u(x)$ for $x > 1.0$ cm. How large must x be for $u(x)$ to differ by less than 20% from the value due to a single point charge of 1.0 μC? (Think carefully about the correct position for the 1-μC charge!)

70. Three 1.0-μC point charges are at the corners of an equilateral triangle of side 1.0 m centered at the origin. Use a spreadsheet program to calculate the electric field components and the electric energy density at points along the x-axis and along parallel lines at $y =$ 0.5 m and $y = 1.5$ m. Choose points with $x = 0, 0.2, 0.4, \ldots, 2$ m. Plot the energy density as a function of x for each y, and comment.

71. File CH27P71 on the supplementary computer disk lists values of dielectric constant as a function of position between the plates of a parallel plate capacitor with plate area 1.0 cm^2. Compute the capacitance. Compare with the value you would obtain by using an average value of the dielectric constant. (The coordinate x varies from $x = 0$ at one of the plates to $x = 1.0$ μm at the other.)

Challenge Problems

72. A capacitor is made from a sheet of material consisting of two layers of aluminum foil and two layers of Mylar (■ Figure 27.35a). The foil thickness is 5×10^{-6} m, and the thickness of the Mylar is to be determined. The foil sheet is $L = 3.0$ cm wide and is rolled into a cylinder 1.0 cm in radius (Figure 27.35b). If the capacitance of the device is to be 0.10 μF, what thickness of Mylar should be used?

73. Model an electron as a uniform sphere of total charge $-e$. What radius must it have for its electrostatic energy to equal its mass-energy?

74. Choose sign conventions for the five unknown capacitors in ■ Figure 27.31. Use Kirchhoff's rules to obtain five equations for the unknown charges. Solve for the charges and find the effective capacitance of the combination. (*Hint:* Make use of the symmetry of the configuration to simplify the calculations.)

75. An arbitrary capacitor has two conductors carrying opposite charges $\pm Q$. Use the following plan to show that the electrostatic energy density inside the capacitor is $u = \frac{1}{2}\epsilon_0 E^2$. Consider a tube of field lines between an element of charge dQ on one plate and an element $-dQ$ on the other. Express the contribution dU of these two elements to the energy stored in the capacitor in terms of dQ and an integral along the field lines between the elements. Use Gauss' law to relate the field E at any point along the tube to the electric field at one of the conductors, and so express dQ in terms of E. Hence, show that:

$$dU = \int_{\text{field tube}} \tfrac{1}{2}\epsilon_0 E^2 \, dV.$$

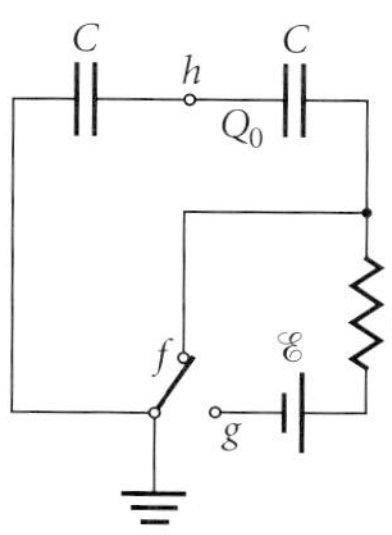

■ FIGURE 27.33

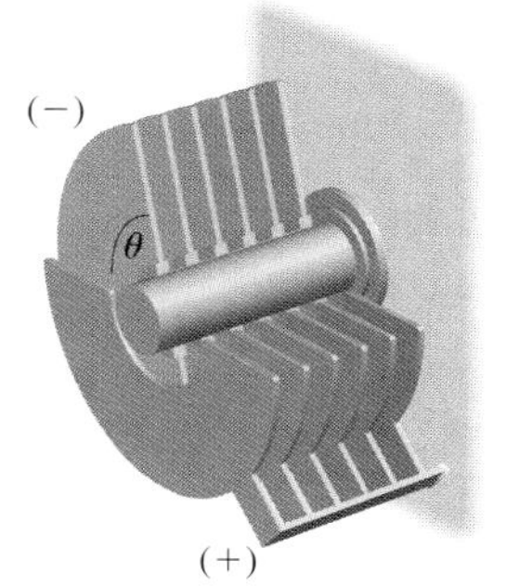

■ FIGURE 27.34

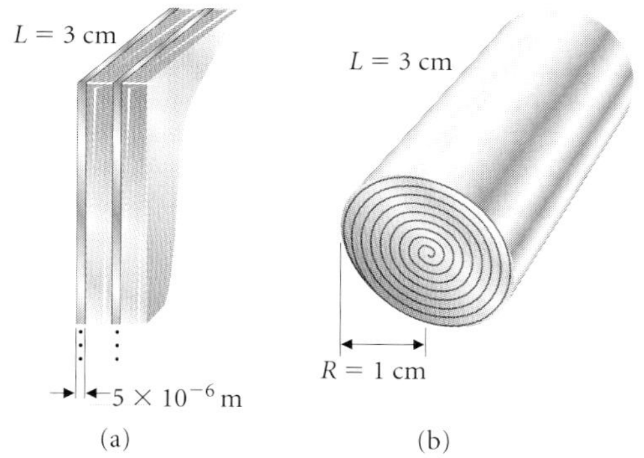

■ FIGURE 27.35

CHAPTER 28
Static Magnetic Fields

Interior of the Princeton Tokamak. The carefully constructed magnetic field of this device has been able to confine a hot plasma long enough for nuclear fusion reactions to occur. In 1994 these reactions lasted long enough to liberate as much energy as it took to operate the machine. This is a major step on the long road toward building a fully functioning nuclear fusion power plant.

CONCEPTS

Magnetic force
Magnetic field
Magnetic moment
Ampère's law
Circulation of magnetic field

GOALS

Be able to:

Apply the Biot-Savart law to find the magnetic field due to a current distribution.

Compute the magnetic moment of a loop.

Apply Ampère's law to calculate magnetic field produced by symmetric systems.

The nation that controls magnetism will control the universe!

CHESTER GOULD, "DICK TRACY"

Magnetic forces are used to control the motion of charged particles in numerous devices, from television sets to controlled nuclear fusion experiments. The fuel in a successful nuclear fusion reactor is so hot that no ordinary container can be used for the experiment chamber. Instead, an elaborately constructed *magnetic bottle* confines the particles. Containing them long enough for the desired nuclear reactions to occur depends on understanding both the particle motion and the design of magnetic fields. In this chapter, we'll investigate how magnetic fields are shaped by their current sources. In the next chapter, we'll study the motion of particles in response to magnetic forces.

VECTOR CROSS PRODUCTS ARE USED EXTENSIVELY IN THIS CHAPTER. YOU MAY WANT TO REVIEW §9.1, ESPECIALLY THE MATH TOOLBOX.

28.1 MAGNETIC FORCE

In the overview of electromagnetism we appealed to your experience with magnets and the forces they exert on each other. But forces between magnets do not provide an operational definition of magnetic field because magnetic poles always occur in pairs, and the force on an isolated pole cannot be measured. Instead, as we did for electric field, we investigate the force exerted on a point *charge*.

NOW WOULD BE A GOOD TIME TO READ THE OVERVIEW AGAIN.

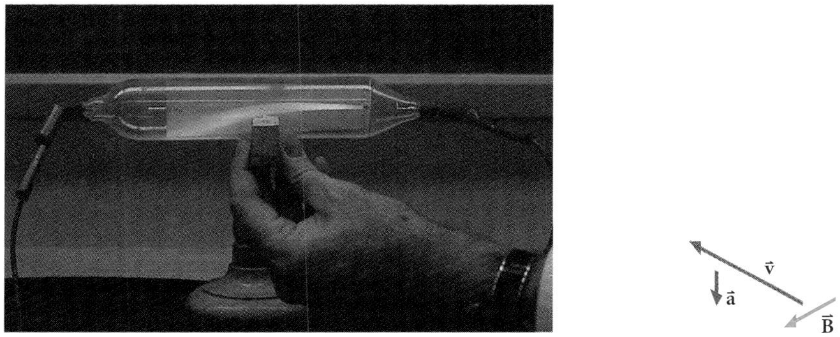

■ FIGURE 28.1
(a) An electron beam. The tube contains a phosphor screen that makes the beam visible. By bringing a bar magnet near the beam, we can investigate the behavior of magnetic force.

(b) The beam bends as the magnet is brought near. The magnetic force is perpendicular to both $\vec{v}$ and $\vec{B}$, so the beam moves downward.

We observe that no magnetic force acts on a stationary charge, whereas a moving charge experiences a force perpendicular to its velocity. ■ Figure 28.1a shows an electron beam moving from right to left in an evacuated tube. When we bring a horizontal bar magnet close to the tube, the beam bends downward. The magnetic force on electrons in the beam is perpendicular to both their velocity and the magnetic field. Further experiments show that the force on a charged particle is proportional to its speed, its charge, and the magnetic field strength:

WE ASSUME THAT WE KNOW THE DIRECTION OF THE MAGNETIC FIELD PRODUCED BY THE BAR MAGNET; SEE, FOR EXAMPLE, FIGURE VI.4 IN THE OVERVIEW.

The magnetic force on a moving point charge equals its charge multiplied by the cross product of its velocity with the magnetic field:

$$\vec{F}_{mag} = q\,\vec{v} \times \vec{B}. \qquad \textbf{(28.1)}$$

Figure 28.1b shows how this rule accounts for the bending of the beam in the evacuated tube.

We can determine the magnitude of the magnetic field by measuring the forces acting on equal point charges moving in different directions. The maximum force acts on a point charge with velocity perpendicular to $\vec{B}$, and we define the magnetic field strength by:

$$B = \frac{F_{max}}{qv}. \qquad (28.2)$$

The direction of $\vec{B}$ is parallel to the velocity of a point charge that is not deflected.

■ FIGURE 28.2
The SI unit of magnetic induction, the tesla, is named after Nikola Tesla (1856–1943), who worked out the basic design of our power transmission system, developed numerous improvements in electrical equipment, and investigated the possibility of wireless power transmission.

TABLE 28.1 **Magnetic Fields**

	Magnetic Field Strength (T)
At the surface of a neutron star	10^8
Near a superconducting magnet	up to 20
In the T-3 Tokamak	3.5
In a high-quality moving-coil loudspeaker	1.5
In the Advanced Light Source at Lawrence Berkeley Laboratory	1.04
In the center of the largest sunspot	0.39
In MRI instruments for medical imaging	0.35
Inside a color TV set	2×10^{-2}
On the Earth's surface, at the equator	3.2×10^{-5}
Due to a power line, on the ground directly below it (This contribution should be added to the Earth's magnetic field.)	10^{-6}
In interstellar space	10^{-10}
In a magnetically shielded experimental environment	10^{-14}

IT IS INFORMALLY CALLED MAGNETIC FIELD, AND WE SHALL CONTINUE TO USE THIS NAME.

The formal name for the field vector $\vec{\mathbf{B}}$ is *magnetic induction.* Its SI unit is the tesla (symbol T) after Nikola Tesla (■ Figure 28.2).

$$1\ \mathrm{T} = 1\ \frac{\mathrm{N \cdot s}}{\mathrm{C \cdot m}} = 1\ \frac{\mathrm{N}}{\mathrm{A \cdot m}}.$$

The cgs unit of magnetic induction, the gauss, is also commonly used: $1\ \mathrm{G} \equiv 10^{-4}\ \mathrm{T}$. The largest magnetic field you are likely to encounter in everyday experience is about 1 T (● Table 28.1).

EXAMPLE 28.1 ♦♦ A particle with a charge of 6.0 μC is used in an experiment to measure magnetic field. The particle moves at 1500 m/s. No force is observed when the particle moves along the x-axis. The maximum force of 7.3 mN is observed when the particle moves along the y-axis. When the particle moves in the positive y-direction, the force it experiences is in the positive z-direction. Find $\vec{\mathbf{B}}$.

MODEL Since the particle experiences no force when moving along the x-axis, the magnetic field is in either the positive or negative x-direction. Since the force is in the $+z$-direction when $\vec{\mathbf{v}}$ is in the $+y$-direction, we conclude that $\vec{\mathbf{B}}$ is in the *negative* x-direction:

$$\hat{\mathbf{j}} \times (-\hat{\mathbf{i}}) = \hat{\mathbf{k}}.$$

SETUP We use eqn. (28.2) to find the field strength:

$$B = \frac{F_{\mathrm{max}}}{qv} = \frac{7.3 \times 10^{-3}\ \mathrm{N}}{(6.0 \times 10^{-6}\ \mathrm{C})(1500\ \mathrm{m/s})} = 0.81\ \mathrm{T}.$$

SOLVE

$$\vec{\mathbf{B}} = -(0.81\ \mathrm{T})\hat{\mathbf{i}}.$$

ANALYZE This is a strong magnetic field, such as might be produced near an electromagnet. ■

EXERCISE 28.1 ❖ ■ Figure 28.3 shows the velocity vectors of several charged particles in a magnetic field represented by the field lines in the figure. What is the direction of the force on each particle? ■

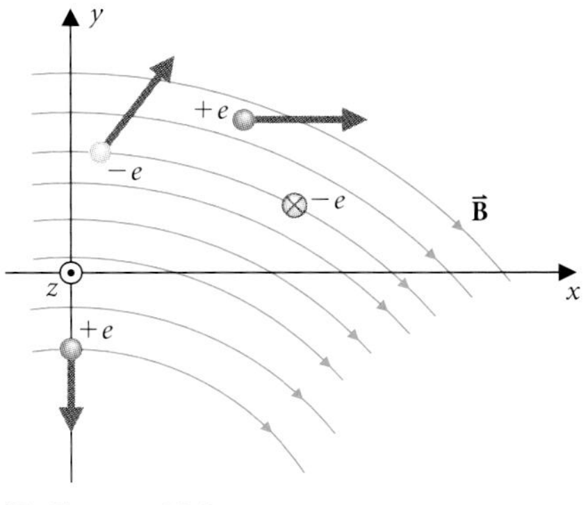

■ FIGURE 28.3
What is the direction of the magnetic force acting on each of these particles?

28.2 CURRENT AS THE SOURCE OF MAGNETIC FIELDS

28.2.1 The Biot-Savart Law

In 1820 Oersted demonstrated that the magnetic field lines produced by a long straight wire form circles centered on the wire. Later the same year, Jean Baptiste Biot (1774–1862) and Felix Savart (1791–1841) showed that the magnetic field strength decreases inversely with distance from the wire. These results provide excellent qualitative images for describing magnetic fields. They were the starting point for finding an exact relation for the magnetic field produced by an arbitrary distribution of current.

WE SHALL REFER TO THIS FACT AS OERSTED'S RULE. FOR A DISCUSSION OF OERSTED'S DISCOVERY, SEE THE OVERVIEW, ESPECIALLY FIGURE VI.16.

The simplest static electric field is that of a point charge. But a point charge in motion does not produce a static magnetic field. As the charge approaches a given point the magnetic field increases to a maximum and then dies away as the particle recedes. Static magnetic field is produced only by steady electric current. The magnetic field is a superposition of contributions from small elements of the current path. Each element $d\vec{\ell}$ is in the direction of the conventional current with magnitude I. A description of the field $d\vec{\mathbf{B}}$ produced by such an element is named after Biot and Savart.

THE ELECTRIC FIELD OF A POINT CHARGE IS GIVEN BY COULOMB'S LAW (EQN. 23.6).

THE BIOT-SAVART LAW (■ FIGURE 28.4):

$$d\vec{\mathbf{B}} = \frac{\mu_0}{4\pi} I \frac{d\vec{\ell} \times \hat{\mathbf{r}}}{r^2}. \quad (28.3)$$

THE BIOT-SAVART LAW IN THIS FORM WAS ANNOUNCED ON OCTOBER 30, 1820, AT A MEETING OF THE FRENCH ACADEMY OF SCIENCES.

In the Biot-Savart law, as in Coulomb's law, the field strength is directly proportional to the source (here $I\,d\vec{\ell}$) and inversely proportional to distance squared. The rule for magnetic field direction, however, is very different from that for electric field. The source $I\,d\vec{\ell}$ is a *vector* quantity, and the magnetic field vector it produces is perpendicular to both the direction of the source and the direction of the position vector $\vec{\mathbf{r}}$ from the source to the field point. Notice that the direction of the vector $d\vec{\ell}$ is determined by the direction of the current.

COMPARE EQN. (28.3) WITH EQN. (23.6). THE DEFINITION OF THE UNIT VECTOR $\hat{\mathbf{r}}$ IS THE SAME IN BOTH CASES.

The constant μ_0 is *defined* to have the exact value:

$$\mu_0 = 4\pi \times 10^{-7}\ \mathrm{N/A^2}.$$

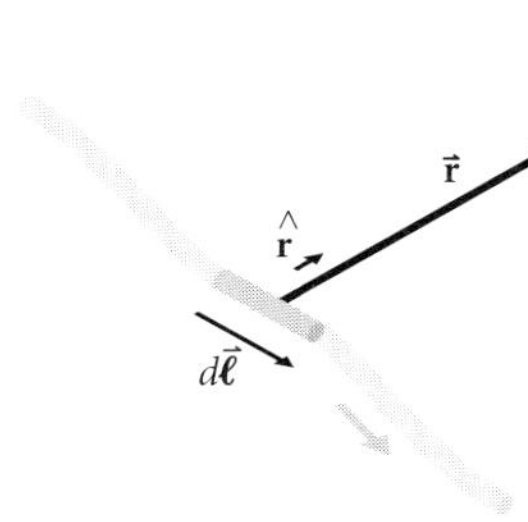

■ FIGURE 28.4
The Biot-Savart law describes the magnetic field produced by a small segment of current-carrying wire. The vector $d\vec{\ell}$ describes the direction of the wire and the sense of the current in it. The vector $\hat{\mathbf{r}}$ points *from* the source point *toward* the field point. [The same definition is used in Coulomb's law, eqns. (23.1) and (23.6).] The direction of the magnetic field at P is perpendicular to both $d\vec{\ell}$ and $\hat{\mathbf{r}}$, in the direction of their cross product.

EXERCISE 28.2 ♦ Show that the source term $I\,d\vec{\ell}$ in the Biot-Savart law equals the number of electrons in the element of wire multiplied by their drift velocity $\vec{\mathbf{v}}_d$ and their charge $(-e)$.

EXERCISE 28.3 ❖ Show that the magnetic field vector $d\vec{\mathbf{B}}$ vanishes at points along the line of the wire element and has maximum magnitude in the plane perpendicular to the element.

EXAMPLE 28.2 ♦♦ A thin wire carrying current $I = 5.0$ A passes through the origin parallel to the x-axis (■ Figure 28.5a). Find the magnetic field at point P on the y-axis at $y = 3.0$ m produced by a 1.0-mm piece of the wire centered at the origin.

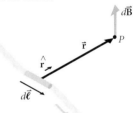

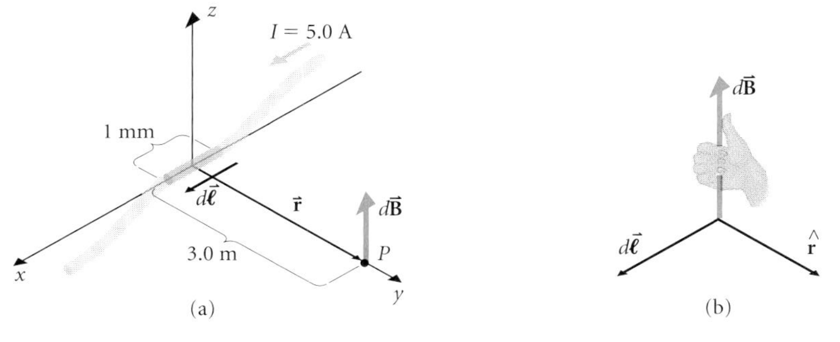

■ FIGURE 28.5
(a) Magnetic field at P due to a segment of wire at the origin. Here $d\vec{\ell}$ is in the x-direction, $\hat{\mathbf{r}}$ is in the y-direction, and $d\vec{\mathbf{B}}$ is in the z-direction. (b) Using the right-hand rule to find the direction of $d\vec{\mathbf{B}}$.

MODEL The piece of wire is very small compared with its distance to P, so we may treat it as a single element and find the field it produces directly from eqn. (28.3).

SETUP The element of wire carries current in the $+x$-direction, so $\vec{\ell} = (1.0\text{ mm})\hat{\mathbf{i}}$. The position vector $\vec{\mathbf{r}} = (3.0\text{ m})\hat{\mathbf{j}}$, and the unit vector $\hat{\mathbf{r}} = \hat{\mathbf{j}}$.

SOLVE

$$\begin{aligned}\vec{\mathbf{B}} &= \frac{\mu_0}{4\pi} I \frac{\vec{\ell} \times \hat{\mathbf{r}}}{r^2} \\ &= (10^{-7}\text{ N/A}^2)(5.0\text{ A})\,\frac{(1.0 \times 10^{-3}\text{ m})(\hat{\mathbf{i}} \times \hat{\mathbf{j}})}{(3.0\text{ m})^2} \\ &= \frac{5.0 \times 10^{-10}}{9.0}\,\frac{\text{N}}{\text{A}\cdot\text{m}}\hat{\mathbf{k}} \\ &= (5.6 \times 10^{-11}\text{ T})\hat{\mathbf{k}}.\end{aligned}$$

ANALYZE The direction is found equally well from formal rules for cross products of unit vectors or from the right-hand rule illustrated in Figure 28.5b. ∎

28.2.2 *The Magnetic Field Produced by a Straight Wire Segment*

The magnetic field due to a longer piece of wire is found from the principle of superposition:

STRICTLY SPEAKING, THE INTEGRAL IS ALWAYS OVER A CLOSED PATH (HENCE $\oint$), SINCE STEADY CURRENT ALWAYS FOLLOWS A CLOSED PATH. IF IT DID NOT, CHARGE WOULD BUILD UP AT THE END OF THE PATH.

PRINCIPLE OF SUPERPOSITION

The magnetic field produced by an extended current distribution is the vector sum of contributions due to its individual elements:

$$\vec{\mathbf{B}} = \oint d\vec{\mathbf{B}} = \oint \frac{\mu_0}{4\pi} I \frac{d\vec{\ell} \times \hat{\mathbf{r}}}{r^2}. \qquad \textbf{(28.4)}$$

We first apply this principle to obtain the field due to current in a straight segment of wire.

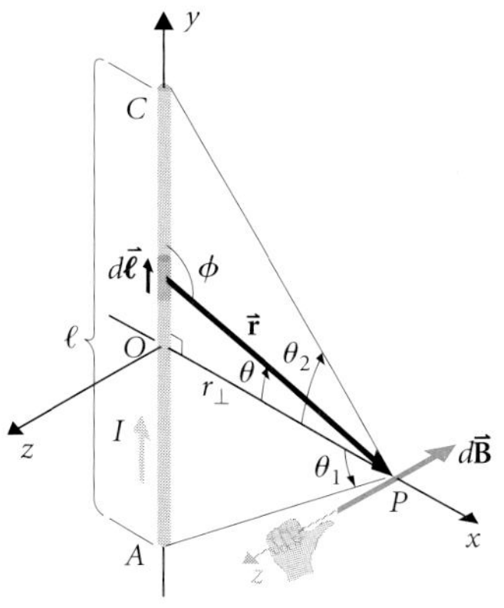

■ FIGURE 28.6
Magnetic field due to a finite segment of wire is the superposition of the contributions due to the individual wire segments. For a straight segment like this, the perpendicular distance $r_\perp = r\sin\phi = x$ is the same for each element. The angles $\theta_1 = \angle OPA$ and $\theta_2 = \angle OPC$ have signs determined by the position of the end of the wire segment with respect to the foot of the perpendicular at O. Going from O to C, we travel in the same direction as the current, so $\angle OPC$ is positive. Similarly, going from O to A, we travel opposite I, so $\angle OPA$ is negative.

EXAMPLE 28.3 ♦♦♦ Find the magnetic field produced by a straight segment of wire of length ℓ carrying current I. Compare your result with the observations of Oersted, Biot, and Savart for a long straight wire ($\ell \to \infty$).

MODEL We are asked for the value of $\vec{\mathbf{B}}$ at an arbitrary point P (■ Figure 28.6). We use the principle of superposition and use the integration procedure from Interlude 2.

SETUP STEPS I, II, AND III We take the y-axis along the wire segment and the x-axis through the field point P. From the Biot-Savart law, the contribution of a typical element of the wire, $d\vec{\ell} = dy\,\hat{\mathbf{j}}$, to the field at P is:

$$d\vec{\mathbf{B}} = \frac{\mu_0}{4\pi} I \frac{d\vec{\ell} \times \hat{\mathbf{r}}}{r^2}.$$

The cross product has magnitude:

$$|d\vec{\ell} \times \hat{\mathbf{r}}| = |d\vec{\ell}||\hat{\mathbf{r}}|\sin\phi = |d\vec{\ell}|\sin(180° - \phi) = |d\vec{\ell}|\frac{x}{r}.$$

The direction of $d\vec{\mathbf{B}}$ is perpendicular to both $\hat{\mathbf{r}}$ and $d\vec{\ell}$. According to the right-hand rule, the field $d\vec{\mathbf{B}}$ at P is in the minus z-direction for every element $d\vec{\ell}$. Since the direction is perpendicular to $\hat{\mathbf{r}}$, it is tangent to the circle through P centered on the wire, in agreement with Oersted's observations. The most convenient coordinate for describing $d\vec{\ell}$ is the angle θ measured from the x-axis. The angle is positive for elements with positive y-coordinate. Then:

$$y = x \tan \theta \qquad \text{and} \qquad r = x \sec \theta.$$

The length of the wire element is:

$$|d\vec{\ell}| = dy = d(x \tan \theta) = x \sec^2 \theta \, d\theta.$$

So:

$$|d\vec{\mathbf{B}}| = \frac{\mu_0 I x^2 \sec^2 \theta \, d\theta}{4\pi (x \sec \theta)^3} = \frac{\mu_0 I}{4\pi x} \cos \theta \, d\theta.$$

WE USED THE SAME ANGLE IN EXAMPLE 24.3 TO FIND THE ELECTRIC FIELD DUE TO A LINE CHARGE.

STEP IV The limits of integration run from θ_1 (negative for the case shown in the figure) to θ_2.

SOLVE
STEP V Since the contributions $d\vec{\mathbf{B}}$ from all elements of the wire are in the same direction, we may pull the constant unit vector $-\hat{\mathbf{k}}$ out of the integral:

$$\vec{\mathbf{B}} = \int d\vec{\mathbf{B}} = \int dB_z \, \hat{\mathbf{k}} = -\hat{\mathbf{k}} \int_{\theta_1}^{\theta_2} \frac{\mu_0 I}{4\pi x} \cos \theta \, d\theta = -\hat{\mathbf{k}} \frac{\mu_0 I}{4\pi x} \sin \theta \Big|_{\theta_1}^{\theta_2}.$$

$$\vec{\mathbf{B}} = -\hat{\mathbf{k}} \frac{\mu_0 I}{4\pi x} (\sin \theta_2 - \sin \theta_1). \qquad (28.5)$$

For a very long wire, ($\ell \gg x$), $\theta_1 \approx -\pi/2$, and $\theta_2 \approx +\pi/2$. Then $\sin \theta_2 = 1$, and $\sin \theta_1 = -1$. The distance x is the perpendicular distance $r_\perp$ of P from the wire segment, so:

$$|\vec{\mathbf{B}}| = \frac{\mu_0 I}{2\pi r_\perp}. \qquad (28.6)$$

ANALYZE Equation (28.6) reproduces Biot and Savart's observations that $|\vec{\mathbf{B}}| \propto 1/r_\perp$ and gives the constant of proportionality $\mu_0/(2\pi)$. The direction of $\vec{\mathbf{B}}$ also follows the subsidiary right-hand rule we observed in the Overview of electromagnetism: with the thumb of your right hand parallel to the current, your fingers curl about the wire in the direction of the magnetic field.

EXERCISE 28.4 ❖ Two long straight parallel wires carry equal currents in opposite directions. What magnetic field do they produce at a point halfway between them?

28.2.3 *The Magnetic Field of Loops and Coils*

In Chapter 24 we found that simple models describe the electric field at a large distance from a system of charges. A system with net charge looks like a point charge, while a system with no net charge looks like a dipole. The magnetic field produced by a current distribution also simplifies at large distances. Experiments show that the simplest magnetic field that occurs is a dipole.

REMEMBER: AN ELECTRIC DIPOLE CONSISTS OF TWO EQUAL AND OPPOSITE POINT CHARGES SEPARATED BY A SMALL DISTANCE (CF. §24.5).

> At large distances from any distribution of current, the magnetic field it produces resembles an ideal dipole field.

As an example we use the Biot-Savart law to compute the field a circular current loop produces at points on its axis. An exact calculation of the field at other points is mathematically intricate and goes beyond our needs. By comparing the field on the axis with the expression computed in Chapter 24 for an electric dipole, we can identify the magnetic dipole moment of the current loop and obtain an excellent approximation to the magnetic field away from the loop axis.

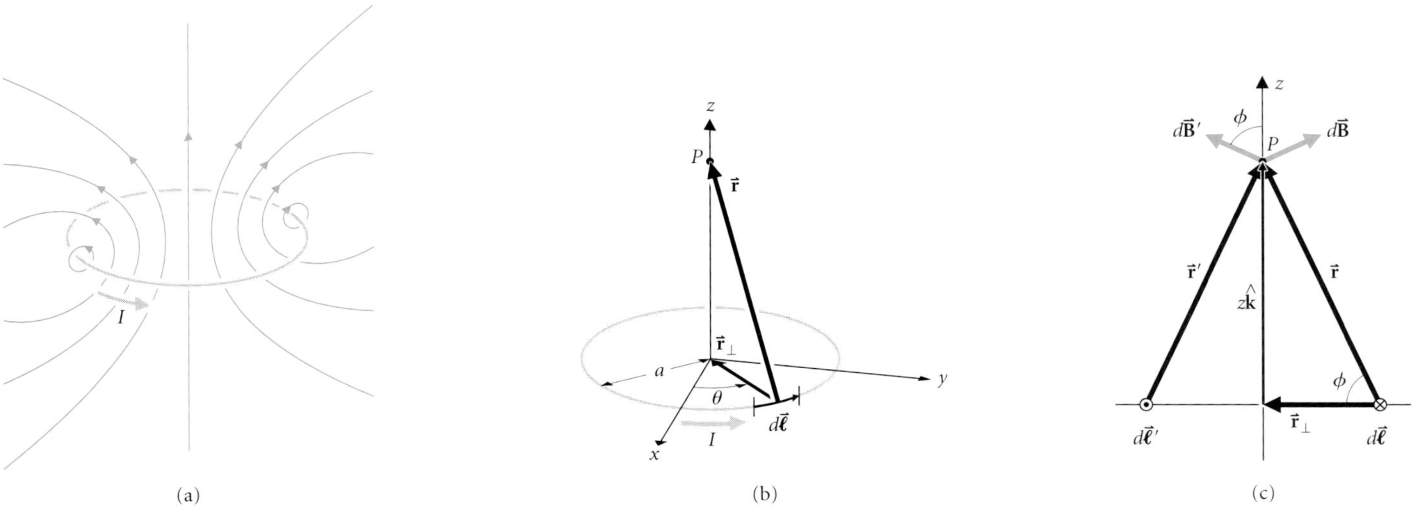

■ **FIGURE 28.7**
(a) The magnetic field on the axis of a circular current loop. Magnetic field lines circle the wire. On the axis, the sum of the contributions from each element of the loop produces a field along the axis. At large distances from the loop, the field pattern resembles that of a dipole.

(b) We describe each element of the loop by its position at angle θ from the x-axis and its length $d\ell = a\,d\theta$. The z-axis is perpendicular to the loop through its center.

(c) View in a plane perpendicular to the loop. Two line elements at opposite ends of a diameter contribute magnetic field elements of equal magnitude. Their z-components add, and their perpendicular components cancel: $d\vec{\mathbf{B}} + d\vec{\mathbf{B}}' = 2\ dB \cos\phi\ \hat{\mathbf{k}} = 2\ dB_z\ \hat{\mathbf{k}}$. The vector $\vec{\mathbf{r}} = \vec{\mathbf{r}}_\perp + z\hat{\mathbf{k}}$.

LIKE ELECTRIC FIELD LINES, MAGNETIC FIELD LINES CANNOT CROSS AT THE CENTER OF THE LOOP.

EXAMPLE 28.4 ♦♦ Find the magnetic field on the axis of a thin circular wire of radius a carrying current I.

MODEL According to Oersted's rule, the magnetic field lines close to the wire circle it, as shown in ■ Figure 28.7a. Along the axis, the magnetic field is perpendicular to the loop.

The magnetic field vector at any point is the sum of the contributions $d\vec{\mathbf{B}}$ from the individual line segments. We use the principle of superposition and evaluate the integral (eqn. 28.4) using the integration process from Interlude 2.

SETUP STEP I To make use of the current loop's symmetry, we choose cylindrical coordinates with the origin at the loop's center, z-axis perpendicular to its plane, and angle θ measured in the sense of the current I (Figure 28.7b). Then the field point has coordinates $(0, 0, z)$.

STEP II A typical current element $d\vec{\ell}$ at position θ measured from the x-axis has length $|d\vec{\ell}| = a\,d\theta$.

STEP III The magnetic field $d\vec{\mathbf{B}}$ produced by $d\vec{\ell}$ at P has a component along the z-axis and a component radially outward, parallel to the x-y-plane (Figure 28.7c). The element $d\vec{\ell}'$ at the opposite end of a diameter makes a contribution $d\vec{\mathbf{B}}'$ with an opposite radial component. Their z-components add. The total magnetic field vector $\vec{\mathbf{B}}$ at P is the sum of contributions from such pairs and is in the z-direction.

For each element $d\vec{\ell}$:

$$d\vec{\ell} \times \hat{\mathbf{r}} = \frac{d\vec{\ell} \times \vec{\mathbf{r}}}{r} = \frac{d\vec{\ell} \times (\vec{\mathbf{r}}_\perp + z\hat{\mathbf{k}})}{r}.$$

Both vectors $d\vec{\ell}$ and $\vec{\mathbf{r}}_\perp$ are in the x-y-plane, so $d\vec{\ell} \times \vec{\mathbf{r}}_\perp$ is in the z-direction. The term $(z/r)d\vec{\ell} \times \hat{\mathbf{k}}$ is the radial component. Since $|\vec{\mathbf{r}}_\perp| = a$, and $d\vec{\ell} \perp \vec{\mathbf{r}}_\perp$:

$$[d\vec{\ell} \times \hat{\mathbf{r}}]_z = \frac{a\ d\ell}{r},$$

where $r = \sqrt{z^2 + a^2}$ is the same for each element. Thus:

$$dB_z = \frac{\mu_0}{4\pi}\frac{a\,d\ell}{r^3}$$

$$= \frac{\mu_0}{4\pi}\frac{a\,d\ell}{(z^2 + a^2)^{3/2}}.$$

SOLVE Both z and a are independent of position on the loop, so:

STEPS IV AND V

$$\vec{\mathbf{B}} = \hat{\mathbf{k}} \oint \frac{\mu_0 a\,d\ell}{4\pi(z^2 + a^2)^{3/2}}$$

$$= \hat{\mathbf{k}} \frac{\mu_0 a}{4\pi(z^2 + a^2)^{3/2}} \oint d\ell.$$

The integral $\oint d\ell$ means the total length around the loop—its circumference, $2\pi a$. So:

$$\vec{\mathbf{B}} = \frac{\mu_0 I a^2}{2(z^2 + a^2)^{3/2}}\hat{\mathbf{k}}. \quad (28.7)$$

ANALYZE Notice how similar this expression is to eqn. (24.5) for the electric field on the axis of a ring of charge. An important difference between the expressions is that $\vec{\mathbf{E}}$ reverses direction on opposite sides of a ring; $\vec{\mathbf{B}}$ does not. ∎

At large distances from the loop, details of its size and shape are no longer important, and its field line diagram looks like that of a dipole (compare Figure 28.7a with Figure VI.4 in the Overview). For $z \gg a$, eqn. (28.7) becomes:

$$B_z \approx \frac{\mu_0 I a^2}{2z^3} = \frac{\mu_0}{4\pi}\frac{2\pi a^2 I}{z^3}.$$

For comparison, the field due to an electric dipole on its axis is (eqn. 24.13):

$$E_z = \frac{2kp}{z^3}.$$

By analogy:

$$B_z = \frac{2k_m m}{z^3},$$

where $k_m = \mu_0/4\pi$, and $\vec{\mathbf{m}} = \pi a^2 I\hat{\mathbf{k}}$ is the *magnetic dipole moment* of the loop.

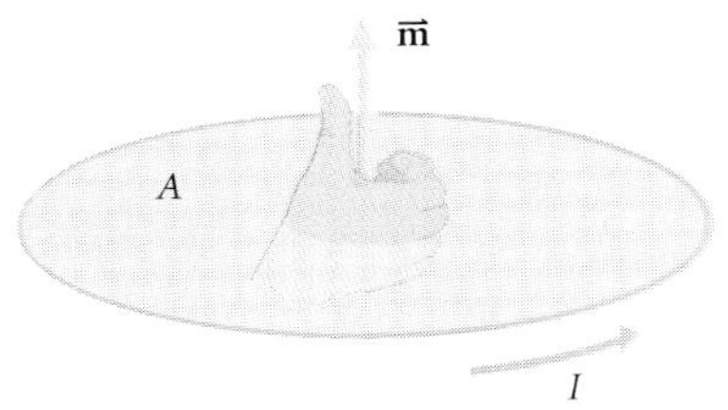

■ FIGURE 28.8
Definition of magnetic moment of a loop. The direction of $\vec{\mathbf{m}}$ is determined by a right-hand rule. Curl the fingers of your right hand around the loop in the direction of I; your thumb gives the direction of $\vec{\mathbf{m}}$. The magnitude $m = IA$.

> A current loop with area A and carrying current I has a *magnetic dipole moment* $\vec{\mathbf{m}}$ of magnitude:
>
> $$m = IA. \quad \mathbf{(28.8)}$$
>
> The direction of $\vec{\mathbf{m}}$ is perpendicular to the loop according to a right-hand rule: when the fingers curl in the direction of the current, the thumb points along $\vec{\mathbf{m}}$ (■ Figure 28.8).

USUALLY $\vec{\mathbf{m}}$ IS CALLED SIMPLY THE MAGNETIC MOMENT, WITH DIPOLE BEING UNDERSTOOD.

Modeling any current loop as a dipole gives an accurate value for the magnetic field at large distances from the loop.

A simple coil is made by wrapping several turns of wire around the same loop. If there are N turns of wire in the coil, it is equivalent to N separate loops producing equal contributions to the field at any point. The coil has a dipole moment with magnitude $m = NIA$ and direction determined as shown in Figure 28.8.

Digging Deeper

Magnetic Moment of an Arbitrary Planar Loop

We use the Biot-Savart law to calculate the magnetic field at a large distance from an arbitrary current loop that lies in a single plane. We choose the origin at a point inside the loop (■ Figure 28.9a), and the z-axis perpendicular to the plane of the loop. Then the magnetic field at a point on the z-axis due to the element $d\vec{\ell}$ is:

$$d\vec{\mathbf{B}} = k_{\mathrm{m}} I \frac{d\vec{\ell} \times \hat{\mathbf{r}}}{r^2} = k_{\mathrm{m}} I \frac{d\vec{\ell} \times \vec{\mathbf{r}}}{r^3}.$$

Writing $\vec{\mathbf{r}} = \vec{\mathbf{r}}_\perp + z\hat{\mathbf{k}}$ and assuming $z \gg |\vec{\mathbf{r}}_\perp|$, we have:

$$d\vec{\mathbf{B}} \approx \frac{k_{\mathrm{m}} I}{z^3} d\vec{\ell} \times (\vec{\mathbf{r}}_\perp + z\hat{\mathbf{k}}).$$

The product $d\vec{\ell} \times \vec{\mathbf{r}}_\perp$ is in the z-direction, while $d\vec{\ell} \times \hat{\mathbf{k}}$ lies in the x-y-plane. To find the loop's dipole moment, we use the z-component of the field. (At this large distance from the loop, B_x and B_y are zero; see Problem 71.)

$$dB_z \approx \frac{k_{\mathrm{m}} I}{z^3} |d\vec{\ell} \times \vec{\mathbf{r}}_\perp|.$$

From Figure 28.9b, $\frac{1}{2}|d\vec{\ell} \times \vec{\mathbf{r}}_\perp|$ is the differential area dA (cf. Chapter 9). Thus:

$$B_z = \int dB_z = \int \frac{k_{\mathrm{m}} I}{z^3} 2\, dA = \frac{2k_{\mathrm{m}} IA}{z^3}.$$

So $m = IA$ is the correct expression for the magnetic dipole moment of an arbitrary current loop.

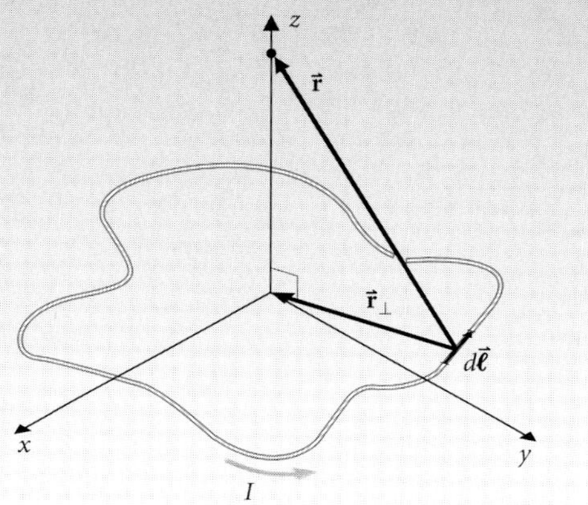

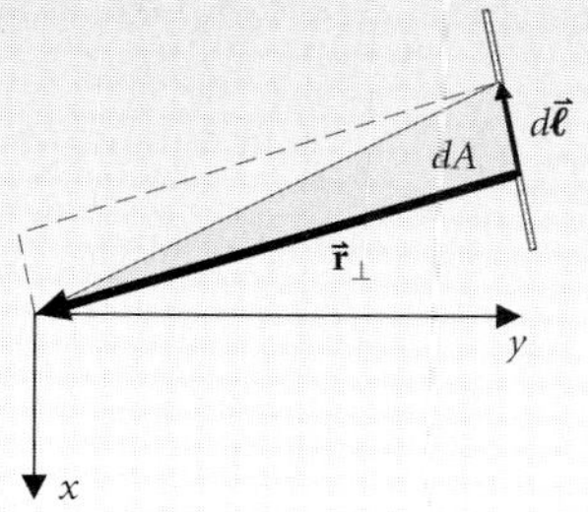

■ **Figure 28.9**
(a) Geometry for computing the magnetic field produced by a plane loop of arbitrary shape. (b) Remember that the cross product of two vectors represents the area of the parallelogram formed by the two vectors. Thus $\frac{1}{2}|d\vec{\ell} \times \vec{\mathbf{r}}_\perp| = dA$, the triangular area element formed by the line element and the center of the loop.

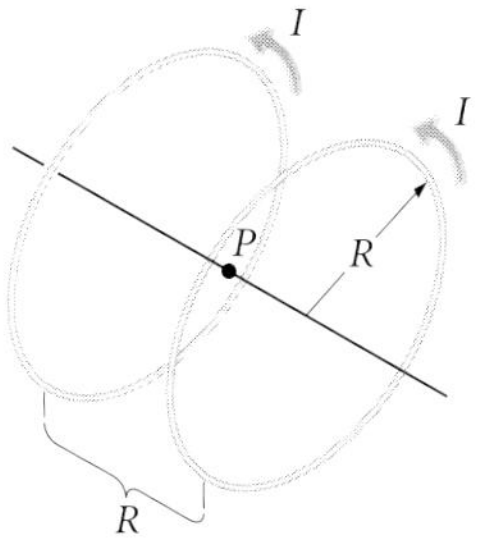

■ **Figure 28.10**
Helmholtz coils (after Hermann von Helmholtz, 1821–1894). Two parallel coils carrying current in the same direction produce a nearly uniform field in the region around point P, halfway between them. (For proof of the uniformity, see Problem 39.)

EXAMPLE 28.5 ♦♦ Helmholtz coils, ■ Figure 28.10, produce a nearly uniform field near the point halfway between their centers. To create a region in the laboratory almost free of magnetic field, a pair of coils with radius $R = 30$ cm and $N = 10^2$ turns of wire each is used to produce a magnetic field equal to and opposite the Earth's magnetic field. If the Earth's field in the laboratory has magnitude $B_e = 0.3$ G, what current is required in the coils?

MODEL The magnetic field on the axis of a coil is the same on either side of the coil. Thus the magnetic field produced by the two Helmholtz coils at their midpoint is just twice that of a single coil.

SETUP The magnetic field of one loop is given by eqn. (28.7) with $z = R/2$ and $a = R$. We have N loops per coil and two coils, so:

$$|\vec{\mathbf{B}}_c| = 2N \frac{\mu_0 IR^2}{2(5R^2/4)^{3/2}} = \frac{8\mu_0 NI}{(5)^{3/2} R}.$$

The field produced by the coils must be opposite the Earth's field and have the same magnitude, so:

$$I = \frac{(5)^{3/2}RB_e}{8\mu_0 N} = \frac{11(0.3\ \text{m})(0.3\ \text{G})(10^{-4}\ \text{T/G})}{8(4\pi \times 10^{-7}\ \text{N/A}^2)(10^2)} = 0.1\ \text{A}.$$

ANALYZE The current is not particularly large and is easily produced in laboratory equipment. Properties of Helmholtz coils are explored further in the problems.

28.3 INTEGRAL LAWS FOR STATIC MAGNETIC FIELDS

Gauss' law for the electric field gives a global relation between the electric field and charge and a useful technique for computing the fields produced by symmetric sources. A relation that plays a similar role for magnetic field is known as Ampère's law. It is based on a different geometric property: magnetic field lines circulate around their sources rather than diverging from them. In this section we shall describe Ampère's law and a version of Gauss' law that applies to magnetic field and give some examples of their use.

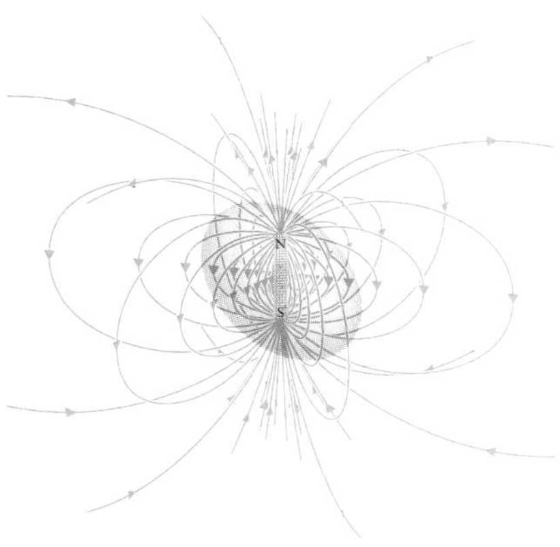

■ FIGURE 28.11
Gauss' law for the magnetic field. The flux through any closed surface is zero: every field line entering the volume through surface S leaves again.

28.3.1 Gauss' Law for the Magnetic Field

Just as electric field lines begin and end on electric charge, we would expect magnetic field lines to begin and end on single magnetic poles. But experiments have repeatedly failed to detect any such poles. Thus magnetic field lines form loops that neither begin nor end. Any field line that enters a volume also leaves it (■ Figure 28.11), and no net magnetic flux emerges. Gauss' law for the magnetic field is quite simple:

> The net magnetic flux Φ_B emerging from any volume is zero:
> $$\Phi_B \equiv \oint \vec{\mathbf{B}} \cdot \hat{\mathbf{n}}\, dA = 0. \qquad \textbf{(28.9)}$$

The integral is taken over the closed surface surrounding the volume.

The currents that are the sources of $\vec{\mathbf{B}}$ do not appear in eqn. (28.9), so Gauss' law is not helpful for calculating magnetic field. But it is very useful for understanding the different behavior of electric and magnetic fields at a boundary. The electric field changes abruptly at the surface of a conductor because surface charge is present. In contrast, since there can be no magnetic charge, the normal component of $\vec{\mathbf{B}}$ is continuous across any boundary.

SEE §25.6.3 FOR THE RELATION BETWEEN σ AND E.

28.3.2 Circulation and Ampère's Law

Flux measures the divergence of electric field lines from their sources. Magnetic field lines form loops around their current sources, and the mathematical quantity that measures this behavior is named *circulation*. Like flux, the term "circulation" is derived from fluid theory, where it describes how the fluid swirls as it flows. To get a rough intuitive idea of circulation, imagine paddling a canoe around a closed curve (■ Figure 28.12). At all points along curve A water flows almost parallel to the boat's path. The water has positive circulation. As you paddle

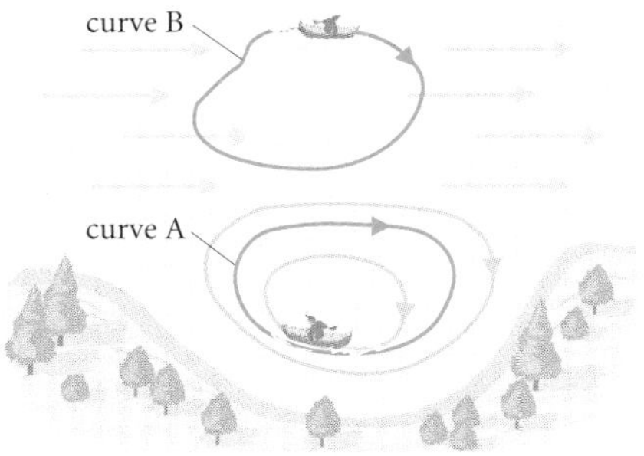

■ FIGURE 28.12
Paddling a canoe around a closed curve, we determine whether the water flow has *circulation* around the curve by noticing whether we mostly paddle downstream ($\mathscr{C} > 0$), paddle upstream ($\mathscr{C} < 0$), or paddle downstream as often as we paddle upstream ($\mathscr{C} = 0$).

STRUGGLING THE OTHER WAY AROUND CURVE A, YOU WOULD GET A CLEAR IMPRESSION OF NEGATIVE CIRCULATION.

around curve B, the water moves in the direction opposite the boat about as often as it moves in the same direction. There is no net circulation.

The circulation $\mathscr{C}$ of a vector field $\vec{\mathbf{v}}$ around a closed curve C is:

$$\mathscr{C} \equiv \oint_C \vec{\mathbf{v}} \cdot d\vec{\ell}. \tag{28.10}$$

The circulation of $\vec{\mathbf{B}}$ around C describes the extent to which field lines follow the direction of the curve.

■ Figure 28.13 shows an application of magnetic circulation. The ammeter measures current when it is clamped around a wire. It does not need to be connected into the circuit because it senses the magnetic field produced by current in the wire. The clamp defines a closed path surrounding the wire, and the circulation of magnetic field around this path provides enough information to determine the current.

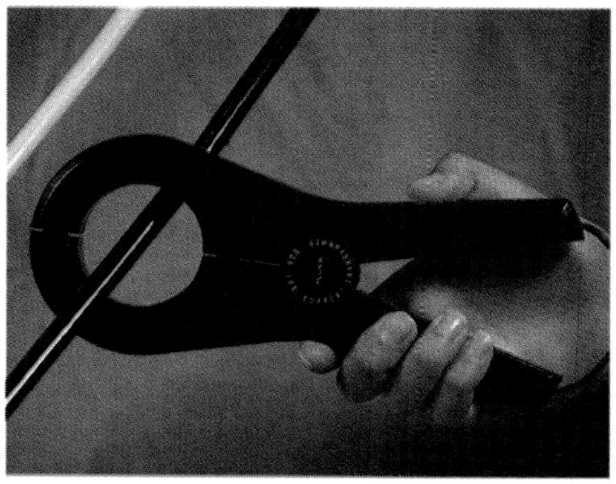

■ FIGURE 28.13
This ammeter is *not* connected into the circuit. Placed *around* a current-carrying wire, it measures the magnetic field.

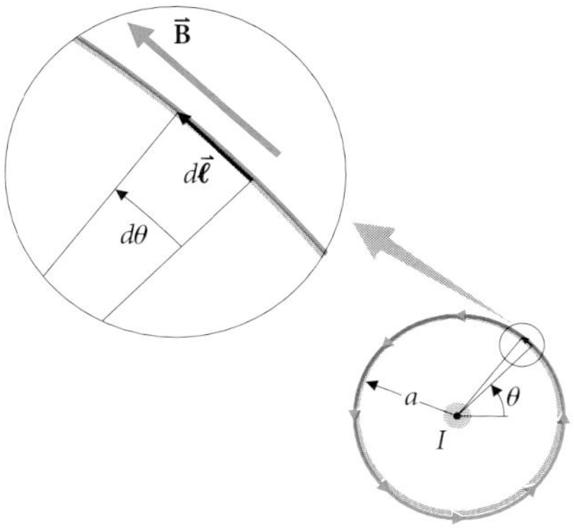

■ FIGURE 28.14
The circulation around the circle $r = a$ equals μ_0 times the current I in the central wire. On the circle, $\vec{\mathbf{B}} \cdot d\vec{\ell} = B\,d\ell$ for each line element.

EXAMPLE 28.6 ♦♦ Find the circulation of $\vec{\mathbf{B}}$ around a circle of radius a centered on a wire carrying current I (■ Figure 28.14).

REMEMBER: THIS IS OERSTED'S RULE.

MODEL The magnetic field lines form circles around the wire, everywhere parallel to the curve C. We choose to go around C counterclockwise, so we expect the circulation to be positive.

SETUP The magnetic field has constant magnitude around the circle and is given by
STEPS I, II, III eqn. (28.6) with $r_\perp = a$:

$$\vec{\mathbf{B}} = \frac{\mu_0 I}{2\pi a}\,\hat{\theta}.$$

We use the angle θ as the coordinate to describe each element of the circle:

$$d\vec{\ell} = a\,d\theta\,\hat{\theta}.$$

Thus:

$$d\mathscr{C} = \vec{\mathbf{B}} \cdot d\vec{\ell} = \frac{\mu_0 I}{2\pi a}\,a\,d\theta = \frac{\mu_0 I}{2\pi}\,d\theta.$$

SOLVE The circulation is:
STEPS IV, V

$$\mathscr{C} = \oint \vec{\mathbf{B}} \cdot d\vec{\ell} = \frac{\mu_0 I}{2\pi}\oint_0^{2\pi} d\theta = \mu_0 I.$$

ANALYZE Note the relation between circulation and current. ■

EXERCISE 28.5 ❖ What is the circulation of a static electric field around any closed curve?

Example 28.6 illustrates the general result known as Ampère's law.

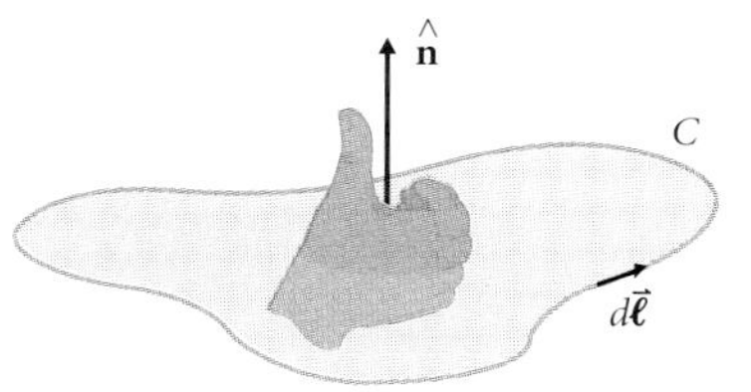

■ FIGURE 28.15
A right-hand rule defines the positive sense of current in Ampère's law. First choose a direction to travel around the curve C. This defines the direction of each element $d\vec{\ell}$. Curl the fingers of your right hand around the curve in the chosen direction; your thumb defines the positive sense for I, or for $\hat{\mathbf{n}}$ in eqn. (28.12).

AMPÈRE'S LAW

The circulation $\mathscr{C}$ of the magnetic field around any closed curve C equals the current through C multiplied by μ_0.

$$\mathscr{C} = \oint_C \vec{\mathbf{B}} \cdot d\vec{\ell} = \mu_0 I. \qquad \textbf{(28.11)}$$

The sign convention for current is determined by a right-hand rule: curl the fingers of your right hand around the curve C in the sense of the vectors $d\vec{\ell}$. The current is positive if it is in the direction of your thumb (■ Figure 28.15). Ampère's law contains the same information as the Biot-Savart law but expressed in a different form.

BY ANALOGY WITH "GAUSSIAN SURFACE," WE SHALL CALL CURVES USED IN AMPÈRE'S LAW "AMPÈRIAN CURVES."

Digging Deeper

DEMONSTRATION OF AMPÈRE'S LAW

A general proof that Ampère's law follows from the Biot-Savart law is mathematically complex. Here we demonstrate the result for a combination of long straight wires. ■ Figure 28.16 shows two arbitrary closed curves near a single long straight wire. Curve (a) surrounds the wire, whereas curve (b) does not. The z-axis is taken along the wire, and polar coordinates r and θ are used in the x-y-plane. A typical element $d\vec{\ell}$ of the curve has components:

$$d\vec{\ell} = dr\,\hat{\mathbf{r}} + r\,d\theta\,\hat{\theta} + dz\,\hat{\mathbf{z}},$$

and the magnetic field produced by the wire is:

$$\vec{\mathbf{B}} = \frac{\mu_0 I}{2\pi r}\hat{\theta}.$$

So, since $\hat{\theta} \cdot \hat{\mathbf{r}}$ and $\hat{\theta} \cdot \hat{\mathbf{z}}$ are both zero,

$$\vec{\mathbf{B}} \cdot d\vec{\ell} = \frac{\mu_0 I}{2\pi}\,d\theta,$$

and the circulation is

$$\mathscr{C} \equiv \oint \vec{\mathbf{B}} \cdot d\vec{\ell} = \frac{\mu_0 I}{2\pi}\oint d\theta.$$

Around curve (a) the angle θ increases uniformly from 0 to 2π:

$$\oint_a d\theta = 2\pi,$$

and

$$\oint_a \vec{\mathbf{B}} \cdot d\vec{\ell} = \mu_0 I.$$

Around curve (b) the angle θ increases from a minimum value to a maximum and back again. The integral is zero.

The circulation around a curve from any combination of long straight wires follows from the principle of superposition. A wire passing through the curve contributes $\mu_0 I$; a wire that does not pass through the curve makes no net contribution.

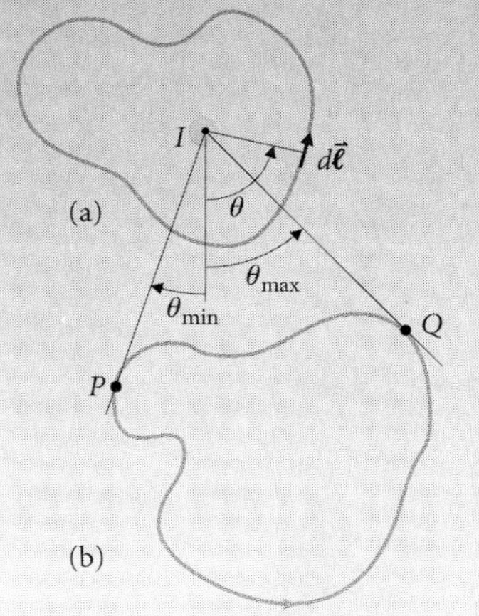

■ FIGURE 28.16
Demonstration of Ampère's law. Around curve a, θ increases from 0 to 2π. Around curve b, θ increases from θ_{min} (which is negative in this case) to θ_{max} and back again as we go from P to Q to P:

$$\oint_b d\theta = 0.$$

The surface is like a soap film attached to a wire loop. As you blow the bubble, the film changes shape but remains attached to the loop. This surface can also have any shape provided that its edge is the chosen curve.

EXERCISE 28.6 ♦ ■ Figure 28.17 shows several long straight wires carrying current perpendicular to the plane of the page. Use Ampère's law to find the circulation around each curve when traversed in the direction indicated.

Sometimes it is useful to express Ampère's law in terms of current density, $\vec{\mathbf{j}}$ = current per unit area (§26.2). Imagine a surface S whose edge is the curve C (■ Figure 28.18). The direction of the normal to the surface at any point is chosen according to the right-hand rule, as above. Then we may write Ampère's law as:

$$\oint_C \vec{\mathbf{B}} \cdot d\vec{\ell} = \mu_0 I = \mu_0 \int_S \vec{\mathbf{j}} \cdot \hat{\mathbf{n}}\, dA. \qquad \textbf{(28.12)}$$

The circulation of $\vec{\mathbf{B}}$ around C is equal to μ_0 multiplied by the flux of $\vec{\mathbf{j}}$ through any surface S spanning the curve C.

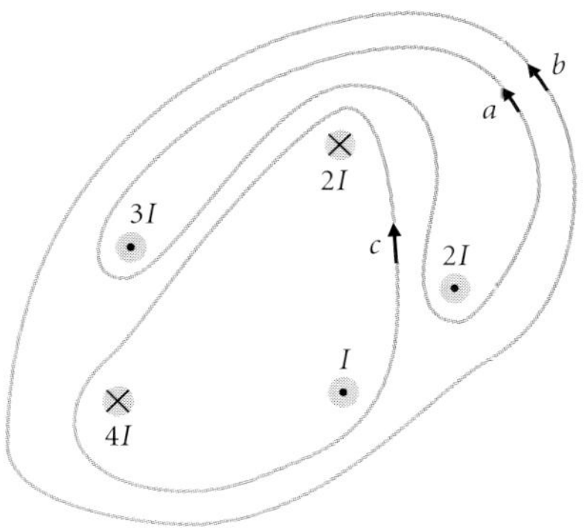

■ Figure 28.17

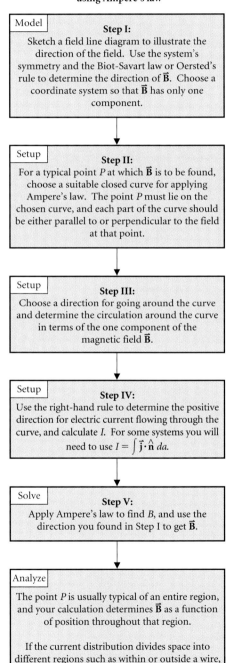

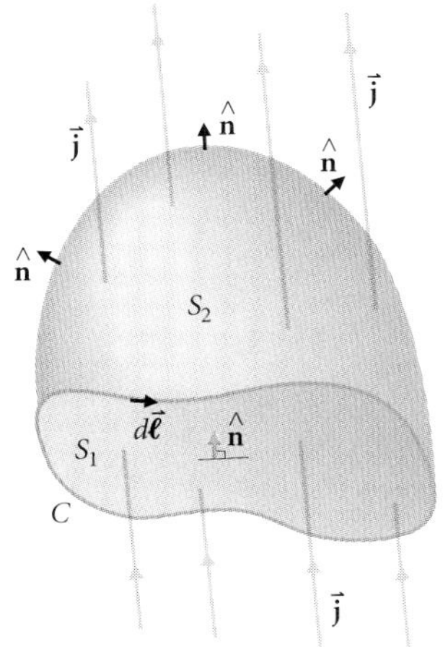

■ Figure 28.18
With steady current, the flux of $\vec{\mathbf{j}}$ is the same through any surface whose edge is the curve C. Two such surfaces are shown here. This result is akin to Kirchhoff's junction rule: if the fluxes were not the same, charge inside the volume would change with time.

Method for finding symmetric magnetic fields using Ampere's law

Model — **Step I:** Sketch a field line diagram to illustrate the direction of the field. Use the system's symmetry and the Biot-Savart law or Oersted's rule to determine the direction of $\vec{\mathbf{B}}$. Choose a coordinate system so that $\vec{\mathbf{B}}$ has only one component.

Setup — **Step II:** For a typical point P at which $\vec{\mathbf{B}}$ is to be found, choose a suitable closed curve for applying Ampere's law. The point P must lie on the chosen curve, and each part of the curve should be either parallel to or perpendicular to the field at that point.

Setup — **Step III:** Choose a direction for going around the curve and determine the circulation around the curve in terms of the one component of the magnetic field $\vec{\mathbf{B}}$.

Setup — **Step IV:** Use the right-hand rule to determine the positive direction for electric current flowing through the curve, and calculate I. For some systems you will need to use $I = \int \vec{\mathbf{j}} \cdot \hat{\mathbf{n}}\, da$.

Solve — **Step V:** Apply Ampere's law to find B, and use the direction you found in Step I to get $\vec{\mathbf{B}}$.

Analyze — The point P is usually typical of an entire region, and your calculation determines $\vec{\mathbf{B}}$ as a function of position throughout that region.

If the current distribution divides space into different regions such as within or outside a wire, you will need to repeat steps II – V for each of those regions.

■ Figure 28.19

28.3.3 Finding Magnetic Fields with Ampère's Law

When the current distribution is symmetric, Ampère's law often provides the most efficient method for calculating the magnetic field. The method is very similar to that for finding electric field from Gauss' law (■ Figure 28.19).

EXAMPLE 28.7 ♦♦ Use Ampère's law to find the magnetic field produced by the two oppositely directed currents I in a coaxial cable (■ Figure 28.20a).

MODEL STEP I The cable has cylindrical symmetry. As in the case of the long straight wire, we expect the magnetic field lines to form circles centered on the symmetry axis.

SETUP STEP II There are two regions to consider: outside the entire system and between the two conductors. For each region we choose as an Ampèrian curve a circle of arbitrary radius r centered on the symmetry axis, so that $\vec{\mathbf{B}}$ is parallel to $d\vec{\ell}$ everywhere on the curve.

STEP III The magnitude of $\vec{\mathbf{B}}$ is constant everywhere on the curve. Going around the curve counterclockwise as shown in Figure 28.20b, the circulation is:

$$\mathscr{C} = \oint \vec{\mathbf{B}} \cdot d\vec{\ell} = \oint B_\theta\, d\ell = B_\theta \oint d\ell = 2\pi r B_\theta. \qquad (28.13)$$

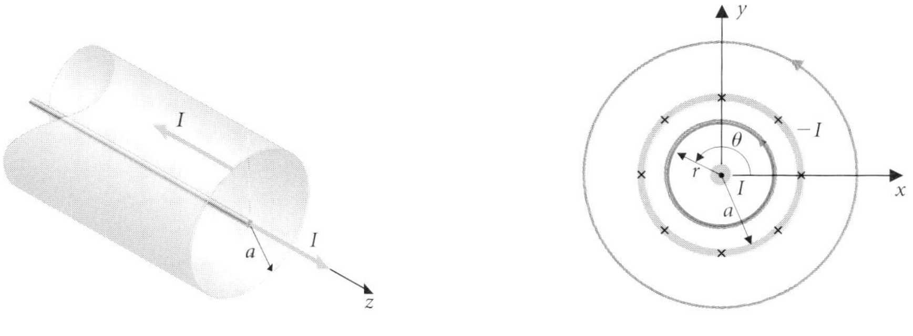

■ FIGURE 28.20
A coaxial cable carries current in opposite directions in the inner wire and the outer cylindrical conducting sheath. From the symmetry of the current distribution, we determine that $\vec{\mathbf{B}}$ has only a θ component. Using Ampère's law, we find that magnetic field is confined to the space between the conductors.

Equation (28.13) gives the circulation in terms of B_θ for a circular curve in any cylindrically symmetric problem.

STEP IV According to the right-hand rule, the positive direction for current is out of the page. The total current through the curve between the conductors is $+I$, while the current through a curve outside the system is $+I - I = 0$.

SOLVE Thus for $r_{\text{wire}} < r < a$:

STEP V

$$2\pi r B_\theta = \mu_0 I \quad \text{and} \quad \vec{\mathbf{B}} = \frac{\mu_0 I}{2\pi r}\hat{\theta},$$

and for $r > a$, $\vec{\mathbf{B}} \equiv 0$.

ANALYZE The current in the outer cylinder does not contribute to the magnetic field inside. Outside the cable, the two currents produce magnetic field in opposite directions, and the net field is zero. ■

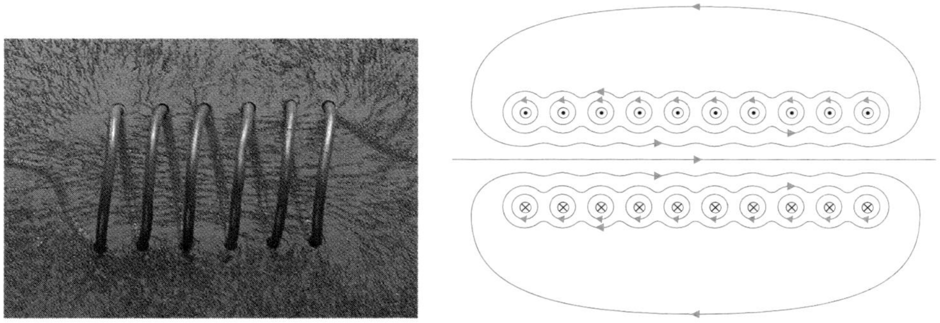

■ FIGURE 28.21
(a) A solenoid is a wire wound into a tight helix.
(b) The field inside the solenoid is almost uniform and parallel to the axis; outside, the field is almost zero.

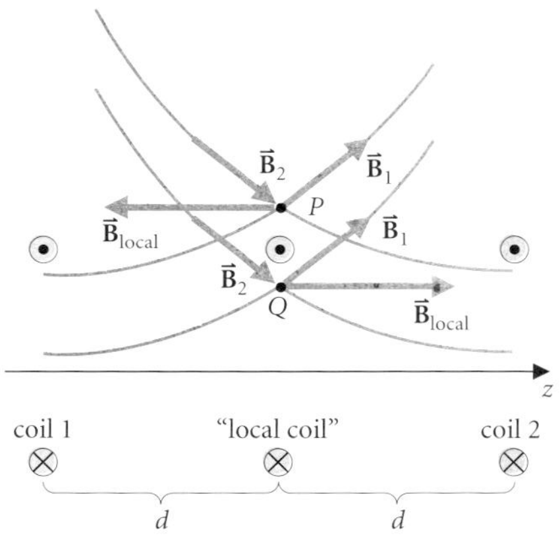

■ FIGURE 28.22
The magnetic field produced by a solenoid at points P and Q just inside and just outside the coil is the sum of contributions from nearby coils ($\vec{\mathbf{B}}_{\text{local}}$) and from pairs of coils at equal distances on either side of P and Q. One such pair is shown here, along with its contributions $\vec{\mathbf{B}}_1$ and $\vec{\mathbf{B}}_2$ whose sum is a vector in the $+z$-direction. The field $\vec{\mathbf{B}}_{\text{local}}$ is in the $+z$-direction inside the solenoid at Q but in the $-z$-direction at P outside. Thus at P the local field opposes the field due to distant pairs of coils, and the sum is almost zero. Inside, at Q, the contributions reinforce each other. (The lengths of vectors $\vec{\mathbf{B}}_1$ and $\vec{\mathbf{B}}_2$ are shown exaggerated for clarity. Many such pairs sum to give $\vec{\mathbf{B}}$ approximately equal to $-\vec{\mathbf{B}}_{\text{local}}$ at P.)

A solenoid is a coil in which the wire is wound in a helix (■ Figure 28.21). The field near a loosely wound helix is quite intricate. Very close to the wire, the field lines form circles around the wire in accordance with Oersted's rule, but at distances comparable to the separation between loops, the lines merge into curves surrounding the entire coil. At great distances, the field lines evolve into the familiar dipole pattern. For long, tightly wound coils, the pattern simplifies: within the coil the field is nearly uniform, while just outside the coil it is nearly zero (■ Figure 28.22). Calculation of the field for a tightly wound coil is correspondingly simplified.

EXAMPLE 28.8 ♦♦ An automobile starter uses a solenoid with radius $a = 2.0$ cm, length $L = 10$ cm, and $N = 500$ turns. Estimate the magnetic field inside the solenoid when it carries a current of 50 A. Assume the solenoid is very long and tightly wound.

MODEL We model the solenoid as infinite and assume that the magnetic field inside is
STEP I parallel to the solenoid's axis, which we choose to be the z-axis. The field outside is nearly zero. The number of turns per unit length is $n = N/L$.

SETUP STEP II A rectangle is a suitable curve (■ Figure 28.23). The magnetic field lies along the side parallel to the solenoid's axis, is perpendicular to the two sides that cut the solenoid, and is zero outside.

STEP III We go around the rectangle clockwise. Along QR and UP, $\vec{\mathbf{B}}$ is perpendicular to $d\vec{\ell}$, so $\vec{\mathbf{B}} \cdot d\vec{\ell} = 0$. Along RS, ST, and TU the magnetic field is almost zero. The only contribution to the circulation comes from side PQ parallel to the axis, where $\vec{\mathbf{B}}$ is parallel to $d\vec{\ell}$ and has constant magnitude:

$$\mathscr{C} = \oint \vec{\mathbf{B}} \cdot d\vec{\ell} = \int_P^Q \vec{\mathbf{B}} \cdot d\vec{\ell} = B_z \ell.$$

STEP IV The positive direction for current is into the page. The number of loops of wire passing through the rectangle is $n\ell$, and each carries current I, so the total current through the rectangle is $In\ell$.

SOLVE STEP V Applying Ampère's law gives:

$$B_z \ell = \mu_0 n \ell I.$$

Thus:

$$\vec{\mathbf{B}} = \mu_0 n I \hat{\mathbf{k}}. \qquad (28.14)$$

For the numbers given:

$$B = (4\pi \times 10^{-7}\ \text{N/A}^2)\frac{500}{(0.1\ \text{m})}(50\ \text{A}) = 0.3\ \text{T}.$$

ANALYZE The arbitrarily chosen length ℓ of the rectangle does not appear in the final expression for $\vec{\mathbf{B}}$, nor does the position of side PQ. Inside the long solenoid, the field has uniform magnitude as well as constant direction. ■

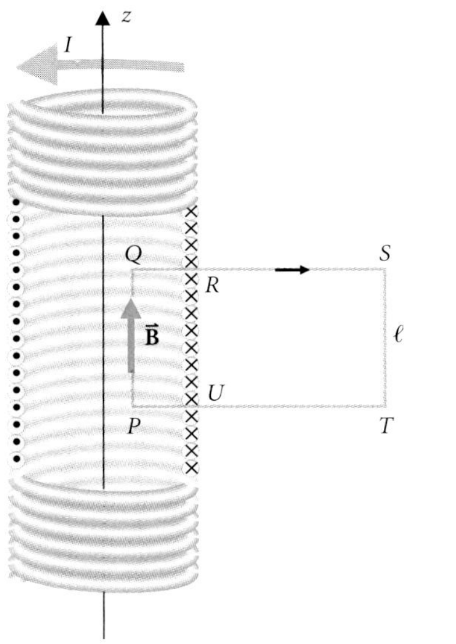

■ FIGURE 28.23
Inside a long, tightly wound solenoid, the magnetic field is parallel to the axis. Outside the solenoid, the field is nearly zero. To apply Ampère's law, we choose a rectangular curve of side ℓ with PQ parallel to the solenoid axis.

EXERCISE 28.7 ◆◆ A coil (called a Rowland ring) is made by wrapping 10^4 turns of wire around a torus (■ Figure 28.24). If the wire carries a current $I = 0.10$ A, apply Ampère's law to find the magnetic field along the center line of the torus.

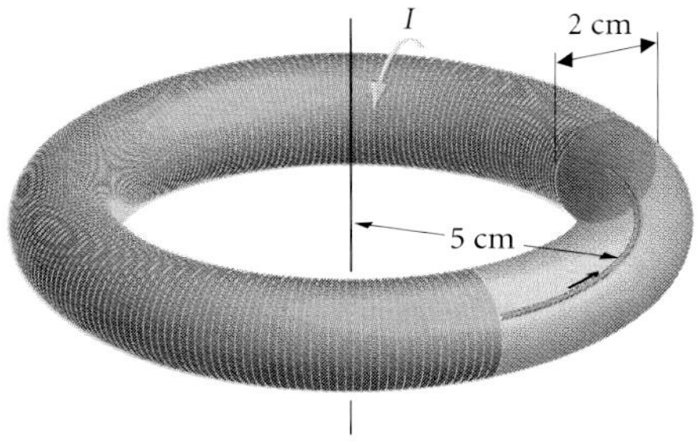

■ FIGURE 28.24

EXAMPLE 28.9 ◆◆◆ Current is not always confined in cylindrical wires. On semiconductor chips, flat conducting strips, much wider than their thickness, carry current between circuit elements. Current sheets also occur where the solar wind meets the Earth's magnetic field. Find the magnetic field produced by an infinite plane current sheet of thickness s that carries a uniform current density $\vec{\mathbf{j}} = j_0 \hat{\mathbf{i}}$ (■ Figure 28.25).

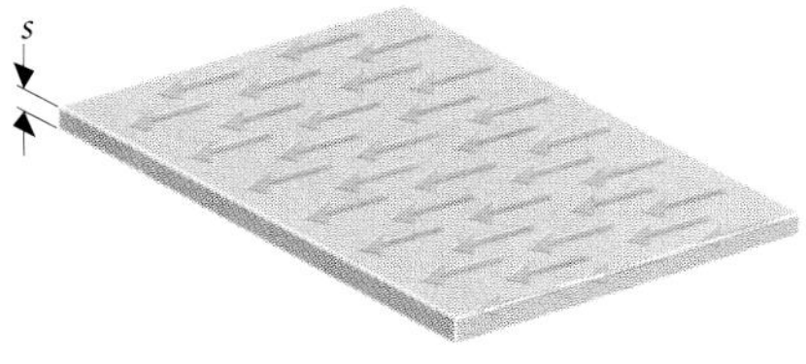

■ FIGURE 28.25
Current in a conducting sheet.

MODEL STEP I We may determine the direction of the magnetic field by modeling the sheet as a set of parallel wires (■ Figure 28.26). We choose the z-axis perpendicular to the sheet. The field directly above a single "wire" points in the $-y$-direction. "Wires" to the right produce magnetic field vectors with negative y- and z-components, while

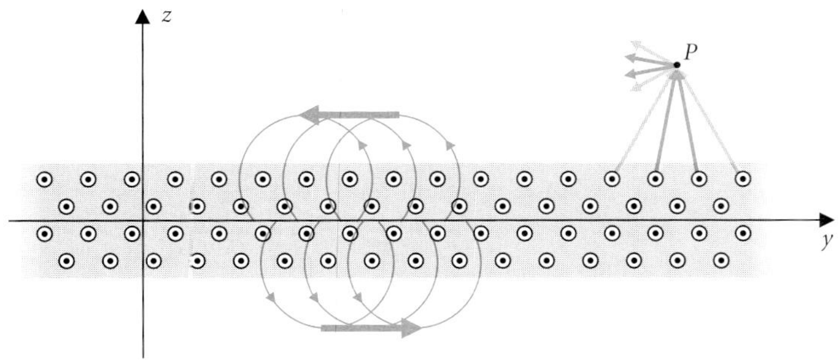

■ FIGURE 28.26
We model the sheet as an infinite set of parallel wires. Each wire produces circular field lines. The superposition of fields due to different wires is parallel to the sheet and perpendicular to I. To be more precise, at any point P, we choose pairs of wires equidistant from P. The magnetic field due to each pair is in the $-y$-direction. The direction of $\vec{\mathbf{B}}$ reverses across the sheet.

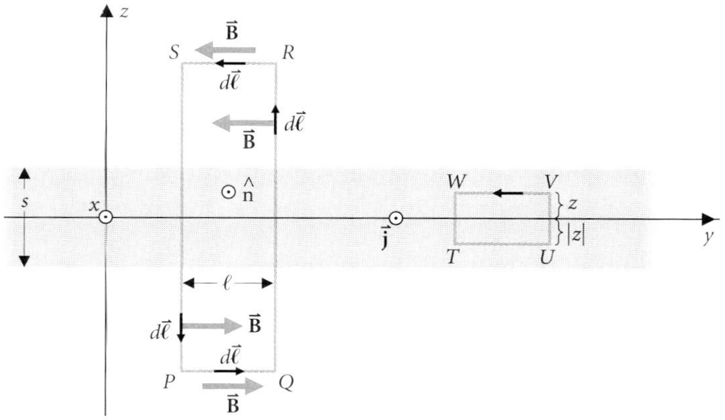

■ FIGURE 28.27
To find the magnetic field due to a uniform current sheet, we choose rectangular curves with two sides parallel to the sheet and the same distance away. The magnetic field is parallel to $d\vec{\ell}$ along PQ, RS, TU, and VW. It is perpendicular to $d\vec{\ell}$ along QR, SP, UV, and WT.

"wires" to the left produce magnetic field vectors with negative y- and positive z-components. In an infinite sheet, every wire on the right has a corresponding wire an equal distance away on the left. The sum of the vectors due to the two wires is in the $-y$-direction. Thus the total magnetic field is in the $-y$-direction. The same argument shows that the magnetic field is in the $+y$-direction below the sheet. At equal distances from the sheet, the field strength is the same both above and below the sheet: B_y (above) $= -B_y$ (below). The same symmetry exists within the sheet.

SETUP STEP II There are two regions: inside the sheet and outside. In each region we choose as Ampèrian curve a rectangle with two sides parallel to the sheet and at equal distances from the midplane (■ Figure 28.27). These two sides are also parallel to the magnetic field. The other two sides are perpendicular to the sheet and perpendicular to the magnetic field.

STEP III First we find the field outside the sheet. We proceed around the curve $PQRS$ counterclockwise. The term $\vec{\mathbf{B}} \cdot d\vec{\ell}$ is zero on the vertical sides. The magnitude of $\vec{\mathbf{B}}$ is constant along the horizontal sides, and:

HERE WE USE $\hat{\mathbf{y}}$ FOR THE UNIT VECTOR IN THE y-DIRECTION, TO AVOID CONFUSION WITH THE CURRENT DENSITY VECTOR $\vec{\mathbf{j}}$.

$$\mathscr{C} = \oint \vec{\mathbf{B}} \cdot d\vec{\ell} = \int_P^Q \vec{\mathbf{B}} \cdot dy\,\hat{\mathbf{y}} + \int_R^S \vec{\mathbf{B}} \cdot dy\,\hat{\mathbf{y}} = B_y(P) \int_P^Q dy + B_y(R) \int_R^S dy$$
$$= \ell[B_y(P) - B_y(R)] = -2\ell B_y(R).$$

NOTE: $d\vec{\ell} = dy\,\hat{\mathbf{y}}$ IS THE CORRECT EXPRESSION ON BOTH ENDS OF THE LOOP. GOING FROM R TO S, $dy < 0$ AND $d\vec{\ell}$ POINTS LEFT.

A similar expression gives the circulation around rectangle $TUVW$ in terms of $B_y(V)$.

STEP IV The direction of $\hat{\mathbf{n}}$ is out of the page. The current through rectangle $PQRS$ is:

$$I = \int \vec{\mathbf{j}} \cdot \hat{\mathbf{n}}\, da = j_0 \int da = j_0 s\ell.$$

SOLVE STEP V Applying Ampère's law gives:

$$-2\ell B_y(R) = \mu_0 j_0 s\ell \Rightarrow B_y(R) = -\tfrac{1}{2}\mu_0 j_0 s.$$

$$\vec{\mathbf{B}} = -\tfrac{1}{2}\mu_0 j_0 s\hat{\mathbf{y}} \text{ above the sheet} \quad \text{and} \quad \vec{\mathbf{B}} = \tfrac{1}{2}\mu_0 j_0 s\hat{\mathbf{y}} \text{ below the sheet.}$$

STEP IV Next we find the magnetic field inside the sheet. The current through rectangle $TUVW$ is:

$$I = 2|z|\ell j_0.$$

SOLVE STEP V Applying Ampère's law gives:

$$-2\ell B_y(V) = 2z\ell j_0.$$
$$\vec{\mathbf{B}} = -\mu_0 j_0 z\hat{\mathbf{y}}.$$

ANALYZE Since z changes sign below the y-axis, one formula works for $\vec{\mathbf{B}}$ everywhere inside the sheet. The solutions for the magnetic field inside and outside the sheet give the same answer at the edge of the sheet, where $z = \pm s/2$. The field outside the sheet is constant, independent of the distance from the sheet. ▮

28.3.4 *Summary of the Integral Laws for Static Fields*

We have now derived flux laws and circulation laws for both electric and magnetic fields. The circulation law for static electric field arises because the Coulomb force is conservative (see Exercise 28.5).

Flux laws:

$$\Phi_E = \oint \vec{\mathbf{E}} \cdot \hat{\mathbf{n}}\, dA = \frac{Q}{\epsilon_0} \quad \textbf{(23.12)} \text{ (Gauss)} \quad \text{and} \quad \Phi_B = \oint \vec{\mathbf{B}} \cdot \hat{\mathbf{n}}\, dA = 0 \quad \textbf{(28.9)}$$

Circulation laws:

$$\oint \vec{\mathbf{E}} \cdot d\vec{\ell} = 0 \quad \textbf{(28.15)} \quad \text{and} \quad \mathscr{C} = \oint \vec{\mathbf{B}} \cdot d\vec{\ell} = \mu_0 I \quad \textbf{(28.11)} \text{ (Ampère)}$$

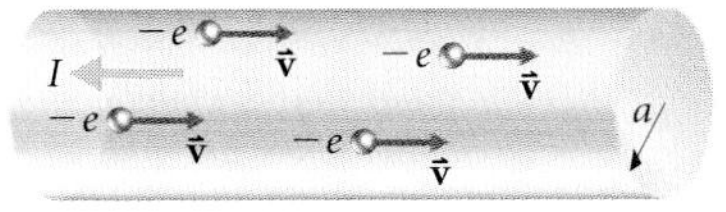

(a)

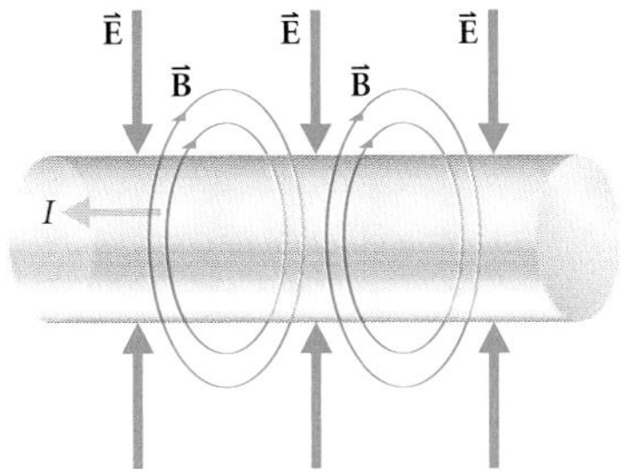

(b)

■ FIGURE 28.28
(a) An electron beam has n electrons per cubic meter, each traveling with velocity $\vec{\mathbf{v}}$ toward the right. The electrons form a current $I = nev$ toward the left. (b) The electric field vector produced by the negatively charged electrons is radially inward. The magnetic field lines form rings surrounding the beam axis, with direction determined by Oersted's right-hand rule.

These relations would be symmetric were it not for the lack of isolated magnetic poles to provide source terms in eqns. (28.9) and (28.15). These equations are equivalent to Coulomb's law and the Biot-Savart law and as such may seem little more than a convenience in a limited class of calculations. However, when fields vary in time (Chapter 30), the circulation laws are profoundly modified, and the four integral laws become our most powerful tool for understanding electromagnetic fields.

Study Problem 18 ♦♦♦ *An Electron Beam*

A beam of electrons has radius R and contains n electrons per cubic meter moving with velocity $\vec{\mathbf{v}}$ along the beam (■ Figure 28.28a). Find the electric and magnetic fields produced by the beam and the resulting force on an electron at the edge of the beam. Does the radius of the beam expand or contract as a result of this force? Assuming the force remains constant, estimate the change in the beam radius during the time the electrons move a distance s along the beam. Is this an important effect in the beam of a TV tube with $s = 15$ cm, $v = 9.4 \times 10^7$ m/s, $R = 0.1$ mm, and $n = 10^{10}$ /m^3?

Modeling the System

A beam that is much longer than its diameter forms a cylindrically symmetric distribution of charge and current. Repulsive electric forces between the electrons tend to disrupt the beam. Electron motion to the right constitutes a current to the left, with magnetic field lines circulating about it (Figure 28.28b). The resulting magnetic force on the electrons is inward. The beam expands if the electric force exceeds the magnetic force. So long as the expansion is slow, the approximations of cylindrical symmetry and static fields remain valid.

Setup

We use Gauss' law and Ampère's law to calculate the fields and then evaluate the force on each electron produced by these fields. We may estimate the beam's expansion by finding the radial displacement of an electron on the edge of the beam in the time s/v that it takes to travel a distance s along the beam.

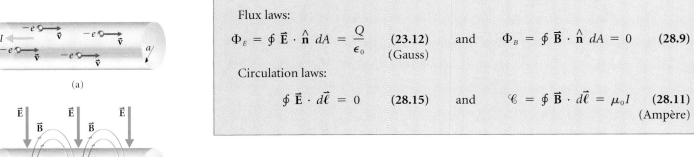

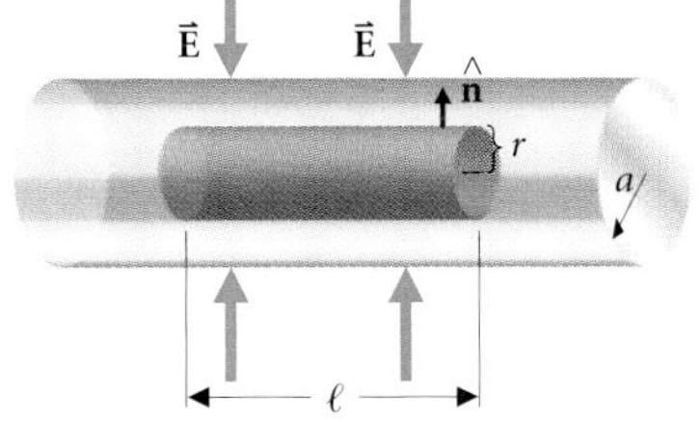

■ FIGURE 28.29
We apply Gauss' law to a cylinder of radius $r < a$ and length ℓ whose axis coincides with the beam axis. The normal to the curved surface is opposite $\vec{\mathbf{E}}$, so the flux through this Gaussian surface is negative, as is the charge enclosed.

STEP I The charge distribution has cylindrical symmetry, so the electric field points radially inward (cf. Example 24.2).

STEP II We apply Gauss' law to a cylinder of length ℓ lying along the beam with its outer surface at radius r (■ Figure 28.29).

STEP III The flux emerging from this cylinder is (cf. eqn. 24.3):

$$\Phi_E = 2\pi r\ell E_r.$$

STEP IV The total charge inside the cylinder is:

$$Q = -ne\pi r^2\ell.$$

Solution

STEP V Applying Gauss' law gives: $\Phi_E = Q/\epsilon_0$.

$$2\pi r\ell E_r = \frac{-ne\pi r^2\ell}{\epsilon_0} \Rightarrow \vec{\mathbf{E}} = -\frac{ner}{2\epsilon_0}\hat{\mathbf{r}}.$$

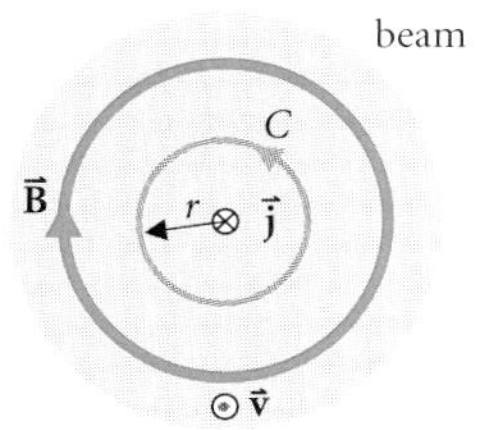

■ FIGURE 28.30
We apply Ampère's law to a circle of radius $r < a$ centered on the beam axis. Going around the circle opposite $\vec{\mathbf{B}}$ (counter-clockwise in this end view), $\hat{\mathbf{n}}$ is out of the page, while $\vec{\mathbf{j}}$ is into the page. The current through the curve is negative.

Setup

STEP I The magnetic field lines are circles around the beam axis.

STEP II We apply Ampère's law to a circle of radius r centered on the beam axis, traversed as shown in ■ Figure 28.30.

STEP III The circulation of $\vec{\mathbf{B}}$ around the circle is (cf. eqn. 28.13):

$$\mathscr{C} = 2\pi r B_\theta.$$

STEP IV The positive direction for current is along the axis parallel to $\vec{\mathbf{v}}$. The current passing through the circle is:

$$I = -\pi r^2 j = -\pi r^2 nev.$$

Solution

STEP V Applying Ampère's law we obtain: $\mathscr{C} = \mu_0 I$.

$$2\pi r B_\theta = -\mu_0\pi r^2 nev \quad \Rightarrow \quad \vec{\mathbf{B}} = \tfrac{1}{2}\mu_0 nevr(-\hat{\theta}).$$

The total force acting on an electron is the sum of the electric and magnetic forces (■ Figure 28.31):

$$\vec{\mathbf{F}} = -e(\vec{\mathbf{E}} + \vec{\mathbf{v}} \times \vec{\mathbf{B}}) = e\left(\frac{ner}{2\epsilon_0} - \mu_0\frac{nerv^2}{2}\right)\hat{\mathbf{r}}$$

$$= \frac{ne^2 r}{2\epsilon_0}(1 - \mu_0\epsilon_0 v^2)\hat{\mathbf{r}}.$$

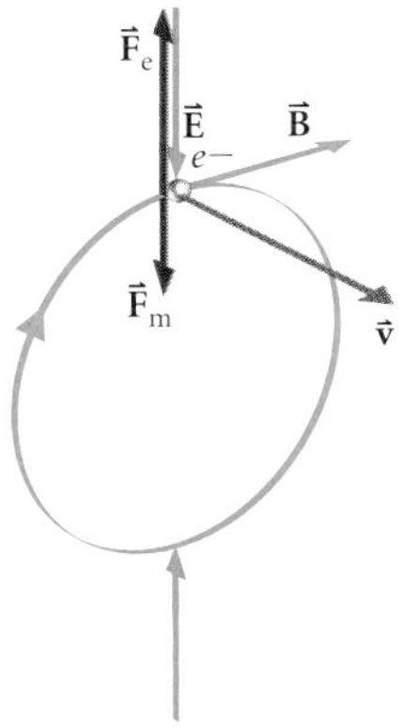

■ FIGURE 28.31
The electric and magnetic forces on an electron in the beam are both radial: $\vec{\mathbf{F}}_e$ is outward, and $\vec{\mathbf{F}}_m$ is inward.

The product $\mu_0\epsilon_0$ has dimensions of (speed)$^{-2}$. The corresponding speed is:

$$\frac{1}{\sqrt{\mu_0\epsilon_0}} = \frac{1}{\sqrt{(4\pi \times 10^{-7}\ \mathrm{N/A^2})(8.85 \times 10^{-12}\ \mathrm{C^2/N\cdot m^2})}}$$
$$= 3 \times 10^8\ \mathrm{m/s},$$

the speed of light! Only if the electrons were moving at the speed of light could the magnetic force equal the electric repulsion. Since $v < c$, the beam expands.

THIS IS NO ACCIDENT! WE'LL SEE WHY IN CHAPTER 33.

To find the displacement of an electron at the edge of the beam, we evaluate the fields at $r = R$. The displacement we want is:

$$s_\perp \sim \tfrac{1}{2}a(\Delta t)^2 = \frac{F}{2m}\left(\frac{s}{v}\right)^2 = \frac{ne^2Rs^2}{4m\epsilon_0 v^2}(1 - \mu_0\epsilon_0 v^2).$$

For the TV beam:

$$s_\perp = \frac{(10^{10}\ /\mathrm{m^3})(1.6 \times 10^{-19}\ \mathrm{C})^2(10^{-4}\ \mathrm{m})(0.15\ \mathrm{m})^2}{4(9.11 \times 10^{-31}\ \mathrm{kg})(8.85 \times 10^{-12}\ \mathrm{C^2/N\cdot m^2})(9.4 \times 10^7\ \mathrm{m/s})^2}\left[1 - \left(\frac{9.4}{30}\right)^2\right]$$
$$= 2 \times 10^{-9}\ \mathrm{m}.$$

Analysis

The expansion of the beam is 2 parts in 10^5 of the beam radius, and is unimportant. (This justifies our use of constant acceleration in calculating the expansion.) In accelerators where beams travel long distances the expansion is significant, and efforts must be made to refocus the beam. When current is carried in an ionized but neutral beam, only the magnetic force acts, and it can *pinch,* or confine, the beam.

In our everyday experience magnetic forces occur in motors, loudspeakers, or between two bar magnets. These forces seem strong compared with the electric forces we encounter, yet the magnetic force is inherently weaker by the ratio v/c. Electric forces only seem to be weaker because of the very small amounts of unbalanced charge, and correspondingly weak fields, that commonly occur.

Chapter Summary

Where Are We Now?

We have discussed methods for calculating the magnetic field produced by an arbitrary distribution of current. Ampère's law gives a global relationship between the magnetic field and its sources that complements the Biot-Savart law. We have now established four integral laws for static electric and magnetic fields.

What Did We Do?

The magnetic force exerted on a moving charge is $\vec{\mathbf{F}} = q\vec{\mathbf{v}} \times \vec{\mathbf{B}}$, perpendicular to both $\vec{\mathbf{v}}$ and $\vec{\mathbf{B}}$. The force law gives an operational definition of $\vec{\mathbf{B}}$.

The Biot-Savart law describes the magnetic field produced by an element $d\vec{\ell}$ of wire carrying current I:

$$d\vec{\mathbf{B}} = I\frac{d\vec{\ell} \times \hat{\mathbf{r}}}{r^2}.$$

Using the principle of superposition, the magnetic field due to a finite current-carrying wire can be found by integration. The magnetic field due to a long straight wire is:

$$\vec{\mathbf{B}} = \left(\frac{\mu_0 I}{2\pi r}, \quad \text{tangent to a circle centered on the wire}\right). \qquad (28.6)$$

At large distances from a current loop, its magnetic field is well approximated by the field of an ideal dipole with magnetic moment $|\vec{\mathbf{m}}| = IA$, where A is the area of the loop. The direction of $\vec{\mathbf{m}}$ is found from a right-hand rule: curl the fingers of your right hand around the loop in the direction of the current, and $\vec{\mathbf{m}}$ lies along your thumb.

The *circulation* of $\vec{\mathbf{B}}$ around a curve C,

$$\mathscr{C} = \oint_C \vec{\mathbf{B}} \cdot d\vec{\ell},$$

describes the extent to which the field lines follow the direction of the curve. Ampère's law relates the circulation of $\vec{\mathbf{B}}$ around a closed curve to the current passing through that curve:

$$\mathscr{C} = \mu_0 I.$$

The positive direction for I is chosen according to a right-hand rule (Figure 28.15). Ampère's law can be used to calculate fields in some important symmetric cases. A solution plan is given in Figure 28.19.

The magnetic field inside a long solenoid, $B = \mu_0 nI$ (eqn. 28.14), is nearly uniform, while the field just outside the solenoid is nearly zero.

Practical Applications

Magnetic fields have numerous practical applications, many of which will be discussed in the next chapter. To design magnetic field structures such as those used for plasma confinement, engineers use both carefully constructed iron pole pieces and electric current in loops and coils of various shapes. In this chapter we have seen how the current path shapes the magnetic field. The current meter mentioned in §28.3.2 is a direct practical application of Ampère's law. Most applications are more subtle, and many require the dynamic relations we shall develop in Chapter 30. Ampère's law is necessary for understanding the behavior of magnetic fields at the boundaries between two different media.

Solutions to Exercises

28.1. See ■ Figure 28.32. Remember that when the charge is negative, the magnetic force is *opposite* $\vec{\mathbf{v}} \times \vec{\mathbf{B}}$.

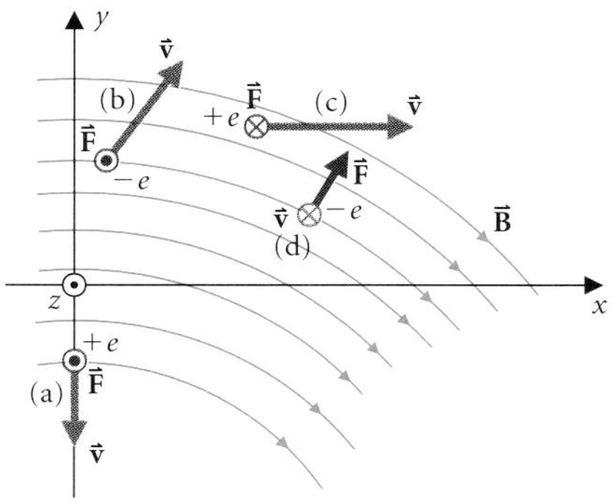

■ FIGURE 28.32
The direction of each magnetic force $\vec{\mathbf{F}}_m = q\vec{\mathbf{v}} \times \vec{\mathbf{B}}$ is determined by the right-hand rule for cross products. The resulting vectors are shown here.

28.2. The current is the charge flowing through a cross section (area A) of the wire per second: $I = nev_dA$. Thus

$$I\,d\ell = nev_d A\,d\ell = ev_d(n\,dV),$$

where $n\,dV$ is the total number of electrons in the element of wire. Electrons moving with velocity $\vec{\mathbf{v}}_d$ produce current in the opposite direction:

$$I\,d\vec{\ell} = -e\vec{\mathbf{v}}_d(n\,dV).$$

28.3. The magnitude of the field produced by an element of wire $d\vec{\ell}$ is proportional to the magnitude of the cross product $|d\vec{\ell} \times \hat{\mathbf{r}}| = d\ell \sin\theta$, where θ is the angle between $d\vec{\ell}$ and $\hat{\mathbf{r}}$. Along the line of $d\vec{\ell}$, $\theta = 0$ and $|d\vec{\mathbf{B}}| = 0$. The maximum value of $|d\vec{\mathbf{B}}|$ occurs for $\theta = 90°$.

28.4. The magnetic field at P (■ Figure 28.33) is the sum of two contributions, one from each wire. According to Oersted's rule, each contribution is into the page. Thus from eqn. (28.6), the net field is:

$$\vec{\mathbf{B}} = 2\left(\frac{\mu_0 I}{2\pi(\ell/2)}\right)\hat{\otimes} = \frac{2\mu_0 I}{\pi\ell}\hat{\otimes}.$$

28.5. The electric potential $\int \vec{\mathbf{E}} \cdot d\vec{\ell}$ between any two points A and B is independent of the path used to evaluate the integral. Thus the change in potential around any closed curve (which equals the circulation of $\vec{\mathbf{E}}$ around the curve) may be written:

$$\mathscr{E}_{SE} = \oint \vec{\mathbf{E}} \cdot d\vec{\ell} = \int_A^B \vec{\mathbf{E}} \cdot d\vec{\ell} + \int_B^A \vec{\mathbf{E}} \cdot d\vec{\ell}$$

$$= \int_A^B \vec{\mathbf{E}} \cdot d\vec{\ell} - \int_A^B \vec{\mathbf{E}} \cdot d\vec{\ell} = 0.$$

The circulation of a static electric field around any closed curve is zero.

28.6. All three curves are traversed counterclockwise, so the positive direction for current is out of the page. All we need do to find the circulations is to compute the current through each curve.

Curve	Net Current	Circulation
a	$3I + 2I = 5I$	$5\mu_0 I$
b	$3I + 2I + I - 2I - 4I = 0$	zero
c	$I - 2I - 4I = -5I$	$-5\mu_0 I$

28.7. The magnetic field inside a solenoid is parallel to its axis. The Rowland ring is a solenoid bent into a circle, and we expect the magnetic field lines to take the form of circles threading the ring. Choosing the circle along the center line as our Ampèrian curve, and going around it counterclockwise, we find that the circulation is $2\pi RB_\theta$, where $R = 5$ cm. The current through the curve is due to all 10^4 turns carrying 0.1 A each. Its direction is downward, so the net current is -10^3 A. Applying Ampère's law, $\oint \vec{\mathbf{B}} \cdot d\vec{\ell} = \mu_0 I$:

$$2\pi(0.05\text{ m})B_\theta = -\mu_0(10^3\text{ A}).$$

$$B_\theta = -\frac{(4\pi \times 10^{-7}\text{ N/A}^2)(10^3\text{ A})}{2\pi(0.05\text{ m})} = -4 \times 10^{-3}\text{ T}.$$

The direction of $\vec{\mathbf{B}}$ is around the circle clockwise.

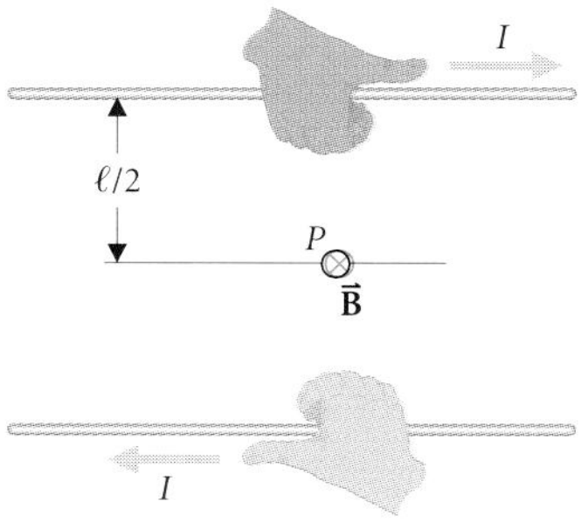

■ FIGURE 28.33

Basic Skills

Review Questions

§28.1 MAGNETIC FORCE

- What magnetic force is exerted on a point charge at rest?
- Describe how the magnetic force exerted on a point charge depends on its velocity.
- If the magnetic field strength is doubled, by how much does the force exerted on a point charge change?
- How could you determine the direction of $\vec{\mathbf{B}}$ by measuring the force exerted on a point charge?
- What is the formal name for $\vec{\mathbf{B}}$? What is its SI unit?

§28.2 CURRENT AS THE SOURCE OF MAGNETIC FIELD

- What did Oersted discover about the magnetic field produced by a long straight wire carrying current?
- State the Biot-Savart law.
- When calculating static magnetic field, why do we not begin with the field due to a point charge?
- How is the Biot-Savart law similar to Coulomb's law? How does it differ?
- What is the principle of superposition for magnetic fields? Why is the integral in eqn. (28.4) written as $\oint$?
- What magnetic field is produced by a straight wire segment of length ℓ carrying current I? How do you determine the angles that appear in this formula?
- What is the *magnetic moment* of a current loop? Under what circumstances is it useful in determining the field produced by the loop?
- What are *Helmholtz coils?*

§28.3 INTEGRAL LAWS FOR STATIC MAGNETIC FIELDS

- State Gauss' law for the magnetic field.
- Define the *circulation* of magnetic field around a curve C.
- State Ampère's law for the static magnetic field.
- Explain how to assign a positive direction for current when using Ampère's law.
- What is the magnetic field inside a solenoid with n turns per unit length?
- What two forces act on charged particles moving in a beam with velocity $\vec{\mathbf{v}}$? Which force is greater? Why?

Basic Skill Drill

§28.1 MAGNETIC FORCE

1. (■ Figure 28.34) What are the directions of the forces acting on the electrons at C and at D?

2. A particle with charge $q = 15$ nC is moving along the x-axis at a speed of 250 m/s. The magnetic field in the region is 3.5×10^{-3} T in the y-direction. What is the force on the particle?

3. The magnetic field in a region of space is in the x-direction and has magnitude 0.48 T. A particle with a charge of 0.65 μC is moving at 375 m/s. For which direction(s) of the particle's velocity will the magnetic force on it be greatest? What is the maximum force?

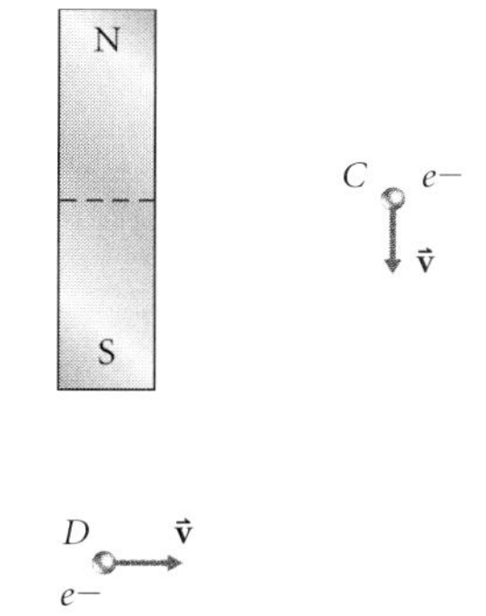

■ FIGURE 28.34

§28.2 CURRENT AS THE SOURCE OF MAGNETIC FIELD

4. A long straight wire carries a current of 15 A. What is the magnetic field 35 cm from the wire?

5. Point P is a perpendicular distance 5.0 cm from the center of a 17-cm-long straight wire segment carrying a current of 1.0 mA. What is the magnetic field at P produced by the wire?

6. You wish to design a simple coil of radius 0.100 m that produces a 6.00×10^{-3}-T magnetic field at its center. If the coil is to carry a current of 1.00 A, how many turns of wire are necessary?

7. Find the magnetic moment of a triangular wire loop with three equal sides 3.5 cm long carrying a 7.7-A current.

§28.3 INTEGRAL LAWS FOR STATIC MAGNETIC FIELDS

8. What is the magnetic flux through the surface of the Earth?

9. Three wires carry currents as shown in ■ Figure 28.35. Find the circulation of $\vec{\mathbf{B}}$ around each of the curves C_1, C_2, and C_3.

10. What is the magnetic field on the axis of a solenoid with 25 turns per centimeter carrying a 15-A current?

11. Find the magnetic field at $r = 0.75$ cm between the conductors of a coaxial cable carrying a current of 5.5 mA. What is the magnetic field outside the cable?

12. The electron beam at the Stanford Linear Accelerator Center (SLAC) has a diameter of 1.9 cm and carries an average current of 25 μA. What is the magnetic field at the edge of the beam? Assume the electrons move at speed $c = 3 \times 10^8$ m/s, and estimate the electron density in the beam. (Ignore relativistic effects.)

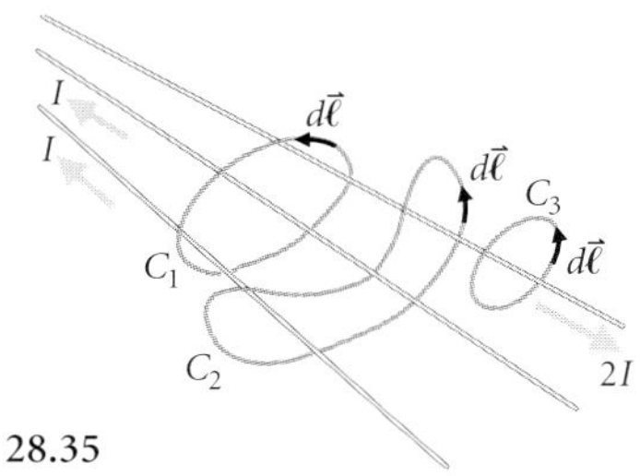

■ FIGURE 28.35

Questions and Problems

§28.1 MAGNETIC FORCE

13. ❖ A charged particle is moving in the x-direction. A bar magnet is brought up with its north pole pointing in the y-direction. The particle accelerates in the $-z$-direction. What is the sign of the charge on the particle?

14. ❖ An electron moving in the $-x$-direction experiences no magnetic force. A second electron moving in the $+y$-direction accelerates in the $+z$-direction. What is the direction of $\vec{\mathbf{B}}$?

15. ❖ A charged particle moving to the right is deflected upward. How could you determine whether the force on the particle is electric or magnetic?

16. ♦ From the operational definition of magnetic field strength, prove the following units relations:

$$\begin{aligned} 1\ \text{T} &= 1\ \text{V}\cdot\text{s/m}^2 \\ &= 1\ \Omega\cdot\text{C/m}^2 \\ &= 1\ \text{kg/C}\cdot\text{s}. \end{aligned}$$

17. ♦ The space shuttle develops a net charge Q of approximately 2 μC during an experiment on plasma beams. The shuttle's speed is 10 km/s, and the Earth's magnetic field B_E at the orbital radius is approximately 3×10^{-5} T. Estimate the magnetic force on the shuttle. Does this effect produce a problem for NASA?

18. ♦ The magnetic field near the surface of a neutron star is about 10^8 T. Estimate the magnetic force on an electron moving at 10^7 m/s perpendicular to $\vec{\mathbf{B}}$.

19. ♦ A particle with a charge of -27 nC moves in a region where the magnetic field is 0.80 T$\hat{\mathbf{k}}$. Find the force acting on the particle when its velocity is:

(a) 3500 m/s$\hat{\mathbf{i}}$.

(b) 2.7×10^4 m/s$\hat{\mathbf{j}}$.

(c) $(1.0 \times 10^5\ \text{m/s})(\hat{\mathbf{i}} + \hat{\mathbf{k}})$.

20. ♦ The magnetic field $\vec{\mathbf{B}}$ is a vector pointing into the page with magnitude 0.050 T. Find the magnitude and direction of the magnetic force acting on:

(a) an electron traveling to the right at $v = 3 \times 10^7$ m/s.

(b) a proton traveling down at $v = 6 \times 10^7$ m/s.

21. ♦ A magnetic field $\vec{\mathbf{B}}$ has magnitude 0.057 T and direction 30.0° upward from the horizontal. An iron nucleus with charge $26e$ is moving horizontally to the right at 1800 m/s. What is the magnetic force exerted on the nucleus?

22. ♦♦ A charged particle with $q = 16\ \mu$C moving at 175 m/s experiences a maximum force of 17×10^{-3} N in the $+z$-direction when it moves at 45° to the $+x$- and $+y$-axes. What is the magnitude of the magnetic field? Can you determine its direction? What force does the particle experience when moving along the y-axis?

§28.2 CURRENT AS THE SOURCE OF MAGNETIC FIELD

23. ❖ Which produces greater magnetic field at the center of a loop with radius R: the loop itself carrying current I or a straight wire segment of length $2\pi R$, tangent to the loop and carrying the same current I? Explain your reasoning, arguing directly from the Biot-Savart law.

24. ❖ The points a, b, and c in ■ Figure 28.36 lie in the same plane as the rectangular current loop. At which point is the magnetic field the greatest? Explain your reasoning.

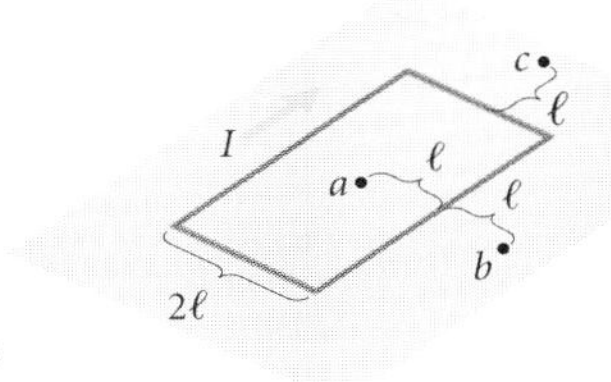

■ FIGURE 28.36

25. ❖ You have a given length of wire and wish to make a device with the largest possible magnetic moment.

(a) What shape loop would you use? Explain why.

(b) Would you wrap the wire in a coil or use a single turn? Explain why.

26. ❖ **(a)** Two parallel long straight wires carry current in the same direction. Draw a diagram showing the pattern of magnetic field lines in a plane perpendicular to the two wires.

(b) Draw a diagram of the field lines if the wires carry current in opposite directions.

27. ♦ An electric power transmission line is 5 m above ground and carries a current of 50 A. What magnetic field does it produce on the ground 10 m from a point directly below the line?

28. ♦ You are using jumper cables to help a friend start his car. The current in the cable is 66 A. What is the magnetic field strength 15 cm from a cable, if it is approximately straight?

29. ♦ If the two jumper cables in Problem 28 form a circle of radius 0.45 m, what is the magnetic field 15 cm above the center of the circle?

30. ♦♦ Two loops of wire with radii $R_2 = 12$ cm and $R_1 = 9.0$ cm are placed concentrically in the x-y-plane (■ Figure 28.37). The outer loop carries current $I_2 = 15$ A, and the inner loop carries current $I_1 = 12$ A. What is the magnetic field at the common center of the loops?

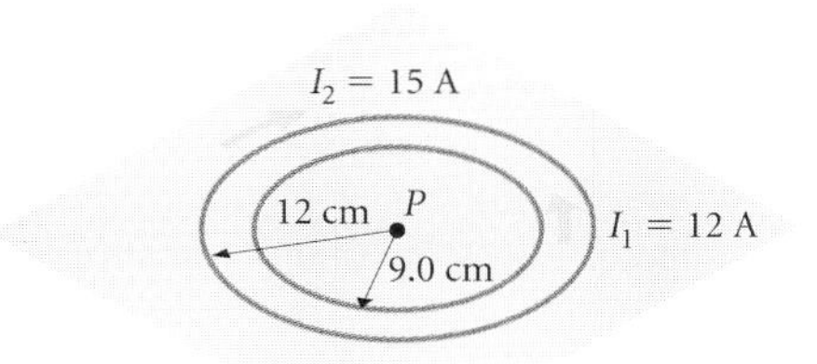

■ FIGURE 28.37

31. ♦♦ A straight wire segment 17 cm long lies along the positive x-axis with one end at the origin. It carries a current $I = 1.0$ mA in the positive x-direction. Find the magnetic field contributed by this segment at the point P with coordinates $x = 8.5$ cm, $y = -5.0$ cm, $z = 0$.

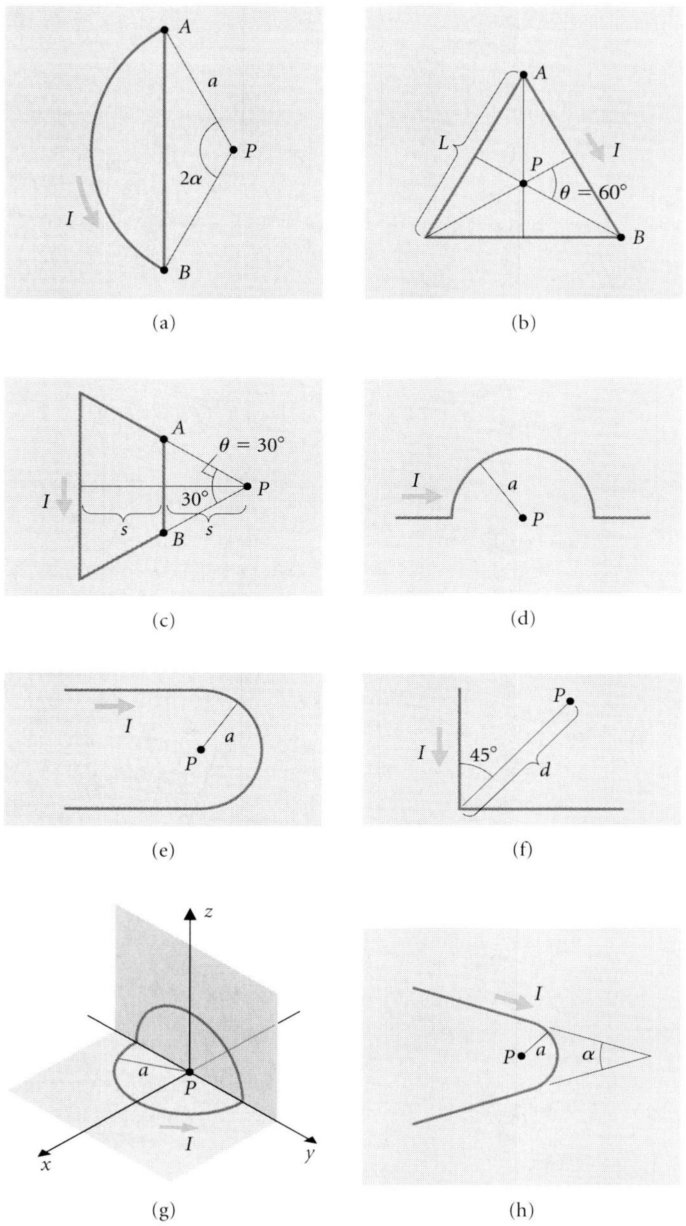

■ FIGURE 28.38

32. ♦♦ ■ Figure 28.38 shows several wires, each carrying current I. In each case, find the magnetic field at point P in the diagram.

33. ♦♦ Two parallel wires are 12 cm apart. Each carries a 25-A current in the same direction. Find the magnetic field midway between the wires and outside the wires at a point in the same plane 6.0 cm from one wire.

34. ♦♦ Assume that the magnetic moment of a finite, tightly wound solenoid equals the number of loops in the solenoid multiplied by the magnetic moment of each loop. Show that the magnetic moment is also given by the formula:

$$m \approx \frac{BV}{\mu_0},$$

where B is the magnetic field and V is the volume inside the solenoid.

35. ♦♦ A thin, uniformly charged ring of radius $a = 0.10$ m carries a total electric charge $Q = 3.3\ \mu$C and rotates with angular speed $\omega = 120$ rad/s about its symmetry axis. Find the magnetic field it produces at its center.

36. ♦♦ A wire loop in the shape of a semicircle of radius 7.5 cm carries a 15-A current. Find: **(a)** the magnetic moment of the loop. **(b)** the magnetic field strength 1.0 m directly above the loop. **(c)** the magnetic field strength in the plane of the loop 1.0 m from the center of the straight side.

37. ♦♦ A square loop of side 3.50 cm carries a 4.77-A current. Find the magnetic field 7.00 cm above the center of the square. Compare your result with the approximate result using the magnetic moment of the loop and the dipole field. How large is the error in the approximate result?

38. ♦♦ Two solenoids are used to produce magnetic field across a cathode-ray tube (■ Figure 28.39). Each coil has 1.0×10^3 turns, is 10.0 cm long, has a radius of 5.0 cm, and carries a 1.1-A current. The CRT is 8.0 cm in diameter. **(a)** Use eqn. (28.14) for the magnetic field inside a solenoid to calculate the magnetic field applied to the CRT. **(b)** Use the result of Problem 41 to calculate the magnetic field at the center of the CRT and at one edge. By how much does each value differ from the result of (a)? By how much do the values differ from each other?

39. ♦♦♦ Helmholtz coils are often used to produce a uniform field in the laboratory. Each of two parallel circular coils of radius R and separated by a distance R carries a current I in the same sense (Figure 28.10). Find the magnetic field at point P and show that $\partial B/\partial z$, $\partial^2 B/\partial z^2$, and $\partial^3 B/\partial z^3$ are all zero at P.

40. ♦♦♦ Find the magnetic field on the axis of a disk of radius $R = 5.0$ cm carrying a uniform charge density $\sigma = 1.0\ \mu\text{C/m}^2$ and spinning at $\omega = 12$ rad/s, at distance $d = 3.0$ cm from the disk.

41. ♦♦♦ Show that the magnetic field on the axis of a finite, tightly wound solenoid (■ Figure 28.40) is given by:

$$\vec{\mathbf{B}} = 2\pi k_m nI(\cos\theta_1 + \cos\theta_2)\hat{\mathbf{k}}.$$

(*Hint:* Consider each element dz of the solenoid to be a circular coil with $n\,dz$ turns, and use eqn. (28.7). Be careful with the limits of integration.) What error is made in Example 28.8 by assuming the solenoid's length to be infinite?

42. ♦♦♦ Use the result of Problem 41 to find how the magnetic field of a finite, tightly wound solenoid varies at large distances along the axis of the solenoid. From the result deduce the magnetic moment of the solenoid. Compare your result with the number of loops in the solenoid multiplied by the magnetic moment of each loop. (*Hint:* Use the binomial theorem to expand the square roots in your expressions for $\cos\theta_1$ and $\cos\theta_2$.)

43. ♦♦♦ A charged ring with $Q = 3.3\ \mu$C and $a = 0.10$ m (cf. Problem 35) is rotating with $\omega = 120$ rad/s about an axis in the plane of

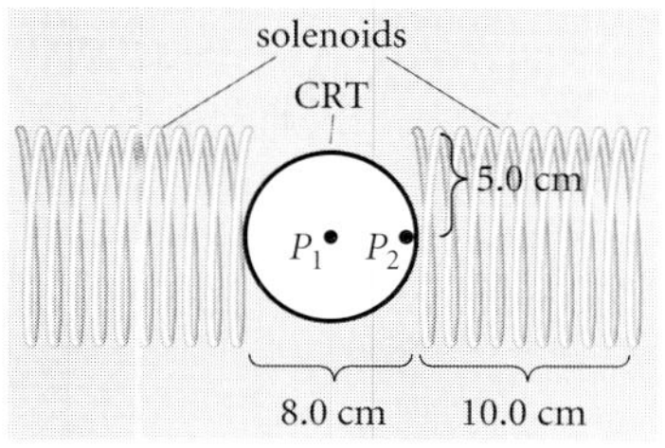

■ FIGURE 28.39

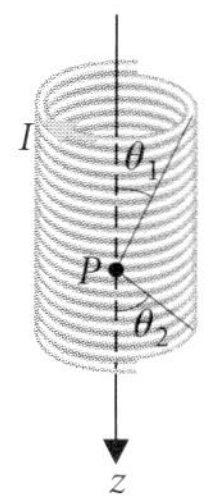

■ FIGURE 28.40

the ring. Find the magnetic field it produces at points on the rotation axis a large distance from the ring.

44. ♦♦♦ A thin, uniformly charged spherical shell of radius $a = 0.10$ m carries a total charge $Q = 3.3\ \mu$C and rotates with angular speed $\omega = 120$ rad/s. Find the magnetic field it produces at points on its rotation axis a large distance from the shell.

§28.3 INTEGRAL LAWS FOR STATIC MAGNETIC FIELDS

45. ❖ A wire carrying current I runs along the axis of a cylindrical box of radius r. What is the magnetic flux through the box?

46. ❖ ■ Figure 28.41 shows a uniform magnetic field between the pole faces of an electromagnet, with zero field elsewhere. Apply Ampère's law to the orange rectangle to prove that the field cannot go abruptly to zero as in the figure.

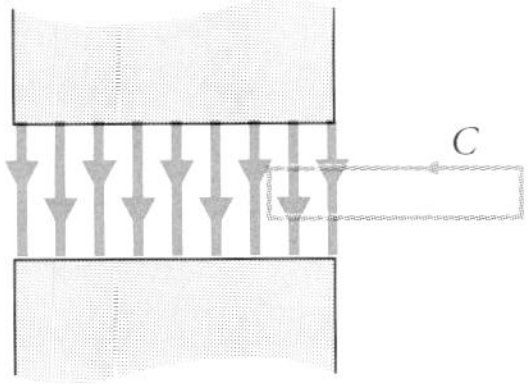

■ FIGURE 28.41

47. ♦♦ A cylinder of radius r has 100 wires closely spaced around its circumference, each running the length of the cylinder and carrying current I. Find the magnetic field: **(a)** inside the cylinder. **(b)** outside the cylinder. In each case consider points at a distance greater than $r/100$ from the surface of the cylinder. What is the purpose of this restriction?

48. ♦♦ A wire of radius 1.5 mm carries a current of 0.22 A. The current density within the wire is uniform. Find the magnetic field inside and outside the wire, and plot $|\vec{\mathbf{B}}|$ as a function of radius.

49. ♦♦ A conducting cylinder of radius b with a hole of radius a bored along its length (■ Figure 28.42) carries a uniform current density $\vec{\mathbf{j}}$. First use Ampère's law to find an expression for the magnetic field within a wire with uniform current density, then use superposition to show that $\vec{\mathbf{B}}$ is constant within the hole.

50. ♦♦♦ A long conducting cylindrical shell carries a uniform current density around its circumference (■ Figure 28.43). Apply Ampère's law to find the resulting magnetic field. (*Hint:* The shell is similar to a solenoid.)

51. ♦♦♦ The current density within a long wire of radius a varies as $j(r) = j_0(r/a)$. Find the magnetic field inside and outside the wire. Plot $B(r)$ as a function of r.

52. ♦♦♦ Uniform surface current in an infinite plane sheet divides evenly between the two halves of a long cylindrical shell then reunites in a second plane sheet diametrically opposite the first (■ Figure 28.44).

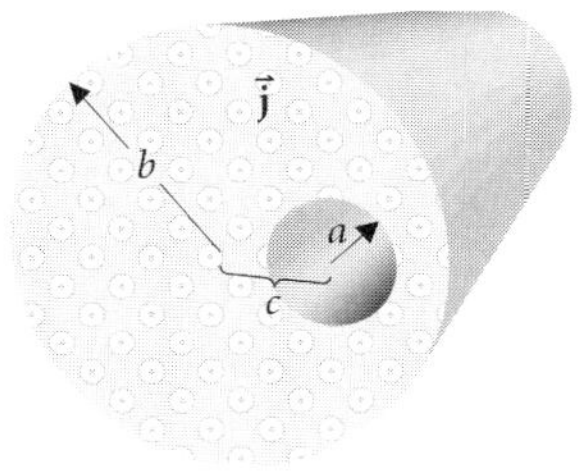

■ FIGURE 28.42

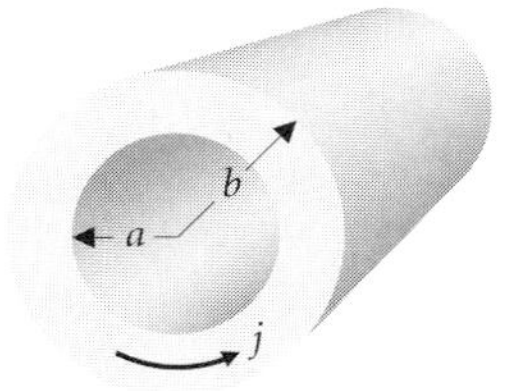

■ FIGURE 28.43

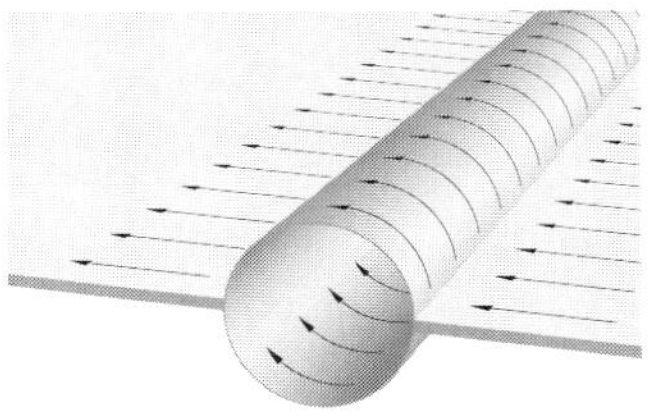

■ FIGURE 28.44

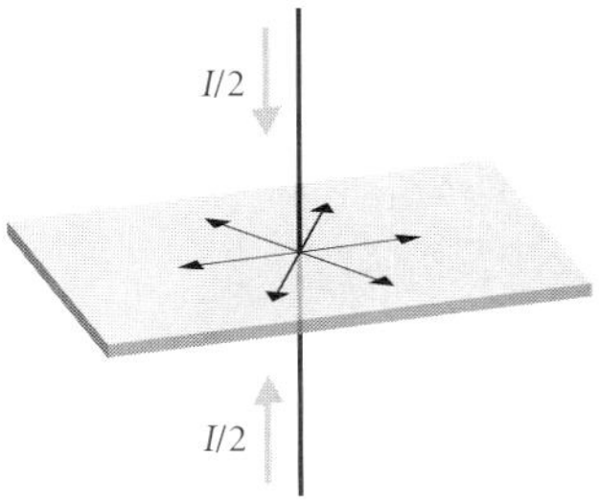

■ FIGURE 28.45

The surface current density in the sheets is K. (K is the product of the current density in each sheet times the very small thickness of the sheet.) **(a)** Describe the symmetry properties of this system and the conclusions you may draw about the symmetry of the magnetic field. **(b)** Determine Ampèrian curves you might use to check each of your conclusions about the field symmetry. **(c)** Show that the magnetic field is zero at any point inside the cylinder. **(d)** Find the magnetic field outside the cylinder.

53. ♦♦♦ **(a)** Each of two long straight wires carries current $I/2$ to an electrode at the center of an infinite conducting sheet (■ Figure 28.45). The current spreads out uniformly along radial lines in the sheet. **(a)** Describe the symmetry of the system. What conclusions can you draw about the symmetry of the magnetic field? **(b)** Describe a curve that allows you to use Ampère's law to find the difference in magnetic induction on opposite sides of the sheet. Does your result support your description of the field in part (a)? **(c)** Further verify your result using curves passing around the wires. **(d)** If the current is delivered to the electrode by a single wire carrying current I, what is the magnetic field on the side of the sheet without the wire? On the side with the wire?

54. ♦♦ A model for a jet that might power an extragalactic radio source is a beam 200 light years in radius carrying a current of 10^{15} A. (A light year is approximately 10^{16} m). **(a)** What is the magnetic field at the edge of the jet? **(b)** If the current were provided by electrons moving at one-tenth the speed of light, what would be the electron density in the jet? **(c)** What is the radial acceleration of an electron at the edge of the beam? If the beam travels 3×10^5 light years, is this a reasonable model for the jet? (*Hint:* Assume that the acceleration is constant, and compare the radial displacement of the electron with the beam radius.)

Additional Problems

55. ❖ A bar magnet is supported at its center of mass by a string so that it may rotate freely. Will it lie horizontally in equilibrium? If not, what might you do to make it lie horizontally?

56. ❖ Two bar magnets are suspended on strings. When the suspension points are close together, the ends of the magnets nearest each other are painted red. When the suspension points are moved far apart, the red end of one magnet points north. In what direction does the red end of the other magnet point?

57. ❖ Is there anywhere on Earth where a compass points toward geographical south?

58. ❖ Ship's compasses are *compensated* by placing magnets near the compass. For what are these magnets compensating?

59. ❖ An electron is moving radially away from a current-carrying wire. Describe the force on the electron.

60. ❖ A positively charged particle is moving parallel to a long straight wire, in the direction opposite the current. Describe the force on the particle.

61. ❖ A charged particle passes through the center of a current loop with velocity perpendicular to the plane of the loop. What force acts on the particle?

62. ❖ In what direction must an electron move near a long straight current-carrying wire to experience no magnetic force? Is this possible?

63. ❖ An electron passing through the plane of a current loop is momentarily unaccelerated. What may you conclude about the direction of the electron's velocity?

64. ❖ Use the fact that the circulation of a static electric field around any closed curve is zero (cf. Exercise 28.5) to show that the electric field in a parallel plate capacitor cannot cut off abruptly at the edge of the plates.

65. ❖ While it is turning, an airplane is banked at an angle to the horizontal. Explain why the airplane's compass reads incorrectly during the turn. Does the error depend on the direction of the turn, on the direction of flight, or on the hemisphere of the Earth where the plane is flying?

66. ♦♦ In a Van de Graaff electric generator, a belt carries a charge density σ and moves with speed v. Find $\vec{\mathbf{E}}$ and $\vec{\mathbf{B}}$ near the surface of the belt.

67. ♦♦ Two plane current sheets are separated by a distance d. Each carries a surface current density K. Find the magnetic field everywhere if the two current densities are **(a)** parallel, **(b)** antiparallel, and **(c)** at right angles.

68. ♦♦ Three wires in a circuit form a Y-shaped junction (■ Figure 28.46). How are the currents in the wires related? If $I_1 = I/3$, find the magnetic field at points A, B, and C, each distance d from the junction.

69. ♦♦ The magnetic field at the Earth's equator ($R = 6400$ km) is 3×10^{-5} T. What is the Earth's magnetic moment? Modeling the current source as a circular loop of radius $R/2$, calculate how large a current would be required to produce the Earth's magnetic field.

70. ♦♦♦ A wire loop is in the form of an ellipse with semimajor axis a and semiminor axis b. Find the magnetic field at a focus, if a current I flows in the loop.

71. ♦♦♦ An arbitrarily shaped current loop lies in the x-y-plane. First, explain why the term $d\vec{\ell} \times (z\hat{\mathbf{k}})$ that we ignored while Digging Deeper integrates to zero. Second, show that for points on the z-axis, $r = \sqrt{z^2 + r_\perp^2}$, so that the approximation $r \approx z$ neglects terms of order $r_\perp^2/z^2$ compared with 1.

72. ♦♦♦ A conducting slab on a silicon chip is 50 μm wide and $a = 1.5$ μm thick. It carries a current density

$$\vec{\mathbf{j}} = (1.0 \times 10^3\ \text{A/m}^2)(z/a)\,\hat{\mathbf{i}}$$

in the region $0 < z < a$. **(a)** Modeling the slab as infinitely long and wide, use Ampère's law to find the magnetic field inside and outside the slab. **(b)** Model the slab as a set of long straight wires each having width dy and carrying current $dI = dy \int j\,dz = dy(0.75 \times 10^{-3}\ \text{A/m})$ in the x-direction. Superpose the contributions of these wires to find the field at a point on the z-axis a distance d above the center of the slab. Estimate the error made by using Ampère's law to compute the field.

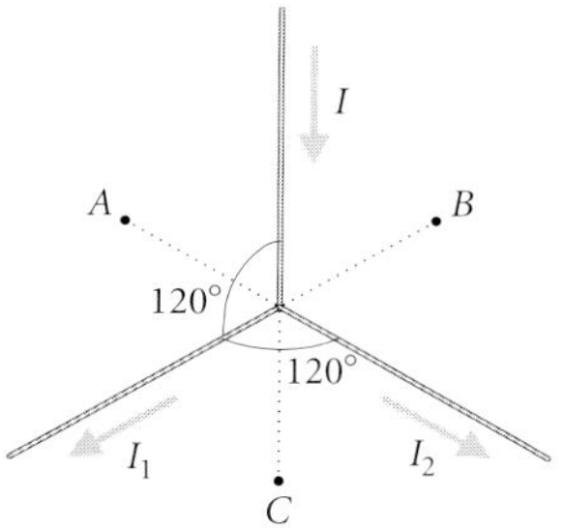

■ FIGURE 28.46

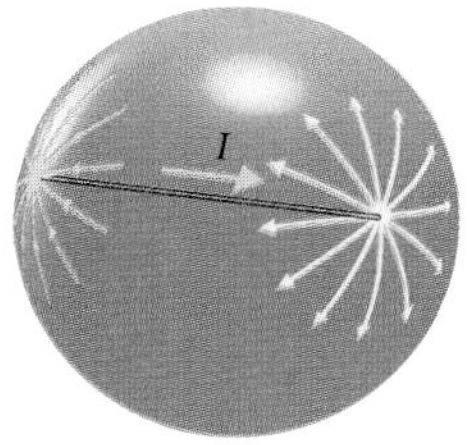

■ FIGURE 28.47

Computer Problems

73. Two Helmholtz coils are parallel to the y-z-plane (Figure 28.10) and each has a 12.00-cm radius. One has its center at the origin, and the other is centered on the x-axis at $x = 12.00$ cm. Each carries a 7.750-A current. Calculate the magnetic field on the x-axis at $x = 0$, $x = 2.00$ cm, $x = 4.00$ cm, and $x = 6.00$ cm. Use your results to plot the field along the axis as a function of position. (See also Problem 39).

74. The magnetic field on the axis of a finite solenoid with a radius a and $n = N/L$ turns per unit length is $B = \mu_0 nI(\cos\theta_1 + \cos\theta_2)/2$. (See Problem 41, Figure 28.40) Express the magnetic field as $B = B_\infty f(z/L)$, where B_∞ is the result for an infinite solenoid. Taking $a = 2.00$ cm and $L = 10.00$ cm, compute values of $B/B_\infty = f(z/L)$ for values of z/L between 0 and 10.00, and plot the results. Compare your answer with the result for an infinite solenoid. Outside the solenoid, its field may be approximated by a dipole field with magnetic moment $m = NIA$, where $A = \pi a^2$ (cf. Problems 34 and 42). Compare your result for B with the dipole field. At what values of z/L is the dipole approximation better than 5%?

Challenge Problems

75. A circular wire loop of radius a and carrying current I lies in the x-y-plane with its center at the origin. Find the magnetic field at a point slightly off axis, with coordinates $(x, 0, z)$, $x \ll z$. **(a)** Use the Biot-Savart law to express $\vec{\mathbf{B}}$ as an integral. **(b)** Use the binomial theorem to expand the $-\frac{3}{2}$ power in the integrand. Assume that x/z is small and neglect terms with powers higher than $(x/z)^1$. **(c)** Perform the integration, and show that your result is consistent with the dipole formula $\vec{\mathbf{B}} = (k_m m/r^3)(\hat{\theta}\sin\theta + 2\hat{\mathbf{r}}\cos\theta)$ for $\theta \ll 1$ and $z \gg a$.

76. Use the Biot-Savart law to write an integral that gives the magnetic field due to a circular current loop of radius a at a point in the plane of the loop a distance d from the axis. **(a)** Assume $d/a \ll 1$, and expand the integrand in powers of d/a, keeping terms through d^2/a^2. Perform the integration to find $\vec{\mathbf{B}}$. **(b)** Repeat part (a) assuming $d/a \gg 1$.

77. An infinitely long wire in the shape of a spiral is described by the formula:

$$\vec{\mathbf{r}}(\theta) = \frac{p\theta}{2\pi}\,\hat{\mathbf{k}} + R(\cos\theta\,\hat{\mathbf{i}} + \sin\theta\,\hat{\mathbf{j}}).$$

If the wire carries a current I flowing toward increasing θ, show that the z-component of the magnetic field at the origin is:

$$B_z(\text{origin}) = \frac{\mu_0 I}{p}.$$

Compare your result with eqn. (28.14) for a tightly wound solenoid.

78. A current I flows in a wire along a diameter of a thin conducting spherical shell of radius R (■ Figure 28.47). The current then flows back to the beginning of the wire through the shell. The current density in the shell is uniformly distributed in angle. Describe the magnetic field produced by the system and find $\vec{\mathbf{B}}$ everywhere.

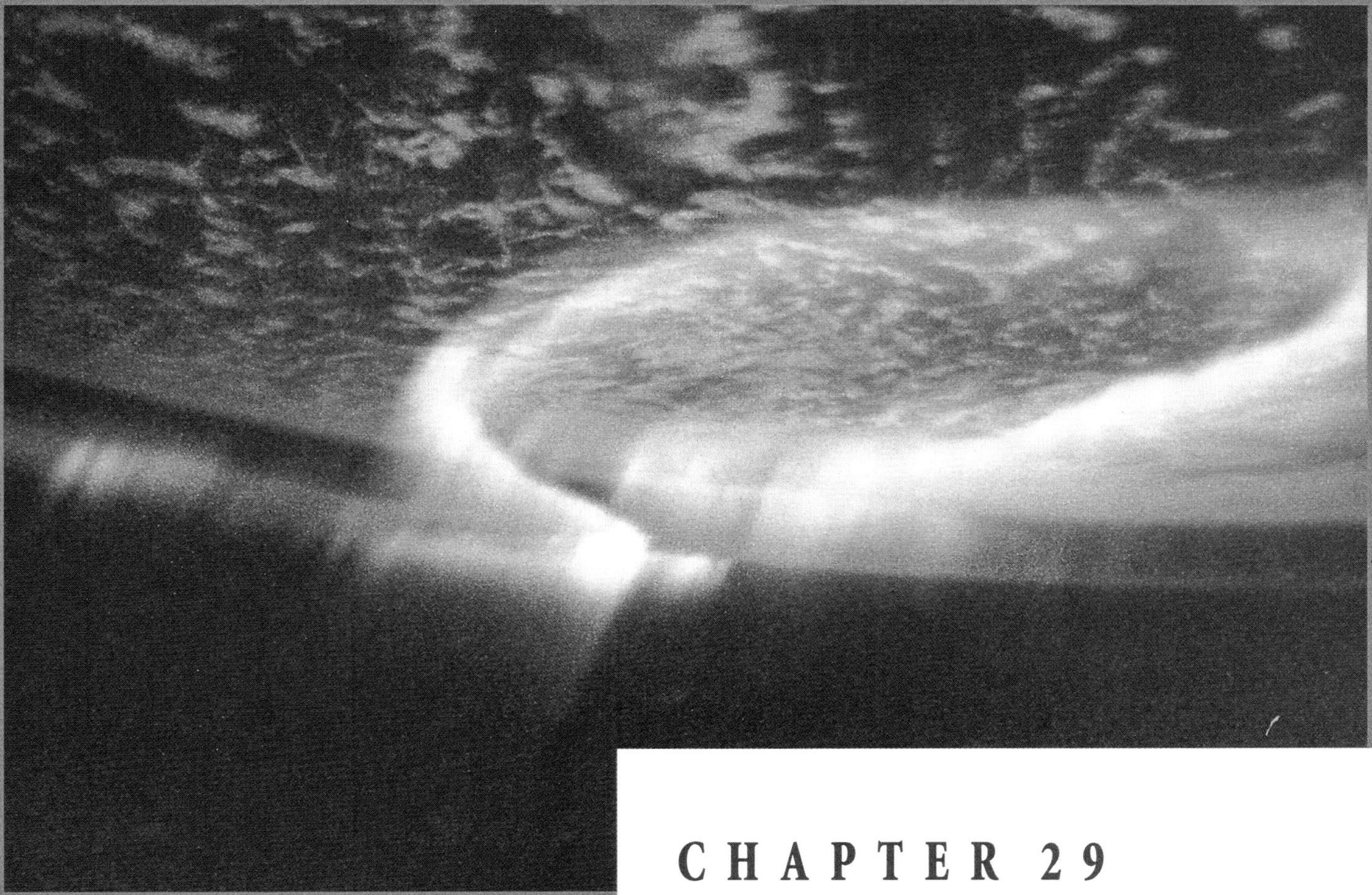

Charged solar wind particles are deflected around the Earth by its magnetic field. Some particles are captured and move in spiral paths along the magnetic field lines. Near the magnetic poles, the particles reach the Earth's upper atmosphere. Particles colliding with air molecules are responsible for the brilliantly colored displays of light called the aurora australis, or southern lights. A similar phenomenon occurs at the North pole.

CHAPTER 29
Static Magnetic Fields: Applications

Concepts

Larmor radius
Cyclotron frequency
Magnetic mirror
Lorentz force
Hall effect
Paramagnetism
Diamagnetism
Ferromagnetism

Goals

Be able to:

Determine the path of a particle moving in a magnetic field.

Compute force and torque exerted by magnetic fields on current-carrying wires.

Understand how atomic behavior results in macroscopic magnetic properties of materials.

She that had all Magnetique force alone,
To draw, and fasten sundred parts in one;

John Donne, The First Anniversary

Our lives are more pleasant because our planet is magnetic. The Sun continually spews out streams of charged particles that would endanger life on the surface of an unprotected planet. The Earth's magnetic field shields us by channeling the flow of particles away from the Earth's surface. When the Sun is particularly active, solar wind particles pose a radiation hazard for astronauts in the shuttle or space station. The aurora is a dramatic result of energetic particles that descend to the upper atmosphere.

Magnetic fields are used extensively in modern society. Magnetic forces help generate the sound from your stereo and start your car. Should you fall seriously ill, magnetic fields help doctors to see inside your body without surgery. A machine in which electrons orbit in a magnetic field can produce x-ray beams for fabricating microcomputer circuits. By accelerating carbon atoms in a similar machine, the different isotopes can be separated to allow dating of historical samples. The high-energy plasmas used in fusion research are confined by carefully designed magnetic fields.

Magnetic field acts on and is produced by electric charge in motion. In this chapter we'll study magnetic forces and look at some of their technological applications.

29.1 Motion of Charged Particles in a Magnetic Field

In Chapter 28 we observed that the force acting on a particle with charge q moving with velocity $\vec{\mathbf{v}}$ in a magnetic field $\vec{\mathbf{B}}$ is (eqn. 28.1):

$$\vec{\mathbf{F}} = q\,\vec{\mathbf{v}} \times \vec{\mathbf{B}}.$$

The force is always perpendicular to the particle's velocity and so does no work on the particle:

$$dW = \vec{\mathbf{F}} \cdot d\vec{\mathbf{s}} = \vec{\mathbf{F}} \cdot \vec{\mathbf{v}}\,dt = 0, \qquad \text{since } \vec{\mathbf{F}} \text{ is perpendicular to } \vec{\mathbf{v}}.$$

Magnetic forces can alter the direction of a particle's motion but cannot alter its energy. Since the force is also perpendicular to $\vec{\mathbf{B}}$, it cannot change a particle's velocity component parallel to the field. The character of a particle's motion depends on the direction of its velocity with respect to the field.

29.1.1 *Motion Perpendicular to a Uniform Magnetic Field*

If a particle's velocity is initially in the plane perpendicular to $\vec{\mathbf{B}}$ (■ Figure 29.1), it remains in that plane, since its velocity component parallel to $\vec{\mathbf{B}}$ remains zero. The particle's speed is constant, and in a uniform field the magnitude of the force acting on the particle is constant. With $\vec{\mathbf{v}}$ perpendicular to $\vec{\mathbf{B}}$:

$$|\vec{\mathbf{F}}| = qvB.$$

A force of constant magnitude always perpendicular to a particle's velocity causes the particle to move in a circle (Figure 29.1; cf. §§3.2 and 4.7). The radius of the circle, called the *Larmor*

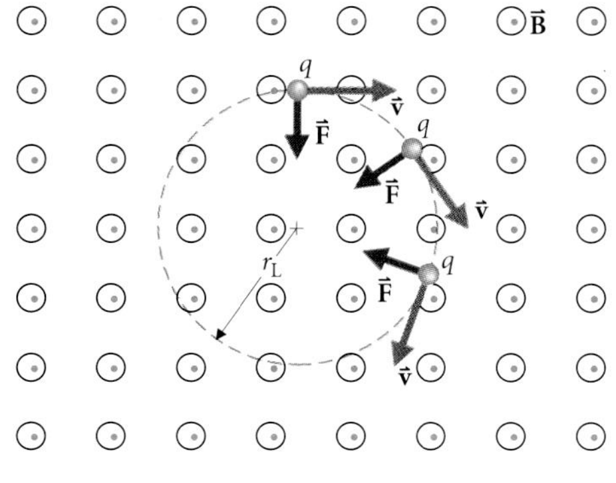

■ FIGURE 29.1
A positively charged particle moves in the plane perpendicular to $\vec{\mathbf{B}}$. The magnetic force is perpendicular to both $\vec{\mathbf{v}}$ and $\vec{\mathbf{B}}$. The particle moves in a circle with radius $r_L = mv/(qB)$.

radius or gyro-radius r_L, is found by recognizing that magnetic force causes the centripetal acceleration v^2/r. Applying Newton's second law, we obtain:

$$\frac{mv^2}{r} = qvB.$$

SEE EQN. 3.11.

Thus the radius of the orbit is: $$r = \frac{mv}{qB} \equiv r_L. \qquad (29.1)$$

The angular speed of a particle circling in a magnetic field is:

$$\omega = \frac{v}{r} = v\frac{qB}{mv} = \frac{qB}{m} \equiv \omega_c. \qquad (29.2)$$

SEE EXAMPLE 29.2 AND *DIGGING DEEPER: MORE ON CYCLOTRONS*.

The frequency is *independent of the particle's speed.* A faster particle moves in a larger circle but requires the same time per orbit. Because this fact is crucial for the design of cyclotron particle accelerators, ω_c is called the *cyclotron frequency.*

STRICTLY, ω_c SHOULD BE CALLED THE CYCLOTRON ANGULAR FREQUENCY, BUT THAT IS AWKWARD AND NOT TRADITIONAL. THE SAME NAME IS USED FOR $f_c = \omega_c/2\pi$: WATCH THE CONTEXT!

EXAMPLE 29.1 ♦ Find the Larmor radius and cyclotron frequency for an electron with kinetic energy 5.0 eV in the Van Allen belts (cf. Figure VI.8), where the Earth's magnetic field is 1.5×10^{-7} T. Assume $\vec{\mathbf{v}}$ is perpendicular to $\vec{\mathbf{B}}$.

MODEL The electron moves on a circle in the plane perpendicular to $\vec{\mathbf{B}}$.

SETUP The electron's kinetic energy is $E = \frac{1}{2}mv^2$, and 1 eV $= 1.6 \times 10^{-19}$ J.

SOLVE The Larmor radius is then (eqn. 29.1):

$$r_L = \frac{mv}{eB} = \frac{m}{eB}\sqrt{\frac{2E}{m}} = \frac{\sqrt{2Em}}{eB}$$

$$= \frac{\sqrt{2(5.0\ \text{eV})(1.6 \times 10^{-19}\ \text{J/eV})(9.1 \times 10^{-31}\ \text{kg})}}{(1.6 \times 10^{-19}\ \text{C})(1.5 \times 10^{-7}\ \text{T})} = 50\ \text{m}.$$

The cyclotron frequency is:

$$\omega_c = \frac{eB}{m} = \frac{(1.6 \times 10^{-19}\ \text{C})(1.5 \times 10^{-7}\ \text{T})}{9.1 \times 10^{-31}\ \text{kg}} = 2.6 \times 10^4\ \text{rad/s}.$$

ANALYZE Compared with the scale of the system (cf. $R_E \approx 6000$ km), the radius of the electron's orbit is tiny. Thus charged particles are constrained to remain close to the magnetic field lines; they are unable to flow freely from the Sun to Earth's surface. ▮

EXERCISE 29.1 ❖ Describe how the orbit of a positron with the same energy would differ from that of the electron. (A positron has the same mass as an electron but the opposite electric charge.)

EXERCISE 29.2 ♦ Find the Larmor radius and cyclotron frequency for a 5.0-eV proton in the Van Allen belts. ▮

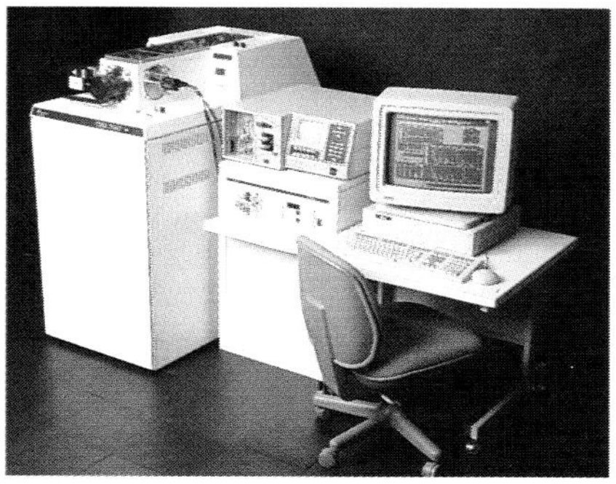

(a)

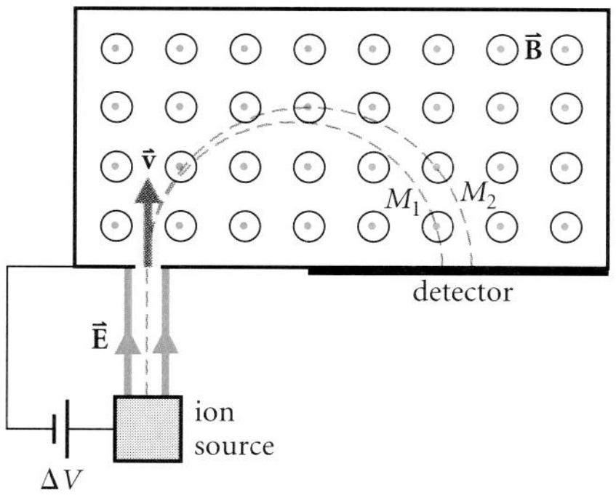

(b)

■ FIGURE 29.2
(a) A mass spectrometer. (b) Atoms are stripped of one or more electrons in the ion source and accelerated by the electric field so that they enter the spectrometer chamber with velocity $\vec{\mathbf{v}}$. The radius of an ion's circular path in the chamber depends on its charge-to-mass ratio q/m. Particles with the same charge and differing mass reach different spots on the detector.

29.1.2 Practical Applications of Circular Particle Motion

■ Figure 29.2 illustrates the operation of a mass spectrometer used for analysis of chemical composition. In the ion source the sample is vaporized and ionized (one or more electrons are removed from each atom). The ions are then accelerated by an electric field and pass through a grid into a region of uniform magnetic field. There the ions travel along circles determined by their Larmor radii and arrive at various points on a position-sensitive detector. In the original mass spectrometer design, photographic film was used as the detector.

An ion with mass m and charge q enters the magnetic field with kinetic energy gained by crossing a potential difference ΔV. Its kinetic energy is $\frac{1}{2}mv^2 = q\,\Delta V$, so its speed is:

IN THIS REGION THE ION IS ACCELERATED BY AN ELECTRIC FORCE.

$$v = \sqrt{\frac{2q\,\Delta V}{m}}.$$

Its Larmor radius in the magnetic field is:

THE CIRCULAR MOTION IS DUE TO A MAGNETIC FORCE.

$$r_L = \frac{mv}{qB} = \frac{(2m\,\Delta V/q)^{1/2}}{B}.$$

For a given ΔV and $\vec{\mathbf{B}}$, an ion's arrival point at the detector depends only on the ion's charge-to-mass ratio, q/m: $\quad r_L \propto \sqrt{m/q}$.

EXERCISE 29.3 ♦ The magnetic field in a mass spectrometer is $B = 0.20$ T, and the accelerating voltage is $\Delta V = 1.0 \times 10^4$ V. At what distance from the beam entry point are singly ionized oxygen ions collected? Where are nitrogen ions collected?

■ FIGURE 29.3
Ernest O. Lawrence (1901–1958) with his original cyclotron.

Acceleration of particles to high energies has found many uses in medicine and industry as well as physics research. Electrostatic accelerators are severely limited by the difficulty of maintaining large potential differences without breakdown. In 1930 Ernest O. Lawrence (■ Figure 29.3) invented the cyclotron, which uses magnetic fields in conjunction with much smaller potential differences.

In a cyclotron (■ Figure 29.4) particles are accelerated by an electric field between two hollow, D-shaped copper electrodes called *dees.* There is no electric field inside the dees, so the particles follow circular orbits in the magnetic field. Regardless of their energy, they require a time $\Delta t = \pi/\omega_c$ between passages across the gap between the dees. During this half-cycle the potential difference between the dees is reversed, so that the particles are accelerated each time they cross the gap. The particles are injected near the center of the dees and spiral outward as they speed up [*Remember:* r increases as v increases (eqn. 29.1)]. The maximum speed is limited by the radius of the dees. Since the particles cross the gap many times, a relatively small potential difference can produce a large energy gain.

EXAMPLE 29.2 ♦♦ In Lawrence's 1939 cyclotron, the dees had a diameter of 60.0 in. The device used a potential difference of 0.20 MV between the dees to produce a beam of singly ionized 14.5-MeV deuterium nuclei. What magnetic field did the device require? How much time did each particle spend in the machine?

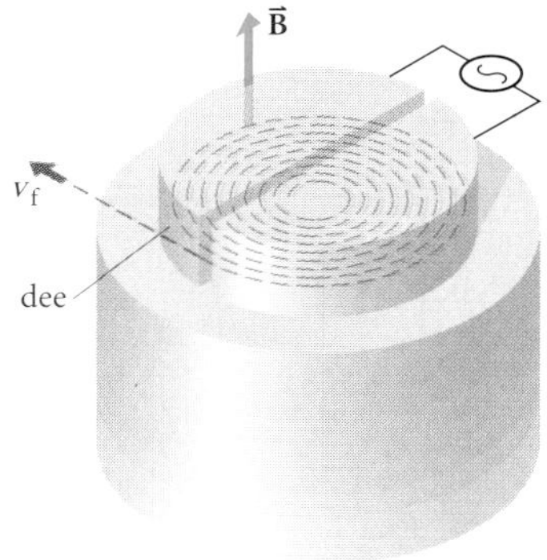

■ FIGURE 29.4
A cyclotron. A potential difference that changes sign at intervals $\Delta t = \pi/\omega_c$ is applied across the dees to accelerate particles spiraling inside. (An electric field exists between the dees.) The diagram shows the acceleration of positively charged particles.

MODEL The magnetic field controls the size of the particle's circular path. When the particle's speed and energy are greatest, its path barely fits inside the machine. This constraint determines the necessary field. The particle completes one revolution each cyclotron period, crosses the gap twice, and gains $2e\,\Delta V = 400$ keV of energy each revolution. Comparing this value with the total energy gain, we can find the number of revolutions, and hence the time.

SETUP The maximum energy E_{max} occurs when the particle's Larmor radius equals the radius of the dees: $r_D = r_L = mv/qB$. So for $E_{max} = 14.5$ MeV, the required field is:

$$B = \frac{mv}{qr_D} = \frac{m\sqrt{2E_{max}/m}}{qr_D} = \frac{\sqrt{2mE_{max}}}{qr_D}.$$

A deuterium nucleus contains one neutron and one proton, and so its mass is 2.0 u. "Singly ionized" means the charge is $+e$. So:

SOLVE
$$B = \frac{[2(2.0)(1.66 \times 10^{-27}\text{ kg})(1.45 \times 10^{7}\text{ eV})(1.6 \times 10^{-19}\text{ J/eV})]^{1/2}}{(1.6 \times 10^{-19}\text{ C})(30.0\text{ in.})(2.54 \times 10^{-2}\text{ m/in.})}$$
$$= 1.0\text{ T}.$$

Every time a particle passes from one dee to the other it gains $\Delta E = 200$ keV. To gain $E_{max} = 14.5$ MeV of kinetic energy in 200-keV steps the particle must pass between the dees $N = E_{max}/\Delta E = (1.45 \times 10^7 \text{ eV})/(2.0 \times 10^5 \text{ eV}) = 73$ times. A particle spends half the cyclotron period in each dee, so the total time for acceleration is:

$$T = 73\left(\frac{1}{2}\frac{2\pi}{\omega_c}\right)$$

$$= 73\pi\frac{m}{qB}$$

$$= \frac{73\pi(2.0)(1.66 \times 10^{-27}\text{ kg})}{(1.6 \times 10^{-19}\text{ C})(1.02\text{ T})}$$

$$= 4.7\ \mu\text{s}.$$

We kept an additional figure in the intermediate result for B.

ANALYZE Since $E_{max} \propto (Br_D)^2$, to obtain more energy per particle requires a larger device or a larger magnetic field.

29.1.3 *Motion in Combined Electric and Magnetic Fields*

If a charged particle moves through a region that contains both electric and magnetic fields, the two fields act independently on the particle, and the total force is the sum of the separate electric and magnetic forces:

$$\vec{\mathbf{F}} = q(\vec{\mathbf{E}} + \vec{\mathbf{v}} \times \vec{\mathbf{B}}). \quad (29.3)$$

This equation is known as the *Lorentz force law.*

After Hendrik A. Lorentz (1853–1928).

To gain intuition about the effects of combined fields, we shall look at some particular examples in which both fields are constant in time and uniform in space. In the simplest case, the electric field is parallel to $\vec{\mathbf{B}}$. It accelerates the particle along the field lines, changing the component of velocity parallel to $\vec{\mathbf{B}}$. Any velocity component perpendicular to $\vec{\mathbf{B}}$ would be influenced only by the magnetic force, resulting in an independent circular motion around the field lines.

Next we'll look at the motion of a particle when the electric and magnetic fields are perpendicular. The magnetic force $q\vec{\mathbf{v}} \times \vec{\mathbf{B}}$ is perpendicular to $\vec{\mathbf{B}}$, so if a charged particle moves with a particular velocity $\vec{\mathbf{v}}_o$, the net Lorentz force on it is zero:

$$0 = \vec{\mathbf{F}} = q(\vec{\mathbf{E}} + \vec{\mathbf{v}}_o \times \vec{\mathbf{B}})$$

$$\Rightarrow \vec{\mathbf{v}}_o \times \vec{\mathbf{B}} = -\vec{\mathbf{E}}.$$

To find $\vec{\mathbf{v}}_o$, we choose the x-direction along $\vec{\mathbf{E}}$ and the y-direction along $\vec{\mathbf{B}}$. Then:

$$\vec{\mathbf{v}}_c \times \vec{\mathbf{B}} = -v_z B\hat{\mathbf{i}} + v_x B\hat{\mathbf{k}}.$$

For this to equal $-\vec{\mathbf{E}} = -E\hat{\mathbf{i}}$, $v_x = 0$ and $v_z = E/B$. Thus:

$$\vec{\mathbf{v}}_o = \frac{E}{B}\hat{\mathbf{k}} = \frac{\vec{\mathbf{E}} \times \vec{\mathbf{B}}}{B^2}. \quad (29.4)$$

A device called a velocity selector uses this effect to produce a beam of particles with nearly identical velocities. The beam must pass through two small holes along a line parallel to $\vec{\mathbf{E}} \times \vec{\mathbf{B}}$. Only particles with velocity very nearly equal to $\vec{\mathbf{v}}_o$ can pass through both holes.

EXERCISE 29.4 ❖ Describe the motion of a particle with initial velocity $\vec{\mathbf{v}} = \vec{\mathbf{v}}_o + v_y\hat{\mathbf{j}}$, where the y-direction is parallel to the magnetic field.

Digging Deeper

More on Cyclotrons

Cyclotrons as archaeological tools

Since accelerating particles cross the dees many times, the potential difference across the dees must vary at precisely the correct cyclotron frequency, otherwise the particles enter the gap when the potential difference acts to decelerate them, and they lose rather than gain energy. With a fixed magnetic field, only particles of a predetermined charge-to-mass ratio are accelerated. Richard Muller of the University of California at Berkeley has taken advantage of this fact to use a cyclotron as a very sensitive mass spectrometer.

Most natural carbon is in the form of ^{12}C, but ^{14}C is present in trace amounts, and it decays at a known rate to ^{14}N (Chapter 36). While alive, organisms take in carbon from the atmosphere or from food and maintain a relatively constant ratio of ^{14}C to ^{12}C. After the organism's death, the proportion of ^{14}C to ^{12}C decreases as the ^{14}C decays. In the technique of carbon dating, the ratio of ^{14}C to ^{12}C is used as a measure of a sample's age. To date very old archaeological samples, a small number of ^{14}C atoms must be detected from among much larger numbers of other atoms. The cyclotron can be tuned to accelerate only singly ionized ^{14}C and ^{14}N, which have almost the same charge-to-mass ratio q/m. The beam exiting the cyclotron then passes through a dilute gas that strips electrons from atoms in the beam. Because of their greater electric charge, the stripped nitrogen nuclei lose energy more rapidly than the carbon nuclei, which are the only species that remains in the beam. Because a very small number of atoms can be counted, this method allows dating of tiny samples of material.

Synchrotrons

According to Einstein's theory of special relativity, the momentum of a particle (§34.4.1) is given by the formula:

$$p = \frac{mv}{(1 - v^2/c^2)^{1/2}} = \gamma mv,$$

where c is the speed of light. The factor γ is not present in Newton's theory, which we have used to analyze the cyclotron (γ is very close to 1 unless v is almost equal to c). Use of the correct relativistic mechanics gives more accurate expressions for the frequency and radius of the particle's motion:

$$\omega = \frac{qB}{\gamma m} \quad \text{and} \quad r = \frac{\gamma mv}{qB}.$$

If a particle in a cyclotron gains enough energy to make γ substantially greater than unity, the particle gets out of phase with the accelerating potential and ceases to gain energy. Most modern accelerators are *synchrotrons,* which use a variable magnetic field to maintain a constant Larmor radius and a variable-frequency accelerating potential that is synchronized

■ Figure 29.5
Dr. Michel Ter-Pogossian with the first medical cyclotron used in a U.S. hospital, at Washington University, St. Louis, MO.

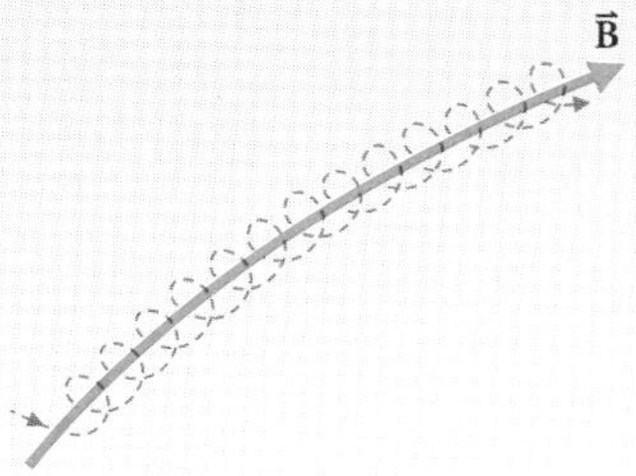

■ Figure 29.6
When a particle has a velocity component along $\vec{B}$, its path is a helix along the field line. This picture shows the path of a positively charged particle.

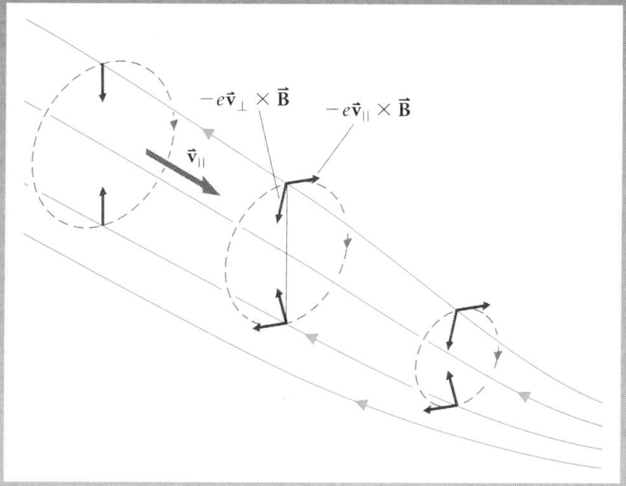

■ FIGURE 29.7
Particle spiraling in a nonuniform field. Here we show a negatively charged electron. The magnetic force has two components due to the two components of $\vec{v}$: $v_\perp$ (the circular motion) and $v_\parallel$ (along the field line). The force component $-e\vec{v}_\perp \times \vec{B}$ does not point exactly at the center of the circle but has a component along the field line, which reduces $v_\parallel$. The force component $-e\vec{v}_\parallel \times \vec{B}$ points along the circumference of the circle and increases $v_\perp$. At the mirror point, $v_\parallel = 0$ and all the particle's energy is in the circular motion.

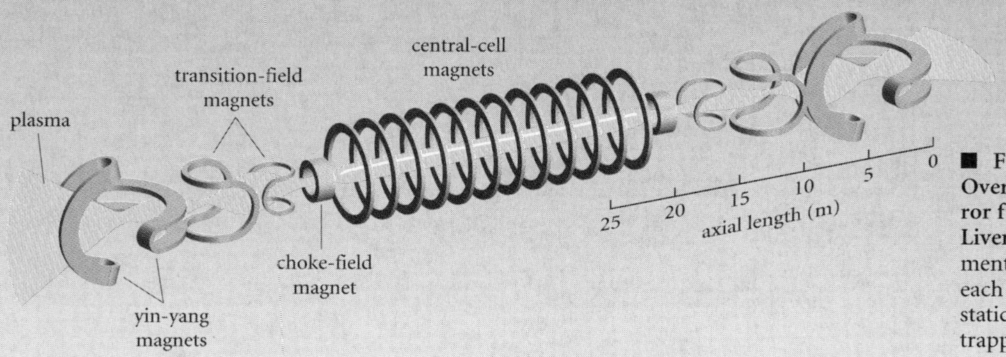

■ FIGURE 29.8
Overall configuration of the tandem mirror fusion reactor developed at Lawrence Livermore Laboratory. Plasma confinement is achieved by the double mirror at each end of the chamber and by electrostatic forces exerted by plasma clouds trapped in the mirror cells.

with the particles as they gain energy. The synchrotron at Fermilab achieves a proton beam with energy of 10^{12} eV ($\gamma \approx 10^3$). Whenever a charged particle is accelerated, it radiates (cf. the Overview §VI.9). Radiation emitted by electrons in synchrotrons is tightly beamed in the direction of the particles' motion. Synchrotrons produce an intense and controllable x-ray beam that can be used in medical and industrial applications (■ Figure 29.5).

The Magnetic Mirror

Particle motion exactly perpendicular to a magnetic field is an ideal case. A particle in a uniform field usually has some velocity component parallel to $\vec{B}$. This constant velocity along the field combines with the particle's circular motion perpendicular to the field to form a helix (■ Figure 29.6).

In a nonuniform magnetic field the field lines converge or diverge. A particle spiraling along converging field lines toward a region of greater $|\vec{B}|$ encounters a force component that reduces the particle's parallel velocity (■ Figure 29.7). In a strongly converging field, this force reverses the particle's velocity parallel to the field, reflecting it from the region of intense field. The converging field lines form a *magnetic mirror.* Only particles originally moving nearly parallel to the field can penetrate the mirror. Particles trapped in the Earth's magnetic field travel along spirals between reflections by magnetic mirrors at the North and South poles. Of those particles that travel nearly parallel to the magnetic field at the equator, a small fraction escapes through the mirror at each pole and enters the Earth's atmosphere to provide power for the polar aurora. Collisions continually replenish the supply of escaping particles and maintain equilibrium between particles captured from the Sun and those escaping to the atmosphere.

Magnetic mirrors have also been used in controlled nuclear fusion experiments such as the Tandem Mirror Fusion Experiment at Lawrence Livermore National Laboratory (■ Figure 29.8). This experiment has now been terminated in favor of the Tokamak design.

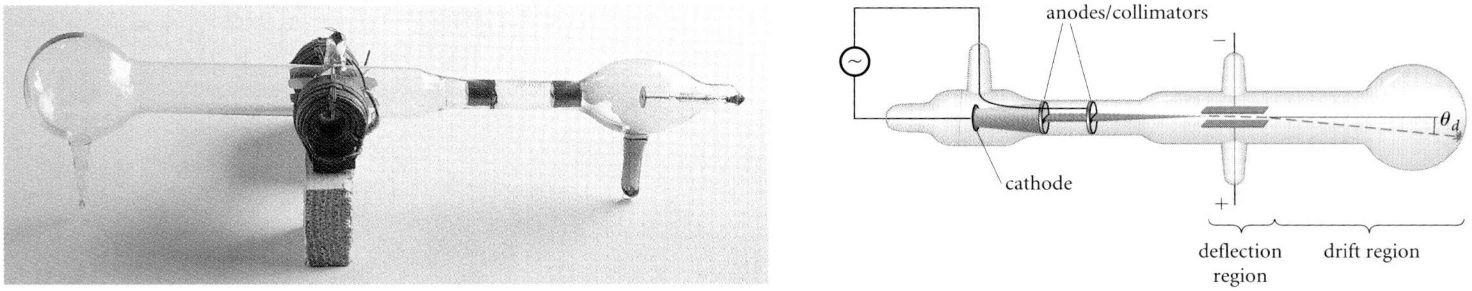

■ FIGURE 29.9
J. J. Thompson used this cathode-ray tube in his 1897 experiment to determine the charge-to-mass ratio e/m for an electron.

In 1897 J. J. Thompson (1856–1940) used the result in eqn. (29.4) to determine the ratio of charge to mass for an electron. He used a cathode-ray tube (■ Figure 29.9) in which a parallel electric field accelerates a beam of electrons along the tube. An electric field $\vec{\mathbf{E}}_d$ between the deflection plates and perpendicular to the tube bends the beam sideways. If the electrons enter the deflection region with speed v, they remain between the plates for a time $t = L/v$ and have an acceleration across the tube $a = F/m = eE_d/m$. They attain a sideways velocity,

$$v_s = at = \frac{eE_d t}{m} = \frac{eE_d L}{mv},$$

and are deflected through a small angle:

REMEMBER: FOR SMALL ANGLES, $\theta \approx \tan\theta$.

$$\theta_d \approx \tan\theta_d = \frac{v_s}{v} = \frac{eE_d L}{mv^2},$$

which is measured to give a relation between the unknowns v and e/m. Helmholtz coils placed on either side of the tube are then used to create a perpendicular magnetic field. Current in the coils is adjusted until the electron beam remains undeflected. Then the electron speed $v = E_d/B$, and:

$$\frac{e}{m} = \frac{E_d\theta_d}{B^2L}.$$

Thompson measured $v = 1.5 \times 10^7$ m/s in his tube, and his value for e/m was within 4% of the modern value of 1.76×10^{11} C/kg.

A particle moving perpendicular to $\vec{\mathbf{B}}$ with a velocity other than $\vec{\mathbf{v}}_o$ undergoes circular motion under the influence of the magnetic force, but the electric field causes the center of the particle's circular orbit to drift in the direction of $\vec{\mathbf{E}} \times \vec{\mathbf{B}}$. ■ Figure 29.10a illustrates this effect for a particle with positive charge q and mass m. Starting at point a, the particle moves in the direction of $\vec{\mathbf{E}}$ and gains speed. After passing point b, the particle moves opposite $\vec{\mathbf{E}}$ and slows down. Since the particle's Larmor radius is proportional to its speed (eqn. 29.1), the radius increases between a and b and decreases between b and d. The speed increases again as the particle moves to point e, and the cycle repeats. The particle's path is not a closed circle, because the curvature $1/r$ is greater where the speed is smaller. Point e is to the left of point a: the particle has drifted to the left, in the direction of $\vec{\mathbf{E}} \times \vec{\mathbf{B}}$. Figure 29.10b shows that the same result is obtained for a negative particle. The drift velocity $\vec{\mathbf{v}}_E$ of the center turns out to equal $\vec{\mathbf{v}}_o$ independent of any of the particle's properties!

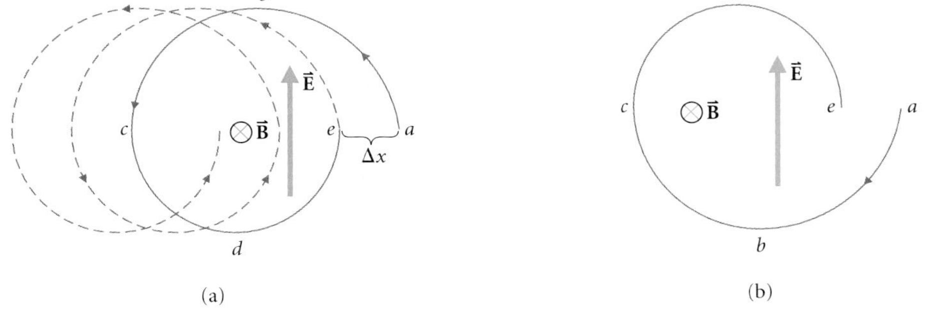

■ FIGURE 29.10
Motion of a particle in a region with $\vec{\mathbf{E}}$ perpendicular to $\vec{\mathbf{B}}$. (a) Positively charged particle. The particle's speed, and hence its Larmor radius, increases as it moves parallel to $\vec{\mathbf{E}}$ (from a to b). The Larmor radius decreases again as the particle moves from b to d and slows down. As a result, the orbit does not close: e is to the left of a. (b) A negatively charged particle orbits in the opposite direction but gains energy when it moves opposite $\vec{\mathbf{E}}$. The resulting drift is the same as for a positively charged particle.

EXAMPLE 29.3 ♦ Near the surface of the Earth at the equator typical values for $\vec{\mathbf{E}}$ and $\vec{\mathbf{B}}$ are (100 V/m, down) and (3×10^{-5} T, north), respectively. If it were not slowed by the atmosphere, with what velocity would a charged particle drift?

MODEL The velocity is in the direction of $\vec{\mathbf{E}} \times \vec{\mathbf{B}}$, or (down) × (north) = east (■ Figure 29.11).

SETUP AND SOLVE We use eqn. (29.4) to find the magnitude $|\vec{\mathbf{v}}_E| = |\vec{\mathbf{v}}_o|$.

$$v_E = \frac{E}{B} = \frac{100\ \text{V/m}}{3 \times 10^{-5}\ \text{T}} = 3 \times 10^6\ \text{m/s}.$$

ANALYZE This speed is surprisingly fast. Neglecting the atmosphere, it would take $2\pi R/v_E$, or approximately 10 s, to circle the Earth! Check the units:

$$\frac{\text{V}}{\text{m}\cdot\text{T}} = \frac{\text{N}}{\text{C}\cdot\text{T}} = \frac{\text{N}\cdot\text{C}\cdot\text{m}}{\text{C}\cdot\text{N}\cdot\text{s}} = \frac{\text{m}}{\text{s}}.$$

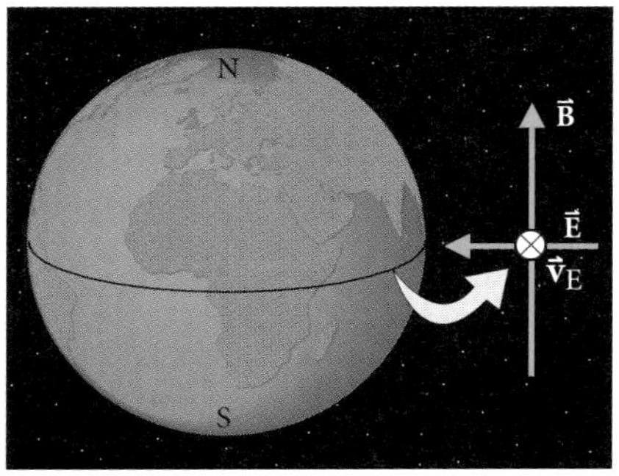

■ FIGURE 29.11
A charged particle near the Earth's surface would circle the Earth in about 10 s due to $\vec{\mathbf{E}} \times \vec{\mathbf{B}}$ drift if there were no atmosphere. In this diagram $\vec{\mathbf{v}}_E$ is into the page.

29.2 FORCES ON CURRENT-CARRYING WIRES

29.2.1 Force on a Wire Segment

Current in a wire arises from the motion of many individual electrons. When a magnetic field is present, the forces acting on the individual electrons add to produce a net force on the wire. We begin by calculating the force on a length $d\ell$ of wire with cross-sectional area A carrying a current I in a magnetic field $\vec{\mathbf{B}}$ (■ Figure 29.12). If a material contains n mobile (*conduction*) electrons per unit volume, the number in a length $d\ell$ of the wire is $nA\ d\ell$, and each drifts with an average velocity $\vec{\mathbf{v}}_d$ (§26.2). The total force on the wire segment is the sum of the forces on each of these electrons:

$$d\vec{\mathbf{F}} = \sum_{\substack{\text{all}\\ \text{moving}\\ \text{electrons}}} -e\vec{\mathbf{v}}_d \times \vec{\mathbf{B}} = -nA\ d\ell\ e\vec{\mathbf{v}}_d \times \vec{\mathbf{B}}. \qquad \text{We used eqn. (28.1).}$$

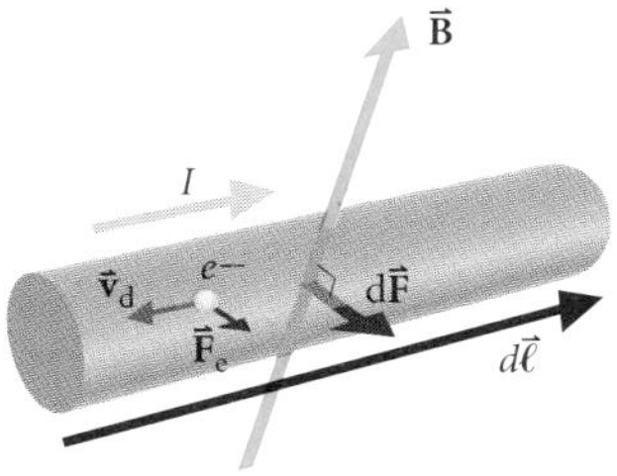

■ FIGURE 29.12
The force on a wire segment is the sum of the forces on all the moving electrons inside. The ions do not move and so do not contribute to the force.

From eqn. (26.5), the current is $I = nAev_d$. Since the vectors $d\vec{\ell}$ and $\vec{\mathbf{v}}_d$ are antiparallel, we find:

$$d\vec{\mathbf{F}} = I\ d\vec{\ell} \times \vec{\mathbf{B}}. \qquad \textbf{(29.5)}$$

REMEMBER: BY CONVENTION, THE DIRECTION OF $d\vec{\ell}$ IS THE DIRECTION OF CONVENTIONAL CURRENT, I.

In §29.3 we'll show how this force is transferred to the wire as a whole.

EXERCISE 29.5 ♦ Find the magnetic force on a 1.0-mm segment of wire carrying a current $I = 5.0$ A in the x-direction if the magnetic field is 0.60 T in the y-direction.

EXAMPLE 29.4 ❖ Show that the net torque about the center of a straight wire segment carrying current I in a uniform field is zero.

MODEL Two symmetrically located elements, equidistant from the center, experience equal forces $d\vec{\mathbf{F}}_1 = d\vec{\mathbf{F}}_2$ (■ Figure 29.13). These forces produce equal and opposite torques $d\vec{\tau}_1 = -d\vec{\tau}_2$ about the center. The total torque is the sum of such balancing pairs and so is zero.

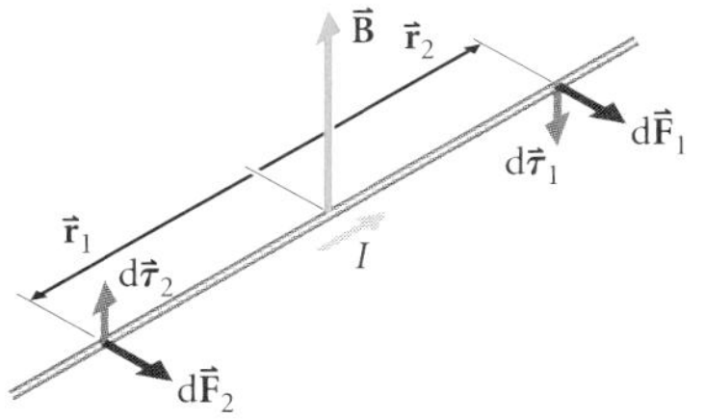

■ FIGURE 29.13
The net torque about the center of a straight wire segment in a uniform magnetic field is zero: the line of action of the net magnetic force passes through the center of the wire.

ANALYZE The total force acts as if it were all applied at the center of the segment.

COMPARE WITH OUR DISCUSSION OF CENTER OF MASS IN §9.3 AND §11.1.

In deriving eqn. (29.5) we assumed that all the electrons have the same drift velocity and experience the same magnetic field. Finite lengths of wire usually are not straight, or they pass through regions of variable field. To calculate forces and torques acting on a finite wire, we model it as a collection of elementary segments and add the forces acting on the individual elements.

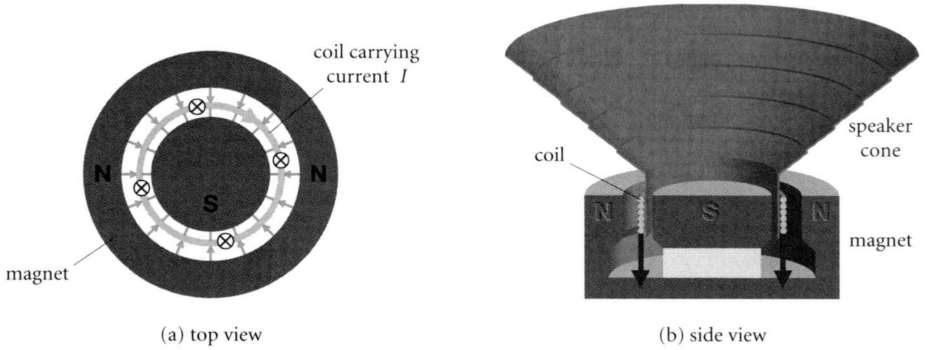

■ FIGURE 29.14
Moving-coil loudspeaker. (a) Top view. The radial magnetic field is produced by pole pieces. A coil attached to the loudspeaker cone carries current I, driven by the amplifier. The force is downward (into the page here) when the current direction is clockwise viewed from above. Reversing the current direction changes the direction of the force. When the current oscillates, so does the speaker cone, and sound waves are produced (cf. Chapter 16). (b) Side view. The gap is shown enlarged for clarity.

EXAMPLE 29.5 ♦♦ A cone-type loudspeaker contains a permanent magnet surrounded by a coil of wire (■ Figure 29.14). Magnetic force on a sinusoidally varying current in the coil oscillates the coil and the attached speaker cone to produce sound waves. If the magnetic field strength at the position of the coil is 1.0 T, the coil has 100 turns of radius 4.0 cm, and the current is 1.0 A, what is the force on the cone?

MODEL The loudspeaker is designed so that the magnetic field is perpendicular to the wire everywhere. With current clockwise around the coil (Figure 29.14a), the direction of the force on each element of the coil, $d\vec{\mathbf{F}} = I\, d\vec{\ell} \times \vec{\mathbf{B}}$, is into the page (downward in Figure 29.14b). We add the contributions due to each element of wire.

SETUP Each element $d\vec{\ell}$ is perpendicular to the field, so:

$$|d\vec{\mathbf{F}}| = I\, d\ell |\vec{\mathbf{B}}|.$$

Finally, $|\vec{\mathbf{B}}|$ is the same at each element, and the total length of the wire is the number N of turns in the coil times the circumference $2\pi r$ of each coil.

SOLVE The total force has magnitude:

$$F = I\ell B = (1.0\text{ A})(100)(2\pi)(4.0 \times 10^{-2}\text{ m})(1.0\text{ T}) = 25\text{ N}.$$
$$\vec{\mathbf{F}} = (25\text{ N, downward in Figure 29.14b}).$$

ANALYZE When the direction of current is reversed, the direction of the force reverses too. To increase the force, we should increase I, ℓ, or B. The cheapest alternative is to increase the length of wire. ■

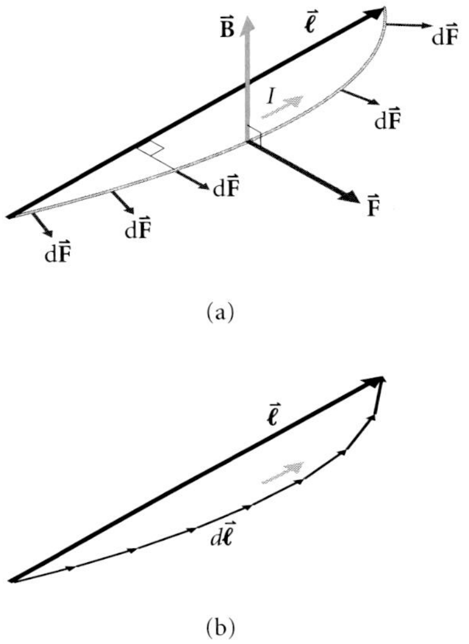

■ FIGURE 29.15
(a) Force on a wire segment in a uniform magnetic field is the sum of the forces on the individual elements of wire: $\vec{\mathbf{F}} = I\vec{\ell} \times \vec{\mathbf{B}}$. (b) The sum of the vectors $d\vec{\ell}$ is the vector $\vec{\ell}$ from one end of the wire to the other.

Next we shall derive a convenient expression for the force acting on a wire segment of arbitrary shape carrying current I in a uniform magnetic field $\vec{\mathbf{B}}$ (■ Figure 29.15a). The force on any differential piece of the wire is given by eqn. (29.5):

$$d\vec{\mathbf{F}} = I\, d\vec{\ell} \times \vec{\mathbf{B}}.$$

The current I and the magnetic field $\vec{\mathbf{B}}$ are the same for each element, so the total force is:

$$\vec{\mathbf{F}} = \int d\vec{\mathbf{F}} = \int I\, d\vec{\ell} \times \vec{\mathbf{B}} = I\left(\int d\vec{\ell}\right) \times \vec{\mathbf{B}}.$$

WE BRING $\vec{\mathbf{B}}$ OUT OF THE INTEGRAL BECAUSE IT IS CONSTANT, BUT WE DO *NOT* CHANGE THE ORDER OF VECTORS IN THE CROSS PRODUCT. THAT WOULD CHANGE THE SIGN OF THE ANSWER.

The integral is the vector sum of the line element vectors and equals the vector $\vec{\ell}$ joining the end points of the segment (Figure 29.15b). Thus:

$$\vec{\mathbf{F}} = I\vec{\ell} \times \vec{\mathbf{B}}. \qquad (29.6)$$

EXAMPLE 29.6 ♦ A semicircular wire segment of radius $a = 2.54$ cm lies in the x-y-plane with its two ends on the x-axis (■ Figure 29.16a). The current in the loop is 1.35 A, and the magnetic field in the region is $\vec{\mathbf{B}} = (2.95 \times 10^{-5}\text{ T})\hat{\jmath} - (4.76 \times 10^{-5}\text{ T})\hat{\mathbf{k}}$. Find the force on the wire segment.

MODEL The vector $\vec{\ell}$ from one end of the segment to the other lies along the x-axis. The magnetic field is uniform, so we may use eqn. (29.6).

SETUP
$$\vec{\mathbf{F}} = I\vec{\ell} \times \vec{\mathbf{B}} = I\ell\hat{\mathbf{i}} \times (B_y\hat{\mathbf{j}} + B_z\hat{\mathbf{k}})$$
$$= I\ell[B_y\hat{\mathbf{k}} + B_z(-\hat{\mathbf{j}})].$$

SOLVE Notice that the given B_z is negative:
$$\vec{\mathbf{F}} = (1.35 \text{ A})(2)(2.54 \text{ cm})[(2.95 \times 10^{-5} \text{ T})\hat{\mathbf{k}} + (4.76 \times 10^{-5} \text{ T})\hat{\mathbf{j}}]$$
$$= (2.02\ \mu\text{N})\hat{\mathbf{k}} + (3.26\ \mu\text{N})\hat{\mathbf{j}}.$$

ANALYZE Check the units:
$$\text{A}\cdot\text{m}\cdot\text{T} = \text{A}\cdot\text{m}\cdot\frac{\text{N}}{\text{A}\cdot\text{m}} = \text{N}.$$

Check the direction using the right-hand rule (Figure 29.16b).

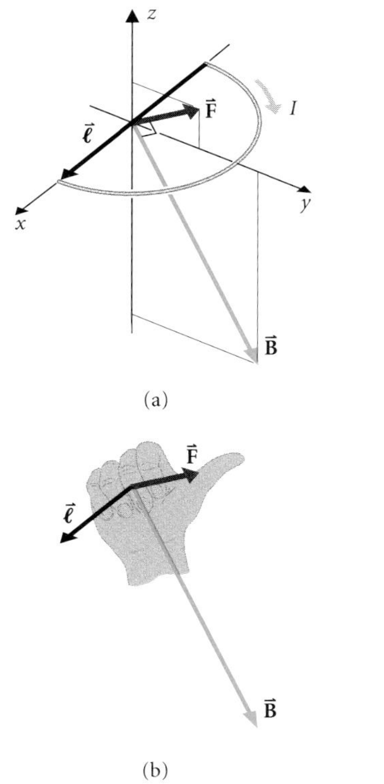

■ FIGURE 29.16
(a) Force on a semicircular piece of wire. The wire lies in the x-y-plane. The magnetic field is uniform with $\vec{\mathbf{B}}$ in the y-z-plane. (b) The right-hand rule shows that the force is also in the y-z-plane, perpendicular to $\vec{\mathbf{B}}$.

Since electric current in wires is a source of magnetic field, one such wire exerts a magnetic force on another. The force between two long parallel wires is used in the SI definition of the ampere (■ Figure 29.17). The wire on the left produces a magnetic field (eqn. 28.6),

$$\vec{\mathbf{B}}_1 = \frac{\mu_0 I_1}{2\pi d}\hat{\mathbf{j}},$$

at the position of the wire on the right. A length $d\ell$ of the second wire experiences a force:

$$d\vec{\mathbf{F}} = I_2\, d\vec{\ell} \times \vec{\mathbf{B}}_1$$
$$= I_2\, d\ell\, \hat{\mathbf{k}} \times \hat{\mathbf{j}}\frac{\mu_0 I_1}{2\pi d}$$
$$= \frac{\mu_0 I_1 I_2}{2\pi d} d\ell(-\hat{\mathbf{i}}).$$

Thus the force per unit length acting on the second wire has magnitude:

$$\frac{dF}{d\ell} = \frac{\mu_0 I_1 I_2}{2\pi d}. \tag{29.7}$$

The force is attractive if the two wires carry current in the same sense, and repulsive if the currents are in opposite directions. If both currents equal 1 A exactly, and $d = 1$ m exactly, then the force per unit length is:

$$\frac{dF}{d\ell} = \frac{2(10^{-7} \text{ N/A}^2)(1 \text{ A})^2}{1 \text{ m}} = 2 \times 10^{-7} \text{ N/m}.$$

The definition of μ_0 (exactly $4\pi \times 10^{-7}$ N/A^2) is effectively the definition of the ampere.

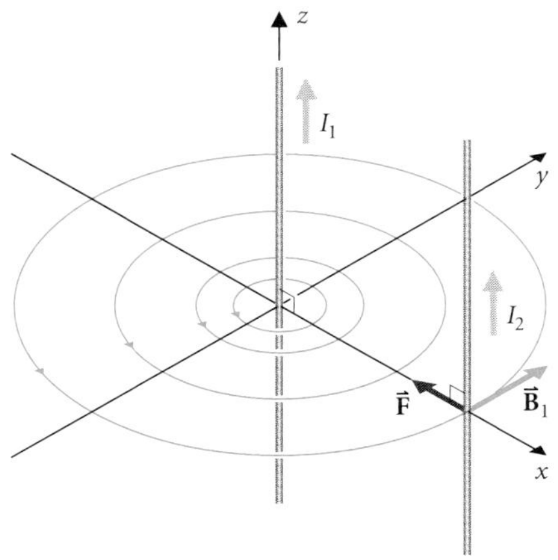

■ FIGURE 29.17
Two parallel wires carrying equal currents exert attractive forces on each other. The wire on the left produces a magnetic field at the position of the wire on the right. The moving charges in the second wire experience a magnetic force toward the first wire.

> If the same current is maintained in two long parallel wires that attract each other with a force of 2×10^{-7} N per meter of wire when the wires are 1 meter apart, then the current in each wire is 1 *ampere* (adopted 1948).

EXERCISE 29.6 ♦ Show that the force per unit length on the first wire is equal to and opposite that on the second.

When calculating forces, a simple model of the current distribution often provides remarkably good approximate results.

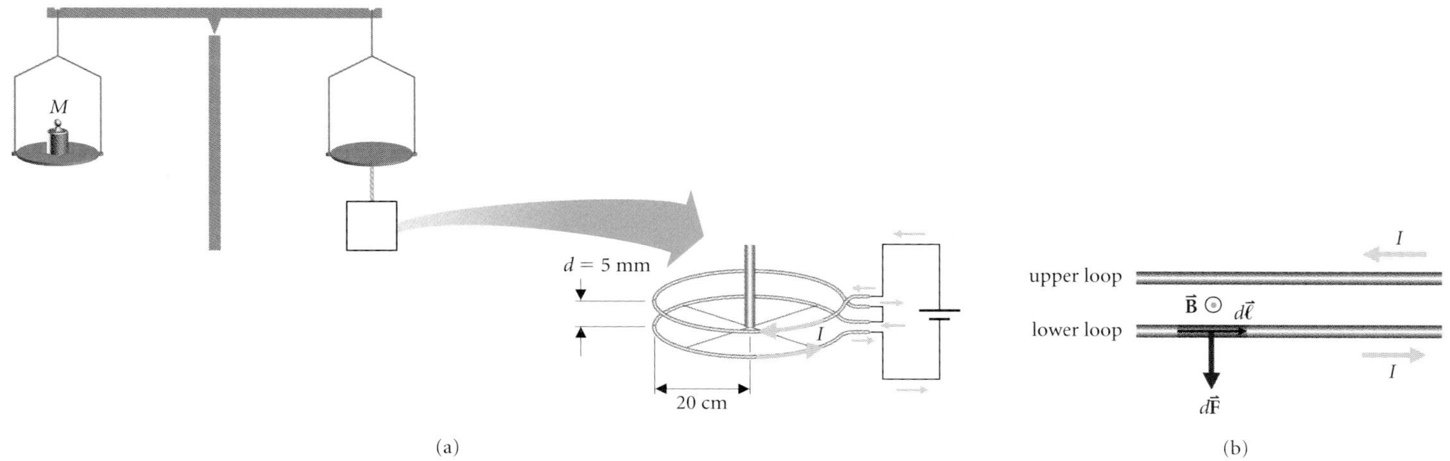

■ **Figure 29.18**
(a) A current balance. Currents are in opposite directions in the two coils. The repulsive force between them is balanced by the gravitational force on the other arm of the balance. (b) The magnetic field at a position on the lower loop may be estimated by modeling the upper loop as a long straight wire: $\vec{\mathbf{B}}$ is out of the page, and $d\vec{\mathbf{F}}$ is downward.

EXAMPLE 29.7 ♦♦♦ ■ Figure 29.18a shows a schematic of a current balance for measuring the forces between parallel current-carrying wires. Two parallel coils of radius $a = 0.20$ m and 150 turns each are connected so that the currents in the coils have the same magnitude but opposite directions. The upper coil is fixed, and the lower coil moves with the balance. In equilibrium, the coils are separated by a distance $d = 5.0$ mm. Estimate the mass M required to balance the system when the current in the coils is 1.0 A.

MODEL The system balances when the weight Mg equals the repulsive force between the two coils. Because the two coils are parallel and close together ($d \ll a$), a reasonable estimate of the field at any position on the lower coil is obtained by modeling the upper coil as a long straight wire tangent to the circle at that point (Figure 29.18b). (This works because the separation of the wires is small compared with their radius of curvature.) The magnetic field lines form circles around the upper wire and so are at right angles to the lower wire.

An exact calculation shows that this method overestimates the force by 0.3%.

SETUP The magnetic field strength produced by the upper coil is the same for each position on the lower coil (eqn. 28.6): $B = \mu_0 NI/(2\pi d)$. Figure 29.18b shows the magnetic field and the resulting force on a typical element of the lower coil:

$$d\vec{\mathbf{F}} = (NI\, d\ell\, B, \quad \text{downward}).$$

Since the force per unit length is the same for each element, the total force has magnitude:

$$F = \int dF = NIB \int_{\substack{\text{lower}\\ \text{coil}}} d\ell = 2\pi aNIB = \frac{a\mu_0 (NI)^2}{d}.$$

We may add the magnitudes because each element of force $d\vec{\mathbf{F}}$ has the *same direction*.

The weight required to balance this force is $Mg = F$.

SOLVE Hence the mass required on the balance is:

$$M = \frac{a\mu_0(NI)^2}{gd} = \frac{(0.20\ \text{m})(4\pi \times 10^{-7}\ \text{N/A}^2)(150)^2(1.0\ \text{A})^2}{(9.8\ \text{N/kg})(5.0 \times 10^{-3}\ \text{m})} = 0.12\ \text{kg}.$$

ANALYZE A similar current balance is used in laboratory equipment that provides the standard for the ampere. ■

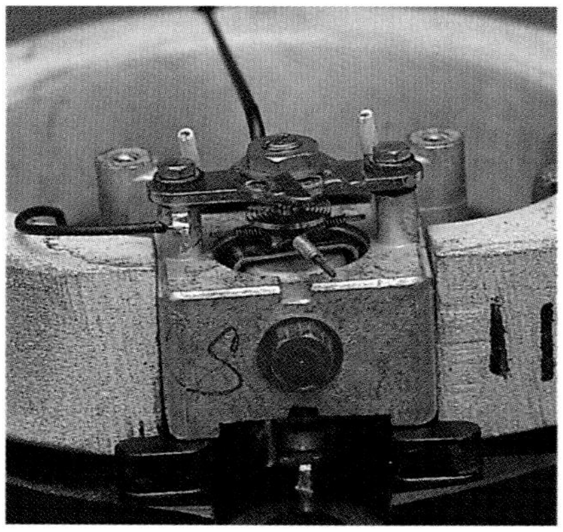

Galvanometer movement. Torque acting on a current-carrying loop causes the pointer movement in this galvanometer.

29.2.2 *Force and Torque on Current Loops*

Many applications of magnetic forces involve rotating machinery in which torques exerted by the field are important. In this section we shall investigate the torque exerted on a current loop. The result is conveniently expressed in terms of magnetic moment.

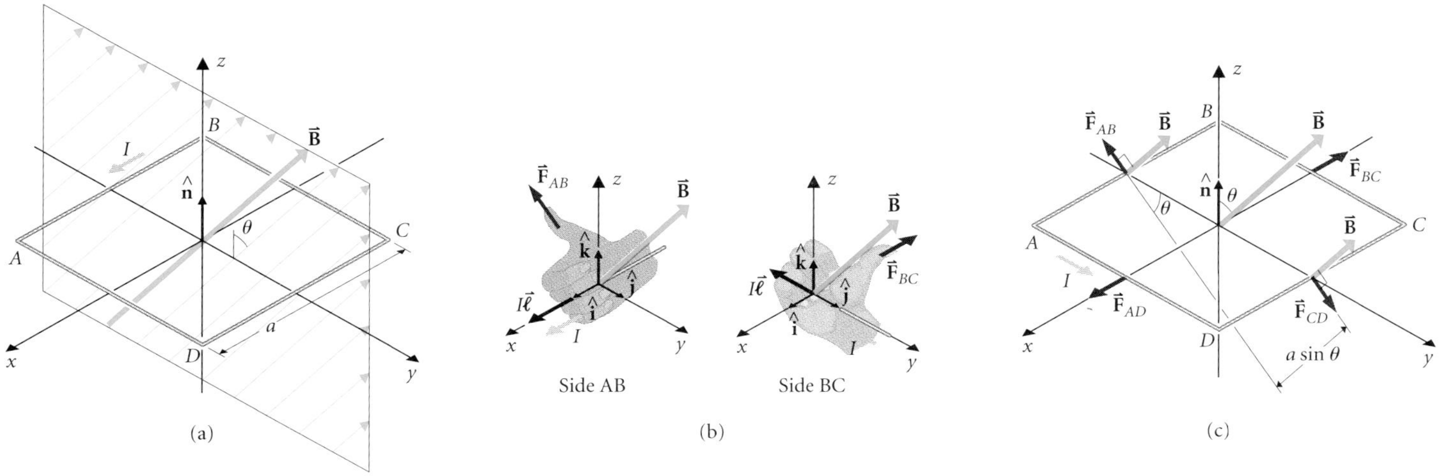

EXAMPLE 29.8 ♦♦ A square current loop is placed in a uniform magnetic field that makes an angle θ with the normal to the loop (■ Figure 29.19a). Find the net force acting on the loop and the net torque about its center.

MODEL We may apply eqn. (29.6) to each side of the loop or to the loop as a whole. If we consider the whole loop as a single wire segment, both its ends are at the same point. The vector $\vec{\ell}$ in eqn. (29.6) is zero, so the net force on the loop is zero. We look at the forces on each side separately to calculate the torque. The line of action of each force is through the center of the corresponding side (Example 29.4).

SETUP We choose the z-axis perpendicular to the loop, with the x- and y-axes parallel to sides AB and BC. Equation (29.6) applies to each side of the square, and $\vec{\mathbf{B}}$ has y- and z-components, but no x-component. Vector $\vec{\ell}$ for side AB is in the x-direction (Figure 29.19b), and thus the force acting on side AB is:

$$\vec{\mathbf{F}}_{AB} = Ia\hat{\mathbf{i}} \times \vec{\mathbf{B}} = Ia\hat{\mathbf{i}} \times (B \sin\theta\, \hat{\mathbf{j}} + B\cos\theta\, \hat{\mathbf{k}})$$
$$= IaB(\sin\theta\, \hat{\mathbf{k}} - \cos\theta\, \hat{\mathbf{j}}).$$

The force on side CD is: $\vec{\mathbf{F}}_{CD} = -Ia\hat{\mathbf{i}} \times \vec{\mathbf{B}} = -IaB(\sin\theta\, \hat{\mathbf{k}} - \cos\theta\, \hat{\mathbf{j}})$.

These forces are equal and opposite. The same is true of the forces on sides BC and AD:

$$\vec{\mathbf{F}}_{AD} = Ia\hat{\mathbf{j}} \times \vec{\mathbf{B}} = Ia\hat{\mathbf{j}} \times (B\sin\theta\, \hat{\mathbf{j}} + B\cos\theta\, \hat{\mathbf{k}}) = IaB(\cos\theta\, \hat{\mathbf{i}}) = -\vec{\mathbf{F}}_{BC}.$$

The net force on the loop is zero.

SOLVE The lines of action of forces $\vec{\mathbf{F}}_{AD}$ and $\vec{\mathbf{F}}_{BC}$ both pass through the center of the loop and so produce no torque about the center of the loop. However, $\vec{\mathbf{F}}_{AB}$ and $\vec{\mathbf{F}}_{CD}$ exert torques of equal magnitude and direction that act to rotate AB up and CD down. The net torque is:

$$\vec{\tau} = \Sigma\, \vec{\mathbf{r}} \times \vec{\mathbf{F}} = -\left(\frac{a}{2}\right)\hat{\mathbf{j}} \times IaB\sin\theta\, \hat{\mathbf{k}} + \left(\frac{a}{2}\right)\hat{\mathbf{j}} \times (-IaB\sin\theta\, \hat{\mathbf{k}})$$
$$= (Ia^2)B\sin\theta(-\hat{\mathbf{i}}).$$

Recognizing $Ia^2\hat{\mathbf{k}} = IA\hat{\mathbf{k}}$ as the magnetic moment of the loop (eqn. 28.8), we have:

$$\vec{\tau} = mB\sin\theta(-\hat{\mathbf{i}}) = \vec{\mathbf{m}} \times \vec{\mathbf{B}}. \quad (29.8)$$

ANALYZE The total force on any closed loop in a uniform $\vec{\mathbf{B}}$ field is zero. According to eqn. (29.6), the total force on any piece of wire is $\vec{\mathbf{F}}_T = I\vec{\ell} \times \vec{\mathbf{B}}$, where $\vec{\ell}$ is the vector between the end points of the wire. For any closed loop the end points are the same, and $\vec{\ell} = 0$. Equation (29.8) describes the torque on any loop in a *uniform* magnetic field, but the proof is left to the problems. The torque tends to align the dipole moment of the loop with the field and the plane of the loop perpendicular to the field (cf. §24.5). ■

■ **Figure 29.19**
(a) The magnetic field makes an angle θ with the normal to the plane of a current loop. The loop lies in the x-y-plane, and $\hat{\mathbf{n}} = \hat{\mathbf{k}}$. (b) Vector diagrams showing the forces on sides AB and BC. (c) All four forces on the loop. The forces on sides BC and AD are equal and opposite and lie along the same line. $\vec{\mathbf{F}}_{AB}$ and $\vec{\mathbf{F}}_{CD}$ are equal and opposite, but their lines of action are separated by a distance $a\sin\theta$. There is a net torque on the loop given by $\vec{\tau} = \vec{\mathbf{m}} \times \vec{\mathbf{B}}$.

$\vec{\mathbf{F}}_{AB}$ and $\vec{\mathbf{F}}_{CD}$ form a couple with lines of action separated by $a\sin\theta$ (Figure 29.19c) and a torque of magnitude $\tau = F_{AB}(a\sin\theta) = (IaB)(a\sin\theta) = (Ia^2)B\sin\theta$ (see §11.1.3).

Compare with eqn. (24.12).

Digging Deeper

Magnetic Forces and Newton's Third Law

■ Figure 29.20a shows two separate elements of wire and the magnetic forces they exert on each other. The forces are not equal and opposite; the magnetic field of element 1 exerts an upward force on element 2, but element 2 exerts no force at all on element 1! The magnetic field due to element 2 is zero at the position of element 1. The mutual magnetic forces on two point charges in motion display the same behavior (Figure 29.20b). What has happened to Newton's third law? In Part II we showed that Newton's third law expresses conservation of linear and angular momentum in a system of particles. The conservation laws are fundamental principles, required by the symmetry of space and time. They must still apply when magnetic forces occur. A proper analysis of momentum conservation for a system of moving charges must include the electric and magnetic fields, which extend over all space. We have already discovered that electric field stores energy (§27.4). The electromagnetic field can also store momentum. The apparent failure of Newton's third law tells us that momentum is being exchanged between particles and fields in the system. Momentum of the entire system of particles and fields *is* conserved. The forces between two finite, closed current loops do obey Newton's third law, since the net momentum exchanged with the fields is zero in that case. Nevertheless, you should always be wary about applying Newton's third law when magnetic forces occur.

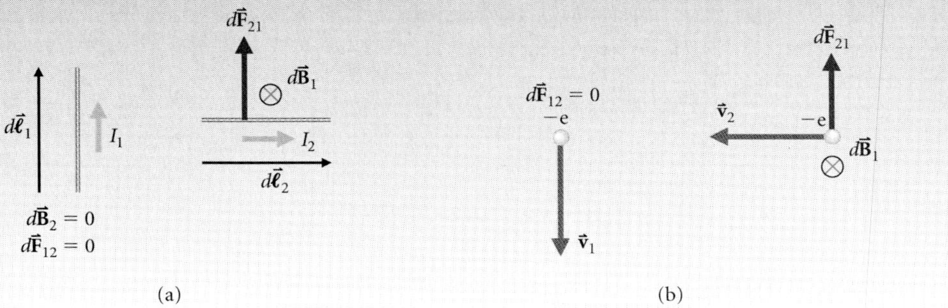

■ Figure 29.20
(a) The forces exerted by two elements of current-carrying wire on each other are not equal and opposite! Here, the force exerted by element 2 on element 1 is zero, while the force exerted by 1 on 2 is not. (b) The same effect occurs for two electrons moving at right angles to each other. These apparent violations of Newton's third law result from exchange of momentum between particles and fields.

EXAMPLE 29.9 ♦ Many satellites use current loops interacting with the Earth's magnetic field to control their orientation in space. A system designed for the *Extreme Ultraviolet Explorer* (EUVE) satellite (■ Figure 29.21) has a coil of area $A = 1\ \text{m}^2$ and $N = 10^3$ turns. What current is required to produce a torque of 0.06 N·m if the Earth's magnetic field is $B = 3 \times 10^{-5}$ T in the plane of the coil?

MODEL We may apply eqn. (29.8) directly.

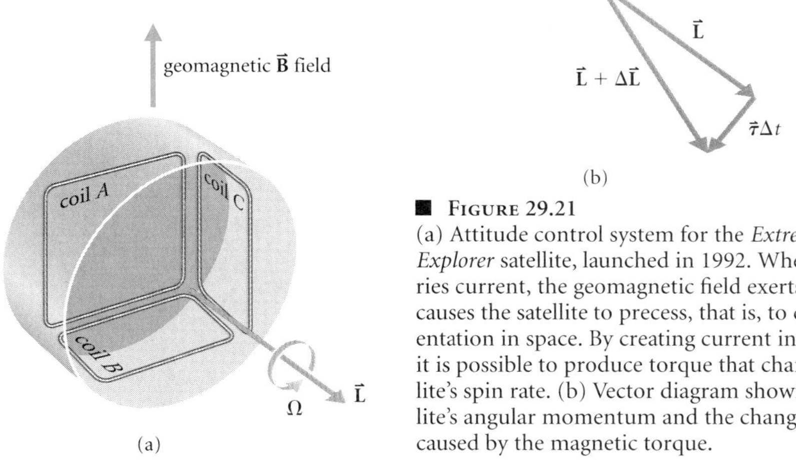

■ Figure 29.21
(a) Attitude control system for the *Extreme Ultraviolet Explorer* satellite, launched in 1992. When coil *A* carries current, the geomagnetic field exerts a torque that causes the satellite to precess, that is, to change its orientation in space. By creating current in coils *B* and *C* it is possible to produce torque that changes the satellite's spin rate. (b) Vector diagram showing the satellite's angular momentum and the change $\Delta\vec{L} = \vec{\tau}\,\Delta t$ caused by the magnetic torque.

SETUP The magnetic moment of the coil has magnitude $m = NIA$ and is perpendicular to the plane of the coil. The torque exerted by the Earth's field is $\vec{\tau} = \vec{m} \times \vec{B}$. If the field is in the plane of the coil, then $\vec{m}$ is perpendicular to $\vec{B}$ and:

$$\tau = mB = NIAB.$$

SOLVE

$$I = \frac{\tau}{NAB} = \frac{0.06 \text{ N}\cdot\text{m}}{1000(1 \text{ m}^2)(3 \times 10^{-5} \text{ T})} = 2 \text{ A}.$$

ANALYZE As designed for *EUVE*, this system requires 24 kg of wire and uses 100 W of power.

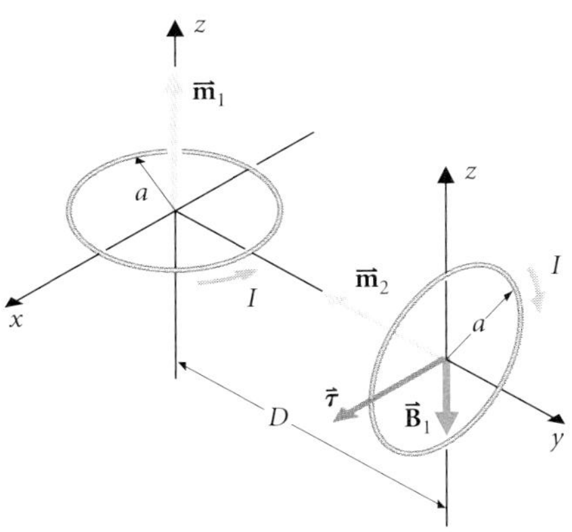

■ FIGURE 29.22
Two current loops lie in orthogonal planes. To find the torque one exerts on the other, we calculate the magnetic field the first loop produces. If the loops are far apart ($D \gg a$), we may use the dipole approximation.

EXAMPLE 29.10 ♦♦ Each of two current loops has N turns, radius a, and area $A = \pi a^2$. One loop lies in the x-y-plane with its center at the origin. The other is parallel to the x-z-plane with its center at $y = D \gg a$. Estimate the torque the first loop exerts on the second if each carries current I.

MODEL ■ Figure 29.22 shows the loops, together with the magnetic field produced by the first loop. We calculate the field at the center of one loop due to the other and then compute the torque using eqn. (29.8). We may model the first loop as a dipole (§28.2.3) in calculating the field it produces.

SETUP We find the magnetic field in the plane of a loop by adapting the result of Exercise 24.6 to a magnetic dipole (i.e., replacing $k\vec{p}$ with $\mu_0\vec{m}/4\pi$). The magnetic moment of the first loop is in the z-direction and has magnitude IA. At the center of the second loop $y = D$:

$$\vec{B}_1 = \frac{\mu_0}{4\pi y^3}(-\vec{m}_1) = \frac{\mu_0 m_1}{4\pi D^3}(-\hat{k}) = \frac{\mu_0 IA}{4\pi D^3}(-\hat{k}).$$

The magnetic moment of the second loop is $\vec{m}_2 = IA(-\hat{j})$, and since $D \gg a$, the magnetic field $\vec{B}_1$ is almost constant over the area of the second loop.

SOLVE Thus the torque is given by eqn. (29.8):

$$\vec{\tau}_{21} = \vec{m}_2 \times \vec{B} = -\frac{\mu_0}{4\pi}\frac{\vec{m}_2 \times \vec{m}_1}{D^3} \qquad (29.9)$$

$$= -\frac{\mu_0}{4\pi}\frac{(NIA)^2}{D^3}(-\hat{j} \times \hat{k}) = \frac{\mu_0}{4\pi}\frac{(NIA)^2}{D^3}\hat{i}.$$

ANALYZE The torque rotates the second loop toward an orientation where the two magnetic moments are antiparallel. Although we derived eqn. (29.9) for a specific orientation of the loops, it is true for any two loops separated by a distance much greater than either's size.

29.3 THE HALL EFFECT

In 1879 E. H. Hall discovered that a conductor carrying current in a magnetic field develops a small electric field perpendicular both to the magnetic field and to the direction of the current. Hall used this observation to prove that current in metals is carried by negative charge.

Current in a metal strip is driven by an applied electric field $\vec{E}$ (■ Figure 29.23). Each conduction electron in a stationary strip has an average drift velocity $\vec{v}_d$. In a region of uniform magnetic field, the magnetic force $-e\vec{v}_d \times \vec{B}$ causes electrons to drift toward one side of the strip (side a in the figure), leaving excess positive charge on side b. These charge concentrations produce an electric field $\vec{E}_H$. The sideways motion ceases when the electric force $-e\vec{E}_H$ on each electron balances the magnetic force $-e\vec{v}_d \times \vec{B}$. The resulting potential difference across the strip,

$$|\Delta V_H| = E_H L = v_d B_\perp L, \qquad (29.10)$$

is known as the *Hall potential.*

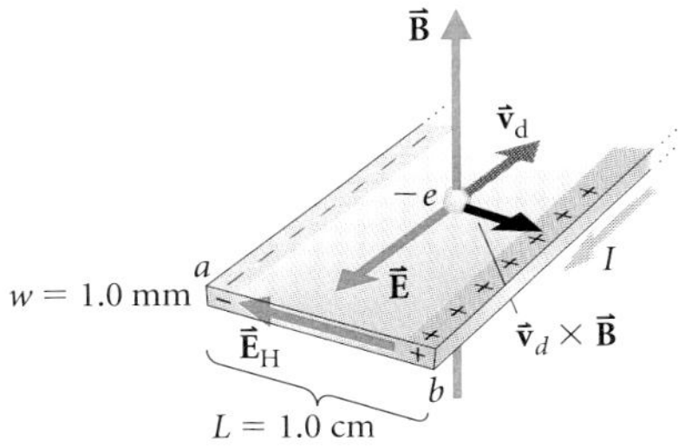

■ FIGURE 29.23
The Hall effect. A strip of conductor carries current I perpendicular to a magnetic field $\vec{B}$. The moving electrons feel a magnetic force that causes them to drift to one edge of the strip. This concentration of charge produces an electric field across the strip. In equilibrium, the electric force component $-eE_H$ on each electron balances the magnetic force component $ev_d B$, and so $\vec{E}_H = -\vec{v} \times \vec{B}$.

$B_\perp$ IS THE COMPONENT OF $\vec{B}$ PERPENDICULAR TO THE DIMENSION L OF THE STRIP.

EXERCISE 29.7 ❖ Show that the sign of the Hall potential would be reversed if current were carried by positive instead of negative charges.

EXAMPLE 29.11 ♦♦ A copper strip has a cross section of 1.0 mm × 1.0 cm (Figure 29.23). The strip carries a current of 5.0 A, and the magnetic field is $|\vec{\mathbf{B}}| = 1.0$ T. What is the Hall potential?

MODEL The number of conduction electrons per unit volume in copper is $n = 8.5 \times 10^{28}$ /m^3 (cf. Example 26.4). As these electrons move, forming the electric current, magnetic force acts on them.

SETUP The drift speed of the electrons is $v_d = j/ne = I/neA$. The Hall potential difference is (eqn. 29.10):

$$|\Delta V_H| = v_d BL = \frac{IBL}{neLw} = \frac{IB}{new}.$$

Thus:

$$|\Delta V_H| = \frac{(5.0\ \text{A})(1.0\ \text{T})}{(8.5 \times 10^{28}\ /\text{m}^3)(1.6 \times 10^{-19}\ \text{C})(1.0 \times 10^{-3}\ \text{m})} = 0.37\ \mu\text{V}.$$

ANALYZE Hall fields are very small because drift speeds are very small (cf. Example 26.4). Nevertheless, they are very important.[1] Since $\Delta V_H \propto (1/ne)$, measurement of the Hall potential with a known magnetic field may be used to determine the sign and number of the charge carriers in a conductor. Alternatively, using a conductor with known properties, the Hall effect may be used to measure $\vec{\mathbf{B}}$.

In equilibrium, the Hall electric force balances the magnetic force on moving electrons, so the component across the strip of the net force on each electron is zero. However, the force on each stationary, positively charged ion is due only to the Hall field and is not balanced by any magnetic force. The Hall electric field transmits the magnetic force to the ions and thus to the strip as a whole.

29.4 Magnetic Materials

29.4.1 *Atomic Model of Magnetization*

SEE THE OVERVIEW OF ELECTROMAGNETISM §VI.5.

Electron motion in individual atoms constitutes a current that can produce magnetic field. The behavior of atoms follows the principles of quantum mechanics, so classical models of magnetism are limited in accuracy. Nevertheless, we can obtain a reasonable, qualitative understanding from a classical approach.

Most materials show only a weak response to an applied external field. They are classified as *paramagnetic* if the field within the material is greater than the applied field and as *diamagnetic* if the internal field is less than the applied field. Paramagnetic materials show a weak attraction toward regions of strong magnetic field, whereas diamagnetic materials are repelled. *Ferromagnetic* materials—primarily iron, cobalt, nickel, and their alloys—show a strong response to applied fields and can form permanent magnets.

Paramagnetism. Each atom in a paramagnetic material has a permanent magnetic moment that is associated with the net angular momentum of its electrons. Normally these atomic moments are randomly oriented and produce no net magnetic field. An external field tends to align the atomic dipoles, which then produce a net field that adds to the applied field. The atoms do not align completely, since thermal motion tends to randomize their directions.

[1] In 1985 the Nobel prize in physics was awarded to Klaus von Klitzing for the discovery of the quantum Hall effect—a related phenomenon that occurs in semiconductors.

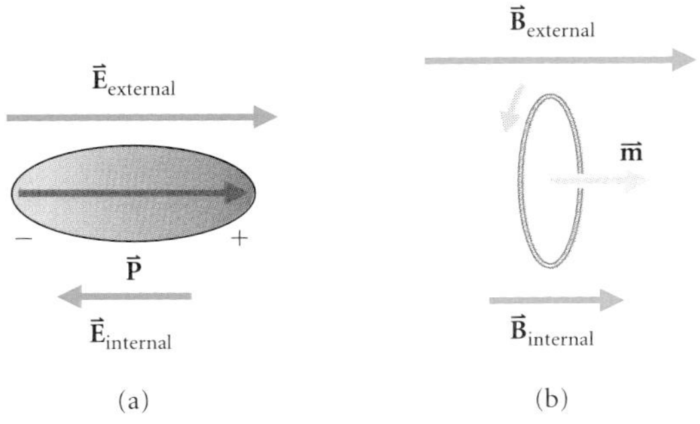

■ FIGURE 29.24
(a) Electric dipoles aligned with an externally applied electric field produce an electric field opposite $\vec{E}_{external}$ within the material (cf. Figure 27.15). (b) Magnetic dipoles aligned with an externally applied magnetic field produce a magnetic field within the material that is in the same direction as $\vec{B}_{external}$.

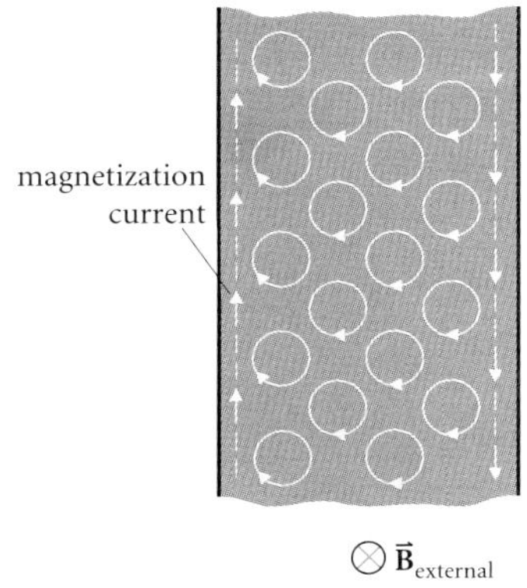

■ FIGURE 29.25
Magnetization current. Each atom behaves like a miniature current loop. Between atoms the currents are in opposite directions, and cancel. At the edge of the material, there is no cancellation. The net current produces magnetic field in the material with direction into the page.

A useful analogy for this effect is a box containing ping-pong balls, each with a small bead of lead attached to its inside surface. The balls tend to settle with the lead bead on the bottom, but if the box is shaken violently, the balls bounce about, and the lead beads appear in random positions. Roughly speaking, greater temperature corresponds to greater shaking of the atoms, and so paramagnetism decreases as temperature increases.

Electric dipoles in a dielectric align with an applied electric field and reduce the applied field. Magnetic dipoles in a paramagnetic material align with an applied magnetic field and increase the applied field (■ Figure 29.24). This reversal of behavior occurs because electric field is produced by charge, and magnetic field is produced by current. ■ Figure 29.25 illustrates what happens when the atoms, each modeled as a small current loop, are aligned. The currents between atoms are in opposite directions and cancel. On the surfaces of the material the atoms produce a net current called *magnetization current* that enhances the external field.

Diamagnetism. In a diamagnetic material with no applied magnetic field the individual atoms have no net magnetic moment; the magnetic effects of the electrons in each individual atom cancel. An external magnetic field changes the motion of the electrons and induces a net magnetic moment in each atom.

THE PATTERN OF ELECTRONS IN AN ATOM DETERMINES ITS MAGNETIC MOMENT. SEE *DIGGING DEEPER.*

An electron in orbit about a nucleus (■ Figure 29.26) forms a current loop with magnetic moment $m = I(\pi r^2)$, where r is the radius of the electron orbit. The current is $I = (e/T) = (ev/2\pi r)$, where T is the period of the orbit and v is the electron's speed. Thus:

REMEMBER: THIS IS A CLASSICAL MODEL.

$$m = \frac{ev}{2\pi r}\pi r^2 = \frac{evr}{2}.$$

WE USED EQN. (28.8) FOR m.

The electric force F_e exerted on the electron of mass M by the atomic nucleus causes the centripetal acceleration:

$$F_e = Mv^2/r.$$

An external magnetic field applied parallel to $\vec{m}$ causes a magnetic force $\vec{F}_m = -e\vec{v} \times \vec{B}$, which is radially outward. Because of the reduced inward force, the centripetal acceleration of the electron decreases. Thus either the orbit radius r increases or the speed decreases. In fact it is the speed that changes. With the new speed $v + dv$, Newton's second law becomes:

A FORCE PARALLEL TO VELOCITY CHANGES THE SPEED. WE'LL SEE IN CHAPTER 30 HOW THIS FORCE ARISES.

$$\frac{M(v + dv)^2}{r} = F_e - evB.$$

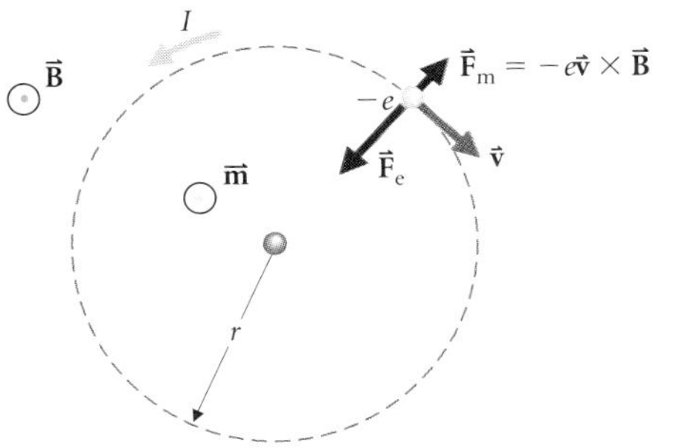

■ FIGURE 29.26
Diamagnetism. The applied field $\vec{B}$ causes a reduction of the net force acting on the electron. The electron's orbital radius remains the same, but its speed is reduced (we'll see why in Chapter 30), and the magnetic moment of the atom changes by $\Delta\vec{m} \propto -\vec{B}$.

Expanding the square and ignoring the term in $(dv)^2$, we have:

$$\frac{Mv^2}{r} + \frac{2Mv\ dv}{r} = F_e - evB.$$

The first term on the left-hand side equals the first term on the right, so:

Remember: Speed decreases, so dv is negative.

$$dv = -\frac{eBr}{2M}.$$

The change in speed is proportional to the applied field B. The magnetic moment changes by an amount:

$$dm = d\left(\frac{evr}{2}\right) = \frac{er}{2}dv = \frac{er}{2}\left(-\frac{eBr}{2M}\right) = -\frac{e^2r^2B}{4M}.$$

The magnetic moment of the electron orbit is reduced by an amount proportional to the applied field B.

EXERCISE 29.8 ♦♦ Show that if $\vec{\mathbf{B}}$ is applied antiparallel to $\vec{\mathbf{m}}$, then m is increased, so that $d\vec{\mathbf{m}}$ is always antiparallel to $\vec{\mathbf{B}}$.

The magnetic moments of the electrons in each atom sum to zero when there is no applied field. But since $d\vec{\mathbf{m}}$ is in the same direction for each electron, the net result of an applied $\vec{\mathbf{B}}$ is to create an induced moment opposite $\vec{\mathbf{B}}$.

The diamagnetic effect occurs in all substances, but the induced magnetic moments are much smaller than the permanent moments of paramagnetic atoms. However, the diamagnetic effect is independent of temperature and becomes the dominant effect at high temperature as the alignment of permanent moments decreases.

TABLE 29.1
Curie Temperatures

Material	Curie Temperature (°C)
Iron	770
Cobalt	1127
Nickel	358
Gadolinium	16
Dysprosium	−168
Chromium dioxide	117
Ferrites:	
$NiFe_2O_4$	585
$CuFe_2O_4$	455
$NiAlFeO_4$	198

After Pierre Curie, 1859–1906.

Note: The degree of magnetization is vastly greater in ferromagnetic materials than in paramagnetic materials.

Ferromagnetism. In ferromagnetic materials a quantum mechanical interaction between electrons in neighboring atoms causes the atomic dipole moments to align with each other even in the absence of an external field. Below a particular temperature, called the *Curie temperature,* the tendency to align overcomes thermal randomization and the material spontaneously *freezes* into *domains* in which all atomic dipoles have the same direction (■ Figure 29.27, ● Table 29.1). Domain formation is similar to the growth of ice crystals as water freezes.

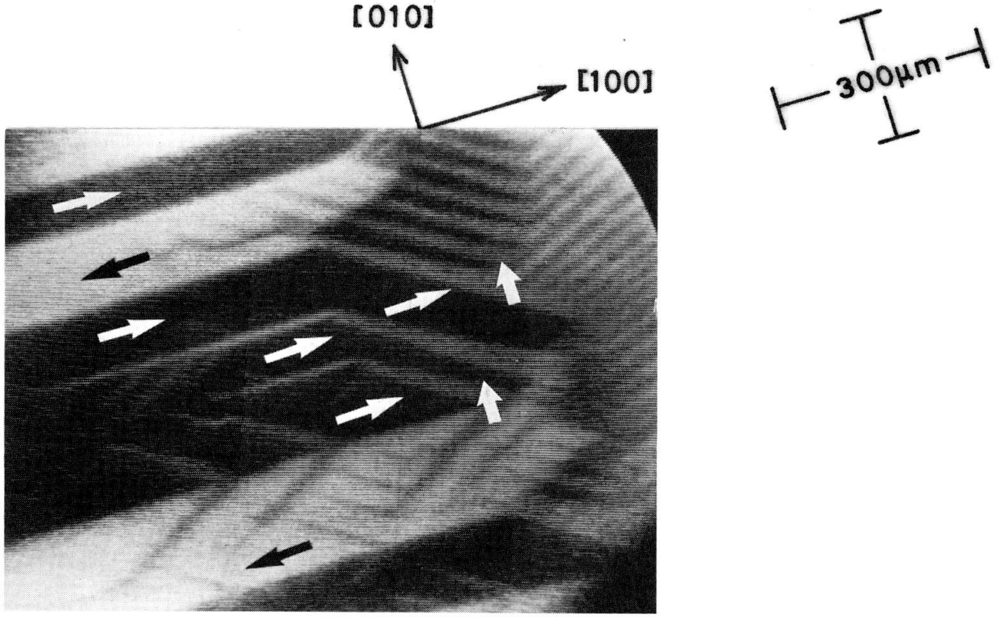

■ Figure 29.27
Ferromagnetism. The magnetic domain structure of silicon iron, as revealed by a scanning electron microscope. The arrows indicate the magnetic axis of the domains.

Digging Deeper

Magnetic Moment and Angular Momentum

The magnetic moment of a particle with mass M and charge e in a circular orbit is:

$$m = \frac{evr}{2} = \frac{e}{2M} L,$$

where $L = Mvr$ is the angular momentum of the particle about the center of the circle. Magnetic moment is also proportional to angular momentum for electrons and nuclei in atoms. Each particle has an intrinsic angular momentum—called spin—as well as its orbital angular momentum, and its magnetic moment is related to L by an empirical factor called g:

$$m = g \frac{e}{2M} L.$$

For an electron, the value of g that relates its intrinsic magnetic moment to its spin is 2.002319304386.

Using the principles of quantum mechanics it is possible to compute the total angular momentum of an atom and hence to compute its magnetic moment m. The same principles determine the chemical behavior of atoms, and so magnetic properties are related to chemical properties. For example, helium is chemically inert, and its magnetic moment is zero. In general, the relations are quite complicated and beyond the scope of this text.

When an external field is applied to a ferromagnetic material, it becomes strongly magnetized. In a weak applied field, domains with magnetization parallel to the applied field grow at the expense of neighboring domains. A strong applied field causes the direction of magnetization within each domain to rotate toward the applied field direction (■ Figure 29.28). In some ferromagnetic materials, especially alloys such as neodymium–iron and Alnico (an alloy of nickel, cobalt, aluminum, and copper), the domains remain aligned when the applied field is removed, forming a permanent magnet. These materials are described as magnetically *hard.* In *soft* materials such as the ferrites (iron oxides combined with iron, nickel, or cobalt), the domains do not remain aligned when the applied field is removed. Pure iron is magnetically soft.

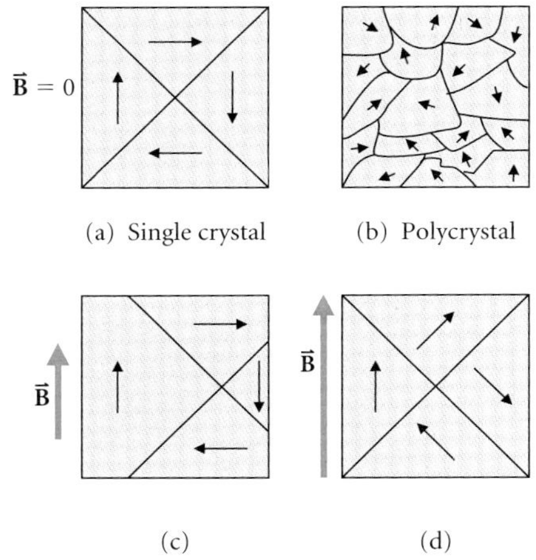

■ Figure 29.28
In an unmagnetized ferromagnetic material (a and b), the magnetic axes of different domains point in random directions, and there is no net magnetic field. A weak applied field (c) causes domains magnetized parallel to $\vec{B}$ to grow. A strong applied field (d) rotates the domains.

⁂ 29.4.2 The Magnetic Field Intensity $\vec{H}$

The magnetic induction $\vec{B}$ at any point within a magnetic material arises both from distant sources and from nearby atomic magnetization. In advanced work, it is useful to separate these two contributions by defining a vector field $\vec{H}$ known as the *magnetic field intensity.* The *magnetization* of the material is described by $\vec{M}$, the magnetic dipole moment per unit volume, and the field intensity is defined by:

$$\vec{H} \equiv \frac{\vec{B}}{\mu_0} - \vec{M}.$$

Frequently, as we saw in the classical model of diamagnetism, the magnetization $\vec{M}$ is proportional to the applied field:

$$\vec{M} = \chi_m \vec{H}.$$

The constant χ_m is called the *magnetic susceptibility.* Then $\vec{B}$ and $\vec{H}$ are also proportional:

$$\vec{B} = \mu_0(1 + \chi_m)\vec{H} \equiv \mu\vec{H}. \qquad (29.11)$$

TABLE 29.2 Magnetic Susceptibilities of Some Common Materials

Paramagnetic Materials (at room temperature)		Diamagnetic Materials	
SUBSTANCE	SUSCEPTIBILITY	SUBSTANCE	SUSCEPTIBILITY
Aluminum	2.1×10^{-5}	Ammonia	-2.6×10^{-6}
Cesium	5.1×10^{-5}	Carbon (graphite)	-1.6×10^{-5}
Iron oxide (FeO)	7.2×10^{-3}	Copper	-9.7×10^{-6}
Lithium	1.4×10^{-5}	Hydrogen (H_2 gas)	-2.1×10^{-9}
Magnesium	1.2×10^{-5}	Lead	-1.6×10^{-5}
Oxygen (O_2 gas)	1.8×10^{-6}	Mercury	-2.8×10^{-5}
Platinum	2.8×10^{-4}	Nitrogen (N_2 gas)	-5.0×10^{-9}
Tungsten	6.8×10^{-5}	Sodium chloride	-1.4×10^{-5}
		Water	-9.1×10^{-6}

SOURCE: *Handbook of Chemistry and Physics*, 64th ed. (Cleveland: Chemical Rubber Co.)

The constant of proportionality μ is called the *magnetic permeability.* (The constant μ_0 is called the permeability of the vacuum, since in vacuum $\mu = \mu_0$.) • Table 29.2 lists magnetic susceptibilities for some common diamagnetic ($\chi_m < 0$) and paramagnetic ($\chi_m > 0$) materials. Note that $|\chi_m|$ is much less than unity for all the materials in the table.

Practical ferromagnetic materials are not well described by eqn. (29.11). For iron, quite reasonable values of H (about 80 A/m) result in complete alignment of all domains at $B =$ 2.15 T. The field is said to be *saturated.* In addition, the domains do not completely return to a random orientation as the external field is removed. The value of B is not uniquely determined by the value of H (■ Figure 29.29). As H is increased and then decreased, B depends on earlier values of H as well as on its present value. This phenomenon is known as *hysteresis,* and the closed curve in a graph of B versus H is called the *hysteresis loop.* The hysteresis loop for soft materials is long and narrow (Figure 29.29a). For hard materials the loop is wider, and a substantial B remains when $H = 0$; such materials are used to make *permanent* magnets (Figure 29.29b).

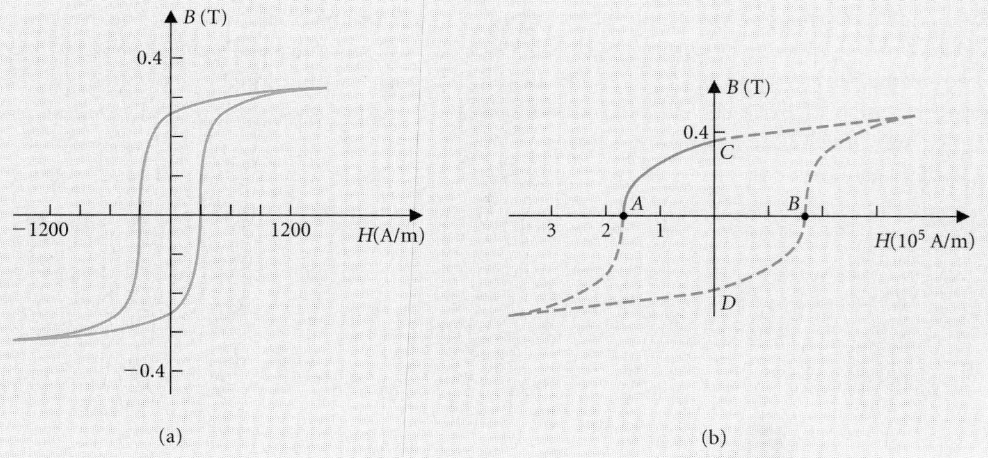

■ FIGURE 29.29
(a) Hysteresis curve for 4C4 Ferrite at 25° C. Ferrite is a *soft* material used in transformer cores. (b) Demagnetization curve for Ferroxdur ($BaFe_{12}O_{18}$), a *hard* magnetic material used for permanent magnets. *Solid line:* data; *dashed line:* schematic. Note how the scale differs from Figure 29.29a. $H = 10^5$ A/m is required to reduce B to zero in Ferroxdur, whereas only 300 A/m is required for Ferrite.

Digging Deeper

Magnetic Resonance Imaging

■ Figure 29.30 shows a view of the interior of a person's head made with the technique of nuclear magnetic resonance. This powerful imaging method uses magnetic field and radio waves to replace the damaging x and γ radiation used by competing techniques. Many atomic nuclei have a net angular momentum and a net magnetic moment. (The magnetic moments of nuclei are several orders of magnitude smaller than the typical moment of a paramagnetic atom.) An external magnetic field $\vec{\mathbf{B}}_o$ causes alignment of the nuclear dipoles and a small bulk magnetization. Individual nuclei act like spinning tops. The external magnetic torque causes the nuclear top to precess like an actual top in a gravitational field (cf. §12.6 and Problem 59). The precession frequency, or *Larmor* frequency, is determined by the external magnetic field and the particular kind of nucleus; it typically falls in the radio frequency range.

If the magnetic field of a radio wave oscillating at the precession frequency is applied at right angles to $\vec{\mathbf{B}}_o$, the nuclei absorb energy from the oscillating field, becoming aligned opposite $\vec{\mathbf{B}}_o$ (cf. Problem 76). This resonant absorption of energy gives the technique its name.

When the applied radio wave is turned off, the nuclei realign with $\vec{\mathbf{B}}_o$ and emit radio waves at the Larmor frequency. Interactions between the magnetic moments of neighboring nuclei (and hence the local chemical properties of the medium) determine the time the nuclei take to realign. Introducing small spatial variations in B_o causes the frequency of the emitted radiation to depend on position. Observations of decay time versus frequency can then be translated into a picture of chemical properties as a function of position—hence the image in Figure 29.30.

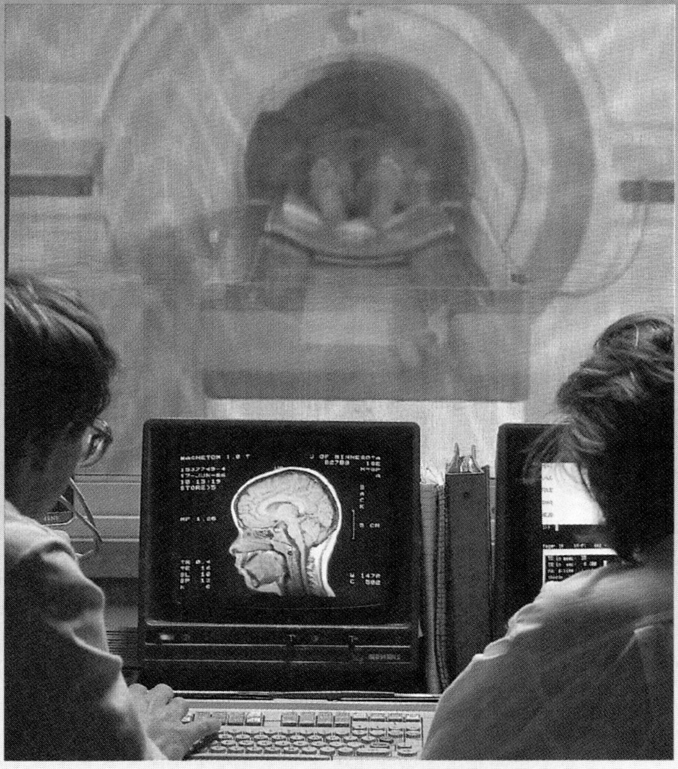

■ FIGURE 29.30
MRI image of a brain. Combinations of proton density and rates at which nuclei realign with $\vec{\mathbf{B}}_0$ are interpreted as different kinds of tissue.

Chapter Summary

Where Are We Now?

We have studied the motion of particles in response to magnetic forces, and the forces and torques on current-carrying wires. These ideas allow us to understand many important technical applications of magnetic fields.

What Did We Do?

The magnetic force exerted on a moving charge is $\vec{\mathbf{F}} = q\vec{\mathbf{v}} \times \vec{\mathbf{B}}$, perpendicular to both $\vec{\mathbf{v}}$ and $\vec{\mathbf{B}}$. A charged particle moving perpendicular to a uniform magnetic field $\vec{\mathbf{B}}$ travels in a circle with an angular frequency $\omega_c = qB/m$ that is independent of its speed. The radius of the circle is the Larmor radius $r_L = mv/qB$. The particle's velocity parallel to $\vec{\mathbf{B}}$ is unaffected by the magnetic field. In a uniform field particles move in regular spirals around the field lines.

Converging field lines form magnetic mirrors that can reverse particle motion along the field lines. This effect causes particles to be trapped in the Earth's Van Allen belts.

The magnetic force on a segment $d\vec{\ell}$ of a current-carrying wire is:

$$d\vec{\mathbf{F}} = I\, d\vec{\ell} \times \vec{\mathbf{B}}.$$

The magnetic force between two wires is used to define the ampere:

> If the same current is maintained in two long parallel wires that attract each other with a force of 2×10^{-7} N per meter of wire when the wires are 1 meter apart, then the current in each wire is 1 ampere.

The net force on a current loop in a uniform magnetic field is zero. The torque on a current loop is:

$$\vec{\tau} = \vec{\mathbf{m}} \times \vec{\mathbf{B}},$$

where $\vec{\mathbf{m}}$ is the magnetic moment of the loop. The force and torque on a wire in a nonuniform field must be found using integration.

A conductor carrying current in a magnetic field develops an electric field perpendicular to both the field and the current. The sign of the resulting Hall potential shows that current in many metals, such as copper, is carried by negative charges. The magnetic force exerted on electrons in the wire is transmitted to the atoms by the Hall field.

When a magnetic field is applied to a medium, it causes two opposing effects: (a) lining up of the individual atomic magnetic moments and (b) induction in each atom of a magnetic moment that opposes the field. These effects give rise to paramagnetic and diamagnetic properties of materials. Ferromagnetism arises from a quantum mechanical interaction between atoms that causes the individual magnetic moments to line up in large regions called domains. * When computing fields in magnetic materials it is useful to define a new field vector $\vec{\mathbf{H}}$, the magnetic field intensity. For linear materials, $\vec{\mathbf{H}}$ is related to $\vec{\mathbf{B}}$ by the magnetic permeability μ:

$$\vec{\mathbf{B}} = \mu\vec{\mathbf{H}} = \mu_0(1 + \chi_m)\vec{\mathbf{H}}.$$

Practical Applications

The spiral motion of particles in magnetic fields is used to advantage in the design of cyclotrons and synchrotrons for accelerating particles to high energy, in mass spectrometers, and in magnetic mirror machines for confining high-temperature plasma. Particles produced in high-energy physics experiments can be identified by the tracks they leave in bubble chambers. A magnetic field causes the tracks to curve (Figure 29.31). Magnetic force is also used to control the electron beam in a TV tube. The magnetic torque on a current loop is used to detect current in some electrical meters. Magnetic forces on current-carrying wires are the basis of electrical motors and moving-coil loudspeakers. Magnetic torques on current loops are used to control the orientation of spacecraft. Nuclear magnetic resonance is used in medical imaging. The direction of magnetization in ferromagnetic materials is used to store information on tapes used in tape recorders, VCRs, and credit cards.

Solutions to Exercises

29.1. The only difference between an electron and a positron is the sign of the charge. The force on the positron is equal in magnitude but opposite in direction. The positron orbits in the opposite sense. ■ Figure 29.31 shows tracks of an electron and a positron in a bubble chamber.

29.2. The proton has a positive charge e and mass 1837 times that of the electron. The cyclotron frequency is thus $\frac{1}{1837}$ that of the electron, or $(2.6 \times 10^4 \text{ rad/s})/1837 = 14$ rad/s. The Larmor radius is proportional to $\sqrt{m}$ and so is $\sqrt{1837} = 43$ times larger than that of the electron and equals 43(50 m), or 2 km.

■ FIGURE 29.31
Tracks of an electron and positron in a bubble chamber. The particles are created when a high-energy γ ray scatters off an atomic electron. The direction of the curvature is determined by the sign of each particle's charge. As they lose energy, the particles follow a spiral of decreasing Larmor radius.

29.3. The mass of an oxygen ion is 16 u. Each singly ionized atom has charge $+e$. The required distance is twice the Larmor radius, so:

$$D = 2r_L = 2\frac{mv}{eB} = 2\frac{(2m\,\Delta V/e)^{1/2}}{B}.$$

For oxygen ions:

$$D_O = 2\frac{[2(16)(1.66\times10^{-27}\text{ kg})(1.0\times10^4\text{ V})/(1.6\times10^{-19}\text{ C})]^{1/2}}{0.20\text{ T}}$$

$$= 0.58\text{ m}.$$

For a nitrogen atom (mass 14 u):

$$D_N = \sqrt{M_N/M_O}\,D_O = \sqrt{14/16}\,(0.58\text{ m}) = 0.54\text{ m}.$$

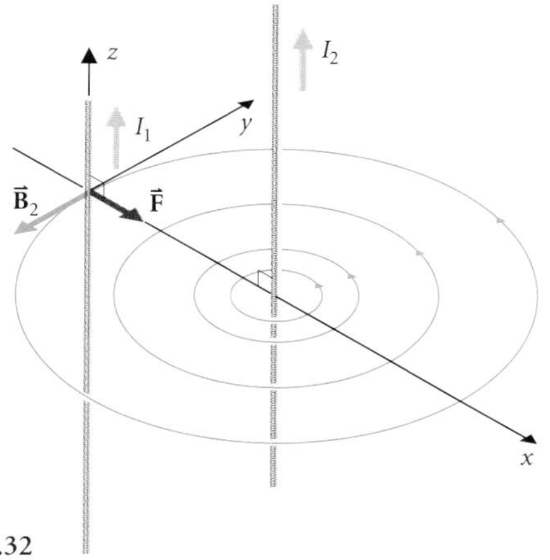

■ FIGURE 29.32

29.4. The Lorentz force on the particle is:

$$\vec{F} = q(\vec{E} + \vec{v}\times\vec{B}) = q(\vec{E} + \vec{v}_o\times\vec{B} + v_y\hat{j}\times\vec{B}).$$

From the definition of $\vec{v}_o$, $\vec{E} + \vec{v}_o\times\vec{B} = 0$, and $\hat{j}\times\vec{B}$ is also zero. The particle's velocity remains constant.

29.5. From eqn. (29.5):

$$\vec{F} = I\,d\vec{\ell}\times\vec{B} = (5.0\text{ A})(1.0\times10^{-3}\text{ m})(0.60\text{ T})(\hat{i}\times\hat{j})$$
$$= 3.0\times10^{-3}\text{ N }\hat{k}.$$

29.6. (■ Figure 29.32) The wire on the right produces magnetic field:

$$\vec{B}_2 = \frac{\mu_0 I_2}{2\pi d}(-\hat{j})$$

at the position of the left-hand wire. So a length $d\ell$ of that wire experiences a force:

$$d\vec{F} = I_1\,d\ell\,\hat{k}\times\frac{\mu_0 I_2}{2\pi d}(-\hat{j}) = \frac{\mu_0 I_1 I_2}{2\pi d}\,d\ell\,\hat{i},$$

which has the same magnitude as but direction opposite the force per unit length on the right-hand wire.

29.7. To produce the same current, the positive charges would have to drift in the opposite direction. With both charge and velocity changing sign, the magnetic force would be in the same direction as with negative charge carriers. Thus positive charge carriers would drift to the same edge as negative carriers do and give rise to Hall electric field in the opposite direction. The sign of the potential difference would be reversed. Certain materials (e.g., beryllium) do have positive charge carriers.

29.8. If $\vec{B}$ is antiparallel to $\vec{m}$, then the magnetic force is radially inward, increasing the centripetal acceleration. The speed increases and so does $|\vec{m}|$.

Basic Skills

Review Questions

§29.1 MOTION OF CHARGED PARTICLES IN A MAGNETIC FIELD

- When a charged particle moves perpendicular to a uniform magnetic field, what is the shape of its path?
- Give a formula for the *Larmor radius.* If a particle's speed doubles, how does the Larmor radius change?
- What is the *cyclotron frequency?* How does it depend on $\vec{B}$? Which properties of the particle determine ω_c?
- How is a cyclotron able to accelerate particles to high energy with a small potential difference?

- If $\vec{\mathbf{E}}$ is perpendicular to $\vec{\mathbf{B}}$, particles with a certain velocity feel no force. What is that velocity? On which properties of the particle does it depend?
- Describe the motion of a particle moving in the plane perpendicular to $\vec{\mathbf{B}}$ when there is a uniform electric field in that plane.

§29.2 FORCES ON CURRENT-CARRYING WIRES

- What force is exerted on a wire segment $d\vec{\ell}$ carrying current I in a region of space with a magnetic field $\vec{\mathbf{B}}$?
- Describe the force on a finite wire segment carrying current I in a region where $\vec{\mathbf{B}}$ is uniform. Explain why the line of action of the net force acts at the center of the segment.
- What is the official SI definition of the ampere?
- What is the net force acting on a closed current loop in a uniform magnetic field?
- What is the net torque on a closed current loop in a uniform magnetic field?

§29.3 THE HALL EFFECT

- Describe the Hall effect. Explain how it allows us to determine the sign of the moving charges in a conductor.

§29.4 MAGNETIC MATERIALS

- Describe the differences between *diamagnetic* and *paramagnetic* materials.
- What is *ferromagnetism?* Name at least one ferromagnetic material.
- What is the Curie temperature?
- ✲ What are the magnetic field intensity $\vec{\mathbf{H}}$ and the magnetization $\vec{\mathbf{M}}$? How are they related?
- What is *hysteresis?*

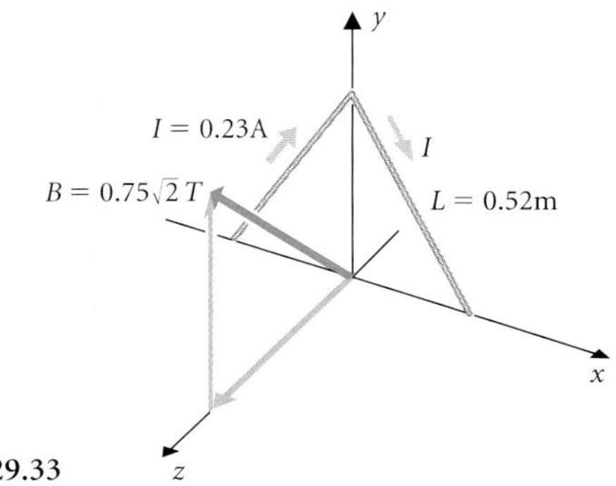

■ FIGURE 29.33

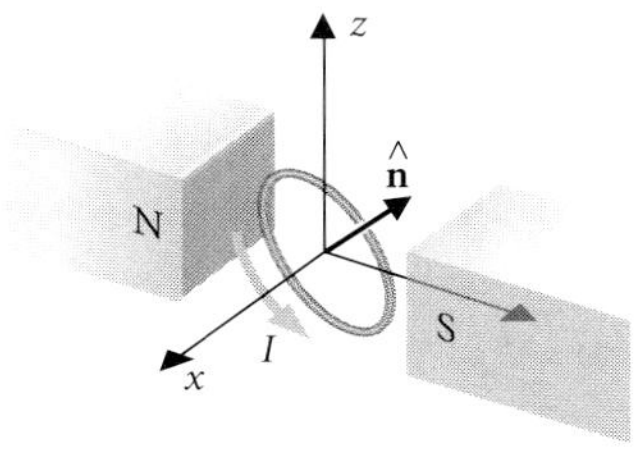

■ FIGURE 29.34

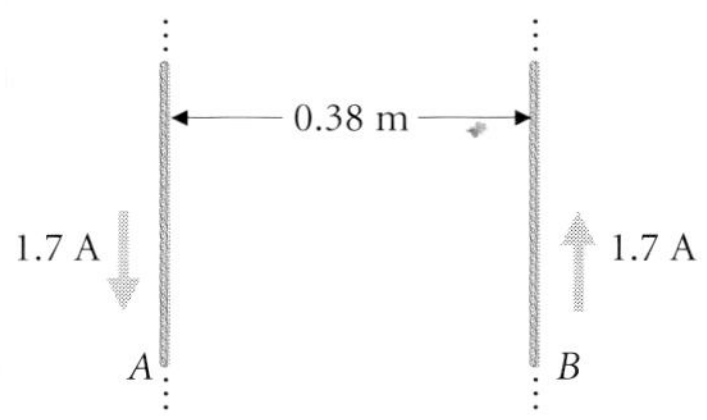

■ FIGURE 29.35

Basic Skill Drill

§29.1 MOTION OF CHARGED PARTICLES IN A MAGNETIC FIELD

1. The magnetic field strength in interstellar space is about 0.3 nT. What are the Larmor radius and cyclotron frequency of an interstellar electron with an energy of 10^{-19} J?

2. Find the Larmor radius and cyclotron frequency for a proton in the solar corona with $v = 1.0 \times 10^5$ m/s and $B = 0.10$ T.

3. The original cyclotron constructed by Lawrence in 1931 was 4 in. in diameter and accelerated H_2 ions to 80 keV. What was the value of the magnetic field?

§29.2 FORCES ON CURRENT-CARRYING WIRES

4. A wire segment of length $d\ell = 1.0 \times 10^{-3}$ m lies along the y-axis and carries current $I = 0.67$ A toward the positive y-direction. The magnetic field at the location of the segment is $\vec{\mathbf{B}} = (1.2\ \text{T})(\hat{\mathbf{i}} - \hat{\mathbf{k}})$. What magnetic force acts on the segment?

5. A segment of wire carrying current $I = 0.23$ A lies along two sides of an equilateral triangle in the x-y-plane (■ Figure 29.33). The third side of the triangle lies along the x-axis. The magnetic field has the uniform value $\vec{\mathbf{B}} = (0.75\ \text{T})(\hat{\mathbf{j}} + \hat{\mathbf{k}})$. What is the net magnetic force acting on the segment?

6. ■ Figure 29.34 shows a current loop that lies between two magnetic pole pieces. The normal $\hat{\mathbf{n}}$ to the loop lies in the y-z-plane. Which of the following statements are correct?
(a) The loop does not move, since the net force acting on it is zero.
(b) The loop rotates so that $\hat{\mathbf{n}}$ points toward the south pole S.
(c) The loop rotates so that $\hat{\mathbf{n}}$ points toward the north pole N.
(d) The loop rotates so that $\hat{\mathbf{n}}$ lies along the x-axis.
(e) The loop rotates clockwise about the x-axis and accelerates along the y-axis.
Explain what is wrong with the answers you didn't choose.

7. Two long straight wires carry a current $I = 1.7$ A each, in opposite directions (■ Figure 29.35). If the separation of the wires is 0.38 m, what is the force per unit length on wire B due to wire A?

8. A compass needle has dipole moment $m = 0.1\ \text{A}\cdot\text{m}^2$. What torque acts on the needle if it is oriented east-west on the Earth's surface at the equator, where the Earth's magnetic field is $B = 3 \times 10^{-5}$ T?

§29.3 THE HALL EFFECT

9. Find the Hall potential across a copper wire 1.0 mm in diameter carrying a 2.5-mA current perpendicular to a magnetic field $B = 1.3$ T.

§29.4 MAGNETIC MATERIALS

10. The magnetic susceptibility of a material is:
(a) necessarily positive.
(b) always small and negative.
(c) always proportional to temperature.
(d) generally $\gg 1$.
(e) none of the above.
Explain what is wrong with the answers you didn't choose.

11. ✲ The magnetic field intensity H in a platinum bar is 1.00000×10^2 A/m. What are the magnetization M and the induction B?

Questions and Problems

§29.1 MOTION OF CHARGED PARTICLES IN A MAGNETIC FIELD

12. ❖ A negatively charged electron moves in vacuum parallel to a long straight wire carrying current as shown in ■ Figure 29.36. The electron is deflected:
(a) left. **(b)** up. **(c)** right. **(d)** down. **(e)** not at all.
Explain why you chose your answer.

13. ❖ An electron is injected with speed v into a region where the magnetic field points into the page (■ Figure 29.37). Describe its subsequent motion.

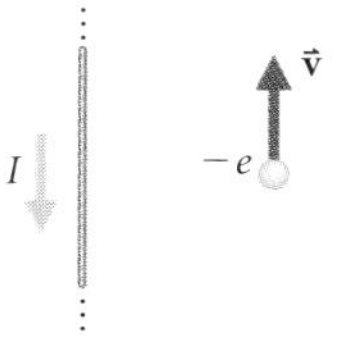

■ FIGURE 29.36

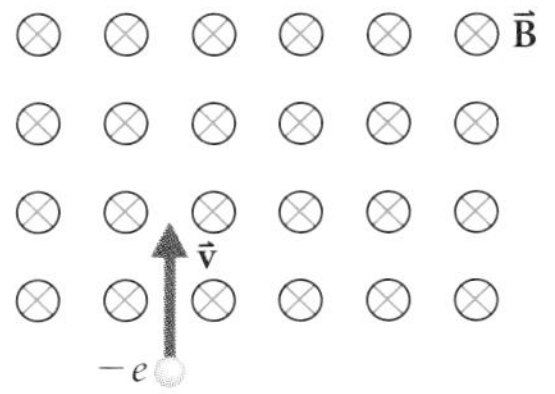

■ FIGURE 29.37

14. ❖ ■ Figure 29.38 shows several possible paths of particles in a uniform magnetic field. Electrons and protons with equal energy enter the system from the right. Which pair of paths represent their motion?
(a) electrons (2) protons (4) **(b)** electrons (4) protons (2)
(c) electrons (1) protons (3) **(d)** electrons (3) protons (1)

15. ❖ Can a cyclotron accelerate particles with either sign of charge? If so, what changes are necessary?

16. ❖ A long straight filament carries a uniform charge density λ, and an electron is moving in a circular orbit about the filament. What happens to the electron's orbit if a current I is established in the filament?

17. ❖ Low-energy cosmic rays reach the Earth's surface in much greater numbers in northern latitudes than near the equator. Assuming these charged particles approach the Earth uniformly from all directions, explain the observed distribution at the Earth's surface.

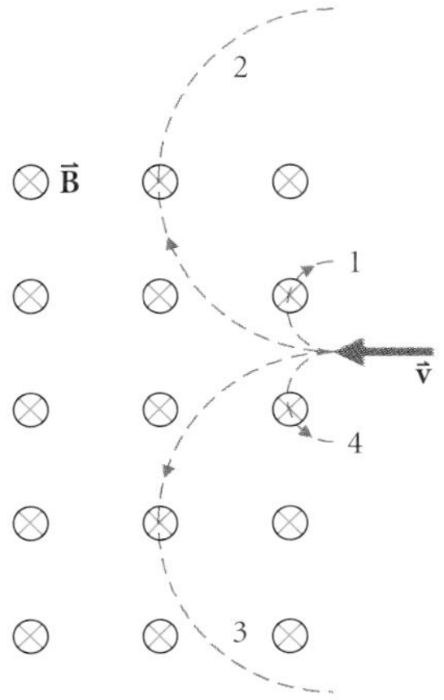

■ FIGURE 29.38
Relative sizes of small and large circles are not to scale.

18. ❖ If an electric field exists parallel to $\vec{B}$, describe the motion of a charged particle with initial velocity **(a)** parallel to $\vec{B}$ and **(b)** perpendicular to $\vec{B}$.

19. ♦ **(a)** Some playful thinkers have imagined the ultimate particle accelerator: a ring around the Earth's equator, using the Earth's magnetic field to hold the particles in orbit. The Earth's field at the equator $B_E \approx 3 \times 10^{-5}$ T. What is the momentum of a proton circling in such an accelerator? Express your results in the physicists' unit GeV/c. **(b)** What proton momentum could such an accelerator achieve if 20-T superconducting magnets were used instead of the Earth's field?

20. ♦ Radiation from the magnetron in your microwave oven is emitted by electrons orbiting in a magnetic field. If the radiation is emitted at the cyclotron frequency f_c, what field strength is necessary to produce microwaves of frequency 3 GHz?

21. ♦♦ Find the Larmor radius and the cyclotron frequency f_c for a proton and an electron in the outer Van Allen belt at 3.4 Earth radii above the equator. Assume each particle has a speed of 3.0×10^5 m/s perpendicular to the magnetic field, and the Earth's magnetic field is a dipole with $B = 3.12 \times 10^{-5}$ T at the equator.

22. ♦♦ A 10-in. diameter cyclotron was built by Lawrence and Livingstone in 1932. It produced 1-MeV protons. What magnetic field was necessary?

23. ♦♦ With magnets that produce a field of 0.13 T, what radius cyclotron would accelerate a beam of protons to an energy of 350 keV? At what frequency should the potential difference across the dees be changed?

24. ♦♦ The synchrotron at Fermi National Laboratory produces a proton beam with a kinetic energy of 10^{12} eV per proton. The radius of the accelerator is 1 km. What magnetic field strength is required? (At this energy, a proton moves with very nearly the speed of light, and its effective mass γm is 1.8×10^{-24} kg.)

25. ♦♦ A mass spectrometer is designed to operate with a magnetic field of 1.3 T and a fixed distance of 0.30 m between the beam entrance and the detector. What potential difference is necessary for the device to detect singly ionized benzene molecules? (The mass of a benzene molecule is 78 u.)

26. ♦♦ A beam of carbon nuclei emerges from an accelerator with kinetic energy $K = 10$ GeV. A "bending magnet" is used to aim the beam toward an experimental chamber. If the beam is to be bent in a circle of 1.0-m radius, what magnetic field is required? Use the relativistic formula for the Larmor radius with $\gamma = 1 + K/mc^2$ (see *Digging Deeper:* More on Cyclotrons), and assume the nuclei have speed $v \approx c$.

27. ♦♦ Deflection of the electron beam in a modern TV set is accomplished by magnetic forces. The electrons are accelerated until they have kinetic energy of 25 keV (cf. Example 24.8). What magnetic field is required to produce a deflection of 3.3 cm if the magnetic field region is 8.0 cm long? (Newtonian mechanics is adequate for this problem. Relativistic effects make about a 3% correction.) If

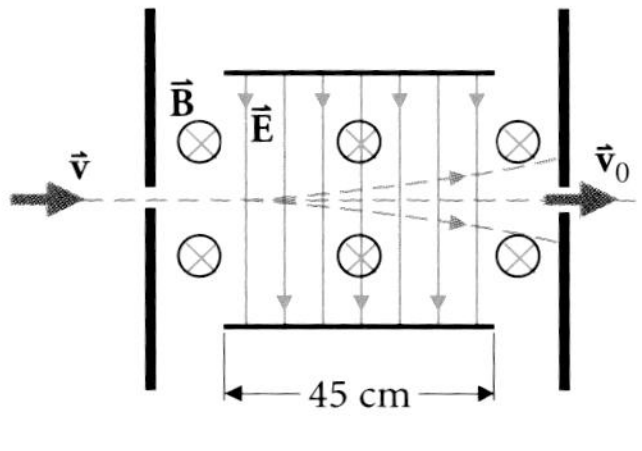

■ FIGURE 29.39

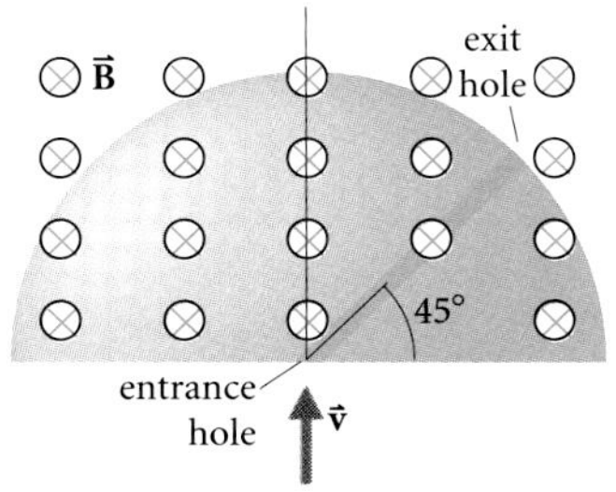

■ FIGURE 29.40

there is a 22-cm gap between the end of the magnetic field region and the end of the tube, how far from the center of the screen does the beam strike?

28. ♦♦ **(a)** A velocity selector for electrons (■ Figure 29.39) has electric and magnetic fields at right angles. If the magnitudes of the fields are $E = 1.5$ kV/m and $B = 0.75$ T, at what speed v_o can particles pass through the selector? **(b)** If a particle's speed is $v_o + \delta v$ and its charge-to-mass ratio is $q/m = \alpha$, what are its acceleration and its displacement perpendicular to its velocity after a time interval Δt? **(c)** If the length of the field region is 45 cm, the exit aperture has radius $r = 0.50$ cm, and all particles enter exactly perpendicular to both $\vec{E}$ and $\vec{B}$, what range of initial speeds exit the selector?

29. ♦♦ A velocity selector for electrons is an evacuated hemisphere within which the magnetic field is parallel to the flat base. Electrons are admitted through a hole at the center of the base and exit through a second hole halfway to the top of the hemisphere (■ Figure 29.40). Find the speed of electrons that exit the hole if the radius of the hemisphere is 0.15 m and $B = 5.4 \times 10^{-4}$ T.

30. ♦♦ Find the separation between the last two orbits in the MIT cyclotron that accelerates deuterons ($m = 2.0$ u, $q = +e$) to an energy of 16 MeV. The potential difference between the dees is 80.0 kV, and the radius of each dee is 18.74 in.

§29.2 FORCES ON CURRENT-CARRYING WIRES

31. ❖ A length of wire carrying current I lies between two magnet pole pieces (■ Figure 29.41). What is the direction of the force on the wire?

32. ❖ A long straight wire runs along the axis of a circular wire loop. If both wires carry current I, what is the net force acting **(a)** on the straight wire and **(b)** on the loop? Justify your answer.

33. ❖ A long straight wire lies along the z-axis and carries current I_1. A small element of wire $d\vec{\ell}$ carrying current I_2 is placed at P, distance d from the long wire, with $d\vec{\ell}$ lying in the x-z-plane. What orientation of the wire segment maximizes the y-component of the force acting on the segment? Explain your reasoning.

34. ❖ A loop carrying a current I has a magnet of dipole moment m at its center (■ Figure 29.42). Both loop and dipole lie in the x-y-plane, and the magnet is fixed. Which of the following statements is correct?
(a) The loop is in equilibrium.
(b) The loop rotates around line BD with C moving out of the page.
(c) The loop rotates around line BD with A moving out of the page.
(d) The loop rotates around line AC with B moving out of the page.
(e) The loop rotates around line AC with D moving out of the page.
Explain how you chose your answer.

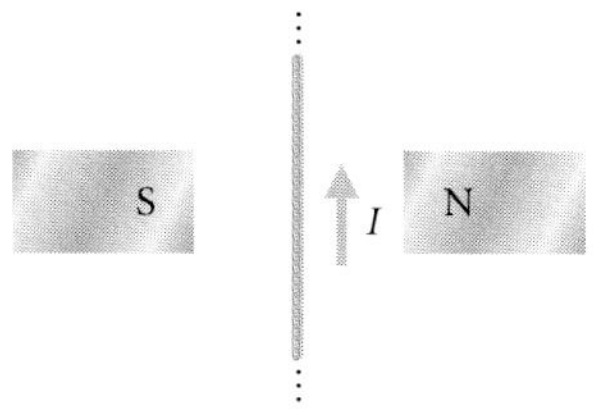

■ FIGURE 29.41

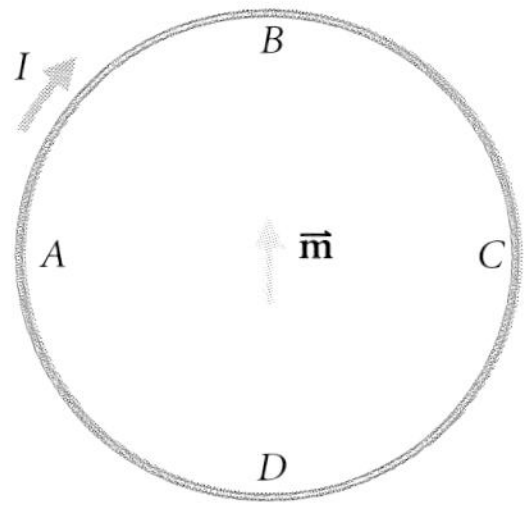

■ FIGURE 29.42

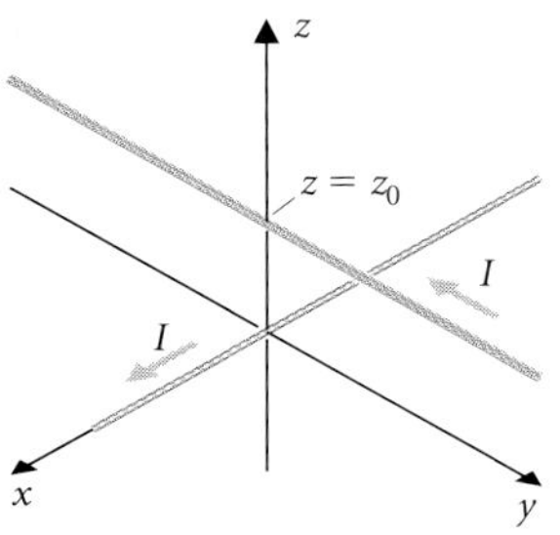

■ FIGURE 29.43

35. ❖ Two long straight wires carry currents of equal magnitude I in the directions shown in ■ Figure 29.43. One lies along the x-axis, and the other is parallel to the y-axis and passes through the point $x = 0$, $y = 0$, $z = z_0$. What is the net force on the second wire due to the first?

36. ❖ Two coils A and B both lie with their axes along the y-axis, and each carries a current I (■ Figure 29.44). What is the direction of the net force on coil B?

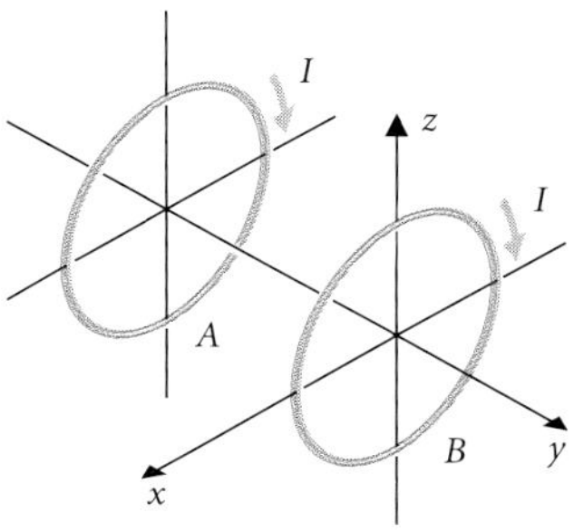

■ FIGURE 29.44

37. ❖ A small current loop lies in the x-y-plane centered at the origin. A long straight wire is parallel to the x-axis and passes through the point $(0, 0, z_0)$. Currents are in the directions shown in ■ Figure 29.45. The loop experiences:
(a) neither a net force nor a net torque.
(b) no torque but a net force in the $+z$-direction.
(c) no net force but a net torque in the $-y$-direction.
(d) no net force but a net torque in the $+x$-direction.
(e) a net force in the $-y$-direction and a net torque in the $+x$-direction.
Explain how you chose your answer.

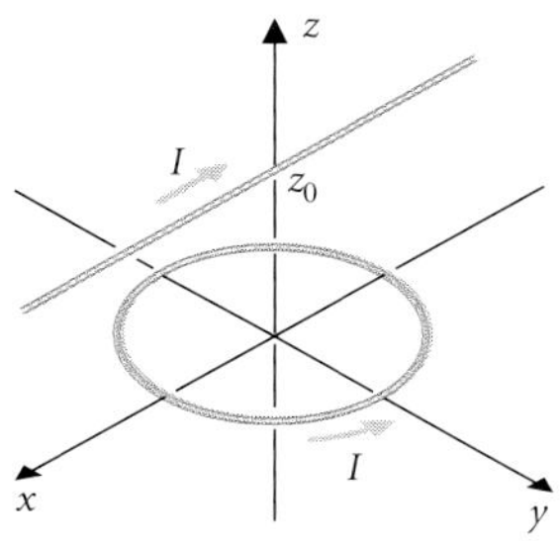

■ FIGURE 29.45

38. ❖ Two concentric wire loops lying in the same plane carry currents in the same direction. Sketch the magnetic field lines produced by the outer loop. What net force and net torque are exerted on the inner loop? Repeat the analysis when the inner loop has been rotated through 90° so that its plane is perpendicular to the plane of the outer loop. Discuss the stability of the first configuration. Is the system stable when the two loops lie in the same plane but have currents in opposite directions?

39. ❖ Two long wires carry opposite currents I parallel to the x-axis at $z = 0$, $y = \pm d$. Describe qualitatively how the net force and net torque on a small square current loop vary as it is moved along the z-axis, along the x-axis, and around a circle centered on the wire at $z = d$. Assume the loop is always oriented parallel to the x-y-plane.

40. ♦ A wire segment 1.2 cm long lies along the x-axis and carries a current of 4.3 mA. The magnetic field is uniform and equals $(0.045\ \text{T})(\hat{\mathbf{i}} + \hat{\mathbf{j}})$. Calculate the force on the wire segment.

41. ♦ A rectangular coil 6.0 cm by 12.0 cm has six turns and carries a 1.1-A current. It lies between the poles of a permanent magnet with its plane parallel to the 0.25-T magnetic field. What is the torque on the loop?

42. (a) ♦ A small current loop of area $A = 1.5\ \text{cm}^2$ carries current $I = 3.6$ mA. What torque acts on it in a lab where its magnetic moment is perpendicular to the Earth's magnetic field of 4.2×10^{-5} T? **(b)** ♦♦ Find its angular acceleration if its rotational inertia about a diameter is $6.2\ \text{g}\cdot\text{cm}^2$.

43. ♦♦ Two parallel, semi-infinite wires are connected by a semicircular wire of radius R (■ Figure 29.46). The magnetic field is uniform, parallel to the straight wires, and perpendicular to the plane of the semicircle. Find the total magnetic force on the object.

44. ♦♦ Find the force and torque on the rectangular loop shown in ■ Figure 29.47.

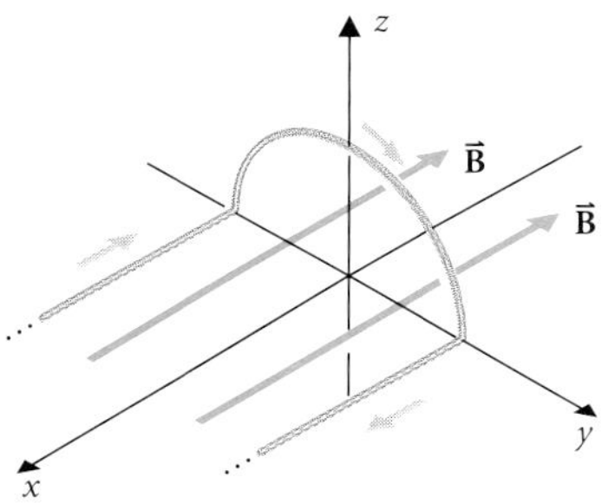

■ FIGURE 29.46

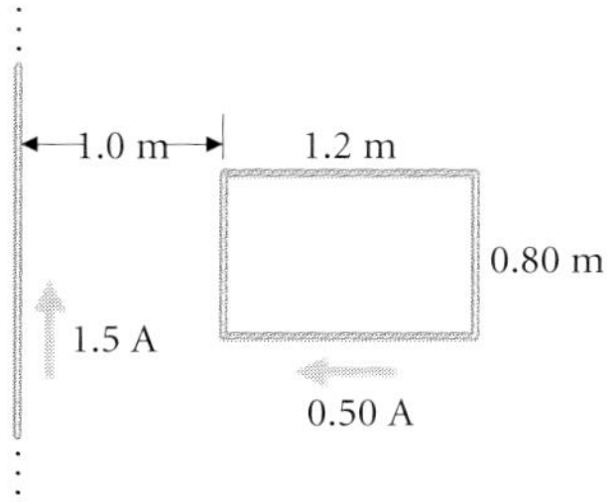

■ FIGURE 29.47

45. ♦♦ A wire loop is in the form of an equilateral triangle. The magnetic field lies in the plane of the triangle, perpendicular to one side. Find the force acting on each side of the triangle, and verify that the net force is zero. Find the torque about the center of the loop, and verify that it equals $\vec{\mathbf{m}} \times \vec{\mathbf{B}}$.

46. ♦♦ A current balance is constructed to measure the force between a solenoid with $n = 1.0 \times 10^4$ turns per meter and a straight

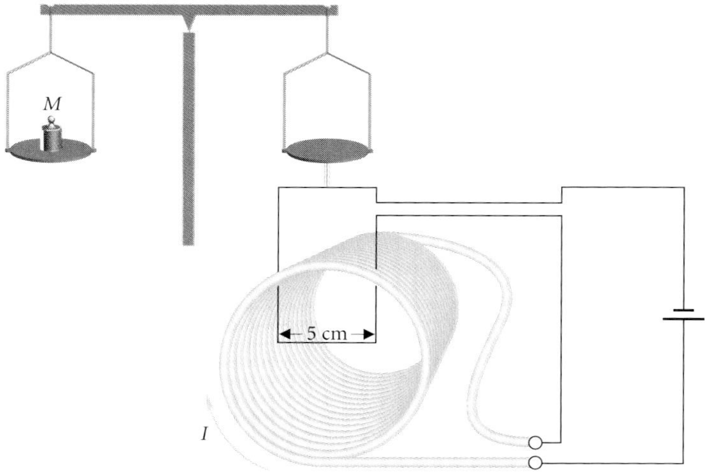

■ FIGURE 29.48

wire segment of length $\ell = 5.0$ cm (■ Figure 29.48). What mass M is necessary for equilibrium when the current in the system is 10.0 A?

47. ♦♦ A galvanometer (cf. §26.5) consists of a square coil of side 2.0 cm with 1.0×10^3 turns lying in the magnetic field produced by a permanent magnet (■ Figure 29.49). The coil experiences a torque when it carries current. Why? Rotation of the coil is opposed by a spring that exerts a torque $\tau = \Gamma\theta$ when the coil is deflected through angle θ (in radians). The coil is attached to a pointer. If the magnetic field at the position of the coil is 0.70 T, find the relation between I and θ. What value of Γ is required if a full-scale deflection of 45° is to correspond to 1.0 mA?

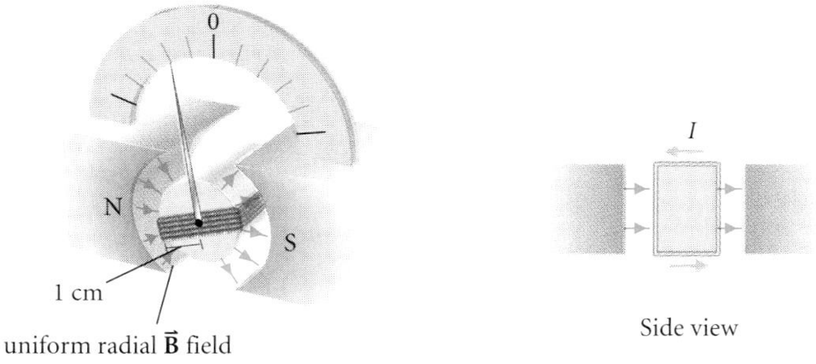

■ FIGURE 29.49

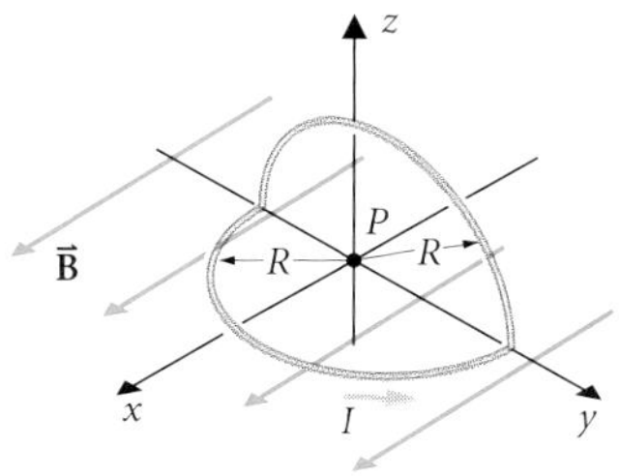

■ FIGURE 29.50

48. ♦♦♦ The wire loop shown in ■ Figure 29.50 carries a current I. If the magnetic field is uniform, $\vec{\mathbf{B}} = B_0\hat{\mathbf{i}}$, compute the total force and torque about the origin acting on the loop.

§29.3 THE HALL EFFECT

49. ❖ A conducting strip has length ℓ and width w perpendicular to a magnetic field B, and thickness d parallel to the field. The material

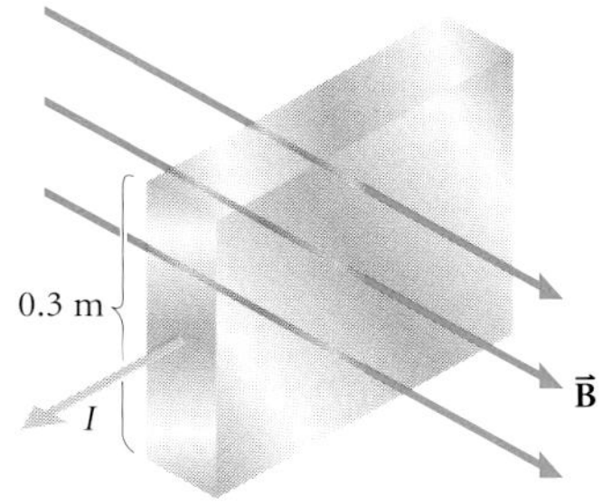

■ FIGURE 29.51

has resistivity ρ and a potential difference V is maintained between the ends of the strip. How does the Hall potential difference depend on each of these parameters?

50. ♦ There is a current along a silver bus bar 0.3 m wide (■ Figure 29.51), immersed in a magnetic field $B = 1$ T. If the drift speed of the electrons required to carry the current is 10^{-5} m/s, compute the potential difference between the top and the bottom of the bus bar required to counteract the magnetic force on conduction electrons. Is the top at positive or negative potential with respect to the bottom?

51. ♦♦ The Hall potential measured across a copper wire 1.17 cm in diameter carrying a 175-A current is 1.05 μV. What is the component of B perpendicular to the current? (Copper has 8.5×10^{28} /m^3 conduction electrons.)

52. ♦♦ Find the current in a copper strip 1.2 cm $\times$ 0.30 cm with a magnetic field of 0.96 T perpendicular to the 1.2-cm side if the Hall potential difference is 0.74 μV. (Density of conduction electrons is 8.5×10^{28} /m^3.)

53. ♦♦ The electric *third rail* in a subway system is made of steel, measures 10.0 cm high by 2.0 cm wide, and carries a 150-A current. Calculate the Hall potential difference across the rail due to the Earth's field with components $B_{\text{horiz}} = 3.0 \times 10^{-5}$ T and $B_{\text{vert}} = 2.7 \times 10^{-5}$ T. (Assume steel has 8×10^{28} /m^3 conduction electrons.)

54. ♦♦♦ An aluminum strip carries a current $I = 30$ A in a magnetic field $B = 1.2$ T. The strip has a rectangular cross section with dimensions 1.0 mm parallel to the magnetic field and 3.00 cm perpendicular to the field. A Hall potential $V_H = 1.2\ \mu$V is observed across the 3-cm dimension. How many electrons per aluminum atom are free to carry current? (data for Al: density = 2700 kg/m^3; mass of Al atom = 27 u)

55. **(a)** ♦♦ To extract energy from the system of Example 29.11, the Hall potential is allowed to drive current I' through each of $N = 100$ shunt resistors (■ Figure 29.52). If each shunt resistance is $r = 1.0$ kΩ, find the power extracted ($L = 1.0$ cm, $w = 1.0$ mm, $B = 1.0$ T, $I = 5.0$ A). **(b)** ♦♦♦ Show that the power output of the battery equals the power dissipated in the bar plus the power dissipated in the shunts.

§29.4 MAGNETIC MATERIALS

56. ❖ What would happen to an iron bar magnet if it were heated to 800° C?

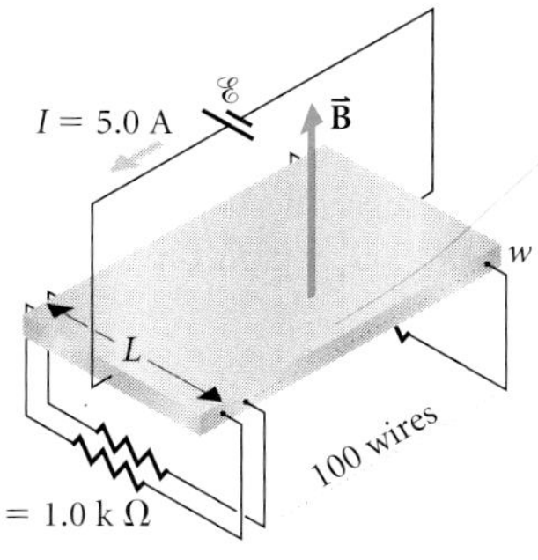

■ FIGURE 29.52 $r = 1.0$ kΩ

57. ❖ A sphere of magnetic material is centered on the axis of a current loop (but not at its center). Model the current loop and the sphere as bar magnets and explain why the loop attracts a paramagnetic sphere and repels a diamagnetic sphere.

58. ♦ The magnetic moment of a finite, tightly wound solenoid may be expressed as (cf. Problem 28.34) $m \approx BV/\mu_0$, where B is the magnetic field and V is the volume inside the solenoid. A cylindrical bar magnet is 10 cm long and has a 5-cm radius. The magnetic induction within the material is $B = 1.3$ T. Use the preceding relation to estimate its magnetic moment.

59. ✲ A magnetic dipole of moment $\vec{\mathbf{m}}$ is spinning about its axis with angular momentum $\vec{\mathbf{L}}$. If the dipole is at an angle θ to a uniform magnetic field $\vec{\mathbf{B}}$, find the precession frequency in terms of L/m. (*Hint:* Refer to Example 12.17.)

Additional Problems

60. ❖ A sphere with both mass and charge distributed uniformly rotates with angular velocity ω. Show that the angular momentum $\vec{\mathbf{L}}$ and the magnetic moment $\vec{\mathbf{m}}$ of the sphere are both parallel to ω.

61. ❖ **(a)** One could design a cyclotron to correct for relativistic variation of the cyclotron frequency by allowing the magnetic field to increase with radius. Explain why. **(b)** This design has an inherent flaw. In a cylindrically symmetric field, the field lines bend toward regions of lesser field strength (■ Figure 29.53). Show that this makes the beam unstable against any deviation from the central plane of the machine.

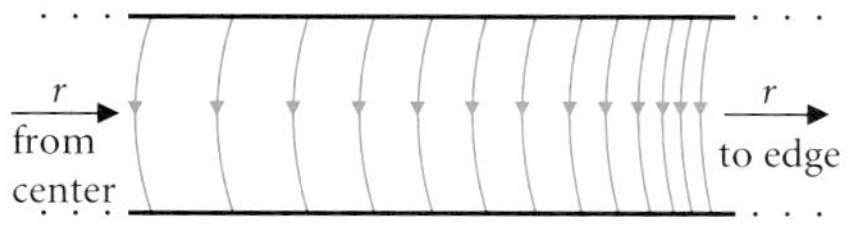

■ FIGURE 29.53

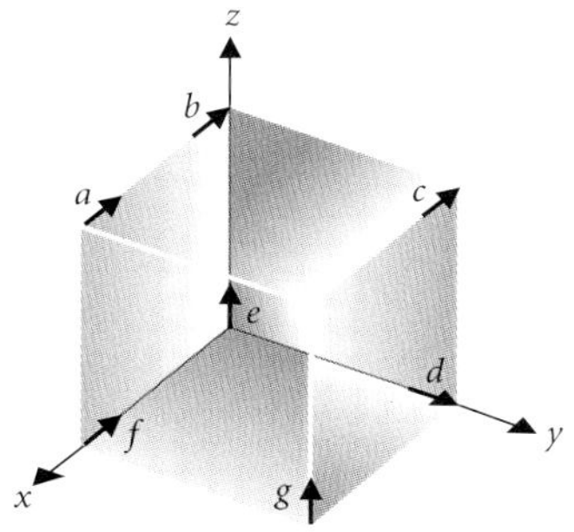

■ FIGURE 29.54

62. ❖ Refer to ■ Figure 29.54. Which pairs of wire elements, if any, exert forces that obey Newton's third law applied to the elements alone?

63. ❖ A long straight wire carries current along the axis of a long, loosely wound solenoid. Is there a net force on either object? What is the effect of the wire on the solenoid?

64. ❖ Two identical coils, each carrying current I, are suspended near each other by threads. When the two coils are in equilibrium, are their currents in the same or in opposite senses? Discuss.

65. ♦ Electrons in Jupiter's magnetosphere radiate cyclotron radiation in the dekameter band. Assuming that the electrons radiate at the cyclotron frequency, estimate Jupiter's magnetic field.

66. ♦ An electron with speed $v = 3 \times 10^5$ m/s is moving parallel to the current in a long straight wire 15 cm away. If the wire carries $I =$ 1.3 mA, what is the acceleration of the electron?

67. ♦ A whimsical engineer has designed a train to use the vertical component of the Earth's magnetic field ($\approx 10^{-5}$ T) for propulsion. What current through the 3-m-long axle is necessary to provide a force of 10^5 N? If the axle has a 1-Ω resistance, what power is necessary? Comment on your result.

68. ♦♦ A small magnetic dipole, $m = 1.0 \times 10^{-4}$ A·m², lies in the x-y-plane at the origin. A current loop of radius $a = 12$ cm carrying a 1.0-mA current also lies in the x-y-plane with its center at the origin. How large is the torque acting on the magnetic dipole?

69. ♦♦ A particle moves in a spiral whose axis is a long straight wire carrying current I. The particle's velocity component parallel to the wire $v_{\parallel}$ is in the direction of I. What is the sign of the particle's charge Q? What relation must hold between Q, I, $v_{\parallel}$, $v_{\perp}$, and r?

70. ♦♦ What is the Larmor radius of a cosmic ray proton with energy $E = 10^9$ GeV and momentum $p = E/c$ in the intergalactic magnetic field of 0.1 μG? If the size of the galaxy is about 3×10^{20} m, what is the maximum energy proton the galaxy can retain?

71. ♦♦ A tangent galvanometer used to measure the strength of the Earth's magnetic field consists of a circular coil with N turns and a small compass needle suspended at its center (■ Figure 29.55). The device is set up with the coil in a north-south plane, so that the compass needle is in the plane of the coil. With a current I in the coil, the needle is deflected by an angle θ. Find the relation between θ, I, the coil's radius a, and the horizontal component of the Earth's field.

72. ♦♦ A charged particle moving in a circle constitutes a current loop. Find the magnetic moment m of the loop in terms of the radius of the circle and the angular frequency. If the circular motion is caused by magnetic force, the appropriate radius is the Larmor radius. Express the magnetic moment in terms of B. When a particle moves in a field of variable strength, its magnetic moment remains constant. (We'll see why in Chapter 30.) This is is another way to understand the magnetic mirror effect. The particle's kinetic energy K also remains constant. (Why?) Express the energy in terms of the components of velocity perpendicular ($v_{\perp}$) and parallel ($v_{\parallel}$) to $\vec{B}$, and use the result to express $v_{\parallel}$ in terms of m and K. At what value of B does the mirror reflection occur?

73. ♦♦♦ A nonuniform field in a region of space has $\vec{B} = (10^{-4}\text{ T})\,\hat{\imath}$, $dB/dx = 10^{-5}$ T/m. Use Gauss' law for the magnetic field applied to the volume shown in ■ Figure 29.56 to find the slope of a field line that is 0.2 m from a line that lies along the x-axis. What is the x-component of acceleration for a proton with Larmor radius 0.2 m moving along the central field line in the $+x$-direction?

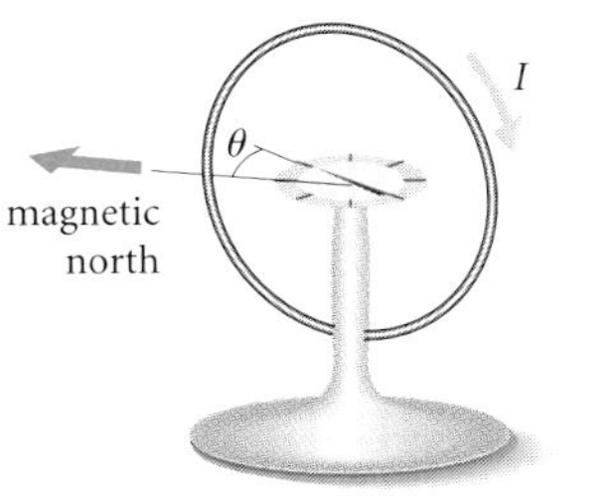

■ FIGURE 29.55

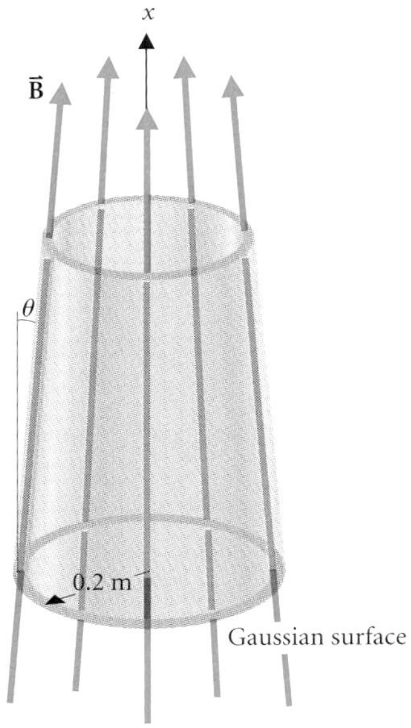

■ FIGURE 29.56

Computer Problem

74. **(a)** Program MAGMOT on your supplementary computer disk computes the motion of a charged particle in a uniform magnetic field. The field is in the z-direction, and the particle starts at $x = -1$, $y = z = 0$, moving in the y-direction. The program asks you to input the initial y-component of v; 1 works well. You also need to input a time step (try 0.01) and how often you want the program to output the particle's position. The program then plots the particle's path in the x-y-plane on the screen. Try changing the initial speed and seeing how the orbit varies. The particle has unit charge and unit mass, and the field strength is also 1 unit. How does the computed orbit radius compare with the expected value? How does the computed period compare with the expected value? **(b)** Program MAGTWO computes the motion of the same particle in a variable magnetic field with components $B_z = 1 + z/5$, $B_x = -(x+1)/10$, and $B_y = -y/10$. The particle starts with $v_z = 0.5$. Choose $v_y = 1$. Where does the particle *mirror?* What happens if you increase v_y? What if you decrease v_y?

Challenge Problems

75. The conclusion that $\vec{E}_H = -\vec{v}_d \times \vec{B}$ is valid in a fixed wire. What correction is necessary to find the Hall electric field in a segment of wire that is accelerated by the magnetic force?

76. **(a)** Find the work required to rotate a magnetic dipole from a direction parallel to the magnetic field to a direction at angle θ to the field. Show that your result can be interpreted in terms of a potential energy of the dipole given by $U = -\vec{m} \cdot \vec{B}$. **(b)** In a nonuniform magnetic field, the potential energy computed in part (a) varies if the dipole is displaced without changing its orientation. This change in energy results from a net force acting on the dipole. Describe how to find the force acting on a dipole in a nonuniform field. Check your idea by repeating Example 29.8 with magnetic field given by $B_0(1 + \alpha x)\hat{k}$. **(c)** Compute the torque exerted by the second loop in Example 29.10 on the first. Note that it is not equal to and opposite the torque exerted by loop 1 on loop 2. **(d)** What couple must act on the two loops if the net torque exerted by the system on itself is zero? **(e)** Use the result of part (b) (force $= \Delta(\vec{m} \cdot \vec{B})$/displacement) to calculate the force acting on each loop, and show that the required couple does act.

77. A coil with $N = 1.0 \times 10^4$ turns and radius $r = 1.0$ cm carries current $I = 10.0$ A. The coil lies in the x-y-plane with its axis along the z-axis. A long straight wire parallel to the x-axis passes through the z-axis at $z = 10.0$ cm. (See also Figure 29.45 and Problem 37). The wire carries current $I' = 5.0$ A toward $-\hat{\imath}$. **(a)** Modeling the coil as a dipole, compute the force and torque it exerts on the wire. **(b)** Compare each with the force and torque the wire exerts on the dipole. (Use the method of Problem 76e.)

78. Use the vector identity

$$\vec{a} \times (\vec{b} \times \vec{c}) = \vec{b}(\vec{a} \cdot \vec{c}) - \vec{c}(\vec{a} \cdot \vec{b})$$

to show that the torque on an arbitrary plane current loop in a uniform field is given by $\vec{\tau} = I \oint (\vec{r} \cdot \vec{B})\, d\vec{\ell}$. (*Hint:* Show that $\oint \vec{r} \cdot d\vec{\ell} \equiv 0$ for a closed loop.) Choose coordinates with the z-axis perpendicular to the loop and the y-axis perpendicular to $\vec{B}$. Let θ be the angle between $\vec{B}$ and the z-axis. Show that the torque is:

$$\vec{\tau} = IB \sin\theta \oint x\, d\vec{\ell}.$$

Show by a geometric argument that:

$$\oint x\, d\vec{\ell} = \hat{\jmath}\,(\text{area of loop}).$$

Thus show that $\vec{\tau} = \vec{m} \times \vec{B}$ for the arbitrary loop.

Part VI Problems

1. ❖ It is possible to increase the magnetic field of a wire loop either **(a)** by increasing the current by a factor N or **(b)** by using N identical loops in a coil. Show that the ratio of power used by the coil to power used by the loop with the larger current is $1/N$. Power consumption is one of the most important considerations in magnet design.

2. ♦♦ A beam of ions is removed from a cyclotron using a deflector (■ Figure VI.25). The deflector is at a high negative potential with respect to the dee so that an outward electric force is applied to the ions. The MIT cyclotron with $R = 18.75$ in. and deflector spacing of 0.3 in. accelerates deuterons (^{2}H ions) to 16 MeV. What potential difference across the deflector is required to increase the orbit radius by 20%?

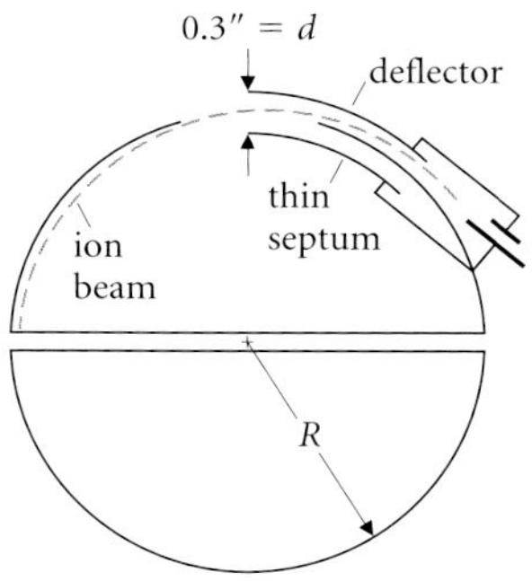

■ FIGURE VI.25

3. ♦♦♦ A crude but effective model of a hydrogen atom consists of a proton attached to an electron by a spring with spring constant $k = 1.6 \times 10^3$ N/m. Find the polarizability of a hydrogen atom, that is, the ratio of its electric dipole moment to the electric field that causes the moment. Estimate the force exerted by a water molecule on a hydrogen atom 10^{-7} m away (cf. Example 24.12 and Exercise 24.6).

4. ♦♦♦ Suppose Kirchhoff's junction rule were not quite obeyed. Specifically, in ■ Figure VI.26 suppose that $I_3 = 0.999(I_1 + I_2)$ and that the remaining 0.1% of the current deposited charge at junction A, 10 cm from each of the two 12-V batteries. How long could this situation persist before the potential at each battery due to the point charge at A exceeded the emf of the battery?

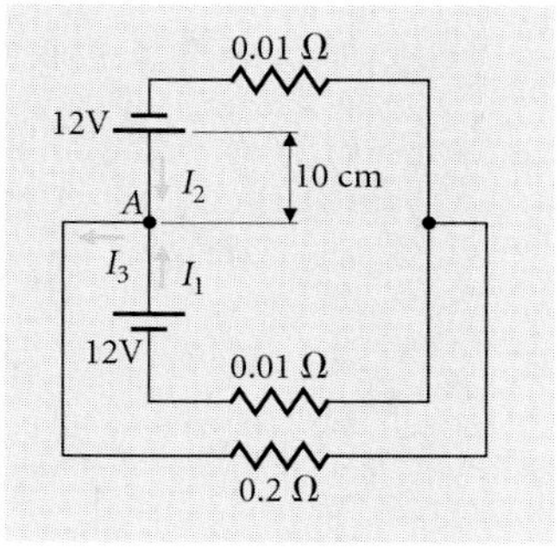

■ FIGURE VI.26

5. ♦♦♦ A parallel plate capacitor has plates of area A and separation d. It is initially connected to a battery whose emf is $\mathcal{E}$ and then disconnected when fully charged. Compute the energy stored in the capacitor. Find the force acting on one plate by computing the change in stored energy as the plate separation is increased by an amount dx. Compute the electric field between the plates. Compare the force on the plate with its charge times the electric field. If the two results are different, explain why. (*Hint:* See §25.6.)

6. (a) ❖ Suppose in Problem 5 that the battery were left connected. Would there be any difference in the forces acting on the capacitor plates? Why or why not? Discuss the significance of the battery in computing the force from a change of potential energy. **(b)** ♦♦♦ Compute the change in stored energy of the system as the plate separation is increased by dx and derive from your result the force acting on one plate.

7. ❖ ■ Figure VI.27 shows a system that we might model as two capacitors in series. Sketch field lines for the distribution of charge shown, and then use the circulation law for static electric field to show that this charge distribution cannot be correct. Show that a consistent field pattern results if the magnitude of charge on the inner plate of the system is slightly less than that on the outer plate. Argue that the capacitance of the series combination of capacitors is slightly greater than the value given by the usual series formula.

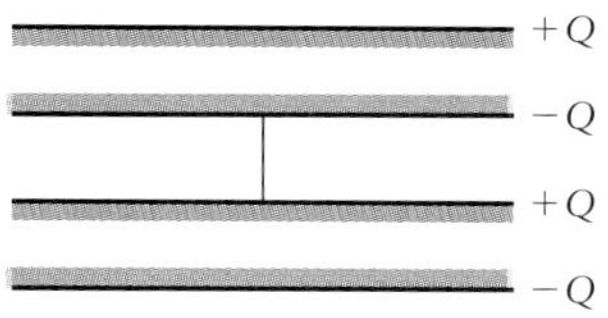

■ FIGURE VI.27

Computer Problem

8. Two charges $+3Q$ and $-Q$ are on the x-axis separated by a distance D with the origin at $+3Q$ (cf. Example 25.11). Compute the potential V_0 at the point on the x-axis at $x = (3 + \sqrt{3})D/2$ where $\vec{\mathbf{E}} = 0$. Find two other points on the x-axis where the potential is V_0. Show that points with potential V_0 on the line through $-Q$ parallel to the y-axis may be found from the equation:

$$-\frac{D}{|y|} + \frac{3}{[1 + (y/D)^2]^{1/2}} = 4 - 2\sqrt{3}.$$

Find the solutions to this equation numerically. Compare your answers with Figure 25.15c, and use them to understand the shape of the equipotential surface through the $E = 0$ point.

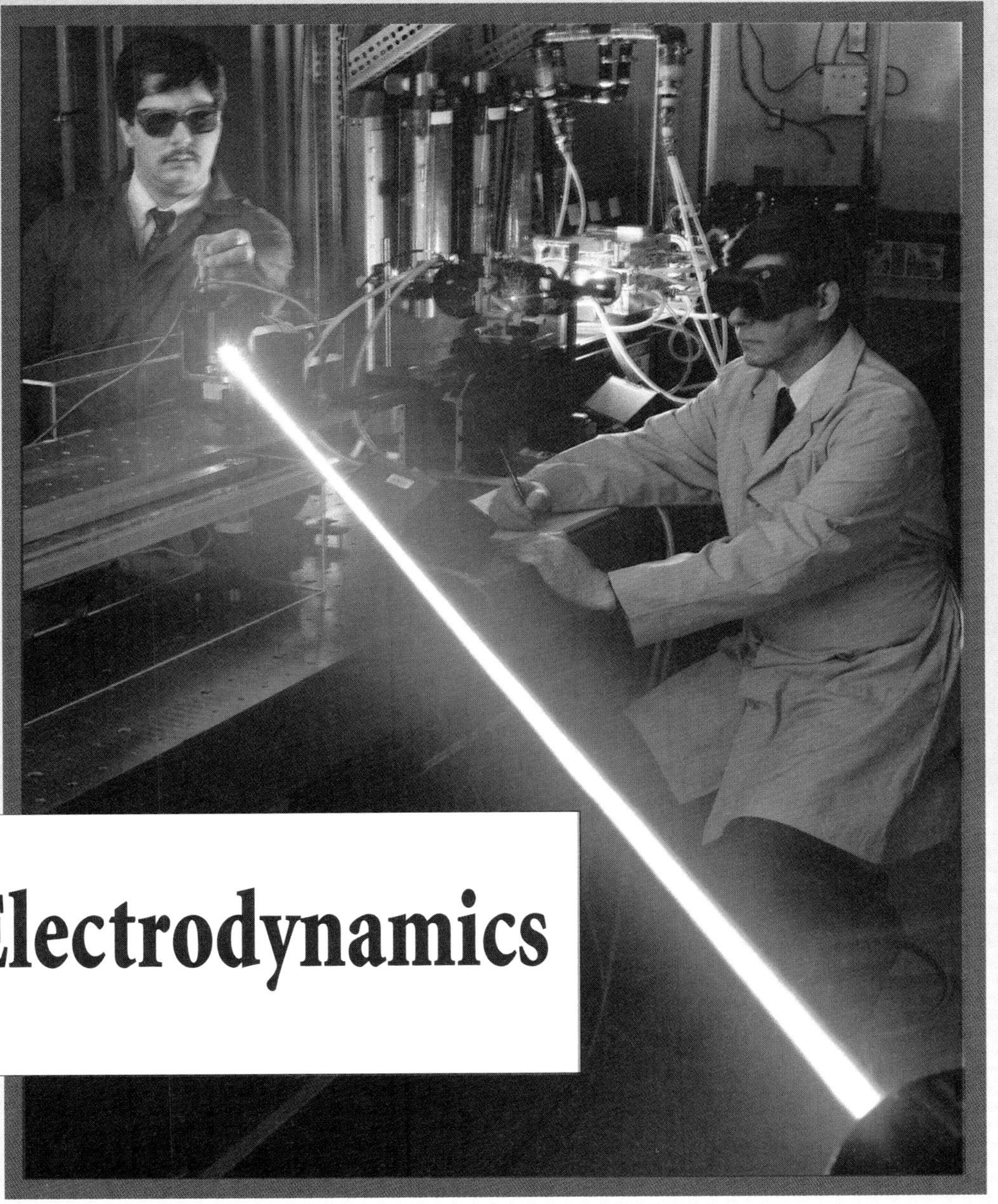

A laser exemplifies the power and beauty of electromagnetic waves. In this part we develop the full set of Maxwell equations and show how they describe the electromagnetic nature of light.

Electrodynamics

PART SEVEN

Static electric fields power your flashlight, and static magnetic fields turn your compass needle toward North. These seem to be unrelated physical effects, but electricity and magnetism appear distinct only when we consider static fields in a single reference frame. Time-varying fields *are* closely related, and the distinction between electric and magnetic depends on the reference frame in which we measure. The subject of this Part is electrodynamics. We'll look at electromagnetic fields that vary in time and learn how to describe electric and magnetic effects in reference frames that move with respect to one another. This study establishes a unity between electric and magnetic phenomena and shows how light waves arise as oscillating fields.

■ FIGURE VII.1
James Clerk Maxwell (1831–1879). Maxwell, a Scot, probably picked up his interest in science from his father. He was educated by his mother until her death, after which time he was sent to school in Edinburgh. He began his undergraduate studies at Edinburgh University. When his mathematical talents became apparent, he moved in 1850 to the university that could offer the best mathematical training, Cambridge, where he graduated second in his class. He began his important work in electromagnetic theory immediately and published his first paper on the topic in 1855. This work culminated in his *Treatise on Electricity and Magnetism,* published in 1873. Maxwell also made important contributions in statistical physics (see §22.7).

■ FIGURE VII.2
Albert Einstein (1879–1955). Einstein was born in Ulm, Germany, on March 14, 1879. He first came to the United States in 1932, and in 1933, he took up a position at the Institute for Advanced Study in Princeton, where he remained until his death.

Einstein made contributions to almost every branch of physics, but he is probably best known for his work on relativity. His first paper on special relativity, published in 1905, was titled *On the electrodynamics of moving bodies.* His first paper on the general theory was published in 1916. In 1921, he was awarded the Nobel Prize for his work on the photoelectric effect (cf. Chapter 35), also published in 1905. Einstein is also remembered for his humanitarian works. In his later years, he campaigned vigorously for a ban on nuclear weapons.

The theory of electrodynamics was put into its modern form by James Clerk Maxwell (■ Figure VII.1) and Albert Einstein (■ Figure VII.2). In this part, we shall study Maxwell's theory and touch on its applications to electric power technology, electronic circuits, optics, and microwaves. Each of the theory's basic equations bears the name of the physicist who is credited with its discovery: Gauss, Ampère, or Faraday. As a set, the equations are named after Maxwell because he was the first to develop them into a complete and consistent theory.

Maxwell's equations relate electromagnetic fields to their sources. We have already studied the two equations that relate the electric and magnetic fields to charge and current, respectively: Gauss' law (Chapters 23 and 24) and the time-independent form of Ampère's law (Chapter 28). Next, we shall see that charge and current are not the only source of these fields. A *changing* magnetic field creates an electric field and, similarly, a changing $\vec{\mathbf{E}}$ creates $\vec{\mathbf{B}}$. In Chapter 30 we study the laws that describe these effects: Faraday's law and Maxwell's extension of Ampère's law. In Chapters 31 and 32 we discuss circuits with time-varying currents, and in Chapter 33 we show how Maxwell's equations describe electromagnetic waves.

Grand Coulee Dam. In a hydroelectric power plant, gravitational potential energy of the water is converted first to kinetic energy as the water falls to the powerhouse at the bottom of the dam, and then to electric energy as the water is made to turn the generators in the powerhouse. The basic principle behind all generators is Faraday's law, the major topic of this chapter. A model of a simple generator is discussed in §30.2.2.

CHAPTER 30

Dynamic Fields

CONCEPTS

Induced electromotive force
Lenz's law
Motional electromotive force
Motor/generator
Induced electric field
Eddy currents
Displacement current

GOALS

Be able to:

Recognize systems in which emf is induced and apply Faraday's law to compute the emf.

Compute induced electric field in symmetric systems.

Recognize systems in which displacement current occurs and compute the current.

. . . creating a new theory . . . is rather like climbing a mountain, gaining new and wider views, discovering unexpected connections between our starting point and its rich environment. But the point from which we started out still exists and can be seen, although it appears smaller and forms a tiny part of our broad view gained by the mastery of the obstacles on our adventurous way up.

ALBERT EINSTEIN AND LEOPOLD INFELD

Hydroelectric power plants are common in the Pacific Northwest. Gravitational potential energy of the water in the reservoir is converted to kinetic energy as the water falls to the powerhouse at the base of the dam, where the water turns generators that produce electric power. On a smaller scale, similar generators provide power to motor homes in remote campsites and to hospitals when the main power source is interrupted. In each of these generators, electric current is created when a coil is forced to rotate in a magnetic field.

WE'LL DISCUSS THESE GENERATORS IN §30.2.2.

In 1820, Oersted's discovery of magnetic effects caused by electric current created a great interest in a search for electric effects produced by magnetic fields—electromagnetic induction. Ten years passed before Michael Faraday (■ Figure 30.1) was able to demonstrate such effects. Ampère had misinterpreted several experiments because he was looking for electric phenomena caused by *static* magnetic fields. Faraday's crucial discovery was the need for *change* in the magnetic field.

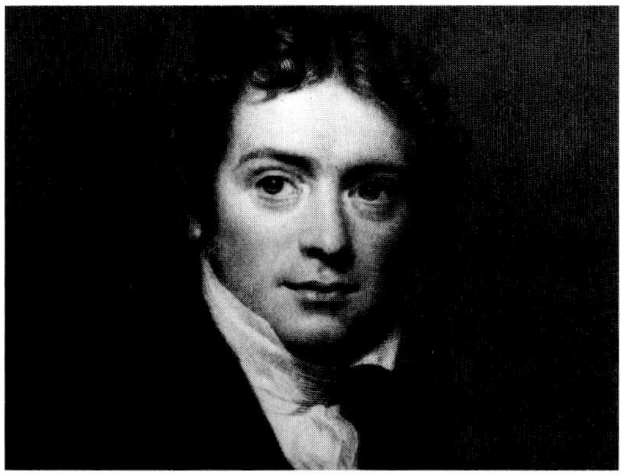

■ FIGURE 30.1
Michael Faraday (1791–1867). The historian and physicist Sir Edmund Whittaker said of Faraday: "Among experimental philosophers Faraday holds by universal consent the foremost place." Faraday made numerous discoveries in physics and chemistry, and is the only scientist to have two units named after him. He also promoted science education for the public.

Faraday's discovery is the basis of the electric power system. Power is transmitted across country at high voltage and low current to minimize ohmic losses in the power lines, yet domestic power is supplied at 110 V. On a pole near your home is a transformer that uses dynamic (i.e., changing) magnetic fields to link the distribution circuit to your household circuitry. Generators at the power plant and electric motors that use the power are equally dependent on dynamic fields.

POWER LINES WITHIN A CITY OPERATE AT 12 KV.

Changing electric fields also produce magnetic fields. This fact was not discovered experimentally, since the effect would have been negligible in the laboratory experiments of the early nineteenth century. It was predicted theoretically by Maxwell during the years 1857–1865 in studies designed to develop a firm mathematical and conceptual basis for electromagnetic theory. He suggested that a changing electric field acts like an additional *displacement current* in Ampère's law. Maxwell's theoretical discovery is the key to understanding the electromagnetic nature of light.

DISPLACEMENT CURRENT IS DISCUSSED IN §30.6.

In this chapter we shall study the two dynamic relations between electric and magnetic fields and arrive at the complete set of Maxwell equations, which form the basis for all of classical electromagnetic theory.

30.1 INDUCED EMF

30.1.1 *Faraday's Law*

In a classroom demonstration of electromagnetic induction (■ Figure 30.2), an ammeter is connected across a coil, and a magnet is moved back and forth along the axis of the coil. So long as the magnet is held fixed, nothing happens, but while the magnet is in motion, the ammeter deflects, indicating the existence of current and an electromotive force in the coil-ammeter circuit:

- A static magnetic field produces no electrical effect.
- A changing magnetic field induces an emf in the coil.

REMEMBER: EMF MEANS ELECTROMOTIVE FORCE. SEE CHAPTER 26.

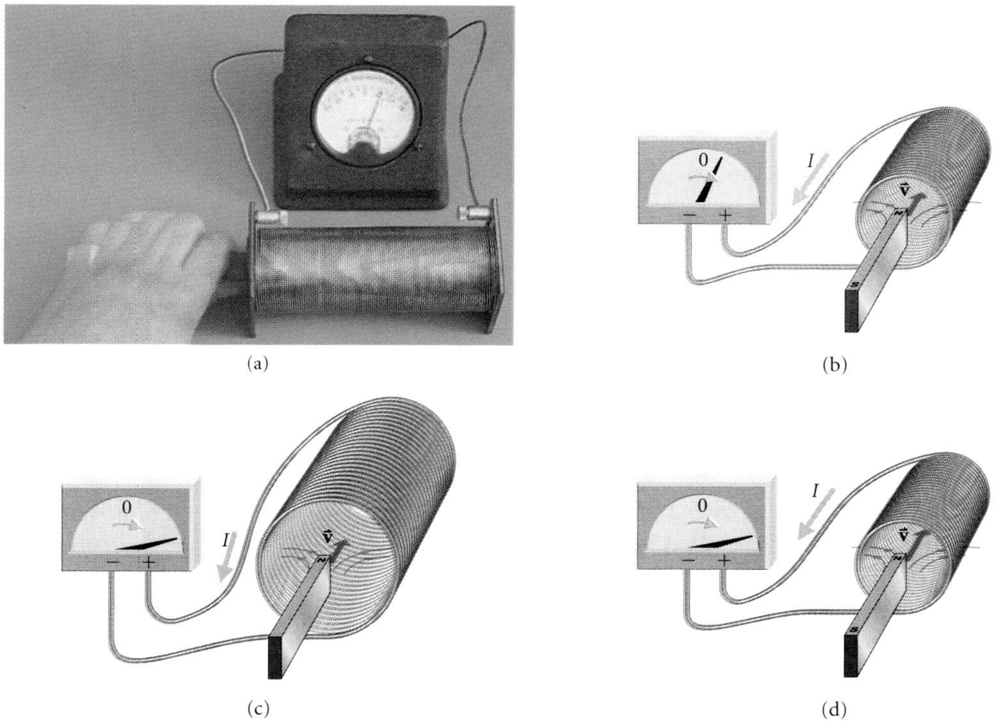

■ FIGURE 30.2
(a) An experiment that demonstrates Faraday's law. As the magnet is moved toward the coil, the ammeter indicates the presence of current in the circuit. (b) Schematic of the experiment shown in the photograph. (c) When the same magnet is used with a coil of greater radius, the ammeter reading indicates that the current is greater. (d) When the magnet is moved more rapidly, the current is again increased.

Several other tests reveal important features of induction. Repeating the experiment with the same magnet but a larger coil results in a larger electromotive force: the emf induced in the coil is proportional to its *area.* It is not change in $\vec{\mathbf{B}}$ itself that is important, but the change in its flux $\Phi_B = \int \vec{\mathbf{B}} \cdot \hat{\mathbf{n}}\, dA$ through the area of the coil.

MAGNETIC FLUX IS DEFINED IN THE SAME WAY AS ELECTRIC FLUX. SEE §23.4.2 AND §28.3.1.

- Induced emf results from change of magnetic flux.

Finally, experiment shows that the ammeter reading is proportional to the number of turns in the coil and to the speed with which the magnet moves.

- The total magnetic flux through a coil with N turns is the sum of the fluxes passing through each of its turns.

$$\Phi_B = \sum_{i=1}^{N} \Phi_{B,\mathrm{i}} = N\Phi_{B,\mathrm{each\,loop}}.$$

- The induced emf $\mathscr{E}$ is proportional to the rate at which flux, Φ_B, changes.

In his experiments, Faraday demonstrated that $\mathscr{E}$ is always *proportional* to $d\Phi_B/dt$: in SI they are equal. Thus we may state *Faraday's law.*

> The magnitude of the emf $\mathscr{E}$ induced in a circuit equals the rate at which magnetic flux through the circuit changes:
>
> $$|\mathscr{E}| = \left|\frac{d\Phi_B}{dt}\right|. \qquad \textbf{(30.1)}$$

The SI unit of magnetic flux is the weber, after Wilhelm Weber (1804–1891):

$$1\ \mathrm{Wb} = 1\ \mathrm{T \cdot m^2}.$$

EXERCISE 30.1 ♦ Show that $1\ \mathrm{Wb/s} = 1\ \mathrm{T \cdot m^2/s} = 1\ \mathrm{V}$.

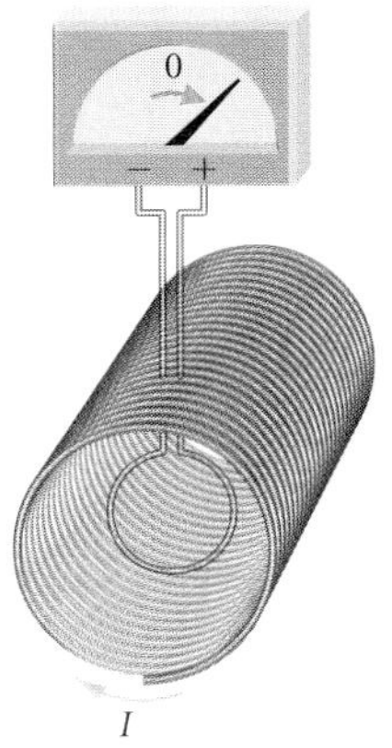

■ FIGURE 30.3
A small circular loop is placed around the axis of a large solenoid. As the current in the solenoid rises from zero, the magnetic field inside the solenoid increases. The increasing flux through the loop generates an emf in the loop.

EXAMPLE 30.1 ♦♦ A small loop of resistance 1 mΩ and radius a = 1 cm lies on the axis of a solenoid with n = 10 000 turns per meter (■ Figure 30.3). The current in the solenoid is increased uniformly from zero to 0.1 A in 10 ms. What is the current through an ammeter connected to the small loop?

MODEL As the current in the solenoid increases, so does the magnetic field it produces. Consequently, magnetic flux through the loop increases, inducing an emf in the loop and producing current in it. We assume the magnetic field inside the solenoid is uniform at any one time (cf. Example 28.8).

SETUP At any instant, the magnetic field inside the solenoid has magnitude $|\vec{\mathbf{B}}| = \mu_0 nI$ (eqn. 28.14). Since $\vec{\mathbf{B}}$ is perpendicular to the loop, the flux through the loop at any time is:

$$\Phi_B = \int \vec{\mathbf{B}} \cdot \hat{\mathbf{n}}\, dA = \pi a^2 |\vec{\mathbf{B}}| = \pi a^2 \mu_0 nI.$$

From Faraday's law, the emf induced in the loop equals the rate of change of this flux. The current I is the only time-dependent variable in Φ_B:

$$|\mathscr{E}| = \left|\frac{d\Phi_B}{dt}\right| = \pi a^2 \mu_0 n \left|\frac{dI}{dt}\right|.$$

SOLVE Finally, the current in the loop is found from the definition of resistance (eqn. 26.2):

$$I_\ell = \frac{|\mathscr{E}|}{R} = \frac{\pi a^2 \mu_0 n}{R}\left|\frac{dI}{dt}\right|$$

$$= \left[\frac{\pi(10^{-2}\text{ m})^2(4\pi \times 10^{-7}\text{ N/A}^2)(10^4\text{ /m})}{10^{-3}\ \Omega}\right]\left(\frac{0.1\text{ A}}{10^{-2}\text{ s}}\right) = 0.04\text{ A}.$$

ANALYZE The current in the loop exists only while the solenoid current is changing.

It is all right to use the result of static field calculations to find $\vec{\mathbf{B}}$ when the current and field are varying as slowly as in this example. In rapidly varying systems, however, the fields propagate through the system as waves at the speed of light. We discuss such systems in Chapter 33. To decide whether the field is slowly varying, ask whether the field changes substantially during the time it takes for light to cross the system. In this example, that time is (1 cm)/(3 × 10^8 m/s) = 3 × 10^{-11} s, which is vastly less than the 10-ms time scale for changes in B and I. ■

30.1.2 *Lenz's Law*

We can learn more from experiments with the magnet, coil, and ammeter (■ Figure 30.4). When the magnet is pulled away from the coil, rather than pushed toward it, the direction of current through the ammeter reverses.

- When the magnetic flux is changed at a fixed rate, the magnitude of induced emf is the same whether flux increases or decreases.
- The sense of induced emf depends on the sense of flux change.

The rule that gives the sense of induced emf is credited to Heinrich Lenz (1804–1865).

> **LENZ'S LAW**
>
> Induced emf acts so as to oppose the change in flux that creates it.

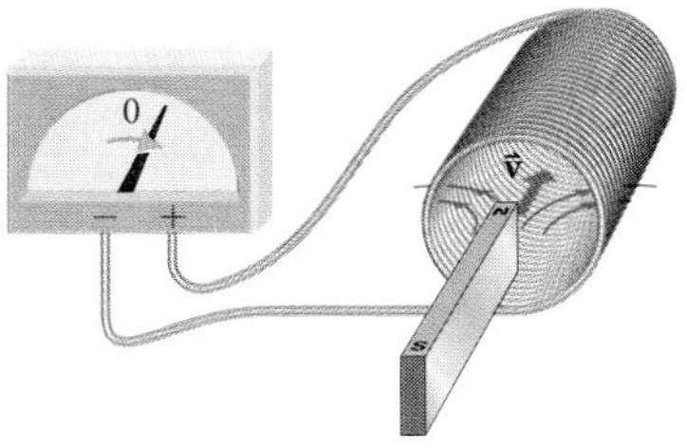

(a)

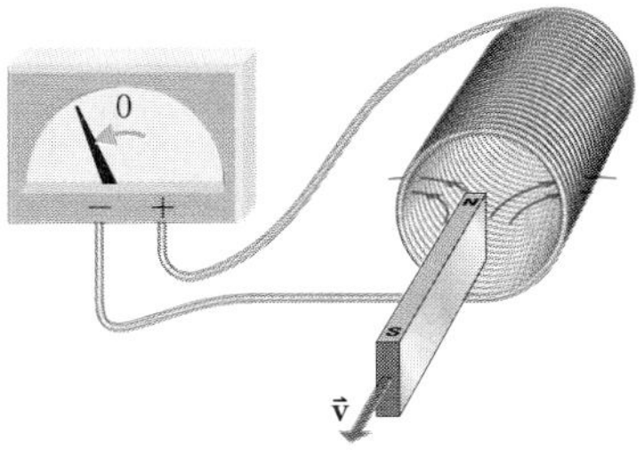

(b)

■ FIGURE 30.4
Lenz's law. Moving the magnet at the same rate in opposite directions generates currents with the same magnitude but opposite directions.

EXAMPLE 30.2 ❖ A magnet lies along the axis of a conducting loop with its N pole pointing toward the loop (■ Figure 30.5). If the magnet is pushed toward the loop, what is the direction of current around the loop?

MODEL According to Lenz's law, the current in the loop opposes any change in magnetic flux caused by the movement of the magnet. The magnetic field strength decreases with distance from the magnet, and its direction is away from the N pole. As the magnet approaches the loop, the magnetic field at the position of the loop increases, and hence the magnetic flux through the loop increases. According to Lenz's law, the current in the loop produces an induced magnetic flux *through the loop* that opposes this change.

The magnetic field vector at the center of the loop is the sum of the contributions from the magnet and from the loop itself: $\vec{\mathbf{B}}_{\text{net}} = \vec{\mathbf{B}}_{\text{magnet}} + \vec{\mathbf{B}}_{\text{loop}}$. As $\vec{\mathbf{B}}_{\text{magnet}}$ increases, the field produced by the loop in response must be in the opposite direction in order to reduce the increase of $\vec{\mathbf{B}}_{\text{net}}$: $|\vec{\mathbf{B}}_{\text{net}}| = |\vec{\mathbf{B}}_{\text{magnet}}| - |\vec{\mathbf{B}}_{\text{loop}}|$. Thus the field lines produced by the loop point toward the magnet and, according to the right-hand rule, the current is counterclockwise around the loop in ■ Figure 30.6.

ANALYZE We do not have to refer to magnetic flux when using Lenz's law. We may view the motion of the magnet toward the loop as the change that induces emf in the loop. Then the current in the loop must cause a force that decelerates the incoming magnet, opposing that change. That is, the magnetic dipole of the current loop has its N pole toward the N pole of the moving magnet. Again, we find that the current is counterclockwise. ■

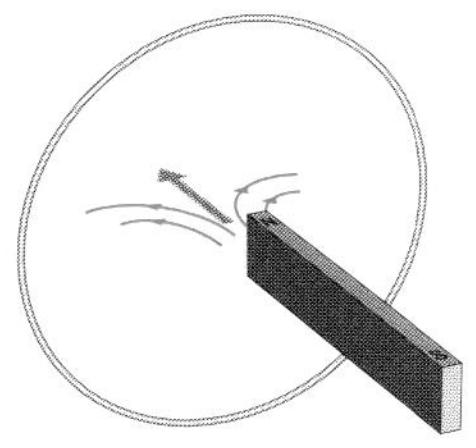

■ FIGURE 30.5
A magnet is pushed, N pole first, toward a conducting loop. If we choose the normal to the plane of the loop to be into the page, magnetic flux through the loop increases because the field at the loop increases as the magnet approaches.

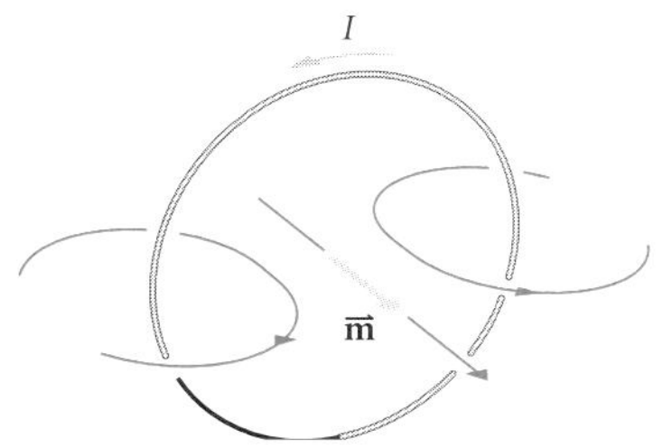

■ FIGURE 30.6
According to Lenz's law, the current in the loop opposes the increase in flux by making magnetic field in the opposite direction. The loop's magnetic moment is opposite that of the incoming magnet, so the loop and magnet repel each other.

EXERCISE 30.2 ❖ Suppose the magnet in Example 30.2 passes through the loop. Show that the current in the loop is clockwise as the magnet recedes.

Lenz's law is necessary for conservation of energy. If the current in Example 30.2 were in the opposite sense, the magnet would be attracted toward the loop and would gain kinetic energy! We could use the increased kinetic energy of the magnet to do work while using the induced emf to operate electrical machinery. Repeating the process would produce endless free energy, but this is, of course, impossible.

Put another way, work must be done on the system to produce energy. If the loop has resistance R, thermal energy is produced in it by the current at a rate I^2R. Thus we have to push the magnet toward the loop *against the opposing force,* doing work at a rate $\vec{\mathbf{F}} \cdot \vec{\mathbf{v}} = I^2R$.

30.1.3 Induced Electric Field

Faraday's law in the form of eqn. (30.1) and Lenz's law together describe the emf arising from a change in flux but do not tell us what produces the emf. Current exists in an ordinary conducting circuit only if work is done on the individual electrons to replace the energy they lose in collisions. There must be an induced electric field in the coil that exerts the necessary force on the electrons (■ Figure 30.7). Induced electric field does not arise directly from electric charge; its source is the changing magnetic field, and its field lines circulate around the source rather than emerging from it. In this way, induced electric field is like magnetic field and unlike static electric field.

WE'LL SAY MORE ABOUT INDUCED ELECTRIC FIELD IN §30.4.

Around a fixed circuit C, the emf may be expressed in terms of the induced electric field:

$$\mathcal{E} = \oint_C \vec{\mathbf{E}}_{\text{ind}} \cdot d\vec{\ell}. \tag{30.2}$$

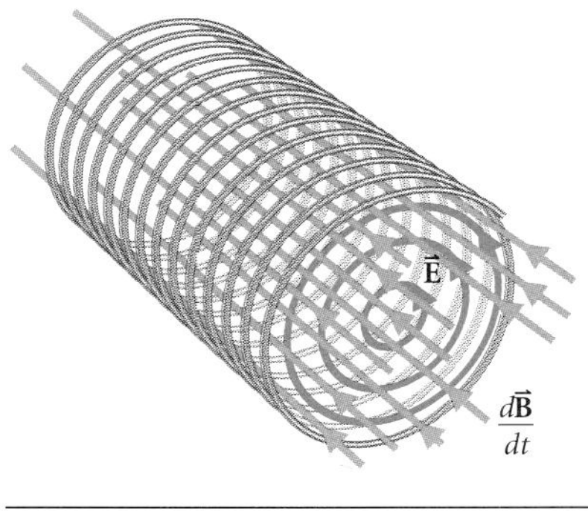

■ FIGURE 30.7
Induced electric field circulates around its source (changing magnetic field). Magnetic field circulates around its source (electric current) in a similar way. There is an important difference between the two cases: $\vec{\mathbf{E}}_{\text{ind}}$ and $\vec{\mathbf{B}}$ circulate in opposite directions around their sources.

30.1.4 Sign Conventions

THIS IS THE SAME CONVENTION USED IN AMPÈRE'S LAW, §28.3, ESPECIALLY FIGURE 28.15.

Faraday's law may be written in a compact form by defining a set of sign conventions. First, we choose the positive direction for current (or equivalently, induced electric field). Then the positive direction for flux is defined from the positive direction for current using a right-hand rule. For example, if we choose current to be positive when it is counterclockwise around the loop in ■ Figure 30.8, then the normal to the area of the loop is out of the page. We use this normal to calculate the flux, which is positive if $\vec{\mathbf{B}}$ is in the same direction as $\hat{\mathbf{n}}$. Once this sign convention has been defined, we can write Faraday's law in an equation that encompasses Lenz's law as well:

$$\mathcal{E} = -\frac{d\Phi_B}{dt}. \tag{30.3}$$

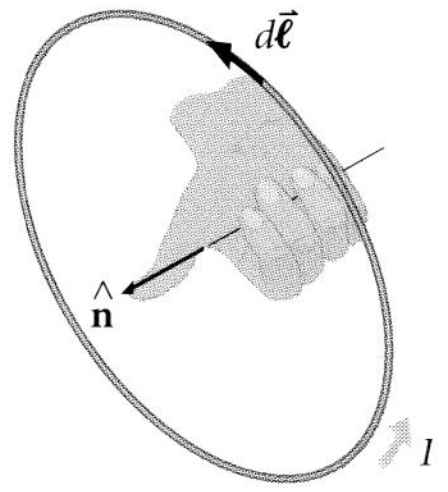

■ FIGURE 30.8
Sign conventions for emf and flux. Choosing the direction for $d\vec{\ell}$ determines the positive direction of $\hat{\mathbf{n}}$, and hence the positive sense of flux. The emf $\mathcal{E}$ is positive if it drives current in the direction of $d\vec{\ell}$, as shown here.

Using this sign convention in Example 30.2, we might define current to be positive if it is counterclockwise in Figure 30.5. Then the normal to the loop is out of the page. The magnet produces magnetic field into the page and hence negative flux through the loop. As the magnet approaches the loop, the field strength increases and the negative flux increases in *magnitude:* $d\Phi_B/dt$ is negative. Using Faraday's law in the form of eqn. (30.3), the emf is positive (counterclockwise), so the current is also counterclockwise as we concluded in the example.

EXAMPLE 30.3 ♦♦ A square loop of wire has sides 3.5 cm long parallel to the x- and y-axes, and a resistance of 0.10 Ω. The magnetic field in the region is increasing with time: $\vec{\mathbf{B}} = B_o(\hat{\mathbf{i}} + \hat{\mathbf{k}})$, where $B_o = (0.050\ \text{T/s}^2)t^2$. Find the current induced in the loop at $t = 1.0$ ms and $t = 0.50$ s.

MODEL Since the magnetic field increases as the square of the time, the rate of change of flux, and hence the current, are also functions of time. We begin by choosing the positive direction around the loop, then we apply eqn. (30.3).

SETUP Choose the counterclockwise direction around the loop to be positive so that $\hat{\mathbf{n}}$ is in the z-direction (■ Figure 30.9). The flux through the loop is:

$$\Phi_B = \int \vec{\mathbf{B}} \cdot \hat{\mathbf{n}}\, dA = \int B_o(\hat{\mathbf{i}} + \hat{\mathbf{k}}) \cdot \hat{\mathbf{k}}\, dA$$
$$= B_o(0 + 1)\int dA = B_o\ell^2.$$

We pulled B_o out of the integral since it does not depend on position. Next we apply Faraday's law:

$$\mathcal{E} = -\frac{d\Phi_B}{dt} = -\frac{dB_o}{dt}\ell^2 = -\frac{d}{dt}[(0.050\ \text{T/s}^2)t^2](3.5\ \text{cm})^2$$
$$= -(0.050\ \text{T/s}^2)2t(3.5\ \text{cm})^2 = -(122\ \mu\text{V/s})t.$$

SOLVE The current in the loop is:

$$I = \frac{\mathcal{E}}{R} = \frac{(-122\ \mu\text{V/s})t}{0.10\ \Omega} = (-1.22\ \text{mA/s})t.$$

At $t = 1.0$ ms, the current is:

$$I = (-1.22\ \text{mA/s})(1.0 \times 10^{-3}\ \text{s}) = -1.2\ \mu\text{A}.$$

And, at $t = 0.50$ s, it is:

$$I = (-1.22\ \text{mA/s})(0.50\ \text{s}) = -0.61\ \text{mA}.$$

ANALYZE The current is negative, which means it is clockwise. It produces a magnetic field in the negative z-direction within the loop, opposing the increase of applied magnetic field. ▮

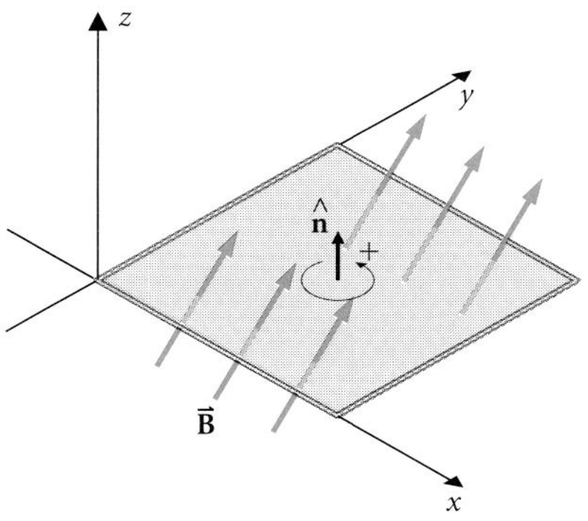

■ FIGURE 30.9
A square loop in the x-y-plane has resistance $R = 0.10\ \Omega$. The magnetic field is at 45° to the x-y-plane and is increasing at a rate that itself increases with time. We choose the positive direction around the loop to be counterclockwise, as shown, so that $\hat{\mathbf{n}} = \hat{\mathbf{k}}$. Then both flux and $d\Phi_B/dt$ are positive for $t > 0$.

30.2 MOTIONAL EMF

30.2.1 EMF in Circuits with Moving Boundaries

Faraday's law applies whenever the magnetic flux through a circuit changes for any reason. Varying the magnetic field is only one way of altering flux.

> *Motional emf* occurs when flux through a circuit varies because the circuit moves or changes its size or shape.

The following example shows how motional emf arises.

EXAMPLE 30.4 ♦♦ Two horizontal conducting rails 1.0 m apart lie in a region where the magnetic field is directly downward with magnitude $B = 0.10$ T (■ Figure 30.10). At one end, the rails are connected by a conducting cross tie. The resistance of rails and cross tie is negligible. A rod is pulled along the rails at constant speed $v = 0.50$ m/s. If the resistance between the two contact points of the rod on the rails is $R = 1.2$ mΩ, what is the current in the circuit? How much power is needed to keep the rod moving?

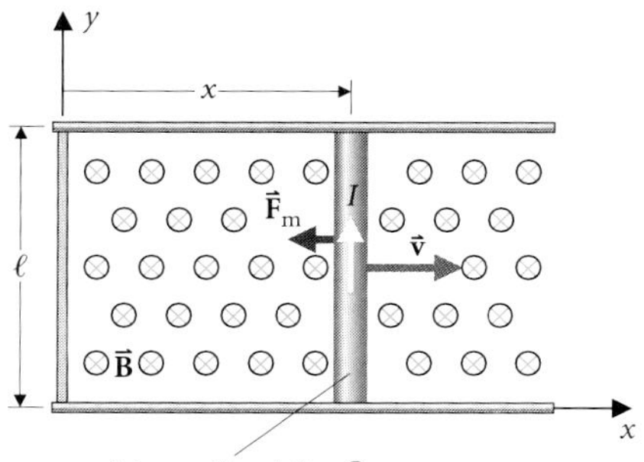

■ FIGURE 30.10
A conducting rod slides in the x-direction on conducting rails. When the rod is pulled at speed v, a current is induced in the circuit. The resulting magnetic force $\vec{\mathbf{F}}_m = I\vec{\ell} \times \vec{\mathbf{B}}$ on the rod opposes its motion.

MODEL The rod, rails, and cross tie form an electric circuit. As the rod moves, the area of the circuit and the magnetic flux through it both increase. Current is established by the resulting emf. From Lenz's law:

1. Magnetic force on the induced current in the rod opposes the rod's motion.
2. Magnetic field lines produced by the current pass upward through the circuit to oppose the increase in downward flux due to the rod's motion.

From either argument, we conclude that the current is counterclockwise.

SETUP Figure 30.10 shows the rod at a distance x from the cross tie. Since $\vec{\mathbf{B}}$ is uniform and perpendicular to the plane containing the rails, the flux through the circuit at this instant is $\Phi_B = BA = B\ell x$. The flux changes at a rate:

$$\left|\frac{d\Phi_B}{dt}\right| = \frac{d}{dt}(B\ell x) = B\ell\frac{dx}{dt} = B\ell v = (0.10\ \text{T})(1.0\ \text{m})(0.50\ \text{m/s}) = 5.0 \times 10^{-2}\ \text{V}.$$

SOLVE We use this result in Faraday's law to find the emf in the circuit. Using eqn. (30.1), the resulting current is:

$$I = \frac{|\mathscr{E}|}{R} = \frac{1}{R}\left|\frac{d\Phi_B}{dt}\right| = \frac{5.0 \times 10^{-2}\ \text{V}}{1.2 \times 10^{-3}\ \Omega} = 42\ \text{A}.$$

Since the current is perpendicular to the magnetic field, the force on the rod has magnitude:

$$|\vec{\mathbf{F}}| = I\ell B = (42\ \text{A})(1.0\ \text{m})(0.10\ \text{T}) = 4.2\ \text{N}.$$

To maintain the rod's speed, an equal and opposite mechanical force must be applied. The required power is:

$$P = Fv = (4.2\ \text{N})(0.5\ \text{m/s}) = 2.1\ \text{W}.$$

ANALYZE The electric power used by the circuit is:

$$P = \mathscr{E}I = (5.0 \times 10^{-2}\ \text{V})(42\ \text{A}) = 2.1\ \text{W}.$$

The mechanical power used to move the rod is dissipated as heat by the electrical resistance. ▮

In Example 30.4 it is a magnetic force that drives current through the resistive rod. The electrons in the moving rod have a velocity component $\vec{\mathbf{v}}$, and thus each experiences a magnetic force component along the rod: $\vec{\mathbf{F}} = -e\vec{\mathbf{v}} \times \vec{\mathbf{B}}$. Motional emf emphasizes two features of Faraday's law:

- The circuits to which the law applies may have moving boundaries.
- Magnetic forces contribute to emf.

AS THE ELECTRONS RESPOND TO THE MAGNETIC FORCE, THEY GAIN A DRIFT VELOCITY $\vec{\mathbf{v}}_d$. WE'LL INVESTIGATE THE CONSEQUENCES OF THIS VELOCITY COMPONENT IN §30.3.

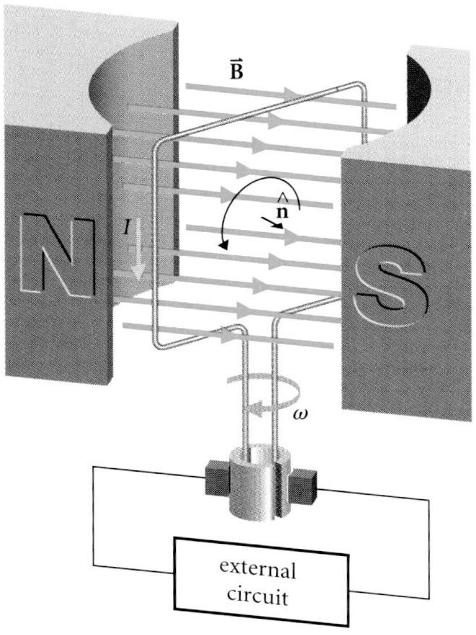

(a)

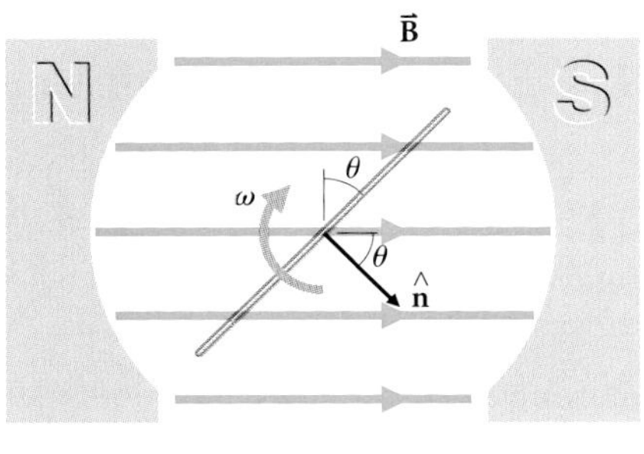

(b)

■ FIGURE 30.11
(a) A simple generator. A wire loop rotates between the poles of a magnet. We define the positive direction around the loop as shown. (b) The loop viewed from above. We assume the magnetic field between the poles is uniform. As the loop rotates, the angle θ between its normal $\hat{\mathbf{n}}$ and the magnetic field vector $\vec{\mathbf{B}}$ increases: $\theta = \omega t$.

30.2.2 Generators and Motors

One of the most important applications of motional emf is in the design of electric generators and motors. A generator converts mechanical work to electrical energy. A motor works the other way, converting electrical energy to mechanical work. Here we look at a simple device that can act as either a generator or a motor.

When a wire loop rotates in a magnetic field as shown in ■ Figure 30.11, there is a changing flux through the loop and hence an induced emf around it. If we define current as positive when it has the direction shown in the figure, then the flux through the loop at the instant shown is:

$$\Phi_B = \int \vec{\mathbf{B}} \cdot \hat{\mathbf{n}}\, dA = BA\cos(\theta) = BA\cos(\omega t),$$

where ω is the angular speed of the loop and we have chosen to start the clock when the loop is perpendicular to the field. As the loop rotates, the flux changes at a rate:

$$\frac{d\Phi_B}{dt} = \frac{d}{dt}[BA\cos(\omega t)] = -BA\omega\sin(\omega t). \tag{30.4}$$

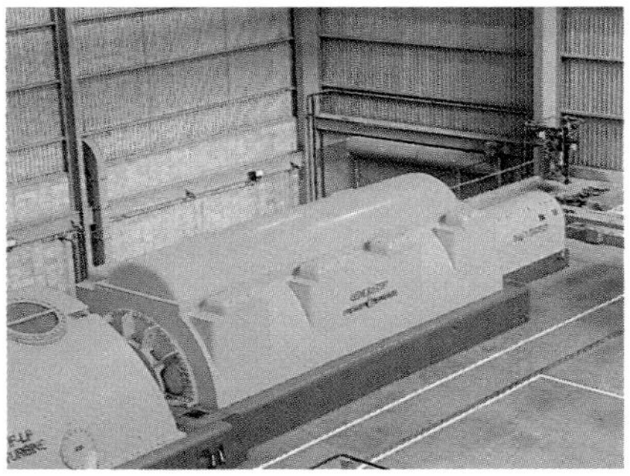

■ FIGURE 30.12
A generator used in a commercial power plant.

The induced emf $\mathscr{E} = BA\omega \sin \omega t$ varies sinusoidally between 0 and $\pm\omega BA$, and the current changes direction twice every period. The device is a simple *generator.* In generators designed for commercial power plants (■ Figure 30.12), $\omega = 2\pi f$ is set by the standard 60-Hz frequency of the power grid. The magnetic field is limited to about 1 T by saturation of the iron magnet cores. Thus the output emf of the generator is determined by the area of the coil and the number of turns of wire.

EXAMPLE 30.5 ♦ A simple generator has a rectangular loop of dimensions 10.0 cm $\times$ 5.0 cm with ten turns rotating in a field of 1.5 T. Find the maximum emf generated when the loop rotates at 60.0 Hz.

MODEL The motional emf is produced by rotation of the loop and varies sinusoidally.

SETUP The flux through the generator coil is ten times the flux through one loop.

SOLVE From eqn. (30.4) and Faraday's law, the maximum emf occurs when $\sin(\omega t) = 1$.

$$\mathscr{E} = NBA\omega = (10)(1.5\ \text{T})(0.10\ \text{m})(0.050\ \text{m})(2\pi)(60.0\ \text{Hz}) = 28\ \text{V}.$$

REMEMBER: $\omega = 2\pi f$.

ANALYZE Slip rings and brushes connect the rotating loop into a stationary circuit. Their design determines whether the current in the circuit is unidirectional (DC) (■ Figure 30.13) or changes direction with the emf (AC). ■

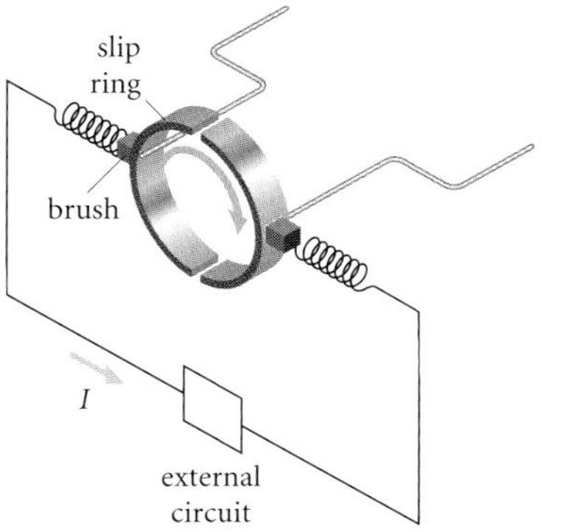

■ FIGURE 30.13
Brushes and a rotating slip ring convert a generator's AC output to DC. Each end of the generator loop is attached to the slip ring. *Brushes* are held against the slip ring and slide along them as they rotate. The brushes transfer current to the stationary wires in the external circuit. In the arrangement shown here, the connections between generator wires and the external circuit are reversed each half rotation as the brushes cross the gaps in the slip ring. The resulting current in the external circuit is always in the same direction. Brushes are typically made of graphite, which both conducts electricity and lubricates the slip rings.

By turning the generator loop, we obtain an electric current. Conversely, if we pass a current through the loop, magnetic forces act on the wire, creating a torque. This device is a simple electric motor.

EXAMPLE 30.6 ♦♦ If a current of 1.0 A is passed through the loop described in Example 30.5 (■ Figure 30.14), find the torque on the loop as a function of the angle θ between the magnetic field and the normal to the loop.

MODEL Since the magnetic field is uniform, we may use the magnetic moment of the loop to compute the torque (see §29.2.2).

SETUP The magnetic moment $\vec{\mathbf{m}} = NIA\hat{\mathbf{n}}$, where $\hat{\mathbf{n}}$ is the normal to the loop and A is its area. Using eqn. (29.8) with the current direction shown in the figure, the torque on the current loop is:

$$\begin{aligned}\vec{\tau} &= \vec{\mathbf{m}} \times \vec{\mathbf{B}} = (NIAB \sin\theta, \text{ counterclockwise as viewed in Figure 30.14b})\\ &= [(1.5\text{ T})(5.0 \times 10^{-3}\text{ m}^2)(10)(1.0\text{ A})\sin\theta, \text{ counterclockwise}]\\ &= [(7.5 \times 10^{-2}\text{ N}\cdot\text{m})\sin\theta, \text{ counterclockwise}].\end{aligned}$$

ANALYZE This torque also varies sinusoidally (Figure 30.14c). To keep the motor rotating in the same direction, the direction of the current must be reversed every half turn. This may be accomplished using brushes and a slip ring. ■

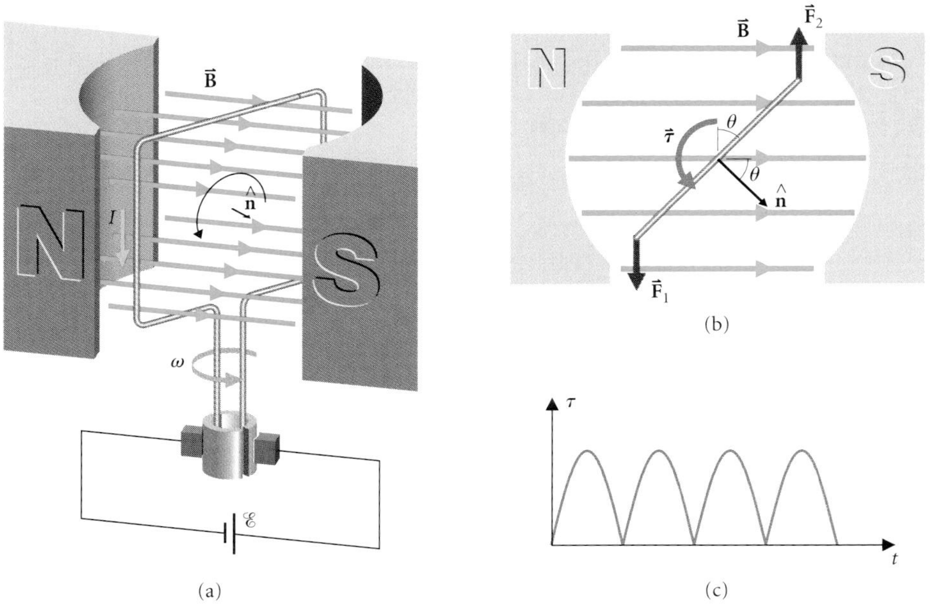

■ FIGURE 30.14
(a) The DC emf drives current through the wire loop. (b) Magnetic force acting on this current develops a net torque on the loop. (c) The net torque varies sinusoidally with the position of the loop. The emf is connected via brushes and slip rings (cf. Figure 30.13) so that the torque has constant direction. Its magnitude is $\tau \propto |\sin(\omega t)|$.

If the coil has a resistance r, then an emf of magnitude Ir is required to drive the current. When the motor is operating, there is also an induced emf of magnitude $\mathcal{E}_{\text{ind}} = BA\omega|\sin \omega t|$ opposing the current, so an emf $\mathcal{E} = Ir + \mathcal{E}_{\text{ind}}$ is needed to maintain the current. Similarly, mechanical force is needed to rotate the loop when it is used as a generator.

EXAMPLE 30.7 ♦♦ Find the power produced by the generator of Example 30.5 when connected in series with a resistance R. Show that it is equal to the mechanical power required to operate the generator.

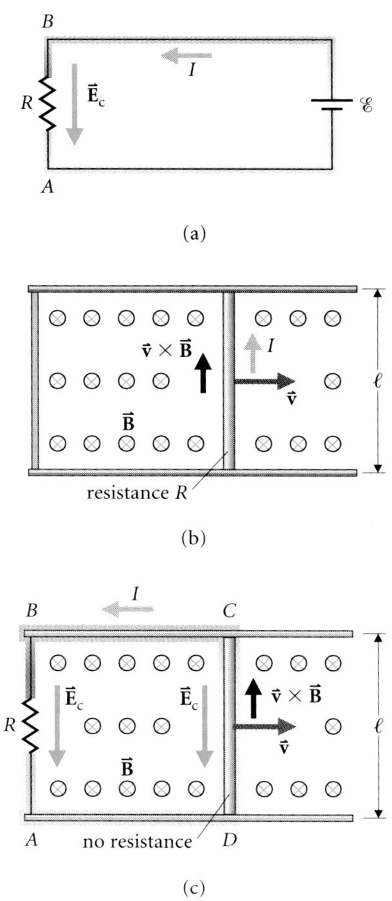

■ FIGURE 30.15
EMF and potential difference in circuits. (a) In a battery-powered circuit, energy is provided by chemical reactions in the battery and is used at the resistor. Charge is distributed throughout the circuit, as indicated schematically by the red and blue colors, and there is a potential difference $V_{BA} = \mathscr{E}$ across the resistor. (b) Conducting rod on rails. The rod has a resistance R. The magnetic force drives electrons through the rod. There are no charge distributions, Coulomb fields, or potential differences. (c) Here the rod is perfectly conducting, but the rails are connected by a *fixed* resistor R. A Coulomb electric force opposes the magnetic force in the perfectly conducting rod. Energy gained by the electrons in the moving rod is transferred to the resistor, where it is used. The potential difference $V_{BA} = V_{CD} = vB\ell$ indicates that the transfer is occurring.

THE ENERGY IS TRANSPORTED BY THE FIELDS, AS WE'LL SEE IN CHAPTER 33.

WE'LL CALL FIELDS DUE TO CHARGE "COULOMB FIELDS" (WITH SUBSCRIPT C) TO DISTINGUISH THEM FROM INDUCED FIELDS.

MODEL The electrical power equals the emf times the current in the circuit (eqn. 26.3). To use this relation, we need to determine the current. The current is the induced emf divided by the total resistance. The mechanical power equals the angular frequency times the torque, $P_m = \omega\tau$ (eqn. 9.14).

SETUP The total resistance in the circuit equals the resistance of the generator coil plus the external resistor in series with it: $R_{tot} = R + r$. The current is:

$$I = \frac{\mathscr{E}}{R + r} = \frac{NBA\omega}{R + r} \sin \omega t.$$

SOLVE The electrical power produced by the generator and used by the circuit is:

$$P_e = I\mathscr{E} = \frac{\mathscr{E}^2}{R + r} = \frac{(NBA\omega)^2 \sin^2 \omega t}{R + r}.$$

From Example 30.6, the magnetic torque exerted on the loop is $\tau = NIAB \sin \omega t$. Then, using our result for I:

$$\tau = NAB \sin \omega t \frac{NBA\omega}{R + r} \sin \omega t = \frac{(NBA)^2\omega}{R + r} \sin^2 \omega t.$$

An equal mechanical torque must be applied to keep the generator turning at constant angular speed. The mechanical power required to exert this torque is:

$$P_m = \omega\tau = \frac{(NBA\omega)^2}{R + r} \sin^2 \omega t.$$

ANALYZE As expected, the required mechanical power input P_m equals the electrical power output of the generator P_e. ▮

EXERCISE 30.3 ♦♦ When the loop is used as a motor, show that the electrical power input equals the mechanical power output plus the joule heat dissipated in the motor loop. ▮

30.3 THE NATURE OF EMF

30.3.1 Potential Difference and EMF

Any circuit has both an energy source and an energy sink. In our study of DC circuits, we used the chemical storage battery as a model for the energy source. The emf of an ideal battery is the energy it provides to the circuit per unit of charge that passes through it, and also equals the potential difference the battery maintains between its terminals. Induced emf is another kind of energy source for circuits, but it is not related so simply to potential difference. We must carefully distinguish the roles of potential difference and energy input.

■ Figure 30.15 shows three circuits, each with the same emf, resistance, and current. Their differences are in the nature of the emf and in the role of static fields.

Circuit (a) is a DC circuit like those we studied in Chapter 26. The battery provides energy that is used *in a different place*—at the resistor. Electric potential provides the bookkeeping system that describes where energy is provided and where it is used. Potential rises across the battery and decreases again across the resistor.

Circuit (b) is the rod-on-rails system we studied in Example 30.4. Here the energy source is the external agent pulling the rod. The motional emf arising from magnetic force on electrons in the rod makes the energy available to the circuit. The energy is dissipated as heat by resistance *in the rod.* In the rest of the circuit, no energy is needed and no fields are required to drive current. Consequently, there is no electric field, or potential difference.

In circuit (c) the energy source and emf are the same as in circuit (b), but here the rod has negligible resistance and the rails are connected by a *stationary* resistor. As in circuit (a), the energy is used in the resistor, distant from the source. To understand this circuit, let's look at what happens as the rod begins to slide. The magnetic force $\vec{F}_m = -e\vec{v} \times \vec{B}$ on conduction electrons in the rod accelerates them rapidly toward end D, but there is no force on the electrons in the stationary resistor to oppose the frictionlike force due to the electrical resistance. The electrons stop and negative charge builds up between A and D, leaving positive charge between B and C. The electric field $\vec{\mathbf{E}}_c$ produced by the charge distribution creates a potential difference of magnitude $E_c\ell$ between BC and AD. Once the electric and magnetic forces on the electrons in the rod balance ($\vec{\mathbf{E}}_c = -\vec{\mathbf{v}} \times \vec{\mathbf{B}}$), the charge buildup ceases. The Coulomb electric force drives current through the resistor, and electrons flow at constant speed through the perfectly conducting rod, where the force on them is now zero. The potential rises by $V_{BA} = vB\ell$ across the rod, indicating where energy enters the circuit, and decreases again across the resistor, indicating where the energy is used.

THIS WHOLE PROCESS HAPPENS VERY RAPIDLY ($\lesssim$1 NS).

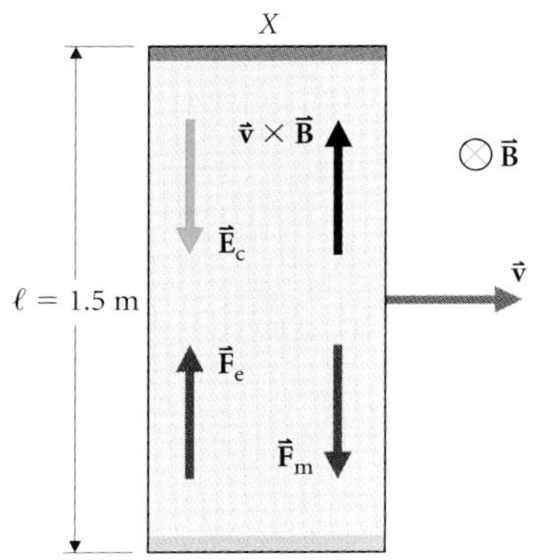

■ FIGURE 30.16
An automobile axle moving through the Earth's magnetic field. The magnetic force due to the vertical component of the Earth's field drives electrons toward end Y of the axle. The resulting Coulomb electric field creates a potential difference between the ends of the axle. In equilibrium, the electric force balances the magnetic force, $\vec{\mathbf{E}}_c = -\vec{\mathbf{v}} \times \vec{\mathbf{B}}$.

EXAMPLE 30.8 ♦♦ An automobile axle is 1.5 m long. If the car is moving at 15 m/s and the vertical component of the Earth's magnetic field is $B_\perp = 2.5 \times 10^{-5}$ T, what is the potential difference between the ends of the axle?

MODEL Initially, electrons move toward end Y of the axle in response to the magnetic force resulting from its motion (■ Figure 30.16). Once the magnetic force is balanced by the electric force due to the charge distribution in the axle, no more electrons move. In equilibrium, the net force on electrons in the axle is zero. The Coulomb electric field produces a potential difference between the ends of the axle.

SETUP In equilibrium, $\vec{\mathbf{F}} = -e(\vec{\mathbf{E}}_c + \vec{\mathbf{v}} \times \vec{\mathbf{B}}) = 0$, so $\vec{\mathbf{E}}_c = -\vec{\mathbf{v}} \times \vec{\mathbf{B}}$. The potential difference between the ends of the axle (eqn. 25.7) is:

$$\Delta V \equiv V_X - V_Y = -\int_Y^X \vec{\mathbf{E}}_c \cdot d\vec{\ell} = |\vec{\mathbf{E}}_c|\ell = vB_{\text{vert}}\ell.$$

SOLVE With the numbers given:

$$\Delta V = (15 \text{ m/s})(2.5 \times 10^{-5} \text{ T})(1.5 \text{ m}) = 0.56 \text{ mV}.$$

ANALYZE We can estimate how much charge is required at each end of the axle to produce the required Coulomb electric field. We approximate the charge distribution with two point charges $\pm q$ at the ends, which produce an electric field:

SEE EXAMPLE 25.14 FOR A SIMILAR APPROXIMATION.

$$2\frac{kq}{(\ell/2)^2} \sim vB \Rightarrow q \sim \frac{vB\ell^2}{8k} = \frac{\ell\,\Delta V}{8k} = \frac{(1.5 \text{ m})(0.56 \times 10^{-3} \text{ V})}{8(9 \times 10^9 \text{ N·m}^2/\text{C}^2)} = 10^{-14} \text{ C}.$$ ■

EXERCISE 30.4 ♦ What potential difference exists between the wing tips of a Boeing 747 aircraft (60-m wingspan) flying at 1000 km/h through a magnetic field 2.5×10^{-5} T perpendicular to the wing?

*30.3.2 *Mathematical Properties of EMF and Potential Difference*

The distinction between emf and potential difference is apparent in their mathematical properties. The potential difference between any two points is given by the integral of the *Coulomb* electric field along any path between the two points:

$$\Delta V \equiv V_B - V_A = -\int_A^B \vec{\mathbf{E}}_c \cdot d\vec{\ell}.$$

THIS IS EQN. (25.7).

If the integral is taken around a closed path starting and ending at A, the potential difference is zero:

$$V_A - V_A = \oint \vec{\mathbf{E}}_c \cdot d\vec{\ell} = 0.$$

This is Kirchhoff's rule—the change of potential around any closed path is zero.

FOR KIRCHHOFF'S RULES, SEE §26.4.

In contrast, Faraday's law tells us that the emf around a closed path is not zero. As we have seen, the emf describes energy sources and results from magnetic effects as well as induced electric fields. The *total* electromagnetic force acting on a charge is $\vec{\mathbf{F}} = q(\vec{\mathbf{E}} + \vec{\mathbf{v}} \times \vec{\mathbf{B}})$.

So:

The *induced emf* acting around a closed path is the work done per unit charge as the charge is moved around the path:

$$\mathcal{E} = \frac{W}{q} = \oint (\vec{\mathbf{E}} + \vec{\mathbf{v}} \times \vec{\mathbf{B}}) \cdot d\vec{\ell}. \qquad (30.5)$$

THE FORCE EXERTED BY THE INDUCED FIELD IS NOT CONSERVATIVE. SEE §8.3.1.

Although the total electric field appears in eqn. (30.5), only the induced electric field contributes to the emf, since the integral of the Coulomb field around a closed curve is zero. On the other hand, potential is defined by the Coulomb field alone; the induced field does not contribute.

*30.3.3 EMF and Choice of Reference Frame

If we push a magnet toward a wire loop at speed v (■ Figure 30.18a), the changing magnetic field induces electric field at the loop, and current flows in the loop. If instead, we push the loop toward the magnet at speed v (Figure 30.18b), magnetic forces produce a motional emf that drives current in the loop. The current is the same in both cases. Einstein, in his famous 1905 paper on special relativity, observed that these two experiments *must* give the same result because they are really the same experiment! Viewed from a reference frame fixed to the magnet (Figure 30.18c), the first experiment looks just like the second experiment. Induced electric field and motional emf are not separate physical effects, but two ways of describing the same effect in different reference frames. A similar correspondence exists between the electric and magnetic fields themselves. The quantity $\vec{\mathbf{E}} + \vec{\mathbf{v}} \times \vec{\mathbf{B}}$ that gives the contribution to emf at a particular element $d\vec{\ell}$ of a moving circuit *is* the electric field $\vec{\mathbf{E}}'$ measured in the reference frame moving with that element.

Digging Deeper

MAGNETIC FORCE AND EMF

Since magnetic force is always perpendicular to the velocity of the charge it acts on, magnetic fields cannot themselves do work on charges. Yet we have used $dW = q(\vec{\mathbf{v}} \times \vec{\mathbf{B}}) \cdot d\vec{\ell}$ in eqn. (30.5). This term is important in some circuits (e.g., the rod on rails). In fact, it is the Hall electric field $\vec{\mathbf{E}}_H = -\vec{\mathbf{v}}_d \times \vec{\mathbf{B}}$ (§29.3) that does the work. The power delivered to a single electron (■ Figure 30.17) is:

$$P_e = -e\vec{\mathbf{E}}_H \cdot \vec{\mathbf{v}}$$
$$= eE_H v = evBv_d.$$

The power delivered to all the electrons in a length $d\ell$ of the rod is:

$$dP = n\ dV\ P_e = nA\ d\ell\, evBv_d$$
$$= neAv_d Bv\ d\ell = I\ d\mathcal{E},$$

with $d\mathcal{E} = Bv\ d\ell$, in agreement with eqn. (30.5).

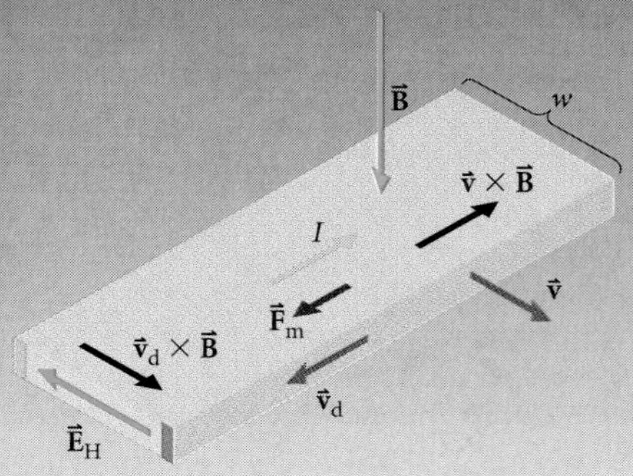

■ FIGURE 30.17
Enlargement of a conducting rod in which a motional emf is developed. When current exists in the rod, electrons move along it with a drift velocity $\vec{\mathbf{v}}_d$. A Hall electric field $E_H = v_d Bw$ develops across the rod. Though $\vec{\mathbf{v}} \times \vec{\mathbf{B}}$ appears in the formula for emf, it is the Hall field that does work on charge.

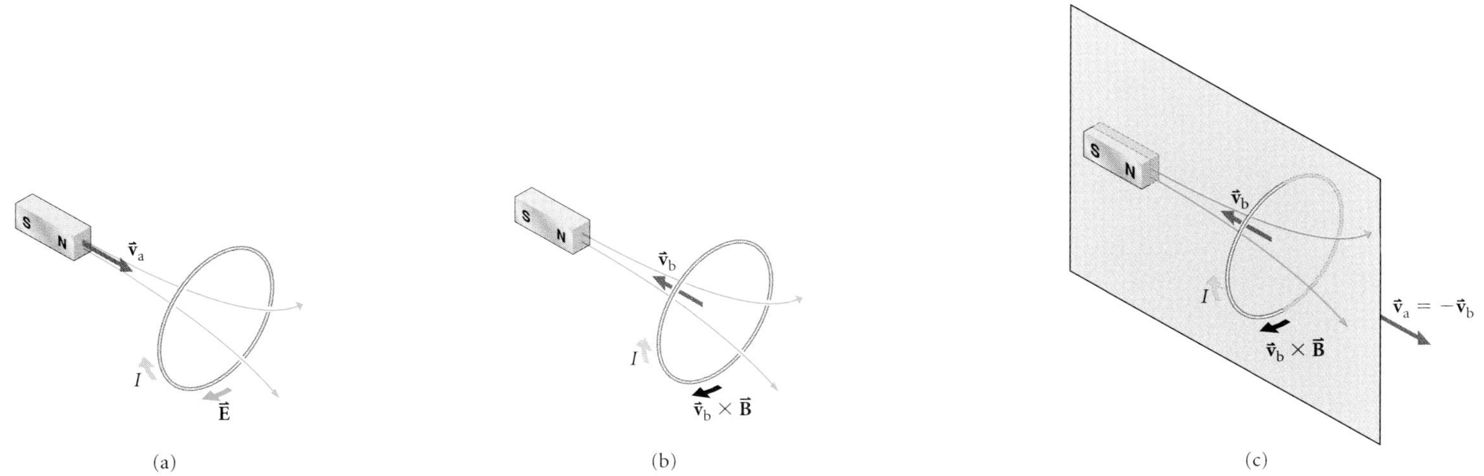

■ FIGURE 30.18
(a) In the lab frame, as we push a magnet toward a wire loop, the increasing flux through the loop creates an induced electric field that drives current. (b) If instead we push the loop toward the magnet, magnetic force $-e\vec{v} \times \vec{B}$ drives the current in the loop. (c) Experiment (a) viewed in a reference frame moving with the magnet is indistinguishable from experiment (b). Thus the electric field $\vec{E}$ in experiment (a) equals $\vec{v} \times \vec{B}$ in experiment (b).

30.4 CALCULATION OF INDUCED ELECTRIC FIELD

In symmetric systems, Faraday's law may be used to calculate the induced electric field. The method, outlined on the left-hand side of ■ Figure 30.19, is similar to that used with Gauss' law and Ampère's law. Faraday's law applies to any curve, whether or not it follows an electric circuit. You should choose a stationary curve that makes full use of the symmetry of the system.

Digging Deeper

THE COMPLETE MATHEMATICAL STATEMENT OF FARADAY'S LAW

A mathematical statement that includes all the features of induction we have discussed is:

$$\mathscr{E} = \oint (\vec{E} + \vec{v} \times \vec{B}) \cdot d\vec{\ell} = -\frac{d}{dt}\oint \vec{B} \cdot \hat{n}\, dA = -\frac{d\Phi_B}{dt}. \qquad (30.6)$$

The emf around any closed curve, regardless of how it is moving, equals the rate of change of flux through the curve. The emf is the work done per unit charge as the charge moves around the curve. The work done as the unit of charge moves along an element $d\vec{\ell}$ of the circuit has an electrical contribution $\vec{E} \cdot d\vec{\ell}$, and there may also be a motional contribution $(\vec{v} \times \vec{B}) \cdot d\vec{\ell}$.

The minus sign expresses Lenz's law. By convention, the choices of direction for the elements $d\vec{\ell}$ and the normal vectors $\hat{n}$ follow the right-hand rule shown in Figures 30.8 and 28.15.

The time derivative outside the integral sign symbolizes the fact that flux change may be caused by change in shape of the circuit (the rod on rails), change in the value of $\vec{B}$ (pushing the magnet through the loop), or change in the orientation of $\hat{n}$ (the generator). If the only time variable quantity is $\vec{B}$ itself, then there is induced electric field but no motional $(\vec{v} \times \vec{B})$ contribution.

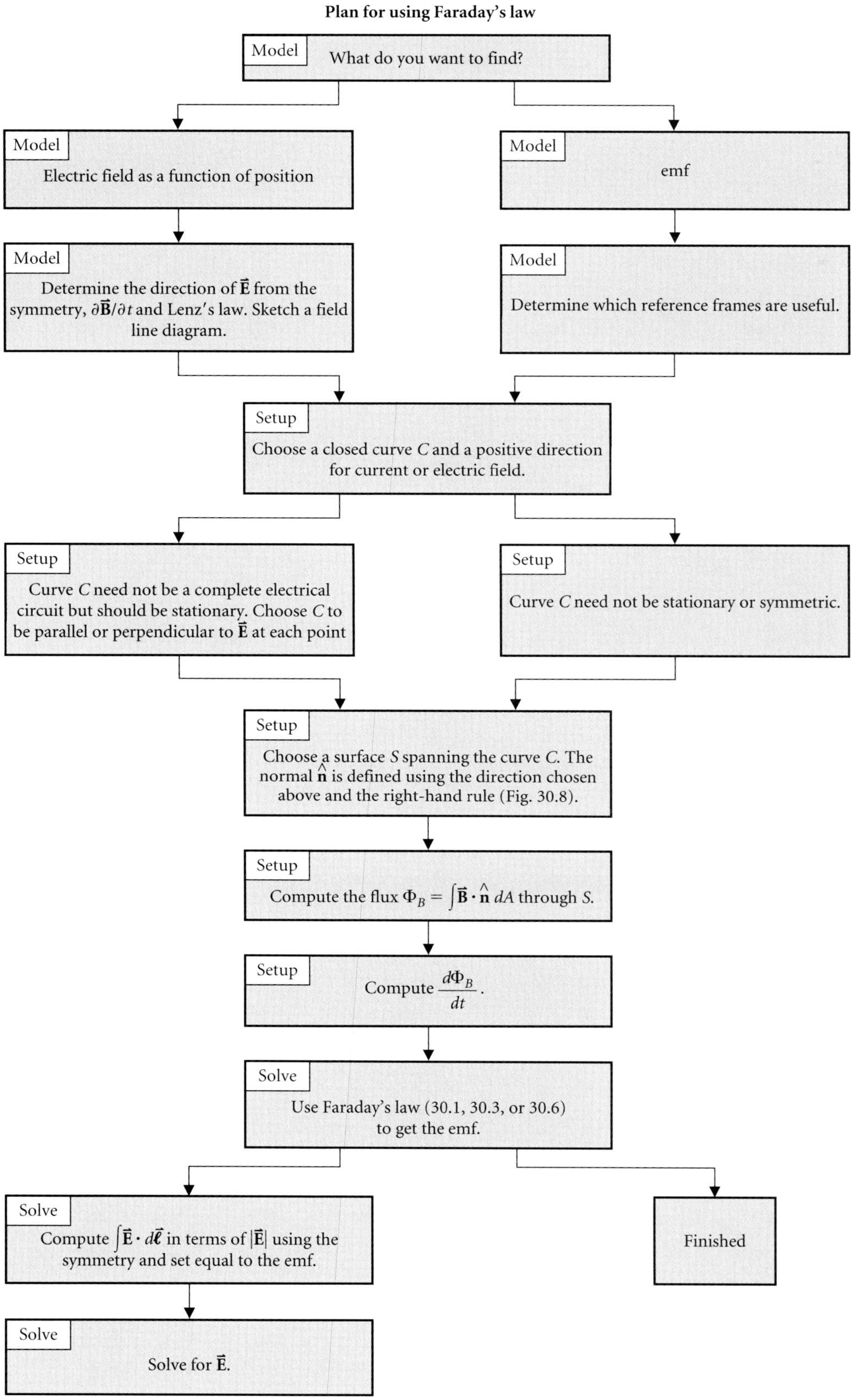
Plan for using Faraday's law
Model
What do you want to find?
Model
Electric field as a function of position
Model
emf
Model
Determine the direction of $\vec{\mathbf{E}}$ from the symmetry, $\partial\vec{\mathbf{B}}/\partial t$ and Lenz's law. Sketch a field line diagram.
Model
Determine which reference frames are useful.
Setup
Choose a closed curve C and a positive direction for current or electric field.
Setup
Curve C need not be a complete electrical circuit but should be stationary. Choose C to be parallel or perpendicular to $\vec{\mathbf{E}}$ at each point
Setup
Curve C need not be stationary or symmetric.
Setup
Choose a surface S spanning the curve C. The normal $\hat{\mathbf{n}}$ is defined using the direction chosen above and the right-hand rule (Fig. 30.8).
Setup
Compute the flux $\Phi_B = \int \vec{\mathbf{B}} \cdot \hat{\mathbf{n}}\, dA$ through S.
Setup
Compute $\frac{d\Phi_B}{dt}$.
Solve
Use Faraday's law (30.1, 30.3, or 30.6) to get the emf.
Solve
Compute $\int \vec{\mathbf{E}} \cdot d\vec{\boldsymbol{\ell}}$ in terms of $|\vec{\mathbf{E}}|$ using the symmetry and set equal to the emf.
Finished
Solve
Solve for $\vec{\mathbf{E}}$.

■ FIGURE 30.19

EXAMPLE 30.9 ♦♦♦ The magnetic field inside a solenoid is increasing at a constant rate $|d\vec{\mathbf{B}}/dt| = \alpha$. Find the induced electric field inside the solenoid.

MODEL The changing magnetic field has cylindrical symmetry, and the electric field lines circulate around this source like the magnetic field lines around a cylindrical current distribution. The electric field lines form circles around the axis of the solenoid. To find the magnitude of the field at any point, we may apply Faraday's law to a circular path passing through that point (■ Figure 30.20). The direction of the field lines in the figure follows from Lenz's law.

IF THERE COULD BE A CURRENT AROUND THE CURVE, IT WOULD OPPOSE THE CHANGE OF FLUX.

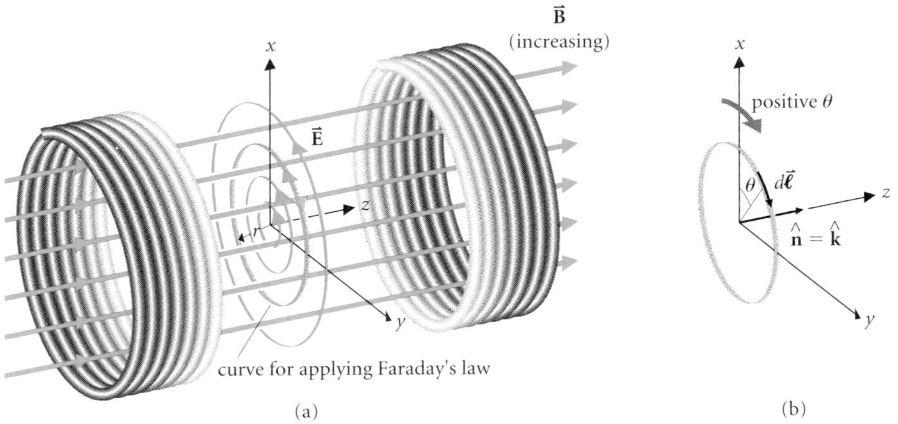

■ FIGURE 30.20
(a) Finding the induced electric field inside a solenoid. We choose the z-axis to be along the axis of the solenoid, parallel to $\vec{\mathbf{B}}$, and apply Faraday's law to a circle of radius r centered on the axis and parallel to the x-y-plane. (b) We choose to go around the circle in the same direction that the angle θ increases, as shown, so that according to the right-hand rule, $\hat{\mathbf{n}} = \hat{\mathbf{k}}$.

SETUP We choose the z-axis to be along the axis of the solenoid, with the circle in the x-y-plane. Using the right-hand rule and choosing $\hat{\mathbf{n}} = \hat{\mathbf{k}}$, we go around the circle as shown in the figure. Since B_z is spatially uniform, the flux passing through the area of the circle is $\Phi_B = \pi r^2 B_z$. Then:

$$\frac{d\Phi_B}{dt} = \pi r^2 \frac{dB_z}{dt} = \pi r^2 \alpha.$$

Here the curve is stationary, so there is no motional contribution to the emf. The electric field has only a θ-component, and its magnitude is the same at each point on the chosen circle. As a result, the integral reduces to a multiplication:

$$\oint \vec{\mathbf{E}} \cdot d\vec{\ell} = \oint E_\theta \, d\ell = E_\theta \oint d\ell = 2\pi r E_\theta.$$

COMPARE EQN. (28.13) FOR THE CIRCULATION OF $\vec{\mathbf{B}}$ AROUND A CIRCLE WHEN THERE IS CYLINDRICAL SYMMETRY.

SOLVE Applying Faraday's law (eqns. 30.2 and 30.3), we find:

$$2\pi r E_\theta = \oint \vec{\mathbf{E}} \cdot d\vec{\ell} = \mathscr{E} = -\frac{d\Phi_B}{dt} = -\pi r^2 \alpha,$$

or

$$E_\theta = -\alpha r/2.$$

ANALYZE The electric field exists whether or not there is a material circuit within the solenoid. If a conducting loop were placed there, the induced field would cause a current. The negative value of E_θ confirms the direction of $\vec{\mathbf{E}}$ shown in the figure. ▮

EXERCISE 30.5 ♦♦ Find the induced electric field outside the solenoid in Example 30.9 if the magnetic field is changing at the same rate. ▮

EXAMPLE 30.10 ♦♦♦ A circular wire loop is perpendicular to the axis of the solenoid (■ Figure 30.21). A small gap of width w is cut in the loop. Assume $w \ll d$, the diameter of the wire, which is in turn much less than the radius R of the loop. What is the electric field within the gap?

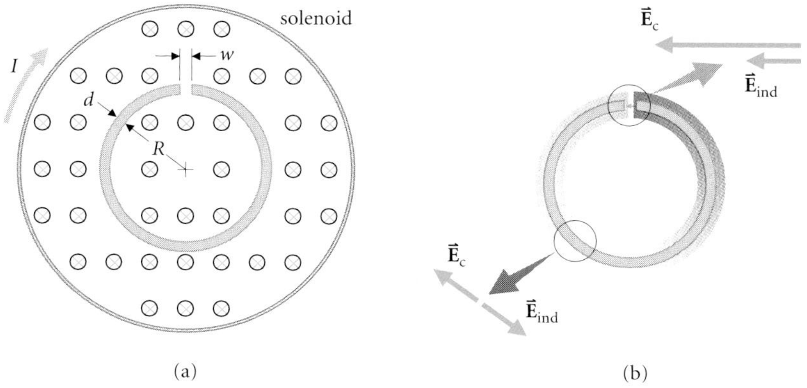

■ FIGURE 30.21
(a) A wire loop inside a solenoid has a small gap of width w. (b) No current can exist because of the gap in the loop. Within the metal, the induced electric field $\vec{E}_{ind}$ is opposed by a Coulomb field $\vec{E}_c$ due to the charge distribution on the loop. Within the gap, $\vec{E}_c$ and $\vec{E}_{ind}$ reinforce each other, creating a large total field.

MODEL The solenoid produces the same induced electric field whether the loop is present or not. Because of the gap, steady current cannot exist, so the net force on any electron in the loop has to be zero. Charge densities develop on the loop, as shown schematically in Figure 30.21b, and produce a Coulomb electric field $\vec{E}_c$ that opposes the induced field within the loop:

$$\vec{E} = \vec{E}_c + \vec{E}_{ind} = 0 \text{ within the metal of the loop.}$$

SETUP From Faraday's law, the circulation of the electric field around the loop is given by:

SEE EXAMPLE 30.9. WE USED THE SAME SIGN CONVENTIONS.

$$\oint \vec{E} \cdot d\vec{\ell} = -\frac{d\Phi_B}{dt} = -\pi R^2 \alpha.$$

The electric field $\vec{E}$ is the total electric field, here a superposition of induced and Coulomb field. The only contribution to the circulation integral comes from the portion of the circle in the gap, which is small enough to be modeled as a single element of the circle:

$$\oint_{\text{circle}} \vec{E} \cdot d\vec{\ell} = \int_{\text{gap}} \vec{E} \cdot d\vec{\ell} \approx w E_{\theta,\text{gap}}.$$

SOLVE Thus the electric field in the gap is:

SEE ALSO PROBLEM 82.

$$E_{\theta,\text{gap}} \approx -\pi R^2 \alpha / w.$$

ANALYZE Because the gap width can be made very small, the electric field in the gap can be very large even when the flux changes slowly. This field is mostly Coulomb field, produced by the charges on the ring. Heinrich Hertz used such a device as an antenna in his experimental discovery of electromagnetic waves. ■

Study Problem 19 ♦♦♦ *The Betatron*

The betatron is designed to accelerate electrons using induced electric field. ■ Figure 30.22b is a schematic drawing of the mechanism. The electrons orbit on a circle of fixed radius within the ceramic *doughnut.* The current in the electromagnet windings oscillates sinusoidally at 60 Hz. Electrons are injected just as the magnetic field is zero, but increasing downward. Show

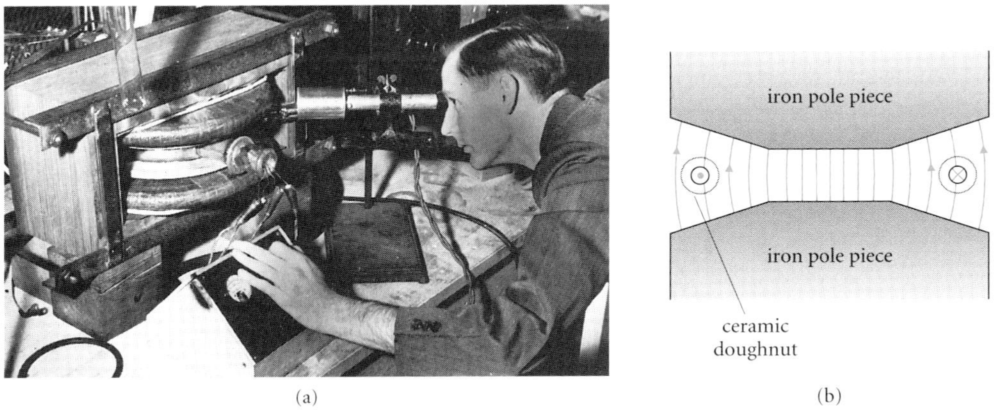

(a) (b)

■ FIGURE 30.22
(a) D. W. Kerst with the first betatron in 1941. This betatron could accelerate electrons to an energy of 2.3 MeV. (b) Schematic of a betatron. The pole pieces are shaped so that the magnetic field at the location of the particles is one-half the average value of the field within the orbit.

that (a) the electron energy increases during the first quarter of an oscillation cycle but would decrease during the second quarter; and (b) for the device to operate, the magnetic field at the radius R of the electron orbit must always be precisely half the average field inside the orbit.

Modeling the System

The magnetic field plays a double role in the betatron. Magnetic force is required to accelerate the electrons along their circular path, and the change of magnetic field in time creates the electric field that increases the electrons' energy. We are to demonstrate that a simple constraint on the field allows it to serve both functions.

Since the magnetic field remains cylindrically symmetric, the electric field lines are circles about the symmetry axis, tangent to the orbit of the electrons. ■ Figure 30.23a shows the field pattern during the first quarter cycle. The electric field accelerates the electrons in the clockwise direction, and the magnetic force points inward. During the second quarter cycle (Figure 30.23b), the direction of $d\vec{\mathbf{B}}/dt$ reverses as the magnetic field begins to decrease. The electric field also reverses and the electrons would lose energy. During the third quarter of the cycle (Figure 30.23c), the electric field would accelerate the electrons in the reverse direction, and in the fourth quarter, $\vec{\mathbf{E}}$ would again produce deceleration.

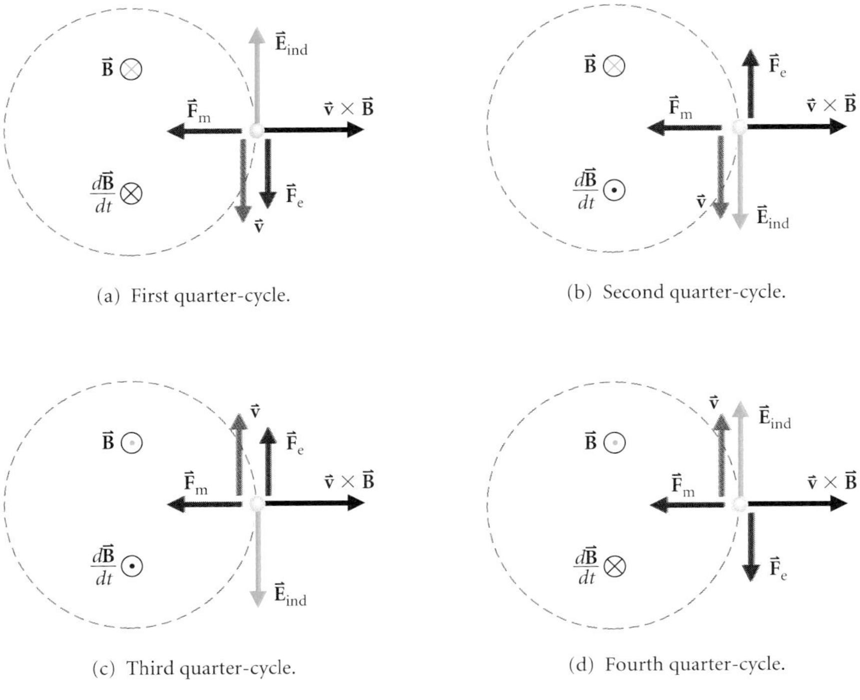

(a) First quarter-cycle. (b) Second quarter-cycle.

(c) Third quarter-cycle. (d) Fourth quarter-cycle.

■ FIGURE 30.23
Forces on an electron in a betatron. In an operating betatron, electrons remain in the machine during the first quarter cycle only. (a) During the first quarter cycle, the magnetic field points into the page and is increasing. The induced electric field is counterclockwise. The electric force increases the electron's speed, while the magnetic force causes the centripetal acceleration v^2/r. (b) During the second quarter cycle, the magnetic field strength is decreasing. The induced electric field is clockwise, and the electric force would decelerate the electrons if they were allowed to remain in the machine. (c) During the third quarter cycle, the magnetic field points out of the page and is increasing. The electrons would be accelerated counterclockwise. (d) During the fourth quarter cycle, the magnetic field strength is decreasing, and the electron energy would be reduced by the electric force opposite $\vec{\mathbf{v}}$.

SEE DIGGING DEEPER: MORE ON CYCLOTRONS IN CHAPTER 29. THERE WE GIVE THE RELATIVISTIC EXPRESSION FOR MOMENTUM.

Because the electrons reach speeds approaching the speed of light, a classical description of their motion should be treated skeptically. In this case, results obtained using classical mechanics and expressed in terms of momentum are the same as the exact relativistic results. Only the relation between the electron velocity and momentum differs.

Setup of Solution

Since magnetic force produces the centripetal acceleration of the electrons, the magnetic field at the orbital radius R is related to the electron speed v:

$$-e\vec{\mathbf{v}} \times \vec{\mathbf{B}} = -evB\hat{\mathbf{r}} = -\frac{mv^2}{R}\hat{\mathbf{r}} \Rightarrow p \equiv mv = eBR. \qquad \text{(i)}$$

The electric force acts parallel to the electron velocity during the first quarter cycle and increases the speed:

REMEMBER: FORCE PARALLEL TO VELOCITY CHANGES SPEED; FORCE PERPENDICULAR TO VELOCITY CHANGES DIRECTION. SEE §2.2 AND §3.2.6.

$$\frac{d|\vec{\mathbf{p}}|}{dt} = |\vec{\mathbf{F}}_e| = eE. \qquad \text{(ii)}$$

We find the electric field by applying Faraday's law to a circle of radius R. We go around the circle shown in Figure 30.23a counterclockwise.

COMPARE THE CALCULATION OF $\vec{\mathbf{E}}$ IN EXAMPLE 30.9.

$$\oint \vec{\mathbf{E}} \cdot d\vec{\ell} = 2\pi R E_\theta = -\frac{d\Phi_B}{dt} \Rightarrow E_\theta = -\frac{1}{2\pi R}\left(\frac{d\Phi_B}{dt}\right). \qquad \text{(iii)}$$

Solution of Equations

Substituting the result (iii) for E into eqn. (ii) gives:

$$\frac{dp}{dt} = eE = \frac{e}{2\pi R}\left|\frac{d\Phi_B}{dt}\right|.$$

Since both the momentum and the flux are approximately zero when the electrons are injected, the solution for p is:

$$p = \frac{e|\Phi_B|}{2\pi R}.$$

Thus, using relation (i) between p and B,

$$eBR = \frac{e|\Phi_B|}{2\pi R}$$

or

$$B = \frac{|\Phi_B|}{2\pi R^2}.$$

The flux within radius R divided by the area πR^2 is just the average magnetic field within radius R. Thus the magnetic field at radius R is half the average value within the orbit, as we were asked to show.

Analysis

With our choice of directions for $d\vec{\ell}$ and $\hat{\mathbf{n}}$, Φ_B is negative and E_θ is positive.

Betatrons built for medical research have produced beams with energies as high as 100 MeV.

EXERCISE 30.6 ♦ The average magnetic field in a betatron reaches a maximum of 1.2 T. If the radius of the electron orbits is 0.60 m, what is the final momentum of electrons accelerated by the device?

30.5 Eddy Currents

When a conductor moves through a magnetic field, Lenz's law predicts a force that opposes the relative motion (see Examples 30.2, 30.4, and 30.7). In each case, the force is due to current, driven by an induced emf, interacting with the magnetic field. A clever design can exploit such forces to drive motors or to build a propulsion system. ■ Figure 30.24 shows a metal plate between two sets of magnet coils. If current is turned on and off in each pair of coils, progressing in sequence from left to right, the coils simulate a magnet moving to the right past the plate. According to Lenz's law, the resulting forces accelerate the plate to keep pace with the moving magnet. This is the principle of the linear induction motor and of mass drivers envisioned as the primary means of launching cargoes from future space colonies. The experimental vehicle shown in ■ Figure 30.25 uses a similar effect to advantage as it "floats" on a cushion of magnetic field using linear synchronous motors in an electrodynamic levitation system.

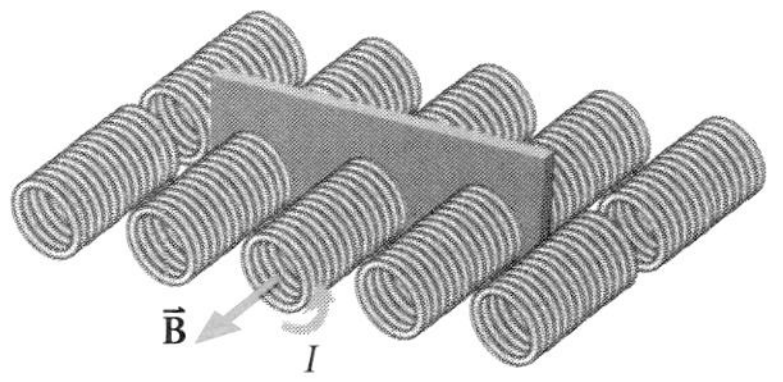

(a) Schematic of a mass driver.

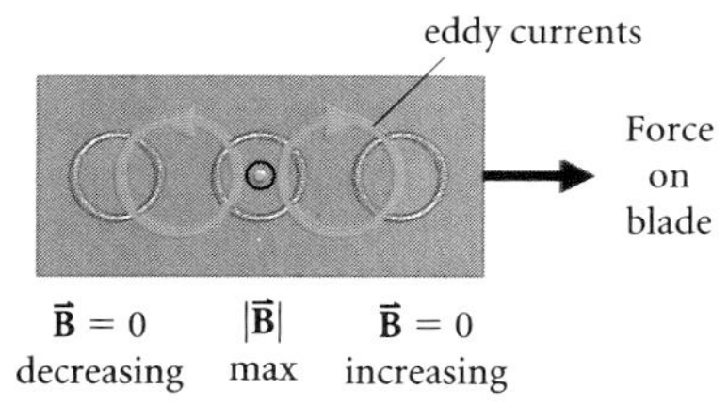

(b) The eddy current pattern and force on the blade.

■ **Figure 30.24**
(a) Professor Gerard O'Neill, late of Princeton University, incorporated mass drivers like this in his dream of colonizing space. Large space colonies in orbit about the Sun would need an economic source of raw materials. A mass driver capable of launching projectiles at speeds of 3 km/s could deliver payloads directly from the Moon's surface to a colony. In this simplified design for either a linear induction motor or a mass driver, a sequence of electromagnets is arranged along a track so that a metal blade passes between their poles in sequence. The current in each magnet coil is increased from zero as the leading edge of the blade approaches, reaches a maximum as the middle of the blade passes, and decreases back to zero as the receding edge of the blade passes. (b) The magnetic field acts like a sequence of magnets always moving faster than the blade and dragging it along. Experimental mass drivers currently achieve an acceleration of 1800 g!

■ **Figure 30.25**
An experimental MAGLEV train. Magnetic forces are used both to support the vehicle and to accelerate it.

In a demonstration of this effect (■ Figure 30.26), a metal pendulum swinging between the poles of a permanent magnet is rapidly brought to a halt. If slots are cut in the metal, interrupting the current path, the pendulum swings more freely. To investigate the force that stops the pendulum, we model the penduium as a large, thin metal sheet moving with velocity

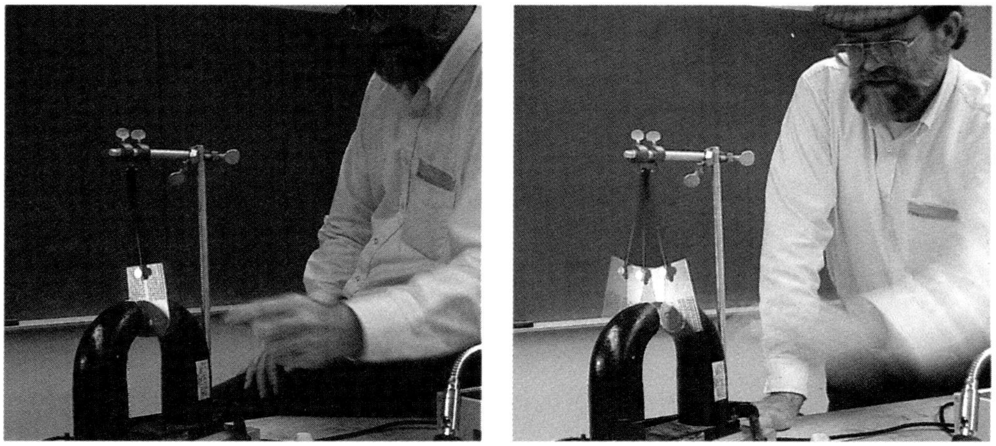

■ **Figure 30.26**
(a) When a metal plate is set swinging between the poles of a magnet, strong damping rapidly brings it to rest on the first swing. (b) The slotted plate swings for a much longer time before coming to rest.

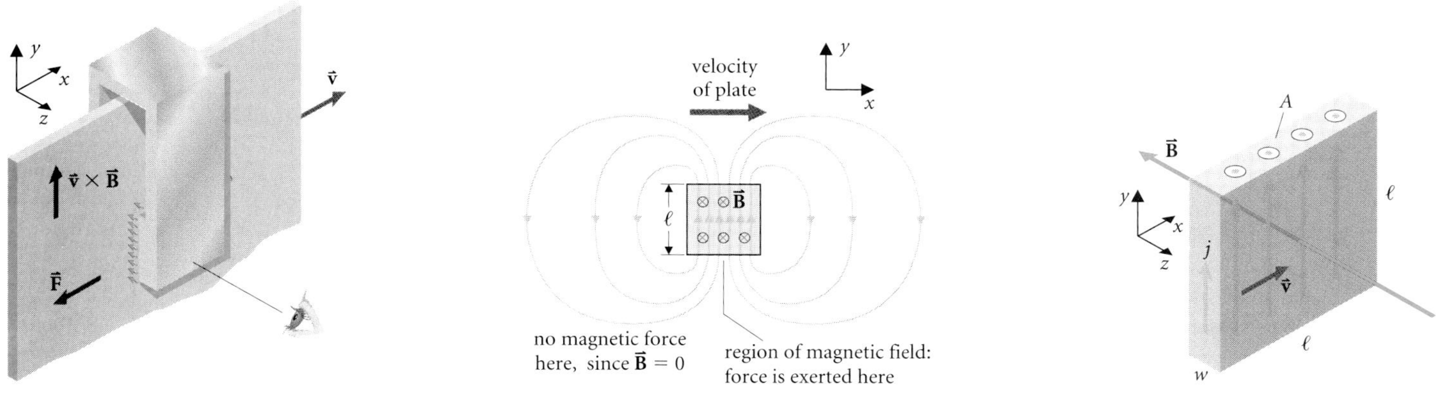

(a) (b) (c)

■ **Figure 30.27**
(a) A very large, thin, flat plate moves between square magnet poles. (b) The plate moves in the x-direction, with $\vec{\mathbf{B}}$ in the $-z$-direction. Thus the magnetic force drives electrons in the $-y$-direction, and the current is in the $+y$-direction between the magnetic poles. The current loops close outside the pole region, driven by a charge distribution on the plate. (c) The plate has thickness w, so the current passes through area $A = \ell w$.

$\vec{\mathbf{v}}$ between the square poles of a magnet (■ Figure 30.27a). Electrons in the metal between the magnet poles experience a magnetic force that drives current in the direction of $\vec{\mathbf{v}} \times \vec{\mathbf{B}}$. The current circulates in the characteristic pattern of loops, or *eddies,* illustrated in Figure 30.27b. The magnetic force on this current is the decelerating force required by Lenz's law.

Eddy currents convert energy to thermal form via joule heating. This dissipation accounts for the kinetic energy lost by the swinging pendulum. It can be a substantial nuisance in electrical machinery. Like the slotted plate, devices such as transformers are built from many separate conducting pieces rather than one continuous piece so as to minimize eddy current effects.

Eddy currents also produce a drag force on magnetically levitated (MAGLEV) vehicles (Figure 30.25). Fortunately, the ratio of drag force to support force decreases as the vehicle's speed increases so that practical designs for high-speed systems may be possible.

Digging Deeper

Forces Due to Eddy Currents

To estimate the magnitude of the force on the swinging pendulum, we approximate the magnetic field as uniform within the square region between the magnet poles, and zero elsewhere. Within that region of the plate, magnetic force acts on the electrons and causes them to drift in the direction of the force. We choose the x-axis to be parallel to $\vec{\mathbf{v}}$, as shown in Figure 30.27b. The current density in the field region is related to the force that drives it:

$$\vec{\mathbf{j}} \sim \frac{\vec{\mathbf{v}} \times \vec{\mathbf{B}}}{\rho} = \frac{v\hat{\mathbf{i}} \times (-B\hat{\mathbf{k}})}{\rho} = \frac{vB}{\rho}\hat{\mathbf{j}},$$

where ρ is the resistivity of the metal in the plate. [This is eqn. (26.9) with $\vec{\mathbf{E}}$ replaced by $\vec{\mathbf{v}} \times \vec{\mathbf{B}}$.] The current emerges uniformly from one face of a rectangular volume (Figure 30.27c), and the total current is:

$$I = jA = \frac{vB}{\rho}\ell w.$$

The magnetic field acts on this current over a length ℓ and exerts a force opposite $\vec{\mathbf{v}}$:

$$\vec{\mathbf{F}} \approx I\vec{\ell} \times \vec{\mathbf{B}} \sim I\ell\hat{\mathbf{j}} \times B(-\hat{\mathbf{k}})$$
$$= \frac{vB^2\ell^2 w}{\rho}(-\hat{\mathbf{i}}).$$

(In fact, charge densities on the plate are necessary to drive current outside the magnetic field region. The resulting electric field opposes the current in the B field region. An exact calculation for circular magnet poles shows that this effect reduces the net force on the plate by a factor of 2.)

In this result, the dependence of eddy current force on the relative speed v, the resistivity ρ, and the square of the magnetic field B is exact. To use the result in other situations, interpret $B^2\ell w$ as an average of B^2 multiplied by the volume V of metal exposed to the field:

$$F \sim \frac{vB_{av}^2 V}{\rho}.$$

30.6 The Ampère–Maxwell Law

Faraday's law and Ampère's law in its static form are oddly asymmetric. A changing magnetic field creates an electric field. Faraday's law relates the circulation of electric field to change in magnetic flux. We might expect that a changing electric field would create a magnetic field, but Ampère's law in its static form relates the circulation of magnetic field only to current. (There is no "current" term in Faraday's law because there are no magnetic currents.) For symmetry, Ampère's law should have a term involving change in electric flux. Maxwell showed that the additional term is not just aesthetic but is necessary to make Ampère's law self-consistent.

The circulation of $\vec{\mathbf{B}}$ around a closed curve is related to the current through a surface bounded by the curve. Mathematically, one may prove that the same relation must hold for *any* surface bounded by the curve, not just the obvious plane surfaces we used in the examples of Chapter 28. In a steady state situation, this mathematical requirement is automatically satisfied. Current through one surface bounded by a curve also passes through any other surface spanning that curve (■ Figure 30.28). However, in a dynamic system this need not be so. Consider charging a capacitor using very long wires (■ Figure 30.29). As the capacitor charges, the current I in the wires produces a magnetic field $\vec{\mathbf{B}}$. Applying Ampère's law to the curve C and the surface S_1, we obtain the usual result: $B = \mu_0 I/(2\pi r)$. But no current passes through the surface S_2. The static form of Ampère's law applied to S_2 predicts $B = 0$—a contradiction. Note, however, that the electric flux between the plates changes as the capacitor charges. If A is the area of the plates and σ is the charge density, then:

Sir George Stokes first published this result as a problem on a Cambridge University prize exam taken by Maxwell!

This is eqn. (28.6).

$$\frac{d\Phi_E}{dt} = A\,\frac{dE_z}{dt} = \frac{A}{\epsilon_0}\left(\frac{d\sigma}{dt}\right) = \frac{1}{\epsilon_0}\left(\frac{dQ}{dt}\right) = \frac{I}{\epsilon_0}.$$

We derived the expression for E_z in §27.1.1.

This thought experiment shows both the problem with Ampère's law and how to fix it. The current may differ through two surfaces with a common boundary, but only if the charge inside changes. The change of electric flux through the surfaces makes up for the difference in current. The quantity

$$I + \epsilon_0\,\frac{d\Phi_E}{dt}$$

is the same for every surface. The second term is called *displacement current.* When charging the capacitor, the conduction current due to the motion of charges through the wires equals the displacement current through the capacitor due to changing electric flux.

Up to this point "conduction current" has been called simply "current." We use this new name to distinguish it from displacement current.

The *displacement current* through a fixed surface is ϵ_0 times the rate of change of electric flux through the surface:

$$I_d \equiv \epsilon_0\,\frac{d\Phi_E}{dt} = \epsilon_0 \int \frac{\partial\vec{\mathbf{E}}}{\partial t}\cdot\hat{\mathbf{n}}\,dA. \qquad (30.7)$$

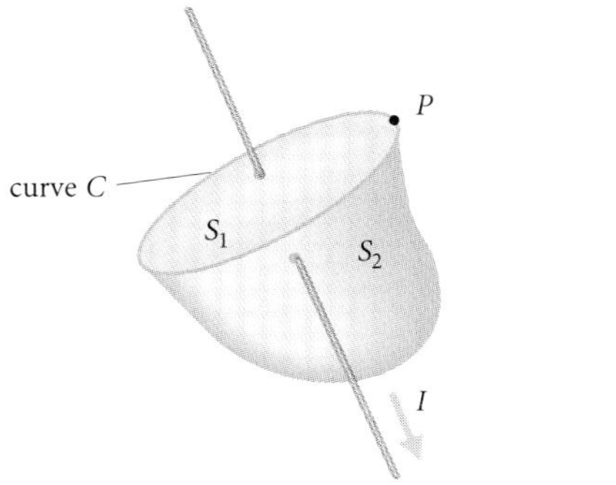

■ **Figure 30.28**
Steady current in a wire. The current is the same through any surface whose edge is the circle through P. For example, I passes through both the plane surface S_1 and the curved surface S_2.

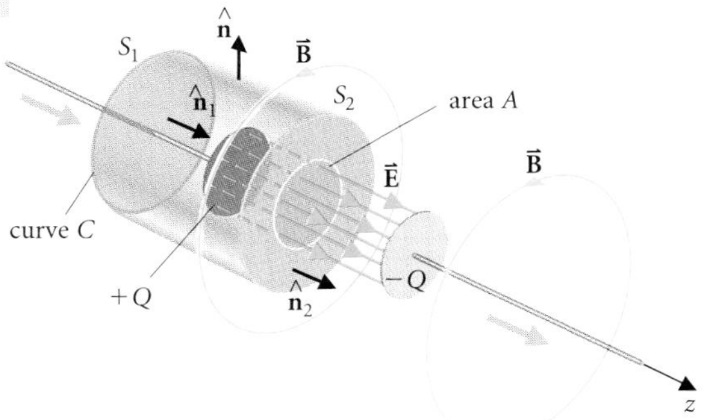

■ **Figure 30.29**
The process of charging a capacitor illustrates the problem with the static form of Ampère's law. We apply Ampère's law to the circle C centered on the wire, and choose the z-axis to be parallel to the wire as shown. Surface S_1 is a plane surface with $\hat{\mathbf{n}}_1 = \hat{\mathbf{k}}$, bounded by curve C. Surface S_2 consists of the curved wall of a cylinder also bounded by C, together with a flat surface parallel to S_1 but between the capacitor plates. On the flat part of S_2, the normal is also equal to $\hat{\mathbf{k}}$. Current I flows through S_1, but there is no conduction current through S_2.

Digging Deeper

Continuity of Total Current

■ Figure 30.30 shows two surfaces S_1 and S_2 with a common boundary curve C. Together they enclose a volume V. If the current I_2 leaving V through S_2 is not equal to the current I_1 entering through S_1, then the charge Q inside the volume changes:

$$I_1 - I_2 = \frac{dQ}{dt}.$$

But, from Gauss' law, Q/ϵ_0 equals the electric flux Φ_E emerging from V. When applying Gauss' law, the normal outward from the volume is always used, but in Ampère's law, the right-hand rule determines the direction of the normal. Thus:

$$\Phi_E = \oint_S \vec{\mathbf{E}} \cdot \hat{\mathbf{n}}\, dA$$

$$= \int_{S_1} \vec{\mathbf{E}} \cdot (-\hat{\mathbf{n}}_1)\, dA + \int_{S_2} \vec{\mathbf{E}} \cdot \hat{\mathbf{n}}_2\, dA$$

$$= \frac{Q}{\epsilon_0} = -\Phi_1 + \Phi_2,$$

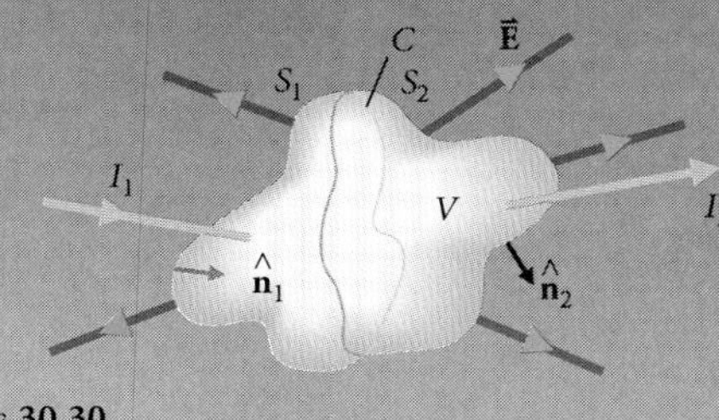

■ FIGURE 30.30
A closed surface S is divided into two pieces S_1 and S_2 by a plane surface. If the current I_1 entering the volume through S_1 does not equal the current I_2 leaving through S_2, then the charge inside the volume changes. Gauss' law relates the electric flux through S to the charge inside. When applying Gauss' law, the normal outward from the volume is always used. Thus $\hat{\mathbf{n}} = -\hat{\mathbf{n}}_1$ on S_1 and $\hat{\mathbf{n}} = +\hat{\mathbf{n}}_2$ on S_2.

where Φ_2 is the flux leaving through S_2, and Φ_1 is the flux entering through S_1. Thus:

$$I_1 - I_2 = \frac{dQ}{dt} = \epsilon_0 \frac{d\Phi_E}{dt} = \epsilon_0 \frac{d\Phi_2}{dt} - \epsilon_0 \frac{d\Phi_1}{dt}$$

$$= I_{d,2} - I_{d,1}.$$

Thus: $\quad I_1 + I_{d,1} = I_2 + I_{d,2}.$

The sum of regular current and displacement current is the same for each surface.

The complete form of the Ampère–Maxwell law includes displacement current on an equal footing with regular current. For a fixed curve C:

To apply this form of Ampère's law, use a fixed curve C.

$$\oint_C \vec{\mathbf{B}} \cdot d\vec{\ell} = \mu_0(I + I_d) = \mu_0 \int_S \left(\vec{\mathbf{j}} + \epsilon_0 \frac{\partial \vec{\mathbf{E}}}{\partial t}\right) \cdot \hat{\mathbf{n}}\, dA. \qquad (30.8)$$

To apply Ampère's law, we may use the method outlined in Figure 28.19, if we include the displacement current. With this change, the plan looks like the one for Faraday's law (Figure 30.19), except that the conduction current through the surface must also be computed.

EXAMPLE 30.11 ♦♦ Find the magnetic field between the circular capacitor plates of radius R in Figure 30.29.

MODEL We expect the magnetic field lines to form circles around the z-axis between the capacitor plates as well as outside them. The electric field between the plates is due to the surface charge already placed there: $\vec{\mathbf{E}} = (\sigma/\epsilon_0)\hat{\mathbf{k}}$. Modeling the charge distribution as uniform, $\sigma = Q/\pi R^2$.

SETUP We choose a circle of radius r centered on the z-axis as our Ampèrian curve. Going counterclockwise in ■ Figure 30.31, the circulation of $\vec{\mathbf{B}}$ around this curve is $2\pi r B_\theta$. The normal to the surface S is in the z-direction, and the electric field is constant over the surface, so the electric flux through the surface is:

$$\Phi_E = \int \vec{\mathbf{E}} \cdot \hat{\mathbf{n}}\, dA = \int (\sigma/\epsilon_0)\hat{\mathbf{k}} \cdot \hat{\mathbf{k}}\, dA = (\sigma/\epsilon_0)(\pi r^2) = (Q/\epsilon_0)(r/R)^2.$$

There is no conduction current through the surface.

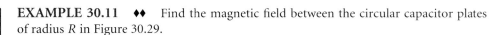

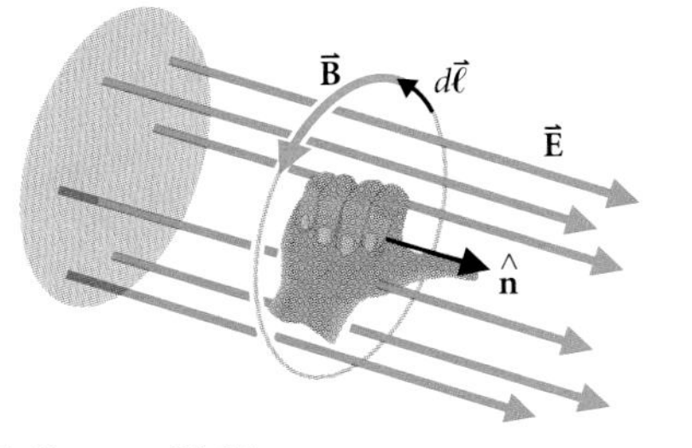

■ FIGURE 30.31
To calculate the magnetic field between the capacitor plates, we choose a circular Ampèrian curve and traverse it counterclockwise, as shown here. Then the electric flux through the curve is positive and increasing.

SOLVE Applying Ampère's law:

$$2\pi r B_\theta = \mathscr{C} = \mu_0(I + I_d) = \mu_0\left(0 + \epsilon_0 \frac{d\Phi_E}{dt}\right)$$

$$= \mu_0\epsilon_0\left(\frac{1}{\epsilon_0}\right)\left(\frac{dQ}{dt}\right)\left(\frac{r^2}{R^2}\right) = \mu_0 I\left(\frac{r}{R}\right)^2.$$

Thus:

$$B_\theta = \frac{\mu_0 I}{2\pi R}\left(\frac{r}{R}\right).$$

ANALYZE At the edge of the plates, the magnetic field reaches the value $B_\theta = \mu_0 I/(2\pi R)$ expected for the current I in the long wire. The capacitor acts like a fat wire with displacement current distributed uniformly across it. The magnetic field increases outward correspondingly. ∎

Chapter Summary

Where Are We Now?

Faraday's law and the Ampère–Maxwell law complete the set of Maxwell's equations. The inclusion of dynamic (i.e., changing) fields allows the discussion of a wide variety of practical systems. Maxwell's equations, together with the Lorentz force law and Newton's laws, are sufficient to solve any problem in classical electrodynamics, although many problems require advanced mathematical techniques.

Maxwell's Equations

(Gauss' law) The electric flux emerging from a volume equals the total charge within the volume divided by ϵ_0:

$$\Phi_E \equiv \oint \vec{\mathbf{E}} \cdot \hat{\mathbf{n}}\, dA = \frac{Q}{\epsilon_0} = \frac{1}{\epsilon_0}\int \rho\, dV.$$

EQN. 23.12, §23.4.3

The magnetic flux emerging from any volume is zero:

$$\Phi_B \equiv \oint \vec{\mathbf{B}} \cdot \hat{\mathbf{n}}\, dA = 0.$$

EQN. 28.9, §28.3.1

(Faraday's law) The electromotive force induced around any closed curve equals the rate of change of the magnetic flux through any surface bounded by the curve. If $\vec{\mathbf{v}}$ is the velocity of the curve element $d\vec{\ell}$, then:

$$\mathscr{E} = \oint (\vec{\mathbf{E}} + \vec{\mathbf{v}} \times \vec{\mathbf{B}}) \cdot d\vec{\ell} = -\frac{d\Phi_B}{dt} = -\frac{d}{dt}\int \vec{\mathbf{B}} \cdot \hat{\mathbf{n}}\, dA.$$

EQN. 30.6, §30.1–30.3

(Ampère–Maxwell law) The circulation of the magnetic field around any fixed closed curve is μ_0 times the total (conduction plus displacement) current through any surface bounded by the curve:

$$\oint \vec{\mathbf{B}} \cdot d\vec{\ell} = \mu_0(I + I_d) = \mu_0\left(I + \epsilon_0\frac{d\Phi_E}{dt}\right) = \mu_0\int\left(\vec{\mathbf{j}} + \epsilon_0\frac{\partial \vec{\mathbf{E}}}{\partial t}\right) \cdot \hat{\mathbf{n}}\, dA.$$

EQN. 30.8, §30.6

THIS PARTICULAR STATEMENT OF AMPÈRE'S LAW IS VALID FOR FIXED, BUT NOT MOVING, CURVES.

What Did We Do?

Faraday's experiments show that a change in magnetic flux through a circuit, for any reason, induces an electromotive force around the circuit. Change in the magnetic field in time induces an electric field that produces the emf.

According to Lenz's law, the response of a system to an induced emf opposes the change in flux that causes the emf. Lenz's law is a consequence of energy conservation; it gives the direction of an induced electric field and predicts the direction of forces acting on a system.

Magnetic forces on electrons in a moving conductor drive current. To find the motional emf, apply Faraday's law to a curve attached to the moving conductor.

Electromotive force describes the energy input to a circuit. Electric potential indicates how energy gained in one part of a circuit is transmitted for use elsewhere. Thus potential differences are intimately related to emf but are not the same thing.

* Electric and magnetic fields are different features of a unified electromagnetic field. In the reference frame moving with a circuit element, the emf arises from the action of the electric field $\vec{\mathbf{E}}'$. In another reference frame, the emf appears to have both electric and motional (magnetic) contributions:

$$\vec{\mathbf{E}}' = \vec{\mathbf{E}} + \vec{\mathbf{v}} \times \vec{\mathbf{B}}.$$

In symmetric systems, induced electric field may be computed from Faraday's law in the same way that magnetic field is found from Ampère's law. (See Figure 30.19.)

Relative motion between a conductor and a magnetic field causes eddies of current within the conductor. Magnetic forces on these eddy currents strongly damp the relative motion.

The static version of Ampère's law is incomplete when applied to nonsteady currents. Maxwell discovered that displacement current must be included with conduction current to obtain a law valid for dynamic fields. The displacement current through a surface is ϵ_0 times the rate of change in the electric flux through the surface.

Practical Applications

One of the most important applications of Faraday's law is the production of electrical energy from mechanical energy. We discussed a simple generator in Examples 30.5 and 30.7. Devices working on the same principle produce electrical energy in commercial power plants. On a smaller scale, magnetos used in piston aircraft engines and lawn mowers use induced emf to produce ignition sparks: the engines run without battery power.

Another important application is the electric motor. The simplest motor is a generator run backwards (Example 30.6). A few more complex designs are explored in the problem set.

Eddy currents resulting from induced emf generate unwanted heat in many electrical devices, especially transformers, which must be designed to suppress the currents. However, mass drivers use eddy current forces to accelerate projectiles. Similar designs have been proposed for transportation systems.

Displacement current is fundamental to the understanding of electromagnetic waves (Chapter 33).

Solutions to Exercises

30.1. From the force law $\vec{\mathbf{F}} = q\vec{\mathbf{v}} \times \vec{\mathbf{B}}$,

$$1\ \text{T} = \frac{1\ \text{N}}{(1\ \text{C})\cdot(1\ \text{m/s})},$$

so
$$1\ \frac{\text{T}\cdot\text{m}^2}{\text{s}} = 1\ \frac{\text{N}\cdot\text{s}\cdot\text{m}^2}{\text{C}\cdot\text{m}\cdot\text{s}} = 1\ \frac{\text{N}\cdot\text{m}}{\text{C}} = 1\ \text{J/C} = 1\ \text{V}.$$

30.2. Argument 1: The force exerted by the loop should decelerate the receding magnet. Thus the N pole of the loop's dipole moment points toward the S pole of the magnet, and so the current is clockwise.

Argument 2: The magnetic flux produced by the magnet is now decreasing, so the response of the loop is to produce field lines in the same direction as those of the magnet. Again, we conclude the current is clockwise.

30.3. The mechanical power output of the motor is its torque multiplied by its angular speed: $P_m = \omega\tau = \omega NBAI \sin\theta$. As the motor rotates, changing flux results in an emf $\mathscr{E}_{\text{ind}} = d\Phi_B/dt = \omega NBA \sin\theta$, and the electrical power supplied is:

$$P_e = \mathscr{E}I = (Ir + \mathscr{E}_{\text{ind}})I = I^2r + \omega NBAI \sin\theta = P_{\text{joule}} + P_m,$$

as required.

30.4. We are given the component of $\vec{\mathbf{B}}$ perpendicular to the (horizontal) airplane wing. We convert the speed to m/s (1 h = 3600 s); then the required potential difference is:

$$\Delta V = B_{\text{vert}}\ell v$$
$$= (2.5 \times 10^{-5}\ \text{T})(60\ \text{m})[(10^6\ \text{m})/(3.6 \times 10^3\ \text{s})]$$
$$= 0.4\ \text{V}.$$

30.5. We apply Faraday's law to a circle of radius $r > R$ centered on the axis of the solenoid (■ Figure 30.32). Here we choose to go around the curve counterclockwise (in the expected direction of $\vec{\mathbf{E}}$), with $\hat{\mathbf{n}}$ out of the page, antiparallel to $\vec{\mathbf{B}}$. All the flux in the solenoid passes through the surface S, and the magnetic field outside the solenoid is approximately 0, so the flux through S is $\Phi_B = -\pi R^2 B$. The emf is:

$$\mathcal{E} = -\frac{d\Phi_B}{dt} = +\pi R^2 \alpha = 2\pi r E_{\text{ind}}.$$

So, $E_{\text{ind}} = R^2\alpha/(2r)$. The direction of $\vec{\mathbf{E}}_{\text{ind}}$ is shown in the figure.

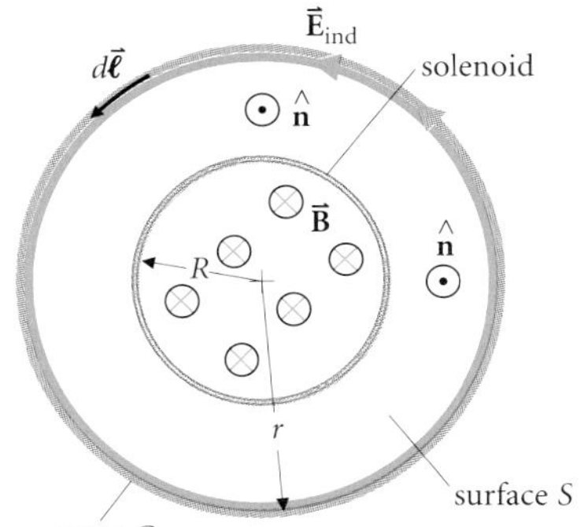

■ FIGURE 30.32

30.6. The field at the electron orbit is $B = \frac{1}{2}B_{\text{av}} = 0.60$ T. So:

$$p = eBR = (1.6 \times 10^{-19}\ \text{C})(0.60\ \text{T})(0.60\ \text{m})$$
$$= 5.8 \times 10^{-20}\ \text{N}\cdot\text{s}.$$

The energy of such an electron is 110 MeV.

Basic Skills

Review Questions

§30.1 INDUCED EMF

- What happens when a bar magnet is pushed N pole first toward a coil connected to an ammeter? How does the effect depend on the size of the coil and the speed of the magnet?
- State Faraday's law.
- Under what conditions is it possible to use static field calculations for time-variable systems?
- State Lenz's law. To which conservation law is it related?
- Describe sign conventions that allow you to combine Faraday's law and Lenz's law into a single mathematical statement. What is that statement?

§30.2 MOTIONAL EMF

- When does motional emf occur?
- Describe the operation of a simple generator. How does the magnetic flux through the loop change? How does the emf generated depend on the angular speed of the loop, its area, and the magnetic field strength?
- How much power is required to operate a simple motor?
- What are *brushes?* In which devices are they used, and why?

§30.3 THE NATURE OF EMF

- Does induced electric field contribute to potential difference? Why or why not?
- Define emf in terms of work.
- How is emf related to the electric and magnetic fields in an arbitrary circuit?
- ✲ If the electric and magnetic fields in a certain reference frame are $\vec{\mathbf{E}}$ and $\vec{\mathbf{B}}$, what is the electric field in a second reference frame moving with velocity $\vec{\mathbf{v}}$ with respect to the first?
- If the magnetic force does no work (cf. §29.1), how does motional emf transform energy?

§30.4 CALCULATION OF INDUCED ELECTRIC FIELD

- Describe how induced electric field is related to its source.

§30.5 EDDY CURRENTS

- What are *eddy currents* and when do they occur?
- Give one example in which eddy currents are useful and one example in which they are a nuisance.

§30.6 THE AMPÈRE–MAXWELL LAW

- Why is the static form of Ampère's law incomplete?
- Define *displacement current.*
- State the complete Ampère–Maxwell law.

Basic Skill Drill

§30.1 INDUCED EMF

1. The magnetic flux through each loop of a coil with 150 turns increases from 1.60 Wb to 2.40 Wb in a time interval of 0.15 s. What is the average emf developed in the coil during that time interval?

2. A loop of wire lies flat on a table with an ammeter connected in it. Also lying on the table is a bar magnet oriented as shown in ■ Figure 30.33. Without being raised from the table, the magnet is turned 180° to reverse the positions of the N and S poles. While this is done, conventional current through the ammeter is:
(a) momentarily to the right.
(b) momentarily to the left.
(c) first to the right and then to the left.
(d) first to the left and then to the right.
(e) essentially zero.
Explain why you chose your answer.

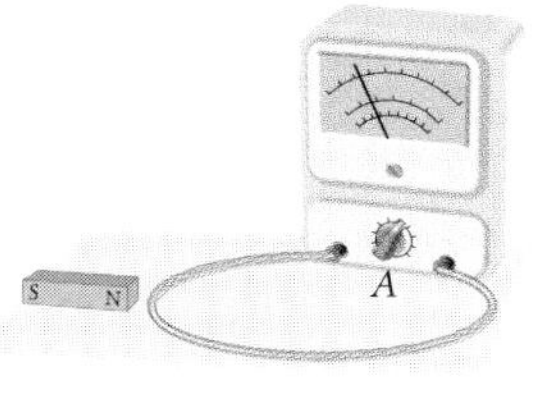

■ FIGURE 30.33

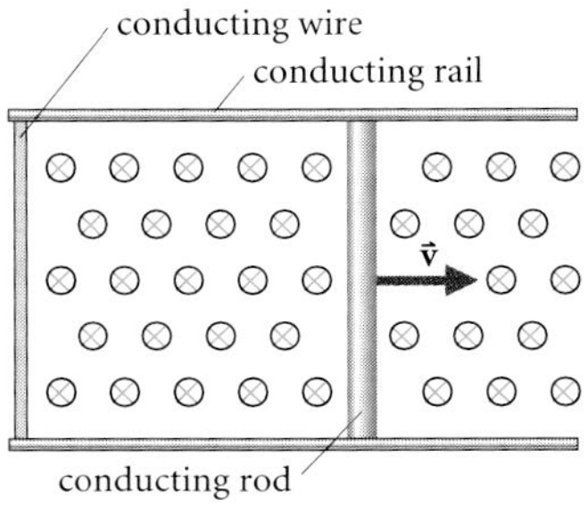

■ FIGURE 30.34

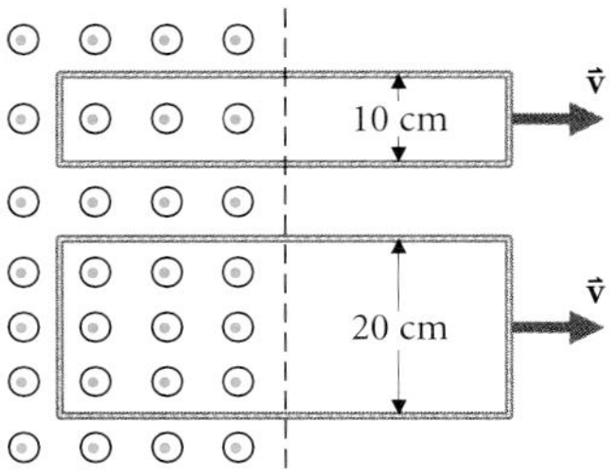

■ FIGURE 30.35

3. The magnetic flux through a loop is given by:

$$\Phi_B(t) = (1\ \text{mWb})(At^4 + \cos \omega t),$$

where t is the time in seconds, $A = 10^{-4}\ /\text{s}^4$, and $\omega = 400$ rad/s. What is the induced emf as a function of time?

§30.2 MOTIONAL EMF

4. A rod moves on metal rails with a velocity $\vec{v}$ as shown in ■ Figure 30.34, in a region where the magnetic field is into the page. The force on the rod is:
(a) zero. **(b)** infinite. **(c)** in the direction of $\vec{v}$.
(d) in the direction opposite $\vec{v}$. **(e)** perpendicular to $\vec{v}$.
Explain why you chose your answer.
5. Two rectangular wire loops each have a resistance of 10 Ω, but one is twice as wide as the other (■ Figure 30.35). They are both pulled out of a strong magnetic field $B = 0.5$ T at the same constant speed $v = 10$ m/s. Use Faraday's law to compare the emfs developed around the two loops and the heat losses in each. On which loop must the greater force act?
6. A simple generator has a rectangular wire loop with dimensions 2.0 cm by 4.0 cm and 150 turns. If it is rotated at $f = 60.0$ Hz in a field of 0.80 T, what is the peak emf produced?
7. A motor has a square coil with 1.0-cm sides and 50 turns that rotates in a magnetic field of 0.75 T. If a current of 2.5 A is supplied to the motor, what is the maximum torque it exerts?

§30.3 THE NATURE OF EMF

8. A railroad supervisor decides to check on his engineers to see whether they are driving the trains at the correct speed. He places his voltmeter leads on either side of the 2.0-m-wide tracks and measures a potential difference 1.3 mV. If the vertical component of the Earth's field at that location is 2.3×10^{-5} T, what is the train's speed?

§30.4 CALCULATION OF INDUCED ELECTRIC FIELD

9. What induced electric field is produced inside a solenoid, 1.0 cm from its axis, if it has 25 turns/cm and the current is increased at 15 A/s?
10. Kerst's 1942 betatron accelerated electrons to 20 MeV at an orbit radius of 19 cm. The momentum of a 20-MeV electron, allowing for relativistic effects, is $39mc$, where m is the electron mass and c is the speed of light. What maximum magnetic field was required?

§30.5 EDDY CURRENTS

11. A copper plate of thickness 0.2 cm moves at 2 m/s between the poles of a magnet. The magnet poles are circular with radius 3 cm, and the magnetic field strength between them is 0.3 T. Estimate the force required to keep the plate moving.

§30.6 THE AMPÈRE–MAXWELL LAW

12. A uniform sheet lying in the x-y-plane has a uniform charge density σ that is increasing at a uniform rate $d\sigma/dt = 16\ \text{mC/m}^2\cdot\text{s}$. Find the displacement current through a circle of radius $r = 2.5$ cm lying parallel to the x-y-plane at $z = z_0 = 50$ cm.
13. A parallel plate capacitor is being charged. At one instant, the surface charge density on the plates is increasing at $35\ \text{mC/m}^2\cdot\text{s}$. Find the magnetic field 1.0 cm from the center of the plates at this instant.

Questions and Problems

§30.1 INDUCED EMF

14. ❖ An ammeter is attached to a circular loop of wire. You grab a bar magnet and push it through the loop N pole end first. Sketch the ammeter reading as a function of time.
15. ❖ What is the direction of current through the ammeter in Example 30.1?
16. ❖ Two loops of wire lie side by side on the table (■ Figure 30.36). One loop includes a battery and a switch. The other includes an ammeter. The switch, initially open, is closed. At the time of closure, current through the ammeter is:
(a) momentarily to the right (The ammeter reading is positive.).
(b) momentarily to the left (The ammeter reading is negative.).

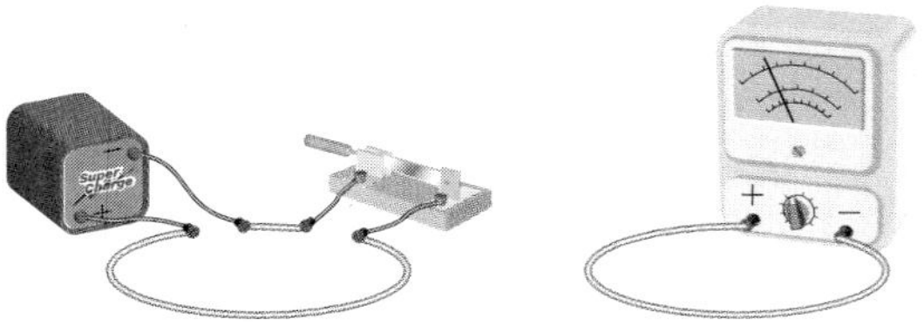

■ FIGURE 30.36

(c) steadily to the right.
(d) steadily to the left.
(e) first to the right and then to the left.
Explain what is wrong with the answers you didn't choose.

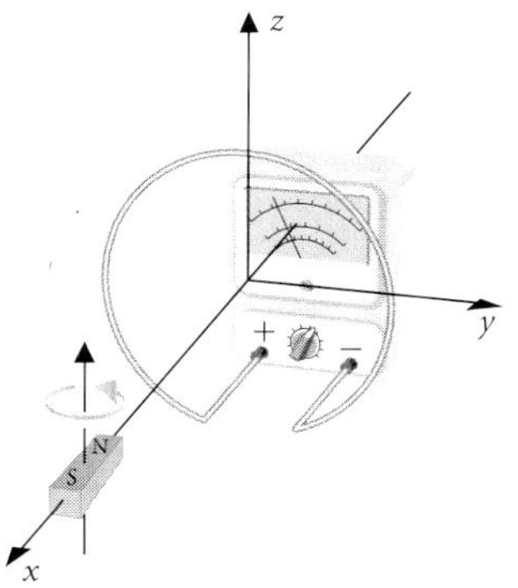

■ FIGURE 30.37

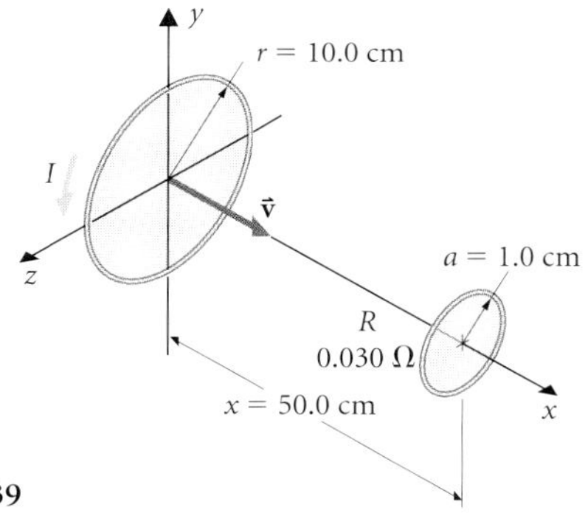

■ FIGURE 30.39

17. ❖ A circular loop of wire lies in the y-z-plane with its center at the origin and is connected to an ammeter that reads positive when the current enters the + terminal. A magnet lying along the x-axis is rotated 180° about an axis parallel to z (■ Figure 30.37), until its N pole points toward the loop. The ammeter reads
(a) first plus then minus.
(b) essentially not at all.
(c) plus while the magnet is rotating.
(d) minus while the magnet is rotating.
(e) first minus then plus.
Explain what is wrong with the answers you didn't choose.

18. ❖ Two conducting loops lie perpendicular to a magnetic field that is increasing at 1.0 mT/s. Loop a is circular with radius 1.5 cm. Loop b is square with 2.5-cm sides. Both have the same resistance. In which loop is the current greater? Justify your answer.

19. ❖ Two identical conducting loops lie perpendicular to a magnetic field. At the position of loop A, the field is 1.5 T and it is increasing at 0.2 T/s. At the position of loop B, the field is 1.2 T and it is decreasing at 0.3 T/s. In which loop is the emf greater? Why?

20. ♦ A coil with 150 turns is connected in series with a voltmeter. If the average voltmeter reading during a 2.7-s interval is 71 V, what change of magnetic flux through the coil occurs during the interval?

21. ♦ A circular loop of radius 2.50 cm lies perpendicular to a magnetic field that is decreasing at 0.305 T/s. What emf is induced in the loop?

22. ♦ A circular wire loop of radius 3.33 cm lies in the y-z-plane with its center at the origin. The magnetic field in the region is given by:

$$\vec{\mathbf{B}} = [0.350\ \mathrm{T} + (0.125\ \mathrm{T/s})t](\hat{\mathbf{i}} - \hat{\mathbf{j}} - \hat{\mathbf{k}}).$$

What is the emf induced in the loop at $t = 0$?

23. ♦♦ The magnetic field in a region of space is:

$$\vec{\mathbf{B}} = (15\ \mathrm{T})\hat{\mathbf{k}}\{[t/(1.0\ \mathrm{s})]^2 - [t/(3.0\ \mathrm{s})]^3\}.$$

A circular loop of radius $r = 3.0$ cm and resistance 0.050 Ω lies in the x-y-plane. **(a)** Find the current in the loop at $t = 0.50$ s, 1.0 s, 2.0 s, and 5.0 s. **(b)** When does the maximum current occur, and what is its value?

24. ♦♦ Two long, tightly wound solenoids have a common axis (■ Figure 30.38). The inner solenoid has radius $a = 2.0$ cm, $N = 10^3$ turns, and a resistance $R = 10.0$ Ω. The outer solenoid has $n = 1.7 \times 10^5$ turns per meter and carries a current that is increasing at a constant rate $dI/dt = 0.60$ A/s. What is the current through an ammeter connected to the inner solenoid? What is its direction?

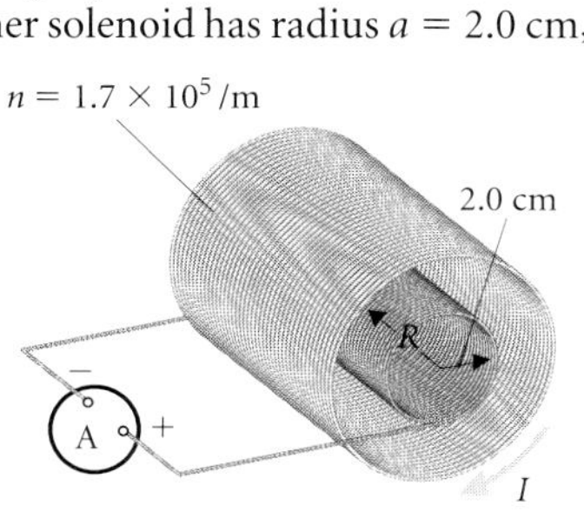

■ FIGURE 30.38

25. ♦♦ (■ Figure 30.39) Two circular loops are parallel to the y-z-plane with their centers on the x-axis. The loop on the right, with resistance $R = 0.030$ Ω and radius $a = 1.0$ cm, is fixed in position. The larger loop, a distance $x = 50.0$ cm to the left, carries current $I = 0.15$ A and has radius $r = 10.0$ cm. It moves directly toward the smaller loop with a velocity of $(2.2\ \mathrm{m/s})\hat{\mathbf{i}}$. Find the magnitude and sense of the current induced in the smaller loop. [*Hint:* Use the result for the magnetic field on the axis of a current loop, eqn. (28.7), assuming $a \ll r \ll x$.]

26. ♦♦ A circuit contains a 10.0-Ω resistor, a 12-V battery, and a 15-cm-radius loop immersed in a 1.0-T magnetic field (■ Figure 30.40). How fast must the magnetic field be changed to prevent current in the circuit? Should it be increased or decreased?

27. ♦♦ A loop of wire is in the form of an equilateral triangle of side $\ell = 3.0$ cm. The wire has resistivity

$$\rho = 1.7 \times 10^{-8}\ \Omega\cdot\mathrm{m}$$

and a diameter of 1.0 mm. What is the resistance of the loop? The loop lies in the x-y-plane and the magnetic field is:

$$\vec{\mathbf{B}} = B_o\hat{\mathbf{k}}[(t/t_o)^2 - (t/t_o)^4],$$

where $B_o = 0.33$ T and $t_o = 1.0$ s. Find the current in the loop as a function of time and plot it. Show that the current reaches a maximum at 0.41 s. What is the power being dissipated in the loop at this time? When, other than at $t = 0$, is the power zero? What happens after that?

§30.2 MOTIONAL EMF

28. ❖ A circular loop is moving at speed v parallel to a diameter (■ Figure 30.41). A uniform magnetic field $\vec{\mathbf{B}}$ is perpendicular to the plane of the loop. Is there a current around the loop? Why or why not?

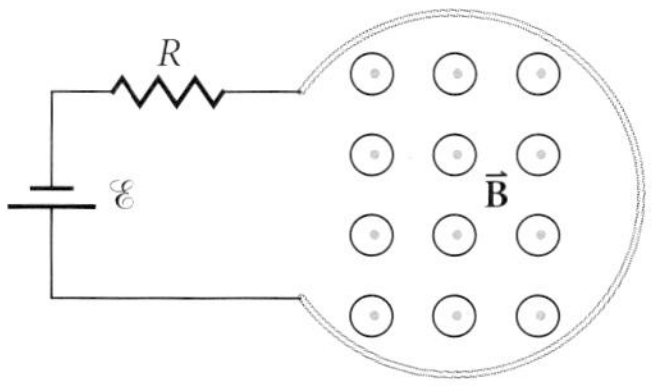

■ FIGURE 30.40

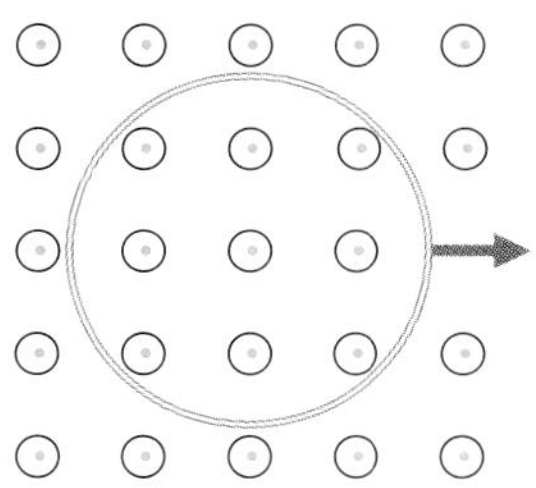

■ FIGURE 30.41

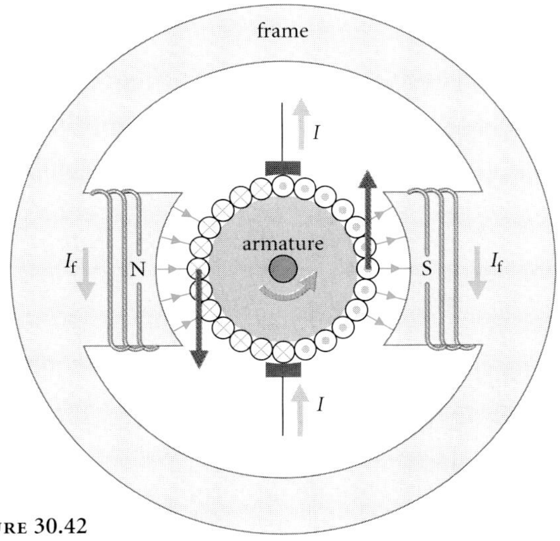

■ FIGURE 30.42

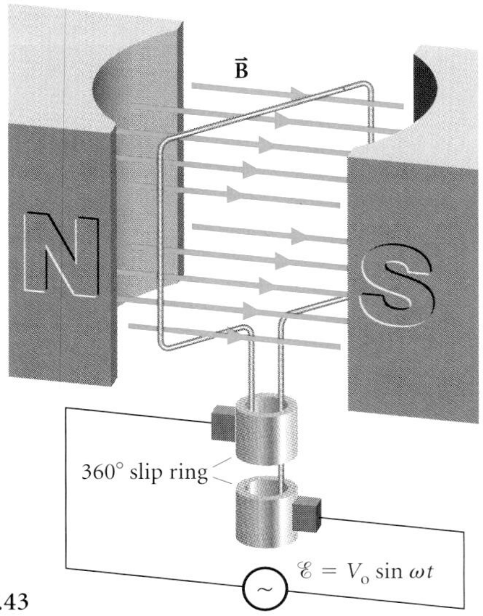

■ FIGURE 30.43

29. ❖ ■ Figure 30.42 shows a schematic of a practical DC motor. Discuss how the output torque varies. Describe the advantages, both mechanical and electrical, of this design over the simple motor discussed in the text.

30. ♦ A simple generator has a coil of area 15.0 cm^2 and 125 turns. It rotates at 60.0 Hz. What magnetic field is required if the peak voltage generated is to be 75.0 V?

31. ♦ A simple generator has a rectangular coil measuring 15.0 cm × 35.0 cm. It rotates at 60.0 Hz in a 1.25-T magnetic field. How many turns are required in the coil for the generator to produce a peak emf of 8.50 kV?

32. ♦♦ A simple generator (cf. Figure 30.11) has a coil with dimensions 0.10 m × 0.20 m, which is rotated at 60.0 Hz in a 0.50-T magnetic field. How many windings are needed if the generator is to develop an average power of 0.10 kW when the total resistance in the circuit is 0.10 kΩ?

33. ♦♦ A simple motor has 250 turns of wire on a coil measuring 12 cm × 4.0 cm. Its resistance is 6.0 Ω. The magnetic field between the pole pieces is 0.95 T. How much current in the coil is needed to produce a maximum torque of 30.0 N·m? What maximum and minimum emf is needed to drive the motor at 60 Hz if the current has constant magnitude?

34. ♦♦ A circular wire loop with resistance $R = 1$ mΩ and radius $a = 5$ cm is placed in a constant magnetic field $B = 1$ T with $\vec{B}$ perpendicular to the plane of the loop. You pull on opposite ends of a diameter and draw the loop into a very narrow rectangle. **(a)** Explain why there is a current in the loop. **(b)** Explain why the total charge that moves past any point on the loop is independent of the time you take to pull the wire taut. **(c)** If you pull the loop closed in $\frac{1}{2}$ s, estimate how much thermal energy is produced in the loop and the average force you have to exert on the wire.

35. ♦♦ A uniform magnetic field $\vec{B} = B_0\hat{k}$ is present in the region $y > 0$. A semicircular wire loop lying in the x-y-plane has radius a, resistance R, and is centered at the origin. It rotates about the $+z$-axis with a constant angular speed ω. The straight diameter of the loop lies along the x-axis at $t = 0$, with the curved side at positive y. Find the emf induced in the loop as a function of time and sketch a graph of the emf versus time for one complete rotation of the loop.

36. ♦♦ A generator is made from a metal rod of length 1.00 m. At one end, a hole in the rod fits smoothly over a metal peg. At the other end, the rod slides smoothly on a circular metal track. The rod rotates about the peg at 50.0 rpm, and a magnetic field of 0.350 T is perpendicular to the plane of rotation. What is the current through a 1.25-kΩ resistor connected between the peg and the track?

37. ♦♦ A rectangular loop of resistance $R = 1.0$ mΩ measuring 12 cm × 3.5 cm is projected with velocity $\vec{v} =$ (7.7 m/s, parallel to its long side) into a region where $\vec{B} =$ (0.65 T, perpendicular to the loop). If the mass of the loop is 38 g, compute the emf around the loop, the current in it, the force on the loop, and its acceleration.

38. ♦♦♦ ■ Figure 30.43 shows a simple design for a synchronous motor. An oscillating potential difference $V(t) = V_0 \sin \omega t$ is applied to the terminals of the rotating loop. **(a)** When it is not driving a mechanical load, the motor rotates at angular speed ω, and there is no current in the loop. Explain why. **(b)** When the motor is running without load, what value of V_0 is required? What is the position of the loop as a function of time? **(c)** If the motor is to produce an average mechanical power P_0 and its windings have resistance r, by how much must the potential difference V_0 be increased?

§30.3 THE NATURE OF EMF

39. ❖ A wooden boat is sailing on a conducting (saltwater) sea at speed $v = 12$ m/s relative to the water. The vertical component of the Earth's field is $B_\perp \approx 10^{-5}$ T. A metal rod 1 m long, lying perpendicular to $\vec{v}$, is on board the boat. An engineer on the boat connects a voltmeter to the two ends of the rod. What does the voltmeter read? Another engineer sitting on a buoy reaches up and touches his voltmeter leads to the ends of the rod as the boat passes. What does his voltmeter read? The engineer on board, believing his voltmeter to be broken, tosses the leads into the water where they land 1 m apart. What does the voltmeter read then? If the engineer on the buoy tosses his leads into the water, what does his meter read?

40. ❖ Suppose, in the system of Example 30.4, that both the sliding rod and the cross tie have equal resistance $R/2 = 0.5$ mΩ. Is the current in the circuit changed? Describe the charge densities that arise on the conducting surfaces (i.e., compare them with the case in which the cross tie has the entire resistance). Describe the fields acting within the rod and the cross tie.

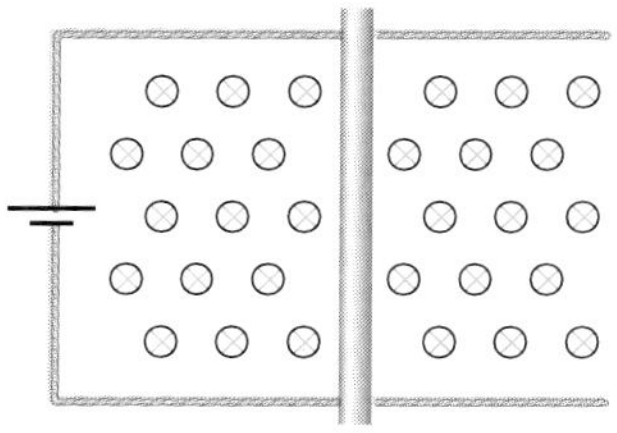

■ FIGURE 30.44

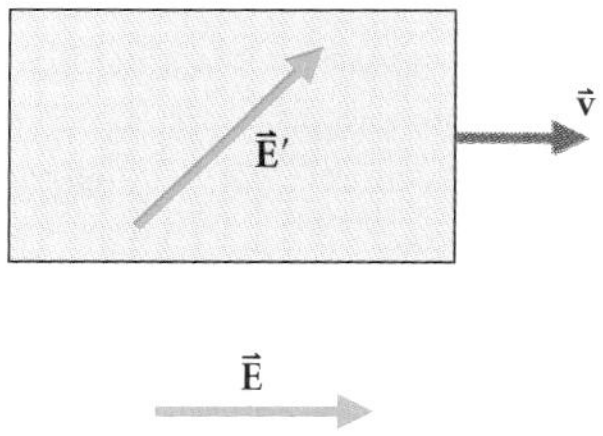

■ FIGURE 30.45

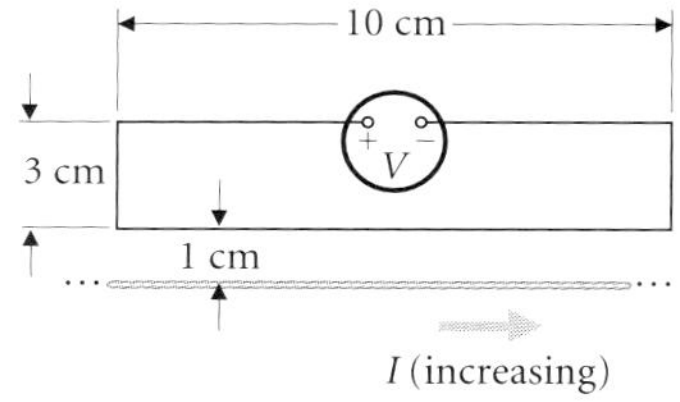

■ FIGURE 30.46

41. ❖ A thin conducting rod rotates about one end with angular speed ω in a region of space with a uniform magnetic field. Looking along the field lines, the rod rotates clockwise. The potential difference between the stationary end A of the rod and its moving end C, $\Delta V = V_A - V_C$, is:
(a) positive. **(b)** negative. **(c)** proportional to ω^2.
(d) inversely proportional to ω. **(e)** zero.
Explain what is wrong with the answers you didn't choose.

42. ❖ At what speed and in what direction must the rod in ■ Figure 30.44 move if there is to be no current in the circuit? Describe qualitatively the surface charges on the system and the fields within the moving rod.

43. ❖ A conducting rod rotates about its center in a region of uniform magnetic field $\vec{B}$ inclined at an angle θ to the rotation axis. Explain why the potential difference between the rod's center and one of its ends is proportional to $\cos\theta$.

44. ✲ ❖ ■ Figure 30.45 shows the direction of the electric field $\vec{E}'$ measured by an observer on a moving cart, and the field $\vec{E}$ measured by an experimenter in the lab. What is the direction of the magnetic field $\vec{B}$ measured in the lab?

45. ♦ A 1-km-long conducting cable tethers a satellite to the space shuttle. The cable moves through the Earth's magnetic field ($B \sim 2 \times 10^{-5}$ T) at a speed of 7 km/s. Estimate the potential difference between the two ends of the tether. The Earth's field near the equator is northward and the space shuttle moves eastward in its orbit. The cable points radially outward from the shuttle to the satellite. Which end of the cable is at higher potential?

46. ✲ ♦ The electric and magnetic fields in a particular region are $\vec{E} = (1.0 \times 10^{-3}\ \text{V/m})\hat{\imath}$ and $\vec{B} = (1.0\ \text{T})\hat{\jmath}$. A particle moves so that the electric field in its rest frame is zero. What is the particle's velocity?

47. ♦♦ A horizontal copper rod of length 0.20 m rotates about one end at angular speed $\omega = 520$ rad/s in a vertical magnetic field $B = 0.25$ T. What is the potential difference between its ends? The rod rotates clockwise when viewed along the field lines, and $\vec{B}$ is parallel to the rotation axis. Which end is at greater potential?

48. ✲ ♦♦ The electric field in a region of space is 125 V/m in the $+x$-direction. The magnetic field is 5×10^{-8} T in the $+z$-direction. Aliens on a spacecraft moving at 2×10^7 m/s in the $+y$-direction measure the electric field in the surrounding space. What is the result of their measurement?

49. ✲ ♦♦ A particle is moving at $(3.00 \times 10^4\ \text{m/s})(\hat{\imath} + \hat{\jmath})$ in an accelerator experiment where the magnetic field in the lab is $(5.50 \times 10^{-3}\ \text{T})(\hat{\jmath} + \hat{k})$. An electric field is applied so that the electric field in a frame moving with the particle is $(0.550\ \text{kV/m})\hat{\imath}$. What is the electric field applied in the lab?

50. ✲ ♦♦♦ A long straight wire near a rectangular circuit carries current that is increasing at a rate $dI/dt = 10^{-3}$ A/s in the sense shown (■ Figure 30.46). The only resistance in the circuit is in the voltmeter. The voltmeter V reads positive when its + terminal is at a higher potential than its − terminal. Compute the reading on the voltmeter. Pay particular attention to the sign of the reading.

§30.4 CALCULATION OF INDUCED ELECTRIC FIELD

51. ❖ Suppose we wanted to use a betatron to slow down electrons. Then when should we inject them into the betatron?

52. ♦ The current in a solenoid with 10^3 turns per meter and radius 1.5 cm is increasing at 6 mA/s. Find the electric field at radii 0.5 cm and 2.5 cm (i.e., inside and outside the solenoid).

53. ♦♦ The poles of an electromagnet have circular cross-sectional area $A = 0.30\ \text{m}^2$. If the field between the poles increases at a rate 0.020 T/s, find the magnitude of the induced electric field at points 0.20 m and 0.50 m from the symmetry axis of the magnet.

54. ♦♦ In Kerst's original betatron, the electrons (mass m) had a final momentum equal to $4.5mc$, where c is the speed of light. Find the maximum induced electric field that acts on the electron in this betatron.

55. ♦♦ The magnetic induction $\vec{B}$ within a long cylindrical shell of magnetic material is increasing at the rate of 10^{-2} T/s (■ Figure 30.47). The magnetic field outside the shell remains negligible. Find the induced electric field as a function of radius from the cylinder axis.

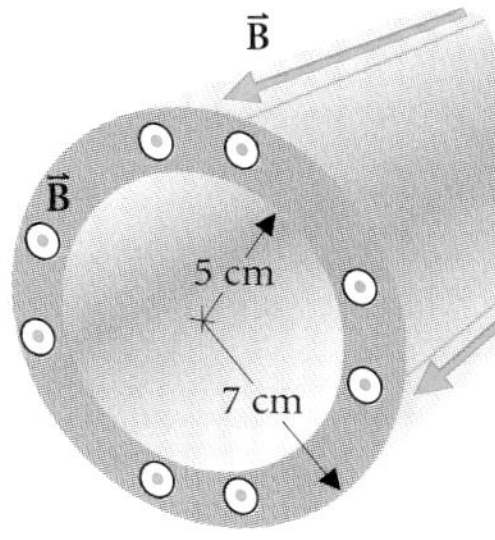

■ FIGURE 30.47

56. ♦♦♦ Two infinite, plane current sheets, each parallel to the x-z-plane, are located at $y = \pm d/2$. The sheets each carry current of magnitude K A/m in the $\pm z$-directions in the top and bottom sheets, respectively. Find the magnetic field produced by the sheets. If K increases with time at a constant rate, find the induced electric field everywhere.

§30.5 EDDY CURRENTS

57. ❖ A uniform magnetic field $\vec{B} = B_0\hat{\imath}$ exists in the region $y < 0$. A thin, rectangular metal plate lying in the y-z-plane measures L in the z-direction and w in the y-direction. It is pulled at constant velocity $\vec{v} = v_0\hat{\jmath}$ out of the magnetic field. Describe how the force on the plate varies with position as it is pulled. Sketch the pattern of current in the plate. Explain why current density is not parallel to $\vec{v} \times \vec{B}$ everywhere.

58. ❖ A thin metal plate is pulled past the poles of a horseshoe magnet so that the field lines penetrate the plate in two patches with opposite polarity. Sketch the resulting pattern of eddy currents. Do the eddy currents caused by the two patches cancel or reinforce each other? Why?

59. ❖ Since the braking force due to eddy currents is proportional to the volume of conductor exposed to magnetic field, wouldn't the largest braking force be achieved by immersing the entire conductor in a uniform magnetic field? Justify your answer.

60. ♦♦ Magnet poles create a rectangular region of magnetic field 3.5 cm × 2.5 cm. The magnetic field strength between the poles is 0.85 T. A sheet of copper 2.5 mm thick moves between the poles at 6.5 m/s. Estimate the power required to keep the plate moving.

61. ♦♦ An aluminum disk with radius $R = 0.50$ m and thickness $t = 1.0$ cm rotates at 1500 rpm about an axle through its center. Magnet poles produce a magnetic field 0.25 T perpendicular to the disk in a circular region 3.0 cm in diameter centered 38 cm from the axle. **(a)** Sketch the eddy current pattern in the disk. **(b)** Calculate the torque acting on the disk. **(c)** Estimate the time for the disk to come to rest.

62. ♦♦ A metal plate swings between the square poles of a permanent magnet as in §30.5. Obtain an approximate expression for the acceleration a of the plate. Estimate the time required for the plate to come to rest by computing v_o/a_o. Estimate the distance traveled as v_o^2/a_o. Obtain numerical values if the mass of the plate is $M = 1$ kg, the magnetic field is $B = 1$ T, the plate thickness is $w = 0.03$ m, the resistivity is $\rho = 10^{-7}\ \Omega\cdot$m, the size of the magnetic field region is $\ell = 0.1$ m, and the initial speed of the plate is $v_o = 10$ m/s.

§30.6 THE AMPÈRE–MAXWELL LAW

63. ♦ A constant current $I = 15$ A charges a parallel plate capacitor with circular plates of radius $R = 3.0$ cm. Use the Ampère–Maxwell law to find the magnetic field between the capacitor plates as a function of r, for $r < R$.

64. ♦♦ The electric field in a region of space is:

$$\vec{\mathbf{E}} = (270\ \text{kV/m}\cdot\text{s})t(\hat{\mathbf{i}} - 2\hat{\mathbf{j}} + 3\hat{\mathbf{k}}).$$

Find the displacement current through a circle of radius 0.65 m in the x-y-plane, and a square with 0.50-m sides in the y-z-plane.

65. ♦♦ The electric field in a region of space is:

$$\vec{\mathbf{E}} = (69.5\ \text{kV/m}\cdot\text{s}^2)t^2(3\hat{\mathbf{i}} + 4\hat{\mathbf{j}}).$$

Find the displacement current through a circle of radius 15 cm in the x-z-plane, centered at the origin, at $t = 0.5$ s and $t = 7.7$ s.

66. (a) ♦♦ Two equal and opposite point charges lie on the x-axis: -0.1 C at $x = 0$ and $+0.1$ C at $x = 10$ cm. Each charge is increasing in magnitude at 10^{-3} C/s. Find the displacement current through a circle centered at the point $x = 5$ cm, $y = 0$, $z = 0$, with radius 4 cm. **(b)** ♦♦♦ Use the result of part (a) to find the magnetic field at the point $x = 5$ cm, $y = 4$ cm, $z = 0$. **(c)** ♦♦ The charges are increasing because there is a constant current I along the x-axis from $+\infty$ to $x = 10$ cm and from $x = 0$ to $-\infty$. Find the magnetic field using the Biot-Savart law (§28.2.1) and show that the result agrees with your answer to part (b).

67. ♦♦♦ A point charge is created by passing a constant current I along a semi-infinite wire as shown in ■ Figure 30.48. Find the magnetic field at point P.

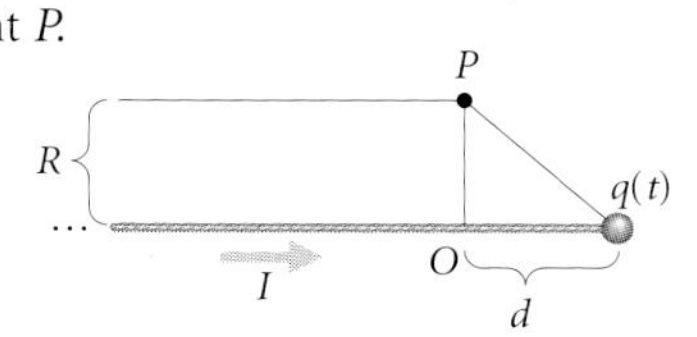

■ FIGURE 30.48

68. ♦♦♦ Two semi-infinite straight wires lie along the positive and negative x-axes. Constant currents I in each wire form a point charge $Q(t)$ at the origin. Find the displacement current through a circle of radius r parallel to the y-z-plane with center on the x-axis at an arbitrary value of x. Apply the Ampère–Maxwell law to find the magnetic field at a point on the circle. Is your result consistent with the Biot-Savart law?

69. ♦♦♦ A coaxial cable lying on the negative x-axis carries current I along its inner conductor of radius r and a return current $I/2$ along the outer sheath at radius $R \gg r$. The cable ends on a conducting sphere of radius R, centered at the origin, which accumulates charge at a rate $dQ/dt = I/2$. Assuming that the charge distribution on the sphere is uniform, calculate the magnetic field at $x = 3R$, $y = 3R$; $x = -3R$, $y = 3R$; and $x = -3R$, $y = R/2$.

Additional Problems

70. ❖ A bar magnet is dropped along the axis of a long, vertical conducting cylinder. The magnetic moment of the magnet lies along the axis of the cylinder. What produces the upward force predicted by Lenz's law? How does the force depend on the speed of the magnet? Describe the resulting motion of the magnet.

71. ❖ The iron cores of transformers are not made of solid iron, but of many thin sheets separated by insulating layers. Why?

72. ♦♦ A wire loop made of copper (resistivity $\rho = 1.7 \times 10^{-8}\ \Omega\cdot$m) has a radius $r = 0.1$ m and is made of wire with cross-sectional area $A = 10^{-5}$ m^2. The loop lies in a horizontal plane and is immersed in a uniform, horizontal magnetic field $B = 0.5$ T. The vertical component of the field is zero but is increasing at a rate $dB/dt = 0.01$ T/s. **(a)** Find the resistance of the wire loop. **(b)** Use Lenz's law to determine the sense of current in the loop. **(c)** Use Faraday's law to find the emf around the loop. **(d)** What is the current in the loop? **(e)** What is the magnetic moment of the loop? **(f)** What is the torque acting on the loop?

73. ♦♦ A long, rectangular wire loop of width 15.0 cm is pulled parallel to its long side at 0.505 m/s into a region where the magnetic field is perpendicular to the loop. The magnetic field strength is decreasing at 0.250 T/s. The loop begins to enter the magnetic field at $t = 0$, when the field strength is 1.500 T. Calculate the rate at which flux through the loop is changing. What is the induced emf at $t = 1.00$ s and $t = 5.00$ s?

74. ♦♦ A small coil with 35 turns of wire and area $A = 6.2 \times 10^{-4}$ m^2 is pulled into a position between the poles of a magnet with its plane perpendicular to the magnetic field. If it takes 2.7 ms from the time that the coil begins to enter the field region until it is in place, and the average emf developed by the coil, as measured by a voltmeter, is $+9.6$ V, what is the magnetic induction between the poles of the magnet? If the coil is then rotated 180° in 3.4 ms, what average emf is measured by the voltmeter? Explain your choice of sign carefully. If the coil is then removed from the magnetic field region in 4.2 ms, what average emf is measured?

75. ♦♦ The magnetic field in a region of space is described by the function:

$$\vec{\mathbf{B}} = B_o(x\hat{\mathbf{i}}/a - y\hat{\mathbf{j}}/a).$$

A square wire loop of side b and resistance R lies parallel to the y-z-plane and moves along the x-axis at constant speed v. It is at the origin at $t = 0$. Find the current in the loop, and evaluate it for $B_o = 0.81$ T, $v = 15$ m/s, $a = 10.0b$, $b = 1.2$ cm, and $R = 0.18\ \Omega$. What force is needed to move the loop? Show that the mechanical power used to move the loop equals the electrical power dissipated in the resistance of the loop.

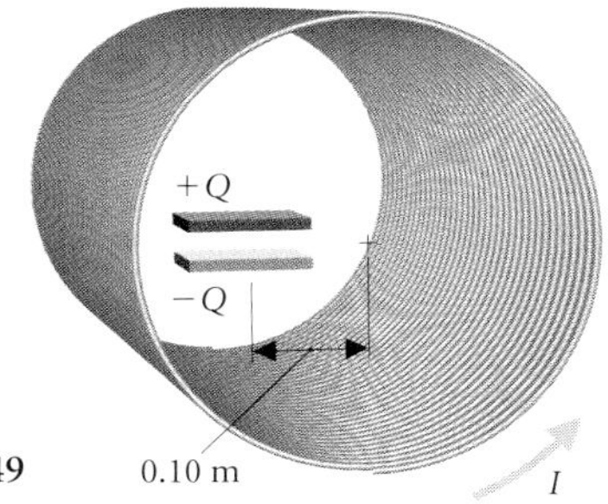

■ FIGURE 30.49

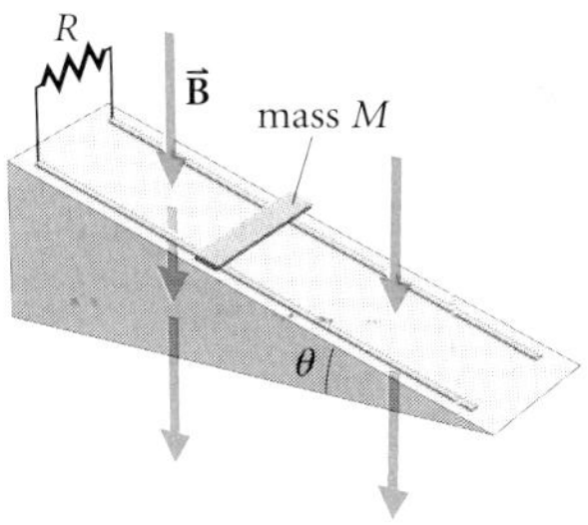

■ FIGURE 30.51

76. ♦♦ A small cylindrical magnet 1.0 cm in diameter, with a mass of 4 g, slides at a constant speed of 3 cm/s down a smooth ramp making a 30° angle with the horizontal. The ramp is made from an aluminum sheet 2.5 mm thick. Estimate the magnetic field produced by the magnet at its surface.

77. ♦♦ A parallel plate capacitor with plate area $A = 2.0 \times 10^{-2}\ \mathrm{m}^2$ and separation $d = 1.0 \times 10^{-4}$ m carries charge $Q = 1.3$ pC. The capacitor is inside a solenoid with $n = 4.0 \times 10^4$ turns/m and current increasing at a rate $dI/dt = 1.0 \times 10^4$ A/s (■ Figure 30.49). The center of the capacitor is at $r = 0.10$ m from the solenoid axis. **(a)** What is the static electric field between the plates? **(b)** What is the potential difference between the plates? **(c)** What is the induced electric field near the center of the capacitor plates? **(d)** What is the total electric field between the plates? **(e)** For what charge Q would the induced field just balance the Coulomb field? Would there still be a potential difference in this case?

78. ♦♦ A perfectly conducting rod slides on frictionless, perfectly conducting rails as shown in ■ Figure 30.50. The magnetic field strength increases at a constant rate $dB/dt = \alpha$. Describe how the rod moves. Find its position as a function of time, if it is initially at rest at $x = x_0$.

79. (a) ♦♦ Two metal rods, each of mass $m = 1.75$ kg and length $\ell = 0.65$ m, are joined at the ends by two metal springs. Each spring has spring constant $k = 9.0 \times 10^{-5}$ N/m and equilibrium length ℓ. The total resistance R in the circuit is 1.5 Ω. A uniform magnetic field $B = 0.37$ T is perpendicular to the plane containing the rods and springs. One rod is fixed and the other is free to move, remaining parallel to the fixed rod. At time $t = 0$, the movable rod is pulled out to a distance 2ℓ from the fixed rod and released. Describe the subsequent motion qualitatively. Calculate the net force on the rod when it is a distance x from the fixed rod and has speed v. **(b)** ♦♦♦ Assume that the position of the rod may be described by the function $x = \ell + A_1 e^{-b_1 t} + A_2 e^{-b_2 t}$. Find A_1, A_2, b_1, and b_2.

80. ♦♦♦ If the rod in Figure 30.44 (Problem 42) has mass M and resistance R, show that it approaches its final speed exponentially: $v = v_f(1 - e^{-\alpha t})$. What is the final speed v_f if there is a coefficient of friction μ between the rod and rails? In that case, show that, when $v = v_f$, the power expended by the battery is equal to the mechanical power needed to overcome friction plus the power dissipated as joule heating in the resistance.

81. ♦♦♦ A perfectly conducting rod of mass M slides on frictionless rails on an inclined plane (■ Figure 30.51) in a region where the magnetic field $\vec{\mathbf{B}}$ is vertical. The rails are joined at the top of the plane by a wire of resistance R. At what speed does the rod slide down the plane? Show that the gravitational energy lost by the rod is dissipated as heat in the resistor.

82. ♦♦♦ Apply Gauss' law to find the charge density on the gap faces of the wire loop in Example 30.10. Estimate the total charge on each gap face and show that the two charges form a dipole with moment p independent of the gap width. Compute the static electric field produced by this dipole at a point P in the loop diametrically opposite the gap. Show that the dipole is inadequate to produce the entire static field known to exist at P. Discuss, qualitatively, how the static field is produced.

83. ♦♦♦ In Example 30.4 the cross tie is replaced by a second rod with resistance $R = 1.0$ mΩ and mass $M = 1.0$ kg. The second rod is free to slide on the rails without friction. At $t = 0$, the two rods are together at $x = 0$, and the one rod is pulled with a constant speed $v = 10.0$ m/s as in the example. Find the velocity and position of the second rod as functions of time. [*Hint:* v is of the form $v = v_f(1 - e^{-\alpha t})$.]

84. ♦♦♦ Between August and November of 1831, Faraday conducted a series of experiments demonstrating electromagnetic induction. In one of them, he rotated a copper disk between the poles of a magnet (■ Figure 30.52) and obtained a steady current in a circuit between the center and edge of the disk. Today, this device is called a *homopolar generator* because it produces DC current. Mercury contacts at the center and the edge of the disk connect it in series with a resistor $R = 15\ \Omega$ and an ammeter. The radius of the disk is $a = 25$ cm, its angular speed is $\omega = 2\pi$ rad/s, and the magnetic field is $B = 1.0$ T. Find the emf developed in the circuit and the reading of the ammeter.

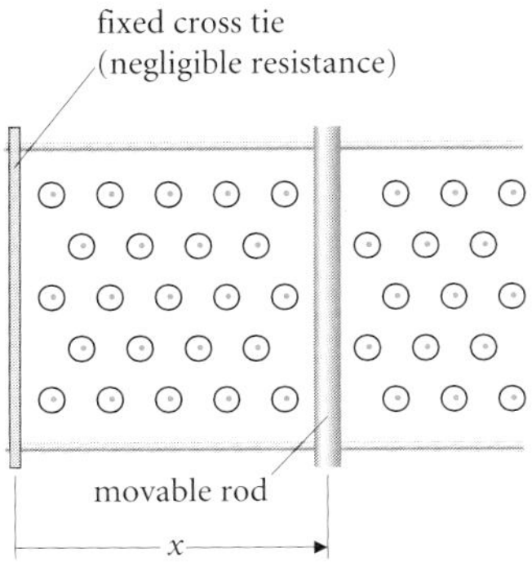

■ FIGURE 30.50

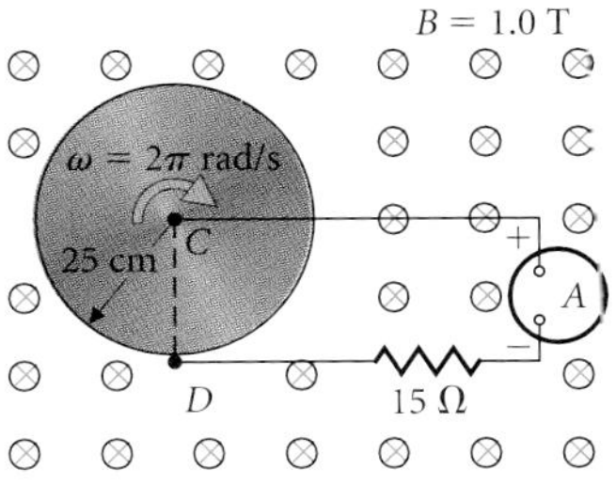

■ FIGURE 30.52

Computer Problems

85. File CH30P85 on the supplementary computer disk lists the emf recorded in a circuit as a function of time. If the flux through the circuit is 0 at $t = 0$, what is the flux through the circuit at $t = 2.0$ s?

86. A metal bar 1.00 m long with mass 0.500 kg slides on rails as shown in Figure 30.44. The emf of the battery is 10.0 V and the resistance of the bar is 1.00 Ω. Find an expression for the acceleration of the bar as a function of its velocity. If the bar starts at the battery at $t = 0$ s, use a numerical method of your choice to compute its velocity and position for $t > 0$. Plot the results and comment. (An analytic solution also exists; compare Problem 80. If you can find it, plot it also and compare with your computed results.)

Challenge Problems

87. A charge Q is at the origin and an equal and opposite charge $-Q$ is at $x = \ell$. There is a current I in a wire joining them along the x-axis. **(a)** Find the electric field at the point $x = \ell/2, y$. The current causes Q to decrease: $I = -dQ/dt$. Use this result to find the displacement current through a circle of radius $\ell/2$ in terms of I. Use the Ampère–Maxwell law to calculate the magnetic field at the point $x = \ell/2, y = \ell/2$. Verify your result using the Biot-Savart law. **(b)** Calculate the potential difference between the two charges if each lies on a sphere of radius $a \ll \ell$. If the connecting wire has a resistance R, how is the current I related to the charge Q? How does Q vary in time? The fields you found in part (a) vary in time as Q does. If fields propagate at the speed of light, under what circumstances do the fields you calculated above give sensible results for the instantaneous fields? (You should be able to state the condition as a minimum value for the resistance R.)

88. A rectangular loop of wire with dimensions $\ell \times w$ has resistance R and mass m. It falls under the influence of gravity into a region of vertical extent w in which the magnetic field $\vec{\mathbf{B}}$ is horizontal and uniform (■ Figure 30.53). At time $t = 0$, the loop is just about to enter the region and its speed is zero. Find the position of the loop as a function of time.

89. **(a)** A magnetic dipole lies along the axis of a circular wire loop with its magnetic moment pointing toward the center of the loop. The magnetic moment of the dipole is increasing. Draw a diagram showing the resulting current and how the force predicted by Lenz's law is produced. What changes occur if the magnetic dipole moment decreases? **(b)** The loop has radius a and is held fixed at a distance z_0 from the dipole. The dipole moment varies sinusoidally as $\vec{\mathbf{m}} = \vec{\mathbf{m}}_0 \sin \omega t$. If the loop has radius R, find the force acting on it as a function of time. Assume that the magnetic field is described by the static dipole formula and that the resistance of the loop obeys Ohm's law. What is the average force on the loop?

90. The magnetic field in a region of space remains uniform, but its strength increases in time. Consider a charged particle moving in the plane perpendicular to the field and assume that its orbit remains circular, though its orbital radius may vary. Show that the following relation holds between the magnetic field and the particle's speed v:

$$\frac{1}{v}\left(\frac{dv}{dt}\right) = \frac{1}{2B}\left(\frac{dB}{dt}\right).$$

Integrate this equation and use the result to show:

$$r^2 B = \text{constant}.$$

The magnetic flux through the particle's orbit remains constant. This is one example of a phenomenon called "flux freezing": conducting matter and magnetic field are strongly coupled. For example, when an ordinary star collapses to form a neutron star 10 km in diameter, the magnetic field lines are dragged inward, resulting in field strengths as high as 10^8 T.

91. ■ Figure 30.54 illustrates the synchronous induction motor. An oscillating current passes through the coils of each electromagnet. The three currents are out of phase with each other by 120°. (Commercial power is usually supplied this way.) Describe the magnetic field produced by the coils. Show that as the currents in the coils vary, the magnetic field pattern they produce in the central region rotates. Hence explain how the field produces a torque on the shaft. Induction motors are particularly useful in applications where sparks are dangerous, since they do not require brushes or slip rings.

92. **(a)** Surface current K spreads charge across the capacitor plates in Problem 63. Show that the surface current at radial distance r from the center of each plate has magnitude:

$$|K(r)| = \frac{I}{2\pi r}\left(1 - \frac{r^2}{R^2}\right).$$

(*Hint:* Show that this distribution of current delivers charge at the required rate to each annular ring of radius r and width dr on the surface of the capacitor plates.) **(b)** Use an Ampèrian curve consisting of circular arcs of radius r on each side of one capacitor plate, connected by lines through and normal to the plate, to show that the surface current is just sufficient to account for the discontinuity of $\vec{\mathbf{B}}$ across the capacitor plate.

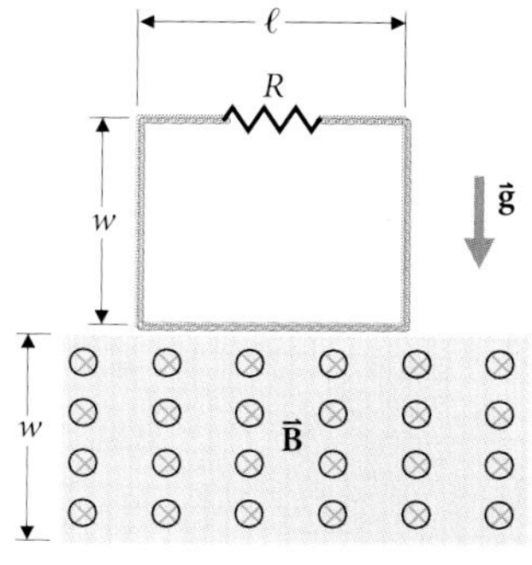

■ FIGURE 30.53

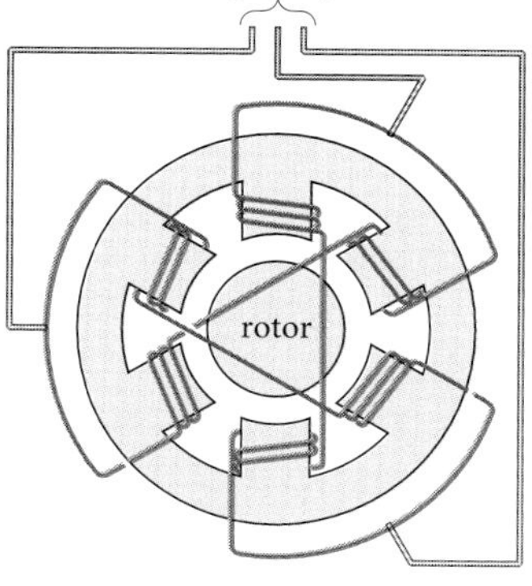

■ FIGURE 30.54

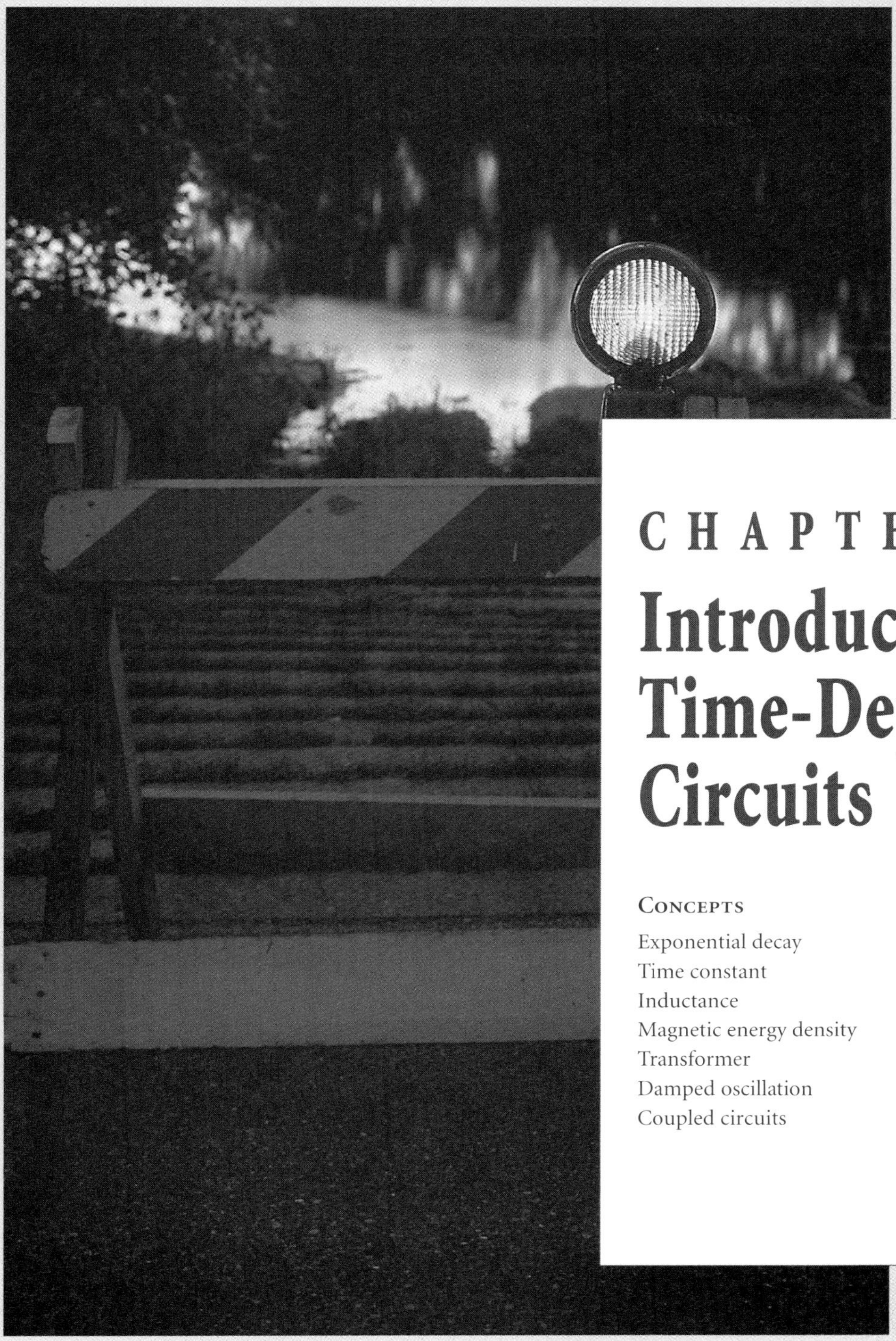

A flashing lamp on a road sign is designed to get your attention. The lamp is part of a circuit that uses a battery and a capacitor. Energy from the battery is stored in the capacitor and periodically released in the bulb's flash. In this chapter we'll investigate several circuits that exhibit similar time-dependent behavior. This circuit is analyzed in Example 31.3.

CHAPTER 31
Introduction to Time-Dependent Circuits

Concepts

Exponential decay
Time constant
Inductance
Magnetic energy density
Transformer
Damped oscillation
Coupled circuits

Goals

Be able to:

Describe qualitatively the behavior of time-dependent circuits.

Use the basic solutions to the exponential and harmonic equations.

Compute the magnetic energy stored in a system.

Recognize systems that exhibit self-induction or mutual induction and estimate their inductance.

"Mine is a long and a sad tale!" said the mouse, turning to Alice, and sighing.
"It is a long tail, certainly," said Alice, looking down with wonder at the mouse's tail; "but why do you call it sad?"

Lewis Carroll

WE'LL DISCUSS MAGNETIC ENERGY IN §31.2.2.

As you drive along the highway at night, the bright flash of a hazard warning catches your eye. Your engine hums and the spark plugs fire in steady rhythm, as you carefully negotiate the danger. Specially designed electric circuits determine the performance of the emergency warning light and the engine's ignition system. Each circuit draws energy slowly from its source, stores it as electric or magnetic field, and then releases it in a rapid burst. Flashguns for photography use a similar circuit. We already have the tools we need to understand the behavior of these circuits: they are the conservation laws for energy and charge, expressed as Kirchhoff's rules (§26.4). In this chapter we'll use these rules to investigate the behavior of circuits in which current changes in time.

31.1 RESISTOR–CAPACITOR CIRCUITS

31.1.1 *Discharging a Capacitor*

The time evolution of charge and current in any circuit follows two characteristic patterns: oscillation and/or exponential decay. We'll learn to recognize the pattern of each circuit that we study before we develop the mathematics. In our first example a resistor and a capacitor are connected in series.

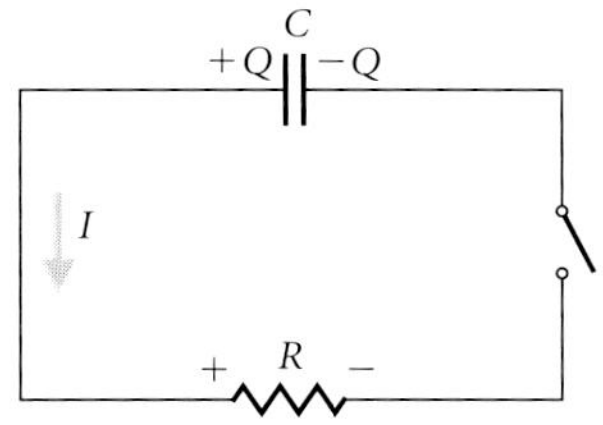

■ FIGURE 31.1
Discharge of a capacitor. When the switch is closed, current through the resistor reduces the charge Q. The arrow defines the positive sense for the current variable $I(t)$.

■ Figure 31.1 is the circuit diagram for a charged capacitor in series with a resistor and a switch. When the switch is closed, the potential difference across the capacitor drives current through the resistor, and the capacitor discharges. As the charge decreases, so do the potential difference and the current. Ultimately, both charge and current decrease to zero. The energy originally stored in the capacitor is converted to thermal energy in the resistor.

The charge on the positive plate of the capacitor is represented by the function $Q(t)$. The current is represented by the function $I(t)$. A positive value for this function represents current in the sense shown by the arrow in the figure. In this case, we expect both functions to be positive at all times and to approach zero as time passes.

REMEMBER: THE DEFINITION OF CAPACITANCE, EQN. (27.2), IS $C = Q/\Delta V$.

To analyze the circuit behavior quantitatively, we apply Kirchhoff's rules. On a counterclockwise path around the circuit, the change in potential across the capacitor is Q/C and across the resistor is $-IR$. From Kirchhoff's loop rule:

$$\frac{Q}{C} - IR = 0.$$

When the function $I(t)$ is positive, current in the circuit reduces the charge $Q(t)$ on the capacitor:

NOTE THE MINUS SIGN: IT'S IMPORTANT! WE'LL DISCUSS SIGNS FURTHER IN THE *MATH TOOLBOX* ON SIGN CONVENTIONS.

$$I = -\frac{dQ}{dt}.$$

Combining these two equations: $$\frac{Q}{C} + \frac{dQ}{dt}R = 0,$$

or $$\frac{dQ}{dt} + \frac{Q}{RC} = 0. \qquad (31.1)$$

Thus the loop rule gives a differential equation for the charge as a function of time. To solve it, we collect all the terms in Q on one side of the equation, leaving t on the other side:

$$\frac{dQ}{Q} = \frac{-1}{RC}\,dt.$$

Next we integrate both sides between $t = 0$, when the switch is closed, and a later time $t > 0$. During this time, Q decreases from its initial value $Q(0) = Q_0$.

WE ARE USING THE DUMMY VARIABLES Q' AND t' IN THE INTEGRANDS TO DISTINGUISH THEM FROM THE LIMITS Q AND t.

$$\int_{Q_0}^{Q} \frac{dQ'}{Q'} = \int_0^t \left(\frac{-1}{RC}\right) dt'.$$

Thus $$\ln Q' \Big|_{Q_0}^{Q} = \ln Q - \ln Q_0 = \ln\left(\frac{Q}{Q_0}\right) = \frac{-t}{RC},$$

or

$$Q(t) = Q_0 e^{-t/RC}. \tag{31.2}$$

Because its solution involves the exponential function, eqn. (31.1) is often called the *exponential equation.* Its characteristic feature is that the derivative of a quantity, here Q, is proportional to the quantity itself.

The current is given by:

$$I(t) = -\frac{dQ}{dt} = \frac{Q_0}{RC} e^{-t/RC}, \tag{31.3}$$

and the potential difference across the capacitor is:

$$\Delta V(t) = \frac{Q(t)}{C} = \frac{Q_0}{C} e^{-t/RC} = \Delta V_0 e^{-t/RC}. \tag{31.4}$$

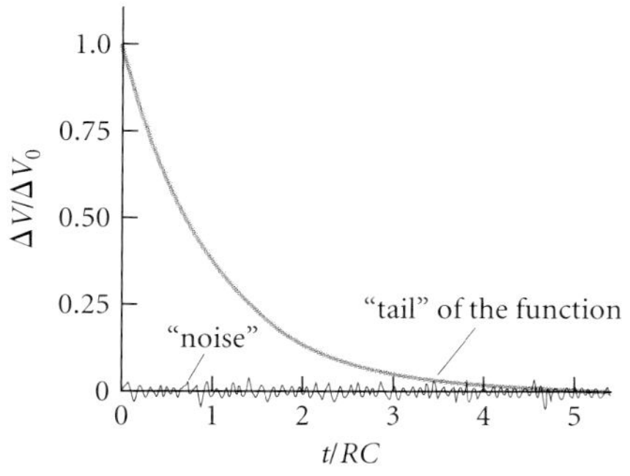

■ **FIGURE 31.2**
Potential difference across the discharging capacitor decays exponentially as a function of time. Random fluctuations (noise) in the circuit exceed the potential difference predicted by the model after a few time constants (here $t \approx 4.5\tau = 4.5RC$), and the potential difference is effectively zero. The part of the function that asymptotically approaches zero ($t \gtrsim 3\tau$) is called the "exponential tail." Current displays the same type of exponential decay.

■ Figure 31.2 is a graph of the potential difference as a function of time. The constant $\Delta V_0 = Q_0/C$ is the potential difference at $t = 0$, when the switch is closed. In a formal mathematical sense, the charge requires infinite time to decay to zero. In practice, once it has decreased to a few percent of its initial value, deviations of the real circuit from the ideal RC model become important. A practical measure of the time for the circuit to discharge is the *time constant*

$$\tau_C \equiv RC. \tag{31.5}$$

At $t = \tau_C$, the charge is reduced by a factor e, and at $t = 3\tau_C$, it is $1/e^3$, or about 5% of its initial value.

REMEMBER $e = 2.718\ldots$, OR ABOUT 3 FOR ORDER OF MAGNITUDE ESTIMATES. SIMILARLY, $e^2 \sim 10$.

THE ARGUMENT OF AN EXPONENTIAL FUNCTION, LIKE THAT OF A LOGARITHM, IS ALWAYS DIMENSIONLESS.

EXERCISE 31.1 ♦ Show that the product RC has dimensions of time. That is, show $1\ \Omega \cdot \text{F} = 1$ s.

EXAMPLE 31.1 ♦♦ A 10.0-μF capacitor is in series with a 2.5-kΩ resistor. The capacitor is initially charged to a potential difference of 120 V. What is the time constant of the circuit? What are the initial current through the resistor and the current after one time constant?

MODEL The capacitor has an initial charge $Q_0 = C\,\Delta V_0$. The results we have just developed describe its discharge through the resistor.

SETUP The time constant (eqn. 31.5) is:

$$\tau_C \equiv RC = (2.5\ \text{k}\Omega)(10.0\ \mu\text{F}) = 2.5 \times 10^{-2}\ \text{s}.$$

The current as a function of time (eqn. 31.3) is:

$$I(t) = \frac{Q_0}{RC} e^{-t/RC}.$$

SOLVE The initial current is:

$$I(0) \equiv I_0 = \frac{Q_0}{RC} = \frac{C\,\Delta V_0}{RC} = \frac{\Delta V_0}{R}$$

$$= \frac{120\ \text{V}}{2.5 \times 10^3\ \Omega} = 0.048\ \text{A}.$$

This is the maximum value of the current. After one time constant,

$$I(\tau_C) = I_0/e = 0.018\ \text{A}.$$

ANALYZE To increase the time constant, we may increase either C or R. At a fixed voltage ΔV_0, increasing R also decreases the initial current in the circuit, whereas increasing C increases the initial charge on the capacitor.

It is the resistor that is responsible for the exponential decay of charge and current in the *RC* circuit because resistors are the only circuit elements that dissipate energy. In its initial state, the charged capacitor contains electrostatic energy that is converted into thermal energy in the resistor as the capacitor discharges. No energy remains stored in the circuit in its final state.

EXAMPLE 31.2 ♦♦♦ Show that when a capacitor is discharged the energy dissipated as heat in the resistor equals the initial energy stored in the capacitor.

MODEL The initial stored energy is $U = Q_0^2/(2C)$ (eqn. 27.6a), and the energy is dissipated in the resistor at a rate $P(t) = I^2R$ (eqn. 26.3b). Since the current decreases with time, the power dissipated by the resistor also decreases, so we integrate the power over time to obtain the total energy dissipated.

SETUP We use eqn. (31.3) for $I(t)$.

SOLVE

$$U = \int_0^\infty P\,dt = \int_0^\infty I^2R\,dt = \int_0^\infty R\left(\frac{Q_0}{RC}e^{-t/RC}\right)^2 dt$$

$$= \int_0^\infty \frac{Q_0^2}{RC^2}e^{-2t/RC}\,dt = \frac{Q_0^2}{RC^2}\left(-\frac{RC}{2}e^{-2t/RC}\right)\Bigg|_0^\infty = \frac{Q_0^2}{2C}.$$

ANALYZE As expected, the total energy dissipated in the resistor equals the initial stored energy. ∎

Math Toolbox

SIGN CONVENTIONS

To analyze a circuit, we describe the physical quantities, charge and current, with mathematical functions, $Q(t)$ and $I(t)$. Both the magnitude and sign of these functions give important information. The function $Q(t)$ means the amount of charge on a specific capacitor plate that we choose. The sign of Q determines which plate is positive. Similarly, the function $I(t)$ means current in a specific direction that we choose: a negative value means current is in the opposite direction. This freedom of definition is similar to our freedom to choose the positive coordinate directions in a mechanics problem. It is an essential freedom, since we cannot always guess in advance the actual behavior of a complicated circuit.

For example, consider two different conventions in the capacitor discharge problem. In our previous discussion, current away from the positive plate of the capacitor was called positive. This choice leads to the relation:

$$I = -\frac{dQ}{dt},$$

and the solution has $I(t) > 0$; charge flows off the capacitor.

If we reverse the definition of I (■ Figure 31.3) so that current *toward* the positive plate of the capacitor is called positive, then:

$$I = +\frac{dQ}{dt}.$$

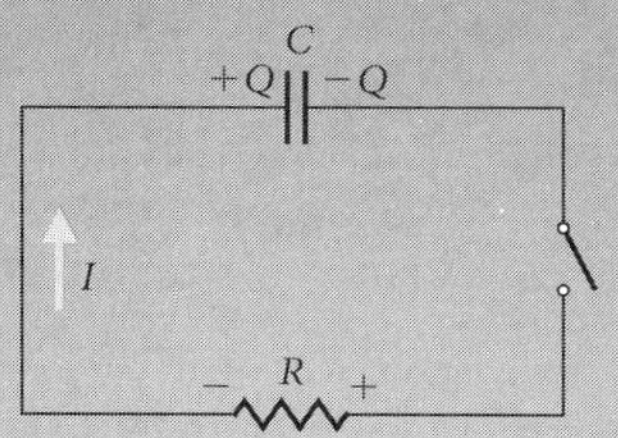

■ FIGURE 31.3
You may define the current variable to have either sense. With I defined as shown here, $I = +dQ/dt$. The solution has the opposite sign from that in eqn. (31.3), but the same physical meaning.

On a counterclockwise path around the circuit, the potential difference across the resistor is $+IR$, and Kirchhoff's rule becomes:

$$\frac{Q}{C} + IR = 0 \Rightarrow \frac{Q}{RC} + \frac{dQ}{dt} = 0.$$

The differential equation for Q and its solution are unchanged. But, in this case,

$$I = +\frac{dQ}{dt} = -\frac{Q_0}{RC}e^{-t/RC},$$

is negative, again indicating that charge flows off the capacitor. The mathematical function has a different sign, but the physical meaning is the same. Some people like to choose I so that $I = +dQ/dt$; other people prefer $I > 0$. You can't have both, so make a decision and be aware of what you have done.

31.1.2 *The Solution Method*

Our analysis of the discharging capacitor illustrates how the behavior of time-dependent circuits depends on relations among charges, currents, and their rates of change; how Kirchhoff's rules applied to these circuits generate differential equations; and how qualitative physical reasoning acts as a guide to their solution. Each kind of system has a typical behavior, which you should learn to recognize. From this first example we have learned that:

> Resistor–capacitor systems evolve exponentially from an initial state to a final state with a characteristic time constant.

In the series *RC* circuit we discussed, the initial state is a charged capacitor, the final state is a discharged capacitor, and the time constant is

$$\tau_C = RC.$$

The method we have to used to solve for $I(t)$ in §31.1.1 is summarized in ■ Figure 31.4.

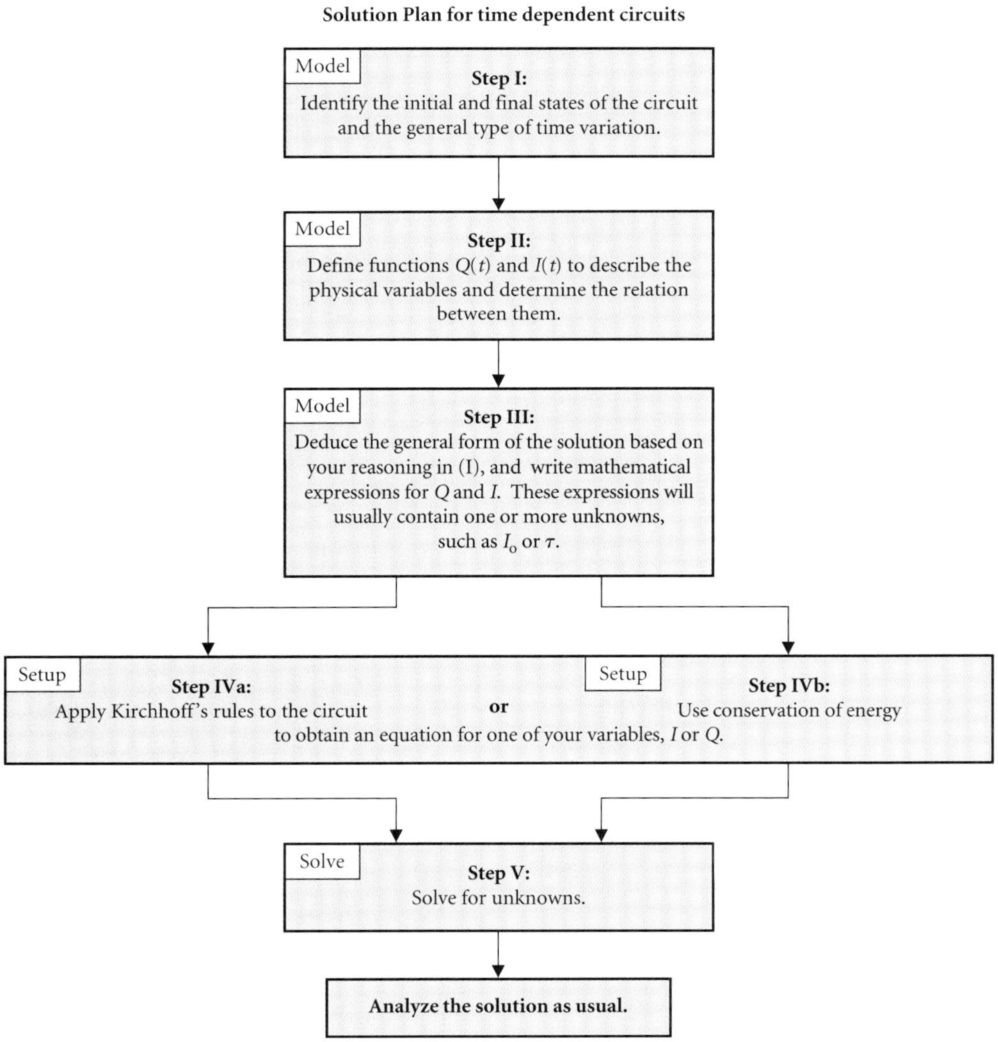

Solution Plan for time dependent circuits
Model
Step I:
Identify the initial and final states of the circuit and the general type of time variation.
Model
Step II:
Define functions Q(t) and I(t) to describe the physical variables and determine the relation between them.
Model
Step III:
Deduce the general form of the solution based on your reasoning in (I), and write mathematical expressions for Q and I. These expressions will usually contain one or more unknowns, such as I0 or τ.
Setup
Step IVa:
Apply Kirchhoff's rules to the circuit
or
Setup
Step IVb:
Use conservation of energy
to obtain an equation for one of your variables, I or Q.
Solve
Step V:
Solve for unknowns.
Analyze the solution as usual.

■ FIGURE 31.4
Solution method for solving time-dependent circuit problems. Compare this with the plan for DC circuits (Figure 26.27).

31.1.3 Charging a Capacitor

THIS IS STEP I OF OUR SOLUTION PLAN, FIGURE 31.4.

■ Figure 31.5 is the diagram of a circuit used to charge a capacitor. Initially, the capacitor is uncharged. At $t = 0$, the switch is closed. The battery drives current through the resistor, building up charge on the capacitor, until the potential difference across the capacitor equals the emf of the battery. Then the potential difference across the resistor and the current in it are zero. The capacitor has a final charge:

$$Q_f = C\mathscr{E}.$$

THE VARIABLES Q AND I ARE DEFINED IN THE DIAGRAM (STEP II).

Following the pattern for RC circuits, the charge evolves exponentially from its initial to its final value. That is, the *difference* $\Delta Q \equiv Q_f - Q(t)$ between the final value of the capacitor charge and its value at time t decays exponentially to zero. ■ Figure 31.6 is a graph of the corresponding potential difference $\Delta V = Q/C$ across the capacitor as a function of time. We expect the time constant to be $\tau_C = RC$. That is, we guess:

$$\Delta Q \equiv Q_f - Q(t) = Q_f e^{-t/RC},$$

THIS IS STEP III.

or

$$Q(t) = C\mathscr{E}(1 - e^{-t/RC}). \qquad (31.6)$$

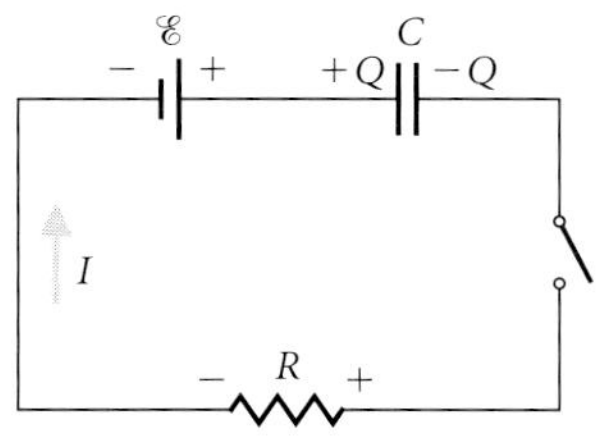

■ FIGURE 31.5
Charging a capacitor. Current increases the capacitor charge until it reaches its final value: with $\Delta V = \mathscr{E}$, $Q_f = \mathscr{E}C$. As the potential difference across the capacitor increases, the current decreases. (Marking Q and I on this figure is Step II of the solution plan.)

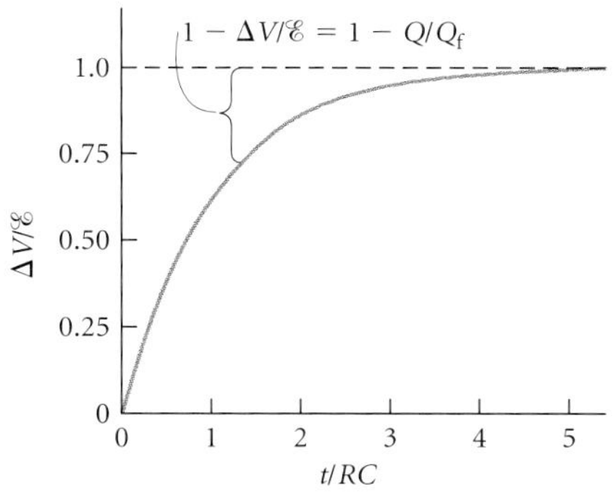

■ FIGURE 31.6
Potential difference across the charging capacitor as a function of time. $\Delta V/\mathscr{E} = Q/Q_f$ approaches unity exponentially. Notice that

$$1 - Q/Q_f = (Q_f - Q)/Q_f = \Delta Q/Q_f \to 0$$

as t increases.

To verify this conclusion, we apply Kirchhoff's rule to obtain a differential equation for the capacitor charge. With the definitions in Figure 31.5, positive current corresponds to increasing capacitor charge:

THIS IS STEP IVA.

$$I(t) = +\frac{dQ}{dt}.$$

Then, traversing the circuit clockwise,

$$0 = \mathscr{E} - \frac{Q}{C} - IR.$$

Combining these two equations, we find:

$$\frac{dQ}{dt} + \frac{Q}{RC} = \frac{\mathscr{E}}{R}. \qquad (31.7)$$

BUT SEE THE *MATH TOOLBOX* IN §31.3.3.

Although we shall not solve this equation directly, it is straightforward to demonstrate that expression (31.6) is the correct solution.

EXERCISE 31.2 ♦♦ Differentiate eqn. (31.6) and substitute the result into eqn. (31.7) to show that we guessed the correct solution.

We find the solution for $I(t)$ by differentiating:

$$I(t) = \frac{dQ}{dt} = \frac{d}{dt}C\mathscr{E}(1 - e^{-t/RC}) = \frac{C\mathscr{E}}{RC}e^{-t/RC} = \frac{\mathscr{E}}{R}e^{-t/RC}.$$

Thus $I(0) = \mathscr{E}/R$, and $I \to 0$ as $t \to \infty$, as we predicted.

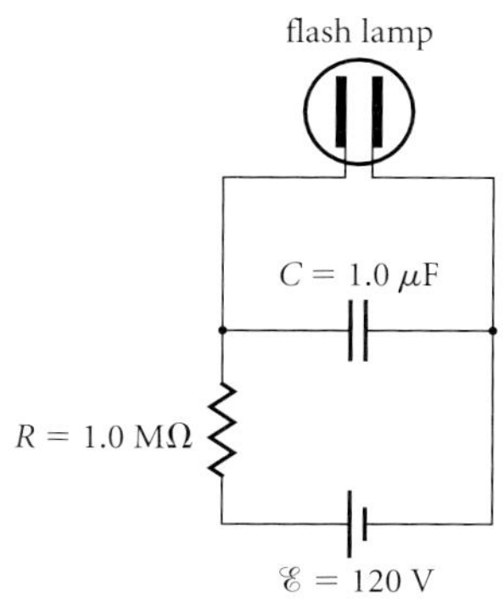

■ **FIGURE 31.7**
Circuit diagram for a highway emergency flasher. The lamp itself has an extremely large resistance until the gas inside ionizes, when its resistance drops to 10 Ω.

EXAMPLE 31.3 ◆◆ A highway emergency flasher uses a 120.0-V battery, a 1.0-MΩ resistor, a 1.0-μF capacitor, and a neon flash lamp in the circuit shown in ■ Figure 31.7. The flash lamp has a resistance of more than 10^7 Ω when the voltage across it is less than 110.0 V. Above 110.0 V, the neon gas ionizes, the lamp's resistance drops to 10 Ω, and the capacitor discharges completely. Find the time between flashes, estimate the duration of each flash, and find the energy released in each.

MODEL Until the capacitor voltage reaches the breakdown voltage $V_b = 110.0$ V, the large resistance of the flash lamp ensures that it draws a negligible current. The capacitor charges as if the lamp were absent. At V_b, however, the lamp resistance quickly becomes negligible, and the capacitor discharges through the lamp as if the battery and series resistor were absent. The time between flashes is the time for the capacitor to charge to V_b. The flash duration is roughly the time for the capacitor to discharge through the lamp, or about three time constants of the capacitor–lamp circuit. The flash energy is the stored energy in the capacitor at 110.0 V.

SEE THE DISCUSSION IN §31.1.1, ESPECIALLY EQN. (31.5).

SETUP Equation (31.6) describes the charging of the capacitor. The flash interval is found by solving for the time when the capacitor voltage is $V_b = 110.0$ V:

$$V_b = Q(t)/C = \mathscr{E}(1 - e^{-t/RC}).$$

So
$$e^{-t/RC} = 1 - \frac{V_b}{\mathscr{E}} = 1 - \frac{110.0\text{ V}}{120.0\text{ V}} = 1 - 0.9167 = 0.0833.$$

SOLVE The time between flashes is:

$$t = -RC\ln(0.0833) = 2.5RC = 2.5(1.0 \times 10^6\ \Omega)(1.0 \times 10^{-6}\text{ F}) = 2.5\text{ s}.$$

The flash duration is:

$$t_f \approx 3\tau_C = 3CR_{\text{lamp}} = 3(1.0 \times 10^{-6}\text{ F})(10\ \Omega) = 30\ \mu\text{s}.$$

The energy in the flash is:

$$U_f = \tfrac{1}{2}CV_b^2 = \tfrac{1}{2}(1.0 \times 10^{-6}\text{ F})(110\text{ V})^2 = 6.1\text{ mJ}.$$

ANALYZE During the flash, the light is as bright as a 200-W light bulb (power $= U_f/t_f$). ▮

31.2 INDUCTANCE

31.2.1 *Self-Inductance*

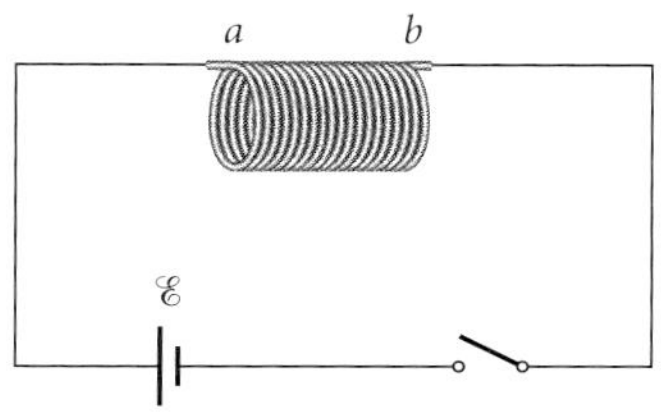

■ **FIGURE 31.8**
The coil is made out of perfectly conducting wire. When the switch is closed, current in the circuit increases from zero.

The simple circuit diagrammed in ■ Figure 31.8 contains a battery (emf $\mathscr{E}$) and a solenoid connected in series with a switch. When the switch is closed, we might expect that a large current $\mathscr{E}/R$ would pass through the low resistance of the wire. Yet even for a perfect conductor ($R = 0$), the current is zero immediately after the switch is closed and increases at a finite rate. As the current in the solenoid increases, the magnetic field inside it also increases, inducing an electric field that opposes the increase in current according to Lenz's law.

REMEMBER: ACCORDING TO FARADAY'S LAW, CHANGING B CREATES AN E.

Let us first assume the solenoid has zero resistance. We'll look at a coil with finite resistance in the next section. The net electric field within the *perfectly conducting* wire, equal to the sum of the Coulomb field produced by the battery and the induced field, must be zero. The current increases at just the right rate to make the net electric field in the wire remain zero:

$$0 = \vec{\mathbf{E}}_{\text{net}} = \vec{\mathbf{E}}_{\text{ind}} + \vec{\mathbf{E}}_c \Rightarrow \vec{\mathbf{E}}_{\text{ind}} = -\vec{\mathbf{E}}_c.$$

THE ELECTRONS IN THE *PERFECT* CONDUCTOR WOULD HAVE INFINITE ACCELERATION IF THE NET FIELD WERE NOT ZERO.

Consequently, the induced emf developed by the inductor equals the potential difference maintained by the battery:

Remember: POTENTIAL DIFFERENCE IS DUE TO THE COULOMB FIELD.

$$\mathscr{E} = \int_{b\ \text{along wire}}^{a} \vec{\mathbf{E}}_{\text{ind}} \cdot d\vec{\ell} = -\int_{b\ \text{along wire}}^{a} \vec{\mathbf{E}}_{\text{c}} \cdot d\vec{\ell} = V_{ab}. \qquad \textbf{(31.8)}$$

Because the induced field is opposite the Coulomb field, the induced emf is often called a *back emf.*

EXERCISE 31.3 ❖ What happens if the switch is suddenly opened when there is current in the coil? Can you explain why switches used to control such devices usually have a capacitor in parallel with the switch?

THIS IS THE MAGNETIC FIELD INSIDE AN INFINITE SOLENOID. WE MAY USE THIS RESULT IF THE SOLENOID IS MUCH LONGER THAN ITS DIAMETER.

We may compute the induced emf of the solenoid from Faraday's law. When there is a current I in the solenoid, the magnetic field inside (eqn. 28.14) is:

$$|\vec{\mathbf{B}}| = \mu_0 nI = \mu_0 NI/\ell,$$

where N is the number of turns in the coil and ℓ is its length. As the current increases, B increases and the induced electric field circulates around the increasing flux (cf. Examples 30.1 and 30.9). The flux through one turn of the coil (area A) is BA, and the flux through the whole solenoid is:

$$\Phi_B = NBA = \frac{\mu_0 N^2 A}{\ell} I.$$

The induced emf equals the rate of change of flux:

$$|\mathscr{E}| = \left|\frac{d\Phi_B}{dt}\right| = \frac{d}{dt}\left|\frac{\mu_0 N^2 AI}{\ell}\right| = \frac{\mu_0 N^2 A}{\ell}\left|\frac{dI}{dt}\right|.$$

The induced emf is proportional to the rate of change of current. The constant of proportionality is called the *inductance* and is given the standard symbol L.

The *inductance* of a circuit element is the ratio of the induced emf in it to the rate of change of current through it:

$$\boldsymbol{L} \equiv \frac{|\mathscr{E}_{\text{ind}}|}{|dI/dt|}. \qquad \textbf{(31.9)}$$

In an ideal solenoid with no resistance, the induced emf equals the potential difference $|V_{ab}|$ across the solenoid (eqn. 31.8):

WE'LL NEED THIS RELATION IN THE NEXT SECTION.

$$|V_{ab}| = \frac{\mu_0 AN^2}{\ell}\left|\frac{dI}{dt}\right| \equiv L\left|\frac{dI}{dt}\right|. \qquad \textbf{(31.10)}$$

Since

$$|\mathscr{E}| = \left|\frac{d\Phi_B}{dt}\right| = L\left|\frac{dI}{dt}\right|,$$

an equivalent definition of inductance is the ratio of flux to current:

$$L = \Phi_B/I. \qquad (31.11)$$

For the solenoid, with $n = N/\ell$ turns per unit length:

WE USED EQN. (28.14) AGAIN.

$$L = \mu_0 AN^2/\ell = \mu_0 n^2 A\ell. \qquad (31.12)$$

Like capacitance, inductance depends on the geometrical structure of a circuit element and is independent of variables such as current or dI/dt.

A circuit element designed to have a particular inductance is called an *inductor.* ■ Figure 31.9 shows the circuit symbol and how the potential changes across an inductor when the current is changing in the direction of the arrow. The sign convention reflects the fact that $\vec{\mathbf{E}}_c$ points in the direction of current change.

The unit of inductance is the henry (symbol H):

$$1\ \text{H} = 1\ \frac{\text{V}\cdot\text{s}}{\text{A}}.$$

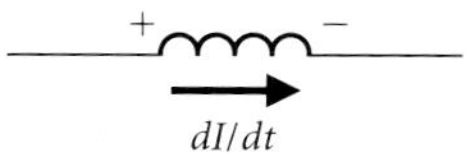

■ FIGURE 31.9
Circuit diagram symbol for an inductor. The sign convention indicates that potential drops across the inductor in the direction that current increases.

AFTER JOSEPH HENRY (1797–1878).

EXAMPLE 31.4 ♦ Find the inductance of a solenoid with 1.0×10^3 turns, length 3.0 cm, and radius 0.50 cm.

MODEL We model the solenoid as *very long* and apply eqn. (31.12).

SETUP The area of the solenoid is $A = \pi r^2$.

SOLVE

$$L = \mu_0 \pi r^2 \frac{N^2}{\ell}$$

$$= (4\pi \times 10^{-7}\ \text{N/A}^2)\pi(0.5 \times 10^{-2}\ \text{m})^2 \left[\frac{(1.0 \times 10^3)^2}{3 \times 10^{-2}\ \text{m}}\right]$$

$$= 3.3\ \text{mH}.$$

ANALYZE Check the units: $\dfrac{\text{N}\cdot\text{m}}{\text{A}^2} = \dfrac{\text{J}}{\text{A}^2} = \dfrac{\text{J}\cdot\text{s}}{\text{A}\cdot\text{C}} = \dfrac{\text{V}\cdot\text{s}}{\text{A}} = \text{H}.$

EXERCISE 31.4 ♦ Verify that 1 H = 1 Wb/A.

31.2.2 *Energy Storage in an Inductor*

Once the switch is closed in Figure 31.8, the constant potential difference maintained by the battery causes current to increase with time. The battery is doing work on charge flowing through it, but no heat is dissipated by resistance. Energy delivered by the battery is stored in the inductor.

REMEMBER: WE ARE STILL ASSUMING THE SOLENOID HAS ZERO RESISTANCE. FOR A REAL INDUCTOR WITH RESISTANCE, THE CURRENT INCREASES LINEARLY FOR $t \ll L/R$. WE'LL DISCUSS THE COMPLETE TIME DEPENDENCE IN THE NEXT SECTION. SEE ALSO EXERCISE 31.10.

The power output of the battery is:

$$P(t) = \mathscr{E}I(t) = \frac{\mu_0 N^2 A}{\ell}\left(\frac{dI}{dt}\right) I = LI\frac{dI}{dt}.$$

The stored energy U at time t equals the total work done to establish the current I in the inductor:

$$U = \int_0^t P\,dt = \int_0^t LI\frac{dI}{dt}\,dt = \int_0^I LI\,dI = \tfrac{1}{2}LI^2\Big|_0^I$$

$$U = \tfrac{1}{2}LI^2. \qquad \textbf{(31.13)}$$

EXERCISE 31.5 ♦ Find the energy stored in a 3.3-mH solenoid when it carries a current of 1.1 A.

Equation (31.13) is similar to the result for energy stored in a capacitor (§27.2): $U = \frac{1}{2}Q^2/C$. In that case, electrostatic energy is stored in the electric field itself throughout the volume of the capacitor. The inductor also stores energy throughout its volume, in the form of magnetic field. For the long solenoid, we have:

$$U = \tfrac{1}{2}LI^2 = \tfrac{1}{2}\mu_0\frac{N^2A}{\ell}I^2 = \tfrac{1}{2}\left(\mu_0\frac{N}{\ell}I\right)^2\left(\frac{\ell A}{\mu_0}\right) = \tfrac{1}{2}\ell AB^2/\mu_0.$$

The stored energy equals the volume ℓA inside the solenoid multiplied by an energy density:

$$u_B = B^2/(2\mu_0). \tag{31.14}$$

Although we derived this expression for the energy density inside the solenoid, it is true generally.

Compare eqn. (31.14) with the corresponding energy density due to electric field ($u_E = \epsilon_0 E^2/2$, eqn. 27.13). When both electric and magnetic fields are present, both store energy and the total energy density is:

$$u \equiv u_E + u_B = \frac{1}{2}\left(\epsilon_0 E^2 + \frac{B^2}{\mu_0}\right). \tag{31.15}$$

EXAMPLE 31.5 ♦♦ Find the magnetic field energy density 6.0 mm from a long straight wire carrying a steady current of 10.0 mA.

MODEL We calculate the magnetic field, and then use eqn. (31.14).

SETUP The magnetic field is $B = \mu_0 I/(2\pi r)$ (eqn. 28.6).

SOLVE So the field energy density is:

$$u_B = \frac{B^2}{2\mu_0} = \frac{(\mu_0 I)^2}{2\mu_0(4\pi^2 r^2)} = \frac{\mu_0 I^2}{8\pi^2 r^2} = (4\pi \times 10^{-7}\ \text{N/A}^2)\left[\frac{(1.0 \times 10^{-2}\ \text{A})^2}{8\pi^2(6.0 \times 10^{-3}\ \text{m})^2}\right]$$

$$= \frac{1}{2\pi(0.36)} 10^{-7}\ \text{N/m}^2 = 4.4 \times 10^{-8}\ \text{J/m}^3.$$

ANALYZE Twice as far from the wire, the energy density is one-quarter as great.

EXERCISE 31.6 ♦♦ How would the answer to Example 31.5 change if the wire also carries a charge density of 5.0 pC/m?

EXAMPLE 31.6 ♦♦♦ The inner conductor of a coaxial cable has radius $a = 0.50$ mm. The radius of the outer sheath is $b = 0.50$ cm (■ Figure 31.10). What is the inductance per unit length of the cable? (Assume the current in the inner conductor is on its outer surface.)

THIS ASSUMPTION IS STRICTLY TRUE ONLY FOR AC CURRENT (CHAPTER 32). WE USE IT HERE TO SIMPLIFY THE CALCULATION.

MODEL The inner conductor and the outer sheath carry equal and opposite currents. Thus, from Ampère's law, the magnetic field is zero everywhere except between the two conductors (cf. Example 28.7). We may compute the inductance of the cable from the stored energy using eqn. (31.13): $L = 2U/I^2$.

SETUP With $\vec{\mathbf{B}}$ given by eqn. (28.6), $B = \mu_0 I/(2\pi r)$, the energy density at radius $r > a$ inside the cable is:

$$u_B(r) = \frac{B^2}{2\mu_0} = \frac{[\mu_0 I/(2\pi r)]^2}{2\mu_0} = \frac{\mu_0 I^2}{8\pi^2 r^2}.$$

The energy density is a function of radius. The energy stored in a cylindrical shell of length ℓ and width dr is:

$$dU = u_B(r)\ dV = u_B(r)(2\pi r\ell\ dr).$$

Thus the energy stored between radii a and b in length ℓ of the cable is:

$$U = \int u_B\ dV = \int_a^b \frac{\mu_0 I^2}{8\pi^2 r^2}(2\pi r\ell)\ dr = \frac{\mu_0 I^2 \ell}{4\pi}\int_a^b \frac{dr}{r} = \frac{\mu_0 I^2 \ell}{4\pi}\ln\left(\frac{b}{a}\right).$$

SOLVE The inductance per unit length is:

$$\frac{L}{\ell} = \frac{2U}{I^2\ell} = \frac{\mu_0}{2\pi}\ln\left(\frac{b}{a}\right) = (2 \times 10^{-7}\ \text{N/A}^2)(\ln 10) = 4.6 \times 10^{-7}\ \text{H/m}.$$

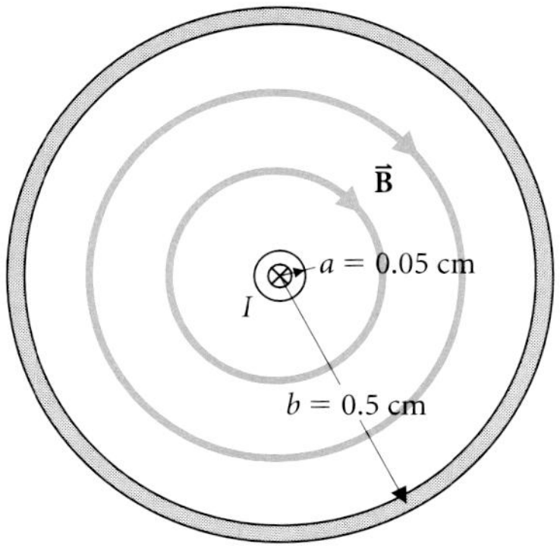

■ FIGURE 31.10
A coaxial cable carrying current I stores magnetic energy between the conductors. The current in the outer sheath is equal and opposite to the current in the inner wire so the net current is zero.

ANALYZE The units calculation is an intermediate result in Exercise 31.4. The result does not depend on the actual values of b and a, but only on their ratio. Does that surprise you? The inductance could also be calculated from eqn. (31.9). ∎

SEE PROBLEM 37.

31.2.3 *Mutual Inductance*

Two circuits may influence each other even if there are no wires connecting them. When a changing magnetic field produced by one circuit passes through another, the resulting induced field produces an emf in the second circuit (cf. Example 30.1).

The *mutual inductance* between two circuit elements is the ratio of emf developed in one element to the rate of change of current in the other element:

$$M \equiv \frac{|\mathcal{E}_2|}{|dI_1/dt|}. \quad \textbf{(31.16)}$$

Conversely, changing current in circuit 2 induces an emf in circuit 1. The corresponding mutual inductance has the same value:

$$M \equiv \frac{|\mathcal{E}_1|}{|dI_2/dt|}.$$

(We omit the proof of this assertion.) The mutual inductance of two circuit elements is usually calculated as the flux through one element divided by the current in the other:

SEE PROBLEM 32 FOR AN EXAMPLE THAT DEMONSTRATES THIS EQUALITY.

$$M = \Phi_1/I_2 = \Phi_2/I_1. \quad (31.17)$$

EXAMPLE 31.7 ♦♦ Find the mutual inductance between a solenoid with 1.0×10^4 turns/m and a single wire loop of radius $r = 1.0$ cm placed inside the solenoid with its plane perpendicular to the axis (■ Figure 31.11).

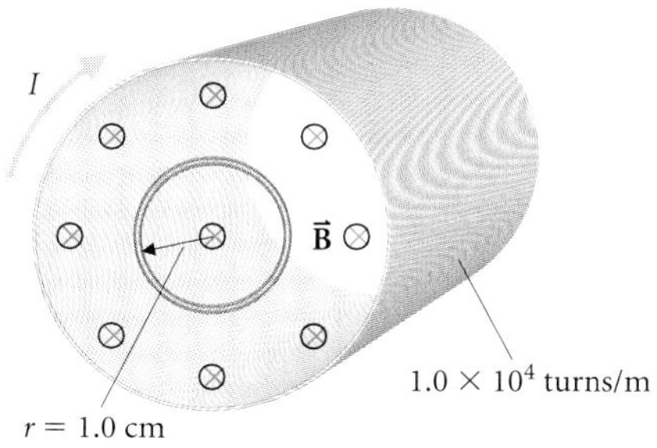

■ **FIGURE 31.11**
A wire loop is placed inside a long solenoid with its plane perpendicular to the solenoid axis.

MODEL We calculate the magnetic flux Φ_2 through the loop due to a current I_1 in the solenoid and use eqn. (31.17).

SETUP The field inside the solenoid is $B = \mu_0 n I_1$, so the flux through the loop is:

$$\Phi_2 = BA = \mu_0 n I_1 (\pi r^2).$$

SOLVE The mutual inductance is:

$$M = \Phi_2/I_1 = \mu_0 n \pi r^2$$
$$= (4\pi \times 10^{-7}\ \text{H/m})(1.0 \times 10^4\ /\text{m})\pi(0.010\ \text{m})^2 = 3.9\ \mu\text{H}.$$

ANALYZE Note that the mutual inductance is independent of I_1. Calculating the flux through the solenoid due to current in the loop is much more difficult. ∎

SEE PROBLEM 90.

One of the most important practical applications of mutual inductance is the electrical transformer. A transformer consists of two wire coils arranged so that flux produced by one links the other. In an air core transformer (■ Figure 31.12a), one coil lies inside the other. Power transformers use an iron core (Figure 31.12b). Nearly all of the magnetic flux remains within the iron and thus links both coils. ■ Figure 31.13 is the circuit symbol for a transformer.

A transformer is frequently used to match the voltage of a power source to that needed by an appliance. A sinusoidally varying potential difference is applied to the *primary* (input) coil, causing a time-varying current in the coil. The resulting variable flux produces an induced electric field that both opposes the applied potential difference and produces an emf in the *secondary* (output) coil.

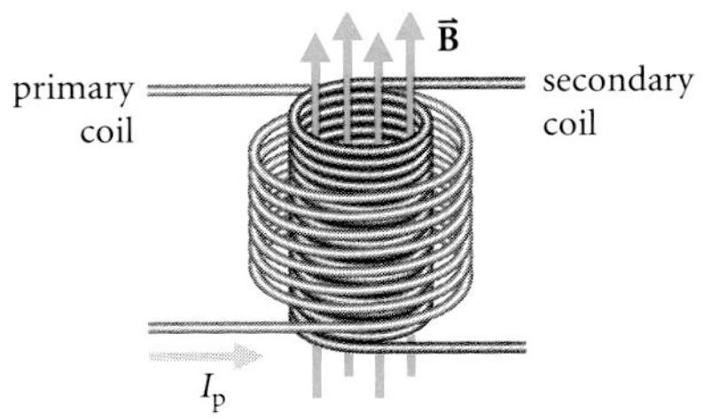

(a) An air core transformer.

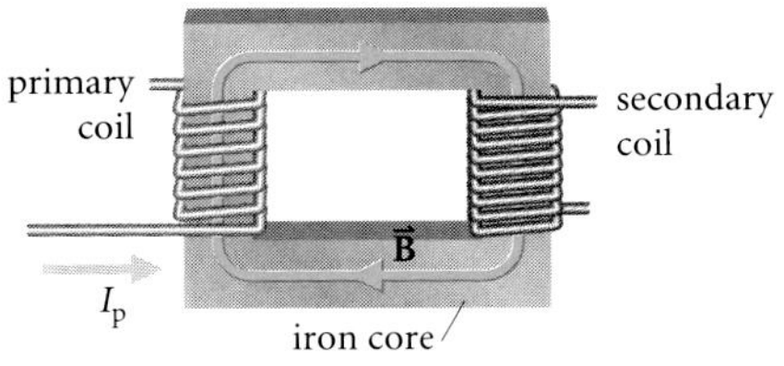

(b) An iron core transformer.

■ **FIGURE 31.12**
(a) In an air core transformer, magnetic flux produced by the primary coil links the secondary. When current changes in the primary, induced electric field drives current in the secondary. (b) An iron core transformer operates on the same principle, but the magnetic flux is trapped inside the core and threads both coils even though they are spatially separated.

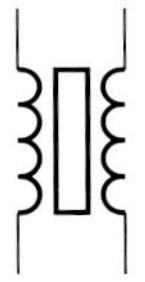

■ **FIGURE 31.13**
Circuit diagram symbol for a transformer.

In a well-designed transformer, resistance in the coils is a relatively small effect that we may neglect here. Then (eqn. 31.8) the induced emf in the primary coil exactly balances the applied potential difference $\mathcal{E}_p$:

$$\mathcal{E}_p = -\mathcal{E}_{\text{induced}} = N_p \frac{d\Phi_B}{dt}.$$

Here N_p is the number of turns in the primary coil and Φ_B is the magnetic flux passing through each turn. The same flux also passes through each turn of the secondary coil, so the induced emf in the secondary has magnitude:

$$|\mathcal{E}_s| = N_s \frac{d\Phi_B}{dt}.$$

Since the flux through each turn Φ_B is the same in each expression, the ratio of output emf to input emf equals the ratio of the numbers of turns in the coils:

$$\left|\frac{\mathcal{E}_s}{\mathcal{E}_p}\right| = \frac{N_s}{N_p}. \tag{31.18}$$

A *step-down* transformer has $N_s < N_p$, and the voltage in the secondary is less than that in the primary. A *step-up* transformer has $N_s > N_p$ and $|\mathcal{E}_s| > |\mathcal{E}_p|$.

The transformer equation (31.18) arises from the fact that the two coils are mutual inductors but does not depend on the value of their mutual inductance. As long as the coil resistances are truly negligible, the result is independent of the current or the power transmitted by the device. The magnitude of the mutual inductance is important when we consider the power transmitted to the secondary circuit, including the effect of resistance.

EXAMPLE 31.8 ♦♦ The line voltage in a city power supply is 12 kV. The voltage supplied to each house is 120 V. What turns ratio is required in the transformer that converts the voltage? What is the ratio of current drawn by the household to current in the power line?

MODEL The turns ratio equals the voltage ratio (eqn. 31.18). The ratio of currents in each coil equals the ratio of fluxes through them (eqn. 31.17).

SETUP AND SOLVE Using eqn. (31.18):

$$\frac{12\,000 \text{ V}}{120 \text{ V}} = \left|\frac{\mathcal{E}_p}{\mathcal{E}_s}\right| = \frac{N_p}{N_s} = 1.0 \times 10^2.$$

Since the flux through each coil is the flux through one turn times the number of turns and the flux through one turn is the same for each coil, then from eqn. (31.17):

$$\frac{I_s}{I_p} = \frac{\Phi_p}{\Phi_s} = \frac{N_p}{N_s} = 1.0 \times 10^2.$$

ANALYZE In an ideal transformer, the power provided by the primary equals the power drawn by the secondary: $I_s\mathcal{E}_s = I_p\mathcal{E}_p$. Transformers used by the Pacific Gas & Electric Company are rated at a few tens of kV·A power, corresponding to a secondary current of up to a few hundred amps. ∎

EXERCISE 31.7 ♦♦ In an ideal transformer, the same flux passes through each turn of the primary and secondary coils. If L_p and L_s are the self-inductances of the two coils, show that their mutual inductance is $M = \sqrt{L_p L_s}$.

31.3 INDUCTOR CIRCUITS

31.3.1 *The LR Circuit*

A real solenoid has some resistance as well as inductance, and we can model it as a pure inductance in series with a resistance (■ Figure 31.14). The behavior of this circuit is similar to the series RC circuit we discussed in §31.1. When the switch has been in position A for a long time, there is a steady current $I_0 = \mathscr{E}/R$ in the circuit, magnetic energy $\frac{1}{2}LI_0^2$ is stored in the inductor, and the potential difference across the resistor equals the battery emf. (Since $dI/dt = 0$, there is no potential difference across the inductor.) When the switch is moved to position B at time $t = 0$, the current decreases exponentially to zero as the energy stored in the inductor is dissipated by the resistor:

$$I(t) = I_0 e^{-t/\tau}. \tag{31.19}$$

COMPARE WITH THE BEHAVIOR OF THE RC CIRCUIT, ESPECIALLY EQN. (31.3).

The time constant τ for the decay is determined by the inductance L and resistance R. The only combination of these with dimensions of time is L/R, so we guess that the time constant for the circuit is $\tau = \tau_L = L/R$.

The potential difference across the inductor due to the changing current (eqn. 31.10) is equal and opposite the potential difference IR across the resistor. As I decreases, so does its rate of change, and both potential differences decrease to zero.

THIS ARGUMENT DOES NOT RULE OUT A DIMENSIONLESS NUMBER LIKE 2 OR π MULTIPLYING L/R. IF SUCH A NUMBER EXISTS, IT WILL SHOW UP IN THE KIRCHHOFF'S RULE ANALYSIS.

EXERCISE 31.8 ♦ Show that the ratio L/R has dimensions of time—that is:

$$(1\ \text{H})/(1\ \Omega) = 1\ \text{s}.$$

We may verify our conclusions about the circuit's behavior by applying Kirchhoff's loop rule. The sign conventions for potential drops across the resistor and inductor are fixed by the positive sense of $I(t)$ chosen in Figure 31.14. (See Figure 31.9). Here we have $I > 0$, and we expect $dI/dt < 0$. Then $\Delta V_R = IR > 0$, and $\Delta V_I = L\, dI/dt < 0$ (eqn. 31.10). Traversing the circuit clockwise with the switch in position B, Kirchhoff's rule gives:

$$0 = \Delta V_I + \Delta V_R = L\frac{dI}{dt} + IR.$$

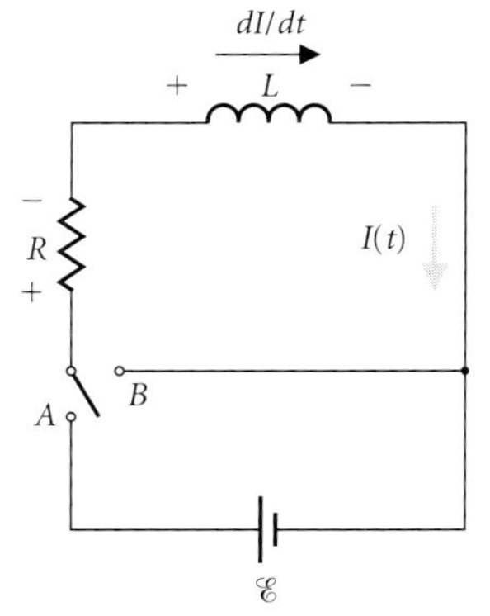

■ FIGURE 31.14
An LR circuit. When the switch has been in position A for a long time, the current in the circuit is $\mathscr{E}/R$. If the switch is then moved to position B, the current decays to zero exponentially.

As expected, this is the exponential differential equation. Substituting expression (31.19), we find:

$$\begin{aligned} L\frac{dI}{dt} + IR &= L\frac{d}{dt}\left(I_0 e^{-t/\tau}\right) + RI_0 e^{-t/\tau} \\ &= LI_0\left(-\frac{1}{\tau}\right)e^{-t/\tau} + RI_0 e^{-t/\tau} \\ &= I_0 e^{-t/\tau}\left(-\frac{L}{\tau} + R\right) = 0. \end{aligned}$$

The differential equation is satisfied for any value of I_0 if:

$$\tau = \tau_L = L/R, \tag{31.20}$$

as we guessed.

THE CURRENT DECREASES TO ABOUT 5% OF ITS INITIAL VALUE IN A TIME $t \approx 3\tau = 3L/R$. WITH A LARGER R, THE CURRENT DECREASES MORE RAPIDLY. COMPARE WITH THE TIME TO DISCHARGE A CAPACITOR (§31.1).

The inductor also regulates the rate at which current increases in the circuit when the switch is returned to position A. When the switch has been left in position B for a time interval $\Delta t \gg \tau_L$, the current is zero. Just after the switch is moved to position A, $I = 0$, all the potential difference of the battery is developed across the inductor and the current increases rapidly. As the current builds, more of the potential drop occurs across the resistor and less

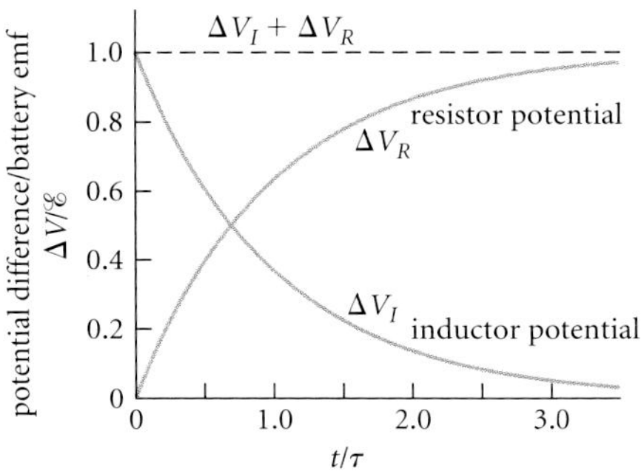

■ **FIGURE 31.15**
Potential differences across the inductor and the resistor in an *LR* circuit as functions of time after the switch is closed. As the current builds, the potential difference across the resistor increases and across the inductor decreases. The sum always equals the battery emf.

across the inductor; the current increases less rapidly. After a long time, the current approaches the final value $\mathscr{E}/R$ and $dI/dt = 0$. All the battery's potential difference is developed across the resistor (■ Figure 31.15). Again, we expect an exponential evolution:

$$I(t) = \frac{\mathscr{E}}{R}(1 - e^{-Rt/L}). \tag{31.21}$$

EXERCISE 31.9 ♦ Apply Kirchhoff's rule to the *LR* circuit (Figure 31.14) to obtain a differential equation for the current after the switch is moved from *B* to *A*. Show that eqn. (31.21) is the correct solution.

EXERCISE 31.10 ♦♦♦ Show that the solution (31.21) predicts $I(t) \propto t$ for $t \ll \tau_L = L/R$. Plot a graph or use the series expansion for the exponential function (Appendix IB). ■

The spark plug in an automobile engine requires a potential difference of 20 kV for a period of a few microseconds during each cycle of the engine (about 0.02 s). The circuit that provides the spark uses an inductor as the energy source and operates somewhat like the emergency flasher in Example 31.3. ■ Figure 31.16 shows a simplified version of the circuit. A rotor synchronized mechanically with the engine closes the switch (the points) and allows current to build through the inductor. When the switch opens, the current decreases rapidly, and a large enough potential difference develops across the inductor for charge to jump the spark plug gap. At any time, there is current in a single loop of the spark coil circuit. In each loop, the inductor and a resistor are in series.

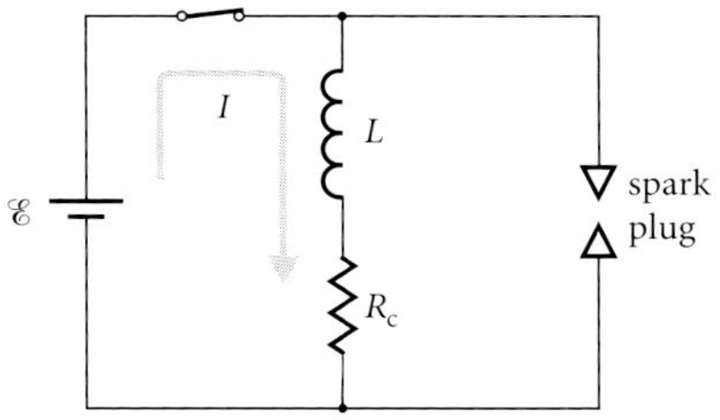

(a) Switch closed.

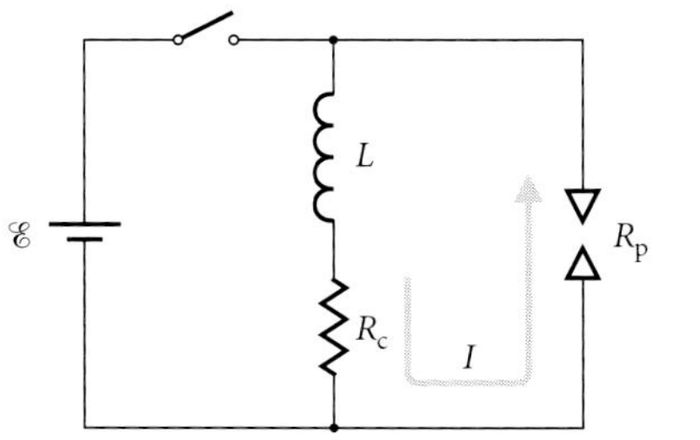

(b) Switch open.

■ **FIGURE 31.16**
Simplified diagram of a spark coil circuit. (a) Switch closed. The battery drives current through the coil, represented by a resistor and inductor in series. Energy is stored as magnetic field in the coil. (b) Switch open. Current in the battery loop is interrupted. The inductor prohibits the current from falling immediately to zero, so a spark jumps the gap across the plug, igniting the fuel.

EXAMPLE 31.9 ♦♦ An automobile spark coil circuit (Figure 31.16) has $\mathscr{E} = 12$ V, $L = 10.0$ mH, $R_c = 10\ \Omega$, and $R_p = 7.0\ \text{k}\Omega$ (when sparking). Estimate: (a) the time the points must remain in contact to build the current, (b) the duration of the spark, and (c) the energy released in the spark.

MODEL With the switch closed, the battery supplies current to the coil in the single loop circuit on the left-hand side of Figure 31.16a. The current builds exponentially according to eqn. (31.21). Once the switch is opened, the inductor prevents the current from dropping immediately to zero, so a spark jumps the gap on the right.

SETUP (a) The time constant for the single loop circuit on the left when the points are in contact is:

$$\tau_L = L/R_c = (1.0 \times 10^{-2}\ \text{H})/(10\ \Omega) = 1\ \text{ms}.$$

SOLVE The points must be closed for $\approx 3\tau_L = 3$ ms to build current in the inductor to within 5% of its maximum value.

ANALYZE This time is much less than the 20-ms rotation period of the crank shaft at 3000 rpm.

SETUP (b) With the points open, current through the spark plug in the right loop decays according to eqn. (31.19). The spark duration is roughly the time constant of a circuit containing the inductor and spark plug in series. The resistance of the coil is negligible compared with the 7-kΩ resistance of the plug: $R = R_c + R_p \approx R_p$.

SOLVE

$$\tau_s \approx L/R_p = (1.00 \times 10^{-2}\ \text{H})/(7.0 \times 10^3\ \Omega) = 1.4\ \mu\text{s}.$$

SETUP (c) The energy in the spark comes from the magnetic energy stored in the inductor (eqn. 31.13). The current in the circuit just before the spark occurs is $I \approx \mathscr{E}/R$.

SOLVE

$$U_s = \tfrac{1}{2}LI^2 \approx \tfrac{1}{2}(10^{-2}\ \text{H})\left(\frac{12\ \text{V}}{10\ \Omega}\right)^2 = 7\ \text{mJ}.$$

ANALYZE The spark duration is much less than the time to build current in the inductor. During the spark, the inductor delivers about (7 mJ)/(1.4 μs) = 5 kW of power. ■

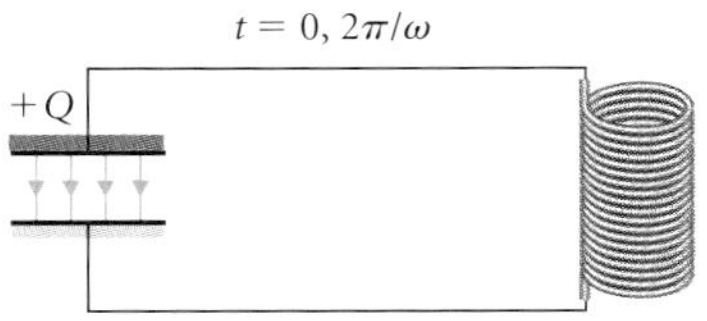

(a) Energy stored in the capacitor.

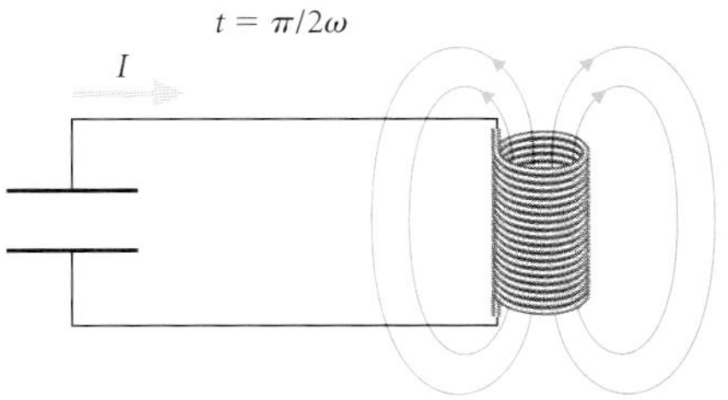

(b) Energy stored in the inductor.

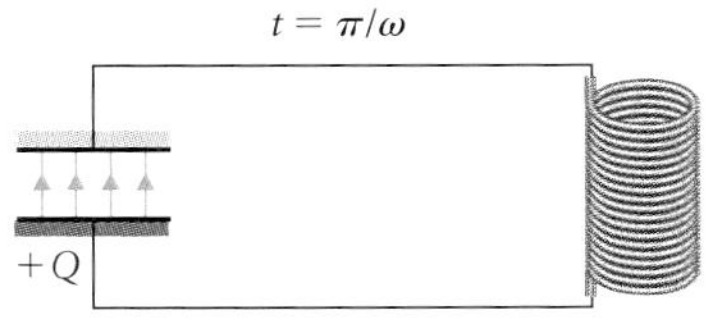

(c) Energy stored in the capacitor again.

■ FIGURE 31.17
Energy in the *LC* circuit. (a) Initially, all the energy is stored in the capacitor as electric field energy. When the switch is closed, the capacitor begins to discharge. The positive charge is on the upper plate.

(b) As current builds in the inductor, the magnetic field energy increases, and energy is transferred from the capacitor to the inductor. When the charge is zero, the current is maximum and all the energy is in the inductor. In this idealized circuit, there are no ohmic losses.

(c) The current begins to charge the capacitor in the opposite sense, transferring energy back to electric form. The current through the inductor decreases until the capacitor is fully charged and the current is zero. The positive charge is then on the lower plate.

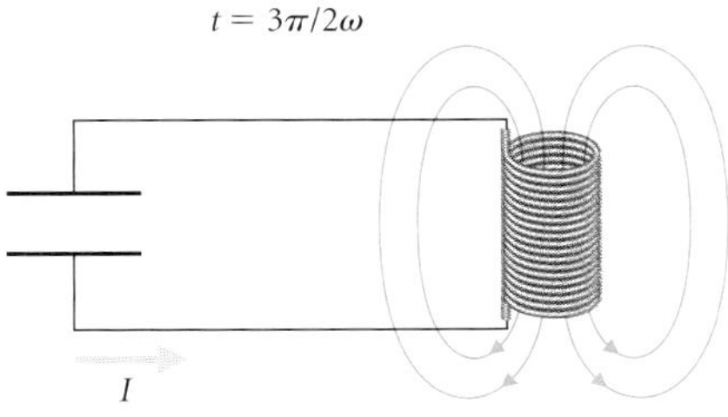

(d) Energy stored in the inductor.

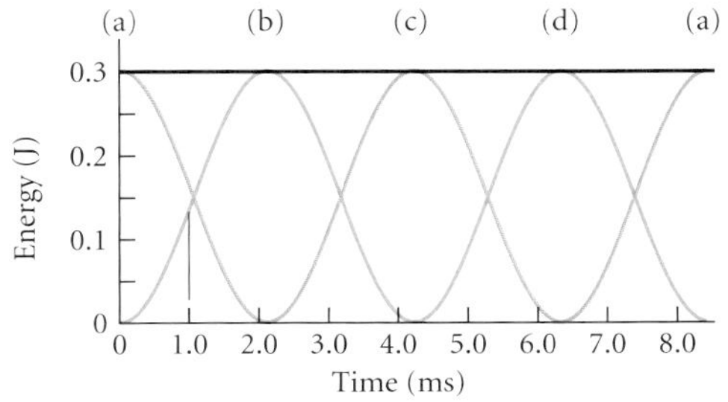

(e) Energy in an *LC* circuit as a function of time.

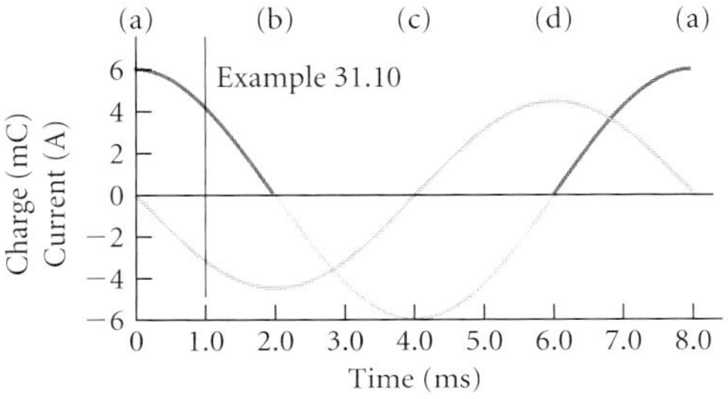

(f) Charge and current in the circuit as a function of time.

(d) The capacitor discharges again, the current increases, and energy is transferred back to the inductor. The current and magnetic field directions are the reverse of their directions in (b). The current then returns the system to state (a), and the cycle repeats indefinitely.

(e) Energy in an *LC* circuit as a function of time. The values were taken from Example 31.10. The corresponding picture, (a) through (d), is marked at the top of the graph. At 1.0 ms (cf. Example 31.10), the energy is almost equally divided between electric and magnetic forms.

(f) Charge and current in the circuit as functions of time. Compare the values at 1.0 ms as computed in Example 31.10 with the values shown here.

31.3.2 *The* LC *Circuit*

An inductor and capacitor in series (■ Figure 31.17) form a circuit with two ways to store energy: as electric energy in the capacitor and as magnetic energy in the inductor. When the capacitor is charged and the switch is closed, energy flows back and forth between the inductor and capacitor. When the inductor stores energy, a current exists that changes the capacitor charge and hence its stored energy. Charge on the capacitor produces a potential difference that causes the inductor current to change. With no resistance to dissipate energy, an ideal *LC* circuit oscillates forever. Such oscillations are the second basic pattern of circuit behavior. The frequency of the oscillation can depend only on L and C, and the combination that has the dimensions of $[T^{-1}]$ is $1/\sqrt{LC}$.

COMPARE WITH THE ENERGY FLOW BETWEEN KINETIC AND POTENTIAL FORMS IN A PENDULUM, §14.3.

EXERCISE 31.11 ♦ Show that $1/\sqrt{LC}$ has the dimensions of frequency—that is, $1\ \text{H}\cdot\text{F} = 1\ \text{s}^2$.

As usual, Kirchhoff's rules give a differential equation that governs the circuit's behavior. We define Q and I as shown in ■ Figure 31.18, and proceed counterclockwise. Kirchhoff's loop rule requires:

$$0 = +L\frac{dI}{dt} + \frac{Q}{C}.$$

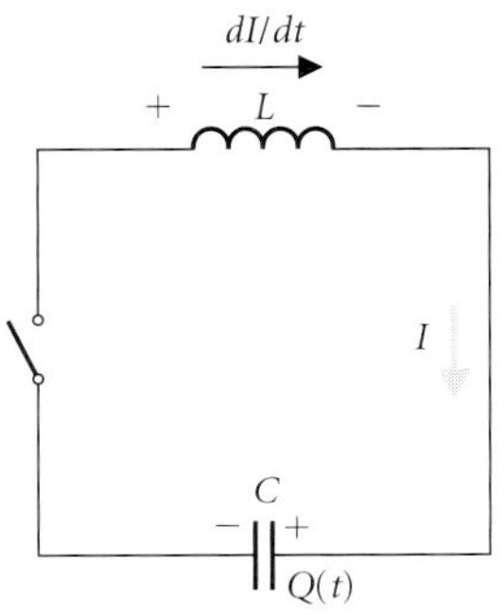

■ FIGURE 31.18
Circuit diagram for an *LC* circuit. In this idealized version, there is no resistance. The current variable has been chosen so that $I = +dQ/dt$.

Positive current $I(t)$ increases the charge $Q(t)$, so:

$$I = +\frac{dQ}{dt}.$$

Substituting:

$$0 = L\frac{d^2Q}{dt^2} + \frac{Q}{C}. \tag{31.22}$$

Rearranging:

$$\frac{d^2Q}{dt^2} = -\frac{1}{LC}Q.$$

THE GENERAL FORM OF THE EQUATION FOR SHM IS $d^2x/dt^2 = -\omega^2 x$.

This is the equation for simple harmonic motion (SHM) (eqn. 14.1). Its general solution (cf. eqn. 14.7) is:

$$Q(t) = Q_0 \cos \omega t + Q_1 \sin \omega t, \tag{31.23}$$

or, equivalently (cf. eqn. 14.3):

$$Q(t) = Q_2 \cos(\omega t + \phi_0). \tag{31.24}$$

The angular frequency is:

$$\omega \equiv \omega_0 = 1/\sqrt{LC}, \tag{31.25}$$

consistent with our argument on dimensional grounds. Equation (31.23) describes any oscillation of the circuit. The constant Q_0 is the charge on the capacitor at $t = 0$ and Q_1 is related to the current at $t = 0$, as we show in the next example. Their values may be found from known values of charge and current at any specific time, or from two values of either charge or current at two different times.

REMEMBER: THE TWO TIMES MUST BE AT DIFFERENT PHASES OF THE OSCILLATION.

EXAMPLE 31.10 ♦♦ In Figure 31.18, $L = 30.0$ mH and $C = 60.0\ \mu$F. At $t = 0$, the capacitor carries a charge $Q_0 = +6.0$ mC, when the switch is closed. (a) Find the charge on the capacitor and the current through the inductor as functions of time. (b) Show that the total energy of the system is constant and find its value. (c) Give numerical values for the capacitor charge and inductor current 1.0 ms after the switch is closed.

MODEL Initially, the capacitor is fully charged but the current is zero. All the stored energy is in the capacitor. The potential difference across the capacitor (and hence across the inductor) causes current through the inductor to increase, discharging the capacitor and storing magnetic energy in the inductor. We may use the results that we have just developed [eqns. (31.23) through (31.25)].

SETUP (a) Just after the switch closes at $t = 0$, the current through the inductor remains zero although its derivative is nonzero.

$$I(0) = \left.\frac{dQ}{dt}\right|_{t=0} = -\omega Q_0 \sin(0) + \omega Q_1 \cos(0) = \omega Q_1.$$

Since $I(0) = 0$, then $Q_1 = 0$.

The frequency of the oscillation is:

$$\omega_0 = \frac{1}{\sqrt{LC}} = \frac{1}{\sqrt{(3.00 \times 10^{-2}\ \text{H})(6.00 \times 10^{-5}\ \text{F})}} = 745\ \text{rad/s}.$$

SOLVE Thus the charge as a function of time is:

$$Q(t) = Q_0 \cos(\omega_0 t) = (6.0\ \text{mC})\cos[(745\ \text{rad/s})t].$$

The current is:

$$I(t) = -Q_0\omega_0 \sin(\omega_0 t) = -(6.0\ \text{mC})(745\ \text{rad/s})\sin[(745\ \text{rad/s})t]$$
$$= -(4.5\ \text{A})\sin[(745\ \text{rad/s})t].$$

IN COMPUTING NUMERICAL VALUES [PART (C)], WE USE THE UNROUNDED AMPLITUDE, 4.47 A.

SETUP (b) The total energy at any time is the sum of the energies stored in the inductor and capacitor.

SOLVE Thus: $$U = \tfrac{1}{2}LI^2 + \tfrac{1}{2}\frac{Q^2}{C} = \tfrac{1}{2}L\left[\frac{Q_0^2}{LC}\sin^2(\omega_0 t)\right] + \frac{1}{2C}[Q_0^2\cos^2(\omega_0 t)]$$

$$= \frac{Q_0^2}{2C}[\sin^2(\omega_0 t) + \cos^2(\omega_0 t)] = \frac{Q_0^2}{2C},$$

and is constant. Its value is: $U = \dfrac{(6.0 \times 10^{-3}\ \text{C})^2}{2(6.00 \times 10^{-5}\ \text{F})} = 0.30\ \text{J}.$

SETUP (c) We obtain values of Q and I from the expressions obtained in (a), with $t = 1.0 \times 10^{-3}$ s.

SOLVE $$Q(1.0\ \text{ms}) = (6.0\ \text{mC})\cos(0.745\ \text{rad}) = 4.4\ \text{mC},$$

and $$I(1.0\ \text{ms}) = -(4.5\ \text{A})\sin(0.745\ \text{rad}) = -3.0\ \text{A}.$$

ANALYZE The minus sign in the answer for I means that the current at $t = 1.0$ ms is opposite the arrow in Figure 31.18. The capacitor is discharging. With no resistance in the circuit, energy is conserved and the oscillation goes on forever. Of course, no real circuit behaves exactly this way. An electrical circuit without resistance of some kind is like a mechanical system without friction—impossible! Even a superconducting LC circuit loses energy by radiating electromagnetic waves (Chapter 33). In the next section we'll see how resistance damps oscillations. ∎

31.3.3 *The* LRC *Circuit*

A pure LC circuit is useful as an idealized illustration of current oscillations, but any real inductor also has a resistance that eventually dissipates the energy. In series with a capacitor, it forms a circuit that exhibits both LC and LR circuit behavior: the current oscillates, but the oscillation damps exponentially.

We can use Kirchhoff's rule to verify this behavior for the LRC circuit. There is a charge Q_0 on the capacitor in ∎ Figure 31.19 when the switch is closed. Defining the capacitor charge and current as shown in the figure,

$$I(t) = -\frac{dQ}{dt},$$

and going counterclockwise around the circuit, Kirchhoff's loop rule gives:

$$-\frac{Q}{C} + L\frac{dI}{dt} + IR = 0.$$

Combining these two equations and rearranging, we find:

$$\frac{Q}{LC} + \frac{R}{L}\left(\frac{dQ}{dt}\right) + \frac{d^2Q}{dt^2} = 0. \quad (31.26)$$

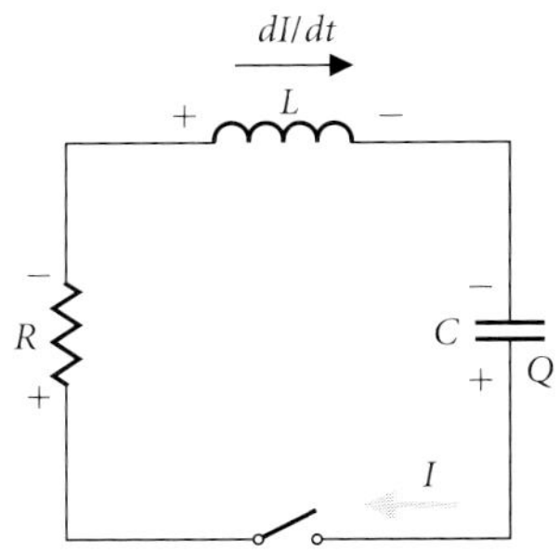

∎ FIGURE 31.19
A series *LRC* circuit. With charge and current variables defined as shown, $I = -dQ/dt$.

This equation combines aspects of the exponential equation (first derivative proportional to Q) and the equation for SHM (second derivative proportional to Q). Its solution is a damped oscillation. As a trial solution, we assume Q has the form:

WE DISCUSSED DAMPED OSCILLATIONS IN §14.4.2.

$$Q(t) = Q_0\cos(\omega t - \phi)e^{-\alpha t},$$

with decay rate $\alpha \equiv 1/\tau$. When the switch is closed, the inductor prevents any instantaneous change of current, so $I(0) = 0$.

$$\left.\frac{dQ}{dt}\right|_{t=0} = -\omega Q_0\sin(\omega t - \phi)e^{-\alpha t} - \alpha Q_0\cos(\omega t - \phi)e^{-\alpha t}\Big|_{t=0}$$

$$= \omega Q_0\sin\phi - \alpha Q_0\cos\phi = 0.$$

Thus $\tan\phi = \alpha/\omega$. Differentiating again, we find:

$$\frac{d^2Q}{dt^2} = Q_0[-\omega^2\cos(\omega t - \phi) + 2\alpha\omega\sin(\omega t - \phi) + \alpha^2\cos(\omega t - \phi)]e^{-\alpha t}$$
$$= Q_0[(\alpha^2 - \omega^2)\cos(\omega t - \phi) + 2\alpha\omega\sin(\omega t - \phi)]e^{-\alpha t}.$$

Substituting the expressions for Q, dQ/dt, and d^2Q/dt^2 into eqn. (31.26), we have:

$$\frac{Q_0}{LC}\cos(\omega t - \phi)e^{-\alpha t} + \frac{R}{L}Q_0[-\omega\sin(\omega t - \phi) - \alpha\cos(\omega t - \phi)]e^{-\alpha t}$$
$$+ Q_0[(\alpha^2 - \omega^2)\cos(\omega t - \phi) + 2\alpha\omega\sin(\omega t - \phi)]e^{-\alpha t} = 0.$$

The expression $Q_0e^{-\alpha t}$ is a factor in each term, and may be divided out. Collecting terms, we have:

$$\left(\frac{1}{LC} - \frac{R}{L}\alpha + \alpha^2 - \omega^2\right)\cos(\omega t - \phi) + \left(-\frac{R}{L}\omega + 2\alpha\omega\right)\sin(\omega t - \phi) = 0. \quad \text{(i)}$$

There are two unknowns: α and ω. We can get two different equations for them by evaluating eqn. (i) at two different times. At $t = \phi/\omega$, $\sin(\omega t - \phi) = 0$, $\cos(\omega t - \phi) = 1$, and

$$1/LC - \alpha(R/L) + \alpha^2 - \omega^2 = 0. \quad \text{(ii)}$$

At $t = (\phi + \pi/2)/\omega$, $\cos(\omega t - \phi) = 0$, $\sin(\omega t - \phi) = 1$, so

$$\omega(R/L - 2\alpha) = 0.$$

Thus:

$$\alpha = \frac{R}{2L} \equiv \frac{1}{\tau_{LRC}}. \quad (31.27)$$

Then from eqn. (ii):

$$\omega^2 = \frac{1}{LC} - \alpha\frac{R}{L} + \alpha^2 = \omega_0^2 - \alpha^2 = \frac{1}{LC} - \frac{R^2}{4L^2} \equiv \omega_1^2. \quad (31.28)$$

So the solution for Q is:

$$Q(t) = Q_0\cos(\omega_1 t - \phi)e^{-Rt/2L}, \quad (31.29)$$

with ω_1 given by eqn. (31.28) and $\phi = \tan^{-1}(\alpha/\omega)$.

The time constant $\tau_{LRC} = 2L/R$ is twice that for the LR circuit. Since the current oscillates (■ Figure 31.20), the average power dissipated (which is proportional to current squared) is one-half what it would be for a steady current, and so the amplitude of the oscillating current decays only half as rapidly as a steady current. The resistance also reduces the frequency of the oscillations compared with an ideal LC circuit.

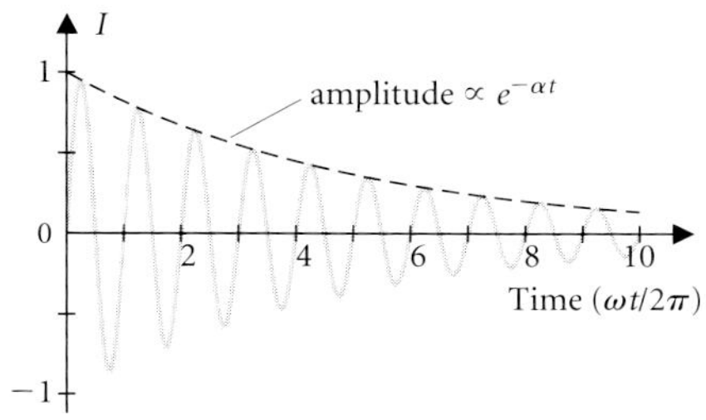

■ FIGURE 31.20
Current in the *LRC* circuit as a function of time. The values of L, R, and C are those in Example 31.11: $L = 4.7$ mH, $R = 1.2$ kΩ, $C = 18$ pF. The current oscillates, but the amplitude decreases (dashed line) as energy is dissipated in the resistor.

COMPARE EQNS. (31.25) AND (31.28).

We determined the constants α and ω by evaluating eqn. (i) at two specific times, $t = \phi/\omega$ and $t = (\phi + \pi/2)/\omega$. Sine and cosine functions vary differently with time, so the only way to satisfy the differential equation at *all* times is to set the coefficients of the sine and cosine terms equal to zero separately. Choosing two specific times at which either $\sin(\omega t - \phi)$ or $\cos(\omega t - \phi)$ is zero emphasizes this point. We'll use this technique often.

EXAMPLE 31.11 ♦ A radio tuning circuit is a series LRC circuit with $C = 18$ pF, $R = 1.2$ kΩ, and $L = 4.7$ mH. Find the natural frequency ω_1 and damping rate α for this circuit.

MODEL We may use eqns. (31.27) and (31.28).

SETUP AND SOLVE The damping rate is:

$$\alpha = \frac{R}{2L} = \frac{1.2 \times 10^3\ \Omega}{2(4.7 \times 10^{-3}\ \text{H})} = 1.3 \times 10^5\ /\text{s}.$$

The natural frequency is:

$$\omega_1^2 = \frac{1}{LC} - \alpha^2 = \frac{1}{(4.7 \times 10^{-3}\ \text{H})(1.8 \times 10^{-11}\ \text{F})} - (1.3 \times 10^5\ /\text{s})^2$$

$$= 1.2 \times 10^{13}\ (\text{rad/s})^2,$$

so: $$\omega_1 = 3.4 \times 10^6\ \text{rad/s}.$$

ANALYZE Equivalently, $f_1 = \omega_1/2\pi = 550$ kHz, in the AM band.

Since the square of the angular frequency is the difference of two terms, it is possible for ω_1^2 to be zero or even negative. This happens when $\alpha \geq \omega_0$ (eqn. 31.28), or equivalently, $R \geq 2\sqrt{L/C}$. With a resistance this large, the energy in the circuit is dissipated sufficiently fast that no oscillations appear. If $\omega_1 = 0$, the circuit is said to be *critically damped;* with $\omega_1^2 < 0$, the circuit is *overdamped.*

An *LRC* circuit is analogous to a mechanical system of a block pushed and pulled through a fluid by a spring. The inductance plays the roll of mass, the capacitance of spring constant, the charge is like displacement, and the current is like velocity. The resistance in the circuit is like drag on the block as it moves through the fluid. If the resistance is small, it doesn't affect the oscillations much, but a large resistance, like strong drag, eliminates the oscillations altogether.

Math Toolbox

How to Solve a Linear Differential Equation

The procedure we have used to solve the differential equations for time-dependent circuits is as follows:

1. Guess the mathematical form of the solution on physical grounds.
2. Plug into the differential equation to find time constants and/or frequencies.
3. Choose values for the adjustable constants in the solution to fit the known initial conditions or other given information.

Using qualitative reasoning in this way is a valuable skill, but what if your guesses don't work? Try this method.

1. Write the equation in standard form with the coefficient of the highest derivative equal to unity, all terms involving Q on the left-hand side (the homogeneous part) and terms not involving Q on the right-hand side (the inhomogeneous part).

   Example: *RC* circuit with battery (eqn. 31.7)

   $$\underbrace{\frac{dQ}{dt} + \frac{1}{RC}Q}_{\text{homogeneous part (H)}} = \underbrace{\frac{\mathcal{E}}{R}}_{\text{inhomogeneous part (I)}}.$$

2. Replace the right-hand side with zero and solve the resulting equation. The text gives a solution to any equation you will encounter in the problems. The solution contains as many adjustable constants as the order of the highest derivative—one for the exponential equation and two for simple harmonic motion.

   Example: $Q_H(t) = \underset{\substack{\uparrow \\ \text{adjustable constant}}}{A}e^{-t/RC}.$

3. Find any function that satisfies the original equation. (This is the hard part.) If the right-hand side is not a function of time, you can get a result by setting all the derivatives on the left-hand side to zero.

   Example: With $dQ/dt = 0$, eqn. (31.7) becomes:

   $$\frac{Q}{RC} = \frac{\mathcal{E}}{R} \Rightarrow Q_I = \mathcal{E}C.$$

   If the right-hand side is a function of time, you must rely on physical intuition, guesswork, experience, or your differential equations text!
4. The complete solution is the sum of the two solutions you have found:

   $$Q(t) = Q_H + Q_I.$$

   Example: $Q(t) = Ae^{-t/RC} + \mathcal{E}C.$
5. Solve for the constants from initial conditions or other given information in the usual way.

   Example: $Q = 0$ at $t = 0 \Rightarrow A + \mathcal{E}C = 0$
   $\Rightarrow A = -\mathcal{E}C.$

The solution is $Q(t) = \mathcal{E}C(1 - e^{-t/RC})$—eqn. (31.6).

EXAMPLE 31.12 ♦♦ If the values of L and R are fixed in the circuit described in Example 31.11, what value of C results in critical damping?

MODEL Critical damping occurs when $\omega_1^2 = 0$.

SETUP AND SOLVE From eqn. (31.28), $\omega_1^2 = 0$ when:

$$0 = \frac{1}{LC} - \frac{R^2}{4L^2} \quad \text{or} \quad C = \frac{4L}{R^2} = \frac{4(4.7 \times 10^{-3}\ \text{H})}{(1.2 \times 10^3\ \Omega)^2} = 1.3 \times 10^{-8}\ \text{F}.$$

ANALYZE With $\omega_1 = 0$ in eqn. (31.29), the charge decays exponentially. ■

EXERCISE 31.12 ♦♦ For the overdamped case, $R > 2\sqrt{L/C}$, assume $Q(t) = Q_0 e^{-\beta t}$ and show that:

$$\beta = \frac{R}{2L} \pm \sqrt{\frac{R^2}{4L^2} - \frac{1}{LC}}.$$

✻ 31.4 Multiloop Circuits

If a circuit has more than one loop, we need to apply Kirchhoff's rules according to the plan in Figure 31.3, and we obtain a set of coupled differential equations. The next example illustrates this technique.

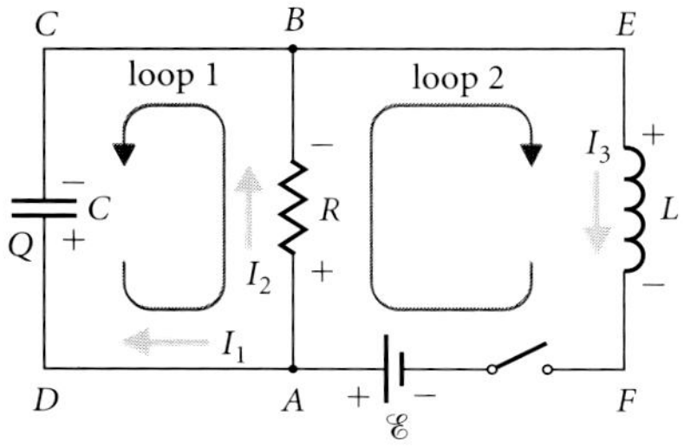

■ **FIGURE 31.21**
We have defined the current variables so that $I_1 = dQ/dt$, and $I_3 = I_1 + I_2$.

EXAMPLE 31.13 ♦♦♦ The capacitor in ■ Figure 31.21 is uncharged at $t = 0$ when the switch is closed. Find the potential difference across the capacitor for $t > 0$.

MODEL STEP I What is the expected behavior? A long time after the switch is closed, the capacitor is fully charged and there is a constant current through the inductor. Because the circuit contains all three kinds of element, we expect a decaying oscillation toward the final equilibrium.

STEP II Defining currents I_1, I_2, and I_3 as shown in the diagram, the expected final state has:

$$I_{1,\text{f}} = 0, \quad I_{2,\text{f}} = I_{3,\text{f}} = \mathcal{E}/R, \quad Q_\text{f}/C = I_{3,\text{f}}R \Rightarrow Q_\text{f} = \mathcal{E}C.$$

Initially, $Q = 0$, and since the inductor does not allow instantaneous change of current, $I_3(0) = 0$ also.

SETUP STEP IV Next we apply Kirchhoff's rules (cf. Figure 26.27):

Junction rule at A: $$I_3 = I_1 + I_2. \tag{i}$$

Loop 1 ($ABCD$): $$-I_2R + Q/C = 0. \tag{ii}$$

Loop 2 ($ABEF$): $$-I_2R - L\frac{dI_3}{dt} + \mathcal{E} = 0. \tag{iii}$$

The relation between I_1 and Q is:

$$I_1 = \frac{dQ}{dt}, \tag{iv}$$

and from eqn. (ii), $I_2 = Q/(RC)$. Then from eqn. (i),

$$I_3 = \frac{Q}{RC} + \frac{dQ}{dt}. \tag{v}$$

Substituting into eqn. (iii), we get:

$$\frac{d^2Q}{dt^2} + \frac{1}{RC}\left(\frac{dQ}{dt}\right) + \frac{Q}{LC} = \frac{\mathcal{E}}{L}. \tag{vi}$$

SOLVE STEPS III AND V This equation can be solved using the method given in the *Math Toolbox*. The homogeneous equation is like the *LRC* circuit eqn. (31.26), with different coefficients [the *LR* circuit decay rate R/L is replaced by the *RC* circuit decay rate $1/(RC)$]. We may adopt that solution (eqn. 31.29) with the appropriate change:

$$Q_H = Q_0 \cos(\omega t - \phi)e^{-\alpha t} \qquad \text{(vii)}$$

with

$$\alpha = \frac{1}{2RC} \quad \text{and} \quad \omega^2 = \frac{1}{LC} - \frac{1}{(2RC)^2}. \qquad \text{(viii)}$$

The solution to the inhomogeneous equation is $Q_I = \mathscr{E}C$. Thus:

$$\begin{aligned} Q(t) &= Q_I + Q_H \\ &= \mathscr{E}C + Q_0 \cos(\omega t - \phi)e^{-\alpha t}. \end{aligned} \qquad \text{(ix)}$$

We complete the solution by finding the constants Q_0 and ϕ. Applying the initial conditions:

$$Q(0) = \mathscr{E}C + Q_0 \cos\phi = 0 \Rightarrow Q_0 = -\mathscr{E}C/\cos\phi.$$

Both I_3 and Q are zero at $t = 0$, so according to eqn. (v), dQ/dt is also zero.

$$\begin{aligned} \left.\frac{dQ}{dt}\right|_{t=0} &= \left. -\omega Q_0 \sin(\omega t - \phi)e^{-\alpha t} - \alpha Q_0 \cos(\omega t - \phi)e^{-\alpha t}\right|_{t=0} \\ &= \omega Q_0 \sin\phi - \alpha Q_0 \cos\phi = 0 \Rightarrow \tan\phi = \alpha/\omega, \end{aligned}$$

as before. Putting these results into the general solution (eqn. ix), with α from eqn. (viii), we find the potential difference across the capacitor, $\Delta V = Q/C$:

$$\Delta V(t) = \mathscr{E}\left[1 - \frac{\cos(\omega t - \phi)}{\cos\phi}e^{-t/2RC}\right].$$

ANALYZE Check that this solution has the expected behavior. For example, $\Delta V \to \mathscr{E}$ as $t \to \infty$. ∎

Chapter Summary

Where Are We Now?

Using Kirchhoff's rules, we have analyzed the behavior of circuits in which current varies with time. These circuits have many practical applications and bring together all of the electromagnetic effects we have studied so far.

What Did We Do?

Coils resist change in the current they carry. The rate of change of current in such an inductor is proportional to the potential difference applied to it:

$$\Delta V = L\, dI/dt,$$

where L is the self-inductance of the coil. The static electric field in the inductor points in the direction of dI/dt. A similar effect occurs when flux produced by one circuit element threads another, giving rise to mutual inductance (eqn. 31.16). Electrical transformers are an application of mutual inductance.

The energy stored in an inductor carrying current I is $\frac{1}{2}LI^2$. The energy is stored in the magnetic field itself, with an energy density $u_B = \frac{1}{2}B^2/\mu_0$. Combining this result with eqn. (27.13) for electric energy, the total energy stored in (electric plus magnetic) fields has density:

$$u = u_E + u_B = \frac{1}{2}(\epsilon_0 E^2 + B^2/\mu_0).$$

We studied four types of single-loop circuit: *RC*, *LR*, *LC*, and *LRC* circuits. These simple circuits demonstrate two important types of behavior: exponential decay and oscillation.

Decay results from energy dissipation in a resistor. The exponential time constants for the *RC* and *LR* circuits are $\tau_C = RC$ and $\tau_L = L/R$.

A system can oscillate only if it has more than one way to store energy. In the *LC* circuit, energy oscillates between electric field energy in the capacitor and magnetic field energy in the inductor. The angular frequency of the oscillation is $\omega_0 = 1/\sqrt{LC}$. The *LRC* circuit exhibits damped oscillations as the resistor dissipates the stored energy. The time constant $\tau_{LRC} = \frac{1}{2}L/R = 1/\alpha$ is one-half the time constant for the *LR* circuit and the oscillation frequency is $\omega_1 = \sqrt{\omega_0^2 - \alpha^2}$. With a large enough resistance, the damping rate $\alpha > \omega_0$ and no oscillations occur. The circuit is overdamped.

Kirchhoff's rules are the tools used to analyze all circuits. They produce linear differential equations for the charge or current. In the *Math Toolbox* we have outlined a method for solving such equations. The first step in any solution is to determine the kind of behavior that is expected and the final equilibrium that is achieved. A solution plan is outlined in Figure 31.4.

* The same principles apply to the solution of multiloop circuits. Kirchhoff's rules produce coupled differential equations. The kinds of behavior—oscillation and decay—are the same.

Practical Applications

Capacitors are frequently used in switching circuits to prevent sparks. Shunt resistors are connected in parallel with capacitor banks to ensure that the stored energy is dissipated safely when the banks are not in use, and *RC* circuits are also used to produce light flashes, either at regular intervals (warning signals) or on demand (camera flash attachments). Inductors find numerous uses in electric circuits, such as smoothing variations in current and storing energy. Inductor circuits provide the spark to automobile engines. Transformers use mutual inductance to couple circuits that transmit power at different voltages. Applications in AC circuits are the subject of the next chapter.

Solutions to Exercises

31.1.
$$(1\ \Omega)\cdot(1\ \text{F}) = \frac{1\ \text{V}}{1\ \text{A}}\cdot\frac{1\ \text{C}}{1\ \text{V}} = \frac{1\ \text{C}}{1\ \text{C/s}} = 1\ \text{s}.$$

31.2. Differentiating eqn. (31.6),

$$\frac{dQ}{dt} = \frac{d}{dt}[C\mathscr{E}(1 - e^{-t/RC})] = -C\mathscr{E}\left(-\frac{1}{RC}\right)e^{-t/RC} = \frac{\mathscr{E}}{R}e^{-t/RC}.$$

Substituting into eqn. (31.7):

$$\frac{\mathscr{E}}{R}e^{-t/RC} + \frac{C\mathscr{E}(1 - e^{-t/RC})}{RC} = \frac{\mathscr{E}}{R}.$$

The two exponential terms cancel and the equation is satisfied.

31.3. The inductor resists change in current. With the circuit broken, it does so by forcing a spark to jump between the blades of the switch as it is opened. Such sparks damage switches; a capacitor in parallel with the switch allows the inductor to charge the capacitor rather than form a spark.

31.4.
$$1\ \text{H} = 1\frac{\text{V}\cdot\text{s}}{\text{A}} = 1\frac{\text{J}\cdot\text{s}}{\text{C}\cdot\text{A}} = 1\frac{\text{N}\cdot\text{m}}{\text{A}^2} = 1\frac{\text{N}}{\text{A}\cdot\text{m}}\cdot\frac{\text{m}^2}{\text{A}}$$
$$= 1\frac{\text{T}\cdot\text{m}^2}{\text{A}} = 1\frac{\text{Wb}}{\text{A}}.$$

31.5. $U = \frac{1}{2}LI^2 = \frac{1}{2}(3.3 \times 10^{-3}\ \text{H})(1.1\ \text{A})^2 = 2.0\ \text{mJ}.$

(*Note:* From Exercise 31.4, $1\ \text{H} = 1\ \text{J/A}^2$.)

31.6. The charge density on the wire produces an electric field at a radius *r* from the wire (eqn. 24.4) $E = \lambda/(2\pi\epsilon_0 r)$, so we must include the electric field energy density:

$$u_E = \frac{1}{2}\epsilon_0 E^2 = \frac{1}{2}\epsilon_0[\lambda/(2\pi\epsilon_0 r)]^2.$$

The electric energy density 6.0 mm from the wire is:

$$u_E = \left(\frac{0.5}{8.85 \times 10^{-12}\ \text{F/m}}\right)\left[\frac{5.0 \times 10^{-12}\ \text{C/m}}{2\pi(6.0 \times 10^{-3}\ \text{m})}\right]^2$$
$$= 9.9 \times 10^{-10}\ \text{J/m}^3.$$

The electric field increases the energy density by 2%. Since both kinds of field decrease as $1/r$, the ratio of their energy densities is independent of distance from the wire.

31.7. The flux through each turn of the primary coil is: $\Phi_p = I_p(L_p/N_p)$. The same flux passes through each turn of the secondary coil, so the total flux through the secondary is:

$$\Phi_s = N_s\Phi_p = I_p L_p N_s/N_p.$$

The mutual inductance is: $M = \Phi_s/I_p = L_p N_s/N_p$. The mutual inductance may also be found from the flux through the primary: $M = \Phi_p/I_s = L_s N_p/N_s$. Multiplying these together:

$$M^2 = \left(L_p\frac{N_s}{N_p}\right)\left(L_s\frac{N_p}{N_s}\right) = L_p L_s.$$

31.8. $$\frac{1\ \text{H}}{1\ \Omega} = \frac{1\ \text{V·s/A}}{1\ \text{V/A}} = 1\ \text{s, as required.}$$

31.9. Traversing the circuit clockwise,

$$0 = \mathcal{E} - IR - L\frac{dI}{dt}.$$

Differentiating eqn. (31.21):

$$\frac{dI}{dt} = \frac{\mathcal{E}}{R}\left(\frac{R}{L}\right)e^{-Rt/L} = \frac{\mathcal{E}}{L}e^{-Rt/L}.$$

Substituting into Kirchhoff's rule:

$$0 \stackrel{?}{=} \mathcal{E} - R\frac{\mathcal{E}}{R}(1 - e^{-Rt/L}) - L\frac{\mathcal{E}}{L}e^{-Rt/L} = 0.$$

The assumed solution satisfies the equation.

31.10. In ■ Figure 31.22 we have plotted the ratio $I/I_f = IR/\mathcal{E}$ of the current I to its final value as a function of the dimensionless time variable t/τ. The straight line is extremely close to the exact solution for $t/\tau < 0.1$. The difference is 4% at $t/\tau = 0.1$.

Alternatively, using the exponential series:

$$e^x = 1 + x + \tfrac{1}{2}x^2 + \cdots,$$

we have: $$1 - e^{-t/\tau} = 1 - [1 - t/\tau + \tfrac{1}{2}(t/\tau)^2 + \cdots] = t/\tau - \tfrac{1}{2}(t/\tau)^2 + \cdots.$$

For $t \ll \tau$, we ignore the squared term and all the higher powers. Then, using eqn. (31.21),

$$I(t) \approx (\mathcal{E}/R)(t/\tau) = (\mathcal{E}/L)t,$$

proportional to time. Compare this result with eqn. (31.9).

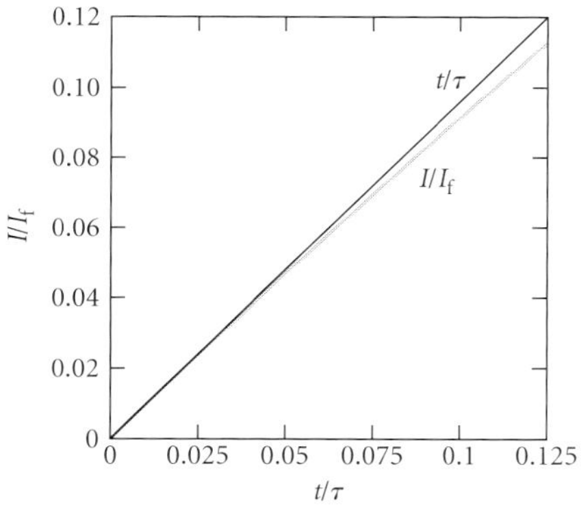

■ FIGURE 31.22

Graph of the current in an inductor as a function of time for $t/\tau < 0.12$. The linear function $I/I_f = t/\tau$ is shown for comparison. The two curves are indistinguishable for $t/\tau < 0.05$ and differ by 4% at $t/\tau = 0.1$.

31.11. $$(1\ \text{H})(1\ \text{F}) = (1\ \text{V·s/A})(1\ \text{C/V}) = \frac{(1\ \text{C})\cdot(1\ \text{s})}{1\ \text{C/s}} = 1\ \text{s}^2.$$

31.12. With the assumed form for Q,

$$-I(t) = \frac{dQ}{dt} = -\beta Q_0 e^{-\beta t} \quad \text{and} \quad -\frac{dI}{dt} = \frac{d^2Q}{dt^2} = \beta^2 Q_0 e^{-\beta t}.$$

Substituting into eqn. (31.26):

$$\beta^2 - \frac{R}{L}\beta + \frac{1}{LC} = 0.$$

Solving the quadratic:

$$\beta_{\pm} = \frac{R}{2L} \pm \sqrt{\frac{R^2}{4L^2} - \frac{1}{LC}},$$

as required. There are two time constants, corresponding to the plus and minus signs. The complete solution has two terms, one with each time constant:

$$Q = Q_1\exp(-\beta_+ t) + Q_2\exp(-\beta_- t).$$

Both β_+ and β_- are positive so both terms decay with time.

Basic Skills

Review Questions

§31.1 RESISTOR–CAPACITOR CIRCUITS

- Which mathematical function describes the potential difference across a discharging capacitor as a function of time?
- What is the time constant for an *RC* circuit?
- Formally, how long does it take for the capacitor to discharge? What is a *practical* measure of the discharge time?
- Is the current variable $I(t)$ always related to the charge variable $Q(t)$ by $I = +dQ/dt$? Why or why not?
- When a capacitor discharges in an *RC* circuit, what happens to the energy that it stored?
- Describe the characteristic behavior of an *RC* circuit as it evolves from its initial state to its final state.

§31.2 INDUCTANCE

- In a circuit containing a superconducting coil, a battery, and a switch, what is the current immediately after the switch is closed? Why isn't it infinite?
- Give two different expressions for the *inductance* of a circuit element.

- What is the SI unit of inductance?
- How does the magnetic energy stored by an inductor depend on the current in it?
- How does magnetic energy density depend on magnetic field strength?
- Define the *mutual inductance* of two circuit elements.
- What is a transformer? Give an example of how one is used.

§31.3 INDUCTOR CIRCUITS

- How does current vary with time in an *LR* circuit? What is the time constant?
- How does current vary with time in an *LC* circuit?
- In mechanical oscillations, energy varies between potential and kinetic forms. What forms of energy occur in an electrical circuit oscillation?
- What happens if resistance is added to an *LC* circuit? How does the oscillation frequency change?
- Does an *LRC* circuit always oscillate? Why or why not?

✲ §31.4 MULTILOOP CIRCUITS

- Which rules are used to analyze multiloop circuits?
- Do any new forms of behavior appear?

Basic Skill Drill

§31.1 RESISTOR–CAPACITOR CIRCUITS

1. Refer to ■ Figure 31.23. If the resistance in the circuit is 6.7 kΩ, and the initial charge on the capacitor is $Q_0 = 0.27$ C, find the meter readings 6.0 s after the switch is closed.

2. A lightning protection system contains a 3.0-mF capacitor connected to ground through a 1.0-Ω resistance. A lightning bolt delivers a charge of 0.70 C to the system. What energy is absorbed by the system, and how long does it take to discharge the energy to ground?

3. A 10.0-V battery, a 56-μF capacitor, a 10.0-kΩ resistor, and a switch are connected in series. What is the final charge on the capacitor a long time after the switch is closed? What is the final current? How much time is required for the charge to reach 95% of its final value? What is the initial current in the circuit?

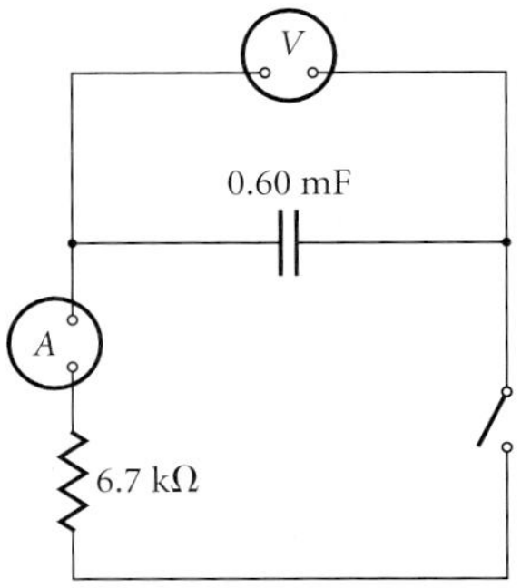

■ FIGURE 31.23

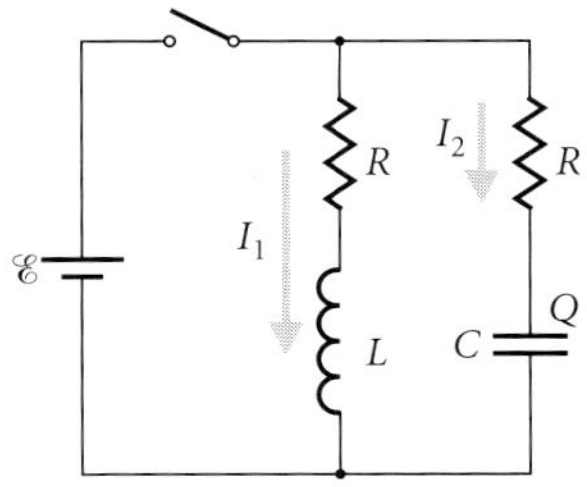

■ FIGURE 31.24

§31.2 INDUCTANCE

4. Find the inductance of a solenoid with 750 turns, length 2.5 cm, and radius 0.45 cm.

5. Find the magnetic energy density at a point 15 cm from a long, straight wire carrying a current of 1.7 A.

6. A transformer with 100 turns of wire on the primary coil and 500 turns on the secondary coil is connected to a potential difference $\Delta V = (100\text{ V})\cos \omega t$. Find the potential difference across the secondary terminals.

§31.3 INDUCTOR CIRCUITS

7. An *LR* circuit contains a 3.0-Ω resistor and a 4.0-H inductor in series with an ammeter. At $t = 4.0$ s, the ammeter reads 0.050 A. What was the current at $t = 0$?

8. When a 0.16-μF capacitor is connected in series with a 12-mH inductor, the circuit is found to be critically damped. What is the resistance of the inductor?

9. A 56-μF capacitor is connected in series with a 3.0-mH inductor. What is the oscillation frequency of the resulting circuit? At $t = 0$, the capacitor is uncharged and the current in the circuit is 17 mA. What is the energy in the inductor? Describe how the current and the total energy in the circuit vary with time.

✲ §31.4 MULTILOOP CIRCUITS

10. The switch in ■ Figure 31.24 is closed at $t = 0$. What are the values of I_1, I_2, and Q immediately after the switch is closed? What are the values a long time later? Do the inductor and capacitor arms of the circuit affect each other's behavior? Why or why not?

Questions and Problems

§31.1 RESISTOR–CAPACITOR CIRCUITS

11. ❖ Explain why the charge on a capacitor cannot change abruptly when the switch is closed in an *RC* circuit.

12. ❖ If the resistance in an *RC* circuit is doubled, how does the time constant change? What if the capacitance is doubled? In each case, give a physical argument to support your answer.

13. ♦ What time is required for the charge on a capacitor in an *RC* circuit to decrease to one-half its original value?

14. ♦ **(a)** What is the time constant of a circuit containing a 32-nF capacitor in series with a 14-kΩ resistor? **(b)** If the charge on the capacitor is initially $Q_0 = 32$ μC, what time interval is required for it to decrease to 1.6 μC? **(c)** What is the initial current in the circuit?

15. ♦ A 16-μF and a 20-μF capacitor are connected in parallel. What resistance must be connected across them for the circuit to have a time constant of 1.0 ms?

16. ♦♦ What is the initial current in an RC circuit (Figure 31.1) if the initial charge on the capacitor is Q_0? Show that if the current were constant, the capacitor would discharge completely in one time constant.

17. ♦♦ Refer to the circuit diagram in Figure 31.5. At $t = 0$, when the switch is closed, the capacitor carries a charge $Q_0 = 2C\mathcal{E}$. Find the charge and current as functions of time after the switch is closed.

18. ♦♦ A 65.0-nF capacitor is connected in series with a resistor. A 12.0-V battery is connected across the capacitor. Immediately after the battery is disconnected, there is a 0.240-A current in the resistor. What is the resistance? Find the current 3.0 μs later.

19. ♦♦ Refer to the circuit diagram in Figure 32.23. At 4.0 s after the switch is closed, the voltmeter reads 150 V and the ammeter reads 30.0 mA. What is the value of the resistance R? What will the meters read 8.2 s later? What was the original charge on the capacitor?

20. ♦♦ Find the charge on each capacitor in ■ Figure 31.25 as a function of time after the switch is closed.

21. ♦♦ You have two capacitors, 56 μF and 22 μF, and two resistors, 18 kΩ and 33 kΩ. You need to build an RC circuit with a time constant of 0.91 s. Can you do it? If so, how?

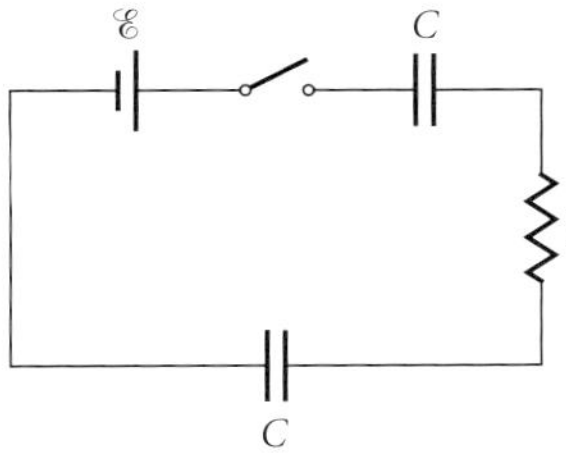

■ FIGURE 31.25

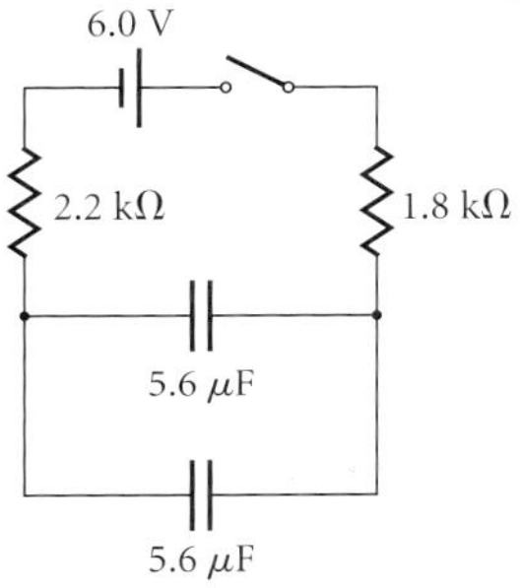

■ FIGURE 31.26

22. ♦♦ A circuit contains two resistors and two capacitors connected as shown in ■ Figure 31.26. What is the charge on each capacitor a long time after the switch is closed? How long does it take to reach that state? How much energy is dissipated in the resistors?

23. ♦♦♦ Two unequal capacitors C_1 and C_2 are connected in series with a resistor R and a switch. Initially, one capacitor has a charge Q_1 and the second is uncharged. Then the switch is closed. Show that the capacitors share the final charge like parallel capacitors, but in the time constant they combine like series capacitors.

§31.2 INDUCTANCE

24. ❖ Two long wires, each carrying current I, are perpendicular to the page, as shown in ■ Figure 31.27. At which point is the magnetic energy density greatest? Explain why you chose your answer.

25. ❖ Compare the total magnetic energy in two solenoids, each with N turns, area A, and current I, but lengths ℓ and 2ℓ.

26. ❖ An inductor is to be made by wrapping a fixed length of wire around a cardboard tube. You have a choice of tubes with the same length but different radii. Does the inductance depend on which tube you choose? If so, how?

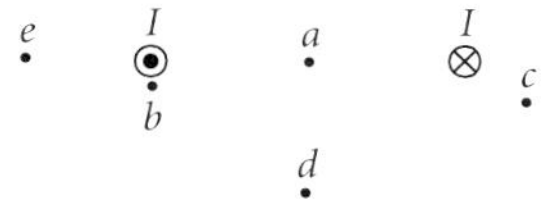

■ FIGURE 31.27

27. ♦ Find the magnetic energy density at a point 5.0 m from a power cable carrying a current of 0.80 kA.

28. ♦ Find the magnetic energy per kilometer of depth inside a large sunspot with area 6×10^{15} m^2 and $B = 4 \times 10^{-3}$ T.

29. ♦ Find the total electromagnetic energy density near the surface of the Earth, where $B = 4 \times 10^{-5}$ T and $E = 100$ V/m. What fraction of the total energy density is electric?

30. ♦♦ Find the magnetic energy density on the axis of a current loop with radius 2.0 cm carrying a current of 6.0 mA, at a distance of 3.0 cm from the center of the loop.

31. ♦♦ A constant current I in a semi-infinite wire lying along the negative z-axis forms a point charge Q at the origin. Find the total electromagnetic energy density at point P with coordinates $z = -d$, $x = R$, with $R = 2.0$ cm, $I = 6.0$ mA, and $d = 1.5$ cm at a time when $Q = 3.0$ pC. Compare Problem 30.67 for the calculation of $B = (\mu_0 I/4\pi R)(1 + d/\sqrt{d^2 + R^2})$.

32. ♦♦ Two solenoids lie along the same axis. The larger has a radius $a = 5.0$ cm, length $\ell = 15$ cm, and $N = 1.3 \times 10^4$ turns. The smaller solenoid has radius $b = 3.0$ cm, length $\ell' = 10.0$ cm, and $N = 5 \times 10^3$ turns. Estimate the mutual inductance of the two solenoids: first by assuming a current I_1 in the large solenoid and computing the flux Φ_2 through the smaller; and second by assuming a current I_2 in the smaller solenoid and calculating Φ_1.

33. ♦♦ A coil consists of 1.5×10^4 turns of wire wrapped around a toroidal (doughnut-shaped) tube. The center of the tube has radius 10.0 cm and the tube's cross section has radius 2.0 cm (cf. Figure 28.24). Estimate the inductance of the ring.

34. ♦♦ **(a)** Estimate the mutual inductance of two concentric, coplanar circular wire loops, one of radius 5 cm and one of radius 0.5 cm. **(b)** Estimate the mutual inductance of the two loops when they are separated by 50 cm (center to center) in their common plane. (*Hint:* Use magnetic moment.)

35. ♦♦♦ Two solenoids of radii a_1 and $a_2 < a_1$ are mounted one inside the other and connected in series so that each carries the same current (■ Figure 31.28). Each has length ℓ and N turns. **(a)** Is the inductance greater if the coils are wound in the same sense (as shown in the figure) or in the opposite sense? Why? **(b)** Compute the inductance of the combination. (You may use the results for long solenoids in your calculations.)

36. ♦♦♦ Use the result $B = \mu_0 nI(\cos\theta_1 + \cos\theta_2)/2$ (Problem 28.41, Figure 28.40) for the magnetic field inside a finite solenoid to improve the estimate of a solenoid's inductance. Assume that the magnetic field is uniform over any cross section of the solenoid though it varies with position along the axis.

37. ♦♦♦ In a coaxial cable, apply Faraday's law to find the emf $\mathcal{E}$ around a rectangular curve, with length ℓ parallel to the cable and whose two ends are radial lines between the two conductors. Find the inductance per unit length of the cable from the relation $L/\ell = (\mathcal{E}/\ell)/(dI/dt)$, (cf. Example 31.6).

38. ♦♦♦ Two parallel wires, each with radius 0.23 mm, carry current in opposite directions. Their centers are separated by 3.0 cm. Calculate the inductance per unit length of the two wires. **(a)** Assume the current is on the outside of the two wires. **(b)** Does it make a difference if the current is uniform within the two wires? If so, how much?

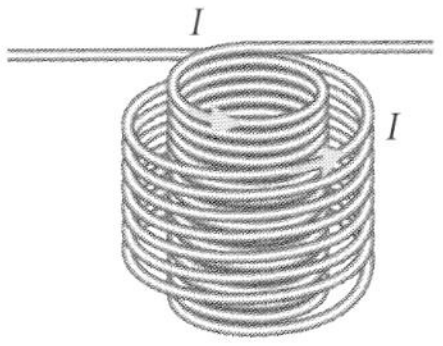

■ FIGURE 31.28

39. ❖ Explain why the current through an inductor cannot change abruptly when the switch is closed in an *LR* circuit in series with a battery.
40. ❖ If the resistance in an *LR* circuit is doubled, how does the time constant change? State the physical reason for the change.
41. ❖ If the resistance in an *LRC* circuit is doubled, how does the time constant change? What happens to the frequency ω_1?
42. ❖ Make an analogy between an *LC* circuit and a particle on a spring (cf. §14.1), and discuss the roles of inductance and capacitance in the circuit.
43. ❖ If the inductance in an *LRC* circuit is doubled, how does this affect the time constant and the frequency ω_1? Give a physical explanation for your answer.
44. ❖ The switch in ■ Figure 31.29 remains in position ② for 0.2 s, then snaps quickly to position ①, remains there for 0.2 s, then snaps back to ②, and so on. Sketch a graph showing the current through the resistor as a function of time. How would the graph change if the oscillation frequency of the switch were increased? If it were decreased?

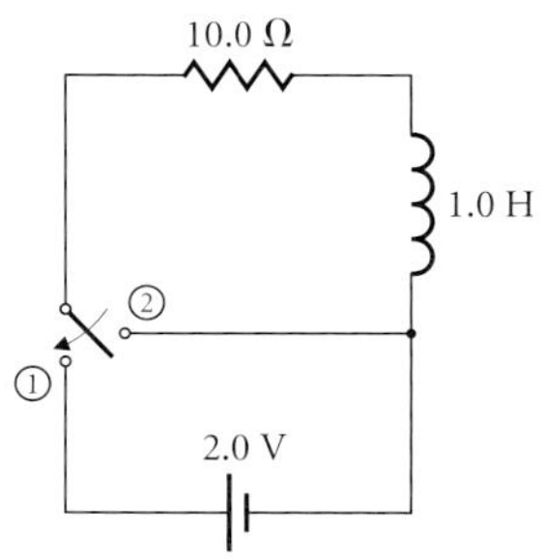

■ FIGURE 31.29

45. ❖ **(a)** Show that the combined inductance of two inductors connected in series is $L_s = L_1 + L_2$. (Assume the inductors are separated widely enough that flux from one does not thread the other.) **(b)** Show that the combined inductance of two inductors connected in parallel is given by $1/L = 1/L_1 + 1/L_2$. (Assume that no flux from one threads the other.)
46. ♦ A circuit contains a 3.0-mH inductor, a 6.0-μF capacitor, and an ammeter. The ammeter has negligible effect on the circuit. If the ammeter reads 0.50 A at $t = 0$ and 0.20 A at $t = 0.21$ ms, find an expression for the current at an arbitrary time.
47. ♦ For a long time prior to $t = 0$, the switch in the circuit of ■ Figure 31.29 has been in position ①. At $t = 0$, the switch is put into position ②. What are the magnitude and direction of current at $t = 0.30$ s?
48. ♦ What turns ratio is needed in a transformer to convert 33-V generator output voltage to 110 V to run a household appliance? How could you build a transformer to do the job? (*Remember:* Turns come in whole numbers!)
49. ♦ A 10.0-V battery, a 10.0-kΩ resistor, and a 0.30-H inductor are connected in series with a switch. The switch is closed at $t = 0$ after remaining open for several seconds. What is the current in the inductor immediately after the switch is closed? What is the final current in the inductor? How much time is required for the current to reach 95% of its final value?
50. ♦ You have a coil with a resistance of 25 Ω and an inductance of 75 mH. What size capacitor should you connect in series with it to make a critically damped circuit?

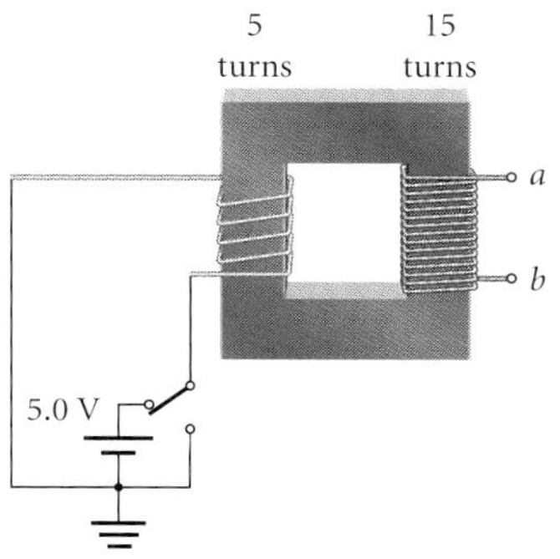

■ FIGURE 31.30

51. ♦♦ The primary coil of the transformer shown in ■ Figure 31.30 is connected to a switching circuit that switches the coil between ground and the positive terminal of a 5.0-V battery at a frequency of 4.0 Hz. Make a sketch showing how the potential difference V_{ab} across the ends of the secondary coil varies in time. How is the behavior affected by the self-inductance of the transformer?
52. ♦♦ An *LRC* circuit contains a 15-μF capacitor and a 23-mH inductor. What resistance is necessary for the circuit to be critically damped ($\omega_1 = 0$)? If the resistance is half this value, find the frequency and time constant for the circuit.
53. ♦♦ You have two capacitors, $C = 5.6\ \mu$F and $C = 2.2\ \mu$F, and two inductors, $L = 22$ mH and $L = 56$ mH. How should you connect them to obtain a circuit that oscillates with a 2.2-ms period?
54. ♦♦ A current I_0 flows initially in a series *LR* circuit. Show that the energy dissipated by the resistor equals the energy initially stored in the inductor.
55. ♦♦ The switch is closed at $t = 0$ in a circuit containing a battery, resistor, and inductor in series. Find the current in the circuit as a function of time. Show that at any time $t > 0$, the energy dissipated by the resistor plus the energy stored in the inductor equals the work done by the battery since $t = 0$.
56. ♦♦ A 56-μF capacitor is connected in series with a 0.30-H inductor. The system oscillates with a total energy of 27 μJ. Compute the maximum values of the capacitor voltage and of inductor current. If the volume within the inductor is 1.9×10^{-4} m^3, estimate the maximum value of the magnetic field in the inductor. If the maximum value of the electric field is 1.0×10^6 V/m and the capacitor is air-filled, find its dimensions.
57. ♦♦ Resistivity of the wire ensures that a real solenoid has both inductance and resistance. If the solenoid is tightly wound with a single layer of wire of length ℓ, show that its *LR* time constant depends only on the resistivity and diameter of the wire and the radius of the coil.
58. ♦♦ A circuit contains a known resistance of 2.7 kΩ, and an unknown inductance. When a 12-V battery is connected, what is the current in the circuit a long time after the switch is closed? If it takes 3.7 ms for the current to reach 95% of this value, what is the inductance in the circuit?
59. ♦♦♦ ■ Figure 31.31 represents the spark coil circuit in an automobile engine. Compare with Figure 31.16. The additional capacitor prevents unwanted sparking across the points. Assume that the switch has been closed for a long time. What are the capacitor charge and the current through the coil? Find the potential difference across the coil as a function of time after the switch is opened but before the plug sparks. What is the maximum voltage achieved? (See also Problem 67.)

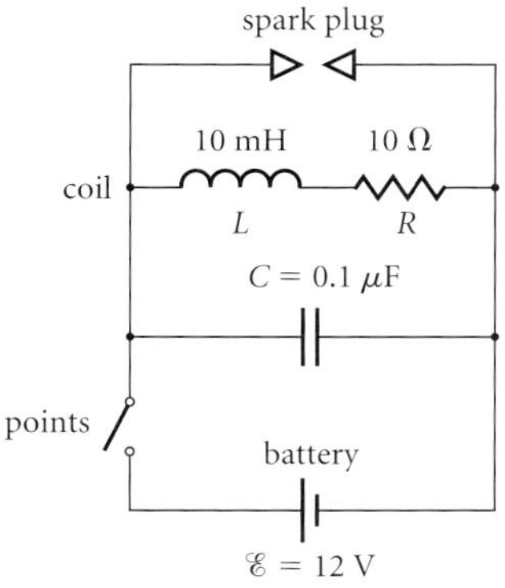

■ FIGURE 31.31

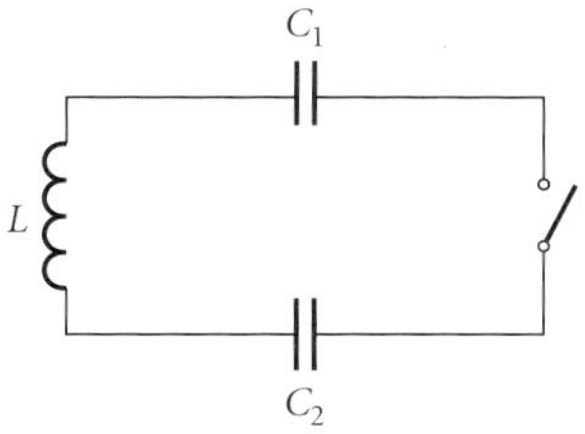

■ FIGURE 31.32

60. ♦♦♦ There is a charge Q_0 on capacitor C_1 in ■ Figure 31.32, and capacitor C_2 is uncharged. At $t = 0$, the switch is closed. What is the current in the circuit immediately after the switch is closed, and a long time after the switch is closed? Find the current in the circuit and the two charges as functions of time for $t > 0$.

61. ♦♦♦ In ■ Figure 31.33, both capacitors are initially uncharged, and all the switches are open. First S_2 is closed, then S_1 is closed. Find the charge on each capacitor. At time $t = 0$, S_1 is opened and S_4 is closed. What is the maximum current through the inductor and when does it occur? At the instant of maximum current, S_2 is opened and S_3 is closed. What is the maximum charge on C_2 and when does it occur?

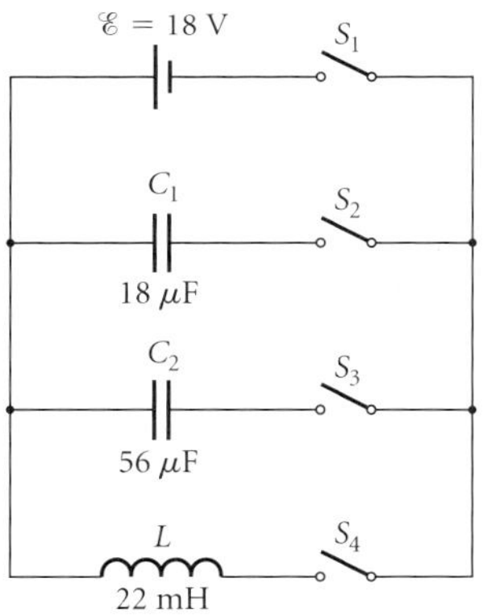

■ FIGURE 31.33

62. ♦♦♦ A 1.2-kΩ resistor, a 3.3-mH inductor, and a 2.5-nF capacitor are connected in series with a 12-V battery. Find the current in the circuit as a function of time after the switch is closed.

63. ♦♦♦ Assume the resistor in a series *LRC* circuit is sufficiently small that the solution for $Q(t)$ is of the form $Q(t) = Q_0 a(t)\sin \omega t$, where $a(t)$ changes negligibly during one cycle of the oscillation. Compute the heat dissipated by the resistor during one cycle, treating $a(t)$ as a constant. Then set the energy dissipated equal to the change in the stored energy $(\frac{1}{2}Q_0^2 a^2/C)$ to obtain an approximate differential equation for $[a(t)]^2$. Solve the equation, and compare the resulting expression for $a(t)$ with the exact solution found in the text.

64. ♦♦♦ An *LRC* circuit contains a 5.0-mH inductor, a 0.10-kΩ resistor, and a 3.0-μF capacitor in series. For a long time prior to $t = 0$, a 6.0-V battery has been connected across the capacitor. Find the charge on the capacitor and the current in the circuit. At $t = 0$, the battery is disconnected from the capacitor. Find the charge and current as functions of time for $t > 0$, and give numerical values at $t = 0.10$ ms.

✲ §31.4 MULTILOOP CIRCUITS

65. ❖ Describe the behavior of the circuit in ■ Figure 31.34 a long time after the switch is closed.

66. ❖ If the resistance in ■ Figure 31.34 is doubled, how is the circuit behavior affected?

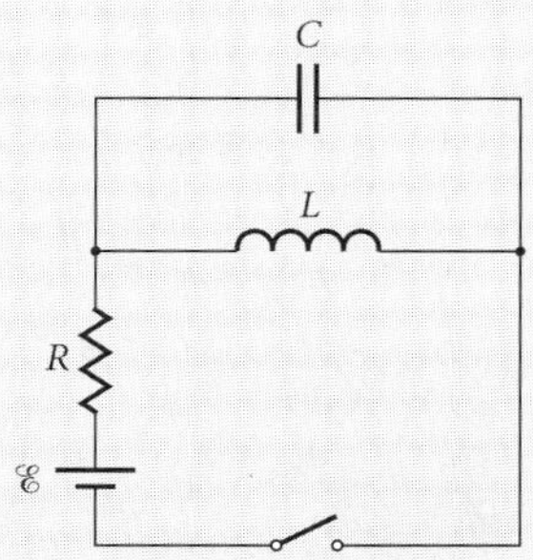

■ FIGURE 31.34

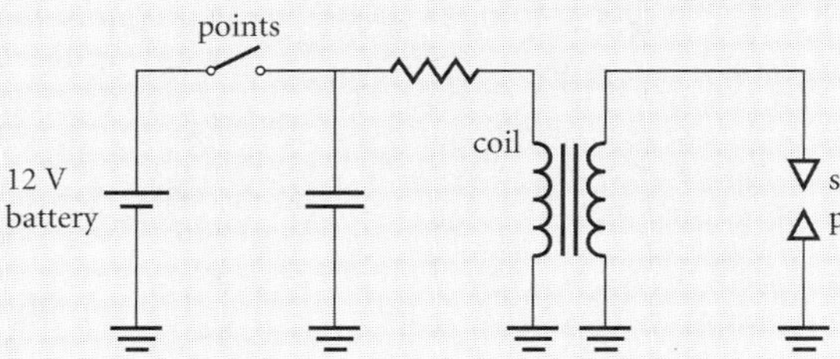

■ FIGURE 31.35

67. ❖ ■ Figure 31.35 is a yet more accurate version of the automotive spark coil circuit. Explain the purpose of the capacitor and of the transformer. (See also Problem 59.)

68. ♦♦ Two identical capacitors, each in series with a resistor R, are connected in parallel (■ Figure 31.36). The combination is connected across a third resistor R. There is a switch between the capacitors, and a charge Q_0 is on one of the capacitors. At $t = 0$, the switch is closed. What is the final state of the system? Find the charge on each capacitor and the current through the third resistor as functions of time.

69. ♦♦ What is the charge on each of the capacitors in ■ Figure 31.37 when the switch has been closed for a long time? Apply Kirchhoff's rules or an energy method to determine the current supplied by the battery as a function of time after the switch is closed. What happens when the switch is opened again?

70. ♦♦♦ The circuit in Example 31.3 should really be treated as a multiloop circuit. Show that when this is done, the time constant becomes $\tau = \frac{10}{11}RC$. How big is the error in the time between flashes computed in the example?

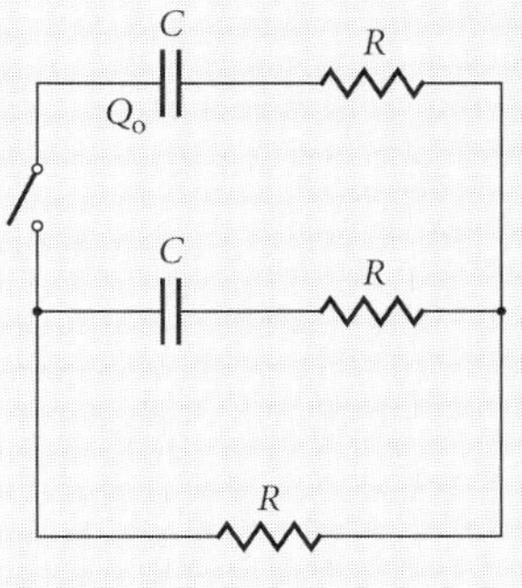

■ FIGURE 31.36

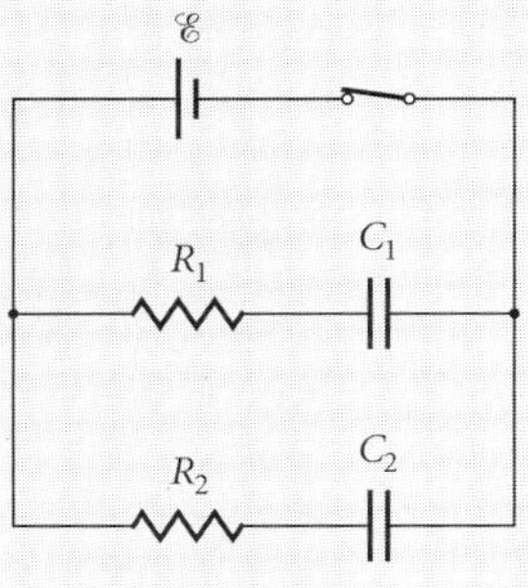

■ FIGURE 31.37

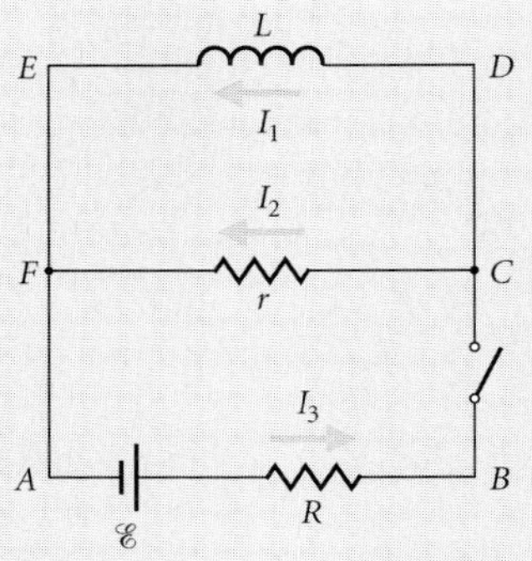

■ FIGURE 31.38

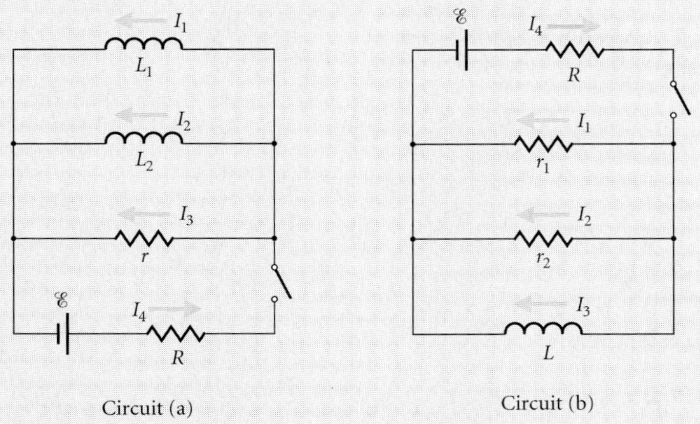

■ FIGURE 31.39

71. ♦♦♦ Apply Kirchhoff's rules to the junction at C and to loops $ABCF$ and $CDEF$ in ■ Figure 31.38 to obtain equations relating I_1, I_2, I_3, and their derivatives. Manipulate these equations to obtain a single equation of the form:

$$\mathcal{E} - L_T \frac{dI_3}{dt} - R_T I_3 = 0.$$

Comment on the value of the time constant. Solve the equation and find I_3, I_1, and I_2 if the switch is closed at $t = 0$.

72. ♦♦♦ Find the charge on the capacitor in the circuit shown in ■ Figure 31.34 as a function of time after the switch is closed. Show that the time constant is $2RC$.

73. ♦♦♦ For each of the circuits in ■ Figure 31.39, find the currents I_1, I_2, I_3, and I_4 as functions of time after the switch is closed. (*Hint:* For circuit (a), use Kirchhoff's loop rule to express dI_2/dt, dI_1/dt, and dI_4/dt in terms of dI_3/dt and I_3. Then differentiate the junction equation and use these expressions to find a single differential equation for I_3. Use a similar method for circuit (b).)

Additional Problems

74. ❖ The switch in the circuit shown in ■ Figure 31.40 has been open a long time. What is the current through the 100-Ω resistor immediately after the switch is closed?

75. ❖ A circuit contains a resistor and a battery in series. When the switch is closed, a current is immediately established in the resistor. The current produces a magnetic field. Where does the magnetic field energy come from? What does "immediately" really mean in this case?

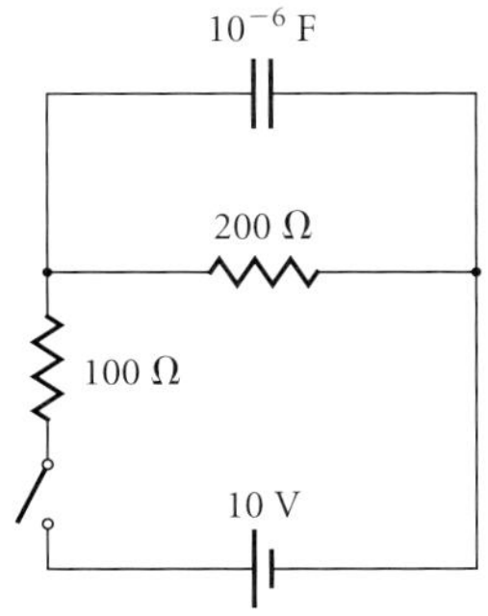

■ FIGURE 31.40

76. ❖ If the resistance in an LR circuit is increased, the time constant decreases. If the resistance in an RC circuit is increased, the time constant increases. Explain the difference.

77. ♦♦ Find the combined inductance of two coils of inductance L_1 and L_2 connected in series if they are close together and have mutual inductance M. Does it make a difference if the two coils wind in the same or opposite senses? Find their combined inductance when they are connected in parallel, again assuming the mutual inductance is M. (Compare with Problem 45.)

78. ♦♦ You have three 220-Ω resistors and a coil with an inductance of 2.5 mH and a resistance of 15 Ω. You wish to construct an LR circuit with a time constant of 15 μs. How may this be done?

79. ♦♦ A circular loop of radius 3.5 cm carries a charge of 2.7 mC/m and spins about an axis through its center at 1500 rpm. What is the total electromagnetic energy density at the center of the loop?

80. ♦♦ A battery of emf ℰ has been connected for a long time in series with a resistor R and a parallel plate capacitor of capacitance C. What is the charge on the capacitor and the current in the circuit? At time $t = 0$, a slab of dielectric with dielectric constant κ is quickly inserted into the capacitor, so that it fills the space between the plates. What is the charge on the capacitor a long time later? Find the current in the circuit and the charge on the capacitor as functions of time for $t > 0$.

81. ✲ ♦♦ Find the current through the battery in ■ Figure 31.41 as a function of time after the switch is closed.

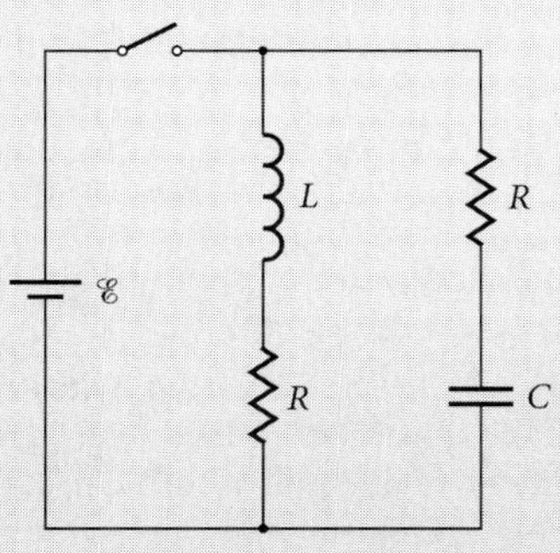

■ FIGURE 31.41

82. ♦♦♦ A loop of radius R is made of wire with radius r. Make two estimates of the loop's self-inductance. **(a)** First assume the flux through the loop equals the area of the loop multiplied by the magnetic field strength at the center. Argue that this is a lower limit for the flux. **(b)** Second, assume the magnetic field in the loop's interior varies with distance from the wire in the same way as for a long straight wire. Do you think this is a better or worse approximation than that in part (a)? Why? **(c)** Evaluate your results for $r = 0.5$ mm and $R = 3.40$ cm. By what factor do they differ?

83. ✲ ♦♦♦ A variable inductor is made by inserting a ferrite plunger partway into a coil. Find the inductance of the coil in terms of the number of turns N, the radius R, the length ℓ, and the distance d that the plunger is inserted. Assume that the ferrite may be described by an effective permeability μ. (Variable inductors of this type are frequently used in tuning circuits for automobile radios.)

Computer Problems

84. The current in an inductor increases linearly for $t \ll \tau_L$. (See Exercise 31.10.) If $L = 1.0$ H and $R = 1.0\ \Omega$, after how long does the actual current differ from the linear increase by 1%? By 10%?

85. A 56-nF capacitor is connected in series with a 12-V battery and a 1.8-kΩ resistor. At $t = 0$, the switch is closed. Plot the energy stored in the capacitor, the total energy dissipated in the resistor, and the total energy drawn from the battery as functions of time. Verify that energy is conserved throughout the charging process.

86. An *LRC* circuit has a 2.2-nF capacitor, a 2.2-mH inductor, and a 780-Ω resistor connected in series across a 12-V battery. The switch is closed at $t = 0$. Modifying eqn. (31.29) to account for the different initial conditions, write down the solution for Q as a function of time, and use it to calculate $I(t)$. Using a spreadsheet or another computer program, calculate the potential difference across the capacitor, the inductor, and the resistor every microsecond from $t = 0$ to $t = 10\ \mu$s. Verify that the sum of the three potential differences equals 12 V at each time.

Challenge Problems

87. Any multiloop circuit containing a resistor has a final state in which currents and capacitor charges are constant in time. True or false? Justify your answer.

88. Two *LR* circuits are coupled by mutual inductance. In one loop, an inductor with $L = 10$ mH is in series with a 100-Ω resistor, a 12-V battery, and a switch. In the other loop, a 20-mH inductor is in series with a 200-Ω resistor. The mutual inductance is 10 mH. Find the current in each loop as a function of time after the switch is closed. What are the values of the two currents at $t = 0.1$ ms?

89. Two *LR* circuits are coupled by their mutual inductance (■ Figure 31.42). If $\mathcal{E} = 12$ V, $L = 2M = 3.0$ mH, and $R = 1.6$ kΩ, find the currents in the circuits as functions of time after the switch is closed. Give numerical values for the currents at $t = 1.0\ \mu$s.

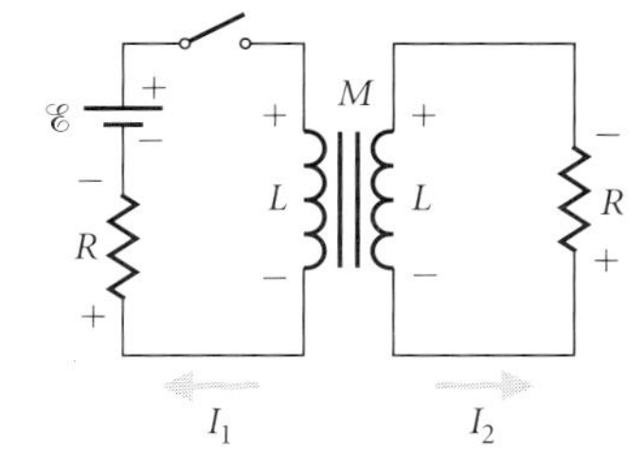

■ FIGURE 31.42

90. **(a)** Calculate the mutual inductance of a small coil carrying current I, with dipole moment $m\hat{\mathbf{k}}$, and a wire loop of radius a that lies in the x-y-plane a distance d from the dipole, with its center on the z-axis. [*Hint:* At distance r from the z-axis, B_z due to the dipole is $B_z = k_m m[-1 + 3z^2/(z^2 + r^2)]/(z^2 + r^2)^{3/2}$.]

(b) Use the result of part (a) to compute the mutual inductance of the dipole and an infinite solenoid with radius a and n turns per meter when the dipole is aligned along the solenoid axis.

(c) Compare your result in part (b) with a calculation of the mutual inductance from the flux produced by the solenoid linking the cross section of the small coil.

91. Analyze the charging and discharging of a capacitor in series with a nonohmic device in which current varies with applied voltage as:

$$I = AV^{\alpha}.$$

Show that for $\alpha < 1$ the process requires a finite time, while for $\alpha \geq 1$ it requires infinite time. In each case, define a time constant and give its value in terms of α, A, and C.

92. The terminals of a superconducting solenoid are connected by a superconducting wire and, initially, there is a current I in the circuit. Imagine an external force squeezing on the solenoid so as to change its radius an amount dr. No emf can develop between the solenoid's terminals without causing infinite current in the superconducting circuit. Use this fact and Faraday's law to find the change in magnetic field within the solenoid. Compute the change in stored energy and the force per unit area needed to compress the solenoid. Your result gives the pressure exerted by a magnetic field. Also find the force per unit area directly from the formula for magnetic force on a current-carrying wire. Compare with the similar result for electric force on a charged, conducting surface (cf. Part VI, Problem 5).

Transformers at an electric substation. In the nation's power grid, power is transmitted at high voltage and low current. In our houses, we use power at much lower voltage but a higher current. This arrangement minimizes ohmic losses in the power grid. A transformer is used to effect the transition to lower voltage. Additional transformers on the power pole outside your home complete the transition to 120-V power.

CHAPTER 32
Introduction to Alternating Current Circuits

CONCEPTS

Phase shift
Reactance
Impedance
Phasor
Power factor
Resonance
Quality

GOALS

Be able to:

Compute the response of a simple circuit to an AC power source.

Determine the power used in a circuit driven by an oscillating emf.

Use phasors to analyze series and parallel AC circuits.

Such harmony is in immortal souls;
But, whilst this muddy vesture of decay
Doth grossly close it in, we cannot hear it.

WILLIAM SHAKESPEARE

In 1882, Nikola Tesla (cf. Chapter 28) demonstrated the great practical advantages of oscillating electrical voltage for the generation, transmission, and use of electrical energy. For example, the transformer in a power company substation reduces the voltage from several hundred thousand volts used in cross-country transmission to the tens of thousands used for transmission within a city. A time-varying potential is essential to the transformer's operation (cf. §31.2.3). Following Tesla's ideas, electric power is now supplied to domestic and industrial users as alternating current (AC). When you plug an appliance into a wall socket in your home, the current in the circuit oscillates sinusoidally in time. Such diverse appliances as industrial motors, vacuum cleaners, and FM radios use AC power. In this chapter we shall study the response of resistors, capacitors, and inductors to oscillating voltage, and develop the fundamentals of AC circuits.

ONCE AGAIN WE SHALL MAKE MAJOR USE OF KIRCHHOFF'S RULES (§26.4) IN ANALYZING CIRCUITS. YOU MAY ALSO WISH TO REVIEW THE MATERIAL ON OSCILLATIONS (CHAPTER 14).

32.1 SINGLE-ELEMENT CIRCUITS

32.1.1 Voltage and Current

EMF. The simplest DC circuit contains a battery to provide energy and a resistor to use it. To build the simplest AC circuit, we replace the battery with an oscillating emf. A single-loop generator (§30.2.2) produces a sinusoidally varying potential difference across its terminals as it rotates. It provides a good mental image for practical generators that operate on the same principles and produce an output of similar form:

DC MEANS DIRECT CURRENT. THE CURRENT IS ALWAYS IN THE SAME DIRECTION.

$$\mathscr{E}(t) = V_0 \cos \omega t. \tag{32.1}$$

THE USE OF A COSINE FUNCTION RATHER THAN A SINE IN EQN. (32.1) IS RELATED TO OUR CHOICE OF ORIGIN FOR TIME. BECAUSE OF THE REPETITIVE NATURE OF THE OSCILLATIONS, SUCH CHOICES DO NOT AFFECT ANY OF THE RESULTS. OUR CHOICES ARE CONSISTENT WITH THOSE IN CHAPTER 14 (SEE, E.G., EQN. 14.3).

■ Figure 32.1 shows the circuit symbol for an ideal AC generator with output given by eqn. (32.1) and with no internal resistance.

Resistance. The currents in a DC network reach a steady state extremely rapidly, in the time required for the electric field to propagate across the circuit at the speed of light. So long as the field and charge distribution that drive current respond much more rapidly than the applied potential varies, resistors behave the same way in an AC circuit as in a DC circuit. This requirement is satisfied for AC power if light can cross the circuit in less than an oscillation period. With 60-Hz commercial power, the circuit could be the size of Texas! Any network made up only of resistors may be analyzed as if it were a DC circuit but with all values of current and potential difference oscillating sinusoidally.

SEE THE ANALYSIS IN EXAMPLE 30.1, WHERE WE ALSO DISCUSSED TIME-VARYING FIELDS.

In the circuit diagrammed in ■ Figure 32.2, the generator develops a sinusoidally varying potential difference that drives current in the resistor:

$$I(t) = \frac{\Delta V(t)}{R} = \frac{V_0}{R} \cos \omega t. \tag{32.2}$$

■ FIGURE 32.1
Circuit diagram symbol for an AC power supply. The symbol illustrates the oscillating form of the potential: $\mathscr{E} = V_0 \cos \omega t$.

The current varies synchronously with the potential difference: it is *in phase.* Its amplitude is

$$I_0 = V_0/R.$$

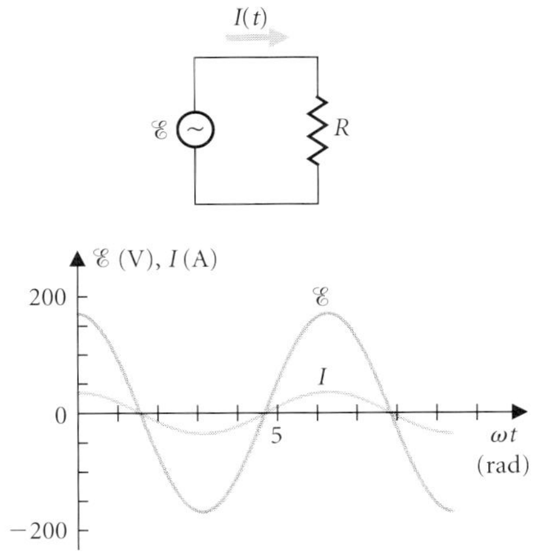

■ FIGURE 32.2
Potential difference across and current in a resistor in series with an AC power supply. The current oscillates in phase with the emf. Its amplitude is $I_0 = V_0/R$. The voltage and current are those for the resistance in Exercise 32.1.

Capacitors. Similarly, when an AC generator is connected to a capacitor, the potential difference across the capacitor, and hence the charge on it, varies in phase with the emf:

$$Q(t) = C\,\Delta V(t) = CV_0 \cos \omega t. \qquad (32.3)$$

Current in the circuit changes the stored charge (■ Figure 32.3):

$$I(t) = \frac{dQ}{dt} = CV_0(-\omega)\sin \omega t = \omega C V_0 \cos\left(\omega t + \frac{\pi}{2}\right). \qquad (32.4)$$

$\cos(\theta + \pi/2) = -\sin\theta$. SEE THE *MATH TOOLBOX* ON HARMONIC FUNCTIONS, CHAPTER 14.

Charge cannot move across a capacitor, so in DC circuits no steady current can exist in a circuit arm containing a capacitor. In AC circuits, however, a capacitor arm can carry oscillating current with constant amplitude $I_0 = \omega C V_0$. The current charges and discharges the capacitor every cycle.

DON'T CONFUSE CONSTANT AMPLITUDE WITH CONSTANT MAGNITUDE! THE MAGNITUDE OF THE CURRENT VARIES SINUSOIDALLY.

Capacitor current is not in phase with the emf. The current is zero when the charge is maximum or minimum, and the current has its maximum value when the charge is zero and is increasing most rapidly. Maximum current occurs in the circuit one-quarter period earlier than the maximum applied voltage (Figure 32.3). We say that the current *leads* the voltage. Equivalently, the voltage follows, or *lags,* the current.

Inductors. When the generator is connected across an inductor (■ Figure 32.4), the potential difference is related to the rate at which current changes (eqn. 31.10):

$$\Delta V(t) = V_0 \cos \omega t = L\frac{dI}{dt}.$$

To obtain the relation between $I(t)$ and $\Delta V(t)$, we integrate:

$$I(t) = \int \frac{\Delta V(t)}{L}\,dt = \frac{V_0}{L}\int \cos \omega t\, dt = \frac{V_0}{\omega L}\sin \omega t.$$

SINCE THERE IS NO DC CURRENT, THE INTEGRATION CONSTANT IS ZERO.

$$I(t) = \frac{V_0}{\omega L}\cos\left(\omega t - \frac{\pi}{2}\right). \qquad (32.5)$$

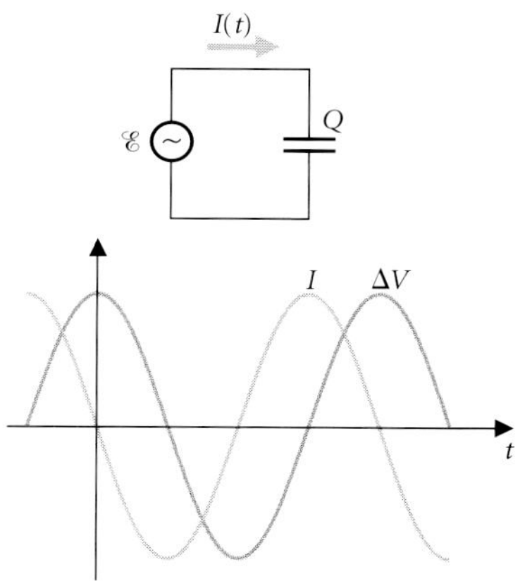

■ FIGURE 32.3
Voltage and current in a capacitor circuit. The current variable is defined as shown in the circuit diagram, so that $I = +dQ/dt$. Current leads the voltage by $\pi/2$ rad (one-quarter period).

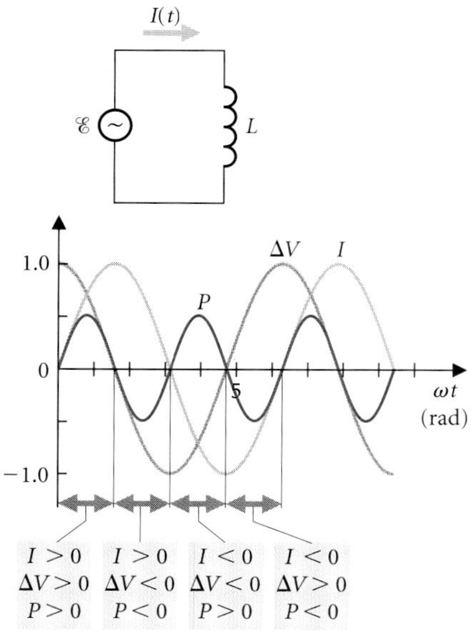

■ FIGURE 32.4
Current and voltage in an inductor circuit. The current lags the voltage by one-quarter period. The power $P(t) = \Delta V(t)I(t)$ oscillates twice as fast. Whenever voltage and current have the *same sign,* positive or negative, the generator provides energy to the circuit. When the two signs are opposite, the circuit returns energy to the generator. With a pure inductance, there is no net energy transfer.

The potential difference is zero when the current is maximum or minimum and has its maximum value when the current is zero and changing most rapidly. Again, there is a phase shift between current and emf, but now the current lags the voltage by $\pi/2$—that is, the voltage leads the current.

32.1.2 Power

Ideal capacitors and inductors store energy in electric or magnetic fields but have no way to convert it to thermal energy. On average, they draw no power from an AC generator. Let's see how the phase shifts between emf and current bring this about.

The instantaneous power input to a circuit element carrying current $I(t)$ is:

$$P(t) = \Delta V(t) I(t).$$

WE USED EQN. (26.3A).

For a resistor in an AC circuit,

$$P(t) = \Delta V(t)\frac{\Delta V(t)}{R} = \frac{V_0^2}{R}\cos^2 \omega t.$$

The average power used by the resistor over a cycle is:

$$P_{av} \equiv \langle P(t) \rangle = \frac{\langle [\Delta V(t)]^2 \rangle}{R} = \frac{V_0^2}{R}\langle \cos^2 \omega t \rangle.$$

Power relations in circuits are usually stated in terms of the root mean square (rms) voltage:

$$V_{rms} \equiv \langle [\Delta V(t)]^2 \rangle^{1/2}. \qquad (32.6)$$

Then the power used by the resistor circuit is:

$$\langle P(t) \rangle = V_{rms}^2/R. \qquad (32.7)$$

Equations (32.6) and (32.7) hold for an arbitrary variation of emf with time.

To evaluate V_{rms} for sinusoidal variations, we need the average value of $\cos^2 \omega t$:

WE FIRST ENCOUNTERED THIS AVERAGE IN §15.3. SEE ALSO PROBLEM 26.

$$\langle \cos^2 \omega t \rangle = \frac{1}{T}\int_0^T \cos^2 \omega t \, dt = \tfrac{1}{2}.$$

Then:

$$V_{rms} = V_0/\sqrt{2}, \qquad (32.8)$$

and:

$$\langle P(t) \rangle = \tfrac{1}{2} V_0^2/R.$$

It is the rms voltage, not the voltage amplitude, that is normally used to rate an AC power source. For standard U.S. household power, $V_{rms} = 120$ V. The corresponding voltage amplitude is $\sqrt{2}\,V_{rms} = 170$ V.

EXERCISE 32.1 ♦ Find the average power input to a 5-Ω room heater coil using commercial power at 60 Hz and $V_{rms} = 120$ V.

The power output of the generator in an inductor circuit is:

$$P(t) = \mathscr{E}(t) I(t) = (V_0 \cos \omega t)\left(\frac{V_0}{\omega L}\sin \omega t\right) = \frac{V_0^2}{2\omega L}\sin(2\omega t).$$

WE USED THE RESULT $\sin(2\theta) = 2\sin\theta\cos\theta$; SEE APPENDIX IB.

The generator provides energy to the circuit when it increases the potential energy of charge flowing through it. With our sign conventions, this occurs when $\mathscr{E}$ and I have the same sign, either both positive or both negative (Figure 32.4). When the signs are opposite, energy returns to the generator. Thus the power output of the generator oscillates at twice the frequency of the potential difference. Energy is drawn from the generator and is stored in the inductor, but the energy flows back into the generator a quarter cycle later. In a capacitor circuit, current leads the voltage rather than lags, but the power it draws from the generator varies in a similar way. In both cases, the generator's power output averages to zero.

RECALL OUR DISCUSSION OF BATTERY CHARGING IN §26.4, ESPECIALLY EXAMPLE 26.12.

EXERCISE 32.2 ♦♦ Show that the power output of the generator in a capacitor circuit is also proportional to $\sin(2\omega t)$.

32.1.3 *Reactance and Phase Shift*

The current in an AC circuit has a *phase shift* ϕ relative to the emf driving it:

$$\mathscr{E}(t) = V_0 \cos \omega t \quad \text{and} \quad I(t) = I_0 \cos(\omega t - \phi). \tag{32.9}$$

With this notation, the phase shift is positive when voltage leads current, as in an inductor, and negative when voltage lags current, as in a capacitor.

COMPARE EQNS. (32.2), (32.4), AND (32.5). *REMEMBER:* WITH A SINGLE INDUCTOR OR CAPACITOR, THE PEAK CURRENT OCCURS A QUARTER CYCLE EARLIER OR LATER THAN THE PEAK VOLTAGE.

Like resistors, where $V_0 = I_0 R$, capacitors and inductors draw AC current with an amplitude proportional to the applied voltage amplitude. Because of the phase shift between voltage and current in capacitors or inductors and the corresponding absence of power dissipation associated with resistance, the constant of proportionality X in the relation $V_0 = I_0 X$ is given a different name: *reactance.*

> The magnitude of the *reactance* $|X|$ of a capacitor or inductor in an AC circuit is the ratio of the applied voltage amplitude to the resulting current amplitude.
>
> $$|X| \equiv V_0/I_0. \tag{32.10}$$

For an inductor, the reactance (eqn. 32.5) is:

$$X_L = \omega L. \tag{32.11}$$

IN ADVANCED WORK, THE REACTANCE IS OFTEN GIVEN A SIGN THAT DESCRIBES THE PHASE SHIFT BETWEEN THE POTENTIAL DIFFERENCE AND THE CURRENT. WE SHALL NOT USE THIS CONVENTION HERE, BUT WE'LL USE ABSOLUTE-VALUE SIGNS ON X_C. THEN ALL OUR EXPRESSIONS ARE CORRECT, NO MATTER WHICH CONVENTION IS USED.

For a capacitor, the reactance (eqn. 32.4) is:

$$|X_C| = 1/(\omega C). \tag{32.12}$$

Since reactance has the same physical dimensions as resistance, it is also measured in ohms. Values of X and ϕ for single-element circuits are summarized in • Table 32.1.

Unlike resistance, reactance depends on the frequency of the applied AC potential. At very low frequencies, flux in an inductor changes slowly, induced electric fields are unimportant, and $X_L \to 0$. At high frequencies, rapidly changing flux produces large induced fields and $X_L \to \infty$. The current required to charge and discharge a capacitor approaches zero at low frequency and $|X_C| \to \infty$. At high frequency, a large current is required to change the charge rapidly, and $X_C \to 0$.

EXERCISE 32.3 ♦ Show that 1 s/F = 1 Ω and 1 H/s = 1 Ω.

TABLE 32.1 **Summary of Phase Relations in a Single-Element Circuit**

Generator voltage $\mathscr{E} = V_0 \cos \omega t$

Element	Current	Resistance/ Reactance	Phase Shift
Resistor	$I = \mathscr{E}/R$	R	0
Capacitor	$I = \omega C V_0 \cos(\omega t + \pi/2)$	$1/\omega C$	$-\pi/2$
Inductor	$I = (V_0/\omega L)\cos(\omega t - \pi/2)$	ωL	$+\pi/2$

EXAMPLE 32.1 ♦ What is the reactance of a 3.5-μF capacitor used in a 60.0-Hz AC circuit? What current amplitude results when the capacitor is connected to a 120-V (rms) power supply?

MODEL We may use relations (32.10) and (32.12).

SETUP The angular frequency ω corresponding to $f = 60$ Hz is $\omega = 2\pi f$.

SOLVE The reactance of the capacitor is:

$$|X_C| = \frac{1}{\omega C} = \frac{1}{2\pi(60\text{ Hz})(3.5 \times 10^{-6}\text{ F})} = 760\ \Omega.$$

The voltage amplitude is $V_0 = \sqrt{2}\,(120\text{ V}) = 170$ V. The current amplitude is:

$$I_0 = V_0/|X_C| = (170\text{ V})/(760\ \Omega) = 0.22\text{ A}.$$

ANALYZE Remember that the current charges and discharges the capacitor. The maximum rate of charge is 0.22 C/s. ■

32.2 Two-Component Circuits

Analysis of an AC circuit network follows essentially the same methods we used for DC circuits. In each case, we expect a current oscillating at the same frequency as the generator and phase shifted with respect to the emf. In this section we'll look at series connections of two circuit elements.

32.2.1 *Steady State Response*

An RC Circuit. ■ Figure 32.5 is a circuit diagram for a capacitor and resistor in series with an AC generator. To analyze the circuit, we use Kirchhoff's loop rule. With I and Q defined as shown in the figure, and proceeding clockwise, we have:

$$V_0 \cos \omega t - I(t)R - \frac{Q(t)}{C} = 0.$$

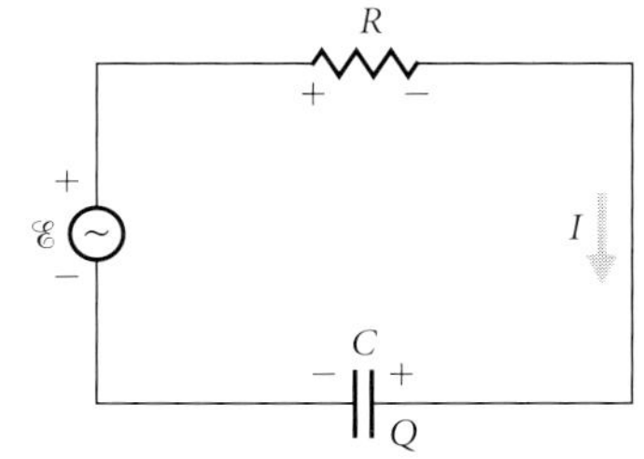

■ Figure 32.5
An *RC* circuit. The current that charges and discharges the capacitor leads the generator voltage by a phase less than 90°.

The results of §32.1 guide us in finding a solution to this equation. The current oscillates with the same frequency as the applied AC voltage but may be out of phase. The resistor alone would give a current in phase with the generator; the capacitor alone would give a current out of phase by $\pi/2$. We expect a combination of both behaviors. Thus we assume the current has the form:

$$I(t) = I_0 \cos(\omega t - \phi).$$

With current and charge defined as in the figure, current increases stored charge, $I = +dQ/dt$, and a similar expression holds for $Q(t)$:

$$Q(t) = Q_0 \sin(\omega t - \phi),$$

with

$$I_0 = \omega Q_0.$$

Substituting this assumed form into the equation relating I, Q, and $\mathcal{E}$:

$$V_0 \cos \omega t - RI_0 \cos(\omega t - \phi) - \frac{I_0}{\omega C} \sin(\omega t - \phi) = 0.$$

This equation must be satisfied at *all* times. In particular, at $t = \phi/\omega$, $\cos(\omega t - \phi) = 1$, $\sin(\omega t - \phi) = 0$, and the equation reduces to:

See §31.3.3, where we argued that the sine and cosine terms vanish separately.

$$V_0 \cos \phi - RI_0 = 0. \qquad \text{(i)}$$

Similarly, at $t = (\phi + \pi/2)/\omega$, $\cos(\omega t - \phi) = 0$, $\sin(\omega t - \phi) = 1$, and the equation becomes:

$$-V_0 \sin \phi - I_0/(\omega C) = 0. \qquad \text{(ii)}$$

Dividing eqns. (i) and (ii), we obtain:

$$\tan\phi = -1/(\omega RC) = -|X_C|/R. \tag{32.13}$$

Compare the discussion in §32.1.3, following eqn. (32.9).

The phase shift is negative and has a magnitude less than 90°. A negative phase shift means that the voltage lags the current. Similarly, squaring and adding, we find I_0:

$$V_0^2 = V_0^2\cos^2\phi + V_0^2\sin^2\phi = (RI_0)^2 + I_0^2/(\omega C)^2.$$

Thus:

$$I_0 = \frac{V_0}{\sqrt{R^2 + 1/(\omega C)^2}} = \frac{V_0}{\sqrt{R^2 + X_C^2}}.$$

The quantity

$$Z = \sqrt{R^2 + X_C^2} \tag{32.14}$$

is called the *impedance* of the circuit. It expresses the combined effect of resistance and reactance on current amplitude. In terms of impedance, the current is:

$$I(t) = \frac{V_0}{Z}\cos(\omega t - \phi), \tag{32.15}$$

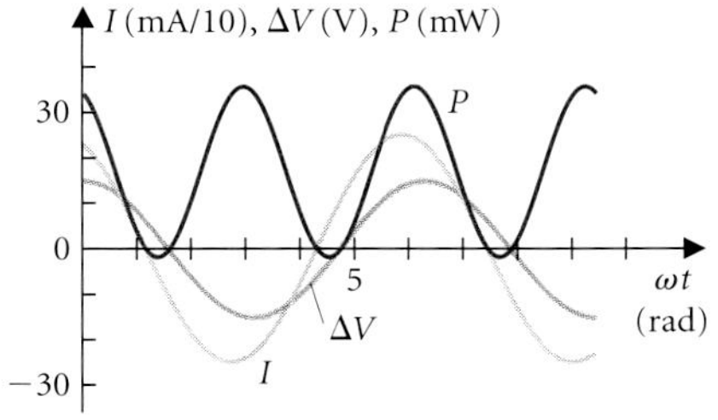

■ Figure 32.6
In this *RC* circuit, voltage lags current by 24°. The power $P(t) = \Delta V(t)I(t)$ oscillates through mostly positive values. On average, the generator transfers energy to the circuit to become thermal energy in the resistor.

and from eqn. (i), we also have: $$\cos\phi = R/Z. \tag{32.16}$$

The effect of adding a resistor to the circuit is to reduce both the current amplitude and the magnitude of the phase shift we'd get with a capacitor alone. The sign of the phase shift is not changed.

EXAMPLE 32.2 ♦ An *RC* circuit contains a 0.82-μF capacitor in series with a 5.6-kΩ resistor and an AC generator with angular frequency $\omega = 5.0 \times 10^2$ rad/s and a peak voltage of 15 V. Find the amplitude of the current in the circuit and the phase shift with respect to the applied voltage.

MODEL We apply relations (32.13) through (32.15).

Remember: Don't round intermediate answers until the end of the calculation.

SETUP The reactance of the capacitor is:

$$|X_C| = 1/\omega C = 1/[(500 \text{ rad/s})(0.82 \times 10^{-6} \text{ F})] = 2.44 \text{ k}\Omega.$$

The impedance in the circuit is:

$$Z = \sqrt{R^2 + X_C^2} = \sqrt{(5.6 \text{ k}\Omega)^2 + (2.44 \text{ k}\Omega)^2} = 6.11 \text{ k}\Omega.$$

SOLVE The current amplitude and phase shift are:

$$I_0 = \frac{V_0}{Z} = \frac{15 \text{ V}}{6.11 \text{ k}\Omega} = 2.5 \text{ mA}.$$

and $$\phi = \tan^{-1}\left(\frac{-|X_C|}{R}\right) = \tan^{-1}\left(\frac{-2.44 \text{ k}\Omega}{5.6 \text{ k}\Omega}\right) = \tan^{-1}(-0.44) = -24°.$$

ANALYZE The voltage lags the current by 24° (see ■ Figure 32.6). ▮

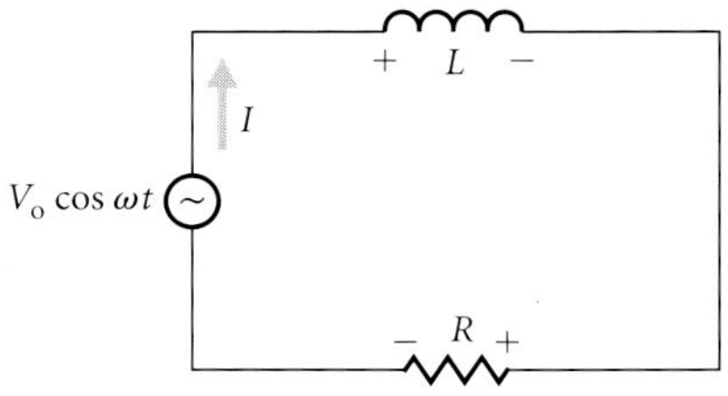

■ Figure 32.7
An *LR* circuit. The current lags the voltage by less than 90°.

An LR Circuit. Next we apply Kirchhoff's loop rule to an *LR* circuit with an AC generator to obtain similar relations for current amplitude and phase. Traversing the loop clockwise in ■ Figure 32.7, we find:

$$V_0\cos\omega t - L\frac{dI}{dt} - IR = 0.$$

Again, we expect the current to be of the form:

$$I(t) = I_0\cos(\omega t - \phi).$$

Substituting this assumed form into the equation above, we have:

$$V_0 \cos \omega t - L[-\omega I_0 \sin(\omega t - \phi)] - I_0 R \cos(\omega t - \phi) = 0.$$

Setting $t = \phi/\omega$:

$$V_0 \cos \phi - I_0 R = 0. \qquad \text{(i)}$$

And, setting $t = (\phi + \pi/2)/\omega$:

$$-V_0 \sin \phi + \omega L I_0 = 0. \qquad \text{(ii)}$$

Dividing eqns. (i) and (ii) gives:

$$\tan \phi = \omega L/R = X_L/R. \qquad (32.17)$$

Squaring and adding:

$$V_0^2 = I_0^2[R^2 + (\omega L)^2] \Rightarrow I_0 = V_0/\sqrt{R^2 + (\omega L)^2}.$$

The complete expression for the current is

$$I(t) = \frac{V_0}{Z} \cos(\omega t - \phi), \qquad (32.18)$$

where the *impedance* of the circuit is:

$$Z = \sqrt{R^2 + X_L^2}. \qquad (32.19)$$

The voltage leads the current ($\phi > 0$) by less than 90°. Equations (32.17) through (32.19) differ from the results for the *RC* circuit (eqns. 32.13–32.15) only in the sign of the reactance terms. Once again, with resistance in the circuit, the magnitude of the phase shift is reduced, but its sign is not changed.

THEY ARE IDENTICAL IF CAPACITIVE REACTANCE IS DEFINED TO BE NEGATIVE.

The impedance of an *LR* circuit is dominated by resistance at low frequencies, $\omega \ll R/L$. At high frequencies, $\omega \gg R/L$, the impedance is dominated by inductance and grows approximately linearly with frequency (■ Figure 32.8). This variation is useful for filtering electrical signals that have a mixture of frequencies. For example, stereo speaker systems usually combine low-frequency "woofers" with high-frequency "tweeters." An *LR* combination in series with the woofer selectively passes the portion of the signal the woofer is designed to reproduce well.

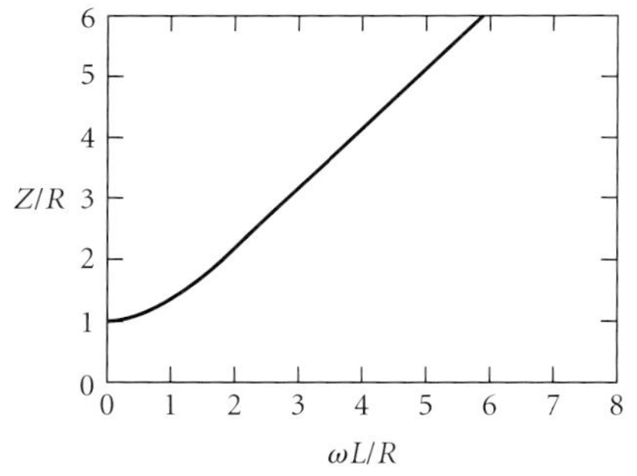

■ FIGURE 32.8
Impedance $Z = R\sqrt{1 + (\omega L/R)^2}$ of an *LR* circuit as a function of angular frequency. The impedance is dominated by the resistance when $\omega \ll R/L$, and by the inductance when $\omega \gg R/L$. Thus Z grows linearly with frequency when $\omega \gg R/L$.

EXAMPLE 32.3 ♦♦ The filter for a woofer is designed to attenuate (i.e., reduce the amplitude of) signals by a factor of 5.0 at 5.0 kHz. If the resistance in the circuit is 8.0 Ω, what inductance is required?

MODEL To attenuate by a factor of 5, the impedance at 5 kHz should be five times the value at zero frequency, where only the resistor contributes.

SETUP Thus the impedance at 5.0 kHz should be $5R = 5.0(8.0\ \Omega) = 40.0\ \Omega$. Using eqn. (32.19), and converting f to $\omega = 2\pi f$:

$$Z^2 = R^2 + X_L^2 = R^2 + (\omega L)^2$$
$$= (8.0\ \Omega)^2 + [(2\pi)(5.0 \times 10^3\ \text{Hz})L]^2.$$

SOLVE We set $Z = 40.0\ \Omega$ and solve for L:

$$L^2 = \frac{(40.0\ \Omega)^2 - (8.0\ \Omega)^2}{[2\pi(5.0 \times 10^3\ \text{Hz})]^2} = 1.6 \times 10^{-6}\ \text{H}^2,$$

and $L = 1.2$ mH.

ANALYZE This value of inductance is readily available at an electrical supply store. Because the ear is insensitive to the relative phase of sounds at different frequencies, the phase shift of the woofer signal is unimportant. ■

32.2.2 Power

The power drawn from the generator in an LR or RC circuit is:

WE USED EQN. (32.18) FOR $I(t)$.

$$P(t) = \mathscr{E}(t)I(t) = V_0 \cos \omega t \left[\frac{V_0}{Z}\cos(\omega t - \phi)\right].$$

The average power output is:

REMEMBER: THE NOTATION $\langle x \rangle$ MEANS THE AVERAGE VALUE OF x; SEE APPENDIX IIC.

$$\langle P(t) \rangle = \frac{V_0^2}{Z}\langle \cos \omega t \cos(\omega t - \phi)\rangle.$$

Expanding $\cos(\omega t - \phi)$ (see Appendix IB) gives:

WE HAVE USED THIS TRIGONOMETRIC IDENTITY BEFORE, IN CHAPTER 14, FOR EXAMPLE.

$$\begin{aligned}\langle \cos \omega t \cos(\omega t - \phi)\rangle &= \langle \cos \omega t(\cos \omega t \cos \phi + \sin \omega t \sin \phi)\rangle \\ &= \cos \phi \langle \cos^2 \omega t\rangle + \sin \phi \langle \cos \omega t \sin \omega t\rangle.\end{aligned}$$

The average value of $\cos^2 \omega t = \frac{1}{2}$, but $\cos \omega t \sin \omega t$ averages to zero, so:

$$\langle P(t) \rangle = \frac{V_0^2}{Z}[\tfrac{1}{2}\cos \phi + (0)\sin \phi].$$

$$\langle P(t) \rangle = \frac{V_0^2}{2Z}\cos \phi = \frac{V_{\text{rms}}^2}{Z}\cos \phi. \tag{32.20}$$

In an AC circuit, impedance plays the same role as resistance in a DC circuit. The *power factor*, $\cos \phi$, arises because current out of phase with the emf does not contribute to the average power.

$$\text{DC circuit: } P = \frac{V^2}{R}; \qquad \text{AC circuit: } \langle P \rangle = \frac{V_{\text{rms}}^2}{Z}\cos \phi.$$

The power factor is $\cos \phi = R/Z$ (eqn. 32.16). No net energy is used unless the circuit contains resistance.

EXAMPLE 32.4 ♦ Find the power used by an LR circuit with $L = 10.0$ mH and $R = 22\ \Omega$ when a 60.0-Hz, 120-V power supply is applied.

MODEL We apply eqn. (32.20). Don't forget to convert f to ω.

SETUP The impedance in the circuit is given by:

$$Z^2 = R^2 + (\omega L)^2 = (22\ \Omega)^2 + [2\pi(60.0\ \text{Hz})(0.010\ \text{H})]^2 = 498\ \Omega^2.$$

(Notice that we don't need to take the square root. The next step involves Z^2, not Z.)

SOLVE The rms voltage is 120 V, so:

$$\langle P \rangle = \frac{V_{\text{rms}}^2}{Z}\left(\frac{R}{Z}\right) = \frac{(120\ \text{V})^2}{498\ \Omega^2}(22\ \Omega) = 640\ \text{W}.$$

ANALYZE For comparison, the power dissipated in a DC circuit with 120 V applied to a 22-Ω resistor would be $\mathscr{E}^2/R = (120\ \text{V})^2/(22\ \Omega) = 650$ W. ∎

Reactance can be a significant nuisance in an electrically powered facility. Current out of phase with the applied voltage delivers no energy to the facility but results in costly resistive losses in the power company's distribution system. The remedy is to reduce the reactance.

An inductor and a capacitor each draw current proportional to $\sin \omega t$ (eqns. 32.4 and 32.5). Thus the current through a series combination of L and C is $I = I_0 \sin \omega t$, and the potential difference across the combination is:

SOME REFERENCES DEFINE X_C TO BE NEGATIVE: $X_C = -|X_C|$. THEN $X_{\text{series}} = X_L + X_C$ AND REACTANCES COMBINE LIKE RESISTANCES.

$$\begin{aligned}\Delta V = \Delta V_L + \Delta V_C &= I_0 \omega L \cos(\omega t) - \frac{I_0}{\omega C}\cos(\omega t) \\ &= I_0\left(\omega L - \frac{1}{\omega C}\right)\cos(\omega t) = I_0(X_L - |X_C|)\cos(\omega t) = V_0 \cos \omega t.\end{aligned}$$

Comparing this with eqn. (32.10), we conclude that the reactance of a series combination equals the inductive reactance minus the capacitive reactance. Since the motors in a factory usually have inductive reactance, a capacitor bank installed in series reduces the total reactance and the corresponding phase shift.

32.2.3 Transient Response

The exponential behavior of the *LR* and *RC* circuits that we studied in Chapter 31 is also present in AC circuits. The general solution for the current is:

$$I(t) = (V_0/Z)\cos(\omega t - \phi) + Ae^{-t/\tau}. \tag{32.21}$$

The exponential term is called the *transient* solution because its magnitude decreases with time, ultimately becoming insignificant compared with the first term (the *steady state* response). In nearly all practical applications, the steady state response is the important result. In what follows we shall neglect transient behavior (but see Problem 38).

32.3 Circuit Analysis Using Phasors

32.3.1 Phasors

In Chapter 14 we discovered that uniform circular motion and simple harmonic motion are closely related. We may describe an oscillating object's position with one component of a position vector that rotates uniformly around a circle. We can also use this analogy in AC circuits; all currents and potentials vary sinusoidally with the same frequency ω, and each may be represented by a rotating vector, or *phasor.*

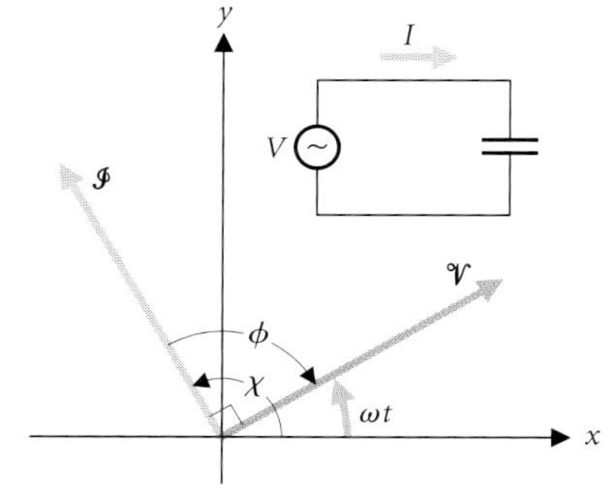

■ Figure 32.9
Voltage and current phasors for a capacitor. Both phasors rotate counterclockwise at angular frequency ω with a fixed angle between them. The current phasor precedes the voltage phasor as they rotate: the current *leads* the voltage by 90°. The measured value of current at any time t equals the x-component of the current phasor. The angle ϕ is drawn from the current phasor to the voltage phasor, with the counterclockwise direction taken as positive. In this diagram, ϕ is negative. (Compare with eqn. 32.4 and Table 32.1.)

We used a similar device when studying interference; see §17.4.2.

The phasor diagram shown in ■ Figure 32.9 represents the relation between current and voltage for a capacitor. The voltage phasor $\mathcal{V}$ has magnitude V_0 and makes an angle ωt with the x-axis. The actual potential difference across the capacitor is $V(t) = V_0 \cos \omega t = |\mathcal{V}|\cos \omega t$, the x-component of the phasor. The y-component of the phasor has no physical meaning. Similarly, the current phasor $\mathcal{I}$ makes an angle $\chi = \omega t + \pi/2$ with the x-axis, and the actual current is the x-component: $I(t) = I_0 \cos(\omega t + \pi/2) = |\mathcal{I}|\cos \chi$. The magnitudes of the current and voltage phasors are related by the capacitive reactance:

$$|\mathcal{V}| = |X_C||\mathcal{I}| = |\mathcal{I}|/(\omega C). \tag{32.22}$$

As time proceeds, both $\mathcal{I}$ and $\mathcal{V}$ rotate counterclockwise with angular frequency ω.

EXERCISE 32.4 ♦♦ Draw a diagram similar to Figure 32.9 showing phasors representing the voltage and current for an inductor. (*Hint:* The current lags the voltage, eqn. 32.5.)

32.3.2 Phasor Representation of a Series Circuit

We may illustrate the phasor method using a series *RC* circuit. In a series circuit, the current is the same in all elements, so the phasor diagram (■ Figure 32.10) has a single current phasor. Since the diagram rotates with time, we may draw the current phasor at any orientation. It is convenient to put it along the x-axis.

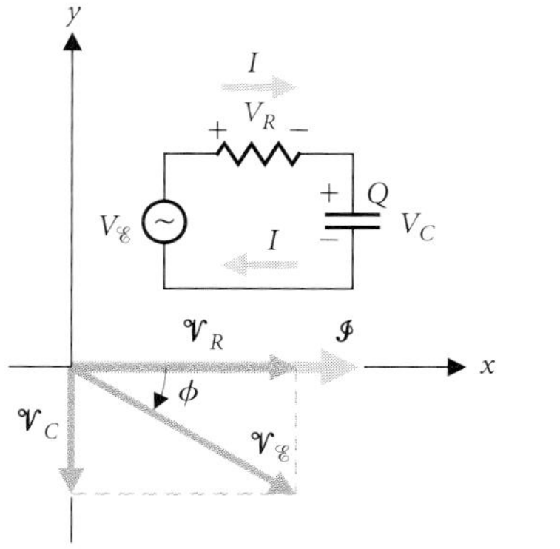

■ Figure 32.10
Phasor diagram for a series *RC* circuit. We may draw the phasor diagram at any orientation. It is convenient to put the single current phasor along the x-axis. The voltage across the resistor is in phase with the current, so it also lies along the x-axis. The voltage across the capacitor lags the current by 90° (see Figure 32.9), so it lies along the negative y-axis. By Kirchhoff's loop rule, the generator voltage is the phasor sum of the resistor and capacitor voltages. It lags the current by angle $|\phi|$, or, equivalently, the current leads the voltage by $|\phi|$. (*Remember:* ϕ is drawn from the current phasor to the voltage phasor. In this diagram, ϕ is negative.)

There are three voltage phasors, corresponding to the potential differences across the resistor, the capacitor, and the generator. The voltage across the resistor has the same phase as the current and is represented by a phasor parallel to the current phasor. The capacitor voltage lags the current by 90° and is represented by a phasor at right angles to the current phasor. The capacitor voltage phasor lies behind (lags) the current phasor as they both rotate counterclockwise in the diagram.

According to Kirchhoff's loop rule, the sum of the potential differences around the circuit is zero:

$$V_0 \cos \omega t - IR - Q/C = 0 \Rightarrow V_0 \cos \omega t = IR + Q/C.$$

REMEMBER: IF $\vec{a} + \vec{b} = \vec{c}$, THEN $a_x + b_x = c_x$.

Since each potential difference is represented by the x-component of the corresponding phasor, the *vector* sum of the resistor and capacitor voltage phasors equals the generator voltage phasor.

> The total potential difference across a series combination is the phasor sum of the individual voltage phasors.

For the resistor and capacitor, the magnitude of the phasor sum is:

$$|\mathcal{V}_{\mathcal{E}}| = \sqrt{|\mathcal{V}_R|^2 + |\mathcal{V}_C|^2} = \sqrt{R^2|\mathcal{I}|^2 + X_C^2|\mathcal{I}|^2} = |\mathcal{I}|\sqrt{R^2 + X_C^2} = |\mathcal{I}|Z.$$

THIS IS THE SAME CONVENTION DESCRIBED ALGEBRAICALLY IN §32.1.3.

By convention, the phase shift is represented by the angle *from* the current phasor *to* the voltage phasor, with the counterclockwise direction taken as positive. When the voltage lags the current, as in this case, the phase shift is negative. Its magnitude is given by:

$$\tan|\phi| = \frac{|\mathcal{V}_C|}{|\mathcal{V}_R|} = \frac{|X_C|}{R}.$$

The current leads the voltage. We have retrieved eqns. (32.13) through (32.15).

EXAMPLE 32.5 ♦♦ A circuit driven harmonically at 60.0 Hz contains a 10.0-μF capacitor and a 0.500-kΩ resistor. What is the current in the circuit if $V_0 = 156$ V?

MODEL The phasor diagram (Figure 32.10) shows the relations we need. The voltage lags the current.

SETUP The angular frequency $\omega = 2\pi f$. The capacitive reactance is:

$$|X_C| = \frac{1}{\omega C} = \frac{1}{2\pi(60.0\text{ Hz})(10.0\ \mu\text{F})} = 265\ \Omega.$$

The total impedance is:

$$Z = \sqrt{R^2 + X_C^2} = \sqrt{(500\ \Omega)^2 + (265\ \Omega)^2} = 566\ \Omega.$$

SOLVE The current amplitude equals the magnitude of the current phasor:

$$I_0 = |\mathcal{I}| = \frac{|\mathcal{V}|}{Z} = \frac{156\text{ V}}{566\ \Omega} = 0.276\text{ A}.$$

The phase angle ϕ is negative. Its magnitude is given by:

$$\tan|\phi| = \frac{|X_C|}{R} = \frac{265\ \Omega}{500\ \Omega} = 0.530 \Rightarrow |\phi| = 27.9°.$$

ANALYZE We might write our answer as $I = (0.276\text{ A})\cos(120\pi t + 0.155\pi)$, where we have converted ϕ to radians. ■ Figure 32.11 is a graph of this function. The voltage lags the current by $0.155\pi/2\pi = 0.08$ of a cycle. The voltage phasors for the resistor and capacitor have magnitude $|\mathcal{V}_R| = |\mathcal{I}|R = (0.276\text{ A})(500\ \Omega) = 138$ V and $|\mathcal{V}_C| = |\mathcal{I}||X_C| = (0.276\text{ A})(265\ \Omega) = 73$ V. These values were used in Figure 32.10. ■

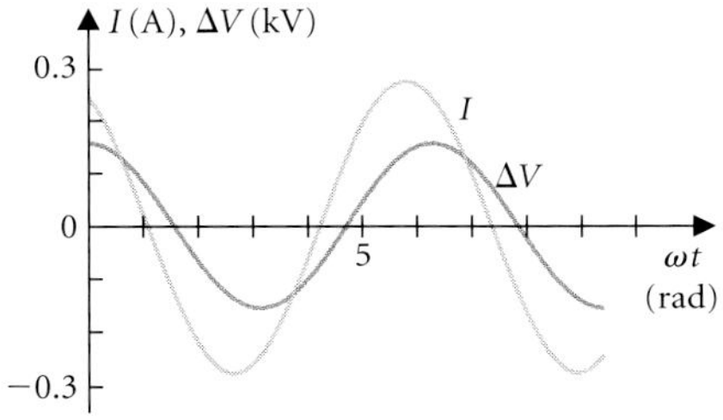

■ FIGURE 32.11
Voltage and current as a function of time. The current leads the voltage by $28° = 0.155\pi$ rad $= 0.08$ cycle.

EXERCISE 32.5 ♦♦ Draw a phasor diagram for a series LR circuit and use it to obtain eqns. (32.17) through (32.19) for the current amplitude and phase.

The expression for average power used by an AC circuit (eqn. 32.20),

$$\langle P(t)\rangle = \tfrac{1}{2}V_0 I_0 \cos\phi,$$

has an elegant form as a dot product of phasors:

$$\langle P(t)\rangle = \tfrac{1}{2}\mathcal{V}\cdot\mathcal{I}. \qquad (32.23)$$

EXERCISE 32.6 ♦ What is the time-averaged power used in the *RC* circuit of Example 32.5?

32.3.3 *Phasor Representation of a Parallel Circuit*

When circuit elements are connected in parallel, the potential difference across each element is the same. For example, in a parallel combination of a resistor and an inductor, there is one voltage phasor, $\mathcal{V}_R = \mathcal{V}_L = \mathcal{V}_\mathcal{E}$. In ■ Figure 32.12, we have drawn the phasor diagram, with the voltage phasor along the x-axis for convenience. The resistor current is in phase with the voltage, while the inductor current lags the inductor voltage by 90°.

> According to Kirchhoff's junction rule, the total current through a parallel combination is the phasor sum of the individual current phasors.

The total current is the vector sum of the individual current phasors $\mathcal{I}_R$ and $\mathcal{I}_L$. The magnitude of $\mathcal{I}$ is:

$$|\mathcal{I}| = \sqrt{|\mathcal{I}_R|^2 + |\mathcal{I}_L|^2} = |\mathcal{V}_\mathcal{E}|\sqrt{(1/R)^2 + (1/X_L)^2}.$$

The voltage leads the current by a phase shift ϕ, where: $\tan\phi = \dfrac{|\mathcal{I}_L|}{|\mathcal{I}_R|} = \dfrac{|\mathcal{V}_\mathcal{E}|/X_L}{|\mathcal{V}_\mathcal{E}|/R} = \dfrac{R}{X_L}.$

As with any circuit, Kirchhoff's rules form the basis of our solution method (■ Figure 32.13). When analyzing an AC circuit, the voltages and currents are represented by phasors and we add them like vectors. In the following example we show how to use the method to analyze a circuit that has both a series and a parallel combination of elements.

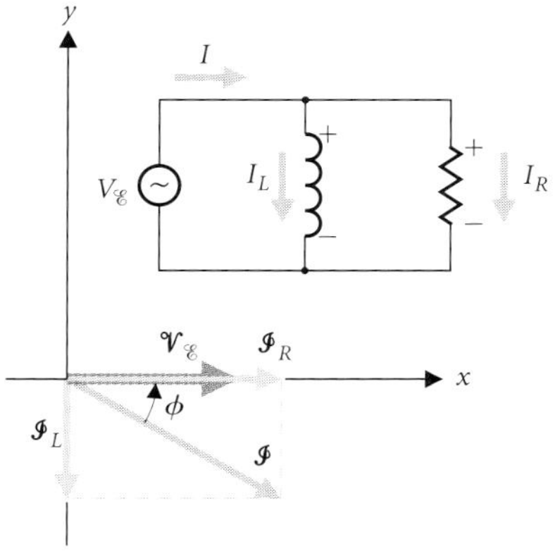

■ FIGURE 32.12
Phasor diagram for a parallel combination of inductor and resistor. The generator and the two circuit elements in parallel have a common voltage phasor, here drawn along the x-axis. According to Kirchhoff's junction rule, the current phasor $\mathcal{I}$ equals the sum of the phasors $\mathcal{I}_L$ and $\mathcal{I}_R$. The resistor current is in phase with the voltage, so we draw it along the x-axis. The inductor current lags the voltage by 90° (cf. Exercise 32.4). The total current lags the voltage by angle ϕ.

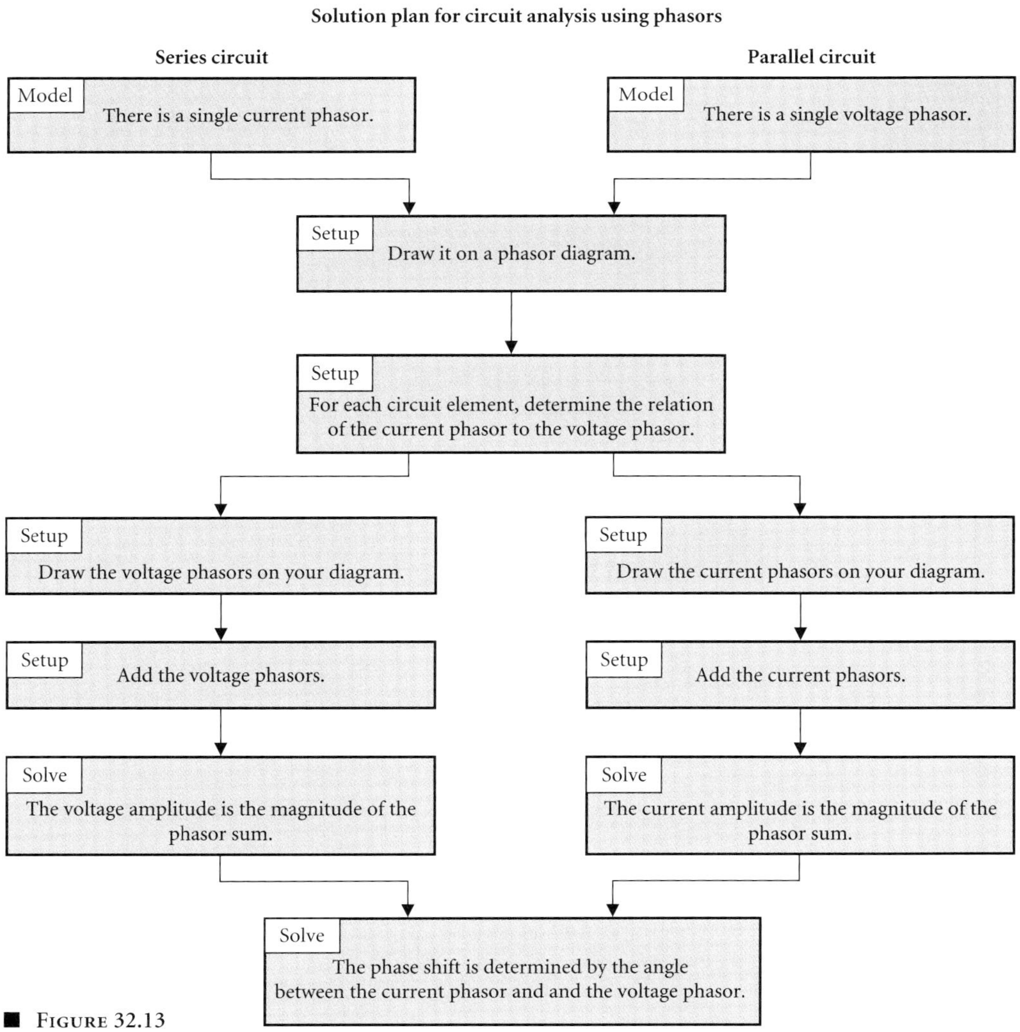

■ FIGURE 32.13

BUT DON'T TRY TO ADD A VOLTAGE TO A CURRENT!

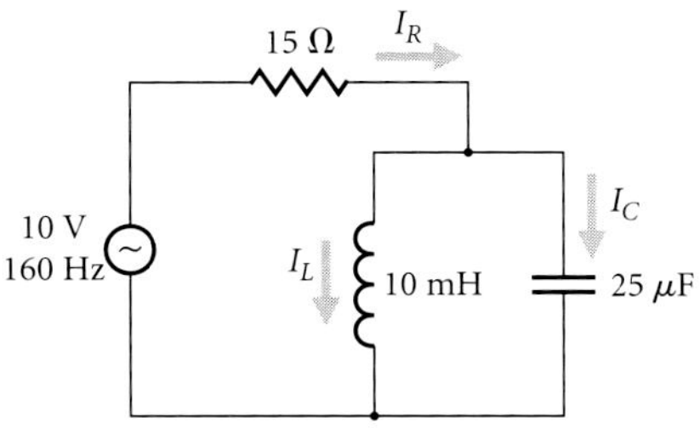

■ FIGURE 32.14
The resistor current divides at the junction with the parallel combination of inductor and capacitor: $I_R = I_L + I_C$.

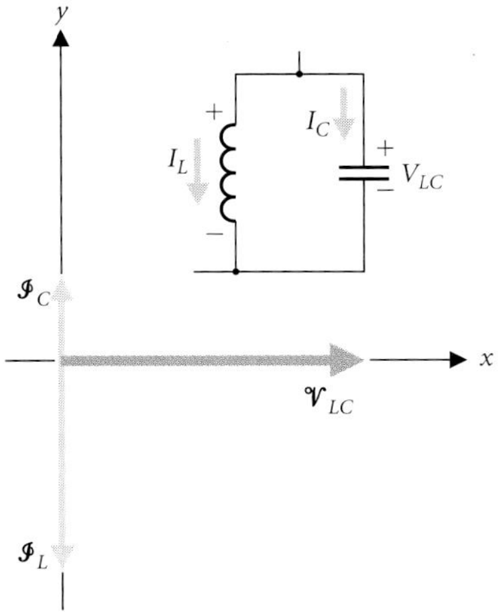

■ FIGURE 32.15
Phasor diagram for the parallel combination of inductor and capacitor. We have chosen to draw the voltage phasor along the x-axis. The capacitor current leads the voltage, while the inductor current lags. These phasors lie along the same line but in opposite directions. The phasor sum has magnitude $|\mathcal{I}| = ||\mathcal{I}_L| - |\mathcal{I}_C||$.

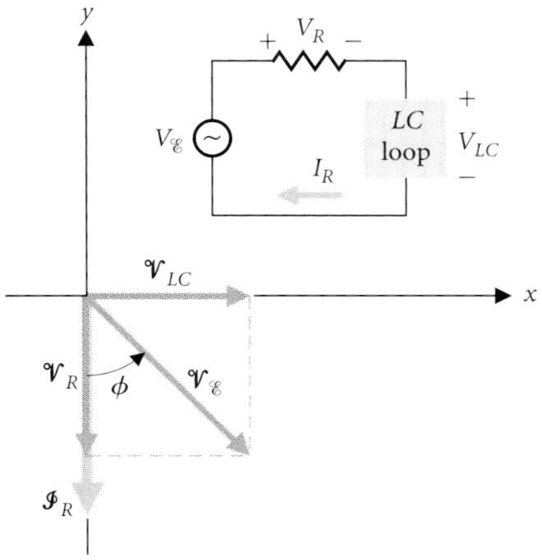

■ FIGURE 32.16
Phasor diagram for the resistor in series with the *LC* combination. The *LC* voltage lies along the x-axis, and the current lags this voltage by 90°, as in Figure 32.15. The resistor voltage is in phase with the current, so it also lies along the negative y-axis. The generator voltage is the phasor sum of $\mathcal{V}_{LC}$ and $\mathcal{V}_R$. The current lags the voltage by $\phi = 42°$.

EXAMPLE 32.6 ♦♦ A 10.0-mH inductor and a 25-μF capacitor are connected in parallel. The combination is in series with a 15-Ω resistor and a 10.0-V, 160-Hz emf (■ Figure 32.14). What is the current through the resistor?

MODEL The generator voltage is the sum of the potential difference across the resistor and the potential difference across the parallel combination of inductor and capacitor. The current through the resistor divides between the inductor and the capacitor. The inductor current lags the voltage and the capacitor current leads. We begin our analysis with the *LC* combination (■ Figure 32.15). Once we have found the voltage phasor for the parallel combination, we may add it to the voltage phasor for the resistor in series (■ Figure 32.16).

SETUP The same potential difference V_{LC} exists across the inductor and the capacitor. The current through each element is related to the potential difference by the reactance:

$$|\mathcal{I}_L| = |\mathcal{V}_{LC}|/X_L \qquad \text{and} \qquad |\mathcal{I}_C| = |\mathcal{V}_{LC}|/|X_C|.$$

The inductor current lags the voltage by $\pi/2$, while the capacitor current leads by $\pi/2$ (eqns. 32.5 and 32.4). The two current phasors are opposite each other (Figure 32.15). The current through the resistor is the phasor sum:

$$\mathcal{I}_R = \mathcal{I}_L + \mathcal{I}_C,$$

with magnitude
$$|\mathcal{I}_R| = |\mathcal{V}_{LC}|\left(\frac{1}{X_L} - \frac{1}{|X_C|}\right) = \frac{|\mathcal{V}_{LC}|}{X_{LC}}, \qquad \text{(i)}$$

THE IMPEDANCE OF A PARALLEL COMBINATION SHOULD REMIND YOU OF THE RULES FOR COMBINING RESISTORS IN PARALLEL.

where
$$\frac{1}{X_{LC}} = \frac{1}{X_L} - \frac{1}{|X_C|} = \frac{1}{\omega L} - \omega C$$
$$= \frac{1}{2\pi(160\text{ Hz})(0.0100\text{ H})} - 2\pi(160\text{ Hz})(25 \times 10^{-6}\text{ F})$$
$$= 0.0743\ /\Omega.$$

So:
$$X_{LC} = 13.5\ \Omega.$$

Figure 32.16 shows the diagram for the series combination of resistor and LC loop. The generator phasor is the vector sum of $\mathcal{V}_R$ and $\mathcal{V}_{LC}$. Each of these phasors is related to the current phasor $\mathcal{I}_R$: $|\mathcal{V}_R| = R|\mathcal{I}_R|$ and, by eqn. (i), $|\mathcal{V}_{LC}| = X_{LC}|\mathcal{I}_R|$. The resistor voltage $\mathcal{V}_R$ is in phase with $\mathcal{I}_R$, but $\mathcal{V}_{LC}$ leads by $\pi/2$ (Figure 32.16). Thus:

$$|\mathcal{V}_\mathcal{E}| = |\mathcal{I}_R|\sqrt{R^2 + X_{LC}^2}. \qquad \text{(ii)}$$

SOLVE Relation (ii) is what we need to solve for $\mathcal{I}_R$:

$$|\mathcal{I}_R| = \frac{|\mathcal{V}_\mathcal{E}|}{\sqrt{R^2 + X_{LC}^2}} = \frac{10.0\ \text{V}}{\sqrt{(15\ \Omega)^2 + (13.5\ \Omega)^2}} = 0.50\ \text{A},$$

and the phase angle is given by

$$\tan\phi = \frac{|\mathcal{V}_{LC}|}{|\mathcal{V}_R|} = \frac{X_{LC}}{R} = \frac{13.5\ \Omega}{15\ \Omega} = 0.90 \Rightarrow \phi = 42°.$$

The current lags the voltage by 42°.

ANALYZE Our final result may be written: $I = (0.50\ \text{A})\cos[(2\pi \times 160)t - 0.23\pi]$, see ■ Figure 32.17.

If capacitive reactance is taken to be negative, then eqn. (i) shows that reactances in parallel combine like resistances in parallel. ■

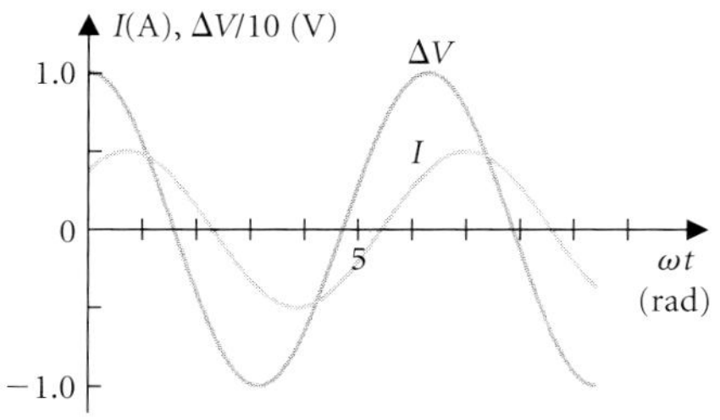

■ FIGURE 32.17
Current and generator voltage as functions of time. The current lags the voltage by 42°.

32.4 THE *LRC* CIRCUIT

In Chapter 31 we found that the transient behavior of a series LRC circuit is a damped oscillation. When the resistance is small, $R \ll 2\sqrt{L/C}$, the circuit oscillates with very nearly the natural frequency $\omega_0 = 1/\sqrt{LC}$. When an AC generator is connected to the circuit, the current undergoes a steady state, forced oscillation at the applied frequency (cf. §14.4). When the applied frequency equals the natural frequency ω_0, the current is in phase with the applied voltage, the circuit efficiently absorbs energy from the generator, and the current amplitude is at a maximum. This phenomenon is known as *resonance*. An LRC circuit can be used to detect weak signals near its resonant frequency or to act as a filter that emphasizes a narrow range of frequencies near ω_0.

WE ARE ASSUMING THAT THE CIRCUIT DAMPING IS LESS THAN CRITICAL; SEE EXAMPLE 31.12.

As in any series circuit, the LRC circuit has a single current phasor $\mathcal{I}$, which we place along the x-axis in ■ Figure 32.18. As time passes, this phasor rotates counterclockwise about the z-axis. The voltage across the resistor is in phase with the current, while the voltages across the inductor and capacitor lead and lag by 90°, respectively. The generator voltage $\mathcal{E} = V_0 \cos\omega t$ equals the sum of the potential differences across the three elements. The phasor sum is:

$$\mathcal{V}_\mathcal{E} = \mathcal{V}_L + \mathcal{V}_C + \mathcal{V}_R.$$

The inductor and capacitor voltage phasors lie along opposite directions, so the magnitude of their sum is the difference of their magnitudes:

$$\left|\mathcal{V}_L + \mathcal{V}_C\right| = \left||\mathcal{V}_L| - |\mathcal{V}_C|\right| = \left|X_L|\mathcal{I}| - |X_C||\mathcal{I}|\right| = \left|X_L - |X_C|\right||\mathcal{I}|.$$

The resistor voltage phasor is at right angles, so we find the magnitude of the sum using the Pythagorean theorem:

$$|\mathcal{V}| = |\mathcal{I}|\sqrt{R^2 + (X_L - |X_C|)^2} = |\mathcal{I}|Z.$$

Thus the impedance of a series LRC circuit is:

$$Z = \sqrt{R^2 + [\omega L - 1/(\omega C)]^2}. \qquad (32.24)$$

The voltage leads the current. The phase shift ϕ is given by:

$$\tan\phi = \frac{X_L - |X_C|}{R} = \frac{\omega L - 1/(\omega C)}{R} = \frac{\omega^2 LC - 1}{\omega RC}. \qquad (32.25)$$

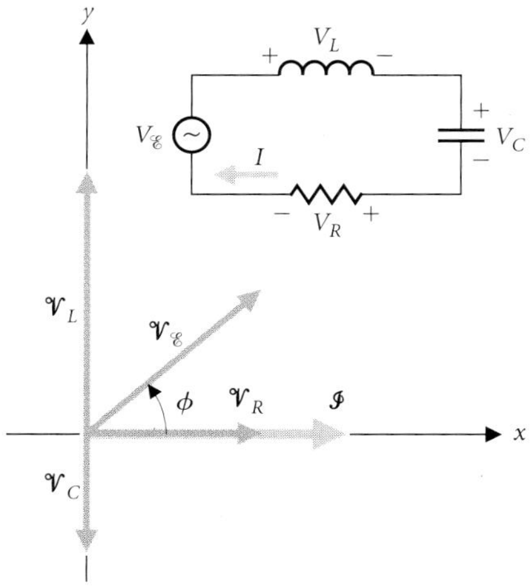

■ FIGURE 32.18
Phasor diagram for a series *LRC* circuit. We have chosen to put the single current phasor along the x-axis. The resistor voltage is in phase with the current (along the x-axis), the inductor voltage leads the current, and the capacitor voltage lags. The generator voltage is the phasor sum of the potential differences across the resistor, capacitor, and inductor. In this example it leads the current.

So the current is: $$I(t) = \frac{V_0}{\sqrt{R^2 + [\omega L - 1/(\omega C)]^2}} \cos(\omega t - \phi). \quad (32.26)$$

We also found this result for X_{LC} in §32.2.2. Compare with the result for a parallel LC combination in Example 32.6.

Results (32.24) through (32.26) are consistent with the relations we developed for the *RC* circuit (eqns. 32.13–32.15) and the *LR* circuit (eqns. 32.17–32.19), and the result that the reactance of a series inductor/capacitor combination is $X_{LC} = \omega L - 1/(\omega C)$.

The current amplitude is a maximum when the impedance Z is a minimum—that is, when the combination of capacitive and inductive reactances is zero. This happens at angular frequency ω_0, where:

$$X_{LC} = \omega_0 L - 1/(\omega_0 C) = 0,$$

or

$$\omega_0^2 = 1/(LC).$$

For $R = 0$, ω_0 is the natural oscillation frequency of the circuit. Compare eqns. (31.25) and (31.28).

If there were no resistance, the current amplitude would become infinite at this *resonant frequency*. Resistance damps the oscillations, keeping the amplitude, $I_0 = V_0/Z$, finite. At the resonant frequency, the phase shift vanishes. The amplitudes of the voltage across the inductor and the capacitor may be each much larger than the generator voltage, but they are equal and opposite.

Look at eqn. (32.25). The numerator is zero at the resonant frequency.

We can investigate the behavior of the circuit at other frequencies by looking at the capacitor voltage amplitude as a function of frequency:

$$V_C(\omega) = I_0/(\omega C) = V_0/(\omega C Z).$$

Remember that Z is itself a function of ω.

At resonance, the impedance Z reduces to the resistance R, and $\omega = \omega_0$:

$$V_C(\omega_0) = \frac{V_0}{\omega_0 CR} = \frac{V_0\sqrt{LC}}{CR} = \frac{V_0}{R}\sqrt{\frac{L}{C}}.$$

The inductor voltage may also be used. It exhibits the same kind of resonant behavior.

The voltage across the capacitor is increased (*amplified*) by the factor:

$$\frac{V_C(\omega_0)}{V_0} \equiv Q \equiv \frac{1}{R}\sqrt{\frac{L}{C}}. \quad (32.27)$$

The *quality factor* Q of the series *LRC* circuit describes its effectiveness as an amplifier of weak signals. It is much larger than unity in most practical circuits. The quality factor also describes how well an *LRC* circuit can discriminate between different applied frequencies. The capacitor voltage amplitude in terms of Q is:

$$V_C(\omega) = \frac{V_0 Q}{\sqrt{(\omega/\omega_0)^2 + Q^2[(\omega/\omega_0)^2 - 1]^2}}. \quad (32.28)$$

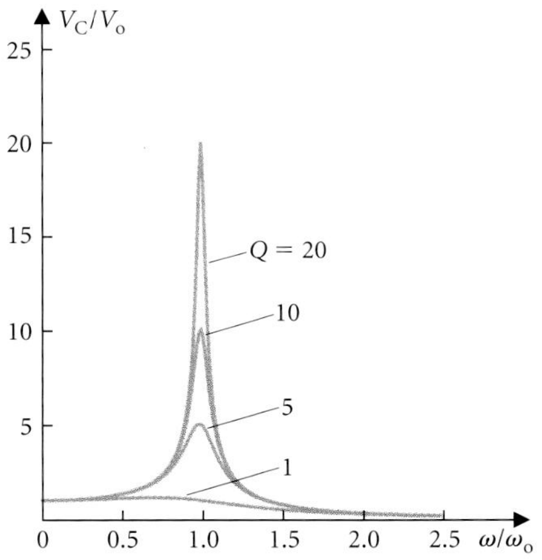

■ **Figure 32.19**
Capacitor voltage as a function of frequency for a series *LRC* circuit. Each curve is labeled with the value of its quality factor Q. As $Q = (1/R)\sqrt{L/C}$ increases, the amplitude peak at resonance becomes taller and narrower. This is a useful feature in circuits designed for frequency selectivity (see Example 32.8).

■ Figure 32.19 shows this function for four different values of Q. As Q increases, the curve reaches a higher maximum value and peaks more sharply. These are desirable characteristics in a circuit designed to respond to a narrow range of frequencies around one selected value.

EXERCISE 32.7 ♦♦ Show that the quality factor may be written $Q = \frac{1}{2}\omega_0\tau$, where τ is the damping time constant of the free *LRC* circuit (eqn. 31.27).

In a radio tuning circuit, a variable capacitor or inductor is used to adjust the resonant frequency. The AC driving voltage is provided by the electric field of radio waves arriving at the antenna.

EXAMPLE 32.7 ♦♦ An AM radio antenna has a resistance of 51.0 Ω. A radio enthusiast wants to build a series *LRC* circuit to tune over a range of frequencies from 0.50 to 2.0 MHz with a Q of 50.0 at 1.00 MHz. If the inductor resistance is 1.0 Ω, what value of L is required? Over what range must the capacitor be adjustable?

MODEL At resonance, the capacitive and inductive reactances are equal.

$$X_L = |X_C| = \sqrt{L/C} \equiv X.$$

From eqn. (32.27), the quality factor is $Q = X/R$ so, with a known resistance in the circuit, we can find the value of X that produces a desired Q value. Once we have X at the given resonant frequency (1.00 MHz), we can find the required values of L and C.

DO NOT CONFUSE THIS Q WITH CHARGE.

Tuning a desired frequency means adjusting the circuit's resonant frequency to that value. The adjustment is made in this circuit by varying the capacitance, with L and R fixed.

SETUP The total resistance in the circuit is the series combination of the antenna and the inductor: $R = 51.0\ \Omega + 1.0\ \Omega = 52.0\ \Omega$. Then the reactance X needed for a Q value of 50.0 is:

$$X = QR = 50.0(52.0\ \Omega) = 2.60\ \text{k}\Omega.$$

With known values for $X = \sqrt{L/C}$ and the resonant frequency $\omega_0 = 2\pi f_0 = 1/\sqrt{LC}$, we can find both L and C.

SOLVE With $f_0 = 1.00$ MHz, the required inductance is:

$$L = \sqrt{\frac{L}{C}}\sqrt{LC} = \frac{X}{\omega_0} = \frac{2.60 \times 10^3\ \Omega}{2\pi(1.00 \times 10^6\ \text{Hz})} = 0.414\ \text{mH}.$$

Since $\omega_0^2 = 1/(LC)$, the capacitance needed to tune a given frequency is:

$$C = \frac{1}{\omega_0^2 L} = \frac{1}{[2\pi(5.0 \times 10^5\ \text{Hz})]^2(0.414\ \text{mH})} = 0.245\ \text{nF at 500 kHz},$$

and $$C = \frac{1}{[2\pi(2.0 \times 10^6\ \text{Hz})]^2(0.414\ \text{mH})} = 15.3\ \text{pF at 2 MHz}.$$

The required value of C ranges from 15 pF for 2 MHz to 245 pF for 500 kHz.

ANALYZE The required capacitance ranges over a factor of 16 to produce a factor of 4 in the frequency range. An alternative design would have a variable inductance. We could have found the capacitance required for a Q of 50 at 1 MHz, and the corresponding range of L. ∎

SEE PROBLEM 27.67, WHICH DESCRIBES A TUNABLE CAPACITOR. PROBLEM 31.83 DESCRIBES A VARIABLE INDUCTOR.

EXAMPLE 32.8 ♦♦ Two radio stations broadcast at 0.810 MHz and 0.850 MHz. When the antenna circuit in Example 32.7 is tuned to 0.810 MHz, it also responds to signals oscillating at 0.850 MHz. If the amplitude of the input signals reaching the antenna is the same at the two frequencies, how large is the unwanted output signal at 0.850 MHz?

MODEL A modern radio uses a second circuit to isolate the desired signal further. The input to the second circuit is the antenna circuit capacitor voltage. In this problem, we want to see how well the antenna circuit itself discriminates, so we compare the peak capacitor voltages at the two frequencies.

SETUP Using the result of Example 32.7, the capacitance required to tune the radio to 0.810 MHz is:

$$C = \frac{1}{\omega_0^2 L} = \frac{1}{[2\pi(8.10 \times 10^5\ \text{Hz})]^2(0.414\ \text{mH})} = 93.25\ \text{pF}.$$

The value of Q at this frequency is:

$$Q = \frac{1}{R}\sqrt{\frac{L}{C}} = \left(\frac{1}{52.0\ \Omega}\right)\sqrt{\frac{4.14 \times 10^{-4}\ \text{H}}{93.25 \times 10^{-12}\ \text{F}}} = 40.52.$$

The capacitor voltage amplitude at the resonant frequency (810 kHz) is related to the input voltage amplitude through eqn. (32.27): $V_C(\omega_0) = V_0 Q$. We obtain the capacitor voltage amplitude at a nearby frequency from eqn. (32.28).

$$\frac{V_C(\omega)}{V_0 Q} = \frac{1}{\sqrt{(\omega/\omega_0)^2 + Q^2[(\omega/\omega_0)^2 - 1]^2}}.$$

For the 850-kHz signal,

$$\frac{\omega}{\omega_0} = \frac{2\pi f}{2\pi f_0} = \frac{f}{f_0} = \frac{0.850 \text{ MHz}}{0.810 \text{ MHz}} = 1.049.$$

SOLVE The ratio of the capacitor voltage at 850 kHz to that at 810 kHz is:

$$\frac{V_C}{V_0 Q} = \frac{1}{\sqrt{1.049^2 + 40.52^2(1.049^2 - 1)^2}} = 0.24.$$

Note: The subtraction reduces the number of significant figures in the result from 3 to 2.

ANALYZE The circuit response to the unwanted signal at 0.850 MHz is 24% of its response to the 0.810-MHz signal. Better selectivity requires a larger Q value. In practice, the second circuit improves the discrimination by as much as a factor of 10. ∎

Chapter Summary

Where Are We Now?

We have extended our analysis of electric circuits to include sinusoidally varying emfs. Such emfs produce alternating current (AC). These circuits are the basis of most household appliances as well as many industrial devices.

What Did We Do?

Circuits, AC as well as DC, are analyzed using Kirchhoff's laws. In an AC circuit the current oscillates at the same frequency as the emf but may be out of phase with the voltage. Thus the power used by the circuit is generally less than the average emf times the average current. The circuit behavior is determined by its resistance R, reactance X, and impedance, $Z = \sqrt{R^2 + X^2}$. The *reactance* is contributed by inductors and capacitors in the circuit. If ω is the angular frequency of the applied emf, the inductor reactance is $X_L = \omega L$ and the capacitor reactance is $|X_C| = 1/(\omega C)$. The reactance of a series combination of inductors and capacitors is the total inductive reactance minus the total capacitive reactance:

$$X_s = X_L - |X_C|.$$

Then: $I = \dfrac{V_0}{Z}\cos(\omega t - \phi)$, $\tan\phi = \dfrac{X_s}{R}$, and $\langle P\rangle = V_{\text{rms}}^2 \dfrac{R}{Z^2}$.

The current through, or voltage across, each component may be described using *phasors*—rotating vectors that describe the magnitudes and phase relations of these quantities. In a series circuit, the voltage phasors add vectorially to give the applied potential; in a parallel circuit, the current phasors add vectorially to give the total current.

A series *LRC* circuit resonates—has peak current amplitude—at frequency $\omega_0 = 1/\sqrt{LC}$. At this frequency, the total reactance is zero and $Z = R$. The breadth of the resonance is determined by the quality factor:

$$Q = \tfrac{1}{2}\omega_0 \tau = \frac{1}{R}\sqrt{\frac{L}{C}}.$$

Practical Applications

The methods we have developed allow us to analyze a large number of practical circuits that do not include solid state devices. Familiar examples we have discussed include radio tuner circuits, loudspeaker crossover circuits, and circuits coupled by transformers.

Solutions to Exercises

32.1 The average power input is:

$$\langle P\rangle = \frac{V_{\rm rms}^2}{R} = \frac{(120\ \text{V})^2}{5.0\ \Omega} = 2.9\ \text{kW}.$$

32.2 Using eqn. (32.4) for $I(t)$, the power output is:

$$P(t) = \mathcal{E}(t)I(t) = V_0\cos\omega t\,(-\omega CV_0)\sin\omega t = -\frac{\omega CV_0^2}{2}\sin(2\omega t).$$

32.3

$$\frac{1\ \text{s}}{1\ \text{F}} = \frac{1\ \text{s}}{1\ \text{C/V}} = \frac{1\ \text{V}}{1\ \text{C/s}} = \frac{1\ \text{V}}{1\ \text{A}} = 1\ \Omega.$$

Similarly, $$\frac{1\ \text{H}}{1\ \text{s}} = \frac{1\ \text{V}}{(1\ \text{A/s})(1\ \text{s})} = \frac{1\ \text{V}}{1\ \text{A}} = 1\ \Omega.$$

32.4 See ■ Figure 32.20. The current lags the voltage by 90°.

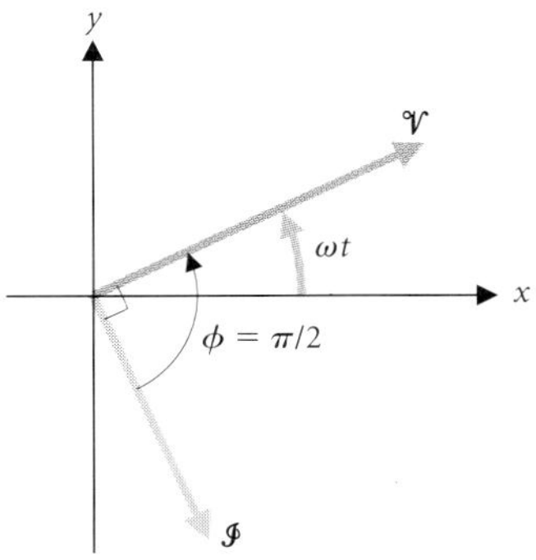

■ FIGURE 32.20
Phasor diagram for an inductor. The phasors rotate counterclockwise with angular speed ω.

32.5 (■ Figure 32.21) The inductor voltage leads the current. The total potential difference is the vector sum of the resistor and inductor voltages:

$$|\mathcal{V}| = \sqrt{|\mathcal{V}_R|^2 + |\mathcal{V}_L|^2} = |\mathcal{I}|\sqrt{R^2 + (\omega L)^2}$$

(cf. eqns. 32.18 and 32.19), and

$$\tan\phi = \frac{|\mathcal{V}_L|}{|\mathcal{V}_R|} = \frac{|\mathcal{I}|\omega L}{|\mathcal{I}|R} = \frac{X_L}{R},\quad \text{which is eqn. (32.17).}$$

32.6 $\langle P\rangle = \frac{1}{2}\mathcal{V}\cdot\mathcal{I} = \frac{1}{2}(156\ \text{V})(0.276\ \text{A})\cos(27.9°) = 19.0\ \text{W}$. The power as a function of time is shown in ■ Figure 32.22.

32.7 The damping time constant is $\tau = 2L/R$, so:

$$\tfrac{1}{2}\omega_0\tau = \frac{1}{2}\left(\frac{1}{\sqrt{LC}}\right)\left(\frac{2L}{R}\right) = \frac{1}{R}\sqrt{\frac{L}{C}} = Q,$$

as required. Qualitatively, Q is π times the number of oscillations the current undergoes in one damping time. Using an acoustic analogy, a bell with high Q rings, but a drum with low Q thuds.

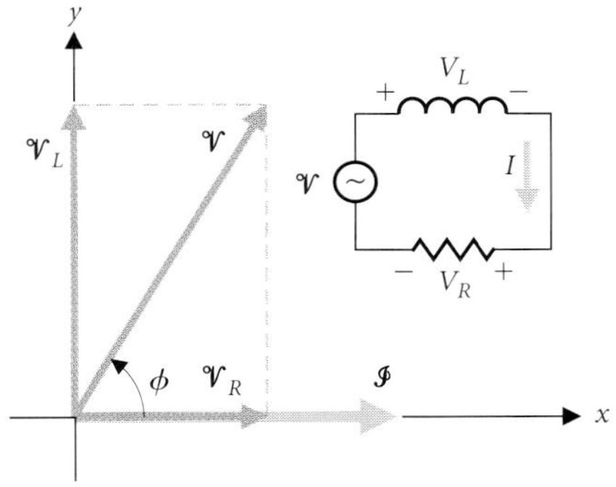

■ FIGURE 32.21
Phasor diagram for a series *LR* circuit. We have chosen to place the current phasor along the x-axis.

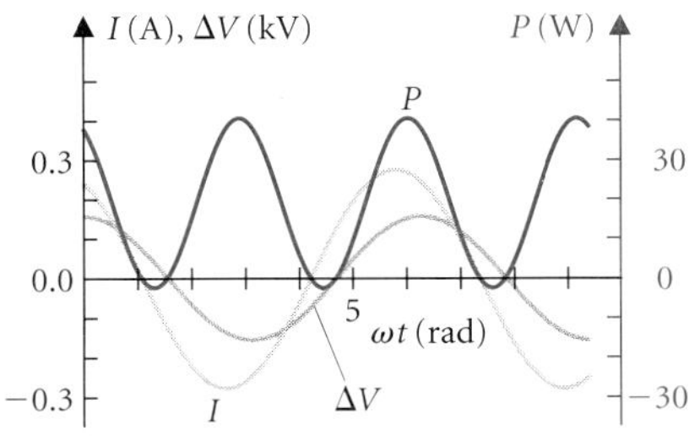

■ FIGURE 32.22
Voltage, current, and power for the *RC* circuit. The power is positive for almost all of the cycle because the phase shift between voltage and current is small.

Basic Skills

Review Questions

§32.1 SINGLE-ELEMENT CIRCUITS

- What constraint exists on the size of a circuit to which you can apply the techniques of this chapter? What is the reason for the constraint?
- What does it mean to say that current *leads* or *lags* voltage?
- Does current in a circuit arm containing a capacitor lead or lag the potential difference across the capacitor?
- Does current in an inductor lead or lag the potential difference across it?
- When AC current passes through a resistor, how does the power dissipated depend on the applied potential difference? What is the time-averaged power?
- Define V_{rms}.
- What is the time-averaged power dissipated when an AC voltage is applied to a capacitor? What about an inductor?
- Define the *reactance* of a capacitor or inductor.
- How does the reactance of an inductor depend on the frequency of the AC power supply?
- How does the reactance of a capacitor depend on the frequency of the AC power supply?

§32.2 TWO-COMPONENT CIRCUITS

- What is *impedance?*
- In a series connection of two circuit elements, how does the impedance depend on resistance and reactance?
- When an AC power supply is connected to a resistor and an inductor in series, how is the current related to the applied voltage? How does the phase shift depend on the reactance? On the resistance?
- What is the *power factor?* How does it depend on resistance?
- Why might you want to reduce the total reactance in a circuit?

§32.3 CIRCUIT ANALYSIS USING PHASORS

- What is a *phasor?*
- In a phasor diagram, how should you measure the phase shift ϕ?
- In a series circuit, how many current phasors are there? How many voltage phasors?
- How do the voltage phasors combine?
- Does it matter at what angle you draw the current phasor? Why or why not?
- Express the power used by a circuit in terms of phasors.
- In a parallel combination of two circuit elements, how many voltage phasors are there? How many current phasors? How do the current phasors combine?

§32.4 THE *LRC* CIRCUIT

- What is the impedance of a series combination of resistor, inductor, and capacitor?
- What is the resonant frequency of a series *LRC* circuit?
- What happens to the current at *resonance?* How are the capacitor and inductor voltages related at resonance?
- What is the quality factor Q for a series *LRC* circuit? What *quality* does it measure?

Basic Skill Drill

§32.1 SINGLE-ELEMENT CIRCUITS

1. What is the reactance of a 5.6-μF capacitor used with a 60.0-Hz power supply?
2. What is the reactance of a 1.2-mH inductor used with a 150-Hz power supply?
3. Find the resistance of a 60-W light bulb designed to operate off the power network (60 Hz, 120 V rms).
4. Find the current amplitude when a 32-μF capacitor is connected to an AC generator providing 25 V rms at 1.0 kHz.

§32.2 TWO-COMPONENT CIRCUITS

5. A coil with an inductance of 5.0 mH and a resistance of 10.0 Ω is connected to an AC power supply providing 10.0-V voltage amplitude at 0.50 kHz. What is the amplitude of the current in the circuit? How much power is used?
6. A 1.0-kΩ resistor and a 2.5-nF capacitor are connected in series to an AC power supply providing 12 V rms at 15 kHz. What is the current amplitude in the circuit? How much power is used?
7. Find the current as a function of time in a circuit with a 2.20-mH inductor and a 5.60-Ω resistor connected to a 60.0-Hz, 120-V power supply.

§32.3 CIRCUIT ANALYSIS USING PHASORS

8. A 120-V, 60-Hz AC generator is connected to a coil with a resistance of 3.0 Ω and an inductance of 2.40 mH. Draw a phasor diagram at $t = 0$, 5 ms, and 10 ms.
9. A 4.8-μF capacitor and a 120-Ω resistor are connected in series with a 120-V, 60.0-Hz power supply. Draw a phasor diagram and use it to find the average power used by the circuit.

§32.4 THE *LRC* CIRCUIT

10. Find the resonant frequency f_0 and Q value for a series *LRC* circuit with $L = 32\ \mu$H, $C = 4.7\ \mu$F, and $R = 6.0\ \Omega$.
11. You have a 56-nF capacitor. You want to make an *LRC* circuit with a resonant frequency at 610 kHz and a Q of 60.0. What values of L and R should you choose? If the input voltage is $V_0 = 0.68$ mV, what is the current amplitude at resonance?

Questions and Problems

§32.1 SINGLE-ELEMENT CIRCUITS

12. ❖ Two capacitors have capacitance C and $2C$. Which has the larger reactance? By what factor? Explain the physical reason for the difference.

13. ❖ Two inductors have inductances L and $2L$. Which has the larger reactance? By what factor? Why?

14. ❖ The frequency of an AC power source is doubled. By how much does the reactance of an inductor in the circuit change? Explain the physical reason for the change.

15. ❖ The frequency of an AC power source is halved. By how much does the reactance of a capacitor in the circuit change? Explain the reason for the change.

16. ❖ What is the rms voltage of a rectified cosine wave

$$V(t) = V_0|\cos \omega t|?$$

17. ❖ Estimate the maximum frequency at which each of the following systems could operate with designs that ignore the response time of electric and magnetic field.

(a) A power grid for the entire United States.

(b) A power grid for a medium-sized city like Denver.

(c) The circuits in a 2-m-diameter supercomputer.

(d) The circuit on a single microprocessor chip 1 cm across in a laptop computer.

18. ♦ A 60.0-mH inductor has a resistance of 5.0 Ω. At what frequency does its resistance equal its reactance?

19. ♦ What is the current amplitude in a circuit with a 15-V rms, 750-Hz power supply connected to a 1.2-nF capacitor?

20. ♦ Commercial power (120 V, 60.0 Hz) is connected to an inductor, and the resulting current amplitude is 1.75 A. What is the inductance?

21. ♦ A hair dryer manufactured for use in the United States is rated at 1200 W. What is its resistance? (Assume it has no inductance or capacitance.)

22. ♦♦ When a power supply is connected to a 5.6-μF capacitor, the current amplitude is 17 A. Connected to a 2.8-mH inductor, the current amplitude is 25 A. What is the frequency of the power supply? How much power is used when the same source is connected to a 5.6-kΩ resistor?

23. ♦♦ What amplitude sine wave produces the same average power in a resistor as a square wave generator with maximum voltage V_0 (■ Figure 32.23)?

24. ♦♦♦ What is the rms voltage of a sawtooth wave (■ Figure 32.24)?

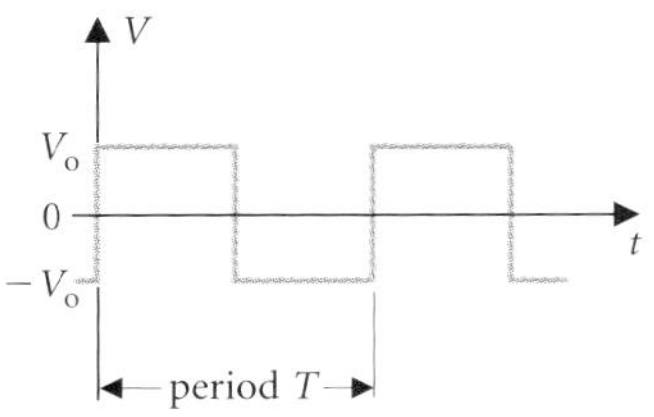

■ FIGURE 32.23

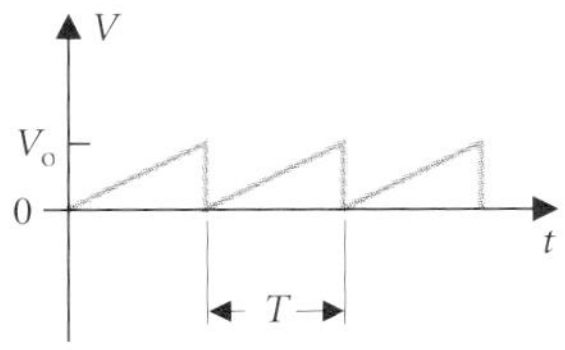

■ FIGURE 32.24

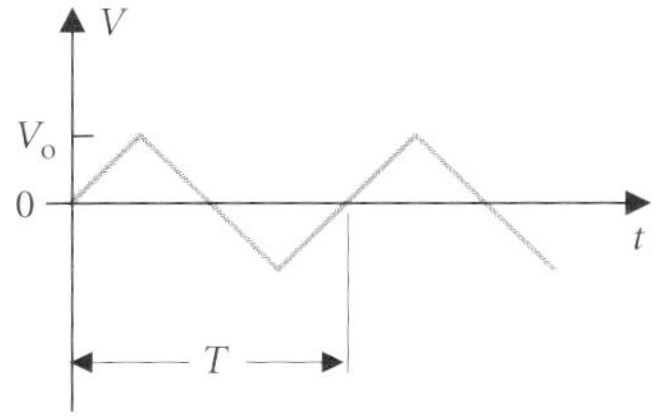

■ FIGURE 32.25

25. ♦♦♦ What is the rms voltage of a triangle wave (■ Figure 32.25)? Based on your results, what would you guess is the rms voltage for a rectified triangle wave?

26. ♦♦♦ Show by direct integration that the average value of $\sin^2 \omega t$ or $\cos^2 \omega t$ over one period is $\frac{1}{2}$.

§32.2 TWO-COMPONENT CIRCUITS

27. ❖ Explain why a resistor uses less power when connected in series with a capacitor or an inductor than when connected alone.

28. ❖ An inductor and a capacitor each have the same reactance X. Each is connected into a circuit with a resistor R and a power supply. Compare the power used by the two circuits.

29. ❖ A resistor is connected in series with an unknown circuit element. If the power consumption of the circuit increases as the frequency of the applied voltage increases, is the unknown element a capacitor, a resistor, or an inductor?

30. ♦ An inductor and a capacitor are connected in series with an AC generator. What is the current in the circuit? Do you believe your solution at $\omega = 1/\sqrt{LC}$? Qualitatively discuss what would happen in a real circuit.

31. ♦ An electric motor has a 75-mH inductance and a 24-Ω resistance. What is the phase shift between current and voltage when it is connected to the commercial power network? What is the average power used?

32. ♦ A coil used in a physics experiment has a 33-mH inductance and an 85-Ω resistance. It is connected to an AC generator with a frequency of $\omega = 1.0 \times 10^3$ rad/s and a peak voltage of 15 V. Find the amplitude of the current through the coil and the phase shift with respect to the applied voltage.

33. ♦ A factory has a total inductance of 8.0 H. What capacitance connected in series is required to make the total reactance at 60.0 Hz zero?

34. ♦♦ Show that the average power used by a series AC circuit is the square of the current amplitude times half the resistance.

35. ♦♦ A coil has resistance $R = 10.0$ Ω and inductance $L = 5.0$ mH. At what frequency is the power used by the coil a maximum? Give a numerical value for the maximum power used when $V_0 = 10.0$ V. Compare with the power used at 60.0 Hz.

36. ♦♦ A 1.2-kΩ resistor and a 2.2-μF capacitor are connected in series across a 120-V, 60-Hz power supply. Find the current in the circuit as a function of time. What is the average power drawn from the power supply?

37. ♦♦ In a stereo crossover network, a 1.9-mH inductor is in series with a 53-μF capacitor. Plot a graph showing the ratio of potential difference across the inductor to the potential difference across the capacitor as a function of frequency. At what frequency is there an equal potential difference across each element? (This is called the crossover frequency.) Across which element should the tweeter be connected, and which the woofer?

38. ♦♦♦ A circuit contains a 33-mH inductor and an 80.0-Ω resistor in series with a 60.0-Hz AC power supply and a switch. The switch is closed at $t = 0$. Find the current in the circuit as a function of time. How many periods of oscillation occur before the transient response damps to 10% of the amplitude of the steady state current?

§32.3 CIRCUIT ANALYSIS USING PHASORS

39. ❖ Does the phasor diagram for a series circuit depend on the order in which circuit elements occur? Why or why not?

40. ❖ Are the voltage phasors due to capacitive and inductive reactance always directly opposite each other in a series circuit? In a parallel circuit?

41. ♦ A 15-V, 25-Hz power supply is connected in series to a 15-kΩ resistor and a 0.56-μF capacitor. Draw the phasor diagram at $t = 0$, 0.01 s, and 0.02 s. What is the current amplitude?

42. ♦ A 35-mH inductor and a 5.6-Ω resistor are connected in parallel across a 44-V, 30.0-Hz power supply. Draw the phasor diagram at $t = 0$, 0.01 s, and 0.02 s. What is the current amplitude?

43. ♦♦ A food mixer has a resistance of 35 Ω and an inductance of 22 mH in series. Draw a phasor diagram for the voltage and current when the mixer is used with home power (120 V, 60.0 Hz). How much power does the mixer draw? How much power would it draw if the inductance were zero?

44. ♦♦ Use a phasor diagram to find the current through the power supply shown in ■ Figure 32.26. Compare and contrast your results with the behavior of a series *RC* circuit (e.g., Example 32.5).

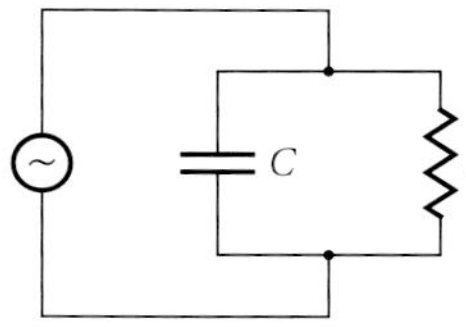

■ FIGURE 32.26

45. ♦♦ Household wiring is arranged so that when different appliances are plugged in, they are in parallel across the power supply. A vacuum cleaner with $R = 55$ Ω and $L = 33$ mH and a television set with $R = 85$ Ω are operating at the same time. Draw a circuit diagram and a phasor diagram. How much power is used? What fraction is used by the vacuum and what fraction by the television?

46. ♦♦ Use the phasor method to find the current through the power supply in the circuit diagramed in ■ Figure 32.27.

47. ♦♦♦ Two coils are connected in parallel across a 75-V, 55-Hz power supply. One coil has an inductance of 15 mH and a resistance of 3.5 Ω; the second has an inductance of 25 mH and a resistance of 6.7 Ω. Draw a phasor diagram and find the currents in the circuit. What is the power absorbed?

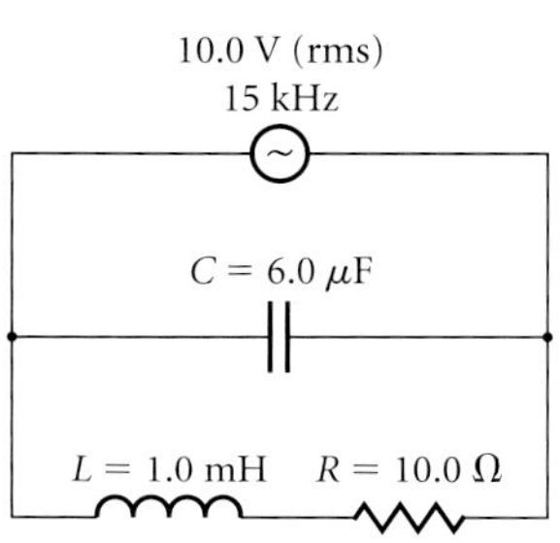

■ FIGURE 32.27

§32.4 THE *LRC* CIRCUIT

48. ❖ If the inductance in a series *LRC* circuit is doubled, how do the resonant frequency, time constant, and quality factor change?

49. ❖ If the capacitor in a series *LRC* circuit is changed for one with four times the capacitance, how do the resonant frequency, time constant, and quality factor change?

50. ❖ If the resistance in a series *LRC* circuit is reduced by a factor of 2, how do the resonant frequency, time constant, and quality factor change?

51. ♦ Find the resonant frequency and *Q* value for a circuit with $R = 18$ Ω, $C = 5.6$ μF, and $L = 33$ mH.

52. ♦ You have a 5.6-mH inductor. You want to build a circuit with a resonant frequency of 2.5 kHz and a *Q* of 45. What values of *C* and *R* should you choose?

53. ♦ You have two inductors, 2.2 mH and 0.75 mH, and two capacitors, 56 pF and 1.0 nF. Each inductor also has 0.95-Ω resistance. Using one inductor and one capacitor, you build a series *LRC* circuit. **(a)** What is the greatest resonant frequency you can achieve? What is the *Q* of that circuit? **(b)** What is the greatest *Q* you can achieve? What is the resonant frequency of that circuit?

54. ♦ Show that the power input to an *LRC* circuit is maximum at resonance.

55. ♦♦ An *LRC* circuit using a 15-μH inductor is tuned to 0.10 MHz. What maximum resistance can the circuit have if an unwanted signal at 90 kHz is amplified by less than 10% of the amplification at 100 kHz?

56. ♦♦ A tuning circuit is adjusted with a variable inductance. If $R = 5.6$ Ω and $C = 1.2$ pF, what range of inductance is required to tune the resonant frequency over the range 75 MHz to 250 MHz? How does the *Q* value for the circuit change over this range? At what values of ω/ω_0 does the capacitor amplitude equal 10% of the peak value at $f_0 = 75$ MHz and 250 MHz?

57. ♦♦ Use a phasor diagram to find the currents in each arm of the parallel *LRC* circuit in ■ Figure 32.28. Find the current amplitude and the average power drawn from the generator. Compare with the behavior of the series *LRC* circuit.

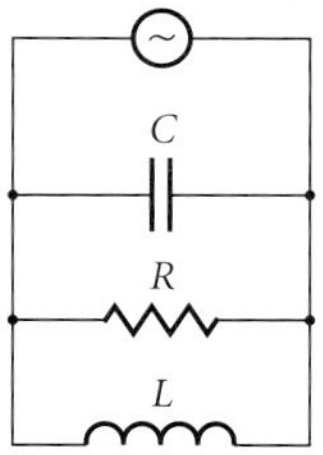

■ FIGURE 32.28

58. ♦♦ Find the potential difference across the inductor in a series *LRC* circuit in terms of ω/ω_0 and *Q*. Plot this function and compare your graph with Figure 32.19. Give a physical reason for each difference.

59. ♦♦ A series *LRC* radio tuning circuit has $L = 0.56$ mH, $R = 2.2$ Ω, and a variable capacitance. What value of *C* is required to tune the circuit to 660 kHz? What is the *Q* value? If another station at 600 kHz has the same amplitude, how large is the unwanted output signal compared with the selected signal?

60. ♦♦ In the FM band, 88–108 MHz, stations may be separated by as little as 200 kHz. In a modern FM tuner, the input signal is converted to an intermediate frequency of 10.7 MHz. Then a secondary circuit discriminates between signals from different stations. What value of Q is required in the secondary circuit if the output signal at 10.5 MHz is 25% of that at 10.7 MHz? Assume equal input amplitudes at the two frequencies. What value of L is required in the circuit if the total resistance is 1.2 Ω? What capacitance is required?

61. ♦♦♦ Show that the maximum amplitude of the capacitor voltage in a series *LRC* circuit occurs at a frequency $\omega = \omega_0\sqrt{1 - 1/(2Q^2)}$ and has the value:

$$V_{\mathrm{C,max}} = \frac{V_0 Q}{\sqrt{1 - 1/(4Q^2)}}.$$

62. ♦♦♦ Show that the maximum amplitude of the inductor voltage in a series *LRC* circuit occurs at a frequency:

$$\omega = \frac{\omega_0}{\sqrt{1 - 1/(2Q^2)}}$$

and has the same value as the maximum capacitor voltage (Problem 61).

63. ♦♦♦ In a series *LRC* circuit driven at frequency ω, show that the ratio of maximum energy stored in the capacitor during a cycle to maximum energy stored in the inductor is:

$$\frac{U_C}{U_I} = \frac{\omega_0^2}{\omega^2}.$$

Show that the difference in the peak energies equals the work done by the AC generator during the quarter cycle in which the capacitor discharges and the inductor reaches peak current, less the thermal energy transformed in the resistor.

64. ♦♦♦ One measure of the width of the resonance curve is the full-width half-maximum, the difference between two angular frequencies at which the circuit response is half its maximum value. Show that, for large Q, the fwhm is approximately $\sqrt{3}\omega_0/Q$.

65. ♦♦♦ In a series *LRC* circuit, show that Q is 2π times the maximum energy stored in the capacitor at resonance divided by the energy dissipated per cycle (at resonance).

Additional Problems

66. ❖ Describe the low- and high-frequency behavior of the circuit shown in ■ Figure 32.29.

67. ❖ Discuss the behavior of the circuit in ■ Figure 32.30. What is the effect, if any, of the capacitor?

68. ❖ A circuit contains a resistor connected to a "black box" and a power supply. As the power supply frequency is increased from zero, you notice that the power absorbed by the circuit decreases. Is the component in the black box a resistor, a capacitor, or an inductor?

69. ❖ How does the rms power depend on the frequency of the AC power source if: **(a)** the circuit contains only resistance? **(b)** the circuit contains a resistor and a capacitor in series?

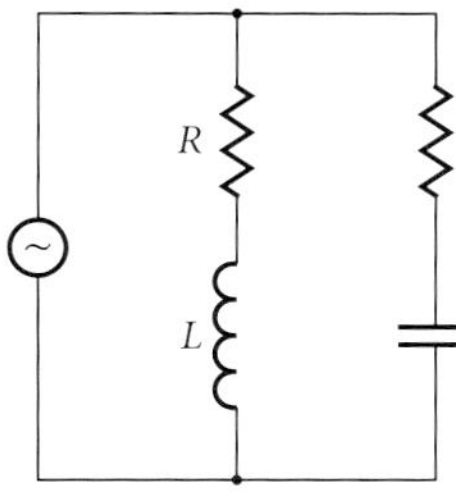

■ FIGURE 32.29

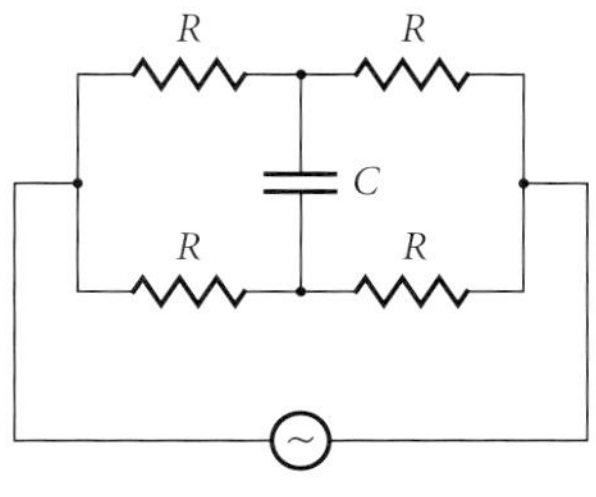

■ FIGURE 32.30

70. ♦♦ Two capacitors are connected in series to an AC power supply. Draw a phasor diagram showing the current and the potential differences across each capacitor. What is the current in the circuit? Do the capacitors combine according to the rules you learned in Chapter 27?

71. ♦♦ A coil with a 20-Ω resistor and a 5-mH inductor is connected in parallel with a 30-μF capacitor. What is the impedance of the combination at 330 Hz?

72. ♦♦♦ Find the current I in the circuit shown in ■ Figure 32.31.

73. ♦♦♦ Use phasor diagrams to analyze the circuit shown in ■ Figure 32.32 if the frequency of each generator is 350 Hz. What is the current in the resistor? How much power is drawn from each emf?

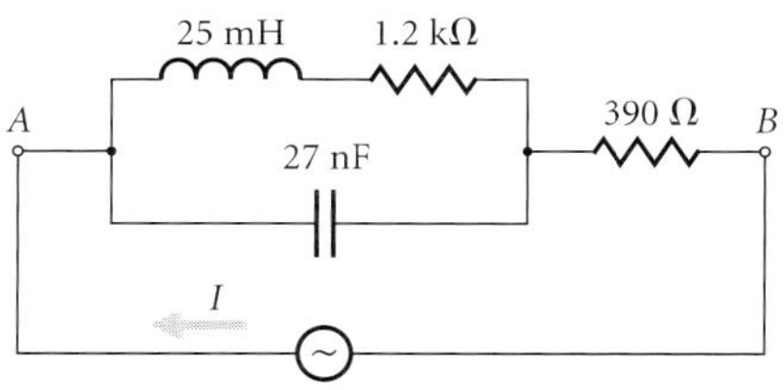

■ FIGURE 32.31 $V_0 = 15$ V, $f = 10.0$ kHz

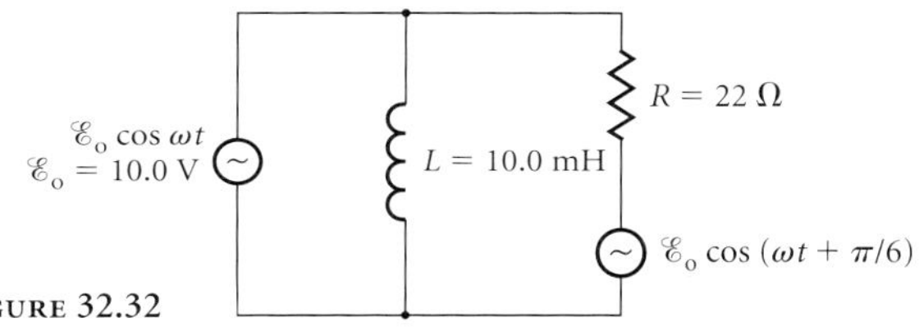

■ FIGURE 32.32

74. ♦♦♦ The quality factor for a parallel antenna circuit (■ Figure 32.33) is the inverse of Q for a series circuit with the same components: $Q_\| = R\sqrt{C/L}$. Derive this expression using the results $Q = \frac{1}{2}\omega_0\tau$ (Exercise 7) and $\tau = 2RC$ (Problem 31.72). Choosing values of L, R, and C, design an antenna circuit with a Q of 5 at 102 MHz. Find the capacitor voltage in terms of Q, and plot $V_C/\mathscr{E}$ versus ω/ω_0 for the circuit you have designed.

75. ♦♦♦ **(a)** A series *LRC* circuit is used in an instrument for measuring the level of liquid in a tank (■ Figure 32.34). The capacitor has parallel plates of dimension $\ell = 20.0$ cm by $w = 5.0$ cm and separation $d = 1.0$ mm. The liquid has dielectric constant $\kappa = 2.3$. The level of liquid is determined by measuring the resonant frequency of the circuit. Show that the liquid level is given by:

$$x = \frac{7.0 \times 10^{14}\ \mathrm{m/s^2}}{\omega_0^2} - 0.15\ \mathrm{m}.$$

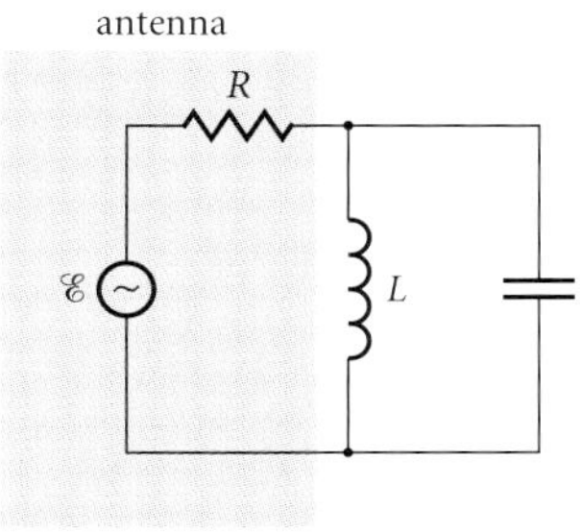

■ FIGURE 32.33

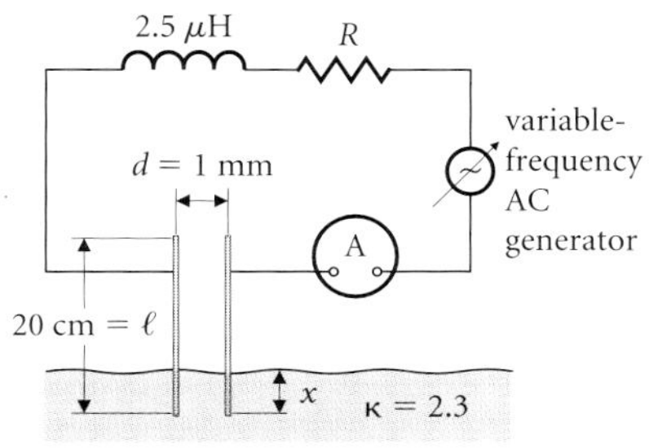

■ FIGURE 32.34

(b) If the ammeter uncertainty in measuring peak current amplitude is 1%, what Q is needed to measure the liquid level to 1% accuracy? What maximum resistance in the circuit will allow this accuracy?

Computer Problems

76. A circuit with $L = 1.2$ mH and $R = 17\ \Omega$ is connected to a 35-V, 2.0-kHz power supply. Use a spreadsheet program to calculate and plot $V(t)$, $I(t)$, and $P(t)$ over two cycles. What is the time-averaged power? (You may calculate the time average analytically or numerically.)

77. A circuit with $R = 56\ \Omega$ and $C = 1.2\ \mu$F is connected to a 5.0-V, 130-kHz power supply. Use a spreadsheet program to plot $V(t)$, $I(t)$, and $P(t)$ over two cycles. What is the time-averaged power?

78. Express the power used by a series LRC circuit as a function of generator frequency in dimensionless form. That is, express the ratio $P \div (V_0^2/2R)$ as a function of the ratio f/f_0. Plot graphs of the function for $Q = 1, 10, 20$. Comment.

Challenge Problems

79. A practical radio antenna uses a parallel LRC circuit (cf. Figure 32.33) in which the incoming signal acts as the emf. Use phasors to find the ratio of capacitor voltage to antenna voltage. The coil has inductance $L = 2.5\ \mu$H *and* resistance $r = 65\ \Omega$. Take $C = 150$ pF, antenna resistance $R = 325\ \Omega$, and the operating frequency of the circuit $f = 0.82$ MHz.

80. Use phasors to find the currents in each arm of the circuit shown in ■ Figure 32.35 and the power drawn from the generator.

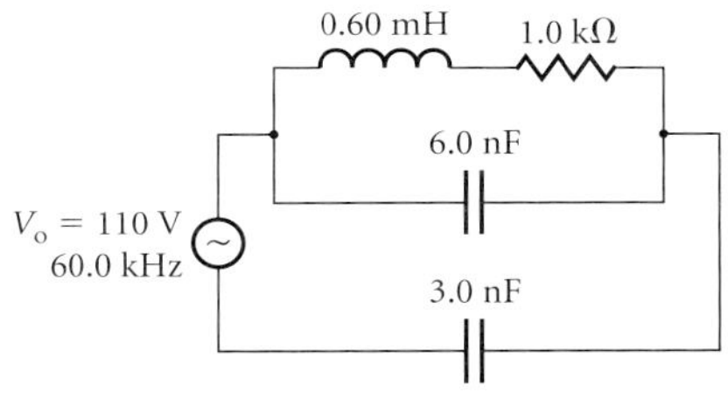

■ FIGURE 32.35

81. **(a)** Show that if the voltage applied to a series LRC circuit is the sum of two oscillating voltages, the voltage across the capacitor is the sum of the two voltages that would result from each of the two inputs applied separately. **(b)** A series LRC circuit with resonant frequency ω_0 and quality factor $Q = 10$ is connected to a signal generator that produces a voltage $V(t) = V_0 \cos(\omega_0 t) + (V_0/3)\cos(3\omega_0 t)$. What is the resulting capacitor voltage as a function of time? **(c)** Fourier's theorem states that any periodic function can be represented as a sum of trigonometric functions. ■ Figure 32.36 shows a square wave produced by alternately switching a battery into and out of a circuit. The square wave is a sum of cosine functions, of which the first two are shown in the figure:

$$S(t) = \frac{4V_0}{\pi}\left[\cos(\omega_s t) - \tfrac{1}{3}\cos(3\omega_s t) + \tfrac{1}{5}\cos(5\omega_s t) + \cdots\right].$$

What capacitor voltage results if the LRC circuit in part (b) is connected to the square wave voltage with $\omega_s = \omega_0$? How good a sine wave generator does this circuit make? **(d)** Describe the capacitor voltage when the frequency of the square wave is (i) $\omega_s = \omega_0/3$, and (ii) $\omega_s = \omega_0/5$. **(e)** When the frequency of the square wave is much

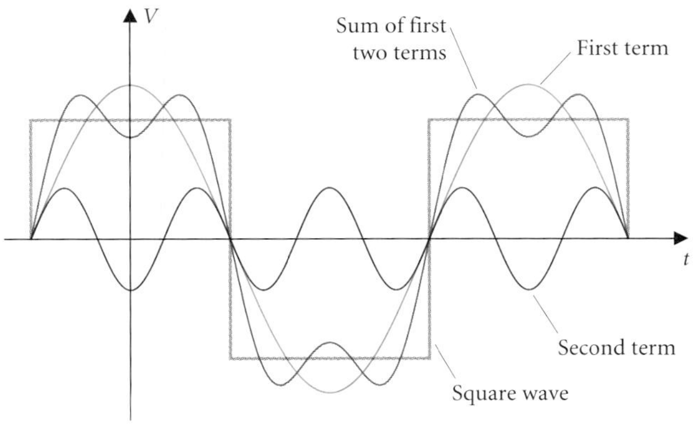

■ FIGURE 32.36

less than the resonant frequency of the LRC circuit, describe the behavior of the capacitor voltage qualitatively, using the methods of Chapter 31. How is this result consistent with parts (a) through (d)?

82. A coil is wound with twin-lead wire (two separate wires bound together so as to remain parallel to each other, but not electrically connected) to form two inductors with the same length ℓ, cross-sectional area A, and number of turns N (■ Figure 32.37). One of the wires is connected to a capacitor C. The other wire is connected in series to a second, identical capacitor C and a signal generator that puts out a voltage $V_0 \cos \omega t$. **(a)** Compute the self-inductance L of one of the wires wrapped around the coil. This will be the self-inductance in each of the two circuits. Also compute the mutual inductance M of the two circuits. How is M related to L? **(b)** With the sign conventions shown in the figure, use Kirchhoff's laws to write differential equations for the two charges $Q_1(t)$ and $Q_2(t)$. Solve for the steady state response of the system, and find its resonant frequency. **(c)** Plot a graph showing the voltage across each capacitor as a function of frequency squared. (Use $\omega^2 LC$ as the frequency variable.) **(d)** Discuss the behavior of the system at $\omega = 1/\sqrt{LC}$.

83. In the jumping rings demonstration (■ Figure 32.38), a metal ring is placed over an iron core that penetrates a solenoid. The ring has resistance R_r and inductance L_r in series. The coils around the central magnet form another LR circuit with inductance L_c and resistance R_c, driven by an AC generator. The mutual inductance is M. Find the current in the ring. Use the result to show that there is a net upward force on the ring.

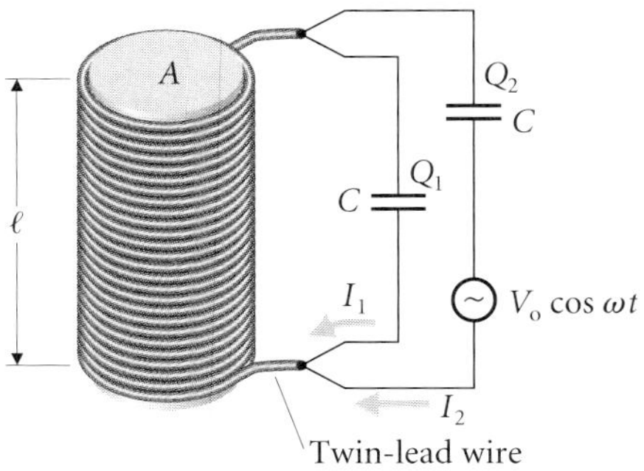

■ FIGURE 32.37

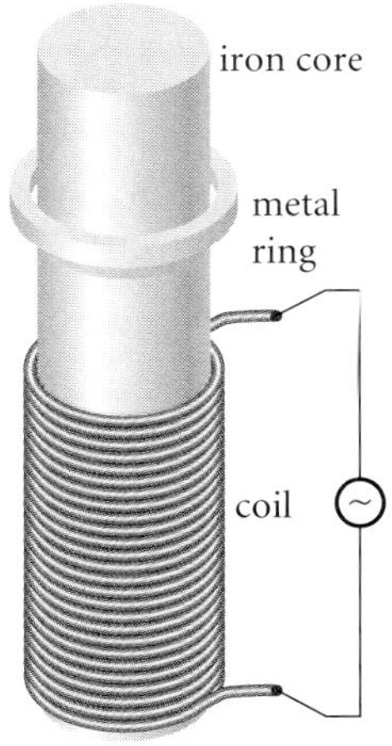

■ FIGURE 32.38

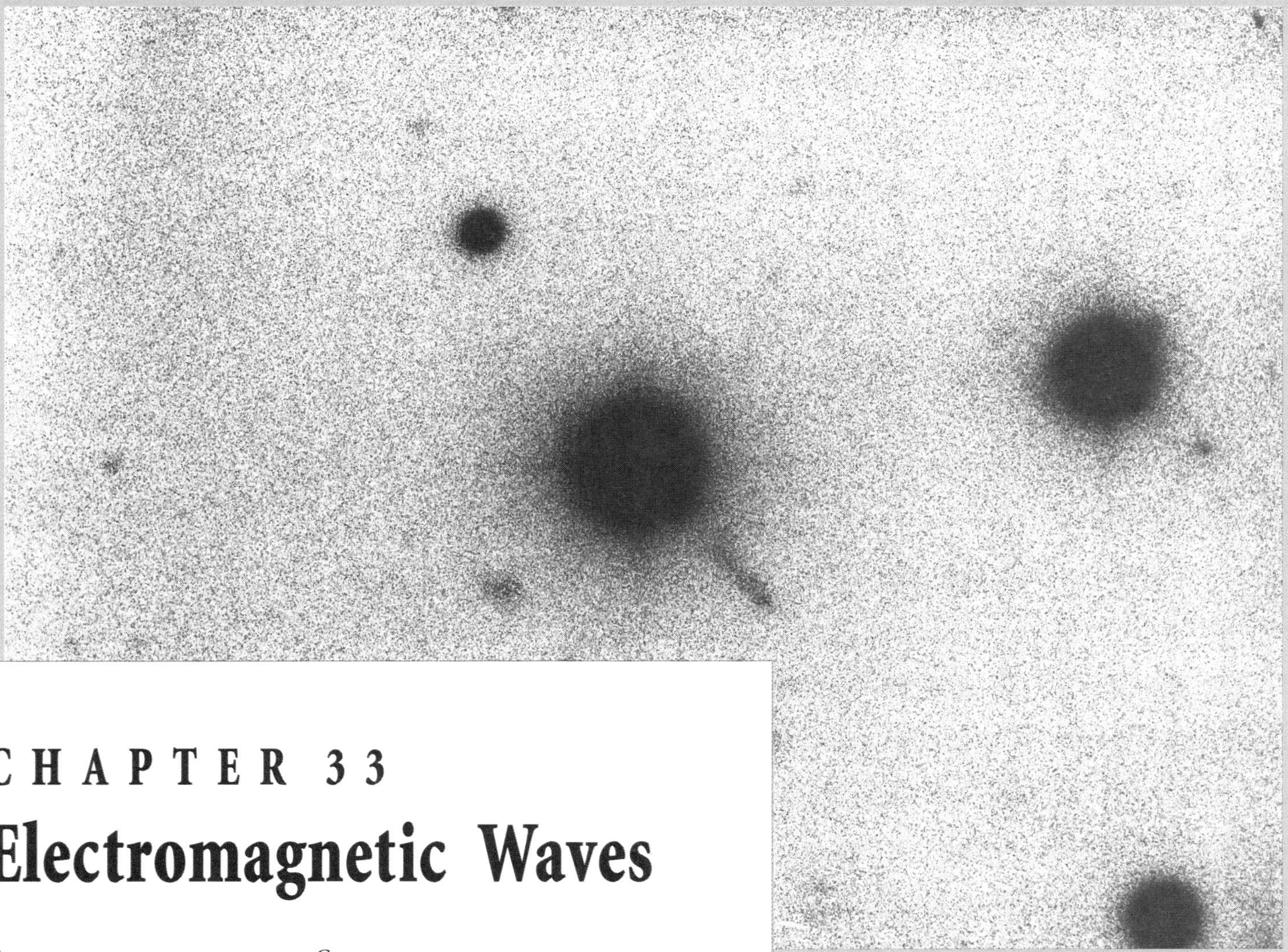

CHAPTER 33
Electromagnetic Waves

Concepts

Electromagnetic wave
Poynting vector
Radiation pressure
Linear polarization
Circular polarization
Cavity oscillator
Waveguide

Goals

Understand:

The configuration of fields in a plane EM wave.

How polarized light is produced.

Why a waveguide has a cutoff frequency.

Be able to:

Compute the radiation pressure exerted by an EM wave.

Compute the wave intensity transmitted by a sequence of polarizing filters.

The quasi-stellar object 3C273. (The name means it is the 273rd object in the third Cambridge catalogue of radio-emitting objects.) Quasi-stellar objects (QSOs), as their name suggests, look like stars on optical photographs, but they are believed to be the bright nuclei of distant galaxies. They are the most distant known objects in the universe. To outshine far closer objects, they must emit enormous amounts of radiation throughout the EM spectrum, from radio waves to γ rays. This radiation is the only information we have about them. By carefully studying these EM waves (their distribution in frequency, time variability, total power, etc.), we can build models of the enigmatic QSOs to learn more about them and their role in the evolution of the universe.

> *I have also a paper afloat, containing an electromagnetic theory of light, which, till I am convinced to the contrary, I hold to be great guns.*
>
> James Clerk Maxwell

Light from a distant quasar has been traveling toward Earth for billions of years. This electromagnetic radiation carries information about the composition and structure of the quasar and also about gas clouds that lie between us and the quasar. Almost everything we know about the universe beyond our solar system we have learned by analysis of electromagnetic (EM) radiation. Maxwell's identification of EM waves with light was one of the most exciting and beautiful discoveries in physics and has been one of the most significant technologically as well. Electromagnetic waves provide the fastest and most efficient way to transmit information, not only across the universe but on Earth as well, through fiber optic cables and satellite links. This chapter is a brief introduction to the theory of EM waves. We hope also to share an appreciation of the elegance and power of Maxwell's theory.

According to special relativity theory (Chapter 34), nothing can travel faster than light does in vacuum.

33.1 Plane Electromagnetic Waves

33.1.1 *Origin and Structure of a Plane EM Wave*

Reread the section on EM waves in the Overview of electromagnetism.

Study Hint: Review the material on waves, especially Chapter 15. We'll encounter the wave equation again in this chapter.

Accelerating charges generate EM waves. In Figure VI.24 we illustrated how an electron that is rapidly displaced from rest and then abruptly stopped again forms a wave pulse. Displacement of the electron produces outward-moving kinks in the electric field lines, and the electron's motion constitutes a current that produces a closely coupled magnetic field pulse. If the electron oscillates, a train of oscillations propagates outward along each electric field line, together with an oscillating magnetic field pattern (■ Figure 33.1). The electric and magnetic fields in the wave are perpendicular to each other and oscillate in phase. The direction of wave propagation, labeled by the wave vector $\vec{k}$, is perpendicular to both $\vec{E}$ and $\vec{B}$ (■ Figure 33.2):

$$\vec{k} \parallel \vec{E} \times \vec{B}. \tag{33.1}$$

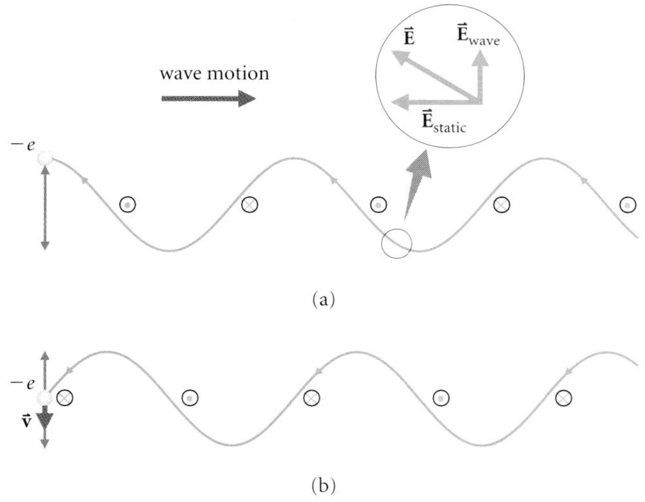

■ Figure 33.1
An oscillating electron produces a train of pulses along each electric field line. Notice that since the electron charge is negative, the electric field direction is toward the electron. The wave itself moves outward. As shown in the enlarged view, the electric field near the electron is a mixture of *static* field, the Coulomb field produced by the electron at rest, and *wave* field, due to the electron's oscillations. The electron in motion constitutes electric current that generates the magnetic field in the wave. Here we show the fields in the direction perpendicular to the electron's motion. The wave amplitude varies with direction, but the field pattern is similar. No radiation is emitted in the direction of the electron's acceleration. (a) At the end of an oscillation, the electron is instantaneously at rest. At this time, both $\vec{E}_{wave}$ and $\vec{B}_{wave}$ are zero at the electron's position (left end of the diagram). (b) One quarter-cycle later, the electron has its maximum speed, and the field line (due to $\vec{E}_{static} + \vec{E}_{wave}$) has its maximum slope. Both E_{wave} and B_{wave} are maximum.

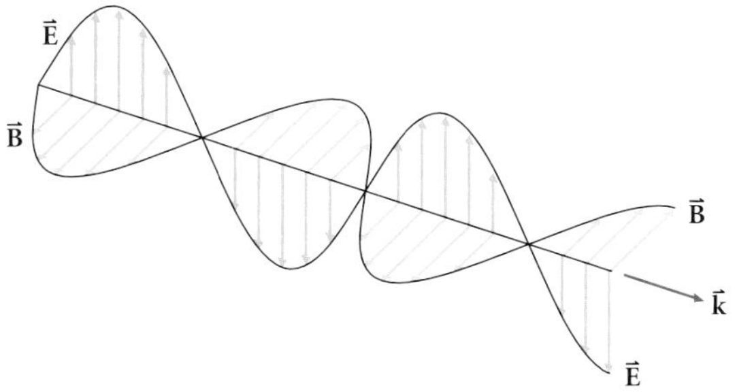

■ FIGURE 33.2
A long way from its source, the static fields have become much smaller than the wave fields, and the wave is approximately plane. This diagram is a snapshot of the wave at a fixed time at points along a single ray. The diagram would be identical along any other ray. The wave vector $\vec{k}$ describes the direction of propagation: it is parallel to $\vec{E} \times \vec{B}$.

The magnitudes of static electric and magnetic fields decrease as one over the distance squared, while, as we shall see in §33.2, the wave fields decrease as one over distance. Thus, at a large distance from the source, the static fields are unimportant and the wave dominates. In this chapter we focus attention on the oscillating (wave) fields. Let's begin by describing the structure of a plane EM wave. In §33.1.2 we shall demonstrate that this structure satisfies Maxwell's equations.

SEE COULOMB'S LAW, §23.2, EQN. (23.6), AND THE BIOT–SAVART LAW, §28.2, EQN. (28.3).

The oscillating electron emits spherical wave fronts. Close to any point P, a great distance d from the accelerating charge, the wave fronts form a set of very nearly parallel planes. We may describe the fields near P as a harmonic plane wave (cf. §16.5.1 and eqns. 15.6–15.8):

HERE "CLOSE TO" MEANS "WITHIN A DISTANCE MUCH LESS THAN d." SEE §16.5.1 FOR FURTHER DISCUSSION OF THE PLANE WAVE APPROXIMATION.

$$\vec{E} = \vec{E}_0 \cos(kx - \omega t) \quad \text{and} \quad \vec{B} = \vec{B}_0 \cos(kx - \omega t). \tag{33.2}$$

The wave has angular frequency ω (equal to $2\pi f$, where f is the frequency of the electron oscillations), wave number $k = |\vec{k}| = 2\pi/\lambda$, and wavelength λ. Its speed is:

$$v = \omega/k = f\lambda. \tag{33.3}$$

THIS IS ALSO EQN. (15.2).

The minus sign in the expression for the wave phase $\phi = kx - \omega t$ indicates that the wave is traveling in the direction of increasing x.

AT A FIXED VALUE OF $\phi = \phi_1$, $x = \phi_1/k + vt$ INCREASES AS t INCREASES. SEE ALSO THE DISCUSSION IN CHAPTER 15.

33.1.2 The Wave Equation for $\vec{E}$ and $\vec{B}$

In this section we show that the structure we described in §33.1.1 for an EM wave in vacuum is consistent with Maxwell's equations. We shall see that, according to Maxwell's theory, the waves travel at the speed of light, and we shall find the relation between $\vec{E}_0$ and $\vec{B}_0$.

We have four equations to satisfy (Chapter 30 summary). Gauss' laws for both the electric and magnetic fields are automatically satisfied by the plane wave in vacuum. The field lines for $\vec{E}$ and $\vec{B}$ are parallel sets of straight lines that neither begin nor end. The flux of either through any closed surface is zero (cf. Example 23.13), regardless of the wave's amplitude or speed.

$\oint \vec{E} \cdot \hat{n}\, dA = Q/\epsilon_0$ AND $\oint \vec{B} \cdot \hat{n}\, dA = 0$.

Faraday's law and Ampère's law describe how changes of each field in time are related to the variation of the other field in space. In order to satisfy these equations, we shall find specific requirements on the wave speed and field amplitudes. We begin by applying Faraday's law (eqn. 30.3):

$$\oint \vec{E} \cdot d\vec{\ell} = -\frac{d}{dt}\int \vec{B} \cdot \hat{n}\, dA.$$

■ Figure 33.3 shows how the wave fields vary in a typical small region at a particular time. Following the plan in Figure 30.19, we choose a curve that is parallel or perpendicular to $\vec{E}$ at each point—in this case, the differential rectangle $GHIJ$ in the x-y-plane with height h and

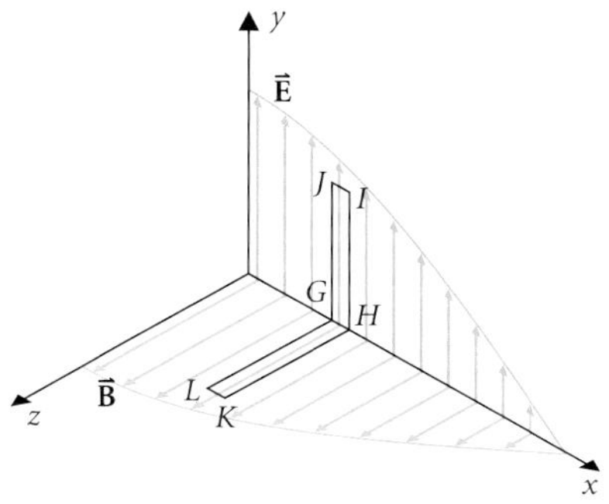

■ FIGURE 33.3 (a)

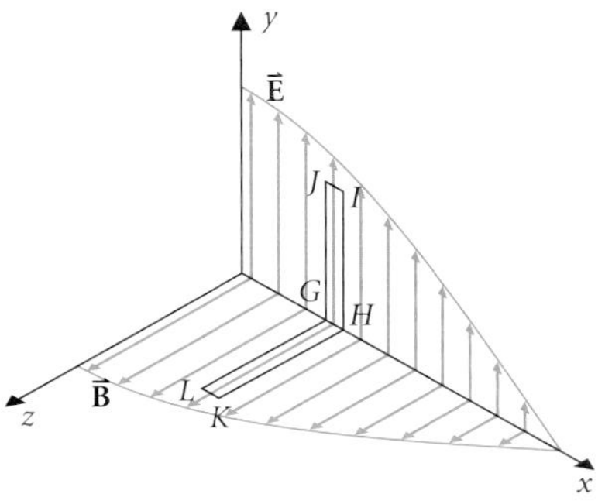

(a) An enlargement of the wave snapshot.

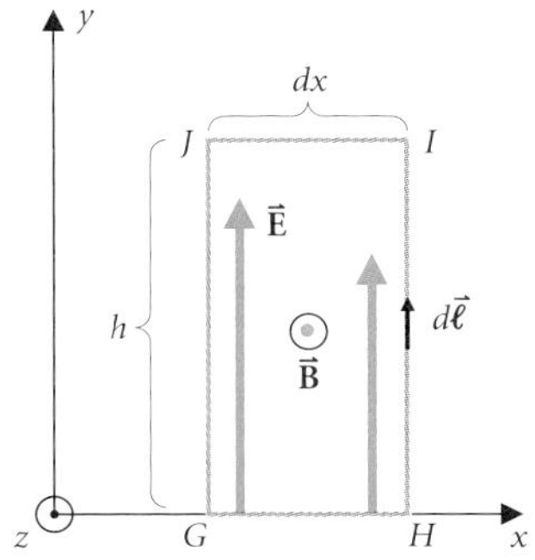

(b) Curve for applying Faraday's law.

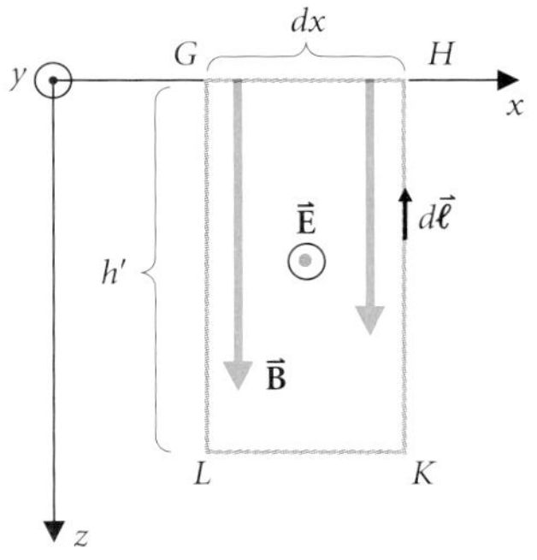

(c) Curve for applying Ampère's law.

■ FIGURE 33.3
(a) An enlargement of the wave snapshot showing the $\vec{\mathbf{E}}$ and $\vec{\mathbf{B}}$ vectors at points along the x-axis. In a plane wave, the field vectors are independent of y and z: the picture would look the same along any line parallel to the x-axis. The wave moves in the $+x$-direction, with $\vec{\mathbf{E}}$ in the y-direction and $\vec{\mathbf{B}}$ in the z-direction. Notice that $\vec{\mathbf{E}} \times \vec{\mathbf{B}} = EB\hat{\mathbf{x}}$. We apply Faraday's law to the rectangle *GHIJ*, traversed counterclockwise. The normal is then in the $+z$-direction. We apply Ampère's law to the rectangle *GLKH*, traversed counterclockwise, with normal in the $+y$-direction. (b) Curve for applying Faraday's law. The fields are shown at a fixed time. (c) Curve for applying Ampère's law.

THROUGHOUT THIS CHAPTER WE USE $\hat{\mathbf{x}}$, $\hat{\mathbf{y}}$, AND $\hat{\mathbf{z}}$ FOR THE UNIT VECTORS, INSTEAD OF $\hat{\mathbf{i}}$, $\hat{\mathbf{j}}$, AND $\hat{\mathbf{k}}$, BECAUSE OF POSSIBLE CONFUSION WITH THE WAVE VECTOR $\vec{\mathbf{k}}$ AND CURRENT DENSITY $\vec{\mathbf{j}}$.

REMEMBER: E IS A FUNCTION OF x; $E(x)$ MEANS "E AT x," *NOT* "E TIMES x."

REMEMBER: THE PARTIAL DERIVATIVE $\partial B/\partial t$ MEANS DERIVATIVE WITH RESPECT TO TIME AT A FIXED PLACE.

width dx. The vectors $d\vec{\ell}$ circulate counterclockwise around the rectangle and the normal vector $\hat{\mathbf{n}} = +\hat{\mathbf{z}}$. The electric field $\vec{\mathbf{E}}$ is parallel to $d\vec{\ell}$ on HI, perpendicular to $d\vec{\ell}$ on *GH* and *IJ*, and antiparallel to $d\vec{\ell}$ on *JG*. Thus:

$$\oint_{GHIJ} \vec{\mathbf{E}} \cdot d\vec{\ell} = hE(x + dx) - hE(x).$$

The magnetic field is parallel to the normal $\hat{\mathbf{n}} = +\hat{\mathbf{z}}$ so:

$$\int \vec{\mathbf{B}} \cdot \hat{\mathbf{n}}\ dA = B(x)h\ dx.$$

We have neglected the small change in B between x and $x + dx$ compared with $B(x)$. The change of flux is caused entirely by change of $\vec{\mathbf{B}}$ with time. So from Faraday's law:

$$h[E(x + dx) - E(x)] = -\frac{d}{dt}[B(x)h\ dx] = -h\ dx\frac{\partial B(x)}{\partial t}.$$

Dividing by $h\,dx$, we find:
$$\frac{E(x + dx) - E(x)}{dx} = -\frac{\partial}{\partial t}B(x).$$

In the limit $dx \to 0$, the left-hand side becomes the derivative of E with respect to x at a fixed time (i.e., in the snapshot shown in Figure 33.3):

$$\frac{\partial E}{\partial x} = -\frac{\partial B}{\partial t}. \tag{33.4}$$

Next we apply the Ampère–Maxwell law (eqn. 30.8) using a similar plan (Figure 28.19). In a region free of moving charge, only displacement current appears:

$$\oint \vec{\mathbf{B}} \cdot d\vec{\ell} = \epsilon_0\mu_0\frac{d}{dt}\int \vec{\mathbf{E}} \cdot \hat{\mathbf{n}}\ dA.$$

We choose the differential rectangle *GLKH* in the x-z-plane with height h' and width dx (Figure 33.3c). The vectors $d\vec{\ell}$ circulate counterclockwise, and the normal vector $\hat{\mathbf{n}} = +\hat{\mathbf{y}}$. The circulation of $\vec{\mathbf{B}}$ around this curve is:

$$\oint_{GLKH} \vec{\mathbf{B}} \cdot d\vec{\ell} = [B(x) - B(x + dx)]h' = -[B(x + dx) - B(x)]h'.$$

The electric field $\vec{\mathbf{E}}$ is parallel to the normal $\hat{\mathbf{n}} = \hat{\mathbf{y}}$, and the electric flux through the rectangle is:

$$\int \vec{\mathbf{E}} \cdot \hat{\mathbf{n}}\, dA = E(x)\ dx\ h'.$$

Therefore, applying Ampère's law:

$$-[B(x + dx) - B(x)]h' = \epsilon_0\mu_0 \frac{\partial E}{\partial t} h'\ dx.$$

$$-\frac{\partial B}{\partial x} = \epsilon_0\mu_0 \frac{\partial E}{\partial t}. \qquad (33.5)$$

Maxwell's equations provide relations between the first derivatives of $\vec{\mathbf{E}}$ and $\vec{\mathbf{B}}$, while the wave equation (15.10) involves second derivatives. This suggests that we should differentiate eqns. (33.4) and (33.5) again. In particular, differentiating eqn. (33.5) with respect to time at a fixed x, we obtain:

$$-\frac{\partial^2 B}{\partial t\, \partial x} = \epsilon_0\mu_0 \frac{\partial^2 E}{\partial t^2}.$$

And, differentiating eqn. (33.4) with respect to x at a fixed time gives us:

$$-\frac{\partial^2 B}{\partial x\, \partial t} = \frac{\partial^2 E}{\partial x^2}.$$

Since the two mixed derivatives of B are equal (Exercise 33.1), we may combine these two equations to obtain:

$$\epsilon_0\mu_0 \frac{\partial^2 E}{\partial t^2} = \frac{\partial^2 E}{\partial x^2}, \qquad (33.6)$$

which is a wave equation for E. The coefficient of the second time derivative determines the wave speed:

SEE §15.2.3 FOR FURTHER DISCUSSION OF THE WAVE EQUATION.

$$v^2 = \frac{1}{\epsilon_0\mu_0} = \frac{1}{(8.85 \times 10^{-12}\ \mathrm{C^2/N\cdot m^2})(4\pi \times 10^{-7}\ \mathrm{H/m})}$$

$$= 8.99 \times 10^{16}\ \mathrm{m^2/s^2}.$$

So the speed v, equal to the speed of light c, is:

$$c \equiv \sqrt{1/(\epsilon_0\mu_0)} = 3.00 \times 10^8\ \mathrm{m/s}. \qquad \mathbf{(33.7)}$$

This remarkable result shows that the speed of light is determined entirely by local electrodynamic properties of space. It not only confirms the electromagnetic nature of light, but it led Einstein to develop the special theory of relativity that profoundly changed our ideas of space and time (see Chapter 34). In accord with that theory, c is now taken to have a fixed value (cf. §1.2.4). Since the value of μ_0 is fixed by the definition of the ampere (§29.2.1), eqn. (33.7) determines the value of ϵ_0.

EXERCISE 33.1 ♦♦ The partial derivatives $\partial^2/\partial x\, \partial t$ and $\partial^2/\partial t\, \partial x$ are equal for any function. Demonstrate this for the harmonic wave function (eqn. 33.2).

EXERCISE 33.2 ♦♦ Differentiate eqn. (33.4) with respect to time and eqn. (33.5) with respect to x to show that B satisfies the same wave equation as E.

The harmonic functions (33.2) are solutions of the wave eqn. (33.6), provided that $\omega/k = c$. The amplitudes E_0 and B_0 are not independent. Using eqn. (33.2):

CHECK THIS BY DIFFERENTIATING EQNS. (33.2) AND SUBSTITUTING INTO EQN. (33.6).

$$\frac{\partial E}{\partial x} = -kE_0 \sin(kx - \omega t) \quad \text{and} \quad \frac{\partial B}{\partial t} = \omega B_0 \sin(kx - \omega t).$$

These relations are consistent with eqn. (33.4), provided that $kE_0 = \omega B_0$, so:

$$B_0 = E_0/c. \tag{33.8}$$

Furthermore, eqns. (33.4) and (33.5) each require that the phases of E and B are the same, as we have assumed. This completes the demonstration that plane waves traveling at the speed of light are a solution of Maxwell's equations in vacuum. Plane waves are not the only solutions, but as we mentioned in Chapter 15, any wave disturbance may be expressed as a sum of harmonic plane waves.

Equations (33.2) are solutions of Maxwell's equations for any value of amplitude and frequency. Thus Maxwell's theory not only explains visible light, but predicts EM waves at all frequencies. Maxwell's theory was triumphantly confirmed in 1887 when Hertz produced and then detected radio waves in the laboratory. By measuring both wavelength and frequency independently, Hertz showed that the radio waves move at the speed of light. We discussed the EM spectrum and the properties that EM waves share with other waves in Chapter 16. In the rest of this chapter, we shall investigate their specific electromagnetic properties.

EXAMPLE 33.1 ♦ An EM wave in the visible part of the spectrum has a wavelength of 550 nm and an electric field amplitude of 670 V/m. Determine the frequency of the wave and the magnetic field amplitude. If the wave travels in the $+x$-direction and its phase is zero when x and t are both zero, write expressions for $E(x, t)$ and $B(x, t)$.

MODEL We use the given information in eqns. (33.2), (33.3), and (33.8).

SETUP AND SOLVE From eqns. (33.3) and (33.7):

$$c = \omega/k = \lambda f = (550\ \text{nm})f = 3.00 \times 10^8\ \text{m/s}.$$

Thus:

$$f = \frac{3.00 \times 10^8\ \text{m/s}}{550 \times 10^{-9}\ \text{m}} = 5.5 \times 10^{14}\ \text{Hz},$$

and

$$\omega = 2\pi f = 3.4 \times 10^{15}\ \text{rad/s}.$$

We use eqn. (33.8) for the magnetic field amplitude:

$$B_0 = \frac{E_0}{c} = \frac{670\ \text{V/m}}{3.0 \times 10^8\ \text{m/s}} = 2.2\ \mu\text{T}.$$

Finally, $k = 2\pi/\lambda = 1.1 \times 10^7$ rad/m. Thus we may apply eqns. (33.2) to this wave.

$$E(x, t) = (670\ \text{V/m})\cos[(1.1 \times 10^7\ \text{rad/m})x - (3.4 \times 10^{15}\ \text{rad/s})t],$$

and

$$B(x, t) = (2.2\ \mu\text{T})\cos[(1.1 \times 10^7\ \text{rad/m})x - (3.4 \times 10^{15}\ \text{rad/s})t].$$

ANALYZE The directions of the electric and magnetic fields are at right angles. We were not given enough information in the problem statement to write vector equations for $\vec{\mathbf{E}}$ and $\vec{\mathbf{B}}$. The magnitude of B is about one-tenth of the Earth's field (cf. Table 28.1). ▮

33.2 Energy and Momentum Transport by EM Waves

33.1.2 *Energy Density and the Poynting Vector*

The warmth of sunshine on a cold winter day and your summer sunburn are both the result of electromagnetic energy transport. Energy is stored in a wave by the electric and magnetic fields, and is transported as the field pattern moves. The EM energy density in the wave (eqn. 31.15) is:

$$u = u_E + u_B = \epsilon_0 \frac{E^2}{2} + \frac{B^2}{2\mu_0}.$$

In an EM wave, the magnitudes of E and B are related ($B = E/c$, eqn. 33.8), so:

$$u_B = \frac{B^2}{2\mu_0} = \frac{\epsilon_0 B^2}{2\epsilon_0\mu_0} = \tfrac{1}{2}\epsilon_0 c^2 \left(\frac{E}{c}\right)^2 = \tfrac{1}{2}\epsilon_0 E^2 = u_E.$$

The electric and magnetic fields have equal energy density, and their total is:

$$u = \epsilon_0 E^2 = \epsilon_0 E_0^2 \cos^2 (kx - \omega t).$$

The distribution of energy in space at a time t is shown in ■ Figure 33.4. At a later time $t + dt$, the distribution has moved to the right a distance $c\,dt$ as the wave carries energy with it. The energy crossing a plane surface of area A perpendicular to the wave in time dt is the energy contained in a volume of thickness $c\,dt$ along the wave:

$$dU = u\,dV = uAc\,dt.$$

The wave *intensity* is the power it transfers per unit area:

WAVE INTENSITY WAS DEFINED IN §16.3.

$$\frac{1}{A}\left(\frac{dU}{dt}\right) = cu = c\epsilon_0 E^2 = c^2\epsilon_0 EB = \frac{EB}{\mu_0}.$$

REMEMBER: $c^2 = 1/(\epsilon_0\mu_0)$.

The energy moves in the direction of $\vec{\mathbf{k}}$—that is, in the direction of $\vec{\mathbf{E}} \times \vec{\mathbf{B}}$. Thus the rate and direction of energy transport per unit area are both described by the *Poynting vector:*

NAMED AFTER JOHN HENRY POYNTING (1852–1914).

$$\vec{\mathbf{S}} \equiv \frac{\vec{\mathbf{E}} \times \vec{\mathbf{B}}}{\mu_0}. \tag{33.9}$$

EXAMPLE 33.2 ◆ Find the average rate at which energy is transported per unit area by a plane EM wave with amplitude $E_0 = 17$ V/m.

MODEL We use eqn. (33.9) together with eqn. (33.8) relating the electric and magnetic field amplitudes.

SETUP The magnitude of the Poynting vector is:

$$|\vec{\mathbf{S}}| = \frac{|\vec{\mathbf{E}} \times \vec{\mathbf{B}}|}{\mu_0} = \frac{|\vec{\mathbf{E}}||\vec{\mathbf{B}}|}{\mu_0} = \frac{|\vec{\mathbf{E}}|^2}{\mu_0 c} = \frac{E_0^2}{\mu_0 c}\cos^2 (kx - \omega t).$$

SOLVE The average value of the cosine squared is $\tfrac{1}{2}$, so:

$$\langle|\vec{\mathbf{S}}|\rangle = \frac{E_0^2}{2\mu_0 c} = \frac{(17\ \text{V/m})^2}{2(4\pi \times 10^{-7}\ \text{H/m})(3.0 \times 10^8\ \text{m/s})} = 0.38\ \text{W/m}^2.$$

ANALYZE Compare this with the intensity of a 60-W light bulb: $\langle S\rangle = (60\ \text{W})/(4\pi r^2) = 0.38$ W/m^2 at $r = 3.5$ m. ■

EXAMPLE 33.3 ◆◆ The Sun is 1.5×10^{11} m from the Earth (on average), and its electromagnetic power (its *luminosity*) is $L = 3.9 \times 10^{26}$ W. What is the mean amplitude of the electric field in radiation from the Sun at the top of the Earth's atmosphere?

MODEL The mean solar energy transported across unit area per unit time is the average magnitude of the Poynting vector at Earth.

SETUP The energy from the Sun passes through a sphere of radius $4\pi R^2$, where R is the Earth–Sun distance (cf. §16.3.3). So:

$$\langle S\rangle = \frac{L}{4\pi R^2} = \frac{3.9 \times 10^{26}\ \text{W}}{4\pi(1.5 \times 10^{11}\ \text{m})^2} = 1.4 \times 10^3\ \text{W/m}^2.$$

But $\langle S\rangle$ is also related to the electric field:

$$\langle S\rangle = E_0^2/(2\mu_0 c).$$

(a)

(b)

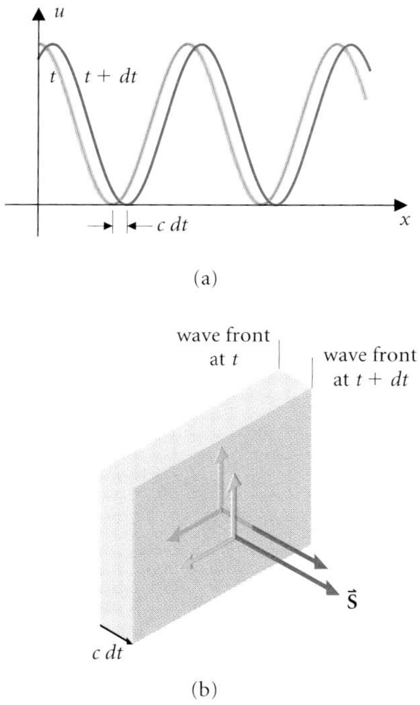

■ **FIGURE 33.4**
(a) Distribution of EM energy in an EM wave propagating in the x-direction. *Blue line:* distribution at time t. *Red line:* distribution at time $t + dt$. As the wave moves, the energy moves with it at speed c in the direction of the wave vector $\vec{\mathbf{k}}$. (b) The energy that passes through a surface of area A in time dt is the energy stored in a volume of thickness $c\,dt$.

$\langle S\rangle = 1400$ W/m^2 IS CALLED THE *SOLAR CONSTANT.*

SOLVE Solving for E_0 and using the calculated value of $\langle S\rangle$:

$$E_0 = \sqrt{2\mu_0 c\langle S\rangle}$$
$$= \sqrt{2(4\pi \times 10^{-7}\ \mathrm{H/m})(3.0 \times 10^{8}\ \mathrm{m/s})(1.4 \times 10^{3}\ \mathrm{W/m^2})}$$
$$= 1.0 \times 10^{3}\ \mathrm{V/m}.$$

ANALYZE This electric field magnitude is about 10^{-3} of the maximum electric field in a capacitor. Note that this is an average electric field amplitude due to all waves at all frequencies. Since $\langle S\rangle \propto 1/R^2$ (the inverse square law, §16.3.3), $E_0 \propto 1/R$. ■

33.2.2 *Momentum Density and Radiation Pressure*

Material particles in motion have linear and angular momentum as well as energy. The same is true of EM waves. Light, absorbed or reflected by a surface, transfers momentum to the surface producing a radiation pressure.

COMPARE WITH THE DEFINITION OF FLUID PRESSURE, §13.2.2, EQN. (13.3).

Radiation pressure is the force per unit area exerted normal to a surface by EM waves.

While basking in the summer sunshine, you absorb nearly a kilowatt of power (cf. Example 33.3), yet you do not feel pushed toward your shadow! Radiation pressure is usually smaller than fluid pressure because radiation has a much smaller ratio of momentum to energy than does an ordinary fluid particle. For a particle with speed v, the ratio is $p/K = mv/(\frac{1}{2}mv^2) = 2/v$. For EM radiation, $p/U = 1/c$.

A wave incident normally on a surface of area A carries momentum across the surface at a rate:

$$\frac{d\vec{\mathbf{p}}}{dt} = A\,\frac{\vec{\mathbf{S}}}{c}. \qquad \mathbf{(33.10)}$$

While radiation pressure is negligible in everyday situations, it is crucial in the atmospheres of very luminous stars and in nuclear fusion research experiments where the pressure of laser light compresses target pellets to a very high density.

EXAMPLE 33.4 ♦ Find the radiation pressure exerted on a mirror by sunlight incident normally.

MODEL The pressure is the force per unit area exerted by the light, and the force is the rate of momentum transfer. If we assume that the mirror is a perfect reflector, then the momentum of the wave is reversed, and $\Delta\vec{\mathbf{p}} = \vec{\mathbf{p}}_{\mathrm{out}} - \vec{\mathbf{p}}_{\mathrm{in}} = -2\vec{\mathbf{p}}_{\mathrm{in}}$. An equal and opposite momentum is transferred to the mirror.

SETUP Thus momentum is transferred to a unit area of the mirror at a rate (eqn. 33.10):

$$\frac{|\Delta\vec{\mathbf{p}}|}{A\,\Delta t} = \frac{|\vec{\mathbf{S}}_{\mathrm{in}} - \vec{\mathbf{S}}_{\mathrm{out}}|}{c} = \frac{2|\vec{\mathbf{S}}_{\mathrm{in}}|}{c}.$$

SOLVE Using the value of S from Example 33.3, the pressure is:

$$P = \frac{2(1.4 \times 10^{3}\ \mathrm{W/m^2})}{3.0 \times 10^{8}\ \mathrm{m/s}}$$
$$= 9.3 \times 10^{-6}\ \mathrm{N/m^2}.$$

ANALYZE Compare this value with typical fluid pressures: 1 atm $\approx 10^5\ \mathrm{N/m^2}$. ■

Digging Deeper

The Momentum of Light

The ratio of momentum to energy in an EM wave is $1/c$. A rigorous proof of this fact requires mathematics beyond the scope of this text, but we can see that it makes sense by looking at a specific example. We calculate the force an EM wave exerts on a weakly conducting medium that contains a low density of charged particles. The wave causes these particles to move and hence to absorb some of the wave's energy and momentum. (A negligible amount of energy and momentum is reflected.) Radiation from the Sun traveling through interplanetary space toward Earth is an example of this situation.

Energy and momentum are both conserved in the interaction of the wave with the medium. Thus the energy gained by the medium is equal to the energy lost by the wave. The same is true of the momentum. By finding the ratio of energy to momentum absorbed, we can deduce the ratio of energy to momentum in the wave.

At a fixed position, $x = 0$, the wave's electric field oscillates in time:

$$\vec{E}(t) = \vec{E}_0 \cos(\omega t).$$

If the conductivity of the medium is σ, the electric field drives a current density:

$$\vec{j}(t) = \sigma\vec{E}(t) = \sigma\vec{E}_0 \cos(\omega t),$$

consisting of n electrons per unit volume, each moving with an average drift velocity (eqn. 26.6, ■ Figure 33.5):

$$\vec{v}_d = \frac{\vec{j}}{-en} = -\frac{\sigma\vec{E}_0}{en}\cos(\omega t).$$

Each electron feels an electric force $\vec{F}_E = -e\vec{E}$. So the power per unit volume delivered to the electrons by the electric field is:

$$\frac{dW}{dt} = n\vec{F}_E \cdot \vec{v}_d = n(-e\vec{E}) \cdot \vec{v}_d$$

$$= -ne\vec{E}_0 \cos(\omega t) \cdot \left[-\frac{\sigma\vec{E}_0}{ne}\cos(\omega t)\right]$$

$$= \sigma E_0^2 \cos^2(\omega t).$$

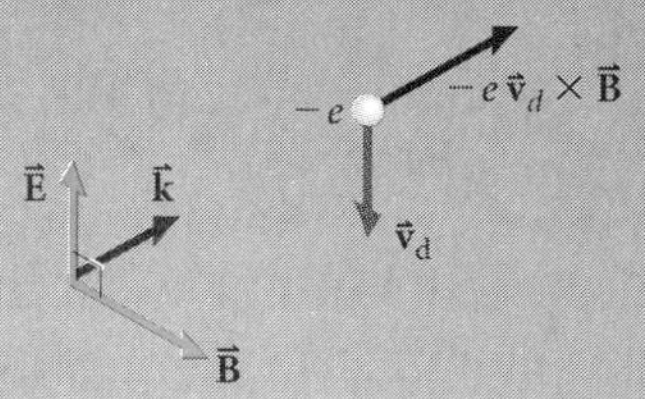

■ Figure 33.5
When an electric field encounters an electron, the electron is accelerated opposite $\vec{E}$ and gains a drift velocity $\vec{v}_d$ through the surrounding material. The magnetic force $-e\vec{v}_d \times \vec{B}$ is in the direction of wave propagation.

The magnetic force perpendicular to $\vec{v}$ does no work, and so delivers no power to the electrons.

Because the electric field oscillates, the net momentum it delivers to the electrons averages to zero. The magnetic field of the waves also exerts force on the drifting electrons. The momentum absorbed by the electrons per unit volume is:

$$\frac{d\vec{p}}{dt} = \vec{F}_B = n(-e\vec{v}_d \times \vec{B})$$

$$= -ne\left[-\frac{\sigma\vec{E}_0}{ne}\cos(\omega t)\right] \times \vec{B}_0\cos(\omega t)$$

$$= \sigma\vec{E}_0 \times \vec{B}_0 \cos^2(\omega t).$$

The rate of momentum transfer divided by the power absorption is:

$$\frac{|d\vec{p}/dt|}{dW/dt} = \frac{E_0 B_0}{E_0^2} = \frac{B_0}{E_0} = \frac{1}{c}.$$

This ratio is independent of the wave amplitude, the conductivity, or any other feature of the specific material. The ratio of momentum to energy gained by the medium from the wave is constant because it is a property of the *wave:*

$$\frac{|\vec{p}_{wave}|}{U_{wave}} = \frac{1}{c}.$$

EXAMPLE 33.5 ♦♦ A laser beam with $S = 1.0 \times 10^6\ \mathrm{W/m^2}$ is incident normally on a sheet of plastic; 70% is reflected and 30% is absorbed. Find the radiation pressure on the plastic.

MODEL When light is absorbed ($\vec{p}_{out} = 0$), the change in momentum is:

$$\Delta\vec{p} = \vec{p}_{out} - \vec{p}_{in} = 0 - \vec{p}_{in} = -\vec{p}_{in},$$

half the result for reflected light. In this example we must treat the reflected light and the absorbed light separately.

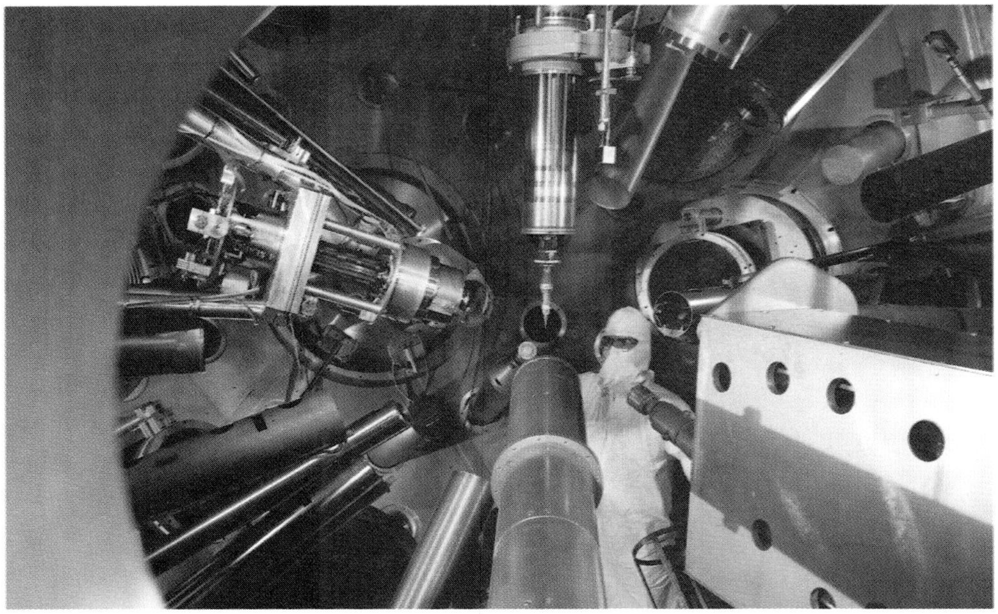

■ FIGURE 33.6
A technician inside the target chamber of the NOVA laser at the Lawrence Livermore Laboratory. A pellet of nuclear fuel is compressed by the radiation pressure due to the laser beams. Ten laser beams with a total power of 10^{14} W converge on the 0.05-mm-diameter pellet.

SETUP For the reflected fraction, $P_r = f_r(2S/c)$.
For the absorbed fraction, $P_a = (1 - f_r)S/c$.

SOLVE The total pressure exerted on the plastic is:

$$P = P_r + P_a = (1 + f_r)\frac{S}{c} = (1.7)\frac{1.0 \times 10^6 \text{ W/m}^2}{3.0 \times 10^8 \text{ m/s}} = 5.7 \times 10^{-3} \text{ N/m}^2.$$

ANALYZE Even lasers don't push very hard.

EXERCISE 33.3 ♦ Find the radiation pressure exerted by the NOVA laser with a power of 10^{14} W focused onto and absorbed by a deuterium/tritium pellet 0.05 mm in diameter (■ Figure 33.6).

33.2.3 *Energy Transport in Circuits*

Energy provided by the battery in a DC circuit is dissipated in resistance, but how does it get from one place to another? The metal wires provide a path for current around the circuit, and one might expect the energy to flow through the metal as well, but that's not the case. The Poynting vector shows where energy flows.

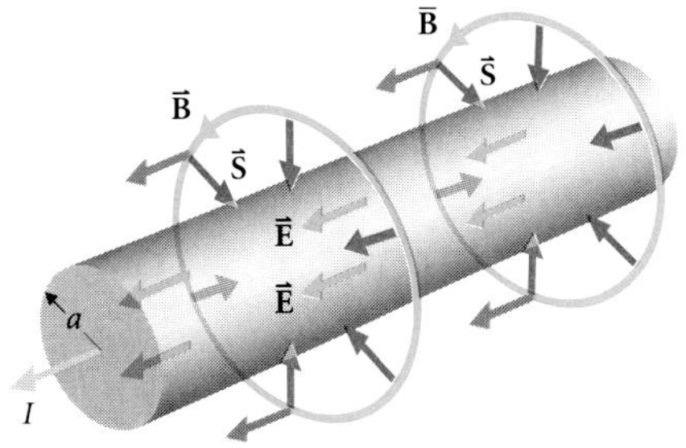

■ FIGURE 33.7
The electric field parallel to a long straight wire drives current in the wire. The magnetic field produced by the current circulates around the wire. The Poynting vector $\vec{S} = \vec{E} \times \vec{B}/\mu_0$ points inward and delivers the energy that heats the wire.

■ Figure 33.7 shows a constant current in a long straight wire with nonzero resistance. A uniform electric field $\vec{E}$ in the wire drives the current, and the current produces a magnetic field that circulates around the wire. The Poynting vector $\vec{S} = (\vec{E} \times \vec{B})/\mu_0$ is perpendicular to the wire's surface and shows that energy flows into the wire from the surrounding space.

The Poynting vector gives the energy flow per unit area per unit time, so we can use it to compute the amount of energy that flows into a length ℓ of the wire with resistance R. Because the fields are cylindrically symmetric around the wire, $|\vec{S}|$ is constant over the surface of the wire, and the energy inflow is $|\vec{S}|$ times the surface area:

NOTE: P HERE MEANS POWER NOT PRESSURE.

$$\frac{dU}{dt} = P = 2\pi a\ell|\vec{S}| = 2\pi a\ell\,\frac{EB}{\mu_0}.$$

This energy is used as joule heating in the wire. To see this, we should express the result in terms of current and resistance. At the surface of the wire,

$$B = \mu_0 I/(2\pi a).$$

Digging Deeper

Oblique Incidence

When light is incident at an oblique angle, the energy reaching an area dA of the surface is the amount crossing area $dA_\perp = dA \cos\theta$ of the wave front (■ Figure 33.8). Thus the rate at which energy reaches the surface is reduced by a factor $\cos\theta$:

$$\frac{dU}{dt}(\text{per unit area}) = |\vec{\mathbf{S}} \cdot \hat{\mathbf{n}}| = S\cos\theta.$$

Only the component of momentum perpendicular to the surface, $p_\perp = p\cos\theta$, contributes to radiation pressure. Including both effects, the radiation pressure is reduced by a factor $\cos^2\theta$.

EXAMPLE 33.6 ♦♦ A microwave beam with an intensity of 98 kW/m^2 is incident on a plane metal surface at a 30° angle and is reflected. What is the radiation pressure exerted on the surface?

MODEL Momentum is transferred to the surface at a rate

$$dp_\perp/dt = 2S\cos^2\theta/c.$$

The factor of 2 appears because the beam is reflected, and the $\cos^2\theta$ accounts for the oblique incidence.

SETUP $P_r = 2S\cos^2\theta/c$

SOLVE $= 2(98 \times 10^3\ \text{W/m}^2)\cos^2(30°)/(3.0 \times 10^8\ \text{m/s})$
$= 4.9 \times 10^{-4}\ \text{N/m}^2.$

ANALYZE The 30° angle of incidence reduces the pressure by 25% compared with normal incidence. ▮

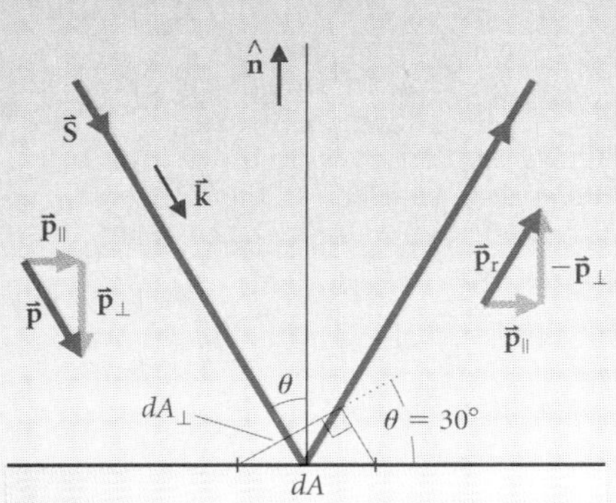

■ **Figure 33.8**
When light is incident at an angle θ, the radiation pressure is reduced by a factor of $\cos^2\theta$:
(i) the energy reaching area A of the surface is the energy crossing area $dA_\perp = dA\cos\theta$ of the wave front; and
(ii) only the normal component $p\cos\theta$ of the momentum contributes to the pressure.

The product $E\ell$ is the potential difference ΔV between the ends of the wire segment, which is related to the segment's resistance R:

Remember: $\vec{\mathbf{E}}$ *is uniform.*

$$E\ell = \Delta V = IR.$$

Thus the power flowing into the wire is:

$$P = 2\pi a(IR)\frac{\mu_0 I}{\mu_0 2\pi a} = I^2R.$$

Some of the thermal energy produced leaves the wire as infrared radiation. Thus $\vec{\mathbf{S}}$ actually consists of a DC inward component and an outward component that oscillates at infrared frequencies.

All of the EM energy converted to thermal form in the wire is brought in from the surrounding space by the fields. The current and the charge distribution are necessary for the fields to exist, but energy is not carried by the charge itself. Remember, the electron drift velocity is so small ($v_d \sim 10^{-3}$ m/s, §26.2) that electrons would take hours to reach a light bulb from the switch across the room, yet the light comes on almost instantaneously.

EXERCISE 33.4 ❖ For any steady current, show that the Poynting vector is everywhere parallel to the equipotential surfaces (cf. §25.4). ▮

EXAMPLE 33.7 ❖ A battery and resistor are connected at opposite ends of a diameter of a circular wire loop. Assuming the wire is perfectly conducting, sketch the Poynting vectors in the plane of the figure.

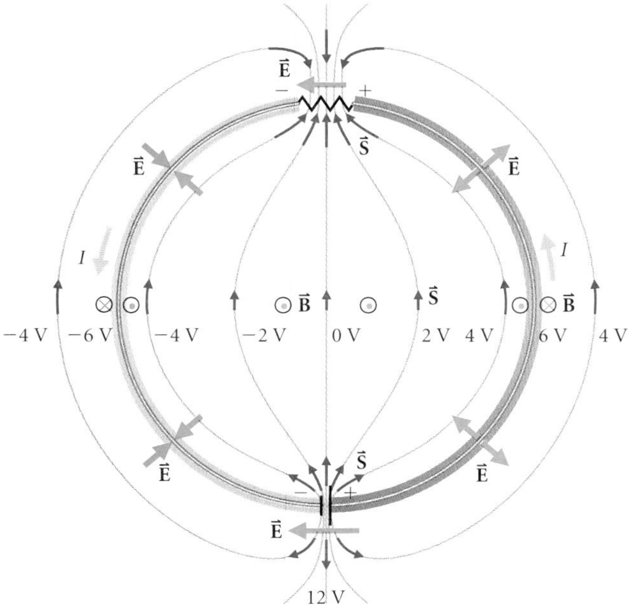

■ **FIGURE 33.9**
Poynting flux in a DC circuit. The connecting wires are assumed to be perfectly conducting. Electric field lines run from right to left across the circuit. Magnetic field lines circle the wire and come up out of the page inside the current loop (cf. Figure 28.7). The Poynting vector runs across the circuit from the battery to the resistor. The energy flow is greatest where E and B are greatest: near but outside the perfectly conducting wire.

MODEL See ■ Figure 33.9. The Poynting vector is parallel to the equipotential surfaces (see Exercise 33.4), so we begin by sketching them. There is no electric field within the perfectly conducting wires: it's not necessary because there is no resistance. Thus each semicircle of wire is at a constant potential. The potential increases across the battery and decreases across the resistor. The equipotential surfaces join points at equal potential and so run across the circuit between the battery and the resistor. The electric field, perpendicular to the equipotentials, crosses the circuit from one side to the other. The electric field is perpendicular to the perfectly conducting wires and is largest close to the wires (cf. §24.2.1, §25.6). Near the wire, the magnetic field forms loops around the wire, and at large distances it forms a dipole pattern. The Poynting flux is along the equipotential surfaces and is strongest near the wire, where the electric and magnetic fields are most intense.

ANALYZE Essentially, the fields transport EM energy over the surface of the wires and not through them. The circuit sets up exactly the right field pattern to deliver the energy where it is needed. ■

33.3 POLARIZATION

33.3.1 Linear Polarization

Sunlight or the light from a desk lamp comes from many atoms, each producing a short wave train. The light is the sum of many waves like the one shown in Figure 33.2. The electric fields in the various wave trains bear no special relation to each other, and the light is said to be *unpolarized* (■ Figure 33.10). By contrast, current in a dipole radio antenna flows back and

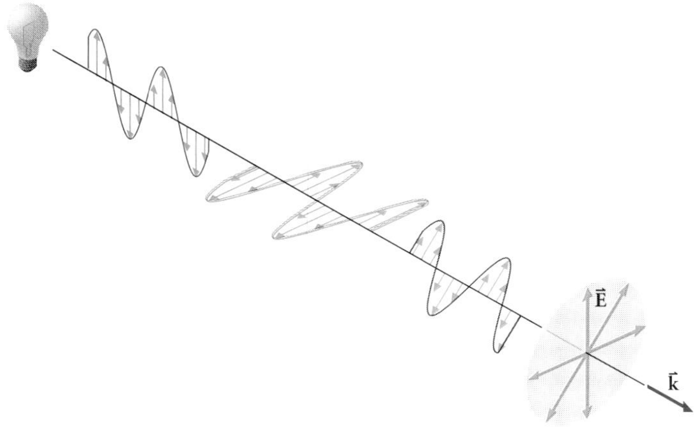

■ **FIGURE 33.10**
Unpolarized light. The electric field vectors in the various wave pulses bear no special relation to each other, except that they are all perpendicular to $\vec{k}$. At any fixed place the direction of $\vec{E}$ varies randomly in time.

forth along a single direction and produces EM waves with electric fields that are all along the same line: the radiation is *linearly polarized.* Certain light sources, such as lasers, also produce linearly polarized light.

Polarized light can be produced from an unpolarized light source by using a *polarizing filter* that transmits only radiation with its electric field along one direction. That direction defines the *transmission axis* of the filter. Unpolarized light falling on the filter emerges linearly polarized but with a reduced intensity. Half the wave energy falling on the filter is absorbed.

To understand how such a filter works, consider microwaves falling on a set of parallel wires (■ Figure 33.11). Waves with electric field parallel to the wires cause current in the wires. Energy is drained from the waves and dissipated as heat. Electric field perpendicular to the wires causes no current, so waves polarized perpendicular to the wires are not absorbed. In an effective filter the separation of the wires is comparable to the wavelength of the radiation, and the wire thickness is much less than the wavelength. A filter for visible light requires spacings of atomic dimensions. A commercial polarizing filter contains molecules of polyvinyl alcohol that are pulled into long parallel lines by stretching a thin sheet. The sheet is then treated with iodine that is absorbed and provides conduction electrons. Each iodized polyvinyl alcohol molecule acts like one of the wires in the microwave filter.

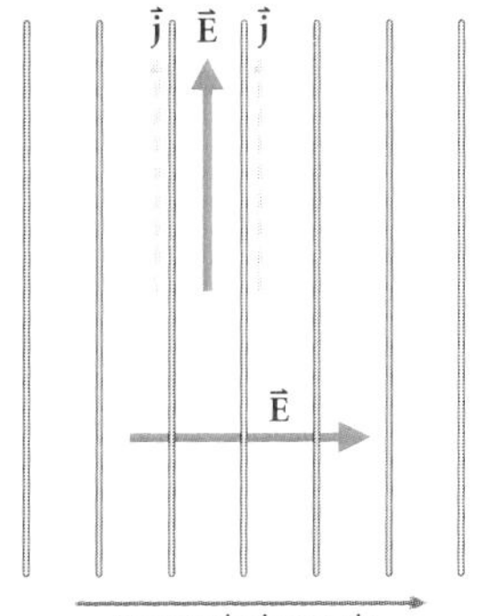

■ FIGURE 33.11
Microwaves incident on a set of parallel wires. Electric field $\vec{E}$ parallel to the wires drives current, and energy is drained from the wave. If $\vec{E}$ is perpendicular to the wires, there is no current and the wave is not absorbed. The transmission axis lies perpendicular to the wires, as shown.

Polarizing filters are also used to analyze the polarization of incident radiation. A wave polarized at an angle to the transmission axis is partially absorbed and partially transmitted. The incident wave's electric field $\vec{E}_i$ has components along, and perpendicular to, the axis (■ Figure 33.12). In effect, it is the sum of two waves, with two perpendicular polarizations whose amplitudes are the two components of $\vec{E}_i$. Only the wave polarized parallel to the filter's axis is transmitted.

If the incident wave is polarized at angle θ to the transmission axis, then the component of E along the axis is $E_{\parallel} = E_i \cos\theta$. The intensity of the waves (given by the Poynting vector) is proportional to the square of the electric field amplitude:

$$I = |\vec{S}| = |\vec{E} \times \vec{B}|/\mu_0 = E^2/\mu_0 c.$$

SEE ALSO §16.3.

So, the transmitted intensity I_t is:

$$I_t = I_i \cos^2\theta, \qquad (33.11)$$

where I_i is the intensity incident on the filter. This relation was discovered by Etienne Malus in 1809; it is known as the law of Malus.

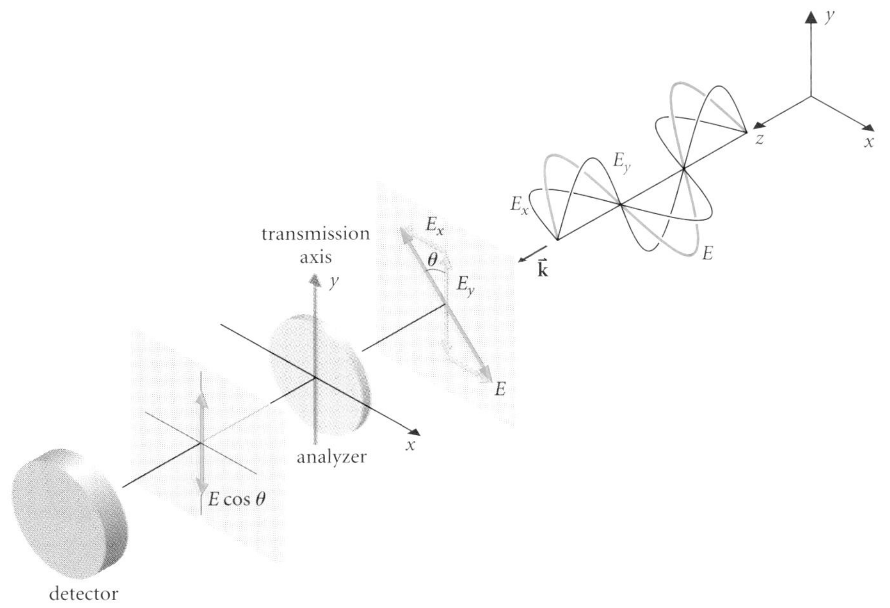

■ FIGURE 33.12
The law of Malus. A polarized wave incident on the analyzer is composed of two waves polarized along, and perpendicular to, the transmission axis. Only the component along the transmission axis is transmitted. Its electric field amplitude is $E \cos\theta$, and the corresponding intensity is $I_i \cos^2\theta$.

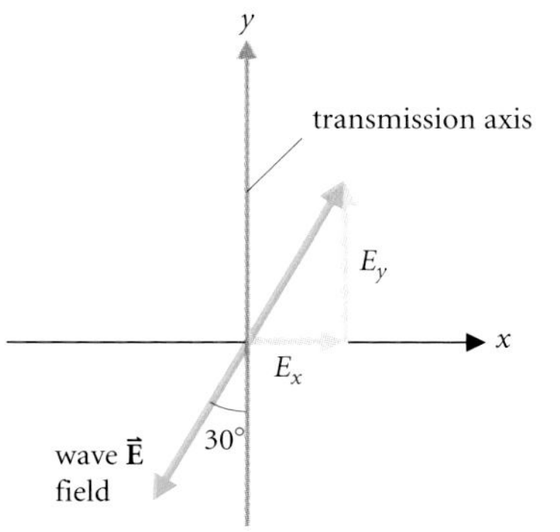

■ FIGURE 33.13
Polarized light is incident on a polarizing filter lying in the plane of the page. Only the component of $\vec{E}$ along the transmission axis is transmitted. We have chosen coordinates with y along the transmission axis.

EXAMPLE 33.8 ♦ Light polarized with $\vec{E}_i$ at 30° to the vertical is incident on a polarizing filter with a vertical transmission axis (■ Figure 33.13). Describe the transmitted wave. What percentage of the incident intensity is transmitted?

MODEL The incident electric field may be resolved into components along (y) and perpendicular (x) to the transmission axis. Only the y-component is transmitted.

SETUP Since $\cos 30° = \sqrt{3}/2$, and $\sin 30° = \frac{1}{2}$: $\vec{E}_i = \frac{E_i}{2}(\hat{x} + \sqrt{3}\,\hat{y})$.

Only the vertical (y) component is transmitted: $\vec{E}_t = \frac{\sqrt{3}}{2} E_i \hat{y}$.

SOLVE The transmitted wave is vertically polarized with an amplitude $\sqrt{3}/2$ of the incident wave. Its intensity is given by:

$$\frac{I_t}{I_i} = \left(\frac{E_t}{E_i}\right)^2 = \frac{3}{4}.$$

Thus 75% of the light is transmitted.

ANALYZE When analyzing the effects of polarizers, always choose components of $\vec{E}$ along and perpendicular to the transmission axis. ■

EXAMPLE 33.9 ♦♦ The transmitted wave in Example 33.8 is incident on a second polarizing filter with its transmission axis at 60° to the first—that is, perpendicular to the wave's original polarization (■ Figure 33.14). What fraction of the original intensity is transmitted by both filters?

MODEL The light emerging from the first filter is vertically polarized. To analyze how it interacts with the second filter, we need to resolve it into the sum of waves polarized perpendicular and parallel to that filter's transmission axis (Figure 33.14b). Only the parallel component is transmitted.

SETUP From Example 33.8, the electric field in the wave incident on the second filter is $\vec{E}_2 = (\sqrt{3}/2)E_i\hat{y}$. Its component parallel to the transmission axis of the second filter is:

$$E_{\parallel,2} = E_2 \cos(60°) = \left(\frac{\sqrt{3}}{2}E_i\right)\frac{1}{2} = \frac{\sqrt{3}}{4}E_i.$$

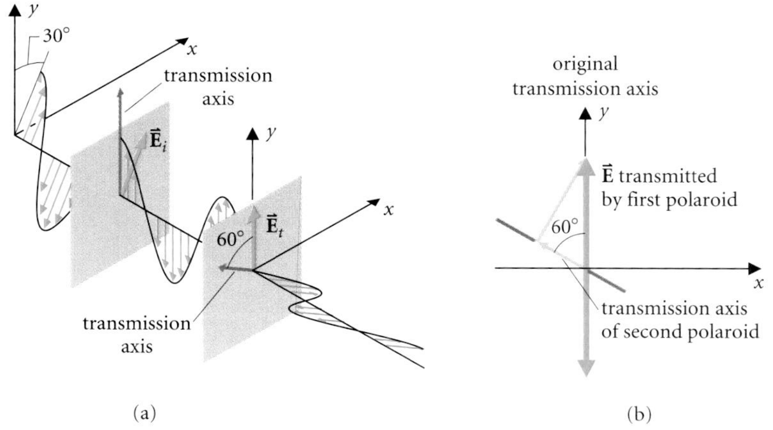

■ FIGURE 33.14
(a) A second filter is placed with its transmission axis at a 60° angle with the first—that is, perpendicular to the polarization of the light incident on the entire system. (b) We resolve the electric field transmitted by the first filter into components along and perpendicular to the transmission axis of the *second* filter. Only the parallel component is transmitted. The beam intensity decreases upon passing through the first filter. Nevertheless, more light passes through the combined system than if the first filter were absent.

SOLVE The transmitted intensity $I_{t,2}$ is proportional to the square of this field:

$$I_{t,2} = \tfrac{3}{16} I_i.$$

ANALYZE The second filter alone would transmit no light, since its transmission axis is perpendicular to the electric field in the wave incident on the entire system. The intermediate filter rotates the polarization of the wave and so allows $\frac{3}{16}$ of the original intensity to pass both filters! ■

EXERCISE 33.5 ♦♦♦ Unpolarized light is incident on a polarizing filter. Use Malus' law to show that 50% of the intensity is transmitted. ■

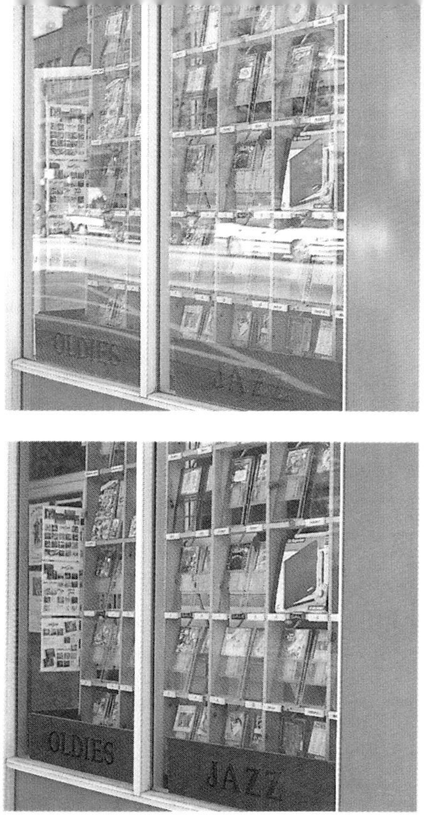

■ FIGURE 33.15
(a) A photograph taken through a window shows unwanted reflections in the glass. (b) A polarizer placed in front of the camera lens reduces the reflections that appear in the photograph because the reflected light is partially polarized.

33.3.2 *Polarization by Reflection*

Reflected light is often highly polarized (■ Figure 33.15). To understand why, we look at what causes reflection and refraction—electrons set oscillating by the wave. These electrons themselves radiate waves polarized parallel to their acceleration. Waves emitted perpendicular to the acceleration have the greatest amplitude; no energy is emitted parallel to the acceleration (cf. Figure 33.1). Thus the direction of the electron motion determines the amplitude and polarization of the reflected wave.

Electrons in the second medium oscillate parallel to the electric field in the transmitted wave. This direction is in turn determined both by the direction of transmitted rays and by the polarization. When the incident wave is polarized parallel to the surface, electrons oscillate parallel to the surface and produce reflected and refracted rays with the same polarization (■ Figure 33.16a). In contrast, when the incident magnetic field is parallel to the surface, the incident electric field makes an angle θ_1 with the surface and the electric field in the transmitted wave makes an angle θ_2 with the surface. The electrons in the second medium oscillate parallel to $\vec{E}_2$, at an angle θ_2 to the surface (Figure 33.16b). The reflected ray makes a small angle ϕ with the direction of electron acceleration, so its amplitude is small.

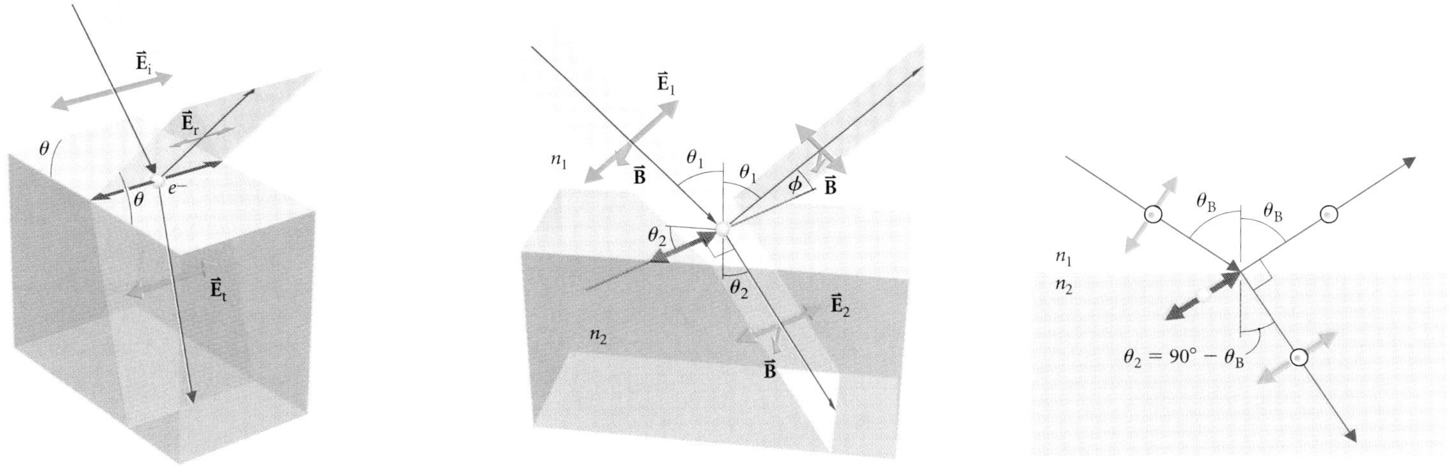

(a) Incident light polarized with $\vec{E}$ parallel to the surface.

(b) Incident light polarized with $\vec{B}$ parallel to the surface.

(c) Reflected ray at right angles to the refracted ray.

■ FIGURE 33.16
Polarization by reflection. (a) Incident light polarized with $\vec{E}$ parallel to the surface. Electron oscillations in the medium are also parallel to the surface, as is the polarization of the reflected and transmitted light. (b) Incident light polarized with $\vec{B}$ parallel to the surface. The electric field lies in the plane of incidence, perpendicular to the ray. Oscillations of electrons in the medium, parallel to $\vec{E}$ in the medium, produced a reflected wave with reduced amplitude. (c) Reflected ray at right angles to the refracted ray. Light polarized with $\vec{B}$ parallel to the surface [cf. (b)] would have to be emitted along the line of the electron acceleration. No radiation is emitted along this line, and there is no reflected ray. Light polarized with $\vec{E}$ parallel to the surface [cf. (a)] is reflected. The incident angle that produces this configuration is called Brewster's angle.

Digging Deeper

Polarization in Nature

Polarization of Skylight

The whole sky looks bright because sunlight is scattered by molecules in the Earth's atmosphere. Because the incident sunlight is unpolarized, electrons in the molecules oscillate in all possible directions in the plane perpendicular to the incident light (■ Figure 33.17). An observer looking at right angles to the Sun is looking along this plane and sees radiation only from electrons oscillating in the single direction at right angles to the line of sight. The observed radiation is polarized (■ Figure 33.18). At other angles, the scattered light is partially polarized. Photographers sometimes use Polaroid filters to reduce the apparent brightness of the sky in their photographs.

Polarization of Starlight

Dust grains in the region between stars scatter and absorb starlight (■ Figure 33.19). Elongated grains tend to align perpendicular to the interstellar magnetic field, and scattered light is polarized with $\vec{\mathbf{E}}$ parallel to the interstellar field direction. By observing the polarization, it is possible to map out the interstellar magnetic field. The observations also allow us to deduce valuable information about the nature of the grains themselves.

■ Figure 33.18
A pair of polarizers with their transmission axes at right angles. The right Polaroid is noticeably darker than the left one, indicating the partial polarization of skylight.

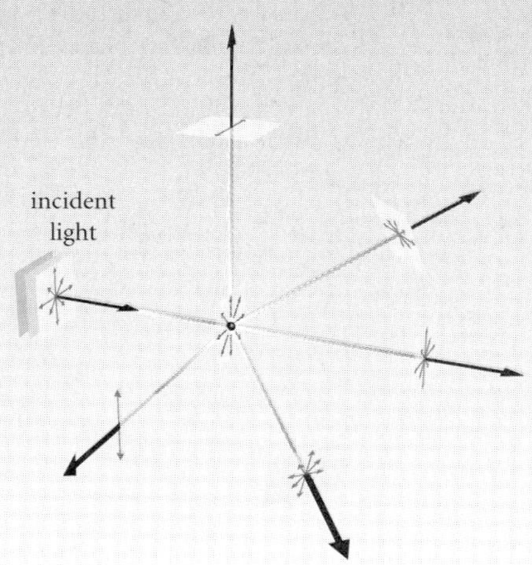

■ Figure 33.17
Light from the Sun is scattered by air molecules to illuminate the whole sky. The molecules oscillate along the electric field direction, perpendicular to the rays, and light emitted in the plane of oscillation is polarized. This is the primary reason why a Polaroid filter on a camera can emphasize clouds, or "cut through" haze.

Interstellar Masers

Water molecules in a giant gas cloud where stars are forming emit very intense, highly polarized microwaves (■ Figure 33.20). The molecules are excited to a high-energy state by an energy source in the cloud, perhaps shock waves, and emit coherent radiation, forming a natural *maser.*

Masers and lasers operate on the same principle though at different frequencies. An electromagnetic wave stimulates the excited molecules to emit radiation in phase with the incident wave and also with the same polarization. Hence the acronym: Microwave Amplification by Stimulated Emission of Radiation.

For one particular incident angle θ_B, known as Brewster's angle, the refracted and reflected rays at are right angles (Figure 33.16c). If the incident light is polarized with $\vec{\mathbf{B}}$ parallel to the surface, the reflected wave would have to be parallel to the oscillating electrons' acceleration. The electrons cannot emit in this direction, so there is no reflected wave. Incident light polarized with $\vec{\mathbf{E}}$ parallel to the surface is reflected as usual. Thus when unpolarized light is reflected from a surface at Brewster's angle, the reflected light is completely polarized parallel to the surface. Light reflected at other angles is partially polarized.

We may obtain an expression for Brewster's angle from Snell's law (eqn. 16.26):

$$n_1 \sin \theta_1 = n_2 \sin \theta_2.$$

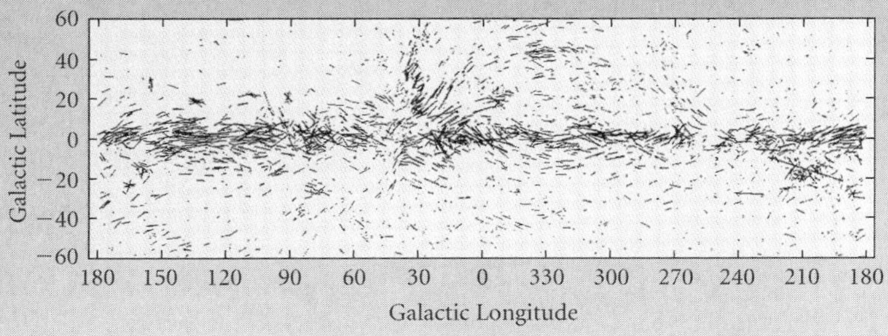

■ FIGURE 33.19

(a) The Horsehead nebula in the constellation of Orion. The dark shape is caused by dust clouds that absorb the light coming from behind them. Elongated dust particles are aligned perpendicular to the local magnetic field.

(b) Light scattered by dust grains, like microwaves scattered by wires (Figure 33.11), is polarized with $\vec{\mathbf{E}}$ perpendicular to the grains and thus parallel to the interstellar magnetic field. Each mark on this diagram shows the measured polarization of light from a single star. Longer lines indicate greater polarization. The diagram maps the direction of the galactic magnetic field. Notice that $\vec{\mathbf{B}}$ tends to lie along the center of the diagram—parallel to the plane of our galaxy (the Milky Way).

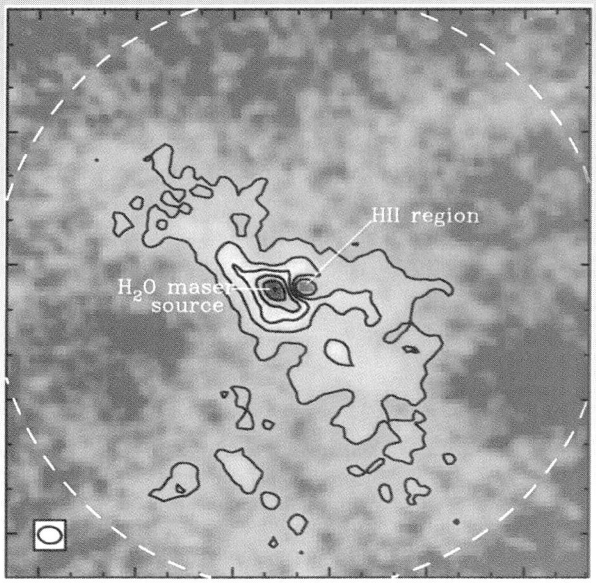

■ FIGURE 33.20

This region of the sky—called W3—is believed to be a birthplace for stars. In the central region, shown in red in this false-color radio map, there is a very bright, highly polarized source of EM waves. Both the intensity of the source and the strong polarization help the identification of this source as a maser.

When the refracted and reflected rays are at right angles (Figure 33.16c), $\theta_2 = 90° - \theta_1$, and Snell's law becomes:

$$n_1 \sin \theta_1 = n_2 \sin(90° - \theta_1) = n_2 \cos \theta_1.$$

Thus, with $\theta_1 = \theta_B$:

$$\tan \theta_B = \frac{n_2}{n_1}. \tag{33.12}$$

Polaroid sunglasses take advantage of the polarization of reflected sunlight. The sunglasses are polarizing filters with a vertical transmission axis. They are particularly effective in reducing the glare from water and highways.

■ FIGURE 33.21
A *smart* antenna array uses 24 helical antennae to receive rocket telemetry data from any direction in any polarization.

EXAMPLE 33.10 ♦ For what angle of incidence is light reflected from water completely polarized?

MODEL The reflected light is completely polarized when the light is incident at Brewster's angle.

SETUP The refractive index of water (Table 16.3) is $n_2 = 1.33$, and $n_1 = 1.00$ is the refractive index of air.

SOLVE Brewster's angle for water is given by eqn. (33.12):

$$\tan\theta_B = n_2/n_1 = 1.33 \qquad \text{and} \qquad \theta_B = 53°.$$

ANALYZE Light incident at other angles is partially polarized, since the reflected light polarized with $\vec{B}$ parallel to the surface has a smaller amplitude than the reflected light polarized with $\vec{E}$ parallel to the surface. ■

33.3.3 *Circular Polarization*

Radio signals transmitted to a satellite from Earth are quite weak by the time they reach its orbit. If the waves were linearly polarized, the receiver antenna would have to be carefully aligned with the wave electric field, a prohibitive requirement for a satellite. Circular polarized waves have a rotating electric field that can be transmitted and received efficiently by a helical antenna (■ Figure 33.21), which is not sensitive to orientation.

Circular polarization can be produced by combining two linearly polarized waves that are 90° out of phase. The electric fields in the two waves are at right angles:

$$\vec{E}_1 = E_0\hat{y}\cos(kx - \omega t) \qquad \text{and} \qquad \vec{E}_2 = E_0\hat{z}\cos(kx - \omega t - \pi/2).$$

The electric field in the superposed wave is:

$$\vec{E} = E_0[\hat{y}\cos(kx - \omega t) + \hat{z}\sin(kx - \omega t)].$$

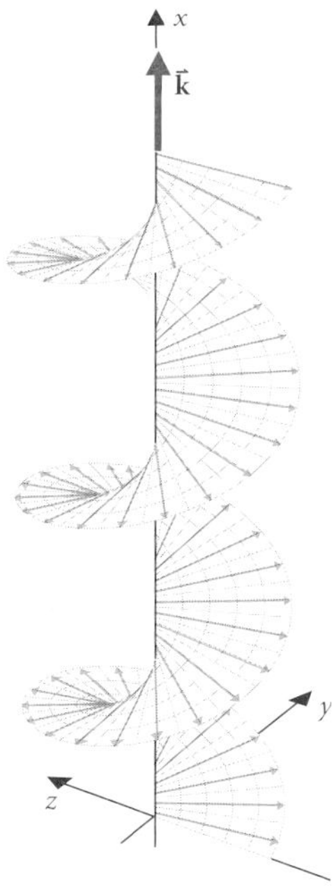

■ FIGURE 33.22
Circular polarization. This picture is a snapshot of the wave at a fixed time. The tips of electric field vectors lie on a helix. In this example the polarization is right circular—with the thumb of your right hand pointing along $\vec{k}$, your fingers indicate the direction of the helix.

■ Figure 33.22 shows how this electric field vector varies in space at one time. The tips of the vectors form a right-handed helix. This polarization is called *right circular.* ■ Figure 33.23 shows how the electric field varies in time at a fixed place. An observer looking toward the source sees the electric field vector rotate clockwise with constant angular speed ω.

If the wave polarized along $\hat{z}$ has a phase lead of 90° instead of a lag, the superposed wave rotates counterclockwise, and the $\vec{E}$ vectors lie on a left-handed helix. Such a wave has *left circular* polarization.

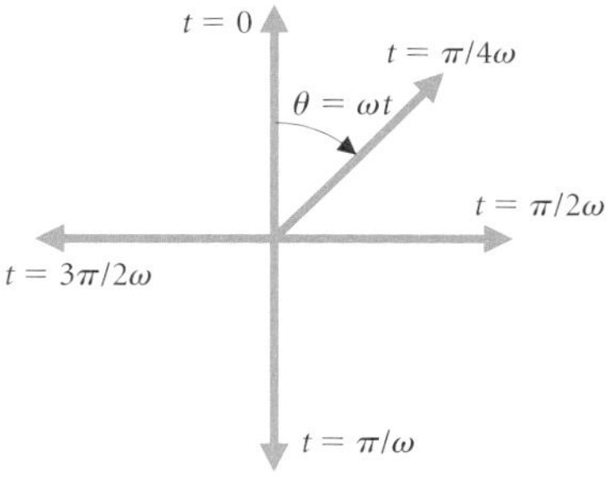

■ FIGURE 33.23
A movie of the wave at a fixed place shows the electric vector rotating. In this right circular wave propagating out of the page toward you, the vector rotates clockwise with angular speed ω. Imagine the helix in Figure 33.22 moving along the axis past you. The electric field vector in the plane of the page changes as the helix moves.

One way to produce circularly polarized light is to pass unpolarized light through a filter in which two linear polarizations travel at different speeds. Calcite is one of a class of substances, called *birefringent,* that have this property. If the calcite crystal has the right thickness, one linear polarization emerges retarded in phase by 90° compared with the other polarization. This filter is called a *quarter-wave plate.*

EXAMPLE 33.11 ❖ Show that a linearly polarized wave is a superposition of right and left circular polarized waves with equal amplitude.

MODEL At a fixed position, the electric field of the right circular wave is rotating clockwise, and $\vec{E}$ in the left circular wave is rotating counterclockwise (■ Figure 33.24). At some time t, both lie along the y-axis. At a later time $t + \Delta t$, they have rotated through equal angles in opposite directions. The sum still lies along the y-axis, although its magnitude is reduced. This is what we expect for a linearly polarized wave.

SETUP AND SOLVE At time $t + \Delta t$, the electric field has magnitude:

$$E = 2E_0 \cos\theta = 2E_0 \cos(\omega\,\Delta t).$$

This is the electric field of a linearly polarized wave with amplitude $2E_0$.

ANALYZE To understand the propagation of a linearly polarized wave through a polarizing filter, we needed to model it as a sum of two waves, each polarized along or perpendicular to the transmission axis (e.g., Example 33.8). Similarly, to understand the propagation of such a wave through a solution of spiral molecules or a magnetized plasma in interstellar space, it is essential to model the wave as a sum of two circular polarizations. (See Problem 51.) This example shows how to do that. Unpolarized light may be modeled as a superposition of linearly or circularly polarized waves.

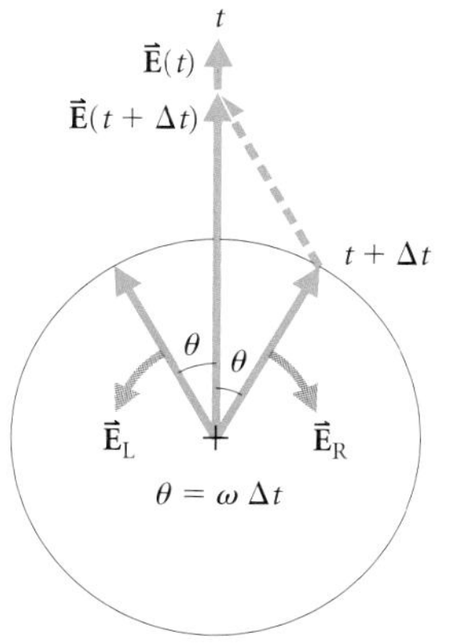

■ FIGURE 33.24
The superposition of a right-circular wave and a left-circular wave, each with the same amplitude E_0 and angular frequency ω, is a linearly polarized wave. In this example the sum of the two electric field vectors always lies along the y-axis and varies between $2E_0\hat{y}$ and $-2E_0\hat{y}$.

⁂ 33.4 ELECTROMAGNETIC OSCILLATIONS AND MICROWAVES

33.4.1 Cavity Oscillators

You need a resonant circuit to detect oscillating radio signals. The *LRC* circuits we discussed in Chapter 32 are effective detectors of FM signals because wavelengths in the FM band, roughly 3 m, are much larger than the size of the circuit. Light can cross the circuit in much less than one oscillation period of the wave and the circuit responds coherently. However, at higher frequencies (shorter wavelengths), this is no longer true. An effective detector or generator for microwaves should be a compact resonant device.

SEE THE DISCUSSION IN §32.1.1.

A cylindrical cavity oscillator (■ Figure 33.25) illustrates how such devices operate. Electromagnetic oscillations can occur in the hollow space surrounded by conducting walls. Like the different standing wave patterns on a string, many different modes of oscillation can occur. Their wavelengths are comparable to the dimensions of the cavity, a few centimeters in the microwave band.

SEE §15.4.2 FOR STANDING WAVE MODES ON A STRING.

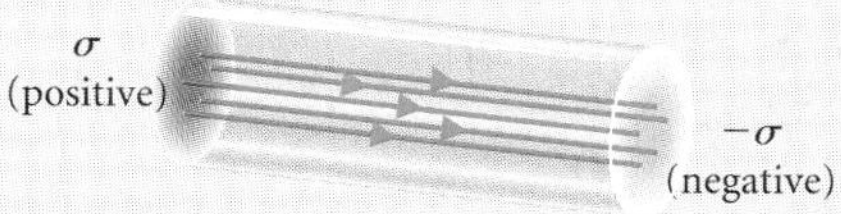

■ FIGURE 33.25
Electromagnetic oscillations in a cylindrical cavity. (a) $t = 0$. Electric field runs along the length of the cavity. Charge density exists on the two ends.

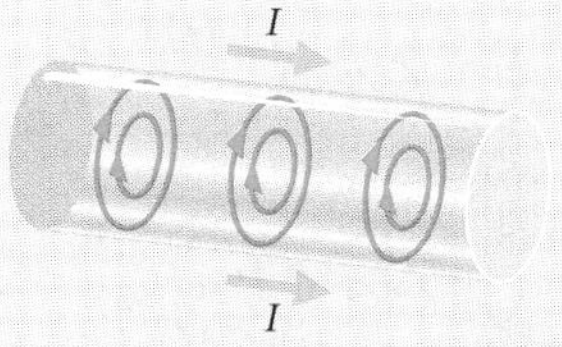

(b) $t = \pi/2\omega = T/4$. As the electric field changes, magnetic field circulates around the displacement current. Current in the conducting walls changes the surface charge density σ.

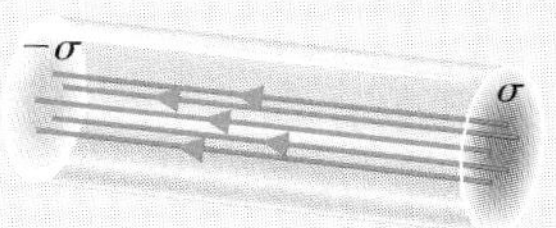

(c) $t = \pi/\omega = T/2$. Electric field and charge density have reversed relative to (a).

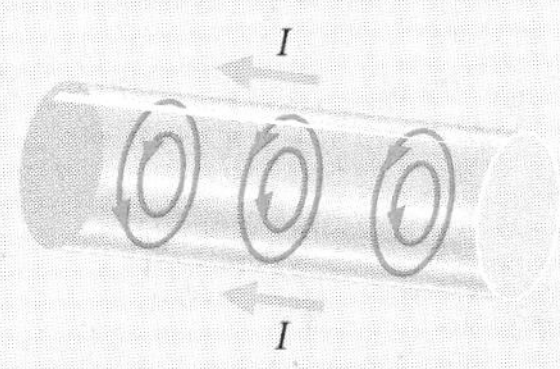

(d) $t = 3\pi/\omega = 3T/4$. Magnetic field circulates around the displacement current: $\vec{B}$ and I are opposite from those in (b).

(a)

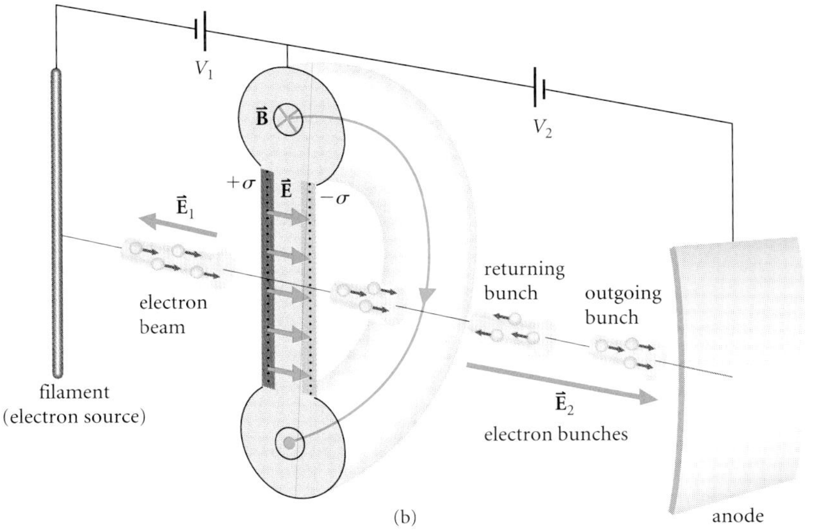

(b)

■ FIGURE 33.26
(a) A Klystron tube. (b) Magnetic field runs around the doughnut-shaped tube, while electric field runs across the doughnut hole. An electron beam passes through a metal grid at the doughnut hole, is reflected by a charged plate, and passes back through the hole. With proper timing of the beam, it transfers its kinetic energy into electromagnetic energy in the cavity.

EACH POSSIBLE MODE OF THE CAVITY HAS A DIFFERENT FREQUENCY, DIFFERENT PATTERN OF FIELDS, AND DIFFERENT EFFECTIVE INDUCTANCE AND CAPACITANCE.

The mode illustrated in Figure 33.25 has electric field lines parallel to the long axis of the cylinder. The electric field is maximum along the axis and zero at the conducting walls. Magnetic field circulates around the displacement current that is produced as the electric field changes. Current flows back and forth along the inner surface of the cavity as the charges on the end faces vary with the electric field. The cavity stores energy in the form of electric and magnetic field, and behaves like an *LC* circuit in which both inductor and capacitor occupy the same volume. Energy loss by the surface current damps the oscillator, just as the resistance in an *LRC* circuit damps its oscillations. An important practical application of resonant cavities is the Klystron tube (■ Figure 33.26), developed as a generator for high-frequency radar signals.

THE MAGNITUDE OF THE ELECTRIC FIELD INSIDE THE CONDUCTOR COMPARED WITH THAT OUTSIDE IS APPROXIMATELY $\sqrt{\omega\epsilon_0/\sigma}$, OR ABOUT $\sqrt{(10^{-19}\text{ s})\omega}$ FOR COPPER.

33.4.2 *Waveguides*

Antennae and transmission lines for microwaves employ waveguides, conducting tubes inside which the microwaves propagate. The structure of EM waves propagating in a guide is different from that of unconfined waves. If the tube's walls were perfectly conducting, the electric field inside any wall would have to be zero. The electric field in the cavity would be exactly normal to the walls (cf. §25.6). Inside ordinary conductors, the field is extremely small and decreases to zero within a small distance from the surface. We shall make very little error if we assume $E = 0$ within the metal. A static magnetic field inside a conductor could exist, but a changing magnetic field would generate an electric field and so cannot occur. Since the normal component of $\vec{B}$ is continuous across any boundary (cf. §28.3.1) and $\vec{B}$ is zero within the conductor, $\vec{B}$ inside the guide is tangential at the surface. Thus the boundary conditions at the surface of the wave guide are:

- $\vec{E}$ is normal to the surface, or zero.
- $\vec{B}$ is tangential to the surface, or zero.

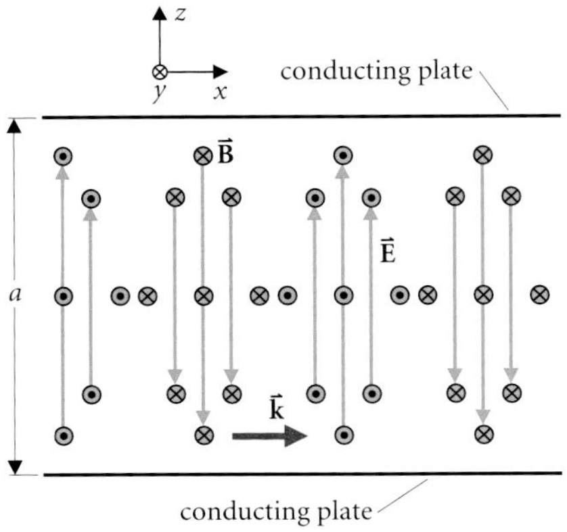

■ FIGURE 33.27
An EM wave propagating between two parallel plates. In the TEM mode, both $\vec{E}$ and $\vec{B}$ are perpendicular to the wave vector, just as in a free-space wave.

We can understand the main features of waveguides by looking at a simple system: parallel metal plates with separation a (■ Figure 33.27). A wave propagating along the guide in the x-direction and polarized with $\vec{E}$ normal to the plates (in the z-direction) and $\vec{B}$ parallel to the plates (the y-direction) satisfies the boundary conditions and so is a possible mode. It is called a *transverse electromagnetic* or TEM mode. A wave polarized with $\vec{E}$ parallel to the plates cannot propagate directly along the guide because $\vec{E}$ is tangential to the surface. Such a *transverse electric* (TE) mode can propagate as a superposition of waves reflecting off the plates as shown in ■ Figure 33.28. Rays (in red) indicate the reflecting waves. Wave fronts, perpendicular to the rays, are indicated by the electric and magnetic field vectors. There is a phase change of π (the electric field reverses direction) at each reflection, which keeps $\vec{E} = 0$ on the guide surface, as required. The components of $\vec{B}$ perpendicular to the plates also sum to zero at the surface, so $\vec{B}$ is tangential. The electric and magnetic fields have maximum strength in the center of the guide (■ Figure 33.29), where the superposition of the magnetic field vectors is perpendicular to the plates.

WE DISCUSSED SUCH PHASE CHANGES IN §15.4 AND §17.1.5.

Only certain reflection angles are possible for a wave with a given frequency because of the way in which waves have to superpose in the guide. Each allowed angle corresponds to a particular wavelength of the disturbance in the guide. The allowed wavelengths are given by:

DIG DEEPER TO SEE HOW THIS WORKS.

$$\lambda_g = \frac{\lambda_f}{\sqrt{1 - (\lambda_f/2a)^2(2m - 1)^2}}, \tag{33.13}$$

where a is the plate separation and m is an integer. The wavelength of any disturbance in the waveguide (λ_g) is greater than the free-space wavelength (λ_f). When $\lambda_f = 2a/(2m - 1)$, the guide wavelength becomes infinite. (This happens when the reflection angle θ is 0.) In our model, this corresponds to the two waves bouncing back and forth in the z-direction, and not traveling in the x-direction at all. If λ_f is greater than $2a$, no mode number m gives a real solution for the wavelength in the guide, and no wave can propagate. The wavelength $\lambda_c = 2a$ is called the *cutoff wavelength* for the guide.

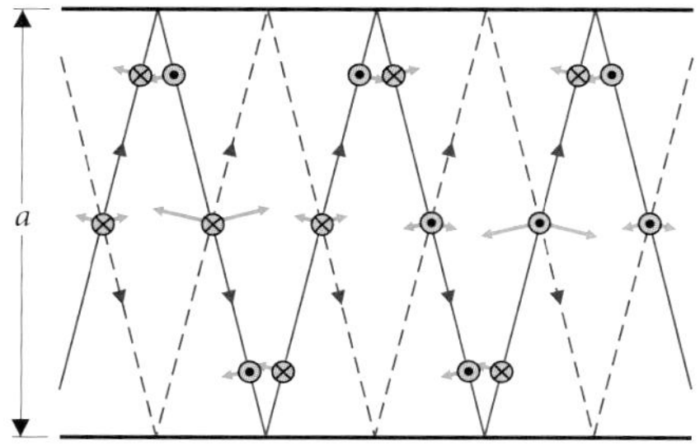

■ FIGURE 33.28
In the TE mode, the electric field is parallel to the plates. Here we model the mode with waves reflecting back and forth off the walls. The electric field vectors sum to zero at each plate, while the magnetic field vectors sum to give $\vec{B}$ parallel to the plates at the surface.

EXAMPLE 33.12 ♦♦ Two parallel conducting slabs are separated by 2.50 cm. What is the cutoff wavelength for this waveguide? What is the wavelength in the guide of radiation with a free-space wavelength of 3.00 cm, propagating in the TE mode?

MODEL The cutoff wavelength is twice the slab separation, or 5.00 cm. Since 3 cm is less than this cutoff wavelength, the 3.00-cm radiation can propagate between the slabs. Allowed wavelengths in the TE mode are given by eqn. (33.13) with integer values of m. However, for $m = 2$, $(\lambda_f/2a)^2(2m - 1)^2 = (\frac{3}{5})^2(3^2) = 3.24 > 1$. For $m \geq 2$, λ_g is not defined, so only the lowest mode ($m = 1$) can propagate.

SETUP We use eqn. (33.13). With $m = 1$, $a = 2.50$ cm, and $\lambda_f = 3.00$ cm:

SOLVE
$$\lambda_g = \frac{3.00 \text{ cm}}{\sqrt{1 - [(3.00 \text{ cm})/(5.00 \text{ cm})]^2}} = 3.75 \text{ cm}.$$

ANALYZE The TEM mode could also propagate. In practice, engineers usually design guides for a single mode. ■

EXERCISE 33.6 ❖ Describe the radiation that emerges when unpolarized radiation with a 6.0-cm wavelength enters the waveguide in Example 33.12.

It is also possible for the reflecting waves in a guide to have $\vec{B}$ parallel to the walls, $\vec{B} = B\hat{y}$. These are called transverse magnetic (TM) modes. In the parallel plate waveguide, the TM modes happen to have the same cutoff frequencies as the TE modes, although in many waveguides there are two different sets of frequencies.

A practical waveguide has two connected sets of plates forming a rectangular cross section. Wave propagation in such a guide shows the same general features that we have discussed for the parallel plates, except that the TEM mode does not propagate. In particular, the TE and TM modes each exhibit cutoffs. Some aspects of these guides are explored in the problems.

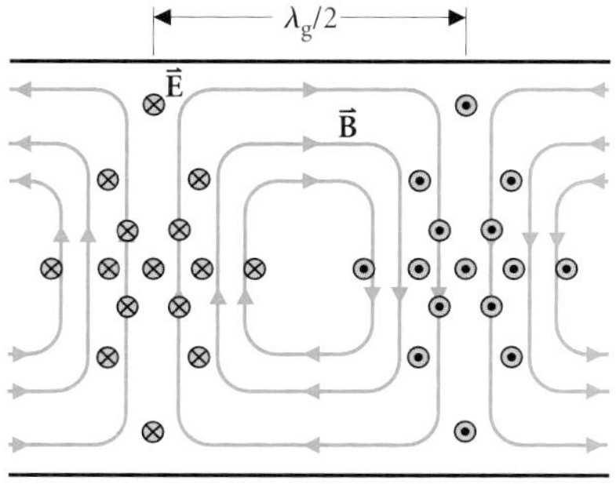

■ FIGURE 33.29
Pattern of electric and magnetic fields in a waveguide. This is the mode described in Figure 33.28. The magnetic field circulates around the displacement current. The magnetic field is tangential at the plates, and the electric field is zero there. The wave propagates to the right. (Check the direction of the Poynting vector.)

Digging Deeper

Superposition of Reflecting Waves

■ Figure 33.30 shows a wave front and a ray of the reflecting wave for an angle of incidence θ that is possible. The electric field must have the same phase everywhere on the wave front PQ. Thus along the ray $APCQ$, the phase difference between P and Q is an integer multiple of 2π. The calculation is easiest if we compute the phase changes from A to P and from A to Q along the ray and then subtract.

The straight-line distance s from A to Q is given by the plate separation a and the angle of incidence θ:

$$s = a \tan \theta.$$

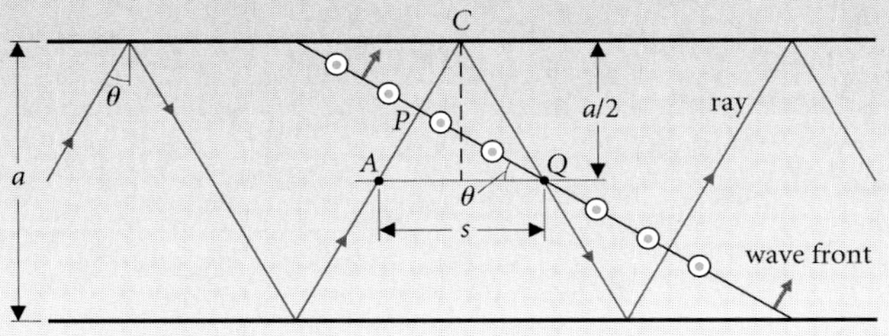

■ **Figure 33.30**
The wave phase is constant along each wave front, such as the wave front PQ shown here. Only for certain angles of incidence θ with $\cos\theta = (2m - 1)\lambda_f/2a$ do the reflecting waves superpose to produce allowed modes in the guide. (Remember that a large number of such wave fronts are superposed in the guide.)

Then the distance AP is $s(\sin\theta) = a(\tan\theta)(\sin\theta)$. The phase difference from A to P is:

$$\phi_P - \phi_A = \frac{2\pi s(\sin\theta)}{\lambda_f} = \frac{2\pi}{\lambda_f}\left(\frac{a \sin^2\theta}{\cos\theta}\right), \qquad \text{(i)}$$

where $\lambda_f = c/f$ is the wavelength of the wave in free space. Now the path length along the ray from A to Q is $a/\cos\theta$, and the phase difference includes a phase shift of π due to reflection at C:

$$\phi_Q - \phi_A = \frac{2\pi}{\lambda_f}\left(\frac{a}{\cos\theta}\right) + \pi. \qquad \text{(ii)}$$

Since P and Q are on the same wave front, separate calculations of the wave phase at the two points can only differ by an integer multiple of 2π:

$$\phi_Q - \phi_P = \frac{2\pi}{\lambda_f}\left(\frac{a}{\cos\theta}\right) + \pi - \frac{2\pi}{\lambda_f}\left(\frac{a\sin^2\theta}{\cos\theta}\right) = 2m\pi,$$

where m is an integer. Then:

$$\lambda_f(2m - 1) = \frac{2a}{\cos\theta}(1 - \sin^2\theta) = 2a\cos\theta,$$

and $$\cos\theta = (2m - 1)\lambda_f/(2a). \qquad \text{(iii)}$$

The corresponding wavelength λ_g of a given mode is the length s along the guide between two points Q and A for which $\phi_Q - \phi_A = 2\pi$. Using eqn. (i) with $s = \lambda_g$:

$$2\pi = 2\pi(\lambda_g/\lambda_f)\sin\theta.$$

So, using eqn. (iii):

$$\lambda_g = \frac{\lambda_f}{\sin\theta} = \frac{\lambda_f}{\sqrt{1 - \cos^2\theta}} = \frac{\lambda_f}{\sqrt{1 - (\lambda_f/2a)^2(2m - 1)^2}}.$$

Chapter Summary

Where Are We Now?

We have completed our exploration of electromagnetism and demonstrated its unity with optics. Throughout this part we have used simple models of the atomic structure of matter to show how electromagnetic theory can explain phenomena. To proceed further, we need the quantum theory of atoms and molecules (Chapter 35) and more advanced mathematical methods.

What Did We Do?

An oscillating electron produces EM waves. At large distances from the electron, the waves are plane, with electric and magnetic fields perpendicular and $\vec{E} \times \vec{B}$ in the direction of wave motion. Maxwell's equations show that the wave speed in vacuum is the speed of light,

$$c = 1/\sqrt{\mu_0\epsilon_0} = 3 \times 10^8 \text{ m/s}.$$

The magnetic field amplitude is related to the electric field amplitude by $E_0 = cB_0$.

The energy flux per unit area in an EM wave or any electromagnetic field is described by the Poynting vector:

$$\vec{\mathbf{S}} = \vec{\mathbf{E}} \times \vec{\mathbf{B}}/\mu_0.$$

When a wave is reflected or absorbed, it exerts a radiation pressure because it transfers momentum. The momentum flux in the wave is $\vec{\mathbf{S}}/c$, and the radiation pressure exerted on a surface is:

$$P = (K/c)|\vec{\mathbf{S}}|\cos^2\theta,$$

where θ is the angle of incidence at the surface and K is a constant that equals 2 for total reflection or 1 for total absorption.

The direction of the electric field vector describes the *polarization* of a wave. When all the $\vec{\mathbf{E}}$ vectors oscillate along a single direction, the wave is linearly polarized. When the $\vec{\mathbf{E}}$ vector rotates, forming a circular helix, the wave is circularly polarized. Most naturally occurring light is unpolarized. Linearly polarized light is produced by passing light through a polarizing filter, or by scattering or reflection. When light is incident on a dielectric surface at Brewster's angle, given by

$$\tan\theta_B = \frac{n_2}{n_1},$$

the reflected light is completely polarized with $\vec{\mathbf{E}}$ parallel to the surface.

* Cavity oscillators and waveguides are used in microwave generators and receivers. The oscillator is a closed cavity containing EM waves. A set of plane waves bouncing back and forth off the walls model wave propagation in a guide. Wavelengths longer than the *cutoff wavelength* cannot propagate in the guide.

Practical Applications

Light and radio waves are used extensively for communication. Microwaves are increasingly used for the transmission of television signals from satellites. Radiation pressure forms comet tails and prevents gravitational collapse of bright stars. Radiation pressure from laser beams is used in nuclear fusion experiments and in traps where atomic properties of cold neutral atoms can be tested. Polarization of reflected and scattered light is used to advantage by people who wear Polaroid sunglasses and by photographers who use polarizing filters. Polarization of transmitted light due to birefringence is used in engineering stress analysis and in minerology.

Solutions to Exercises

33.1. The first derivatives with respect to x and t are calculated in the text. The second derivatives are:

$$\frac{\partial^2}{\partial t\,\partial x}[\cos(kx - \omega t)] = \frac{\partial}{\partial t}[-k\sin(kx - \omega t)] = \omega k\cos(kx - \omega t),$$

and:

$$\frac{\partial^2}{\partial x\,\partial t}[\cos(kx - \omega t)] = \frac{\partial}{\partial x}[\omega\sin(kx - \omega t)] = k\omega\cos(kx - \omega t).$$

These are equal, as required.

33.2. The derivative of eqn. (33.4) with respect to time is:

$$\frac{\partial^2 E}{\partial t\,\partial x} = -\frac{\partial^2 B}{\partial t^2}.$$

And, the derivative of eqn. (33.5) with respect to x is:

$$-\frac{\partial^2 B}{\partial x^2} = \epsilon_0\mu_0\frac{\partial^2 E}{\partial x\,\partial t}.$$

Combining these gives:

$$\frac{\partial^2 B}{\partial t^2} = \frac{1}{\epsilon_0\mu_0}\frac{\partial^2 B}{\partial x^2}, \quad \text{as required.}$$

33.3. The Poynting flux in the laser beam is the total power divided by the area onto which it is focused, $A = \pi d^2$:

$$S = \frac{10^{14}\ \text{W}}{\pi(0.05 \times 10^{-3}\ \text{m})^2} = 1.3 \times 10^{22}\ \text{W/m}^2.$$

Since the radiation is totally absorbed, the radiation pressure is:

$$P = \frac{S}{c} = \frac{1.3 \times 10^{22}\ \text{W/m}^2}{3 \times 10^8\ \text{m/s}} = 4 \times 10^{13}\ \text{Pa}.$$

A big enough laser focused finely enough can be pushy.

33.4. The equipotential surfaces are perpendicular to the Coulomb field lines. (There is no induced field if the current is steady.) Since $\vec{\mathbf{S}} \propto \vec{\mathbf{E}} \times \vec{\mathbf{B}}$, $\vec{\mathbf{S}}$ is also perpendicular to $\vec{\mathbf{E}}$, so $\vec{\mathbf{S}}$ runs along equipotential surfaces.

33.5. Unpolarized light is the superposition of individual wave pulses, each polarized at a random angle θ ($0 < \theta < \pi$) to the transmission axis. At a particular time, any value of this angle is equally likely. Averaged over time, the incident intensity with polarization in a range of angles $d\theta$ depends only on the range $d\theta$, and not on the value of the angle.

$$\frac{dI}{d\theta} = \frac{I_0}{\pi}.$$

According to Malus' law, a fraction $\cos^2\theta$ of this intensity passes the filter. The total transmitted intensity is found by summing up the contributions from all the incident waves:

$$I = \int dI_{\text{transmitted}} = \frac{I_0}{\pi}\int_0^{\pi}\cos^2\theta\, d\theta = \frac{I_0}{2}.$$

33.6. Since $2a = 5.0$ cm, the free-space wavelength is greater than the cutoff. Waves polarized parallel to the plates cannot propagate. The TEM mode can propagate, so the emerging radiation is linearly polarized perpendicular to the plates.

Basic Skills

Review Questions

§33.1 PLANE ELECTROMAGNETIC WAVES

- How are electromagnetic waves generated?
- Describe the configuration of electric and magnetic field vectors in a plane EM wave.
- Write the wave equation for the electric field.
- What is the relation between the amplitudes of the electric and magnetic fields in an EM wave?
- Describe the phase relation between $\vec{\mathbf{E}}$ and $\vec{\mathbf{B}}$ in an EM wave.
- Why do we believe that light is an EM wave?

§33.2 ENERGY AND MOMENTUM TRANSPORT BY EM WAVES

- What is the *Poynting vector?* How is it related to $\vec{\mathbf{E}}$ and $\vec{\mathbf{B}}$?
- Which physical property of an EM wave is described by the Poynting vector?
- What is the relation between energy and momentum in an EM wave?
- Define *radiation pressure.*
- How could you calculate the radiation pressure on a surface when a light wave is incident on it?
- How is energy transported from the battery to the resistor in a simple circuit?

§33.3 POLARIZATION

- Describe the electric field vectors in *linearly polarized* light.
- How does a polarizing filter work?
- State the law of Malus.
- How is light polarized by reflection?
- What is Brewster's angle?
- Describe a left circularly polarized wave.
- Discuss how a circularly polarized wave can result from the superposition of two linearly polarized waves.
- What kind of wave results from the superposition of a right and a left circular wave, each with the same amplitude and phase?

✲ §33.4 ELECTROMAGNETIC OSCILLATIONS AND MICROWAVES

- Describe the configuration of fields in one mode of oscillation for a cylindrical cavity oscillator.
- What boundary conditions apply to the electric and magnetic fields in a waveguide?
- What is the TEM mode?
- Sketch the configuration of fields in the TE mode between two parallel conducting plates.
- Why is there a cutoff wavelength below which the TE mode cannot propagate?

Basic Skill Drill

§33.1 PLANE ELECTROMAGNETIC WAVES

1. An electromagnetic wave has a magnetic field amplitude of 1.56×10^{-8} T. What is the electric field amplitude?
2. The electric field amplitude in an EM wave is $(0.45\ \text{V/m})\hat{\mathbf{z}}$. The magnetic field is in the $-x$-direction. What is the magnetic field amplitude, and what is the direction of propagation?
3. The electric field in a plane EM wave propagating in the z-direction at a certain place and time is $\vec{\mathbf{E}} = (6.0\ \text{V/m})(\hat{\mathbf{x}} + \hat{\mathbf{y}})$. Find $\vec{\mathbf{B}}$.
4. The magnetic field in an EM wave is given by:

$$\vec{\mathbf{B}} = (4.3 \times 10^{-10}\ \text{T})\hat{\mathbf{z}}\cos\{[\pi/(1.0\ \text{mm})](x - ct)\}.$$

Find $\vec{\mathbf{E}}$.

§33.2 ENERGY AND MOMENTUM TRANSPORT BY EM WAVES

5. The power reaching the Earth from a distant star is $2 \times 10^{-8}\ \text{W/m}^2$. Find E_0 in the radiation from the star.
6. A 1.5-mW laser beam produces light with an electric field amplitude of 0.50 kV/m. What is the diameter of the beam?
7. A plane EM wave has $E_0 = 3.0$ V/m. What is the time-averaged magnitude of $\vec{\mathbf{S}}$? At what rate is momentum carried over a 1.5-m^2 surface perpendicular to $\vec{\mathbf{k}}$?
8. A 15-W light beam 3.0 mm in diameter, incident normally on a human hand, is absorbed. What is the radiation pressure on the hand?

§33.3 POLARIZATION

9. Light with intensity $I_0 = 7.75\ \text{W/m}^2$ polarized along the y-axis falls on a polarizing filter with its transmission axis at 20° to $\hat{\mathbf{y}}$. What is the intensity of the transmitted light?

10. Linearly polarized light traveling in the x-direction is incident on a polarizing filter with its transmission axis in the y-direction. The transmitted intensity is 45% of the incident intensity. What is the angle of polarization in the incident beam?
11. Find Brewster's angle for diamond in air (cf. Table 16.3).
12. A laser beam is incident from air on a plane surface. You test the reflected ray with a polarizing filter and determine that the reflected beam is completely polarized when the angle of incidence is 55°. What is the refractive index of the material?

✲ §33.4 ELECTROMAGNETIC OSCILLATIONS AND MICROWAVES

13. Two parallel conducting plates are 3.56 cm apart. What is the cut-off frequency for waves propagating in the TE mode in this waveguide?
14. Radio waves are passed between two parallel plates. The emerging radiation is unpolarized when $\lambda < 1.25$ cm but is polarized with $\vec{E}$ perpendicular to the plates when $\lambda > 1.25$ cm. What is the separation of the plates?

Questions and Problems

§33.1 PLANE ELECTROMAGNETIC WAVES

15. ❖ An EM wave propagating in the x-direction has $\vec{B}$ in the y-direction at a certain place and time. What is the corresponding direction of $\vec{E}$?
16. ❖ Show that a plane wave with $\vec{E}$ parallel to $\vec{B}$ cannot satisfy Maxwell's equations.
17. ❖ Show that a plane wave with $\vec{E}$ parallel to $\vec{k}$ and $\vec{B}$ perpendicular to $\vec{k}$ cannot satisfy Maxwell's equations.
18. ♦ In a wave propagating in the y-direction, the magnetic field amplitude is: $\vec{B}_0 = (1.2 \times 10^{-7}\ \text{T})(\hat{x} + \sqrt{3}\,\hat{z})$. Find $\vec{E}_0$.
19. ♦ A 16.5-MHz radio wave is propagating in the x-direction. The electric field amplitude is 0.792 V/m. Determine the wave vector $\vec{k}$ and the magnetic field amplitude.
20. ♦♦ An EM wave with a frequency of 1.80 GHz has a magnetic field amplitude of $(4.75\ \text{nT})\hat{z}$ and propagates in the y-direction. The magnetic field is maximum in the positive z-direction at $y = 0.250$ m, $t = 0.650$ ns. Write expressions for $\vec{E}$ and $\vec{B}$ in the wave.
21. ♦♦ Two EM waves with equal electric field amplitudes of 0.44 V/m are propagating in the x-direction. One wave has $\vec{E}$ in the y-direction, and the other has $\vec{E}$ in the z-direction. Find the magnetic field vector amplitude in the superposed wave.
22. ♦♦ The direction of the electric field vector in an EM wave is $(1/\sqrt{6})(\hat{x} + \sqrt{3}\,\hat{y} + \sqrt{2}\,\hat{z})$, and the direction of the magnetic field is $(0.29)(2\hat{x} + \hat{y} - 2.6\hat{z})$. What is the direction of propagation of the wave?
23. ♦♦ An EM wave has wave vector $\vec{k} = (\hat{x} + 2\hat{y})$ rad/m. The magnetic field amplitude is $\vec{B}_0 = (1.43 \times 10^{-10}\ \text{T})(2\hat{x} - \hat{y} + \hat{z})$. Determine the electric field vector amplitude and the frequency of the wave.

§33.2 ENERGY AND MOMENTUM TRANSPORT BY EM WAVES

24. ❖ **(a)** Two plane EM waves propagating in the same direction have angular frequencies of ω and 2ω. They have the same electric field amplitude. Compare the energy transported by each wave. **(b)** Two plane EM waves have the same frequency and direction, but the electric field amplitude of one is twice the amplitude of the other. Compare the energy transported by each wave.
25. ❖ **(a)** A power supply drives current through the central wire of a perfectly conducting coaxial cable, through a resistor at the far end, and back through the cable's outer cylinder. Sketch the electric and magnetic fields in the region between the conductors, and show that the Poynting flux flows along the cable in the direction of the inner current. **(b)** Describe the pattern of the Poynting vector in the coaxial cable if the conductors have a small resistivity.
26. ❖ Discuss how the flow of energy from a battery to a resistor changes as the circuit is made larger in diameter.
27. ♦♦ A solar collector on a satellite is proposed to provide 10^9 W of power for the city of Chicago. Assuming perfect absorption and 1% efficiency in conversion to useful power, what area should the solar collector have? The power is beamed down to a perfectly efficient collector on the surface of the Earth. If the collector's area is $2.5 \times 10^4\ \text{m}^2$, what microwave intensity is required? What is the electric field amplitude in the microwave beam at the Earth's surface?
28. ♦♦ Find the average electric and magnetic field amplitudes in EM waves 1 m from a 100-W light bulb.
29. ♦♦ A laser beam reflecting normally off a surface with an area of $3.0\ \text{cm}^2$ exerts a force of 2.4 N on it. What is the electric field amplitude in the laser beam?
30. ♦♦ A possible spacecraft design uses a solar sail made of aluminized plastic. As sunlight reflects off the sail, radiation pressure drives the spacecraft outward. If the sail material has a density of $700\ \text{kg/m}^3$, what is the maximum thickness of the sail for which the force due to radiation pressure exceeds the gravitational force on the sail? If the sail area is $10^6\ \text{m}^2$, its thickness is 1 μm, and the craft carries a 100-kg payload, what is its acceleration at the radius of Earth's orbit?
31. ♦♦ A scheme for powering an interstellar starship envisions using a laser on the Moon for propulsion. What laser power is needed to propel a 30 000-kg starship with an acceleration of $0.01g$ if the laser beam is completely absorbed by the starship?
32. ♦♦ An aircraft tracking radar radiates 1 kW of power into a fan-shaped beam that is thin in the horizontal direction and spreads through an angle of 45° in the vertical direction. At a distance of 10 km, the beam's horizontal dimension is 200 m. Assuming perfect reflection, what radiation pressure does the beam exert on an aircraft 10 km away?
33. ♦♦ A satellite has an average power consumption of 350 W. Assuming 5.5% efficiency for conversion of solar energy to electrical energy, what area is required for the solar panels?
34. ♦♦ A 1.0-kW laser with a beam area of $1.0\ \text{cm}^2$ is incident on a mirror with an angle of incidence equal to 15°. Assuming perfect reflection, what is the force due to radiation pressure on the mirror?

35. ♦♦♦ Perfectly absorbing interplanetary dust grains have density $\rho = 3 \times 10^3$ kg/m^3. Show that spherical grains with radii less than a critical value R_c are blown out of the solar system. Calculate R_c.

36. ♦♦♦ A plane EM wave with $\vec{\mathbf{E}} = E_0\hat{\mathbf{y}}\cos(kx - \omega t)$ is traveling along the x-axis. How much EM energy U exists at time t in a rectangular volume of dimensions dy by dz between x and $x + dx$? Compute the time rate of change of energy in this volume as the wave passes. Show that dU/dt equals the energy transported into the volume at x by the Poynting vector minus the energy transported out at $x + dx$. Show that:

$$\frac{\partial u}{\partial t} + \frac{\partial S}{\partial x} = 0,$$

where u is the energy density in the wave. This is the differential form of the energy conservation law for EM field.

37. ♦♦♦ Show that the total power transported down a coaxial cable is $I\,\Delta V$, where ΔV is the potential difference between the core and the sheath (cf. Problem 25).

§33.3 POLARIZATION

38. ❖ A beam of light passes through two polarizing filters with vertical transmission axes. A third filter is placed between the other two with its transmission axis at an angle θ to the vertical. For what angle θ does the transmitted beam have maximum intensity? For what angle is it minimum? How do the answers change if the two outer filters have perpendicular transmission axes? Explain your reasoning.

39. ❖ You wish to photograph a reflection of trees in a lake. Can you use a polarizing filter to make the reflection appear brighter? How?

40. ❖ A beam of light is a superposition of unpolarized light with intensity I_u and a plane-polarized wave with intensity I_p. Describe how the intensity transmitted by a polarizing filter varies as the filter rotates in the plane perpendicular to the light beam.

41. ♦ Laser light is linearly polarized with its electric field vector along the x-axis. What is the intensity of the beam, as a fraction of its initial intensity, after passing through: **(a)** A polarizing filter with its axis at 37.0° to the x-axis? **(b)** A polarizing filter with its axis at 77.0° to the x-axis?

42. ♦ Find Brewster's angle for a water–glass interface if the glass has a refractive index of 1.6.

43. ♦ Light shines on a glass surface at an angle of incidence equal to 58°. The reflected light is 100% polarized. What is the refractive index of the glass?

44. ♦♦ An EM wave with amplitude $\vec{\mathbf{E}}_0 = (340\text{ V/m})(2\hat{\mathbf{x}} - \hat{\mathbf{y}})$ passes through a polarizing filter with its axis parallel to the vector $\hat{\mathbf{x}} + 3\hat{\mathbf{y}}$. What is the intensity of the transmitted light?

45. ♦♦ Laser light of intensity 1.6 W/m^2, polarized so that its electric field vector makes an angle 45° with the x-axis, passes through two polarizing filters. The transmission axis of the first is along the x-axis, and the second filter's axis is parallel to the incident polarization. What is the intensity of the transmitted light? How would the result differ if the second transmission axis were perpendicular to the initial polarization?

46. ♦♦ An unpolarized light beam of intensity I_0 passes along the z-axis through three polarizing filters. The first filter's transmission axis is parallel to $\hat{\mathbf{x}}$. The second filter's axis makes an angle of 25° with the y-axis, and the third's axis is parallel to $\hat{\mathbf{y}}$. What is the intensity of the transmitted light?

47. ♦♦ A beam of unpolarized light of intensity I_0 is incident at Brewster's angle on a slab of glass with a refractive index of 1.5. The reflected beam intensity is 9.5% of I_0. Assume that there is no absorption. **(a)** Calculate Brewster's angle for the glass in air. **(b)** The percent polarization of a light beam is:

$$p_{\%} = 100\left|\frac{I_1 - I_2}{I_1 + I_2}\right|,$$

where I_1 and I_2 are the intensities in two orthogonal polarizations. Find the percent polarization of the reflected ray. **(c)** Find the percent polarization of the transmitted ray.

48. ♦♦ Compute the y- and z-components of the electric field vector in the superposition of two waves with electric field vectors given by:

$$\vec{\mathbf{E}}_1 = (25.0\text{ V/m})\hat{\mathbf{y}}\cos[(6.0\text{ rad/m})x - (1.8 \times 10^9\text{ rad/s})t],$$

and

$$\vec{\mathbf{E}}_2 = (30.0\text{ V/m})\hat{\mathbf{z}}\sin[(6.0\text{ rad/m})x - (1.8 \times 10^9\text{ rad/s})t],$$

at $x = 0$ and $t = 0$, $T/4$, and $T/2$. Draw the vectors. How would you describe the polarization of the superposition?

49. ♦♦♦ A quarter-wave plate produces a phase shift of 90° between waves polarized at angles of ±45° with the vertical. Unpolarized light of intensity I_0 passes through a polarizing filter with its transmission axis vertical, then through the quarter-wave plate, and finally through another filter with a horizontal transmission axis. What is the intensity of the light emerging from the system?

50. ♦♦♦ The transmission axes of two polarizing filters make an an angle θ_0. A third filter may be rotated between the other two. Prove that the three filters transmit maximum intensity when the middle filter's axis makes an angle $\theta_0/2$ with each of the other two.

51. ♦♦♦ *Faraday rotation.* A linearly polarized EM wave with its electric field vector along the z-axis propagates into an ionized medium. In this medium, right-circular waves propagate faster than left-circular waves. The difference in speed is $\Delta v/c = 7 \times 10^{-5}$, and the wave frequency is 15 GHz. What is the direction of polarization after the wave has traveled 25 m? (*Hint:* Describe the linearly polarized wave as the superposition of two circularly polarized waves, as in Example 33.11. Write the y- and z-components of $\vec{\mathbf{E}}$ after the wave has propagated a distance x, remembering that k_R and k_L are different, and calculate the angle E makes with the z-axis from $\tan\theta = E_y/E_z$.)

* §33.4 ELECTROMAGNETIC OSCILLATIONS AND MICROWAVES

52. ❖ Use the Poynting vector to describe the energy flow in the cylindrical cavity oscillator discussed in §33.4.1.

53. ❖ At high frequencies, a coaxial cable acts as a waveguide. Show that a TEM mode can propagate along the cable, and sketch the configuration of the fields. Is there a cutoff?

54. ❖ Use the Poynting vector to describe the energy flow in the TE mode shown in Figure 33.29.

55. ❖ Sketch the field lines for the $m = 1$ TM mode in a parallel plate waveguide.

56. ❖ The electric field of the $m = 2$ TE mode in a parallel plate waveguide is proportional to $\sin(2\pi z/a)$. Sketch the field configuration in the guide.

57. ❖ A microwave oven is a rectangular resonant cavity. Describe the configuration of fields inside in the lowest mode. What disadvantages would the lowest mode have for cooking?

58. ♦ You wish to use two parallel conducting plates to make a waveguide for 2.7-cm wavelength TE microwaves, but 3.0-cm wavelength waves are not to propagate. What should the separation of the plates be?

59. ♦ Calculate the cutoff frequency for the TE mode between two parallel plates with a 15-cm separation.

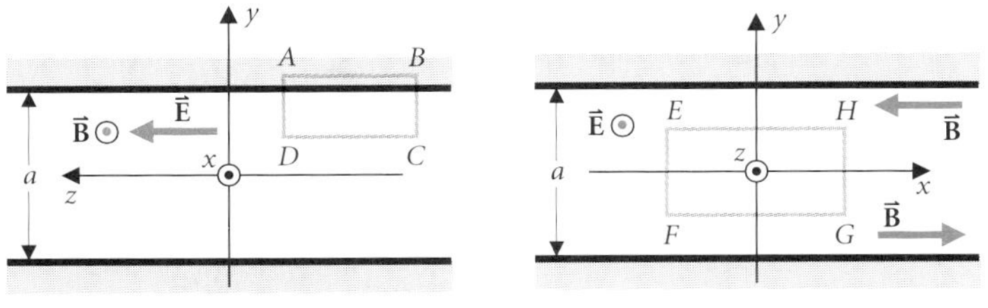

(a) Circuit ABCD. (b) Circuit EFGH.

■ FIGURE 33.31

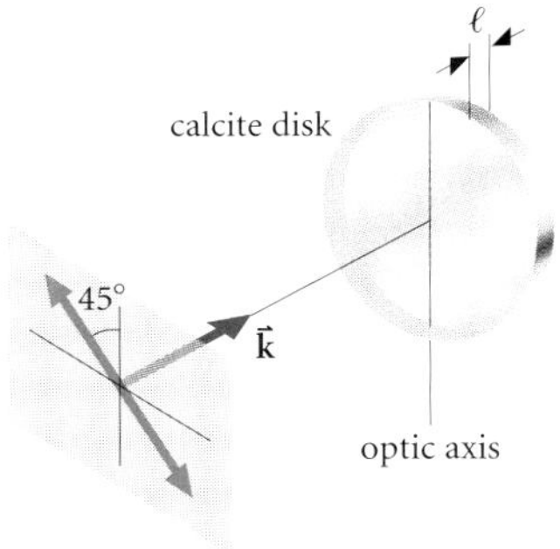

■ FIGURE 33.32

60. ♦ What separation of conducting plates is required to make a waveguide with 165-MHz cutoff frequency in the TE mode?

61. ♦♦♦ A waveguide consists of two conducting plates with a 2.0-cm separation. It is excited so that the $m = 1$ TE mode and the TEM mode have equal amplitude. The TM mode is not excited. The radiation has free-space wavelength $\lambda_f = 3.0$ cm. What length of guide is necessary if the emerging wave is to be circularly polarized?

62. ♦♦♦ A cavity oscillator is formed by two infinite, parallel, and perfectly conducting metal plates with separation a. The magnetic field inside the cavity is given by:

$$\vec{\mathbf{B}} = -B_0 \sin(\pi y/a)\cos(\omega t)\hat{\mathbf{x}}.$$

(a) Assuming that $\vec{\mathbf{E}}$ is in the z-direction, apply Faraday's law to circuit *ABCD* (■ Figure 33.31a) to find the electric field inside the cavity. **(b)** Compute the displacement current everywhere inside the cavity. **(c)** Using the result of (b), apply Ampère's law to circuit *EFGH* (Figure 33.31b) to obtain another expression for $\vec{\mathbf{B}}$ in the cavity. (*Hint:* Use the symmetry of $\vec{\mathbf{B}}$). **(d)** What value of ω makes the result of (c) the same as the given expression for $\vec{\mathbf{B}}$?

63. ♦♦♦ Two infinite, flat, parallel, and perfectly conducting metal plates lie in the planes $y = \pm a/2$. The space between the plates acts as a cavity oscillator in which the electric field is given by:

$$\vec{\mathbf{E}} = E_0 \cos(\pi y/a)\sin(\omega t)\hat{\mathbf{z}}.$$

(a) Assuming that $\vec{\mathbf{B}}$ is in the x-direction, compute the displacement current density and apply Ampère's law to a rectangular circuit inside the cavity to obtain B everywhere inside. **(b)** What is $\vec{\mathbf{B}}$ within the conducting plates? Use Ampère's law to compute the current per meter on the metal surfaces, and compare with the displacement current per meter between the plates.

Additional Problems

64. ❖ Assuming that the amplitude of the ripple in a field line of an oscillating electron is proportional to the amplitude of its acceleration component perpendicular to the field line, argue that the intensity radiated by the electron is proportional to $\sin^2\theta$, where θ is the angle between the field line and the direction of oscillation.

65. ♦ What charge on the Sun would make the static electric field at the Earth's orbit equal the electric field amplitude (about 1 kV/m, Example 33.3) in sunlight?

66. ♦♦ What current around the Sun's equator would produce a magnetic field at the Earth's orbit equal to the average magnetic field amplitude in radiation reaching the Earth? (*Hint:* Use magnetic moment.)

67. ♦♦ Lasers are increasingly being used for surgery. The laser cauterizes and seals the wound as it cuts. To get a rough idea of how the process works, use the following model. Model flesh as water at 300 K and imagine that flesh is removed when the water boils. Assuming total absorption, what power laser is required to remove 1 mm of flesh per second over an area 1 mm in diameter? How much force is exerted on the wound as a result of this process?

68. ♦♦♦ Calcite is a mineral in which the speed of an EM wave depends on its polarization. Electrons in calcite oscillate more freely along certain crystal planes than perpendicular to them. The direction perpendicular to the planes is called the optic axis. Waves polarized parallel to the optic axis travel more rapidly than waves polarized perpendicular to it. Light with wavelength λ_0 and polarized at 45° to the optic axis falls on the calcite disk shown in ■ Figure 33.32. If $\Delta n = n_\perp - n_\parallel$ is the difference in the refractive index for the two polarizations, describe the polarization of the emerging wave if the thickness ℓ of the disk is **(a)** $\lambda_0/(4\,\Delta n)$, **(b)** $\lambda_0/(2\,\Delta n)$, and **(c)** $\lambda_0/(3\,\Delta n)$.

Computer Problems

69. Use a spreadsheet program to compute the electric field vectors

$$\vec{\mathbf{E}}_1 = (3\text{ V/m})\cos[(5\text{ rad/s})t]\hat{\mathbf{x}},$$

and

$$\vec{\mathbf{E}}_2 = (3\text{ V/m})\sin[(5\text{ rad/s})t]\hat{\mathbf{y}}$$

every 0.04 s for one complete cycle. Compute the components of the superposition $\vec{\mathbf{E}}_1 + \vec{\mathbf{E}}_2$, and plot the vector. How would you describe the polarization? What happens if the magnitude of $\vec{\mathbf{E}}_2$ is 2 V/m?

70. In Example 33.9 we showed how a polarizer placed between two others with their axes perpendicular allows some light to be transmitted. Take the y-axis along the first polarizer's transmission axis, and the x-axis along the final polarizer's axis. Set up a spreadsheet program to calculate the amplitude of the transmitted wave when the intermediate polarizer's axis makes an angle θ with the y-axis. For what value of θ is the most light transmitted? Now add another polarizer between the second and the last, with its axis at 2θ to the y-axis. What value of θ maximizes the transmitted light? How does the transmitted intensity compare with the previous case? Now, add another polarizer and repeat the calculations. Can you draw any general conclusions?

Challenge Problems

71. An electron moves uniformly in a circle. In the spirit of §33.1.1, describe the field line emerging from the electron perpendicular to the plane of its motion. Show that the corresponding wave is circularly polarized and that its Poynting vector is directed away from the electron. Describe how the polarization of waves emitted by the electron depends on direction.

72. A nonreflective coating for TV screens consists of a polarizing filter on top of a quarter-wave plate. Explain how it works.

73. Consider a rectangular guide measuring $a = 3$ cm by $b = 5$ cm. Argue that the $m = 1$ TE mode can propagate in this guide as well as one with $b = \infty$. What is the cutoff? What cutoff results from the finite value of b? If a 4-GHz wave is introduced into the guide, what is the polarization of the emerging wave?

74. A uniformly, positively charged filament lies along the y-axis with its center at the origin. A small coil with dipole moment $\vec{\mathbf{m}} = m\hat{\mathbf{z}}$ is also at the origin. Sketch a diagram showing field lines for the electric and magnetic fields in the x-z plane. From your diagram, determine the direction of the Poynting vector at points in the x-z plane. From your results, argue that the EM field carries angular momentum in the z-direction. Assuming the filament is free to rotate about the z-axis, describe what happens when the current in the coil is reduced to zero. Where does the angular momentum go? Use Faraday's law to show how the necessary force is exerted on the filament.

75. A plane EM wave propagating into a material with resistivity ρ has the form:

$$\vec{\mathbf{E}}(z, t) = E_0\hat{\mathbf{y}}e^{-\alpha z}\cos(kz - \omega t),$$

and

$$\vec{\mathbf{B}}(z, t) = -B_0\hat{\mathbf{x}}e^{-\alpha z}\cos(kz - \omega t + \phi_0).$$

Apply Maxwell's equations and the relation $\vec{\mathbf{E}} = \rho\vec{\mathbf{j}}$ to find α, k, ϕ_0 and the ratio B_0/E_0 as functions of ω.

Part VII Problems

1. ♦♦ A coil is made by wrapping 24 m of 18 gauge copper wire (157 feet per ohm at 68° F, diameter 4.1 mm) around a cardboard tube 2 in. in diameter. Calculate the impedance of the coil when connected to 240-Hz, 2.2-V AC power. What is the current in the coil?

2. ♦♦♦ An electron is held in a circular orbit about an atomic nucleus by the attractive Coulomb force between them. The electron forms a current loop with magnetic moment $\vec{\mathbf{m}}$. A magnetic field $\vec{\mathbf{B}}$ is applied to the atom, parallel to $\vec{\mathbf{m}}$. Assuming cylindrical symmetry about the center of the atom, find the induced electric field in terms of dB/dt, as the field is changed from 0 to $\vec{\mathbf{B}}$. Show that the change in the electron's speed as a result of the applied $\vec{\mathbf{B}}$ is $\Delta v = -(er/2m)B$ (cf. §29.4.1).

3. ♦♦♦ A current $I = 1.0$ A is charging a circular capacitor with plates of radius 4.0 cm. Compute the electric and magnetic energy densities at a point 3.0 cm from the axis of symmetry when the charge density on the plates is 10.0 nC/m^2. In considering the energy used to charge a capacitor, we neglected magnetic energy density. Is this justified?

4. ♦♦♦ A parallel plate capacitor filled with a dielectric is connected to a resistor and a battery. If you suddenly pull the dielectric slab out of the capacitor, what happens to the charge on the capacitor? What happens to the energy? Taking into account the battery, the resistor, and the work you do, calculate the energy input and output from the circuit. Verify that energy is conserved.

5. ♦♦♦ **(a)** A star has mass M, luminosity L, and radius R. Find the average value of the Poynting flux at the surface of the star.
(b) If an electron acts like an absorbing surface with effective area $\sigma = 6.7 \times 10^{-29}$ m^2, find the force exerted on an electron at the surface by absorption of radiation from the star.
(c) If the star's luminosity is large enough, the force due to radiation pressure blows off the star's outer layers. Because of the electric force between them, each electron drags a proton along with it. Find the luminosity—called the Eddington luminosity—for which the force on an electron at the star's surface due to radiation pressure balances the gravitational force on a proton. What is the Eddington luminosity for a star with the mass of the Sun?
(d) If the luminosity of a star is proportional to the 3.45 power of its mass, what is the largest mass a star can have and still have a luminosity less than its Eddington limit?

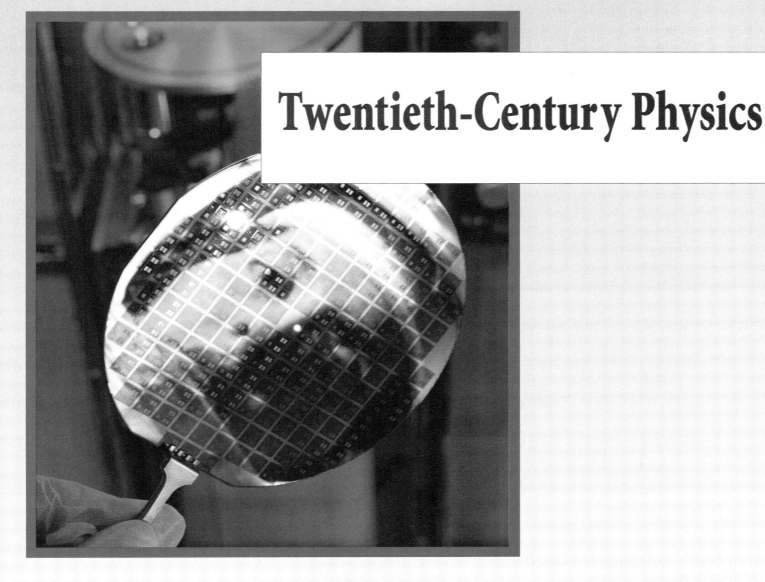

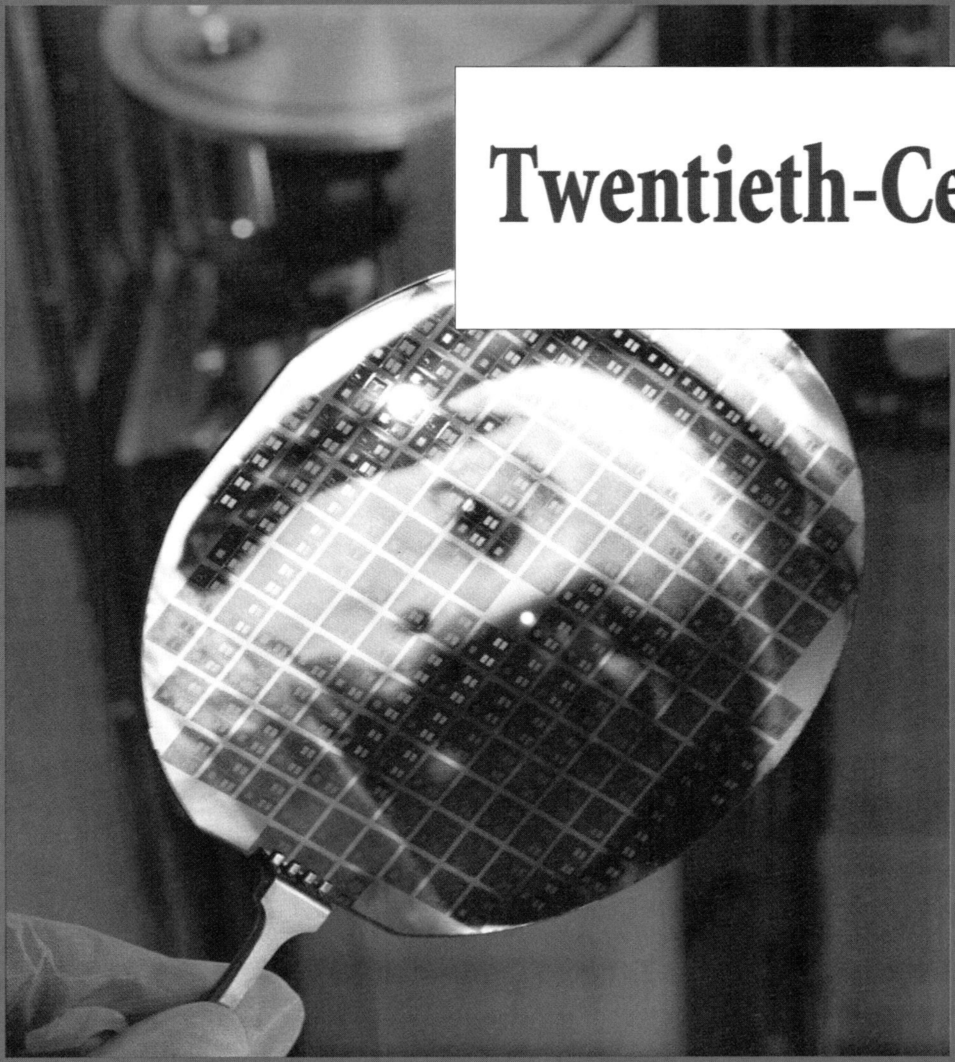

Twentieth-Century Physics

PART EIGHT

Physicists could perhaps be excused for feeling smug in the years following Maxwell's work. Newton's theory of mechanics and Maxwell's electromagnetic theory seemed to be completely accurate and capable, in principle, of describing all known phenomena. Astronomers understood the universe as a roughly spherical collection of stars with the Sun near its center. The same laws of mechanics, light, and chemistry could be observed at work in the stars and on Earth. With the basic rules governing the universe known, all that remained was to work out the details.

But it's never wise to be smug! In the years following 1890, this neat picture began to come apart, and our view of the universe underwent a complete revolution. As a result of several careful experiments dealing with the propagation of light, the theory of electromagnetic waves had to be modified and several ad hoc assumptions were required—the theory became ugly. Einstein instead suggested a complete revision of Newtonian ideas about space and time. Max Planck's explanation of cavity radiation and Einstein's explanation of how electrons are ejected from illuminated metal surfaces both showed that light acts like a stream of particles as well as like an electromagnetic wave. The discovery of radioactivity introduced a new kind of process, neither gravitational nor electromagnetic, and provided the particles Ernest Rutherford used to prove that atoms are mostly vacuum—very small heavy nuclei, each surrounded by a cloud of electrons.

While physicists were revolutionizing our understanding of atomic phenomena, astronomers were making equally profound discoveries about the nature of the universe. The visible sphere of stars is just a small region near the edge of a much larger system that is itself but one of millions. Edwin Hubble showed that the entire universe is expanding. Aleksandr Friedmann and others used Einstein's general theory of relativity to explain the observations in terms of a universe that abruptly came into existence 10–20 billion years ago in an extremely hot, dense state.

The modern conception of the physical world is a four-dimensional composite of space and time in which a sea of elementary particles forms and dissolves. The behavior of such particles can only be predicted on average. Because the physics of everyday events involves very large numbers of particles, these averages are well defined, and the Newtonian picture we have used thus far is an extremely good approximation to this average behavior. Modern physics does not replace Newtonian physics, but extends it.

In this part we introduce the exciting ideas of twentieth-century physics. In Chapter 34 we describe Einstein's ideas of relativity and space-time. In Chapter 35 we explore how the wave/particle nature of both light and matter leads to an understanding of the structure of atoms and their chemical behavior. In Chapters 36 and 37 we describe the structure of atomic nuclei and the principles that underlie the behavior of subnuclear particles. We conclude with an epilogue on current attempts to unify all of physics and how these ideas relate the smallest scale structure of matter with the structure and evolution of the universe.

The solar eclipse of 1991. Bright streamers of the solar corona become visible when the Moon blocks the disk of the Sun. The star Delta Geminorum is visible between the streamers to the right (west) of the Sun. By comparing a star's position measured during an eclipse with its position when not near the Sun, astronomers can observe the effect of the Sun's gravity on light from the star. The first such measurements were made by Arthur Eddington and Andrew Crommelin. Their results were announced at a meeting of the Royal Astronomical Society on November 6, 1919. The next day The Times *carried the headline "Revolution in Science." These observations provided the first confirmation of the general theory of relativity.*

CHAPTER 34

Relativity and Space-Time

Concepts

Relative quantity
Invariant
Simultaneity
Event
Space-time
World line
Time dilation
Length contraction
Lorentz transformation
Rest mass energy
Energy-momentum invariant

Goals

Be able to:

Compute the effects of time dilation and length contraction.

Describe the world lines of particles and light rays in a space-time diagram.

Relate the energy and momentum of a particle in different reference frames.

Henceforth space by itself and time by itself are doomed to fade away into mere shadows, and only a kind of union of the two will preserve an independent reality.

Hermann Minkowski

Before starting this chapter, review the material on relatively moving reference frames in §3.3.

Einstein first predicted the bending of light in 1907. It wasn't until 1915 that he was able to predict its magnitude correctly.

In May 1919, two groups of British scientists, led by Arthur Eddington and Andrew Crommelin, traveled halfway around the world to observe a solar eclipse. Their goal was to test the prediction of a young German physicist, Albert Einstein, that the Sun's gravity bends light rays. Beginning in 1905, Einstein had published a series of papers that forever changed the way we look at the world. Eddington and Crommelin verified his predictions, and his name became a household word.

As a young man, Einstein asked himself how a light beam would appear to someone traveling with it. To such an observer, the sinusoidal pattern of electric and magnetic fields in the wave would appear fixed in space. That is a mystery, for such a stationary pattern of fields violates Maxwell's equations. The mystery is resolved by Einstein's special theory of relativity, published in 1905. Convinced that fields are a more fundamental concept than particles, Einstein assumed that Maxwell's equations are exactly correct for all observers, and he showed that the Newtonian concepts of space and time needed fundamental revision. A consequence of Einstein's theory is that no observer can travel at the speed of light!

Since 1905, the predictions of special relativity have been verified in many experiments. While most often used for exotic applications in particle physics and astronomy, the theory also touches our lives through nuclear power plants, aircraft navigation, and synchrotron x-ray sources that are used in the production of computer chips or in medical applications such as the study of cells in the human body (■ Figure 34.1) and the diagnosis of heart disease. In §34.1–34.3 we shall study the relativistic view of space and time, and in §34.4 we'll look at the implications of Einstein's theory for particle dynamics.

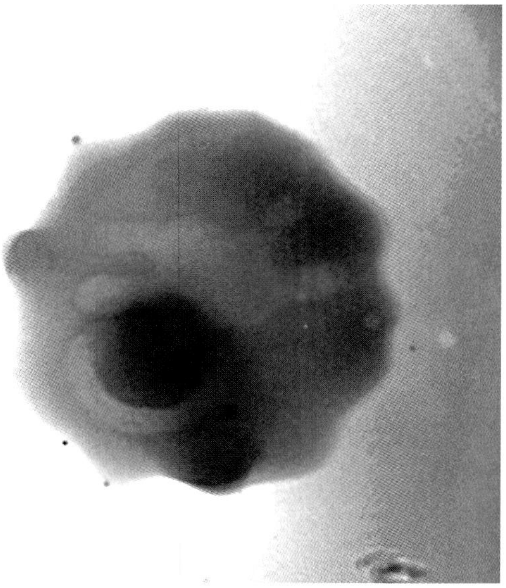

■ Figure 34.1
A human red blood cell infected with the malaria parasite. The parasite is in the ring-structured phase of its development. The image was taken with an x-ray microscope using 2.4 nm x rays from the Advanced Light Source at the Ernest O. Lawrence Berkeley National Laboratory. The short wavelength x rays allow higher resolution than would be possible with visible light. The ALS, completed in 1993, is a synchrotron light source that produces the brightest beams of soft-x-ray and ultraviolet light.

34.1 Special Relativity

34.1.1 What Is a Relativity Theory?

We discussed Galileo's thought experiment in §0.3.

In the seventeenth century, Galileo reasoned that a passenger in a ship's cabin cannot tell from the behavior of things on board whether the ship is anchored in harbor or moving at constant velocity. We verify Galileo's conclusion continuously as we revolve around the Sun with a speed of 30 km/s and fall toward the Virgo cluster of galaxies at 600 km/s, without any sense of our motion. All physical phenomena proceed as they would in the absence of such motions. A theory of relativity describes *how* the laws of physics agree with this observation. We shall illustrate this idea using *Galilean* relativity, which is valid only for relative speeds much less than the speed of light. In §34.1.2 we shall see how Einstein extended the idea to develop a theory valid for all possible speeds.

But we accelerate around the Sun, you might say. The Earth is an accelerated reference frame! However, the Earth falls freely under the influence of gravity. According to general relativity, it is therefore an inertial frame. (See the essay following this chapter.)

Newton's second law exhibits the important features of any relativity theory. Suppose a person on a train moving at constant velocity $\vec{\mathbf{v}}_{t,E}$ throws a baseball upward so that it falls back into the same car it started from (■ Figure 34.2). A person standing beside the tracks (i.e., on the Earth) describes the ball's velocity at a particular time with a vector $\vec{\mathbf{u}}_E$. A passenger on the train describes the ball's velocity at the same time with a different vector $\vec{\mathbf{u}}_t$. The ball's velocity is an example of a relative quantity.

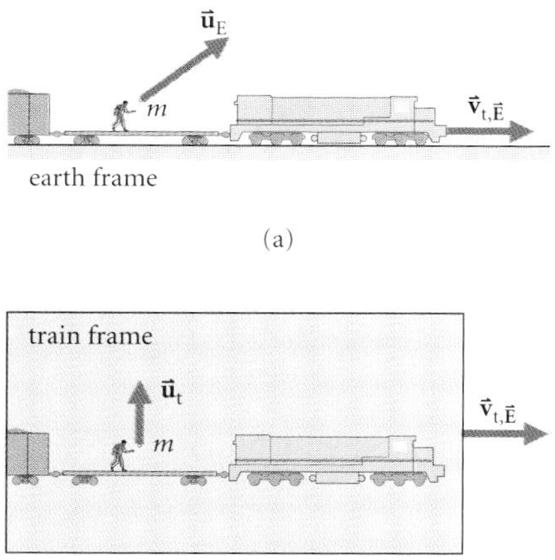

■ FIGURE 34.2
Galilean relativity. The mass of the ball is the same in the Earth frame (a) and in the train frame (b). It is an invariant. The ball's velocity at any instant differs in the two frames: $\vec{\mathbf{u}}_t = \vec{\mathbf{u}}_E - \vec{\mathbf{v}}_{t,E}$. It is a relative quantity.

> A *relative quantity* is one whose value depends on the reference frame in which it is measured.

Relative quantities are said to *transform* between reference frames. The Galilean transformation law for the baseball's velocity is:

$$\vec{\mathbf{u}}_t = \vec{\mathbf{u}}_E - \vec{\mathbf{v}}_{t,E}. \tag{34.1}$$

The mass of the baseball and the gravitational force on it have the same value in both reference frames; they are *invariant.*

> An *invariant* is a quantity that is independent of reference frame.

The most important invariant in this example is Newton's second law itself. Its form, $\vec{\mathbf{F}} = m\vec{\mathbf{a}}$, is the same in both frames. That is, in the Earth reference frame $\vec{\mathbf{F}}_E = m\vec{\mathbf{a}}_E$, and in the train reference frame $\vec{\mathbf{F}}_t = m\vec{\mathbf{a}}_t$. Let's see why this is true.

In the Earth frame,

$$\vec{\mathbf{F}}_E = m\vec{\mathbf{a}}_E = m\frac{d\vec{\mathbf{u}}_E}{dt}. \tag{i}$$

Since the force is invariant, the force measured in the train frame equals the force measured in the Earth frame: $\vec{\mathbf{F}}_t = \vec{\mathbf{F}}_E$. The mass is also invariant. Using the transformation law (eqn. 34.1), we may write $\vec{\mathbf{u}}_E$ in terms of $\vec{\mathbf{u}}_t$ to find the force in the train frame in terms of $\vec{\mathbf{a}}_t$.

TO CONVINCE YOURSELF THAT FORCE IS AN INVARIANT, LOOK AT THE EXPRESSIONS FOR, SAY, GRAVITATIONAL FORCE OR COULOMB FORCE. THE FORCES DEPEND ON FUNDAMENTAL PROPERTIES OF PARTICLES, LIKE MASS, AND ON PARTICLE SEPARATION, WHICH IS ALSO A GALILEAN INVARIANT.

$$\begin{aligned}\vec{\mathbf{F}}_t &= \vec{\mathbf{F}}_E = m\frac{d}{dt}(\vec{\mathbf{u}}_E)\\ &= m\frac{d}{dt}(\vec{\mathbf{u}}_t + \vec{\mathbf{v}}_{t,E}) = m\left(\frac{d\vec{\mathbf{u}}_t}{dt} + \frac{d\vec{\mathbf{v}}_{t,E}}{dt}\right).\end{aligned}$$

Since $\vec{\mathbf{v}}_{t,E}$ is constant, $d\vec{\mathbf{v}}_{t,E}/dt = 0$, and so:

$$\vec{\mathbf{F}}_t = m\frac{d\vec{\mathbf{u}}_t}{dt} = m\vec{\mathbf{a}}_t. \tag{ii}$$

NOTE: THIS DERIVATION SHOWS THAT THE ACCELERATION IS ALSO INVARIANT.

Equation (ii) relates the force to the acceleration in the train frame in the same way that eqn. (i) relates those vectors in the Earth frame. Newton's second law thus illustrates the *special principle of relativity:*

> The laws of physics have the same form in any two reference frames in uniform relative motion.

The laws that govern the behavior of physical systems and the fundamental properties of objects (like charge) are invariant.[1] Practical computations are usually carried out in one particular reference frame and involve relative quantities. Thus a complete theory of relativity

[1] Many popular expositions of relativity miss this point entirely, leading to misguided claims that science justifies relativism in philosophy.

Table 34.1 Some Invariant and Relative Quantities in Galilean Relativity

(Velocity of prime frame as measured by an observer in the unprime frame is $\vec{v}$.)

Invariants

Mass	m
Time	t
Distance	s
Acceleration	$\vec{a}$
Force	$\vec{F}$
Charge	Q

Relative Quantities		Transformation Law	Text Reference
Position	$\vec{r}$	$\vec{r}' = \vec{r} - \vec{v}t$	§3.3
Velocity	$\vec{u}$	$\vec{u}' = \vec{u} - \vec{v}$	§3.3
Momentum	$\vec{p} = m\vec{u}$	$\vec{p}' = \vec{p} - m\vec{v}$	§9.3.4
Kinetic energy	$K = \frac{1}{2}mu^2$	$K' = K + \frac{1}{2}mv^2 - m\vec{u}\cdot\vec{v}$	§9.3.4
Electric field	$\vec{E}$	$\vec{E}' = \vec{E} + \vec{v} \times \vec{B}$	§30.3.3

states which physical quantities are relative and which are invariant, gives the transformation laws for relative quantities, and describes physical laws in an invariant manner. In our discussion of classical physics we have developed the set of rules for Galilean relativity, some of which are collected in ● Table 34.1.

34.1.2 Einstein's Postulates

Maxwell's equations defy Galilean relativity: they predict a speed $c = 3 \times 10^8$ m/s for electromagnetic waves. According to the Galilean velocity transformation law (eqn. 34.1), this prediction can hold in only one reference frame. Thus, either the Galilean transformation laws are incorrect or, as physicists believed in the nineteenth century, the special principle of relativity does not apply to electromagnetism. At that time, physicists thought of electromagnetic waves as disturbances in a medium known as the ether, and Maxwell's equations were considered valid only in the frame in which the ether is at rest. Toward the end of the nineteenth century, measurements of the refractive index of a moving liquid, stellar aberration (see Figure 16.14), and Michelson's attempts to detect motion with respect to the ether (§17.2) had forced physicists, led by H. A. Lorentz, to modify the ether theory and to introduce theoretically unmotivated assumptions such as ether drag.

In his later years, Einstein could not recall that Michelson and Morley's result had influenced his thinking.

Einstein's approach was quite different. He assumed the special principle of relativity to be valid for all physical laws, discarded the ether concept, and took Maxwell's equations as the proper, invariant description of electromagnetism. With these assumptions, Galilean relativity can be only an approximation.

In his 1905 paper on relativity, Einstein stated his premises in the form of two postulates. The first of these is the special principle of relativity, and the second is a consequence of Maxwell's equations.

We defined an inertial reference frame for Newtonian physics in §4.8.1. Einstein modified the definition.

Postulates of special relativity

1. The laws of physics take the same form in all inertial reference frames.
2. In any given inertial frame, the speed of light c is observed to be the same no matter what the velocity of observer or emitter.

Invariance of the speed of light seems at odds with common sense. If you run toward a baseball, it approaches your glove faster than if you run away. A flash of light approaches you at the same speed whether you run toward the source or away from it! Since we have no personal experience with things that move almost as fast as light, we must build intuition by applying Einstein's postulates in special situations to derive relations that we can check against experiments.

34.1.3 *Time Dilation*

Our first situation involves a clock that uses light pulses to define a time interval. Using it, we find that time is a relative quantity.

This clock uses the reflection of light between two parallel mirrors to measure equal intervals of time (■ Figure 34.3). In one cycle of the clock, a light pulse is emitted at A, reflects from the upper mirror at B, and returns to A, where it is detected. In one frame of reference (#1), the clock is at rest (Figure 34.3a). The pulse travels a distance $2L$ straight up and down and takes a time $\Delta t_1 = 2L/c$ to complete one cycle.

A second frame moves to the left, parallel to the mirrors, at speed v with respect to the first. According to observers in frame 2, the clock moves to the right at a constant speed v, and the light pulse follows the triangular path $A'B'D'$ shown in Figure 34.3b. According to Einstein's second postulate, the speed of the pulse is c, the same as in frame 1, so the time interval Δt_2 between emission and reception of the light pulse is $\Delta t_2 = 2\ell/c$. Since the light has to keep pace with the moving clock, $\ell > L$, and $\Delta t_2 > \Delta t_1$. The clock cycle takes more time in a reference frame where the clock is in motion. To find how much more, we note that ℓ depends on Δt_2. Applying the Pythagorean theorem to triangle $A'B'C'$:

$$\left(\frac{c\,\Delta t_2}{2}\right)^2 = \ell^2 = L^2 + \left(\frac{v\,\Delta t_2}{2}\right)^2.$$

So,

$$\Delta t_2 = \frac{2L}{c}\left(\frac{1}{\sqrt{1 - v^2/c^2}}\right) = \frac{\Delta t_1}{\sqrt{1-\beta^2}} \equiv \gamma\,\Delta t_1,$$

where we have introduced the standard abbreviations:

$$\beta \equiv \frac{v}{c} \quad \text{and} \quad \gamma \equiv \frac{1}{\sqrt{1 - v^2/c^2}} = \frac{1}{\sqrt{1-\beta^2}}. \tag{34.2}$$

The time interval between the emission and detection of the light pulse is a relative quantity, different for different observers. This effect is an example of *time dilation*. The concept of time interval is inseparably linked to specific pairs of events and to the observers measuring the interval.

Note: A' means event A (light pulse is emitted) as observed in frame 2. The light pulse is received at location D', different from A', because the clock is moving.

The effect is independent of the construction of the clock or its direction of motion. For another example, see Problem 27.

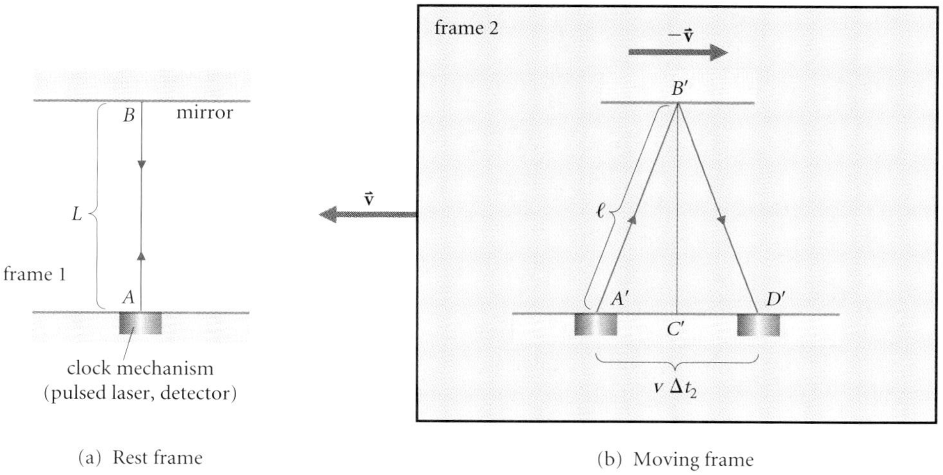

■ Figure 34.3
A clock measures the time interval between the emission and reception of a light pulse. (a) Rest frame. In one cycle the pulse travels from A to B and back to A, taking time $\Delta t = 2L/c$. (b) Moving frame. The clock moves at speed v and the light pulse follows a triangular path $A'B'D'$, still at speed c. The time for the light pulse to travel the longer path is greater than the interval Δt measured in the rest frame.

The smallest time interval Δt_1 between two events is measured in the reference frame in which both events occur at the same place. In another frame moving at speed v with respect to the first, the measured time interval is:

$$\Delta t_2 = \gamma \, \Delta t_1. \tag{34.3}$$

Cesium clocks are accurate enough to measure the time dilation effect between two similar clocks, one on the Earth's surface and one carried in a jet aircraft. Substantial time dilations occur for objects moving close to the speed of light and are observed in the interactions of subatomic particles.

EXAMPLE 34.1 ♦♦ Laboratory experiments produce low-speed μ^- mesons (muons), which then decay spontaneously with an average lifetime of $\tau = 2.197\ \mu s$. Cosmic rays striking the Earth's atmosphere at an altitude $h = 25.00$ km also produce muons that are later detected at the surface. What speed is necessary for an average muon to reach the surface before decaying?

MODEL The muon's average lifetime is determined by its physical structure. Crudely speaking, in the reference frame in which it is at rest, each muon carries a clock with a definite period. The events of muon production and detection occur within 2.197 μs (on average) measured in this muon rest frame, where they occur at the same place—the position of the muon. The production and detection events occur at different places in the Earth reference frame, and the time interval between them is longer: $\Delta t_E = \gamma\tau$.

SETUP The required speed of the muons is:

$$v = h/\Delta t_E = h/(\gamma\tau).$$

Using eqn. (34.2) for γ in terms of β:

$$\beta \equiv \frac{v}{c} = \frac{h}{\gamma\tau c} = \frac{h}{\tau c}\sqrt{1 - \beta^2}.$$

SOLVE

$$\beta^2 = \left(\frac{h}{\tau c}\right)^2 (1 - \beta^2) \Rightarrow \beta^2\left[1 + \left(\frac{h}{\tau c}\right)^2\right] = \left(\frac{h}{\tau c}\right)^2.$$

Putting in the numbers:

$$\frac{h}{\tau c} = \frac{25.00 \times 10^3 \text{ m}}{(2.197 \times 10^{-6} \text{ s})(2.998 \times 10^8 \text{ m/s})} = 37.96.$$

Thus:

$$\beta = \frac{37.96}{\sqrt{37.96^2 + 1}} = 0.9997.$$

ANALYZE By contrast, naively using Newtonian theory without time dilation would give a speed $v = (25 \text{ km})/\tau = 38c$!

The detection of muons in cosmic ray showers (in 1937) occurred long after special relativity had been accepted. Nevertheless, it constitutes one of the earliest and most forceful demonstrations of time dilation. ∎

34.1.4 *Length Contraction*

In the Earth reference frame, time dilation explains how muons can reach detectors on the Earth's surface. However, in the muon's reference frame, the detector has only 2.2 μs to reach the muon. At a speed of $0.9997c$, the detector cannot move 25 km in that time. The resolution of this dilemma is that the detector doesn't have to travel 25 km; its distance from the muon is reduced by a factor $1/\gamma$!

EXERCISE 34.1 ♦ Calculate γ for $\beta = 0.9997$ and show that the detector must travel 660 m in the muon's frame. Show that it can just do this in 2.2 μs.

Like time intervals, distances depend on the measuring procedure. Let's look at what happens when each observer computes distance as speed multiplied by time interval. An observer on the Earth sees the muon travel a distance ℓ_E in a time $t_E = \gamma\tau$ and computes the muon's speed as $v = \ell_E/t_E = \ell_E/\gamma\tau$. Conversely, the muon "sees" the detector travel a distance ℓ_μ in a time $t_\mu = \tau$ and computes the detector's speed as $v = \ell_\mu/t_\mu = \ell_\mu/\tau$. But the relative speed of Earth and muon is the same in each of their reference frames. Thus $\ell_\mu/\tau = \ell_E/\gamma\tau$, and so $\ell_\mu = \ell_E/\gamma$.

NOTICE THAT WHEN WE COMPUTE THE SPEED OF SOMETHING THIS WAY WE MUST MEASURE BOTH ℓ AND t IN THE *SAME FRAME*.

EXAMPLE 34.2 ♦♦ A rocket 1.0×10^2 m long (measured by engineers on board) flies past a space station at $v = 0.80c$ (■ Figure 34.4). How long does the rocket take to pass an astronaut on the space station? What is the length of the rocket as measured by the astronaut?

MODEL The astronaut measures the time interval between two events: E_1—nose of rocket passes astronaut, and E_2—tail of rocket passes astronaut. The astronaut determines the rocket's length from its speed times the time interval. We compare these measurements with those of the rocket engineers to see how the rocket lengths in the two frames compare.

SETUP First, we calculate the time as observed by the rocket engineers:

$$\Delta t_R = \frac{\ell_R}{v} = \frac{100\ \text{m}}{0.80(3.0 \times 10^8\ \text{m/s})} = 0.42\ \mu\text{s}.$$

The two events E_1 and E_2 occur at the same place in the astronaut's frame, so the time interval in that frame is the shorter. From eqn. (34.3):

$$\Delta t_A = \Delta t_R/\gamma = \ell_R/(v\gamma) = (0.42\ \mu\text{s})\sqrt{1 - 0.80^2}$$
$$= (0.42\ \mu\text{s})(0.60) = 0.25\ \mu\text{s}.$$

SOLVE The length of the rocket in the astronaut's frame is:

$$\ell_A = v\,\Delta t_A = v\,\Delta t_R/\gamma = \ell_R/\gamma = (100\ \text{m})(0.60) = 60\ \text{m}.$$

ANALYZE It is important to identify clearly the *pair of events* that defines the measurement of any length.

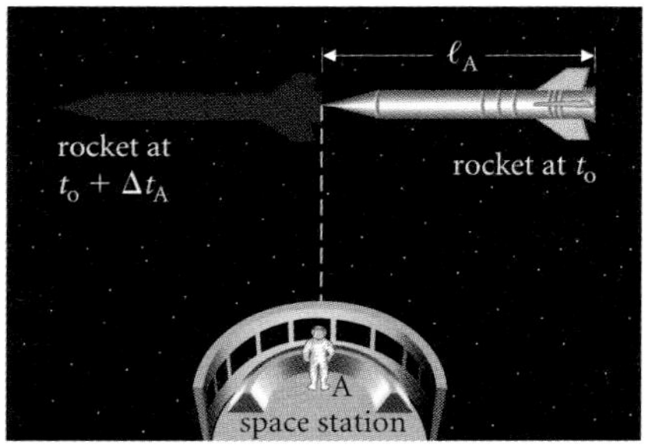

■ FIGURE 34.4
A rocket flies past an astronaut A in the space station, who measures the rocket's length as $\ell = v\,\Delta t_A$.

The length of the rocket and the distance traveled by the muon are relative quantities. This effect is called *length contraction*.

> The length ℓ of an object is longest in the reference frame in which it is at rest. In another frame moving *parallel* to the object, its length is:
>
> $$\ell' = \ell/\gamma. \qquad \textbf{(34.4)}$$

Like time dilation, length contraction depends on the value of γ, which is greater than or equal to 1 and becomes infinite as v approaches c. This is our first indication that physical objects cannot exceed the speed of light.

34.1.5 Simultaneity

When events happen before our eyes, we can usually tell whether they happen at the same time or not. But how do we know about the timing of separated events? In action thrillers, the team meets to synchronize watches, or waits for a radio message and then follows a set

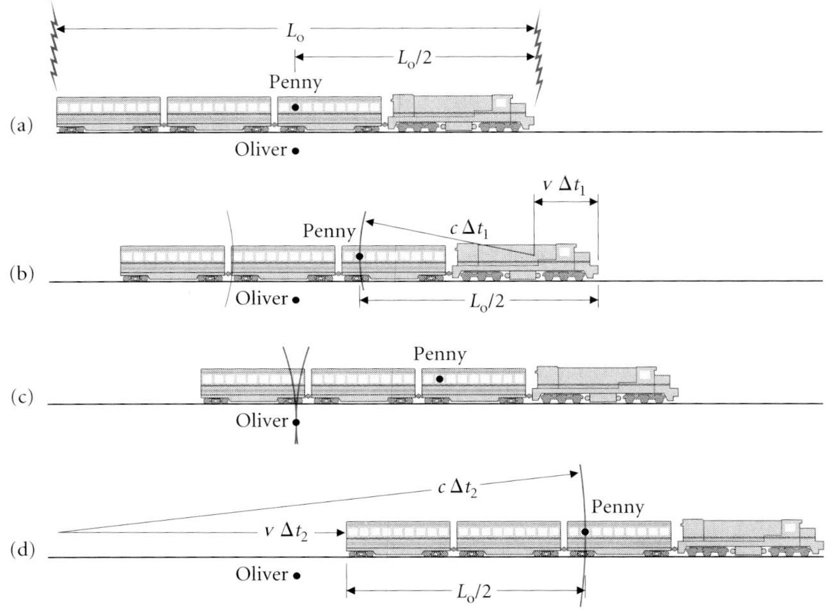

■ FIGURE 34.5
Simultaneity. Lightning bolts strike each end of the train at the same time, as measured by the observer Oliver. The sequence of events in Oliver's frame is: (a) Lightning bolts strike the train as Penny passes Oliver. (b) Light signal from front lightning bolt (bolt 1) reaches Penny after a time Δt_1. (c) Light signals from both bolts reach Oliver. (d) Light signal from rear bolt (bolt 2) reaches Penny a time Δt_2 after (a).

schedule. We do the same in everyday life. One of the most intriguing consequences of Einstein's postulates is that this method of synchronizing events depends on the reference frame. To see why, let's study one of Einstein's thought experiments.

A train, the Relativistic Flyer, moves to the right at a large speed v. A passenger P (Penny) sits in the middle of the train, and an observer O (Oliver) stands beside the tracks. Two lightning bolts strike the ends of the train. ■ Figure 34.5 shows the sequence of events in Oliver's reference frame.

OLIVER CONCLUDES THAT BOTH FLASHES OCCURRED AT THE SAME TIME BECAUSE THEY REACH HIM FROM EQUAL DISTANCES AT THE SAME TIME.

Oliver's interpretation: The two bolts strike at the same time, just as Penny passes him. Penny is moving to the right and encounters the light flash from bolt 1 before it reaches him. Next, both flashes arrive together at Oliver's position. Finally, the light flash from the second bolt reaches Penny.

Penny's interpretation of these events is different. The two flashes start at equal distances from her, travel at the same speed, and thus require the same time interval to reach her. The only way the flashes can arrive one after the other is for them to have occurred at different times. According to Penny, bolt 1 strikes before Oliver passes by and bolt 2 strikes later.

EXERCISE 34.2 ❖ Illustrate the sequence of events in Penny's reference frame, beginning with bolt 1 striking. (Assume the relative velocity of Oliver and Penny is such that bolt 2 strikes before Penny sees the flash from bolt 1.)

EXAMPLE 34.3 ♦♦♦ If the train's length is $L_P = 5.0 \times 10^2$ m (measured by Penny) and it is moving at $4c/5$, what time difference does Penny observe between the occurrence of the two bolts?

MODEL Since the light from each bolt travels at the same speed c in Penny's frame, the difference in the arrival time of the flashes equals the difference between the times the bolts struck. That difference is most readily calculated in Oliver's reference frame and then transformed to Penny's reference frame using the time dilation formula.

SETUP Since the train is at rest in Penny's frame, her measurement of its length is greater than Oliver's. So its length in his reference frame is $L_O = L_P/\gamma$ (eqn. 34.4). In Oliver's frame, a time interval Δt_1 elapses between the bolts striking and Penny seeing the first flash. During this interval Penny travels a distance $v\ \Delta t_1$ and the flash travels $c\ \Delta t_1$. The sum of these distances is $L_O/2$ (Figure 34.5):

$$(c + v)\Delta t_1 = L_O/2 = L_P/(2\gamma).$$

Similarly, Penny sees the second flash after an interval Δt_2, in Oliver's frame, where:

$$c\ \Delta t_2 = L_O/2 + v\ \Delta t_2,$$

or

$$(c - v)\Delta t_2 = L_P/(2\gamma).$$

The time interval in Oliver's frame between the two flashes reaching Penny is:

$$\delta t_O = \Delta t_2 - \Delta t_1 = \frac{L_P}{2\gamma}\left(\frac{1}{c-v} - \frac{1}{c+v}\right) = \frac{L_P v}{\gamma(c^2 - v^2)} = \frac{\gamma L_P v}{c^2}.$$

SOLVE The two flash receptions occur at different positions in Oliver's frame but at the same position in Penny's frame. Thus the time interval measured by Penny is the shorter:

$$\delta t_P = \frac{\delta t_O}{\gamma} = \frac{L_P v}{c^2} = \left(\frac{500\ \text{m}}{3.0 \times 10^8\ \text{m/s}}\right)\left(\frac{4}{5}\right) = 1.3 \times 10^{-6}\ \text{s}.$$

ANALYZE Like time dilation and length contraction, the disagreement among observers over simultaneity of events is appreciable only when the relative speed of the observers is close to that of light. ∎

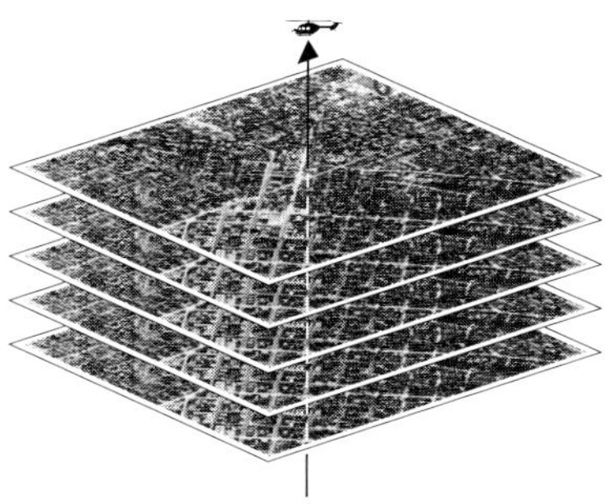

■ FIGURE 34.6
The helicopter parable. Photographs taken from a helicopter hovering over 42nd and Broadway are stacked to form a record of the history of New York City. The intersection's image is always at the center of the photographs.

34.2 SPACE-TIME

Time dilation, length contraction, and relativity of simultaneity are closely related phenomena. For example, length measurements involve the distance between two positions *at the same time* so that length contraction is just another form of disagreement over simultaneity. Hermann Minkowski realized in 1905 that these phenomena have a natural interpretation if we abandon space and time as separate concepts and think of them as unified in a single, four-dimensional *space-time.* To visualize this concept, let us imagine that an artist decides to record a history of New York City by making a sequence of aerial photographs from a helicopter hovering high above the city. Each photograph is labeled with the time it is taken, and the history is constructed by stacking the photographic plates vertically (■ Figure 34.6). This marvelous work of art shows all details of the city's life, including the surging crowds at 42nd and Broadway and poor Mr. Smith, who trips while running to catch the bus. The artist has constructed a *space-time diagram.*

BECAUSE THE SPEED OF LIGHT IS FINITE, A PHOTOGRAPH USING LIGHT CAN ONLY BE A METAPHOR FOR A RECORD OF SIMULTANEOUS EVENTS. IN §34.2.1 WE SHALL DESCRIBE A METHOD THAT ELIMINATES THE INACCURACIES OF A PHOTOGRAPH.

34.2.1 *Representation of Space-time in a Single Reference Frame*

Every occurrence (e.g., Mr. Smith trips) has a location in space and time. An ideal *event* occurs at a point in space and at an instant in time. It is described by the time when it happens and the three Cartesian coordinates of its position (t, x, y, z). Space-time is the four-dimensional collection of all events. Space evolves through time. A point, such as 42nd and Broadway, has a history, a sequence of events, represented by a line in space-time called a *world line.* An alternative view of space-time is the collection of world lines corresponding to each spatial point.

The most common convention for drawing space-time is to place the time axis vertically. All three spatial dimensions are perpendicular to the time axis, so at least one has to be left out of the diagram! Then simultaneous events occur in parallel planes perpendicular to the time-axis (■ Figure 34.7). It is convenient to use ct as the coordinate along the time axis so that the units on all axes are meters. Equivalently, we could put (distance/c) on each space-axis and use units of seconds.

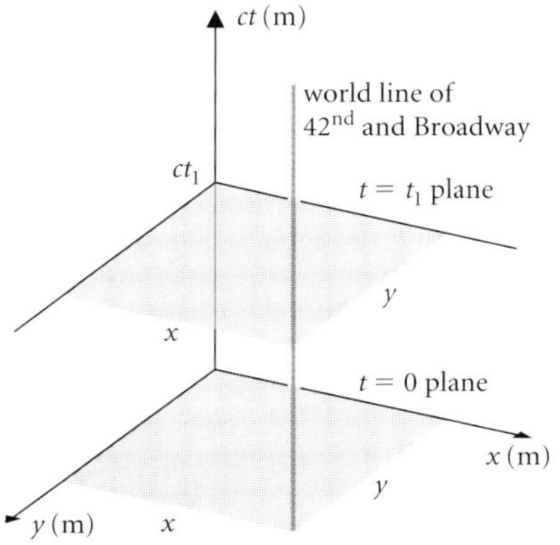

■ FIGURE 34.7
Space-time diagram with two spatial coordinates, and time (coordinate ct) vertical. Simultaneous events occur in parallel planes. Two such planes are shown here. A spatial point has a history, represented by a line in the diagram called a *world line.* The world line of 42nd and Broadway is vertical because the street intersection does not move in this reference frame.

Digging Deeper

Defining Coordinates in a Reference Frame

To establish a reference frame, we need a way to compare the times of events on different world lines. We can use standard clocks to measure time but, because of time dilation, they may not move with respect to each other. Thus we need many clocks and many observers to read them. In real space-time, there are usually only a few spacecraft or pieces of laboratory equipment available to receive and respond to signals. So for thought experiments, we use an ideal space-time that is filled with observers, one at each spatial point. One particular observer's world line is the time-axis of the frame, and an event along this world line is chosen as the origin. The method for synchronizing clocks and assigning spatial coordinates to the observers closely mimics measurement procedures in a real experiment. Each observer must be equipped with a device for sending EM waves (e.g., radar).

An observer, A, first measures her distance from the origin by sending a radar signal as shown in ■ Figure 34.8. The observer at the origin sends a reply immediately upon receiving the signal. According to Einstein's postulate, the signals travel at speed c, so if the first signal is emitted at time t_e and the reply is received at time t_r, the signal travels a total distance $c(t_r - t_e) = 2d$, and so observer A's measured distance from the origin is:

$$d = \tfrac{1}{2}c(t_r - t_e).$$

The observers check that they are mutually stationary by repeating the measurements and ensuring that $d =$ constant.

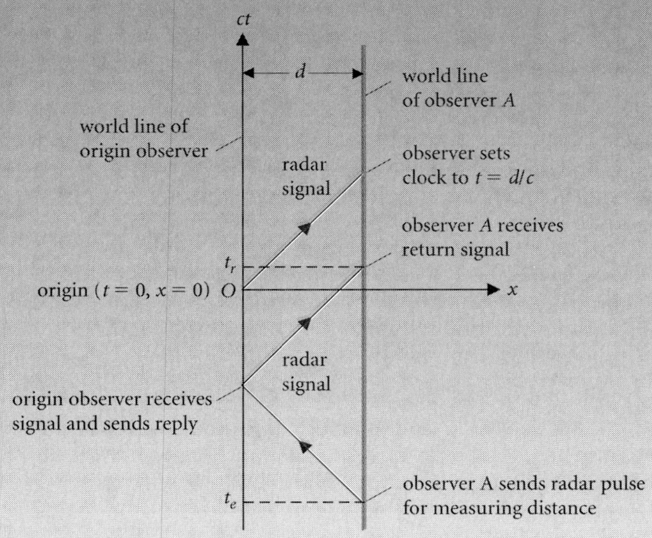

■ **Figure 34.8**
Defining coordinates. A radar signal is sent at time t_e from observer A to the observer at the origin, and back. The reply is received at time t_r. The distance d is determined to be $\frac{1}{2}c(t_r - t_e)$. Once A knows the distance from O, A's clock is synchronized using a signal sent from the origin.

Next, they synchronize their clocks using a light flash emitted at the origin. (*Remember:* The origin means $t = 0$ as well as $x = y = z = 0$.) Knowing the signal travels at speed c, observer A, distance d away, sets her clock to $t = d/c$ as the light flash passes. Now the observers are ready to conduct experiments.

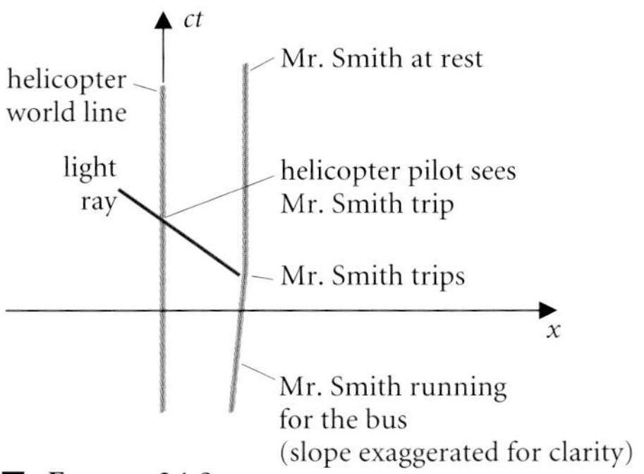

■ **Figure 34.9**
Events in the helicopter parable shown in a space-time diagram. Only one space dimension is shown. The part of Mr. Smith's world line where he is running has a barely discernible slope ($v \ll c$) even though we have exaggerated it enormously. After he trips, his world line is vertical, indicating that he is at rest in this frame.

EXAMPLE 34.4 ❖ Draw a space-time diagram illustrating the events in the helicopter parable.

MODEL As the helicopter hovers over the city, it remains for an extended time in one spatial position that we choose to call $x = 0$, $y = 0$, $z = 0$. Thus there is a sequence of events with increasing time coordinates but fixed space coordinates that describe the helicopter. This is the helicopter's *world line.* Since its spatial position is the spatial origin, this line is also the time-axis. We pick one event on this line as the origin ($t = 0$, $x = 0$, $y = 0$, $z = 0$).

Mr. Smith, running for the bus at speed v, occupies a sequence of positions

$$x = x_o + vt = x_o + (v/c)ct = x_o + \beta(ct).$$

His world line is the set of points with coordinates $(t, x, y, z) = (t, x_o + \beta ct, 0, 0)$—that is, a line with slope $1/\beta$ (■ Figure 34.9). When he trips, his speed becomes zero. His world line changes direction and moves vertically ($x =$ constant), indicating that he is at rest.

The helicopter pilot sees Mr. Smith by observing a light wave that propagates from Mr. Smith to the helicopter. It travels at speed c ($\beta = 1$), so its world line has slope 1 and is therefore at 45° to each axis.

ANALYZE We have to exaggerate the slope of Mr. Smith's world line in the diagram to notice it at all. For any reasonable estimate of the speed at which he can run, β would be tiny. ∎

34.2.2 Space-time Interval

Since space and time coordinates are not invariant, familiar concepts like past and future need careful definition. There is an invariant quantity, called the space-time interval, that allows us to order events in space-time and provide those definitions.

Suppose a flash of light is emitted in all directions at an event A. The expanding wave front comprises the events labeled *light cone* in ∎ Figure 34.10. For example, light moving in the positive x-direction defines events with coordinates t, $x = x_A + c(t - t_A)$. These coordinates describe the straight line shown in the diagram. Light moving in all possible directions in the x-y-plane gives a set of straight lines forming the surface of a cone, hence the name.

WE CANNOT EASILY DRAW THE RESULT OF LIGHT MOVING IN ALL THREE SPATIAL DIRECTIONS.

The light cone from A divides space-time into several regions. Suppose event A is you at this moment reading this page. Your future is all the events you can influence. For example, by shining a flashlight on a piece of photographic film, you can expose it. The flashlight beam travels at the speed of light, and the event *film is exposed* would lie on the surface of the light cone. If you throw a baseball through a window, the event *window breaks* would be well inside your future light cone since the ball travels at a speed much less than c. Anything leaving A carrying energy or information moves at a speed $\leq c$ and has a world line on or within the light cone: this defines the future of A.

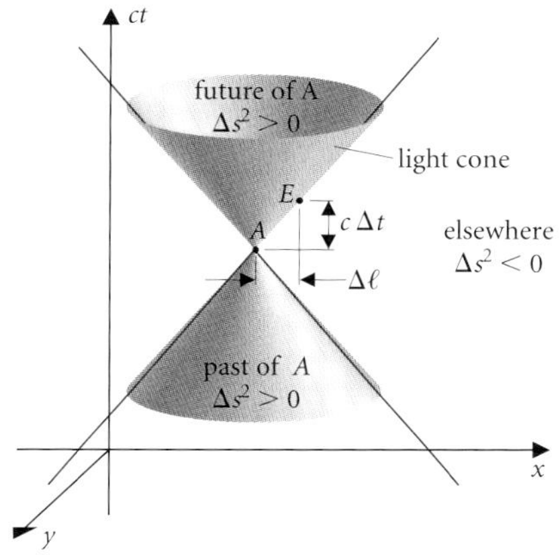

∎ FIGURE 34.10
The past and future. Two spatial dimensions are shown here. The past and future of event A are defined by the *light cone:* the set of events corresponding to a flash of light emitted in all directions at A. At E, $\Delta\ell/\Delta t = c$. The space-time interval between events on the light cone is zero. Events not in the past or future of A are *elsewhere.*

> The *future* of an event A is the set of events that could receive a signal from A—that is, the events within the light cone that begins at A.

Event A also has a past light cone. For example, your sunburned nose is a result of light from the Sun reaching your nose. Your strained shoulder muscle is a result of your walk to the library carrying your books. All these events lie within the light cone that reaches event A from the past.

> The *past* of an event A is the region within the past light cone—that is, the set of events from which a signal could reach A.

Events that are in the past or future of A are said to have a *timelike* separation from A. Events that can neither send a signal to, nor receive a signal from, A are *elsewhere* and are said to have a *spacelike* separation from A.

If one event causes another, we would expect that relation to be independent of reference frame. A quantity called the *space-time interval* describes such relations between events. The correct expression for interval is most easily discovered by looking at events on the same light cone. The time interval Δt and the spatial distance $\Delta\ell$ between two such events A and E are related: $\Delta\ell = c\,\Delta t$ (Figure 34.10). Thus:

$$\Delta s^2 \equiv c^2(\Delta t)^2 - (\Delta\ell)^2$$

is zero for every event E on the light cone. According to Einstein's postulate, the speed of light is the same in every reference frame so that $\Delta s^2 = 0$ is an invariant description of the light cone.

Next, we notice that for events in the past or future of A, $\Delta\ell < c\,\Delta t$, so:

$$\Delta s^2 = c^2(\Delta t)^2 - (\Delta\ell)^2 > 0,$$

and for events elsewhere, $\Delta s^2 < 0$. These distinctions are also invariant. Combining these results, we may deduce the basic geometric property of space-time.

COMPARE WITH THE PYTHAGOREAN THEOREM: $\Delta\ell^2 = \Delta x^2 + \Delta y^2 + \Delta z^2$. BUT NOTE THAT THE MINUS SIGN IN EQN. (34.5) IS NOT A TYPOGRAPHICAL ERROR.

The square of the space-time interval,

$$\Delta s^2 \equiv c^2(\Delta t)^2 - (\Delta\ell)^2, \qquad (34.5)$$

between any two events in space-time is an invariant.

This interval may be used to define invariant quantities that replace the Newtonian concepts of time interval and length.

THE PROPER TIME BETWEEN TWO EVENTS IS ONLY DEFINED FOR EVENTS WITH A TIMELIKE SEPARATION.

The *proper time* interval $\Delta\tau$ between two events with a timelike separation is the space-time interval between them divided by c:

$$\Delta\tau = \Delta s/c = \sqrt{(\Delta t)^2 - (\Delta\ell)^2/c^2}. \qquad (34.6)$$

The proper time interval $\Delta\tau$ equals the coordinate time interval Δt in a reference frame where the events occur at the same place ($\Delta\ell = 0$). If the two events are on the world line of a particle, this is the reference frame in which the particle is at rest. Proper time measures the rate of all physical processes—the decay of muons or the rate at which your heart beats. Two events in your life occur at the same place in your reference frame. Observers in any other reference frame measure a spatial separation between those events and a different time interval, but the proper time they calculate is the interval shown by your watch.

THERE IS NO SUCH THING AS A PROPER DISTANCE BETWEEN EVENTS WITH A TIMELIKE SEPARATION.

The *proper distance* between two events with a spacelike separation is the magnitude of the space-time interval between them:

$$L = \sqrt{-\Delta s^2} = \sqrt{(\Delta\ell)^2 - c^2(\Delta t)^2}. \qquad (34.7)$$

In the reference frame where an object is at rest, the locations of its ends at some time t define two events with $\Delta t = 0$. The proper distance between these events is called the proper length or *rest length* of the object.

EXAMPLE 34.5 ♦ A sigma particle produced in an accelerator experiment is captured in a detector 0.150 m away 0.520 ns later. Find the speed of the particle and the proper time interval between its production and detection.

MODEL We are given the laboratory coordinate time interval. Ordinary kinematics applies within a given reference frame to determine the speed. To obtain the proper time, we apply eqn. (34.6).

SETUP AND SOLVE The particle's speed is:

$$v = \Delta\ell/\Delta t = (0.150 \text{ m})/(0.520 \times 10^{-9} \text{ s}) = 2.88 \times 10^8 \text{ m/s}.$$

The proper time interval is:

$$\Delta\tau = \sqrt{(\Delta t)^2 - (\Delta\ell)^2/c^2} \qquad \text{(i)}$$

$$= \sqrt{(0.520 \times 10^{-9} \text{ s})^2 - (0.150 \text{ m})^2/(3.00 \times 10^8 \text{ m/s})^2} = 0.143 \text{ ns}.$$

ANALYZE As we saw in the muon experiment (Example 34.1), it is time in the particle's rest frame—the proper time—that determines whether a particle is likely to reach a detector before it decays. The mean lifetime of a sigma particle—0.15 ns—is longer than the proper time we calculated, consistent with its capture in the detector. Einstein's theory is verified daily by such events in accelerators worldwide.

WE'LL HAVE MORE TO SAY ABOUT PARTICLE DECAY IN CHAPTERS 36 AND 37.

Since $\Delta\ell = v\,\Delta t$, eqn. (i) may be written $\Delta\tau = \Delta t/\gamma$, so it is just the time dilation formula, eqn. (34.3). ∎

34.3 The Lorentz Transformation

34.3.1 *Coordinate Transformation*

Frequently, we need to compare representations of space-time in different reference frames. For this we require the transformation equations relating the coordinates of an arbitrary event in one frame to its coordinates in the other frame. Relative motion is the important difference between the two frames, so we avoid irrelevant complications by choosing both x-axes to be along the direction of the relative velocity and placing the origin at the same event in each frame. The coordinates in the second, prime, frame are labeled t', x', y', and z'. Thus the *prime frame* moves with velocity $\vec{\mathbf{v}} = v\hat{\mathbf{i}}$ with respect to the *unprime frame.* Dig Deeper (p. 1082) to see that the transformation is given by:

$$ct' = \gamma(ct - \beta x), \qquad x' = \gamma(x - \beta ct),$$
$$y' = y, \qquad z' = z. \tag{34.8}$$

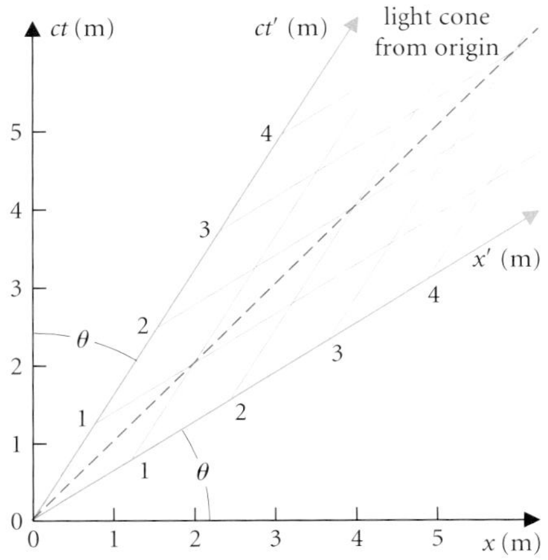

■ Figure 34.11
The Lorentz transformation. The prime axes are shown in the unprime frame. Both axes are rotated toward the light cone from the origin, with $\tan\theta = \beta$. Keep in mind when using this diagram that the scales on the two sets of axes are not the same!

Using the Lorentz transformation (eqns. 34.8), we can determine how the axes of the prime frame appear in a space-time diagram drawn by unprime observers (■ Figure 34.11). Along the t'-axis, x', y', and z' equal zero. At points on this line, $x = \beta ct$. (Use eqns. 34.8 with $x' = 0$). This is the equation of a particle (the observer at the prime origin) moving to the right at speed βc. Similarly, along the x'-axis, t', y', and z' equal zero. At points along this line, $t' = 0 \Rightarrow ct = \beta x$, which describes a line making angle $\theta = \tan^{-1}\beta$ with the x-axis. Both the time- and space-axes are rotated toward the light cone. In the limit as $v \rightarrow c$, both axes coincide along a 45° line. This is a second indication that a relative speed greater than c cannot occur since, if it did, the space- and time-axes would trade places.

Note that $\beta c \equiv v$. As $\beta \rightarrow 0$, $\gamma \rightarrow 1$, and eqns. (34.8) reduce to the Galilean transformation in Table 34.1.

EXERCISE 34.3 ♦♦ In the prime frame, the unprime frame moves with velocity $\vec{\mathbf{v}} = -v\hat{\mathbf{i}}$. Use this fact to write down the Lorentz transformation from prime to unprime coordinates. Locate the unprime axes in a space-time diagram drawn by prime observers.

EXAMPLE 34.6 ♦♦ A rocket leaves Earth traveling at $v = 4c/5$. After 1 day, a radio signal is sent to the rocket from Earth. The rocket immediately sends a reply. Find the following: (a) The time at which the signal is received by the rocket, as perceived by the rocket crew. (b) The distance of the rocket from Earth at that time. (c) The time at which the return signal is received on Earth, as measured by both Earth and rocket observers.

MODEL We begin by drawing the space-time diagram in the Earth frame (■ Figure 34.12) with the events E_0 (signal sent to rocket), E_1 (signal received at rocket), and E_2 (return signal received at Earth). We use the given information to find the coordinates of these events in the Earth frame and then use the Lorentz transformation to find the coordinates in the rocket frame. We choose the x-axis to be parallel to the rocket's velocity, with the origin at the event *rocket leaves Earth.*

SETUP E_0: $x_0 = 0$, $t_0 = 1$ day.

E_1: E_1 occurs at the intersection of the rocket's world line and the light cone from E_0. The rocket moves at speed v and thus its x-coordinate at time t is $x = vt$. The radio signal travels at speed c. At time t, it has been traveling for $\Delta t = t - t_0$ and its x-coordinate is $x = c(t - t_0)$. The radio signal meets the rocket when their x-coordinates are the same—that is, at t_1 when:

$$vt_1 = c(t_1 - t_0),$$

or

$$t_1 = \frac{ct_0}{c - v} = \frac{t_0}{1 - \beta} = \frac{1 \text{ day}}{1 - 4/5} = 5 \text{ days.}$$

Then $x_1 = vt_1 = 4$ light-days. [This is the answer to (b).]

E_2: The light signal takes 4 days to reach the rocket and an equal time to come back.

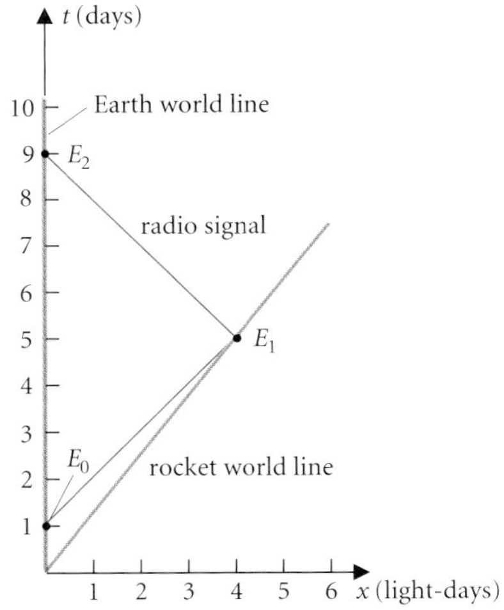

■ Figure 34.12
Space-time diagram for a rocket leaving Earth. The rocket launch is taken as the origin. Events E_0 (radio signal sent from Earth), E_1 (signal received at rocket), and E_2 (reply received at Earth) are shown. The distance unit is a light-day: the distance traveled by light in 1 day.

Digging Deeper

Derivation of the Lorentz Transformation

Since both the prime and unprime frames are inertial, the transformation equations should be linear in the coordinates so that uniform motion in one frame is uniform in the other. When the relative speed of the frames is much less than c, the transformation reduces to the Galilean result (Table 34.1). These requirements are satisfied by a transformation of the form:

$$ct' = Act + Bx, \qquad x' = Dct + Ex, \qquad y' = Fy, \qquad z' = Fz, \tag{i}$$

where the coefficients A, B, D, E, and F depend only on the relative velocity. The same coefficient F occurs for y' and z' because with both directions perpendicular to the relative velocity, relative motion should affect them in the same way.

To determine the coefficients, we use the invariance of the space-time interval: Δs^2 between any two events is the same in both frames. We choose the origin and another event with $y = z = 0$ but arbitrary values of x and t. Then also $y' = z' = 0$, and the space-time interval between the two events is:

$$\Delta s^2 = c^2t^2 - x^2 = c^2(t')^2 - (x')^2.$$

Applying the transformation (i) to the prime coordinates, we find:

$$c^2t^2 - x^2 = (Act + Bx)^2 - (Dct + Ex)^2.$$

Gathering up like terms, we have:

$$c^2t^2(A^2 - D^2 - 1) + 2xct(AB - DE) + x^2(B^2 - E^2 + 1) = 0.$$

Since this equation holds for arbitrary values of x and t, the three coefficients each vanish separately:

$$A^2 - D^2 - 1 = 0, \tag{ii}$$

$$AB - DE = 0, \tag{iii}$$

and

$$B^2 - E^2 + 1 = 0. \tag{iv}$$

From eqn. (iii), $AB = DE$, so we can eliminate B from eqn. (iv):

$$0 = \frac{D^2E^2}{A^2} - E^2 + 1 = E^2\left(\frac{D^2 - A^2}{A^2}\right) + 1.$$

Then using eqn. (ii), we find: $E^2\left(\frac{-1}{A^2}\right) + 1 = 0$ or $A^2 = E^2$.

Both coefficients A and E are positive, since t and t' increase together, as do x and x'. So $A = E$, and it follows that $D = B$.

Next we consider a particular event whose coordinates we know. The spatial origin of the prime frame ($x' = 0$, $y' = 0$, $z' = 0$) moves to the right with speed v in the unprime frame. Thus at any time T in the unprime frame, its coordinates are: $x = vT$, $t = T$ and $x' = 0$, $t' = T'$.

Using the second of eqns. (i) to relate these coordinates, we have:

$$x' = 0 = Dct + Ex = DcT + EvT,$$

or

$$D = -(v/c)E = -\beta E = -\beta A.$$

Then from eqn. (ii),

$$(1 - \beta^2)A^2 - 1 = 0 \Rightarrow A = \frac{1}{\sqrt{1 - \beta^2}} \equiv \gamma.$$

With $E = A$ and $D = -\beta A$, the resulting Lorentz transformation is:

$$ct' = \gamma(ct - \beta x) \quad \text{and} \quad x' = \gamma(x - \beta ct).$$

We complete the derivation by noting that F equals 1 (■ Figure 34.13 and Exercise 34.4).

EXERCISE 34.4 ♦♦ By considering an event with $y, z \neq 0$, show that $F = 1$.

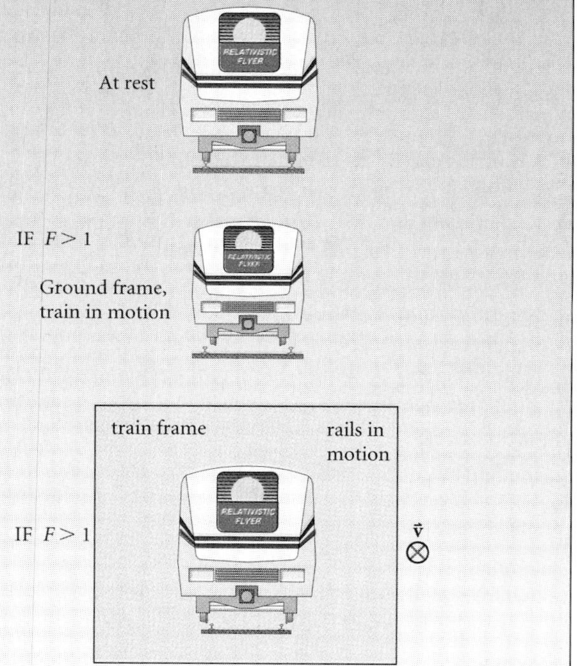

■ Figure 34.13
What if rapidly moving objects were to contract in the direction perpendicular to their motion—that is, if the coefficient F in the Lorentz transformation were not unity? The Relativistic Flyer is designed to fit exactly on the rails when at rest. For $F > 1$, an engineer on the ground would observe the train falling between the tracks as it sped up. An engineer on the train would observe the moving rails contracting between the wheels of the train. These two opposite disasters cannot both happen. A similar argument rules out width expansion ($F < 1$). This argument relies on the fact that both engineers agree on the simultaneity of the events that they use to measure widths (see Problem 16). They can agree because the separation of the events is perpendicular to the velocity.

SOLVE $x_2 = 0$, $t_2 = 5 \text{ days} + 4 \text{ days} = 9 \text{ days}$.

SETUP It remains to find t_1' and t_2'. From eqn. (34.8):

$$t_1' = \gamma(t_1 - \beta x_1/c),$$

with

$$\gamma = \frac{1}{\sqrt{1-\beta^2}} = \frac{1}{\sqrt{1 - 16/25}} = \frac{5}{3}.$$

SOLVE

$$t_1' = \tfrac{5}{3}[5 \text{ days} - \tfrac{4}{5}(4 \text{ days})] = (\tfrac{5}{3})(\tfrac{9}{5}) \text{ days} = 3 \text{ days}.$$

Similarly:

$$t_2' = \tfrac{5}{3}(9 \text{ days}) = 15 \text{ days}.$$

ANALYZE Notice that for two events occurring at the same place in the rocket frame (*rocket leaves Earth* and *rocket receives signal*), the rocket interval of 3 days is shorter than the Earth interval of 5 days. For events that occur at the same place in the Earth frame, the reverse is true. ∎

EXERCISE 34.5 ❖ Draw these events in the rocket frame.

Study Problem 20 ♦♦♦ *The Student's Revenge*

A chemistry professor is a passenger on the Relativistic Flyer, a train that travels at $0.80c$. A disgruntled physics student who failed the chemistry course notices that the train, length $\ell = 150$ m, must pass through a tunnel $L = 0.10$ km long. The student knows the train is contracted to a length $\ell/\gamma = \frac{3}{5}(150 \text{ m}) = 90$ m and arranges to close gates at both ends of the tunnel when the train is inside, thus trapping the professor (■ Figure 34.14). The professor is not worried. Knowing the tunnel is contracted to a length of $\frac{3}{5}(100 \text{ m}) = 60$ m, the professor concludes that the train will never be completely inside the tunnel and the student's scheme will fail. Who is right?

Modeling the System

The professor is certainly correct in that, from his point of view, the tunnel is shorter than the train. However, he has forgotten that events that are simultaneous in the student's frame, such as closing gates at the two ends of the tunnel, are not simultaneous in his frame. If the tunnel exit is closed first, the train can continue to enter the shortened tunnel (crumpling as it goes) until it is completely inside. Then the entrance closes and the professor is trapped.

To demonstrate the truth of this conjecture, we identify the four relevant events and look at their coordinates in both reference frames.

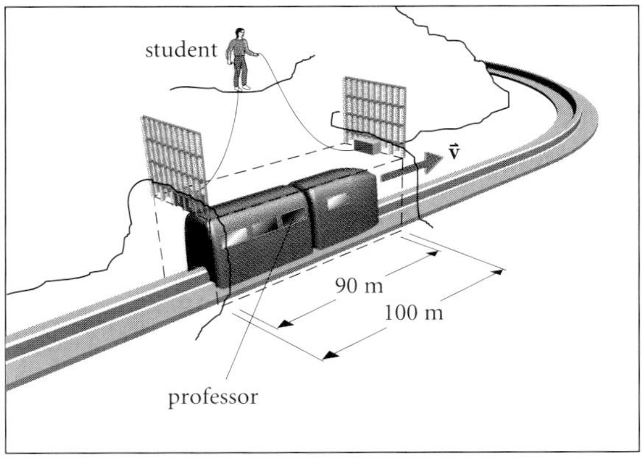

(a) Student's picture.

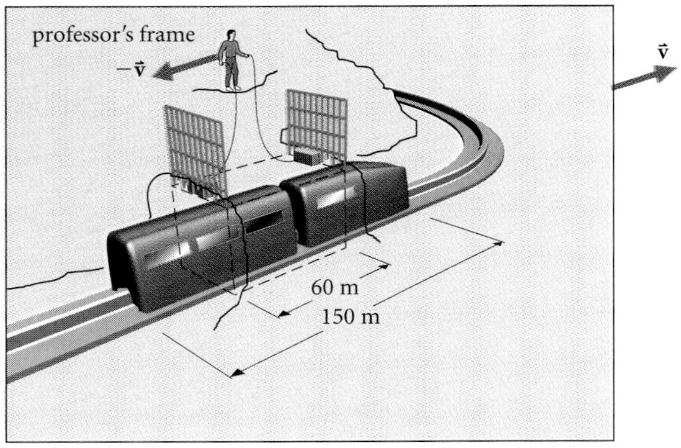

(b) Professor's belief.

■ FIGURE 34.14
(a) The disgruntled student is poised to close the gates and trap the Lorentz-contracted train inside the tunnel.

(b) The chemistry professor believes that, in his frame, the train will never be completely inside the Lorentz-contracted tunnel and thinks that he is safe.

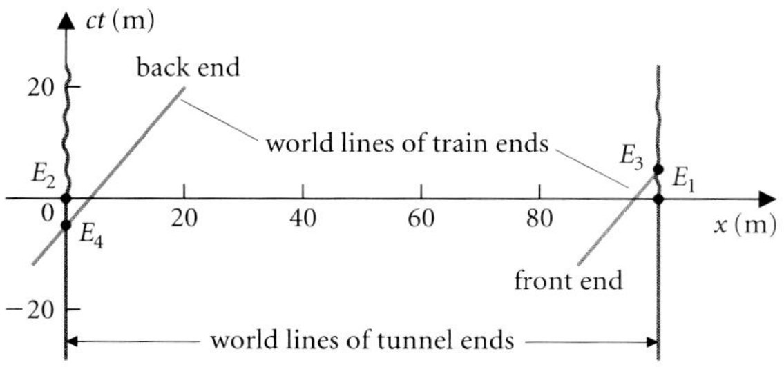

■ FIGURE 34.15
Space-time diagram for the events in the student's frame. The rest length of the tunnel is $L = 100$ m and the length of the train is $\ell/\gamma = 3(150 \text{ m})/5 = 90$ m. The world lines of both ends of the train and the tunnel are shown. Events E_1 and E_2, gates closing at the tunnel entrance and exit, are simultaneous in this frame.

Setup of Solution

The four events are as follows.

E_1: Tunnel exit closes.
E_2: Tunnel entrance closes.
E_3: Front end of train reaches tunnel exit.
E_4: Back end of train reaches tunnel entrance.

We choose the origin to be at the tunnel entrance, at the instant the student closes the gate. We assume the train is centered in the tunnel at $t = 0$ in the student's (unprime) frame. The tunnel has length L (■ Figure 34.15), so the first two events have coordinates:

$$x_1 = L,\ ct_1 = 0 \qquad \text{and} \qquad x_2 = 0,\ ct_2 = 0.$$

In the student's frame, the train has length ℓ/γ. At $t = 0$, its front end is a distance $(L - \ell/\gamma)/2$ from the tunnel exit and its back end is an equal distance from the tunnel entrance. Thus the front reaches the exit at time $ct_3 = (L - \ell/\gamma)/(2\beta)$ and the back end was at the entrance at time $ct_4 = -\ ct_3$. Thus events 3 and 4 have coordinates:

$$x_3 = L,\ ct_3 = \frac{\text{L} - \ell/\gamma}{2\beta} \qquad \text{and} \qquad x_4 = 0,\ ct_4 = \frac{-(L - \ell/\gamma)}{2\beta}.$$

Next, we use eqns. (34.8) with $\beta = \frac{4}{5}$, $\gamma = \frac{5}{3}$ to find the time coordinates in the professor's (prime) frame:

NOTE: E_2 IS THE ORIGIN IN BOTH FRAMES.

$$ct_1' = \gamma(ct_1 - \beta x_1) = \gamma(0 - \beta L) = -\gamma\beta L.$$
$$ct_2' = \gamma(ct_2 - \beta x_2) = \gamma(0 - 0) = 0.$$
$$ct_3' = \gamma(ct_3 - \beta x_3) = \gamma\left(\frac{L - \ell/\gamma}{2\beta} - \beta L\right) = \frac{\gamma L(1 - 2\beta^2) - \ell}{2\beta}.$$
$$ct_4' = \gamma(ct_4 - \beta x_4) = \gamma\left[\frac{-(L - \ell/\gamma)}{2\beta} - 0\right] = \frac{-\gamma L + \ell}{2\beta}.$$

Solution

We evaluate the times of the four events in the professor's frame:

$$ct_1' = -(\tfrac{5}{3})(\tfrac{4}{5})(100 \text{ m}) = -133 \text{ m}.$$
$$ct_2' = 0.$$
$$ct_3' = \frac{(5/3)(100 \text{ m})(1 - 32/25) - 150 \text{ m}}{8/5} = -123 \text{ m}.$$
$$ct_4' = -\frac{(5/3)(100 \text{ m}) - (150 \text{ m})}{8/5} = -10.4 \text{ m}.$$

In this frame the sequence of events is as follows. First, the tunnel exit is blocked [E_1 at $t' = -(133 \text{ m})/c = -0.44\ \mu\text{s}$]. Next, the front of the train smashes into the closed tunnel exit [E_3

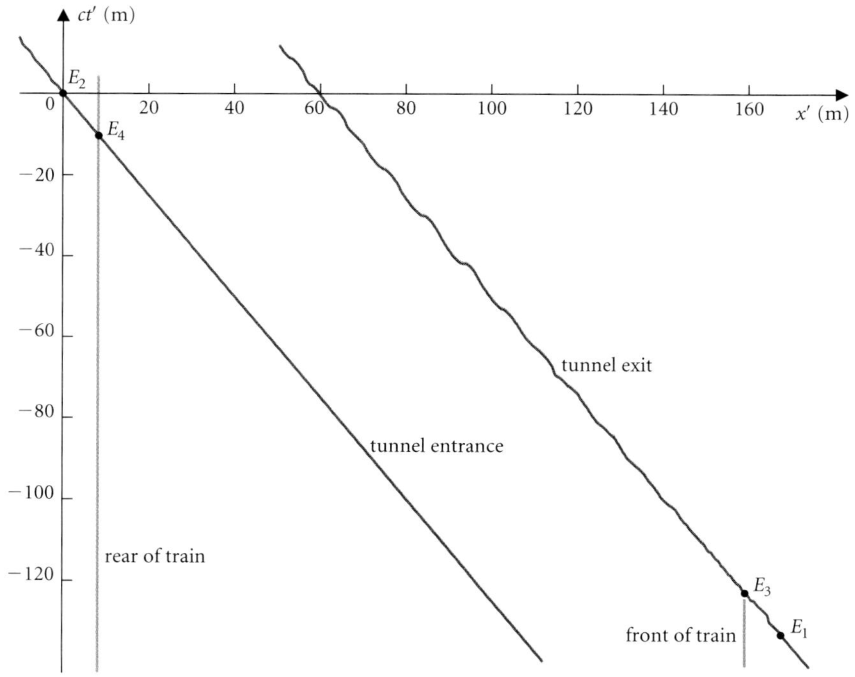

■ FIGURE 34.16
The tunnel viewed in the professor's frame. The tunnel has length $L/\gamma = 3(100\text{ m})/5 = 60$ m and the train has length 150 m. Events E_1 and E_2 are not simultaneous in this frame: E_1 occurs first. However, E_1 still precedes E_3, and E_4 precedes E_2. Compare with Figure 34.15.

at $t' = (-123\text{ m})/c = -0.41\ \mu\text{s}$]. Then the back end of the train enters the tunnel [E_4 at $t' = (-10.4\text{ m})/c = -0.035\ \mu\text{s}$]. Finally, the tunnel entrance is closed and the professor is trapped (E_2 at $t' = 0$). The events in the professor's frame are shown in ■ Figure 34.16.

Analysis

The sequence of cause and effect is preserved by the Lorentz transformation, even though events that are simultaneous in one frame are not simultaneous in another. Notice that in both frames E_1 precedes E_3 (the tunnel exit is blocked before the train runs into it) and E_4 precedes E_2 (the back end of the train enters the tunnel before it is blocked). However, the sequence of events that do not affect each other (E_1 and E_2, or E_3 and E_4) can change.

The moral of this story is never to jump to conclusions when solving problems in special relativity! And, of course, the student could (and should) have found a better way to handle her frustrations!

34.3.2 Velocity Transformation

From the Lorentz transformation for coordinates, we may derive rules for transforming an object's velocity components between frames. The definition of instantaneous velocity involves events separated by a small time interval along the world line of an object. Suppose two such events are separated by a displacement $d\vec{\mathbf{r}}'$ and a time dt' in the prime frame. The velocity in that frame is then:

$$\vec{\mathbf{u}}' = \frac{d\vec{\mathbf{r}}'}{dt'} = \frac{dx'\hat{\mathbf{i}} + d\vec{\mathbf{r}}'_{\perp}}{dt'},$$

where $d\vec{\mathbf{r}}_{\perp} = d\vec{\mathbf{r}}'_{\perp}$ is the component of $d\vec{\mathbf{r}}$ in the y-z-plane (■ Figure 34.17). The prime frame has velocity $\vec{\mathbf{v}} = \beta c\hat{\mathbf{i}}$ relative to the unprime frame. By differentiating eqns. (34.8), we can write both $d\vec{\mathbf{r}}'$ and dt' in terms of unprime coordinates:

$$\vec{\mathbf{u}}' = \frac{\gamma(dx - \beta c\ dt)\hat{\mathbf{i}} + d\vec{\mathbf{r}}_{\perp}}{\gamma(dt - \beta\ dx/c)},$$

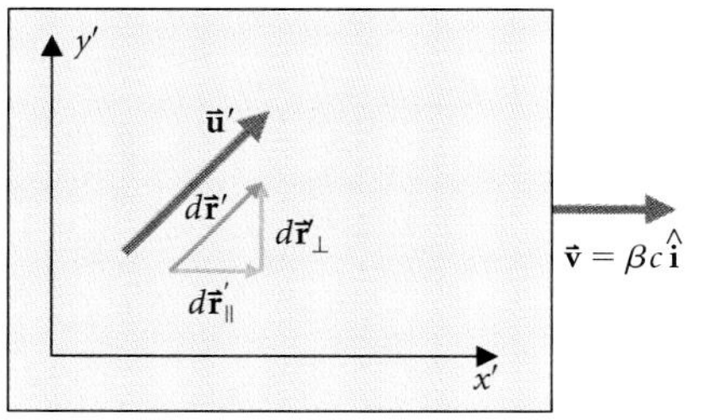

(a) Prime frame.

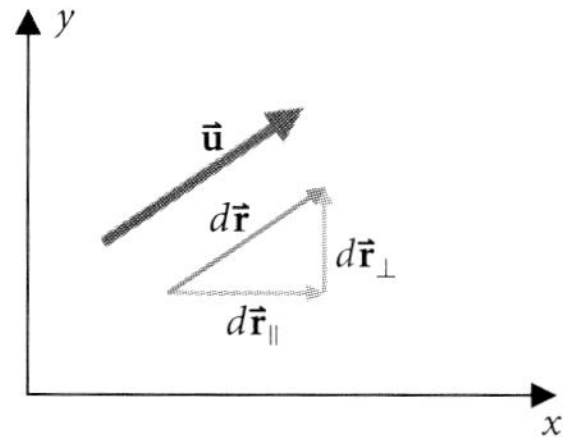

(b) Unprime frame.

■ FIGURE 34.17
Lorentz transformation for velocities. Here we are looking at the velocity vectors in the x-y-plane. In the prime frame (a), a particle undergoes a displacement $d\vec{\mathbf{r}}'$ in time dt'. In the unprime frame (b), its displacement is $d\vec{\mathbf{r}}$ in time dt.

IN THIS SECTION WE USE $\vec{\mathbf{v}}$ FOR THE VELOCITY OF THE PRIME FRAME MEASURED BY OBSERVERS IN THE UNPRIME FRAME, AND $\vec{\mathbf{u}}$ AND $\vec{\mathbf{u}}'$ FOR THE VELOCITY OF AN OBJECT MEASURED WITHIN EACH FRAME. DISTINGUISH THEM CAREFULLY!

Dividing top and bottom by dt, we find:

$$\vec{\mathbf{u}}' = \frac{(dx/dt) - \beta c}{1 - (\beta/c)(dx/dt)}\hat{\mathbf{i}} + \frac{d\vec{\mathbf{r}}_\perp/dt}{\gamma[1 - (\beta/c)(dx/dt)]}.$$

In components:

When using eqns. (34.9), be sure to specify clearly what defines each frame and which object has the velocity $\vec{\mathbf{u}}$.

$$u'_x = \frac{u_x - \beta c}{1 - \beta u_x/c} \quad \text{and} \quad \vec{\mathbf{u}}'_\perp = \frac{\vec{\mathbf{u}}_\perp}{\gamma(1 - \beta u_x/c)} \tag{34.9}$$

In the limit $\beta \ll 1$, the denominators in each expression tend to 1, and we recover the Galilean results (Table 34.1). In the special case $\vec{\mathbf{u}} = 0$, we find $\vec{\mathbf{u}}' = -\beta c\hat{\mathbf{i}}$, as expected. In general, the velocity components parallel and perpendicular to $\vec{\mathbf{v}}$ transform differently. The γ factors cancel in u'_x because both time dilation and length contraction influence the ratio dx'/dt'. Length contraction has no effect on the ratio $d\vec{\mathbf{r}}'_\perp/dt'$ as lengths perpendicular to $\vec{\mathbf{v}}$ do not change. The factor $1 - \beta u_x/c$ that appears in both expressions arises from the disagreement about simultaneity between the frames.

EXERCISE 34.6 ♦ A rocket moving with speed v directly away from a planet emits a flash of light. Use eqns. (34.9) to find the speed of the flash as measured by an observer on the planet.

EXAMPLE 34.7 ♦ As a rocket passes a space station, the inhabitants of the station measure the rocket's speed to be $v = 0.65c$. At the instant the rocket passes, they launch a supply package at speed $0.78c$ in the same direction. How fast does the supply package approach the rocket, as measured by the rocket crew?

MODEL The space station defines the unprime frame and the rocket defines the prime frame. The package has velocity $\vec{\mathbf{u}}$ in the unprime (space station) frame and $\vec{\mathbf{u}}'$ in the prime (rocket) frame. The velocities of rocket and package are parallel.

SETUP To convert to the rocket frame, we choose the x-axis to be along the rocket's path and apply eqns. (34.9).

SOLVE We have $u_x = 0.78c$ and $\beta = 0.65$. Thus:

$$u'_x = \frac{0.78c - 0.65c}{1 - (0.65)(0.78)} = \frac{0.13c}{0.49} = 0.26c.$$

ANALYZE Compare with the Galilean result of $0.78c - 0.65c = 0.13c$. The supply package approaches the rocket twice as fast as the Galilean result would predict. ■

EXAMPLE 34.8 ♦♦ A second rocket is approaching the space station at right angles to the track of the first, as seen from the space station, with speed $0.45c$. Describe its velocity as seen by the first rocket's crew (■ Figure 34.18).

MODEL Again, we choose the space station to define the unprime frame and the first rocket to define the prime frame, with the x- and x'-axes along the first rocket's path. In the space station frame, the second rocket moves in the y-direction. In the prime (first rocket's) frame, the second rocket's velocity has both x'- and y'-components.

SETUP We use eqns. (34.9) with $\beta = 0.65$, $u_x = 0$, and $u_y = 0.45c$. Then:

$$\gamma = \frac{1}{\sqrt{1 - 0.65^2}} = 1.32.$$

SOLVE

$$u'_x = \frac{u_x - \beta c}{1 - \beta u_x/c} = \frac{0 - 0.65c}{1 - 0} = -0.65c,$$

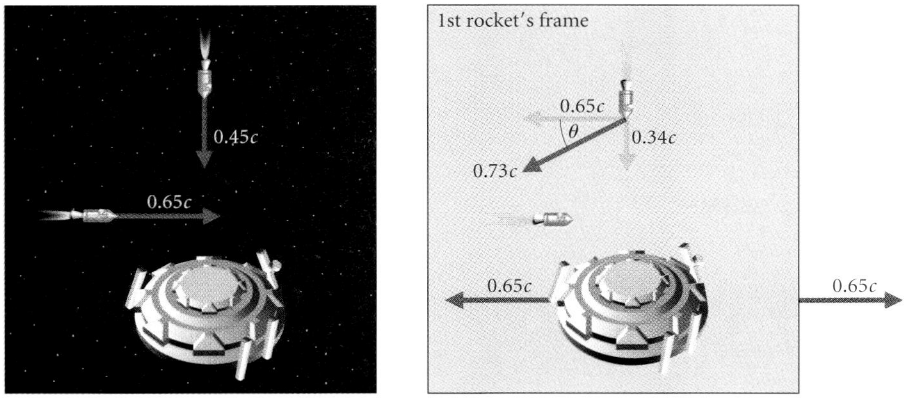

(a) (b)

■ FIGURE 34.18
(a) View in the space station frame as the second rocket approaches. (b) The situation in the first rocket's frame. The second rocket is approaching from in front.

and
$$u'_y = \frac{u_y}{\gamma(1 - \beta u_x/c)} = \frac{0.45c}{1.32(1 - 0)} = 0.34c.$$

Equivalently, the second rocket's speed is:
$$u' = c\sqrt{(0.65)^2 + (0.34)^2} = 0.73c,$$

and it is moving at an angle
$$\theta = \tan^{-1}(0.34/0.65) = 28° \text{ to the first rocket's path.}$$

ANALYZE Compare with the Galilean result: $\theta_G = \tan^{-1}(0.45/0.65) = 35°$. The relativistic expression for $\tan\theta$ is smaller by a factor γ. As $u \to c$, everything appears to approach the rocket from almost directly in front, and the geometry of space appears distorted. ■

FOR MORE ON THIS KIND OF DISTORTION, SEE PROBLEM 56.

34.3.3 *Acceleration in Special Relativity*

Special relativity is *special* because it describes the relations between *uniformly moving* reference frames. Acceleration involves a change of reference frame that we must describe carefully. A famous thought experiment considers the dilemma of the traveling twin.

THIS THOUGHT EXPERIMENT IS KNOWN AS THE *TWIN PARADOX*.

One twin (mobile Melia) volunteers for the first trip to Sirius (8.6 light-years away) and her sister (domestic Delia) stays home. The expedition accelerates rapidly, cruises at $\beta = \frac{4}{5}$, turns around quickly at Sirius after encountering its hostile inhabitants, and returns at $\beta = \frac{4}{5}$. On Earth, 21.5 years have passed when the expedition returns, but the expedition members have aged only $\Delta t' = \Delta t/\gamma = 12.9$ years. Melia, the traveling twin, is 8.6 years younger than her sister.

Melia is perplexed. After all, according to rocket observations, the Earth sped away and later returned. Sister Delia moves with the Earth, so why doesn't she end up younger?

Acceleration at Sirius provides the answer (■ Figure 34.19). Mobile Melia occupies one inertial reference frame G while going to Sirius and a second frame R while returning. With special relativity, she correctly calculates the time intervals LB and CR. The time interval BC is left out. Event B is simultaneous with the rocket firing (event O) in Melia's outbound frame, and event C is simultaneous with O in her inbound frame. Events B and C are certainly not simultaneous from domestic Delia's point of view! Because Melia accelerates and so changes from one inertial frame to another, she cannot successfully apply results from special relativity that are valid only when the observer remains in a single, inertial reference frame.[2]

In 1960, Chalmers Sherwin pointed out that iron nuclei could be used as the twins in an experimental test of round-trip time dilation, and R. V. Pound and G. A. Rebka were able to

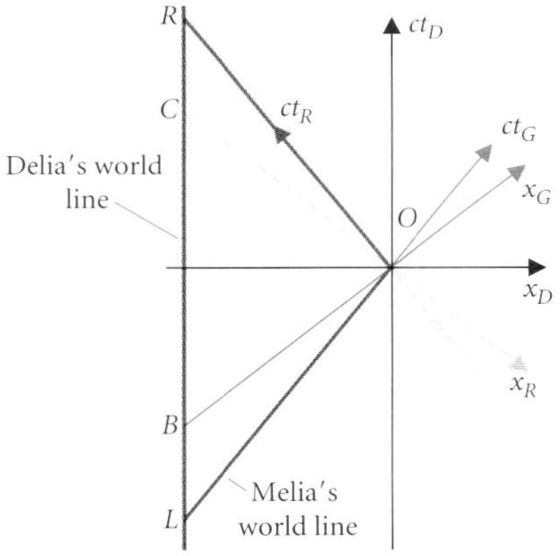

■ FIGURE 34.19
Space-time diagram for the twins Melia and Delia. Origin is chosen at the event *rocket engines fire at Sirius.* Melia's reference frame changes from G to R as a result of the acceleration caused by the rocket engines. Using special relativity to compute the time intervals in her rest frame, she misses the time interval BC due to the change of frame: O and B are simultaneous in frame G; O and C are simultaneous in frame R.

[2] In spite of the name, there is no real paradox here. Note carefully that Melia's youth is not a *dynamic* effect due to forces acting on her as she turns around. It is a purely *kinematic* effect due to the change of reference frame, as experiment confirms.

In Problems 60–62 we ask you to investigate further aspects of the twin paradox.

For thermal motion, see Chapter 19, especially §19.2. For a detailed comparison of the experiment with theory, see Problem 83.

measure the effect. Because of its thermal motion, an iron nucleus undergoes periods of constant velocity with short periods of abrupt acceleration that change its velocity, rather like Melia on her rocket trip. The average speed of the traveling atom increases as temperature increases. The iron nucleus also carries an internal clock whose period can be measured by looking at the frequency of light emitted by the nucleus. Pound and Rebka found that an iron sample 1 K hotter than another had a lower effective frequency, $\Delta f/f = -2 \times 10^{-15}$. The lower frequency corresponds to the lesser aging of the traveling twin.

34.4 Relativistic Dynamics

34.4.1 Momentum

In this section we shall continue the convention of using $\vec{\mathbf{v}}$ for the velocity of a frame and $\vec{\mathbf{u}}$ for the velocity of a particle in a frame.

When a constant force acts on a particle, according to Newtonian theory, its speed and its momentum $|\vec{\mathbf{p}}_N| \equiv mu$ increase uniformly with time. According to special relativity, the speed is limited by the finite value $u = c$, and so the Newtonian momentum also approaches a limit mc, apparently contradicting Newton's second law. According to Einstein's first postulate, if Newton's second law in the form $\vec{\mathbf{F}} = d\vec{\mathbf{p}}/dt$ is invariant, the Newtonian expression for momentum cannot be exact.

Although $\vec{\mathbf{p}}$ cannot remain proportional to speed when $u \sim c$, we have no reason to doubt its dependence on mass or that its direction is parallel to $\vec{\mathbf{u}}$. So we suppose that the correct relativistic expression is of the form:

$$\vec{\mathbf{p}} = mf(u)\vec{\mathbf{u}},$$

In eqn. (34.10), $\gamma = (1 - u^2/c^2)^{-1/2}$. Previously we used $\gamma = (1 - v^2/c^2)^{-1/2}$, where v is the relative speed of two reference frames. You must use context to distinguish the two γs. When both appear in the same expression, we use γ_u for the value of γ that pertains to a particle with speed u.

where $f(u)$ corrects the dependence on speed. Since Newton's expression is correct for $u \ll c$, we know that $f(u) \to 1$ as $u \to 0$. *Digging Deeper,* we show that $f(u) = \gamma$. The correct relativistic expression for momentum is:

$$\vec{\mathbf{p}} = \gamma m\vec{\mathbf{u}}. \qquad \textbf{(34.10)}$$

As the particle's speed approaches the speed of light, a large increase in momentum corresponds to a negligible increase in speed.

Sometimes eqn. (34.10) is interpreted by saying that a particle's mass increases with speed according to the formula $m = \gamma m_o$, where m_o is the *rest mass.* We prefer to reserve the word "mass" and the symbol m for rest mass, which is an invariant property of a particle or system. This is the convention used in eqn. (34.10) and throughout this book.

EXAMPLE 34.9 ♦ A cosmic ray electron of mass $m = 9.11 \times 10^{-31}$ kg is traveling with speed $u = 0.995c$. What is its momentum?

MODEL With a speed almost equal to the speed of light, we should use the relativistic formula for momentum, eqn. (34.10).

SETUP First we calculate γ using the electron's speed:

$$\gamma = (1 - u^2/c^2)^{-1/2} = (1 - 0.995^2)^{-1/2} = 10.0.$$

SOLVE The momentum has magnitude:

$$\begin{aligned} p &= \gamma mu \\ &= (10.0)(9.11 \times 10^{-31}\ \text{kg})(0.995)(3.00 \times 10^8\ \text{m/s}) \\ &= 2.72 \times 10^{-21}\ \text{kg}\cdot\text{m/s}. \end{aligned}$$

ANALYZE The momentum is ten times the Newtonian value. *Remember:* When using the expression γmu, use the particle's speed u to compute γ.

EXERCISE 34.7 ♦ At what speed does an electron have twice the momentum predicted by the Newtonian formula?

Digging Deeper

Relativistic Momentum

We can determine the function $f(u)$ from a thought experiment about colliding particles, analyzed in two different reference frames. Two identical particles each of mass m undergo the elastic collision shown in ■ Figure 34.20a. The first particle moves with speed u at a small angle α to the x-axis, while the second moves with speed $w \ll c$ along the y-axis. The particles collide, and the y-component of each particle's velocity is reversed.

Let's look at the momentum of the two particles before and after the collision:

	Before	After
p_x	$muf(u)\cos\alpha + 0$	$muf(u)\cos\alpha + 0$
p_y	$-muf(u)\sin\alpha + mwf(w)$	$muf(u)\sin\alpha - mwf(w)$

Next we apply conservation of momentum. The x-components are already equal. For the y-components to be equal, we need:

$$wf(w) = uf(u)\sin\alpha. \qquad \text{(i)}$$

Now let's look at the same collision in a frame where particle 1 moves parallel to the y-axis. This (prime) frame moves to the right with speed $v = u\cos\alpha$. We use eqns. 34.9 with $\beta = (u/c)\cos\alpha$ and $\gamma = [1 - (u/c)^2\cos^2\alpha]^{-1/2}$ to Lorentz-transform the velocities.

For the first particle:

$$u'_{1,y} = \gamma u \sin\alpha \equiv u' \quad \text{and} \quad u'_{1,x} = 0.$$

Its speed is u'.

For the second particle:

$$u'_{2,x} = w'_x = -u\cos\alpha \quad \text{and} \quad u'_{2,y} = w'_y = w/\gamma.$$

Its speed is given by $(w')^2 = (w'_x)^2 + (w'_y)^2$:

$$w' = \left[u^2\cos^2\alpha + w^2\left(1 - \frac{u^2\cos^2\alpha}{c^2}\right)\right]^{1/2}$$

$$= u\left[\cos^2\alpha\left(1 - \frac{w^2}{c^2}\right) + \frac{w^2}{u^2}\right]^{1/2}.$$

As in the unprime frame, we apply conservation of momentum in the y-direction:

	Before	After
p'_y	$-mu'f(u') + mw'_yf(w')$	$mu'f(u') - mw'_yf(w')$

Set before = after:

$$u'f(u') = w'_yf(w'). \qquad \text{(ii)}$$

We find $f(u)$ by considering the limit $w/c \to 0$. We may neglect quantities of order w^2/c^2 (or equivalently α^2) in eqns. (i) and (ii). As $w \to 0$, $f(w) \to 1$, and:

$$\cos\alpha = \sqrt{1 - \sin^2\alpha} \to 1.$$

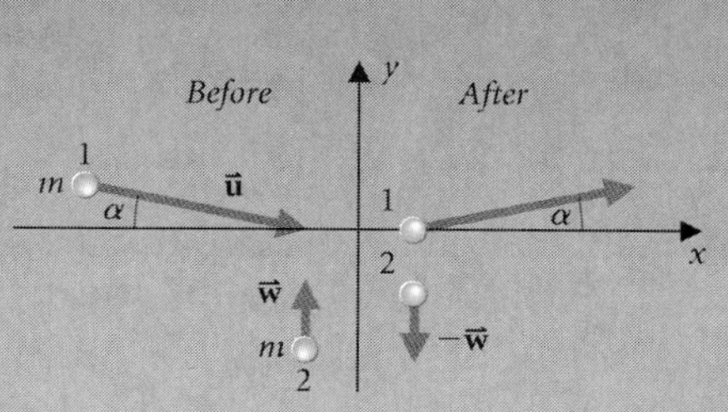

(a) Unprime frame.

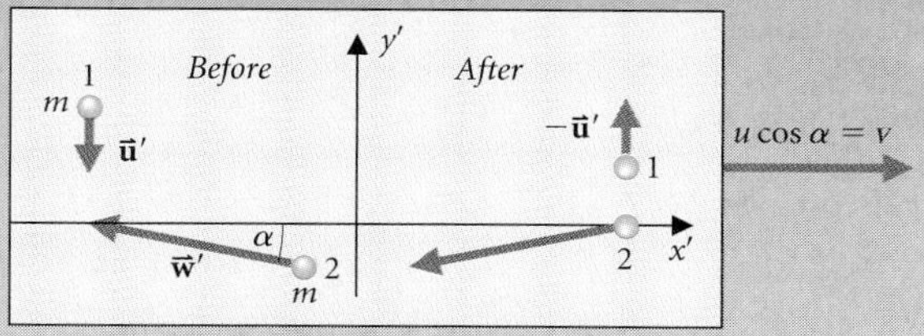

(b) Prime frame.

■ Figure 34.20
(a) Two identical particles collide. One particle has a large speed u but makes a small angle α with the x-axis. The other particle has a small speed $w \ll c$ and travels parallel to the y-axis. During the collision, each particle's y-component of momentum is reversed. (b) The same collision in the prime frame. This frame moves at speed $v = u\cos\alpha$ in the x-direction with respect to the unprime frame. The first particle moves parallel to the y-axis with a small speed u', and the second particle moves at a large speed w', making a small angle with the x'-axis. When $w \ll c$, this diagram is the mirror image of (a).

So:

$$\gamma \to \frac{1}{\sqrt{1 - u^2/c^2}} \equiv \gamma_u,$$

$$w' \to u,$$

$$w'_y \to w/\gamma_u$$

$$u' \to u\gamma_u \sin\alpha.$$

Then eqn. (i) becomes:

$$w = u\sin\alpha f(u), \qquad \text{(iii)}$$

and eqn. (ii) becomes:

$$u\gamma_u \sin\alpha f(u\gamma_u\sin\alpha) = (w/\gamma_u)f(u). \qquad \text{(iv)}$$

We know that $f \to 1$ as its argument approaches 0. On the left-hand side of eqn. (iv), f multiplies $\sin\alpha$, which is already small, so any correction to $f = 1$ gives a term that is even smaller and may be neglected. Thus eqn. (iv) becomes:

$$u\gamma_u \sin\alpha = (w/\gamma_u)f(u). \qquad \text{(v)}$$

Combining eqns. (iii) and (v) to eliminate $\sin\alpha$, we have:

$$f(u) = \gamma_u.$$

So, the correct relativistic formula for momentum is:

$$\vec{\mathbf{p}} = \gamma m\vec{\mathbf{u}}.$$

In Chapter 6 we expressed Newton's second law in its most fundamental form: $\vec{\mathbf{F}} = d\vec{\mathbf{p}}/dt$. In this form, Newton's law remains valid for particles whose speed approaches c, provided that we use the proper relativistic expression for $\vec{\mathbf{p}}$ and work in a single reference frame. To find the corresponding relation between force and acceleration, we differentiate expression (34.10).

$$\vec{\mathbf{F}} = \frac{d\vec{\mathbf{p}}}{dt} = \frac{d}{dt}(\gamma m\vec{\mathbf{u}}) = m\left(\gamma\frac{d\vec{\mathbf{u}}}{dt} + \frac{d\gamma}{dt}\vec{\mathbf{u}}\right).$$

Since $$\frac{d\gamma}{dt} = \frac{d}{dt}\left(\frac{1}{\sqrt{1 - u^2/c^2}}\right) = \frac{(-\frac{1}{2})(-2u/c^2)}{(1 - u^2/c^2)^{3/2}}\left(\frac{du}{dt}\right) = \gamma^3\left(\frac{u}{c^2}\right)\left(\frac{du}{dt}\right),$$

$$\vec{\mathbf{F}} = \gamma m\left[\frac{d\vec{\mathbf{u}}}{dt} + \gamma^2\left(\frac{u}{c^2}\right)\left(\frac{du}{dt}\right)\vec{\mathbf{u}}\right] = \gamma m\left[\vec{\mathbf{a}} + \gamma^2\left(\frac{u}{c^2}\right)\left(\frac{du}{dt}\right)\vec{\mathbf{u}}\right].$$

Now we resolve $\vec{\mathbf{a}}$ into components parallel and perpendicular to $\vec{\mathbf{u}}$. The parallel component changes the particle's speed: $a_\parallel = du/dt$. Choose the x-axis to be along $\vec{\mathbf{u}}$. Then $\vec{\mathbf{a}}_\parallel = a_\parallel\hat{\mathbf{i}}$, and:

$$\vec{\mathbf{F}} = \gamma m\left(\vec{\mathbf{a}}_\perp + a_\parallel\hat{\mathbf{i}} + \gamma^2\frac{u}{c^2}a_\parallel u\hat{\mathbf{i}}\right) = \gamma m\left[\vec{\mathbf{a}}_\perp + \vec{\mathbf{a}}_\parallel\left(1 + \gamma^2\frac{u^2}{c^2}\right)\right].$$

The quantity multiplying $\vec{\mathbf{a}}_\parallel$ simplifies:

$$1 + \gamma^2\beta^2 = 1 + \beta^2/(1 - \beta^2) = 1/(1 - \beta^2) = \gamma^2.$$

So:

$$\vec{\mathbf{F}} = \gamma m\vec{\mathbf{a}}_\perp + \gamma^3 m\vec{\mathbf{a}}_\parallel. \tag{34.11}$$

Since the coefficients of $\vec{\mathbf{a}}_\perp$ and $\vec{\mathbf{a}}_\parallel$ are different, the acceleration is in general not parallel to the force. It is more difficult to increase the particle's speed than to change its direction of motion. The Newtonian result $\vec{\mathbf{F}} = m\vec{\mathbf{a}}$ is the limit of eqn. (34.11) as $u \to 0$ and $\gamma \to 1$.

34.4.2 *Mass and Energy*

One of Einstein's most famous discoveries is that mass is a form of energy. As a consequence, mass may be converted into other forms of energy, and particles may be annihilated and created as these changes occur. To show how Einstein's relation arises, we calculate the work done on a particle and, as in Newtonian physics, interpret it as an increase in energy. Suppose, for example, that an accelerator applies a constant electric force $\vec{\mathbf{F}}$ to an electron. Using eqn. (34.11) for the component of $\vec{\mathbf{F}}$ parallel to $\vec{\mathbf{u}}$, the force does work at a rate:

ALTHOUGH WE HAVE NOT PROVED IT, THE EXPRESSION $P = \vec{\mathbf{F}}\cdot\vec{\mathbf{u}}$ REMAINS VALID AS $u \to c$.

$$P = \vec{\mathbf{F}}\cdot\vec{\mathbf{u}} = \gamma^3 mu\frac{du}{dt} = mc^2\left[\frac{(1/2)(d/dt)(u^2/c^2)}{(1 - u^2/c^2)^{3/2}}\right] = \frac{d}{dt}(\gamma mc^2).$$

According to the work-energy theorem, the work done on the electron increases the quantity γmc^2, which is the electron's energy:

FOR THE WORK-ENERGY THEOREM, SEE §7.1.

$$E = \gamma mc^2. \tag{34.12}$$

For a particle at rest, $\gamma = 1$ and its energy is $E_0 = mc^2$, the *rest energy* discovered by Einstein. The kinetic energy—that is, the energy due to motion—is given by:

$$K = E - E_0 = (\gamma - 1)mc^2. \tag{34.13}$$

EXERCISE 34.8 ♦♦ Expand γ in powers of u^2/c^2 and show that the first two terms in eqn. (34.13) are the Newtonian kinetic energy and a relativistic correction:

$$K \approx \tfrac{1}{2}mu^2\left[1 + \frac{3}{4}\left(\frac{u^2}{c^2}\right)\right] \qquad (\text{for } u \ll c).$$

EXAMPLE 34.10 ♦♦ An electron in a television tube has a kinetic energy of about 35 keV. Cosmic ray electrons reach energies of many GeV. What is the rest energy of an electron? Find the speed of an electron in the television tube and of a 1-GeV cosmic ray electron. How accurate is the Newtonian kinetic energy in each case?

MODEL In each case, we compare the kinetic energy (eqn. 34.13) with the Newtonian value $K = \frac{1}{2}mu^2$.

SETUP The mass of an electron is 9.110×10^{-31} kg.

SOLVE Its rest energy is:

$$E_0 = mc^2 = \frac{(9.11 \times 10^{-31}\ \text{kg})(2.998 \times 10^8\ \text{m/s})^2}{1.602 \times 10^{-19}\ \text{J/eV}} = 511\ \text{keV}.$$

SETUP The electron's kinetic energy is $(\gamma - 1)mc^2$. We find its speed from γ:

$$\gamma^2 = \frac{1}{1 - \beta^2} \Rightarrow \beta = \sqrt{1 - \frac{1}{\gamma^2}}.$$

SOLVE In the television tube:

$$\gamma - 1 = K/mc^2 = (35\ \text{keV})/(511\ \text{keV}) = 0.0685.$$

With $\gamma = 1.0685$, we find β:

$$\beta = \sqrt{1 - \frac{1}{1.0685^2}} = 0.352.$$

For small β, the correction term to the Newtonian K (Exercise 34.8) is:

$$\tfrac{3}{4}\beta^2 = 0.093.$$

For the cosmic ray, we have:

$$\gamma = E/mc^2 = (10^6\ \text{keV})/(511\ \text{keV}) = 2 \times 10^3.$$

The corresponding speed is:

$$\beta = \sqrt{1 - (2 \times 10^3)^{-2}} = 0.99999987,$$

which is indistinguishable from 1, to the significance we are working.

ANALYZE For the electron in the television set, the Newtonian kinetic energy is accurate to 9%, but the Newtonian formula underestimates the cosmic ray kinetic energy by a factor of almost 4000. ∎

Unlike the arbitrary constants we encountered with potential energies, the rest energy has real physical significance. It is unchanged when a particle is accelerated, but it can be released when the particle combines with another and changes form. This is precisely what happens in nuclear reactions—for example, in power plants and the Sun. Because of this, particle physicists often write masses in energy units: the mass of a proton is written

$$m_p = 938\ \text{MeV}/c^2.$$

EXAMPLE 34.11 ♦♦ The principal energy-producing reaction in the Sun fuses protons into a helium nucleus. Four protons, each with mass 1.0079 u, form a helium nucleus with mass 4.00260 u ($1\ \text{u} = 1.66 \times 10^{-27}$ kg). How many reactions occur in the Sun each second? Estimate how long the Sun can continue to shine.

MODEL First we find the energy liberated in each reaction. Then the number of reactions per second equals the total energy radiated per second divided by the energy per reaction. We can obtain an upper limit to the Sun's life by calculating the time for it to convert all its hydrogen to helium.

THE SUN'S MASS AND LUMINOSITY ARE GIVEN INSIDE THE FRONT COVER.

SETUP In each reaction the total rest mass decreases by

$$\Delta m = 4m_{\mathrm{H}} - m_{\mathrm{He}} = (4 \times 1.0079 - 4.0026)\ \mathrm{u} = 0.029\ \mathrm{u},$$

releasing an energy:

$$\Delta E = \Delta mc^2 = 0.029(1.66 \times 10^{-27}\ \mathrm{kg})(3.0 \times 10^8\ \mathrm{m/s})^2 = 4.3 \times 10^{-12}\ \mathrm{J}.$$

The Sun's luminosity is 3.9×10^{26} W, so the number of reactions per second is:

SOLVE

$$\frac{dn}{dt} = \frac{L}{\Delta E} = \frac{3.9 \times 10^{26}\ \mathrm{W}}{4.3 \times 10^{-12}\ \mathrm{J}} = 9.0 \times 10^{37}\ /\mathrm{s}.$$

SETUP The mass of helium produced per second is:

$$\frac{dM}{dt} = \frac{dn}{dt} m_{\mathrm{He}}.$$

And, the time to convert the entire Sun to helium is:

$$T \sim \frac{M}{dM/dt} = \left(\frac{M}{m_{\mathrm{He}}}\right)\left(\frac{1}{dn/dt}\right).$$

SOLVE The Sun's mass is 2.0×10^{30} kg, so:

$$T \sim \left(\frac{2.0 \times 10^{30}\ \mathrm{kg}}{4(1.7 \times 10^{-27}\ \mathrm{kg})}\right)\left(\frac{1}{9.0 \times 10^{37}\ /\mathrm{s}}\right) = 3.3 \times 10^{18}\ \mathrm{s} \approx 10^{11}\ \mathrm{y}.$$

ANALYZE This crude estimate is about ten times too large. When its inner 10% is converted to helium, the Sun will undergo structural changes and destroy the Earth. Of its 10-billion-year lifetime, about 4.5 billion years have passed since the Sun's formation. ∎

As a particle is accelerated towards the speed of light, its energy and momentum increase dramatically. To reach light speed would require infinite energy. Again, we see that no particle with $m > 0$ can travel at (or above) the speed of light.[3]

34.4.3 *The Energy-Momentum Invariant*

Neither energy nor momentum is an invariant quantity since they both depend on a relative quantity: the particle's speed. However, the combination $E^2 - p^2c^2$ is invariant:

$$E^2 - p^2c^2 = (\gamma mc^2)^2 - (\gamma mu)^2c^2 = \gamma^2 m^2 c^4\left(1 - \frac{u^2}{c^2}\right).$$

$$E^2 - p^2c^2 = m^2c^4. \tag{34.14}$$

This energy-momentum invariant has important similarities to the interval invariant (§34.2.2). The location of an event in space-time has four components: time and the three components of the position vector. The invariant is:

$$\text{Invariant} = (\text{time component})^2 - (\text{length of spatial vector})^2.$$

Equation (34.14) has the same form if we identify energy as the time component that goes with the momentum vector. Thus energy and momentum form a four-dimensional vector whose invariant length is determined by the mass. Three Newtonian concepts—mass, energy, and momentum—are unified into a single quantity that obeys a four-dimensional vector conservation law.

[3] It has been suggested that tachyons, particles with imaginary mass, travel with speed $> c$. Such particles could never go slower than light! There is no evidence that such particles actually exist.

■ FIGURE 34.21
The Bevatron at the University of California, Berkeley, shortly after its construction. This machine was designed to accelerate protons to an energy sufficient to generate antiprotons (see Example 34.12).

In 1954, the Bevatron accelerator was constructed at the University of California, Berkeley (■ Figure 34.21). A major purpose of the machine was to create antiprotons—particles with the same mass as a proton but with opposite values of charge and certain other physical properties. Accelerated protons would collide with stationary protons to create pairs of protons and antiprotons out of kinetic energy:

$$2 \text{ protons} + \text{kinetic energy} \rightarrow 3 \text{ protons} + 1 \text{ antiproton}.$$

THE TOTAL MOMENTUM OF A SYSTEM WAS DISCUSSED IN CHAPTER 6.

To design the Bevatron, it was important to determine how much kinetic energy would be needed. To see how this was done, we apply eqn. (34.14) to the total energy and momentum of the system. This step is valid, although we omit the proof here.

EXAMPLE 34.12 ♦♦♦ Compute the minimum energy necessary in the Bevatron beam to produce antiprotons.

MODEL If all the particles were at rest in the laboratory after the collision, $2mc^2$ of kinetic energy would suffice. However, the momentum of the protons in the beam is conserved: the reaction products cannot be at rest in the lab frame (■ Figure 34.22). As in nonrelativistic mechanics, the description is simplest in a frame where the system has zero momentum. [The name is changed from the center of mass frame to the center of energy (CE) frame in relativistic discussions.] In the CE frame, both the beam and the target protons have kinetic energy that is converted into the particle pair. At the minimum beam energy, there is no energy left in kinetic form: the final state has four stationary particles in the CE frame.

To find the minimum energy, we note that the energy-momentum invariant has the same value in the CE frame after the collision as in the lab frame before the collision. In any one frame, energy and momentum are separately *conserved,* and therefore so is their invariant combination. The invariant has the same value before and after the collision. Because it is *invariant,* its value is the same in both the lab and CE reference frames.

(a) Lab frame.

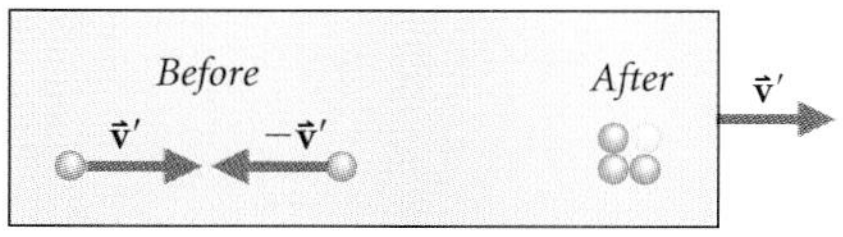

(b) CE frame.

■ FIGURE 34.22
(a) Proton–proton collision in the lab frame results in three protons and one antiproton. All four particles have the same mass. (b) Here the four particles are at rest in the CE frame after the collision. The initial energy of the proton in the lab frame is then the minimum necessary for production of an antiproton.

A BeV (for billion elecron volts) is an old name for GeV—hence the Bevatron's name.

SETUP We use the standard conservation law plan. In the lab frame before the collision, there is one moving proton ($E = \gamma mc^2$, $p_x = \gamma mu$) and one proton at rest ($E = mc^2$, $p_x = 0$). In the CE frame after the collision, there are four particles, each with equal mass and energy ($E = mc^2$), all at rest ($p_x = 0$).

$$\begin{aligned} & \qquad\quad \text{CE AFTER} \qquad\qquad\qquad \text{LAB BEFORE} \\ E^2 - p^2c^2 &= (4mc^2)^2 - 0 = (mc^2 + \gamma mc^2)^2 - (\gamma mu)^2c^2. \\ 16m^2c^4 &= m^2c^4[(\gamma + 1)^2 - \gamma^2u^2/c^2] \\ &= m^2c^4[\gamma^2(1 - u^2/c^2) + 2\gamma + 1] \\ &= 2m^2c^4(\gamma + 1). \end{aligned}$$

SOLVE Thus $\gamma + 1 = 8$. To produce antiprotons requires a minimum γ factor of 7 and a minimum kinetic energy:

$$K_{\text{min}} = (\gamma - 1)mc^2 = 6(0.938\text{ GeV}) = 5.6\text{ GeV}.$$

ANALYZE The Bevatron was designed to provide 6.4 GeV per proton. In general, the probability that a reaction will occur is exceedingly small at the minimum energy. The extra 0.8 GeV provided in the design ensures a reasonable production rate. The Bevatron was a proton synchrotron (see *Digging Deeper,* §29.1). ■

Chapter Summary

Where Are We Now?

According to the special principle of relativity, physical laws have the same form in any inertial reference frame. Einstein developed his special theory of relativity so that Maxwell's electromagnetic theory should be consistent with this principle. Together with his later revision of gravitational theory (general relativity), these theories require that we recognize a four-dimensional space-time as the arena in which physical events occur. We have developed a basic description of space-time and studied the kinematics of special relativity. We have learned how expressions for a particle's momentum and energy are modified when it moves at speeds approaching that of light, and how the structure of space-time leads us to define a single energy-momentum invariant that is independent of reference frame.

What Did We Do?

A relativity theory classifies physical quantities as relative or invariant, describes how to express relative quantities in different reference frames and how to express physical laws in invariant form.

Physics is described by events occurring in a four-dimensional space-time. A particle follows a world line through space-time that consists of the sequence of events in the particle's history. Space-time diagrams provide a convenient geometrical representation of the relations among events and world lines. The Lorentz transformation shows how to convert relative quantities from one frame to another. When the prime frame moves with velocity $\vec{v} = v\hat{\mathbf{i}}$ as measured in the unprime frame, the coordinates are related by:

$$ct' = \gamma(ct - \beta x), \qquad x' = \gamma(x - \beta ct), \qquad y' = y, \qquad \text{and} \qquad z' = z.$$

The space-time interval $\Delta s^2 = c^2\,\Delta t^2 - \Delta\ell^2$ is invariant.

Special relativity is valid in inertial reference frames. The inertial frame in which an object's velocity is zero is called its rest frame. While an object accelerates, its rest frame changes continuously. Calculations involving acceleration may require the use of several reference frames.

TABLE 34.2 Invariant and Relative Quantities in Special Relativity *(Relative velocity of frames is $\vec{v} = \beta c\hat{i}$.)*

Invariants

Rest mass of a particle	m
Charge of a particle	Q
Space-time interval	$\Delta s^2 = c^2\,\Delta t^2 - \lvert\Delta\vec{x}\rvert^2$
Energy-momentum invariant	$m^2c^4 = E^2 - p^2c^2$
Electromagnetic field invariants*	$\vec{E}\cdot\vec{B}$
	$\lvert\vec{E}\rvert^2 - c^2\lvert\vec{B}\rvert^2$

Relative Quantities		Transformation Law	Text Reference
Time coordinate	t	$ct' = \gamma(ct - \beta x)$	34.8
Spatial coordinate	x	$x' = \gamma(x - \beta ct)$	34.8
Velocity of a particle	$\vec{u}$	$u'_x = \dfrac{u_x - \beta c}{1 - \beta u_x/c}$	34.9
		$\vec{u}'_\perp = \dfrac{\vec{u}_\perp}{\gamma(1 - \beta u_x/c)}$	34.9
Energy	$E = \gamma_u mc^2$	$E' = \gamma(E - \beta p_x c)$	Problem 77
Momentum	$\vec{p} = \gamma_u m\vec{u}$	$cp'_x = \gamma(cp_x - \beta E)$	
		$\vec{p}'_\perp = \vec{p}_\perp$	
Electric field*	$\vec{E}$	$\vec{E}'_\perp = \gamma(\vec{E}_\perp + c\vec{\beta}\times\vec{B})$,	$E'_x = E_x$
Magnetic field*	$\vec{B}$	$\vec{B}'_\perp = \gamma(\vec{B}_\perp - \dfrac{\vec{\beta}}{c}\times\vec{E})$,	$B'_x = B_x$

*Not derived in the text. Note that the Lorentz transformation for $\vec{E}$ differs from the transformation in Table 34.1 only by the factor γ in the expression for $\vec{E}'_\perp$.

As a particle approaches the speed of light, its momentum, $\vec{p} = \gamma m\vec{u}$, increases without bound. In any reference frame, no material object can have a speed equal to or greater than c. The relativistic expression for energy, $E = \gamma mc^2$, shows that each particle has a rest energy mc^2. When particles interact, energy may be transformed between rest energy and other forms.

We have listed some important invariant and relative quantities in • Table 34.2.

Practical Applications

Newtonian physics is an excellent approximation in everyday situations, but relativity theory must be used in calculations involving objects moving near the speed of light. Examples include analyzing the results of accelerator experiments or studying cosmic rays. A recent application is the free-electron laser. Relativistic effects are important in the production of synchrotron radiation by accelerator beams. Synchrotron radiation is finding increasing use in manufacturing, diagnostic, and healthcare applications.

Einstein's relation $E = \gamma mc^2$ is the basis for the operation of nuclear reactors and also allows us to understand energy generation in stars. We'll explore these topics further in Chapter 36.

The relativistic expression for momentum is essential in understanding the behavior of particles in accelerators (cf. §29.1.2, §30.4) and in cosmic rays.

Solutions to Exercises

34.1 From Example 34.1,

$$\gamma = h/(\beta\tau c) = 38/\beta = 38/0.9997 \approx 38.$$

At the muon's birth, the detector's distance is:

$$d = (25 \text{ km})/\gamma = (25 \text{ km})/38 = 0.66 \text{ km}.$$

The detector is just able to travel this distance in the 2.2-μs lifetime of the muon: $v\tau \approx c\tau = (3.0 \times 10^8 \text{ m/s})(2.2 \times 10^{-6} \text{ s}) = 660$ m.

34.2 See ■ Figure 34.23. **(a)** Bolt 1 strikes. **(b)** Bolt 2 strikes. **(c)** Flash from bolt 1 reaches Penny. **(d)** Flashes from both bolts reach Oliver. **(e)** Flash from bolt 2 reaches Penny.

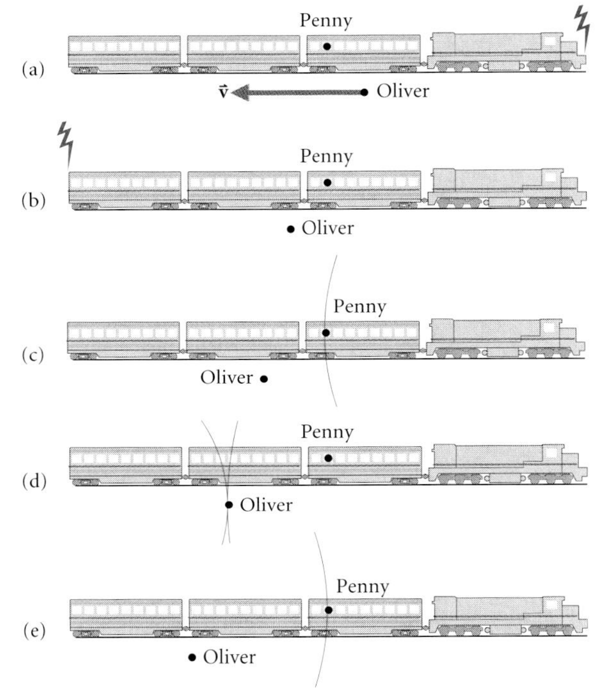

■ FIGURE 34.23
The lightning bolts in Penny's reference frame. Remember, the light from the bolts moves at speed c, and Oliver moves with speed v in this frame.

34.3 The reverse Lorentz transformation is given by eqns. (34.8), with the prime and unprime switched, and $\beta \to -\beta$:

$$ct = \gamma(ct' + \beta x') \qquad \text{and} \qquad x = \gamma(x' + \beta ct').$$

(You could also derive these results by solving eqns. (34.8) for x and ct in terms of x' and ct'.)

Along the x-axis, $t = 0$. Thus $ct' = -\beta x'$. This line has a negative slope and makes an angle θ with the x'-axis (■ Figure 34.24), where:

$$\tan\theta = |ct'|/x' = \beta.$$

Along the t-axis, $x = 0$ so that $x' = -\beta ct'$; again, the angle between

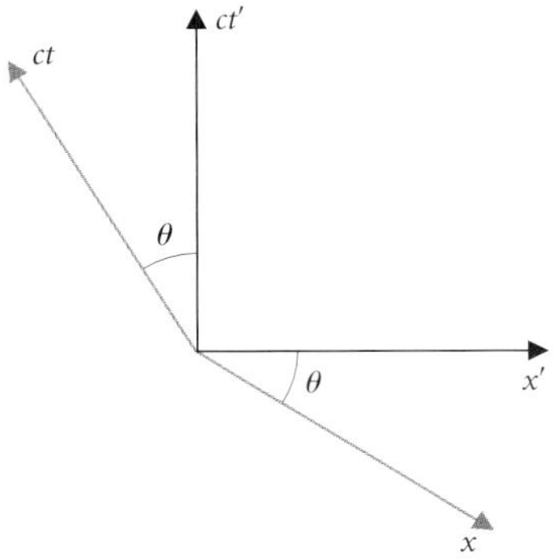

■ FIGURE 34.24
The unprime frame has velocity $-\beta c\hat{\mathbf{i}}$ with respect to the prime frame. If the prime axes are drawn perpendicular, then the unprime axes are at an angle $90° + 2\tan^{-1}\beta$. The relative positions and the angle between the prime and unprime axes are the same no matter which axes we choose to draw perpendicular.

the axes is $\theta = \tan^{-1}\beta$. The angles between the axes are the same as in Figure 34.11.

34.4 The interval between the origin and the new event is:

$$s^2 = c^2t^2 - x^2 - y^2 - z^2 = c^2(t')^2 - (x')^2 - (y')^2 - (z')^2.$$

We have already shown that $c^2t^2 - x^2 = c^2(t')^2 - (x')^2$, so we must have:

$$y^2 + z^2 = (y')^2 + (z')^2 = F^2(y^2 + z^2) \Rightarrow F^2 = 1.$$

The argument for choosing the positive root is the same as that given for A.

34.5 See ■ Figure 34.25. In the rocket frame the Earth moves with $\beta = 0.8$. The light signal travels a much smaller distance to the rocket than it does returning to earth.

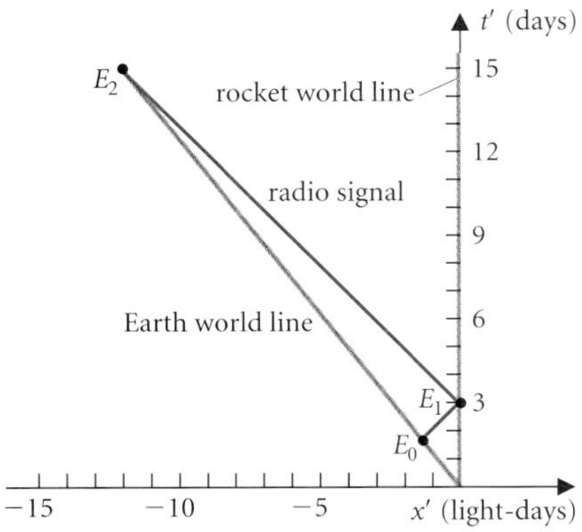

■ FIGURE 34.25
The events of Example 34.6 in the rocket frame. The signal emitted from Earth to the rocket travels a short distance, but the signal from the rocket has to chase the receding Earth: it travels a much larger distance.

34.6 The planet moves at speed $v = \beta c$ with respect to the rocket. The speed of the flash is c in the rocket frame. In the planet frame, it is:

$$c' = \frac{c - \beta c}{1 - \beta c/c} = c, \text{ as required by Einstein's postulates.}$$

34.7 With $\gamma = 2$, the momentum is twice that given by the Newtonian formula. Then $1/\gamma^2 = 1 - u^2/c^2 = 1/4$, and so $u = (\sqrt{3}/2)c = 0.87c$.

34.8

$$\gamma = \left(1 - \frac{u^2}{c^2}\right)^{-1/2} = 1 + \frac{1}{2}\left(\frac{u^2}{c^2}\right) + \left(-\frac{1}{2}\right)\left(-\frac{3}{2}\right)\frac{1}{2}\left(\frac{u^4}{c^4}\right) + \cdots.$$

Therefore: $$(\gamma - 1)mc^2 = \frac{1}{2}mc^2\frac{u^2}{c^2} + \frac{3}{8}mc^2\frac{u^4}{c^4} + \cdots$$
$$\approx \frac{1}{2}mu^2\left[1 + \frac{3}{4}\left(\frac{u}{c}\right)^2\right].$$

Basic Skills

Review Questions

§34.1 SPECIAL RELATIVITY

- What is a *relative quantity?* Give an example of a relative quantity in Galilean relativity and in special relativity.
- What is an *invariant?* Give an example of an invariant in Galilean relativity and in special relativity.
- What are the features of a complete theory of relativity?
- List Einstein's postulates for special relativity. Which of them is known as the *special principle of relativity?*
- What is *time dilation?*
- How can you tell which observer measures the shortest time interval between two events?
- Describe *length contraction.* Are all dimensions contracted? If not, which one or ones?
- If you determine that two events occur at the same time, do all observers agree that they are simultaneous? Why or why not?

§34.2 SPACE-TIME

- What is *space-time?* How might you represent it on paper?
- Describe a procedure for setting up a coordinate system in space-time.
- How is the *future* of an event defined? How is the *past* defined?
- What is the space-time *interval* between two events?
- What is a *spacelike* interval? A *timelike* interval?
- Define *proper time* and *proper distance.*

§34.3 THE LORENTZ TRANSFORMATION

- What assumptions were made in deriving the Lorentz transformation for the coordinates in the form (34.8)?
- Describe how to locate the coordinate axes of a relatively moving reference frame in a space-time diagram.
- What three effects are important in the Lorentz transformation for velocities? Do they all play a role in transforming each component of $\vec{\mathbf{u}}$?

§34.4 RELATIVISTIC DYNAMICS

- How do we know that the Newtonian expression for momentum cannot be correct as $u \to c$?
- What is the correct relativistic expression for a particle's momentum?
- What is a particle's *rest energy?* Can it ever be extracted?
- What is the *energy-momentum invariant?* What relation among mass, energy, and momentum does it suggest?
- Using energy and momentum as examples, carefully state the difference between conserved and invariant quantities.

Basic Skill Drill

§34.1 SPECIAL RELATIVITY

1. A rocket ship with length 0.10 km in its own rest frame is moving at $\beta = 0.70$ toward an asteroid. What is the length of the ship as measured by an observer on the asteroid?
2. Neutral pions (a kind of subatomic particle) have an average lifetime of 8.4×10^{-17} s in their own rest frame. In a lab experiment pions are produced with a speed $\beta = 0.995$. What is the average pion lifetime in the lab? How far do these pions travel, on average?
3. The captain of a spaceship en route from Aldebaran to Sirius sends signals to Earth every 12 h as she passes the solar system. Earth Central receives signals every 12 h 3 min. What is the speed of the spaceship?

§34.2 SPACE-TIME

4. Draw the following events in a space-time diagram: a rocket leaves Earth at $0.3c$; after departure, the rocket sends a signal to Earth; Earth receives the signal 17 h after the rocket departure. Use your diagram to determine when and where the signal was sent (as measured in the Earth frame). Calculate the space-time interval between the events *rocket leaves Earth* and *rocket sends signal.*
5. Event E_1 occurs at $t = 0$, $x = 3.50$ m. Event E_2 occurs at $t = 1.3$ ns, $x = -0.75$ m. Calculate the space-time interval between the events. Is the interval spacelike or timelike?

§34.3 THE LORENTZ TRANSFORMATION

6. An event occurs at $t = 0.27$ s, $x = 7900$ km. Find the coordinates of this event in a reference frame moving at speed $v = 1.4 \times 10^8$ m/s along the x-axis with respect to the first frame.
7. Two rockets leave a space station along parallel paths. The space station personnel record their speeds as $v_1 = 0.85c$ and $v_2 = 0.73c$. What is the speed of the first rocket, as measured by the crew on the second?

8. A positron is the antiparticle of an electron; it has the same mass but opposite charge. When a positron and an electron collide, they annihilate, producing two photons each with the same energy. If the particles approach with speed $v \ll c$, what is the photon energy? (*Note:* γ rays with this energy have been observed arriving from the center of the galaxy.)

9. A proton has kinetic energy $K = 450$ MeV. What is the value of γ for the proton? What is its momentum? What is its energy-momentum invariant?

Questions and Problems

34.1 SPECIAL RELATIVITY

10. ❖ Decide whether each of the following statements is true or false, and explain your answers. **(a)** The wave fronts propagating outward from a flash of light are spherical according to any observer in an inertial reference frame. **(b)** The wave fronts remain centered on their source according to every inertial observer.

11. ❖ For each transformation listed, decide whether each quantity is relative or invariant. Justify your answers.

Quantity	Transformation
Speed	Galilean transformation
Velocity	translation of coordinates
Length	reflection of y-coordinate axis in x-z-plane
Mass	rotation of coordinates
Position	Lorentz transformation

12. ❖ If a perfectly rigid rod AB is struck at one end, the whole rod accelerates simultaneously (■ Figure 34.26). Show that such a rigid rod cannot exist because, for some observers, the far end B would accelerate before the rod is struck.

13. ❖ A space navy cruiser of rest length L passes directly by a disabled space freighter at speed β (■ Figure 34.27). Two astronauts leap from the front and the back of the cruiser so as to pass simultaneously (according to the freighter crew) through two entry ports distance L/γ apart on the freighter. How is their procedure described by the crew of the cruiser?

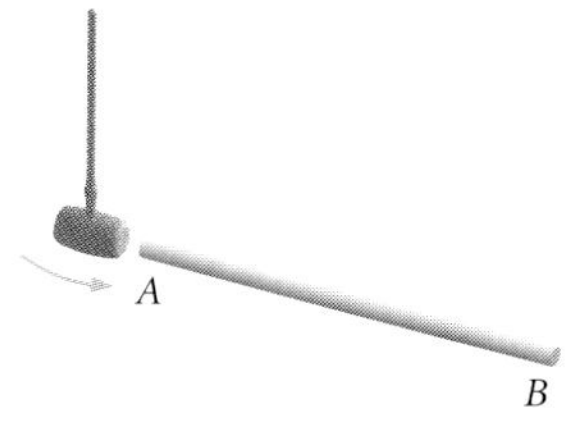

■ FIGURE 34.26

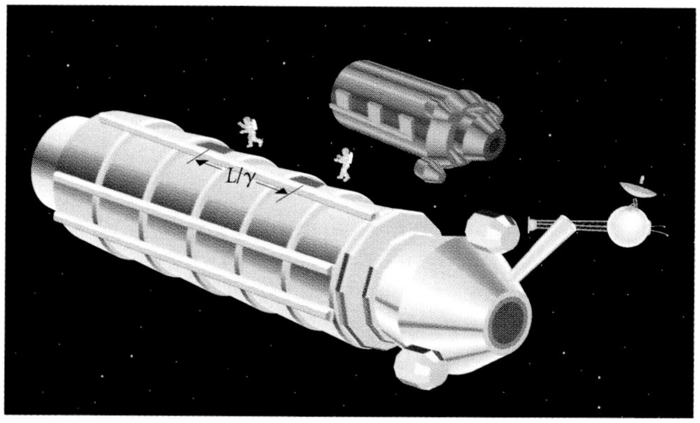

■ FIGURE 34.27

14. ❖ A rod of rest length L slides at speed $\beta \approx 1$ parallel to its length on a frictionless surface. A groove of width L is cut in the surface perpendicular to the rod's path. While crossing the groove, the rod accelerates downward and strikes the far edge of the groove in a fearful collision (■ Figure 34.28). How does the system appear in the rest frame of the rod? What must happen to the rod if events in this reference frame are to be consistent with those described in the groove rest frame?

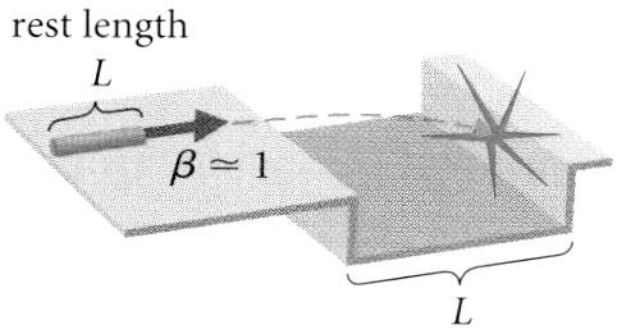

■ FIGURE 34.28

15. ❖ A champion runner who runs at nearly the speed of light holds a mirror in front of herself as she runs. Can she see her reflection in the mirror? Use the special principle of relativity in your answer. Would the answer be different in Galilean relativity? Why or why not?

16. ❖ Consider the following variation on the tale of Oliver, Penny, and the lightning bolts. Oliver is standing in the middle of a bridge that crosses over the railroad tracks at right angles. Two lightning bolts strike the ends of the (fortunately well-grounded) bridge, simultaneously in Oliver's frame, just as Penny, on the train, passes beneath. Show that Penny and Oliver agree that the bolts strike simultaneously.

17. ❖ What is the minimum number of observers needed to test time dilation if each observer can make measurements only at her own location? Specify the number of observers needed in each reference frame, and explain your reasoning.

18. ❖ There once was a young fencer named Fisk
Whose swordplay was exceedingly brisk.
So fast was his action
The Fitzgerald contraction
Reduced his épée to a disk.

Is this possible? If so, describe the motion of the épée relative to the observer.

19. ♦ A space tug is towing a square space station panel past a space colony at one-quarter the speed of light. What is the shape of the panel in the reference frame of a colonist?

20. ♦ Which has the greater length according to lab frame observers, a rod of rest length 1.25 m moving at $\beta = 0.60$ or a rod of rest length 1.333 m moving at $\beta = 0.80$? Assume that each rod's velocity is parallel to its length.
21. ♦ Lambda particles have an average lifetime $\tau = 2.9 \times 10^{-10}$ s. If in a collision, lambda particles are produced with speed $v = 0.95c$ in the lab frame, how far do they travel, on average, before decaying?
22. ♦ A transmitter that emits radar pulses at intervals of 7.600 μs is mounted on an alien spacecraft that passes Earth at a speed $\beta = 0.9975$. At what frequency are the pulses received on Earth just as the spacecraft passes?
23. ♦ An omega particle moving at $\beta = 0.9990$ traveled 0.8386 m in the lab before decaying. How long did it exist in its own rest frame before it decayed?
24. ♦♦ Two particles emerge from a collision experiment. The first has speed $\beta = 0.975$ and decays after 26.39 ns in its own rest frame. The second particle has speed $\beta = 0.880$ and decays after 34.79 ns in its rest frame. Which particle survives longer in the lab frame? By how much?
25. ♦♦ While mining an asteroid, you observe a rocket pass by. You measure its length to be 417 m, and it takes 3.0 μs to pass you. What is the speed of the rocket and its length as measured by its crew?
26. ♦♦ How regularly would your heart have to beat to detect the effects of time dilation when you ride on a jet aircraft traveling at 250 m/s?
27. ♦♦ Suppose the light clock in §34.1.3 is in motion parallel to the direction of the light beams. Show that the same time dilation formula holds.
28. ♦♦ *The transverse Doppler effect.* An atom radiates light at frequency f_o as measured by an observer at rest with respect to the atom. What is the frequency measured by an observer moving at speed v perpendicular to the line from observer to atom?
29. ♦♦ A centrifuge with rotor arms 10 cm long spins with an angular speed $\omega = 10^5$ rad/s. By what fraction $\Delta\ell/\ell$ is the length of a molecule at the rim of the centrifuge contracted with respect to a similar molecule at rest in the laboratory?
30. ♦♦♦ *The Doppler effect for light.* Consider an atom radiating light of frequency f_o and moving directly away from the observer with speed $v = \beta c$. By considering the time interval between the arrival of successive wave fronts, derive eqn. (16.24):

$$f/f_o = \sqrt{(1 - \beta)/(1 + \beta)}.$$

§34.2 SPACE-TIME

31. ❖ Is an interval between two events on a particle's world line spacelike or timelike? Explain.
32. ❖ Two rocket ships depart Earth 20 min apart heading in the same direction, with $\beta = 0.8$. The trailing rocket emits light flashes at regular intervals $\Delta t'$ in its frame. Show that the interval between flashes as observed by the leading rocket is also $\Delta t'$.
33. ❖ Two events P and Q have a spacelike separation. Show that both the future and past light cones of P and Q overlap. This result shows that such events may have a common cause in the past and may mutually influence events in their common future. Can the past of one intersect the future of the other?
34. ❖ The light cone is the set of events with zero interval ($\Delta s = 0$) from the origin. Describe the set of all events that have the same spacelike separation ($\Delta s^2 = \Delta s_o^2 < 0$) from the origin. Also describe the set of events with the same timelike separation ($\Delta s^2 = \Delta s_1^2 > 0$).
35. ♦ (■ Figure 34.29) Calculate the space-time interval between events A and B, and between events C and D. Assume $y = z = 0$ for all the events. Decide whether each interval is spacelike or timelike.

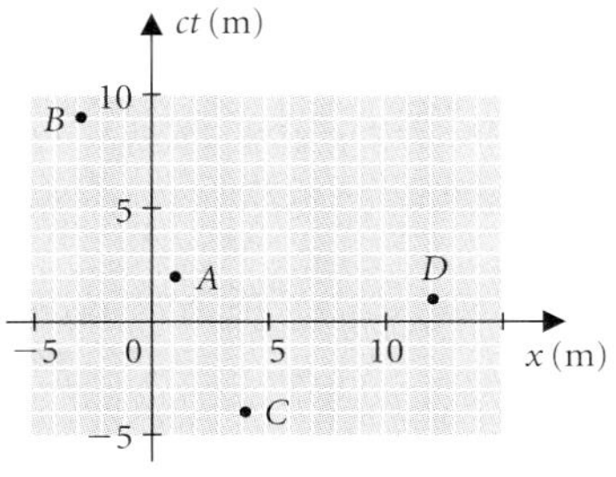

■ FIGURE 34.29

36. ♦ The coordinates of two events are $ct_1 = 0.70$ m, $x_1 = 2.0$ m, and $ct_2 = 10.37$ m, $x_2 = 35.6$ m. Calculate the proper distance between the events. Does proper time between the two events have any meaning? Why or why not?
37. ♦ The coordinates of two events are $ct_1 = 0.45$ m, $x_1 = 16.29$ m, and $ct_2 = 37.56$ m, $x_2 = 23.49$ m. Calculate the proper time between the two events. Does proper length between the events have any meaning? Why or why not?
38. ♦♦ Draw the following events on a space-time diagram. At $t = 0$, a rocket leaves a space station at $0.67c$ enroute to Earth, 7.5 light-hours away. After 0.5 hours the space station sends a radio message to Earth announcing the departure; 0.5 hours after receiving the signal, Earth sends a message to the rocket clearing it to land. Assume the space station is at rest with respect to Earth. Use your diagram to determine when (in Earth-time) the rocket receives clearance to land. How far is it from Earth then? How much proper time has elapsed on board the rocket since it departed the space station?

§34.3 THE LORENTZ TRANSFORMATION

39. ❖ Construct a space-time diagram that represents two reference frames moving at speed β with respect to each other and using different events as origin. How are the equations of the Lorentz transformation modified to describe this case?
40. ❖ Using a space-time diagram, show that for any two events separated by a timelike interval, there is a reference frame with both events on its time-axis.
41. ❖ Using a space-time diagram, show that for any two events separated by a spacelike interval, there is a reference frame with both events on its space-axis.
42. ❖ Suppose that (just in case the student is correct) the chemistry professor (Study Problem 20) plans to look for light from the crash as the front end of the train hits the tunnel's blocked exit, and escape by rocket if a crash occurs. Can the professor escape the tunnel by this method? Use a space-time diagram to justify your answer.
43. ♦ What error is made by using the Newtonian velocity addition formula to find the relative speed of two jet aircraft, both with a speed of 1800 km/h, approaching each other head on.
44. ♦ Consider two events on the time-axis of the prime reference frame, separated by a time interval $\Delta t'$. Use the Lorentz transformation to find the time interval between the events in the unprime frame and so derive the time dilation formula.
45. ♦ Two signaling lasers are pointed directly at each other. Show that the relative speed of two light pulses, one from each laser, is c.
46. ♦ A space navy cruiser moving parallel to the x-axis of a space station at $\beta = 0.80$ launches a rescue pod in the y-direction with speed $\beta = 0.90$ in the cruiser reference frame. What is the pod's speed relative to observers on the space station?
47. ♦♦ Refer to Figure 34.29. Determine the velocity of a reference frame with origin O in which event A occurs on the time-axis. Find the coordinates of events A, B, C, and D in this frame. Calculate the

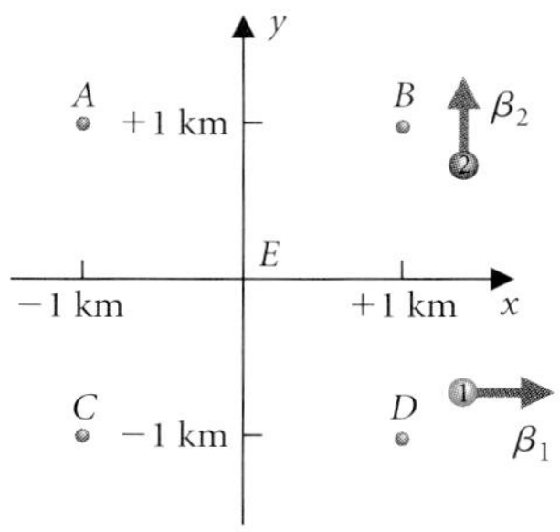

■ FIGURE 34.30

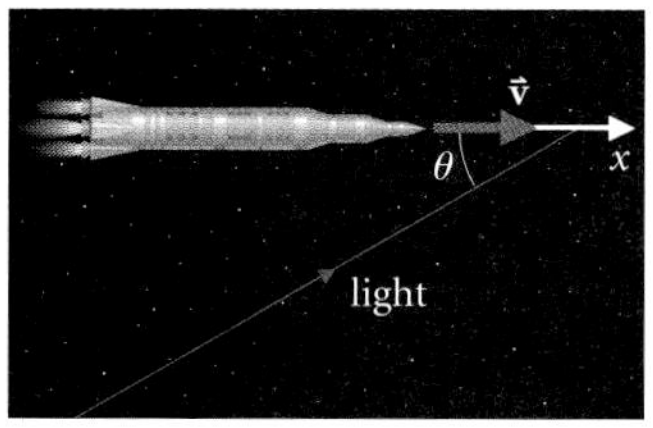

■ FIGURE 34.31

space-time intervals between event pairs *AB* and *CD*, and verify that they are the same as those calculated in the original frame. (See also Problem 35.)

48. ♦♦ At $t = 0$, a rocket departs Earth at $0.90c$ on a journey to a colony 12 light-years away. After 1 year, a radio signal is sent from the Earth to the rocket and after 2 years a signal is sent from the colony to the rocket. Draw these events on a space-time diagram. Which signal arrives first? Find the arrival times of the signals in both the Earth and rocket frames.

49. ♦♦ Events *A*, *B*, *C*, *D*, and *E* occur simultaneously according to observers at rest in the reference frame shown in ■ Figure 34.30. Observers 1 and 2 each move at speed $\beta = 0.800$ with respect to that frame, observer 1 along the *x*-axis and observer 2 along the *y*-axis. Determine the sequence of events in each observer's frame. Assuming both observers choose event *E* as the origin, compute the coordinates of events *A*–*D* in each frame.

50. ♦♦ In the Earth reference frame, flashes occur simultaneously at evenly spaced points 1.0 km apart along the *x*-axis. A rocket ship with velocity $(0.80c)\hat{\mathbf{i}}$ is at $x = 0$ when the flash occurs there. Find the time interval between the rocket's reception of flashes from $x = +N$ km and $x = -N$ km. What is this time interval in the rocket's frame? What time of occurrence does the rocket record for each flash?

51. ♦♦ A hedgehog-class automated interstellar exploration module approaches the Betelgeuse system at speed $\beta = 0.80$ along the *x*-axis. The hedgehog splits into six submodules that depart with speed $\beta' = 0.60$, one each along the directions $\pm x$, $\pm y$, and $\pm z$. **(a)** What is the speed of each submodule with respect to the Betelgeuse system? **(b)** What is the relative speed of each pair of submodules with opposite directions? **(c)** With perpendicular directions?

52. ♦♦ Use the velocity transformation law to show that a light pulse traveling along the *y*-axis (i.e., perpendicular to $\vec{\mathbf{v}}$) in the unprime frame also has speed *c* in the prime frame. (As usual, the prime frame is moving at $\vec{\mathbf{v}} = v\hat{\mathbf{i}}$ with respect to the unprime frame.)

53. ♦♦ In Figure 34.11, suppose that 1 cm represents a time $t = 1\ \mu$s ($ct = 300$ m) along the time-axis and a length 300 m on the space-axis. What length represents 1 μs on the t'-axis? Show that the same length represents 300 m on the x'-axis.

54. ♦♦ Two light pulses are moving in the positive *x*-direction. In the unprime frame, they are separated by distance *d*. Show that in a prime frame moving at velocity $u\hat{\mathbf{i}}$ with respect to the unprime frame, they are separated by distance $d' = d\sqrt{(c + u)/(c - u)}$. From this result, derive the Doppler effect for light (eqn. 16.24). (See also Problem 30).

55. ♦♦ Suppose the prime frame could have speed *c* with respect to the unprime frame. Show that all objects moving with speed less than *c* in the unprime frame would move at speed *c* in the prime frame.

56. ♦♦ A rocket is moving at speed *v* in a certain reference frame (■ Figure 34.31), and a beam of light is traveling at an angle θ to the *x*-axis in the same frame. Use the velocity transformation law to show that in the rocket frame the speed of light is still *c* but that its angle of approach to the rocket is changed. Find an expression for $\tan\theta'$. Discuss qualitatively the appearance of the sky as seen by observers on a rocket traveling directly away from Earth toward a star-filled universe.

57. ♦♦ Using the Lorentz transformation, derive the following law for the transformation of a position vector $\vec{\mathbf{r}}$ into a reference frame (prime) moving with velocity βc with respect to the unprime frame:

$$\vec{\mathbf{r}}' = \vec{\mathbf{r}} + (\gamma - 1)(\vec{\mathbf{r}}\cdot\hat{\beta})\hat{\beta} - \gamma\vec{\beta}ct.$$

From this, show that the velocity transformation law is:

$$\vec{\mathbf{w}}' = \frac{\vec{\mathbf{w}} + (\gamma - 1)(\vec{\mathbf{w}}\cdot\hat{\beta})\hat{\beta} - \gamma\vec{\beta}c}{\gamma(1 - \vec{\beta}\cdot\vec{\mathbf{w}}/c)}.$$

Show that this expression reduces to eqns. (34.9) when $\vec{\beta} = \beta\hat{\mathbf{i}}$.

58. ♦♦♦ A quantity called rapidity, $\xi \equiv \tanh^{-1}\beta$, is sometimes used as a measure of an object's speed. If an object moves parallel to the x'-axis and has rapidity ξ_1 measured in the prime frame, and if the prime frame moves parallel to the *x*-axis and has rapidity ξ_2 with respect to the unprime frame, show that the object's rapidity in the unprime frame is $\xi = \xi_1 + \xi_2$.

59. ♦♦♦ Derive the Lorentz transformation for acceleration by differentiating the velocity transformation.

60. ❖ We have supposed that acceleration at Sirius has no direct effect on proper time as experienced by mobile Melia. With the cooperation of a Sirian scientist, Melia could test the hypothesis. How?

61. ♦♦♦ To investigate the Case of the Youthful Twin, each twin transmits a radio pulse every 2 y, as measured in her own frame. Draw a space-time diagram in domestic Delia's reference frame showing mobile Melia's world line (constant speed out, instantaneous acceleration, same speed back), and the world lines of the radio pulses sent by each. Show that by counting received pulses, each twin may conclude that Delia is the older upon reunion.

62. ♦♦♦ Reversing her velocity at the destination, mobile Melia switches from the outgoing inertial frame *G* to the returning inertial frame *R*. In these two reference frames, different events E_1 and E_2 on Delia's world line are simultaneous with the turnaround event. Find the time interval in Delia's frame between E_1 and E_2. Show that when this extra time is added to Delia's 7.7 y of aging computed in Melia's rest frame, the result is consistent with the computation in Delia's frame. (This calculation resolves the paradox by correctly computing proper time along Delia's world line. Notice that we must use Delia's frame during the time that Melia is accelerating. We can add the results because proper time is invariant.)

§34.4 RELATIVISTIC DYNAMICS

63. ❖ Modern accelerators such as LEP (Linear Electron Positron Collider) use two beams of particles that approach each other with equal speeds before colliding. What is the advantage of this design?

64. ❖ The energy-momentum invariant of a system of particles is observed to be M^2c^4 in the lab frame. What is its value in a frame moving at speed $\beta = 0.977$ in the laboratory? If the system remains isolated but undergoes a reaction that reduces the number of particles in the system, how does the invariant change? Describe how the invariant would change if the reaction also produces energetic γ rays that are lost from the system.

65. ♦ In nuclear fusion experiments, deuterium (mass 2.01410 u) and tritium (3.01605 u) combine to form helium (4.0026 u) and a neutron (1.00867 u). How many reactions per second are needed to provide 1.0 kW of power?

66. ♦ In nuclear power plants, a nucleus of ^{235}U (mass 235.044 u) combines with a neutron (mass 1.0087 u) to produce a krypton nucleus (mass 91.926 u), a barium nucleus (141.916 u), and two more neutrons. How much uranium is processed per second in a 1.00-GW power plant?

67. ♦ When a proton and an electron combine to form neutral hydrogen, 13.6 eV of energy is released. By what fraction is the mass of the system decreased because of this energy release? (The 13.6 eV is called the binding energy of the system.)

68. ♦ What value of γ would be required for a single proton to have the same kinetic energy as a 10^3-kg car traveling at 20 m/s? How would the momentum of the proton compare with that of the car?

69. ♦ What is the energy-momentum invariant for an electron with a kinetic energy of 75.0 eV? Of 6.9 GeV? What magnitude of momentum does each electron have?

70. ♦ A proton has a total energy of 2.75 TeV. Find its energy-momentum invariant and its momentum.

71. ♦♦ In the later stages of stellar evolution, the primary energy-producing reaction is the triple-α process in which three helium nuclei (mass 4.0026 u) combine to form a carbon nucleus (mass 12.000 u). How much energy is produced per reaction? If a star in this phase has 400 times the luminosity of the Sun and 6.3 times the Sun's mass, how long would it take to change 1% of its mass from helium to carbon?

72. ♦♦ How many years' worth of the Earth's current power usage (roughly 10 TW) would be required to accelerate a 70-kg person to a speed of $0.997c$ (assuming perfect efficiency)?

73. ♦♦ A cosmic ray proton with an energy of 10^8 GeV is absorbed by a stray pea (mass 5.0 g) floating in the space shuttle galley. What is the resulting speed of the pea?

74. ♦♦ A particle moving at $0.85c$ along the x-axis collides inelastically with another identical particle at rest. After the collision, one particle moves off along the x-axis in the same direction as the incoming particle at $\beta = 0.74$. Find the velocity of the other particle after the collision. How much of the initial kinetic energy was converted to other forms in the collision?

75. ♦♦♦ Two particles of mass m_1 and m_2 and speeds u_1 and u_2 approach each other at an angle α and collide. During the collision, the particles coalesce, forming a new particle of mass m. Show that:

$$m^2 = m_1^2 + m_2^2 + 2m_1m_2\gamma_1\gamma_2(1 - \beta_1\beta_2 \cos \alpha).$$

Show that $m > m_1 + m_2$ for any values of u_1, u_2, and α. Explain this result. (*Hint:* Compare expressions for the system's energy-momentum invariant before and after the collision.)

76. ♦♦♦ Cosmic rays are often described by their rigidity, $R = pc/(Ze)$, where p is the momentum and Ze the charge of the nucleus. Show that the Larmor radius (§29.1, eqn. 29.1) of the particle is $r_L = R/(Bc)$, so particles of high rigidity deviate little from straight-line paths.

77. ♦♦♦ **(a)** Use the Lorentz velocity transformation (eqns. 34.9) to show that the particle's γ factor transforms according to the rule: $\gamma_{u'} = \gamma_u\gamma(1 - \beta u/c)$, where γ (without a subscript) $= (1 - v^2/c^2)^{-1/2}$, $\gamma_u = (1 - u^2/c^2)^{-1/2}$, and similarly for $\gamma_{u'}$. **(b)** Use this result to verify the transformation laws for energy and momentum given in Table 34.2.

Additional Problems

78. ❖ We have tacitly assumed that relatively moving observers agree on their relative speed. Describe an experiment that would test this hypothesis.

79. ❖ When studying the Newtonian dynamics of collisions, we discovered that when two identical particles collide elastically, their speeds in the CM frame are unchanged, although the direction of the velocity may change. Is the same true if the particle speed is relativistic? Why or why not?

80. ❖ In one reference frame, a plane rectangular wire loop is at rest. A uniform magnetic field is perpendicular to the plane of the loop and increasing in time. A second reference frame moves with speed v parallel to one side of the loop. Show that the magnetic field in the second frame is nonuniform as well as time variable. Use the motional emf in the second frame to argue that a Faraday-type law must exist.

81. ❖ Evaluate the electromagnetic field invariants (Table 34.2) for an EM wave propagating in vacuum. Use the results to show that an EM wave has the same structure (as given in §33.1) in any two inertial frames.

82. ♦♦ Two lamps that flash once every 50 ns (each in its rest frame) are mounted one on the tunnel entrance and one on the back end of the train in Study Problem 20. Each lamp emits a flash just as they pass each other. Locate the next flash of each light in a space-time diagram in both the tunnel and train frames. Show that observers in each reference frame observe time dilation of the other's light flashes. Explain why there is no paradox in the results.

83. ♦♦ How would you expect the emitted frequency of a particular spectral line to vary with temperature because of the speed of the emitting particles? Find the fractional change per kelvin, $(1/f)\ df/dT$, for iron nuclei. [*Hint:* Use the relation between temperature and rms speed for an ideal gas of nuclei (§19.2), and use time dilation to relate the time intervals (a) between the emission of two wave fronts, and (b) between the reception of two wave fronts. Compare with the results of the Pound–Rebka experiment (§34.3.3).]

84. ♦♦ Draw the following events on a space-time diagram. A rocket pirate ship leaves its base for a mineral-rich asteroid, 1.375 light-years away, traveling at $0.800c$. Light from the pirates' rocket thrust at departure is observed by the space patrol at its port, 0.170 light-years from the asteroid and 1.545 light-years from the pirate base. At what speed must the space patrol travel to reach the asteroid before the pirates? How much does the patrol age during its journey?

85. ♦♦♦ As the space patrol approaches the asteroid (Problem 84) at $\beta = 0.978$, what is its speed as observed by the pirates who approach the asteroid from the opposite direction at $\beta = 0.800$? If the pirates observe the space patrol's rocket burn as it leaves port, how much time do they have to plan a retreat between receiving the signal and the arrival of the patrol?

86. ♦♦♦ A cylinder lying along the x-axis is rotating about its axis with angular velocity ω. Show that if the cylinder is viewed from a frame moving in the x-direction with speed βc, the cylinder is twisted by an amount $\Delta\phi = \omega\beta/c$ per unit length.

87. ♦♦♦ **(a)** The helicopter that we imagined photographing Manhattan in §34.2.1 is 1 km above the ground and records an area 1 km

on a side. Explain why the events shown in a photograph are not exactly simultaneous, and find the maximum time difference between events shown in a single photograph. **(b)** As the photographs were taken, an airplane was flying past the helicopter at a speed of 600 km/h. In the airplane's frame, what is the maximum time difference between events that are simultaneous for the helicopter pilot? **(c)** Is the travel time of light or the aircraft's motion the more important source of error in constructing a space-time diagram from photographs?

88. ♦♦♦ *The symmetric space-time diagram.* The representation of prime and unprime frames in Figure 34.11 is asymmetric and disguises the equivalence of the two frames for describing physics. For example, representation of length contraction is unconvincing because the scales on the two sets of axes are different. We can construct a symmetric representation by using a third reference frame (labeled *) in which the original two frames move with equal speed in opposite directions (■ Figure 34.32). In the asterisk frame, the unprime frame moves with speed $c\beta_*$. **(a)** Explain why the prime and unprime axes make equal angles ϕ with the * axes, as shown in Figure 34.32. Explain why the x- and ct'-axes are perpendicular, as are the x'- and ct-axes. **(b)** Consider the prime origin as a particle with velocity $\vec{u}_* = \beta_* c\hat{i}$ in the asterisk frame and $\vec{u} = \beta c\hat{i}$ in the unprime frame. Use the velocity transformation law to prove that

$$\beta_* = (\beta - \beta_*)/(1 - \beta\beta_*).$$

(c) Show that $\beta_* = (1 - \sqrt{1 - \beta^2})/\beta$. Explain why you may reject the + sign when solving the quadratic. **(d)** Use the result of (b) to express β in terms of β_*. Then substitute $\tan\phi = \beta_*$ (cf. Figure 34.11) to show that $\sin 2\phi = \beta$.

89. ❖ Explain why the length and time scales are identical on all four axes in the symmetric space-time diagram (Problem 88, Figure 34.32). Describe how to locate **(a)** all events that are simultaneous with a given event in either reference frame, and **(b)** all events that occur at the same place in either reference frame.

90. ♦♦ In a symmetric space-time diagram (Figure 34.32, Problem 88), construct the world lines of the ends of a meter stick that moves with speed $\beta = \frac{4}{5}$ along the x-axis. Identify events corresponding to length measurements in the prime and unprime frames to show that, in this diagram, the rod is clearly shortened in the frame in which it moves.

91. ♦♦♦ Physical quantities in special relativity are expressed most simply in four dimensions as functions of proper time (§34.2). For example, the four-dimensional position vector $\boldsymbol{r} = (ct, \vec{r})$ has components equal to the coordinates of an event. The four-dimensional velocity vector of a particle is the rate of change of its four-dimensional position vector with respect to proper time: $\boldsymbol{u} = d\boldsymbol{r}/d\tau$. Find the time and space components of $\boldsymbol{u}$ in terms of the three-dimensional position vector $\vec{u}$. Show that the invariant magnitude [(time component)^2 − (spatial vector)^2] of $\boldsymbol{u}$ equals c^2, and also show that the relativistic momentum-energy vector (§34.4) is $\boldsymbol{p} = m\boldsymbol{u}$.

92. ♦♦♦ An observer living on a rotating disk decides to measure the radii and circumferences of circles centered on the rotation axis, using a standard meter stick. Show that the observer finds the circumference of a circle of radius r to be $2\pi r/\sqrt{1 - \omega^2 r^2/c^2}$.

93. ♦♦♦ Use the transformation laws for $\vec{E}$ and $\vec{B}$ given in Table 34.2 to compute $\vec{E}' \cdot \vec{B}'$ and $(E')^2 - c^2(B')^2$ in terms of $\vec{E}$ and $\vec{B}$. Show that each quantity is invariant.

Computer Problems

94. At $t = 0$, a rocket, the *Interstellar Ramjet,* leaves Earth with a constant acceleration a' as measured by accelerometers on board. At time t, the rocket has speed $v(t)$. In a "prime" frame moving at speed v, the rocket has speed $dv' = a'\ dt'$ after a time dt' measured in the prime frame. Show that the corresponding time interval in the Earth frame is $dt = dt'/\gamma$. Use the velocity transformation to show that the speed in the Earth frame at time $t + dt$ is:

$$v(t + dt) = \frac{v(t) + a'\ dt'}{1 + v(t)a'\ dt'/c^2}.$$

Use a spreadsheet or other computer program to compute dt' and $v(t + dt)$, given $v(t)$ and dt. You may find it convenient to use dimensionless variables $\beta = v/c$ and time variable t/t_o, where $t_o \equiv c/a'$. If the acceleration a' has magnitude $g = 9.8\ \text{m/s}^2$, after how long is the rocket's speed = 75% of c? What is the rocket's speed after 2 y? Plot the world line of the *Interstellar Ramjet* on a space-time diagram. Show that a light signal emitted in the x-direction from an event at $ct = 0$ and $x < -c^2/a'$ never reaches the *Ramjet.*

95. Compute the magnitude of the momentum and the kinetic energy of a particle with speed v and compare with the Newtonian values. Make plots of p/p_N and K/K_N as functions of β. For what values of β is each ratio equal to 2? To 10?

Challenge Problems

96. A distant observer takes a photograph of a cube with sides of proper length ℓ moving at speed v perpendicular to the line of sight as shown in ■ Figure 34.33. The photograph is made with light that crosses line AB in the observer's frame simultaneously. Show that light from face PS contributes to the image, which looks like a rotated cube. What is the apparent angle of rotation?

97. In this problem we demonstrate that the existence of electric field and two signs of charge implies the existence of magnetic field. A long, straight wire is electrically neutral and carries a current I produced by electrons moving at speed v. A single electron outside the wire also moves parallel to the wire at speed v. **(a)** Show that the wire is electrically charged in a reference frame moving with the electrons. (*Hint:* Consider two lengths of wire, one containing N electrons and one containing N positive charges. The wire is electrically neutral because these two lengths are the same in the lab frame. How do they compare in the electron frame?) **(b)** Compare the electric force on the isolated electron computed in the electron frame with the magnetic force computed in the laboratory.

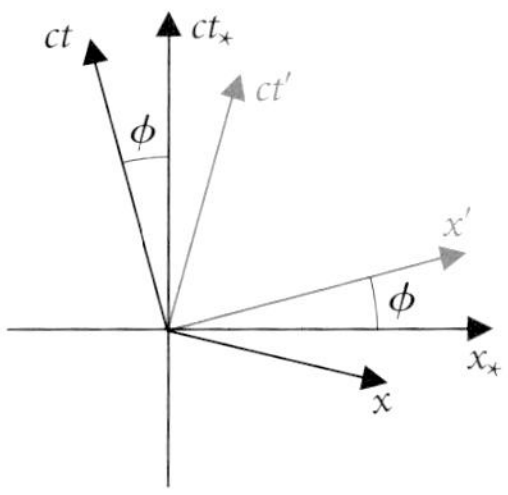

■ FIGURE 34.32

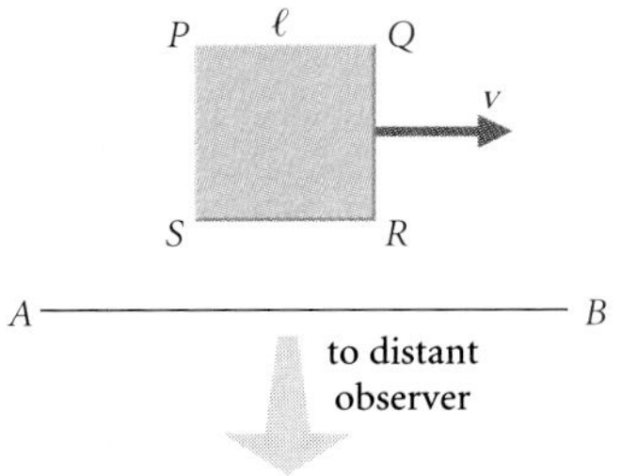

■ FIGURE 34.33

This is news distinctly shocking and apprehensions for the safety of confidence even in the multiplication table will arise . . .

NEW YORK TIMES EDITORIAL, NOVEMBER 11, 1919

ESSAY 8
General Relativity: A Geometric Theory of Gravity

E8.1 GRAVITY AS AN INERTIAL FORCE

When a bus decelerates rapidly, loose packages slide, briefcases fall over, and passengers stand at an angle as if, for a moment, gravity had been tilted toward the front of the bus. In the accelerated reference frame of the bus, each object is subject to an apparent force that overturns the briefcases and throws the passengers forward.

A physics student on the sidewalk interprets events differently; for example, the sliding packages move at constant velocity while the bus around them decelerates. In this reference frame, the apparent force doesn't exist at all; it is an example of an *inertial force.*

> An *inertial force* is an apparent force needed to account for the acceleration of free bodies with respect to a noninertial reference frame.

Inertial forces mimic gravitational force. The acceleration resulting from the noninertial nature of the reference frame is the same for all free objects. The inertial force on an object equals its mass multiplied by this acceleration. For example, the inertial force on an object of mass m in the bus is:

$$\vec{\mathbf{F}}_{\text{inertial}} = -m\vec{\mathbf{a}},$$

where $\vec{\mathbf{a}}$ is the acceleration of the bus with respect to the sidewalk.

Prototype of the Laser Interferometer Gravitational-Wave Observatory (LIGO). General relativity relates the structure of space to the matter within it, and predicts the existence of gravitational waves: ripples in space-time. Interference between laser beams reflected from mirrors at the ends of the two arms measures the path difference to within a fraction of a wavelength. Passage of a gravitational wave (§E8.4) alters these path lengths and is detectable as a change in the interference pattern.

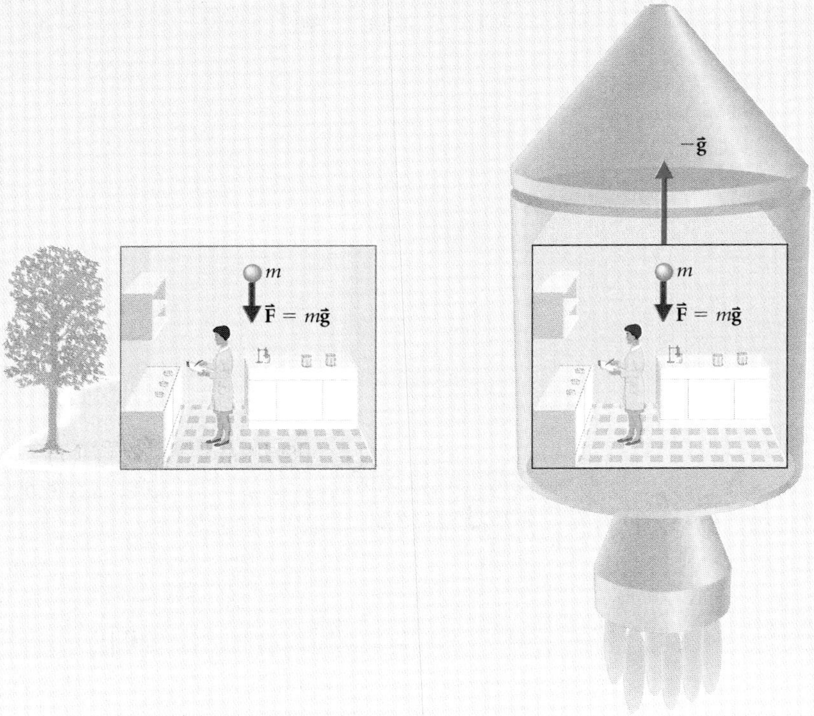

■ FIGURE E8.1
The equivalence principle. The experimenter cannot determine whether the lab is in a uniform gravitational field or in an accelerating rocket.

Einstein realized that gravity and inertial forces are more than just similar; gravity *is* an inertial force. Consider two identical laboratories (■ Figure E8.1), one on Earth in which there is a gravitational acceleration $\vec{g}$, and one far from any astronomical object but accelerating at $-\vec{g}$ due to rocket thrust. Each piece of equipment in the rocket lab is subject to an inertial force equal to the gravitational force on the corresponding equipment in the Earth lab. The result of any mechanical experiment is the same in each lab. Einstein generalized this conclusion in the *equivalence principle.*

> No experiment of any kind can distinguish a uniform gravitational field from a uniform acceleration of the reference frame.

Gravitational force appears in frames that are *not* inertial. Conversely, the local inertial frame is one in which gravity disappears—that is, a freely falling frame. Practical examples are the interior of the orbiting space shuttle, or NASA's *Vomit Comet* flown on a parabolic path.

Einstein's concept of an inertial frame differs from the definition we have used in Newtonian physics. Previously, we described gravity as a force in the same class as spring or electric forces. An inertial frame meant one in which every acceleration is produced by an identifiable force. In Einstein's picture, the gravitational force is not on the force list at all: it is a fictitious force due to acceleration of the reference frame. Inertial frames are defined to be ones in which there is no gravitational force and no gravitational acceleration, and all *other* accelerations are produced by identifiable forces.

Physics in these inertial frames is described by special relativity. Using special relativity with the equivalence principle allows us to deduce some remarkable effects of gravity.

Gravity causes time dilation. To demonstrate this effect, imagine dropping a clock to Earth from a great distance (■ Figure E8.2). Since it remains in an inertial frame, the clock measures the rate proper time passes in the absence of gravitational effects. We compare it with another identical clock at rest with respect to the Earth and at a distance R from the center. The freely falling clock passes the fixed clock at speed $v = \sqrt{2GM/R} = \sqrt{-2\Phi}$, where $\Phi = -GM/R$ is the gravitational potential at radius R. We use the special relativistic time dilation formula to compare the time interval between ticks of the earthbound clock (in the prime frame) and the freely falling clock (in the unprime frame). The time interval in the falling frame is the shorter, since the earthbound clock ticks do not occur at the same place in *any* inertial reference frame. Thus the period of the earthbound clock $\Delta t'$ is:

$$\Delta t' = \gamma\,\Delta t = \frac{\Delta t}{\sqrt{1 - v^2/c^2}} = \frac{\Delta t}{\sqrt{1 - 2GM/Rc^2}},$$

where Δt is the period of the freely falling standard clock. In general, if $\Delta t'$ is the time interval between clock ticks where the gravitational potential is Φ,

$$\Delta t'\sqrt{1 + 2\Phi/c^2} = \text{constant.} \tag{E8.1}$$

Figure E8.2 also compares the readings on three stationary clocks at one instant. All the clock readings were the same when the moving clock was released. Stationary clocks near the Earth read earlier times than the falling one. Clocks closer to the Earth show a greater difference.

To see how large the effect is, let's compare the rate of aging of twin brothers, Joe, who stays at his home at sea level, and Stan, who climbs to the top of Mount Everest (8848 m above sea level). Because the gravitational potential is different at the two locations, the brothers' internal clocks tick at different rates.

Stan is 8848 m farther from the center of the Earth than Joe. The Earth's mass is 6.0×10^{24} kg and its radius is 6.4×10^3 km. The quantity

$$\frac{GM}{Rc^2} = \frac{(6.7 \times 10^{-11}\ \text{m}^3/\text{kg}\cdot\text{s}^2)(6.0 \times 10^{24}\ \text{kg})}{(6.4 \times 10^6\ \text{m})(3.0 \times 10^8\ \text{m/s})^2} = 7.0 \times 10^{-10}$$

is so small that we can approximate the square root using the binomial expansion (Appendix IB): $(1 + x)^{1/2} \approx 1 + \frac{1}{2}x$.

The intervals between the twins' heartbeats are in the ratio (eqn. E8.1):

$$\frac{\Delta t_J}{\Delta t_S} = \frac{\sqrt{1 - 2GM/(R_S c^2)}}{\sqrt{1 - 2GM/(R_J c^2)}} \approx \frac{1 - GM/(R_S c^2)}{1 - GM/(R_J c^2)}.$$

We use the binomial expansion again, $(1 + x)^{-1} \approx 1 - x$:

$$\frac{\Delta t_J}{\Delta t_S} = 1 - \frac{GM}{c^2}\left(\frac{1}{R_S} - \frac{1}{R_J}\right) = 1 - \frac{GM}{c^2}\left(\frac{R_J - R_S}{R_S R_J}\right) = 1 + \frac{GM}{Rc^2}\left(\frac{R_S - R_J}{R}\right),$$

where we ignore the difference between the two radii except when they are subtracted.

$$\frac{\Delta t_J}{\Delta t_S} = 1 + (7.0 \times 10^{-10})\left(\frac{8848\ \text{m}}{6.4 \times 10^6\ \text{m}}\right) = 1 + 9.6 \times 10^{-13}.$$

The difference between their rates of aging is 1 part in 10^{12}. To make the total difference equal to 1 min, Stan would have to stay on Mt. Everest for a time T, where $10^{-12}\,T = 1$ min, or $T = 2$ million years!

Since Δt_J is greater than Δt_S, Joe's heart beats slower than Stan's, and Joe ages less. The difference should not dissuade Stan from making the trip. Only very carefully designed experiments can measure relativistic effects on the Earth.

■ FIGURE E8.2
Gravitational time dilation. A freely falling clock measures proper time between events. A clock at rest in the Earth's frame is initially synchronized with an identical clock that is dropped from a great height. The falling clock registers a greater elapsed time than the stationary clock it is passing.

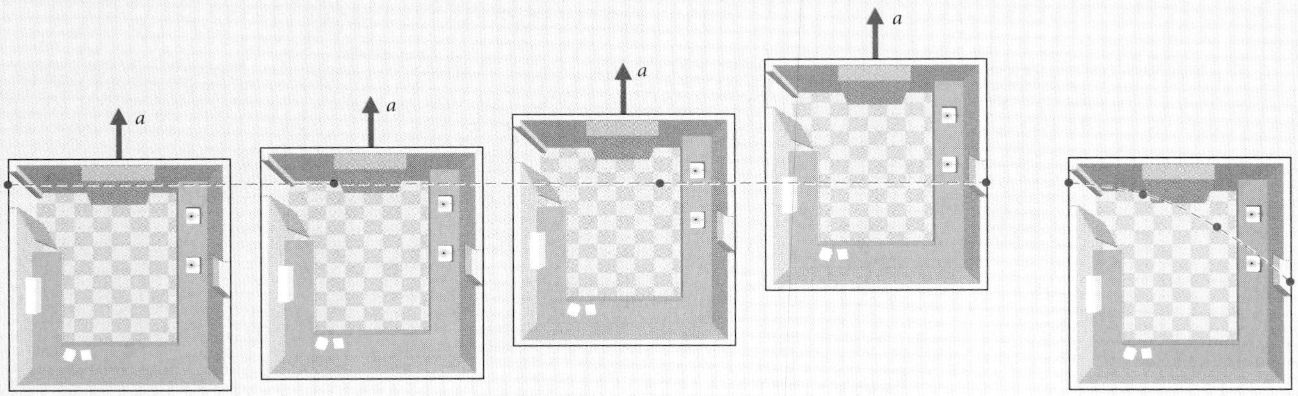

(a) (b)

■ FIGURE E8.3
Gravitational bending of light.
(a) Light pulse crossing an accelerated laboratory. The light follows a straight line in an unaccelerated (inertial) frame.
(b) Relative to the laboratory, the light follows a curved path.

Digging Deeper

WHY SHOULD WE BELIEVE THE EQUIVALENCE PRINCIPLE?

The equivalence principle has been extensively tested. If it is correct, then everything should fall at the same rate, even though nuclear and electromagnetic energy contribute in different proportions to the masses of different chemical substances. ■ Figure E8.4 shows an experiment that demonstrates that gravity and inertial forces have identical effects on these forms of energy. The experiment was performed by Roland von Eötvös in 1889.[1] Two objects of equal mass but different chemical composition are at the ends of a bar suspended by a thin fiber. Both gravity and inertial force due to the Earth's rotation act on the objects. Any difference in the gravitational accelerations would cause a torque on the fiber. Eötvös found no difference to within 1 part in 10^9. More recent experiments have improved the accuracy to 1 part in 10^{13}.

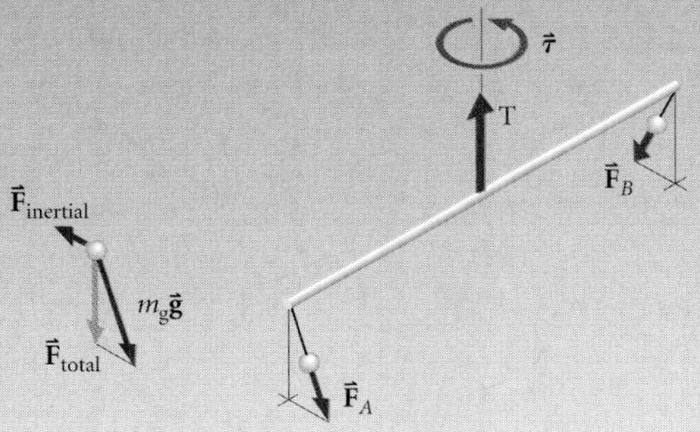

(a) (b)

■ FIGURE E8.4
The Eötvös experiment. Two objects of equal inertial mass are suspended from a bar. (a) The objects hang at the local vertical, defined by the gravitational force plus the inertial force in the rotating frame of the Earth. (b) If the two gravitational masses are not quite equal, the objects hang at different angles and a torque is exerted on the bar. (The angles are shown exaggerated for clarity.)

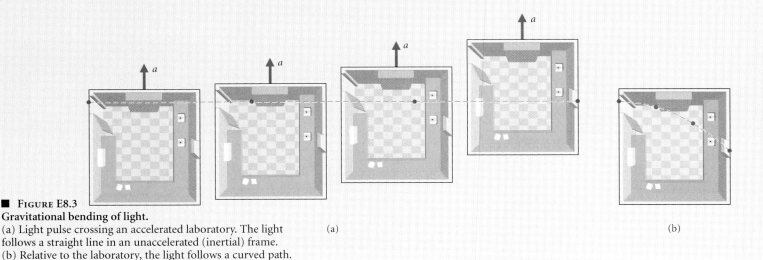

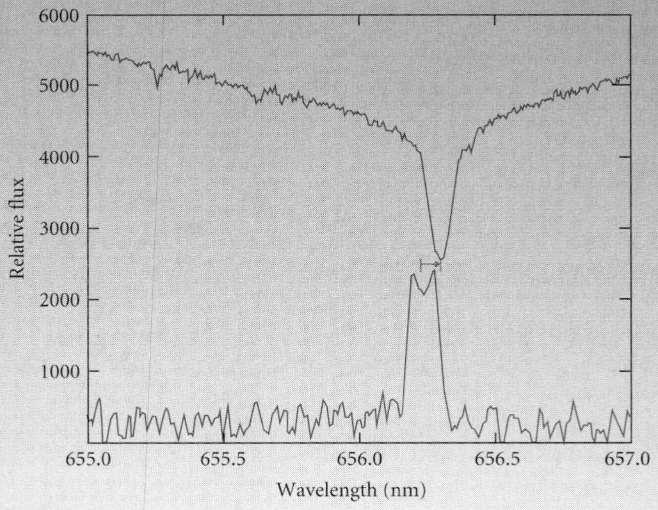

■ FIGURE E8.5
The spectrum of the white dwarf star 40 Eridani B shows a gravitational redshift. The upper spectrum shows the Balmer α absorption line of hydrogen (§35.2.2) in the white dwarf spectrum. The lower spectrum shows the same line in emission in the red dwarf binary companion 40 Eridani C. There is an obvious offset between the centers of the two lines of about 0.06 nm, of which 7% is a Doppler shift (§16.4) due to the motion of the two stars, and the rest is a gravitational redshift. The mass of the white dwarf is about 0.5 times the sun's mass, but its radius is only about 8000 km so GM/Rc^2 is about 10^{-4}.

[1] Eötvös' aim was to show, in the context of Newtonian theory, that the mass that produces gravitational force is the same as the mass that responds to force. This must be true if the equivalence principle holds.

A second consequence of the equivalence principle is that gravity accelerates light as well as matter. Since gravity is an effect of the reference frame, everything in the frame is accelerated equally. ■ Figure E8.3 shows a light pulse moving through an accelerated laboratory. The light moves in a straight line as seen by an observer in an inertial frame but traces a curved path across the accelerated frame. In a gravitational field, light travels along straight lines in a freely falling frame. Relative to the Earth's surface, light falls downward, pulled by gravity.

E8.2 CURVATURE OF SPACE-TIME

A real gravitational field is uniform only in small regions, and a freely falling inertial frame can only be defined *locally.* We need different inertial frames to discuss physics on board the space shuttle and in New Orleans. To construct a reference frame that covers the entire solar system, we must imagine a patchwork of many local reference frames. The extended reference frame so constructed describes a curved space-time. Navigation on the Earth's surface provides a good example of what this means.

When light is emitted by atoms close to a massive body, gravitational time dilation causes a distant observer to record the passage of wave fronts at a rate slower than that for waves emitted by identical atoms near the observer. This effect, called the gravitational red shift, has been observed in the spectra of certain stars (■ Figure E8.5). It has also been tested on Earth. Iron nuclei emit γ rays with an energy of 14.4 keV. In 1960, Pound and Rebka measured the difference in frequency between γ rays emitted at the bottom of a 22.5-m-high tower and the γ rays absorbed at the top. In 1965, Pound and Snider improved the experiment, measuring a frequency shift $\Delta f/f = (2.56 \pm 0.03) \times 10^{-15}$, in agreement with the predictions of general relativity.

Gravitational bending of light by the Sun was first detected during the solar eclipse of 1919 (■ Figure E8.6). More accurate results are obtained using radio waves emitted by distant quasars. The results confirm Einstein's predictions to within 1%. ■ Figure E8.7 shows multiple images formed as the light from a single quasar is deflected by a massive cluster of galaxies between us and the quasar. There are several known examples of such gravitational lenses.

■ FIGURE E8.6
Photograph of the May, 1919 solar eclipse taken by Eddington and Cottingham at Principe Island. Measurement of star positions on this photograph gave the first evidence of gravitational bending of light rays. Unfortunately, because of the age of the photograph, the star images are no longer distinguishable.

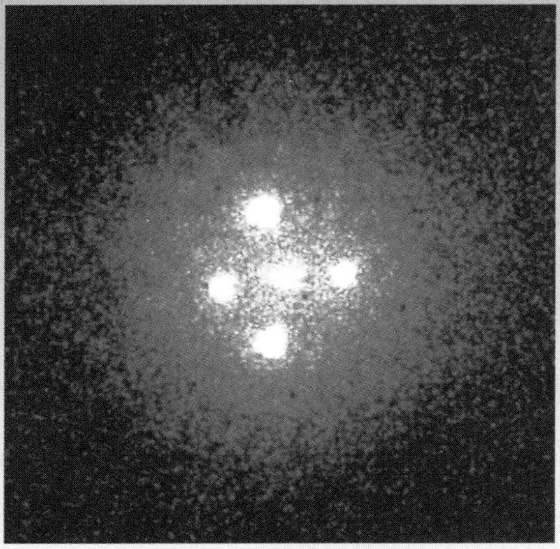

■ FIGURE E8.7
The *Einstein cross:* a gravitational lens. Light from a distant quasar (approximately 8 billion light-years away) is bent by the gravitational field of an intervening galaxy (the diffuse central object, approximately 400 million light-years away) to form four distinct images of the quasar. The number and position of the images formed depends on the precise alignment of background and foreground objects.

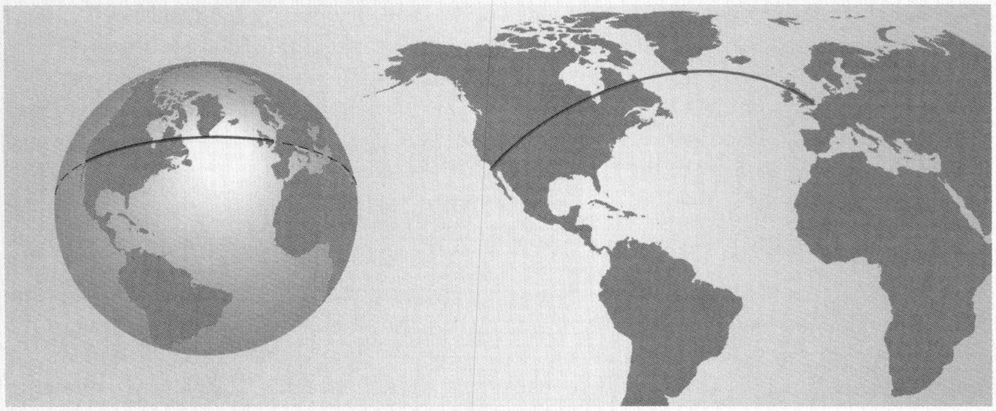
(a) (b)

■ FIGURE E8.8
(a) The airline route from Los Angeles to London is part of a great circle around the globe. (b) When drawn on a flat piece of paper, the route looks curved.

An airplane traveling between Los Angeles and London flies over northern Canada and Greenland, heading first north and then south (■ Figure E8.8). The plane's path is the shortest distance from Los Angeles to London, a straight line on the curved surface of the Earth. It *seems* to change direction because the rectangular grid of north-south-east-west is laid down *locally* on small patches of the Earth's surface and does not describe the whole Earth well. On a map showing the whole Earth on a flat piece of paper, the *straight line* route from Los Angeles to London *looks* curved.

In an inertial frame, a particle moves in a straight line unless acted upon by a force. Its world line is the path with minimum proper time between any two events. This principle also holds in a global reference frame. The nearly circular orbit of the Earth and the highly elliptical orbit of a comet around the Sun are both straight world lines in curved space-time. They appear curved both because we are looking at projections into three-dimensional space, and because the Newtonian description of space distorts its geometry, like the flat map of the Earth's surface. Like an ant crawling in a straight line over a plum, the Earth's path curves right around and comes back to the same place in space (although not to the same event in space-time)!

As in the Newtonian model, mass causes gravity. The mass of the Sun distorts space-time, stretching and curving it (■ Figure E8.9). In the solar system, differences between Einstein's geometric theory of gravity (general relativity) and Newtonian theory are small. Among the successful predictions of Einstein's theory are a rotation of Mercury's elliptical orbit by 43 arc seconds per century and corrections of the order of 100 μs in timing of radar signals reflected from Mercury and Venus. Near compact, massive stars or in the dynamics of the universe as a whole, space-time curvature produces dramatic effects.

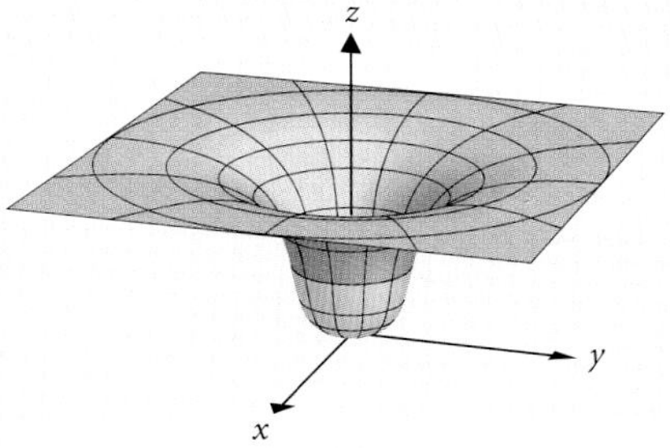

■ FIGURE E8.9
In the curved space around a star, circles of radius r have a circumference less than $2\pi r$. This diagram helps us visualize the effect. Distances in a plane through the equator of a star are represented by distances on a sheet that curves in an imaginary third dimension. This diagram corresponds to a very compact star: a neutron star. The orange region is the interior of the star. For a black hole, the central region would extend infinitely *downward,* forming a hole in the sheet.

E8.3 BLACK HOLES

In 1798, Pierre Simon de Laplace imagined a star so massive that the escape velocity from its surface exceeds the speed of light. No light could escape from such a star: it would be a *black hole,* swallowing things up through its gravitational pull but invisible to an outside observer.

A particle leaving the surface of a sphere just escapes to infinity if its total (kinetic plus potential) energy $E = \frac{1}{2}mv^2 - GMm/R$ is zero (cf. Example 8.8). It is said to have escape speed v_{esc}, where $v_{esc} = 2GM/R$. The escape speed equals the speed of light if the sphere's radius is:

$$R = 2GM/c^2 \equiv R_S. \quad \text{(E8.2)}$$

This radius is called the Schwarzschild radius of the star, after Karl Schwarzschild, who first calculated the results of Einstein's theory for a point mass.

The mass of the Sun is 2.0×10^{30} kg (front cover), so its Schwarzschild radius is:

$$R_S = \frac{2GM}{c^2} = \frac{2(6.7 \times 10^{-11}\ \text{m}^3/\text{kg}\cdot\text{s}^2)(2.0 \times 10^{30}\ \text{kg})}{(3.0 \times 10^8\ \text{m/s})^2} = 3.0\ \text{km}.$$

The Sun is much larger than this ($R = 7 \times 10^{10}$ m), but stars called neutron stars exist with radii equal to a few Schwarzschild radii.

Although we used Newtonian theory to calculate the Schwarzschild radius, eqn. (E8.2) is also the correct result according to Einstein's theory. The entire future of an event within R_S lies within the black hole, and there can be no communication with the outside. Any world line that passes within R_S remains inside.

From the outside, the object is invisible.[2] Nothing escapes from within R_S, and gravitational time dilation, which is infinite at R_S, ensures that any signal from just outside takes infinite time to reach an outside observer. We would never actually see anything fall in. Far from the hole, its gravitational field is the same as for any other star of the same mass, so it might be observed by its influence on the motion of nearby visible objects.

Normal astronomical objects are supported by internal pressure at radii much greater than their Schwarzschild radii. However, some exploding stars may compress their central regions enough to form a black hole. A few astronomical x-ray sources have been interpreted as black holes accreting gas from binary companions in orbit about the holes. The x rays are emitted by hot infalling gas swirling around the hole before being swallowed (■ Figure E8.10). The central regions of a star cluster in a galactic nucleus could collapse to form a black hole with a billion times the mass of the Sun. Observations of our own galactic center show the violent motions that we would expect near such a large point mass. Hubble telescope observations of the centers of some distant galaxies show dense concentrations of rapidly moving gas and stars (■ Figure E8.11): exactly what would be expected in the vicinity of massive black holes.

■ FIGURE E8.10
Artist's conception of a galactic x-ray source. Gas pulled from a companion star forms a disk of matter spiraling onto a black hole at the disk's center. The x rays are emitted by hot gas at the inner edge of the disk.

E8.4 GRAVITATIONAL WAVES

Mass causes gravity by curving space-time. Accelerating mass can produce oscillating distortions—that is, gravitational waves—that propagate at the same speed as light. Such waves are extremely difficult to detect directly, and experimental attempts have so far proved unsuccessful. However, astronomers have discovered a remarkable system in which two very condensed stars orbit each other. Observed changes in the system are consistent with the predicted loss of energy and momentum to gravitational waves. In 1993, Richard Hulse and Joe Taylor received the Nobel prize for this discovery.

E8.5 THE HISTORY OF THE UNIVERSE

Observations of distant galaxies show a uniform distribution of matter on the largest scales (> 300 million light-years). According to general relativity, such a universe expands from an initial state of infinite density. Whether the expansion continues, or stops and is followed by contraction, depends on the specific values of the expansion velocity and the density of the universe. Distant galaxies do have relative speed proportional to separation, as predicted by the theory. The expansion rate is somewhat uncertain, but probably in the range 15–30 km/s per million light-years, corresponding to an age of the universe between 10 and 20 billion years. The observed mass density of the universe is more uncertain, since it is likely that only a small fraction of the total material emits light. According to the best current estimates, the universe will expand forever.

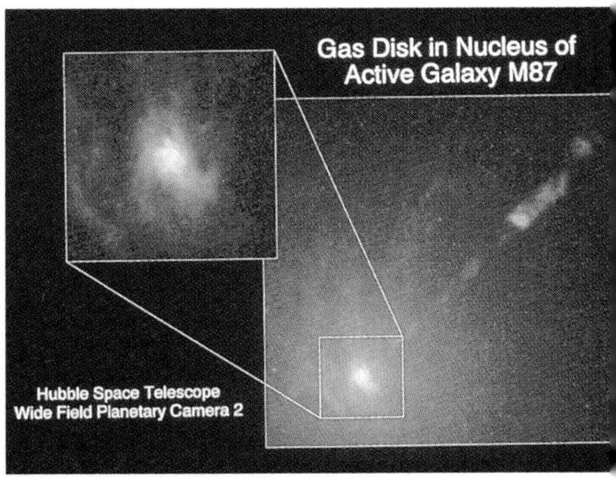

■ FIGURE E8.11
A disk of gas in the core of the giant galaxy called M87 (the 87th object in Messier's catalogue). Doppler-shifted emission lines (cf. §16.4.4 and Chapter 35) show that the gas is moving rapidly (about 550 km/s) in opposite directions on opposite sides of the disk. This rapid rotation is expected if there is a large mass in a small volume at the galaxy's center—a black hole.

[2] Quantum physics modifies this result. The hole emits blackbody radiation at a temperature inversely proportional to R_S^2. This effect is negligible for a hole with the mass of a star.

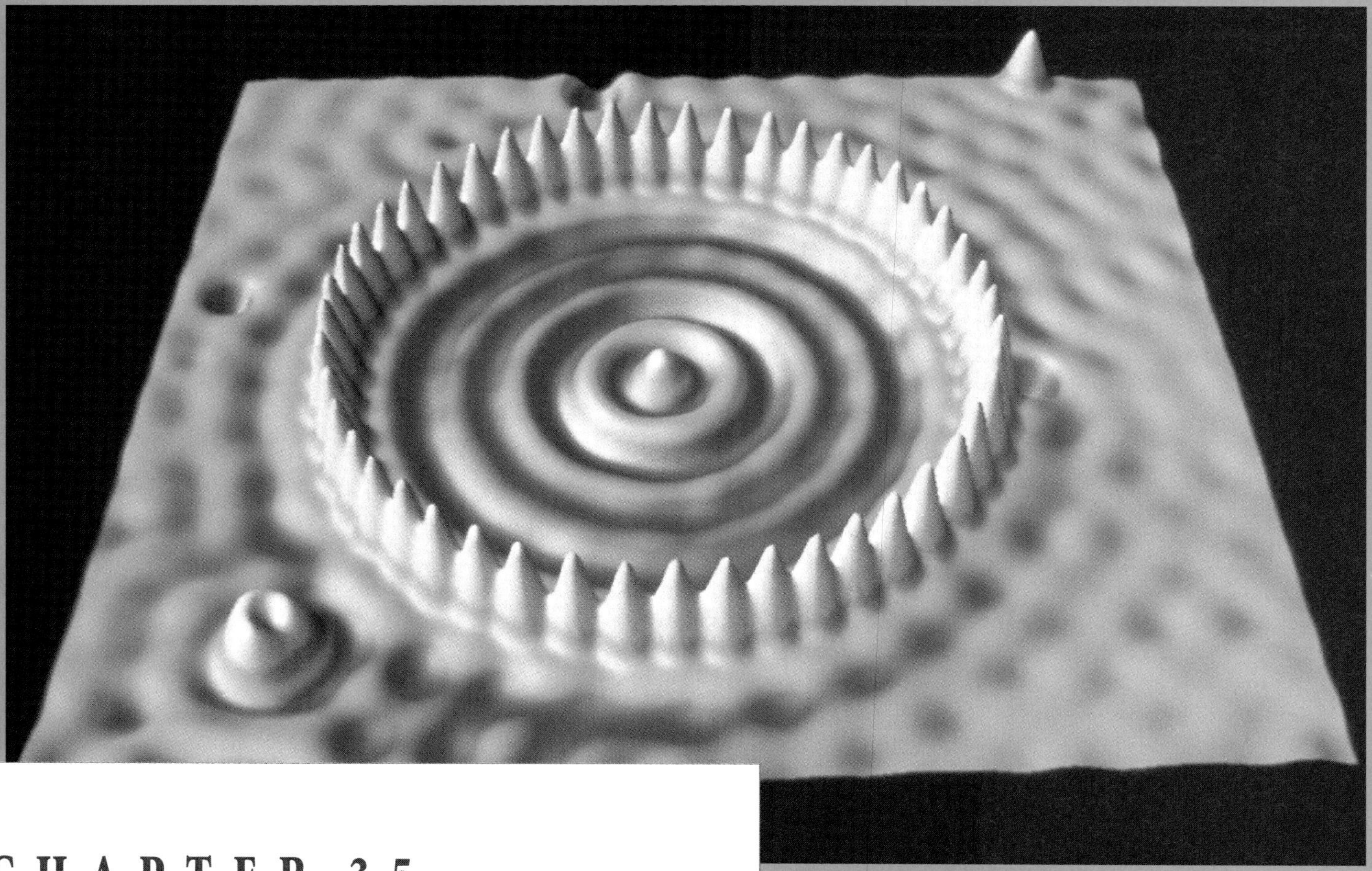

CHAPTER 35
Light and Atoms

Scanning tunneling microscope image of a copper surface. Researchers at IBM's Almaden Research Center built a corral of iron atoms, carefully positioned on the surface. The electrons on the surface within the corral form a wave with circular "ripples." The system is cooled to 4 K to prevent the iron atoms from moving over the surface.

Concepts

Photon
Quantization
Correspondence principle
De Broglie waves
Pauli principle
Uncertainty principle

Goals

Be able to:

Relate energy and momentum of particles to wavelength and frequency of the corresponding waves.

Compute the frequencies of photons emitted during transitions in hydrogenic atoms.

Understand:

Why the photoelectric effect requires a particle description of light.

How a wave description of matter explains atomic stability.

How wave and particle descriptions complement each other.

I don't think anyone understands quantum mechanics.

Richard P. Feynman

With a scanning tunneling microscope, we can directly observe how electrons behave on the surface of a metal. Confined within a *corral* of iron atoms arranged on a copper surface, the electrons form a standing wave that looks like ripples on a pond. This is an odd picture that seems to conflict with our previous model of electrons as particles—pointlike concentrations of mass and charge.

In classical physics, the particle model provides a valuable description of objects, even atoms. For example, we used it to study the thermal properties of gases (Chapter 19). In Part IV we found that waves transport energy and momentum through a system. They describe the collective behavior of the particles or fields that comprise the system. At the beginning of the twentieth century, physicists discovered that the best way to understand the structure of matter was to combine elements of both the particle and wave pictures. Light is a wave in the electromagnetic field, but light interacts with atoms in particlelike bundles called photons. Similarly, an electron beam shows wave diffraction when reflected from a crystal.

The picture that emerges is bizarre by classical standards. Its philosophical interpretation is still hotly debated, but its predictions are verified with incredible precision in many experiments. The electronic switches that operate your pocket calculator or route your telephone call are but one development based on the principles of atomic physics. The great chemical diversity of the world, important not only for technology but for our very existence, is a consequence of these principles, which determine the patterns of electrons in atoms. In this

Digging Deeper

The Origins of the Quantum Idea

Quantum theory is based on the concept that certain physical properties of systems—energy, momentum, and angular momentum—are not truly continuous variables, but occur in small chunks called quanta. In Newtonian theory, we ignore this granularity because the systems we study contain enormous numbers of quanta. A spinning ice skater, for example, has about 10^{35} quanta of angular momentum. Using Newtonian physics is like modeling charge distributions as continuous while knowing that charge comes in integer multiples of e.

Max Planck first suggested the idea of energy quanta in December 1900, as "an act of desperation." Planck had worked for decades attempting to understand thermal radiation from hot objects (§§21.3 and 35.1.3). In September 1900, fresh laboratory data led him to the precise empirical formula we know as Planck's law, but he could not yet explain why the law should hold. He knew that radiation was emitted and absorbed by atoms and molecules (oscillators) in the object, and he had obtained a relation between the electromagnetic energy at frequency f and the energy of the oscillators at the same frequency. Planck made two further, astonishing assumptions. The first was the *quantum postulate:*

> The energy of each oscillator in the body is an integer times a basic energy unit E:
>
> $$U = NE.$$

The second step was a novel procedure for determining the number of possible oscillator energy states (§22.7), which implied that the basic energy unit is proportional to frequency:

$$E = hf.$$

Both steps were bizarre by the standards of classical physics, and Planck himself considered them tentative.

As in many other problems of the time, it was Einstein who saw the significance of the quantum hypothesis. In 1905, he introduced the quantum of energy in radiation. He was motivated by his derivation of the entropy of radiation, which he based on the observed frequency dependence of thermal radiation at high frequencies. He immediately used his new idea to explain the photoelectric effect (§35.1.1), and his theories were fully confirmed in 1916 by Millikan.

In 1906, Einstein realized that "Planck's theory makes implicit use of the light-quantum hypothesis." In 1909, Einstein first used the word "pointlike" in connection with the light quanta. The idea that the light quanta carry momentum was first stated by Johannes Stark in the same year. In 1916, Einstein established the connection between Bohr's theory of atomic spectra (§35.2) and Planck's formula, and explicitly added momentum to his quantum theory. The photon was born.

chapter we study how a particle description of light and a wave description of electrons fuse into a model of atomic structure. We then look briefly at the meaning of this wave/particle description and the fundamental changes it requires in relations of cause and effect.

35.1 Photons

35.1.1 *The Photoelectric Effect*

Since its discovery, the photoelectric effect (■ Figure 35.1) has become an important part of modern technology. In a television camera, light falls on a semiconductor surface and liberates electrons, converting an image to an electronic signal. In a photomultiplier tube (■ Figure 35.2), used to detect extremely faint light sources, light falls on the metal cathode surface and ejects electrons. Each electron ejected from the cathode creates a current that is amplified within the tube and detected by an electronic circuit connected to it.

The ejected electrons are called *photoelectrons*.

Heinrich Hertz was the first to observe the photoelectric effect. In 1887, while performing an experiment to confirm Maxwell's theory of electromagnetic waves, he noticed that ultraviolet light from sparks generated by his transmitting coil influenced the sensitivity of his receiving coil. Careful experiments by Hertz and others over the next 18 years revealed several important characteristics of the photoelectric effect:

- The rate of electron emission is proportional to light intensity.
- The energy of the ejected electrons is independent of light intensity.
- For any given metal, there is a minimum frequency f_0 of light that can eject electrons.
- The energy of electrons ejected by light of frequency f is proportional to the frequency difference $f - f_0$.
- Electrons are emitted immediately (within 10^{-9} s) after light exposure begins, regardless of intensity.

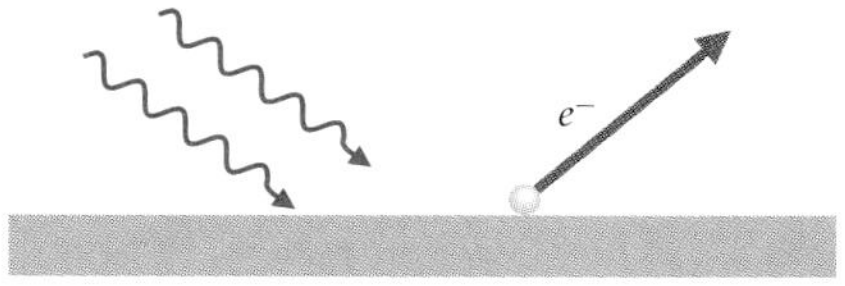

■ Figure 35.1
The photoelectric effect. Light waves incident on a surface eject electrons.

Of these features, only the first is easily understood using classical electromagnetic theory: the rate electrons absorb energy from the light is proportional to the rate energy is delivered to the metal surface. The next three features are puzzling, since most classical models predict a much more complicated frequency dependence. Classical theory cannot account for the rapid start of electron flow, as we illustrate in the following example.

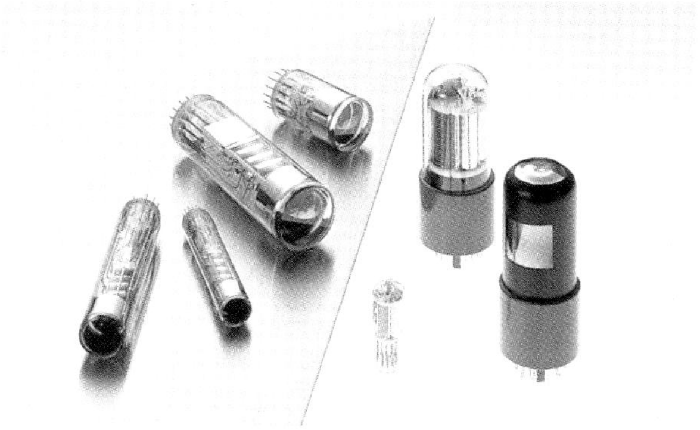

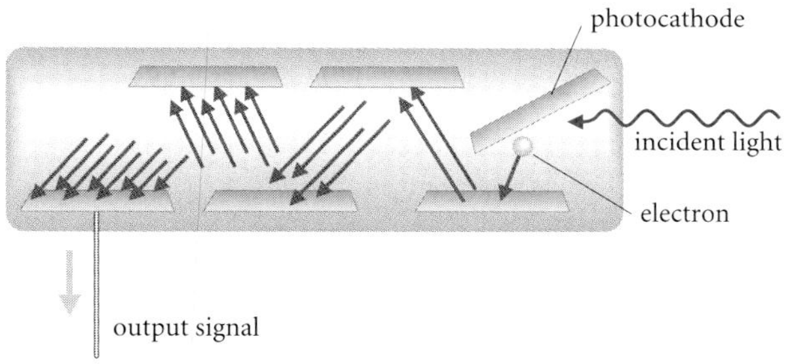

(b)

■ Figure 35.2
(a) Commercial photomultiplier tubes make use of the photoelectric effect.

(b) Incident light releases electrons at the photocathode. Cathodes inside the tube, called dynodes, are held at increasingly positive potentials so that the electrons are accelerated from dynode to dynode. At each dynode, the electrons eject more electrons, increasing the total current in the tube. A measurable current results from a tiny amount of incident light.

EXAMPLE 35.1 ♦♦ Light of intensity $I = 10\ \text{W/m}^2$ falling on a metal surface ejects electrons, each with 3 eV of energy. According to classical theory, how much time is required for an atom of radius 10^{-10} m to receive the necessary 3 eV of energy?

MODEL In a metal, there are at most a few free electrons per atom, so the cross section of an atom is a reasonable guess for the absorbing area per electron. To estimate the shortest time before electrons can be ejected, we also suppose that the energy is absorbed with perfect efficiency:

SETUP
$$\Delta t_{\text{classical}} = \frac{\text{observed energy}}{\text{absorption rate}} = \frac{\text{observed energy}}{\text{intensity} \times \text{area}}.$$

SOLVE
$$\Delta t_{\text{classical}} = \frac{(3\ \text{eV})(1.6 \times 10^{-19}\ \text{J/eV})}{(10\ \text{W/m}^2)(\pi)(10^{-10}\ \text{m})^2} = 2\ \text{s}.$$

ANALYZE The classical prediction is vastly greater than the observed time required for electron ejection. ∎

In 1905, Einstein gave the correct explanation of the photoelectric effect: though propagating as a wave according to Maxwell's theory, the electromagnetic field delivers energy to electrons in the metal in chunks, or *quanta*. The magnitude of an energy quantum is proportional to the wave frequency:

$$E = hf. \tag{35.1}$$

The constant of proportionality h is Planck's constant:

$$h = 6.626 \times 10^{-34}\ \text{J}\cdot\text{s}.$$

Planck first obtained this value in 1901 by fitting a theoretical function to the experimentally determined blackbody radiation spectrum. Millikan obtained it independently from his photoelectric data (■ Figure 35.3).

PLANCK'S 1901 VALUE WAS 6.55×10^{-34} J·s.

The incident light is a stream of energy quanta in the form of particles, or *photons*, each with energy hf and each with an equal probability of ejecting an electron. Thus the number of electrons ejected per second is proportional to the number of photons per second reaching the surface, and so is proportional to the light intensity. Electrons can be ejected as soon as the first photon strikes the metal.

Einstein's model directly explains the minimum frequency f_0 and the proportionality of the observed electron energy and frequency. Electrons are free to move within the metal but are bound by the attractive force of the metal atoms with a potential energy called the *work function*.

The *work function* ϕ of a metal is the minimum energy required to remove a single electron from the metal.

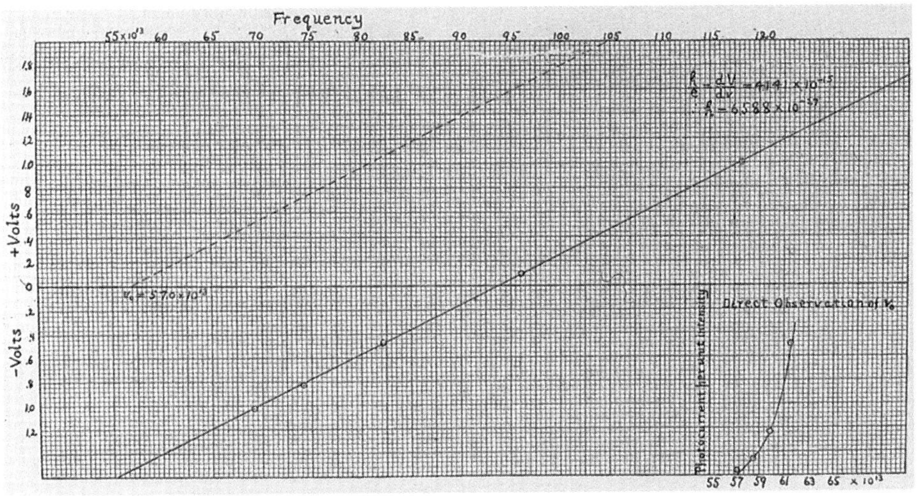

■ FIGURE 35.3
Millikan's original data on photoemission from lithium. The potential of the anode that is just sufficient to suppress the photocurrent is plotted against the wavelength of light. Millikan calculated h from the slope of the line, obtaining the value 6.588×10^{-34} J·s. The value written on his graph is in cgs units of erg·s.

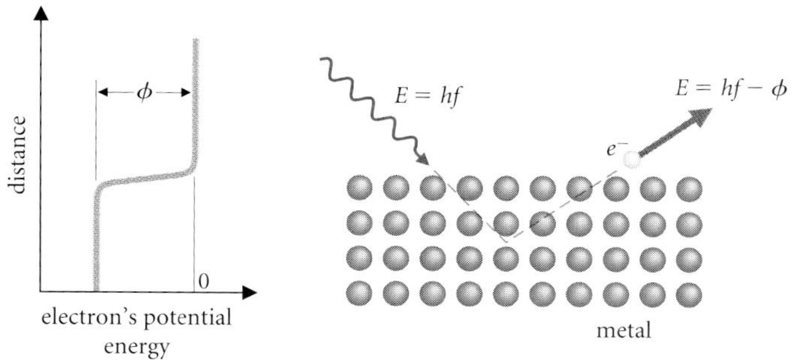

■ FIGURE 35.4
The work function. An incident photon is absorbed by an electron beneath the surface of a metal, where the electron's potential energy is $-\phi$. The electron escapes from the metal. Outside the metal, the electron has zero potential energy, and its kinetic energy is the amount hf absorbed from the photon less the amount ϕ of potential energy gained.

When the incident light wave delivers a quantum of energy to an electron, a minimum amount ϕ is used to escape from the metal. So electrons emerge with a range of kinetic energy up to a maximum (■ Figure 35.4):

$$\tfrac{1}{2}mv^2 = hf - \phi. \tag{35.2}$$

The minimum frequency f_0 is that for which the energy of a quantum just equals the work function: $hf_0 = \phi$. Not every photon ejects an electron.

> The fraction of photons reaching a given metal surface with sufficient energy that actually eject an electron is the *quantum efficiency* of the metal.

Typical metals (like aluminum, copper, and gold) have work functions of 4–5 eV and peak quantum efficiency of 10–20% at about $\lambda = 100$ nm. ■ Figure 35.5 is a plot of quantum efficiency for a gold photodiode as a function of photon wavelength. Few materials are efficient photoemitters in the visible. Cesium antimonide (CsSb) is a notable exception, with a peak quantum efficiency of 15% at 450 nm.

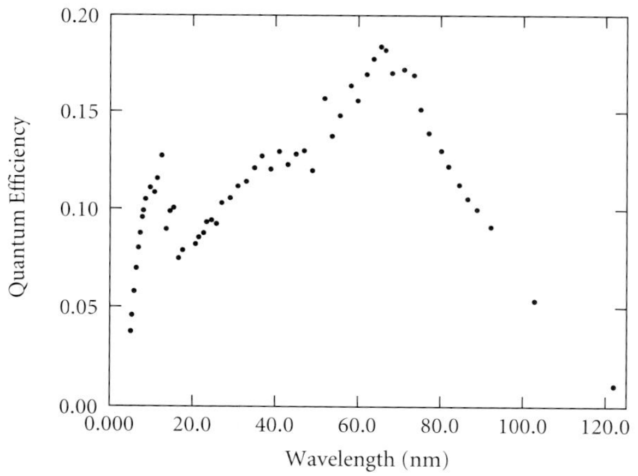

■ FIGURE 35.5
Quantum efficiency of a gold photodiode as a function of frequency. The systematic uncertainty in the measurements is about 10%.

MONOCHROMATIC (SINGLE COLOR) MEANS HAVING A NARROW RANGE OF FREQUENCIES.

EXAMPLE 35.2 ♦ A monochromatic light beam falls on a potassium surface. If the photoelectrons emerge with a maximum kinetic energy of 2.0 eV, what is the frequency of the light?

MODEL The energy of an individual light quantum equals the work function plus the kinetic energy of the photoelectrons. The frequency of the light is determined directly from the energy of a quantum using eqn. (35.1).

SETUP We take the work function for potassium from ● Table 35.1.

SOLVE From eqn. (35.2),

$$f = \frac{\tfrac{1}{2}mv^2 + \phi}{h} = \frac{(2.0\text{ eV} + 2.3\text{ eV})(1.6 \times 10^{-19}\text{ J/eV})}{6.6 \times 10^{-34}\text{ J}\cdot\text{s}} = 1.0 \times 10^{15}\text{ Hz.}$$

TABLE 35.1 Photoelectric Properties of Selected Metals

Metal		Work Function (eV)	Spread in Measured Values
Aluminum	Al	4.25	±0.13
Barium	Ba	3	*
Calcium	Ca	2.9	*
Cerium	Ce	2.9	*
Copper	Cu	4.6	±0.16
Gold	Au	5.3	±0.13
Iron	Fe	4.7	±0.11
Lithium	Li	3	*
Mercury	Hg	4.5	*
Nickel	Ni	5.15	±0.12
Platinum	Pt	5.85	±0.25
Potassium	K	2.3	*
Silver	Ag	4.6	±0.09
Sodium	Na	2.75	*
Tungsten	W	4.6	±0.32

*Only one source available.

ANALYZE This corresponds to a wavelength $\lambda = c/f = 290$ nm, which is in the ultraviolet.

EXAMPLE 35.3 ♦♦ A star known to be as bright as the Sun (luminosity $\approx 4 \times 10^{26}$ W) is 3×10^{18} m from Earth. A photoelectric detector with a quantum efficiency of $e_q = 0.1$ and a 1-cm² area is used to observe that star. Estimate the photon count rate.

MODEL Without much more detailed information both about the star and about the detector, we can only obtain a rough estimate. So we assume that the given quantum efficiency is an appropriate average over the star's spectrum and that there is no absorption of starlight along its path to Earth. Then, if the star is at distance R, the power per unit area reaching the detector is the star's luminosity L divided by the area of a sphere of radius R. The power incident on the detector of area A is:

$$P = AL/(4\pi R^2).$$

SETUP This power (= energy/time) equals the number of photons per second reaching the detector times the energy per photon:

$$P = \frac{dn}{dt}hf \Rightarrow \frac{dn}{dt} = \frac{P}{hf}.$$

A fraction e_q of these photons actually eject an electron, so the count rate is:

$$r = \frac{e_q P}{hf} = \frac{e_q AL}{4\pi R^2 hf}.$$

For a star like the Sun, an appropriate average wavelength is 500 nm, and $f = c/\lambda$.

SOLVE $$r = \frac{(0.1)(10^{-4}\ \text{m}^2)(4 \times 10^{26}\ \text{W})(5 \times 10^{-7}\ \text{m})}{4\pi(3 \times 10^{18}\ \text{m})^2(6.6 \times 10^{-34}\ \text{J}\cdot\text{s})(3 \times 10^8\ \text{m/s})} = 90\ /\text{s}.$$

ANALYZE This is quite a large count rate by astronomical standards. Multiplication devices can increase the electron yield considerably. An astronomer could easily observe the star with this detector.

EXERCISE 35.1 ♦♦ A light beam of intensity 1.0 W/m^2 and wavelength 100 nm falls on a gold photodiode (Figure 35.5). Find the photoelectron current emitted per unit area of the metal surface.

To determine the work function for a metal, we could use an experimental photocell (■ Figure 35.6). Incident light with a fixed, known frequency liberates electrons at a surface called the *cathode.* Electrons collected on another surface called the *anode* produce current in the external circuit. If the anode is held at a negative potential with respect to the cathode, the resulting electric force on the ejected electrons decelerates them. With a great enough negative potential, no electrons reach the anode at all. The potential difference that is just great enough to stop the electrons is called the *stopping potential.* If both cathode and anode are made of the same material, the stopping potential measures the work function of that material.

ODDLY, WITH DIFFERENT MATERIALS YOU GET THE WORK FUNCTION OF THE ANODE. SEE PROBLEM 98.

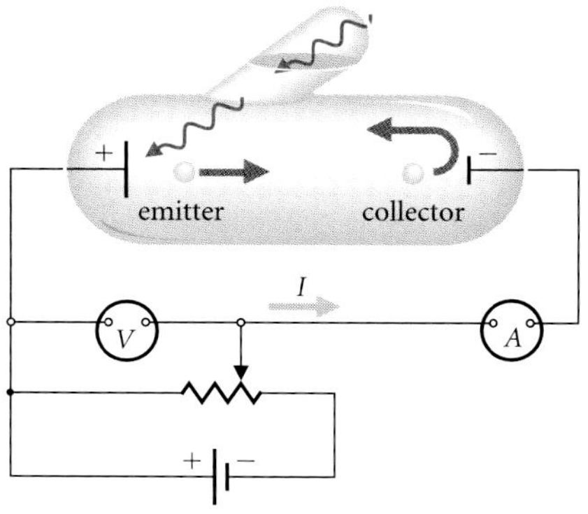

■ FIGURE 35.6
Apparatus for measuring the work function. Incident photons eject electrons from the emitter, or cathode. The collector, or anode, is maintained at a negative potential with respect to the emitter so that the resulting electric field decelerates the emitted electrons. When V is set to the stopping potential, no electrons reach the collector, and the measured current in the circuit drops to zero.

EXAMPLE 35.4 ♦♦ In a photoelectric experiment, light of wavelength 253.5 nm falls on a lithium cathode. The electron curent is suppressed when the lithium anode is held at a potential of -1.85 V with respect to the cathode. Find the work function for lithium.

MODEL Electrons leave the cathode with kinetic energy that is converted to electric potential energy at the anode. With a stopping potential of about 2 V, the electrons are nonrelativistic. Einstein's relation (35.2) determines the work function.

SETUP The electron kinetic energy may be found using the usual conservation law methods. As the potential difference between anode and cathode is made more negative, the potential energy of electrons arriving at the anode is increased and their kinetic energy is decreased. With the anode at the stopping potential, the electrons have zero kinetic energy there.

	BEFORE (at cathode)		AFTER (at anode)
Kinetic energy	$\frac{1}{2}mv^2$		0
Potential energy	$-eV_{\text{cathode}}$		$-eV_{\text{anode}}$
Set totals equal:	$\frac{1}{2}mv^2 - eV_{\text{cathode}}$	$=$	$-eV_{\text{anode}}.$

Use eqn. (35.2) to find ϕ:

$$-e(V_{\text{anode}} - V_{\text{cathode}}) = \tfrac{1}{2}mv^2 = hf - \phi = h\frac{c}{\lambda} - \phi.$$

IN PROBLEMS LIKE THIS, THE COMBINATION $hc = 1.24 \times 10^3$ eV·nm IS OFTEN USEFUL.

SOLVE
$$\phi = h\frac{c}{\lambda} + e\,\Delta V = \frac{(6.63 \times 10^{-34}\ \text{J·s})(3.00 \times 10^8\ \text{m/s})}{(1.60 \times 10^{-19}\ \text{J/eV})(253.5 \times 10^{-9}\ \text{m})} - 1.85\ \text{eV}$$
$$= (4.90 - 1.85)\ \text{eV} = 3.05\ \text{eV}.$$

ANALYZE From Table 35.1, the work function for lithium is 3 eV, in good agreement with this result.

If stopping potential is measured as a function of wavelength, the value of Planck's constant may be found from the slope of a graph of ΔV versus $f = c/\lambda$ and the work function from its intercept. Millikan measured the slope (Figure 35.3) and showed it to be the same for several different metals.

IN THE EXAMPLE WE FOUND $\Delta V = -(h/e)f + \phi/e$, A LINE WITH SLOPE $-h/e$.

35.1.2 *The Compton Effect*

In 1922, Arthur H. Compton obtained experimental evidence that photons behave like particles. Compton shone a beam of x rays with wavelength $\lambda = 0.07$ nm onto a sample of graphite and observed that the wavelength of x rays emerging at 90° from their original direction was increased by 0.002 nm (■ Figure 35.7). The energy of the x-ray photons,

$$E = hc/\lambda \approx 18\ \text{keV},$$

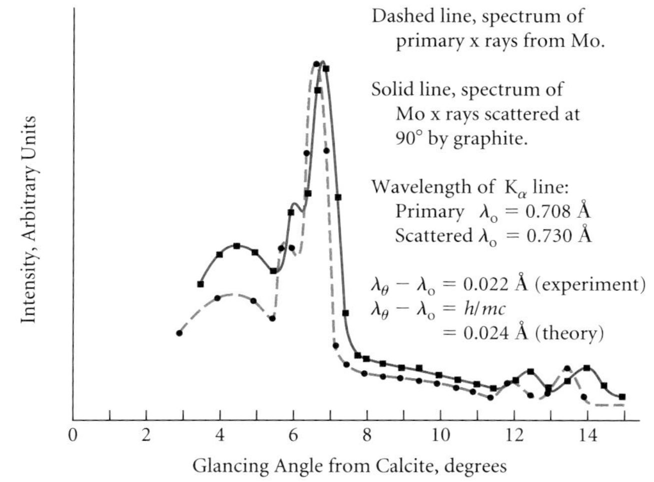

■ **FIGURE 35.7**
Compton's spectrum of x rays from molybdenum. Dashed line—primary spectrum. Solid line—spectrum of x rays scattered through 90° by graphite. The x-ray wavelength was determined by diffraction using a calcite crystal (§17.5). Large glancing angles correspond to longer wavelengths. The uniform increase in wavelength is evident. Redrawn from Figure 4, Physical Review second series, vol. 21, #5, May 1923.

is much larger than the few eV binding energy of electrons to the carbon atoms in the graphite block. In effect, the electrons act like free particles exposed to the x-ray beam. Classically, such free electrons would scatter an electromagnetic wave with no change in wavelength. Compton explained the observed wavelength change as the result of elastic collisions between x-ray photons and stationary free electrons. The effect is now named after him.

Before a Compton collision, the incident photon has frequency f_1, and the electron is assumed to be at rest (■ Figure 35.8). After the collision, the photon has frequency f_2 and travels at angle θ to the incident beam. The electron recoils at angle ϕ. Electromagnetic radiation carries momentum given by its energy divided by the speed of light. Thus, the initial momentum of the photon is:

COMPARE THE DISCUSSION IN §33.2.2.

$$p = E_1/c = hf_1/c. \tag{35.3}$$

THUS COMPTON SCATTERING ALSO TESTS THE RELATIVISTIC EXPRESSIONS FOR ENERGY AND MOMENTUM (§34.4).

The electron recoils at high speed and must be described using special relativity. Treating the encounter as a particle collision, we apply conservation of energy and momentum:

	BEFORE		AFTER	
Energy	$hf_1 + mc^2$	$=$	$hf_2 + \gamma mc^2.$	(i)
x-component of momentum	hf_1/c	$=$	$(hf_2/c)\cos\theta + \gamma mv\cos\phi.$	(ii)
y-component of momentum	0	$=$	$(hf_2/c)\sin\theta - \gamma mv\sin\phi.$	(iii)

For comparison with Compton's data, we need a relation among the two photon wavelengths and the scattering angle θ. First, we use the two momentum equations, (ii) and (iii), to eliminate ϕ. (Recall $\beta \equiv v/c$ and $\gamma \equiv 1/\sqrt{1-\beta^2}$.)

$$(\gamma mv)^2(\cos^2\phi + \sin^2\phi) = (h/c)^2[(f_1 - f_2\cos\theta)^2 + (f_2\sin\theta)^2],$$

or,

$$\gamma^2\beta^2 = \gamma^2 - 1 = (h/mc^2)^2(f_1^2 + f_2^2 - 2f_1f_2\cos\theta). \tag{iv}$$

From the energy equation (i):

$$(\gamma - 1)^2 = (h/mc^2)^2(f_1 - f_2)^2. \tag{v}$$

Subtracting (v) from (iv) and using the energy equation again:

$$(h/mc^2)^2[2f_1f_2(1 - \cos\theta)] = 2(\gamma - 1) = 2(h/mc^2)(f_1 - f_2).$$

Dividing by $2hf_1f_2/(mc^3)$:

$$\frac{h}{mc}(1 - \cos\theta) = \frac{c(f_1 - f_2)}{f_1f_2}.$$

The factor,

$$h/(mc) \equiv \lambda_C = 2.43 \times 10^{-3}\text{ nm}, \tag{35.4}$$

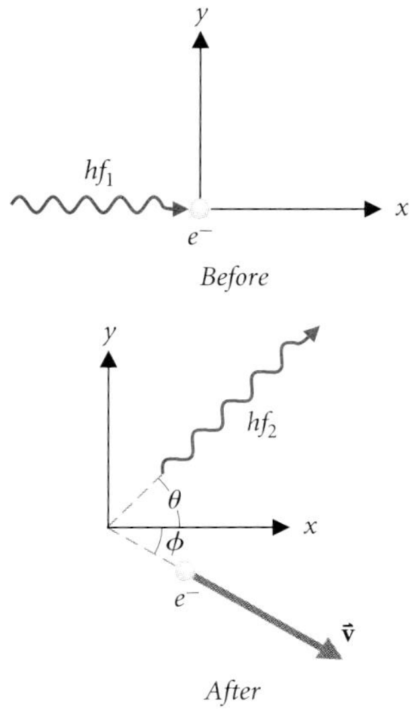

■ **FIGURE 35.8**
Compton scattering. We have chosen the x-axis to be along the direction of the incoming photon. The electron is initially at rest. After the collision, the photon has frequency f_2 and travels at an angle θ to its incoming direction. The electron recoils with speed v at angle ϕ to the x-axis.

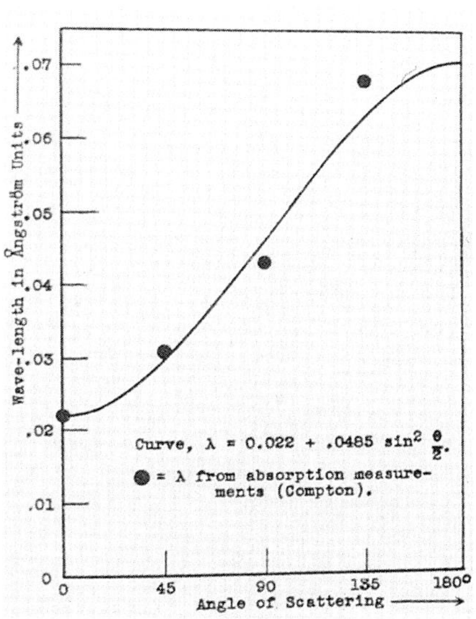

■ FIGURE 35.9
Compton's original data, published in 1923, show the variation in wavelength of 0.22-nm γ rays scattered by iron, aluminum, and paraffin. The incident γ rays were from a radium source (see Chapter 36). Compton determined the scattered frequency by absorbing the γ rays in lead, whose absorption properties were well known. The measurements are in good agreement with the theoretical curve (eqn. 35.5).

is known as the *Compton wavelength* of the electron. Converting from frequency to wavelength, $f = c/\lambda$, we find:

$$\lambda_2 - \lambda_1 = \lambda_C(1 - \cos\theta). \tag{35.5}$$

For photons scattered at 90° ($\cos\theta = 0$), eqn. (35.5) predicts a wavelength increase $\Delta\lambda = \lambda_C$, in agreement with Compton's results. Compton performed a second experiment in which 565-keV γ rays from a radium source were scattered at various angles from iron, aluminum, and paraffin, and he measured the wavelength of the scattered photons as a function of angle. Equation (35.5) provided an excellent fit to his measured values (■ Figure 35.9).

Compton's experiment also exhibits the simultaneous validity, or *duality*, of wave and particle pictures in the description of light. While the effect being measured is understood to result from collisions of x-ray particles with electrons, it is observed as the difference in two wavelengths that are measured from wave interference effects in a crystal spectrometer (cf. §17.5).

EXAMPLE 35.5 ♦ If x rays of wavelength 0.12 nm are scattered from free electrons in a plasma sample, what change in wavelength is observed for x rays scattered at 45° to the incident beam? (A plasma is a gas in which the particles are charged.)

MODEL The assumption of zero initial electron velocity is reasonable so long as the electrons are nonrelativistic.

SETUP We use eqn. (35.5) with $\cos 45° = \sqrt{2}/2$.

SOLVE $\lambda_2 - \lambda_1 = (2.43 \times 10^{-3}\text{ nm})(1 - \sqrt{2}/2) = 7.12 \times 10^{-4}\text{ nm}$.

ANALYZE The wavelength increases by 7.1×10^{-4} nm, or 0.59%. ■

Using the relation between photon wavelength and energy, $\lambda = c/f = hc/E$, eqn. (35.5) becomes:

$$\frac{\lambda_2 - \lambda_1}{\lambda_1} \equiv \frac{\Delta\lambda}{\lambda} = \frac{E_1}{mc^2}(1 - \cos\theta). \tag{35.6}$$

The fractional change in λ is determined primarily by the ratio of photon energy to the electron's rest mass energy, $mc^2 = 0.511$ MeV. Substantial effects occur for very high energy photons (γ rays).

EXERCISE 35.2 ♦♦ Find the fractional change in energy when 1.0-MeV γ rays are scattered through 90°.

✲ 35.1.3 *The Planck Radiation Law*

Max Planck introduced the idea of energy quanta in 1900 to explain blackbody radiation. Planck's formula for the electromagnetic energy per unit volume du between the frequencies f and $f + df$ in a cavity at temperature T is:

SEE § 21.3 FOR A DISCUSSION OF BLACKBODY RADIATION.

$$du = \frac{8\pi h f^3}{c^3}\frac{df}{\exp(hf/kT) - 1}. \tag{35.7}$$

We can see why photons are necessary to explain blackbody radiation by considering a special case: the radiation inside a perfectly reflecting, cubical box. In equilibrium, according to classical thermodynamics, there is, on average, $\frac{1}{2}kT$ of energy in each possible mode of energy storage. An oscillator has two modes of energy storage: kinetic and potential energy. So each of a system's possible oscillation modes has an energy kT (cf. §19.7). Thus, for a classical calculation, we need only determine the number of modes of oscillation within the cube. The result was derived independently by Rayleigh, Jeans, and Einstein, and is called the *Rayleigh–Jeans law* for the energy density in blackbody radiation:

$$du_{\text{R-J}} = \frac{8\pi f^2 df}{c^3} kT. \tag{35.8}$$

The Rayleigh–Jeans law gives the correct energy density in the low-frequency limit but fails disastrously at high frequencies. Because the number of possible oscillation modes increases without limit as frequency increases, classical theory predicts an infinite energy density and correspondingly infinite energy flow from any source of radiation. Sitting before your fireplace, you should be vaporized by an infinite flux of γ radiation!

THIS NONSENSICAL RESULT IS CALLED THE "ULTRAVIOLET CATASTROPHE."

The flaw in classical reasoning lies not in the counting of modes, but in the assumption of energy equipartition among them. Because the energy is bundled into photons, a mode of frequency f cannot have an energy less than hf. When this quantum is much larger than kT, according to Boltzmann's law (§22.7), it is improbable that a mode contains any energy at all, much less an equal share with all the other modes. At high frequencies, the number of modes at frequency f should be multiplied not by the equipartition energy kT, but by a product of the photon energy hf and a Boltzmann factor $\exp(-hf/kT)$, which gives the probability of the mode being excited.

35.2 BOHR'S ATOMIC MODEL

35.2.1 The Structure of Atoms

At the beginning of the twentieth century, three experimental facts about atoms were known:

- The volume occupied by a single atom in solids is approximately a sphere with a radius of a few times 10^{-10} m.
- Atoms contain negatively charged electrons that can be separated from and recombined with the main, positively charged body of the atom.
- An atom of a specific element emits and absorbs specific frequencies of light (■ Figure 35.10).

The plum pudding model of the atom (■ Figure 35.11) had been proposed based on these properties. The plum pudding atom was a uniform sphere of positively charged material with a fixed pattern of electrons embedded in it. In 1910, Ernest Rutherford and his students Hans Geiger and Ernest Marsden devised an experiment to test this model. They bombarded a thin gold foil with a beam of α-particles emitted by a radioactive substance. The α-particles, which

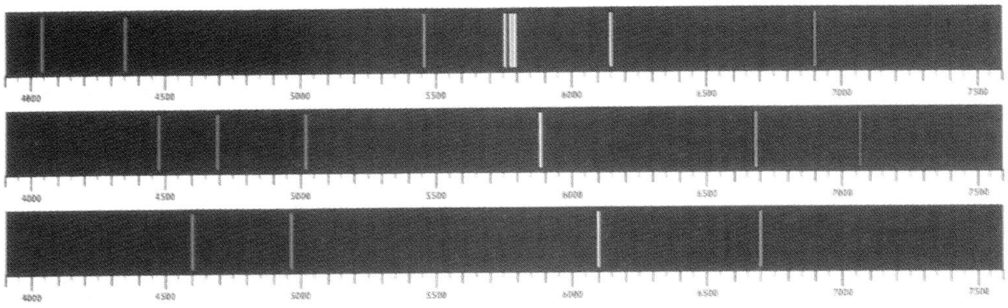

■ FIGURE 35.10
Spectra of light from atoms is emitted at a set of discrete frequencies. This photograph shows the light emitted by mercury (top), helium, and lithium (bottom).

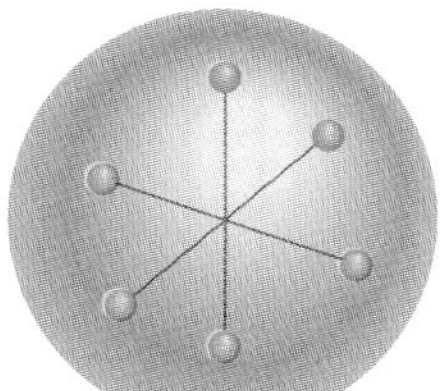

■ FIGURE 35.11
Plum pudding model of the atom. In this model, positively charged material forms a uniform sphere of radius $\approx 10^{-10}$ m, the observed size of the atom. Pointlike electrons are embedded in the sphere in a regular lattice. While this model can explain the simplest properties of atoms, Rutherford's experiment showed that it can't be correct.

we now know are bare nuclei of helium atoms, are positively charged and are deflected by electrical forces encountered while passing near, or through, atoms in the foil. Rutherford measured the directions at which the α-particles emerged (■ Figure 35.12).

EXAMPLE 35.6 ♦♦♦ Use the plum pudding model to predict the results of Rutherford's experiment. Estimate the maximum deflection of a 1-MeV α-particle (assumed pointlike) caused by a uniform sphere of radius 4×10^{-10} m with the total positive charge of a gold nucleus ($+79e$).

MODEL The force exerted on the α-particle at any distance r from the center of a sphere with uniform charge density ρ is:

$$F \propto \frac{\text{charge within radius } r}{r^2} = \frac{q(r)}{r^2}.$$

For paths that pass within the sphere, $r < R$, $q(r) = \rho V = \frac{4}{3}\pi r^3 \rho$, and $F \propto r$. For paths that pass outside the sphere, $r > R$, $q(r)$ equals the total charge of the particle, and $F \propto 1/r^2$. Thus the force is maximum at the surface of the sphere. A particle that just skims the surface is deflected the most (■ Figure 35.13). The particle gains momentum Δp perpendicular to its original velocity and is deflected through a small angle $\theta \approx \tan\theta = \Delta p/p$. If the deflection is small, we may ignore any deviation of the path from a straight line *during* the interaction.

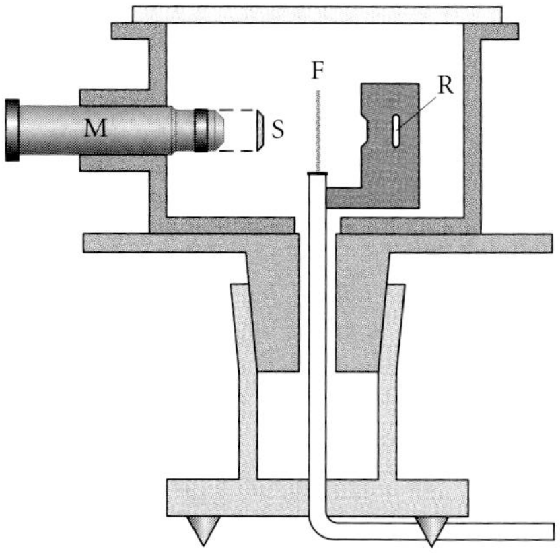

■ FIGURE 35.12
Rutherford's apparatus. Helium nuclei (α-particles) from a radioactive source R pass through a thin gold foil F. Particles striking screen S produce flashes that are observed through telescope M. The telescope is rotated and the rate α-particles strike the screen is observed as a function of telescope position.

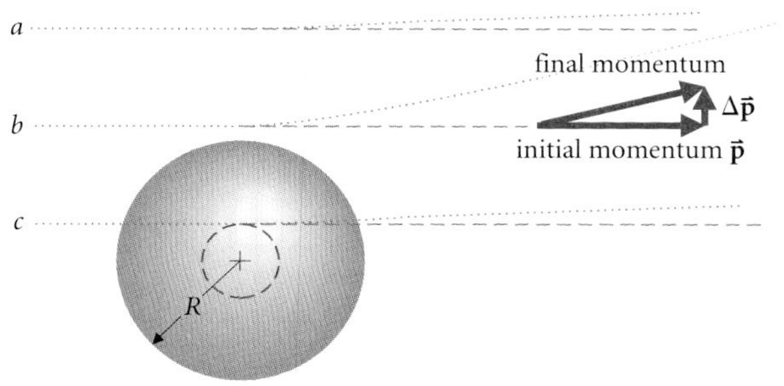

■ FIGURE 35.13
Deflection of charged particles by a uniform sphere of charge (the *plum pudding* model). The deflection is greatest when the particle just skims the surface (track b). Particles that pass farther (track a) from the sphere experience a lesser force because of the inverse square Coulomb force law. Particles that pass through the sphere (track c) experience a lesser force because a smaller amount of charge (that within the dashed sphere) produces force on the particle.

SETUP The particle travels at speed v and is near the sphere for a time $\Delta t \approx 2R/v$. Assuming for simplicity that the maximum force acts for a time Δt, it produces a change in momentum:

$$\Delta p \approx F_{\max}\,\Delta t = \frac{kQ_\alpha Q_{\mathrm{Au}}}{R^2}\left(\frac{2R}{v}\right).$$

In terms of the kinetic energy $E_\alpha = \frac{1}{2}mv^2$ of the α-particle, the deflection angle is:

$$\begin{aligned}\theta &\approx \frac{\Delta p}{p} = \frac{2kQ_\alpha Q_{\mathrm{Au}}/Rv}{mv} = \frac{kQ_\alpha Q_{\mathrm{Au}}}{RE_\alpha}\\ &= \frac{(9 \times 10^9\ \mathrm{N \cdot m^2/C^2})(2)(79)(1.6 \times 10^{-19}\ \mathrm{C})^2}{(4 \times 10^{-10}\ \mathrm{m})(10^6\ \mathrm{eV})(1.6 \times 10^{-19}\ \mathrm{J/eV})}\\ &= 6 \times 10^{-4}\ \mathrm{rad}.\end{aligned}$$

ANALYZE This calculation overestimates the deflection caused by a plum pudding atom because it neglects the effects of the negatively charged electrons. Thus Rutherford expected to see his α-particles undergo small deflections. ■

To Rutherford's amazement, some of the particles were scattered to large angles, and some even returned in the direction they had come, contradicting the predictions of the plum pudding model. Rutherford showed that his data are explained exactly by a model in which the positive charge of the atom and almost all its mass are concentrated in a point at the center (■ Figure 35.14). Modern scattering experiments show that a typical atomic nucleus is a few times 10^{-15} m in radius.

SEE CHAPTER 36 FOR MORE DETAILS OF NUCLEAR STRUCTURE.

EXERCISE 35.3 ♦♦ How close can a 1-MeV α-particle get to a pointlike gold nucleus if initially aimed directly at it? What happens to the α-particle subsequently?

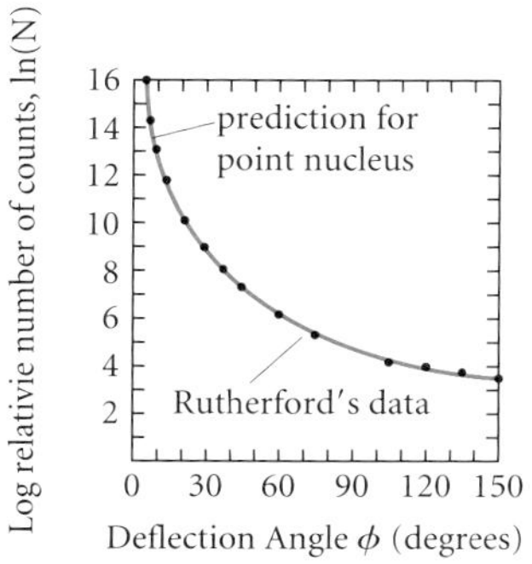

■ FIGURE 35.14
Rutherford-scattering data. The dots represent scattering rates of α-particles passing through gold foil, as measured by Rutherford's students, Geiger and Marsden. The solid line is the theoretical expectation for scattering by a point nucleus.

35.2.2 *Balmer's Spectrum and Bohr's Atom*

Rutherford's experiment leaves no doubt that solid material is made up of small particles surrounded by empty space. Yet this discovery posed impossible problems for classical physics. Electrons could orbit an atomic nucleus to form an atom but, because they are accelerating, the electrons should radiate electromagnetic waves, lose energy, and plunge into the nucleus within 10^{-8} s. Material objects should immediately collapse by a factor of 10^{15} in volume while emitting a tremendous burst of radiation!

In 1913, Niels Bohr showed how the photon model of light could explain the stability of atoms. He assumed that an electron orbiting an atomic nucleus cannot radiate energy continuously; it emits or absorbs radiation one photon at a time, making abrupt transitions between orbits with different energies. Furthermore, for a given atom, only specific photon frequencies are possible.

The frequencies of light emitted or absorbed by a hydrogen atom can be described using a formula discovered by Johann Balmer in 1885:

$$f_{nn'} = f_o\left(\frac{1}{n^2} - \frac{1}{(n')^2}\right), \tag{35.9}$$

ACTUALLY, BALMER'S FORMULA HAD $n = 2$. LATER WORKERS GENERALIZED HIS RESULT TO ARBITRARY VALUES OF n.

where n and $n' > n$ are integers and $f_o = 3.288 \times 10^{15}$ Hz. The spectral lines corresponding to $n = 2$, called the Balmer series, lie in the visible (■ Figure 35.15). The Lyman series, $n = 1$, is in the ultraviolet. The Paschen series, $n = 3$, and the Brackett series, $n = 4$, lie in the infrared.

In Bohr's model, a photon with energy $hf_{nn'}$ is emitted when an electron makes a transition from orbit n' to orbit n (■ Figure 35.16). Since only certain photon energies are emitted,

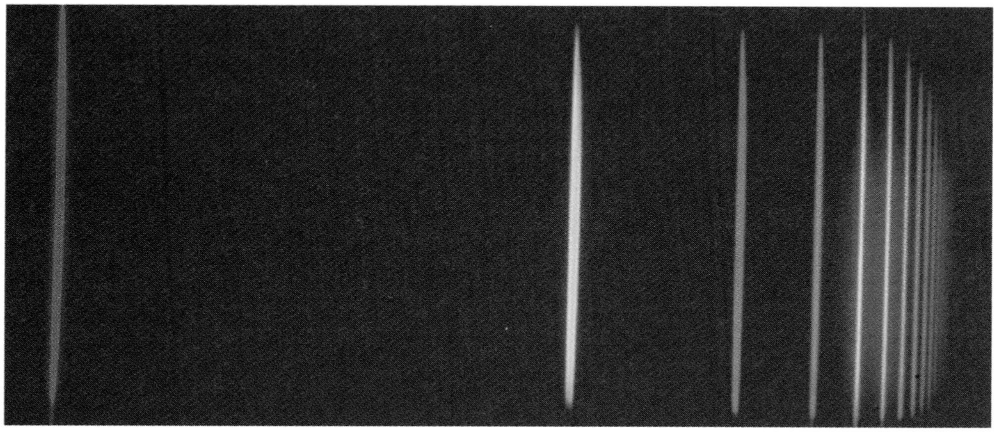

■ FIGURE 35.15
The hydrogen spectrum. The lines that appear in the visible are called the Balmer sequence: the brightest line, at the red end of the spectrum, is Balmer alpha. It corresponds to $n = 2$, $n' = 3$ in eqn. (35.9). The rest of the lines are in the blue part of the spectrum.

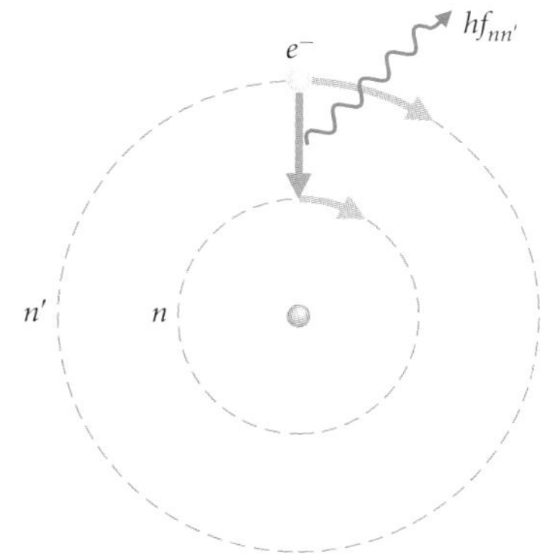

■ FIGURE 35.16
Photon emission in Bohr's model of the hydrogen atom corresponds to an electron jumping from one allowed orbit, labeled by integer n', down to another, labeled by integer n.

the electron orbits in hydrogen are also restricted. Balmer's formula suggests that the possible electron orbits in a hydrogen atom can also be labeled by an integer n. Since each orbit corresponds to a specific value of the electron's angular momentum, Bohr assumed that the angular momentum is quantized.

THE SYMBOL $\hbar$ IS READ "H BAR." ITS VALUE IS $\hbar = h/(2\pi) = 1.0546 \times 10^{-34}$ J·s.

The angular momentum of an electron in its orbital motion is an integer multiple of Planck's constant divided by 2π.

$$L = nh/(2\pi) \equiv n\hbar. \tag{35.10}$$

Working from this bold assumption, Bohr used *classical* (Newtonian) physics to investigate the properties of the atom. Each electron moves in a circular orbit of radius r_n with speed v_n. Bohr's principle (35.10) imposes a relation between the radius of the nth orbit and the electron's speed v_n in that orbit (■ Figure 35.17):

WE USE NONRELATIVISTIC EXPRESSIONS FOR L AND K. WE'LL CHECK THE VALIDITY OF THIS LATER.

$$L = mv_n r_n = nh/(2\pi) \equiv n\hbar.$$

The electric (Coulomb) force due to the positively charged nucleus causes the electron's acceleration as it moves in a circle:

$$ke^2/r_n^2 = mv_n^2/r_n. \tag{i}$$

Combining these two relations, we find:

$$ke^2 = \frac{(mv_n r_n)^2}{mr_n} = \frac{(n\hbar)^2}{mr_n}.$$

Solving for the orbit radius:

$$r_n = n^2 \frac{\hbar^2}{mke^2}. \tag{35.11}$$

Using eqn. (i), the kinetic and potential energies of the electron are:

$$K_n \equiv \tfrac{1}{2}mv_n^2 = \tfrac{1}{2}ke^2/r_n \qquad \text{and} \qquad U_n = -ke^2/r_n.$$

So, the total energy of the electron is:

$$E_n \equiv K_n + U_n = -\tfrac{1}{2}ke^2/r_n. \tag{35.12}$$

Using eqn. (35.11) for r_n, the total energy of the electron is:

$$E_n = -\frac{m(ke^2)^2}{2\hbar^2}\left(\frac{1}{n^2}\right). \tag{35.13}$$

Thus using the assumption that angular momentum is quantized together with classical ideas about motion, Bohr obtained expressions for both the orbital radius r_n and the energy in that orbit E_n that depend solely on fundamental constants (e, m, k, $\hbar$) and an integer n.

When an electron drops from orbit n' to orbit $n < n'$, its energy decreases. A photon is emitted and carries away the energy lost by the electron:

$$hf_{nn'} = E_{n'} - E_n = \frac{m(ke^2)^2}{2\hbar^2}\left[\frac{1}{n^2} - \frac{1}{(n')^2}\right]. \tag{35.14}$$

Thus Balmer's empirical formula for the hydrogen spectrum arises from the principle that angular momentum exists in multiples of a fundamental quantum $\hbar$.

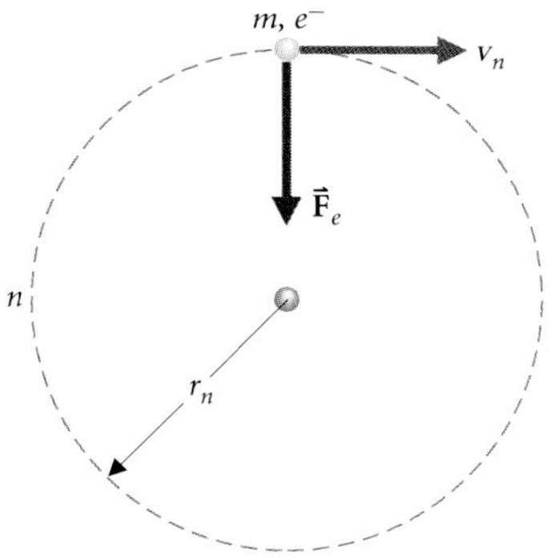

■ FIGURE 35.17
Calculation of allowed orbits in Bohr's model for the hydrogen atom. The electron follows a circular orbit under the influence of the Coulomb force due to the nucleus. Bohr assumed that angular momentum $L_n = mv_n r_n$ is quantized, $L_n = n\hbar$, and thus there is a specific relation between the radius r_n of the nth orbit and the electron's speed v_n in that orbit.

We may also write eqn. (35.14) as:

$$E_{n'} - E_n = E_o \left[\frac{1}{n^2} - \frac{1}{(n')^2} \right],$$

with

$$E_o = \frac{m(ke^2)^2}{2\hbar^2} = \frac{m(ke^2)^2(2\pi)^2}{2h^2} = \frac{me^4}{8\epsilon_0^2 h^2} \qquad (35.15)$$

$$= \frac{(9.109 \times 10^{-31}\ \text{kg})(1.602 \times 10^{-19}\ \text{C})^4}{8(8.854 \times 10^{-12}\ \text{F/m})^2(6.626 \times 10^{-34}\ \text{J}\cdot\text{s})^2}$$

$$= 2.180 \times 10^{-18}\ \text{J} = 13.6\ \text{eV}.$$

Each possible state of a hydrogen atom is labeled by the integer n of the electron orbit and is represented in an *energy-level diagram* (■ Figure 35.18). The vertical axis represents the energy of the atom, while the horizontal axis is used only for convenience in representing processes with the diagram. The lowest possible energy ($n = 1$) occurs in the *ground state* of the atom. States with $n > 1$ are called *excited* states. When an electron makes a transition between two states, $n_i \rightarrow n_f$, it is accompanied by emission ($n_f < n_i$) or absorption ($n_f > n_i$) of a photon. States with finite n have negative energy; the electron is bound to the atom. A state with positive energy represents a free electron passing by an isolated proton with positive kinetic energy and zero potential energy. The energy required to raise an electron from the ground state ($n = 1$) to $n' = \infty$ is called the *ionization potential* $\chi = E_o$. When a photon with energy $hf > \chi$ is absorbed by an atom, an electron is ejected, and the atom is said to be *ionized.*

Compare with our discussions of gravitational (§8.2.2) and electrical (Chapter 25) potential energy.

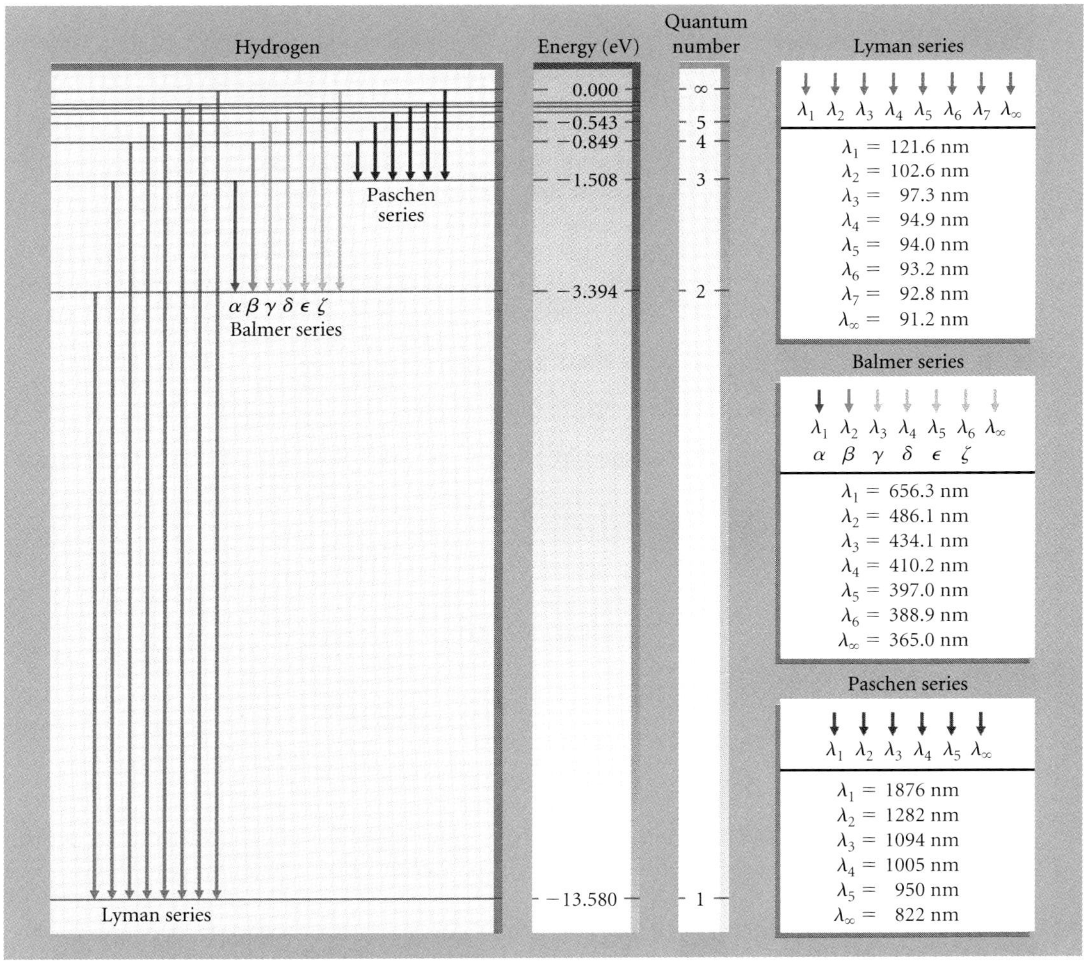

■ Figure 35.18
Energy-level diagram for hydrogen, according to the Bohr model, showing the transitions responsible for some of the observed lines. The levels corresponding to large n are closer together than those corresponding to small n. Transitions between closely spaced levels with large n values produce lines in the radio range.

NOTE: r_1 IS OFTEN WRITTEN a_o AND IS CALLED THE *BOHR RADIUS.*

EXAMPLE 35.7 ♦♦ Find the radius of the first Bohr orbit ($n = 1$) for hydrogen and the speed of an electron in that orbit.

MODEL Equation (35.11) gives the radius. We expect the answer to be of the order of 10^{-10} m. Knowing the radius, we apply Bohr's condition on angular momentum to get the speed.

SETUP Setting $n = 1$ in eqn. (35.11), we have:

$$r_1 = \frac{\hbar^2}{mke^2}$$

SOLVE
$$r_1 = \frac{(1.0546 \times 10^{-34}\ \text{J}\cdot\text{s})^2}{(9.1094 \times 10^{-31}\ \text{kg})(8.9876 \times 10^9\ \text{N}\cdot\text{m}^2/\text{C}^2)(1.6022 \times 10^{-19}\ \text{C})^2}$$
$$= 0.5292 \times 10^{-10}\ \text{m}.$$

The speed is:

$$v_1 = \frac{\hbar}{mr_1} = \frac{1.05 \times 10^{-34}\ \text{J}\cdot\text{s}}{(9.11 \times 10^{-31}\ \text{kg})(0.529 \times 10^{-10}\ \text{m})}$$
$$= 2.18 \times 10^6\ \text{m/s}.$$

ANALYZE Since this speed is less than 1% that of light, we are justified in ignoring special relativity in the Bohr theory for hydrogen. ▮

EXERCISE 35.4 ♦ Verify that to three significant figures Bohr's theory gives the correct numerical value for the constant f_o in Balmer's formula.

EXERCISE 35.5 ♦ The dimensionless number $\alpha \equiv ke^2/(\hbar c) = \frac{1}{137}$, called the *fine structure constant,* is useful in comparing values of atomic parameters. Show that the energy E_o of an electron in the first Bohr orbit is $\frac{1}{2}\alpha^2 mc^2$, and $v_1 = \alpha c$. ▮

EXAMPLE 35.8 ♦ Clouds of hydrogen in the Milky Way typically appear red (■ Figure 35.19). Find the wavelength of light emitted when the electron in a hydrogen atom makes a transition from the $n = 3$ state to the $n = 2$ state (*Balmer* α), and use your answer to explain the color.

MODEL We use the Bohr model of the atom.

SETUP First, we convert Balmer's formula (eqn. 35.9) to an expression for wavelength:

$$f_{nn'} = f_o\left[\frac{(n')^2 - n^2}{(nn')^2}\right].$$

So:

$$\lambda_{nn'} = \frac{c}{f_{nn'}} = \frac{c}{f_o}\left[\frac{(nn')^2}{(n')^2 - n^2}\right].$$

SOLVE With $n = n_f = 2$ and $n' = n_i = 3$,

$$\lambda_{23} = \frac{c}{f_o}\left[\frac{(nn')^2}{(n')^2 - n^2}\right] = \frac{(2.998 \times 10^8\ \text{m/s})(2^2)(3^2)}{(3.288 \times 10^{15}\ \text{Hz})(3^2 - 2^2)} = 656\ \text{nm}.$$

ANALYZE This wavelength is at the red end of the visible range. Balmer α is a prominent emission line from hydrogen gas at temperatures of a few thousand kelvin, and accounts for the red color.

The constant $f_o/c = R = 1.097 \times 10^7$ /m is called the Rydberg. It is useful when using the Balmer formula to compute wavelengths. ▮

Niels Bohr, 1885–1962

EXAMPLE 35.9 ♦ What is the minimum frequency of a photon that can ionize a hydrogen atom from the $n = 3$ state?

MODEL The photon would just raise the atom to the $n' = \infty$ state; we find its frequency from Balmer's formula.

SETUP AND SOLVE

$$f_{min} = f_o\left[\frac{1}{3^2} - \frac{1}{\infty^2}\right] = \frac{f_o}{9} = 3.65 \times 10^{14}\text{ Hz}.$$

ANALYZE The light would be described as infrared (cf. Table 16.2). The corresponding wavelength is $c/f = 0.82\ \mu\text{m}$.

An atom from which all electrons but one have been stripped is called a *hydrogenic atom*. Bohr's model is easily modified to apply to any such atom. If the charge on the atomic nucleus is Ze, then the force between the nucleus and electron is given by $F = kZe^2/r^2$. All other results are also modified by replacing the factor ke^2 by kZe^2.

UPON BEING TOLD THAT BOHR'S MODEL CORRECTLY PREDICTS E_o FOR HELIUM, EINSTEIN REMARKED, "THEN IT IS ONE OF THE GREATEST DISCOVERIES."

EXERCISE 35.6 ♦♦ Repeat the calculations of the Bohr model for a hydrogenic atom with nuclear charge Ze. Show that the radius of the nth orbit is reduced by a factor Z and the binding energy of the atom is multiplied by Z^2.

EXAMPLE 35.10 ♦ Hot gas in distant clusters of galaxies emits an x-ray spectral line with a photon energy of 6.7 keV. If the emission is from the $n' = 2$ to $n = 1$ state in a hydrogenic atom, what is the nuclear charge of the atom?

MODEL We use the expression for photon energy from Bohr's model, multiplied by the square of the nuclear charge (see Exercise 35.6).

SETUP Thus:

$$Z^2E_o\left(\frac{1}{1^2} - \frac{1}{2^2}\right) = 6.7\text{ keV},$$

$$Z^2E_o\left(\tfrac{3}{4}\right) = 6.7\text{ keV}.$$

SOLVE So:

$$Z = \left[\frac{4(6.7 \times 10^3\text{ eV})}{3(13.6\text{ eV})}\right]^{1/2} = 26.$$

ANALYZE This is the charge of an iron nucleus. Astronomers use observations of the 6.7-keV line to estimate the amount of iron in the gas. That information provides important clues about the origin of the gas in the cluster.

■ FIGURE 35.19
The central portion of the Eta Carinae Nebula (also known as the Keyhole Nebula). The red color is due to emission from the Balmer α line of hydrogen.

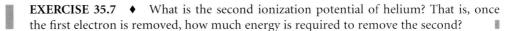

EXERCISE 35.7 ♦ What is the second ionization potential of helium? That is, once the first electron is removed, how much energy is required to remove the second?

35.2.3 *The Correspondence Principle*

According to quantum theory, atoms undergo abrupt transitions between states as they absorb or emit photons, while classical theory considers continuous fields and continuous change in macroscopic systems. Both theories are very successful, and both describe the behavior of charged objects responding to electromagnetic forces. Bohr realized that classical theory must be a limiting case of a complete quantum theory; the two should give corresponding results for systems that are large on the atomic scale, though small on the macroscopic scale, or for systems in which the energy of a single photon is negligible.

We have seen an example of the correspondence principle in the spectrum of thermal radiation. In the low-frequency limit, $hf \ll kT$, a single photon's energy is much less than the typical thermal energy, and the classical Rayleigh–Jeans formula gives the correct spectrum. A second example is provided by radiation from electrons in Bohr orbits with very large quantum numbers n. The frequency of a photon emitted in a transition from $n' = n + 1$ to n is given by:

$$f_{n+1,n} = f_o\left[\frac{1}{n^2} - \frac{1}{(n+1)^2}\right] = f_o\frac{2n+1}{n^2(n+1)^2} \approx \frac{2f_o}{n^3} \quad \text{for } n \gg 1.$$

Classically, the radiation frequency should equal the frequency of the electron in its orbit: $f_c = v/(2\pi r_n)$. Using Bohr's principle to express v in terms of n and r_n, and eqn. (35.11) for r_n:

$$f_c = \frac{n\hbar}{2\pi m r_n^2} = \frac{n\hbar}{2\pi m}\left(\frac{mke^2}{n^2\hbar^2}\right)^2.$$

Next, we substitute eqn. (35.15) for $E_o = hf_o = 2\pi\hbar f_o$:

$$f_c = \frac{(ke^2)^2 m}{2\pi\hbar^3 n^3} = \frac{E_o}{\pi n^3\hbar} = \frac{2f_o}{n^3}.$$

The classical and quantum results correspond when n is large.

Atoms having electrons in large n states (called *highly excited* atoms) are big ($r \gg r_1$) and fragile. They do not occur naturally on Earth. In 1965, astronomers detected radiation from hydrogen atoms in interstellar space undergoing transitions between levels with n near 100. Values of n as large as 350 have since been discovered. Large values of n (10–100) can be obtained in the laboratory using a laser to excite the atoms. These atoms provide an excellent vehicle for testing the correspondence principle.

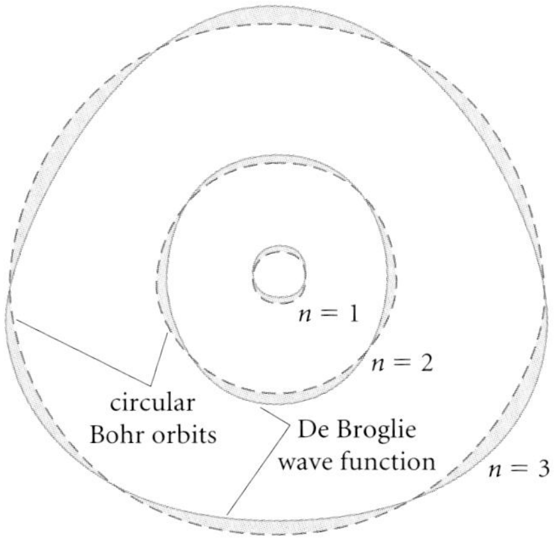

■ FIGURE 35.20
De Broglie waves and the Bohr theory. Exactly n de Broglie waves fit into the circumference of the nth Bohr orbit. If the wavelength were slightly longer or shorter, the wave would suffer destructive interference as it wrapped around the orbit. This picture serves as a useful image of how quantization occurs: it is not an accurate model. Historically, it served as a transition between the Bohr and Schrödinger descriptions.

35.3 ELECTRON WAVES

35.3.1 *De Broglie's Hypothesis*

COMPARE §15.4.2 AND §33.4.

Bohr's atomic model can explain the stability of atoms and connect atomic structure with the photon description of light. But what constrains the angular momentum of an electron? The answer, discovered by Louis de Broglie in 1923, is that electrons, like photons, have both a wave and a particle nature. When a wave is confined in a system with finite dimensions, a standing wave pattern is formed. Certain wavelengths are reinforced by reflections at the boundaries of the system, while other wavelengths interfere destructively. Each state of a hydrogen atom corresponds to a different standing-wave pattern (■ Figure 35.20).

The energy and momentum of a photon are related to the frequency and wavelength of the corresponding electromagnetic wave (eqn. 35.1):

$$E = hf$$

and

$$p = E/c = hf/c = h/\lambda.$$

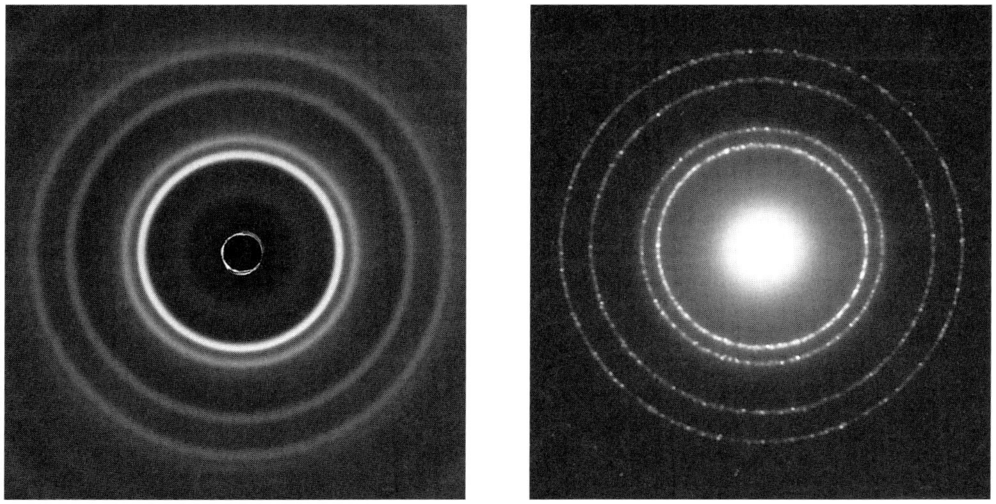

(a) (b)

■ FIGURE 35.21
The wave nature of electrons is confirmed by the diffraction pattern produced when electrons scatter off crystals. (a) The x-ray diffraction pattern of aluminum. A lead disk was placed over the center of the film for most of the exposure to avoid over-exposing the film in these bright regions. (b) Diffraction pattern of aluminum using electrons. Both patterns show the characteristic diffraction rings.

The frequency and wavelength of a particle's associated de Broglie wave are related to the particle's total energy $E = \gamma mc^2$ and momentum $p = \gamma mv$ in the same way:

$$f = E/h \quad \text{and} \quad \lambda = h/p. \tag{35.16}$$

When an electron beam scatters from a crystal surface, a diffraction pattern is observed (■ Figure 35.21). Since diffraction is purely a wave phenomenon and cannot be explained with particles, this observation provides direct experimental confirmation of de Broglie's hypothesis.

EXERCISE 35.8 ♦ Show that the wave speed $v_d = \lambda f$ of a de Broglie wave $= c^2/v$.

We may use relations (35.16) to show that exactly n de Broglie wavelengths fit into the nth Bohr orbit. With $p = mv_n$, and L from eqn. (35.10):

$$\frac{2\pi r_n}{\lambda_n} = \frac{2\pi r_n}{h/mv_n} = \frac{2\pi}{h} mv_n r_n = \frac{L}{\hbar} = \frac{1}{\hbar}\hbar n = n.$$

EXAMPLE 35.11 ♦ What is the wavelength of electrons in a television set that have a kinetic energy of 35 keV?

MODEL Since the kinetic energy is much less than the 511-keV rest energy of an electron, we may use the Newtonian formula for the electron momentum in eqn. (35.16).

SETUP
$$K = \tfrac{1}{2}mv^2, \quad \text{and} \quad p = mv = m\sqrt{2K/m} = \sqrt{2mK}.$$

SOLVE Then:

$$\begin{aligned}\lambda &= \frac{h}{p} = \frac{h}{\sqrt{2mK}}\\ &= \frac{6.63 \times 10^{-34}\ \text{J}\cdot\text{s}}{[2(9.11 \times 10^{-31}\ \text{kg})(3.5 \times 10^{4}\ \text{eV})(1.6 \times 10^{-19}\ \text{J/eV})]^{1/2}}\\ &= 6.6 \times 10^{-12}\ \text{m}.\end{aligned}$$

ANALYZE This is some 10^5 times smaller than the wavelengths of visible light. Electron microscopes can achieve much greater resolution than is available with light microscopes, since wave diffraction is no longer a limiting factor.

DIFFRACTION AND ITS EFFECTS ON RESOLUTION ARE DISCUSSED IN §17.3.

35.3.2 *Schrödinger's Picture of the Hydrogen Atom*

The work of Bohr and de Broglie led to a qualitative understanding of atomic structure and provided quantitative results for simple atoms. Bohr's model of the atom explained both Rutherford's data and the hydrogen spectrum, and de Broglie's waves offered a believable mechanism for quantization. However, careful examination of the spectra of certain elements such as sodium showed that some of the spectral lines are split—the line is actually two lines with a very small separation in frequency (■ Figure 35.22). Bohr's theory could not account for this

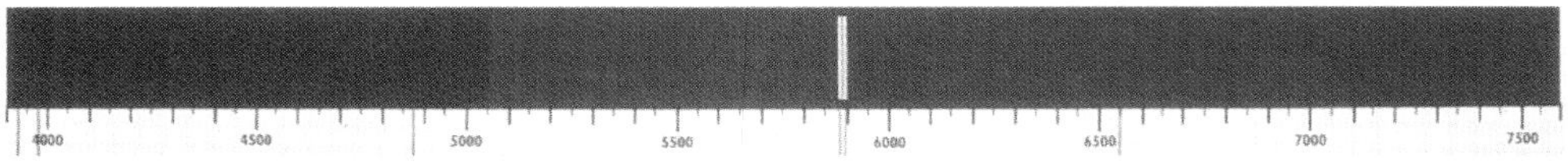

■ FIGURE 35.22
The bright sodium line in the yellow part of the spectrum is actually two very closely spaced lines. Such line *splitting* is not explained by Bohr's model.

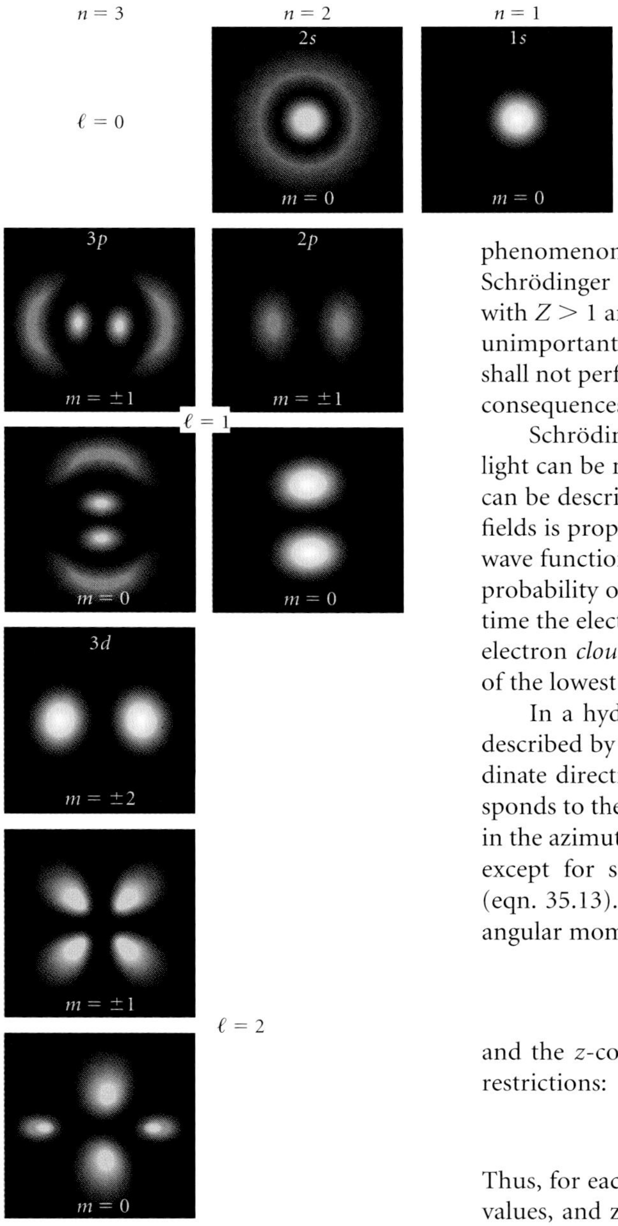

■ FIGURE 35.23
Electron wave functions in hydrogen. The $n = 1$ state has only one possible value of ℓ—0. This is called the 1*s* state. The wave function is spherically symmetric and $|\psi|^2$ is greatest at the origin. When $n = 2$, there are two possible values of ℓ. The 2*s* state has $\ell = 0$ and the 2*p* state has $\ell = 1$. With $\ell = 1$, there are three possible values of m: 0 and ± 1. All $\ell = 0$ states are spherically symmetric. Note that as ℓ increases, the distribution becomes less centrally concentrated. In general, ψ is a complex-valued function; its magnitude is real-valued.

phenomenon, nor could it explain the structure of atoms with many electrons. In 1926, Erwin Schrödinger produced a theory capable of explaining the distribution of electrons in atoms with $Z > 1$ and many details of atomic spectra. His theory is valid where relativistic effects are unimportant, and it is used for fundamental calculations in chemistry and atomic physics. We shall not perform any calculations with Schrödinger's theory, but we can discuss its qualitative consequences.

Schrödinger's equations describe the behavior of the electron wave function ψ. A pulse of light can be modeled with photons or EM waves and, similarly, a particle such as an electron can be described by a matter (de Broglie) wave. Just as the energy density in electromagnetic fields is proportional to the square of the field amplitudes, it is the squared magnitude of the wave function that describes electron density. One may view the density $|\psi|^2$ at a point as the probability of an electron being in a unit volume surrounding the point, or as the fraction of time the electron spends in such a unit volume, or as a description of the charge density in an electron *cloud* surrounding the atomic nucleus. ■ Figure 35.23 illustrates this cloud for a few of the lowest energy states in hydrogen.

In a hydrogen atom, the electron is modeled with a three-dimensional standing wave, described by three integers that give the number of nodes of the pattern in each of the coordinate directions. In spherical coordinates, Bohr's n, the *principal* quantum number, corresponds to the radial coordinate. Nodes in the polar (θ) direction are described by ℓ and nodes in the azimuthal (ϕ) direction are described by m. The energy of each state depends on n and, except for small corrections we'll discuss later, is the same as that predicted by Bohr (eqn. 35.13). In Schrödinger's theory, ℓ and m are related to angular momentum. The total angular momentum of the electron is:

$$L = \sqrt{\ell(\ell + 1)}\hbar, \tag{35.17}$$

and the z-component of the angular momentum is $m\hbar$. The numbers ℓ and m satisfy the restrictions:

$$0 \le \ell \le n - 1 \quad \text{and} \quad -\ell \le m \le \ell. \tag{35.18}$$

Thus, for each value of ℓ, there are $2\ell + 1$ possible values of m (ℓ negative values, ℓ positive values, and zero). For $n = 1$ (the ground state), both ℓ and m are zero. Thus $L = 0$ in the ground state, whereas in Bohr's approximate model $L = \hbar$. The set of possible states in Schrödinger's theory is much richer than in the Bohr model it replaced, but the basic concept—quantization of energy and angular momentum—remains fundamental. Schrödinger found that quantization is not an ad hoc assumption, but a direct consequence of the electron's wave nature.

■ Figure 35.24 is an energy-level diagram for hydrogen, including all the levels found in the Schrödinger theory. The diagram is very similar to that for the Bohr theory, except that for $n > 1$, each level is a composite of states with different values of ℓ and m. Careful measurements of the hydrogen spectrum show that states with different ℓ actually have slightly different energies.

This *fine structure* in the spectrum arises from relativistic motion and electron spin. Modeling the electron as a small, rotating, charged ball is a useful visualization so long as you do not take it too literally. Such a ball has a magnetic moment and behaves like a small magnetic dipole. The electron's magnetic moment is parallel or antiparallel to its orbital angular momentum. Thus an individual electron has a spin angular momentum described by the quantum number $s = \frac{1}{2}$ and with z-component of spin described by the quantum number m_s, which takes only the values $\pm\frac{1}{2}$. In a reference frame where the electron is at rest at any instant, the positively charged nucleus is in motion. The moving nucleus acts as an electric

SEE § 29.4.1 FOR THE RELATION BETWEEN MAGNETIC MOMENT AND ANGULAR MOMENTUM.

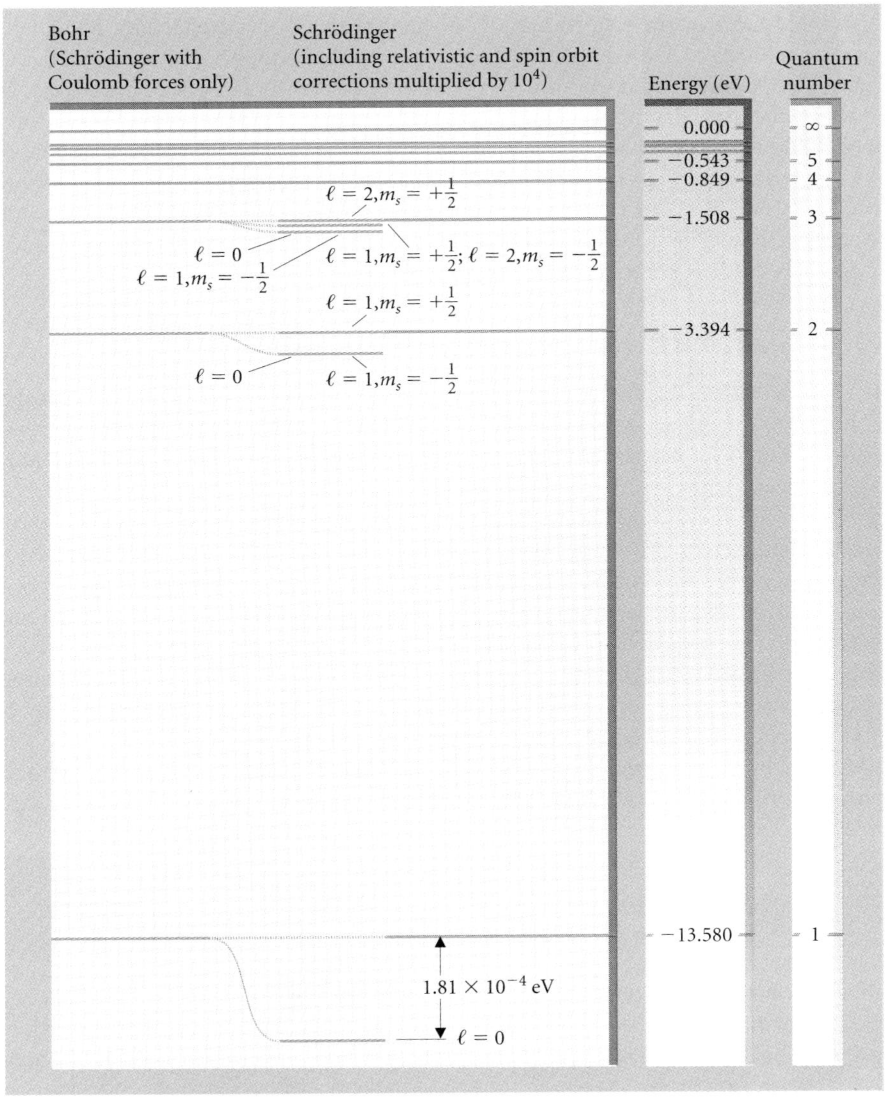

■ FIGURE 35.24
Energy-level diagram for hydrogen. The Bohr and Schrödinger theories make identical predictions when relativity is ignored and only Coulomb forces are considered (left column). Using the correct relativistic expression for the electron's energy and momentum, and including the electron's magnetic properties in Schrödinger's theory, we find small shifts and splittings of the levels, as illustrated in the right column of the diagram for $n \leq 3$. These effects are about 10^4 times smaller than the energy differences between Bohr levels and are shown greatly exaggerated for clarity. These predictions of Schrödinger's theory are in excellent agreement with measurements.

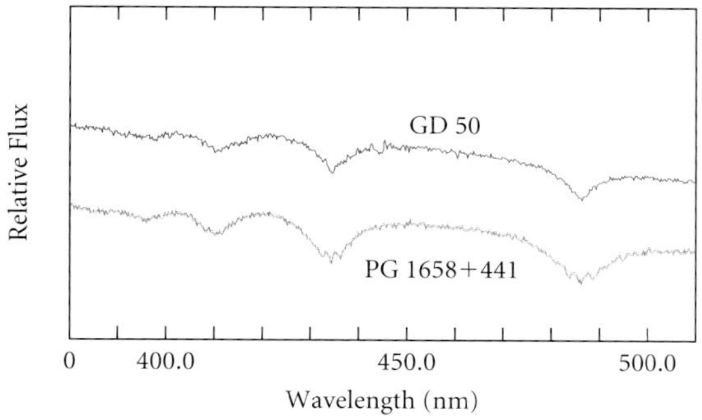

■ FIGURE 35.25
Zeeman splitting of hydrogen absorption lines in the spectrum of a white dwarf star. Compare the Balmer ($n = 2$) absorption features at 435 and 485 nm in the two spectra shown. The strong magnetic field (about 300 T) of the white dwarf PG 1658+441 causes different m states to have slightly different energies and results in the observed triple line structure. *Note:* These stars are known only by their numbers in specialized astronomical catalogues.

current and produces a magnetic field. As a result, the spinning electron has a small amount of magnetic potential energy. Energy differences caused by this spin–orbit interaction are of the order of 10^{-4} eV. In a spectroscope, the result is an apparent splitting of a spectral line into several closely spaced components.

Hyperfine structure due to the spin of the nucleus causes line splittings roughly 100 times smaller still. The electron energy changes when its spin flips between states parallel and antiparallel to the nuclear spin. Low-energy photons emitted as a result of such spin flips in the hydrogen ground state are very important in astronomy. The resulting radio waves with wavelength $\lambda = 21$ cm can traverse thousands of light-years of interstellar gas, allowing radio astronomers to observe distant regions of our galaxy that cannot be seen in visible light.

States with different quantum numbers respond differently to outside influences. Different values of m can be distinguished if an external magnetic field is applied to the atom. The magnetic field distinguishes one direction—parallel to the field—from all others. The orbital motion of the electron gives the atom as a whole a magnetic moment and a magnetic potential energy ($\approx 10^{-3}$ eV/T), which depends on the component of angular momentum parallel to the external field and hence on m. The resulting energy differences cause spectral line splittings known as the *Zeeman effect* (■ Figure 35.25).

AFTER PIETER ZEEMANN (1865–1943).

35.3.3 The Pauli Exclusion Principle and Chemistry

THE PERIODIC TABLE IS INSIDE THE BACK COVER.

In the periodic table, elements are organized into groups with similar chemical properties. Using the Schrödinger theory, we can understand why the elements form these groups. Carbon differs from silver and resembles silicon because of the ways electrons are arranged in atoms of these elements.

Hydrogen has the simplest atomic structure, since it has but one electron. The next simplest case is helium with a nuclear charge of $2e$ and two electrons. On average, the two electrons are farther from each other than either is from the nucleus. The electrical forces of the electrons on each other are less important than their individual interactions with the nucleus. Then each electron is described by a wave function very similar to the ground state wave function of hydrogen, though the orbital radius is smaller and the binding energy larger because of the greater charge on the nucleus. *Both* electrons in helium are held tightly by the nucleus and helium is a chemically unreactive (noble or inert) gas.

Even in helium, however, the electrons do not entirely behave as if each were alone. A new physical principle with no classical counterpart comes into play. It is the *Pauli exclusion principle.*

AFTER WOLFGANG PAULI (1900–1958).

> No more than one electron in a system can occupy any quantum state.

BECAUSE OF ENRICO FERMI'S BASIC WORK ON THE RELATION BETWEEN A PARTICLE'S SPIN AND ITS THERMODYNAMIC BEHAVIOR, PARTICLES THAT SHARE THIS PROPERTY OF ELECTRONS ARE CALLED *FERMIONS.*

The quantum state of an electron in an atom is described by its quantum numbers n, ℓ, m, and m_s. According to the Pauli principle, no two electrons in an atom can have the same values of all four numbers. The two electrons in helium may have the same principal quantum number $n = 1$ and the same angular momentum quantum numbers $\ell = m = 0$, but if they do, their spins *must* be oriented in opposite directions: $m_s = +\frac{1}{2}$ and $m_s = -\frac{1}{2}$. These quantum numbers describe the ground state of helium.

In atoms with three or more electrons, the exclusion principle forces electrons to be in higher energy states. In lithium, as in helium, the first two electrons fill the $n = 1$ state with opposite spins. The third electron must enter a much less tightly bound state with $n = 2, \ell = m = 0$. Lithium is a highly reactive metal that readily shares the one loosely bound electron in chemical bonds. Beryllium's four electrons fill the $n = 2, \ell = m = 0$ state. It is less reactive than lithium, but much more so than helium. Six electrons can be put in the three states with $n = 2, \ell = 1$, and $m = -1, 0$, or $+1$. These correspond to the elements from boron through neon (● Table 35.2). Neon, with all the $n = 2$ states filled is, like helium, an inert gas. When all the quantum states in a certain n level are filled in this way, we say the atom has a *closed shell.* Fluorine, lacking one electron from such a complete shell, is a highly corrosive gas that, in reactions, takes electrons from atoms such as lithium.

The entire periodic table can be understood in terms of the quantum numbers of the electrons. After the third period, however, the mutual electrical forces among the electrons are important in determining the order in which the states are filled. We use the distributions shown in Figure 35.23 to see how this happens.

In states with $\ell = 0$, the wave function is spherically symmetric and has large values close to the origin. When ℓ is not zero, the wave function vanishes at the origin and is concentrated at a distance r, which increases with ℓ. In a multielectron atom, the presence of other electrons

TABLE 35.2

Element	Z	Electron Distribution						
		$n = 1$	$n = 2$		$n = 3$			$n = 4$
		$\ell = 0$	$\ell = 0$	$\ell = 1$	$\ell = 0$	$\ell = 1$	$\ell = 2$	$\ell = 0$
H	1	1						
He	2	2	$n = 1$ shell filled					
Li	3	2	1					
Be	4	2	2					
B	5	2	2	1				
C	6	2	2	2				
N	7	2	2	3				
O	8	2	2	4				
F	9	2	2	5				
Ne	10	2	2	6	$n = 1$ and $n = 2$ shells filled			
Na	11	2	2	6	1			
Mg	12	2	2	6	2			
Al	13	2	2	6	2	1		
Si	14	2	2	6	2	2		
P	15	2	2	6	2	3		
S	16	2	2	6	2	4		
Cl	17	2	2	6	2	5		
Ar	18	2	2	6	2	6		
K	19	2	2	6	2	6		1
Ca	20	2	2	6	2	6		2
Sc	21	2	2	6	2	6	1	2

has the effect of shielding the nucleus, reducing its effective charge. This shielding effect is greater for states with higher ℓ since they are concentrated farther from the center of the distribution. Electron states with greater ℓ see a smaller effective nuclear charge and have a correspondingly smaller binding energy. In the fourth period of the periodic table, this effect becomes sufficiently pronounced that the $n = 4$, $\ell = 0$ states fill before the $n = 3$, $\ell = 2$ states.

Since 1926, Schrödinger's picture has been developed into a complete theory of chemical bonding. Computational complexity is now the main limit to understanding chemical reactions from first principles.

35.4 Quantum Mechanics

The success of the photon and electron wave models in explaining atomic phenomena leaves no doubt that dual wave/particle behavior is a fundamental aspect of nature. That duality at the atomic level does not contradict our experience of distinct particles and waves on the macroscopic scale, but the duality introduces new principles that we don't encounter in the everyday world. We conclude our discussion of atomic physics by showing that wave mechanics imposes fundamental limits on measurement accuracy and on our ability to predict the behavior of physical systems.

35.4.1 *The Heisenberg Uncertainty Principle*

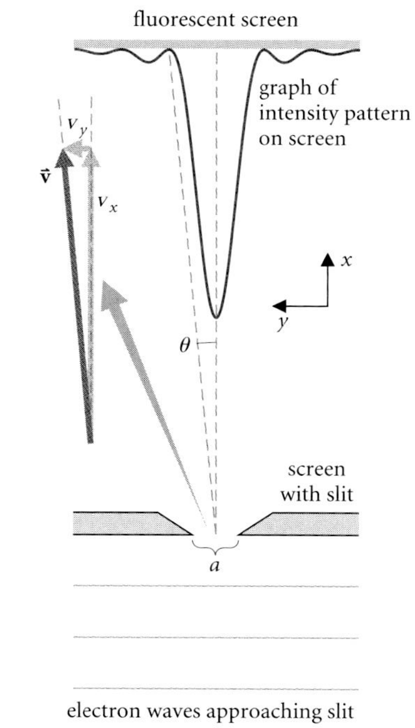

■ **Figure 35.26**
Measuring the position of an electron beam with a horizontal slit of width *a*. The uncertainty in the position is $\delta y = a$. A diffraction pattern forms on the screen. In order to reach a point on the screen at angle θ from the slit, the y-component of the electron's velocity must satisfy $v_y/v = \sin\theta$. The uncertainty in the electron's position within the diffraction maximum corresponds to an uncertainty $\delta p_y = mv_y = mv\sin\theta = p\lambda/a = h/a$ in the electron's momentum.

Suppose we attempt to measure the y-coordinate of electrons by passing a beam through a long, horizontal slit (■ Figure 35.26). The experiment determines the position of each electron in the beam with an uncertainty equal to the width a of the slit: $\delta y = a$. At the same time, each electron's wave function is diffracted by the slit and forms the usual single-slit diffraction pattern (cf. §17.3.1). Almost all the electrons strike the screen within the central diffraction maximum. This means that each individual electron leaves the slit at some unknown angle between 0 and θ, where, for electrons of wavelength λ, θ is given by $\sin\theta = \lambda/a$. Uncertainty in the y-direction of the electron's motion corresponds to uncertainty in the y-component of its momentum, given by:

$$\delta p_y \sim p\sin\theta = p\lambda/a.$$

Using de Broglie's formula for the electron wavelength, $\lambda = h/p$, we find:

$$\delta p_y \sim h/a,$$

or

$$\delta y\, \delta p_y \sim h.$$

In principle, the accuracy of the position measurement can be improved indefinitely by decreasing the width of the slit. However, there is a penalty for increased accuracy; the uncertainty in momentum increases in proportion as the central diffraction maximum expands. Improving our knowledge of the electron's position inevitably reduces the information available about its momentum.

Werner Heisenberg showed in 1927 that this limit is not a peculiar feature of any particular experiment but a basic feature of wave/particle duality. To see why, let's investigate how waves superpose to form a localized, particlelike disturbance. A single plane wave has a well-defined wavelength λ, but spreads uniformly throughout space (■ Figure 35.27a). Superposing a second wave of slightly different wavelength results in constructive interference where the two waves are in phase and destructive interference where they are out of phase (Figure 35.27b). Because the probability of finding an electron at a point is proportional to the wave amplitude squared, such a wave superposition describes a somewhat localized particle. By superposing a large number of individual waves with a range of wavelengths, we can arrange to have constructive interference in one small region with destructive interference everywhere else (Figure 35.27c). This superposition represents a particle located within a range of positions δx and represented by waves with a range of momenta δp. The localization in space can be achieved only by superposing waves with a spread in wavelength. According to Fourier's theorem of wave superpositions, the range of wavelengths is related to the spatial localization by:

$$\delta x \, \delta(1/\lambda) \geq 1/(4\pi).$$

Since $p = h/\lambda$, $\delta(1/\lambda) = \delta p/h$, and so:

$$\delta x \, \delta p \geq \hbar/2. \qquad \textbf{(35.19)}$$

This analysis shows that eqn. (35.19) is more than just a limit on the accuracy of measurement; it is a fundamental relation between actual physical properties of a system.

A similar general relation holds between uncertainties in energy and time measurements. If we observe the wave representation of a localized particle passing a fixed observation point, the time we would record for its passage is uncertain by:

$$\delta t \sim \delta x/v,$$

where $v = p/m$ is the speed of the particle. The particle's energy is uncertain by:

$$\delta E = \delta(p^2/2m) = (p\,\delta p)/m.$$

So, the product of uncertainties in energy and time is:

$$\delta E \, \delta t = \frac{p\,\delta p}{m}\left(\frac{\delta x}{v}\right) = \delta p \, \delta x \geq \frac{\hbar}{2}. \qquad (35.20)$$

EXAMPLE 35.12 ♦♦ Spectral lines from atoms are not infinitely sharp. Atoms typically remain in one state for about 10^{-8} s before emitting a photon. Use Heisenberg's uncertainty relations to estimate the uncertainty in the energy of an emitted photon. Compare this uncertainty with the fine structure splitting.

MODEL The uncertainty in the energy of the photon is related to the uncertainty in the time of its emission, represented here by the typical time the atom remains in one state before emitting a photon (the *lifetime* of the state).

SETUP The uncertainty in the energy is given by eqn. (35.20):

$$\delta E_{\text{photon}} \sim \frac{\hbar}{2\,\delta t} = \frac{1.06 \times 10^{-34}\ \text{J}\cdot\text{s}}{2(10^{-8}\ \text{s})(1.6 \times 10^{-19}\ \text{J/eV})} = 3 \times 10^{-8}\ \text{eV}.$$

ANALYZE This is a few tenths of a percent of the fine structure splitting (§35.3.2) and is roughly comparable with hyperfine effects. ■

EXERCISE 35.9 ♦ Excited iron nuclei emit the 14.4-keV x rays used in the Pound–Rebka experiment (cf. Essay 8: General Relativity). The excited state has a mean lifetime of 10^{-7} s. What is the uncertainty of the photon energy? Compare with the energy shift $\Delta E/E \sim 2 \times 10^{-15}$ that Pound and Rebka were trying to measure.

35.4.2 *The Meaning of the Wave Function*

Atomic phenomena are statistical: experiments with a single atom suspended by laser radiation show transitions between states occurring at randomly distributed times. Quantum theory accurately predicts the average time interval but cannot predict when an individual transition will occur. This strange situation is inconsistent with classical physics. If we knew the position and velocity of every particle, we could imagine using Newton's laws to compute the complete future of the universe. In quantum mechanics, the uncertainty principle denies us accurate knowledge of both position and velocity, and only allows us to find the probability that any future state of the universe will occur.

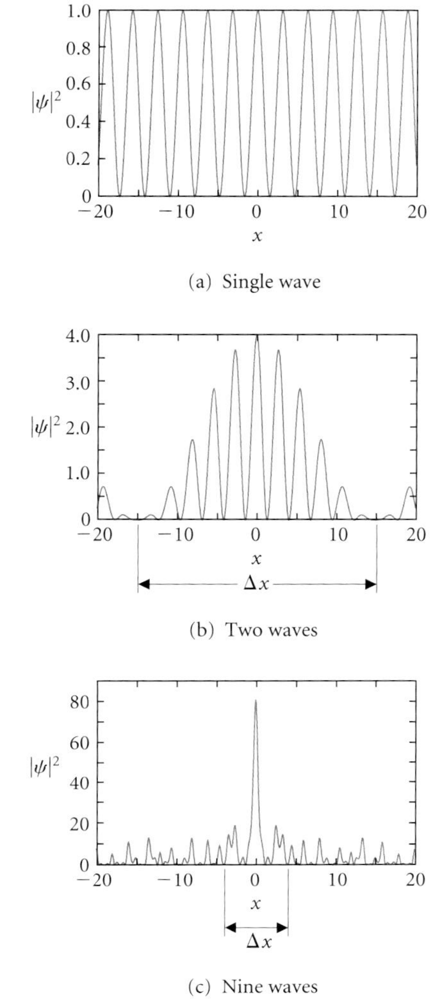

■ FIGURE 35.27
Wave superposition. Each plot shows $|\psi|^2$ as a function of x. (a) A wave with a single precisely known wavelength is not localized at all in space. (b) The superposition of two waves with slightly different wavelengths produces a modulated wave: here $\Delta\lambda$ is one unit and the wave function is concentrated in the region $-15 \lesssim x \lesssim +15$. The pattern repeats periodically in both positive and negative x-directions. (c) Superposition of nine waves ($\Delta\lambda \approx 8$) produces a wave packet that is localized to a fairly small region about the origin. Again, the pattern repeats to left and right. An infinite number of waves must be superposed to eliminate this repetition and localize the particle at a single spot.

To see how this interpretation follows from wave/particle duality, consider the interference of electrons passing through a screen with two small holes (■ Figure 35.28) and striking a television screen. An intense electron beam produces a two-source interference pattern on the screen. If the intensity is reduced so that electrons arrive one at a time, isolated flashes occur randomly on the screen but never at minima of the diffraction pattern. How can we interpret such behavior? The experiment tells us that electric charge leaves the electron gun in a bundle and that sometime later a flash occurs on the screen, but only our theoretical description tells us what it is that propagates from the gun to the screen. It can't be a classical particle, which would have to pass through one hole or the other and couldn't show interference at all. What propagates is Schrödinger's wave function ψ, which undergoes two-source interference in the standard way. But the wave strikes the entire screen and the flash occurs at an isolated spot! The wave represents not a substance, in any sense we are used to, but the probability of an event occurring. The event—*electron leaves gun*—will be followed by a flash; a total probability of unity leaves the gun, propagates as a wave, and spreads out in an interference pattern to strike the screen. Once the event—*flash*—occurs somewhere, it has no further chance of occurring elsewhere. According to the standard interpretation of quantum mechanics, the wave function abruptly ceases to exist everywhere else.

The standard interpretation of quantum mechanics provides extremely accurate methods of calculation and has satisfied every experimental test yet devised, but it has provoked fierce philosophical debate. Einstein was never able to accept it as final, because it seems to deny that any objective real thing connects events such as *charge leaves gun* and *flash.* Measurement—locating the flash on the screen—occurs as a discontinuous event involving collapse and regeneration of the wave function describing the electron. The role of conscious observers in such measurements has sparked particularly lively debate. As an amusing and provocative introduction to such topics, we close the chapter with a parable devised by Schrödinger, which assumes the quantum theory applies to macroscopic systems as well as microscopic.

A cat is placed in an airtight box with an oxygen supply and with a glass vial containing cyanide gas to be released if a radiation detector is struck by a particle from a radioactive sample. The radioactive sample is a quantum system for which we can accurately compute the *probability* a particle will be released in a given time interval. We propose to place the cat in the box just long enough for the probability to be one-half, then open the box, and see what has happened.

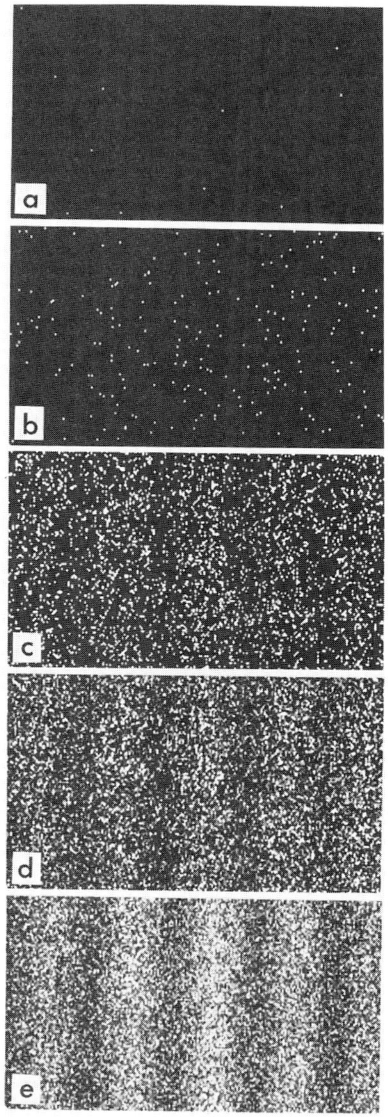

■ FIGURE 35.28
Electron interference. In this two-slit experiment, the electrons are separated by electric forces. Electrons leave their source at so slow a rate that two electrons are not present between the source and the detector at the same time. Thus each electron interferes with itself. Individual electrons striking the screen produce flashes of light. All the flashes, taken together, produce a two-source interference pattern. Most electrons strike the screen near the interference maxima. (a) 10 electrons have reached the screen; their positions appear completely random. (b) 200 electrons; an overall pattern is detectable if you know what to look for. (c) 3000 electrons; a striped pattern is easily visible in the distribution of impact points. The contrast between interference maxima and minima becomes more pronounced as the number of electrons increases. (d) 20 000 electrons. (e) 70 000 electrons.

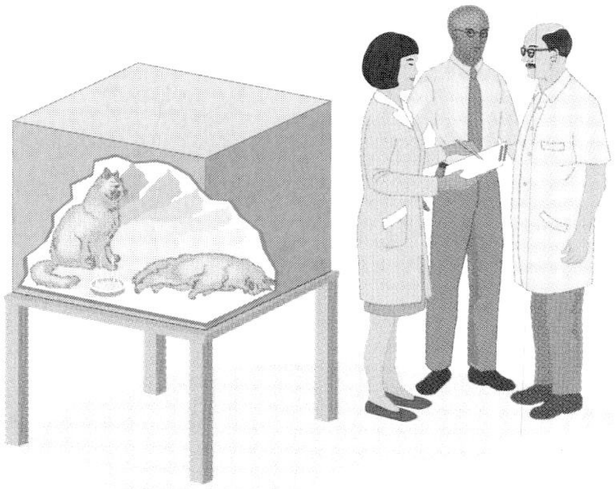

Schrödinger's cat. The cat has a 50% chance of surviving the experiment. According to quantum mechanics, the cage contains neither a live nor a dead cat, but an even mixture of live-cat and dead-cat states—whatever that means!

Erwin Schrödinger, 1887–1961

According to quantum mechanics, what is in the box just before you open it is neither a dead cat nor a live cat but a wave function describing equal probabilities for finding a live cat or a dead cat. When you open the box, the wave function collapses and you observe one or the other state. While you are pondering how strange this description is, your research supervisor approaches the lab room knowing that it contains a wave function describing equal probabilities of finding a researcher who will report a live cat and one who will report a dead cat . . .

Chapter Summary

Where Are We Now?

We have seen that a dual wave/particle model is needed to explain atomic phenomena. By applying the model in simple cases, we have been able to extract some fundamental results such as the uncertainty principle. In the next two chapters we shall use this model to discuss atomic nuclei and the many kinds of subatomic particle discovered in the quest for the ultimate building blocks of nature.

What Did We Do?

Light incident on certain metal surfaces ejects electrons with energy that depends on the light frequency. Einstein explained this *photoelectric effect* by assuming electromagnetic energy is absorbed as photons, particles whose energy is related to their wave frequency by:

$$E = hf.$$

When the energy of a photon is completely absorbed, an electron may be ejected with kinetic energy:

$$K = hf - \phi.$$

The work function ϕ is the energy required to remove an electron from the metal's surface. The probability that an electron is ejected is the quantum efficiency of the metal at that frequency.

The scattering of x rays by stationary electrons—the *Compton effect*—is best understood as elastic particle collisions between photons and free electrons. The momentum of a photon is:

$$p = E/c = hf/c = h/\lambda.$$

In collisions between photons and stationary electrons, the photon wavelength is increased as its energy is decreased:

$$\lambda_2 - \lambda_1 = \lambda_C(1 - \cos\theta),$$

where the Compton wavelength $\lambda_C = h/mc$. For an electron, $\lambda_C = 2.43 \times 10^{-3}$ nm.

✲ Classical electromagnetism and thermodynamics predict a thermal radiation spectrum that becomes ever more intense with increasing frequency. The photon picture of radiation leads to the correct spectrum, which decreases exponentially at high frequency.

Rutherford's α-particle-scattering experiment shows that the positive charge in an atom is concentrated in a nucleus vastly smaller than the atom as a whole. According to classical theory, such atoms should collapse as orbiting electrons radiate electromagnetic waves. In Bohr's atomic model, electrons orbit the nucleus and can only radiate energy in the form of photons as they make abrupt transitions between orbits. The frequency of the radiation is given by the Balmer formula:

$$f_{nn'} = f_o\left(\frac{1}{n^2} - \frac{1}{(n')^2}\right), \quad \text{where } f_o = 3.288 \times 10^{15}\ \text{Hz}.$$

The angular momentum of an electron in its orbit is quantized:

$$L = nh/2\pi \equiv n\hbar.$$

The correspondence principle states that classical and quantum predictions must be the same in situations involving many quanta (e.g., $n \gg 1$ in the expressions above).

Electrons, like photons, have wave as well as particle properties. The frequency and wavelength of a particle are related to its energy and momentum in the same way as for photons:

$$E = hf \qquad \text{and} \qquad p = h/\lambda.$$

An integral number of electron wavelengths fit into a Bohr orbit, indicating that atomic states result from standing-wave patterns. A proper three-dimensional calculation of standing-wave patterns in hydrogen follows from Schrödinger's theory of wave mechanics. The sequential filling of these standing-wave states, which occurs as the number of electrons in an atom is increased, explains the observed regularities in the chemical behavior of the elements.

Wave/particle duality places fundamental limits on the accuracy of measurement. According to Heisenberg's uncertainty principle, the product of uncertainties in position and momentum measurements or of energy and time measurements has a minimum value:

$$\delta x\, \delta p_x \geq \tfrac{1}{2}\hbar \qquad \text{and} \qquad \delta E\, \delta t \geq \tfrac{1}{2}\hbar.$$

Atomic phenomena appear to be inherently statistical; current theory is limited to predicting the probability of future events. Schrödinger's wave function represents the probability of an event occurring. The wave function collapses abruptly when the event occurs. This description is a subject of lively philosophical debate.

Practical Applications

Photoelectric phenomena are used to detect light in applications ranging from TV cameras to astronomical research. Wave mechanics of electrons is essential for the design of microwave generators and any of the semiconductors used in electronics. Chemical research relies on quantum theory to understand chemical bonding and to predict the molecular structure and properties of new chemicals.

Solutions to Exercises

35.1 The rate electrons (charge e) are emitted per unit area equals the quantum efficiency e_q multiplied by the rate photons strike the surface (intensity I divided by the energy per photon). We take the quantum efficiency for gold at 100 nm from Figure 35.5. The photocurrent per unit area j is:

$$j = e\frac{I}{hf}e_q = e\frac{I\lambda}{hc}e_q$$

$$= \frac{(1.6 \times 10^{-19}\ \text{C})(1.0\ \text{W/m}^2)(100 \times 10^{-9}\ \text{m})}{(6.6 \times 10^{-34}\ \text{J}\cdot\text{s})(3.0 \times 10^{8}\ \text{m/s})}(0.06)$$

$$= 5 \times 10^{-3}\ \text{A/m}^2.$$

35.2 With $\theta = 90°$, $\cos\theta = 0$, and eqn. (35.6) gives:

$$\frac{\lambda_2 - \lambda_1}{\lambda_1} = \frac{E_1}{mc^2} = \frac{1.0\ \text{MeV}}{0.511\ \text{MeV}} = 2.0.$$

Using $\lambda = hc/E$, we find:

$$\frac{(1/E_2) - (1/E_1)}{1/E_1} = \frac{E_1 - E_2}{E_2}$$

$$= 2.0$$

Thus $$E_1 = 3.0E_2$$

and so $$(E_1 - E_2)/E_1 = 2/3.$$

35.3 This is a conservation law problem. At the minimum distance d, the gold nucleus and the α-particle are at rest with respect to each other. Since the gold nucleus is much more massive than the α, the two are nearly stationary in the lab frame as well, and we may ignore their kinetic energy. Then, their electrical potential energy equals the initial kinetic energy K_α of the α:

$$kQ_\alpha Q_{\text{Au}}/d = K_\alpha.$$

Then:

$$d = \frac{kQ_\alpha Q_{Au}}{K_\alpha} = \frac{(9 \times 10^9\ \text{N}\cdot\text{m}^2/\text{C}^2)(2)(79)(1.6 \times 10^{-19}\ \text{C})^2}{(1 \times 10^6\ \text{eV})(1.6 \times 10^{-19}\ \text{J/eV})}$$

$$= 2 \times 10^{-13}\ \text{m}.$$

The directly aimed α-particle penetrates to about 0.002 of the atomic diameter, which is still approximately 50 nuclear diameters from the nucleus. Subsequently, the α-particle reaccelerates back along its original path and appears as a particle deflected by 180°.

35.4 Comparing Balmer's empirical formula with Bohr's theoretical result gives:

$$f_o = \frac{E_o}{h} = \frac{2.180 \times 10^{-18}\ \text{J}}{6.626 \times 10^{-34}\ \text{J}\cdot\text{s}} = 3.290 \times 10^{15}\ \text{Hz},$$

as expected. (The difference in the fourth significant figure is due to the assumption that the proton remains stationary.)

35.5 The quantity $ke^2/\hbar$ that appears in the expression for E_o equals αc. Thus:

$$E_o = \frac{m(ke^2)^2}{2\hbar^2} = \frac{m}{2}(\alpha c)^2 = \tfrac{1}{2}\alpha^2 mc^2.$$

Similarly: $$v_1 = \frac{\hbar}{mr_1} = \frac{\hbar}{m}\left(\frac{mke^2}{\hbar^2}\right) = \frac{ke^2}{\hbar} = \alpha c.$$

35.6 Equating electric force Zke^2/r^2 to mass times acceleration and using quantization of angular momentum gives:

$$Zke^2 = mv_n^2 r_n = (n\hbar)^2/(mr_n).$$

Then,

$$r_n = \frac{n^2}{Z}\left(\frac{\hbar^2}{mke^2}\right) = \frac{n^2}{Z}a_o,$$

as required. The kinetic, potential, and total energies are:

$$K_n = \tfrac{1}{2}mv_n^2 = \tfrac{1}{2}kZe^2/r_n, \quad U_n = -kZe^2/r_n,$$

and $$E_n \equiv K_n + U_n = -\tfrac{1}{2}kZe^2/r_n.$$

Using the result above for r_n, we obtain:

$$E_n = -\frac{m(ke^2)^2}{2\hbar^2}\left(\frac{Z^2}{n^2}\right) = -E_o\frac{Z^2}{n^2}, \quad \text{as required.}$$

35.7 Since $Z = 2$ for helium, the energy of the electron in the ground state of singly ionized helium is $Z^2 = 4$ times the corresponding energy in hydrogen. Thus the second ionization potential of helium is:

$$4hf_o = 4(13.6\ \text{eV}) = 54.4\ \text{eV}.$$

35.8 $v_d = \lambda f = (h/p)(E/h) = E/p = \gamma mc^2/(\gamma mv) = c^2/v$.

35.9 Heinsenberg's uncertainty principle (eqn. 35.20) gives the energy spread directly:

$$\delta E = \frac{\hbar}{2\,\delta t} = \frac{1.05 \times 10^{-34}\ \text{J}\cdot\text{s}}{2(1.6 \times 10^{-19}\ \text{J/eV})(10^{-7}\ \text{s})} = 3 \times 10^{-9}\ \text{eV},$$

and $$\frac{\delta E}{E} = \frac{3 \times 10^{-9}\ \text{eV}}{14.4 \times 10^3\ \text{eV}} = 2 \times 10^{-13}.$$

The line width is more than 100 times the expected shift.

Basic Skills

Review Questions

§35.1 PHOTONS

- Describe the *photoelectric effect,* and state five experimental facts about it. Which fact is most devastating for classical theory?
- How is the energy of a light quantum related to its frequency?
- What is the *work function* of a metal?
- Define *quantum efficiency.* Is it a constant for a given metal?
- What is *stopping potential?*
- Describe the *Compton effect.* How does it differ from the classical description of light scattered from electrons?
- Which length determines the wavelength change for scattered light at any specific angle of scattering?
- * What explains the exponential decrease toward high frequency in Planck's radiation law?

§35.2 BOHR'S ATOMIC MODEL

- What experimental evidence suggests that atomic nuclei are very small?
- How did Balmer describe the frequencies of light emitted by a hydrogen atom?
- What hypotheses did Bohr make in his model of the hydrogen atom?
- Roughly how large is the first Bohr orbit for hydrogen?
- How much larger is the second Bohr orbit as compared with the first?
- What is an *energy-level diagram?* Sketch the diagram for hydrogen.
- What is the *ionization potential* of an atom?
- What is the *correspondence principle?*

§35.3 ELECTRON WAVES

- How does the idea of electron waves help clarify the Bohr atomic model?
- What physical property of a particle determines its de Broglie wavelength?
- Bohr's n is the *principal* quantum number in Schrödinger's model of the atom. What are the others? What properties do they represent? How are they related to fine structure and hyperfine structure in the energy-level diagram?
- What is the Pauli *exclusion principle?* Explain how it determines the structure of the periodic table.
- What is a *closed shell?* What is its significance for an atom's chemical properties? Describe the chemical properties of atoms with nearly closed shells.

§35.4 QUANTUM MECHANICS

- State two forms of the Heisenberg uncertainty principle.
- What is the standard interpretation of a particle's *wave function?*

Basic Skill Drill

§35.1 PHOTONS

1. What are the wavelength and frequency of a 1.0-keV photon?
2. An FM station broadcasts a power of 250 kW at 102.2 MHz. At what rate does the station emit photons?
3. What is the energy of electrons emitted from a silver surface when UV photons with an energy of 15.0 eV shine on the surface?
4. A photon of wavelength 0.135 nm scatters from an electron and returns along its original path. What is the wavelength of the scattered photon?

§35.2 BOHR'S ATOMIC MODEL

5. What wavelength photon is emitted when a hydrogen atom undergoes a transition from $n' = 5$ to $n = 2$?
6. Compare the radius of the second Bohr orbit to the wavelength of a photon emitted in an $n' = 2$ to $n = 1$ transition.
7. Find the energy required to remove the eighth electron from an oxygen atom, leaving a totally ionized atom.

§35.3 ELECTRON WAVES

8. At what speed would an electron have a de Broglie wavelength of 1.0 m?
9. How many different angular momentum states are there in the third shell ($n = 3$)? What is the maximum number of electrons an atom can have in the third shell?
10. Would you expect oxygen and fluorine to form a stable chemical compound? Why or why not? What about sodium and fluorine?

§35.4 QUANTUM MECHANICS

11. A certain apparatus can measure the momentum of an electron to 1 part in 10^3. If the speed of the electron is 1.5×10^8 m/s, what is the minimum uncertainty in its position after the measurement?
12. An electron is known to be between $x = 0$ and $x = 1.5$ nm. What is the minimum uncertainty in the x-component of its velocity?
13. An electron remains in the $n = 4$ state of a nitrogen atom for 6×10^4 s, on average, before it transitions to $n = 2$. What is the uncertainty in the energy of the photon emitted? (Lines with such long lifetimes are observed in space but not on Earth.)

Questions and Problems

§35.1 PHOTONS

14. ♦ Compare the momentum of a 10-keV photon with that of an electron that has 10 keV of kinetic energy.
15. ♦ A certain organic molecule dissociates into two smaller molecules when it absorbs 7.4 eV. What is the maximum-wavelength photon that can dissociate the molecule?
16. ♦ What is the binding energy of a molecule that can be dissociated by photons with wavelengths less than 555 nm?
17. ♦ What is the energy of a photon whose wavelength is 425 nm?
18. ♦ Show that the Compton wavelength of the electron equals the wavelength of a photon with energy $E = mc^2$.
19. ♦ What is the energy of photoelectrons emitted from a sodium surface when the incident light has wavelength $\lambda = 415$ nm?
20. ♦ What is the longest wavelength of light that can eject photoelectrons from a platinum surface?
21. ♦ The Rayleigh–Jeans law is valid when $hf < kT$. Determine whether it is adequate to describe the following situations: **(a)** Cosmic radio waves with temperature $T = 2.7$ K at wavelength $\lambda = 3$ cm. **(b)** The flux of "green" light ($\lambda = 550$ nm) from the Sun ($T \approx 6000$ K). **(c)** Infrared radiation from the Sun at wavelength $\lambda = 10\ \mu$m.
22. ♦ Light of wavelength 550 nm has intensity $|\vec{\mathbf{S}}| = 1.5$ W/m^2. What is the area of a surface that receives 1 photon/s, on average?
23. ♦♦ A photon with wavelength $\lambda = 4.00$ nm is Compton scattered through a 90° angle. What is the fractional change in wavelength?
24. ♦♦ A radio station broadcasts 150 kW at a frequency of 95.0 MHz. Assuming the station radiates uniformly in all directions, how many photons per unit area and per unit time strike a surface 1.5 km from the station? A detector with area 10 cm^2 and 1.5 km from the station can count individual photons at rates up to 10^{12} /s. Can it distinguish whether the station emits photons rather than a continuous wave?
25. ♦♦ One hundred days after light from the explosion first reached Earth, Supernova 1987a in the Large Magellanic Cloud was observed to emit visible radiation at a rate of 10^{35} W. How many photons per second would be collected by a telescope 5 m in diameter? (The Large Magellanic Cloud is 1.7×10^5 light-years from Earth.)
26. ♦♦ The intensity of light from a barely visible object is 1.5×10^{-11} W/m^2. At what minimum rate must photons enter the eye for an object to be visible? Take the diameter of the pupil to be 0.70 cm and the photon wavelength to be 560 nm.
27. ♦♦ A celestial x-ray source emits 10^{26} W of 6-keV x rays. If the source is 3×10^{19} m away, at what rate do photons arrive at a 900-cm^2 detector in low Earth orbit?
28. ♦♦ A 1.3-Mev x ray makes a head-on collision with a stationary electron and is scattered through an angle of 180°. What are the final velocity of the electron and the frequency of the scattered photon?
29. ♦♦ What is the final speed of an electron that has just scattered a 10.0-keV photon through an angle of 45°? A 1.0-MeV photon?
30. ♦♦ An x ray of wavelength 3.00 nm is Compton scattered through an angle of 60.0°. What is the recoil energy of the electron?
31. ♦♦ In a Compton-scattering experiment, both the scattered photon and the recoil electron are detected. If the electron has 61 keV of kinetic energy and the scattered photon has an energy of 150 keV, what is the initial wavelength of the x rays?
32. ♦♦ What is the maximum energy that can be transferred to electrons by Compton scattering of x rays with an initial wavelength of 5.0 nm?
33. ♦♦ The maximum kinetic energy transferred to electrons as they scatter x rays is observed to be 6.0 keV. What is the wavelength of the incident x rays?
34. ♦♦ What is the maximum energy a photon of UV light ($\lambda = 400$ nm) can transfer to an electron in a Compton collision? Could such a collision eject the electron from a metal with a work function of 1.0 eV? (*Note:* Your result proves that Compton collisions are *not* the mechanism of the photoelectric effect.)

35. ♦♦ Photoelectrons ejected from a surface by light of wavelength 4.0×10^2 nm are stopped by a potential difference of 0.80 V. What potential difference is required to stop electrons ejected by light of wavelength 3.0×10^2 nm?

36. ♦♦ A telescope on a satellite receives light over a 0.78-m^2 area and focuses it onto a gold photodiode. The system is used to observe a star's ultraviolet light. When light with wavelength in the range $\lambda =$ 85 nm–95 nm falls on the detector, 7 electrons per second are liberated. What is the observed intensity of the star's radiation in this wavelength range?

37. ♦♦ You want to build a photodetector for blue light ($\sim$ 420 nm). Which of the materials in Table 35.1 would make a suitable detector? Justify your choice.

38. ✲ ♦♦ In a cavity with volume 1.0 m^3 at temperature 300 K, find the number of photons in the frequency ranges $f =$ (3.00 to 3.03) $\times$ 10^{13} Hz, $f =$ (2.00 to 2.02) $\times$ 10^{14} Hz, and $f =$ (3.00 to 3.03) $\times$ 10^{14} Hz. What does a photon number less than unity mean?

39. ♦♦ Both the emitting and collecting surfaces of a photocell (Figure 35.6) are made of cerium. What electrical potential difference V_s (stopping potential) is required to eliminate the flow of photoelectrons between the surfaces if the incident light has wavelength $\lambda =$ 320 nm? What is the change in stopping potential if the wavelength is decreased to 300 nm?

40. ♦♦♦ Show that the directions of a Compton-scattered photon and the recoil electron are related by $\cot \phi = (1 + hf/mc^2)\tan(\theta/2)$.

41. ♦♦♦ Use conservation of energy and momentum to show that a free electron cannot completely absorb a photon. How can an electron in metal absorb a photon in the photoelectric effect?

42. ♦♦♦ Energy transfers to or from an oscillating system are quantized according to the same rule as photons: $\Delta E = nhf$, where n is an integer. What is the smallest energy that can be transferred to an object of mass 5 g suspended by a spring with a spring constant of 20 N/m? How many energy quanta are needed for this system to oscillate with an amplitude of 1.0 cm?

§35.2 BOHR'S ATOMIC MODEL

43. ❖ On the basis of Bohr's theory, which has more lines in the optical portion of its spectrum, hydrogen or doubly ionized lithium?

44. ❖ Museums often exhibit mineral samples that are drab when seen by ordinary room lighting but glow with brilliant colors when bathed in ultraviolet (*black*) light. This phenomenon is called fluorescence. The electrical discharge in a fluorescent light produces ultraviolet radiation, which causes a coating material on the inside surface of the tube to fluoresce. Use the concept of atomic energy levels to explain how fluorescence works.

45. ❖ Chemical isotopes are atoms with the same nuclear charge but different mass. For example, deuterium (*heavy hydrogen*) has an extra neutron, giving its nucleus a mass about twice that of a proton. Do all isotopes of a given element have the same spectrum; that is, are their energy levels the same? Why or why not?

46. ♦ Use Balmer's formula to determine the wavelengths of visible lines in the hydrogen spectrum.

47. ♦ What transition in hydrogen produces higher-frequency photons, $n' = 5$ to $n = 3$, or $n' = 20$ to $n = 4$?

48. ♦ For what initial Bohr states is it possible for a 950-nm photon to ionize hydrogen?

49. ♦ What is the binding energy of an electron in the $n = 3$ state of hydrogen?

50. ♦♦ Find the short-wavelength limits ($n' = \infty$) of the Lyman, Balmer, Paschen, and Brackett series for hydrogen. In what part of the spectrum does each limit lie?

51. ♦♦ Is it possible for the short-wavelength limit of a spectral series with $n = N + 1$ to have a shorter wavelength than the α line ($n' = N + 1$ to $n = N$) of the spectral series with $n = N$? If so, for which series is it possible?

52. ♦♦ Through what minimum potential difference must an electron be accelerated so that it can transfer energy to a hydrogen atom in its ground state via an inelastic collision?

53. ♦♦ An ultraviolet photon of wavelength 85 nm is absorbed by a hydrogen atom. What is the kinetic energy of the emerging electron?

54. ♦♦ A photon emitted when an electron makes a transition from the $n' = 2$ state to $n = 1$ state in hydrogenic helium is later absorbed by a hydrogen atom in its ground state. What is the kinetic energy of the resulting electron?

55. ♦♦ An electron typically remains in the $n = 2$ state of hydrogen for 10^{-8} s. How many revolutions does a typical electron make according to the Bohr theory before a transition occurs?

56. ♦♦ According to classical physics, an electron in circular motion at frequency f radiates EM waves at frequencies f, $2f$, $3f$, and so on. Consider the allowed transitions between energy levels at high n (cf. the calculation in §35.2.3) to show that Bohr's model makes this same prediction in the limit of large n.

57. ♦♦ Compute the first four energy levels for hydrogenic oxygen. Determine the energy and wavelength of spectral lines corresponding to $n' = 2$, $n = 1$ and $n' = 3$, $n = 2$. In which part of the spectrum do these lines lie?

58. ♦♦ Find the angular momentum quantum number for the orbit of the Moon around the Earth. If the angular momentum of the Moon were quantized, could you tell? Why or why not? What about a bicycle wheel?

59. ♦♦ What is the radius of the electron in the $n = 350$ orbit of hydrogen? Compare the atom's diameter with the average separation between atoms in a gas with $p = 1$ atm, $T = 300$ K and in the interstellar medium, where the density is about 10^3 m^{-3} and $T = 100$ K.

60. ♦♦♦ Show that the frequency of the photon emitted in a transition from the nth to the $(n - 1)$th Bohr orbit lies between the two revolution frequencies of the electron in those two orbits.

61. ♦♦♦ The classical formula for electromagnetic power radiated by an electron is $P = e^2a^2/(6\pi\epsilon_0 c^3)$, where a is the magnitude of the electron's acceleration. Evaluate the classical power radiated by an electron in the nth Bohr orbit, and use the result to estimate the time for transition to the $(n - 1)$th orbit. For what values of n do you expect your value to be correct? How does your time for $n = 2$ compare with the observed time of about 10^{-8} s?

62. ♦♦♦ Neutral hydrogen atoms in a gas discharge tube, originally in the ground state, are excited by collisions with free electrons with 12.2 eV of kinetic energy. What possible wavelengths can subsequently be emitted by the hydrogen atoms?

§35.3 ELECTRON WAVES

63. ❖ A particle is confined between two rigid plane surfaces, one at $x = 0$ and one at $x = L$. Schrödinger's wave function ψ for the lowest energy state of the particle is proportional to $\sin(\pi x/L)$. If you observe at a random time, where is the particle most likely to be found?—Least likely to be found? Compare with the corresponding predictions for a classical particle bouncing (elastically) back and forth between the surfaces. Higher energy states have wave functions $\psi \propto \sin(n\pi x/L)$. For large n, describe where the particle is most likely to be found. Do the quantum and classical descriptions correspond in the limit of very large n? Why or why not?

64. ❖ Can an electron and a photon have both the same wavelength and the same frequency? Why or why not?

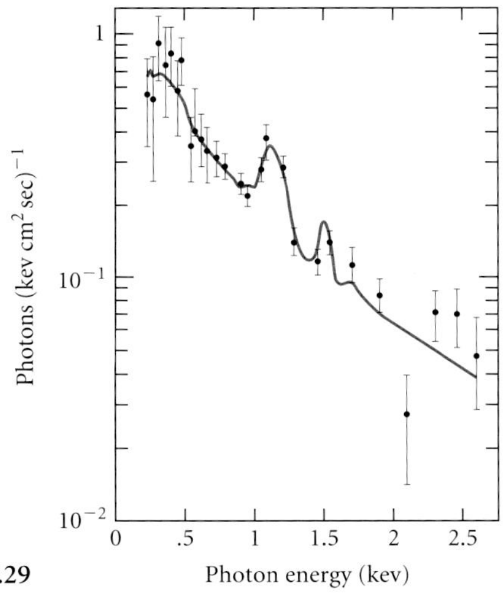

■ FIGURE 35.29

65. ❖ An absorption line at 6.4 keV is observed in the spectra of galactic x-ray sources. It is identified as the transition from $n = 1$ to $n' = 2$ in iron. Why does the value not agree with that from the Bohr formula (Example 35.10)? What is the effective charge internal to the transitioning electron?
66. ♦ Find the de Broglie wavelength of a 0.22 rifle bullet with a mass of 0.010 kg and a speed of 150 m/s.
67. ♦ Compare the de Broglie wavelengths of an electron, a proton, and a 3.0-g pea, each with speed 100 km/s.
68. ♦♦ Find the de Broglie wavelength of a cosmic ray proton with a kinetic energy of 3.0 GeV.
69. ♦♦ Compute the first four Bohr energy levels for hydrogenic iron (i.e., an iron nucleus with one electron around it). A spectral line at 1.15 keV is observed in the spectrum of the Virgo cluster of galaxies (■ Figure 35.29). Can you identify which transition this is? What is the effective charge on the iron nucleus?
70. ♦♦ What accelerating voltage is necessary to produce an electron beam with a de Broglie wavelength equal to the 0.15-nm spacing between atoms in a crystal target?
71. ♦♦ A radium atom ejects an α-particle (helium nucleus) with 5.87 MeV of kinetic energy. Compare the de Broglie wavelength of the α-particle with the radius of the radium nucleus (2×10^{-14} m).
72. ♦♦ What energy electrons are required in an electron microscope with a 1-nm resolution? (Assume the electron wavelength should be less than one-tenth the desired linear resolution.)
73. ♦♦ The diameter of the ring of iron atoms shown on the chapter title page is 144 nm. Estimate the wavelength of the electron wave and hence the electron's kinetic energy.
74. ♦♦ What is the de Broglie wavelength of an electron in the $n = 2$ orbit of helium? Compare with the orbital radius.
75. ♦♦ At what energy does an electron's de Broglie wavelength equal its Compton wavelength? (*Hint:* The electron is relativistic.)
76. ♦♦♦ An electron is confined between two infinite walls at $x = 0$ and $x = 1.0\ \mu$m, as in Problem 63. What is the de Broglie wavelength of the electron in its lowest energy state? What is the electron's energy? How many electrons can have this energy? If there are six electrons, what is the total energy of the system in its lowest energy state?

§35.4 QUANTUM MECHANICS

77. ❖ Electrons in highly excited states of a hydrogen atom ($n \gg 1$) can remain in those states for rather long times (a millisecond or longer). Use the uncertainty principle to explain why.
78. ♦ If the speed of an electron in the x-direction is measured with an accuracy of 10^{-6} m/s, to what accuracy can its position along the x-axis be located?
79. ♦♦ Transitions from a certain atomic energy level to the ground state occur on average in 10^{-8} s. If the energy released in the transition is 2.48 eV, what is the minimum uncertainty in the wavelength of the resulting photon?
80. ♦♦ An electron is confined between two plane surfaces a distance L apart. From Heisenberg's uncertainty principle, estimate the minimum kinetic energy the electron can have. Evaluate the result when L is the size of an atom, about 10^{-10} m, and when L is the size of an atomic nucleus, about 10^{-15} m.
81. ♦♦ Estimate from Heisenberg's uncertainty principle the minimum momentum of an electron confined in a region equal in size to the diameter of the first Bohr orbit. Compare the result with the momentum deduced from Bohr's theory.
82. ♦♦ An electron is confined in a cubical box with 1.0-mm sides. What is the uncertainty in the electron's energy?

Additional Problems

83. ♦♦ What error is made in computing the de Broglie wavelength of a 35-keV electron using the Newtonian formula for its momentum? Of a 1-MeV electron?
84. ♦♦ Show that the ratio of a particle's Compton wavelength to its de Broglie wavelength is $\lambda_C/\lambda_d = \sqrt{\gamma^2 - 1}$.
85. ♦♦ Calculate the de Broglie wavelength λ_d of the recoiling electron in Compton scattering with $\theta = 90°$. Examine the three cases: **(a)** $\lambda_i \ll \lambda_C$, **(b)** $\lambda_i \approx \lambda_C$, and **(c)** $\lambda_i \gg \lambda_C$. In each case, show that $\lambda_d \approx \lambda_i$.
86. ♦♦ When a hydrogen atom emits a photon, the atom recoils to conserve momentum. Show that this effect causes a change in wavelength of an emitted photon that is essentially independent of photon energy. Compute the value of the wavelength change.
87. ♦♦ Show that the classical electron radius (Example 27.12), the Compton wavelength, and the radius of the first Bohr orbit are approximately in the ratio $\alpha^2:\alpha:1$.
88. ♦♦♦ A photoemitting metal surface with work function $\phi =$ 2.30 eV is in a region of uniform magnetic field strength $B = 0.635$ T. Find the maximum radius of the paths traveled by photoelectrons as a function of the wavelength of incident light.
89. ♦♦♦ The first step in forming an image on an ordinary photographic plate is the dissociation of silver bromide (AgBr) molecules by incident light. The dissociation energy of AgBr is measured to be 1.00×10^5 J/mol. Find the energy in electron volts required to dissociate a single molecule of AgBr and estimate the maximum wavelength to which the photographic plate is sensitive.
90. ♦♦♦ A silver sphere (work function 4.6 eV) is suspended in a vacuum chamber by an insulating thread. Ultraviolet light of wavelength 0.20 μm shines on the sphere. What is the resulting electric potential of the sphere?
91. ♦♦♦ In the Bohr model for hydrogen, we assumed that the proton remains at rest. In a better model, both electron and proton orbit about their common center of mass. Taking this into account, assume that the total angular momentum of the system is quantized and calculate E_o. Show that the result is given by eqn. (35.15) if we replace m with the *reduced mass* $\mu = m_e m_p/(m_e + m_p)$. How large is the correction to the wavelength of Balmer α?
92. ♦♦♦ Use the result of Problem 91 to calculate the frequency difference between Balmer α lines produced by hydrogen and deute-

rium. What resolution is necessary for a spectrograph to resolve lines from the two isotopes? (Compare §17.4.4 and Problem 17.61.)

93. ♦♦♦ Calculate a Bohr model for positronium—an electron in orbit about a positron—assuming that the total angular momentum of the system is quantized. Be careful to include the motion of both particles (cf. Problem 91). What is the separation of the particles in the ground state? What is the frequency of light emitted in a transition from $n' = 2$ to $n = 1$?

94. ♦♦♦ A beam of electrons with an energy of 5 keV passes through a powder of silver crystals. If the interatomic spacing of silver atoms in a crystal is 0.408 nm, find the angular radius of the first-order diffraction ring from the principal Bragg planes (cf. §17.5).

Computer Problem

95. In 1916, Millikan published measurements of the potential difference required to suppress the photocurrent when a freshly cut sodium surface is illuminated by each of six different spectral lines from a mercury vapor lamp. (See file CH35P95 on the supplementary computer disk.) Graph these data and determine Planck's constant h from the slope of the graph.

Challenge Problems

96. A sample of hydrogen gas with density $\rho = 0.065\ \text{kg/m}^3$ is heated by an electric discharge to a temperature of several thousand kelvin. Explain why only a few Balmer lines are visible in the light emission from the sample, even though from the temperature alone one would expect a sufficient fraction of atoms to be in levels with very high n. Use Bohr's theory to estimate the number of lines that can be observed. (*Hint:* How big can an atom get without interfering with nearby atoms?)

97. Compute the ratio of magnetic moment to angular momentum for a ring of mass M and charge Q rotating about an axis perpendicular to the ring. Assuming this same ratio holds for electron orbital angular momentum and spin, find the orbital and spin magnetic moments of an electron in the $n = 1$ Bohr orbit of hydrogen. (This assumption makes an error of about a factor of 2 in the spin magnetic moment.) To estimate the energy of interaction between the electron's orbital and spin magnetic moments (spin-orbit coupling), treat the orbital magnetic moment as at the center of the atom and parallel to the spin magnetic moment, which is located at the electron. Estimate the magnitude of the hyperfine energy assuming the same ratio to find the nuclear magnetic moment from its spin of $\frac{1}{2}\hbar$.

98. Draw a schematic graph of the potential energy of electrons versus position in the circuit shown in Figure 35.6 if emitter and collector are made of different metals. From your graph, argue that the stopping potential is given by:

$$V = hf/e - \phi_c,$$

where ϕ_c is the work function of the collector.

99. Model a particle as the superposition of two waves with wavelength λ_1 and λ_2. Show that the superposition moves with speed:

$$v_s = \frac{f_1 - f_2}{(1/\lambda_1) - (1/\lambda_2)}.$$

Express v_1 and v_2 as $v \pm \Delta v/2$ and show that $v_s = v$, ignoring terms of order $(\Delta v/c)^2$. (See Exercises 17.2 and 35.8.)

100. **(a)** A μ meson is an elementary particle with the same charge as an electron but with 207 times the mass. Its average lifetime before decay into lighter particles is 2.22 μs. Rework the Bohr theory for a μ meson in orbit about a nucleus of charge $+Ze$ to find the radii and energies of the stationary states of such a μ-mesic atom. What is the energy of a photon emitted in the $n' = 2$ to $n = 1$ state for a μ-mesic atom ($Z = 82$). **(b)** The measured value of the photon energy in an $n' = 2$ to $n = 1$ transition in μ-mesic lead is 6.0 MeV, which disagrees with the result of part (a). The radius of a lead nucleus is 7.1×10^{-15} m. Show that the first Bohr orbit lies inside the nucleus. Find a corrected radius and energy of the first Bohr orbit, taking this fact into account, and recompute the photon energy. (*Hint:* How do the electric force on the μ meson and its potential energy depend on distance from the center of the nucleus, when the meson is inside the nucleus?) How much better is this approximation?

ESSAY 9
The Scanning Tunneling Microscope

Shirley Chiang
Department of Physics
University of California at Davis

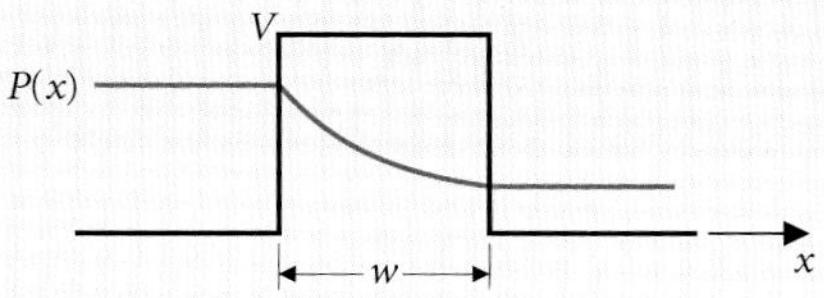

■ **Figure E9.1**
$P(x)$ indicates the probability of an electron incident on the square barrier with height V being at position x. Note that the probability does not include the reflected electron wave at the barrier but gives a qualitative idea of the exponential decrease of the probability with the width w of the barrier.

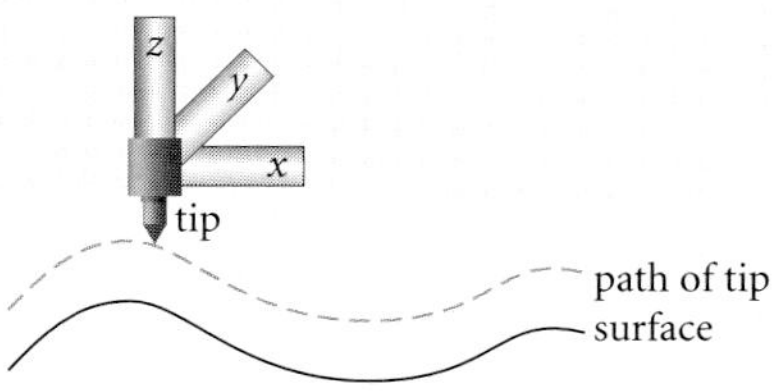

■ **Figure E9.2**
Schematic diagram showing operation of Scanning Tunneling Microscope. The tunneling tip is shown moving at a constant distance above the surface, as its height is regulated by a constant current feedback loop. x, y, and z indicate the piezoelectric elements which allow the movement of the tip in three dimensional space with respect to the sample.

The scanning tunneling microscope (STM) is a remarkable instrument which permits the measurement of the topography of a surface down to the atomic level. It was invented by Gerd Binnig and Heinrich Rohrer in 1982, and they shared half of the 1986 Nobel Prize in Physics for this achievement. The operation of this instrument is based on the phenomenon of quantum mechanical tunneling of electrons from a sharp tip to a solid surface through a vacuum gap when a voltage is applied between the tip and the surface. Classically, a potential barrier exists between the tip and the surface, and an electron whose energy is smaller than the barrier height could not exist within the barrier, nor would it have enough energy to surmount the barrier. But because electrons also have wave properties, for a barrier of finite width and height, a quantum mechanical calculation shows that the electron has a finite probability of being within the barrier, and a finite probability of "tunneling" through the barrier and appearing on the other side. (■ Figure E9.1). The probability that the electron can tunnel through the barrier increases as the barrier becomes narrower.

In the STM, the distance between the tip and sample is on the order of a few atomic diameters, and the quantum mechanical tunneling current I has the following dependence: $I \propto \exp(-Az)$, where A is a variable which can depend on the materials used for tip and sample, and z is the distance between the tip and sample. For a particular set of tip and sample materials, A can usually be considered to be constant. When z is in Ångstroms ($1\ \text{Å} = 10^{-8}$ cm, and a typical lattice spacing in a solid is $2.5 - 3$ Å), $I \propto 10^{-z}$, i.e. the current changes by one order of magnitude when the distance changes 1 Å. This tunneling current is measured and kept constant by a feedback circuit. The tip in the STM is moved over the surface in three dimensions (■ Figure E9.2) using piezoelectric crystals, which are materials which expand or contract by an amount proportional to the voltage which is applied to their electrodes. As the tip is scanned in a raster pattern in the x-y plane over the surface, the feedback circuit moves the tip to and from the surface so as to keep the current constant. Thus, the tip traces a path which is a constant distance away from the surface. A computer is normally used to make a plot of the voltages applied to the x, y, and z piezoelectric elements, which can be calibrated to yield a three-dimensional (3D) topographic map of the surface. Because of the exponential dependence of the current on the distance, only the atoms closest to the end of the tip contribute significantly to the tunneling current. When the tip is sufficiently sharp, perhaps with a single atom at its apex, one can obtain atomically resolved images of the surface for some materials.

The STM has been used to measure many types of materials, including metals, semiconductors, and molecules. ■ Figure E9.3 shows a 3D view of individual atoms of a gold (Au(111)) surface, which are arranged in a close-packed hexagonal array. For many metals, the STM can resolve individual atoms which are adjacent to one another. Many surfaces have reconstructed top layers, in which the equilibrium structure of the top layer of the surface is different from the bulk crystal below it. One such case is the top layer of a clean silicon (Si(111)) surface, in which the top layer of the surface has a rhombus-shaped unit cell which is 7 times as long along each axis as the unit cell in the rest of the sample. Therefore, this structure is called a 7×7 reconstruction. Although the structure of the surface layer is now understood, it is quite complicated, and ■ Figure E9.4 shows only the 12 Si "adatoms," which sit above the

■ FIGURE E9.3
Three-dimensional view of STM image of gold (Au(111)) surface, in which atoms are arranged in a close-packed hexagonal array.

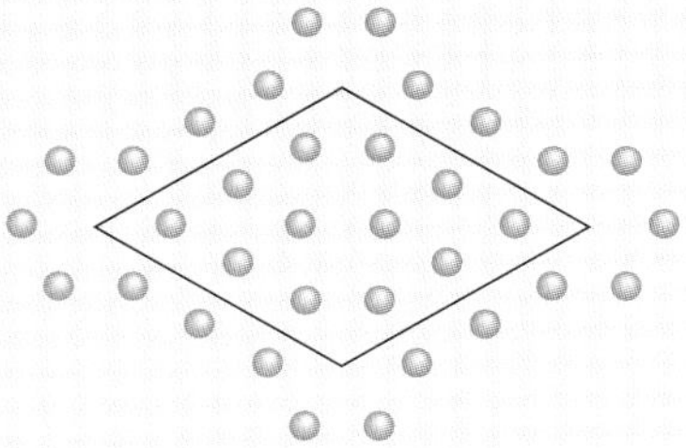

■ FIGURE E9.4
Schematic diagram showing the 12 Si "adatoms" that sit above the surface in the Si(111) 7 × 7 reconstruction. These are the atoms which are observed in the STM image in Figure E9.5. The surface unit cell is indicated by the rectangle. Note that the structure has 6 adatoms around holes in the corners of the unit cell.

surface, in each unit cell. As these atoms are actually higher than other structures in the surface layer, the STM images only these atoms. ■ Figure E9.5 shows a 3D view of a Si(111) 7 × 7 surface as measured in an STM image. Distinctive features in the structure are not only the 12 adatoms in a unit cell, but also the arrangement of 6 adatoms around what appear to be holes in the corners of the unit cell. This particular image also shows an atomic step on the surface, so that one sees the heights of two different layers on the surface simultaneously. Most STM images of semiconductors, however, cannot be interpreted so simply as pictures of individual atoms, as the STM often measures the electronic orbitals of atoms of the surface rather than the positions of individual atoms.

The STM is an instrument which has developed very quickly over the last decade and is now used actively in many fields of research, ranging from physics and chemistry to materials science and biology. It can be used in air, in liquids, and in vacuum environments. It has often been used to study metal and semiconductor substrates, some of which have atoms and molecules adsorbed on top. In addition, it has been used to study such widely varying samples as electrodes in electrochemical reactions, superconducting and layered samples, molecular and metallic thin films, as well as some biological materials.

THE PICTURE ON PAGE VI IN THE PROLOGUE SHOWS BENZENE MOLECULES.

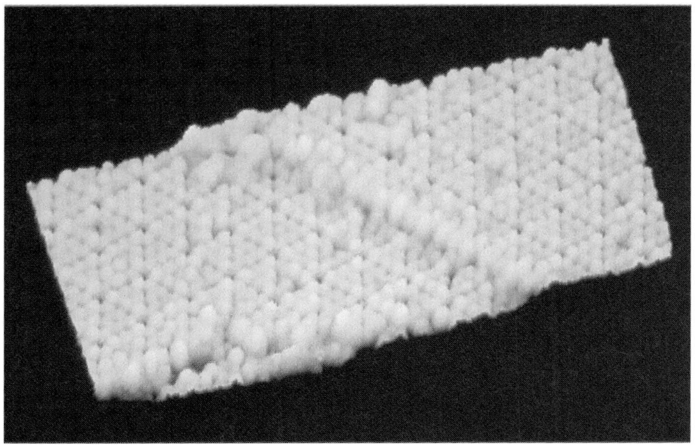

■ FIGURE E9.5
Three-dimensional view of an STM image of Si(111) 7 × 7, showing the 12 adatoms per unit cell which are sketched in Figure E9.4. This image also shows clearly an atomic step on the surface.

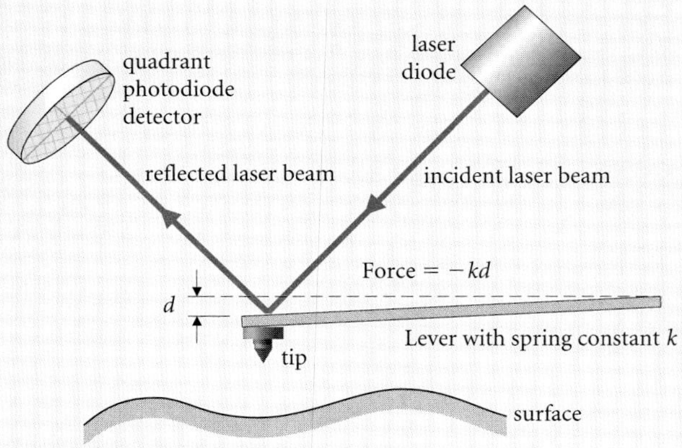

■ **Figure E9.6**
Schematic diagram showing operation of an atomic force microscope.

In addition, many other studies on non-conducting samples are now performed using an atomic force microscope (AFM), also called a scanning force microscope, a close cousin of the STM which was an invention of Gerd Binnig, together with Calvin Quate and Christoph Gerber in 1985. This instrument senses the amount of deflection of a small lever near the surface to measure the interaction force between a tip mounted on the end of the lever and the surface (see ■ Figure E9.6). The most common method of sensing this deflection is to reflect a laser diode beam from the end of the lever. The deflection of the beam is measured using a position sensitive detector—a photodiode divided into 2 or 4 parts so that signals can be compared among the parts. Such difference signals indicate the movement of the beam across the face of the detector and therefore the deflection of the lever. Using this technique, many types of forces have been measured, including atomic repulsion, Van der Waals, electric, magnetic, and frictional forces. Surface topography can often be measured in the contact mode, in which the tip is pushed directly against the surface, while the deflections of the lever are measured. In addition, images can be obtained at a constant force or a constant force derivative, permitting surface topographic imaging without actually having contact between the tip and surface. Such contact can damage soft samples. AFM images are less likely to have atomic resolution than STM images and are more dependent on the actual tip shape. The AFM has been used to measure samples ranging from magnetic and compact disks to soft biological materials. In addition to research applications, AFM's are now being used in development labs and for characterization of manufacturing problems.

This nuclear power plant is operated by Pacific Gas and Electric Company at Diablo Canyon, California. Unlike fossil fuels, nuclear fuel is sufficient to last for centuries. Yet the use of nuclear power has become a politically charged issue. The safety of such power plants has been questioned since the accident at Chernobyl, Ukraine, even though most plants used worldwide are of a safer design. Radioactive wastes must be safely stored for millennia.

CHAPTER 36
Atomic Nuclei

Concepts

Atomic number
Atomic mass number
Atomic mass unit
Strong nuclear force
Weak nuclear force
Isotope
Radioactive series
Antiparticles
Quantum numbers

Goals

Be able to:

Write reaction equations for α, β, and γ decays.

Use mass tables to determine whether a particular nucleus is unstable against α or β decay.

Relate observed radiation rates to the number of radioactive atoms in a sample.

A new thing had just been born; a new control; a new understanding of man, which man had acquired over nature.

I. I. Rabi

Nuclear power plants operating normally have excellent safety records and emit much less pollution (including radiation!) than coal-fired plants. Smaller reactors contribute to public health through life-saving medical therapies (■ Figure 36.1) and effective long-term sterilization of food products. But they also raise the specters of radioactive waste products that must be stored for centuries, reactor accidents such as occurred at Chernobyl, Ukraine, in 1986, causing worldwide nuclear pollution, and the diversion of plutonium for weapons use. Nuclear energy has become one of the major technical and political issues of our time. France, for example, has embarked on an ambitious nuclear power program, while in the United States, the nuclear power industry is in decline.

Both the benefits and risks of the nuclear age stem from fundamental research accomplished between 1890 and 1940. In this chapter we describe the model of nuclear structure that emerged from that research and discuss some applications of the model in technology, medicine, and astrophysics.

36.1 Basic Nuclear Structure

36.1.1 *Charge and Mass*

IF ATOMS WERE NOT *EXACTLY* NEUTRAL, BECAUSE OF A SMALL DIFFERENCE BETWEEN THE SIZES OF THE PROTON AND ELECTRON CHARGES, THE RESULTING ELECTRICAL FORCES WOULD CAUSE ALL PARTS OF THE UNIVERSE TO ACCELERATE AWAY FROM EACH OTHER. ASTRONOMICAL OBSERVATION SHOWS NO SUCH ACCELERATIONS, AND LIMITS ANY CHARGE IMBALANCE TO 1 PART IN 10^{30}.

An ordinary hydrogen atom has the simplest nucleus: a particle called a *proton* (■ Figure 36.2). Each proton has a positive electric charge equal in magnitude to the electron charge; its mass is about 2000 times larger than the electron mass:

$$q_p = +e = 1.60218 \times 10^{-19}\ \mathrm{C} \qquad \text{and} \qquad M_p = 1.67265 \times 10^{-27}\ \mathrm{kg} \approx 1836 m_e .$$

Because the proton and electron have opposite charges, a hydrogen atom is electrically neutral. All neutral atoms have a similar structure: the electric charge on any nucleus is a whole multiple Z of the proton charge ($q = +Ze$) and is exactly balanced by the charge on the surrounding electrons.

> The *atomic number Z* of an element equals the number of protons in an atomic nucleus of that element.

As we saw in Chapter 35, the chemical properties of an element are determined by the number of electrons surrounding each atom.

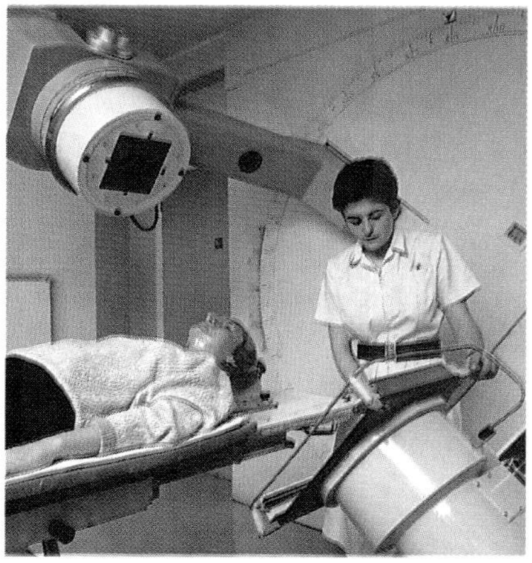

■ FIGURE 36.1
A patient being treated with a beam of radiation from radioactive cobalt nuclei prepared in a nuclear reactor. In modern medicine, radioactive nuclei are used in both diagnosis and treatment of disease.

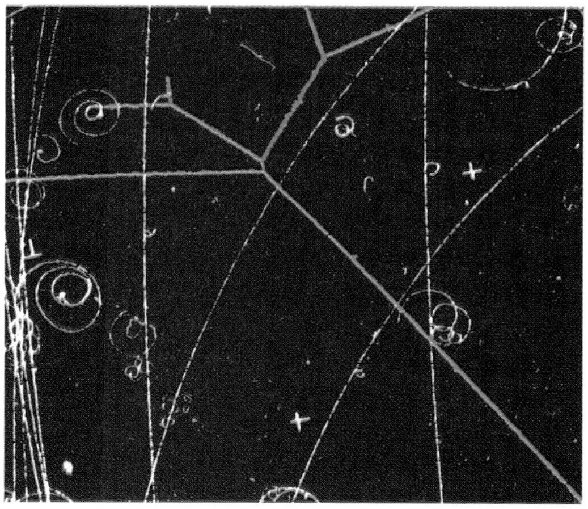

■ FIGURE 36.2
Track formed by a proton passing through a bubble chamber. Liquid hydrogen is maintained at its boiling point so that bubbles form around atoms ionized by the passing proton. The proton undergoes a collision with a stationary proton—the nucleus of a hydrogen atom in the liquid. The two protons subsequently undergo further collisions. Notice that the outgoing tracks are at 90° (cf. Example 10.3).

The *atomic number Z* of an element is also the number of electrons in a neutral atom of the substance.

The masses of atomic nuclei may be measured directly with a mass spectrometer (§29.1.2). Such measurements show that individual nuclear masses are nearly integral multiples of the proton mass:

$$M \approx AM_p.$$

The integer A is the *atomic mass number* of the nucleus. Since the chemical properties of an atom depend only on atomic number, nuclei with the same Z form atoms of the same element regardless of the value of A.

Nuclei with the same atomic number but different atomic mass form different *isotopes* of the same element.

A naturally occurring sample of an element is often a mixture of several isotopes. Atomic masses determined by chemical methods give the average atomic mass number for the mixture, which is usually not an integer. Chemical atomic masses are the values most often found in tables.

CHEMICAL MASSES ARE LISTED IN THE PERIODIC TABLE INSIDE THE BACK COVER.

Nuclei with particular values of A and Z belong to a *nuclear species*. The symbol for a species denotes the atomic number by the name of the corresponding chemical element. The atomic mass is written as a superscript preceding the name:

$$^{A}(\text{Name}).$$

Examples are ^{4}He for the helium ($Z = 2$) isotope with mass number $A = 4$, and ^{56}Fe for the iron ($Z = 26$) isotope with $A = 56$. The atomic mass number A and atomic number Z of a nucleus are generally different. The only exception is hydrogen, for which both equal unity. For light nuclei, A is roughly twice Z (■ Figure 36.3).

SOME AUTHORS PREFER TO INCLUDE THE ATOMIC NUMBER AS A SUBSCRIPT: $^{4}_{2}$He AND $^{56}_{26}$Fe.

The atomic mass number only gives an approximate value for the mass of a nucleus. Precise mass values are tabulated in terms of the atomic mass unit:

One *atomic mass unit* (u) is one-twelfth the mass of a neutral ^{12}C atom.

$$1\ \text{u} = 1.66054 \times 10^{-27}\ \text{kg}.$$

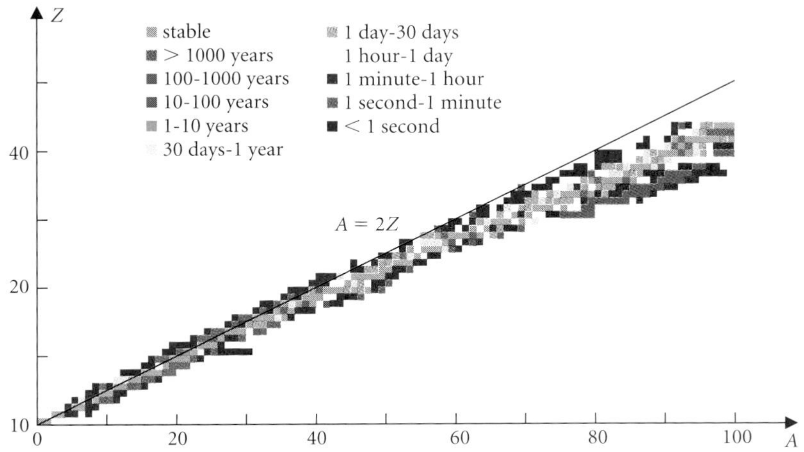

■ FIGURE 36.3
Light, stable nuclei occur along a line in a Z/A diagram with $A \sim 2Z$. As Z increases, A tends to be slightly larger than $2Z$. In this diagram the gray boxes indicate stable nuclei. The colors indicate the lifetimes of unstable nuclei.

TABLE 36.1 **Masses of Selected Atoms** *(in u)*

Z\A	2Z − 3	2Z − 2	2Z − 1	2Z	2Z + 1	2Z + 2	2Z + 3	2Z + 4	2Z + 5	2Z + 6	2Z + 7	2Z + 8
1 H			1.007825	2.01410	3.01605							
2 He			3.01603	4.00260	5.0122							
3 Li			5.0125	6.01512	7.016005							
4 Be		6.01973	7.01693	8.005305	9.01218							
5 B			9.01333	10.01294	11.00931	12.01435						
6 C				12.000000	13.00336	14.00324	15.01060					
7 N			13.00574	14.00307	15.00011	16.00610						
8 O	13.02481	14.00860	15.00307	15.99492	16.99913	17.99916						
9 F		16.01148			18.998405							
10 Ne				19.99244	20.99385	21.99139						
11 Na					22.98977							
12 Mg			22.99413	23.98504	24.98584	25.98259						
13 Al					26.98154	27.98191						
14 Si				27.97693	28.97650	29.97377						
15 P					30.97376							
16 S				31.97207	32.97146	33.96787		35.96708				
17 Cl					34.96885		36.96590					
18 Ar				35.96755	36.96677	37.96273		39.96238				
19 K					38.96371	39.96400	40.96183					
20 Ca				39.96259		41.95863	42.95878	43.95549		45.95369		47.95253
21 Sc							44.95592					
22 Ti						45.95263	46.95177	47.94795	48.94787	49.94478		
23 V								49.94716	50.94396			
24 Cr						49.94605		51.94051	52.94065	53.93888		
25 Mn									54.93805			
26 Fe						53.93961		55.93493	56.93639	57.93328		
27 Co									58.93319	59.93388		
28 Ni						57.93534		59.93078	60.93105	61.92834		63.92796
29 Cu									62.92959		64.92779	

Masses are often given in energy units using Einstein's relation; $m = E/c^2$:

$$1 \text{ u} = 931.49 \text{ MeV}/c^2.$$

Tabulated masses are those of neutral atoms. To obtain the mass of an isolated nucleus, you should subtract the mass of Z electrons. The binding energy of the electrons to the atom also affects the mass of the atom but is typically a few keV/c^2 or less and is negligible in nearly all practical calculations. We list the masses of some selected nuclei in ● Table 36.1.

EXAMPLE 36.1 ♦♦ Assuming natural chlorine (chemical atomic mass 35.453) consists entirely of the isotopes ^{35}Cl and ^{37}Cl, estimate the proportion of ^{35}Cl in the natural mixture.

MODEL The two isotopes contribute to the measured chemical mass in direct proportion to their fraction in the natural mixture.

SETUP Letting f be the fraction of ^{35}Cl,

$$(\text{Chemical atomic mass}) = f(\text{mass of } ^{35}\text{Cl}) + (1 - f)(\text{mass of } ^{37}\text{Cl}).$$

Table 36.1 (*Continued*)

Z		A	mass	A	mass	A	mass	A	mass	A	mass	A	mass	A	mass	A	mass
30	Zn	64	**63.92914**	66	**65.92604**	67	**66.927132**	68	**67.924848**	70	**69.925325**						
35	Br	79	**78.918332**	81	**80.91629**	87	86.9199										
36	Kr	78	**77.92040**	80	**79.91638**	82	**81.91348**	83	**82.91413**	84	**83.91151**	86	**85.91062**	92	91.9262		
56	Ba	130	**129.90628**	132	**131.90449**	134	**133.90449**	135	**134.90567**	136	**135.90456**	137	**136.9058**	138	**137.90524**	141	140.9141
57	La	138	**137.90716**	139	**138.9064**	143	142.916										
58	Ce	136	**135.90718**	138	**137.90603**	140	**139.90548**	142	**141.9093**	144	143.91359						
59	Pr	141	**140.9077**														
60	Nd	141	141.90777	143	**142.90986**	144	143.91013	145	**144.9126**	146	**145.91315**	148	**147.9169**	150	**149.92092**		
62	Sm	144	**143.91207**	146	145.9131	147	146.9149	148	147.91485	149	148.91721	150	**149.9173**	152	**151.91976**	154	**153.92222**
80	Hg	196	**195.96582**	198	**197.9668**	199	**198.96828**	200	**199.96832**	201	**200.9703**	202	**201.97064**	204	**203.9735**		
81	Tl	203	**202.97235**	205	**204.97444**												
82	Pb	204	**203.97305**	206	**205.97448**	207	**206.9759**	208	**207.97666**								
83	Bi	209	**208.9804**	211	210.98729	212	211.99129	213	212.99439	214	213.99873						
84	Po	203	202.9813	204	203.9802	205	204.9810	206	205.9804	207	206.98161	208	207.98125	209	208.98243	210	209.98288
		211	210.98666	212	211.98887	213	212.99286	214	213.99521	215	214.99945	216	216.00192	217	217.0064	218	218.0090
85	At	205	204.9860	206	205.9865	207	206.9857	208	207.9865	209	208.98617	210	209.9870	211	210.98751	212	211.99074
		213	212.99294	214	213.99634	215	214.99865	216	216.00243	217	217.00472	218	218.00871	219	219.0113		
86	Rn	207	206.9906	208	207.9895	209	208.9902	210	209.98956	211	210.99062	212	211.99072	213	212.99389	214	213.99537
		215	214.99875	216	216.00028	217	217.00394	218	218.00562	219	219.00951	220	220.0114	221	221.0155	222	222.01761
87	Fr	209	208.9959	210	209.9962	211	210.9955	212	211.9962	213	212.99619	214	213.99887	215	215.00036	216	216.0032
		217	217.00464														
88	Ra	211	211.0008	212	211.9997	213	213.0002	214	213.99997	215	215.00274	216	216.0035	217	217.00632	218	218.00715
		219	219.01008	220	220.0110	221	221.01393	222	222.01539	223	223.01853	224	224.02021	225	225.0236	226	226.02544
92	U	233	233.03965	234	234.0410	235	235.04394	236	236.04558	238	238.05082						
93	Np	231	231.0383	233	233.0408	235	235.04408	237	237.04819								
94	Pu	233	233.0430	234	234.0433	235	235.0453	236	236.04605	237	237.04843	238	238.04958	239	239.05218	240	240.05383
		241	241.05687	242	242.05877	243	243.062	244	244.06423								

boldface figures = stable

Taking the masses of ^{35}Cl and ^{37}Cl from Table 36.1, we have:

$$35.453 = f(34.969) + (1 - f)(36.966).$$

SOLVE

$$f = \frac{36.966 - 35.453}{36.966 - 34.969} = 0.7576.$$

ANALYZE Natural chlorine is about 76% ^{35}Cl and 24% ^{37}Cl.

36.1.2 *The Size of Nuclei*

Although they contain almost all the mass, nuclei occupy a tiny fraction of the volume of matter. Ernest Rutherford and his students first carried out experiments that established the very small size of atomic nuclei (cf. §35.2.1). Using the Coulomb force law to explain the results of their scattering experiments, they determined that the radii of gold, silver, and copper nuclei were less than 32, 20, and 12 fm, respectively. Scattering α-particles (helium nuclei) from aluminum nuclei, they discovered deviations from the Coulomb model when the α-particles came within 8 fm of the center of the atom. This was the first indication of a distinct *nuclear* force.

The SI symbol for 10^{-15} m, 1 fm = 1 femtometer, is also an abbreviation for Enrico Fermi's last name. Fermi made numerous early discoveries in nuclear physics, and this unit is often called a fermi in his honor.

Remember: The electron wave has wavelength $\lambda = h/p$, §35.3. Diffraction is discussed in §17.3 and §17.4.5.

When relativistic electrons are used as the scattering particles, they are diffracted by the nucleus. The diffraction pattern provides an accurate measurement of the nuclear size (■ Figure 36.4). The results of these experiments show that the radii of nuclei are described by the formula:

$$R \approx (1.2 \text{ fm})A^{1/3}. \qquad (36.1)$$

The volume of a nucleus is roughly proportional to the mass number A so that nuclei all have about the same mass density.

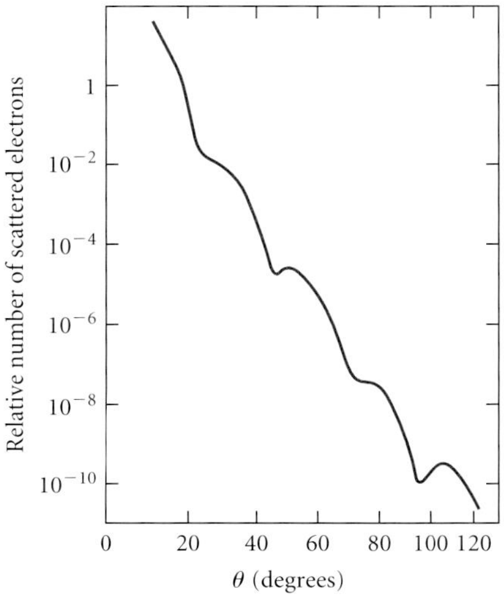

■ **Figure 36.4**
In an experiment to measure the size of a nucleus, 450-MeV electrons were scattered by ^{58}Ni nuclei. The electron wavelength ($\lambda \approx 3$ fm) is small enough to probe the nuclear size. Diffraction minima should occur at $\sin\theta = m\lambda/d$ (§17.3). Here the experimenters have plotted the number of electrons scattered in each direction. The positions of the minima indicate a nuclear radius equal to 4.1 fm. Adapted from a figure by Ingo Sick, Institut für Physik der Basel. Used with permission.

36.1.3 Nucleons

The word "atom" was introduced by ancient Greek philosophers to mean a particle that cannot be cut into smaller pieces. We now think of atoms as composites of nuclei and electrons. The electrons may be removed by photon absorption or by collisions, and they may be shared among atoms in chemical reactions. The concept of a chemical element is useful because these electron reactions do not alter the atomic nuclei. However, much more energetic reactions can change the structure of nuclei (■ Figure 36.5), indicating that nuclei are themselves composite structures of more fundamental particles.

The electric charges of nuclei are easily explained by supposing that a nucleus of atomic number Z contains Z protons, but these protons account for only a part of the nuclear mass. The rest is contributed by a number $N = A - Z$ of electrically neutral particles called *neutrons* (■ Figure 36.6), each with nearly the same mass as a proton. Neutrons are unstable and cannot exist outside a nucleus for more than about 10 min. Their role in explaining nuclear properties was the first clue to their existence. The reality of neutrons was established in 1932 by James Chadwick, who was able to produce a beam of free neutrons and observe their collisions with protons in hydrogen-containing materials (■ Figure 36.7). The properties of neutrons are summarized, along with those of protons and electrons, in ● Table 36.2. (We discuss the decay of neutrons in §36.2.3.)

Atomic structure is remarkably simple: all of the ordinary matter we observe in the universe is made up of protons, neutrons, and electrons. Protons and neutrons are both referred to as *nucleons*. Next, we describe how the nucleons interact to form atomic nuclei.

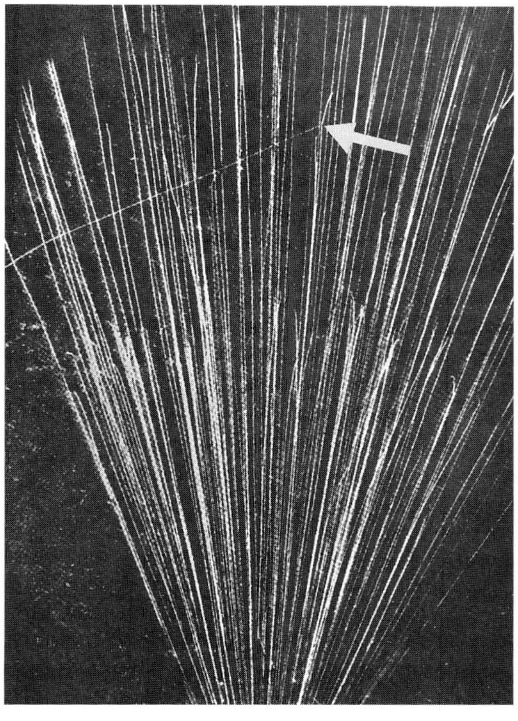

■ **Figure 36.5**
This cloud chamber photograph, taken in 1925 by Blackett and Lees, shows one of the first nuclear reactions observed visually. An α-particle, moving upward in the photograph, collides with a nitrogen atom in the chamber, producing an oxygen nucleus and a proton. The proton recoils backward and to the left. The cloud chamber contains supersaturated gas. Moisture condenses on charged ions produced by the passage of the fast particles. Cloud chambers were important in the early days of nuclear physics but have been replaced by more modern technologies.

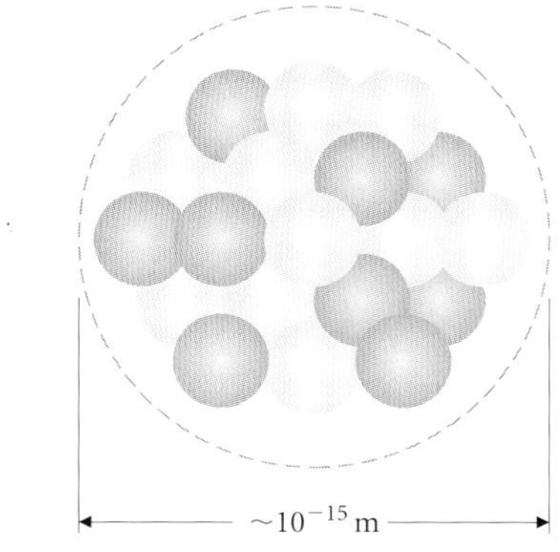

■ **Figure 36.6**
A nucleus is composed of Z protons (colored red in the diagram) and $N = A - Z$ neutrons (colored tan) bound together in a region with a diameter of the order of 10^{-15} m $\equiv$ 1 fm. This diagram represents a nucleus with mass number $A = 19$, atomic number $Z = 9$, and $N = A - Z = 10$ neutrons—that is, ^{19}F.

Table 36.2 Properties of Atomic Constituents					
	Mass			Electric Charge	Spin
	(10^{-27} kg)	(u)	(MeV)	(e)	($\hbar$)
Electron	9.109×10^{-4}	5.49×10^{-4}	0.5110	−1	$\frac{1}{2}$
Nucleons					
Proton	1.67262	1.00727	938.272	+1	$\frac{1}{2}$
Neutron	1.67493	1.00866	939.566	0	$\frac{1}{2}$

36.1.4 *Nuclear Forces*

Electrical repulsion between its positively charged protons would rapidly destroy a nucleus unless some other force acts to hold it together. This force is called the *strong nuclear force.* Collision experiments show that the strong force between protons overwhelms their electrical repulsion when the protons are less than about 10^{-15} m apart but decreases much more rapidly than electrical forces for distances greater than 10^{-15} m. The strong force doesn't distinguish between the two kinds of nucleon; it acts in the same way between a pair of neutrons or between a neutron and a proton as it does between a pair of protons. There is no

Digging Deeper

Why the Neutron is Necessary

The proton is named after Proteus, a mythical Greek figure who could take on any form at will. The name was given when an atomic nucleus was thought to consist of A protons and $N \equiv A - Z$ electrons bound together to give the proper charge and mass. Such an economical description of all matter is very attractive. However, it has some major problems that can be solved by including neutrons in the theory.

- According to the uncertainty principle, an electron confined within a nucleus must have a large momentum. The resulting kinetic energy is ten times greater than the observed energies of electrons emitted by radioactive nuclei (§36.2).
- The spin angular momentum of nuclei is observed to depend only on the mass number A. According to quantum mechanics, spin is determined by the total number of particles in a system, which depends on both A and Z in the proton + electron model.
- Electrons have magnetic moments 600 times larger than do protons, while nuclear magnetic moments are comparable to that of a proton.

To address these observations, we might suppose an improbable cancellation of magnetic moments and abandon the rules of angular momentum so crucial to understanding chemistry and atomic spectra. Such ideas are not logically faulty but have an ugliness that in physics is often a symptom of error. By contrast, neutrons explain all three observations in a straightforward way. Chadwick believed in the existence of neutrons long before he discovered them experimentally.

Chadwick's discovery illustrates an important theme of modern physics. Given the choice between accepting an ugly theory or a new kind of particle, physicists tend to bet on the new particle. Experiment has often confirmed this approach.

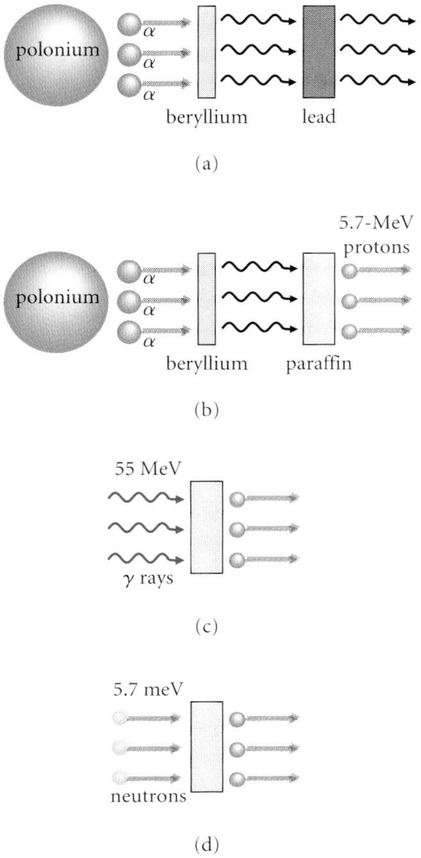

■ Figure 36.7
James Chadwick, a protégé of Rutherford, conducted experiments that established the existence of neutrons. Chadwick's colleagues described him as "the personification of the ideal experimentalist." (a) Radioactive polonium emits α-particles that strike a beryllium target and produce radiation that penetrates a thick lead shield. (b) The particles from the interaction in beryllium eject 5.7-MeV protons from a paraffin target. (c) To eject 5.7-MeV protons from paraffin using γ rays, the γ rays would have to have at least 55 MeV of energy, much more than is available from the nuclear reaction in beryllium. (d) Chadwick showed that neutral particles with about the same mass as the proton need only about 5.7 MeV of energy to eject the protons via elastic collisions.

simple formula, like Coulomb's law, that completely describes the strong force exerted by two nucleons on each other. In this chapter we shall rely on approximate models of the strong force.

The short range of the strong force limits the size of nuclei. In a nucleus very much larger than 1 fm, protons near the surface would feel the strong attraction of relatively few nearby nucleons while feeling the electrical repulsion of every other proton (■ Figure 36.8). In large nuclei, the number of neutrons N exceeds the number of protons Z. The additional neutrons increase the effects of the strong force without increasing electrical repulsion, thus tending to hold a nucleus together. The largest known stable nuclei have $A \approx 240$. Larger nuclei can be produced in carefully designed experiments, but they quickly reduce their size by expelling nucleons in the form of smaller nuclei.

A second kind of nuclear interaction due to the *weak nuclear force* causes a neutron to disrupt, or β-decay, into a proton and an electron (see §36.2.3). Inside a nucleus, β decay is inhibited by the strong force. In a nucleus with too many neutrons, the weak decay process overwhelms the strong force, and the nucleus reduces its neutron number by emitting electrons. Either too few or too many neutrons cause a nucleus to be unstable. That is why stable nuclei cluster near a line in a plot of Z versus A (Figure 36.3).

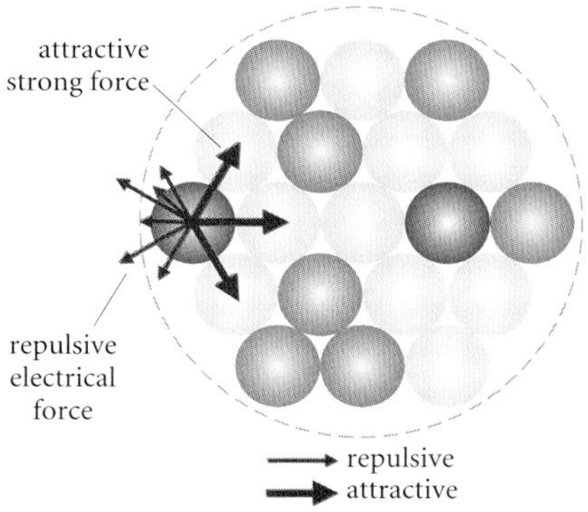

■ FIGURE 36.8
A nucleus is bound together by the attractive *strong* force its nucleons exert on each other. Electric forces between the protons are repulsive and tend to disrupt the nucleus. In a large nucleus, the short-range strong force acts primarily between a nucleon and its nearest neighbors. A proton, however, feels the long-range electrical repulsion of every other proton in the nucleus. For this reason, nuclei with $A \gtrsim 240$ are unstable.

36.1.5 *Binding Energy*

The strong force pulls nucleons together and binds them into a nucleus. As with the electrons in an atom, this binding can be described either by the force itself or by the corresponding potential energy. The potential energies of nuclei are large enough to be observed in measurements of nuclear masses; a stable nucleus has less mass than its nucleons would if they could be separated as free particles. According to Einstein's mass-energy relation (§34.4.2), this mass difference corresponds to the energy needed to separate the nucleons—the binding energy B of the nucleus.

THE BINDING ENERGY OF THE NUCLEUS IS THE *NEGATIVE* OF ITS POTENTIAL ENERGY. THUS WHEN BINDING ENERGY INCREASES, SOME OTHER FORM OF ENERGY ALSO INCREASES CORRESPONDINGLY.

The *mass defect* of a nucleus equals the difference between the total mass of its individual nucleons and the mass of the assembled nucleus:

$$\Delta M(A, Z) = Z \times (\text{proton mass}) + N \times (\text{neutron mass}) - (\text{nuclear mass}).$$

Data are usually tabulated for neutral atoms. For example, in Table 36.1,

$$M_{\text{H}} = 1.0078 \text{ u} = 1.0073 \text{ u} + 0.0005 \text{ u} = M_{\text{p}} + m_{\text{e}}.$$

Thus a formula for the mass defect in terms of tabulated data must include the mass of the electrons:

$$\Delta M(A, Z) = ZM_{\text{H}} + NM_{\text{n}} - M(A, Z). \tag{36.2}$$

NOTICE THAT THE ELECTRON MASS CANCELS OUT IN EQN. (36.2). THAT ISN'T TRUE IN ALL THE EQUATIONS WE SHALL WRITE, SO CHECK CAREFULLY!

The *binding energy* of a nucleus equals its mass defect expressed in energy units:

$$B = c^2 \, \Delta M. \tag{36.3}$$

The binding energy per nucleon (B/A) in a nucleus determines how easy it is to break up the nucleus. When B/A is large, the nucleus is tightly bound; if B/A is small, the nucleus is more easily split.

The short range of the strong force is evident in a plot of B/A versus atomic mass number for stable nuclei (■ Figure 36.9). Adding a nucleon to an element less massive than iron causes each nucleon to be more tightly bound to the larger nucleus. Adding a nucleon to atoms more massive than iron increases the size of the already large nucleus and decreases the attraction between nucleons already present. The shape of the binding energy curve has great importance for nuclear technology. Energy may be released either by breaking apart large nuclei, as in fission reactors using uranium, or by fusing together lighter nuclei. Fusion reactors under development would operate by fusing deuterium and tritium into helium.

IN BOTH CASES, THE TOTAL BINDING ENERGY OF THE NUCLEI PRODUCED IS GREATER THAN THAT OF THE ORIGINAL PARTICLES.

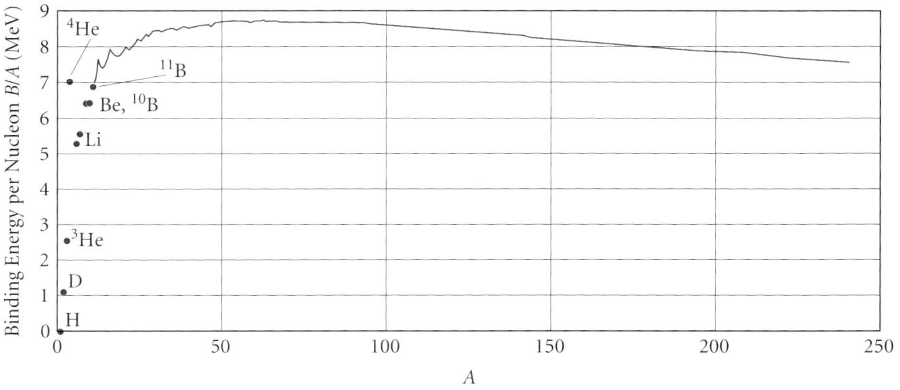

■ FIGURE 36.9
Binding energy per nucleon versus mass number *A* for stable nuclei. The curve has a maximum at elements near iron. Notice also that helium is more tightly bound than its neighbors.

EXAMPLE 36.2 ♦ Find the binding energy of a ^{4}He nucleus. Compare the binding energy per nucleon with the value in Figure 36.9.

MODEL A helium nucleus contains two neutrons and two protons. We use eqns. (36.2) and (36.3).

SETUP From Table 36.1, the mass of a ^{4}He atom is: $M(^4\text{He}) = 4.00260$ u. So, using values for M_p, m_e, and M_n from Table 36.2, the mass defect of ^{4}He is:

$$\Delta M(4, 2) = 2(M_p + m_e) + 2M_n - M(^4\text{He})$$
$$= [2(1.00727 + 0.00055 + 1.00866) - 4.00260]\ \text{u} = 0.03036\ \text{u}.$$

SOLVE This corresponds to a binding energy of:

$$B = c^2\,\Delta M = c^2(0.03036\ \text{u})(931.49\ \text{MeV}/c^2\cdot\text{u}) = 28.28\ \text{MeV}.$$

ANALYZE The binding energy per nucleon is $B/A = (28.28\ \text{MeV})/4 = 7.07$ MeV/nucleon, as compared with a value of 7.1 MeV/nucleon read directly from the figure. ■

EXERCISE 36.1 ♦♦ Through a sequence of reactions in stars, six helium nuclei fuse into one magnesium nucleus. Use Figure 36.9 to estimate the energy released.

36.2 NATURAL RADIOACTIVITY

36.2.1 Conservation Laws and Quantum Numbers

The nuclear age began in 1896, when Becquerel discovered the spontaneous emission of energetic particles by samples of uranium. In 1898, Marie and Pierre Curie (■ Figure 36.10) discovered similar radiation from thorium and radium, and gave the phenomenon its modern name: radioactivity. Three kinds of radiation are common in nature. They were originally named according to their *range*—the distance they can penetrate into surrounding material:

- α rays (bare helium nuclei), which typically penetrate a few centimeters of air;
- β rays (electrons) capable of penetrating several meters of air;
- γ rays (photons) whose range depends strongly on energy but is typically much greater than for α or β rays.

In each case, the emission of a particle changes a *parent* nucleus into a different, *daughter* nucleus. We typically have accurate descriptions of the parent nucleus and of the daughter products. The intermediate processes by which decay occurs are only now beginning to be understood in detail. Whenever we wish to compare initial and final states without reference

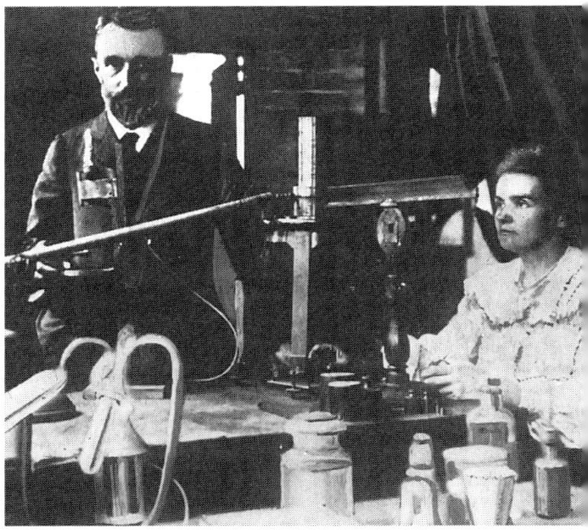

■ FIGURE 36.10
Marie and Pierre Curie in their laboratory. With Henri Becquerel, they were awarded the Nobel prize in 1903 for their discovery of radioactivity. Marie Curie won a second Nobel prize, in chemistry, for her later work with radium. Their daughter, Irene Joliot-Curie, also won a Nobel prize for her work on the production of radioactive substances by bombardment with α-particles.

to intermediate states, conservation laws provide the most powerful tools. The familiar conservation laws for energy, linear and angular momentum, and electric charge remain valid in nuclear processes. In addition, nuclear decays obey two conservation laws that are not apparent on a larger scale.

Technical names for the conserved quantities are *lepton* (light particle) *number* and *baryon* (heavy particle) *number.* We have already discussed one kind of lepton, the electron, and two kinds of baryon, the nucleons. As such new and unfamiliar particle properties enter our tale, it is useful to think of each in analogy with electric charge. Each particle has a charge number, and its charge equals that number times the electron charge e. Conservation of electric charge was first established experimentally and only later, with the discovery of Maxwell's equations, found to be required by basic principles. Lepton and baryon numbers also describe chargelike quantities; they are examples of *quantum numbers.* We don't yet know if their conservation is required by fundamental principles.

Next let us examine the three types of natural radioactive decay to see how these conservation laws apply.

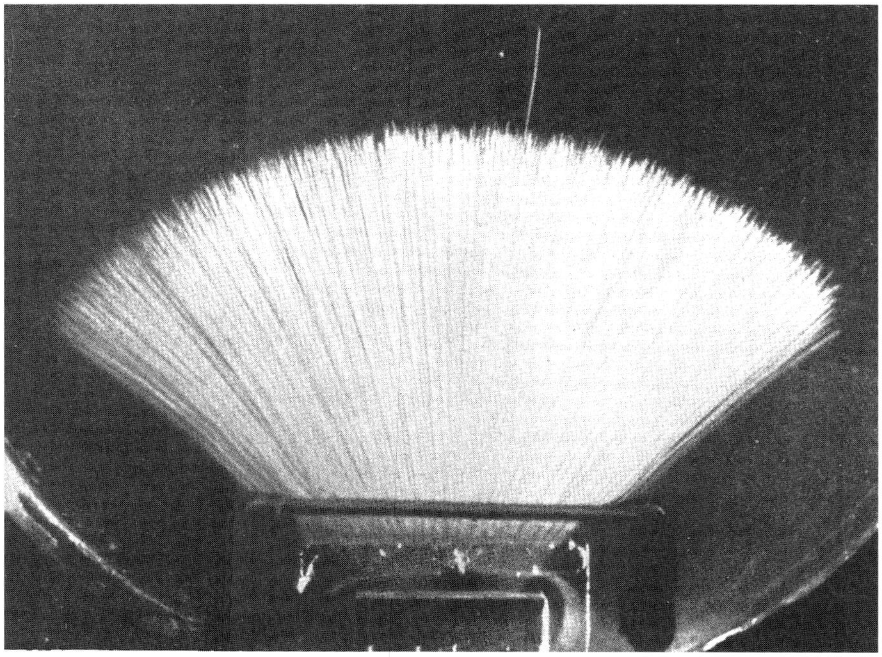

■ FIGURE 36.11
Cloud chamber photograph of particles produced in the α decay of ^{214}Po to ^{210}Pb. Note the almost constant range of the particles. Each α has an energy of 7.69 MeV. The one long track is due to a rare decay from an excited state of ^{214}Po.

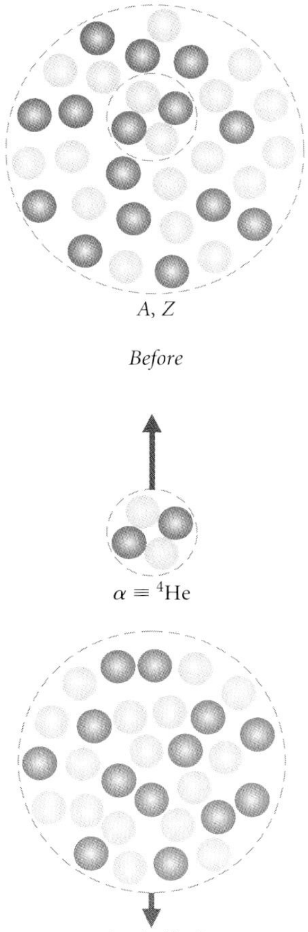

■ FIGURE 36.12
An α decay. Four nucleons bind together and escape as an α-particle. Baryons are conserved, so the total mass number A of the two daughter nuclei equals that of the parent. Electric charge is also conserved, so the sum of the atomic numbers of the daughters equals that of the parent. Kinetic energy of the daughters arises from a reduction in the total mass.

36.2.2 α Decay

In α decay, two neutrons and two protons within a parent nucleus bind together to form a helium nucleus, which then escapes (■ Figures 36.11, 36.12, and 36.13). The potential energy released in forming the α-particle is converted to kinetic energy, which is shared by the daughter particles. The daughter particles have a smaller total mass than the parent nucleus; the loss of rest energy (traditionally given the symbol Q) equals the gain of kinetic energy:

$$K = Q = (\text{mass of parent nucleus} - \text{total mass of daughter particles})c^2.$$

The Q values in α decay are always much less than the rest energy of a single nucleon. This observation is a consequence of baryon conservation. Nucleons cannot disappear in the reaction; they can only be bound more tightly. Thus the total atomic mass number of the daughter particles is the same as that of the parent nucleus. Electrons play no role in α decay.

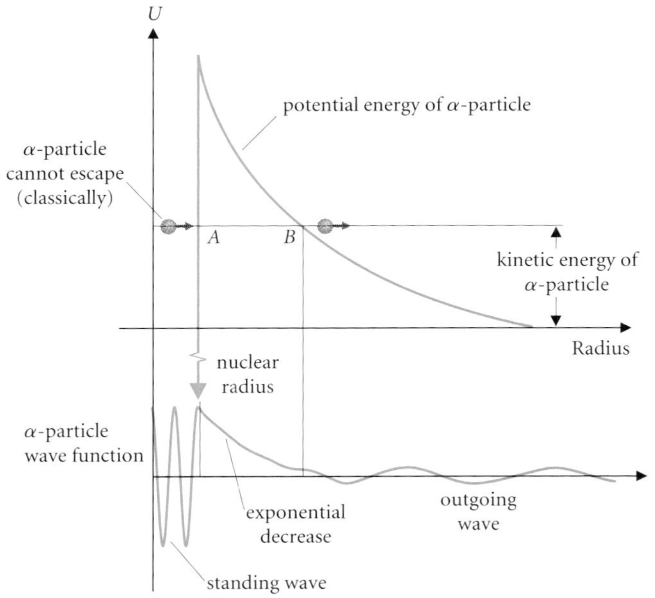

■ FIGURE 36.13
α-decay. Four nucleons associate into an α-particle and escape the nucleus together. Such combinations of nucleons continually form and redissolve within a nucleus. Because α-particles are particularly tightly bound (Figure 36.9), they form with sizable kinetic energy and have a particularly large probability of escape compared with other small substructures that might form. The potential energy function of an α-particle is large and negative from the center of the nucleus to a radius of a few fermis due to the attractive strong forces exerted by the rest of the nucleons. Farther out, its potential energy is positive because of repulsive Coulomb forces. A classical particle could not exist in the region between points A and B on the graph, where it would have to have negative kinetic energy. However, the wave function (cf. §35.3 and §35.4) for an α-particle moving outward is partially transmitted through this region. The outgoing wave that results corresponds to a small probability that the α-particle will escape. (The particle is said to *tunnel* through the classical barrier.) Calculations based on this tunneling model are in excellent agreement with measured α-decay rates.

Conservation of electric charge requires that the number of protons be the same before and after the reaction. That is, the total atomic number of the daughters is the same as that of the parent nucleus.

A general equation for α decay that expresses these conservation laws (■ Figure 36.14) is:

$$^{A}Z \rightarrow {}^{(A-4)}(Z - 2) + {}^{4}\text{He} + Q. \quad (36.4)$$

For example:

$$^{144}\text{Nd} \rightarrow {}^{140}\text{Ce} + {}^{4}\text{He} + Q.$$

(Neodymium decays to cerium plus helium, releasing energy Q.)

The Q value of a reaction can be calculated from tabulated masses. Decay is possible when Q is positive; otherwise, the parent nucleus is stable.

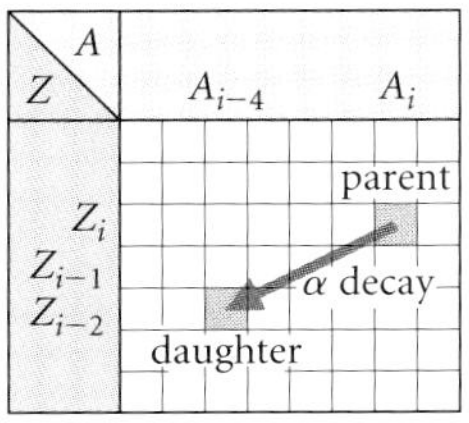

■ FIGURE 36.14
An α decay displayed on a table of atomic number Z versus mass number A. The daughter nucleus is displaced by 4 in mass number A and 2 in atomic number Z.

EXAMPLE 36.3 ♦♦ Find the energy released in the α decay of ^{144}Nd.

MODEL The Q value is determined by the decrease of total *nuclear* mass, while tabulated data give atomic masses, which include electrons. However, as many electrons are included in the atomic mass of the parent as in the total atomic mass of the daughters. So, the electron masses cancel in the calculation of Q. The total binding energy of the electrons differs somewhat between parent and daughter atoms, resulting in errors of at most a few keV.

SETUP AND SOLVE We obtain the masses from Table 36.1. Thus:

$$\begin{aligned} Q &= c^2(M_{\mathrm{Nd}} - M_{\mathrm{Ce}} - M_{\mathrm{He}}) \\ &= c^2(143.9101 - 139.9055 - 4.0026)\ \mathrm{u} \\ &= c^2(0.0020\ \mathrm{u})(931.49\ \mathrm{MeV}/c^2\cdot\mathrm{u}) \\ &= 1.9\ \mathrm{MeV}. \end{aligned}$$

ANALYZE Notice how many significant figures are lost in the subtraction! ■

36.2.3 β Decay

In β decay, the parent nucleus emits an energetic electron. To conserve charge, the daughter nucleus must have one more proton than the parent and, since the total number of baryons cannot change, one fewer neutron. In effect, a single neutron within the nucleus decays to produce a proton and an electron (■ Figure 36.15). Free neutrons undergo just such a reaction.

A tentative formula for β decay (■ Figure 36.16) is:

$$^{A}Z \rightarrow {}^{A}(Z + 1) + e^{-} + Q + X.$$

The reaction product X must be included; otherwise, nearly all the conservation laws are violated. For example, the measured kinetic energies of emitted electrons spread over a continuous range from 0 to Q (■ Figure 36.17). Without X, this could happen only if individual decays violate energy conservation. In cloud chamber photographs (■ Figure 36.18), the electron and daughter nucleus leave tracks that are not in opposite directions. Without X, this would indicate a violation of linear momentum conservation. Nor can the reaction conserve angular momentum. One neutron with spin $\frac{1}{2}\hbar$ is replaced with a proton and electron, each with spin $\frac{1}{2}\hbar$. According to quantum mechanics, these spins either cancel to zero or add to a spin of $1\hbar$. Orbital angular momentum comes in whole multiples of $\hbar$ and so cannot account for the difference of $\frac{1}{2}\hbar$ between final and initial spin angular momentum. Finally, total lepton number would not be the same if X is not present. The parent has zero lepton number (no electrons) while the electron daughter has lepton number 1. Thus X must have lepton number -1. To understand β decay, we must either believe in a new kind of particle or abandon conservation laws altogether.

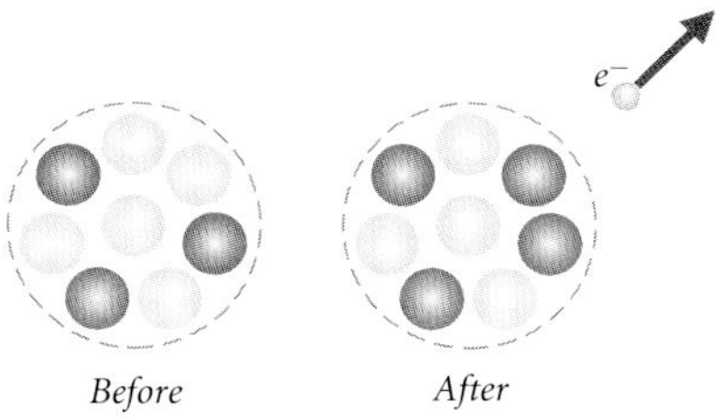

■ FIGURE 36.15
A β decay. The daughter nucleus has the same atomic mass number A. It has one fewer neutron but one more proton, so its charge is increased by 1. An energetic electron is ejected from the nucleus.

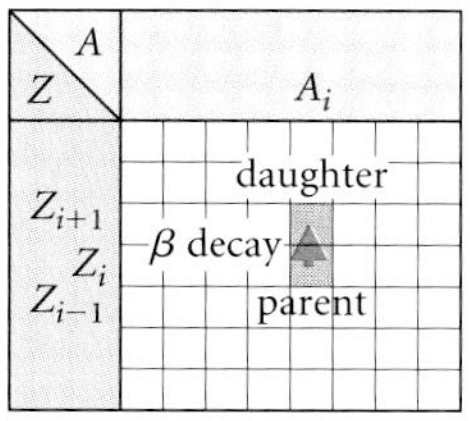

■ FIGURE 36.16
A β decay displayed on a table of atomic number Z versus mass number A. The daughter nucleus has the same mass number A but is displaced by 1 in atomic number Z.

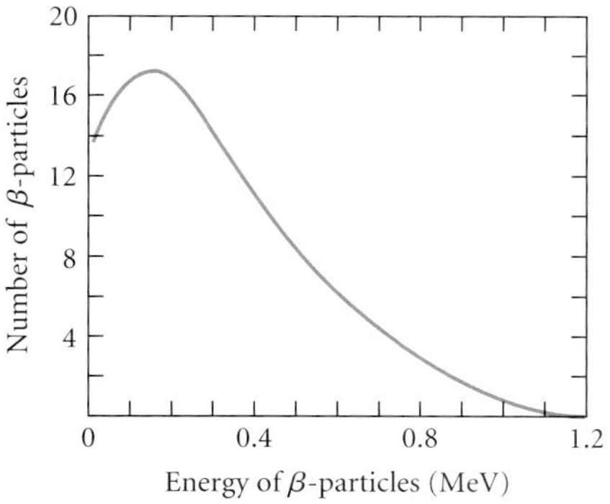

■ FIGURE 36.17
Distribution of energy of electrons emitted in the β decay of ^{210}Bi. Because the electrons have a continuous distribution of energies, there must be another particle that shares the released energy with the electron.

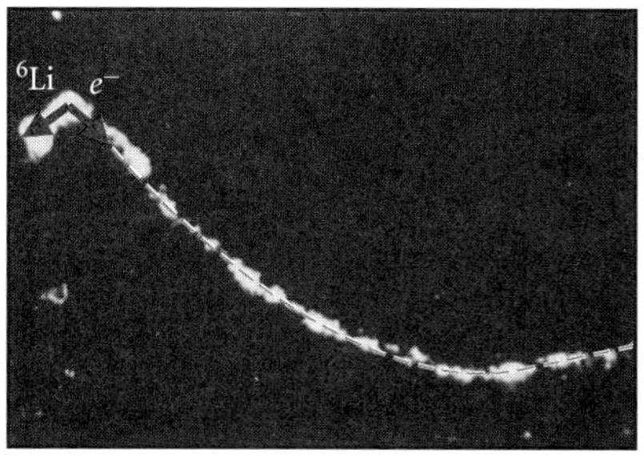

■ FIGURE 36.18
Cloud chamber photograph of β decay of ^{6}He. The short, thick track is the ^{6}Li daughter nucleus; the thinner, curved track is the electron. The two particles do not move opposite each other but instead have a net downward momentum in this picture, indicating the presence of a third, upward-moving, particle. Since it leaves no track, the third particle is electrically neutral. The photograph was taken in 1957 by S. Szalay and J. Csikay.

In summary, the new particle has spin $\frac{1}{2}\hbar$, no electric charge, and lepton number -1. It is a *neutrino*—a name given by Enrico Fermi meaning "little neutral one" in Italian. It is believed to have zero—or at least a very small—mass. This specific type is called an *electron antineutrino,* symbolized $\bar{\nu}_e$. Neutrinos have a very small probability of interacting with other particles and remained undetected by experiment for 20 years after Fermi proposed their existence. They were first observed in 1953 by Cowan and Reines, who detected reactions of chlorine atoms with the intense neutrino flux from a nuclear reactor.

The general equation for β decay, including the antineutrino, is

$$^{A}Z \rightarrow {}^{A}(Z+1) + e^- + \bar{\nu}_e + Q. \qquad (36.5)$$

The Q value for a β decay may also be computed using tabulated atomic masses. The atomic mass of the daughter element *includes* the mass of the emitted electron.

EXERCISE 36.2 ♦♦ What kinetic energy is shared by the electron and neutrino in the β decay of ^{12}B?

36.2.4 *Antiparticles and Positron (β^+) Decay*

For each kind of particle, a corresponding *antiparticle* exists that has equal mass but opposite values of electric charge and baryon or lepton number. In 1927, Paul Dirac predicted that such antiparticles must exist if both quantum mechanics and special relativity are correct. Antielectrons (*positrons*) were first observed by Carl Anderson in 1932 in a cloud chamber photograph of cosmic rays (■ Figure 36.19). Antiprotons and antineutrons were observed in accelerator experiments during the 1950s (■ Figure 36.20). Since then, Dirac's prediction has been completely verified experimentally.

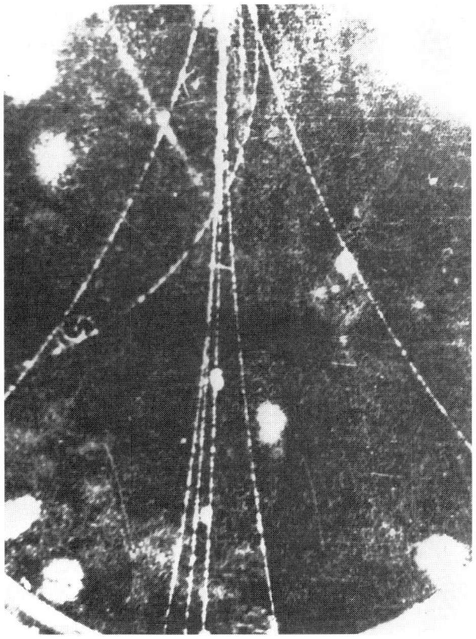

■ FIGURE 36.19
One of the first pictures of a cosmic ray electron shower passing through a cloud chamber. A magnetic field causes the charged particles to follow curved paths (§29.1). Tracks curving to both right and left indicate the presence of both electrons and positrons.

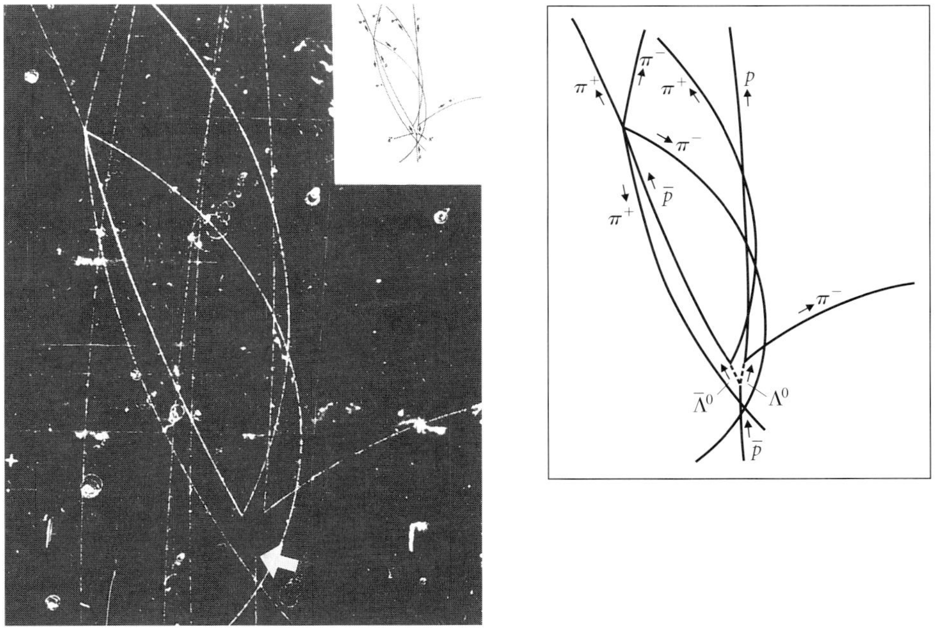

■ FIGURE 36.20
Track of an antiproton in a bubble chamber. Upon colliding with a proton (at arrow), the pair annihilates, producing two neutral Λ-particles that leave no tracks. The Λ^0, on the right, decays into a proton and a π^{-1}. The antiparticle $\bar{\Lambda}^0$ decays into the exact antiparticles, an antiproton and a π^+. The Λ- and π-particles are discussed in Chapter 37.

Digging Deeper

NEUTRINOS

Neutrinos interact only very weakly with ordinary matter. While most kinds of particle are stopped by a few centimeters thickness of solid material, a typical neutrino could pass through several light-years of solid iron!

Cosmic Gall*

Neutrinos, they are very small
  They have no charge and have no mass
And do not interact at all.
  The earth is just a silly ball
To them, through which they simply pass,
  Like dustmaids down a drafty hall
Or photons through a sheet of glass.
  They snub the most exquisite gas,
Ignore the most substantial wall,
  Cold shoulder steel and sounding brass,
Insult the stallion in his stall,
  And, scorning barriers of class,
Infiltrate you and me! Like tall
  And painless guillotines, they fall
Down through our heads into the grass.
  At night they enter at Nepal
And pierce the lover and his lass
  From underneath the bed—you call
It wonderful; I call it crass.

John Updike

The Sun seems to be producing only a third as many neutrinos as expected from the nuclear reactions that fuel its energy output. For more than 20 years, Raymond Davis has been observing neutrino reactions with the chlorine atoms in a 400 000-L tank of perchlorethylene cleaning fluid (■ Figure 36.21). The tank is kept in a deep gold mine to shield it from particles that cannot pass through a few thousand feet of earth. Neutrino reactions produce 30 argon atoms per month, which are flushed from the tank every 3 months and counted! The argon production rate is too small, and we don't understand why. Neutrino detectors using water (■ Figure 36.22) and gallium also fail to detect enough neutrinos.

■ FIGURE 36.21
Tank of perchlorethylene used in a neutrino detection experiment. Located deep underground at the Homestake gold mine near Lead, South Dakota, the tank is covered with water that provides even more shielding against stray particles that might mimic neutrino reactions. A small fraction of the neutrinos entering the tank undergo reactions that convert chlorine nuclei to argon. Periodically, the argon is flushed from the system and its amount measured.

Neutrinos may be able to change form during their 8-min flight from the Sun; models of the nuclear reactions inside the Sun may be wrong in some unknown way. Some astronomers

*From *Telephone Poles and Other Poems* by John Updike. Copyright © 1963 by John Updike. Reprinted by permission of Alfred A. Knopf Inc.

The particle and antiparticle have exactly opposite values of electric charge, lepton number, and baryon number, so the total value of each is zero. A system consisting of a particle and its antiparticle possesses only mass/energy and linear and angular momentum, precisely the conserved quantities possessed by a photon. Thus, the system can *annihilate* to form two photons. For example:

$$e^+ + e^- \rightarrow 2\gamma.$$

Two photons are required in order to conserve the system's momentum.

■ FIGURE 36.22
The Kamioka detector. The detector is at a depth of 2700 m in a lead-zinc mine, about 300 km west of Tokyo. The detector contains a total of 1000 photomultipliers (cf. §35.1) that detect photons produced by interactions of particles with the water in the detector.

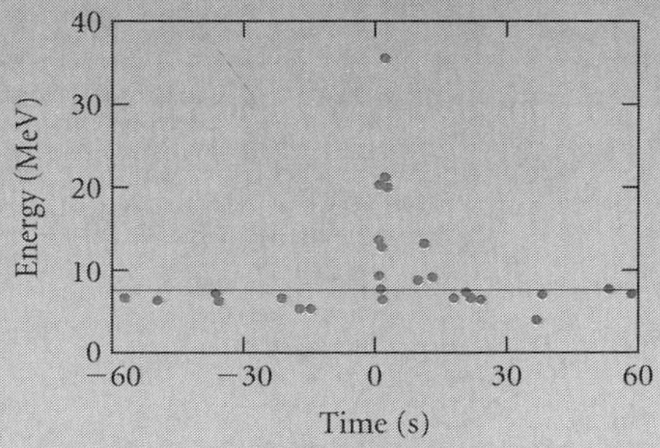

■ FIGURE 36.23
Neutrino events observed in the Kamiokande detector on February 23, 1987. The time of the detection coincides with the arrival of light from supernova 1987a in the Large Magellanic Cloud. (This detection was an important confirmation of supernova theory.)

have proposed that the Sun may turn off its energy production periodically. Because the photons produced inside the Sun take about 10 million y to work their way out, we would have to wait that long to detect a reduction in heat and light.

Neutrinos from outside the solar system were detected in early 1987 (■ Figure 36.23), when two laboratories detected a total of 16 neutrinos at the same time as astronomers first observed light from the explosion of supernova SN1987a in a nearby galaxy (■ Figure 36.24). The explosion compressed the star's core, forcing protons and electrons together in reverse β decay, converting the entire core to neutrons. The number of neutrinos observed is consistent with the theoretical expectation.

■ FIGURE 36.24
The region of the Large Magellanic Cloud surrounding supernova 1987a before (top) and after the supernova outburst. The arrow indicates the star prior to the explosion.

The inverse reaction is called *pair production.* An energetic photon disappears and produces a particle-antiparticle pair: $\gamma \to e^- + e^+$. The energy of the photon must exceed the rest energy of the two electrons:

$$E_\gamma > 2m_e c^2 = 1.022 \text{ MeV}.$$

Pair production by a single photon alone does not conserve momentum, but the process can occur near a charged particle or in an intense electromagnetic field, where momentum can be

SEE PROBLEM 76.

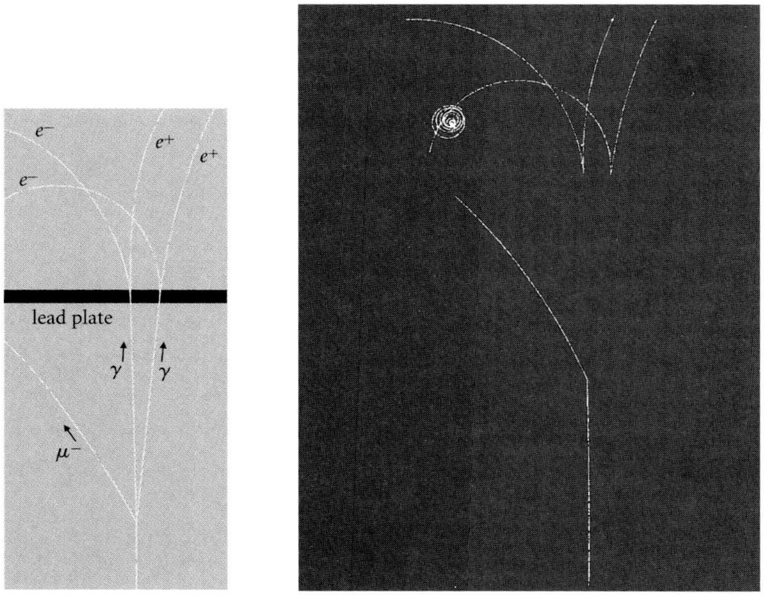

■ **FIGURE 36.25**
Pair production. The K meson entering the bubble chamber at the bottom decays into an antineutrino, a μ meson, and two γ rays. The γ rays pass through a lead plate placed across the chamber, where they produce electron-positron pairs. (Notice that the tracks curve in opposite directions, indicating the opposite signs of charge.) The antineutrino does not interact at all and its track remains invisible.

transferred to the particle or the field. Pair production is the principal energy loss mechanism for high-energy γ rays passing through matter (■ Figure 36.25).

Some nuclei are unstable against positron emission. The daughter nucleus has one more neutron and one fewer proton than its parent:

THIS PROCESS IS CALLED β^+ DECAY.

$$^{A}Z \rightarrow {}^{A}(Z - 1) + e^{+} + \nu_e + Q. \quad (36.6)$$

Here ν_e represents an electron neutrino, which is the antiparticle of the electron antineutrino produced in β decay.

EXERCISE 36.3 ❖ What are the baryon number and lepton number of each particle involved in a positron emission?

36.2.5 γ Decay

In γ decay, a nucleus emits a photon, which carries neither baryon number, lepton number, nor electric charge. Neither the atomic number nor the atomic mass number of the parent nucleus changes in the decay. The parent and daughter in this case are just different energy states of the same nuclear species (■ Figure 36.26). The process is very similar to the emission of photons by transitions of electrons between atomic levels, except that γ emission is only crudely modeled by transitions involving a single nucleon. The general formula for γ emission is:

$$^{A}Z^{*} \rightarrow {}^{A}Z + \gamma, \quad (36.7)$$

where * denotes a nucleus excited from its ground state. For example:

$$^{57}Fe^{*} \rightarrow {}^{57}Fe + \gamma.$$

To conserve momentum, the daughter nucleus recoils and carries away a small kinetic energy that is usually, but not always, unimportant.

If a photon of just the right energy strikes an iron nucleus, γ emission can go in reverse. The γ emission by $^{57}Fe^{*}$ is the transition used by Pound and Rebka in their search for the gravitational redshift. With even a small nuclear recoil, an emitted photon would not have enough energy to cause the inverse reaction, and the Pound–Rebka experiment would be impossible. In iron, however, individual atoms are bound in a crystal that absorbs the recoil

SEE THE ESSAY ON A GEOMETRIC THEORY OF GRAVITY.

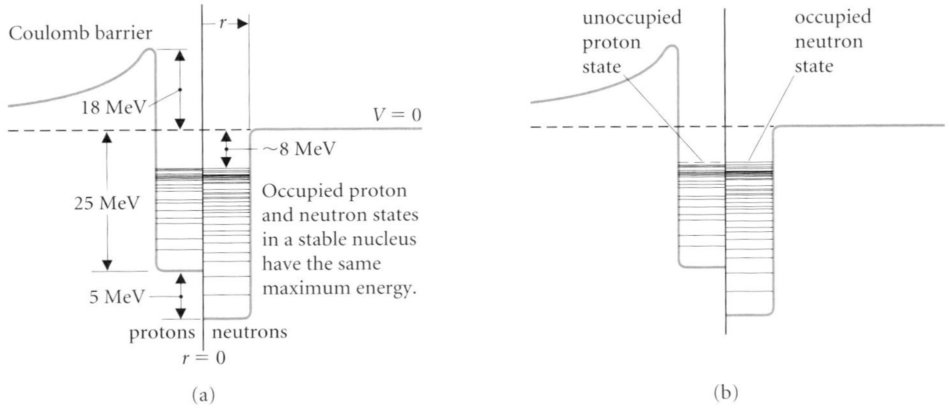

■ **FIGURE 36.26**
The shell model. Individual nucleons move rapidly throughout the volume of the nucleus influenced by an average potential energy function due to all the other nucleons. Neutrons experience only the strong force, whose potential is represented in the right half of the figure. Protons have, in addition, positive electric potential energy. The energy values shown here are estimates for a heavy nucleus like uranium. The energy states for a particle in this collective potential are represented by the horizontal lines in the figure. Individual nucleons fill these states in a manner similar to that of the electron states in a neutral atom. The Pauli exclusion principle allows only two nucleons in each level so neutrons and protons must occupy ever higher quantum number states in larger nuclei.

Gamma ray emission by nuclei occurs for the same reason that electrons in atoms emit photons. A nucleon, excited to a state above the lowest unoccupied state emits a γ ray during transition to a lower energy state. Absorption of a γ ray results in transition of a nucleon to a higher state. (a) In stable nuclei, neutrons and protons occupy states at nearly the same energy. Neutrons cannot decay because the resulting proton would have to enter a state with higher quantum numbers and would require energy exceeding the mass difference between the neutron and proton. (b) A nucleus with an unoccupied proton state close in energy to an occupied neutron state is unstable against β emission.

momentum with negligible change in energy. The resulting photons are precisely tuned to be reabsorbed and so provide a tool for measuring small frequency shifts due to relative motion or different height in a gravitational field.

EXERCISE 36.4 ♦♦ What is the kinetic energy of a free ^{57}Fe nucleus after the emission of the 14.4-keV x ray? If the half-life of the γ decay is 10^{-7} s, compare the energy of the nucleus with the uncertainty in the x-ray energy obtained from Heisenberg's principle. How large a crystal must the iron nucleus couple with so that the recoil energy loss is less than the uncertainty in the energy?

36.2.6 The Law of Decay

All three kinds of radiation (α, β, and γ) follow a simple statistical law.

RADIOACTIVE DECAY IS GOVERNED BY QUANTUM MECHANICS, WHICH ONLY ALLOWS US TO KNOW THE PROBABILITY THAT ANY FUTURE STATE WILL OCCUR (CF. CHAPTER 35). THUS THE LAW OF DECAY IS A DIRECT CONSEQUENCE OF QUANTUM MECHANICS.

The number of nuclei per second that decay—that is, emit a particle—is proportional to the number N of nuclei present in the sample:

$$R = \lambda N.$$

The decay rate R equals the rate of decrease in the number of parent nuclei:

$$\frac{dN}{dt} = -R = -\lambda N. \qquad (36.8)$$

WE'VE SEEN THIS EQUATION BEFORE. SEE, FOR EXAMPLE, §§13.3.4, 14.4, 31.1.

In a given time interval, every parent nucleus in a sample has an equal probability of decaying. In a large sample, the number of nuclei that actually decay equals the number present multiplied by the chance each has of decaying. The constant of proportionality λ, the probability of decay per unit time, depends on the particular parent nucleus and on the particular type of decay. Values of λ are usually measured experimentally and range over more than 40 orders of magnitude. The solution of the differential equation (36.8) is:

$$N(t) = N_o e^{-\lambda t}, \qquad (36.9)$$

where N_o is the number of parent nuclei present in the sample at $t = 0$. The rate of particle emission also decreases exponentially:

$$R(t) \equiv -\frac{dN}{dt} = \lambda N_o e^{-\lambda t}. \qquad (36.10)$$

Equation (36.10) expresses the characteristic experimental feature of radioactive decay: exponential decrease of the decay rate.

Nuclear decay may also be described by the half-life of a parent nucleus.

The *half-life* $T_{1/2}$ of a nuclear species is the time required for half of a large sample to decay.

According to eqn. (36.9), the number of parent nuclei remaining after one half-life is:

$$N(T_{1/2}) \equiv N_o/2 = N_o \exp(-\lambda T_{1/2}).$$

Canceling the common factor of N_o and taking natural logarithms, we find:

$$-\ln 2 = -\lambda T_{1/2},$$

or

$$T_{1/2} = (\ln 2)/\lambda. \qquad (36.11)$$

EXERCISE 36.5 ♦♦ Show that the law of decay may be expressed in the form:

$$N(t) = N_o 2^{-t/T_{1/2}}.$$

EXAMPLE 36.4 ♦♦ A sample of ^{214}Pb nuclei (half-life = 3.05 min) initially emits 352 β-particles per second. When will the emission rate have decreased to 10 /s?

MODEL The decay rate is proportional to the number of parent nuclei.

SETUP We use the result of Exercise 36.5. At $t = 0$, $R(0) = R_o = \lambda N_o$. Then:

$$\frac{R}{R_o} = \frac{N}{N_o} = 2^{-t/T_{1/2}}.$$

SOLVE Taking logarithms:

$$\ln\left(\frac{R}{R_o}\right) = -\frac{t}{T_{1/2}} \ln 2.$$

So,

$$t = T_{1/2} \frac{-\ln(R/R_o)}{\ln 2}$$

$$= (3.05 \text{ min})\frac{\ln(35.2)}{0.69315} = 15.7 \text{ min}.$$

ANALYZE The emission rate decreases by a factor of 2 per half-life. In five half-lives, the rate decreases by $2^5 = 32$.

The decay rate R of a radioactive sample is called its *activity*. The SI unit for activity is named after Henri Becquerel.

Digging Deeper

γ-Ray Imaging

Doctors can examine a patient's internal organs by injecting a radioactive substance that is absorbed by the organ and then detecting γ rays from the absorbed substance. The most popular type of detector is the Anger camera (■ Figure 36.27), which measures the location of light pulses (scintillations) produced when γ rays are absorbed in a sodium iodide crystal. The resulting image of the organ is displayed on a television screen (■ Figure 36.28).

The radioisotope most commonly used is ^{99m}Tc, an excited state of ^{99}Tc with a half-life of 6 h. The 140-keV γ ray emitted by ^{99m}Tc is strongly absorbed by NaI, allowing use of a thin detector crystal and a correspondingly light and versatile camera. Furthermore, technetium enters a variety of nontoxic chemical compounds designed for absorption by different organs.

The medical goal of imaging is to make an accurate diagnosis while minimizing the radiation dose to the patient. Trade-offs are inevitable. For example, the radiation dose would be minimized by choosing a radioisotope with a half-life of about 70% of the time for the target organ to absorb the isotope. In thyroid examinations, 2 mCi of sodium pertechnetate ($Na^{99m}TcO_4$) are injected 2 h before the image is taken. The 6-h half-life is not the optimum value, but this disadvantage is much less important than the advantages of efficient camera design and effective medical chemistry.

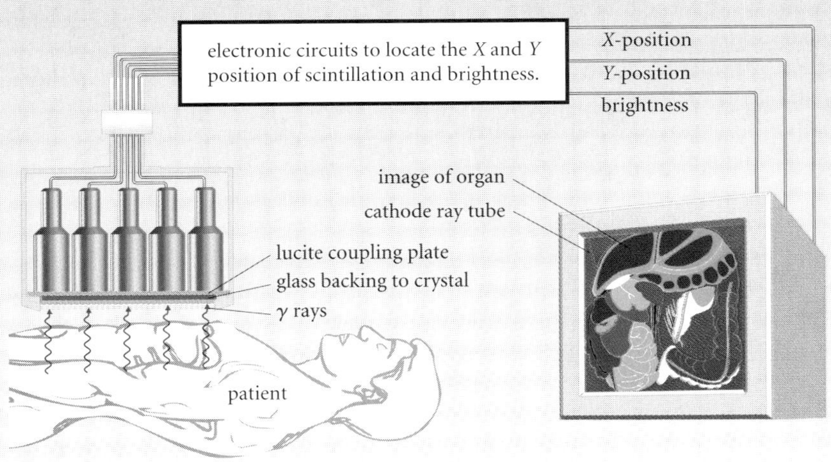

■ FIGURE 36.27
Schematic of an Anger-type camera. The γ rays from nuclear decay interact with the sodium iodide crystal, producing visible light that is detected by the array of photomultiplier tubes (cf. §35.1) and converted to an image on the cathode ray tube.

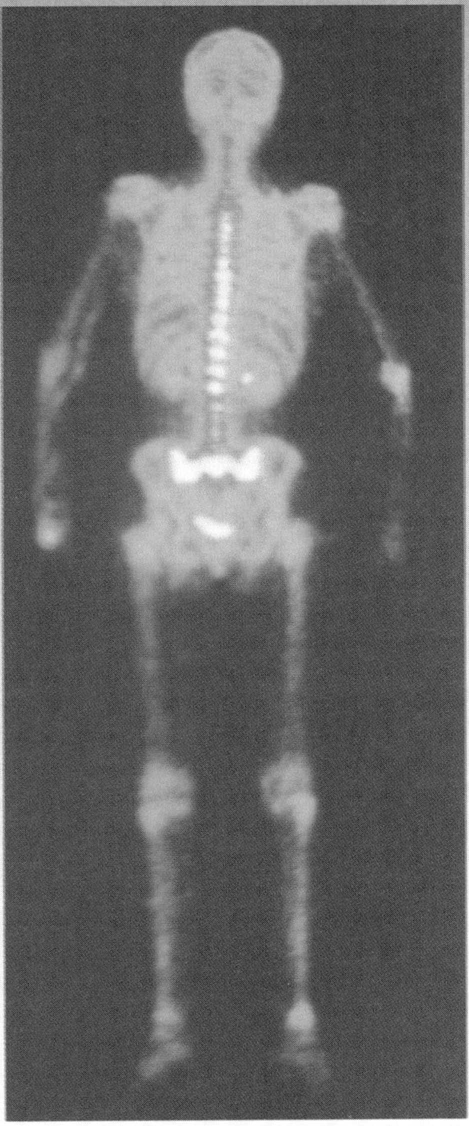

■ FIGURE 36.28
Whole skeleton image produced by an Anger camera using ^{99}Tc.

One *becquerel* (1 Bq) is an activity of one disintegration per second.

$$1 \text{ Bq} \equiv 1 \text{ s}^{-1}.$$

A unit still in common use is named after Marie Curie. The curie (Ci), originally the activity of 1 gram of pure radium metal, is now defined in terms of the becquerel:

$$1 \text{ Ci} \equiv 3.7 \times 10^{10} \text{ Bq}.$$

■ FIGURE 36.29
Radioactive series. (a) The $4n$ series begins with ^{232}Th, which has a half-life of 14 billion years. It passes through elements with mass number $A = 4n$, $n = 58-52$, ending with ^{208}Pb, which is stable. The series is a sequence of α decays (orange lines) and β decays (green lines). In some instances, two possible paths are available. For example, ^{212}Bi α-decays to ^{208}Tl 34% of the time and β-decays to ^{212}Po 66% of the time. The half-life of each element in the series is indicated on the chart.

(b) The $4n + 1$ series begins with ^{237}Np. Its half-life of 2 million years is so short that there is essentially none left in nature. The series ends at the stable isotope ^{209}Bi.

(c) The $4n + 2$ series begins with ^{238}U and ends with ^{206}Pb; ^{234}Th β-decays to an excited state of ^{234}Pa, which can itself β-decay or γ-decay to the ground state, which subsequently β-decays.

(d) The $4n + 3$ series begins with ^{235}U and ends with ^{207}Pb.

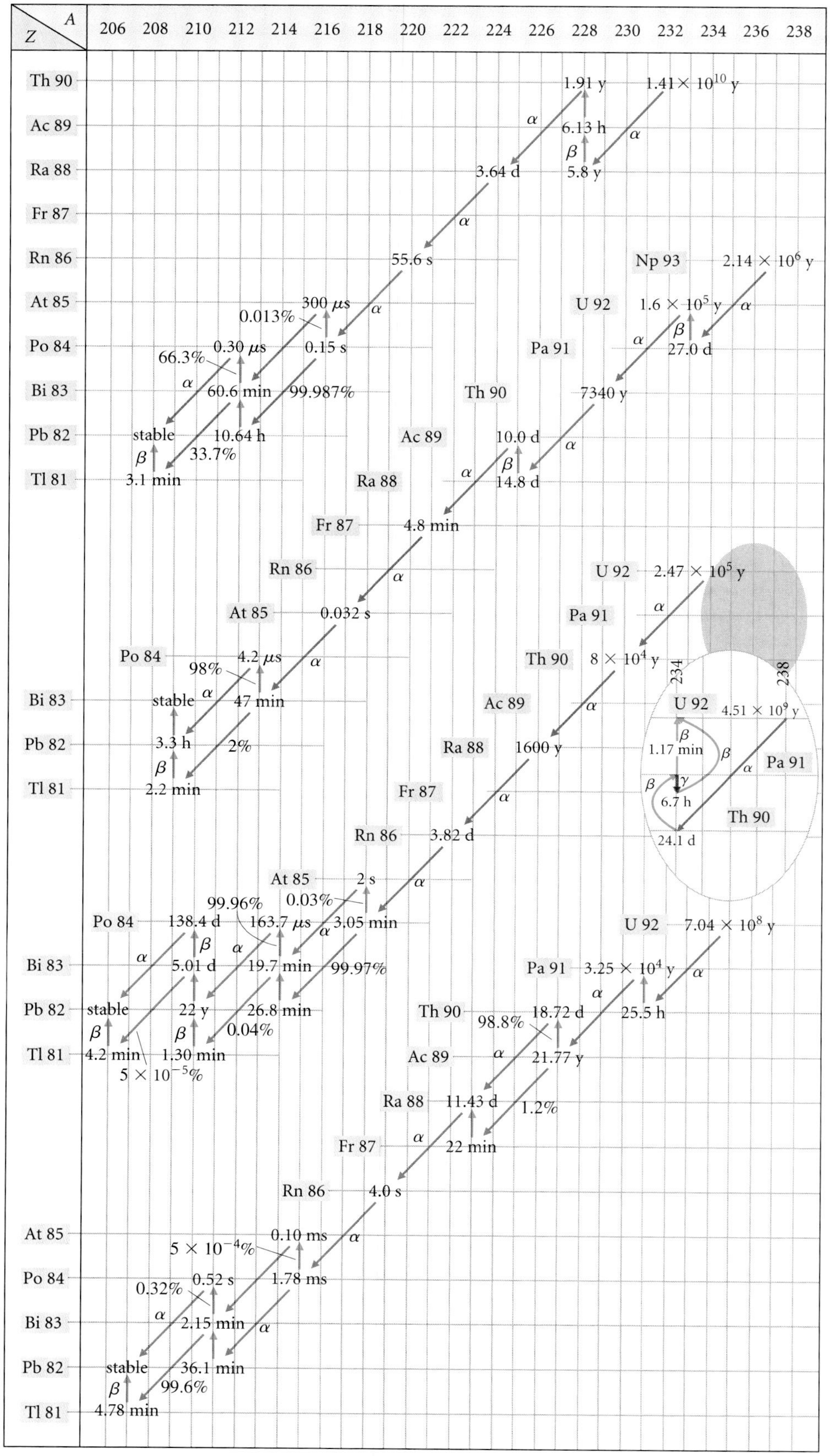

36.2.7 *Radioactive Series*

Among the heavier nuclei, radioactive decay frequently results in an unstable daughter nucleus, which then itself decays to form granddaughter nuclei, and so forth, until at last a stable form is produced. In such a series, β decays do not alter the atomic mass number and α decays each reduce A by 4. So all the nuclear species in a series have mass numbers that differ from each other by whole multiples of 4, and four distinct series exist (■ Figure 36.29). Three of the series are found in nature. In each case, the heaviest isotope has a half-life of a billion or so years. These heavy elements are thought to have been produced by exploding stars shortly before the solar system formed, some 4.5 billion years ago. Elements heavier than the series leader have long since decayed. The fourth possible series begins with the isotope ^{237}Np, whose half-life is too short for a noticeable quantity to remain in nature.

The elements lighter than the series leader have much shorter half-lives. They are continually formed from earlier elements in the series and decay into later members. In a particular sample of material, the series reaches an equilibrium in which each species is formed at the same rate it decays. The series is like a sequence of lakes connected by a mountain stream. Water melts from the parent snowpack and flows through the lakes. The level of each lake adjusts until water flowing in is balanced by water flowing out (■ Figure 36.30).

Radioactive series provide a useful tool for dating geological samples. As ^{238}U in the rock decays, ^{206}Pb is produced at the end of the series. Similarly, ^{235}U decays to ^{207}Pb. Assuming that no lead was present originally, the ratios of ^{238}U/^{206}Pb or ^{235}U/^{207}Pb determine the age of the sample via the law of decay. Of the naturally occurring lead isotopes, only ^{204}Pb is not a product of radioactivity. This isotope may be used to check the assumption of zero initial lead.

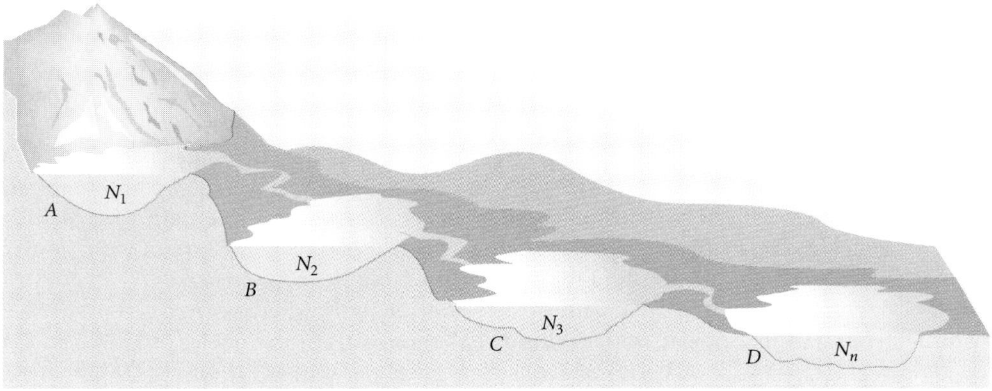

■ FIGURE 36.30
Water analogy for radioactive series. The decay of the parent element is like water pouring from a large reservoir, through a small outlet into a series of lakes and streams. The level of each lake adjusts until water flowing out of it just balances the water flowing in. The water ultimately enters a reservoir with no outlet, which corresponds to the stable end product of the series.

EXAMPLE 36.5 ♦♦ A 1.0-g sample of granite contains 0.16 μg of ^{238}U and 0.13 μg of ^{206}Pb. No ^{204}Pb was found. Estimate the age of the rock.

MODEL The ^{206}Pb is produced by the decay of ^{238}U. Because there is no ^{204}Pb, we assume there was no lead in the rock when it was formed.

SETUP If the sample initially contained N_o atoms of uranium, after a time t, $N(t) = N_o e^{-\lambda t}$ would remain. We assume all the decayed atoms have reached the end of the series. Thus:

$$\frac{N_{Pb}}{N_U} = \frac{N_o(1 - e^{-\lambda t})}{N_o e^{-\lambda t}} = e^{\lambda t} - 1.$$

The ratio of masses is:

$$\frac{M_{Pb}}{M_U} = \frac{A_{Pb}N_{Pb}}{A_U N_U} = \frac{206}{238}(e^{\lambda t} - 1).$$

SOLVE

$$e^{\lambda t} = 1 + \frac{238 M_{Pb}}{206 M_U} = 1 + \frac{238(0.13\ \mu g)}{206(0.16\ \mu g)} = 1.94.$$

$$\lambda t = \frac{\ln(2)}{T_{1/2}} t = \ln(1.94).$$

$$t = T_{1/2}\frac{\ln(1.94)}{\ln(2)} = (4.51 \times 10^9\ y)(0.96) = 4.3 \times 10^9\ y.$$

ANALYZE The result is independent of N_o, the original amount of ^{238}U present. A check on the answer could be obtained from the ratio $^{235}U/^{207}Pb$ in the same sample. ∎

Uranium is present in small quantities in the ground in most parts of the country. One of the products of uranium decay is radon gas (^{222}Rn). With a half-life of about 4 d, radon can diffuse out of the ground before it decays. The gas mixes into the atmosphere and disperses. However, if the gas diffuses through the foundations into a modern home that has been tightly sealed for fuel efficiency, radon can build up within the house, forming an air pollutant that can be a health hazard. Not all radon is of natural origin. Uranium was used in a bright orange paint during the 1930s and 1940s. Some dishes made with this paint have become collector's items that can contribute to the radon level within a home. The hazard can be readily avoided by ensuring that your home is adequately ventilated.

36.3 Nuclear Reactions

Interactions between nuclei depend strongly on the energy available. At energies of ≤ 1 MeV, the nuclei collide elastically, while at energies of several GeV per nucleon, they disintegrate into a spray of small fragments. The subject of reactions in general is much too vast to be dealt with here, so we shall discuss only a few specific reactions of particular interest. The same conservation laws we used in the study of radioactive decay apply to nuclear reactions, although the reactions often involve more complex initial and final states than occur in simple decay.

36.3.1 *Transmutation by Neutron Bombardment*

Because neutrons have no electric charge and are not repelled electrically by a nucleus, even neutrons moving with thermal speeds can penetrate a nucleus and cause a reaction. Once it gets close enough, the neutron is attracted by the strong force of the other nucleons and forms a *compound nucleus* with the same Z but larger A than the original target nucleus. Absorption of the neutron typically releases about 8 MeV of binding energy, leaving the compound nucleus in an excited state that then undergoes γ decay. Decays continue until the resulting product is stable. For example, the resulting nucleus may β-decay to form an atom of one higher Z. The original element is said to be *transmuted* into the new element.

This kind of reaction is used in neutron activation analysis, where a sample is bombarded with neutrons and its original composition is determined from the nuclei produced in the bombardment. One use of the technique is in testing whether valuable paintings are authentic. The induced radiation is short lived, and the painting need not be damaged to remove a sample for chemical analysis. Many of the chemical elements in the solar system were formed by neutron-capture reactions in the interior of earlier stars.

EXAMPLE 36.6 ♦♦♦ Neglecting any kinetic energy of the incident neutron, what are the energies of the γ ray and the electron + neutrino produced in the transmutation of ^{13}C into ^{14}N?

MODEL When the neutron is absorbed by the carbon nucleus, its mass number increases by 1 to form ^{14}C, which then β-decays to ^{14}N, releasing energy Q. The formation of the compound nucleus and its γ decay follow eqn. (36.7). Equation (36.5) describes the subsequent β decay. We neglect the initial kinetic energy of the neutron and any recoil kinetic energy of the ^{14}C nucleus.

SETUP Applying eqn. (36.7):

$$\mathrm{n} + {}^{13}\mathrm{C} \to {}^{14}\mathrm{C}^* \to {}^{14}\mathrm{C} + \gamma.$$

and eqn. (36.5):

$$^{14}\mathrm{C} \to {}^{14}\mathrm{N} + e^- + \bar{\nu}_e + Q.$$

Then, the energy of the photon is:

$$E_\gamma = c^2(\text{neutron mass} + \text{mass of } {}^{13}\mathrm{C} - \text{mass of } {}^{14}\mathrm{C}).$$

SOLVE We take the mass values from Tables 36.1 and 36.2.

$$E_\gamma = c^2(1.00866 + 13.00336 - 14.00324)(\text{u})(931.5\ \text{MeV}/c^2\cdot\text{u})$$
$$= 8.2\ \text{MeV}.$$

Remember: The tabulated masses are atomic masses, so the extra electron is included in the mass of ^{14}N.

SETUP Neglecting recoil of the daughter nucleus, the energy Q released in the β decay is transformed into kinetic energy of the electron and the neutrino.

SOLVE

$$Q = c^2(\text{mass of } ^{14}\text{C} - \text{mass of } ^{14}\text{N})$$
$$= c^2(14.00324 - 14.00307)(\text{u})(931.5\ \text{MeV}/c^2\cdot\text{u})$$
$$= 0.16\ \text{MeV}.$$

ANALYZE Once again, subtraction of numbers that are almost equal greatly reduces the number of significant figures in the results. Note that ^{14}N is ordinary nitrogen, the major constituent of air. ∎

36.3.2 Energy Generation in Stars

A star has to have a central temperature of millions of kelvin so that gas pressure can support its material against its own gravitational attraction. At the same time, the star cools by radiation so that an intense energy source is needed to maintain its temperature. Nuclear reactions provide that energy source. Stars form out of material that is primarily hydrogen, and the most important reaction is the fusion of hydrogen to form helium. In a star like the Sun, the reaction begins with the fusion of two protons to form deuterium:

Arthur Eddington was the first to suggest that nuclear reactions could power the Sun. Critics asserted that the Sun was not hot enough, to which he replied "Then find me a hotter place!"

The Proton–Proton Cycle

Reaction	Energy Release (MeV)
$p + p \rightarrow {}^2\text{H} + e^+ + \nu_e$	1.442
${}^2\text{H} + p \rightarrow {}^3\text{He} + \gamma$	5.493
${}^3\text{He} + {}^3\text{He} \rightarrow {}^4\text{He} + 2p$	12.859

Stars more massive than the Sun have higher central temperatures at which a sequence of reactions catalyzed by carbon, nitrogen, and oxygen synthesize helium more rapidly than in the proton–proton cycle.

Even in the Sun, some He is synthesized via the CNO cycle.

The CNO Cycle

Reaction	Energy Release (MeV)
$p + {}^{12}\text{C} \rightarrow {}^{13}\text{N} + \gamma$	1.944
${}^{13}\text{N} \rightarrow {}^{13}\text{C} + e^+ + \nu_e$	2.221
$p + {}^{13}\text{C} \rightarrow {}^{14}\text{N} + \gamma$	7.550
$p + {}^{14}\text{N} \rightarrow {}^{15}\text{O} + \gamma$	7.293
${}^{15}\text{O} \rightarrow {}^{15}\text{N} + e^+ + \nu_e$	2.761
$p + {}^{15}\text{N} \rightarrow {}^{12}\text{C} + {}^4\text{He}$	4.965

The carbon, nitrogen, and oxygen nuclei are not consumed by the reaction cycle so that a relatively small number of these nuclei can process a large number of the star's protons into helium.

Because a proton needs a large amount of kinetic energy to overcome the Coulomb repulsion between nuclei, both of these reaction cycles are extremely temperature-sensitive. Under conditions in stellar interiors, the rate of the proton–proton cycle is proportional to T^{12} and the rate of the CNO cycle to T^{20}. It is much easier for two protons to come close enough to react than for a proton to approach a more heavily charged CNO nucleus. However, a compound nucleus formed from two protons is much more likely to separate again rather than become ^{2}H via β decay. The compound nuclei in the CNO cycle tend to β-decay rather than dissociate. Consequently, the CNO is the more efficient cycle at temperatures where it occurs at all.

Digging Deeper

We Are Children of the Stars

During their lives, stars serve as factories for fusing hydrogen into heavier elements (■ Figure 36.31a). Helium formation releases the greatest binding energy and lasts for the greatest part of the star's lifetime (about 10^{10} y for the Sun but as short as 10^{6} y for the biggest and brightest stars). To maintain energy production as its core is converted to helium, the star contracts, causing its temperature to increase and the zone of hydrogen fusion to move outward in a shell surrounding the helium core. Gravitational contraction raises the central temperature until helium can fuse into carbon:

Triple α Process

$$^{4}\text{He} + {}^{4}\text{He} \rightarrow {}^{8}\text{Be} + \gamma.$$

$$^{4}\text{He} + {}^{8}\text{Be} \rightarrow {}^{12}\text{C} + \gamma.$$

This reaction stabilizes the star for a few thousand years until a carbon core builds up, requiring further contraction.

In small stars like the Sun, the core stops contracting because its electrons cannot move closer together without sharing quantum numbers and thus violating the Pauli exclusion principle. The core no longer has to remain hot to support itself, and the star ends its life as a slowly cooling object about the size of the Earth. In larger stars, the core becomes hot enough to start a carbon-burning reaction. Eventually, such stars build an onionlike structure, ■ Figure 36.31b, adding layers ever faster until an iron core forms at the center.

At this point, the star runs out of nuclear resources. Very high temperatures are required to overcome the Coulomb repulsion between large nuclei, and a large flux of γ rays is associated with matter at high temperatures (see §35.1.3). These γ rays drive inverse nuclear reactions that dissociate the large nuclei. The balance of fusion and fission tends to produce a majority of ^{56}Fe nuclei, which have the maximum binding energy per nucleon. The continuing loss of energy to neutrinos, which escape the star, causes its ultimate demise. Contraction proceeds rapidly, and catastrophe occurs in a few hours as the temperature of the core continues to increase and dissociation of iron nuclei by γ rays dominates fusion.

The star is forced to return the nuclear energy it has used for millions of years. The core collapses in less than a millisecond, squeezing itself into a ball of neutrons and emitting a flash of some 10^{57} neutrinos (Figure 36.23). Unsupported by the collapsed core, the outer layers fall inward and heat up. Nuclear reactions in the still fresh outer material, together with the neutrino flash, produce a vast explosion that reverses the infall and hurls the outer layers back into space. This material, much of it now processed into heavy elements, mixes with the interstellar gas. Millions of years later, these elements join in the formation of new stars and provide the stuff to make planets and people. We are the product of several generations of stars that preceded the Sun.

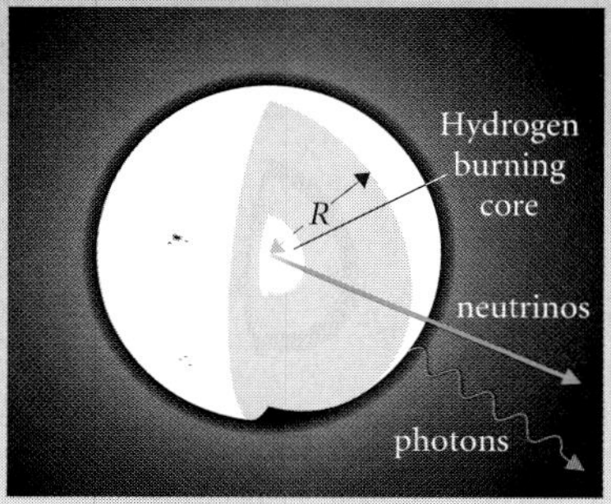

(a)

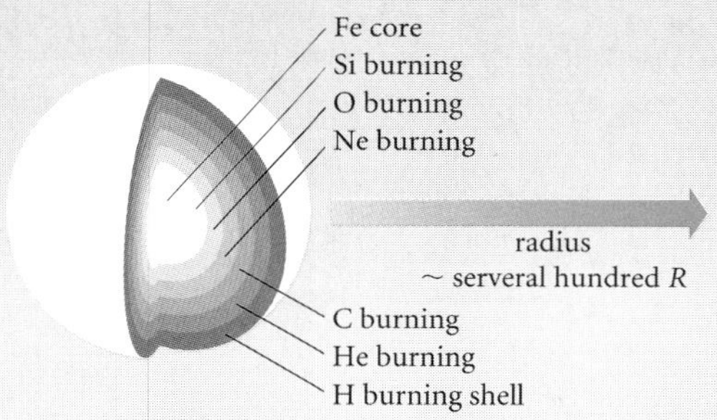

(b)

■ Figure 36.31
Structure of a star at various stages of its evolution. (a) During most of its lifetime, a star burns H to He in its inner core. (b) Late red giant phase. Shells have developed around the core. In each shell, a different nuclear reaction occurs, the heavier elements being produced closer to the center. The process terminates when an iron core is formed because elements heavier than iron are less tightly bound (Figure 36.9). If the star is sufficiently massive, it collapses violently, followed by a supernova explosion (Figure 36.24). Heavy elements can be formed by neutron bombardment during the explosion.

The difficulty in forming deuterium (^{2}H) from protons is equally significant for fusion power research, where the chief problem is to confine a dense, hot plasma long enough for the plasma nuclei to form helium. Current designs use deuterium and tritium (^{3}H) as fuel and avoid the troublesome proton-fusion step.

36.3.3 *Fission*

Nuclear *fission* is the breakup of a nucleus into smaller fragments.

Fission reactions are possible because the binding energy per nucleon for elements heavier than iron decreases with increasing mass number. Massive nuclei can thus release potential energy by reorganizing into smaller nuclei (■ Figure 36.32). Spontaneous fission, a common mode of decay for artificially produced nuclei more massive than uranium, also occurs in uranium, though at a much lower rate than α decay. In nuclear power plants, induced fission results when a free neutron is absorbed by a uranium (235) nucleus. The 8 MeV released when the neutron is absorbed leaves an excited compound nucleus that has a high probability of fission.

The fission reaction is useful because it is a messy reaction that releases neutrons as well as daughter nuclei. If these neutrons strike other uranium nuclei and result in further fissions, it is possible for a *chain reaction* to release energy from every atom in a macroscopic quantity of material. The uranium sample need only be large enough to give a daughter neutron a high probability of causing fission before escaping the sample or being absorbed.

The probability of a neutron causing fission is described by the cross section, or effective target area, σ, of a uranium nucleus. We showed in §22.7 that a molecule traveling through a gas with n molecules per cubic meter, each with cross-sectional area σ, travels an average distance $\ell = 1/n\sigma$ before striking another molecule. A similar formula describes the average distance a neutron travels through a sample of uranium before causing a fission reaction. Thus ℓ gives the size of a uranium sample that can chain-react. The criterion for a chain reaction is usually described by a critical mass. Roughly speaking, this is the mass in a sphere of radius ℓ. The critical mass depends on the concentration of uranium in the sample, but it is typically of the order of a kilogram.

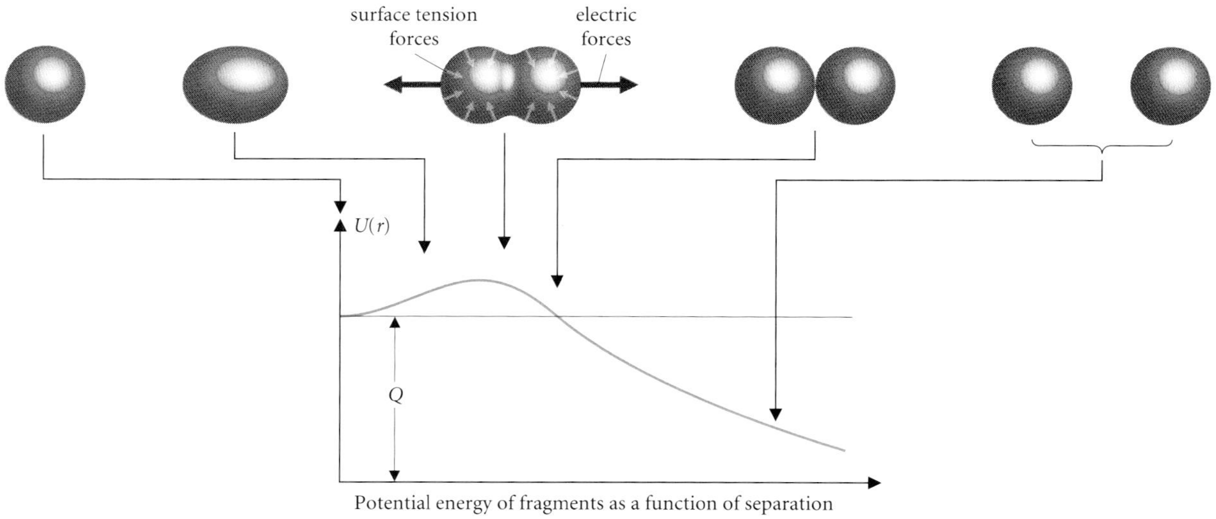

■ FIGURE 36.32
The liquid drop model. Nuclear fission may be understood with a model that, in contrast to the shell model, ignores the behavior of individual nucleons and treats the nucleus as a drop of liquid. The nucleons play the role of molecules of the liquid, and the nuclear forces are treated like the surface tension in a liquid. The fission instability arises when the energy released by absorption of a neutron leaves the nucleus rotating and thus with a larger diameter perpendicular to the rotation axis. Electrical repulsion tends to stretch the nucleus. Nuclear forces are less effective in the distorted, rotating nucleus. As the two parts of the nucleus separate, the surface tension tends to pull them into separate spheres rather than opposing the electric force and holding them together.

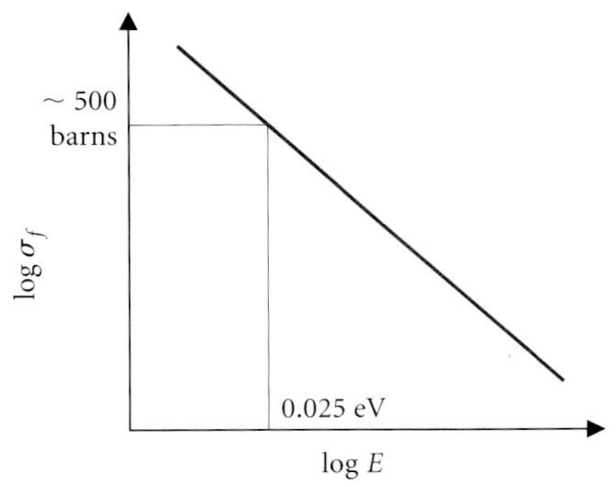

■ FIGURE 36.33
Fission cross section of ^{235}U as a function of neutron energy. The unit of nuclear cross section, the *barn,* is equal to 10^{-28} m^2.

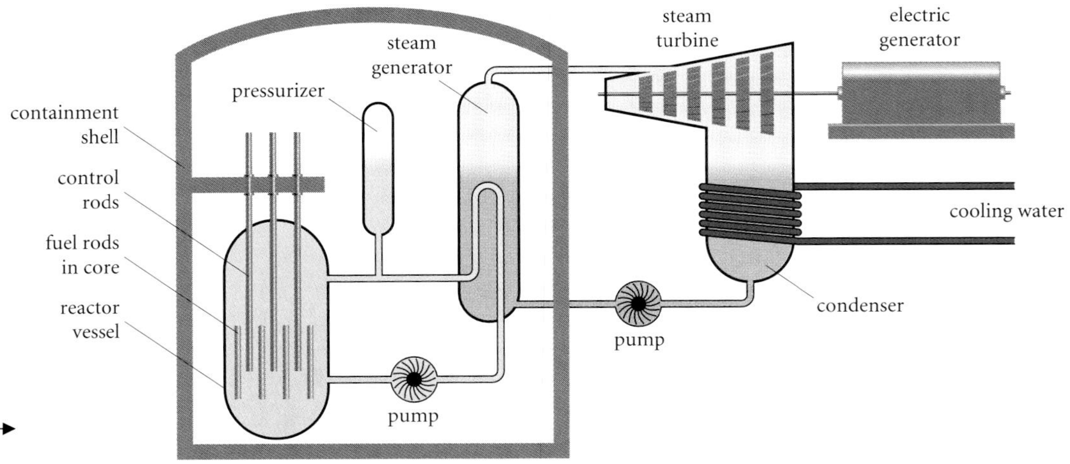

■ FIGURE 36.34
Schematic of a pressurized water reactor. Pellets of uranium fuel are placed in fuel rods and immersed in water, which serves both as coolant and moderator of neutrons. Control rods absorb neutrons and stop the reaction when inserted into the reactor core.

PLUTONIUM AND THORIUM ARE ALSO USED. THORIUM IS A NATURALLY OCCURRING ELEMENT THAT PLAYS A MINOR ROLE IN TECHNOLOGY. PLUTONIUM IS PRODUCED IN URANIUM REACTORS AND THEN ITSELF USED AS FUEL.

The fission cross section of the fissionable isotope of uranium, ^{235}U, decreases rapidly at high neutron energy (■ Figure 36.33). For the neutrons released in fission of uranium, the cross section is so low that an impractical amount of material would be needed for a reactor, unless the neutrons can be slowed down, or *moderated.* ■ Figure 36.34 shows the schematic of a typical reactor operated in the United States. The uranium is concentrated in fuel pellets that are much too small to sustain a chain reaction. The water surrounding the pellets serves both to transport the heat released by the reactor and to moderate the neutrons. Collisions with the nearly equal-mass hydrogen nuclei in the water efficiently transfer the neutrons' kinetic energy to the water. The control rods are usually made of cadmium, which absorbs neutrons efficiently. The reactor is designed so that, with the control rods completely inserted, any neutron is almost certainly absorbed before causing a fission. In operation, the rods are adjusted so that the neutrons from each fission, on average, produce precisely one new fission.

EXAMPLE 36.7 ♦♦ An induced fission reaction is described by:

$$n + {}^{235}U \rightarrow {}^{236}U^* \rightarrow {}^{141}Ba + {}^{92}Kr + 3n + Q.$$

Estimate the energy of each neutron produced in this reaction.

MODEL Since the neutrons are much less massive than the two nuclei that are produced, nearly all the energy released is shared by the three neutrons. For this estimate, we may assume that, on average, they share the energy equally.

SETUP The energy released is due to the mass difference between the reacting particles and the fission products.

$$\begin{aligned} Q &= c^2[\text{mass of } {}^{235}U - (\text{mass of } {}^{141}Ba + {}^{92}Kr + 2n)] \\ &= c^2[235.0439 - 140.9141 - 91.9262 - 2(1.0087)](u)(931.49 \text{ MeV}/c^2\cdot u) \\ &= (0.1862)(931.49) \text{ MeV} \\ &= 173.4 \text{ MeV}. \end{aligned}$$

SOLVE The energy per neutron is $Q/3 = 57.8$ MeV.

ANALYZE This energy is off scale in Figure 36.33. The cross section for one of these unmoderated neutrons to produce another fission is tiny. ■

Chapter Summary

Where Are We Now?

In this chapter we have delved into the atomic nucleus to find that it is composed of more fundamental particles—nucleons—and that these particles can be split apart and rearranged in nuclear reactions, thus changing one chemical species into another. In our efforts to understand the reactions between the nucleons, we have identified the strong and weak nuclear forces and the quantum mechanical quantities of lepton number and baryon number. In the next chapter we shall pursue our quest to even smaller sizes and observe that the nucleons themselves are composite structures.

What Did We Do?

Atomic nuclei are made up of nucleons: neutrons and protons. A nuclear species is described by its atomic number Z and atomic mass number A. A nucleus contains Z protons and $N \equiv A - Z$ neutrons. The two kinds of nucleon have similar masses. Protons each have electric charge $+e$ and neutrons are electrically neutral.

Besides the electrical force we have encountered previously, two kinds of force act between nucleons: the strong force and the weak force. The strong force has short range and cannot bind nuclei larger than a few fermis in radius. Stable nuclei generally contain more neutrons than protons since neutrons enhance the stabilizing strong force but do not share the electrical repulsion of the protons. The weak force is responsible for reactions that change neutrons to protons (β decay), and vice versa. Nuclei with too large an excess of neutrons are unstable against β decay.

There are three principal categories of natural radioactivity. In α decay, a parent nucleus decays to a daughter with the emission of an α-particle (helium nucleus). Both the charge and mass of the nucleus are changed.

$$\alpha \text{ decay: } {}^{A}Z \rightarrow {}^{(A-4)}(Z - 2) + {}^{4}\text{He} + Q.$$

In β decay, the weak force causes the decay of a neutron within the nucleus, changing the charge of the nucleus but not its mass number A.

$$\beta \text{ decay: } {}^{A}Z \rightarrow {}^{A}(Z + 1) + e^{-} + \bar{\nu}_e + Q.$$

In γ decay, a nucleus in an excited state decays to a lower energy state with the emission of a photon (γ ray). All such reactions may be analyzed using conservation laws for energy, charge, lepton and baryon number, and linear and angular momentum. To conserve energy, nuclei always decay to a combination of particles whose total mass is less than or equal to that of the original nucleus. Applying the conservation laws to β decay, we find that a particle called the neutrino must exist. The neutrino is a lepton with zero charge and zero (or very small) mass.

The rate of particle emission by a sample of radioactive material decreases exponentially according to the *law of decay:*

$$\frac{dN}{dt} = -\lambda N.$$

The half-life of a nuclear species is the time required for half of a large sample to decay.

Two important classes of nuclear reactions are fusion, in which elements lighter than iron combine to form more massive nuclei with the release of energy, and fission, in which nuclei heavier than iron split into smaller nuclei, again releasing energy. Fusion of hydrogen to helium provides the energy source in the Sun and other stars, while fission of uranium and related elements is used in nuclear power plants.

Practical Applications

Nuclear power plants have become a major source of energy worldwide and are also used to power some ships and submarines. Nuclear decay has applications in the diagnosis and treatment of various ailments, including thyroid disease and cancer. Radioactive decay provides a reliable method for finding the ages of archaeological and geological samples. Nuclear fusion of deuterium and tritium into helium is being developed as a possible power source for the next millennium.

Solutions to Exercises

36.1 From Figure 36.9, the binding energy per nucleon for $A = 24$ (^{24}Mg) is $B/A \approx 8.3$ MeV, while $B/A \approx 7.1$ MeV at $A = 4$. The energy released by fusing six helium nuclei into one ^{24}Mg nucleus is:

$$\Delta E = 24(8.3 \text{ MeV}) - 24(7.1 \text{ MeV}) = 24(1.2 \text{ MeV}) = 29 \text{ MeV}.$$

Equivalently, 1.2 MeV/nucleon equals:

$$\frac{(1.2 \text{ MeV/nucleon})(1.6 \times 10^{-13} \text{ J/MeV})}{1.7 \times 10^{-24} \text{ g/nucleon}} = 1.1 \times 10^{11} \text{ J/g}.$$

36.2 The equation for β decay of ^{12}B is:

$$^{12}\text{B} \rightarrow {}^{12}\text{C} + e^- + \bar{\nu}_e + Q.$$

The tabulated mass for ^{12}C includes a carbon nucleus and six electrons, while the tabulated value for ^{12}B includes a nucleus and only five electrons. The neutrino is assumed massless. Therefore, the energy released in the reaction is:

$$Q = [M(^{12}\text{B}) - M(^{12}\text{C})]c^2$$
$$= [(12.0144 - 12.0000) \text{ u}](931.49 \text{ MeV/u}) = 13.4 \text{ MeV}.$$

36.3 In the general case of positron emission, the two nuclei ^AZ and $^A(Z-1)$ have zero lepton number and baryon number A. The positron and the electron neutrino have zero baryon number. The positron, being the antiparticle of the electron, has the opposite lepton number of the electron—that is, -1. The total lepton number equals that of the parent nucleus, zero; so the neutrino has lepton number $+1$.

36.4 The photon has momentum $p_\gamma = E/c$. The momentum of the nucleus is equal and opposite to that of the γ, and its kinetic energy is given by the nonrelativistic expression:

$$E_{\text{Fe}} = \frac{p^2}{2M}$$
$$= \frac{[(14.4 \text{ keV})(1.60 \times 10^{-16} \text{ J/keV})/(3.00 \times 10^8 \text{ m/s})]^2}{2(57)(1.66 \times 10^{-27} \text{ kg})}$$
$$= 3.12 \times 10^{-22} \text{ J}$$
$$= 1.95 \times 10^{-3} \text{ eV}.$$

From Heisenberg's principle, the uncertainty in the photon's energy is:

$$\Delta E = \frac{\hbar}{\Delta t} = \frac{1.05 \times 10^{-34} \text{ J}\cdot\text{s}}{(10^{-7} \text{ s})(1.6 \times 10^{-19} \text{ J/eV})} \approx 7 \times 10^{-9} \text{ eV}.$$

For the photon to be reabsorbed by another iron nucleus, its energy (in the rest frame of the absorbing nucleus) must not be shifted by more than the Heisenberg uncertainty. The recoil energy of the nucleus is large enough to prevent reabsorption. To suppress the recoil energy to an amount less than ΔE requires that M be increased by a factor of 3×10^5. This can be achieved if the nucleus couples to a crystal containing 3×10^5 atoms.

36.5 Using eqn. (36.11) in eqn. (36.9), we have:

$$N(t) = N_o \exp[-(\ln 2)t/T_{1/2}].$$

Since $e^{ab} = (e^a)^b$, we may write the law of decay as:

$$N(t) = N_o(e^{\ln 2})^{-t/T_{1/2}}$$
$$= N_o 2^{-t/T_{1/2}}.$$

Basic Skills

Review Questions

§36.1 BASIC NUCLEAR STRUCTURE

- What is the *atomic number* Z of an element?
- What is the *atomic mass number* A of an element? What rule gives the relation of A to Z?
- Which nuclear property determines the chemical behavior of an atom? How do *isotopes* of an element differ?
- How is the *atomic mass unit* defined?
- How is the radius of a nucleus related to its mass number A?
- Approximately how big is a carbon nucleus?
- What are the three basic constituents of atoms?
- What are *nucleons*?

- What are the two kinds of nuclear force? Which is responsible for holding nuclei together?
- What is β decay? Which nuclear force is responsible for this process?
- What is the *binding energy* of a nucleus?

§36.2 NATURAL RADIOACTIVITY

- List three types of natural radioactivity.
- Which quantities are conserved in radioactive decay?
- Describe α decay. How are the mass and charge of the daughter nucleus related to those of the parent?
- What particles emerge from a nucleus that undergoes β decay?
- Which nuclear quantities change as a result of γ decay?
- What is a *positron?* Which of its parameters differ from those of an electron?
- What happens when a particle and its antiparticle collide?
- What is the *law of decay?*
- What is the *half-life* $T_{1/2}$ of an atomic species? How is it related to the decay probability λ?
- What is the *activity* of a radioactive sample? In which units is it measured?
- How many radioactive series are there? How many are found in nature?

§36.3 NUCLEAR REACTIONS

- What happens when neutrons bombard a relatively small nucleus like carbon? What happens if the target nucleus is large, like uranium?
- What is the proton–proton cycle? Why is it important?
- Why does a nuclear reactor need water to moderate the neutrons?

Basic Skill Drill

§36.1 BASIC NUCLEAR STRUCTURE

1. Write the standard symbols for nuclei of yttrium containing 90 nucleons and for aluminum nuclei with 27 nucleons.
2. Natural lithium has a chemical atomic mass of 6.941 u. It consists of isotopes with atomic masses of 6.015123 u and 7.016005 u. Find the percentages of the two isotopes in the natural mixture.
3. Calculate the binding energy of a deuterium nucleus (^{2}H).
4. Some nuclear theorists speculate that a nucleus with atomic number $Z = 110$ and atomic mass number $A = 294$ might be stable. What would be the radius of such a nucleus?

§36.2 NATURAL RADIOACTIVITY

5. Write a formula for the α decay of ^{183}Au.
6. Write a formula for the β decay of ^{8}Li.
7. What is the energy of γ rays produced in the annihilation of a proton/antiproton pair?
8. By how much does the mass of a ^{60}Co nucleus change when it emits a 58.6-keV x ray?
9. ^{23}Ne has a half-life of 37.6 s against β decay. What is the activity of 15 mg of ^{23}Ne?
10. A natural sample of ^{235}U ore contains 0.056 mg of ^{227}Th. How much ^{235}U is in the sample?

§36.3 NUCLEAR REACTIONS

11. What nucleus is produced when a ^{59}Co nucleus captures a neutron? What is the Q value for this reaction?
12. How much energy is liberated in the triple α process in which three α-particles combine to produce a ^{12}C nucleus?
13. When ^{235}U absorbs a neutron, one possible reaction produces ^{143}La and ^{87}Br. How many neutrons are produced? How much energy is liberated?

Questions and Problems

§36.1 BASIC NUCLEAR STRUCTURE

14. ❖ A stable isotope is known to have a mass number $A = 45$. Within roughly what range of atomic number Z would you expect the isotope to be?
15. ❖ Discuss the factors that place lower and upper limits on the size of nuclei with a given atomic number. Which nucleus would you expect to be more stable, ^{7}Li or ^{8}Li?
16. ♦ Natural boron is a mixture of ^{10}B and ^{11}B and has a chemical atomic mass of 10.82 u. What is the isotope ratio of natural boron?
17. ♦ Natural lead contains the isotopes ^{204}Pb (1.4%), ^{206}Pb (24.1%), ^{207}Pb (22.1%), and ^{208}Pb (52.4%). Calculate the chemical atomic mass, and compare with the value in the periodic table.
18. ♦ Calculate the binding energy of a ^{12}C nucleus.
19. ♦ Use eqn. (36.1) to estimate the size of a californium nucleus ($Z = 98$) with $A = 245$. (This artificially produced nucleus has a lifetime of 8500 y.)
20. ♦♦ A deuteron (^{2}H) consists of a neutron and proton confined within a distance of approximately 1.5 fm. Use Heisenberg's uncertainty principle to estimate the kinetic energy of the proton and neutron. Compare your result with the binding energy of the deuteron.
21. ♦♦ Calculate the binding energy per nucleon for tritium (^{3}H), and compare with the result for ^{3}He.

§36.2 NATURAL RADIOACTIVITY

22. ❖ Explain what is wrong with each of the following reaction equations. There may be more than one error in each.

$$^{227}\text{Th} \rightarrow {}^{223}\text{Ca} + \alpha.$$

$$^{15}\text{C} \rightarrow {}^{15}\text{B} + e^+ + \bar{\nu}_e.$$

$$^{3}\text{H} \rightarrow {}^{3}\text{He} + e^+ + \nu_e.$$

23. ❖ It is never possible for the activity of a radioactive substance to increase with time. Is this statement true or false? Explain your reasoning.

24. ❖ A great, great . . . granddaughter species in a radioactive series can never have an activity greater than that of the parent species. Is this statement true or false? Explain your reasoning.
25. ❖ Under what circumstances can a great, great . . . granddaughter species in a radioactive series have an activity less than that of the parent species?
26. ❖ The oxygen isotopes ^{14}O and ^{19}O both decay by β emission. Which do you expect to emit an electron, and which a positron? Why?
27. ❖ State the rule for computing Q, the energy released by a positron decay, in terms of tabulated atomic masses. Show explicitly how the rule accounts for the extra electrons included in the atomic masses.
28. ❖ Recent observations of the central regions of our galaxy show strong γ-ray emission at an energy of 0.511 MeV. How do you suppose these photons may be produced?
29. ❖ Suppose the reaction $n \rightarrow \gamma$ were possible, in violation of baryon conservation. Would it be possible to have nuclear species heavier than hydrogen? Why or why not?
30. ❖ Since antielectrons, antiprotons, and antineutrons all exist, why do we not encounter antinuclei, or even antipeople? (*Hint:* What would be the major experimental problem with maintaining a macroscopic quantity of antimatter?)
31. ❖ Referring to Figure 36.26, explain why the shell model requires large, stable nuclei to contain more neutrons than protons.
32. ♦ What energy photons would result from the annihilation of a neutron and an antineutron?
33. ♦ Write a formula for the α decay of each of the following nuclear species: ^{224}Th, ^{211}Bi, ^{243}Cm, ^{221}Fr, ^{215}At.
34. ♦ Write a formula for the β decay of each of the following nuclear species. Determine from Table 36.1 and/or Figure 36.3 which are electron emitters and which are positron emitters: ^{12}B, ^{34}Cl, ^{140}La, ^{209}Pb, ^{138}Pr.
35. ♦♦ Compute the activity of a 1.0-g sample of ^{226}Ra, whose half-life is 1622 y.
36. ♦♦ Use tabulated data to show that ^{56}Fe is stable against α decay.
37. ♦♦ A ^{60}Co source has a half-life of 5.2 y and an initial activity of 1100 Ci. Express its activity in becquerels. What is its activity after 40 y? How many cobalt atoms does it contain initially?
38. ♦♦ A sample of ^{131}I has a half-life of 8.06 d and an initial activity of 2.5×10^8 Bq. What will be the total number of disintegrations from the sample?
39. ♦♦ The mass of ^{64}Cu is 63.929759 u. It decays by positron emission to ^{64}Ni, which has mass 63.92796 u. What is the maximum energy of positrons emitted by a ^{64}Cu sample?
40. ♦♦ Find the number of γ rays emitted inside a patient given a 2-mCi standard dose of ^{99m}Tc ($T_{1/2} = 6$ h) for a thyroid examination.
41. ♦♦ What is the energy of α-particles emitted by ^{152}Sm?
42. ♦♦ The half-life of ^{239}Pu is 2.439×10^4 y. By how much is the activity of a sample decreased after 500 y?
43. ♦♦ Find the energy released in the α decay of ^{212}Rn.
44. ♦♦ Find the energy released in the β decay of ^{16}N.
45. ♦♦ A ^{226}U nucleus undergoes five successive α decays before reaching a stable form. What is the final nucleus?
46. ♦♦ The ^{235}U in a sample of ore has an activity of 3.4×10^5 Bq. How much ^{227}Th is in the sample?
47. ♦♦ A sample of thorium ore emits 1.0×10^7 Bq of β radiation. How much thorium is contained in the sample?
48. ♦♦ A sample of ^{65}Ni decays by β emission with a half-life of 25.20 h. At $t = 0$, a sample of ^{78}As ($T_{1/2} = 1.515$ h) has 4.00 times the β activity of the nickel sample. At what time will the two samples have equal activities?
49. ♦♦♦ A technician wishes to assay a prepared sample of ^{112}Ag but knows it is contaminated with ^{111}Pd. Both nuclides undergo β decay, the silver with a half-life of 3.13 h and the palladium with a half-life of 22 min. The initial activity of the sample is 6.35×10^7 Bq. After 1.00 h, the activity has decreased to 4.50×10^7 Bq. What are the initial activities of the silver and the palladium? How long must the technician wait for the palladium contaminant to contribute less than 0.1% of the activity?
50. ♦♦♦ A subdivision is to be built on land where the waste rock from a uranium mine was discarded 10 y ago. Assume that all of the original 30 $g \cdot m^{-3}$ of uranium was removed and all of the gaseous daughter elements escaped from the waste. Estimate the remaining activity per cubic meter of the waste rock.
51. ♦♦♦ A geochemist wants to know the age of a granite rock sample. Analysis shows the sample to contain 600 g of ^{206}Pb for every kilogram of ^{238}U. We may assume the sample has been undisturbed since its formation, at which time it contained no ^{206}Pb. How old is the sample?
52. ♦♦♦ A sample initially contains a mixture of two isotopes with decay constants λ_1 and λ_2 and initially equal activities. Find an expression for the ratio of the activities of the two isotopes at later times. After what time will the shortest-lived isotope contribute less than 1% of the total activity?
53. ♦♦♦ Show that the minimum number of parent nuclei N_o required to obtain a given decay rate R after a time t_a occurs when the half-life of each nucleus is $T_{1/2} = t_a \ln 2$. Find the minimum number N_o. (This result has application in medicine, where a sufficient decay rate R for a procedure should be obtained with a minimum total radioactive dose to the patient.)

§36.3 NUCLEAR REACTIONS

54. ❖ Give reasons why the following reactions cannot occur:

$$^{23}\text{Na} + \text{n} \rightarrow {}^{19}\text{F} + \alpha.$$

$$^{3}\text{H} + {}^{2}\text{H} \rightarrow {}^{4}\text{He} + \text{p}.$$

55. ❖ Suppose you wish to investigate the amounts of copper and zinc in the paint of a supposed masterpiece using neutron activation analysis. What radioactive elements would you expect to produce? (Use Figure 36.3.)
56. ❖ Using Table 36.3, determine which elements are likely to be produced when ^{238}U is exposed to an intense neutron flux in a reactor. Do you see why plutonium is important in nuclear technology?
57. ❖ Consider some of the fission products that are likely to be produced in the fission reactions $^{235}\text{U} \rightarrow X + Y + 3\text{n}$. Are the products stable? Do you see why a fission reactor continues to require a large flow of cooling water even after the fission reaction is shut down?
58. ♦ Calculate the energy released in forming a ^{16}O nucleus from a ^{12}C nucleus plus an α-particle.
59. ♦ A neutron is absorbed by an unknown target atom. After a γ decay and a β^+ decay, the product is observed to be ^{65}Cu. What was the target species?
60. ♦ What is the unknown product X in each of the following reactions?

$$^{14}\text{N} + \alpha \rightarrow X + \text{n}.$$

$$^{12}\text{C} + \gamma \rightarrow X + \alpha.$$

$$^{7}\text{Li} + \text{p} \rightarrow X + \text{n}.$$

$$^{31}\text{P} + {}^{2}\text{H} \rightarrow X + \text{p}.$$

61. ♦ Find the energy Q released by the fusion reaction:

$$^{2}\text{H} + {}^{2}\text{H} \rightarrow {}^{3}\text{He} + \text{n} + Q.$$

TABLE 36.3 **Half-lives of Heavy Elements Near ^{238}U**

Z \ A	237	238	239	240	241	242
91 Pa	8.7 min	2.3 min				
92 U	6.75 d	4.5×10^9 y	23.5 min	14.1 h		
93 Np	2×10^6 y	2.12 d	2.35 d	7.5 min	16.0 min	
94 Pu	1.1 μs	87.8 y	8 μs	6540 y	23 μs	4×10^5 y
95 Am	75 min	0.06 s	11.9 h	0.9 ms	1.5 μs	14 ms

62. ♦♦ Each of the following reactions leads to the same compound nucleus X. What is that nucleus?

$$^{17}\text{O} + {}^{3}\text{He} \to X.$$
$$^{14}\text{N} + {}^{6}\text{Li} \to X.$$
$$^{10}\text{B} + {}^{10}\text{B} \to X.$$

What is the remaining reaction product when nucleus X emits each of the following: α, ^{2}H, ^{3}H, ^{9}Be?

63. ♦♦ What compound nucleus results when an α-particle is incident on a ^{19}F nucleus? What is the remaining reaction product when the compound nucleus emits each of the following: α, ^{2}H, ^{3}H, ^{9}Be?

64. ♦♦ Assume that a star consuming helium by the triple α process produces ten times the power of the Sun and can consume 0.3 solar masses of helium. How long will the star's helium-burning phase last? You will need the result of Problem 12.

65. ♦♦ What is the minimum antineutrino energy necessary to cause the reaction:

$$\bar{\nu}_e + \text{p} \to \text{n} + \text{e}^+?$$

66. ♦♦ What is the minimum-energy photon necessary to remove a neutron from magnesium in the reaction $^{24}\text{Mg} + \gamma \to {}^{23}\text{Mg} + \text{n}$? Neglect any kinetic energy of the reaction products.

67. ♦♦ Find the value of Q for the reaction $^8\text{Be} \to {}^4\text{He} + {}^4\text{He}$. Compare your result with the electrostatic energy of two α-particles, modeled as uniformly charged spheres barely in contact.

68. ♦♦ The wear of automobile parts is frequently tested by neutron activation. For example, a 30-g steel piston ring is exposed to neutrons to form ^{59}Fe with a half-life of 45 d and an activity of 15 μCi. The ring is then installed in a car, and the oil is removed after 30 d of testing. If the oil then shows an activity of 6.3 Bq, how much steel has worn from the ring?

69. ♦♦ The remains of a hominid woman, named Lucy by her discoverer, were found in Africa in 1974, forcing anthropologists to reevaluate their theories of human origins. The bones were embedded in rock containing potassium. A fraction $f = 1.18 \times 10^{-5}$ of the potassium was ^{40}K, which decays to ^{40}Ar with a half-life of 1.28×10^9 y. For each 0.1 g of potassium, the rock contained 2.39×10^{-9} g of ^{40}Ar. Contamination by the atmosphere accounts for 7.25×10^{-10} g of the argon. What is the age of the fossils?

70. ♦♦ How much energy is released when an α-particle combines with ^{52}Cr to form ^{56}Fe? What is the Q value for the reaction $^{54}\text{Cr} + {}^{55}\text{Mn} \to {}^{109}\text{In}$ (mass 108.9071 u)?

71. ♦♦ Find the energy released in the reaction $^{12}\text{C} + {}^{12}\text{C} \to {}^{24}\text{Mg}$.

72. ♦♦ Calculate the energy per gram released in the fusion reaction sequence:

$$^2\text{D} + {}^2\text{D} \to {}^3\text{T} + \text{p},$$
$$^3\text{T} + {}^2\text{D} \to {}^4\text{He} + \text{n}.$$

Compare your answer with the energy per gram released in a fission reaction such as that in Example 36.7.

73. ♦♦♦ At what temperature is the mean kinetic energy of two protons equal to their electrical potential energy at a separation of 2.0 fm? This greatly overestimates the temperature required for fusion, because the protons can tunnel through the Coulomb barrier in a reverse version of α decay. (See Figure 36.13.)

74. ♦♦♦ The neutron's mass cannot be measured in a mass spectrometer, and so it is obtained indirectly. From the following data find the neutron mass:

$$^1\text{H} + {}^1\text{H} \to {}^2\text{H} + \text{e}^+ + \nu_e + 1.444\ \text{MeV}$$
$$^2\text{H} + {}^2\text{H} \to {}^3\text{H} + {}^1\text{H} + 4.031\ \text{MeV}$$
$$^2\text{H} + {}^2\text{H} \to {}^3\text{He} + \text{n} + 3.269\ \text{MeV}$$
$$^3\text{H} \to {}^3\text{He} + \text{e}^- + \bar{\nu}_e + 0.0186\ \text{MeV}$$
$$^1\text{H} + \text{n} \to {}^2\text{H} + \gamma \text{ with } E_\gamma = 2.225\ \text{MeV}$$
$$M(^1\text{H})c^2 = 938.78\ \text{MeV}.$$

Additional Problems

75. ❖ Explain why a free neutron can undergo β decay while a free proton can*not* emit a positron. Why *can* a proton emit a positron when it is inside a nucleus? (Refer to Figure 36.26.)

76. ❖ By comparing initial and final states of the pair production and annihilation reactions in the CM reference frame of an $\text{e}^+\,\text{e}^-$ pair, show that a single photon cannot by itself decay into a pair, nor can a pair annihilate to form a single photon.

77. ♦ In neutron stars, which contain approximately one solar mass (2×10^{30} kg) within a radius of 10 km, gravitation is intense enough to prevent β decay and allow the existence of a single nucleus containing some 10^{57} neutrons! Compute the density of a typical nucleus and compare your result with the density of a neutron star.

78. ♦♦ Compare the de Broglie wavelength of an α-particle emitted by ^{217}Fr with the diameter of the nucleus.

79. ♦♦ Natural uranium is 0.72% ^{235}U and 99.28% ^{238}U. If the two isotopes were equally abundant when the Earth was formed, how old is the Earth?

80. ♦♦ In the decay of ^{144}Nd (Example 36.3), what fraction of the total energy release is in the form of kinetic energy of the cerium daughter nucleus?

81. ♦♦♦ A 5.000-MeV γ ray dissociates a stationary deuteron (binding energy 2.226 MeV). Find the kinetic energies of the resulting neutron and proton if the neutron emerges at an angle of 45° to the direction of the γ ray.

82. ♦♦♦ Compare the fission cross section of ^{235}U (Figure 36.33) with **(a)** the geometrical cross section of the uranium nucleus, and **(b)** a circle with radius equal to the de Broglie wavelength of the neutron. In this case, compare both the magnitude and dependence on neutron energy.

83. ♦♦♦ The unstable nucleus ^{243}Am emits an α-particle with kinetic energy 5.3 MeV. Assuming the α has the same kinetic energy inside the nucleus as outside, compute the rate at which the α undergoes collisions with the inner surface of the nucleus. If the half-life for the decay is 8000 y, what is the probability of escape at each collision?

84. ♦♦♦ A patient's body excretes a particular chemical element with a biological half-life T_b of 2 d and is given a dose of a radioisotope of that element with a decay half-life $T_{1/2} = 8$ h. Derive a formula for the fraction of the original dose remaining in the patient's body as a function of time and compute the fraction remaining after 1.5 d.

85. ♦♦♦ A natural sample of ^{238}U ore emits 1.26×10^4 α-particles per second from decays to ^{234}Th. How many ^{238}U nuclei are contained in the sample? How many decays per second of ^{226}Ra are expected? What mass of radium is present? How many β decays of ^{210}Tl are expected? (Refer to Figure 36.29c.)

86. ♦♦♦ Assuming a neutron loses one-half its kinetic energy in each collision with a thermal proton, what number N of such collisions is necessary to reduce a 2-MeV neutron to thermal energies of 0.04 eV? Assuming the cross section for such collisions equals the geometrical cross section of a proton (radius 1.2 fm), what is the average distance ℓ the neutron travels between collisions in water? Neglect any collisions with oxygen nuclei. Estimate the root mean square distance the neutron travels during this process, taking $d_{rms} \approx \ell\sqrt{N}$.

87. ♦♦♦ Use the results of Problem 90 to show: If $T_2 \ll t \ll T_1$, then N_2 attains a value proportional to N_1 such that both decay with the same activity. Show that, in the case $T_1 \ll t \ll T_2$, species 1 has decayed away, that nearly all the atoms are daughters, and that almost none are granddaughters. Discuss how these results relate to the hydrodynamic analogy for radioactive series (Figure 36.30).

Computer Problems

88. The activity of a sample of ^{55}Cr after successive intervals of 5 min is found to be 9.60, 3.57, 1.33, 0.495, and 0.185 mCi. Plot the logarithm of the activity versus time. What is the half-life of ^{55}Cr?

89. One of the fission products of uranium is ^{149}Nd, which decays with a 1.8-h half-life to ^{149}Pm. It subsequently decays with a 54-h half-life to ^{149}Sm, which is stable. Assuming there is 1 g of Nd at $t = 0$, use a spreadsheet program to calculate the activity of the Nd, the Pm, and the whole sample as a function of time. Plot your results.

Challenge Problems

90. Suppose that initially we have a pure sample containing N_o atoms of a radioactive species with half-life T_1. The daughter species decays with half-life T_2 to a stable granddaughter species. Show that the numbers of atoms of the three species are given by:

$$N_1(t) = N_o 2^{-t/T_1},$$

$$N_2(t) = N_o\left(\frac{T_2}{T_1 - T_2}\right)(2^{-t/T_1} - 2^{-t/T_2}),$$

and

$$N_3(t) = N_o\left(\frac{T_2 2^{-t/T_2} - T_1 2^{-t/T_1}}{T_1 - T_2} + 1\right).$$

91. Use the results of Problem 90 to find the time when the maximum number of atoms are in the daughter form and the value of that maximum number. A medical radiology department receives a "source" containing 100 mCi of ^{99}Mo Monday morning at 10:00 A.M. The molybdenum decays to ^{99m}Tc with a half-life of 66.02 h. The technetium decays with a half-life of 6.02 h. When should technicians milk the source to obtain the maximum quantity of ^{99m}Tc? All the technetium is used for scheduled tests, but a new patient urgently requires a test 3 h after the previous milking. How many curies of technetium are available for the test?

92. Use Heisenberg's uncertainty principle and the observed Q value of the reaction $n \rightarrow p + e^+ + \nu_e$ to argue that we *cannot* think of a neutron as a bound state of proton, electron, and neutrino that somehow splits up. How does the reaction $p + e^- \rightarrow n + \nu_e$ support this conclusion?

93. After a long time has passed and essentially all the parent nuclei in a large sample have decayed, show that the average time a parent nucleus survived before decaying is λ^{-1}. [*Hint:* Calculate the average as $\int tN(t)\, dt/N_o$, where $N(t)$ is the number of nuclei that reach age t and then decay before $t + dt$, and N_o is the initial number of nuclei.] If a given atom has two distinct modes of decay with decay constants λ_1 and λ_2, find an expression for the number of atoms in a sample remaining after time t, and show that the mean lifetime of an atom is $t_{mean} = (\lambda_1 + \lambda_2)^{-1}$.

94. In Chadwick's original experiment, 5.30-MeV α-particles from a polonium source strike ^{9}Be nuclei. Supposing that the resulting reaction is: $^9\text{Be} + \alpha \rightarrow {}^{13}\text{C} + \gamma$, what is the energy of the emerging γ rays? Assuming that these hypothetical γ rays Compton-scatter from the protons in Chadwick's target, what is the minimum-energy γ necessary to produce the observed 5.7-MeV recoil protons? Supposing instead that the reaction is $^9\text{Be} + \alpha \rightarrow {}^{12}\text{C} + n$, what is the energy of the resulting neutron? What energy would it transfer to a stationary proton in a head-on elastic collision?

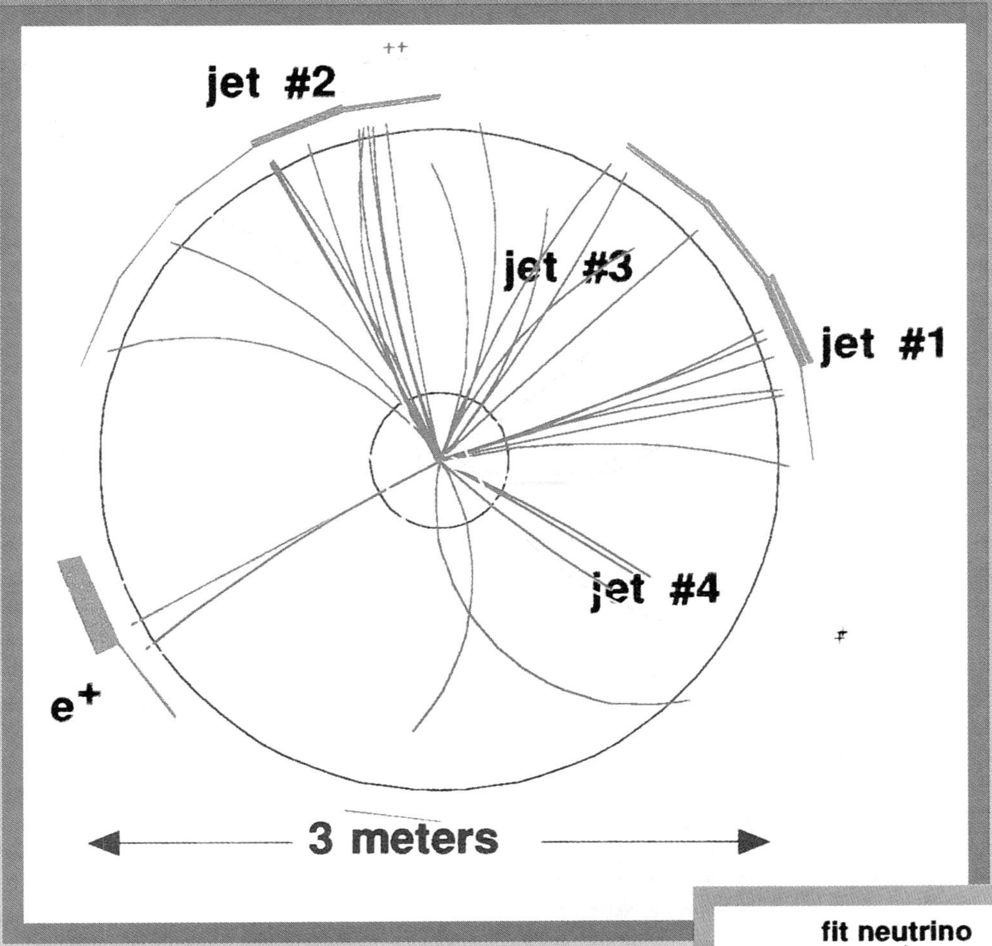

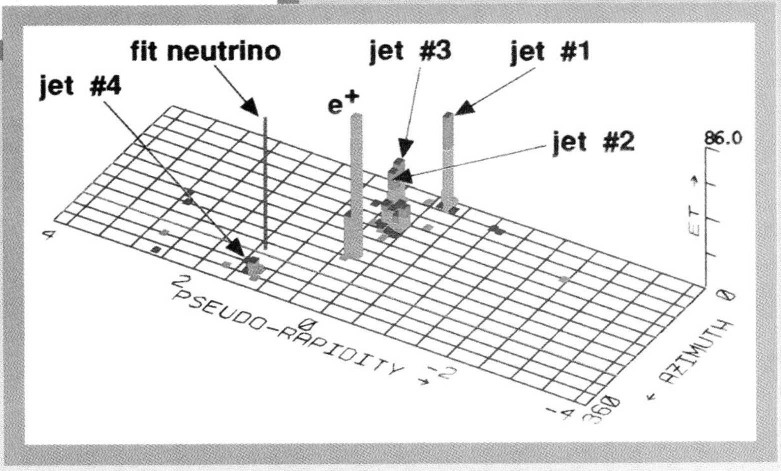

Result of a proton/antiproton collision in the Tevatron at Fermi National Accelerator Laboratory. A top quark and a top antiquark were created in the collision. Each quark decayed into a W boson and a bottom quark. One W boson decayed into a positron and a neutrino, while the other decayed into a light quark and a light antiquark. The particle tracks form four jets, *showing the tracks of particles formed in the decay of the four quarks (two bottom quarks, and the light quark/antiquark pair). The positron caused the red track at the left. The neutrino left no track. Its presence is inferred from momentum conservation. The bottom panel shows the energy deposited in detectors* (calorimeters) *around the experiment chamber, as a function of angle. The position and energy of the neutrino are deduced, not directly observed.*

. . . three quarks for Muster Mark . . .

James Joyce

CHAPTER 37
Particle Physics

Concepts

Virtual particles
Quarks
Color charge
Unification of forces

Goals

Be able to:
Describe qualitatively the electroweak and quantum chromodynamic models of fundamental particles.

THE EUROPEAN LABORATORY FOR PARTICLE PHYSICS, RECENTLY RENAMED, IS STILL KNOWN BY THE INITIALS OF ITS PARENT ORGANIZATION, CONSEIL EUROPÉEN DE RECHERCHE NUCLÉAIRE (EUROPEAN COUNCIL FOR NUCLEAR RESEARCH)

REMEMBER: THE SYMBOL T STANDS FOR TERA $= 10^{12}$ (TABLE 1.1).

Richard Feynman once remarked that doing particle physics is like studying Swiss watches by smashing them together and looking at the broken pieces that fly out. The structure of nuclei and of particles once called "elementary" is determined by accelerating them to high energy, smashing them together, and looking at the resulting debris. Understanding the smallest constituents of matter requires enormous energies and correspondingly gigantic machines. Since the 1950s, the size and cost of particle accelerators have grown rapidly, and they are now major national facilities such as the Fermi National Accelerator Laboratory (■ Figure 37.1) or international collaborations such as the CERN laboratory in Geneva, Switzerland.

As the energy available for experimental reactions has increased from a few GeV to several TeV per particle, a wealth of subnuclear particles has been revealed. Between 1940 and 1970 physicists gradually developed a new concept of force and discovered the principles that govern the subnuclear world. The resulting *standard model* explains subnuclear behavior with a small number of *fundamental* particle types and three rules of behavior. In this chapter we describe the standard model, which takes us to the frontiers of research. It currently provides our best answers to the questions the ancient Greeks asked at the birth of physics: What are the constituents of the universe and what are the rules they follow?

■ FIGURE 37.1
Aerial photograph of Fermilab. The circular structure in the photograph follows the path of the underground tunnel where particles are accelerated. Counterrotating beams of particles meet in experiment chambers where they collide within elaborate detectors. Computer analysis of the detector readings (see frontispiece) determines which particles are produced in the collisions.

37.1 PARTICLE CREATION AND FUNDAMENTAL FORCES

37.1.1 Creation and Destruction

We are used to a stable world. Everyday objects neither appear nor disappear spontaneously. Even in nuclear reactions the particles convert only a small fraction of their mass to energy. The subnuclear world, however, could not be more different. An energetic photon disappears and its energy reappears as a positron and electron, or a neutron is destroyed and a proton, an electron, and an antineutrino are created in its place. Such creation and destruction of subnuclear particles occurs incessantly and is limited only by the need to satisfy conservation laws. This chaos is not apparent in our macroscopic world because the electron and proton appear to be stable. If an electron were to decay, the resulting particles would have to have a total mass less than m_e, total charge $-e$, and lepton number $+1$. We know of no such set of particles. Proton decay, if it occurs at all, has a half-life greater than 10^{31} y.

DESPITE INTENSIVE SEARCHES, PROTON DECAY HAS NEVER BEEN OBSERVED.

37.1.2 *Virtual Particles and Fundamental Forces*

Particle creations may appear to violate energy conservation, but only for limited periods of time! For example, a photon with energy E can come into existence for a time Δt limited by the Heisenberg uncertainty principle (■ Figure 37.2):

$$\Delta t \approx \hbar/2E.$$

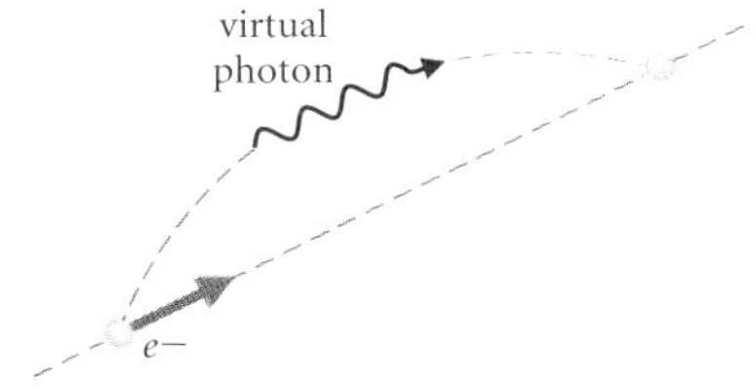

■ FIGURE 37.2
An electron emits and reabsorbs a virtual photon. According to Heisenberg's uncertainty principle, a photon of energy E can exist for a time $\Delta t \approx \hbar/(2E)$ before being reabsorbed.

Such a particle, existing for a limited time on borrowed energy, is called a *virtual particle.*

Virtual particles are the mechanism by which forces are transmitted (■ Figure 37.3). For example, when a virtual photon emitted by one electron is absorbed by another, the photon transfers momentum between the electrons. The continuous exchange of such photons produces a continuous momentum transfer—the repulsive Coulomb force between the electrons. We introduced the idea of fields, such as the electric field (Chapter 23), to explain the interaction of separated particles. On the subatomic level, we recognize the electromagnetic field as a swarm of virtual photons.

During its brief life, a virtual photon can travel a maximum distance:

$$c\,\Delta t \sim \frac{\hbar c}{2E} = \frac{\hbar c}{2hf} = \frac{c}{4\pi f} = \frac{\lambda}{4\pi}.$$

Because photons have no rest mass, they can have arbitrarily small energies, and the distance $c\,\Delta t$ can be arbitrarily large. For this reason, the Coulomb force acts over distances much larger than atomic dimensions.

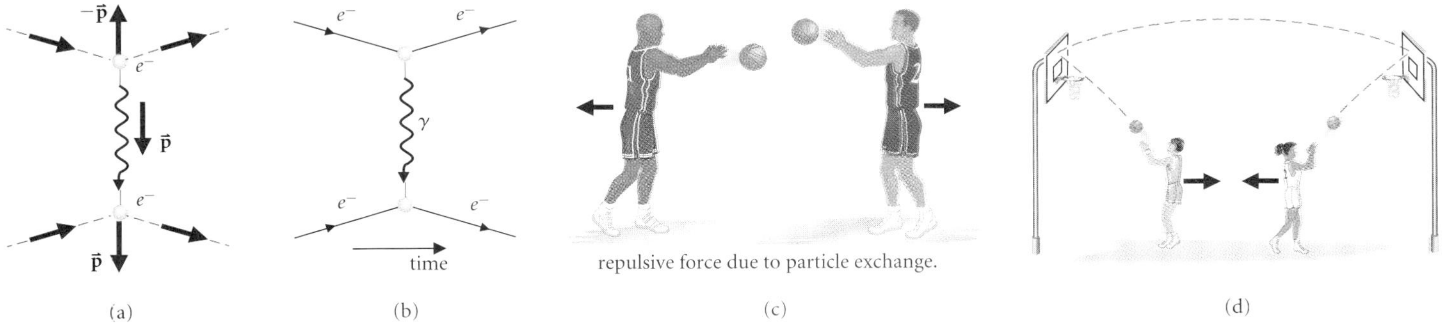

■ FIGURE 37.3
(a) A virtual photon transmits electromagnetic force between two electrons. An electron emits a photon with momentum $\vec{\mathbf{p}}$, changing its own momentum by $-\vec{\mathbf{p}}$. The photon is reabsorbed by a second electron, which gains momentum $\vec{\mathbf{p}}$. A continual exchange of photons results in the Coulomb repulsion of the electrons. (b) A diagram showing these processes is called a Feynman diagram. The photon is labeled γ. (c) In a classical analog of particle exchange, two students standing on ice throw basketballs back and forth. As the student on the left throws his ball, he gives it momentum $\vec{\mathbf{p}}$ to the right and he recoils with momentum $-\vec{\mathbf{p}}$ to the left. Upon catching the ball thrown by his partner, he gains additional leftward momentum. Similarly, the student on the right gains rightward momentum. (d) A similar mechanism produces an attractive Coulomb force between oppositely charged particles, but there is no obvious classical analog. If the two students stand back to back, they can bounce their balls off walls before catching them, and hence be pushed together. But there are no walls in the particle world.

37.1.3 *Feynman Diagrams*

Richard Feynman (1918–1988) invented a diagram that illustrates the role of virtual particles. A *Feynman diagram* (Figure 37.3b) is a schematic of the paths of the real and virtual particles involved in an interaction. Particles whose paths cross the boundaries of the diagram are real. A point where the paths join is called a vertex. At each vertex all the conservation laws we have discussed so far must be obeyed: the sum of energy, momentum, charge, lepton number, and baryon number of the particles entering the vertex must equal the sum for the particles leaving. In this example time increases from left to right. It is conventional to draw antiparticles moving backward in time.

THESE DIAGRAMS ARE USEFUL GUIDES FOR CALCULATING THE OUTCOME OF A PARTICLE INTERACTION.

SEE §36.2.1 FOR DEFINITIONS OF THESE QUANTUM NUMBERS.

The theory of quantum electrodynamics (QED), based on virtual photons, allows us to calculate interactions between charged particles with great accuracy. Experimental tests currently match theory to better than 1 part in 10^7. Because of this success, physicists now believe that all fundamental forces act by the exchange of virtual particles. Thus to understand subatomic physics we need a catalog of the particles that participate in the interactions that govern the behavior of matter.

37.1.4 The π Meson

In 1935 Hideki Yukawa (■ Figure 37.4) proposed that a virtual particle carries the strong force (§36.1.4). This Yukawa particle is now known as the π meson or *pion.* The name *meson,* or middle particle, reflects the fact that the pion mass is between those of electrons and of nucleons. Yukawa predicted its mass from the short range (10^{-15} m) of the strong force. The energy of a virtual Yukawa particle is at least as great as its rest energy mc^2. This minimum energy corresponds to the maximum time the particle may exist and so to the maximum distance it can travel:

NOTE THAT THE DISTANCE $c\,\Delta t \sim \hbar/mc$ IS APPROXIMATELY THE PION COMPTON WAVELENGTH.

$$c\,\Delta t_{\max} = \hbar c/(2mc^2) \approx \text{range of strong force} \approx 10^{-15}\text{ m}.$$

So:

$$mc^2 \approx \frac{\hbar c}{2\times 10^{-15}\text{ m}} = \frac{(1.0\times 10^{-34}\text{ J}\cdot\text{s})(3\times 10^8\text{ m/s})}{(2\times 10^{-15}\text{ m})(1.6\times 10^{-19}\text{ J/eV})} = 100\text{ MeV}.$$

MESON NOW MEANS ANY STRONGLY INTERACTING PARTICLE WITH SPIN ANGULAR MOMENTUM EQUAL TO AN INTEGER TIMES $\hbar$.

This estimate is very close to the measured value of 135 MeV for the neutral pion. Pions were among the first of the new kinds of particle detected in accelerator experiments in the 1950s (■ Figure 37.5).

■ FIGURE 37.4
Hideki Yukawa won the Nobel Prize for physics in 1949 for predicting the existence of mesons.

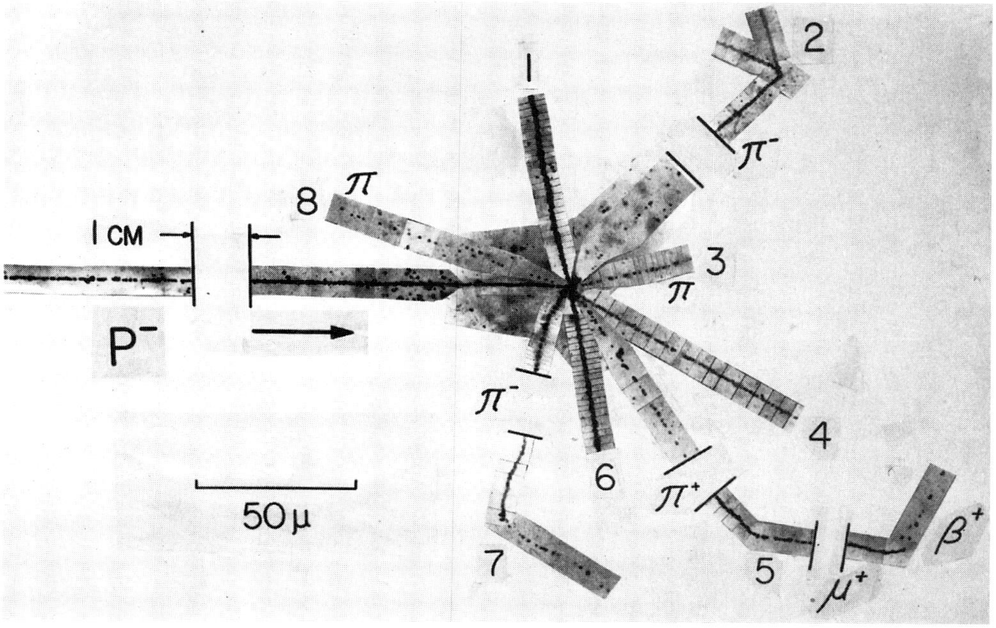

■ FIGURE 37.5
Proton/antiproton collision observed at the Bevatron, University of California, Berkeley, in 1956. The two particles annihilate, forming a spray of pions. One pion at the lower right later decays into a positively charged muon and a neutral particle that leaves no track. Several other decays are also visible.

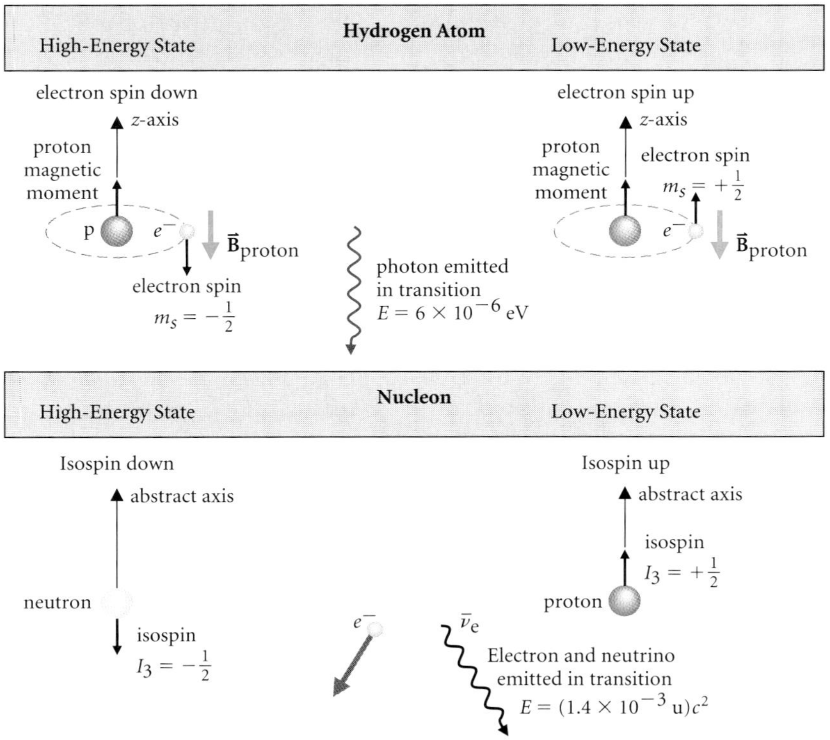

■ FIGURE 37.6
Analogy between electron spin and the isotopic spin of a nucleus. An atom emits a photon as the electron makes a transition from the high-energy, "spin-down" state to the low-energy, "spin-up" state. The neutron ("isospin-down" state) is the high-energy state of the nucleon. In the transition to the low-energy, "isospin-up" state (the proton), an electron (e^-) and an antineutrino ($\overline{\nu}_e$) are emitted.

37.1.5 Isospin

Discovery of the pion confirmed the idea that forces occur because of particle exchange, but also introduced a new puzzle. The π meson occurs in three forms (π^0, π^+, and π^-) with different electric charges (0, $+e$, and $-e$) but very nearly the same mass and identical strong interactions. Neutrons and protons behave similarly: they have nearly equal masses and experience the strong force in exactly the same way. In an imaginary world where only strong nuclear forces act, neutrons and protons would be identical. We could think of them as being two different quantum states of one particle, the nucleon. The observed differences in mass and electric charge arise because the weak and electric forces act differently on the two quantum states.

The two possible spin states of the electron in a hydrogen atom (§35.3) provide an analogy for the two nucleon states (■ Figure 37.6). If there were no magnetic fields, both electron spin states would have the same energy and be indistinguishable. In the actual world, magnetic fields cause the two spin states to have different energies. This analogy between spin and nucleon states is not just superficial: as we shall see, it proves very important in understanding the strong force (§37.2). Because of this analogy and because *iso*topes differ in neutron number, the property that distinguishes protons from neutrons was named *isospin.*

The neutron and proton have different names because they have different charges and respond differently to electromagnetic fields. But electric charge is irrelevant to the strong force. Had we understood atoms from the inside out instead of from the outside in, we would have recognized a single particle, the nucleon, with two isospin states.

Isospin is described by a quantum number I. States with different isospin have different energies (i.e., different masses) and different electric charges. For nucleons $I = \frac{1}{2}$, and there are two charge states. The pions have $I = 1$, and three charge states. In both cases, the number of charge states equals $2I + 1$.

THIS NUMERICAL RELATION HOLDS FOR PARTICLES CONTAINING FIRST-GENERATION QUARKS. SEE §37.2 AND §37.3.

37.2 SUBNUCLEAR PARTICLES AND THE QUARK MODEL

37.2.1 The Population Explosion

Machines capable of accelerating proton beams to energies of several GeV per particle were first built in the 1950s. Experiments with these machines soon discovered evidence for numerous kinds of strongly interacting particle whose existence was unexpected from experience with nuclear reactions (■ Figure 37.7). These particles exist for very short times before decaying into less massive products (■ Figure 37.8). If their lifetime is too short, the particles do not travel a detectable distance from their origin. The existence of such particles is inferred from a phenomenon called *resonance:* The rate at which particles interact in a collision experiment suddenly increases when their total energy in the CM frame is equal to the new particle's rest

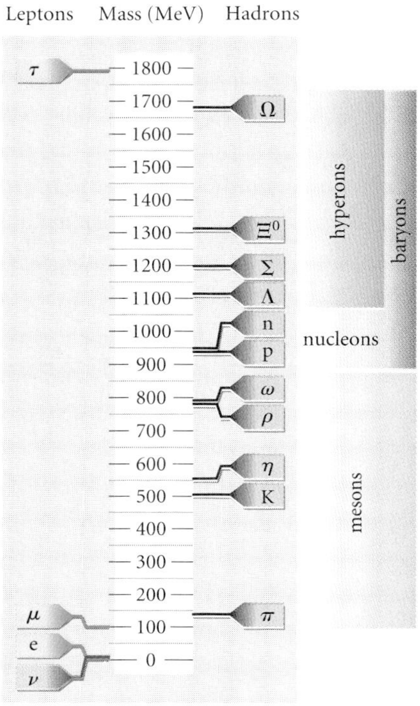

■ FIGURE 37.7
Some of the particles that have been discovered at high-energy particle accelerators. The initial classification scheme was based on particle mass: *leptons,* or light particles; *mesons,* intermediate mass particles; and *baryons,* or heavy particles. As more particles were discovered, the mass scheme was found to be inappropriate—notice the mass of the τ—and was replaced by a classification in terms of particle spin and response to forces. Leptons do not respond to the strong force, and they have spin $\frac{1}{2}$. Hadrons—mesons and baryons—do respond to the strong force; mesons have integer spin, hyperons and nucleons have half-integer spin. The forces felt by each particle are indicated by the color code. Red: strong, electromagnetic, and weak. Green: electromagnetic and weak. Blue: weak only.

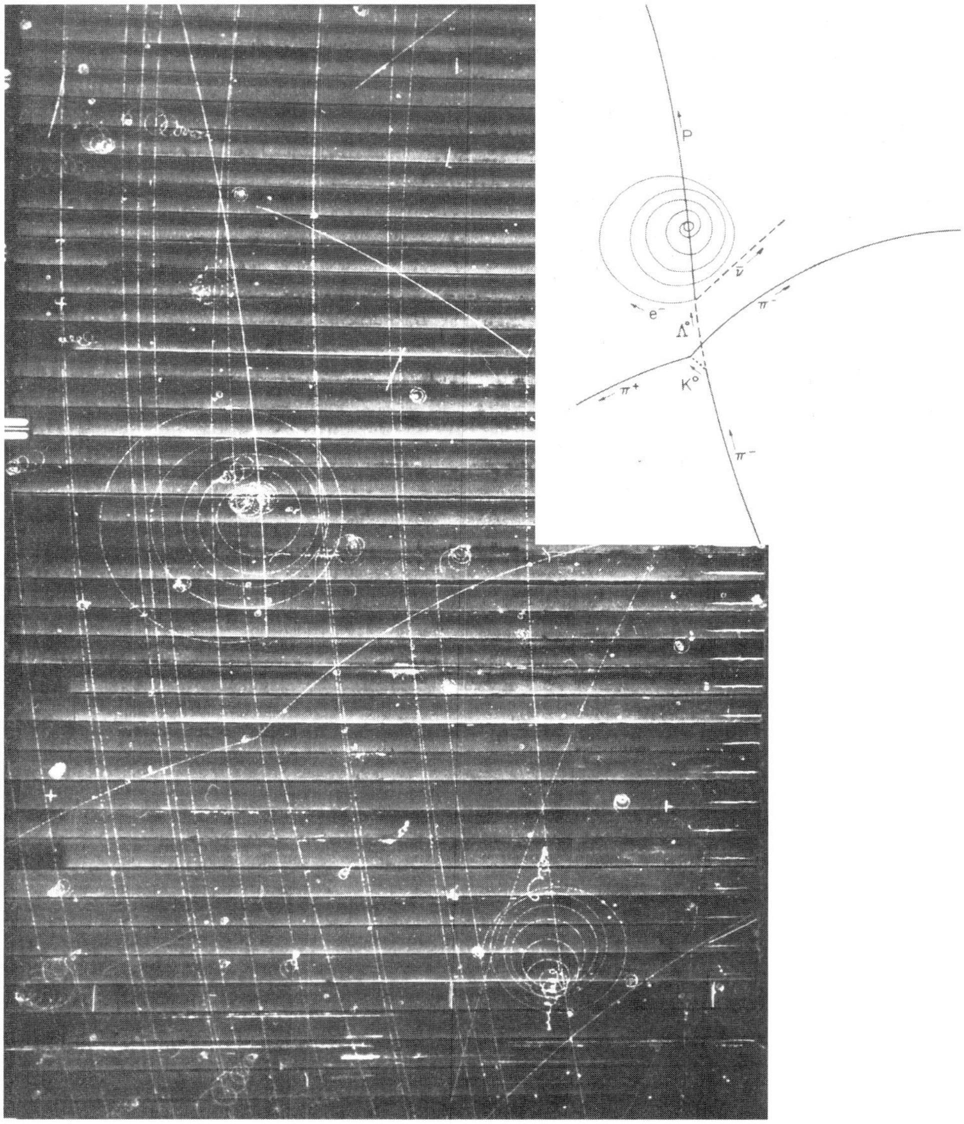

■ FIGURE 37.8
Several particle decays are evident in this bubble chamber photograph. A π^- collides with a proton in the chamber, producing a Λ^0 and a K^0. The K^0 subsequently decays into a π^- and a π^+. The Λ^0 decays into a proton, an electron, and an antineutrino, which leaves no track. The distance traveled by each particle between production and decay is indicative of its lifetime. The magnetic field in the chamber causes charged particle tracks to curve: note the spiral followed by the electron.

energy (■ Figure 37.9). The colliding particles form the new particle, which then decays into particles of lesser mass. The population explosion of such particles demolished the tidy explanation of matter as electrons, protons, and neutrons exerting forces via virtual photons and pions. The new kinds of particle seemed to have no role in nuclear phenomena, yet their existence demanded explanation.

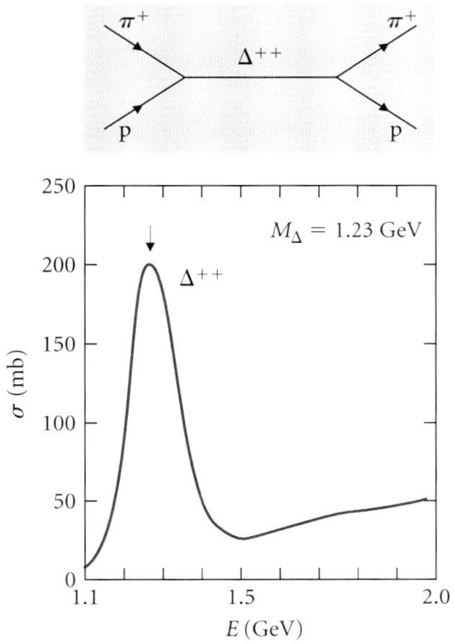

■ FIGURE 37.9
A Δ^{++} particle with mass 1.23 GeV/c^2 is formed in collisions between protons and pions. The particle itself decays rapidly and no track is seen. Its existence is inferred from its effect on the scattering of the π^+ and p. The rate of interactions (measured by the cross section) shows a marked peak as the total beam energy approaches 1.23 GeV. (1 mb $= 10^{-31}$ m^2, cf. Figure 36.33.)

37.2.2 Strangeness and Quarks

Our understanding of subnuclear particles has come with the recognition of several new properties that are important in subnuclear reactions but not apparent in nuclear or atomic structure. The first of these properties, recognized in 1953, is called *strangeness* because particles possessing it seemed at the time to behave strangely—they decay too slowly. Nuclei and other strongly interacting systems have radii of the order of 10^{-15} m. Strong interactions require about 10^{-22} s to occur, roughly the time for light to travel this typical distance of 10^{-15} m. The strange particles ultimately decay to stable particles but require the much longer time, about 10^{-10} s, typical of weak interactions.

Apparently the strange particles are unable to decay via strong interactions, but can decay via weak interactions. This behavior can be understood if strong interactions obey a conservation law that weak interactions may violate. Strangeness, the new partially conserved quantity, occurs only in multiples of a basic unit, just as electric charge occurs in multiples of the electron charge. Thus particles are assigned an integer S, which gives the number of strangeness units they possess. The Λ^0 in Figure 37.8, for example, has a strangeness of -1. Strangeness may change by one unit during weak interactions.

The recognition of strangeness led to a powerful classification scheme for strongly interacting particles. ■ Figure 37.10 shows how particles of roughly similar mass fall into groups that show great regularity in their properties. When this classification was proposed, the

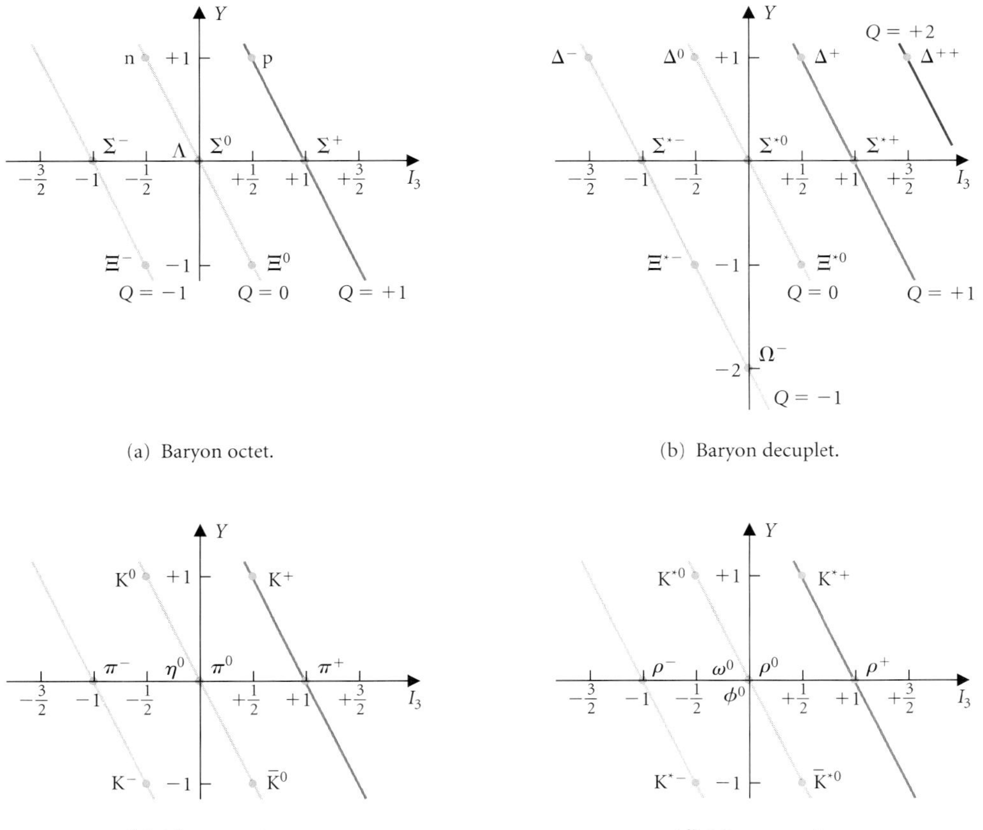

(a) Baryon octet.

(b) Baryon decuplet.

(c) Meson octet.

(d) Meson nonet.

■ FIGURE 37.10
Isospin/hypercharge diagrams. Hypercharge, $Y = B + S$, is the sum of baryon number plus strangeness. When we plot particle properties in a diagram of Y versus I_3, the z-component of isospin, an amazing regularity emerges. Particle families of roughly similar mass form regular geometrical figures. Lines of constant electric charge are diagonal lines in these plots. The notation K* denotes an excited state of the K particle. These patterns correspond to symmetries in the underlying mathematical description of the particles (§37.4).

■ **Figure 37.11**
Discovery of the Ω^-. The K^- incident at the bottom of the photograph collides with a proton in the chamber, producing an Ω^-, a K^0, and a K^+. The Ω^- subsequently decays into a Ξ^0 and a π^-.

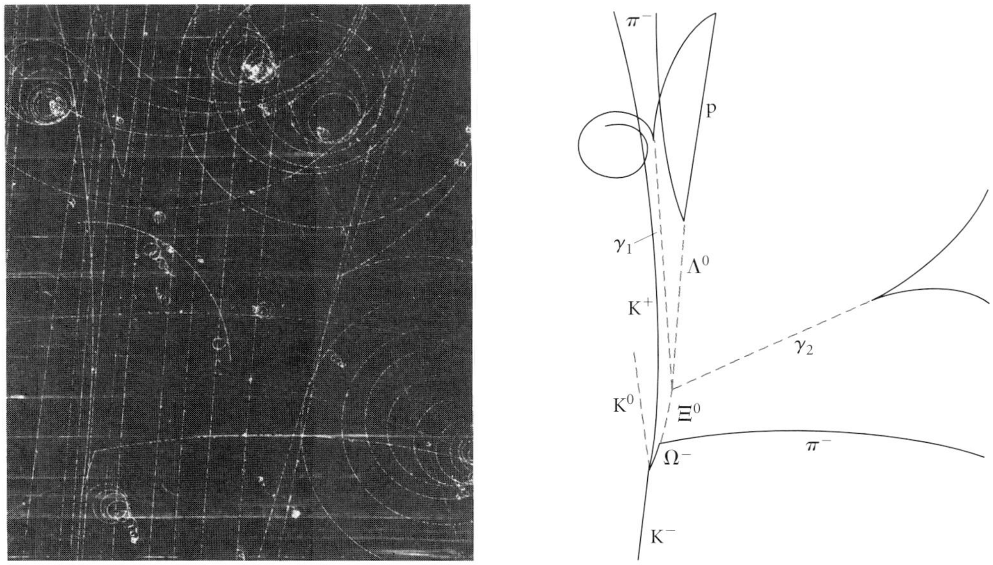

Ω^- particle in Figure 37.10b had not been observed. Its subsequent discovery (■ Figure 37.11) is convincing evidence for the reality of strangeness.

Just as the periodic table of chemical elements reflects the behavior of electrons in atoms, this classification of subnuclear particles hints at an underlying structure. In 1964, Murray Gell-Mann and Stephan Zweig pointed out that the classification is explained if the strongly interacting particles are composites of three kinds of particles they dubbed *quarks.* Each quark has spin $\frac{1}{2}$, baryon number $\frac{1}{3}$, and charge that is a multiple of $e/3$. The different kinds of quark are referred to as *flavors.* The first two flavors, *up* and *down,* refer to the isotopic spin: up

The name is German for milk curds and is taken from a passage in *Finnigan's Wake* by James Joyce. (See chapter quote.)

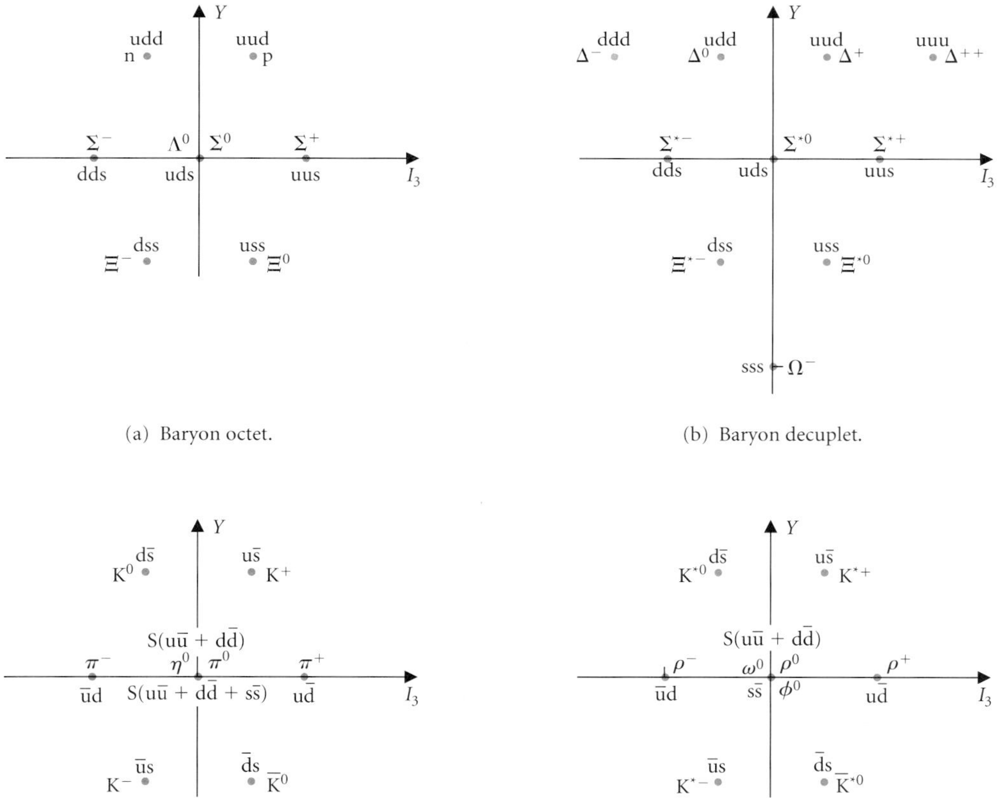

(a) Baryon octet.

(b) Baryon decuplet.

(c) Meson octet.

(d) Meson nonet.

■ **Figure 37.12**
Quark theory explains the hypercharge/isospin diagram. Each particle is described by the *quarks* that compose it. The flavorless particles at the centers of diagrams (c) and (d) are superpositons of combinations of quark/antiquark pairs. As in Figure 37.10, the * denotes an excited state.

quarks have isospin up ($I_3 = +\frac{1}{2}$), while the down quark has isospin down ($I_3 = -\frac{1}{2}$). The strange quark is assigned $S = -1$ and $I = 0$. ■ Figure 37.12 illustrates how the model explains particle properties. As more massive particles were discovered, additional quarks with additional quantum numbers were needed to explain them. The other three flavors are distinguished by quantum numbers called C (charm), T (topness), and B (bottomness) (● Table 37.1). In ● Table 37.2 we list properties of some mesons and baryons, including their masses and quark

Remember: You might find it useful to think of these new quantum numbers in analogy with electric charge (cf. §36.2.1). Topness is sometimes called truth and bottomness, beauty.

Table 37.1 Properties of Quarks

Flavor	Symbol	Electric Charge	Isospin	I_3	S	C	B	T	Mass[a] (MeV)
Up	u	$2e/3$	$\frac{1}{2}$	$+\frac{1}{2}$	0	0	0	0	330
Down	d	$-e/3$	$\frac{1}{2}$	$-\frac{1}{2}$	0	0	0	0	333
Strange	s	$-e/3$	0	0	-1	0	0	0	486
Charm	c	$2e/3$	0	0	0	$+1$	0	0	1650
Bottom	b	$-e/3$	0	0	0	0	-1	0	4500
Top	t	$2e/3$	0	0	0	0	0	$+1$	180 000

[a]Masses are not directly observable. Values are derived from an analysis of hadron masses.

Table 37.2 A Selection of Mesons and Baryons

Name	Valence Quarks	Mass (MeV)	Isospin	I_3
π^+	$u\bar{d}$	139.57	1	$+1$
π^-	$\bar{u}d$	139.57	1	-1
π^0	$S(\bar{u}u, \bar{d}d)$[a]	134.97	1	0
K^+	$u\bar{s}$	493.65	$\frac{1}{2}$	$\frac{1}{2}$
K^-	$\bar{u}s$	493.65	$\frac{1}{2}$	$-\frac{1}{2}$
K^0	$d\bar{s}$	497.7	$\frac{1}{2}$	$-\frac{1}{2}$
η	$S(\bar{u}u, \bar{d}d, \bar{s}s)$[a]	549	0	0
p	uud	938.272	$\frac{1}{2}$	$+\frac{1}{2}$
n	udd	939.566	$\frac{1}{2}$	$-\frac{1}{2}$
Λ	uds	1115.6	0	0
Σ^+	uus	1189.4	1	$+1$
Σ^0	uds	1192.6	1	0
Σ^-	dds	1197.4	1	-1
Δ^{++}	uuu	1230	$\frac{3}{2}$	$+\frac{3}{2}$
Δ^+	uud	1230	$\frac{3}{2}$	$+\frac{1}{2}$
Δ^0	udd	1230	$\frac{3}{2}$	$-\frac{1}{2}$
Δ^-	ddd	1230	$\frac{3}{2}$	$-\frac{3}{2}$
Ξ^0	uss	1315	$\frac{1}{2}$	$+\frac{1}{2}$
Ξ^-	dss	1321	$\frac{1}{2}$	$-\frac{1}{2}$
Ω^-	sss	1672	0	0
Λ_c^+	udc	2285	0	0
B^+	$u\bar{b}$	5271	$\frac{1}{2}$	$\frac{1}{2}$
B^0	$d\bar{b}$	5275	$\frac{1}{2}$	$-\frac{1}{2}$

[a]The notation $S(x, y)$ means that the particle is a superposition of states x and y.

TABLE 37.3 Leptons

Name		Mass (MeV)	Charge (e)
Electron	e	0.511	−1
Muon	μ	105.658	−1
Tau	τ	1784 ± 3	−1
Electron neutrino	ν_e	$< 1.8 \times 10^{-5}$	0
Muon neutrino	ν_μ	< 0.25	0
Tau neutrino	ν_τ	< 35	0

content. All the mesons are quark/antiquark pairs. Each baryon contains three quarks. In • Table 37.3 we list the leptons. Each particle has an antiparticle with opposite quantum numbers.

37.2.3 *Proton Structure and the Reality of Quarks*

IN 1977, WILLIAM FAIRBANK AND CO-WORKERS AT STANFORD UNIVERSITY REPORTED OBSERVING FRACTIONALLY CHARGED NIOBIUM SPHERES. THE RESULT REMAINS UNCONFIRMED AND IN CONTRADICTION WITH SEVERAL OTHER EXPERIMENTS.

THESE SUBPARTICLES WERE ORIGINALLY NAMED PARTONS TO AVOID PREMATURE IDENTIFICATION WITH THE QUARKS.

No experiment has found evidence for an isolated quark, and experiments fail to detect objects with fractional charge. A quark, it seems, cannot be separated from another quark by more than the 10^{-15}-m scale typical of strong interactions. Since they cannot be isolated, we must look elsewhere for proof of their existence. Rutherford detected atomic nuclei by showing that charged particles are scattered by something much smaller than a single atom. If protons really consist of three quarks, we should be able to observe scattering by individual quarks. Doing such an experiment is like using a microscope; one must use a wavelength small enough to resolve what one wants to see. By 1970, the Stanford Linear Accelerator Center had produced electron beams with de Broglie wavelengths of the order of 10^{-16} m, and had obtained definite evidence of pointlike particles within protons. The continued experimental and theoretical success of the quark model has overcome initial skepticism among physicists, who now generally accept it as the working model of reality.

EXERCISE 37.1 ♦♦ What is the energy of an electron that has a de Broglie wavelength of 10^{-16} m?

EXAMPLE 37.1 ♦♦ One possible decay mode of the Λ^0 is: $\Lambda^0 \rightarrow p + \pi^-$ (Figure 36.24). Verify that the masses of the particles permit this decay, and determine which quantum numbers are conserved.

MODEL This is a conservation law problem for which we may use the standard plan.

SETUP The particle properties are given in Table 37.2 and Figure 37.12, and the quantum numbers for the quarks are in Table 37.1. The charge of the Λ^0 is the sum of the charges of the quarks that compose it: $q = 2e/3 - e/3 - e/3 = 0$. The baryon number is 1 for each baryon and zero for each meson or lepton.

	BEFORE	AFTER
	Λ^0	$p + \pi^-$
Mass (MeV)	1115.6	$938.3 + 139.6 = 1077.9$
Charge	0	$+e + -e = 0$
Baryon number	+1	$+1 + 0 = +1$
Strangeness	−1	$0 + 0 = 0$
Isospin I_3	0	$\frac{1}{2} - 1 = -\frac{1}{2}$

SOLVE The mass of the two decay products is less than that of the original particle, as required. The Q value for the reaction is 1115.6 MeV − 1077.9 MeV = 37.7 MeV. Charge and baryon number are conserved; strangeness and isospin are not.

ANALYZE The Λ^0 is composed of the quarks u, d, and s. The proton is uud and the pion is $d\bar{u}$. Thus in this interaction the strange quark s converted to a down quark d with the production of a $u\bar{u}$ pair. Since the strangeness changes by one unit, this decay is governed by the weak interaction. We may conclude that isospin, like strangeness, is not conserved under the weak interaction. ∎

37.3 THE STANDARD MODEL

The quark model of the 1960s offered no explanation of the forces that hold quarks together or of why the composites they form interact with each other. Since then, the idea behind quantum electrodynamics—that forces are transmitted by virtual particles—has been applied successfully to the strong force and to a unified version of the weak and electromagnetic forces known as the electroweak interaction. Together, these ideas form the standard model, which we describe in this section.

37.3.1 Electroweak Unification

In a weak interaction, the charge of a particle may change (as in neutron decay) or it may remain unchanged (■ Figure 37.13). Thus three different species of virtual particle with different electric charges are necessary to transmit weak forces. These particles are given the names $W^{\pm}$ and Z^0, with the superscripts indicating their electric charges. Furthermore, under weak interactions, the electron and the neutrino act as different states of the same kind of particle—just as the neutron and proton are the same when only strong forces are involved. To accommodate these features, the electroweak theory requires *four* force-transmitting particles, called *gauge bosons,* instead of three (● Table 37.4). The fourth has neither mass nor electric charge, and transmits forces between electric charges—it is the photon. Weak and electromagnetic

WE INTRODUCED THE NEUTRINO IN THE LAST CHAPTER; SEE §36.2.3. THINK OF THIS IN ANALOGY WITH THE ENERGY STATES OF AN ATOM. THE STATE OF THE ATOM CHANGES WHEN A PHOTON IS EMITTED OR ABSORBED. SIMILARLY, THE STATE OF A PARTICLE (AND ITS NAME!) CHANGE WHEN ONE OF THE GAUGE BOSONS IS EMITTED OR ABSORBED.

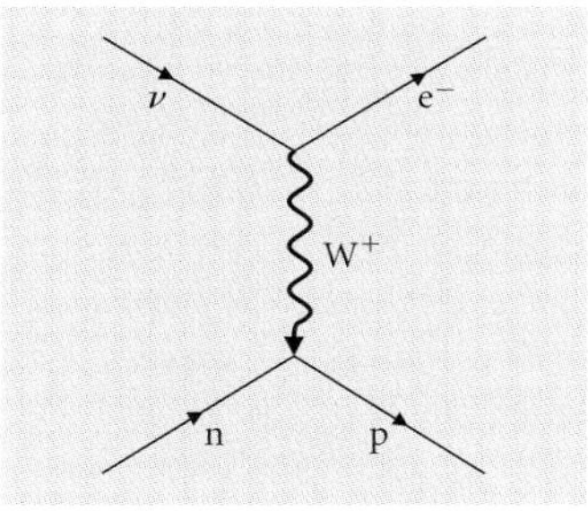

(a) $n + \nu_e \rightarrow p + e^-$.

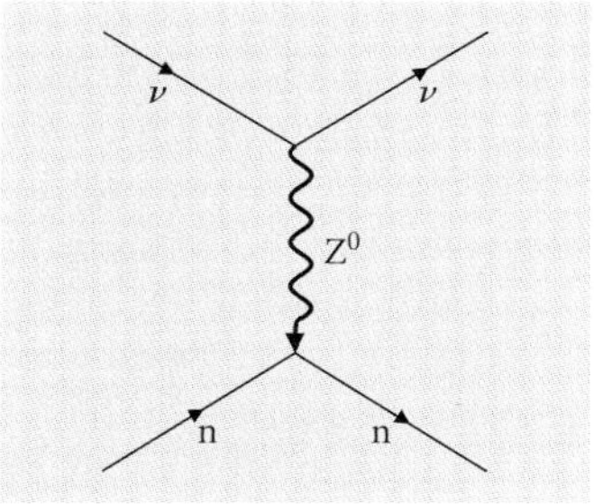

(b) Elastic collision between a ν and n.

■ FIGURE 37.13
Examples of weak interactions. (a) A neutrino emits a W^+ and becomes an electron. A neutron absorbs the W^+ and becomes a proton. (b) Elastic collision between a neutrino and a neutron. The Z^0 exchanges energy and momentum between the particles but does not alter the character of the neutron or the neutrino.

TABLE 37.4 Gauge Bosons

Name	Force	Mass (MeV)	Charge
Photon γ	electromagnetic	0	0
	weak		
W^+	weak	80	$+e$
W^-	weak	80	$-e$
Z	weak	91	0

GAUGE REFERS TO A MATHEMATICAL FEATURE OF THE THEORY. BOSONS, AFTER SATYENDA NATH BOSE (1894–1974), ARE PARTICLES WITH INTEGER SPIN. THE QUARKS AND LEPTONS ARE FERMIONS, WITH HALF-INTEGER SPIN.

■ FIGURE 37.14
The UA1 detector at CERN, Switzerland. This detector was used in the detection of the Z^0 (Figure 10.16).

ELECTRICITY AND MAGNETISM OFFER A SIMILAR EXAMPLE OF UNIFICATION. IN SPECIAL RELATIVITY, THE DISTINCTION BETWEEN ELECTRIC AND MAGNETIC FIELDS BREAKS DOWN UNDER A CHANGE OF REFERENCE FRAME. THE FIELDS ARE TWO ASPECTS OF ONE EFFECT.

forces turn out to be the same interaction acting under different conditions. The great apparent difference between the two arises because massless virtual photons are easily produced and can act over macroscopic distances, while the W and Z bosons, with masses of about 80 GeV/c^2, are produced rarely and can act only over very short distances. In an (imaginary!) experiment in which 80 GeV were a negligible particle energy, the W, Z, and γ particles would produce forces of equal strength. This electroweak theory, developed by Sheldon Glashow, Steven Weinberg, and Abdus Salaam in the late 1960s, was confirmed in 1983 by the production of free W and Z bosons at the CERN accelerator (■ Figure 37.14).

The quarks are also subject to weak interactions. For example, ■ Figure 37.15 illustrates the quark model of the neutron–neutrino interaction shown in Figure 37.13a. The interaction changes a down quark into an up quark and also changes a neutrino into an electron. The weak interaction sees up and down quarks as different states of the same particle, as it does the electron/neutrino pair. The strange quark alone doesn't fit this scheme, since it doesn't have a pair, so a fourth quark flavor, dubbed *charm,* proves necessary. The last significant skepticism about the quark model faded quickly after the simultaneous discovery at the Stanford and Brookhaven accelerators of both ground and excited states of charmonium—a charmed and anticharmed quark in orbit about each other! (See ■ Figure 37.16.)

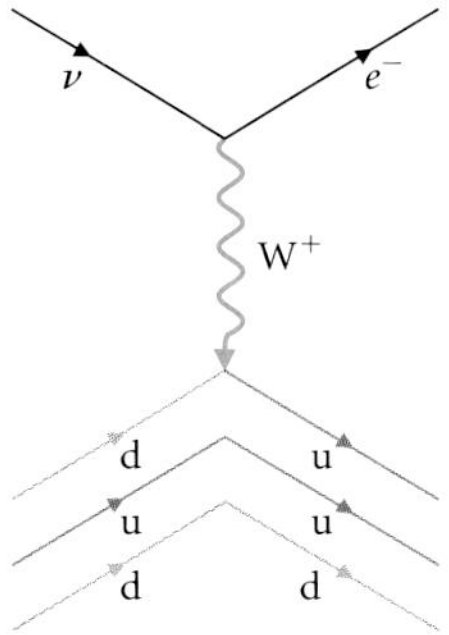

■ FIGURE 37.15
Quark model of the reaction n + $\nu_e \rightarrow$ p + e^-. The W^+ boson emitted by the neutrino is absorbed by one of the d quarks in the neutron, converting it into a u quark. The resulting combination of one d and two u quarks forms a proton.

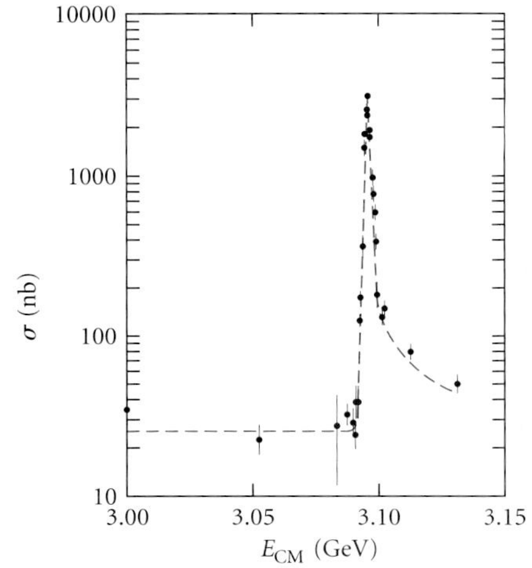

■ FIGURE 37.16
Discovery of the charmed quark. The graph shows the probability for an electron/positron pair to annihilate and produce strongly interacting particles. The peak is a resonance due to the formation of a charm/anticharm pair (J/ψ particle), which then decays to other particles. This peak occurs when the total energy of the interacting particles (in the CM frame) equals 3.1 GeV, corresponding to the 3.1 GeV/c^2 mass of the ψ. (One nanobarn is an effective cross-sectional area of 10^{-37} m^2. The probability of a reaction, and hence the number of observed events, is proportional to this effective area.)

TABLE 37.5 The World According to the Weak Interaction

Family	Leptons		Quarks	
1	electron	e^-, ν_e	u, d	up, down
2	muon	μ^-, ν_μ	s, c	strange, charm
3	tau	τ^-, ν_τ	t, b	top, bottom

The weak interaction organizes the fundamental particles into the families, or *generations,* shown in ● Table 37.5. Particles from family 1 appear to be stable for indefinite periods under present conditions and form the macroscopic world we observe about us. Family 2 is involved in the particle reactions we have discussed so far. Recent discovery of the τ lepton and the top and bottom species of quark suggests a complete third family, though the τ neutrino has not been observed as of 1996. We have no theory that explains why there should be more than one generation nor how many generations there should be altogether.

The W boson can change a quark within a generation or between generations. Such interactions, involving a charged W boson, are called charged-current interactions. The Z boson participates in neutral current interactions and does not change quark flavor.

PERHAPS THE BEST STATEMENT OF OUR PUZZLEMENT OVER THE PARTICLE FAMILIES WAS I. I. RABI'S COMMENT ON THE DISCOVERY OF THE MUON: "WHO ORDERED THAT?" COSMOLOGY PROVIDES A LIMIT ON THE MENU'S SIZE. THE AMOUNT OF ^{4}He PRODUCED IN THE EARLY UNIVERSE DEPENDS STRONGLY ON THE NUMBER OF NEUTRINO SPECIES. THE EXISTENCE OF MORE THAN FOUR SPECIES IS INCONSISTENT WITH THE OBSERVED AMOUNT OF ^{4}HE.

EXAMPLE 37.2 ❖ Draw a diagram like Figure 37.15 showing the decay $\Lambda^0 \to p + \pi^-$ discussed in Example 37.1.

MODEL From Table 37.2, the quark compositions are:

$$\Lambda^0 = \text{uds}, \qquad p = \text{uud}, \qquad \text{and} \qquad \pi^- = \bar{\text{u}}\text{d}.$$

Thus this is a weak decay in which a strange quark is converted to a down plus a $\text{u}\bar{\text{u}}$ pair. The Z boson cannot change quark flavors, so the interaction involves a W.

SETUP The s (charge $-e/3$, Table 37.1) converts to a u (charge $2e/3$) by emitting a W^- with charge $-e$: $-e/3 - (-e) = 2e/3$. The W^- then decays to a $\bar{\text{u}}$ and a d.

SOLVE ■ Figure 37.17 is the diagram for this interaction.

ANALYZE We shall call these diagrams *quark line diagrams* because they trace the paths of the quarks as they rearrange into new particles. It is customary to draw the arrows on antiparticle lines showing them moving backward in time, as we have done in Figure 37.17 for the $\bar{\text{u}}$. ▮

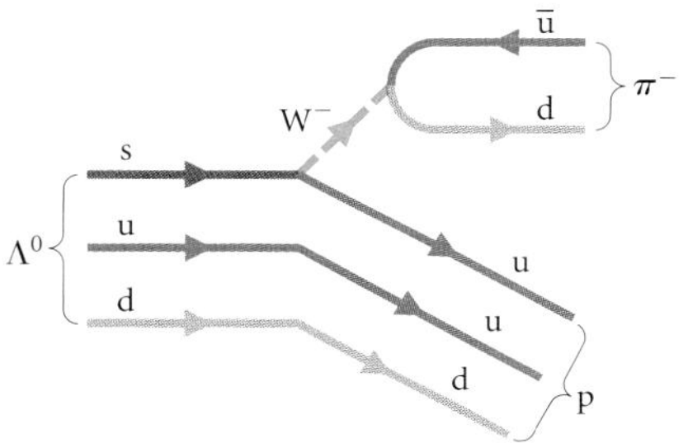

■ FIGURE 37.17
Feynman diagram for the weak hadronic decay of a Λ^0 to a proton plus π^-. The up and the down quark remain unchanged. The strange quark converts to a u by emitting a W^-, which then decays to the $\bar{\text{u}}$ and d comprising the π^-. Check the conservation of charge at each vertex of the diagram.

37.3.2 Quantum Chromodynamics

Quarks are also bound together by force-transmitting particles, named *gluons* in the whimsical tradition of high-energy physics names. The same scattering experiments that detect quarks inside protons also show that they do not carry all of a proton's momentum. Something else—the gluons—must exist inside a proton to carry the remaining momentum. A proton, then, is a little ball containing three quarks together with a cloud of virtual quark/antiquark pairs and virtual gluons which continually transmit the forces that hold the ball together.

In quantum electrodynamics, the electromagnetic field is represented by a cloud of virtual photons. We also know that the electromagnetic field is produced by, and acts on, the electric charge carried by particles. We say that the photons *couple with* electric charge. By analogy, we would expect quarks to possess a property, a strong force charge, by which they couple with the gluons. None of the observed particles, which are combinations of quarks, show any net amount of this strong force charge, just as neutral atoms contain positive and negative electric charge with a total of zero.

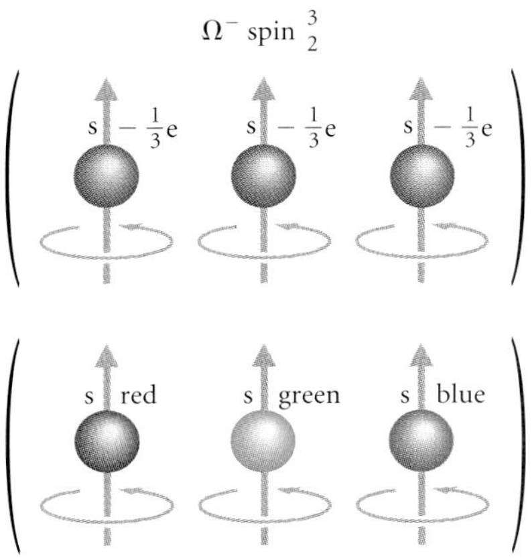

■ FIGURE 37.18
The Ω^- consists of three apparently identical quarks, which seems to contradict the Pauli exclusion principle. The introduction of color allows us to distinguish the quarks and preserve the principle.

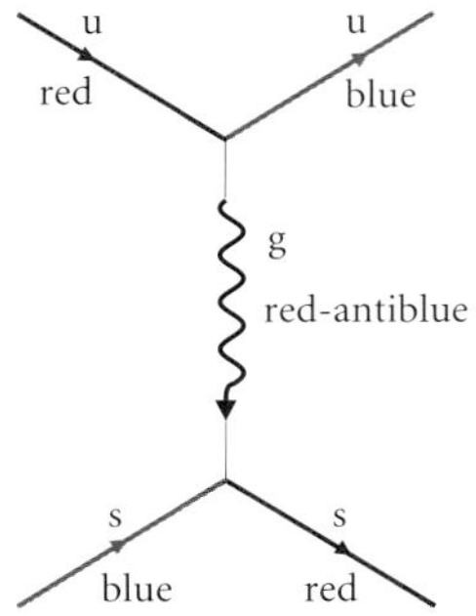

■ FIGURE 37.19
The chromodynamic force is transmitted by gluons. Each gluon carries a combination of color charges. In this example a red, up quark emits a red-antiblue gluon. The up quark changes from red to blue. The blue, strange quark absorbs the gluon and becomes red.

We can pull atoms apart and observe electric charge directly, but since we cannot isolate individual quarks, evidence for the existence of a strong force charge is indirect. The Ω^- particle gives an example of such evidence and shows that strong force charge has to have three different forms. The quark model of an Ω^- (■ Figure 37.18) consists of three s quarks with parallel spins to account for the particle's total spin of $3\hbar/2$. According to the Pauli exclusion principle (§35.3), no two quarks can be in identical states. Each of the three quarks in the Ω^- must have a different value of some quantum number, which we recognize as the value of their strong force charge. The accepted name for strong force charge is *color* and the three forms are *red, green,* and *blue.* Each of the observed particles has zero *net* color. Any names could have been chosen, but these have the virtue of being easy to remember. Of course, color charge has no relation to light, which is an electromagnetic wave produced by electric charge.

A GOOD REASON FOR USING THESE NAMES IS THE ANALOGY WITH PRIMARY COLORS OF LIGHT. IF RED, GREEN, AND BLUE LIGHT BEAMS WITH EQUAL INTENSITY SHINE ON A REFLECTING SURFACE, THE RESULT IS SEEN AS WHITE. THIS MIMICS THE FACT THAT EQUAL AMOUNTS OF THE THREE COLOR CHARGES COMBINE TO BE NEUTRAL.

A theory modeled on quantum electrodynamics but involving *colors* of strong charge is called *quantum chromodynamics.* ■ Figure 37.19 illustrates a typical chromodynamic interaction and shows the crucial new feature that the gluons themselves carry color charge.

A TOTAL OF EIGHT GLUONS ARE NEEDED IN THE THEORY. IN ADDITION TO THE SIX COMBINATIONS LIKE BLUE-ANTIRED, THERE ARE TWO THAT CARRY MIXTURES SUCH AS BLUE-ANTIBLUE WITH RED-ANTIRED.

■ Figure 37.20 shows why isolated quarks cannot exist. When quarks are separated by distances of the order of 10^{-15} m, one can think of the chromodynamic force between them, like the electric force, in terms of field lines. The field lines represent gluon fields that attract each other and so concentrate in a tube between the quarks, rather than spreading out as electric field lines do. As a result, the chromodynamic force between quarks does not decrease at large separations. Separating two quarks would require an arbitrarily large amount of work, which would be stored as energy in the gluon field. Once this stored energy exceeds the rest energy of a quark/antiquark pair, such a pair materializes, producing two mesons rather than two separated quarks.

37.3.3 Conservation Laws for the Strong and the Weak Forces

• Table 37.6 shows the quantum numbers we have discussed, and their conservation properties. The traditional conserved quantities—energy, momentum, angular momentum, and charge—are also conserved in particle interactions. Quark number and baryon number are also absolutely conserved in the standard model. Certain theories that have been developed in an attempt to unify the strong and the electroweak forces predict nonconservation of quark and baryon number. At this time (1996), there is no experimental support for this idea. Quark flavor is conserved by the strong and electromagnetic forces but not by the weak force.

TABLE 37.6 The Conservation Laws of Particle Physics			
Quantity	**Strong**	**Weak**	**Electromagnetic**
Energy	✓	✓	✓
Momentum	✓	✓	✓
Angular momentum	✓	✓	✓
Electric charge	✓	✓	✓
Quark number	✓	✓	✓
Baryon number	✓	✓	✓
Lepton number	✓	✓	✓
Quark flavor	✓	×	✓
Isotopic spin	✓	×	×
Quark color	×	✓	✓
Lepton generation	✓	✓	✓

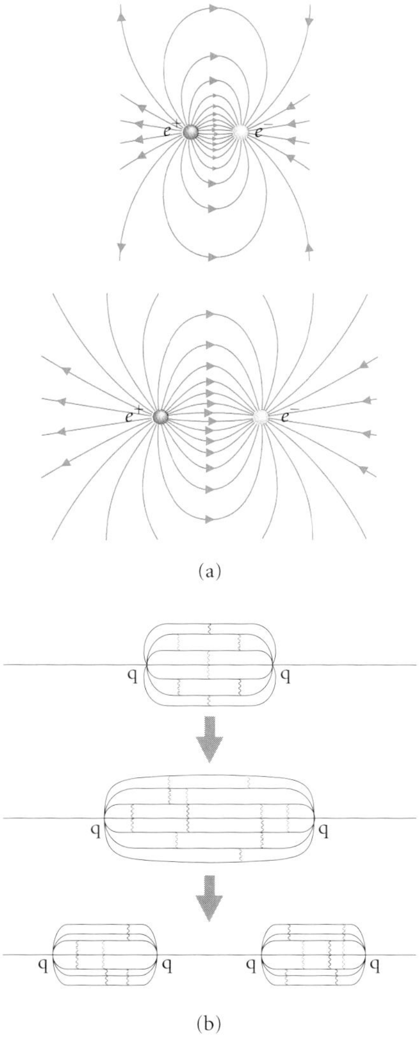

■ FIGURE 37.20
Isolated quarks do not exist because the color force does not decrease with distance. (a) Electric field lines between oppositely charged particles spread out as the separation of the charges increases, illustrating the decrease of the Coulomb force with distance (cf. §23.1 and §23.2). (b) In contrast, color force lines form a tubelike structure and do not separate as the quarks separate. Eventually, when the applied force has done enough work, a single tube splits into two, forming a quark/antiquark pair. The result is two mesons rather than isolated quarks.

EXAMPLE 37.3 ♦♦ Compare the reactions $\pi^- + p \rightarrow K^0 + \Lambda^0$ (Figure 37.8) and $\Lambda^0 \rightarrow p + e^- + \bar{\nu}_e$. Use the conservation laws to determine whether each reaction involves the strong or the weak interaction.

MODEL The conserved quantities that distinguish strong hadronic interactions from weak ones are quark flavor and isotopic spin (I_3). (*Remember:* The hadrons are the baryons plus the mesons. See Figure 37.7.)

SETUP We investigate the values of these two quantities before and after the interaction. We find the quark composition of each particle from Table 37.2.

$$\pi^-: \bar{u}d, \qquad \pi^+: u\bar{d}, \qquad p: uud, \qquad K^0: d\bar{s}, \qquad \Lambda^0: uds.$$

	BEFORE		AFTER	BEFORE		AFTER
	$\pi^- + p$	$\rightarrow$	$K^0 + \Lambda^0$	Λ^0	$\rightarrow$	$p + e^- + \bar{\nu}_e$
Quark number	$0 + 3 = 3$		$0 + 3 = 3$	3		3
Quark flavor	$\bar{u}d + uud = udd$		$d\bar{s} + uds = udd$	uds		uud
I_3	$-1 + \frac{1}{2} = -\frac{1}{2}$		$-\frac{1}{2} + 0 = -\frac{1}{2}$	0		$\frac{1}{2}$
Lepton number				0		$0 + 1 - 1 = 0$

SOLVE Neither quark flavor nor isospin change in the $p + \pi^-$ interaction, but both change in the decay of the Λ^0. Thus the first interaction is a strong interaction and the second is a weak interaction.

ANALYZE Both interactions conserve quark number, lepton number, and charge. Production of leptons in the Λ decay is a telltale sign of a weak process. ■

EXERCISE 37.2 ♦♦ Is the decay $\rho^0 \rightarrow \pi^+ + \pi^-$ governed by the strong or weak interaction? What about $K^0 \rightarrow \pi^+ + \pi^-$?

37.3.4 *Limitations of the Standard Model*

The standard model has been completely successful in explaining qualitative features of subnuclear particles. Quantitative calculations with the model are often extremely intricate, but those that have been done are in good agreement with experiment. Nevertheless, no one accepts the model as a final description of nature.

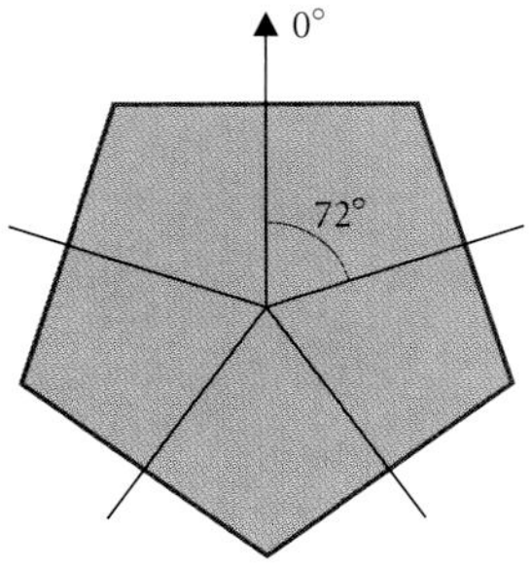

■ FIGURE 37.21
A pentagon illustrates the classical idea of symmetry. Rotation of the figure through any multiple of 72° leaves its appearance unchanged.

- Its own great success argues against it.
  The model describes electroweak and chromodynamic interactions by very similar theories. If such theories show weak and electromagnetic forces to be unified, should we not look for a more comprehensive model that includes the strong force and perhaps even gravity in the unified theory?
- Important phenomena lie beyond the scope of the theory.
  Why, for example, are there three apparently different types of force rather than some other number? What determines the number of generations of particles? Why do the numerical parameters in the theory have the values they do?

In the last section of this chapter, we describe some of the ideas that led to discovery of the standard model and that are guiding attempts to extend it.

37.4 CHARACTERISTICS OF MODERN PARTICLE THEORIES

37.4.1 Symmetries and Groups

The pentagon in ■ Figure 37.21 illustrates the classical, geometrical idea of symmetry—a regularity of shape that is pleasing to the eye. A precise statement of this regularity is that a rotation of the figure through 72° results in an identical shape. This statement describes the pentagon's symmetry in a way that generalizes to much more abstract *symmetries* of physical systems: a set of operations (rotation by 72°, 144°, etc.), and a property (shape) that remains unchanged when any of the operations is performed. A closely related example, important in physics, is the fact that all the laws of physics remain unchanged if the coordinate system is rotated in any way. Equivalently, the behavior of a physical system does not depend on spatial orientation.

FOR EXAMPLE, A PENDULUM SWINGING IN EARTH'S GRAVITY IS PART OF A SYSTEM THAT INCLUDES THE EARTH. THE PENDULUM'S BEHAVIOR DOESN'T VARY DURING THE COURSE OF A DAY, AS THE EARTH ROTATES.

Symmetries are closely related to the mathematical idea of groups and to the physical idea of conservation laws. Any set of symmetry operations corresponds to the elements of a group, and every symmetry of physical laws corresponds to a conserved physical property. For example, the symmetry of physical law under rotations corresponds to conservation of angular momentum.

THIS RESULT, WHILE TRUE, IS CERTAINLY NOT OBVIOUS AND WE DO NOT CLAIM TO HAVE PROVED IT.

The group properties can be satisfied only by special sets of elements, which in turn means that only special kinds of symmetry are possible for physical laws. As a result, the mathematics of groups has proved a powerful guide in the discovery of the unified electroweak and quantum chromodynamic theories. One approach to further unification is the attempt to determine which group describes the symmetry of the unified physical laws.

The power of group theory is well illustrated by isotopic spin. We could imagine changing each proton in the universe into a neutron, and vice versa. This operation is a symmetry of the strong force, since it leaves the strong forces acting on any object unchanged. The group of such particle-exchange symmetries of the strong force is the same as the group of spatial rotations. Thus the corresponding quantity, conserved by strong interactions, should follow the same rules as angular momentum. This quantity is the isotopic spin.

WE SAY THAT THE ROTATIONS AND THE STRONG FORCE SYMMETRIES ARE DIFFERENT REPRESENTATIONS OF THE SAME GROUP.

Math Topic

GROUP PROPERTIES ILLUSTRATED BY AN EXAMPLE

PROPERTIES OF A GROUP	ROTATIONS OF A PENTAGON
A group is collection of elements with an operation *.	Rotations through 0°, 72°, 144°, 216°, 288° (360° ≡ 0°) follow one rotation by another.
If a and b are in the group, then so is $a*b$.	72°*144° = rotation through 144° + 72° = 216°.
There is an identity element e.	e = 0°—no rotation.
$e*a = a$ for every element a in the group.	0°*144° = rotation through 144° + 0° = 144°.
For any element a in the group, there is an inverse element a^{-1} such that $a^{-1}*a = e$.	$(144°)^{-1}$ = 360° − 144° = 216° 216°*144° = rotation through 144° + 216° = 360° ≡ 0°.

■ FIGURE 37.22
An electron is surrounded by a sea of virtual electron/positron pairs. Like the molecules in a dielectric, the pairs are polarized by the central charge and shield the *bare* charge from outside observers. The charge we observe and call e is the shielded charge. Both the bare charge and the shielding are infinite. Fortunately, the shielded charge can be calculated. Predictions made using this theory are extremely well supported by experiment.

37.4.2 Renormalization

In Chapter 25 we found that the total electrostatic energy of a point charge is infinite. Though startling, this infinite energy does not cause trouble, since we only observe energy changes. In the classical theory of the electron, we neglect the self-energy, which does not change during classical electrical processes. Such neglect of constant, infinite quantities is an example of *renormalization.*

Infinities appear in the quantum theory via the existence of virtual particles. An electron, ■ Figure 37.22, produces a cloud of virtual photons. According to the Heisenberg principle, the most energetic particles have the shortest time of existence and so remain close to the electron. These virtual photons can themselves produce virtual electron/positron pairs. It seems we can't have a single isolated particle; particles are necessarily surrounded by a virtual particle/photon cloud. The cloud possesses energy and, because the central electron attracts virtual positrons and repels virtual electrons, contributes to the total observed charge. The effects of the cloud are infinite. The observed mass and charge of the electron are the renormalized differences between infinite *bare* mass and charge and the infinite effects of the virtual cloud. Renormalization must result in finite quantities if a quantum theory is to make sensible statements about observed reality. This is a trivial requirement for classical electromagnetism but is a strong restriction on quantum theories. Indeed, one of the strongest reasons for believing the electroweak and quantum chromodynamic theories is that they are renormalizable.

37.4.3 Spontaneous Symmetry Breaking

The symmetry of a physical theory is often disguised by the phenomena we observe. One example is the magnetism of an iron sample. The magnetization of the sample points in some specific direction, yet Maxwell's equations show no preference for any particular direction over another. The theory is symmetric, but the iron sample is not. If the sample is heated above its Curie temperature T_C, the iron atoms no longer align, the magnetization disappears, and the sample regains the symmetry of electromagnetic theory. This behavior of the iron sample is called *spontaneous symmetry breaking.* The rotational symmetry of electromagnetism is *broken* by the iron sample once the typical energy of an atom becomes less than approximately kT_C. The breaking is spontaneous in that random fluctuations in the sample will cause it to magnetize even in the absence of external fields. Furthermore, the loss of symmetry is not complete: the direction in which an isolated sample will magnetize is completely random.

The differences between electromagnetism and the weak force are another example of symmetry breaking. The energy at which the breaking occurs is set by the masses of the W and Z bosons that transmit the weak force. The quest for further unification of physics presumes that electroweak and chromodynamic forces differ only because some, as yet undemonstrated, symmetry remains broken at energies currently achievable by accelerators.

37.5 Conclusion

Quantum chromodynamics and the unified electroweak theory are as far as we can go in describing fundamental particle phenomena, based on correspondence between theory and experimental data. Together with general relativity, they form a working set of answers to the ancient Greek questions about the constituents of the universe and the rules it follows. Three rules (for the three fundamental forces) and 24 basic constituents suffice. Furthermore, each of the three theories shows a high degree of mathematical elegance and economy of physical assumptions. We have every reason to be pleased with the achievements of physics in explaining the world, but no excuse for being smug about them. It remains for the physicists of the future to show how quantum principles apply to gravity and to continue the quest for unity in physical law.

Six quarks, six leptons, and their antiparticles. We have not included the force-carrying gauge bosons in this count.

Chapter Summary

Where Are We Now?

We have completed our survey of physical theory, including results discovered centuries ago and those discovered only shortly before the publication of this book. Theories of atomic and subatomic physics involve complex and abstract mathematics, which we have described more qualitatively than our treatment of classical ideas. The quest for knowledge goes on, and the physical models we have described will be modified and refined as research progresses.

What Did We Do?

In the standard model of particle physics, matter is composed of 24 fundamental quantities: 6 quarks, 6 leptons, and their antiparticles. Quarks have a property, akin to electric charge, called color charge, which comes in three kinds: red, green, and blue. All forces result from the exchange of particles: exchange of photons causes the electromagnetic force between electrically charged particles. Similarly, the weak force is due to the exchange of W and Z bosons, and the strong force results from the exchange of gluons between particles possessing color charge. All known particles are accounted for as either individual leptons or as combinations of 2 or 3 of the 12 quark types, bound by a cloud of gluons.

Quarks have fractional electromagnetic charge, in multiples of $\frac{1}{3}e$. Familiar nuclear particles such as the proton are combinations of quarks with an integer total electric charge and no net color.

Conserved quantities in physics are related to underlying symmetries in the physical laws. For example, space-time is invariant under spatial rotations; all forces respect this symmetry, and the corresponding conserved quantity is angular momentum. Symmetry is described by the mathematical idea of a group. The groups describing the symmetries of electromagnetic and weak interactions are combined into one larger group to describe the unified electroweak force. Physicists hope to find a theory that unifies the strong force with the electroweak force.

Practical Applications

The theory presented in this chapter is so new that practical applications, if any, remain for the future.

Solutions to Exercises

37.1 A particle's de Broglie wavelength is given by its momentum: $\lambda = h/p$. For so small a wavelength as we desire, we expect to require highly relativistic electrons, each with an energy much greater than its rest mass. So:

$$p^2c^2 = E^2 - m_0^2c^4 \approx E^2.$$

Then,
$$E \approx pc = \frac{hc}{\lambda} = \frac{(6.64 \times 10^{-34}\ \text{J}\cdot\text{s})(3 \times 10^8\ \text{m/s})}{10^{-16}\ \text{m}} = 2 \times 10^{-9}\ \text{J} = 10\ \text{GeV}.$$

37.2 The ρ^0, at the center of the meson nonet (Figure 37.12d), is composed of equal numbers of quarks and their corresponding antiquarks, and has no net quark flavor. The π^+ and π^-, being the antiparticle of each other, also have no net quark flavor. $I_3 = 0$ for the ρ^0, -1 for the π^-, and $+1$ for the π^+. Quark flavor and the total value of I_3 are both zero before and after the decay, which is therefore governed by the strong interaction.

The K^0 has quark flavor $d\bar{s}$, so quark flavor is not conserved in its decay to two pions. Additionally, I_3 changes from $-\frac{1}{2}$ to 0. This is a weak decay.

Basic Skills

Review Questions

§37.1 PARTICLE CREATION AND FUNDAMENTAL FORCES

- Why does the macroscopic world appear stable?
- What is a *virtual particle?* How long can it exist?
- Which particle did Yukawa first propose? Why?
- To what does *isospin* refer? Explain why the name is appropriate.

§37.2 SUBNUCLEAR PARTICLES AND THE QUARK MODEL

- What were the three original classes of particles? What distinguished the classes?
- Why are *strange* particles strange?
- What is a *quark?* What is the fundamental unit of charge for quarks?
- How many original quark flavors were there?

§37.3 THE STANDARD MODEL

- Which particles transmit the weak force? Are any of them charged? If so, which?
- Which bosons can change quark flavor?
- How many generations of particles are there?
- What name is given to the strong force charge? How many kinds are there and what are their names?
- How does the chromodynamic force vary with quark separation?
- State two limitations of the standard model.

§37.4 CHARACTERISTICS OF MODERN PARTICLE THEORIES

- Which mathematical structure is important for understanding particle theory?
- State a symmetry of the strong force and the corresponding conserved quantity.
- What is *renormalization?*
- What is *spontaneous symmetry breaking?*

Basic Skill Drill

§37.1 PARTICLE CREATION AND FUNDAMENTAL FORCES

1. For how long can a virtual electron/positron pair exist?
2. What is the energy of each photon produced in the decay of a π^0: $\pi^0 \to \gamma + \gamma$? (The two photons have equal energy in the CE frame. Do you recall why?)

§37.2 SUBNUCLEAR PARTICLES AND THE QUARK MODEL

3. Find the Q value for the reaction:

$$p + \bar{p} \to \pi^+ + \pi^+ + \pi^- + \pi^-.$$

4. Draw a quark line diagram for the reaction:

$$\pi^- + p \to \Lambda^0 + K^0.$$

What is the Q value for this reaction?

§37.3 THE STANDARD MODEL

5. Can the reaction

$$K^0 + p \to K^+ + n$$

proceed via the strong interaction? Why or why not?
6. Draw a Feynman diagram for the interaction:

$$K^- \to \mu^- + \bar{\nu}_\mu.$$

Which boson is involved?

§37.4 CHARACTERISTICS OF MODERN PARTICLE THEORIES

7. The Δ particle has isospin $I = -\frac{3}{2}$. How many different Δs are there? What are the corresponding values of I_3?

Questions and Problems

§37.1 PARTICLE CREATION AND FUNDAMENTAL FORCES

8. ❖ Discuss the reasons why one must increase the particle energy available from accelerators in order to study phenomena on smaller spatial scales or on shorter time scales.

9. ♦ Calculate the minimum energy of a photon that can cause the reaction:

$$\gamma + p \rightarrow \pi^0 + p.$$

10. ♦ Calculate the minimum energy in the CE frame for each incident proton in the reaction:

$$p + p \rightarrow p + p + \Lambda + \bar{\Lambda}.$$

11. ♦♦ For an electric motor to function according to Maxwell's equations, virtual photons must be able to travel at least as far as the dimensions of the motor. Derive a value for the maximum possible rest mass of a photon from the dimensions of a large industrial motor, approximately 1 m. What limit on the photon rest mass follows from the fact that Maxwell's equations describe the Earth's magnetic field over distances of the order of 10^7 m?

§37.2 SUBNUCLEAR PARTICLES AND THE QUARK MODEL

12. ❖ Strangeness is conserved in some processes and not in others. In what sense is this similar to your experience with *liquid* water. In what ways is it different?

13. ❖ Mesons have integer spin and zero baryon number. Explain why only combinations of a quark and an antiquark can form a meson. Similarly, explain why only combinations of three quarks may form a baryon.

14. ❖ The baryon octet in Figures 37.10 and 37.12 consists of spin $\frac{1}{2}$ particles, while the baryon decuplet contains spin $\frac{3}{2}$ particles. How are the quarks arranged differently in an octet particle and in a decuplet particle with the same isospin and hypercharge?

15. ♦♦ Determine which quantum numbers are conserved in the decay:

$$\overline{\Sigma^+} \rightarrow \bar{p} + \pi^0. \qquad (\overline{\Sigma^+} \text{ is the antiparticle of the } \Sigma^+.)$$

§37.3 THE STANDARD MODEL

16. ❖ Draw diagrams showing the exchange of bosons in each of the following reactions:

(a) $p + \bar{\nu}_e \rightarrow n + e^+$. **(b)** $e^- + \nu_\mu \rightarrow \nu_e + \mu^-$.
(c) $d + e^- \rightarrow u + \nu_e$. **(d)** $\nu_e + \nu_\mu \rightarrow \nu_e + \nu_\mu$.
(e) $\bar{\nu}_e + p \rightarrow e^+ + \Lambda$.

17. ❖ Determine which of these reactions cannot occur at all and which cannot occur by the strong interaction. In each case, indicate the reasons for your decision.

(a) $K^- + p \rightarrow \overline{K^0} + n$. **(b)** $\pi^- + p \rightarrow K^- + \Sigma^+$.
(c) $\pi^- + p \rightarrow K^+ + \Sigma^0 + \pi^-$. **(d)** $K^- + p \rightarrow \Sigma^+ + n + \pi^-$.
(e) $K^- + p \rightarrow K^0 + \pi^+ + e^-$. **(f)** $\pi^- + p \rightarrow \pi^0 + \Sigma^+$.

Draw quark line diagrams for those reactions that can occur by the strong interaction.

18. ♦♦ Draw a quark line diagram for the reaction:

$$p + \pi^- \rightarrow n + \pi^0.$$

What is the Q value for this reaction?

19. ♦♦ Draw a quark line diagram for the reaction:

$$\Xi^- + p \rightarrow \Lambda + \Lambda.$$

Find the Q value.

§37.4 CHARACTERISTICS OF MODERN PARTICLE THEORIES

20. ✲ ❖ Do the positive integers (0, 1, 2, 3, . . .) form a group? If so, what are the operation and the identity element of the group? Answer the same question for the following sets:

- All integers (0, ±1, ±2, ±3, . . .).
- All rational fractions (a/b, where a and b are positive integers).
- Translations of the origin of coordinates along a single axis.

21. ❖ At what temperature would you expect the weak interaction to attain comparable strength to electromagnetism? Explain your reasoning.

Additional Problems

22. ♦♦ Given that the Λ_c^+ charmed baryon can decay weakly into $\Delta^{++} + K^-$ or $p + \overline{K^0} + \pi^+ + \pi^-$, determine a limit to the mass of Λ_c^+.

23. ♦♦ Calculate the Q value for the weak decay:

$$\Omega^- \rightarrow \Lambda + K^-.$$

Verify that quark number is conserved.

24. ♦♦ The Δ^0 and the Λ both decay to $p + \pi^-$, but their half-lives are very different. Explain why. Which half-life is the shorter? (The values are approximately 10^{-23} s and 3×10^{-10} s, respectively.)

25. ♦♦ Both the ρ^0 and the K^0 decay to $\pi^+ + \pi^-$, but the ρ^0 has a half-life of about 10^{-23} s, while the half-life of the K^0 is about 10^{-10} s. Explain the difference. What can you conclude about the quark flavor of the ρ^0? Determine the Q value for each decay [$M(\rho^0) = 770$ MeV].

26. ♦♦ What is the minimum energy of each proton in the CE frame for the reaction

$$p + p \rightarrow \Lambda^0 + K^0 + p + \pi^+$$

to occur? Determine whether this is a strong or weak interaction and draw a quark line diagram or a Feynman diagram for it.

27. ♦♦ Draw a Feynman diagram at the quark level for the following reactions:

(a) $\Omega^- \rightarrow \Xi^0 + \pi^-$. **(b)** $\Lambda_c^+ \rightarrow p + \overline{K^0}$.

Find the Q value for each reaction.

Part VIII Problems

1. ❖ Tachyons are imaginary particles that move faster than light. Just as ordinary particles cannot move faster than light, tachyons cannot go slower. On a space-time diagram, draw the world line of a tachyon emitted from the origin in the positive x-direction at speed $2.2c$, toward an observer moving in the x-direction at speed $0.8c$. Upon receiving the signal, the observer transmits a tachyon signal at $2.2c$ (in his own frame) in the negative x'-direction. Show that this signal arrives at the unprime time-axis *before* the original signal was transmitted. (See Greg Benford's novel *Timescape* for some consequences of tachyon communication!)

2. ❖ Explain, using the shell model (Figure 36.26), why the spins of individual nucleons do *not* add to produce nuclei with large total spin.

3. ♦♦ Show that when a particle's energy is much greater than its rest energy, its de Broglie wavelength is approximately the same as the wavelength of a photon with the same energy.

4. ♦♦ In an accelerator experiment, a K^0 particle decays into two π^0 mesons. In the lab frame, the π^0 mesons' velocities make equal angles $\theta = 27°$ with the direction of motion of the K^0. Find the speed of the K^0. Mass of $K^0 = 497.71$ MeV; mass of $\pi^0 = 134.97$ MeV. (*Hint:* It is easiest to work in the rest frame of the K^0 and then transform to the lab frame.)

5. ♦♦♦ A quasar 3×10^9 light-years away emits a blob of luminous gas moving with $\gamma = 10$ at an angle of 0.01 rad from the line of sight to the Earth. Show that the blob will *appear* to move away from the quasar at greater than the speed of light. Such *superluminal motion* is observed in many quasars. (*Hint:* Assume the blob emits light pulses at regular intervals, and find the separation of the pulses in space and time as seen from Earth. Only distance perpendicular to the line of sight can be measured!)

6. ♦♦♦ In 1959, Pound and Rebka measured the effects of gravitational time dilation on γ rays propagating from the top to the bottom of a 22.5-m-high tower. In the experiment, nuclei in a sample of ^{57}Co emit γ rays with precisely determined frequency as they decay to ^{57}Fe. The γ rays are absorbed in a sample of ^{57}Fe at the bottom of the tower. The detector absorbs the γ rays only when the Doppler effect due to its motion cancels the frequency shift due to time dilation. What fractional change in frequency of the γ rays results from gravitational time dilation? What velocity of the detector results in maximum absorption?

7. ♦♦♦ In an inertial reference frame at rest with respect to the center of a rotating disk, events A and B occur simultaneously at the opposite ends of a diameter. At the same time observers O and O' on the disk arrive at opposite ends of the perpendicular diameter (■ Figure VIII.1). Observers O and O' do not agree that A and B are simultaneous in their respective instantaneous rest frames. They do not even agree as to which event occurs first. By how much time do they disagree? This illustrates that observers in a noninertial reference frame, such as the surface of the Earth, cannot consistently synchronize their clocks. How big is the effect on Earth? Atomic clocks are stable to 1 part in 10^{13}. How much time must pass before the uncertainty in the clock reading is larger than the synchronization error?

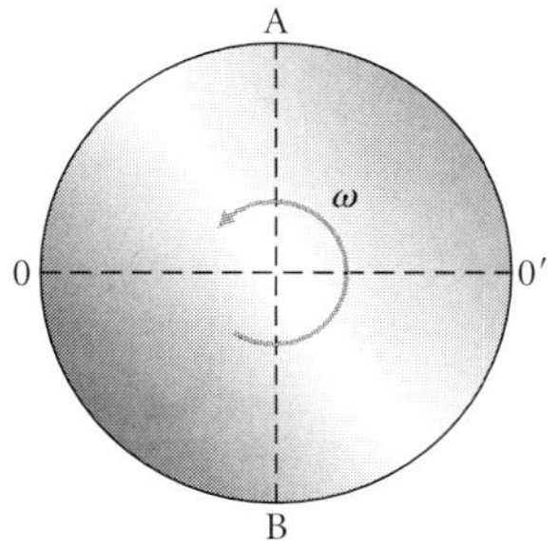

■ FIGURE VIII.1

8. ♦♦♦ An astronaut jokes that NASA should pay overtime to space shuttle crews because of relativistic effects on the experience of time. Which is the more important effect in a space shuttle orbit, special relativistic time dilation due to the shuttle's speed or general relativistic effects due to the different gravitational potential in orbit as compared with the Earth's surface? Should the astronauts be paid overtime or have their pay docked due to the dominant effect?

9. ♦♦♦ Use Heisenberg's uncertainty principle to estimate the minimum kinetic energy of an electron confined within a nucleus with radius 4.2×10^{-15} m. If the nucleus is a uniformly charged sphere with atomic number $Z = 14$, *estimate* the electric potential energy of the electron. From your results, is it plausible that electric forces can bind the electron within the nucleus?

10. ♦♦♦ Consider a 10-MeV proton incident on a ^{12}C nucleus. Compare the speed of the proton with the speed of an electron in the $n = 1$ Bohr orbit. Also compute the maximum energy the proton can transfer to the electron in a head-on collision. Comment on why the chemical state of an atom is unimportant in understanding its nuclear reactions.

11. ♦♦♦ **(a)** Apply Bohr's atomic theory to the orbital motion of an electron and a positron around each other. Show that the energy of the nth orbit is given by: $E_n = -mc^2\alpha^2/(4n^2)$, where $\alpha = ke^2/(\hbar c) \approx \frac{1}{137}$ is the fine-structure constant, which measures the strength of the electromagnetic force. **(b)** An early model for charmonium assumed that the force between a c and a $\bar{c}$ quark follows an inverse square law, but with strength α_s instead of α. Assume that the J/Ψ particle (mass = 3.096 GeV) and the Ψ' particle (mass = 3.687 GeV) are the $n = 1$ and $n = 2$ Bohr states of charmonium. Assume also that the mass of a charmed quark is $m_c \approx 1.5$ GeV. Find the ratio α_s/α.

Challenge Problems

12. A photon with initial frequency f Compton-scatters from an electron moving relativistically in the same direction as the photon. Find the frequency of the scattered photon as a function of scattering

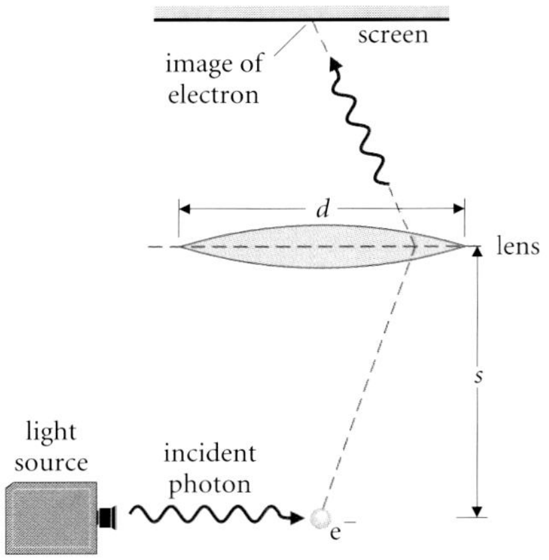

■ FIGURE VIII.2

angle and the initial momentum of the electron. Repeat the calculation for an electron moving in the opposite direction from the photon. Comment on the reasons why your results differ from the Compton formula for a stationary electron. A beam of 400-keV x rays is scattered by a gas of singly ionized helium atoms. Estimate the error made in assuming the electrons are stationary rather than in orbit about the helium nuclei.

13. Suppose we attempt to measure the position of an electron with the apparatus shown in ■ Figure VIII.2. A photon from the light source scatters from the electron and is focused by the lens on a photographic plate. The accuracy of the measured position is limited by the resolution of the lens. Show that the uncertainty in the measured position is $\delta x \sim s\lambda/d$. In principle, the accuracy of the measurement can be improved indefinitely by decreasing the wavelength of the photons used. However, the photon transfers momentum to the electron as it scatters. Show that the resulting uncertainty in the momentum of the electron after the measurement is $\delta p_x = (h/\lambda)(d/s)$ so Heisenberg's uncertainty principle is satisfied.

This is not the end.
It is not even the beginning of the end.
It is perhaps, the end of the beginning.

WINSTON CHURCHILL

EPILOGUE

If history is a guide, today's fundamental understanding will ultimately be recognized as an approximation. The standard model succeeds in explaining physics with only three interactions and a small number of particle types. Yet, as we noted in Chapter 37, it contains many arbitrary features. Following the spirit of Einstein, physicists now have appetites whetted for a theory with a single interaction in which all details arise as necessary consequences. With little experimental evidence, the search for such a theory is highly speculative, and we shall take only a quick glance at current ideas.

The logic of the standard model suggests that the electroweak and strong interactions appear different only because we cannot yet observe phenomena in which their unity is apparent. Pursuing this logic leads to a class of Grand Unified Theories, or GUTs, in which the photons, W and Z bosons, and gluons of the standard model are seen as special cases of particles that carry the grand unified interaction. The quest for unity is also a quest for higher experimental energy. Energies of 80 GeV are required to produce the W and Z bosons that carry the weak interaction; much higher energies are needed to test GUTs. The Superconducting Supercollider would have produced particle energies of approximately 40 TeV, at a cost of billions. That massive an enterprise is needed to search for phenomena predicted by GUTs.

THE EARLIEST AND SIMPLEST GUT MODELS PREDICTED DECAY OF PROTONS WITH A LIFETIME OF $10^{32\pm2}$ y. SEVERAL EXPERIMENTS HAVE RULED OUT A LIFETIME THIS SHORT.

THE SUPERCONDUCTING SUPERCOLLIDER, ALREADY UNDER CONSTRUCTION IN TEXAS, WAS CANCELED BY CONGRESS IN 1994.

The universe provides the ultimate high-energy physics laboratory. It is expanding away from an immensely hot, dense origin some 10–20 billion years ago. Near this beginning, temperatures were high enough to produce any kind of elementary particle. Cosmologists have shown that the current state of the universe may be the result of particle reactions in the first fraction of the first second. An exciting consequence of some GUT models is that universes arise spontaneously from a quantum vacuum. Fundamental particle dynamics may explain the existence of our universe!

Beyond GUTs comes the unification of gravity with the other interactions. Gravity is intimately related to the geometry of space-time, and abstract geometrical notions play a large role in such theories. For example, *superstring* theories involve a space-time with ten dimensions. Four are the usual space and time dimensions. Space in the remaining six dimensions is curved so strongly that the entire six-dimensional space we usually think of as a single space-time event is approximately 10^{-35} m in extent. In these theories, the fundamental type of object is not a point particle, but a one-dimensional *string* that wraps around the *compactified* six-space. The possible vibrations of such strings give rise to the different kinds of particle observed at the level of the standard model.

We do not know now which, if any, of these speculations will prove correct. The only certain thing is that physics will remain exciting and continue to stretch our imagination for some time to come.

APPENDIXES

Appendix I

Mathematics

A. Scientific Notation
B. Arithmetic, Algebra, Geometry and Trigonometry
C. Vectors
D. Coordinate Systems
E. Derivatives
F. Integrals
G. Rules of Probability

Appendix II

Symbols

A. The Greek Alphabet
B. Symbols Used in the Text
C. Mathematical Symbols
D. Symbols for Units
E. Roman Numerals

Appendix III

Answers to Selected Problems

APPENDIX I

Mathematics

A: SCIENTIFIC NOTATION

In problems of interest to physicists, quantities may vary over a very large range of magnitudes. For example, consider the concept of length; sizes much smaller than single atoms and as large as the visible universe may be important in the same discussion. Scientific discussions thus require a convenient and efficient way to describe these magnitudes and to calculate with the very large or very small numerical ratios that arise from comparing them.

Common decimal notation certainly allows the expression of such numbers. A ratio of one-millionth can be written 0.000001, but already the number of zeros makes the expression difficult to read quickly. Most people would be unable to read a much smaller number directly and would have to resort to counting zeros. Scientific notation is a way of using exponents to express numbers by counting zeros and reporting the count rather than writing the zeros. Consider the following examples.

$$0.001 = 1/1000 = 1/(10 \times 10 \times 10) = (1/10)^3 = 10^{-3}$$

Or, $$1000. = 10 \times 10 \times 10 = 10^3.$$

If the leading digit is to the right of the decimal, the exponent is negative; if the leading digit is to the left, the exponent is positive.

$$33\,760\,000 = 3.376 \times 10\,000\,000 = 3.376 \times 10^7.$$

$$0.0000462 = 4.62 \times 0.00001 = 4.62 \times 10^{-5}.$$

The number that multiplies the power of 10 is chosen by convention. One convention, which we generally use in this text, is to express numbers greater than 1000 or less than 0.001 as a value between 0.1 and 10 multiplied by the appropriate power of 10. Another convention, popular among engineers, is to make the exponent of 10 be a multiple of 3. Thus we might write the numbers above as:

$$3.376 \times 10^7 = 0.3376 \times 10^8$$

and $$4.62 \times 10^{-5} = 0.462 \times 10^{-4},$$

while the engineering convention would be

$$33.76 \times 10^6 \quad \text{and} \quad 46.2 \times 10^{-6}.$$

Multiplication and division of numbers in scientific notation rely on the distributive law and the rules for multiplying or dividing exponents. For example:

$$\begin{aligned}(3.376 \times 10^7) \times (4.62 \times 10^{-5}) &= (3.376 \times 4.62) \\ &\quad \times (10^7 \times 10^{-5}) \\ &= 15.6 \times 10^{7-5} \\ &= 15.6 \times 10^2 \\ &= 1.56 \times 10^3.\end{aligned}$$

$$\frac{3.376 \times 10^7}{4.62 \times 10^{-5}} = \left(\frac{3.376}{4.62}\right)\left(\frac{10^7}{10^{-5}}\right) = 0.731 \times 10^{12}.$$

To add and subtract numbers in scientific notation, first express the numbers as multiples of the *same* power of 10. Then the numbers can be added or subtracted in the normal manner:

$$\begin{aligned}3.376 \times 10^7 + 6.417 \times 10^5 &= 3.376 \times 10^7 + 0.06417 \times 10^7 \\ &= 3.440 \times 10^7.\end{aligned}$$

$$\begin{aligned}4.62 \times 10^{-5} - 3.71 \times 10^{-4} &= 0.462 \times 10^{-4} - 3.71 \times 10^{-4} \\ &= -3.25 \times 10^{-4}.\end{aligned}$$

B: Arithmetic, Algebra, Geometry, and Trigonometry

Arithmetic and Algebra

Associative law of multiplication:	$a(b + c) = ab + ac$
Distributive law of addition:	$(a + b) + c = a + (b + c)$
Distributive law of multiplication:	$(ab)c = a(bc)$
Logarithm of a product:	$\log(ab) = \log(a) + \log(b)$
Change of base of a logarithm:	$\log_a x = (\log_a b)(\log_b x)$

Conversion of the natural logarithm to base 10:	$\log_e 10 \approx 2.303$
Conversion of base 10 to the natural logarithm:	$\log_{10} e \approx 0.434$
Multiplication of exponents:	$(a^x)(a^y) = a^{x+y}$
Powers of exponents:	$(a^x)^y = a^{xy}$
Solution of a quadratic, $ax^2 + bx + c = 0$:	$x = \dfrac{-b \pm \sqrt{b^2 - 4ac}}{2a}$

Factors: $a^2 + 2ab + b^2 = (a + b)^2$

$a^2 - b^2 = (a - b)(a + b)$

Factorial: $n! = n(n - 1)(n - 2) \cdots (2)1$

Taylor series:

$$f(x_o + x) = f(x_o) + f'(x_o)x + f''(x_o)x^2/2! + \cdots$$

Binomial expansion:

$$(1 + x)^n = 1 + nx + n(n - 1)x^2/2! + \cdots$$

Series expansions of common functions:

$$e^x = 1 + x + x^2/2! + x^3/3! + \cdots$$
$$\ln(1 + x) = x - x^2/2 + x^3/3 + \cdots$$
$$\sin(x) = x - x^3/3! + x^5/5! + \cdots$$
$$\cos(x) = 1 - x^2/2! + x^4/4! + \cdots$$
$$\tan(x) = x + x^3/3 + 2x^5/15 + \cdots$$

Geometry

Circumference of circle of radius *r*:	$2\pi r$
Area of circle of radius *r*:	πr^2
Surface area of a sphere of radius *r*:	$4\pi r^2$
Volume of a sphere of radius *r*:	$4\pi r^3/3$
Volume of a circular cylinder of radius *r* and height *h*:	$\pi r^2 h$
Area of a triangle:	$bh/2$

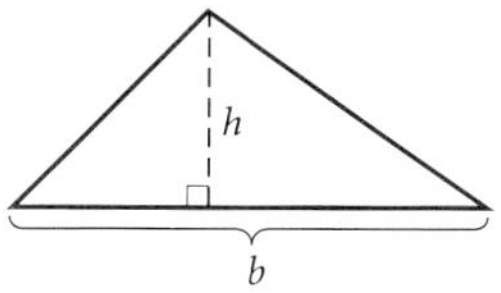

Pythagorean theorem: $a^2 + b^2 = c^2$

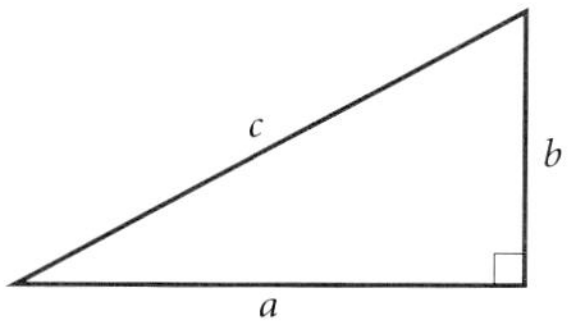

Straight line: $y = mx + c$

$m = \tan\theta = \text{rise/run}$

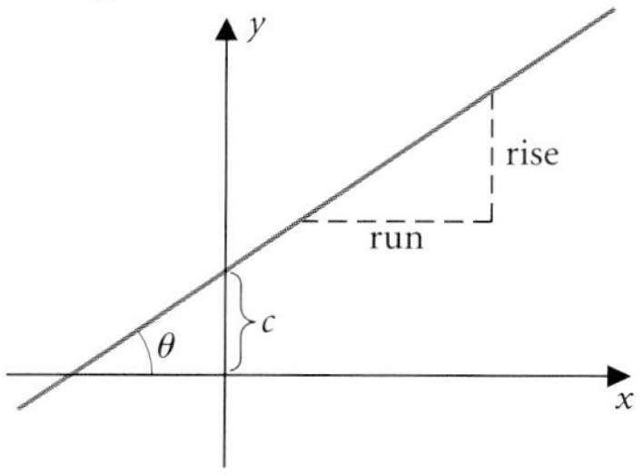

Circle: $(x - x_o)^2 + (y - y_o)^2 = r^2$

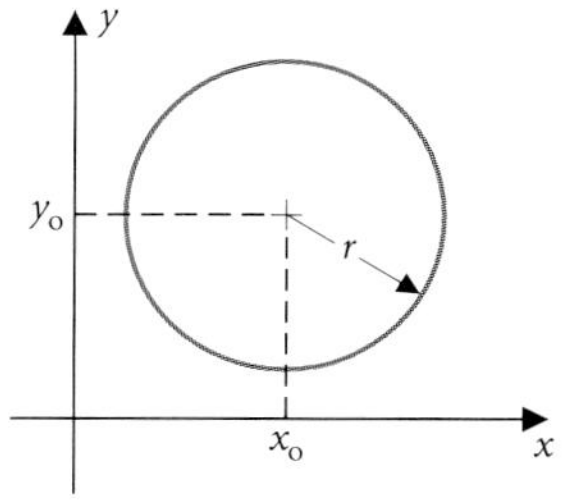

Ellipse: $\dfrac{(x - x_o)^2}{a^2} + \dfrac{(y - y_o)^2}{b^2} = 1$

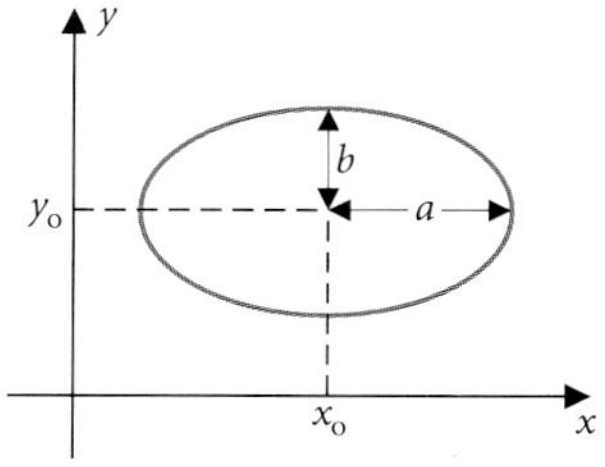

Hyperbola: $\dfrac{(x - x_o)^2}{a^2} - \dfrac{(y - y_o)^2}{b^2} = 1$

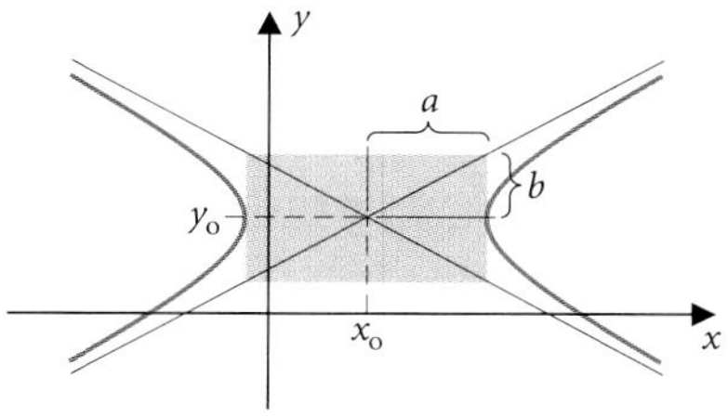

Parabola: $(y - y_o)^2 = 4a(x - x_o)$

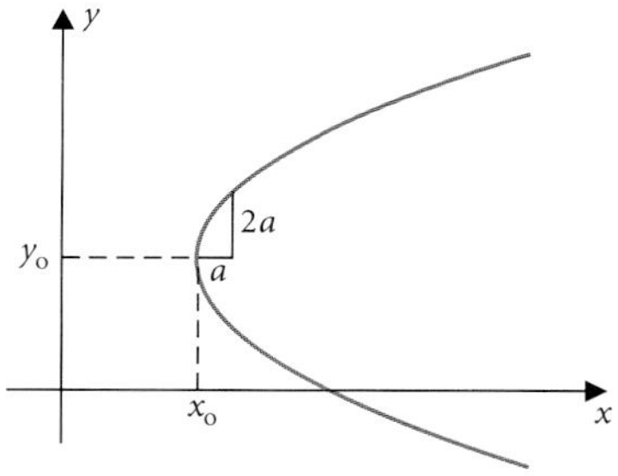

Trigonometry

$$\sin\theta = \frac{b}{c} \qquad \csc\theta = \text{cosec}\,\theta = \frac{1}{\sin\theta} = \frac{c}{b}$$

$$\cos\theta = \frac{a}{c} \qquad \sec\theta = \frac{1}{\cos\theta} = \frac{c}{a}$$

$$\tan\theta = \frac{b}{a} \qquad \cot\theta = \text{cotan}\,\theta = \frac{1}{\tan\theta} = \frac{a}{b}$$

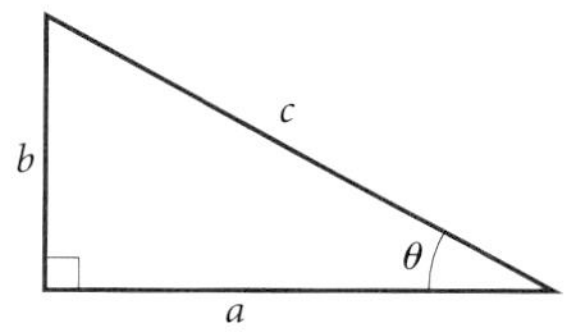

$$\sin^2\theta + \cos^2\theta = 1 \qquad 1 + \tan^2\theta = \sec^2\theta$$

$$\sin(2\theta) = 2\sin\theta\cos\theta \qquad \sin(\theta/2) = \sqrt{(1-\cos\theta)/2}$$

$$\cos(2\theta) = \cos^2\theta - \sin^2\theta \qquad \cos(\theta/2) = \sqrt{(1+\cos\theta)/2}$$

$$= 2\cos^2\theta - 1$$

$$= 1 - 2\sin^2\theta$$

$$\tan(2\theta) = \frac{2\tan\theta}{1-\tan^2\theta} \qquad \tan(\theta/2) = \frac{\sin\theta}{1+\cos\theta} = \frac{1-\cos\theta}{\sin\theta}$$

$$\sin A + \sin B = 2\sin\left(\frac{A+B}{2}\right)\cos\left(\frac{A-B}{2}\right)$$

$$\cos A + \cos B = 2\cos\left(\frac{A+B}{2}\right)\cos\left(\frac{A-B}{2}\right)$$

$$\sin(A+B) = \sin A\cos B + \cos A\sin B$$

$$\cos(A+B) = \cos A\cos B - \sin A\sin B$$

$$\tan(A+B) = \frac{\tan A + \tan B}{1 - \tan A\tan B}$$

Triangle Rules

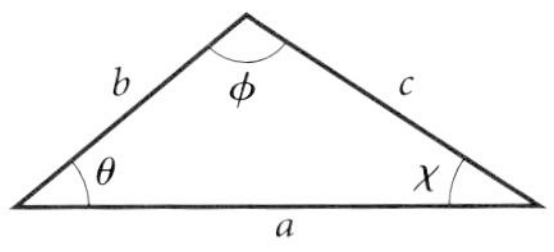

Sine rule: $\dfrac{\sin\theta}{c} = \dfrac{\sin\phi}{a} = \dfrac{\sin\chi}{b}$

Cosine rule: $c^2 = a^2 + b^2 - 2ab\cos\theta$

C: Vectors

Name of vector: $\vec{\mathbf{v}}$

Magnitude of vector: $|\vec{\mathbf{v}}|$ or v

$|\vec{\mathbf{v}}| = \sqrt{v_x^2 + v_y^2 + v_z^2}$

Components of vectors: $\vec{\mathbf{v}} = (v_x, v_y, v_z)$

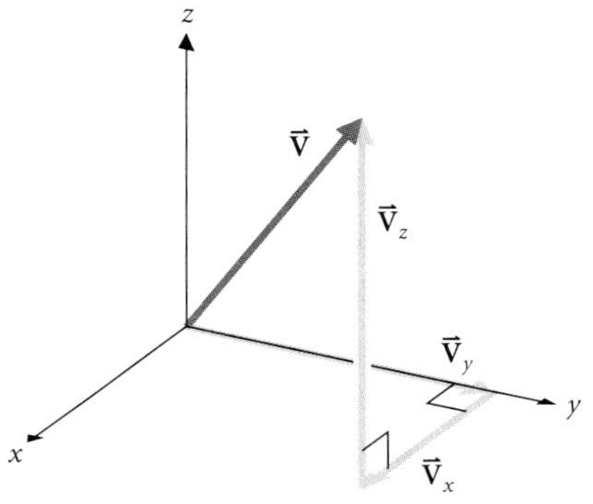

Unit vectors: $\hat{\mathbf{V}} \equiv \dfrac{\vec{\mathbf{V}}}{|\vec{\mathbf{V}}|}$ and $\vec{\mathbf{V}} = \vec{\mathbf{V}}_x + \vec{\mathbf{V}}_y + \vec{\mathbf{V}}_z = V_x\hat{\mathbf{i}} + V_y\hat{\mathbf{j}} + V_z\hat{\mathbf{k}}$

Addition of vectors: $S_x = a_x + b_x$; $S_y = a_y + b_y$; $S_z = a_z + b_z$

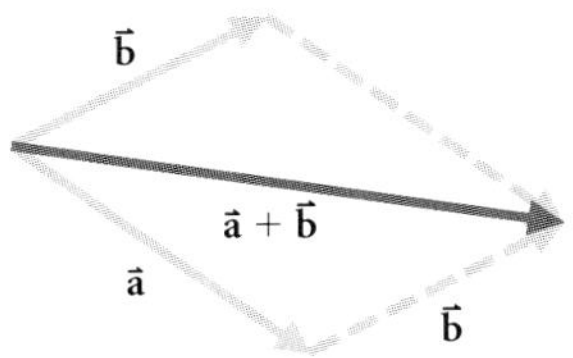

Subtraction of vectors: $D_x = a_x - b_x$; $D_y = a_y - b_y$; $D_z = a_z - b_z$

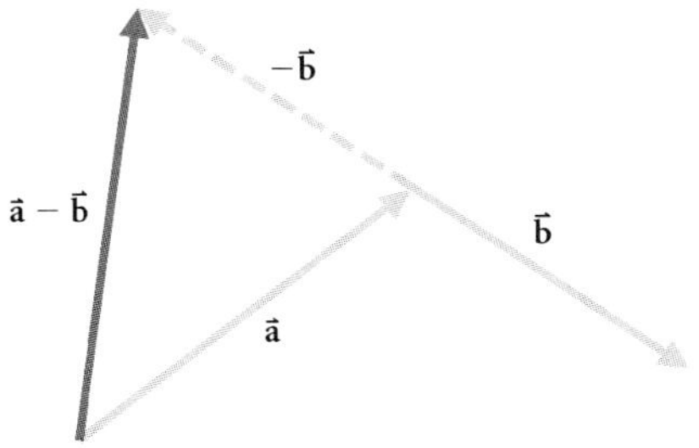

Dot product of two vectors: $\vec{\mathbf{a}}\cdot\vec{\mathbf{b}} = ab\cos\theta = \vec{\mathbf{b}}\cdot\vec{\mathbf{a}}$

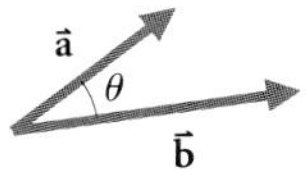

Distributive rule for dot products:

$$\vec{\mathbf{a}}\cdot(\vec{\mathbf{b}}+\vec{\mathbf{c}}) = \vec{\mathbf{a}}\cdot\vec{\mathbf{b}} + \vec{\mathbf{a}}\cdot\vec{\mathbf{c}}$$

Dot product in terms of components:

$$\vec{\mathbf{a}}\cdot\vec{\mathbf{b}} = a_xb_x + a_yb_y + a_zb_z$$

Cross product of two vectors:

$|\vec{\mathbf{a}}\times\vec{\mathbf{b}}| = ab\sin\theta$, direction given by right-hand rule

$$\vec{\mathbf{a}}\times\vec{\mathbf{b}} = -\vec{\mathbf{b}}\times\vec{\mathbf{a}}$$

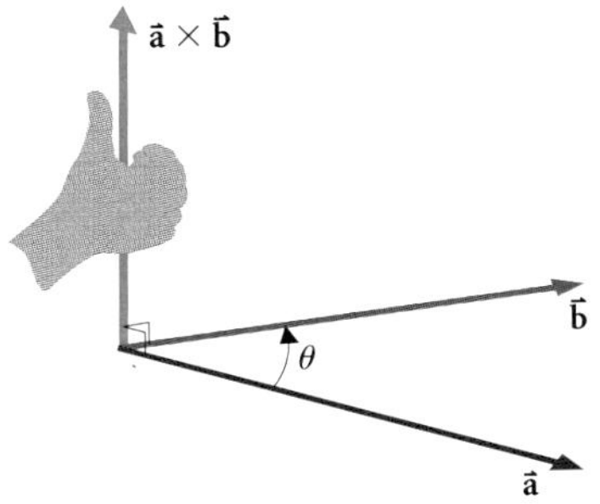

Distributive rule for cross products:

$$\vec{\mathbf{a}}\times(\vec{\mathbf{b}}+\vec{\mathbf{c}}) = \vec{\mathbf{a}}\times\vec{\mathbf{b}} + \vec{\mathbf{a}}\times\vec{\mathbf{c}}$$

Cross product in terms of components:

$$\vec{\mathbf{a}}\times\vec{\mathbf{b}} = \hat{\mathbf{i}}(a_yb_z - a_zb_y) + \hat{\mathbf{j}}(a_zb_x - a_xb_z) + \hat{\mathbf{k}}(a_xb_y - a_yb_x)$$

Identities:

$$\vec{\mathbf{a}}\cdot(\vec{\mathbf{b}}\times\vec{\mathbf{c}}) = \vec{\mathbf{b}}\cdot(\vec{\mathbf{c}}\times\vec{\mathbf{a}}) = \vec{\mathbf{c}}\cdot(\vec{\mathbf{a}}\times\vec{\mathbf{b}})$$

$$\vec{\mathbf{a}}\times(\vec{\mathbf{b}}\times\vec{\mathbf{c}}) = (\vec{\mathbf{a}}\cdot\vec{\mathbf{c}})\vec{\mathbf{b}} - (\vec{\mathbf{a}}\cdot\vec{\mathbf{b}})\vec{\mathbf{c}}$$

$$(\vec{\mathbf{a}}\times\vec{\mathbf{b}})\cdot(\vec{\mathbf{c}}\times\vec{\mathbf{d}}) = (\vec{\mathbf{a}}\cdot\vec{\mathbf{c}})(\vec{\mathbf{b}}\cdot\vec{\mathbf{d}}) - (\vec{\mathbf{a}}\cdot\vec{\mathbf{d}})(\vec{\mathbf{b}}\cdot\vec{\mathbf{c}})$$

$$\vec{\mathbf{a}} = (\vec{\mathbf{a}}\cdot\hat{\mathbf{b}})\hat{\mathbf{b}} + \hat{\mathbf{b}}\times(\vec{\mathbf{a}}\times\hat{\mathbf{b}})$$

$$|\vec{\mathbf{a}}\times\vec{\mathbf{b}}|^2 = a^2b^2 - (\vec{\mathbf{a}}\cdot\vec{\mathbf{b}})^2$$

D: Coordinate Systems

Rectangular: (x, y, z)

Unit vectors: $\hat{\mathbf{i}} = \hat{\mathbf{x}}$; $\hat{\mathbf{j}} = \hat{\mathbf{y}}$; $\hat{\mathbf{k}} = \hat{\mathbf{z}}$

For a right-handed system: $\hat{\mathbf{i}} \times \hat{\mathbf{j}} = \hat{\mathbf{k}}$

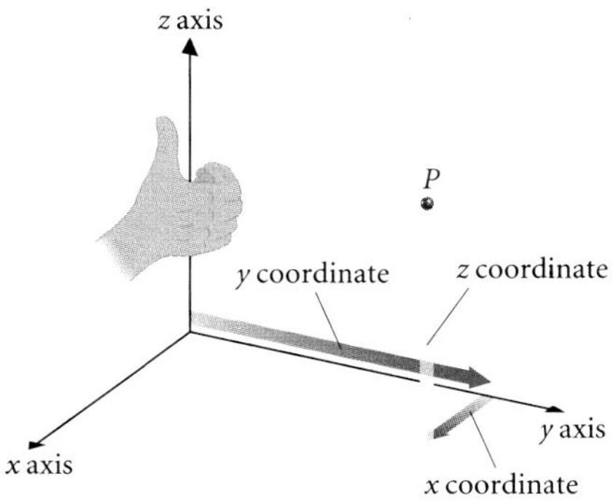

Plane polar: (r, θ)

Unit vectors: $\hat{\mathbf{r}}$; $\hat{\boldsymbol{\theta}}$

$$x = r\cos\theta \qquad y = r\sin\theta$$

$$r = \sqrt{x^2 + y^2} \qquad \theta = \tan^{-1}(y/x)$$

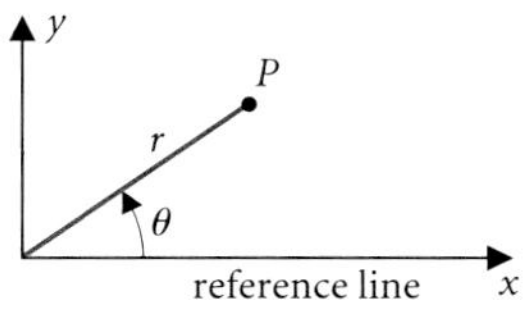

Cylindrical polar: (r, ϕ, z)

Unit vectors: $\hat{\mathbf{r}}$; $\hat{\boldsymbol{\phi}}$; $\hat{\mathbf{z}}$

$$\hat{\mathbf{r}} \times \hat{\boldsymbol{\phi}} = \hat{\mathbf{z}}$$

$$x = r\cos\phi; \qquad y = r\sin\phi; \qquad z = z$$

$$r = \sqrt{x^2 + y^2} \qquad \phi = \tan^{-1}(y/x)$$

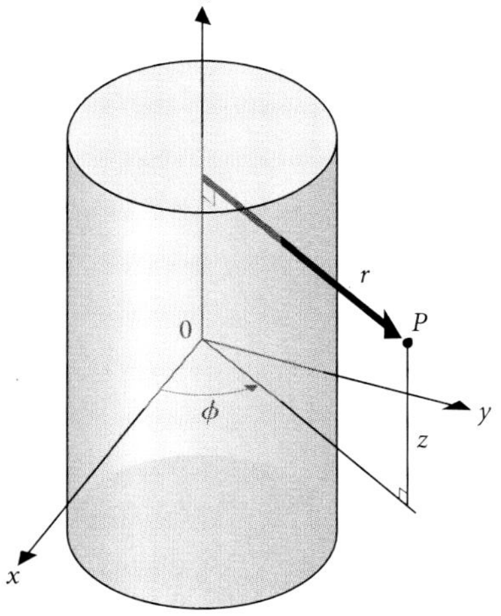

Spherical polar: (r, θ, ϕ)

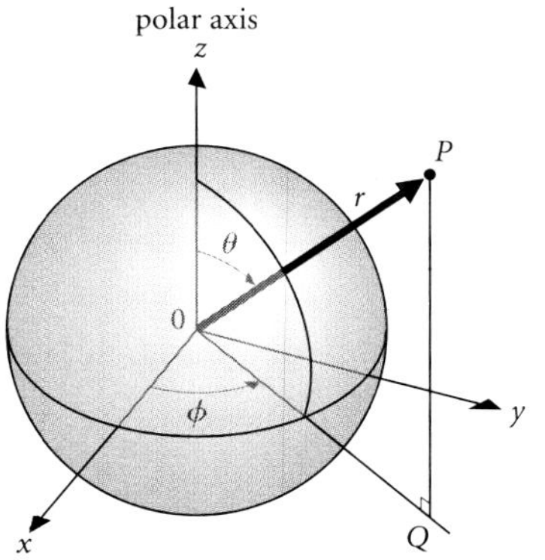

Note: Mathematics texts often define θ and ϕ to have reverse meanings. We have given the definitions that are standard in physics.

Unit vectors: $\hat{\mathbf{r}}$; $\hat{\boldsymbol{\theta}}$; $\hat{\boldsymbol{\phi}}$

$$\hat{\mathbf{r}} \times \hat{\boldsymbol{\theta}} = \hat{\boldsymbol{\phi}}$$

$$x = r\sin\theta\cos\phi; \qquad y = r\sin\theta\sin\phi; \qquad z = r\cos\theta$$

$$r = \sqrt{x^2 + y^2 + z^2};$$

$$\theta = \cos^{-1}\left[z/\sqrt{x^2 + y^2 + z^2}\right];$$

$$\phi = \tan^{-1}(y/x)$$

E: Derivatives

Let f be a function of variable x.

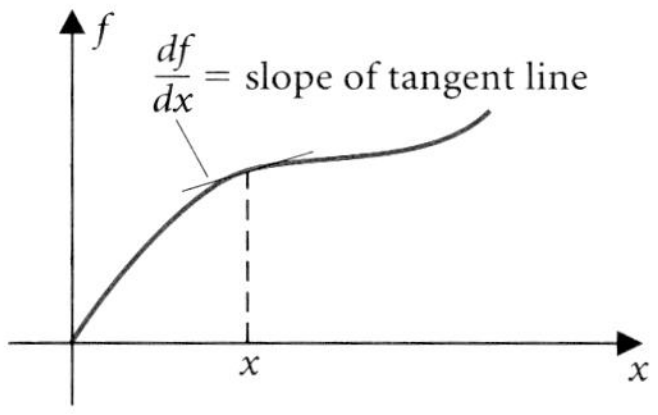

$$\frac{df}{dx} = \lim_{dx\to 0} \frac{f(x + dx) - f(x)}{dx} = \text{slope of curve } f(x)$$

f	df/dx
x^n	nx^{n-1}
e^{ax}	ae^{ax}
$\sin(kx)$	$k\cos(kx)$
$\cos(kx)$	$-k\sin(kx)$
$\tan(kx)$	$k\sec^2(kx)$
$\sec(kx)$	$k\sec(kx)\tan(kx)$
$\ln(kx)$	$\frac{1}{x}$

Chain rule: Let g be a function of $f(x)$, then:

$$\frac{dg}{dx} = \left(\frac{dg}{df}\right)\left(\frac{df}{dx}\right)$$

Example: $g = \ln(\sin x)$

$$\frac{dg}{dx} = \frac{1}{\sin x}\cos x = \cot x$$

Partial derivatives: Let $f = f(x, t)$, then:

$$\frac{\partial f}{\partial x} = \frac{df(x, t)}{dx}, \quad \text{holding } t \text{ constant}$$

Example: $f = \sin(kx - \omega t) \qquad \partial f/\partial x = k \cos(kx - \omega t)$

Product rule: Let $h(x) = f(x)g(x)$, then:

$$\frac{dh}{dx} = f(x)\frac{dg}{dx} + \frac{df}{dx}g(x)$$

Let $h(x) = \dfrac{f(x)}{g(x)}$, then:

$$\frac{dh}{dx} = -\left(\frac{f(x)}{g(x)^2}\right)\left(\frac{dg}{dx}\right) + \left(\frac{1}{g(x)}\right)\left(\frac{df}{dx}\right)$$

$$= \frac{g(x)\; df/dx - f(x)\,dg/dx}{g(x)^2}$$

Inverse functions: $\dfrac{dg}{dx} = \dfrac{1}{dx/dg}$

Example: $\dfrac{d(\cos^{-1} x)}{dx} = \dfrac{d\theta}{d(\cos\theta)}$ (letting $x = \cos\theta$)

$$= \frac{1}{d(\cos\theta)/d\theta} = \frac{1}{-\sin\theta} = \frac{-1}{\sqrt{1 - x^2}}$$

F: Integrals

$$\int_a^b f(x)\; dx = \text{the area under the curve of } f(x) \text{ between } a \text{ and } b$$

$$= \lim_{\Delta x_i \to 0} \sum_a^b f(x_i)\; \Delta x_i$$

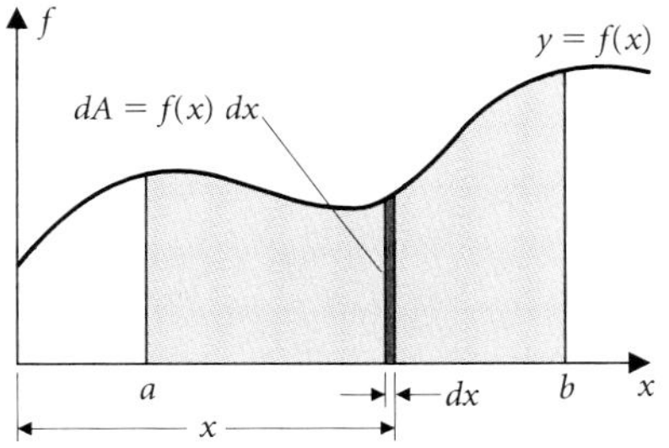

Integration is the inverse of differentiation:

$$\int_a^b \frac{df}{dx}\,dx = f(b) - f(a) \qquad \text{and} \qquad \frac{d}{dx}\int_a^x f(x')\; dx' = f(x)$$

Substitution

Consider $f(x)$ where $x = g(u)$. Then $dx = (dg/du)\; du.$

$$\int_a^b f(x)\; dx = \int_{g^{-1}(a)}^{g^{-1}(b)} f(g(u))\,\frac{dg}{du}\,du$$

Example: Evaluate

$$\int_0^a \frac{dx}{\sqrt{a^2 - x^2}}.$$

Letting $u = \theta$, $g(u) = a \sin\theta$, $dx = a \cos\theta\; d\theta$, the limits then become:

$$g^{-1}(0) = \sin^{-1}(0) = 0,\; g^{-1}(a) = \sin^{-1}(1) = \pi/2.$$

So, performing the indicated substitutions, we get:

$$\int_0^{\pi/2} \frac{a\cos\theta}{\sqrt{a^2 - a^2\sin^2\theta}}\,d\theta = \int_0^{\pi/2} d\theta = \frac{\pi}{2}.$$

Integration by Parts

Let $h(x) = f(x)\,\dfrac{dg}{dx}$

$$\int_a^b h(x)\; dx = f(x)g(x)\bigg|_a^b - \int_a^b \frac{df}{dx}g(x)\; dx$$

Some Indefinite Integrals

f	$\int f(x)dx$
$x^p\ (p \neq -1)$	$\dfrac{x^{p+1}}{p+1}$
$\sin(kx)$	$\dfrac{-\cos(kx)}{k}$
$\dfrac{1}{x+a}$	$\ln(x + a)$
$\dfrac{1}{1+x^2}$	$\tan^{-1}(x)$
$\sin^2(x)$	$\dfrac{x}{2} - \dfrac{\sin(2x)}{4}$
$\tan^2(x)$	$\tan(x) - x$
$\dfrac{1}{\sqrt{1-x^2}}$	$\sin^{-1}(x)$
e^{ax}	$\dfrac{e^{ax}}{a}$
$\cos(kx)$	$\dfrac{\sin(kx)}{k}$
$\tan(kx)$	$\dfrac{-\ln[\lvert\cos(kx)\rvert]}{k}$
$\sec(x)$	$\ln[\lvert\sec(x) + \tan(x)\rvert]$
$\cos^2(x)$	$\dfrac{x}{2} + \dfrac{\sin(2x)}{4}$
$\sec^2(x)$	$\tan(x)$
$\sqrt{1-x^2}$	$\frac{1}{2}\left[x\sqrt{1-x^2} + \sin^{-1}(x)\right]$

Some Definite Integrals

$$\int_0^{2\pi} \sin^2(x)\, dx = \int_0^{2\pi} \cos^2(x)\, dx = \pi$$

$$\int_0^{2\pi} \sin(x)\, dx = \int_0^{2\pi} \cos(x)\, dx = 0$$

$$\int_0^{\infty} e^{-a^2x^2}\, dx = \frac{\sqrt{\pi}}{2a}$$

$$\int_0^{\infty} x^{2n+1} e^{-a^2x^2}\, dx = \frac{n!}{2a^{2(n+1)}}$$

$$\int_0^{\infty} e^{-ax}\, dx = \frac{1}{a} \qquad (a > 0)$$

$$\int_0^{\infty} x^2 e^{-a^2x^2}\, dx = \frac{\sqrt{\pi}}{4a^3}$$

$$\int_0^{\infty} x^4 e^{-a^2x^2}\, dx = \frac{3\sqrt{\pi}}{8a^5}$$

G: Rules of Probability

If events A and B are independent, then:

$$\begin{aligned} p(A \text{ and } B) &= p(A)p(B) \\ p(A \text{ or } B) &= 1 - p(\text{not } A \text{ and not } B) \\ &= 1 - [1 - p(A)][1 - p(B)] \\ &= p(A) + p(B) - p(A)p(B) \end{aligned}$$

Appendix II
Symbols

A: The Greek Alphabet

Α	α	alpha
Β	β	beta
Γ	γ	gamma
Δ	δ	delta
Ε	ϵ	epsilon
Ζ	ζ	zeta
Η	η	eta
Θ	θ	theta
Ι	ι	iota
Κ	κ	kappa
Λ	λ	lambda
Μ	μ	mu
Ν	ν	nu
Ξ	ξ	xi
Ο	o	omicron
Π	π	pi
Ρ	ρ	rho
Σ	σ	sigma
Τ	τ	tau
Υ	υ	upsilon
Φ	ϕ	phi
Χ	χ	chi
Ψ	ψ	psi
Ω	ω	omega

B: Symbols Used in the Text

The entry under "Chapter" is the chapter in which the symbol is first used.

Chapter	Symbol	Definition
0	a	semimajor axis of an ellipse
2	$\vec{\mathbf{a}}$	acceleration
2	$\vec{\mathbf{a}}_o$	specific constant value of acceleration
17	a	slit width
20	a	van der Waals constant
35	a_o	Bohr radius
1	A	area
14	A	amplitude
1	AB	straight line distance between A and B
1	$\overrightarrow{\mathbf{AB}}$	vector between points A and B
2	$\overset{\frown}{\mathrm{AB}}$	path length along a circular arc from A to B
28	$\vec{\mathbf{B}}$	magnetic field (magnetic induction)
36	B	binding energy
37	B	baryon number
37	B	bottomness or beauty
14	b	damping constant
20	b	van der Waals constant
37	b	bottom quark
1	c	speed of light in vacuum
12	c	compression of a spring
37	c	charmed quark
27	C	capacitance
37	C	charm
0	C	a constant
19	c_p	specific heat at constant pressure
19	c_p'	molar specific heat at constant pressure
19	c_v	specific heat at constant volume
19	c_v'	molar specific heat at constant volume
28	$\mathscr{C}$	circulation
0	d, D	distance
3	d, D	diameter
17	d	slit separation
37	d	down quark
2	$\vec{\mathbf{D}}$	displacement = change in position
27	$\vec{\mathbf{D}}$	electric displacement
17	D	dispersion
36	D	deuterium
2	dx	differential amount of x
0	e	eccentricity of an ellipse

Chapter	Symbol	Definition
13	e	2.7182818
7	e	efficiency
23	e	electron charge $= 1.6 \times 10^{-19}$ C
24	e, e^-	electron
36	e^+	positron
1	E	east
8	E	energy
23	$\vec{\mathbf{E}}$	electric field
26	$\mathcal{E}$	electromotive force
3	f	frequency (Hz)
4	$\vec{\mathbf{f}}$	friction force
4	$\vec{\mathbf{F}}$	force
0	F_1, F_2	foci of an ellipse
29	g	electron g factor = 2.002 . . .
2	$\vec{\mathbf{g}}$	acceleration due to gravity
E2	$\vec{\mathbf{g}}$	gravitational field strength
5	G	gravitational constant
0	h, H	height
21	H	rate of heat flow
29	$\vec{\mathbf{H}}$	magnetic field vector
35	h	Planck's constant
35	$\hbar = h/(2\pi)$	Planck's constant
6	$\vec{\mathbf{I}}$	impulse
12	I	rotational inertia
16	I	intensity
26	I	electric current
30	I_d	displacement current
37	I	isospin
37	I_3	z-component of isospin
1	$\hat{\mathbf{i}}$	unit vector in the x-direction
26	$\vec{\mathbf{j}}$	current density
1	$\hat{\mathbf{j}}$	unit vector in the y-direction
1	$\hat{\mathbf{k}}$	unit vector in the z-direction
4	k	spring constant
15	k	wave number $= 2\pi/\lambda$
33	$\vec{\mathbf{k}}$	wave vector
19	k	Boltzmann's constant
21	k	thermal conductivity
23	k	electric force constant
28	k_m	$= \mu_o/4\pi$, magnetic force constant
7	K	kinetic energy
22	K	coefficient of performance
2	ℓ	distance
3	ℓ	length
35	ℓ	angular momentum quantum number
1	L	length
1	L	dimension of length
31	L	(self) inductance
9	$\vec{\mathbf{L}}$	angular momentum
15	m	an integer
35	m	quantum number for z-component of angular momentum
19	m	mass of a molecule
28	$\vec{\mathbf{m}}$	magnetic moment
1	M, m	mass
1	M	dimension of mass
31	M	mutual inductance
29	$\vec{\mathbf{M}}$	magnetization
4	$\vec{\mathbf{n}}$	normal force
23	$\hat{\mathbf{n}}$	unit normal vector
13	n	number density (particles per unit volume)
16	n	refractive index
15	n	any integer
36	n	neutron
28	n	number of turns per unit length in a coil
1	N	north
19	N	number of molecules of substance
28	N	integer
19	N_A	Avogadro's number
19	$\mathcal{N}$	number of moles of substance
1	O	origin of coordinates
36	p	proton
6	$\vec{\mathbf{p}}$	momentum
24	$\vec{\mathbf{p}}$	electric dipole moment
1	P	a point
7	P	power
13	P	pressure
33	P	radiation pressure
16	P_*	pressure amplitude in a sound wave
6	$\vec{\mathbf{P}}$	momentum of a system
27	$\vec{\mathbf{P}}$	electric polarization
19	Q	heat transfer
23	Q, q	electric charge
32	Q	quality factor for an *LRC* circuit
36	Q	energy produced in a nuclear or particle reaction
3	r, R	radius
1	$\vec{\mathbf{r}}$	position vector
3	$\hat{\mathbf{r}}$	unit vector in the radial direction
3	R	range of a projectile
17	R	resolving power
19	R	ideal gas constant
21	R	thermal resistance
26	R	electric resistance
21	R_f	R factor for thermal resistance = *RA*
35	R	Rydberg constant
4	$\vec{\mathbf{s}}$	displacement
1	s	arc length
4	s	stretch or compression of a spring
35	s	spin
37	s	strange quark
1	S	south
2	S	speed

Chapter	Symbol	Definition
22	S	entropy
37	S	strangeness
33	$\vec{\mathbf{S}}$	Poynting vector
2	t	time variable
37	t	top quark
1	T	dimension of time
0	T	period
4	T	tension
19	T	temperature
36	T	tritium
27	u	energy density
37	u	up quark
10	$\vec{\mathbf{u}}$	velocity
8	U	potential energy
8	U	internal energy
2	$\vec{\mathbf{v}}$	velocity
26	$\vec{\mathbf{v}}_{\mathrm{d}}$	drift velocity
2	$\vec{\mathbf{v}}_{\mathrm{i}}$	initial velocity
2	$\vec{\mathbf{v}}_{\mathrm{f}}$	final velocity
2	$\vec{\mathbf{v}}_{\mathrm{o}}$	specific constant value of velocity
6	V	volume
25	V	electric potential
25	V_{AB}	potential difference between A and B
1	$\vec{\mathbf{V}}$	arbitrary vector
1	V_x, V_y, V_z	components of $\vec{\mathbf{V}}$
4	$\vec{\mathbf{W}}$	weight
7	W	work
1	W	west
32	X	reactance
1	x	cartesian coordinate
33	$\hat{\mathbf{x}}$	unit vector in the x-direction
1	y	cartesian coordinate
28	$\hat{\mathbf{y}}$	unit vector in the y-direction
37	Y	hypercharge
1	z	cartesian coordinate
28	$\hat{\mathbf{z}}$	unit vector in the z-direction
32	Z	impedance
10	α	alpha-particle or helium nucleus
15	α	angle
20	α	coefficient of linear thermal expansion
35	α	fine structure constant
3	$\vec{\alpha}$	angular acceleration
16	β	v/c
21	β	coefficient of volume thermal expansion
36	β	beta-particle, or electron
19	γ	ratio of specific heats
29	γ	$1/\sqrt{1-\beta^2}$, relativistic speed factor
36	γ	photon
2	δx	small amount of x, or uncertainty in x
1	Δx	change in x = final value − initial value

Chapter	Symbol	Definition
37	Δ	delta-particle
10	ϵ	coefficient of restitution
21	ϵ	emissivity
23	ϵ_0	$= 1/(4\pi k)$, permittivity of the vacuum
1	Θ, θ	angle
28	$\hat{\boldsymbol{\theta}}$	unit vector in θ-direction
27	κ	dielectric constant
3	λ	latitude
5	λ	mass per unit length
15	λ	wavelength
22	λ	mean free path
24	λ	charge per unit length
35	λ_{C}	Compton wavelength
36	Λ	lambda particle
15	μ	mass per unit length
34	μ, μ^-	muon
4	μ_{s}	coefficient of static friction
4	μ_{k}	coefficient of kinetic friction
28	μ_{o}	permeability of the vacuum
29	μ	magnetic permeability
36	ν	neutrino
36	$\bar{\nu}$	antineutrino
1	π	3.14159
36	π	π meson, pion
6	ρ	mass density
24	ρ	charge density
26	ρ	resistivity
37	ρ	rho particle
11	σ	surface mass density
21	σ	Stephan–Boltzmann constant
22	σ	collision cross section
24	σ	surface charge density
26	σ	conductivity
37	Σ	sigma particla
14	τ	time interval, characteristic time
37	τ	tau-particle
9	$\vec{\tau}$	torque
1	ϕ	angle
14	ϕ	phase
E2	ϕ	gravitational potential
23	Φ_E	electric flux
28	Φ_B	magnetic flux
27	χ	electric susceptibility
29	χ_{m}	magnetic susceptibility
4	ψ, Ψ	angle
35	ψ	Schrödinger wave function
3	ω, Ω	angular frequency (rad/sec)
12	$\vec{\omega}$	angular velocity
12	Ω	precession frequency
22	Ω	number of microstates
37	Ω	omega-particle

C: Mathematical Symbols

$=$	equals; set equal to	$\gtrsim$	greater than or about equal to	$\angle$	angle
$\approx$	is approximately equal to	$\lesssim$	less than or about equal to	$\frac{d}{dx}$	derivative with respect to x
$\neq$	is not equal to	$\ll$	much less than	$\frac{\partial}{\partial x}$	partial derivative with respect to x
$\sim$	of the order of	$\gg$	much greater than	$\int$	integral
$\equiv$	is defined as; is equivalent to	$\pm$	plus or minus	$\oint$	integral over a closed path or surface
$\Rightarrow$	implies	$\lvert x \rvert$	absolute value of x	lim	limit
$\propto$	proportional to	$\lvert \vec{\mathbf{x}} \rvert$	magnitude of the vector $\vec{\mathbf{x}}$	$\sum_{i=n}^{m} x_i$	sum over x_i, $i = n$ through m, that is, $x_n + x_{n+1} + x_{n+2} + \cdots + x_m$
$\rightarrow$	tends to, becomes	$[x]$	physical dimensions of x		
$<$	less than	$<x>$	average value of x		
$>$	greater than	$\parallel$	parallel		
$\geq$	greater than or equal to	$\perp$	perpendicular		
$\leq$	less than or equal to	$n!$	n factorial		

D: Symbols for Units

kg	kilogram (Fundamental SI units)	ft	foot	N	newton
m	meter (Fundamental SI units)	g	gram	nm	nautical mile = 1.15 statute mile
s	second (Fundamental SI units)	G	gauss	Oe	oersted
A	ampere (Fundamental SI units)	gal	gallon	oz	ounce
Å	angstrom	H	henry	Pa	pascal
atm	atmosphere	h	hour	qt	quart
AU	astronomical unit	hp	horse power	sr	steradian
Bq	bequerel	Hz	hertz	rad	radian
Btu	British thermal unit	in.	inch	rev	revolution
C	coulomb	J	joule	rpm	revolutions per minute
°C	degree celsius	K	kelvin	T	tesla
cal	calorie	kt	knot = nautical mile per hour	ton	ton
Cal	Calorie = kilocalorie	kWh	kilowatt-hour	u	atomic mass unit
cm Hg	centimeters of mercury	lb	pound	V	volt
Ci	curie	ly	light-year	W	watt
d	day	L	liter	Wb	Weber
dyn	dyne	mi	mile	y	year
erg	erg	min	minute	Ω	ohm
eV	electron volt	mol	mole	°	degree
°F	degree Fahrenheit	mph	miles per hour	′	minute of arc
F	farad	$M_{\odot}$	solar mass	″	second of arc

E: Roman Numerals

i	I	1	xi	XI	11
ii	II	2	xii	XII	12
iii	III	3	xiii	XIII	13
iv	IV	4	xiv	XIV	14
v	V	5	xv	XV	15
vi	VI	6	xvi	XVI	16
vii	VII	7	xvii	XVII	17
viii	VIII	8	xviii	XVIII	18
ix	IX	9	xix	XIX	19
x	X	10	xx	XX	20

APPENDIX III

Answers to Selected Problems

CHAPTER 0

0.1 Brand X **0.3** 7.2 cm; 31.2 cm **0.5** The comet's speed is greatest at point *A* and smallest at point *C*. **0.7** straight down **0.9** all the way to the bottom **0.11** *GH* **0.15** No; The difference is only 0.2 mm

CHAPTER 1

1.1 ± 0.5 mm
1.3(a) 5×10^{-8} s; 4 significant figures
(b) 50 kg; 1 significant figure
(c) 5×10^{-6}; 3 significant figures
(d) 50 μm; 3 significant figures
1.5 $10^{18} \mu$s
1.7(a) $\dfrac{M}{L^3}$ **(b)** dimensionless
(c) L^2 **(d)** L^3 **1.9** The vector would have a magnitude of 15 m and point north.
1.11

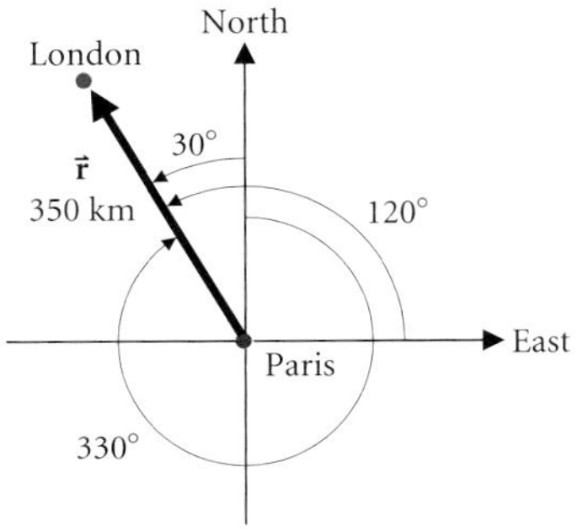

1.13 $a_x = -1,\ a_y = 2$
$b_x = 3,\ b_y = 2$
$c_x = 1,\ c_y = -3$
$d_x = 2,\ d_y = 2$
$e_x = -2,\ e_y = -1$
1.15 (2, 3) **1.19(a)** ± 0.5 mm; four decimal places **(b)** the uncertainty is higher; ± 0.5 mm **(c)** ± 0.5 sec; 4 digits
(d) same as part (c); precision of the measurement is totally unsuited for its purpose **1.23** 4×10^4 km
1.27(a) ± 0.8 m **(b)** unnecessary
(c) 0.458° **(d)** Ridiculous **(e)** no
1.31 67 in. = 1.7 m
1.35 1.8×10^{12} furlong/fortnight
1.39 $\dfrac{1}{(\text{length})^2}$; $1.5 \times 10^7\ \text{m}^{-2}$; no
1.43(a) $3x^2(\Delta x)$ **(b)** $|\cos(x)|(\Delta x)$
(c) $\left| -\dfrac{22x}{(2x^2-1)^2} \right| (\Delta x)$;
6.6×10^{-2}; 1.0×10^{-2}; 0.32 **1.47** (38 m, 23° from the ground) **1.51** (e)
1.55(a) $|\vec{\mathbf{s}}| = 2|\vec{\mathbf{b}}| \cos\left[\dfrac{1}{2}(\theta_b - \theta_a)\right]$
(b) $|\vec{\mathbf{d}}| = 2|\vec{\mathbf{a}}| \sin\left(\dfrac{\theta_b - \theta_a}{2}\right)$;
its direction from north is $\dfrac{\theta_a + \theta_b}{2} - \dfrac{\pi}{2}$
1.59 (53 Nmi, 5° N of E) **1.63** $\dfrac{1}{2}\vec{\mathbf{A}} + \vec{\mathbf{B}}$: (1 m, 35 m); (35 m, 1.6° from *y*-axis);
$\vec{\mathbf{B}} - \vec{\mathbf{A}}$: (−68 m, −4 m); (68 m, 3° from *x*-axis)
1.67 $\sqrt{14}$ m; $\sqrt{12}$ m; $\sqrt{2}$ m
1.71 6.62 cm^3 $\pm$ 0.17 cm^3; 6.62 $\pm$ 0.05 cm^3; Implied uncertainty is $\approx \frac{1}{3}$ true uncertainty.
1.75 7600 s; 4×10^{-8} sec/day; 5×10^{-13}
1.79 9×10^7 y

CHAPTER 2

2.1 85 km/h **2.3** −2.6 cm/s; +1.5 cm/s
2.5 *A:* (1.1 m/s, parallel to the *x*-axis)
B: (1.8 m/s, parallel to the *y*-axis)
C: (0.77 m/s, toward 30° clockwise from the *y*-axis) **2.7** b **2.9** only when $v_i = 0$
2.11

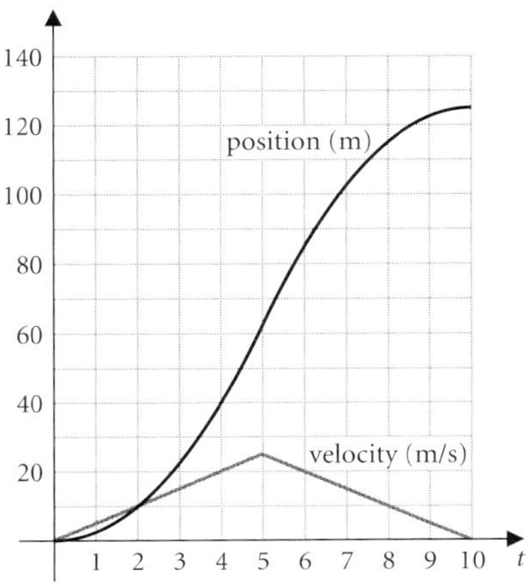

2.13 (9.3×10^{-2} m/s^2, in the direction from *D* to A) **2.17** (10 m/s, 30° north of west) **2.21** (8.8 m/s, 59° north of east)
2.25 5.0 m/s^2 **2.29** a
2.33 Slope $= \dfrac{\Delta \text{ position}}{\Delta \text{ time}} =$ velocity; no
2.37 (290 km/h, parallel to the track)
2.41 4.00 m/s^2 **2.45** 0.39 s; 3.1 m/s
2.49 24 s **2.53** 78 m; 16 m
2.57(a) 1.2 m; **(b)** (0.29 m/s, downward)
(c) $\alpha = 0.90$ m/s^4 **2.61** 18 m/s
2.65 Modeling the spacecraft as a particle is an excellent approximation
2.69 $g = 980.6$ cm/s^2; $\alpha = 3.5 \times 10^2$ cm/s^4
2.73 $\theta = 57°$

CHAPTER 3

3.1 d **3.3(a)** 6.28 m/s; 0.100 Hz; 0.628 rad/s; **(b)** 3.95 m/s^2
3.5 $\vec{\mathbf{a}}(C) = 0$ m/s^2; $\vec{\mathbf{a}}(A) = 20$ m/s$^2\,\hat{\mathbf{i}}$
$\vec{\mathbf{a}}(B) = -20$ m/s$^2\,\hat{\mathbf{j}}$
$\vec{\mathbf{a}}(D) = 20$ m/s$^2\,\hat{\mathbf{j}}$
$\vec{\mathbf{a}}(E) = -20$ m/s$^2\,\hat{\mathbf{i}}$
3.7(a) (4.4 mi) $\hat{\mathbf{j}}$ + (1.4 mi) $\hat{\mathbf{i}}$
(b) (3.7 mi) $\hat{\mathbf{j}}$ − (0.71 mi) $\hat{\mathbf{i}}$

(c) $|\vec{\mathbf{D}}_{rel}| = (0.71\text{ mi})(-3\hat{\mathbf{i}} - \hat{\mathbf{j}})$
3.9 23° **3.13** 15°, or 75°
3.17 4.5 s; 45 m away; yes
3.21 19.8 m/s
3.25(a) $y = -\frac{1}{2}gt^2$;
$x = v_0 t$
(b) $+\frac{gD^2}{2v_0^2}; -g\frac{D}{v_0}$
(c) $2D - v_0\sqrt{\frac{2h}{g}}; D\sqrt{\frac{2g}{h}}$ **3.29** 52°
3.33 $\left(\frac{v_e}{v_i}\right)^2\left[\frac{2\sqrt{3}}{3} \pm \sqrt{\frac{4}{3} - \frac{1}{3}\left(\frac{v_i}{v_e}\right)^2\left(1 + 3\left(\frac{v_i}{v_e}\right)^2\right)}\right]$;
yes **3.37** $\Delta\vec{\mathbf{v}} = \vec{\mathbf{v}}_f - \vec{\mathbf{v}}_i = 0$; therefore $\vec{\mathbf{a}}_{avg}$ must be 0 **3.41** b **3.45** 7.3 km
3.49 286 m/s; close to, but not quite equal to, the speed of sound **3.53** 2.00
3.57 Its speed *relative to the air* is greater.
3.61 1.2 km; 0.67 km; $\frac{1}{3}$ h **3.65** 3.14 h
3.69 She's cutting it close; her round trip will take 1.9 hours, just 6 minutes short of her limit. **3.73** speed = 4.0 m/s
3.77 $(7.1\text{ m/s})\sqrt{1 - \sin[(10\text{ rad/s})t]}$
3.81 13.9 m/s²; 17.1 m/s²; 17.3 m/s²
3.85 $2h\sin 2\phi\,[\cos 2\phi + \sqrt{\cos^2 2\phi + 1}]$
3.89 Camelot **3.93** Air resistance reduces both maximum height and range. Air resistance does not noticeably affect the path unless $\alpha > 0.01$ **3.97(a)** 35°
(b) 4.8 m **(c)** 5.2 m
(d) $\vec{\mathbf{a}}_{AV,\text{land}}$ = (53 m/s², 28° right of upward); $\vec{\mathbf{a}}_{AV,\text{horse}}$ = (33 m/s², 5° right of upward);
(e) Leap to horse: 1.5 rad/s; Horse to landing: 1.6 rad/s

CHAPTER 4

4.1 rock climber: normal forces, frictional forces, gravitational force; buoy: tension force, drag force, gravitational force, buoyant force
4.3 (5.6 N, 60° below the x-axis)
4.5 force; 1×10^{28} N **4.7** 15 boxes
4.9 If I weigh 840 N, I am able to exert half of this on the lever: 420 N. Then s = 15 cm;
$\vec{\mathbf{F}}_{\text{on spring}} = +(420\text{ N})\,\hat{\mathbf{i}}; = +k\vec{\mathbf{s}}$
$\vec{\mathbf{F}}_{\text{spring on you}} = -(420\text{ N})\,\hat{\mathbf{i}}; = -k\vec{\mathbf{s}}$
$\vec{\mathbf{F}}_{\text{on mounting}} = +(420\text{ N})\,\hat{\mathbf{i}}$
4.11 $f_s = F$ **4.13** $F_{\text{bump}} = Ma$
4.15 $n = mg = 1.5 \times 10^4$ N; friction = (2.1×10^3 N, inward) **4.19** $2M$; $2Mg\sin\theta$ **4.23** 1.8×10^3 N
4.27 6.00 m/s² **4.31** k = (1 lb-force)/[(1 lb-mass)(32 ft/s²)]; 1 slug = 32 lb-mass = 14.5 kg **4.35** $n = Mg$; the normal force increases by an amount mg.
4.39 The springs all have the same spring constant. **4.43** $\frac{2k}{3}$ **4.47** By pulling harder on the string, the person reduces the normal force. If the person releases the tension in the string, n increases to mg.
4.51(a) 640 N **(b)** 770 N
(c) 510 N **(d)** 640 N **(e)** zero
4.55 19 N **4.59** 1900 N
4.63 F_D = 120 N; 0.34 m/s²
4.67 Frictional force; (1.22×10^4 N, opposite the car's velocity) **4.71** $\frac{mg}{\mu_s d}$; $\frac{2mg}{\mu_s d}$; mg **4.75** $\frac{1}{2}\ell\left(1 + \frac{Mg}{k\ell}\right)$ below the ceiling **4.79** the plane's path will curve upward **4.83** 1.0×10^5 N
4.87(a) 29 m/s **(b)** 7.0×10^1 m/s; 63 m/s **(c)** If $v < v_1$, friction acts up the slope; there is a minimum speed if $\tan\theta > \mu$. **4.93** If the rope doesn't stretch the climber suffers a major injury.
4.97(a) 830 N **(b)** 2800 N
4.101 $F = \frac{\mu mg}{\cos\theta + \mu\sin\theta}$; 22°; 0.40 kN; the slope doesn't change the value of θ at the limit; 850 N **4.105** 57°

CHAPTER 5

5.1 1700 N; 3400 N **5.3** When the cable does not hang in a curve there is no force to balance the cable's weight.
5.5 Assume you and your friend have a mass of 65 kg and 75 kg respectively, then;
4×10^{-8} N; $2 \times 10^{10} = \frac{W}{Fg}$
5.7 2.3×10^{13} kg; 1.5×10^{-3} m/s²; 400 m from the asteroid's surface
5.9 Mg; $2Mg$; $3Mg$ **5.13(a)** Mg
(b) $\frac{Mg}{k} + \ell$ **5.17** $s = \frac{M_2 g}{k}$; $g(M_1 + M_2) = T$ **5.21** 3.13 m/s
5.25 the blocks do slide with respect to each other; 2.9 m/s²; 5.3 m/s²
5.29 $a = g\sin\theta - \frac{g}{2}\cos\theta(\mu_1 + \mu_2)$;
$s = \frac{Mg\cos\theta}{2k}(\mu_1 + \mu_2)$ **5.33** Since the measured tension differs from the ideal value by much more than the accuracy of the data, we conclude that the pulley system is not ideal. **5.37** $\frac{2Mg}{5}$ **5.41** $\frac{1}{4}Mg$
5.45 $\theta = \phi$; $\sin\theta = \frac{1}{1+\alpha}$;
$M = \frac{2k\ell}{g}\alpha\sqrt{1 - \left(\frac{1}{1+\alpha}\right)^2}$; $\theta = 49° = \phi$;
$M = 7.7 \times 10^3$ kg **5.49** The jar would remain right by her hand when released.
5.53 2.0×10^{30} kg **5.57** 1.4 h
5.61 The length of the Lunar beanstalk is about $\frac{1}{4}$ of the distance from the Earth to the Moon; the lunar beanstalk will not function properly. **5.65(a)** The monkey accelerates at 1.96 m/s² downward while the bananas accelerate at 1.96 m/s² upward.
(b) The bananas accelerate at 4.9 m/s²
(c) The bananas accelerate at 9.8 m/s²
(d) The monkey will minimize the time it takes for him to reach the bananas by accelerating up the rope; a, 2.26s; b, 2.02s; c, 1.75s **(e)** no **5.69** $\frac{1}{3}Mg\sin\theta$
5.73 3.70 m/s²; Mars
5.77(a) $g(r) = \frac{g_0 r}{R_E}$ **5.81(a)** 1.0×10^5 N
(b) $T = 9.3 \times 10^4$ N **(c)** 9.0×10^4 N
5.85 $\frac{dy}{dx} = \tan\theta = \frac{g\lambda x}{C}$; $y(x) = \frac{g\lambda x^2}{2C}$; sag $\propto D^2$

CHAPTER 6

6.1 $\frac{p_1}{p_2} = \frac{3}{8}$ **6.3** c
6.5 2×10^2 N·s **6.7** 12×10^3 N·s$\hat{\mathbf{i}}$
6.9 12 m/s **6.11** 0.15 m/s
6.13 $\frac{p_{ms}}{p_{vw}} = 0.20$ **6.17** (50 N·s, "down-alley") **6.21** (12.5 N·s, opposite the ball's initial velocity) **6.25** 30 m/s
6.29 $(3\text{ m/s})\hat{\mathbf{i}} + (8\text{ m/s})\hat{\mathbf{j}}$
6.33 3×10^4 N **6.37(a)** (6 N·s, upward) **(b)** 3.0×10^3 N **6.41** zero
6.45 less than $m_{\text{exhaust}}v_{ex}$ **6.49** It slides off the ice in a direction opposite the dumb dog. **6.53** astronaut: 0.72 m/s; toolkit: 4.3 m/s **6.57** 8.3×10^{-2} m/s
6.61 $(0.14\text{ N·s})\hat{\mathbf{i}}$; 2.9 m/s **6.65** 58 N
6.69 $-0.69\,v_{ex}$ **6.73** The speed will be increased by a factor of 1.6
6.77

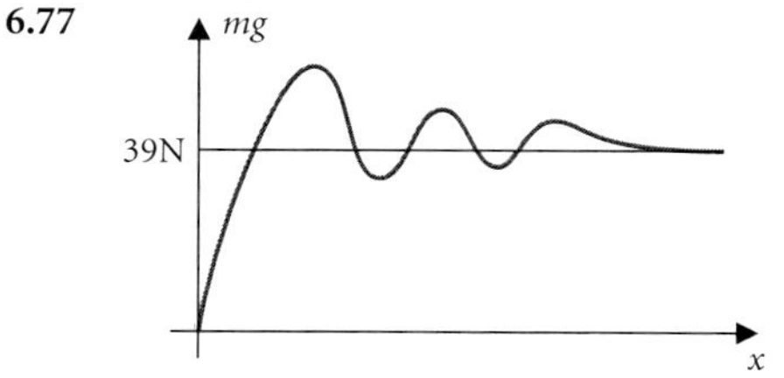

6.81 $\frac{5h}{4}$ **6.85(a)** v_1 = 0.27 m/s, v_2 = 0
(b) v_1 = 0.27 m/s, v_2 = 0.26 m/s
(c) The skater's gliding speeds are not changed. But the velocity components perpendicular to the gliding direction are as given in (a) and (b) above.
6.89 $\frac{v_0^2}{2\mu_r g}$; Back of car #2 is $\frac{v_0^2}{2\mu_r g} - \frac{7L}{4}$ from point A; $2.3\sqrt{\mu_r gL}$ **6.93** $\pi F_0 t_0\hat{\mathbf{i}}$

6.97

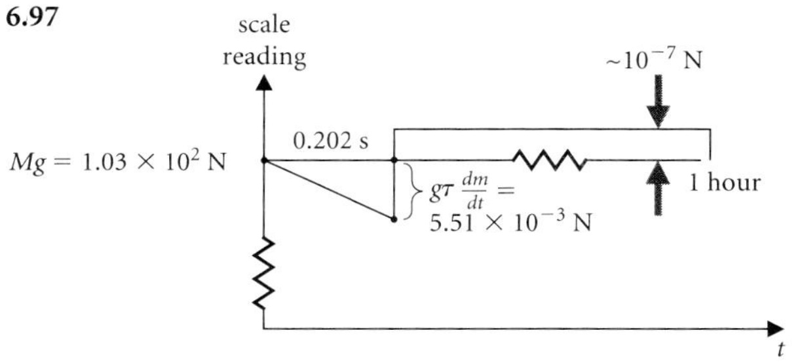

CHAPTER 7

7.1 $K_{\text{car 1}} = 7.2 \times 10^4$ J; $K_{\text{truck}} = 7.2 \times 10^5$ J; $\frac{K_{\text{car 1}}}{K_{\text{truck}}} = 10^{-1}$; $\frac{K_{\text{car 1}}}{K_{\text{car 2}}} = 0.12$
7.3 $W_{\text{haystack on car}} = -4.7 \times 10^5$ J; the car does $+4.7 \times 10^5$ J of work on the hay.
7.5 $\vec{\mathbf{a}} \cdot \vec{\mathbf{b}} = -3.1$ **7.7** zero
7.9 2.8 kN **7.13** 100 J **7.17** d
7.21 98 J **7.25** 9.0×10^1 J
7.29 0 **7.33** 26 **7.37** 70°
7.41(a) $W_{\text{pirate}} = 3 \times 10^{11}$ J; $W_{\text{carrier}} = -2 \times 10^{11}$ J; $v_f = 450$ m/s
(b) Yes; -2×10^{10} J
7.43(a) 1.1×10^6 m/s
(b) $\frac{\Delta v}{v} = 4 \times 10^{-3}$ **7.49** 2×10^9 J; the same **7.53(a)** $\hat{\mathbf{i}}_2 = \cos\theta\,\hat{\mathbf{i}}_1 + \sin\theta\,\hat{\mathbf{j}}_1$; $\hat{\mathbf{j}}_2 = -\sin\theta\,\hat{\mathbf{i}}_1 + \cos\theta\,\hat{\mathbf{j}}_1$
7.55(a) $\mu Mg2\pi R$ **7.59** By turning East, the rocket is thrusting in the same direction that it's already moving in—increasing its K as efficiently as possible.
7.63 10; 50 N **7.67** 5; $\frac{5}{6}$
7.71 9.2×10^5 W **7.75** 11
7.79 0.857; 50 J **7.83** True
7.87(a) 16 m/s **(b)** 78 km/h
(c) 32 m/s^2 **(d)** fine for backs of envelopes but not OK for a detailed accident report. **(e)** 2.0×10^5 J **7.91** $2\sin\theta$
7.95(a) $s_1 = \frac{v_0^2}{2\mu g}\left(1 - \left(\frac{\alpha}{1+\alpha}\right)^2\right)$; $s_2 = \frac{v_0^2}{2\mu g}\frac{\alpha}{(1+\alpha)^2}$; $W = -\frac{1}{2}\frac{mv_0^2}{(1+\alpha)}$
(b) $\sqrt{2\mu gL(1+\alpha)} = v_{\max}$;
$v_m = v_0\frac{\alpha}{1+\alpha} + \frac{\sqrt{v_0^2 - 2\mu gL(1+\alpha)}}{1+\alpha}$
$v_M = \frac{\alpha}{1+\alpha}v_0 - \frac{\alpha}{1+\alpha}\sqrt{v_0^2 - 2\mu gL(1+\alpha)}$;
$W = -\mu_k mgL$

CHAPTER 8

8.1 $\frac{U_1}{U_2} = \frac{1}{4}$ **8.3** 3 cm **8.5** Reference level; floor; $U_i = 1400$ J; $U_f = 0$; Reference level: initial position; $U_i = 0$ J; $U_f = -1400$ J; Reference level: highest shelf; $U_i = -4200$ J; $U_f = -5600$ J; $\Delta U = U_f - U_i = -1400$ J; $v = 7.9$ m/s
8.7 choice d) is correct **8.9** 1×10^7 J
8.11 700 W **8.13** The spring energy decreases by a factor of 2.
8.17 $\frac{mg}{\mu(\ell - L)}$; $\frac{mg(\ell - L)}{2\mu}$
8.21 Same; The smaller mass will have most of the energy. **8.25** Our exact formula applies to all planets. The practical formula applies to all planets as long as you use the correct value of the acceleration due to gravity.
8.29(a) 4×10^4 W **(b)** The KE given to the people requires an additional power of 7 W **8.33** 10^{14} kg/s
8.37 $\frac{3\ell}{2} + \frac{2Mg}{k}$ **8.41** 4.1 kJ
8.45 42 m/s **8.49** $x = \frac{\ell[k_1 + 3k_2]}{2(k_1 + k_2)}$
8.53 The decrease of TME is due to friction forces within the car's brakes dissipating energy as heat. **8.57** normal force; none; forces that occur in the muscles in your legs.
8.61 Normal and weight forces don't do any work on the truck. Friction and tension do work. The work Isaac does ends up as thermal energy **8.65** 2.4 J; normal and weight force, tension; none do work **8.69** 5 m/s; probably a large overestimate **8.73** No
8.77 $\Delta E = -40$ J **8.81** 2.0×10^5 J; 240 N **8.85** $U = -GMm\left[\frac{1}{R+h} + \frac{1}{4R+h}\right]$; Yes; $g = \frac{17GM}{16R^2}$ for $h \ll R$
8.89 $s_0 = \sqrt{\frac{5mRg}{k}}$
8.93 $k = 1140$ N/m, (stretch ≤ 0.9 m). For larger stretches, the two estimates differ increasingly with $k_1 > k_2$.
8.97
$$v = \sqrt{\frac{2g(d - \ell)\sin\theta}{(1+\alpha)}\frac{[1 + (1+\alpha)^2\tan^2\theta]}{[1 + (1+\alpha)\tan^2\theta]}}$$

CHAPTER 9

9.1(a) 1.5×10^7 J·s **(b)** $\frac{L_{O,\text{car}}}{L_{O,\text{truck}}} = 0.20$;
9.3 $-(22\text{ m}^2)\hat{\mathbf{j}}$ **9.5** $-(200\text{ N}\cdot\text{m})\hat{\mathbf{k}}$
9.7 The remaining person's weight force will exert a torque on the bench that will cause the bench to rotate. **9.9** from above
9.11 $\vec{\mathbf{r}}_{cm} = (30\text{ cm})\hat{\mathbf{i}}$ **9.13** $(-1.0\text{ J}\cdot\text{s})\hat{\mathbf{k}}$
9.15 $\frac{L_{O,1}}{L_{O,2}} = 0.73$ **9.19(a)** for any point 0 on a line through the particle's position parallel to its velocity. **(b)** points O in the x-y plane above the $\vec{\mathbf{L}} = 0$ line **(c)** O must be in the x-y plane below the $\vec{\mathbf{L}} = 0$ line **(d)** O on the line parallel to the z-axis through the particle's position **(e)** a straight path at constant velocity
9.25 $[(2.9)(-\hat{\mathbf{k}}) + (1.6)\hat{\mathbf{i}} + (6.1)\hat{\mathbf{j}}]$ J·s; $\vec{\mathbf{C}}$ describes an angular momentum vector.
9.29 $-3\hat{\mathbf{k}}$ **9.33** 3.14×10^{43} J·s; Jupiter contributes the most
9.37 31 m^2; 126° **9.41** $m(\vec{\mathbf{r}}_o \times \vec{\mathbf{v}}_o)$; $|\vec{\mathbf{L}}_O| = mr_o v_o$; $t_\perp = -\frac{\vec{\mathbf{r}}_o \cdot \hat{\mathbf{v}}_o}{v_o}$; $|\vec{\mathbf{L}}_O| = mv_o\,r(t_\perp)$ **9.45** $(5 \times 10^3\text{ J}\cdot\text{s})\hat{\mathbf{k}}$
9.49 20 N·m **9.53** $F = \frac{rMg}{R}$; $P = rMg\omega$ **9.57(a)** $\vec{\mathbf{L}}_i = 2(\ell^2 m\omega_0\hat{\mathbf{k}})$; $\vec{\tau} = -\ell\mu mg\hat{\mathbf{k}}$; $t = \frac{\ell\omega_0}{\mu_k g}$; **(b)** $m\omega_0^2\ell^2$; same time **9.61** The observed motion could result from the star along with a massive companion traveling through space and orbiting about the system CM as they go.
9.65 You should support the stick at its center.
9.69 $\vec{\mathbf{r}}_{\text{CM}} = \left(\frac{7}{4}\text{ m}\right)\hat{\mathbf{i}} + \left(\frac{7}{4}\text{ m}\right)\hat{\mathbf{j}}$; $\vec{\mathbf{v}}_{\text{CM}} = \left(\frac{1}{2}\text{ m/s}\right)\hat{\mathbf{i}} + \left(\frac{1}{4}\text{ m/s}\right)\hat{\mathbf{j}}$; $K_{\text{int}} = 22.9$ J; $\vec{\mathbf{L}}_{\text{int}} = (8.8\text{ J}\cdot\text{s})\hat{\mathbf{k}}$
9.73 $\frac{1}{18}L$ below the square's geometric center.
9.77 $-40\text{ m/s}\,\hat{\mathbf{i}} + 60\text{ m/s}\,\hat{\mathbf{j}} + 0.4\text{ m/s}\,\hat{\mathbf{k}}$
9.81 The satellite begins to rotate. The astronaut rotates in the opposite direction.
9.85 $(4.9\text{ m/s})\hat{\mathbf{i}}$; $-(3.1\text{ m/s})\hat{\mathbf{i}}$
9.89 The other half ends up with twice its initial velocity and escapes from the Earth.
9.93(a) 0.14 rad/s; 2.5×10^{10} J·s
(b) The velocity of the CM is $-\frac{\omega\ell}{6}\hat{\mathbf{j}}$; The rotational speed about the CM is ω
9.101 The period is 34.3 in scaled units.
9.103 $\frac{\sqrt{2}\,\alpha\, m^2}{36g_o}$

Chapter 10

10.3 The initial velocity of the target ball is 0.1 m/s in the direction of the incident ball. **10.5** $\vec{v}_{CM} = (1\ m/s)\hat{i}$; $\vec{v}_{astr} = 5\ m/s\,\hat{i}$; $\vec{v}_{sat} = -1\ m/s\ \hat{i}$; Both zero **10.7** inelastic; $\frac{Mv_0^2}{4}$ **10.9** No **10.11** only if the particles have equal mass **10.15** 0.28 m/s; 6.2 m/s **10.19** 330 kg **10.23** Yes; $\frac{1}{9}h$

10.27 $\vec{v}_f = -\frac{v_0}{2}[2.5\hat{i} + \hat{j}]$; 8.8 mv_0^2

10.31 $\frac{1}{2}f$ **10.35** 450 m/s

10.39 43 kg·m/s; 79 J
10.43 $v_f = 0.36v_i$

$\vec{u} = (0.49v_i$ at 66° to original axis of molecule)

$\omega = 0.89\frac{v_i}{\ell}$

where ℓ = length of molecule
$\vec{\omega}$ is clockwise in Fig. 10.31

10.47(a) $\vec{v}_{CM} = \frac{\Delta\vec{p}}{2M}$; $\omega = \frac{\Delta p}{M\ell}$
(b) The dumbbell travels along a straight path without rotating. Its CM velocity is the same as in (a). **(c)** $\frac{(\Delta p)^2}{2M}$; $\frac{(\Delta p)^2}{4M}$
10.51 570 m/s; 2400 J **10.55** the outgoing velocities are perpendicular;

$\vec{v}_1 = \left(\frac{\sqrt{3}}{2}v_0,\ \text{at } 30°\right)$

$\vec{v}_2 = \left(\frac{v_0}{2},\ \text{at } 60°\right)$

10.59 $\frac{1}{3}\vec{v}_0$; $\vec{\omega} = \left(\frac{v_0}{4\ell},\ \text{clockwise}\right)$; $\frac{5}{16}mv_0^2$

10.63(b) $\sqrt{v_n^2 - \mu_k gd}$ **(e)** u_{n+1} increases if $v_1 > 3.0$ m/s and u_{n+1} decreases if $v_1 < 3.0$ m/s; 6 **10.67** No; Yes

Chapter 11

11.1 b, c, and d **11.3** 0.21 m
11.5 The CM is higher **11.7** No; No **11.9** No **11.11** $(0, \frac{3}{2}\ell)$

11.13 590 N **11.15** $\frac{1}{2\sqrt{3}}L$ from the bottom of the crate. **11.19** 4.0×10^2 N;

$\ell_{Mike} = \frac{M_{Jen}}{M_{Mike}}\ell_{Jen}$; Mike must sit at a maximum distance 0.94 m from the center.
11.23 1.2×10^5 N; 1.2×10^4 N·m; (1.6×10^5 N, rightward); (0.4×10^5 N, leftward) **11.27** 9 **11.33** Vertical force = 95 N. Horizontal force = 110 N. Put bottom screws in first. **11.37(a)** at her CM **(b)** None **(c)** 15° **(d)** 6° **(e)** 0.25 **11.41** We want a larger force to be exerted by the shear's blades than the tailor's blades. **11.45** 20 N
11.49 0.80 g **11.53** Equilibrium is not possible unless θ_1 and $\theta_2 \le 45°$ and $M_1 \sin\theta_1 = M_2 \sin\theta_2$. **11.57** 4500 N;
Force on each front tire = 4.0×10^3 N
Force on each back tire = 8.0×10^3 N
11.61 Top barrel: $\sqrt{2}Mg$ on middle barrel, Mg on wall; Middle barrel: $2Mg\sqrt{2}$ on lowest barrel, $3Mg$ on wall; Bottom barrel: $3Mg$ on floor, $2Mg$ on wall; the value of μ_s is irrelevant

11.65 $m_{max}(\theta) = \frac{Mr_1}{2r_2} \times \frac{[\cos^2\theta(1+\mu^2) + 4\mu^2\sin^2\theta + 4\mu\sin\theta\cos\theta]^{1/2}}{\cos\theta + 2\mu\sin\theta}$;
$m_{min}(\theta)$ can be found by replacing μ by $-\mu$ in the above formula. **11.69** 56 kg

11.73 at its geometric center **11.77** $\frac{4R}{3\pi}$ from straight edge **11.81** 8.5×10^4 N
11.87 2.5×10^3 N; 1.2×10^4 N
11.91 (670 N @ 66° from horizontal)
11.95 11.3 cm **11.99(b)** 0.53 m
11.103 26.1%

Chapter 12

12.1 centered on module so that the whole station rotates on the $\hat{k}$ axis
12.3 8.0×10^5 **12.5** 60 J·s
12.7 7.6 N **12.9** 8.96 rad/s; 94.1 J
12.11 0.31 rad/s **12.19** 0.8 rad/s; 0.3 m/s **12.23** The square has the greatest rotational inertia, the rod has the smallest
12.27 490 W **12.31** 5.4 N; 4.9 W

12.35 $\sqrt{\frac{3L}{g}}$; $\sqrt{\frac{4gL}{3}}$ downward; $\frac{1}{R}\sqrt{\frac{4gL}{3}}$; $\frac{2}{3}MgL$; $\frac{1}{3}MgL$; MgL **12.39** 57.7 N; 57.0 N **12.43** at the bottom of the barrel; At angle $\sin^{-1}\left[+\frac{2\alpha_{barrel,z}R}{5g}\right]$ from the bottom; $\frac{5\mu g}{2R\sqrt{1+\mu^2}}$ **12.49** 2.7 m/s
12.53 $2.7R - 1.7r$; $2.5R - 1.5r$

12.57 $\frac{2\pi}{1 + \frac{I_c}{I_y}}$; Only if $\frac{I_c}{I_y}$ is irrational

12.61 No; When the object rotates about the z-axis $\vec{L}$ and $\vec{\omega}$ are ||; Neither

12.65 $\alpha = \frac{\alpha}{4} + \frac{1}{8}$ **12.69** $\frac{3MR^2}{2}$

12.73 $\frac{1}{2}M\ell^2$; $\frac{1}{4}M\ell^2$ **12.77** 5.5×10^{22} N·m; 1.4×10^{22} kg $\approx 2 \times 10^{-3} M_E$
12.79 vertically
12.83(a) $\vec{\omega} = (5.2\ rad/s)\hat{k}$
(b) $-(0.12\ rad/s^2)\hat{k}$
(c) 420 J; $(160\ J\cdot s)\hat{k}$
(d) $-(3.8\ N\cdot m)\hat{k}$; 18; -420 N·m
(e) $-0.87\ rad/s\hat{k}$; 18 J **12.87** $\frac{2g}{5}$; $\frac{2g}{5R}$ **12.91** $3m\ell^2\omega$; $\frac{3}{2}m\ell^2\omega^2$
12.95 8.3 cm from the junction, on the steel side; $I_{CM} = 0.91$ kg·m²; I(steel end) = 3.5 kg·m²; I(Al end) = 6.0 kg·m²
12.99 0.31 m **12.103(a)** No
(b) $N_i = (23.87\ s)\omega(t_i)$; $b = 3.79 \times 10^{-5}$/s
12.107(a)
$\vec{v}(t) = [\omega \sin\omega t(v_O t - R_O) - v_O\cos\omega t]\hat{i} + [\omega\cos\omega t(R_O - v_O t) - v_O\sin\omega t]\hat{j}$;
$\vec{a}(t) = [\omega^2\cos\omega t(v_O t - R_O) + 2\omega v_O\sin\omega t]\hat{i} - [\omega^2\sin\omega t(R_O - v_O t) + 2\omega v_O\cos\omega t]\hat{j}$

(b) $\frac{1}{2}m\omega^2[R_O^2 - r^2]$ **(d)** $\sqrt{7}\,\omega R_O/2$

Chapter 13

13.1 The beam is primarily supported by shear. **13.3** 37 atm; 6.3×10^8 N
13.5 the larger piston; 110 N
13.7 3×10^5 Pa **13.9** 750 m³
13.11 5.85 m/s **13.13** 14 m
13.17 No; No **13.21** .2 kg
13.25 3 atm **13.29** .0207 m²
13.33 1×10^{-3} m² **13.37** No changes need to be made. **13.41** 13 m
13.45 1400 m **13.49** 0.5 N
13.53 neither **13.59** 41.5%
13.63 448 bags **13.67** 1.8×10^3 kg/m³
13.73 pushed **13.77** greater
13.81 60 m **13.85** 9.7 atm; 0.98 MW
13.89 2.1 cm **13.91** $\frac{3}{2}\rho v_0^2 LW$
13.95 greater; the same
13.97 1.02×10^5 Pa **13.99** angular momentum **13.103** 9.3 km; 140 m/s
13.105 5.05×10^7 N; 1.68×10^8 N·m; 6.67 m below the surface of the water
13.109 6.26 rad/s
13.113(a) 4×10^5 Pa **(b)** a parabola

Chapter 14

14.3 $x(t) = -(7.5\ cm)\cos[(2.6\ rad/s)t]$; $v(t) = (19.5\ m/s)\sin[(2.6\ rad/s)t]$
14.5 The period of pendulum A is $\sqrt{2}$ times the period of pendulum B.
14.7 $\theta(t) = 3.0°\cos([3.6\ rad/s]t)$
14.9 8.7° **14.11** 0.3 m **14.13** 0.7 s
14.17 $x(t) = (+1.1\ m)\cos\left([2.6\ /s]t + \frac{3\pi}{2}\right)$
14.21 4.5 s **14.25** 0.55 m
14.29 towards the support

14.33 2.005 s
14.37 $\theta(t) = (0.081 \text{ rad}) \sin[(1.98 \text{ rad/s})t]$
14.41 0.64 Hz **14.45** 1.9 s
14.49 $\frac{E_{2^\circ}}{E_{1^\circ}} = 4$ **14.53** 8.0 rad/s; 0.056 m/s
14.57 3.3×10^3 N/m
14.61 $Z_{res} = -(0.08 \text{ m}) \cos[(6 \text{ rad/s})t]$
14.65 5.0 s; 0.11 rad/s
14.69 $2\pi\sqrt{\frac{m}{k_1 + k_2}}$
14.73 .023 m/s; $+\frac{1}{2}\pi r^2 \rho_w g z_0^2$
14.77 2.10 s, 5% larger than for the simple pendulum. **14.81** $7.8\sqrt{\frac{r}{g}}$

Chapter 15

15.1 $\sqrt{3}$ **15.3** 440 Hz
15.5 0.48 s **15.7** $\sqrt{\frac{7}{5}\frac{P}{\rho}} = 330$ m/s at 0° C **15.9** 2:1 **15.11** 43 W
15.13
$y(x, t) = (5.0 \text{ cm}) \cos\left[\begin{matrix}(3.0 \text{ rad/m})x \\ -(5.0 \text{ rad/s})t + \pi\end{matrix}\right]$
For a free boundary, the phase difference would be zero rather than π
15.15 9.9 g/m **15.19** No.
15.23 440 Hz **15.27** $v = 5.00$ m/s in $-x$ direction; $\lambda = 0.100$ m; $T = 0.0200$ s; $f = 50.0$ Hz; A = 1.00 mm **15.31** $y(x, t) = (0.750 \text{ mm}) \times \cos[(31.4 \text{ rad/m})x - (314 \text{ rad/s})t + \pi]$ **15.35** 0.6 μm
15.39 No; Yes **15.43** 2.5 mm
15.47 The wave displacements will not always be equal and opposite at *any* point.
15.51(a) $\frac{1}{2}$ wavelength **(b)** $\frac{1}{2}$ wavelength **(c)** $\frac{1}{4}$ wavelength **15.55** 13 cm
15.59 increase by 2.3% **15.63** $13\frac{\text{g}}{\text{m}}$; 0.060 mm **15.67** The path of an element is elliptical. **15.69** $\frac{m}{2\ell}\sqrt{\frac{T}{\mu}}$, m = integer
15.73 One pulse will appear to stand still and the other will race around at $2v$.
15.77 The superposition is a pulse that travels at 330 m/s, the speed of each individual wave.
15.83 $P = \frac{A^2 Tk\omega}{4} \sin(2\pi x) \cos(400\pi t)$

Chapter 16

16.1 0.75 m; 0.19 m **16.3** 6 displacement nodes; 7 pressure nodes; 1.0 kHz
16.5 3.0×10^8 m/s **16.7(a)** 33 ns **(b)** 0.13 s **(c)** 4 y
16.9 $3 \times 10^{-9} \frac{\text{W}}{\text{m}^2}$ **16.11** 440 Hz
16.13 7×10^{-5} **16.15** 7.0 m
16.17 24.4° **16.19** 131 Hz
16.23(a) 5.2 m **(b)** 2.6 m
16.27 49 Hz; 97 Hz **16.31** 8.3 min
16.35 No **16.39** 3.2
16.43 33 W/m^2; 140 dB **16.49** 3×10^8 m/s **16.53** -9.17×10^{-4}; blue-shifted **16.57** $f = 9.0 \times 10^2$ Hz
16.65 You should point the laser at the apparent position of the object. **16.69** at 0° to the normal of side BC **16.73** θ_i; $\frac{2t}{\sqrt{\left(\frac{n_{glass}}{n_{air}} \csc \theta_i\right)^2 - 1}}$ **16.79(a)** 33μs
(b) 1.3 ms; No; estimates v as $\frac{\Delta d}{\Delta t}$
16.83(a) 3 **(b)** 2.5; a node at each wall
16.87 7.4×10^3 m/s; Yes
16.91 $\theta_1 = \sin^{-1}\frac{d}{r}$, $\theta_2 = \sin^{-1}\left(\frac{d}{nr}\right)$; maximum deflection $\phi \approx 138°$ occurs for $\frac{d}{r} \sim 0.9°$ **16.95(a)** 1.3×10^{-5}
(b) Yes; No

Chapter 17

17.1 1.20 mm; 2.40 mm; 3.60 mm; 1.20 mm; The separation is not constant except near the center of the screen.
17.3 0.10 μm **17.5** green; 17.7 mm
17.7(a) θ = 6.89°, 13.9°, 21.1°, 28.7°, 36.9° . . . 73.7° **(b)** θ = 36.9°
17.9 $y = (3.9 \text{ mm}) \times \cos\left[\left(3\frac{\text{rad}}{\text{m}}\right)x - \left(10^3\frac{\text{rad}}{\text{s}}\right)t + 0.14 \text{ rad}\right]$
17.11 $\frac{2\pi}{7}$ **17.13** $\frac{0.215 \text{ nm}}{m}$, $m = 1, 2, 3 \ldots$ **17.15** The film deforms into a *wedge* shape that is thicker at the bottom. **17.19** orange **17.23** 19; 6.1×10^{-5} m from apex **17.27** $y = 0$, ± 1.3 mm, ± 2.5 mm, ± 3.8 mm etc; 1.52
17.31 67 mm; 44 mm **17.35** The 600 kHz signals, with their longer wavelengths are bent further into the tunnel.
17.41 $y = 0.74$ m **17.45** 0.02″
17.49 $d = 2a = 4.0$ μm
17.53 Maximum intensity is 9 times that of the fork. Minimum intensity equals that of the fork. **17.57** $\theta \simeq$ 3.2°, 6.4°, 9.6°, 12.8°, 16.1°, 22.9°, 26.4°, 30° etc.; θ = 0, 19.5°, 41.8°, 90°; 4.5°, 8°, 11°, 14.5° etc.
17.61(b) 1.2×10^5; 1.7×10^{-6} m; 1.2×10^6 m^{-1} **17.65** 4 Hz; Maximum intensity is $\approx$ 7.5 times that of the fork. Minimum intensity is $\approx$ 0.54 that of the fork.
17.69 2.6 nm; 5.1 nm; 7.2 nm
17.73 $\lambda = \frac{2t \sin \theta}{m}$; $\theta + 2\tan^{-1}\left(\frac{t}{5d}\right)$
17.79(a) 530 nm green, 410 nm violet
(b) 580 nm yellow, 490 nm blue, 425 nm violet **(c)** 456 nm blue, 638 nm red
17.83 $\langle u \rangle_{short} = \frac{2I_0}{v_s}[1 + \cos(\Delta kx - \Delta\omega t)]$; $\frac{2\pi}{\Delta k}$; $\frac{1}{f_b}$; $\langle u \rangle_{long} = \frac{2I_0}{v_s} = 2\langle u \rangle_1$
17.91 $d = \frac{\lambda L}{\ell}$; $\frac{N_A}{N_B} = 3$

Chapter 18

18.1

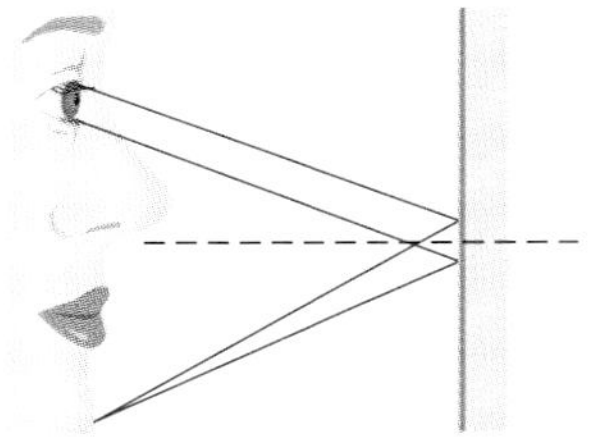

18.3 3 m **18.5** $x_i = +6.7$ cm; real image; $x_i = -12$ cm; virtual image
18.7(a) Virtual image 2.7 m behind the glass.
(b) Virtual image 0.79 m behind glass.
18.9 2.0 **18.11** A virtual image is formed 17 cm from the lens, on the same side as the object. **18.13** 10.1 cm
18.15 $x_i = +40$ cm; real; m = -4.0; inverted **18.17** 52 **18.19** 6.137 cm @ 480.0 nm; 6.246 cm @ 656.3 nm
18.21 5 **18.25** twice the separation of the 2 mirrors **18.29** 0.686 cm
18.31 $\theta = 2\phi$ **18.35** No
18.39 real, inverted image, 0.33 m from the mirror **18.43** 5.2 cm behind the surface of the bottle. **18.47** 5.13 cm behind the lens. **18.53** 0.70 m from the outside surface of the window. **18.55** The image is virtual and is located inside the bubble, 0.92 cm from the center, on the same side as the crystal. The bubble behaves like a diverging lens. **18.59** $f = \frac{Rn}{2(n - 1)}$
18.63 $x_o > 0$ **18.67** -11 cm; $x_i = -6.4$ cm; $m = -0.40$; virtual
18.71 $m = 0.63$; on axis magnification would be 0.60 **18.73** same; converging
18.77 image moves 20 cm away from object
18.81 +1.0 **18.85(a)** $x_i = \frac{5f}{4}$
(b) $x_i = f$ **(c)** $x_i = +\frac{f}{2}$
(d) Image formed at ∞. **(e)** $x_i = -f$
(f) $x_i = -\frac{5}{6}f$ **(g)** $x_i = \frac{13}{8}f$
(h) $x_i = \frac{5}{3}f$ **(i)** $x_i = 1.7f$
18.89 0.6 cm **18.93** The travel time of light along any path through the lens

from O to I is the same. **18.97** The mirror should be at 45°; parallel light
18.99 33 cm **18.103** The object must be virtual and closer to the surface than $\frac{Rn_1}{(n_2 - n_1)}$. **18.111** 10 km

CHAPTER 19

19.1 Yes; No **19.3** 351.7 K; 173° F
19.5(a) 6.2×10^{-3} mol **(b)** 4
(c) helium **19.7** no
19.9(a)

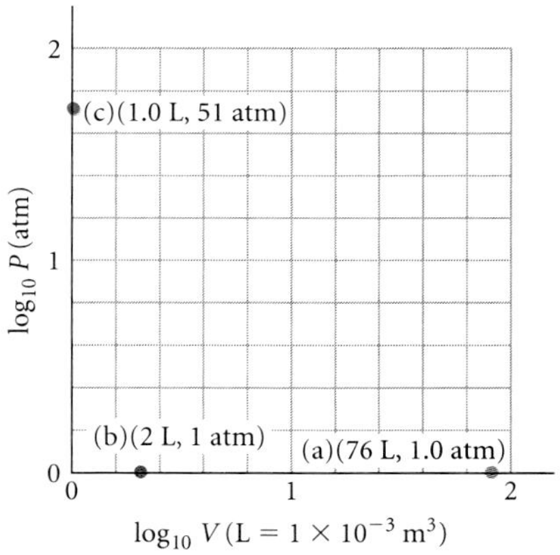

19.11 1030 J/kg·K **19.13** 1.1 atm
19.15 $\frac{7}{5}$ **19.17** 740 J/kg·K
19.19(a) Yes. Time scale is months to years.
(b) Yes. Time scale is on the order of hours.
(c) No. **(d)** Yes. Minutes time scale.
19.23 −37.97° F; 234.28 K **19.27** each molecule has the same kinetic energy; less massive molecules have larger rms speed
19.31 The 2 L container
19.35(a) 1.6×10^6 m/s
(b) 6.7×10^7 m/s **19.39** 740 J
19.41(a) 24 J **(b)** +63 J **(c)** The block rises 5.3 m. **19.45(a)** 1.8 mol
(b) 0.063 atm **(c)** No
19.49 0.080 mol **19.53** 1010 J/kg·K; 1680 J/kg·K; $\gamma = 1.66$; no change
19.57 330 m/s; $v_{rms} = 490$ m/s
19.61 $W = 1300$ J **19.65(a)** 0.20 mol
(b) 0.028 mol **(c)** $n\,V_p$
(d) $n = 22.5$; 0.64 mol **(e)** 475 K
(f) 3.2×10^5 Pa **19.69** 1850 J/K·kg
19.73 8×10^{-23} J; No
19.77 $\frac{v_{rms}\text{ electron}}{v_{rms}\text{ ion}} = 205$;
The contributions to the pressure are equal.
19.81 1.47 cm **19.83** 245 K
19.87 No; Yes; work is done on the system
19.91 6.2×10^{-5} m

CHAPTER 20

20.1 7.5×10^5 Pa **20.3** 2.89×10^6 J
20.5 1.9×10^{-5} K^{-1}
20.7 Heat capacity of water is much greater
20.9 48° C **20.13** 1.06×10^6 Pa
20.17 0.59 mol **20.21** 2.89×10^5 kJ
20.25 No **20.29** 0.2%
20.33 3.8×10^{-4} K^{-1}
20.37 $T = \frac{C_A T_A + C_B T_B}{C_A + C_B}$
20.41 $246 \frac{\text{J}}{\text{kg} \cdot \text{K}}$ **20.43** 13.5° C
20.47 5.012 J/K; 1.22×10^5 Pa; 232 K; The thermometer *lowered* the system temperature by ~1.5 K **20.51** 0.59 kg
20.55 7.6 kg **20.59** 5.2° C
20.63 9×10^{-3} *R above* the equator.

CHAPTER 21

21.1 3100 W **21.3** concrete: 7.3 W/m^2·K; oak: 4.9 W/m^2·K; oak
21.5 b; c **21.7** 8.85 μm **21.9** 8 min
21.13 skewer conducts heat to interior of potato **21.17** $R_{plywood} = 3.4 \times 10^{-3}$ K/W; $R_{window} = 0.058$ K/W; $R_{wall} = 3.2 \times 10^{-3}$ K/W
21.21 360° C; 410 W
21.25 $H = \frac{k\pi rR}{h(T_2 - T_1)}$; The heat flow is unchanged. **21.29** Buoyant force accelerates the hot gases upward; no.
21.33 It is very small.
21.37 3×10^{-5} m^2; No **21.39** 10 min
21.43 The stove **21.47(a)** 47 W
(b) 7 W; $H_{conv} < H_{rad}$ **21.51** $\alpha \approx 0.01$

CHAPTER 22

22.1 melting of ice cubes; free expansion of gas; an inelastic collision
22.3(a)

Process:	Q(J)	W(J)	ΔU(J)
AB	3030	0	3030
BC	10100	4040	6060
CD	−6060	0	−6060
DA	−5050	−2020	−3030
Cycle	2020	2020	0

(b) 0.154 **22.5** 1.4×10^3 J; 1.6×10^3 J **22.7** No
22.9 $e < 50\%$ **22.11** 2.4×10^{-5}
22.13 1×10^{14} m
22.21(a) $T_A = \frac{P_A V_A}{\mathcal{N} R}$; $T_B = \frac{1}{8}\frac{P_A V_A}{\mathcal{N} R}$
(b) $W_{AB} = -\frac{7}{8} P_A V_A$; $Q_{AB} = -\frac{35}{16} P_A V_A$;
(c) $T_C = 4\frac{P_A V_A}{\mathcal{N} R}$; $Q_{BC} = \frac{93}{16} P_A V_A$
(d) $W_{CA} = \frac{9}{2} P_A V_A$ **(e)** 0.6
22.25 The pressure and volume throughout the cycle increase; decrease **22.29** by a factor of 2.25 or by 95 W
22.37 $\frac{dS}{dt} = -6.2 \times 10^{22}$ W/K
22.39 850 J/K **22.43** 4.0×10^2 K
22.47 decreases if $\frac{v}{v_{mp}} < 1.02$; increases for $\frac{v}{v_{mp}} > 1.02$
22.51 $\lambda_2 = \frac{1}{2}\lambda_1$; $\frac{\tau_1}{\tau_2} = \sqrt{2}$
22.55 40 nm **22.59** The temperature dependence of the entropy is given by: $\frac{3}{2} Nk \ln T$ or $3Nk \ln v_{rms}$.
22.63 $10^{-3 \times 10^8}$; $N(10^{-3 \times 10^8})$; $10^{3 \times 10^8}$ y
22.67 0° C; $\Delta S_{ice} = 12.26$ J/K; $\Delta S_{water} = -10.76$ J/K; $\Delta S_{net} = +1.5$ J/K
22.71 $e_B = 1 - \frac{T_A}{T_B}$; $e_C = 1 - \frac{T_A}{T_C}$; $e_C > e_B$
22.75(a) 0.42 **(b)** 6×10^{-6}
22.79(a) $\frac{V_G}{V_F} = \left[\frac{\Delta P}{P_C} + 1\right]^{\frac{1}{\gamma - 1}}$
22.83(c) $\frac{P_A + P_B}{2}(V_B - V_B) + \frac{P_B V_B + P_A V_A}{\gamma - 1}$
(d) $\frac{P_A + P_D}{2}(V_D - V_A) + \frac{P_D V_D - P_A V_A}{\gamma - 1}$
(e) 0.366 **(f)** $1.691 V_A$; $(1.201)\frac{P_A V_A}{\mathcal{N} R}$
(g) 0.5997

CHAPTER 23

23.1 zero **23.3** positive charge on the sweater and negative charge on the cotton T-shirt **23.5** $9.0 \times 10^9 \frac{\text{N}}{\text{C}}$ towards Q
23.7 a
23.9

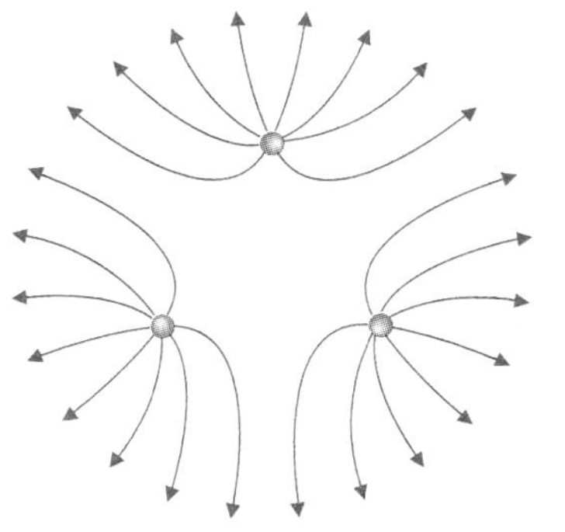

23.11 $3.00 \times 10^6 \frac{\text{N} \cdot \text{m}^2}{\text{C}}$; $1.00 \times 10^6 \frac{\text{N} \cdot \text{m}^2}{\text{C}}$; $-2.00 \times 10^6 \frac{\text{N} \cdot \text{m}^2}{\text{C}}$
23.15 2×10^{-8} N **23.19** $\left(\frac{kQ^2}{k_s}\right)^{1/3}$
23.23 0.18 nC

23.27 $\vec{\mathbf{E}} = \left(5.58 \times 10^{-11}\ \frac{\text{N}}{\text{C}}, \text{downward}\right)$

23.31 At R, $\vec{\mathbf{E}}$ is upward. At P, $\vec{\mathbf{E}}$ is to the right.

23.35

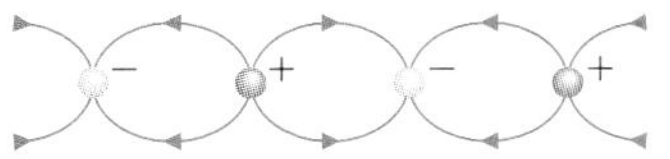

23.39 zero; directly away from the empty corner **23.43(a)** 3.4 mN **(b)** 4.8 mN **23.47(a)** zero **(b)** $\left(\frac{kQ}{2a^2}(1 + 2\sqrt{2}), \text{outward along the diagonal}\right)$ **(c)** $\left(\frac{16}{25}\sqrt{5}\frac{kQ}{a^2}, \perp \text{ to the side}\right)$

23.51 $\frac{2\sqrt{2}kq}{a^2}\hat{\mathbf{i}}$ **23.55** $\frac{Q}{6\epsilon_0}$; No

23.59 47 N·m²/C **23.63** Unstable; same; no **23.67** $y = 3.3D$

23.71 $y = \pm 1.3\ L$

23.75 $\frac{dy}{dx} = \frac{y}{x + a\left(\frac{r_-^3 - r_+^3}{r_-^3 + r_+^3}\right)}$ where $r_- = \sqrt{(x - a)^2 + y^2}$ is the distance from the charge at $+a$, and r_+ is defined similarly.

Chapter 24

24.1 $(3.3 \times 10^2\ \text{N/C})\hat{\mathbf{r}}$

24.3 (6.3×10^6 N/C, radially outward from axis along which filament lies)

24.5 (2.6×10^8 N/C, normal to the plate)

24.7 $9.47 \times 10^{14}\ \text{m/s}^2$; 1.45×10^{-8} s

24.9 0.17 μC **24.13** −.5 MC; −.9 nC/m² **24.17** zero

24.21 4.7×10^4 N/C; 2.7×10^4 N/C

24.25 $\frac{k\lambda\ell}{x_0^2 - \frac{1}{4}\ell^2}\hat{\mathbf{i}}$ **24.29** $-\frac{\pi k\lambda_0}{2R}\hat{\mathbf{j}}$

24.33 **b** **24.37** 240 N/C

24.41 $\vec{\mathbf{E}}_{\text{net}}$ will always point 45° from each plane at every point in space. $|\vec{\mathbf{E}}_{\text{net}}| = \frac{\sigma\sqrt{2}}{2\epsilon_0}$

24.45 $\frac{2\pi k\sigma_0 z\hat{\mathbf{k}}}{a^2}\left[\frac{2z^2 + a^2}{\sqrt{z^2 + a^2}} - 2z\right]$

24.49 $\frac{\rho}{3\epsilon_0}$ **24.53** **e**

24.57 $5.2 \times 10^{-2}\ \text{m/s}^2$

24.61 −56 nC **24.65** 0.42 m/s

24.71 at points A and C the field points in the $-y$-direction; The field is in the y-direction at both points B and D **24.73** 1.0 N/C

24.77 zero; $\frac{kQ}{9D^2}$ **24.81** $\frac{L^2}{mke^2}$

24.85 the rod hangs at 50° with the vertical. 11 N **24.89(a)** $F_x = -9.3\ \mu$N; $F_y = 38\ \mu$N; **(b)** $F_x = -59\ \mu$N; $F_y = -28\ \mu$N; **(c)** $F_x = 0$; $F_y = -86\ \mu$N

24.93 $\frac{\pi k\sigma_0 q}{2g}$; Yes

Chapter 25

25.1 4.6×10^{-20} J **25.3** 600 J

25.5 (a)

25.7

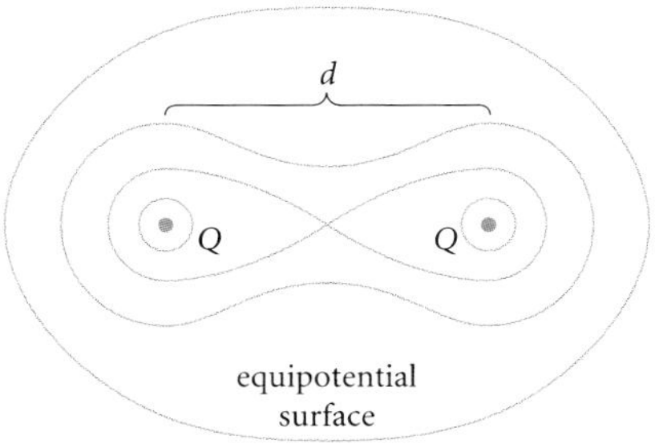

25.9 $\frac{kQ}{a}$ **25.11** (c) **25.13** (b)

25.15 zero **25.19** 2.0×10^{-11} m

25.23 same **25.25** zero

25.29 −54 V **25.33** -4.91×10^4 V

25.37 10 **25.41** $-\frac{kQ^2}{a}$

25.45 $2\frac{kQq}{\ell}$ **25.47** $k\frac{p\cos\theta}{r^2}$

25.51

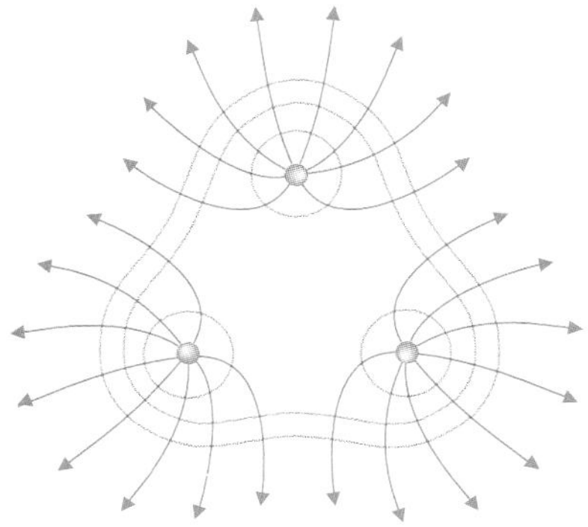

25.55(a) $y = \frac{2}{3}$ cm, −2.0 cm **(b)** $y = -4.8$ cm; -1.2×10^5 V

25.59 0.3 μC

25.63(a) $V(y) = 2k\lambda \ln\left(\frac{y_0}{y}\right)$ where y_0 is a suitably chosen reference point

(b) $V(y) = \frac{k\lambda L}{y}$

(c) $V(y) = k\lambda \ln\left(\frac{\sqrt{(L^2/4) + y^2} + L/2}{\sqrt{(L^2/4) + y^2} - L/2}\right)$

25.67 $V = \pi k\rho(b^2 - a^2)$

25.71(a) a **(b)** d

25.75

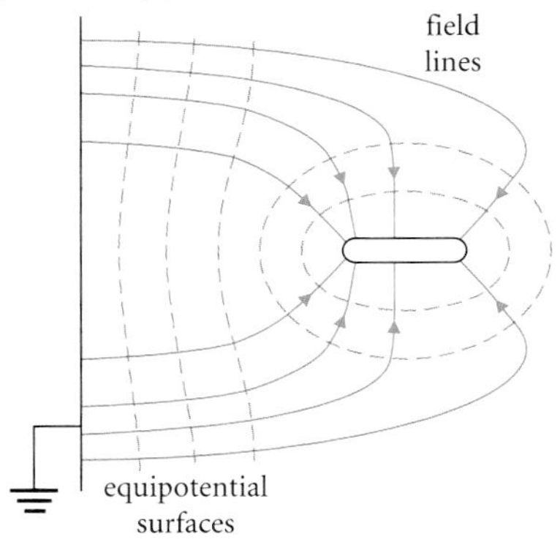

25.79 88 kV/m; The field strength becomes larger **25.81** 44 V

25.83 163×10^6 C; 900 **25.87** 190 V

25.91 39 s; 51 W

25.95 $V(P_1) = 1.45 \times 10^5$ V
$V(P_2) = -2.0 \times 10^5$ V

25.99(a) Point outside: $\frac{kQ}{d}$ where d = distance from center of sphere; Point inside: $V(P) = \frac{kQ}{2}\frac{(3R^2 - d^2)}{R^3}$

(b) $\frac{3}{5}\frac{kQ^2}{R}$ **25.103** $\frac{\pi}{4q}\sqrt{\frac{m\ d^3}{k}}$

Chapter 26

26.1 1200 C **26.3** 3 A; 1.7 Ω

26.5 $\frac{1}{3}$ **26.7** 4.4 kΩ **26.9** 11.2 V

26.11 2.26 Ω **26.13** $\lambda\omega r$

26.17 1.44 V **26.21** 1.9 V

26.25 ≥ 3.3 mm **26.29** 1.5 mm; 1.7×10^{-4} V/m **26.33** closed

26.37 21.7 kΩ **26.41** 0.4 Ω; 49.0 V

26.45 16 Ω; 0.50 A; 0.33 A top arm; 0.17 A through 8 Ω resistor; 0.057 A through each 12 Ω resistor; $V_A = 6.7$ V; $V_B = 6.6$ V; $V_C = 6.0$ V **26.49** both decrease

26.53 14 A through 3 Ω resistor; 4 A through 6 Ω resistor; 18 A through 24 V battery

26.59 top arm: $I = \mathscr{E}/R$ to the left,
2nd arm: $I = 0$,
3rd arm: $I = 0$,
Bottom arm: $I = \mathscr{E}/R$ to the right

26.63 2.5 A; $\frac{2}{3}$ A; 1.8 A; 1.3 V

26.67 2.49 V; 2.50 V **26.71** 0.500 mA; less sensitive; $R_2 = 0.20\ \Omega$; $R_3 = 0.020\ \Omega$

26.75 0.03 Ω **26.79** 2R

26.83 0.13 A

26.87 $i_1 = \frac{(R_3 + R_2)\mathscr{E}_1 - R_3\mathscr{E}_2}{R_1R_2 + R_1R_3 + R_2R_3}$; $i_2 = \frac{[\mathscr{E}_2(R_1 + R_3) - R_3\mathscr{E}_1]}{(R_1R_2 + R_1R_3 + R_2R_3)}$

26.91 $I_N = 25.3$ A; $R_N = 0.26\ \Omega$

26.95 5.00 V; 9.99 Ω; 5.00 V; 0.477 A; Case A gives the closer answer for the resistance. Case B gives the closer answer for the emf.

Chapter 27

27.1 120 pC **27.3** $d = 3.1$ mm

27.5 7.0×10^{-8} J **27.7** 6.5 pF

27.9 33 pF **27.11** $2.1 \times 10^{-6}\ \text{J}\cdot\text{m}^{-3}$

27.15 (e) **27.19** 0.90 kV/m

27.23 $V_1/V_2 = \frac{1}{3}$; $U_1/U_2 = \frac{1}{3}$

27.27 3 μF **27.29** in series; $\frac{8}{7}\ \mu$F

27.33 $Q_i = C\mathcal{E}$; $U = \frac{3}{2}C\mathcal{E}^2$; $3\mathcal{E} = \Delta V_{tot}$;

$C\mathcal{E} = Q_f$; $U_f = \frac{3}{2}C\mathcal{E}^2$

27.37 $|\sigma_{bound}| < |\sigma_{free}|$; (b)
27.41 20 pF **27.45** 2.2×10^{-11} F
27.47 $C_1 = 22$ pF, $C_2 = 4.7$ pF, $Q = 320$ pC, $U = 2.0$ nJ, $C_1' = 7.1$ pF, $C_2' = 8.6$ pF, $Q_1 = 0.14$ nC, $Q_2 = 0.18$ nC, $U' = 3.3$ nJ, Work ($W = 1.3$ nJ) is done *on* the system.
27.51 $U_E = 100\ \mathrm{J/m^3}$ **27.55** $r = 2R$

27.57(a) $\frac{1}{2\pi}kQ^2\frac{(\ell^2 + x^2)^2}{(\ell^2 - x^2)^4}$

(b) $\frac{2}{\pi}kQ^2\frac{(x\ell)^2}{(x^2 - \ell^2)^4}$; $\frac{8}{81\pi}\frac{kQ^2}{\ell^4}$

27.61 The Forsterite-filled capacitor can hold 2.7 times as much charge as the paraffin.
27.65 $V_{min} = 10\ \mathrm{m^3}$
27.69 $|x| > 8.25$ cm

27.73 $R = \frac{3}{5}\frac{ke^2}{mc^2} = 1.7 \times 10^{-15}$ m

CHAPTER 28

28.1 *C*: zero; *D*: into the page
28.3 along the *y*-axis or the *z*-axis; 1.2×10^{-4} N **28.5** 3.4×10^{-9} T
28.7 $4.1 \times 10^{-3}\ \mathrm{A \cdot m^2}$ **28.9** 0; $\mu_0 I$; $2\mu_0 I$ **28.11** 1.5×10^{-7} T; zero
28.13 negative **28.17** 6×10^{-7} N
28.21 (2.2×10^{-16} N, out of the page)
28.25(a) a circle **(b)** a single turn
28.29 7.9×10^{-5} T **28.33** 0; 1.1×10^{-4} T **28.37** 3.02×10^{-6} T; using magnetic moment $B_z = 3.41 \times 10^{-6}$ T which gives a 13% error
28.41 7% **28.45** zero

28.49 $B_o = \frac{1}{2}\mu_0 jr$; Within the hole, $\vec{\mathbf{B}}$ is perpendicular to the line from the center of the cylinder to the center of the hole and has magnitude $\frac{1}{2}\mu_0 jc$. **28.53(b),(c)** $B = \frac{\mu_0 I}{4\pi r}$

(d) 0; $\frac{\mu_0 I}{2\pi r}$ **28.57** yes **28.61** zero

28.67(a) above both sheets $\vec{\mathbf{B}} = (\mu_0 K$, perpendicular to $\vec{\mathbf{K}}$); between the sheets $\vec{\mathbf{B}} = 0$; below the sheets $\vec{\mathbf{B}} = (\mu_0 K$, opposite $\vec{\mathbf{B}}_{above}$) **(b)** Above and below the sheets $\vec{\mathbf{B}} = 0$; Between the sheets $\vec{\mathbf{B}} = (\mu_0 K$, perpendicular to $\vec{\mathbf{K}}$) **(c)** Above both sheets

$\vec{\mathbf{B}} = \frac{\mu_0}{2}K(\hat{\mathbf{i}} - \hat{\mathbf{j}})$ where $\hat{\mathbf{i}}$ and $\hat{\mathbf{j}}$ are the directions of the two currents; Between the sheets

$\vec{\mathbf{B}} = \frac{-\mu_0}{2}K(\hat{\mathbf{i}} + \hat{\mathbf{j}})$; Below the sheets

$\vec{\mathbf{B}} = \frac{\mu_0}{2}K(-\hat{\mathbf{i}} + \hat{\mathbf{j}})$

28.69 $7.9 \times 10^{22}\ \mathrm{A \cdot m^2}$; 2×10^9 A
28.73

x(cm)	B(x) (T)
0.0	5.49×10^{-5}
2.0	5.73×10^{-5}
4.0	5.80×10^{-5}
6.0	5.81×10^{-5}

CHAPTER 29

29.1 9×10^3 m; 53 rad/s
29.3 1.1 T **29.5** $(0.090\ \mathrm{N})(\hat{\mathbf{k}} - \hat{\mathbf{j}})$
29.7 (1.5 μN/m, to the right)
29.9 0.30 nV **29.11** 0.028 A/m; 1.25699×10^{-4} T **29.13** clockwise circle
29.17 At the poles, an incoming charged particle will be undeflected. **29.21** Proton 3.9 km, 12 Hz; Electron 2.2 m, 22 kHz
29.25 23 kV **29.29** 1.0×10^7 m/s
29.33 There is no *y*-component of force regardless of orientation. **29.37** e
29.41 $1.2 \times 10^{-2}\ \mathrm{N \cdot m}$

29.45 $\vec{\mathbf{F}}_1 = \frac{1}{2}ILB\hat{\mathbf{k}}$ ($\hat{\mathbf{k}}$ is $\perp$ to plane of loop)

$\vec{\mathbf{F}}_2 = \frac{1}{2}ILB\hat{\mathbf{k}}$

$\vec{\mathbf{F}}_3 = -ILB\hat{\mathbf{k}}$

$\vec{\mathbf{F}}_{net} = 0$;

$\vec{\tau} = \left(\frac{\sqrt{3}\,L^2\,IB}{4}\right.$, in plane of triangle and $\perp$ to $\vec{\mathbf{B}}\Big)$ **29.49** ΔV_H is proportional to *V*, *w*, and *B*; ΔV_H is inversely proportional to ρ and ℓ; ΔV_H has no dependence on *d*
29.53 corner to corner, 21 p*V*.

29.55 1.4×10^{-14} W **29.59** $\frac{B \sin\theta}{(L/m)}$

29.63 No; Forces attempt to expand the solenoid. **29.67** 3×10^9 A; 10^{19} *W*

29.71 $\tan\theta = \frac{\mu_0 NI}{2a\,B_E}$

29.75 $\frac{-e(\vec{\mathbf{v}}_d \times \vec{\mathbf{B}})\ (\text{electrons per atom})}{(\text{mass per atom})}$

For copper: The correction is 2 parts in 10^5

CHAPTER 30

30.1 0.80 kV
30.3 $(4\ \mathrm{mV})\{(100)\sin[(400\ \mathrm{rad/s})t] - (10^{-4}/\mathrm{s}^3)t^3\}$ **30.5** Smaller loop: 0.5 V, 0.025 W. Larger loop: 1 V, 0.1 W; The larger loop. **30.7** $9.4 \times 10^{-3}\ \mathrm{N \cdot m}$
30.9 0.24 mV/m **30.11** 60 N
30.13 0.22 nT **30.17** c
30.21 0.599 mV **30.25** 1.0×10^{-9} A in the sense opposite the current in the large loop.
30.27 $(0.13\ \mathrm{A})\{[t/(1.0\ \mathrm{s})] - 2[t/(1.0\ \mathrm{s})]^3\}$; 2.5 μW; 0.71 s **30.31** 344 turns

30.35 $\frac{\omega a^2}{2}B \quad 0 < t < \frac{\pi}{\omega}$;

$\frac{-\omega a^2}{2}B \quad \frac{\pi}{\omega} < t < \frac{2\pi}{\omega}$

30.39 zero; 0.1 mV; 0.1 mV; zero
30.41 (b) **30.45** 140 V; satellite end
30.49 $(385\ \mathrm{V/m})\hat{\mathbf{i}} + (165\ \mathrm{V/m})(\hat{\mathbf{j}} - \hat{\mathbf{k}})$
30.53 2.0 mV/m; 1.9 mV/m
30.57 $y > 0$ $\vec{\mathbf{F}} = 0$;
$y < 0$ $\vec{\mathbf{F}} = -qvB\hat{\mathbf{k}}$
30.61(b) 400 N·m **(c)** 1 ms
30.65 0.17 μA; 2.7 μA

30.69 $\frac{\mu_0 I}{24\pi R}\left(1 - \frac{1}{\sqrt{2}}\right)\hat{\theta}$;

$\frac{\mu_0 I}{24\pi R}\left(1 + \frac{1}{\sqrt{2}}\right)\hat{\theta}$; $0.32\,\mu_0 I\,\hat{\theta}$

30.73 $Wv\left[B(0) + 2t\frac{dB}{dt}\right]$; 75.8 mV (clockwise); 75.8 mV (counterclockwise)
30.77(a) 7.3 V/m **(b)** 7.3×10^{-4} V **(c)** 25 V/m **(d)** 18 V/m

(e) 4.4 pC, Yes **30.81** $\frac{mgR\sin\theta}{B^2\ell^2\cos^2\theta}$

30.85 -7.80 Wb

30.89 $\vec{\mathbf{F}} = -\hat{\mathbf{k}}\frac{\mu_0^2\omega}{16R}\frac{m_0^2 a^3 \sin 2\omega t \sin 2\theta}{(z_0^2 + a^2)^3}$

where $\hat{\mathbf{k}}$ is parallel to $\vec{\mathbf{m}}$; zero

CHAPTER 31

31.1 $V = 0.10$ kV; $I = 15$ mA
31.3 For large *t* $Q = 0.56$ mC, $I = 0$; 1.7 s; $I_{initial} = 1.0$ mA **31.5** 2.0 $\mu\mathrm{J/m^3}$
31.7 1.0 A **31.9** 390 Hz; 0.43 μJ
31.13 0.69 *RC*
31.17 $Q(t) = C\mathcal{E}(1 + e^{-t/(RC)})$;

$I(t) = -\frac{\mathcal{E}}{R}e^{-t/RC}$

31.21 capacitors in parallel, resistors in parallel **31.25** solenoid with length 2ℓ will have half the magnetic energy of one with length ℓ **31.29** 0.6 mJ/m³; 7×10^{-5}
31.33 0.57 H **31.35(a)** same

(b) $\frac{N^2\pi\mu_0}{\ell}(a_2^2 + 3a_1^2)$

31.41 Doubling the resistance halves the time constant; ω_1 decreases
31.43 doubles; ω_1 will increase if $L < \frac{3}{8}R^2C$ otherwise, ω_1 decreases
31.47 1.0×10^{-2} A clockwise
31.49 0; 1.00 mA; 0.090 ms
31.53 all in series or all in parallel

31.57 $\tau = \frac{\mu_0\pi rd}{8\rho}$ **31.61** 0.32 mC, 0; 0.51 A, 1.0 ms; 0.57 mC, 2.7 ms
31.65 steady current $\mathcal{E}/R$ through the resistor; capacitor is uncharged

31.69 $Q_1 = \mathcal{E} C_1$; $Q_2 = \mathcal{E} C_2$; $\frac{\mathcal{E}}{R_1} e^{-t/(R_1C_1)} + \frac{\mathcal{E}}{R_2} e^{-t/(R_2C_2)}$; Nothing

31.73(a) $\tau = \frac{L_1 L_2 (R + r)}{(L_1 + L_2) R r}$; $I_1 = \frac{\mathcal{E}}{R} \frac{L_2}{(L_1 + L_2)} (1 - e^{-t/\tau})$; $I_2 = \frac{\mathcal{E}}{R} \left(\frac{L_1}{L_1 + L_2}\right) (1 - e^{-t/\tau})$; $I_3 = \frac{\mathcal{E}}{R + r} e^{-t/\tau}$; $I_4 = \frac{\mathcal{E}}{R}\left[1 - \frac{r}{R + r} e^{-t/\tau}\right]$

(b) $\tau = L\left(\frac{1}{r_1} + \frac{1}{r_2} + \frac{1}{R}\right)$; $I_1 = \frac{\mathcal{E} r_2}{(r_2 R + r_1 R + r_1 r_2)} e^{-t/\tau}$; $I_2 = \frac{\mathcal{E} r_1}{r_2 R + r_1 R + r_1 r_2} e^{-t/\tau}$; $I_3 = \frac{\mathcal{E}}{R}(1 - e^{-t/\tau})$; $I_4 = \frac{\mathcal{E}}{R}\left[1 - \frac{r_1 r_2}{R(r_1 + r_2) + r_1 r_2} e^{-t/\tau}\right]$

31.77 $L_1 + L_2 \pm 2M$; Yes; $\frac{L_1 L_2 - M^2}{L_1 + L_2 \mp 2M}$

31.81 $\frac{\mathcal{E}}{R}(1 + e^{-t/RC} - e^{-tR/L})$

31.85

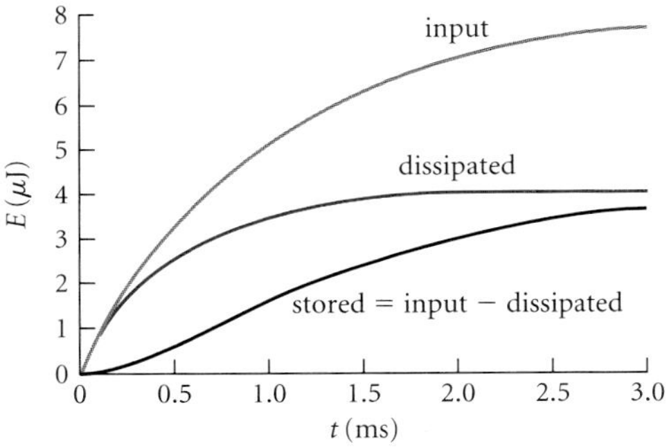

Stored: $\mathcal{E}^2 C\left(\frac{1}{2} + \frac{1}{2} e^{-2t/(RC)} - e^{-t/(RC)}\right)$; dissipated: $\frac{1}{2}\mathcal{E}^2 C(1 - e^{-2t/(RC)})$; drawn from the battery: $\mathcal{E}^2 C(1 - e^{-t/(RC)})$

31.89 $\frac{\mathcal{E}}{R}\left[1 - \frac{1}{2} e^{-2Rt/3L} - \frac{1}{2} e^{-2Rt/L}\right]$, $\frac{\mathcal{E}}{2R}[e^{-2Rt/L} - e^{-2Rt/3L}]$; 3.6 mA, −1.3 mA

Chapter 32

32.1 470 Ω **32.3** 240 Ω
32.5 0.54 A; 1.4 W
32.7 (30 A) cos (ωt − 0.147 rad). The amplitude has 2 significant figures.

32.9 5.4 W

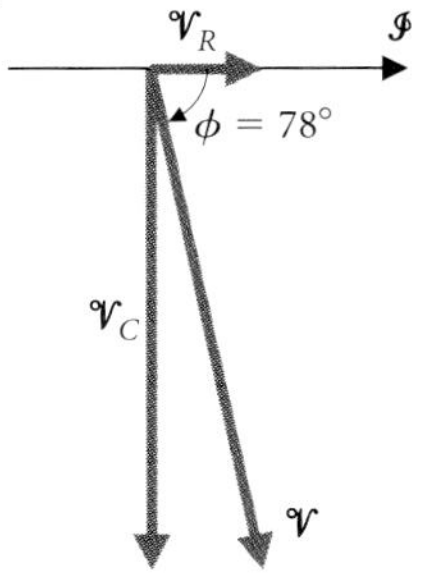

32.11 0.078 Ω; 1.2 μH; 8.7 mA
32.13 The inductance 2*L* has the larger reactance by a factor of 2. **32.17** United States, 60 Hz; Denver, 10 kHz; supercomputer, 150 MHz; chip, 30 GHz
32.21 12 Ω **32.25** $V_{rms} = \frac{V_o}{\sqrt{3}}; \frac{V_o}{\sqrt{3}}$
32.29 capacitor **32.33** 0.88 μF
32.37 f = 0.50 kHz; The woofer should be connected to the capacitor and the tweeter should be connected to the inductor
32.41 I_o = 1.1 mA

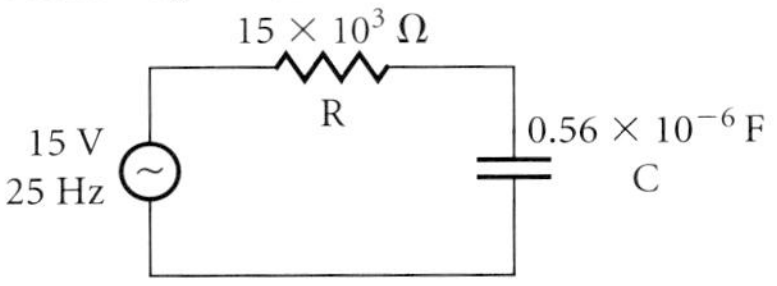

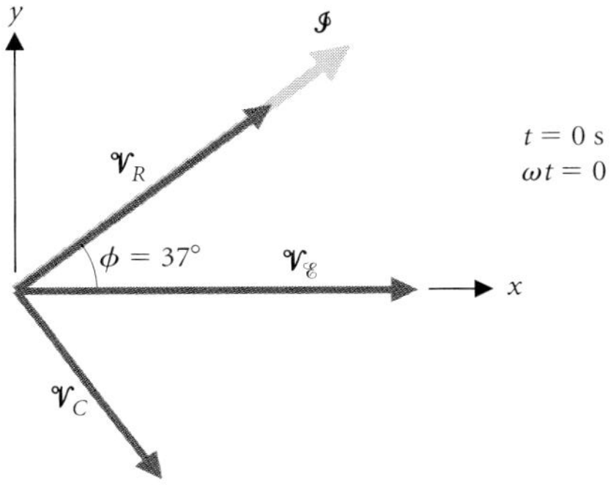

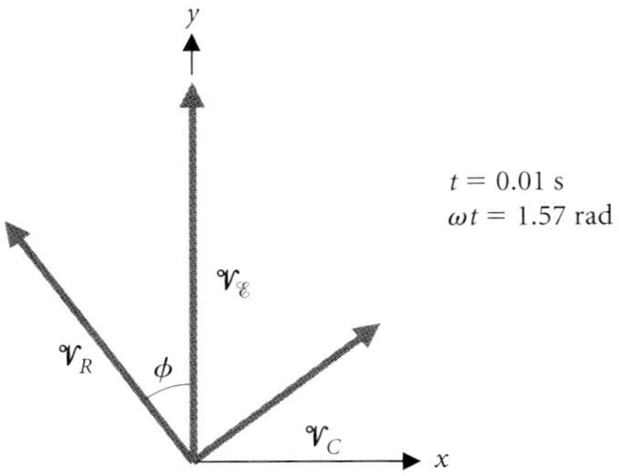

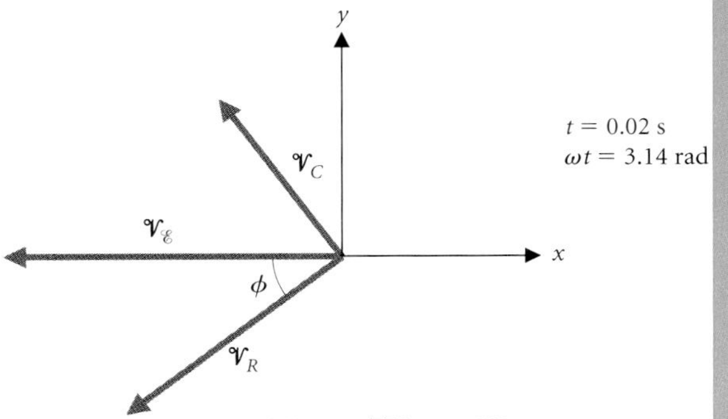

32.45 420 W; 40% TV set; 60% vacuum cleaner

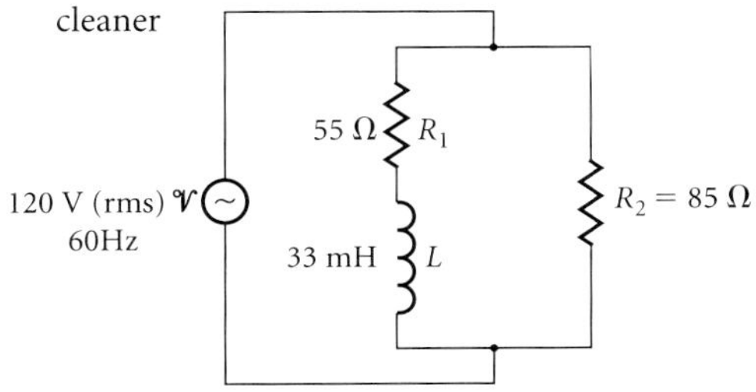

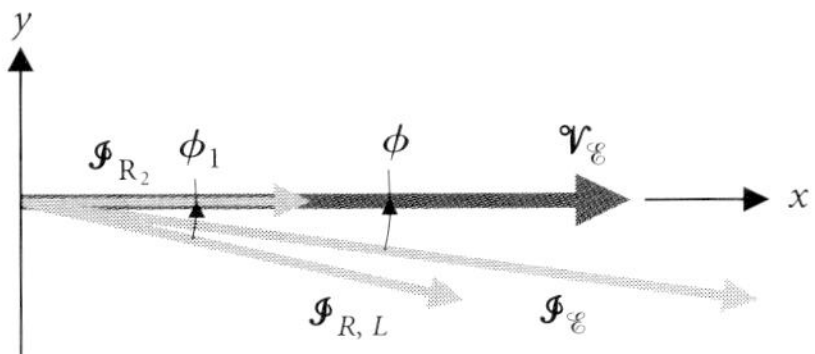

$\phi_1 = 13°$, $\phi = 8°$

32.49 ω_o and Q are each halved, τ is unchanged
32.53(a) 0.78 MHz, 3900; **(b)** 6600, 0.45 MHz **32.55** $R \leq 0.18$ Ω
32.59 1.04×10^{-10} F; 1100; 0.55%
32.63 $W_{generator} = \frac{V_0^2}{\omega Z}\left[\frac{\pi}{4}\cos\phi + \frac{1}{2}\sin\phi\right]$; The first term is the work dissipated by the resistor. The second term is $U_{L,max} - U_{C,max}$
32.67 In the steady state no potential difference occurs across the capacitor.
32.71 20 Ω **32.75(a)** The capacitance is $C \equiv \frac{1}{\omega_o^2 L} = \frac{\epsilon_o w}{d}[\ell + (\kappa - 1)x]$
(b) Q_{min} = 7.1; R_{max} = 24 Ω
32.79 V_C / V_{ant} = 0.17
32.83 $I_r = \frac{X_M I_o}{\sqrt{R_r^2 + (\omega L_r)^2}} \cos\left[\omega t + \frac{\pi}{2} - \tan^{-1}\left(\frac{\omega L_r}{R_r}\right)\right]$; $\langle F_{up}\rangle \propto \frac{R_r}{R_r^2 + \omega^2 L_r^2}$

Chapter 33

33.1 4.68 V/m
33.3 $\vec{\mathbf{B}} = (2.8 \times 10^{-8}\ \text{T})\frac{(-\hat{\mathbf{x}} + \hat{\mathbf{y}})}{\sqrt{2}}$
33.5 $E_0 = 4 \times 10^{-3}$ V/m
33.7 6.0×10^{-11} N **33.9** 6.84 W/m²

33.11 67.5° **33.13** 4.21×10^9 Hz
33.15 the $-z$-direction
33.19 $\hat{\mathbf{k}} = (0.346 \text{ rad/m})\,\hat{\mathbf{x}}$
$B_0 = 2.64 \times 10^{-9}$ T
33.23 $\vec{\mathbf{E}}_0 = (1.92 \times 10^{-2} \text{ V/m})(-2\hat{\mathbf{x}} + \hat{\mathbf{y}} + 5\hat{\mathbf{z}})$; $f = 107$ MHz **33.27** 7×10^7 m^2; 4×10^4 W/m^2; 5.5×10^3 V/m
33.31 9×10^{11} W
33.35 $R_c = 0.19\ \mu$m **33.39** Yes; set the filter's transmission axis horizontal
33.43 1.6 **33.47(a)** 56°
(b) 100% **(c)** 10.5% **33.51** 16°
33.55

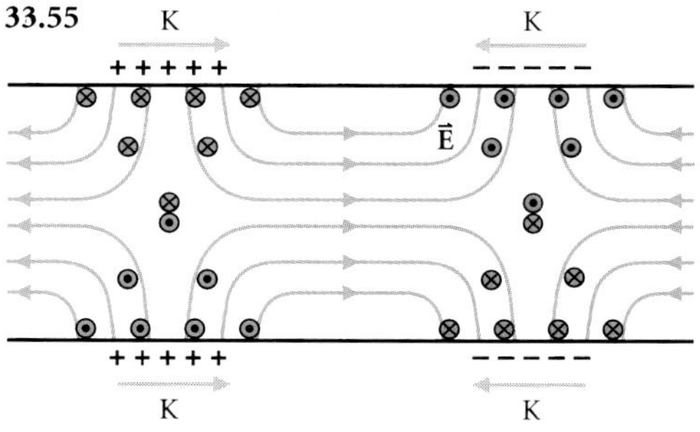

33.59 1 GHz
33.63(a)

$$B_x(y) = -\mu_0\epsilon_0 E_0 \frac{\omega a}{\pi} \cos(\omega t) \sin\left(\frac{\pi y}{a}\right)$$

(b) $\vec{\mathbf{B}}$ is zero within the conductor; The surface current per unit length is $2\epsilon_0 E_0 \frac{\omega a}{\pi} \cos \omega t$ in the $-z$-direction. The displacement current is equal in magnitude to, but opposite, the surface current.
33.67 $\frac{1}{2}$ W; 2×10^{-9} N **33.73** 6 cm; 10 cm; polarized parallel to the 3-cm sides

33.75
$$\alpha = \frac{\mu_0 c}{\rho\sqrt{2}}\left[1 + \sqrt{1 + (\epsilon_0\omega\rho)^{-2}}\right]^{-1/2}$$

$$k = \frac{\omega}{c\sqrt{2}}\left[1 + \sqrt{1 + (\epsilon_0\omega\rho)^{-2}}\right]^{1/2}$$

$$\phi_0 = \tan^{-1}\left[\frac{1}{\epsilon_0\omega\rho + \sqrt{(\epsilon_0\omega\rho)^2 + 1}}\right]$$

$$\frac{B_0}{E_0} = \frac{1}{c}[1 + (\epsilon_0\omega\rho)^{-2}]^{1/4}$$

Chapter 34

34.1 71 m **34.3** 2.7×10^7 m/s
34.5 -18 m^2; spacelike **34.7** $0.32\,c$
34.9 1.48; 1020 MeV/c; (938 MeV)^2
34.13 The astronaut at the front must leap first, $\frac{\gamma\beta x}{c}$ seconds earlier. **34.17** The minimum number of observers is three. To test the symmetry of time-dilation four observers are needed. **34.21** 0.26 m
34.25 1.4×10^8 m/s, 470 m
34.29 $\Delta\ell/\ell_0 = -6 \times 10^{-10}$
34.33 No **34.37** $\Delta s = 36.4$ m; No. In no frame do both events occur at the same time.
34.43 The fractional error is 2.8×10^{-10}%
34.47 $x'_A = 0$; $x'_B = -5\sqrt{3}$ m;
$x'_C = 4\sqrt{3}$ m; $x'_D = \frac{23\sqrt{3}}{3}$ m.
$ct'_A = \sqrt{3}$ m; $ct'_B = 7\sqrt{3}$ m;
$ct'_C = -4\sqrt{3}$ m; $ct'_D = \frac{-10\sqrt{3}}{3}$ m.
$\Delta s_{AB} = \sqrt{33}$ m; $\Delta s_{CD} = \sqrt{-39}$ m.
34.49 $x_{1A} = x_{1C} = -\frac{5}{3}$ km, $x_{1B} = x_{1D} = \frac{5}{3}$ km; $t_{1A} = t_{1C} = 4.4 \times 10^{-6}$ s, $t_{1B} = t_{1D} = -4.4 \times 10^{-6}$ s; $y_{2A} = y_{2B} = \frac{5}{3}$ km;
$y_{2C} = y_{2D} = -\frac{5}{3}$ km; $y_{2E} = 0$;
$t_{2A} = t_{2B} = -4.44 \times 10^{-6}$ s;
$t_{2C} = t_{2D} = +4.44 \times 10^{-6}$ s
34.53 1.5 cm; To represent a given space time interval requires a geometrical length on the prime axes that is longer by a factor $\sqrt{(1 + \beta^2)/(1 - \beta^2)}$.

34.59
$$a'_{\parallel} = \frac{a_{\parallel}}{\gamma^3(1 - \beta u_{\parallel}/c)^3}$$
$$a'_{\perp} = \frac{[(1 - \beta u_{\parallel}/c)a_{\perp} + (\beta u_{\perp}/c)a_{\parallel}}{\gamma^2(1 - \beta u_{\parallel}/c)^3}$$

34.63 All of the energy in two colliding particles is available for producing massive reaction products. **34.67** The fractional change $= 1.44 \times 10^{-8}$ **34.69** 6.72×10^{-27} J^2, the same, 4.68×10^{-24} kg·m/s, 3.7×10^{-18} kg·m/s **34.73** 1.0×10^{-8} m/s **34.79** Yes. The same argument applies as in the non-relativistic case.
34.83 -2.6×10^{-15} K^{-1}
34.87 Light requires different time intervals to arrive from different events. 7.5×10^{-7} s, 5.6×10^{-4} s, the aircraft's motion
34.91 $u_0 = \gamma c$; $\tilde{\mathbf{u}} = \gamma\vec{\mathbf{v}}$
34.95

	2.000	10.000
p/p_N	0.866	0.995
K/K_N	0.786	0.985

Chapter 35

35.1 1.2 nm; 2.4×10^{17} Hz
35.3 10.4 eV **35.5** 0.434 μm
35.7 870 eV **35.9** 9; 18
35.11 0.3 nm **35.13** $\geq 5 \times 10^{-21}$ eV
35.15 170 nm **35.19** 0.24 eV
35.23 6.08×10^{-4} **35.27** 0.8 photons/s
35.31 5.9×10^{-3} nm **35.35** 1.8 V
35.39 0.98 V; +0.2 V **35.43** lithium
35.47 $n^1 = 5 \rightarrow n = 3$ **35.51** Yes; $n \geq 3$
35.55 8×10^6 **35.59** 6480 nm; $d = 3$ nm; $d_i = 10^8$ nm; $r_{350} \gg d$, $r_{350} \ll d_i$ **35.65** $z_{\text{eff}} = 25$
35.67 7 nm; 4×10^{-12} m; 2×10^{-36} m
35.71 5.9×10^{-15} m, about $\frac{1}{3}$ the nuclear radius **35.75** 0.723 MeV **35.79** 7 fm
35.83 7%; 200% **35.89** 1 eV/molecule; 1.2 μm **35.93** 0.106 nm; 1.24×10^{15} Hz
35.97 $\frac{Q}{2M}$; 4.63×10^{-24} A·m^2; 9.26×10^{-24} A·m^2; 1.8×10^{-4} eV; $E_{\text{hyperfine}} \approx 5 \times 10^{-8}$ eV

Chapter 36

36.1 ^{90}Y; ^{27}Al **36.3** 1.8 MeV
36.5 $^{183}\text{Au} \rightarrow {}^{179}\text{Ir} + {}^4\text{He} + Q$
36.7 γ ray has energy 938 MeV
36.9 7.2×10^{18} Bq **36.11** ^{60}Co; 7.42 MeV **36.13** 6; 153.5 MeV
36.17 207.241 u **36.21** 2.82 MeV; 2.57 MeV **36.25** If sample has been decaying undisturbed for less than $t \sim 1/(\lambda d)$
36.29 No **36.33** $^{224}\text{Th} \rightarrow {}^{220}\text{Ra} + {}^4\text{He} + Q$; $^{211}\text{Bi} \rightarrow {}^{207}\text{Tl} + {}^4\text{He} + Q$; $^{243}\text{Cm} \rightarrow {}^{239}\text{Pu} + {}^4\text{He} + Q$; $^{221}\text{Fr} \rightarrow {}^{217}\text{At} + {}^4\text{He} + Q$; $^{215}\text{At} \rightarrow {}^{211}\text{Bi} + {}^4\text{He} + Q$
36.37 4.1×10^{13} Bq; 2×10^{11} Bq; 9.6×10^{21}
36.41 0.186 MeV **36.45** ^{206}Pb
36.49 silver 5.45×10^7 Bq; palladium 0.90×10^7 Bq; 3.1 h **36.53** $\frac{eRT_{1/2}}{\ln 2}$
36.59 ^{64}Zn **36.61** 3.27 MeV
36.65 1.805 MeV **36.69** 2.6×10^6 y
36.73 2.8×10^9 K **36.77** $\rho_* \approx 5 \times 10^{17}$ kg/m^3; $\rho_{\text{nuc}} = 2.3 \times 10^{17}$ kg/m^3
36.81 neutron 1.479 MeV; proton 1.295 MeV
36.85 2.59×10^{21}; 1.26×10^4 decays/sec; 3.4×10^{-18} kg; 10^{18}
36.89

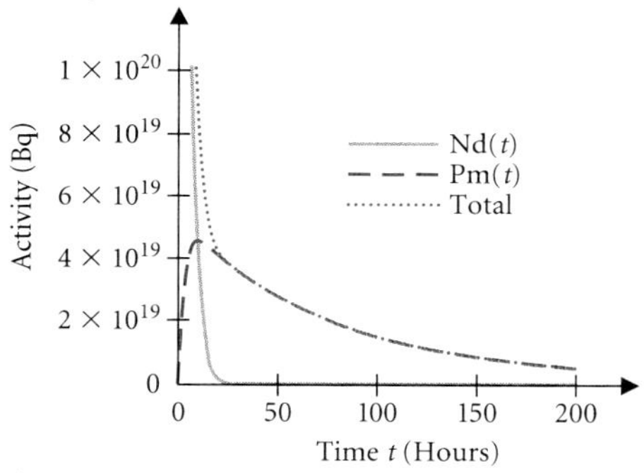

36.91 $\left(\frac{T_1 T_2}{T_1 - T_2}\right) \ln\left(\frac{T_1}{T_2}\right) \frac{1}{\ln 2}$;
$N_0 \frac{T_2}{T_1 - T_2}\{2^{-[T_2/(T_1 - T_2)] \ln (T_1/T_2)/\ln 2} - 2^{-[T_1/(T_1 - T_2)] \ln (T_1/T_2)/\ln 2}\}$; Tues at $8^{\text{h}}53^{\text{m}}$ am; 22.6 mCi

Chapter 37

37.1 3×10^{-22} s **37.3** 1.3183 GeV
37.5 Yes; quark flavor and isospin are unchanged **37.7** 4; $-\frac{3}{2}$, $-\frac{1}{2}$, $+\frac{1}{2}$, $+\frac{3}{2}$
37.11 10^{-7} eV; 10^{-14} eV **37.15** charge, baryon number, energy, linear momentum, angular momentum **37.19** 27 MeV
37.23 62 MeV **37.27(a)** 217 MeV
(b) 849 MeV

INDEX

Major discussions and definitions are in boldface.

D

F

Q

R

X

Y

Z

PHOTO CREDITS

PROLOGUE xxv Brookhaven National Laboratory; **xxviii (top left)** © Brian Yarvin 1994/Photo Researchers, Inc.; **(top right)** © Renee Lynn/Photo Researchers, Inc.; **(bottom)** © David Nunuk/Science Photo Library/Photo Researchers, Inc.; **xxix (top left & top middle)** NASA; **(bottom left)** NOAO; **(bottom middle)** Painting by Alfred Kamajian, from "How the Milky Way Formed" by Sidney Van den Bergh and James E. Hesser. © 1993 by Scientific American, Inc. All Rights Reserved; **(bottom right)** Alan Dressler (Carnegie Institution) and NASA; **xxx (top)** M. J. Geller, J. P. Huchva and V. deLapparent (graphics by M. J. Geller and E. E. Falco) © Smithsonian Astrophysical Observatory; **(bottom left)** © A. B. Dowsett/Science Photo Library/Photo Researchers, Inc.; **(bottom right)** IBM Corp., Research Division, Almaden Research Center.

PART I 1 Rockwell International/NASA; **(inset)** NASA.

CHAPTER 0 2 NASA; **3 (top)** Agence Photographique de la Réunion des Musées Nationaux; **(bottom)** © English Heritage/Photographic Library; **5** Erich Lessing/Art Resource, New York; **6** © John W. Allen; **7 (top)** The Bettmann Archive, Inc.; **(left)** Science Photo Library/Photo Researchers, Inc.; **(right)** © Owen Gingerich. By permission of the Houghton Library, Harvard University; **8** © Owen Gingerich. By permission of the Houghton Library, Harvard University; **9** Deutsche Forschungsanstalt für Luft-und Raumfahrt e.V.; **10** Science Photo Library/Photo Researchers, Inc.; **13 (top left)** © Owen Gingerich, 1981, by permission of the Houghton Library, Harvard University; **13 (top right)** Authors; **(bottom left & right)** Scala/Art Resource, NY.

CHAPTER 1 18 By Chesley Bonestell/Space Art International; **20** The Granger Collection, New York; **22 (top)** © Charles M. Falco/Photo Researchers, Inc.; **(bottom)** © Courtesy of the National Institute of Standards and Technology; **23 (both)** © Courtesy of the National Institute of Standards and Technology; **24** Photo courtesy of the design consulting firm Greenhorne & O'Mara, Inc.; **26** National Institute of Standards; **29 (top)** NASA; **(bottom)** AIP Emilio Segrè Visual Archives; **47** Appeared in "Why Round the Answer?" by D. L. Mathieson, *The Physics Teacher,* Oct. 1990, p. 471; **48** © Dorval/Explorer/Photo Researchers, Inc.

CHAPTER 2 51 © Jo Lillini, Agence Vandystadt/Photo Researchers, Inc.; **59 (left)** Smithsonian; **(right)** © Deborah Davis/PhotoEdit; **65** James E. Faller/NIST; **67** © 1994 Ken M. Johns/Photo Researchers, Inc.; **71** © The Harold E. Edgerton 1992 Trust, courtesy of Palm Press, Inc.

CHAPTER 3 83 © Jean Marc Barey/Agence Vandystadt/Photo Researchers, Inc.; **84 (top)** © Richard Megna/Fundamental Photographs; **(bottom)** David Young Wolff/PhotoEdit; **93 (top)** © Mike Fizer. All rights reserved; **(bottom)** Tom Nebbia/Palm Press, Inc.; **99** Authors/Scottish Heavyweight Champion Francis Brebner throwing the hammer; **102** © 1990, Richard Megna/Fundamental Photographs.

CHAPTER 4 125 © Classic Bike/EMAP Archives; **126** Courtesy of the Trustees of the Portsmouth Estates; **129** © Stephen Dalton/NHPA; **132** AP/Wide World Photos; **133** NASA, Johnson Space Center; **135** NIST; **136 (both)** NASA; **139** F. P. Bowden, D. Tabor, *Friction and Lubrication of Solids,* Oxford University Press © 1950. By permission of Oxford University Press; **140** Reuters/Bettmann.

ESSAY 1 162 Courtesy of the Royal Ontario Museum.

CHAPTER 5 172 **(left)** Jonathan Blair, © National Geographic Society; **(right)** © Mirror Syndication International; **173** West Publishing; **179** Bettmann Archive; **191** © Earl Zwicker, from *The Physics Teacher,* January 1987, p. 59; **196** Michael Lampton.

PART II 197 © The Stock Market/Globus Brothers, 1989.

CHAPTER 6 198 Roger Ressmeyer, © 1990 Corbis; **(inset)** © 1986 Max-Planck-Institut für Aeronomie, Lindau/Harz, Germany, by courtesy of Dr. H. U. Keller; **200** © The Harold E. Edgerton 1992 Trust, courtesy of Palm Press, Inc.; **212** NASA; **215** NOAO; **220** © Dennis MacDonald/PhotoEdit.

CHAPTER 7 224 Boeing Aircraft Corp.; **228** © Tony Freeman/PhotoEdit; **235** © The Granger Collection, New York; **237** © Tom Pantages; **242** NASA.

CHAPTER 8 257 © Busch Entertainment Corp.; **274** © 1992 Richard Megna, Fundamental Photographs.

ESSAY 2 290 © Giraudon/Art Resource, NY; **292** USGS.

CHAPTER 9 294 © David Madison 1994; **295** © David Madison 1994; **302** John Burke; **310** © 1993 Barry Rosenthal/FPG; **324** © Vandystadt/Photo Researchers, Inc.

ESSAY 3 329 NASA; **331** NASA.

CHAPTER 10 335 © 1995 Comstock; **336** © Bernard Asset, Agence Vandystadt/Photo Researchers, Inc.; **345** Photograph of model of a solid, assembled for "inelastic" collisions. Figure 2 in "Elastic and Inelastic Collisions" by Uri Ganiel, *The Physics Teacher,* January 1992, p. 19; **347 (top)** Photo courtesy of CERN, Geneva; **(bottom)** Pekka Parvianen/Science Photo Library/Photo Researchers, Inc.; **359** English Heritage; **360** Space Telescope Science Institute/NASA.

PART III **361** © Tad Stamm. All Rights Reserved. Picturesque.

CHAPTER 11 **362** West Publishing; **(inset)** © Tom Pantages; **363** Susan Lea; **382** © Reuters/Bettmann.

CHAPTER 12 **390** © 1990 Comstock; **394** © FPG International 1936; **407** Permission granted by the Courier-Journal, Louisville, KY; **413** © Tom Pantages; **414 (left)** © Jean Kugler/1986 FPG International; **(right)** Reprinted by permission from page 24 of *The Dynamic Universe: An Introduction to Astronomy,* 4/e, by Theodore P. Snow, Copyright © 1991 by West Publishing Company, All rights reserved; **424** © David Young-Wolff/PhotoEdit.

ESSAY 4 **426** © Scott Markewitz 1992/FPG International; **427 (top)** © FPG International; **(middle)** © C. Michael Lewis; **(bottom)** Peter Marlow/SYGMA; **428** © Scott Markewitz 1991/FPG International.

CHAPTER 13 **431** © Merpool/Photo Researchers, Inc.; **432** © The Granger Collection, New York; **437** Photo Deutsches Museum Munchen; **446** © Hubertus Kanus/Photo Researchers, Inc.; **449 (top)** © Maarten Rutgers and Xiao-lun Wu **(bottom)** ONERA; **454** U.S. Navy; **455 (left)** P. Bradshaw, Stanford University; **(right)** Courtesy Thomas J. Mueller/University of Notre Dame; **456** ONERA; **462 (top)** © Tom Pantages; **(bottom)** © Robert Brenner/PhotoEdit; **465 (both)** ONERA.

PART IV **469** © Farley Lewis/Photo Researchers, Inc.

CHAPTER 14 **470** © Tom Pantages; **482** © Reuters/Bettmann; **486** UPI/Bettmann; **488** © Tom Pantages; **493** © Deborah Davis, PhotoEdit.

CHAPTER 15 **495** © Tony Freeman/Photo Edit; **496** Alan Oddie/PhotoEdit; **498** From *Inquiry Into Physics,* by Ostdiek and Bord, © 1995 West Publishing Company, All rights reserved; **500** © Tom Pantages; **511** © Tom Pantages; **512 (both)** © Tom Pantages.

CHAPTER 16 **523** Jane Preston; **524** Shuleen Martin; **532** From "The Physics of Organ Pipes" by N. H. Fletcher and S. Thwaites. *Scientific American* January 1983, p. 97; **533** The Granger Collection, New York; **536** The Smith-Kettlewell Eye Research Institute pamphlet; **537** Alan Oddie/PhotoEdit; **538** © Jack Stein Grove/PhotoEdit; **540** A. J. Hudspeth and R. A. Jacobs; **546** Alan Bridle/NRAO; **547** © Erich Schrempp/Photo Researchers, Inc.; **548** © Tom Pantages; **549** © Tom Pantages; **551 (top)** © 1990 Richard Megna/Fundamental Photographs; **(bottom)** Amnon Yariv; **552** © David Parker/Science Photo Library/Photo Researchers, Inc.; **558** © 1994 Comstock, Inc.

CHAPTER 17 **562** © Raymond F. Newell; **563** © 1986 Richard Megna/Fundamental Photographs; **569** © Tom Pantages; **571** © 1987 Ken Kay/Fundamental Photographs; **573** Picture from article by G. Ramme: "Colors on Soap Films—An Interference Phenomenon" in *The Physics Teacher,* October 1990, p. 480; **574** Special Collections Division, Nimitz Library, United States Naval Academy, Courtesy Hale Observatories; **576 (all)** © Tom Pantages; **578** © Tom Pantages; **579 (all)** © Tom Pantages; **583 (both)** © Tom Pantages; **585** © Tom Pantages; **587** PSSC PHYSICS, 2/e, 1965; D. C. Heath & Company with Educational Development Center, Inc., Newton, MA. By permission of D. C. Heath & Company; **588 (left)** Kodak RL/Visuals Unlimited; **(right)** © C. Canizares, MIT; **592** © Dr. Brian J. Thompson, Provost Emeritus; Distinguished University Professor; **594** The Observatories of the Carnegie Institute of Washington; **595 (left)** R. J. Nelmes and M. I. McMahon, "High-Pressure Powder Diffraction on Synchrotron Sources", *Journal of Synchrotron Radiation,* Vol. I, Part 1, October 1994, p. 71, Figure 2; **(right)** NRAO.

CHAPTER 18 **597** NASA; **598 (both left)** © Tom Pantages; **(right)** NRAO; **599** © Jeff Isaac Greenberg/Photo Researchers, Inc.; **600** © Tom Pantages; **603** © Tom Pantages; **604** © Tom Pantages; **605 (left)** © Tom Pantages; **(right)** © 1993 Richard Megna/Fundamental Photographs; **613** © 1992 Martin Hough/Fundamental Photographs; **620** © Tom Pantages; **623** From *Science Year,* The 1969 World Book Science Annual. © Field Enterprises Educational Corporation. By permission of World Book, Inc.; **624** Roger Ressmeyer, © 1992 Corbis; **629** © Robert Greenler/Robert Greenler Sky Photos.

ESSAY 5 **634** NASA.

PART V **640** © Bonnie Sue Rauch/Photo Researchers, Inc.; **641 (top)** Courtesy of Central Scientific Company; **(bottom)** The Bettmann Archive; **642** © The Granger Collection, New York.

CHAPTER 19 **643** © Will & Demi McIntyre. All Rights Reserved. Picturesque; **644** © Tom Pantages; **648** Science Photo Library/Photo Researchers, Inc.; **654** The Science Museum/Science & Society Picture Library; **667** From "Brownian Motion" by Bernard H. Lavenda, *Scientific American,* February 1985, p. 71.

CHAPTER 20 **676** © Uniphoto Inc.; **684** © Tom Pantages; **686** © Bill Terr. All Rights Reserved. Picturesque Stock Photos; **691** Courtesy of the National Institute of Standards and Technology; **692** Authors.

ESSAY 6 **698 (top)** Jim Lockhart; **(bottom)** R. Plambeck, Radio Astronomy Lab, U.C. Berkeley; **699 (right)** after data courtesy of Lakeshore Cryotronics.

CHAPTER 21 **700** © Tom Pantages; **702 (left)** © 1993 James L. Amos, All Rights Reserved; **(right)** © Tom Pantages; **707 (top)** NASA; **(bottom left)** © Hans Namuth/Science Photo Source/Photo Researchers; **(bottom middle)** Courtesy of E. L. Kohlschmieder; **(bottom right)** NASA; **716** © Tom Pantages.

CHAPTER 22 **718** Tadashi Hachimandai; **723** © Photo Deutsches Museum Munchen.

ESSAY 7 **751 (both)** © Tom Pantages.

PART VI **754** © Murray & Associates, Inc. 1987, Picturesque Stock Photo; **756** © Tom Pantages; **757** NASA; **758 (left)** © The Granger Collection, New York; **(right)**; Tom Pantages; **759 (all)** Authors; **760** © Tom Pantages; **761** Authors; **762** Reprinted from *The Physics Teacher,* Dec. 1978, p. 631. Illustration from Faraday's "Diary" Vol. I. London, 1932.

CHAPTER 23 **764** © 1994 Comstock; **768** Courtesy of IBM Corp., Research Division, Almaden Research Center; **772** Dr. Roger Bland, San Francisco State University; **781** © The Bettmann Archive.

CHAPTER 24 **790** AP/Wide World Photos; **(inset)** © Tom Pantages; **793** © Tom Pantages; **797** © Tom Pantages; **801 (left)** Jim Lockhart, San Francisco State University; **(right)** © 1980 Kip Peticolas/Fundamental Photographs.

CHAPTER 25 **813** © 1993 Peter Menzel Photography/Boston Museum of Science; **826** Christopher Johnson, The University of Utah.

CHAPTER 26 **843** © Tom Pantages; **844 (left)** © Tom Pantages; **(right)** Authors; **846** The Bettmann Archive; **848 (left)** © Tom Pantages; **(right)** © Kip Peticolas/Fundamental Photographs; **850** © Tom Pantages; **852** © Tom Pantages; **862** © Tom Pantages.

CHAPTER 27 **875** Lawrence Berkeley Laboratory; **883** Lawrence Berkeley Laboratory; **884** © Tom Pantages.

CHAPTER 28 **898** Courtesy of Princeton University; **899** Authors; **900** © The Bettmann Archive; **908** Chip Kozy; **911** © Tom Pantages.

CHAPTER 29 **923** Mark Marten/Science Source/Photo Researchers, Inc.; **925** Courtesy of Finnigan MAT, San Jose, CA; **926** Lawrence Berkeley Laboratory; **928** Courtesy of Mallinckrodt Institute of Radiology at Washington University; **930** The Science Museum/Science & Society Picture Library; **934** © Tom Pantages; **940** K. Koike and K. Hayakawa from *Applied Physics Letters* vol. 45, No. 5, Sept. 1984; **943** Paul Shambroom/Photo Researchers, Inc.; **945** Lawrence Berkeley Laboratory.

PART VII **953** Courtesy of GE Research and Development Center; **954 (left)** The Bettmann Archive; **(right)** Courtesy of the Archives, California Institute of Technology.

CHAPTER 30 **955** Bureau of Reclamation; **956** AIP Emilio Segrè Visual Archives; **957** © Tom Pantages; **962** © 1996 by Joseph Somsel. Photograph courtesy of Pacific Gas and Electric Company; **971** University of Illinois/Courtesy AIP Emilio Segrè Visual Archives; **973 (right)** Courtesy of the Japanese Railway Technical Research Institute; **(both on bottom)** Authors.

CHAPTER 31 **987** © Tom Pantages.

CHAPTER 32 **1016** © 1996 by Joseph Somsel. Photograph courtesy of Pacific Gas and Electric Company.

CHAPTER 33 **1039** Maarten Schmidt/California Institute of Technology; **1048** Courtesy of Lawrence Livermore National Laboratory; **1053 (both)** © Tom Pantages; **1054** © Tom Pantages; **1055 (top)** Royal Observatory, Edinburgh/Science Photo Library/Photo Researchers, Inc.; **(bottom)** Dr. W. J. Welch, Radio Astronomy Lab, U.C. Berkeley; **1056** Russ Underwood, Lockheed Martin Missiles & Space; **1058** © Tom Pantages.

PART VIII **1067** photo courtesy of Lockheed Corporation.

CHAPTER 34 **1069** © courtesy of Roland and Marjorie Christen, taken with Astro-Physics 130mm StarFire EDT refractor; **1070** Werner Meyer-Ilse, Center for X-ray Optics, Lawrence Berkeley National Laboratory; **1091** © 1977 by Sidney Harris—"What's So Funny About Science?", William Kaufmann, Inc; **1093** Lawrence Berkeley Laboratory.

ESSAY 8 **1103** The LIGO Project, California Institure of Technology; **1106** From data courtesy of Iain F. Reid, California Institute of Technology, taken at the Keck Telescope; **1107 (left)** © Royal Astronomical Society Library; **(right)** NASA/STSI; **1109 (top)** Dana Berry/STScI (Hubble Space Telescope Science Institute); **(bottom)** NASA/STSI.

CHAPTER 35 **1110** IBM Corporation, Research Division, Almaden Research Center; **1112** Hamamatsu Corporation; **1113** Permission to copy granted from *Physical Review* Vol. VII, No. 3, 2nd series, March 1916; **1118** Permission to copy granted from *Physical Review* Vol. VII, 2nd series, March 1916; **1119** © 1994 Wabash Instrument Corp./Fundamental Photographs; **1121** David Gauntt and A. O. Schawlow/Stanford University; **1124** Bettmann Archive; **1125** NOAO; **1127** PSSC Physics 2/e, 1965 D. C. Heath and Company Educational and Development Center; **1128** © 1994 Wabash Instrument Corp./Fundamental Photographs; **1130** From data courtesy of Gary D. Schmidt, Steward observatory, University of Arizona; **1134 (left)** Akira Tonomura/Hitachi, Ltd.; **(right)** Bettmann Archive.

ESSAY 9 **1142** Shirley Chiang; **1143 (both)** courtesy of IBM Corp., Reasearch Division, Almaden Research Center.

CHAPTER 36 **1145** © 1996 by Joseph Somsel. Photograph courtesy of Pacific Gas and Electric Company; **1146 (left)** © Tim Bedow/Science Photo Library/Photo Researchers, Inc.; **(right)** Lawrence Berkeley National Laboratory; **1150** Blackett and Lees, Plate 7 from the Proceeding of the Royal Society, A136, pp. 325–338, 1932; **1153** © The Bettmann Archive; **1154** K. Phillipp, Naturwissenschaft, Vol. 14, p. 1204 (1926). © Springer-Verlag; **1156** copyright Società Italiana di Fisica; **1157 (left)** © California Institute of Technology; **(right)** Lawrence Berkeley Laboratory; **1158** photo courtesy of Brookhaven National Laboratory; **1159 (both left)** Kamiokande; **(right)** © Anglo-Australian Observatory/Royal Observatory Edinburgh. Photographs made from UK Schmidt plates by David Malin; **1160** Lawrence Berkeley Laboratory; **1163** © 1994 Peter Berndt, M.D., P.A. All Rights Reserved/Custom Medical Stock Photo.

CHAPTER 37 **1177** CDF Collaboration; **1178** Fermi National Accelator Laboratory/Science Photo Library/Photo Researchers, Inc.; **1180 (left)** AP/Wide World Photos; **(right)** Lawrence Berkeley Laboratory; **1182** Lawrence Berkeley Laboratory; **1184** Courtesy Brookhaven National Laboratory; **1188** Photo courtesy of CERN Geneva.

List of Tabulated Data

Solution Plans